University Physics Volume 1

SENIOR CONTRIBUTING AUTHORS

SAMUEL J. LING, TRUMAN STATE UNIVERSITY

JEFF SANNY, LOYOLA MARYMOUNT UNIVERSITY

WILLIAM MOEBS, PHD

OpenStax
Rice University
6100 Main Street MS-375
Houston, Texas 77005

To learn more about OpenStax, visit https://openstax.org.
Individual print copies and bulk orders can be purchased through our website.

PRINT BOOK ISBN-10	1-938168-27-5
PRINT BOOK ISBN-13	978-1-938168-27-7
PDF VERSION ISBN-10	1-947172-20-4
PDF VERSION ISBN-13	978-1-947172-20-3
ENHANCED TEXTBOOK ISBN-10	1-947172-15-8
ENHANCED TEXTBOOK ISBN-13	978-1-947172-15-9
Revision Number	UP1-2016-002(05/18)-MJ
Original Publication Year	2016

Printed in Indiana, USA

OpenStax

OpenStax provides free, peer-reviewed, openly licensed textbooks for introductory college and Advanced Placement® courses and low-cost, personalized courseware that helps students learn. A nonprofit ed tech initiative based at Rice University, we're committed to helping students access the tools they need to complete their courses and meet their educational goals.

Rice University

OpenStax, OpenStax CNX, and OpenStax Tutor are initiatives of Rice University. As a leading research university with a distinctive commitment to undergraduate education, Rice University aspires to path-breaking research, unsurpassed teaching, and contributions to the betterment of our world. It seeks to fulfill this mission by cultivating a diverse community of learning and discovery that produces leaders across the spectrum of human endeavor.

Foundation Support

OpenStax is grateful for the tremendous support of our sponsors. Without their strong engagement, the goal of free access to high-quality textbooks would remain just a dream.

Laura and John Arnold Foundation (LJAF) actively seeks opportunities to invest in organizations and thought leaders that have a sincere interest in implementing fundamental changes that not only yield immediate gains, but also repair broken systems for future generations. LJAF currently focuses its strategic investments on education, criminal justice, research integrity, and public accountability.

The William and Flora Hewlett Foundation has been making grants since 1967 to help solve social and environmental problems at home and around the world. The Foundation concentrates its resources on activities in education, the environment, global development and population, performing arts, and philanthropy, and makes grants to support disadvantaged communities in the San Francisco Bay Area.

Calvin K. Kazanjian was the founder and president of Peter Paul (Almond Joy), Inc. He firmly believed that the more people understood about basic economics the happier and more prosperous they would be. Accordingly, he established the Calvin K. Kazanjian Economics Foundation Inc, in 1949 as a philanthropic, nonpolitical educational organization to support efforts that enhanced economic understanding.

Guided by the belief that every life has equal value, the Bill & Melinda Gates Foundation works to help all people lead healthy, productive lives. In developing countries, it focuses on improving people's health with vaccines and other life-saving tools and giving them the chance to lift themselves out of hunger and extreme poverty. In the United States, it seeks to significantly improve education so that all young people have the opportunity to reach their full potential. Based in Seattle, Washington, the foundation is led by CEO Jeff Raikes and Co-chair William H. Gates Sr., under the direction of Bill and Melinda Gates and Warren Buffett.

The Maxfield Foundation supports projects with potential for high impact in science, education, sustainability, and other areas of social importance.

Our mission at The Michelson 20MM Foundation is to grow access and success by eliminating unnecessary hurdles to affordability. We support the creation, sharing, and proliferation of more effective, more affordable educational content by leveraging disruptive technologies, open educational resources, and new models for collaboration between for-profit, nonprofit, and public entities.

The Bill and Stephanie Sick Fund supports innovative projects in the areas of Education, Art, Science and Engineering.

Table of Contents

PREFACE

Welcome to *University Physics*, an OpenStax resource. This textbook was written to increase student access to high-quality learning materials, maintaining highest standards of academic rigor at little to no cost.

About OpenStax

OpenStax is a nonprofit based at Rice University, and it's our mission to improve student access to education. Our first openly licensed college textbook was published in 2012 and our library has since scaled to over 25 books used by hundreds of thousands of students across the globe. OpenStax Tutor, our low-cost personalized learning tool, is being used in college courses throughout the country. The OpenStax mission is made possible through the generous support of philanthropic foundations. Through these partnerships and with the help of additional low-cost resources from our OpenStax partners, OpenStax is breaking down the most common barriers to learning and empowering students and instructors to succeed.

About OpenStax's resources

Customization

University Physics is licensed under a Creative Commons Attribution 4.0 International (CC BY) license, which means that you can distribute, remix, and build upon the content, as long as you provide attribution to OpenStax and its content contributors.

Because our books are openly licensed, you are free to use the entire book or pick and choose the sections that are most relevant to the needs of your course. Feel free to remix the content by assigning your students certain chapters and sections in your syllabus in the order that you prefer. You can even provide a direct link in your syllabus to the sections in the web view of your book.

Instructors also have the option of creating a customized version of their OpenStax book. The custom version can be made available to students in low-cost print or digital form through their campus bookstore. Visit your book page on OpenStax.org for more information.

Errata

All OpenStax textbooks undergo a rigorous review process. However, like any professional-grade textbook, errors sometimes occur. Since our books are web based, we can make updates periodically when deemed pedagogically necessary. If you have a correction to suggest, submit it through the link on your book page on OpenStax.org. Subject matter experts review all errata suggestions. OpenStax is committed to remaining transparent about all updates, so you will also find a list of past errata changes on your book page on OpenStax.org.

Format

You can access this textbook for free in web view or PDF through OpenStax.org, and for a low cost in print.

About *University Physics*

University Physics is designed for the two- or three-semester calculus-based physics course. The text has been developed to meet the scope and sequence of most university physics courses and provides a foundation for a career in mathematics, science, or engineering. The book provides an important opportunity for students to learn the core concepts of physics and understand how those concepts apply to their lives and to the world around them.

Due to the comprehensive nature of the material, we are offering the book in three volumes for flexibility and efficiency.

Coverage and scope

Our *University Physics* textbook adheres to the scope and sequence of most two- and three-semester physics courses nationwide. We have worked to make physics interesting and accessible to students while maintaining the mathematical rigor inherent in the subject. With this objective in mind, the content of this textbook has been developed and arranged to provide a logical progression from fundamental to more advanced concepts, building upon what students have already learned and emphasizing connections between topics and between theory and applications. The goal of each section is to enable students not just to recognize concepts, but to work with them in ways that will be useful in later courses and future careers. The organization and pedagogical features were developed and vetted with feedback from science educators dedicated to the project.

Pedagogical foundation

Throughout *University Physics* you will find derivations of concepts that present classical ideas and techniques, as well as modern applications and methods. Most chapters start with observations or experiments that place the material in a context of physical experience. Presentations and explanations rely on years of classroom experience on the part of long-time physics professors, striving for a balance of clarity and rigor that has proven successful with their students. Throughout the text, links enable students to review earlier material and then return to the present discussion, reinforcing connections between topics. Key historical figures and experiments are discussed in the main text (rather than in boxes or sidebars), maintaining a focus on the development of physical intuition. Key ideas, definitions, and equations are highlighted in the text and listed in summary form at the end of each chapter. Examples and chapter-opening images often include contemporary applications from daily life or modern science and engineering that students can relate to, from smart phones to the internet to GPS devices.

Assessments that reinforce key concepts

In-chapter **Examples** generally follow a three-part format of Strategy, Solution, and Significance to emphasize how to approach a problem, how to work with the equations, and how to check and generalize the result. Examples are often followed by **Check Your Understanding** questions and answers to help reinforce for students the important ideas of the examples. **Problem-Solving Strategies** in each chapter break down methods of approaching various types of problems into steps students can follow for guidance. The book also includes exercises at the end of each chapter so students can practice what they've learned.

Conceptual questions do not require calculation but test student learning of the key concepts.

Problems categorized by section test student problem-solving skills and the ability to apply ideas to practical situations.

Additional Problems apply knowledge across the chapter, forcing students to identify what concepts and equations are appropriate for solving given problems. Randomly located throughout the problems are **Unreasonable Results** exercises that ask students to evaluate the answer to a problem and explain why it is not reasonable and what assumptions made might not be correct.

Challenge Problems extend text ideas to interesting but difficult situations.

Answers for selected exercises are available in an **Answer Key** at the end of the book.

Additional resources

Student and instructor resources

We've compiled additional resources for both students and instructors, including Getting Started Guides, PowerPoint slides, and answer and solution guides for instructors and students. Instructor resources require a verified instructor account, which you can apply for when you log in or create your account on OpenStax.org. Take advantage of these resources to supplement your OpenStax book.

Community Hubs

OpenStax partners with the Institute for the Study of Knowledge Management in Education (ISKME) to offer Community Hubs on OER Commons – a platform for instructors to share community-created resources that support OpenStax books, free of charge. Through our Community Hubs, instructors can upload their own materials or download resources to use in their own courses, including additional ancillaries, teaching material, multimedia, and relevant course content. We encourage instructors to join the hubs for the subjects most relevant to your teaching and research as an opportunity both to enrich your courses and to engage with other faculty.

To reach the Community Hubs, visit **www.oercommons.org/hubs/OpenStax (https://www.oercommons.org/hubs/OpenStax)** .

Partner resources

OpenStax partners are our allies in the mission to make high-quality learning materials affordable and accessible to students and instructors everywhere. Their tools integrate seamlessly with our OpenStax titles at a low cost. To access the partner resources for your text, visit your book page on OpenStax.org.

About the authors
Senior contributing authors

Samuel J. Ling, Truman State University

Dr. Samuel Ling has taught introductory and advanced physics for over 25 years at Truman State University, where he is currently Professor of Physics and the Department Chair. Dr. Ling has two PhDs from Boston University, one in Chemistry and the other in Physics, and he was a Research Fellow at the Indian Institute of Science, Bangalore, before joining Truman. Dr. Ling is also an author of *A First Course in Vibrations and Waves*, published by Oxford University Press. Dr. Ling has considerable experience with research in Physics Education and has published research on collaborative learning methods in physics teaching. He was awarded a Truman Fellow and a Jepson fellow in recognition of his innovative teaching methods. Dr. Ling's research publications have spanned Cosmology, Solid State Physics, and Nonlinear Optics.

Jeff Sanny, Loyola Marymount University

Dr. Jeff Sanny earned a BS in Physics from Harvey Mudd College in 1974 and a PhD in Solid State Physics from the University of California–Los Angeles in 1980. He joined the faculty at Loyola Marymount University in the fall of 1980. During his tenure, he has served as department Chair as well as Associate Dean. Dr. Sanny enjoys teaching introductory physics in particular. He is also passionate about providing students with research experience and has directed an active undergraduate student research group in space physics for many years.

William Moebs, Formerly of Loyola Marymount University

Dr. William Moebs earned a BS and PhD (1959 and 1965) from the University of Michigan. He then joined their staff as a Research Associate for one year, where he continued his doctoral research in particle physics. In 1966, he accepted an appointment to the Physics Department of Indiana Purdue Fort Wayne (IPFW), where he served as Department Chair from 1971 to 1979. In 1979, he moved to Loyola Marymount University (LMU), where he served as Chair of the Physics Department from 1979 to 1986. He retired from LMU in 2000. He has published research in particle physics, chemical kinetics, cell division, atomic physics, and physics teaching.

Contributing authors

Stephen D. Druger
Alice Kolakowska, University of Memphis
David Anderson, Albion College
Daniel Bowman, Ferrum College
Dedra Demaree, Georgetown University
Edw. S. Ginsberg, University of Massachusetts
Joseph Trout, Richard Stockton College
Kevin Wheelock, Bellevue College
David Smith, University of the Virgin Islands
Takashi Sato, Kwantlen Polytechnic University
Gerald Friedman, Santa Fe Community College
Lev Gasparov, University of North Florida
Lee LaRue, Paris Junior College
Mark Lattery, University of Wisconsin
Richard Ludlow, Daniel Webster College
Patrick Motl, Indiana University Kokomo
Tao Pang, University of Nevada, Las Vegas
Kenneth Podolak, Plattsburgh State University

Reviewers

Salameh Ahmad, Rochester Institute of Technology–Dubai
John Aiken, University of Colorado–Boulder
Raymond Benge, Terrant County College

Gavin Buxton, Robert Morris University
Erik Christensen, South Florida State College
Clifton Clark, Fort Hays State University
Nelson Coates, California Maritime Academy
Herve Collin, Kapi'olani Community College
Carl Covatto, Arizona State University
Alejandro Cozzani, Imperial Valley College
Danielle Dalafave, The College of New Jersey
Nicholas Darnton, Georgia Institute of Technology
Ethan Deneault, University of Tampa
Kenneth DeNisco, Harrisburg Area Community College
Robert Edmonds, Tarrant County College
William Falls, Erie Community College
Stanley Forrester, Broward College
Umesh Garg, University of Notre Dame
Maurizio Giannotti, Barry University
Bryan Gibbs, Dallas County Community College
Lynn Gillette, Pima Community College–West Campus
Mark Giroux, East Tennessee State University
Matthew Griffiths, University of New Haven
Alfonso Hinojosa, University of Texas–Arlington
Steuard Jensen, Alma College
David Kagan, University of Massachusetts
Sergei Katsev, University of Minnesota–Duluth
Jill Leggett, Florida State College–Jacksonville
Alfredo Louro, University of Calgary
James Maclaren, Tulane University
Ponn Maheswaranathan, Winthrop University
Seth Major, Hamilton College
Oleg Maksimov, Excelsior College
Aristides Marcano, Delaware State University
James McDonald, University of Hartford
Ralph McGrew, SUNY–Broome Community College
Paul Miller, West Virginia University
Tamar More, University of Portland
Farzaneh Najmabadi, University of Phoenix
Richard Olenick, The University of Dallas
Christopher Porter, Ohio State University
Liza Pujji, Manakau Institute of Technology
Baishali Ray, Young Harris University
Andrew Robinson, Carleton University
Aruvana Roy, Young Harris University
Gajendra Tulsian, Daytona State College
Adria Updike, Roger Williams University
Clark Vangilder, Central Arizona University
Steven Wolf, Texas State University
Alexander Wurm, Western New England University
Lei Zhang, Winston Salem State University
Ulrich Zurcher, Cleveland State University

1 | UNITS AND MEASUREMENT

Figure 1.1 This image might be showing any number of things. It might be a whirlpool in a tank of water or perhaps a collage of paint and shiny beads done for art class. Without knowing the size of the object in units we all recognize, such as meters or inches, it is difficult to know what we're looking at. In fact, this image shows the Whirlpool Galaxy (and its companion galaxy), which is about 60,000 light-years in diameter (about 6×10^{17} km across). (credit: modification of work by S. Beckwith (STScI) Hubble Heritage Team, (STScI/AURA), ESA, NASA)

Chapter Outline

1.1 The Scope and Scale of Physics

1.2 Units and Standards

1.3 Unit Conversion

1.4 Dimensional Analysis

1.5 Estimates and Fermi Calculations

1.6 Significant Figures

1.7 Solving Problems in Physics

Introduction

As noted in the figure caption, the chapter-opening image is of the Whirlpool Galaxy, which we examine in the first section of this chapter. Galaxies are as immense as atoms are small, yet the same laws of physics describe both, along with all the rest of nature—an indication of the underlying unity in the universe. The laws of physics are surprisingly few, implying an underlying simplicity to nature's apparent complexity. In this text, you learn about the laws of physics. Galaxies and atoms may seem far removed from your daily life, but as you begin to explore this broad-ranging subject, you may soon come to realize that physics plays a much larger role in your life than you first thought, no matter your life goals or career choice.

1.1 | The Scope and Scale of Physics

Physics is devoted to the understanding of all natural phenomena. In physics, we try to understand physical phenomena at all scales—from the world of subatomic particles to the entire universe. Despite the breadth of the subject, the various subfields of physics share a common core. The same basic training in physics will prepare you to work in any area of physics and the related areas of science and engineering. In this section, we investigate the scope of physics; the scales of length, mass, and time over which the laws of physics have been shown to be applicable; and the process by which science in general, and physics in particular, operates.

The Scope of Physics

Take another look at the chapter-opening image. The Whirlpool Galaxy contains billions of individual stars as well as huge clouds of gas and dust. Its companion galaxy is also visible to the right. This pair of galaxies lies a staggering billion trillion miles $(1.4 \times 10^{21} \, \text{mi})$ from our own galaxy (which is called the *Milky Way*). The stars and planets that make up the Whirlpool Galaxy might seem to be the furthest thing from most people's everyday lives, but the Whirlpool is a great starting point to think about the forces that hold the universe together. The forces that cause the Whirlpool Galaxy to act as it does are thought to be the same forces we contend with here on Earth, whether we are planning to send a rocket into space or simply planning to raise the walls for a new home. The gravity that causes the stars of the Whirlpool Galaxy to rotate and revolve is thought to be the same as what causes water to flow over hydroelectric dams here on Earth. When you look up at the stars, realize the forces out there are the same as the ones here on Earth. Through a study of physics, you may gain a greater understanding of the interconnectedness of everything we can see and know in this universe.

Think, now, about all the technological devices you use on a regular basis. Computers, smartphones, global positioning systems (GPSs), MP3 players, and satellite radio might come to mind. Then, think about the most exciting modern technologies you have heard about in the news, such as trains that levitate above tracks, "invisibility cloaks" that bend light around them, and microscopic robots that fight cancer cells in our bodies. All these groundbreaking advances, commonplace or unbelievable, rely on the principles of physics. Aside from playing a significant role in technology, professionals such as engineers, pilots, physicians, physical therapists, electricians, and computer programmers apply physics concepts in their daily work. For example, a pilot must understand how wind forces affect a flight path; a physical therapist must understand how the muscles in the body experience forces as they move and bend. As you will learn in this text, the principles of physics are propelling new, exciting technologies, and these principles are applied in a wide range of careers.

The underlying order of nature makes science in general, and physics in particular, interesting and enjoyable to study. For example, what do a bag of chips and a car battery have in common? Both contain energy that can be converted to other forms. The law of conservation of energy (which says that energy can change form but is never lost) ties together such topics as food calories, batteries, heat, light, and watch springs. Understanding this law makes it easier to learn about the various forms energy takes and how they relate to one another. Apparently unrelated topics are connected through broadly applicable physical laws, permitting an understanding beyond just the memorization of lists of facts.

Science consists of theories and laws that are the general truths of nature, as well as the body of knowledge they encompass. Scientists are continuously trying to expand this body of knowledge and to perfect the expression of the laws that describe it. **Physics**, which comes from the Greek *phúsis*, meaning "nature," is concerned with describing the interactions of energy, matter, space, and time to uncover the fundamental mechanisms that underlie every phenomenon. This concern for describing the basic phenomena in nature essentially defines the *scope of physics*.

Physics aims to understand the world around us at the most basic level. It emphasizes the use of a small number of quantitative laws to do this, which can be useful to other fields pushing the performance boundaries of existing technologies. Consider a smartphone (**Figure 1.2**). Physics describes how electricity interacts with the various circuits inside the device. This knowledge helps engineers select the appropriate materials and circuit layout when building a smartphone. Knowledge of the physics underlying these devices is required to shrink their size or increase their processing speed. Or, think about a GPS. Physics describes the relationship between the speed of an object, the distance over which it travels, and the time it takes to travel that distance. When you use a GPS in a vehicle, it relies on physics equations to determine the travel time from one location to another.

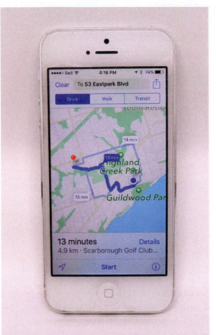

Figure 1.2 The Apple iPhone is a common smartphone with a GPS function. Physics describes the way that electricity flows through the circuits of this device. Engineers use their knowledge of physics to construct an iPhone with features that consumers will enjoy. One specific feature of an iPhone is the GPS function. A GPS uses physics equations to determine the drive time between two locations on a map. (credit: Jane Whitney)

Knowledge of physics is useful in everyday situations as well as in nonscientific professions. It can help you understand how microwave ovens work, why metals should not be put into them, and why they might affect pacemakers. Physics allows you to understand the hazards of radiation and to evaluate these hazards rationally and more easily. Physics also explains the reason why a black car radiator helps remove heat in a car engine, and it explains why a white roof helps keep the inside of a house cool. Similarly, the operation of a car's ignition system as well as the transmission of electrical signals throughout our body's nervous system are much easier to understand when you think about them in terms of basic physics.

Physics is a key element of many important disciplines and contributes directly to others. Chemistry, for example—since it deals with the interactions of atoms and molecules—has close ties to atomic and molecular physics. Most branches of engineering are concerned with designing new technologies, processes, or structures within the constraints set by the laws of physics. In architecture, physics is at the heart of structural stability and is involved in the acoustics, heating, lighting, and cooling of buildings. Parts of geology rely heavily on physics, such as radioactive dating of rocks, earthquake analysis, and heat transfer within Earth. Some disciplines, such as biophysics and geophysics, are hybrids of physics and other disciplines.

Physics has many applications in the biological sciences. On the microscopic level, it helps describe the properties of cells and their environments. On the macroscopic level, it explains the heat, work, and power associated with the human body and its various organ systems. Physics is involved in medical diagnostics, such as radiographs, magnetic resonance imaging, and ultrasonic blood flow measurements. Medical therapy sometimes involves physics directly; for example, cancer radiotherapy uses ionizing radiation. Physics also explains sensory phenomena, such as how musical instruments make sound, how the eye detects color, and how lasers transmit information.

It is not necessary to study all applications of physics formally. What is most useful is knowing the basic laws of physics and developing skills in the analytical methods for applying them. The study of physics also can improve your problem-solving skills. Furthermore, physics retains the most basic aspects of science, so it is used by all the sciences, and the study of physics makes other sciences easier to understand.

The Scale of Physics

From the discussion so far, it should be clear that to accomplish your goals in any of the various fields within the natural sciences and engineering, a thorough grounding in the laws of physics is necessary. The reason for this is simply that the laws of physics govern everything in the observable universe at all measurable scales of length, mass, and time. Now, that is easy enough to say, but to come to grips with what it really means, we need to get a little bit quantitative. So, before surveying the various scales that physics allows us to explore, let's first look at the concept of "order of magnitude," which we use to come to terms with the vast ranges of length, mass, and time that we consider in this text (**Figure 1.3**).

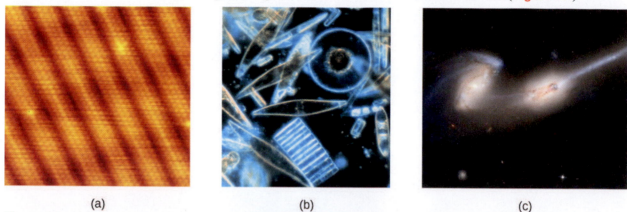

(a)   (b)   (c)

Figure 1.3 (a) Using a scanning tunneling microscope, scientists can see the individual atoms (diameters around 10^{-10} m) that compose this sheet of gold. (b) Tiny phytoplankton swim among crystals of ice in the Antarctic Sea. They range from a few micrometers (1 μm is 10^{-6} m) to as much as 2 mm (1 mm is 10^{-3} m) in length. (c) These two colliding galaxies, known as NGC 4676A (right) and NGC 4676B (left), are nicknamed "The Mice" because of the tail of gas emanating from each one. They are located 300 million light-years from Earth in the constellation Coma Berenices. Eventually, these two galaxies will merge into one. (credit a: modification of work by "Erwinrossen"/Wikimedia Commons; credit b: modification of work by Prof. Gordon T. Taylor, Stony Brook University; NOAA Corps Collections; credit c: modification of work by NASA, H. Ford (JHU), G. Illingworth (UCSC/LO), M. Clampin (STScI), G. Hartig (STScI), the ACS Science Team, and ESA)

Order of magnitude

The **order of magnitude** of a number is the power of 10 that most closely approximates it. Thus, the order of magnitude refers to the scale (or size) of a value. Each power of 10 represents a different order of magnitude. For example, 10^1, 10^2, 10^3, and so forth, are all different orders of magnitude, as are $10^0 = 1$, 10^{-1}, 10^{-2}, and 10^{-3}. To find the order of magnitude of a number, take the base-10 logarithm of the number and round it to the nearest integer, then the order of magnitude of the number is simply the resulting power of 10. For example, the order of magnitude of 800 is 10^3 because $\log_{10} 800 \approx 2.903$, which rounds to 3. Similarly, the order of magnitude of 450 is 10^3 because $\log_{10} 450 \approx 2.653$, which rounds to 3 as well. Thus, we say the numbers 800 and 450 are of the same order of magnitude: 10^3. However, the order of magnitude of 250 is 10^2 because $\log_{10} 250 \approx 2.397$, which rounds to 2.

An equivalent but quicker way to find the order of magnitude of a number is first to write it in scientific notation and then check to see whether the first factor is greater than or less than $\sqrt{10} = 10^{0.5} \approx 3$. The idea is that $\sqrt{10} = 10^{0.5}$ is halfway between $1 = 10^0$ and $10 = 10^1$ on a log base-10 scale. Thus, if the first factor is less than $\sqrt{10}$, then we round it down to 1 and the order of magnitude is simply whatever power of 10 is required to write the number in scientific notation. On the other hand, if the first factor is greater than $\sqrt{10}$, then we round it up to 10 and the order of magnitude is one power of 10 higher than the power needed to write the number in scientific notation. For example, the number 800 can be written in scientific notation as 8×10^2. Because 8 is bigger than $\sqrt{10} \approx 3$, we say the order of magnitude of 800 is $10^{2+1} = 10^3$. The number 450 can be written as 4.5×10^2, so its order of magnitude is also 10^3 because 4.5 is greater than 3. However, 250 written in scientific notation is 2.5×10^2 and 2.5 is less than 3, so its order of magnitude is 10^2.

The order of magnitude of a number is designed to be a ballpark estimate for the scale (or size) of its value. It is simply a way of rounding numbers consistently to the nearest power of 10. This makes doing rough mental math with very big and very small numbers easier. For example, the diameter of a hydrogen atom is on the order of 10^{-10} m, whereas the diameter of the Sun is on the order of 10^9 m, so it would take roughly $10^9/10^{-10} = 10^{19}$ hydrogen atoms to stretch across the diameter of the Sun. This is much easier to do in your head than using the more precise values of $1.06 \times 10^{-10}\,\mathrm{m}$ for a hydrogen atom diameter and $1.39 \times 10^{9}\,\mathrm{m}$ for the Sun's diameter, to find that it would take 1.31×10^{19} hydrogen atoms to stretch across the Sun's diameter. In addition to being easier, the rough estimate is also nearly as informative as the precise calculation.

Known ranges of length, mass, and time

The vastness of the universe and the breadth over which physics applies are illustrated by the wide range of examples of known lengths, masses, and times (given as orders of magnitude) in **Figure 1.4**. Examining this table will give you a feeling for the range of possible topics in physics and numerical values. A good way to appreciate the vastness of the ranges of values in **Figure 1.4** is to try to answer some simple comparative questions, such as the following:

- How many hydrogen atoms does it take to stretch across the diameter of the Sun?
  (Answer: $10^{9}\,\mathrm{m}/10^{-10}\,\mathrm{m} = 10^{19}$ hydrogen atoms)

- How many protons are there in a bacterium?
  (Answer: $10^{-15}\,\mathrm{kg}/10^{-27}\,\mathrm{kg} = 10^{12}$ protons)

- How many floating-point operations can a supercomputer do in 1 day?
  (Answer: $10^{5}\,\mathrm{s}/10^{-17}\,\mathrm{s} = 10^{22}$ floating-point operations)

In studying **Figure 1.4**, take some time to come up with similar questions that interest you and then try answering them. Doing this can breathe some life into almost any table of numbers.

Length in Meters (m)	Masses in Kilograms (kg)	Time in Seconds (s)
10^{-15} m = diameter of proton	10^{-30} kg = mass of electron	10^{-22} s = mean lifetime of very unstable nucleus
10^{-14} m = diameter of large nucleus	10^{-27} kg = mass of proton	10^{-17} s = time for single floating-point operation in a supercomputer
10^{-10} m = diameter of hydrogen atom	10^{-15} kg = mass of bacterium	10^{-15} s = time for one oscillation of visible light
10^{-7} m = diameter of typical virus	10^{-5} kg = mass of mosquito	10^{-13} s = time for one vibration of an atom in a solid
10^{-2} m = pinky fingernail width	10^{-2} kg = mass of hummingbird	10^{-3} s = duration of a nerve impulse
10^{0} m = height of 4 year old child	10^{0} kg = mass of liter of water	10^{0} s = time for one heartbeat
10^{2} m = length of football field	10^{2} kg = mass of person	10^{5} s = one day
10^{7} m = diameter of Earth	10^{19} kg = mass of atmosphere	10^{7} s = one year
10^{13} m = diameter of solar system	10^{22} kg = mass of Moon	10^{9} s = human lifetime
10^{16} m = distance light travels in a year (one light-year)	10^{25} kg = mass of Earth	10^{11} s = recorded human history
10^{21} m = Milky Way diameter	10^{30} kg = mass of Sun	10^{17} s = age of Earth
10^{26} m = distance to edge of observable universe	10^{53} kg = upper limit on mass of known universe	10^{18} s = age of the universe

Figure 1.4   This table shows the orders of magnitude of length, mass, and time.

 Visit **this site (https://openstaxcollege.org/l/21scaleuniv)** to explore interactively the vast range of length scales in our universe. Scroll down and up the scale to view hundreds of organisms and objects, and click on the individual objects to learn more about each one.

Building Models

How did we come to know the laws governing natural phenomena? What we refer to as the laws of nature are concise descriptions of the universe around us. They are human statements of the underlying laws or rules that all natural processes follow. Such laws are intrinsic to the universe; humans did not create them and cannot change them. We can only discover and understand them. Their discovery is a very human endeavor, with all the elements of mystery, imagination, struggle, triumph, and disappointment inherent in any creative effort (**Figure 1.5**). The cornerstone of discovering natural laws is observation; scientists must describe the universe as it is, not as we imagine it to be.

(a) Enrico Fermi   (b) Marie Curie

Figure 1.5 (a) Enrico Fermi (1901–1954) was born in Italy. On accepting the Nobel Prize in Stockholm in 1938 for his work on artificial radioactivity produced by neutrons, he took his family to America rather than return home to the government in power at the time. He became an American citizen and was a leading participant in the Manhattan Project. (b) Marie Curie (1867–1934) sacrificed monetary assets to help finance her early research and damaged her physical well-being with radiation exposure. She is the only person to win Nobel prizes in both physics and chemistry. One of her daughters also won a Nobel Prize. (credit a: modification of work by United States Department of Energy)

A **model** is a representation of something that is often too difficult (or impossible) to display directly. Although a model is justified by experimental tests, it is only accurate in describing certain aspects of a physical system. An example is the Bohr model of single-electron atoms, in which the electron is pictured as orbiting the nucleus, analogous to the way planets orbit the Sun (**Figure 1.6**). We cannot observe electron orbits directly, but the mental image helps explain some of the observations we can make, such as the emission of light from hot gases (atomic spectra). However, other observations show that the picture in the Bohr model is not really what atoms look like. The model is "wrong," but is still useful for some purposes. Physicists use models for a variety of purposes. For example, models can help physicists analyze a scenario and perform a calculation or models can be used to represent a situation in the form of a computer simulation. Ultimately, however, the results of these calculations and simulations need to be double-checked by other means—namely, observation and experimentation.

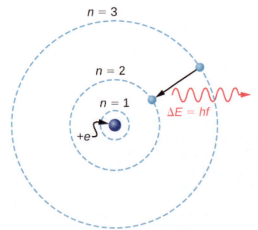

Figure 1.6 What is a model? The Bohr model of a single-electron atom shows the electron orbiting the nucleus in one of several possible circular orbits. Like all models, it captures some, but not all, aspects of the physical system.

The word *theory* means something different to scientists than what is often meant when the word is used in everyday conversation. In particular, to a scientist a theory is not the same as a "guess" or an "idea" or even a "hypothesis." The phrase "it's just a theory" seems meaningless and silly to scientists because science is founded on the notion of theories. To a scientist, a **theory** is a testable explanation for patterns in nature supported by scientific evidence and verified multiple times by various groups of researchers. Some theories include models to help visualize phenomena whereas others do not. Newton's theory of gravity, for example, does not require a model or mental image, because we can observe the objects directly with our own senses. The kinetic theory of gases, on the other hand, is a model in which a gas is viewed as being composed of atoms and molecules. Atoms and molecules are too small to be observed directly with our senses—thus, we picture them mentally to understand what the instruments tell us about the behavior of gases. Although models are meant only to describe certain aspects of a physical system accurately, a theory should describe all aspects of any system that falls within its domain of applicability. In particular, any experimentally testable implication of a theory should be verified. If an experiment ever shows an implication of a theory to be false, then the theory is either thrown out or modified suitably (for example, by limiting its domain of applicability).

A **law** uses concise language to describe a generalized pattern in nature supported by scientific evidence and repeated experiments. Often, a law can be expressed in the form of a single mathematical equation. Laws and theories are similar in that they are both scientific statements that result from a tested hypothesis and are supported by scientific evidence. However, the designation *law* is usually reserved for a concise and very general statement that describes phenomena in nature, such as the law that energy is conserved during any process, or Newton's second law of motion, which relates force (F), mass (m), and acceleration (a) by the simple equation $F = ma$. A theory, in contrast, is a less concise statement of observed behavior. For example, the theory of evolution and the theory of relativity cannot be expressed concisely enough to be considered laws. The biggest difference between a law and a theory is that a theory is much more complex and dynamic. A law describes a single action whereas a theory explains an entire group of related phenomena. Less broadly applicable statements are usually called principles (such as Pascal's principle, which is applicable only in fluids), but the distinction between laws and principles often is not made carefully.

The models, theories, and laws we devise sometimes imply the existence of objects or phenomena that are as yet unobserved. These predictions are remarkable triumphs and tributes to the power of science. It is the underlying order in the universe that enables scientists to make such spectacular predictions. However, if experimentation does not verify our predictions, then the theory or law is wrong, no matter how elegant or convenient it is. Laws can never be known with absolute certainty because it is impossible to perform every imaginable experiment to confirm a law for every possible scenario. Physicists operate under the assumption that all scientific laws and theories are valid until a counterexample is observed. If a good-quality, verifiable experiment contradicts a well-established law or theory, then the law or theory must be modified or overthrown completely.

The study of science in general, and physics in particular, is an adventure much like the exploration of an uncharted ocean. Discoveries are made; models, theories, and laws are formulated; and the beauty of the physical universe is made more sublime for the insights gained.

1.2 | Units and Standards

Learning Objectives
By the end of this section, you will be able to: • Describe how SI base units are defined. • Describe how derived units are created from base units. • Express quantities given in SI units using metric prefixes.

As we saw previously, the range of objects and phenomena studied in physics is immense. From the incredibly short lifetime of a nucleus to the age of Earth, from the tiny sizes of subnuclear particles to the vast distance to the edges of the known universe, from the force exerted by a jumping flea to the force between Earth and the Sun, there are enough factors of 10 to challenge the imagination of even the most experienced scientist. Giving numerical values for physical quantities and equations for physical principles allows us to understand nature much more deeply than qualitative descriptions alone. To comprehend these vast ranges, we must also have accepted units in which to express them. We shall find that even in the potentially mundane discussion of meters, kilograms, and seconds, a profound simplicity of nature appears: all physical quantities can be expressed as combinations of only seven base physical quantities.

We define a **physical quantity** either by specifying how it is measured or by stating how it is calculated from other measurements. For example, we might define distance and time by specifying methods for measuring them, such as using a meter stick and a stopwatch. Then, we could define *average speed* by stating that it is calculated as the total distance traveled divided by time of travel.

Measurements of physical quantities are expressed in terms of **units**, which are standardized values. For example, the length of a race, which is a physical quantity, can be expressed in units of meters (for sprinters) or kilometers (for distance runners). Without standardized units, it would be extremely difficult for scientists to express and compare measured values in a meaningful way (**Figure 1.7**).

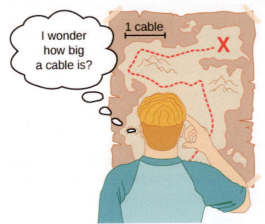

Figure 1.7 Distances given in unknown units are maddeningly useless.

Two major systems of units are used in the world: **SI units** (for the French *Système International d'Unités*), also known as the *metric system*, and **English units** (also known as the *customary* or *imperial system*). English units were historically used in nations once ruled by the British Empire and are still widely used in the United States. English units may also be referred to as the *foot–pound–second* (fps) system, as opposed to the *centimeter–gram–second* (cgs) system. You may also encounter the term *SAE units*, named after the Society of Automotive Engineers. Products such as fasteners and automotive tools (for example, wrenches) that are measured in inches rather than metric units are referred to as *SAE fasteners* or *SAE wrenches*.

Virtually every other country in the world (except the United States) now uses SI units as the standard. The metric system is also the standard system agreed on by scientists and mathematicians.

SI Units: Base and Derived Units

In any system of units, the units for some physical quantities must be defined through a measurement process. These are called the **base quantities** for that system and their units are the system's **base units**. All other physical quantities can then be expressed as algebraic combinations of the base quantities. Each of these physical quantities is then known as a **derived quantity** and each unit is called a **derived unit**. The choice of base quantities is somewhat arbitrary, as long as they are independent of each other and all other quantities can be derived from them. Typically, the goal is to choose physical quantities that can be measured accurately to a high precision as the base quantities. The reason for this is simple. Since the derived units can be expressed as algebraic combinations of the base units, they can only be as accurate and precise as the base units from which they are derived.

Based on such considerations, the International Standards Organization recommends using seven base quantities, which form the International System of Quantities (ISQ). These are the base quantities used to define the SI base units. **Table 1.1** lists these seven ISQ base quantities and the corresponding SI base units.

ISQ Base Quantity	SI Base Unit
Length	meter (m)
Mass	kilogram (kg)
Time	second (s)

Table 1.1 ISQ Base Quantities and Their SI Units

ISQ Base Quantity	SI Base Unit
Electrical current	ampere (A)
Thermodynamic temperature	kelvin (K)
Amount of substance	mole (mol)
Luminous intensity	candela (cd)

Table 1.1 ISQ Base Quantities and Their SI Units

You are probably already familiar with some derived quantities that can be formed from the base quantities in **Table 1.1**. For example, the geometric concept of area is always calculated as the product of two lengths. Thus, area is a derived quantity that can be expressed in terms of SI base units using square meters $(\text{m} \times \text{m} = \text{m}^2)$. Similarly, volume is a derived quantity that can be expressed in cubic meters (m^3). Speed is length per time; so in terms of SI base units, we could measure it in meters per second (m/s). Volume mass density (or just density) is mass per volume, which is expressed in terms of SI base units such as kilograms per cubic meter (kg/m^3). Angles can also be thought of as derived quantities because they can be defined as the ratio of the arc length subtended by two radii of a circle to the radius of the circle. This is how the radian is defined. Depending on your background and interests, you may be able to come up with other derived quantities, such as the mass flow rate (kg/s) or volume flow rate (m^3/s) of a fluid, electric charge $(\text{A} \cdot \text{s})$, mass flux density $[\text{kg/}(\text{m}^2 \cdot \text{s})]$, and so on. We will see many more examples throughout this text. For now, the point is that every physical quantity can be derived from the seven base quantities in **Table 1.1**, and the units of every physical quantity can be derived from the seven SI base units.

For the most part, we use SI units in this text. Non-SI units are used in a few applications in which they are in very common use, such as the measurement of temperature in degrees Celsius $(^\circ\text{C})$, the measurement of fluid volume in liters (L), and the measurement of energies of elementary particles in electron-volts (eV). Whenever non-SI units are discussed, they are tied to SI units through conversions. For example, 1 L is 10^{-3} m^3.

 Check out a comprehensive source of information on **SI units (https://openstaxcollege.org/l/21SIUnits)** at the National Institute of Standards and Technology (NIST) Reference on Constants, Units, and Uncertainty.

Units of Time, Length, and Mass: The Second, Meter, and Kilogram

The initial chapters in this textbook are concerned with mechanics, fluids, and waves. In these subjects all pertinent physical quantities can be expressed in terms of the base units of length, mass, and time. Therefore, we now turn to a discussion of these three base units, leaving discussion of the others until they are needed later.

The second

The SI unit for time, the **second** (abbreviated s), has a long history. For many years it was defined as 1/86,400 of a mean solar day. More recently, a new standard was adopted to gain greater accuracy and to define the second in terms of a nonvarying or constant physical phenomenon (because the solar day is getting longer as a result of the very gradual slowing of Earth's rotation). Cesium atoms can be made to vibrate in a very steady way, and these vibrations can be readily observed and counted. In 1967, the second was redefined as the time required for 9,192,631,770 of these vibrations to occur (**Figure 1.8**). Note that this may seem like more precision than you would ever need, but it isn't—GPSs rely on the precision of atomic clocks to be able to give you turn-by-turn directions on the surface of Earth, far from the satellites broadcasting their location.

Figure 1.8 An atomic clock such as this one uses the vibrations of cesium atoms to keep time to a precision of better than a microsecond per year. The fundamental unit of time, the second, is based on such clocks. This image looks down from the top of an atomic fountain nearly 30 feet tall. (credit: Steve Jurvetson)

The meter

The SI unit for length is the **meter** (abbreviated m); its definition has also changed over time to become more precise. The meter was first defined in 1791 as 1/10,000,000 of the distance from the equator to the North Pole. This measurement was improved in 1889 by redefining the meter to be the distance between two engraved lines on a platinum–iridium bar now kept near Paris. By 1960, it had become possible to define the meter even more accurately in terms of the wavelength of light, so it was again redefined as 1,650,763.73 wavelengths of orange light emitted by krypton atoms. In 1983, the meter was given its current definition (in part for greater accuracy) as the distance light travels in a vacuum in 1/299,792,458 of a second (**Figure 1.9**). This change came after knowing the speed of light to be exactly 299,792,458 m/s. The length of the meter will change if the speed of light is someday measured with greater accuracy.

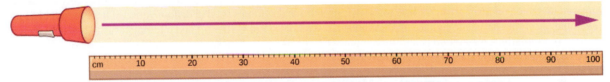

Light travels a distance of 1 meter
in 1/299,792,458 seconds

Figure 1.9 The meter is defined to be the distance light travels in 1/299,792,458 of a second in a vacuum. Distance traveled is speed multiplied by time.

The kilogram

The SI unit for mass is the **kilogram** (abbreviated kg); it is defined to be the mass of a platinum–iridium cylinder kept with the old meter standard at the International Bureau of Weights and Measures near Paris. Exact replicas of the standard kilogram are also kept at the U.S. National Institute of Standards and Technology (NIST), located in Gaithersburg, Maryland, outside of Washington, DC, and at other locations around the world. Scientists at NIST are currently investigating two complementary methods of redefining the kilogram (see **Figure 1.10**). The determination of all other masses can be traced ultimately to a comparison with the standard mass.

 There is currently an effort to redefine the SI unit of mass in terms of more fundamental processes by 2018. You can explore the history of mass standards and the contenders in the quest to devise a new one at the **website (https://www.nist.gov/pml/productsservices/redefining-kilogram)** of the Physical Measurement Laboratory.

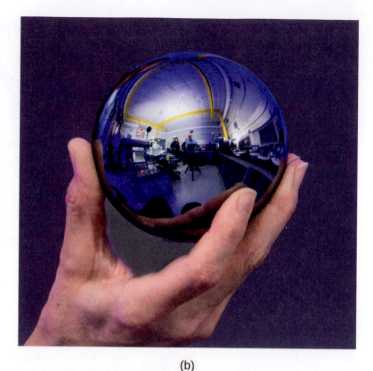

(a)   (b)

Figure 1.10 Redefining the SI unit of mass. Complementary methods are being investigated for use in an upcoming redefinition of the SI unit of mass. (a) The U.S. National Institute of Standards and Technology's watt balance is a machine that balances the weight of a test mass against the current and voltage (the "watt") produced by a strong system of magnets. (b) The International Avogadro Project is working to redefine the kilogram based on the dimensions, mass, and other known properties of a silicon sphere. (credit a and credit b: modification of work by National Institute of Standards and Technology)

Metric Prefixes

SI units are part of the **metric system**, which is convenient for scientific and engineering calculations because the units are categorized by factors of 10. **Table 1.2** lists the metric prefixes and symbols used to denote various factors of 10 in SI units. For example, a centimeter is one-hundredth of a meter (in symbols, 1 cm = 10^{-2} m) and a kilometer is a thousand meters (1 km = 10^3 m). Similarly, a megagram is a million grams (1 Mg = 10^6 g), a nanosecond is a billionth of a second (1 ns = 10^{-9} s), and a terameter is a trillion meters (1 Tm = 10^{12} m).

Prefix	Symbol	Meaning	Prefix	Symbol	Meaning
yotta-	Y	10^{24}	yocto-	y	10^{-24}
zetta-	Z	10^{21}	zepto-	z	10^{-21}
exa-	E	10^{18}	atto-	a	10^{-18}
peta-	P	10^{15}	femto-	f	10^{-15}
tera-	T	10^{12}	pico-	p	10^{-12}
giga-	G	10^9	nano-	n	10^{-9}
mega-	M	10^6	micro-	μ	10^{-6}
kilo-	k	10^3	milli-	m	10^{-3}
hecto-	h	10^2	centi-	c	10^{-2}
deka-	da	10^1	deci-	d	10^{-1}

Table 1.2 Metric Prefixes for Powers of 10 and Their Symbols

The only rule when using metric prefixes is that you cannot "double them up." For example, if you have measurements in petameters (1 Pm = 10^{15} m), it is not proper to talk about megagigameters, although $10^6 \times 10^9 = 10^{15}$. In practice, the only time this becomes a bit confusing is when discussing masses. As we have seen, the base SI unit of mass is the kilogram (kg), but metric prefixes need to be applied to the gram (g), because we are not allowed to "double-up" prefixes. Thus, a thousand kilograms (10^3 kg) is written as a megagram (1 Mg) since

$$10^3 \, \text{kg} = 10^3 \times 10^3 \, \text{g} = 10^6 \, \text{g} = 1 \, \text{Mg}.$$

Incidentally, 10^3 kg is also called a *metric ton*, abbreviated t. This is one of the units outside the SI system considered acceptable for use with SI units.

As we see in the next section, metric systems have the advantage that conversions of units involve only powers of 10. There are 100 cm in 1 m, 1000 m in 1 km, and so on. In nonmetric systems, such as the English system of units, the relationships are not as simple—there are 12 in. in 1 ft, 5280 ft in 1 mi, and so on.

Another advantage of metric systems is that the same unit can be used over extremely large ranges of values simply by scaling it with an appropriate metric prefix. The prefix is chosen by the order of magnitude of physical quantities commonly found in the task at hand. For example, distances in meters are suitable in construction, whereas distances in kilometers are appropriate for air travel, and nanometers are convenient in optical design. With the metric system there is no need to invent new units for particular applications. Instead, we rescale the units with which we are already familiar.

Example 1.1

Using Metric Prefixes

Restate the mass $1.93 \times 10^{13} \, \text{kg}$ using a metric prefix such that the resulting numerical value is bigger than one but less than 1000.

Strategy

Since we are not allowed to "double-up" prefixes, we first need to restate the mass in grams by replacing the prefix symbol k with a factor of 10^3 (see **Table 1.2**). Then, we should see which two prefixes in **Table 1.2** are closest to the resulting power of 10 when the number is written in scientific notation. We use whichever of these two prefixes gives us a number between one and 1000.

Solution

Replacing the k in kilogram with a factor of 10^3, we find that

$$1.93 \times 10^{13} \, \text{kg} = 1.93 \times 10^{13} \times 10^3 \, \text{g} = 1.93 \times 10^{16} \, \text{g}.$$

From **Table 1.2**, we see that 10^{16} is between "peta-" (10^{15}) and "exa-" (10^{18}). If we use the "peta-" prefix, then we find that $1.93 \times 10^{16} \, \text{g} = 1.93 \times 10^1 \, \text{Pg}$, since $16 = 1 + 15$. Alternatively, if we use the "exa-" prefix we find that $1.93 \times 10^{16} \, \text{g} = 1.93 \times 10^{-2} \, \text{Eg}$, since $16 = -2 + 18$. Because the problem asks for the numerical value between one and 1000, we use the "peta-" prefix and the answer is 19.3 Pg.

Significance

It is easy to make silly arithmetic errors when switching from one prefix to another, so it is always a good idea to check that our final answer matches the number we started with. An easy way to do this is to put both numbers in scientific notation and count powers of 10, including the ones hidden in prefixes. If we did not make a mistake, the powers of 10 should match up. In this problem, we started with $1.93 \times 10^{13} \, \text{kg}$, so we have 13 + 3 = 16 powers of 10. Our final answer in scientific notation is 1.93×10^1 Pg, so we have 1 + 15 = 16 powers of 10. So, everything checks out.

If this mass arose from a calculation, we would also want to check to determine whether a mass this large makes any sense in the context of the problem. For this, **Figure 1.4** might be helpful.

 1.1 **Check Your Understanding** Restate $4.79 \times 10^5 \, \text{kg}$ using a metric prefix such that the resulting number is bigger than one but less than 1000.

1.3 | Unit Conversion

Learning Objectives

By the end of this section, you will be able to:

- Use conversion factors to express the value of a given quantity in different units.

It is often necessary to convert from one unit to another. For example, if you are reading a European cookbook, some quantities may be expressed in units of liters and you need to convert them to cups. Or perhaps you are reading walking directions from one location to another and you are interested in how many miles you will be walking. In this case, you may need to convert units of feet or meters to miles.

Let's consider a simple example of how to convert units. Suppose we want to convert 80 m to kilometers. The first thing to do is to list the units you have and the units to which you want to convert. In this case, we have units in *meters* and we want to convert to *kilometers*. Next, we need to determine a conversion factor relating meters to kilometers. A **conversion factor** is a ratio that expresses how many of one unit are equal to another unit. For example, there are 12 in. in 1 ft, 1609 m in 1 mi, 100 cm in 1 m, 60 s in 1 min, and so on. Refer to **Appendix B** for a more complete list of conversion factors. In this case, we know that there are 1000 m in 1 km. Now we can set up our unit conversion. We write the units we have and then multiply them by the conversion factor so the units cancel out, as shown:

$$80 \; \cancel{m} \times \frac{1 \; km}{1000 \; \cancel{m}} = 0.080 \; km.$$

Note that the unwanted meter unit cancels, leaving only the desired kilometer unit. You can use this method to convert between any type of unit. Now, the conversion of 80 m to kilometers is simply the use of a metric prefix, as we saw in the preceding section, so we can get the same answer just as easily by noting that

$$80 \; m = 8.0 \times 10^1 \; m = 8.0 \times 10^{-2} \; km = 0.080 \; km,$$

since "kilo-" means 10^3 (see **Table 1.2**) and $1 = -2 + 3$. However, using conversion factors is handy when converting between units that are not metric or when converting between derived units, as the following examples illustrate.

Example 1.2

Converting Nonmetric Units to Metric

The distance from the university to home is 10 mi and it usually takes 20 min to drive this distance. Calculate the average speed in meters per second (m/s). (*Note:* Average speed is distance traveled divided by time of travel.)

Strategy

First we calculate the average speed using the given units, then we can get the average speed into the desired units by picking the correct conversion factors and multiplying by them. The correct conversion factors are those that cancel the unwanted units and leave the desired units in their place. In this case, we want to convert miles to meters, so we need to know the fact that there are 1609 m in 1 mi. We also want to convert minutes to seconds, so we use the conversion of 60 s in 1 min.

Solution

1. Calculate average speed. Average speed is distance traveled divided by time of travel. (Take this definition as a given for now. Average speed and other motion concepts are covered in later chapters.) In equation form,

$$\text{Average speed} = \frac{\text{Distance}}{\text{Time}}.$$

2. Substitute the given values for distance and time:

$$\text{Average speed} = \frac{10 \; mi}{20 \; min} = 0.50 \; \frac{mi}{min}.$$

3. Convert miles per minute to meters per second by multiplying by the conversion factor that cancels miles and leave meters, and also by the conversion factor that cancels minutes and leave seconds:

$$0.50 \, \frac{\cancel{\text{mile}}}{\cancel{\text{min}}} \times \frac{1609 \, \text{m}}{1 \, \cancel{\text{mile}}} \times \frac{1 \, \cancel{\text{min}}}{60 \, \text{s}} = \frac{(0.50)(1609)}{60} \, \text{m/s} = 13 \, \text{m/s}.$$

Significance

Check the answer in the following ways:

1. Be sure the units in the unit conversion cancel correctly. If the unit conversion factor was written upside down, the units do not cancel correctly in the equation. We see the "miles" in the numerator in 0.50 mi/min cancels the "mile" in the denominator in the first conversion factor. Also, the "min" in the denominator in 0.50 mi/min cancels the "min" in the numerator in the second conversion factor.

2. Check that the units of the final answer are the desired units. The problem asked us to solve for average speed in units of meters per second and, after the cancellations, the only units left are a meter (m) in the numerator and a second (s) in the denominator, so we have indeed obtained these units.

 1.2   **Check Your Understanding** Light travels about 9 Pm in a year. Given that a year is about 3×10^7 s, what is the speed of light in meters per second?

Example 1.3

Converting between Metric Units

The density of iron is $7.86 \, \text{g/cm}^3$ under standard conditions. Convert this to kg/m^3.

Strategy

We need to convert grams to kilograms and cubic centimeters to cubic meters. The conversion factors we need are $1 \, \text{kg} = 10^3 \, \text{g}$ and $1 \, \text{cm} = 10^{-2} \, \text{m}$. However, we are dealing with cubic centimeters $(\text{cm}^3 = \text{cm} \times \text{cm} \times \text{cm})$, so we have to use the second conversion factor three times (that is, we need to cube it). The idea is still to multiply by the conversion factors in such a way that they cancel the units we want to get rid of and introduce the units we want to keep.

Solution

$$7.86 \, \frac{\cancel{g}}{\cancel{\text{cm}}^3} \times \frac{\text{kg}}{10^3 \, \cancel{g}} \times \left(\frac{\cancel{\text{cm}}}{10^{-2} \, \text{m}} \right)^3 = \frac{7.86}{(10^3)(10^{-6})} \, \text{kg/m}^3 = 7.86 \times 10^3 \, \text{kg/m}^3$$

Significance

Remember, it's always important to check the answer.

1. Be sure to cancel the units in the unit conversion correctly. We see that the gram ("g") in the numerator in $7.86 \, \text{g/cm}^3$ cancels the "g" in the denominator in the first conversion factor. Also, the three factors of "cm" in the denominator in $7.86 \, \text{g/cm}^3$ cancel with the three factors of "cm" in the numerator that we get by cubing the second conversion factor.

2. Check that the units of the final answer are the desired units. The problem asked for us to convert to kilograms per cubic meter. After the cancellations just described, we see the only units we have left are "kg" in the numerator and three factors of "m" in the denominator (that is, one factor of "m" cubed, or "m^3"). Therefore, the units on the final answer are correct.

 1.3   **Check Your Understanding** We know from **Figure 1.4** that the diameter of Earth is on the order of 10^7 m, so the order of magnitude of its surface area is $10^{14} \, \text{m}^2$. What is that in square kilometers (that is, km^2)? (Try doing this both by converting 10^7 m to km and then squaring it and then by converting $10^{14} \, \text{m}^2$ directly to square kilometers. You should get the same answer both ways.)

Unit conversions may not seem very interesting, but not doing them can be costly. One famous example of this situation was seen with the *Mars Climate Orbiter*. This probe was launched by NASA on December 11, 1998. On September 23, 1999, while attempting to guide the probe into its planned orbit around Mars, NASA lost contact with it. Subsequent investigations showed a piece of software called SM_FORCES (or "small forces") was recording thruster performance data in the English units of pound-seconds (lb-s). However, other pieces of software that used these values for course corrections expected them to be recorded in the SI units of newton-seconds (N-s), as dictated in the software interface protocols. This error caused the probe to follow a very different trajectory from what NASA thought it was following, which most likely caused the probe either to burn up in the Martian atmosphere or to shoot out into space. This failure to pay attention to unit conversions cost hundreds of millions of dollars, not to mention all the time invested by the scientists and engineers who worked on the project.

 1.4   Check Your Understanding Given that 1 lb (pound) is 4.45 N, were the numbers being output by SM_FORCES too big or too small?

1.4 | Dimensional Analysis

Learning Objectives

By the end of this section, you will be able to:

* Find the dimensions of a mathematical expression involving physical quantities.
* Determine whether an equation involving physical quantities is dimensionally consistent.

The **dimension** of any physical quantity expresses its dependence on the base quantities as a product of symbols (or powers of symbols) representing the base quantities. **Table 1.3** lists the base quantities and the symbols used for their dimension. For example, a measurement of length is said to have dimension L or L^1, a measurement of mass has dimension M or M^1, and a measurement of time has dimension T or T^1. Like units, dimensions obey the rules of algebra. Thus, area is the product of two lengths and so has dimension L^2, or length squared. Similarly, volume is the product of three lengths and has dimension L^3, or length cubed. Speed has dimension length over time, L/T or LT^{-1}. Volumetric mass density has dimension M/L^3 or ML^{-3}, or mass over length cubed. In general, the dimension of any physical quantity can be written as $L^a M^b T^c I^d \Theta^e N^f J^g$ for some powers a, b, c, d, e, f, and g. We can write the dimensions of a length in this form with $a = 1$ and the remaining six powers all set equal to zero: $L^1 = L^1 M^0 T^0 I^0 \Theta^0 N^0 J^0$. Any quantity with a dimension that can be written so that all seven powers are zero (that is, its dimension is $L^0 M^0 T^0 I^0 \Theta^0 N^0 J^0$) is called **dimensionless** (or sometimes "of dimension 1," because anything raised to the zero power is one). Physicists often call dimensionless quantities *pure numbers*.

Base Quantity	Symbol for Dimension
Length	L
Mass	M
Time	T
Current	I
Thermodynamic temperature	Θ
Amount of substance	N
Luminous intensity	J

Table 1.3 Base Quantities and Their Dimensions

Physicists often use square brackets around the symbol for a physical quantity to represent the dimensions of that quantity. For example, if r is the radius of a cylinder and h is its height, then we write $[r] = L$ and $[h] = L$ to indicate the dimensions of the radius and height are both those of length, or L. Similarly, if we use the symbol A for the surface area of a cylinder and V for its volume, then $[A] = L^2$ and $[V] = L^3$. If we use the symbol m for the mass of the cylinder and ρ

for the density of the material from which the cylinder is made, then $[m] = \text{M}$ and $[\rho] = \text{ML}^{-3}$.

The importance of the concept of dimension arises from the fact that any mathematical equation relating physical quantities must be **dimensionally consistent**, which means the equation must obey the following rules:

- Every term in an expression must have the same dimensions; it does not make sense to add or subtract quantities of differing dimension (think of the old saying: "You can't add apples and oranges"). In particular, the expressions on each side of the equality in an equation must have the same dimensions.

- The arguments of any of the standard mathematical functions such as trigonometric functions (such as sine and cosine), logarithms, or exponential functions that appear in the equation must be dimensionless. These functions require pure numbers as inputs and give pure numbers as outputs.

If either of these rules is violated, an equation is not dimensionally consistent and cannot possibly be a correct statement of physical law. This simple fact can be used to check for typos or algebra mistakes, to help remember the various laws of physics, and even to suggest the form that new laws of physics might take. This last use of dimensions is beyond the scope of this text, but is something you will undoubtedly learn later in your academic career.

Example 1.4

Using Dimensions to Remember an Equation

Suppose we need the formula for the area of a circle for some computation. Like many people who learned geometry too long ago to recall with any certainty, two expressions may pop into our mind when we think of circles: πr^2 and $2\pi r$. One expression is the circumference of a circle of radius r and the other is its area. But which is which?

Strategy

One natural strategy is to look it up, but this could take time to find information from a reputable source. Besides, even if we think the source is reputable, we shouldn't trust everything we read. It is nice to have a way to double-check just by thinking about it. Also, we might be in a situation in which we cannot look things up (such as during a test). Thus, the strategy is to find the dimensions of both expressions by making use of the fact that dimensions follow the rules of algebra. If either expression does not have the same dimensions as area, then it cannot possibly be the correct equation for the area of a circle.

Solution

We know the dimension of area is L^2. Now, the dimension of the expression πr^2 is

$$[\pi r^2] = [\pi] \cdot [r]^2 = 1 \cdot \text{L}^2 = \text{L}^2,$$

since the constant π is a pure number and the radius r is a length. Therefore, πr^2 has the dimension of area. Similarly, the dimension of the expression $2\pi r$ is

$$[2\pi r] = [2] \cdot [\pi] \cdot [r] = 1 \cdot 1 \cdot \text{L} = \text{L},$$

since the constants 2 and π are both dimensionless and the radius r is a length. We see that $2\pi r$ has the dimension of length, which means it cannot possibly be an area.

We rule out $2\pi r$ because it is not dimensionally consistent with being an area. We see that πr^2 is dimensionally consistent with being an area, so if we have to choose between these two expressions, πr^2 is the one to choose.

Significance

This may seem like kind of a silly example, but the ideas are very general. As long as we know the dimensions of the individual physical quantities that appear in an equation, we can check to see whether the equation is dimensionally consistent. On the other hand, knowing that true equations are dimensionally consistent, we can match expressions from our imperfect memories to the quantities for which they might be expressions. Doing this will not help us remember dimensionless factors that appear in the equations (for example, if you had accidentally conflated the two expressions from the example into $2\pi r^2$, then dimensional analysis is no help), but it does help us remember the correct basic form of equations.

1.5   Check Your Understanding Suppose we want the formula for the volume of a sphere. The two expressions commonly mentioned in elementary discussions of spheres are $4\pi r^2$ and $4\pi r^3/3$. One is the volume of a sphere of radius r and the other is its surface area. Which one is the volume?

Example 1.5

Checking Equations for Dimensional Consistency

Consider the physical quantities s, v, a, and t with dimensions $[s] = L$, $[v] = LT^{-1}$, $[a] = LT^{-2}$, and $[t] = T$. Determine whether each of the following equations is dimensionally consistent: (a) $s = vt + 0.5at^2$; (b) $s = vt^2 + 0.5at$; and (c) $v = \sin(at^2/s)$.

Strategy

By the definition of dimensional consistency, we need to check that each term in a given equation has the same dimensions as the other terms in that equation and that the arguments of any standard mathematical functions are dimensionless.

Solution

a. There are no trigonometric, logarithmic, or exponential functions to worry about in this equation, so we need only look at the dimensions of each term appearing in the equation. There are three terms, one in the left expression and two in the expression on the right, so we look at each in turn:

$$[s] = L$$
$$[vt] = [v] \cdot [t] = LT^{-1} \cdot T = LT^0 = L$$
$$[0.5at^2] = [a] \cdot [t]^2 = LT^{-2} \cdot T^2 = LT^0 = L.$$

All three terms have the same dimension, so this equation is dimensionally consistent.

b. Again, there are no trigonometric, exponential, or logarithmic functions, so we only need to look at the dimensions of each of the three terms appearing in the equation:

$$[s] = L$$
$$[vt^2] = [v] \cdot [t]^2 = LT^{-1} \cdot T^2 = LT$$
$$[at] = [a] \cdot [t] = LT^{-2} \cdot T = LT^{-1}.$$

None of the three terms has the same dimension as any other, so this is about as far from being dimensionally consistent as you can get. The technical term for an equation like this is *nonsense*.

c. This equation has a trigonometric function in it, so first we should check that the argument of the sine function is dimensionless:

$$\left[\frac{at^2}{s}\right] = \frac{[a] \cdot [t]^2}{[s]} = \frac{LT^{-2} \cdot T^2}{L} = \frac{L}{L} = 1.$$

The argument is dimensionless. So far, so good. Now we need to check the dimensions of each of the two terms (that is, the left expression and the right expression) in the equation:

$$[v] = LT^{-1}$$
$$\left[\sin\left(\frac{at^2}{s}\right)\right] = 1.$$

The two terms have different dimensions—meaning, the equation is not dimensionally consistent. This equation is another example of "nonsense."

Significance

If we are trusting people, these types of dimensional checks might seem unnecessary. But, rest assured, any textbook on a quantitative subject such as physics (including this one) almost certainly contains some equations with typos. Checking equations routinely by dimensional analysis save us the embarrassment of using an incorrect equation. Also, checking the dimensions of an equation we obtain through algebraic manipulation is a great way to make sure we did not make a mistake (or to spot a mistake, if we made one).

 1.6 **Check Your Understanding** Is the equation $v = at$ dimensionally consistent?

One further point that needs to be mentioned is the effect of the operations of calculus on dimensions. We have seen that dimensions obey the rules of algebra, just like units, but what happens when we take the derivative of one physical quantity with respect to another or integrate a physical quantity over another? The derivative of a function is just the slope of the line tangent to its graph and slopes are ratios, so for physical quantities v and t, we have that the dimension of the derivative of v with respect to t is just the ratio of the dimension of v over that of t:

$$\left[\frac{dv}{dt}\right] = \frac{[v]}{[t]}.$$

Similarly, since integrals are just sums of products, the dimension of the integral of v with respect to t is simply the dimension of v times the dimension of t:

$$\left[\int v\,dt\right] = [v] \cdot [t].$$

By the same reasoning, analogous rules hold for the units of physical quantities derived from other quantities by integration or differentiation.

1.5 | Estimates and Fermi Calculations

Learning Objectives
By the end of this section, you will be able to: • Estimate the values of physical quantities.

On many occasions, physicists, other scientists, and engineers need to make *estimates* for a particular quantity. Other terms sometimes used are *guesstimates, order-of-magnitude approximations, back-of-the-envelope calculations,* or *Fermi calculations.* (The physicist Enrico Fermi mentioned earlier was famous for his ability to estimate various kinds of data with surprising precision.) Will that piece of equipment fit in the back of the car or do we need to rent a truck? How long will this download take? About how large a current will there be in this circuit when it is turned on? How many houses could a proposed power plant actually power if it is built? Note that estimating does not mean guessing a number or a formula at random. Rather, **estimation** means using prior experience and sound physical reasoning to arrive at a rough idea of a quantity's value. Because the process of determining a reliable approximation usually involves the identification of correct physical principles and a good guess about the relevant variables, estimating is very useful in developing physical intuition. Estimates also allow us perform "sanity checks" on calculations or policy proposals by helping us rule out certain scenarios or unrealistic numbers. They allow us to challenge others (as well as ourselves) in our efforts to learn truths about the world.

Many estimates are based on formulas in which the input quantities are known only to a limited precision. As you develop physics problem-solving skills (which are applicable to a wide variety of fields), you also will develop skills at estimating. You develop these skills by thinking more quantitatively and by being willing to take risks. As with any skill, experience helps. Familiarity with dimensions (see **Table 1.3**) and units (see **Table 1.1** and **Table 1.2**), and the scales of base quantities (see **Figure 1.4**) also helps.

To make some progress in estimating, you need to have some definite ideas about how variables may be related. The following strategies may help you in practicing the art of estimation:

 • *Get big lengths from smaller lengths.* When estimating lengths, remember that anything can be a ruler. Thus, imagine breaking a big thing into smaller things, estimate the length of one of the smaller things, and multiply to get the length of the big thing. For example, to estimate the height of a building, first count how many floors it has. Then, estimate how big a single floor is by imagining how many people would have to stand on each other's shoulders to reach the ceiling. Last, estimate the height of a person. The product of these three estimates is your estimate of the height of the building. It helps to have memorized a few length scales relevant to the sorts of problems you find yourself solving. For example, knowing some of the length scales in **Figure 1.4** might come in handy. Sometimes it also helps to do this in reverse—that is, to estimate the length of a small thing, imagine a bunch of them making up a bigger thing. For example, to estimate the thickness of a sheet of paper, estimate the thickness of a stack of paper and then divide by the number of pages in the stack. These same strategies of breaking big things into smaller things or aggregating smaller things into a bigger thing can sometimes be used to estimate other physical quantities, such as masses and times.

- *Get areas and volumes from lengths.* When dealing with an area or a volume of a complex object, introduce a simple model of the object such as a sphere or a box. Then, estimate the linear dimensions (such as the radius of the sphere or the length, width, and height of the box) first, and use your estimates to obtain the volume or area from standard geometric formulas. If you happen to have an estimate of an object's area or volume, you can also do the reverse; that is, use standard geometric formulas to get an estimate of its linear dimensions.

- *Get masses from volumes and densities.* When estimating masses of objects, it can help first to estimate its volume and then to estimate its mass from a rough estimate of its average density (recall, density has dimension mass over length cubed, so mass is density times volume). For this, it helps to remember that the density of air is around 1 kg/m^3, the density of water is 10^3 kg/m^3, and the densest everyday solids max out at around 10^4 kg/m^3. Asking yourself whether an object floats or sinks in either air or water gets you a ballpark estimate of its density. You can also do this the other way around; if you have an estimate of an object's mass and its density, you can use them to get an estimate of its volume.

- *If all else fails, bound it.* For physical quantities for which you do not have a lot of intuition, sometimes the best you can do is think something like: Well, it must be bigger than this and smaller than that. For example, suppose you need to estimate the mass of a moose. Maybe you have a lot of experience with moose and know their average mass offhand. If so, great. But for most people, the best they can do is to think something like: It must be bigger than a person (of order 10^2 kg) and less than a car (of order 10^3 kg). If you need a single number for a subsequent calculation, you can take the geometric mean of the upper and lower bound—that is, you multiply them together and then take the square root. For the moose mass example, this would be

$$\left(10^2 \times 10^3\right)^{0.5} = 10^{2.5} = 10^{0.5} \times 10^2 \approx 3 \times 10^2 \text{ kg}.$$

The tighter the bounds, the better. Also, no rules are unbreakable when it comes to estimation. If you think the value of the quantity is likely to be closer to the upper bound than the lower bound, then you may want to bump up your estimate from the geometric mean by an order or two of magnitude.

- *One "sig. fig." is fine.* There is no need to go beyond one significant figure when doing calculations to obtain an estimate. In most cases, the order of magnitude is good enough. The goal is just to get in the ballpark figure, so keep the arithmetic as simple as possible.

- *Ask yourself: Does this make any sense?* Last, check to see whether your answer is reasonable. How does it compare with the values of other quantities with the same dimensions that you already know or can look up easily? If you get some wacky answer (for example, if you estimate the mass of the Atlantic Ocean to be bigger than the mass of Earth, or some time span to be longer than the age of the universe), first check to see whether your units are correct. Then, check for arithmetic errors. Then, rethink the logic you used to arrive at your answer. If everything checks out, you may have just proved that some slick new idea is actually bogus.

Example 1.6

Mass of Earth's Oceans

Estimate the total mass of the oceans on Earth.

Strategy

We know the density of water is about 10^3 kg/m^3, so we start with the advice to "get masses from densities and volumes." Thus, we need to estimate the volume of the planet's oceans. Using the advice to "get areas and volumes from lengths," we can estimate the volume of the oceans as surface area times average depth, or $V = AD$. We know the diameter of Earth from **Figure 1.4** and we know that most of Earth's surface is covered in water, so we can estimate the surface area of the oceans as being roughly equal to the surface area of the planet. By following the advice to "get areas and volumes from lengths" again, we can approximate Earth as a sphere and use the formula for the surface area of a sphere of diameter d—that is, $A = \pi d^2$, to estimate the surface area of the oceans. Now we just need to estimate the average depth of the oceans. For this, we use the advice: "If all else fails, bound it." We happen to know the deepest points in the ocean are around 10 km and that it is not uncommon for the ocean to be deeper than 1 km, so we take the average depth to be around $(10^3 \times 10^4)^{0.5} \approx 3 \times 10^3$ m.

Now we just need to put it all together, heeding the advice that "one 'sig. fig.' is fine."

Solution

We estimate the surface area of Earth (and hence the surface area of Earth's oceans) to be roughly

$$A = \pi d^2 = \pi (10^7 \, \text{m})^2 \approx 3 \times 10^{14} \, \text{m}^2.$$

Next, using our average depth estimate of $D = 3 \times 10^3$ m, which was obtained by bounding, we estimate the volume of Earth's oceans to be

$$V = AD = (3 \times 10^{14} \, \text{m}^2)(3 \times 10^3 \, \text{m}) = 9 \times 10^{17} \, \text{m}^3.$$

Last, we estimate the mass of the world's oceans to be

$$M = \rho V = (10^3 \, \text{kg/m}^3)(9 \times 10^{17} \, \text{m}^3) = 9 \times 10^{20} \, \text{kg}.$$

Thus, we estimate that the order of magnitude of the mass of the planet's oceans is 10^{21} kg.

Significance

To verify our answer to the best of our ability, we first need to answer the question: Does this make any sense? From **Figure 1.4**, we see the mass of Earth's atmosphere is on the order of 10^{19} kg and the mass of Earth is on the order of 10^{25} kg. It is reassuring that our estimate of 10^{21} kg for the mass of Earth's oceans falls somewhere between these two. So, yes, it does seem to make sense. It just so happens that we did a search on the Web for "mass of oceans" and the top search results all said 1.4×10^{21} kg, which is the same order of magnitude as our estimate. Now, rather than having to trust blindly whoever first put that number up on a website (most of the other sites probably just copied it from them, after all), we can have a little more confidence in it.

 1.7   Check Your Understanding Figure 1.4 says the mass of the atmosphere is 10^{19} kg. Assuming the density of the atmosphere is 1 kg/m^3, estimate the height of Earth's atmosphere. Do you think your answer is an underestimate or an overestimate? Explain why.

How many piano tuners are there in New York City? How many leaves are on that tree? If you are studying photosynthesis or thinking of writing a smartphone app for piano tuners, then the answers to these questions might be of great interest to you. Otherwise, you probably couldn't care less what the answers are. However, these are exactly the sorts of estimation problems that people in various tech industries have been asking potential employees to evaluate their quantitative reasoning skills. If building physical intuition and evaluating quantitative claims do not seem like sufficient reasons for you to practice estimation problems, how about the fact that being good at them just might land you a high-paying job?

 For practice estimating relative lengths, areas, and volumes, check out this **PhET (https://openstaxcollege.org/l/21lengthgame)** simulation, titled "Estimation."

1.6 | Significant Figures

Learning Objectives
By the end of this section, you will be able to: • Determine the correct number of significant figures for the result of a computation. • Describe the relationship between the concepts of accuracy, precision, uncertainty, and discrepancy. • Calculate the percent uncertainty of a measurement, given its value and its uncertainty. • Determine the uncertainty of the result of a computation involving quantities with given uncertainties.

Figure 1.11 shows two instruments used to measure the mass of an object. The digital scale has mostly replaced the double-pan balance in physics labs because it gives more accurate and precise measurements. But what exactly do we mean by *accurate* and *precise*? Aren't they the same thing? In this section we examine in detail the process of making and reporting a measurement.

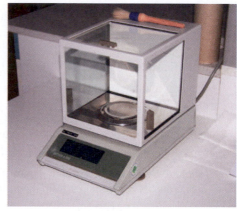

(a)   (b)

Figure 1.11   (a) A double-pan mechanical balance is used to compare different masses. Usually an object with unknown mass is placed in one pan and objects of known mass are placed in the other pan. When the bar that connects the two pans is horizontal, then the masses in both pans are equal. The "known masses" are typically metal cylinders of standard mass such as 1 g, 10 g, and 100 g. (b) Many mechanical balances, such as double-pan balances, have been replaced by digital scales, which can typically measure the mass of an object more precisely. A mechanical balance may read only the mass of an object to the nearest tenth of a gram, but many digital scales can measure the mass of an object up to the nearest thousandth of a gram. (credit a: modification of work by Serge Melki; credit b: modification of work by Karel Jakubec)

Accuracy and Precision of a Measurement

Science is based on observation and experiment—that is, on measurements. **Accuracy** is how close a measurement is to the accepted reference value for that measurement. For example, let's say we want to measure the length of standard printer paper. The packaging in which we purchased the paper states that it is 11.0 in. long. We then measure the length of the paper three times and obtain the following measurements: 11.1 in., 11.2 in., and 10.9 in. These measurements are quite accurate because they are very close to the reference value of 11.0 in. In contrast, if we had obtained a measurement of 12 in., our measurement would not be very accurate. Notice that the concept of accuracy requires that an accepted reference value be given.

The **precision** of measurements refers to how close the agreement is between repeated independent measurements (which are repeated under the same conditions). Consider the example of the paper measurements. The precision of the measurements refers to the spread of the measured values. One way to analyze the precision of the measurements is to determine the range, or difference, between the lowest and the highest measured values. In this case, the lowest value was 10.9 in. and the highest value was 11.2 in. Thus, the measured values deviated from each other by, at most, 0.3 in. These measurements were relatively precise because they did not vary too much in value. However, if the measured values had been 10.9 in., 11.1 in., and 11.9 in., then the measurements would not be very precise because there would be significant variation from one measurement to another. Notice that the concept of precision depends only on the actual measurements acquired and does not depend on an accepted reference value.

The measurements in the paper example are both accurate and precise, but in some cases, measurements are accurate but not precise, or they are precise but not accurate. Let's consider an example of a GPS attempting to locate the position of a restaurant in a city. Think of the restaurant location as existing at the center of a bull's-eye target and think of each GPS attempt to locate the restaurant as a black dot. In **Figure 1.12**(a), we see the GPS measurements are spread out far apart from each other, but they are all relatively close to the actual location of the restaurant at the center of the target. This indicates a low-precision, high-accuracy measuring system. However, in **Figure 1.12**(b), the GPS measurements are concentrated quite closely to one another, but they are far away from the target location. This indicates a high-precision, low-accuracy measuring system.

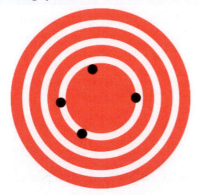 

(a) High accuracy, low precision   (b) Low accuracy, high precision

Figure 1.12   A GPS attempts to locate a restaurant at the center of the bull's-eye. The black dots represent each attempt to pinpoint the location of the restaurant. (a) The dots are spread out quite far apart from one another, indicating low precision, but they are each rather close to the actual location of the restaurant, indicating high accuracy. (b) The dots are concentrated rather closely to one another, indicating high precision, but they are rather far away from the actual location of the restaurant, indicating low accuracy. (credit a and credit b: modification of works by "DarkEvil"/Wikimedia Commons)

Accuracy, Precision, Uncertainty, and Discrepancy

The precision of a measuring system is related to the **uncertainty** in the measurements whereas the accuracy is related to the **discrepancy** from the accepted reference value. Uncertainty is a quantitative measure of how much your measured values deviate from one another. There are many different methods of calculating uncertainty, each of which is appropriate to different situations. Some examples include taking the range (that is, the biggest less the smallest) or finding the standard deviation of the measurements. Discrepancy (or "measurement error") is the difference between the measured value and a given standard or expected value. If the measurements are not very precise, then the uncertainty of the values is high. If the measurements are not very accurate, then the discrepancy of the values is high.

Recall our example of measuring paper length; we obtained measurements of 11.1 in., 11.2 in., and 10.9 in., and the accepted value was 11.0 in. We might average the three measurements to say our best guess is 11.1 in.; in this case, our discrepancy is 11.1 – 11.0 = 0.1 in., which provides a quantitative measure of accuracy. We might calculate the uncertainty in our best guess by using the range of our measured values: 0.3 in. Then we would say the length of the paper is 11.1 in. plus or minus 0.3 in. The uncertainty in a measurement, A, is often denoted as δA (read "delta A"), so the measurement result would be recorded as $A \pm \delta A$. Returning to our paper example, the measured length of the paper could be expressed as 11.1 ± 0.3 in. Since the discrepancy of 0.1 in. is less than the uncertainty of 0.3 in., we might say the measured value agrees with the accepted reference value to within experimental uncertainty.

Some factors that contribute to uncertainty in a measurement include the following:

- Limitations of the measuring device

- The skill of the person taking the measurement

- Irregularities in the object being measured

- Any other factors that affect the outcome (highly dependent on the situation)

In our example, such factors contributing to the uncertainty could be the smallest division on the ruler is 1/16 in., the person using the ruler has bad eyesight, the ruler is worn down on one end, or one side of the paper is slightly longer than the other.

At any rate, the uncertainty in a measurement must be calculated to quantify its precision. If a reference value is known, it makes sense to calculate the discrepancy as well to quantify its accuracy.

Percent uncertainty

Another method of expressing uncertainty is as a percent of the measured value. If a measurement A is expressed with uncertainty δA, the **percent uncertainty** is defined as

$$\text{Percent uncertainty} = \frac{\delta A}{A} \times 100\%.$$

Example 1.7

Calculating Percent Uncertainty: A Bag of Apples

A grocery store sells 5-lb bags of apples. Let's say we purchase four bags during the course of a month and weigh the bags each time. We obtain the following measurements:

- Week 1 weight: 4.8 lb

- Week 2 weight: 5.3 lb

- Week 3 weight: 4.9 lb

- Week 4 weight: 5.4 lb

We then determine the average weight of the 5-lb bag of apples is 5.1 ± 0.2 lb. What is the percent uncertainty of the bag's weight?

Strategy

First, observe that the average value of the bag's weight, A, is 5.1 lb. The uncertainty in this value, δA, is 0.2 lb. We can use the following equation to determine the percent uncertainty of the weight:

$$\text{Percent uncertainty} = \frac{\delta A}{A} \times 100\%. \qquad (1.1)$$

Solution

Substitute the values into the equation:

$$\text{Percent uncertainty} = \frac{\delta A}{A} \times 100\% = \frac{0.2 \text{ lb}}{5.1 \text{ lb}} \times 100\% = 3.9\% \approx 4\%.$$

Significance

We can conclude the average weight of a bag of apples from this store is 5.1 lb ± 4%. Notice the percent uncertainty is dimensionless because the units of weight in $\delta A = 0.2$ lb canceled those in $A = 5.1$ lb when we took the ratio.

 1.8 Check Your Understanding A high school track coach has just purchased a new stopwatch. The stopwatch manual states the stopwatch has an uncertainty of ±0.05 s. Runners on the track coach's team regularly clock 100-m sprints of 11.49 s to 15.01 s. At the school's last track meet, the first-place sprinter came in at 12.04 s and the second-place sprinter came in at 12.07 s. Will the coach's new stopwatch be helpful in timing the sprint team? Why or why not?

Uncertainties in calculations

Uncertainty exists in anything calculated from measured quantities. For example, the area of a floor calculated from measurements of its length and width has an uncertainty because the length and width have uncertainties. How big is the uncertainty in something you calculate by multiplication or division? If the measurements going into the calculation have small uncertainties (a few percent or less), then the **method of adding percents** can be used for multiplication or division. This method states *the percent uncertainty in a quantity calculated by multiplication or division is the sum of the percent* *uncertainties in the items used to make the calculation.* For example, if a floor has a length of 4.00 m and a width of 3.00 m, with uncertainties of 2% and 1%, respectively, then the area of the floor is 12.0 m^2 and has an uncertainty of 3%. (Expressed as an area, this is 0.36 m^2 [12.0 m^2 × 0.03], which we round to 0.4 m^2 since the area of the floor is given to a tenth of a square meter.)

Precision of Measuring Tools and Significant Figures

An important factor in the precision of measurements involves the precision of the measuring tool. In general, a precise measuring tool is one that can measure values in very small increments. For example, a standard ruler can measure length to the nearest millimeter whereas a caliper can measure length to the nearest 0.01 mm. The caliper is a more precise measuring tool because it can measure extremely small differences in length. The more precise the measuring tool, the more precise the measurements.

When we express measured values, we can only list as many digits as we measured initially with our measuring tool. For example, if we use a standard ruler to measure the length of a stick, we may measure it to be 36.7 cm. We can't express this value as 36.71 cm because our measuring tool is not precise enough to measure a hundredth of a centimeter. It should be noted that the last digit in a measured value has been estimated in some way by the person performing the measurement. For example, the person measuring the length of a stick with a ruler notices the stick length seems to be somewhere in between 36.6 cm and 36.7 cm, and he or she must estimate the value of the last digit. Using the method of **significant figures**, the rule is that *the last digit written down in a measurement is the first digit with some uncertainty.* To determine the number of significant digits in a value, start with the first measured value at the left and count the number of digits through the last digit written on the right. For example, the measured value 36.7 cm has three digits, or three significant figures. Significant figures indicate the precision of the measuring tool used to measure a value.

Zeros

Special consideration is given to zeros when counting significant figures. The zeros in 0.053 are not significant because they are placeholders that locate the decimal point. There are two significant figures in 0.053. The zeros in 10.053 are not placeholders; they are significant. This number has five significant figures. The zeros in 1300 may or may not be significant, depending on the style of writing numbers. They could mean the number is known to the last digit or they could be placeholders. So 1300 could have two, three, or four significant figures. To avoid this ambiguity, we should write 1300 in scientific notation as 1.3×10^3, 1.30×10^3, or 1.300×10^3, depending on whether it has two, three, or four significant figures. *Zeros are significant except when they serve only as placeholders.*

Significant figures in calculations

When combining measurements with different degrees of precision, *the number of significant digits in the final answer can be no greater than the number of significant digits in the least-precise measured value.* There are two different rules, one for multiplication and division and the other for addition and subtraction.

1. *For multiplication and division, the result should have the same number of significant figures as the quantity with the least number of significant figures entering into the calculation.* For example, the area of a circle can be calculated from its radius using $A = \pi r^2$. Let's see how many significant figures the area has if the radius has only two—say, $r = 1.2$ m. Using a calculator with an eight-digit output, we would calculate

$$A = \pi r^2 = (3.1415927...) \times (1.2 \text{ m})^2 = 4.5238934 \text{ m}^2.$$

But because the radius has only two significant figures, it limits the calculated quantity to two significant figures, or

$$A = 4.5 \text{ m}^2,$$

although π is good to at least eight digits.

2. *For addition and subtraction, the answer can contain no more decimal places than the least-precise measurement.* Suppose we buy 7.56 kg of potatoes in a grocery store as measured with a scale with precision 0.01 kg, then we drop off 6.052 kg of potatoes at your laboratory as measured by a scale with precision 0.001 kg. Then, we go home and add 13.7 kg of potatoes as measured by a bathroom scale with precision 0.1 kg. How many kilograms of potatoes do we now have and how many significant figures are appropriate in the answer? The mass is found by simple addition and subtraction:

$$7.56 \text{ kg}$$
$$-6.052 \text{ kg}$$
$$\frac{+13.7 \text{ kg}}{15.208 \text{ kg}} = 15.2 \text{ kg}.$$

Next, we identify the least-precise measurement: 13.7 kg. This measurement is expressed to the 0.1 decimal place, so our final answer must also be expressed to the 0.1 decimal place. Thus, the answer is rounded to the tenths place, giving us 15.2 kg.

Significant figures in this text

In this text, most numbers are assumed to have three significant figures. Furthermore, consistent numbers of significant figures are used in all worked examples. An answer given to three digits is based on input good to at least three digits, for example. If the input has fewer significant figures, the answer will also have fewer significant figures. Care is also taken that the number of significant figures is reasonable for the situation posed. In some topics, particularly in optics, more accurate numbers are needed and we use more than three significant figures. Finally, if a number is *exact*, such as the two in the formula for the circumference of a circle, $C = 2\pi r$, it does not affect the number of significant figures in a calculation. Likewise, conversion factors such as 100 cm/1 m are considered exact and do not affect the number of significant figures in a calculation.

1.7 | Solving Problems in Physics

Learning Objectives

By the end of this section, you will be able to:

- Describe the process for developing a problem-solving strategy.
- Explain how to find the numerical solution to a problem.
- Summarize the process for assessing the significance of the numerical solution to a problem.

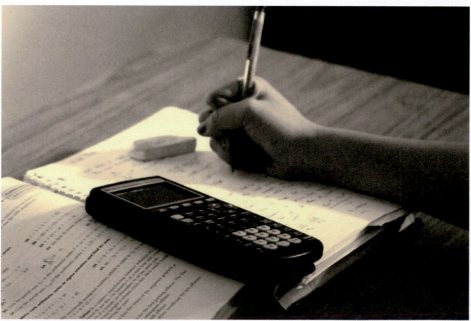

Figure 1.13 Problem-solving skills are essential to your success in physics. (credit: "scui3asteveo"/Flickr)

Problem-solving skills are clearly essential to success in a quantitative course in physics. More important, the ability to apply broad physical principles—usually represented by equations—to specific situations is a very powerful form of knowledge. It is much more powerful than memorizing a list of facts. Analytical skills and problem-solving abilities can be applied to new situations whereas a list of facts cannot be made long enough to contain every possible circumstance. Such analytical skills are useful both for solving problems in this text and for applying physics in everyday life.

As you are probably well aware, a certain amount of creativity and insight is required to solve problems. No rigid procedure works every time. Creativity and insight grow with experience. With practice, the basics of problem solving become almost automatic. One way to get practice is to work out the text's examples for yourself as you read. Another is to work as many end-of-section problems as possible, starting with the easiest to build confidence and then progressing to the more difficult. After you become involved in physics, you will see it all around you, and you can begin to apply it to situations you encounter outside the classroom, just as is done in many of the applications in this text.

Although there is no simple step-by-step method that works for every problem, the following three-stage process facilitates problem solving and makes it more meaningful. The three stages are strategy, solution, and significance. This process is used in examples throughout the book. Here, we look at each stage of the process in turn.

Strategy

Strategy is the beginning stage of solving a problem. The idea is to figure out exactly what the problem is and then develop a strategy for solving it. Some general advice for this stage is as follows:

- *Examine the situation to determine which physical principles are involved*. It often helps to *draw a simple sketch* at the outset. You often need to decide which direction is positive and note that on your sketch. When you have identified the physical principles, it is much easier to find and apply the equations representing those principles. Although finding the correct equation is essential, keep in mind that equations represent physical principles, laws of nature, and relationships among physical quantities. Without a conceptual understanding of a problem, a numerical solution is meaningless.

- *Make a list of what is given or can be inferred from the problem as stated (identify the "knowns")*. Many problems are stated very succinctly and require some inspection to determine what is known. Drawing a sketch can be very useful at this point as well. Formally identifying the knowns is of particular importance in applying physics to real-world situations. For example, the word *stopped* means the velocity is zero at that instant. Also, we can often take initial time and position as zero by the appropriate choice of coordinate system.

- *Identify exactly what needs to be determined in the problem (identify the unknowns)*. In complex problems, especially, it is not always obvious what needs to be found or in what sequence. Making a list can help identify the unknowns.

- *Determine which physical principles can help you solve the problem*. Since physical principles tend to be expressed in the form of mathematical equations, a list of knowns and unknowns can help here. It is easiest if you can find equations that contain only one unknown—that is, all the other variables are known—so you can solve for the unknown easily. If the equation contains more than one unknown, then additional equations are needed to solve the problem. In some problems, several unknowns must be determined to get at the one needed most. In such problems it is especially important to keep physical principles in mind to avoid going astray in a sea of equations. You may have to use two (or more) different equations to get the final answer.

Solution

The solution stage is when you do the math. *Substitute the knowns (along with their units) into the appropriate equation and obtain numerical solutions complete with units*. That is, do the algebra, calculus, geometry, or arithmetic necessary to find the unknown from the knowns, being sure to carry the units through the calculations. This step is clearly important because it produces the numerical answer, along with its units. Notice, however, that this stage is only one-third of the overall problem-solving process.

Significance

After having done the math in the solution stage of problem solving, it is tempting to think you are done. But, always remember that physics is not math. Rather, in doing physics, we use mathematics as a tool to help us understand nature. So, after you obtain a numerical answer, you should always assess its significance:

- *Check your units*. If the units of the answer are incorrect, then an error has been made and you should go back over your previous steps to find it. One way to find the mistake is to check all the equations you derived for dimensional consistency. However, be warned that correct units do not guarantee the numerical part of the answer is also correct.

- *Check the answer to see whether it is reasonable. Does it make sense?* This step is extremely important: –the goal of physics is to describe nature accurately. To determine whether the answer is reasonable, check both its magnitude and its sign, in addition to its units. The magnitude should be consistent with a rough estimate of what it should be. It should also compare reasonably with magnitudes of other quantities of the same type. The sign usually tells you about direction and should be consistent with your prior expectations. Your judgment will improve as you solve more physics problems, and it will become possible for you to make finer judgments regarding whether nature is described adequately by the answer to a problem. This step brings the problem back to its conceptual meaning. If you can judge whether the answer is reasonable, you have a deeper understanding of physics than just being able to solve a problem mechanically.

- *Check to see whether the answer tells you something interesting. What does it mean?* This is the flip side of the question: Does it make sense? Ultimately, physics is about understanding nature, and we solve physics problems to learn a little something about how nature operates. Therefore, assuming the answer does make sense, you should always take a moment to see if it tells you something about the world that you find interesting. Even if the answer to this particular problem is not very interesting to you, what about the method you used to solve it? Could the method be adapted to answer a question that you do find interesting? In many ways, it is in answering questions such as these that science progresses.

CHAPTER 1 REVIEW

KEY TERMS

accuracy the degree to which a measured value agrees with an accepted reference value for that measurement

base quantity physical quantity chosen by convention and practical considerations such that all other physical quantities can be expressed as algebraic combinations of them

base unit standard for expressing the measurement of a base quantity within a particular system of units; defined by a particular procedure used to measure the corresponding base quantity

conversion factor a ratio that expresses how many of one unit are equal to another unit

derived quantity physical quantity defined using algebraic combinations of base quantities

derived units units that can be calculated using algebraic combinations of the fundamental units

dimension expression of the dependence of a physical quantity on the base quantities as a product of powers of symbols representing the base quantities; in general, the dimension of a quantity has the form $L^a M^b T^c I^d \Theta^e N^f J^g$ for some powers a, b, c, d, e, f, and g.

dimensionally consistent equation in which every term has the same dimensions and the arguments of any mathematical functions appearing in the equation are dimensionless

dimensionless quantity with a dimension of $L^0 M^0 T^0 I^0 \Theta^0 N^0 J^0 = 1$; also called quantity of dimension 1 or a pure number

discrepancy the difference between the measured value and a given standard or expected value

English units system of measurement used in the United States; includes units of measure such as feet, gallons, and pounds

estimation using prior experience and sound physical reasoning to arrive at a rough idea of a quantity's value; sometimes called an "order-of-magnitude approximation," a "guesstimate," a "back-of-the-envelope calculation", or a "Fermi calculation"

kilogram SI unit for mass, abbreviated kg

law description, using concise language or a mathematical formula, of a generalized pattern in nature supported by scientific evidence and repeated experiments

meter SI unit for length, abbreviated m

method of adding percents the percent uncertainty in a quantity calculated by multiplication or division is the sum of the percent uncertainties in the items used to make the calculation.

metric system system in which values can be calculated in factors of 10

model representation of something often too difficult (or impossible) to display directly

order of magnitude the size of a quantity as it relates to a power of 10

percent uncertainty the ratio of the uncertainty of a measurement to the measured value, expressed as a percentage

physical quantity characteristic or property of an object that can be measured or calculated from other measurements

physics science concerned with describing the interactions of energy, matter, space, and time; especially interested in what fundamental mechanisms underlie every phenomenon

precision the degree to which repeated measurements agree with each other

second the SI unit for time, abbreviated s

SI units the international system of units that scientists in most countries have agreed to use; includes units such as meters, liters, and grams

significant figures used to express the precision of a measuring tool used to measure a value

theory testable explanation for patterns in nature supported by scientific evidence and verified multiple times by various groups of researchers

uncertainty a quantitative measure of how much measured values deviate from one another

units standards used for expressing and comparing measurements

KEY EQUATIONS

Percent uncertainty $\text{Percent uncertainty} = \frac{\delta A}{A} \times 100\%$

SUMMARY

1.1 The Scope and Scale of Physics

- Physics is about trying to find the simple laws that describe all natural phenomena.

- Physics operates on a vast range of scales of length, mass, and time. Scientists use the concept of the order of magnitude of a number to track which phenomena occur on which scales. They also use orders of magnitude to compare the various scales.

- Scientists attempt to describe the world by formulating models, theories, and laws.

1.2 Units and Standards

- Systems of units are built up from a small number of base units, which are defined by accurate and precise measurements of conventionally chosen base quantities. Other units are then derived as algebraic combinations of the base units.

- Two commonly used systems of units are English units and SI units. All scientists and most of the other people in the world use SI, whereas nonscientists in the United States still tend to use English units.

- The SI base units of length, mass, and time are the meter (m), kilogram (kg), and second (s), respectively.

- SI units are a metric system of units, meaning values can be calculated by factors of 10. Metric prefixes may be used with metric units to scale the base units to sizes appropriate for almost any application.

1.3 Unit Conversion

- To convert a quantity from one unit to another, multiply by conversions factors in such a way that you cancel the units you want to get rid of and introduce the units you want to end up with.

- Be careful with areas and volumes. Units obey the rules of algebra so, for example, if a unit is squared we need two factors to cancel it.

1.4 Dimensional Analysis

- The dimension of a physical quantity is just an expression of the base quantities from which it is derived.

- All equations expressing physical laws or principles must be dimensionally consistent. This fact can be used as an aid in remembering physical laws, as a way to check whether claimed relationships between physical quantities are possible, and even to derive new physical laws.

1.5 Estimates and Fermi Calculations

- An estimate is a rough educated guess at the value of a physical quantity based on prior experience and sound physical reasoning. Some strategies that may help when making an estimate are as follows:
  - Get big lengths from smaller lengths.
  - Get areas and volumes from lengths.
  - Get masses from volumes and densities.
  - If all else fails, bound it.

- One "sig. fig." is fine.
- Ask yourself: Does this make any sense?

1.6 Significant Figures

- Accuracy of a measured value refers to how close a measurement is to an accepted reference value. The discrepancy in a measurement is the amount by which the measurement result differs from this value.
- Precision of measured values refers to how close the agreement is between repeated measurements. The uncertainty of a measurement is a quantification of this.
- The precision of a measuring tool is related to the size of its measurement increments. The smaller the measurement increment, the more precise the tool.
- Significant figures express the precision of a measuring tool.
- When multiplying or dividing measured values, the final answer can contain only as many significant figures as the value with the least number of significant figures.
- When adding or subtracting measured values, the final answer cannot contain more decimal places than the least-precise value.

1.7 Solving Problems in Physics

The three stages of the process for solving physics problems used in this book are as follows:

- *Strategy*: Determine which physical principles are involved and develop a strategy for using them to solve the problem.
- *Solution*: Do the math necessary to obtain a numerical solution complete with units.
- *Significance*: Check the solution to make sure it makes sense (correct units, reasonable magnitude and sign) and assess its significance.

CONCEPTUAL QUESTIONS

1.1 The Scope and Scale of Physics

1. What is physics?

2. Some have described physics as a "search for simplicity." Explain why this might be an appropriate description.

3. If two different theories describe experimental observations equally well, can one be said to be more valid than the other (assuming both use accepted rules of logic)?

4. What determines the validity of a theory?

5. Certain criteria must be satisfied if a measurement or observation is to be believed. Will the criteria necessarily be as strict for an expected result as for an unexpected result?

6. Can the validity of a model be limited or must it be universally valid? How does this compare with the required validity of a theory or a law?

1.2 Units and Standards

7. Identify some advantages of metric units.

8. What are the SI base units of length, mass, and time?

9. What is the difference between a base unit and a derived unit? (b) What is the difference between a base quantity and a derived quantity? (c) What is the difference between a base quantity and a base unit?

10. For each of the following scenarios, refer to **Figure 1.4** and **Table 1.2** to determine which metric prefix on the meter is most appropriate for each of the following scenarios. (a) You want to tabulate the mean distance from the Sun for each planet in the solar system. (b) You want to compare the sizes of some common viruses to design a mechanical filter capable of blocking the pathogenic ones. (c) You want to list the diameters of all the elements on the periodic table. (d) You want to list the distances to all the stars that have now received any radio broadcasts sent from Earth 10 years ago.

1.6 Significant Figures

11. (a) What is the relationship between the precision and the uncertainty of a measurement? (b) What is the relationship between the accuracy and the discrepancy of a measurement?

1.7 Solving Problems in Physics

12. What information do you need to choose which equation or equations to use to solve a problem?

13. What should you do after obtaining a numerical answer when solving a problem?

PROBLEMS

1.1 The Scope and Scale of Physics

14. Find the order of magnitude of the following physical quantities. (a) The mass of Earth's atmosphere: 5.1×10^{18} kg; (b) The mass of the Moon's atmosphere: 25,000 kg; (c) The mass of Earth's hydrosphere: 1.4×10^{21} kg; (d) The mass of Earth: 5.97×10^{24} kg; (e) The mass of the Moon: 7.34×10^{22} kg; (f) The Earth–Moon distance (semimajor axis): 3.84×10^8 m; (g) The mean Earth–Sun distance: 1.5×10^{11} m; (h) The equatorial radius of Earth: 6.38×10^6 m; (i) The mass of an electron: 9.11×10^{-31} kg; (j) The mass of a proton: 1.67×10^{-27} kg; (k) The mass of the Sun: 1.99×10^{30} kg.

15. Use the orders of magnitude you found in the previous problem to answer the following questions to within an order of magnitude. (a) How many electrons would it take to equal the mass of a proton? (b) How many Earths would it take to equal the mass of the Sun? (c) How many Earth–Moon distances would it take to cover the distance from Earth to the Sun? (d) How many Moon atmospheres would it take to equal the mass of Earth's atmosphere? (e) How many moons would it take to equal the mass of Earth? (f) How many protons would it take to equal the mass of the Sun?

For the remaining questions, you need to use **Figure 1.4** to obtain the necessary orders of magnitude of lengths, masses, and times.

16. Roughly how many heartbeats are there in a lifetime?

17. A generation is about one-third of a lifetime. Approximately how many generations have passed since the year 0 AD?

18. Roughly how many times longer than the mean life of an extremely unstable atomic nucleus is the lifetime of a human?

19. Calculate the approximate number of atoms in a bacterium. Assume the average mass of an atom in the bacterium is 10 times the mass of a proton.

20. (a) Calculate the number of cells in a hummingbird assuming the mass of an average cell is 10 times the mass of a bacterium. (b) Making the same assumption, how many cells are there in a human?

21. Assuming one nerve impulse must end before another can begin, what is the maximum firing rate of a nerve in impulses per second?

22. About how many floating-point operations can a supercomputer perform each year?

23. Roughly how many floating-point operations can a supercomputer perform in a human lifetime?

1.2 Units and Standards

24. The following times are given using metric prefixes on the base SI unit of time: the second. Rewrite them in scientific notation without the prefix. For example, 47 Ts would be rewritten as 4.7×10^{13} s. (a) 980 Ps; (b) 980 fs; (c) 17 ns; (d) 577 μs.

25. The following times are given in seconds. Use metric prefixes to rewrite them so the numerical value is greater than one but less than 1000. For example, 7.9×10^{-2} s could be written as either 7.9 cs or 79 ms. (a) 9.57×10^5 s; (b) 0.045 s; (c) 5.5×10^{-7} s; (d) 3.16×10^7 s.

26. The following lengths are given using metric prefixes on the base SI unit of length: the meter. Rewrite them in scientific notation without the prefix. For example, 4.2 Pm would be rewritten as 4.2×10^{15} m. (a) 89 Tm; (b) 89 pm; (c) 711 mm; (d) 0.45 μm.

27. The following lengths are given in meters. Use metric prefixes to rewrite them so the numerical value is bigger than one but less than 1000. For example, 7.9×10^{-2} m could be written either as 7.9 cm or 79 mm. (a) 7.59×10^7 m; (b) 0.0074 m; (c) 8.8×10^{-11} m; (d) 1.63×10^{13} m.

28. The following masses are written using metric prefixes on the gram. Rewrite them in scientific notation in terms of the SI base unit of mass: the kilogram. For example, 40 Mg would be written as 4×10^4 kg. (a) 23 mg; (b) 320 Tg; (c) 42 ng; (d) 7 g; (e) 9 Pg.

29. The following masses are given in kilograms. Use metric prefixes on the gram to rewrite them so the numerical value is bigger than one but less than 1000. For example, 7×10^{-4} kg could be written as 70 cg or 700 mg. (a) 3.8×10^{-5} kg; (b) 2.3×10^{17} kg; (c) 2.4×10^{-11} kg; (d) 8×10^{15} kg; (e) 4.2×10^{-3} kg.

1.3 Unit Conversion

30. The volume of Earth is on the order of 10^{21} m³. (a) What is this in cubic kilometers (km³)? (b) What is it in cubic miles (mi³)? (c) What is it in cubic centimeters (cm³)?

31. The speed limit on some interstate highways is roughly 100 km/h. (a) What is this in meters per second? (b) How many miles per hour is this?

32. A car is traveling at a speed of 33 m/s. (a) What is its speed in kilometers per hour? (b) Is it exceeding the 90 km/h speed limit?

33. In SI units, speeds are measured in meters per second (m/s). But, depending on where you live, you're probably more comfortable of thinking of speeds in terms of either kilometers per hour (km/h) or miles per hour (mi/h). In this problem, you will see that 1 m/s is roughly 4 km/h or 2 mi/h, which is handy to use when developing your physical intuition. More precisely, show that (a) 1.0 m/s = 3.6 km/h and (b) 1.0 m/s = 2.2 mi/h.

34. American football is played on a 100-yd-long field, excluding the end zones. How long is the field in meters? (Assume that 1 m = 3.281 ft.)

35. Soccer fields vary in size. A large soccer field is 115 m long and 85.0 m wide. What is its area in square feet? (Assume that 1 m = 3.281 ft.)

36. What is the height in meters of a person who is 6 ft 1.0 in. tall?

37. Mount Everest, at 29,028 ft, is the tallest mountain on Earth. What is its height in kilometers? (Assume that 1 m = 3.281 ft.)

38. The speed of sound is measured to be 342 m/s on a certain day. What is this measurement in kilometers per hour?

39. Tectonic plates are large segments of Earth's crust that move slowly. Suppose one such plate has an average speed of 4.0 cm/yr. (a) What distance does it move in 1.0 s at this speed? (b) What is its speed in kilometers per million years?

40. The average distance between Earth and the Sun is 1.5×10^{11} m. (a) Calculate the average speed of Earth in its orbit (assumed to be circular) in meters per second. (b) What is this speed in miles per hour?

41. The density of nuclear matter is about 10^{18} kg/m³. Given that 1 mL is equal in volume to cm³, what is the density of nuclear matter in megagrams per microliter (that is, Mg/μL)?

42. The density of aluminum is 2.7 g/cm³. What is the density in kilograms per cubic meter?

43. A commonly used unit of mass in the English system is the pound-mass, abbreviated lbm, where 1 lbm = 0.454 kg. What is the density of water in pound-mass per cubic foot?

44. A furlong is 220 yd. A fortnight is 2 weeks. Convert a speed of one furlong per fortnight to millimeters per second.

45. It takes 2π radians (rad) to get around a circle, which is the same as 360°. How many radians are in 1°?

46. Light travels a distance of about 3×10^8 m/s. A light-minute is the distance light travels in 1 min. If the Sun is 1.5×10^{11} m from Earth, how far away is it in light-minutes?

47. A light-nanosecond is the distance light travels in 1 ns. Convert 1 ft to light-nanoseconds.

48. An electron has a mass of 9.11×10^{-31} kg. A proton has a mass of 1.67×10^{-27} kg. What is the mass of a proton in electron-masses?

49. A fluid ounce is about 30 mL. What is the volume of a 12 fl-oz can of soda pop in cubic meters?

1.4 Dimensional Analysis

50. A student is trying to remember some formulas from geometry. In what follows, assume A is area, V is volume, and all other variables are lengths. Determine which formulas are dimensionally consistent. (a) $V = \pi r^2 h$; (b) $A = 2\pi r^2 + 2\pi rh$; (c) $V = 0.5bh$; (d) $V = \pi d^2$; (e) $V = \pi d^3/6$.

51. Consider the physical quantities s, v, a, and t with dimensions $[s] = L$, $[v] = LT^{-1}$, $[a] = LT^{-2}$, and $[t] = T$. Determine whether each of the following equations is dimensionally consistent. (a) $v^2 = 2as$; (b) $s = vt^2 + 0.5at^2$; (c) $v = s/t$; (d) $a = v/t$.

52. Consider the physical quantities m, s, v, a, and t with dimensions $[m] = M$, $[s] = L$, $[v] = LT^{-1}$, $[a] = LT^{-2}$, and $[t] = T$. Assuming each of the following equations is dimensionally consistent, find the dimension of the quantity on the left-hand side of the equation: (a) $F = ma$; (b) $K = 0.5mv^2$; (c) $p = mv$; (d) $W = mas$; (e) $L = mvr$.

53. Suppose quantity s is a length and quantity t is a time. Suppose the quantities v and a are defined by $v = ds/dt$ and $a = dv/dt$. (a) What is the dimension of v? (b) What is the dimension of the quantity a? What are the dimensions of (c) $\int v\, dt$, (d) $\int a\, dt$, and (e) da/dt?

54. Suppose $[V] = L^3$, $[\rho] = ML^{-3}$, and $[t] = T$. (a) What is the dimension of $\int \rho\, dV$? (b) What is the dimension of dV/dt? (c) What is the dimension of $\rho(dV/dt)$?

55. The arc length formula says the length s of arc subtended by angle $\cup$ in a circle of radius r is given by the equation $s = r\Theta$. What are the dimensions of (a) s, (b) r, and (c) Θ?

1.5 Estimates and Fermi Calculations

56. Assuming the human body is made primarily of water, estimate the volume of a person.

57. Assuming the human body is primarily made of water, estimate the number of molecules in it. (Note that water has a molecular mass of 18 g/mol and there are roughly 10^{24} atoms in a mole.)

58. Estimate the mass of air in a classroom.

59. Estimate the number of molecules that make up Earth, assuming an average molecular mass of 30 g/mol. (Note there are on the order of 10^{24} objects per mole.)

60. Estimate the surface area of a person.

61. Roughly how many solar systems would it take to tile the disk of the Milky Way?

62. (a) Estimate the density of the Moon. (b) Estimate the diameter of the Moon. (c) Given that the Moon subtends at an angle of about half a degree in the sky, estimate its distance from Earth.

63. The average density of the Sun is on the order 10^3 kg/m^3. (a) Estimate the diameter of the Sun. (b) Given that the Sun subtends at an angle of about half a degree in the sky, estimate its distance from Earth.

64. Estimate the mass of a virus.

65. A floating-point operation is a single arithmetic operation such as addition, subtraction, multiplication, or division. (a) Estimate the maximum number of floating-point operations a human being could possibly perform in a lifetime. (b) How long would it take a supercomputer to perform that many floating-point operations?

1.6 Significant Figures

66. Consider the equation 4000/400 = 10.0. Assuming the number of significant figures in the answer is correct, what can you say about the number of significant figures in 4000 and 400?

67. Suppose your bathroom scale reads your mass as 65 kg with a 3% uncertainty. What is the uncertainty in your mass (in kilograms)?

68. A good-quality measuring tape can be off by 0.50 cm over a distance of 20 m. What is its percent uncertainty?

69. An infant's pulse rate is measured to be 130 ± 5 beats/min. What is the percent uncertainty in this measurement?

70. (a) Suppose that a person has an average heart rate of 72.0 beats/min. How many beats does he or she have in 2.0 years? (b) In 2.00 years? (c) In 2.000 years?

71. A can contains 375 mL of soda. How much is left after 308 mL is removed?

72. State how many significant figures are proper in the results of the following calculations: (a) $(106.7)(98.2)/(46.210)(1.01)$; (b) $(18.7)^2$; (c)

$$\left(1.60 \times 10^{-19}\right)(3712)$$

73. (a) How many significant figures are in the numbers 99 and 100.? (b) If the uncertainty in each number is 1, what is the percent uncertainty in each? (c) Which is a more meaningful way to express the accuracy of these two numbers: significant figures or percent uncertainties?

74. (a) If your speedometer has an uncertainty of 2.0 km/h at a speed of 90 km/h, what is the percent uncertainty? (b) If it has the same percent uncertainty when it reads 60 km/h, what is the range of speeds you could be going?

75. (a) A person's blood pressure is measured to be 120 ± 2 mm Hg. What is its percent uncertainty? (b) Assuming the same percent uncertainty, what is the uncertainty in a blood pressure measurement of 80 mm Hg?

76. A person measures his or her heart rate by counting the number of beats in 30 s. If 40 ± 1 beats are counted in 30.0 $\pm$ 0.5 s, what is the heart rate and its uncertainty in beats per minute?

77. What is the area of a circle 3.102 cm in diameter?

78. Determine the number of significant figures in the following measurements: (a) 0.0009, (b) 15,450.0, (c) 6×10^3, (d) 87.990, and (e) 30.42.

79. Perform the following calculations and express your answer using the correct number of significant digits. (a) A woman has two bags weighing 13.5 lb and one bag with a weight of 10.2 lb. What is the total weight of the bags? (b) The force F on an object is equal to its mass m multiplied by its acceleration a. If a wagon with mass 55 kg accelerates at a rate of 0.0255 m/s^2, what is the force on the wagon? (The unit of force is called the *newton* and it is expressed with the symbol N.)

ADDITIONAL PROBLEMS

80. Consider the equation $y = mt + b$, where the dimension of y is length and the dimension of t is time, and m and b are constants. What are the dimensions and SI units of (a) m and (b) b?

81. Consider the equation $s = s_0 + v_0 t + a_0 t^2/2 + j_0 t^3/6 + S_0 t^4/24 + ct^5/120$, where s is a length and t is a time. What are the dimensions and SI units of (a) s_0, (b) v_0, (c) a_0, (d) j_0, (e) S_0, and (f) c?

82. (a) A car speedometer has a 5% uncertainty. What is the range of possible speeds when it reads 90 km/h? (b) Convert this range to miles per hour. Note 1 km = 0.6214 mi.

83. A marathon runner completes a 42.188-km course in 2 h, 30 min, and 12 s. There is an uncertainty of 25 m in the distance traveled and an uncertainty of 1 s in the elapsed time. (a) Calculate the percent uncertainty in the distance. (b) Calculate the percent uncertainty in the elapsed time. (c) What is the average speed in meters per second? (d) What is the uncertainty in the average speed?

84. The sides of a small rectangular box are measured to be 1.80 ± 0.1 cm, 2.05 ± 0.02 cm, and 3.1 ± 0.1 cm long. Calculate its volume and uncertainty in cubic centimeters.

85. When nonmetric units were used in the United Kingdom, a unit of mass called the pound-mass (lbm) was used, where 1 lbm = 0.4539 kg. (a) If there is an uncertainty of 0.0001 kg in the pound-mass unit, what is its percent uncertainty? (b) Based on that percent uncertainty, what mass in pound-mass has an uncertainty of 1 kg when converted to kilograms?

86. The length and width of a rectangular room are measured to be 3.955 ± 0.005 m and 3.050 ± 0.005 m. Calculate the area of the room and its uncertainty in square meters.

87. A car engine moves a piston with a circular cross-section of 7.500 ± 0.002 cm in diameter a distance of 3.250 ± 0.001 cm to compress the gas in the cylinder. (a) By what amount is the gas decreased in volume in cubic centimeters? (b) Find the uncertainty in this volume.

CHALLENGE PROBLEMS

88. The first atomic bomb was detonated on July 16, 1945, at the Trinity test site about 200 mi south of Los Alamos. In 1947, the U.S. government declassified a film reel of the explosion. From this film reel, British physicist G. I. Taylor was able to determine the rate at which the radius of the fireball from the blast grew. Using dimensional analysis, he was then able to deduce the amount of energy released in the explosion, which was a closely guarded secret at the time. Because of this, Taylor did not publish his results until 1950. This problem challenges you to recreate this famous calculation. (a) Using keen physical insight developed from years of experience, Taylor decided the radius r of the fireball should depend only on time since the explosion, t, the density of the air, ρ, and the energy of the initial explosion, E. Thus, he made the educated guess that $r = kE^a \rho^b t^c$ for some dimensionless constant k and some unknown exponents a, b, and c. Given that $[E] = ML^2T^{-2}$, determine the values of the exponents necessary to make this equation dimensionally consistent. (*Hint*: Notice the equation implies that $k = rE^{-a}\rho^{-b}t^{-c}$ and that $[k] = 1$.

) (b) By analyzing data from high-energy conventional explosives, Taylor found the formula he derived seemed to be valid as long as the constant k had the value 1.03. From the film reel, he was able to determine many values of r and the corresponding values of t. For example, he found that after 25.0 ms, the fireball had a radius of 130.0 m. Use these values, along with an average air density of 1.25 kg/m^3, to calculate the initial energy release of the Trinity detonation in joules (J). (*Hint*: To get energy in joules, you need to make sure all the numbers you substitute in are expressed in terms of SI base units.) (c) The energy released in large explosions is often cited in units of "tons of TNT" (abbreviated "t TNT"), where 1 t TNT is about 4.2 GJ. Convert your answer to (b) into kilotons of TNT (that is, kt TNT). Compare your answer with the quick-and-dirty estimate of 10 kt TNT made by physicist Enrico Fermi shortly after witnessing the explosion from what was thought to be a safe distance. (Reportedly, Fermi made his estimate by dropping some shredded bits of paper right before the remnants of the shock wave hit him and looked to see how far they were carried by it.)

89. The purpose of this problem is to show the entire concept of dimensional consistency can be summarized by the old saying "You can't add apples and oranges." If you have studied power series expansions in a calculus course, you know the standard mathematical functions such as trigonometric functions, logarithms, and exponential functions can be expressed as infinite sums of the form

$$\sum_{n=0}^{\infty} a_n x^n = a_0 + a_1 x + a_2 x^2 + a_3 x^3 + \cdots, \qquad \text{where}$$

the a_n are dimensionless constants for all $n = 0, 1, 2, \cdots$ and x is the argument of the function. (If you have not studied power series in calculus yet, just trust us.) Use this fact to explain why the requirement that all terms in an equation have the same dimensions is sufficient as a definition of dimensional consistency. That is, it actually implies the arguments of standard mathematical functions must be dimensionless, so it is not really necessary to make this latter condition a separate requirement of the definition of dimensional consistency as we have done in this section.

2 | VECTORS

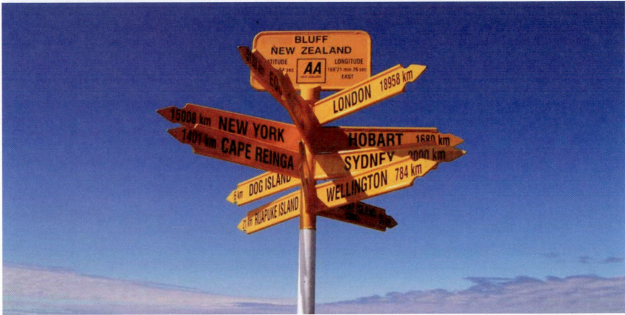

Figure 2.1 A signpost gives information about distances and directions to towns or to other locations relative to the location of the signpost. Distance is a scalar quantity. Knowing the distance alone is not enough to get to the town; we must also know the direction from the signpost to the town. The direction, together with the distance, is a vector quantity commonly called the displacement vector. A signpost, therefore, gives information about displacement vectors from the signpost to towns. (credit: modification of work by "studio tdes"/Flickr, thedailyenglishshow.com)

Chapter Outline

Introduction

Vectors are essential to physics and engineering. Many fundamental physical quantities are vectors, including displacement, velocity, force, and electric and magnetic vector fields. Scalar products of vectors define other fundamental scalar physical quantities, such as energy. Vector products of vectors define still other fundamental vector physical quantities, such as torque and angular momentum. In other words, vectors are a component part of physics in much the same way as sentences are a component part of literature.

In introductory physics, vectors are Euclidean quantities that have geometric representations as arrows in one dimension (in a line), in two dimensions (in a plane), or in three dimensions (in space). They can be added, subtracted, or multiplied. In this chapter, we explore elements of vector algebra for applications in mechanics and in electricity and magnetism. Vector operations also have numerous generalizations in other branches of physics.

2.1 | Scalars and Vectors

Many familiar physical quantities can be specified completely by giving a single number and the appropriate unit. For example, "a class period lasts 50 min" or "the gas tank in my car holds 65 L" or "the distance between two posts is 100 m." A physical quantity that can be specified completely in this manner is called a **scalar quantity**. Scalar is a synonym of "number." Time, mass, distance, length, volume, temperature, and energy are examples of **scalar** quantities.

Scalar quantities that have the same physical units can be added or subtracted according to the usual rules of algebra for numbers. For example, a class ending 10 min earlier than 50 min lasts $50 \text{ min} - 10 \text{ min} = 40 \text{ min}$. Similarly, a 60-cal serving of corn followed by a 200-cal serving of donuts gives $60 \text{ cal} + 200 \text{ cal} = 260 \text{ cal}$ of energy. When we multiply a scalar quantity by a number, we obtain the same scalar quantity but with a larger (or smaller) value. For example, if yesterday's breakfast had 200 cal of energy and today's breakfast has four times as much energy as it had yesterday, then today's breakfast has $4(200 \text{ cal}) = 800 \text{ cal}$ of energy. Two scalar quantities can also be multiplied or divided by each other to form a derived scalar quantity. For example, if a train covers a distance of 100 km in 1.0 h, its speed is 100.0 km/1.0 h = 27.8 m/s, where the speed is a derived scalar quantity obtained by dividing distance by time.

Many physical quantities, however, cannot be described completely by just a single number of physical units. For example, when the U.S. Coast Guard dispatches a ship or a helicopter for a rescue mission, the rescue team must know not only the distance to the distress signal, but also the direction from which the signal is coming so they can get to its origin as quickly as possible. Physical quantities specified completely by giving a number of units (magnitude) and a direction are called **vector quantities**. Examples of vector quantities include displacement, velocity, position, force, and torque. In the language of mathematics, physical vector quantities are represented by mathematical objects called **vectors** (**Figure 2.2**). We can add or subtract two vectors, and we can multiply a vector by a scalar or by another vector, but we cannot divide by a vector. The operation of division by a vector is not defined.

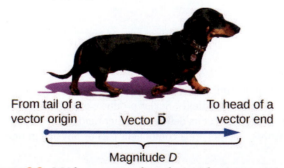

From tail of a vector origin   Vector $\vec{D}$ To head of a vector end

Magnitude D

Figure 2.2 We draw a vector from the initial point or origin (called the "tail" of a vector) to the end or terminal point (called the "head" of a vector), marked by an arrowhead. Magnitude is the length of a vector and is always a positive scalar quantity. (credit "photo": modification of work by Cate Sevilla)

Let's examine vector algebra using a graphical method to be aware of basic terms and to develop a qualitative understanding. In practice, however, when it comes to solving physics problems, we use analytical methods, which we'll see in the next section. Analytical methods are more simple computationally and more accurate than graphical methods. From now on, to distinguish between a vector and a scalar quantity, we adopt the common convention that a letter in bold type with an arrow above it denotes a vector, and a letter without an arrow denotes a scalar. For example, a distance of 2.0 km, which is a scalar quantity, is denoted by $d = 2.0$ km, whereas a displacement of 2.0 km in some direction, which is a vector quantity, is denoted by $\vec{d}$.

Suppose you tell a friend on a camping trip that you have discovered a terrific fishing hole 6 km from your tent. It is unlikely your friend would be able to find the hole easily unless you also communicate the direction in which it can be found with respect to your campsite. You may say, for example, "Walk about 6 km northeast from my tent." The key concept here is that you have to give not one but *two* pieces of information—namely, the distance or magnitude (6 km) *and* the direction (northeast).

Displacement is a general term used to describe a *change in position*, such as during a trip from the tent to the fishing hole. Displacement is an example of a vector quantity. If you walk from the tent (location A) to the hole (location B), as shown in **Figure 2.3**, the vector $\vec{D}$, representing your **displacement**, is drawn as the arrow that originates at point A and ends at point B. The arrowhead marks the end of the vector. The direction of the displacement vector $\vec{D}$ is the direction of the arrow. The length of the arrow represents the **magnitude** D of vector $\vec{D}$. Here, D = 6 km. Since the magnitude of a vector is its length, which is a positive number, the magnitude is also indicated by placing the absolute value notation around the symbol that denotes the vector; so, we can write equivalently that $D \equiv \left| \vec{D} \right|$. To solve a vector problem graphically, we

need to draw the vector $\vec{D}$ to scale. For example, if we assume 1 unit of distance (1 km) is represented in the drawing by a line segment of length $u = 2$ cm, then the total displacement in this example is represented by a vector of length $d = 6u = 6(2\,\text{cm}) = 12\,\text{cm}$, as shown in **Figure 2.4**. Notice that here, to avoid confusion, we used $D = 6$ km to denote the magnitude of the actual displacement and $d = 12$ cm to denote the length of its representation in the drawing.

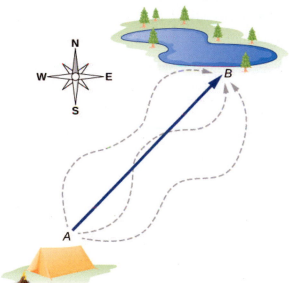

Figure 2.3 The displacement vector from point A (the initial position at the campsite) to point B (the final position at the fishing hole) is indicated by an arrow with origin at point A and end at point B. The displacement is the same for any of the actual paths (dashed curves) that may be taken between points A and B.

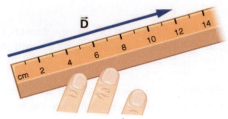

Figure 2.4 A displacement $\vec{D}$ of magnitude 6 km is drawn to scale as a vector of length 12 cm when the length of 2 cm represents 1 unit of displacement (which in this case is 1 km).

Suppose your friend walks from the campsite at *A* to the fishing pond at *B* and then walks back: from the fishing pond at *B* to the campsite at *A*. The magnitude of the displacement vector $\vec{D}_{AB}$ from *A* to *B* is the same as the magnitude of the displacement vector $\vec{D}_{BA}$ from *B* to *A* (it equals 6 km in both cases), so we can write $D_{AB} = D_{BA}$. However, vector $\vec{D}_{AB}$ is *not* equal to vector $\vec{D}_{BA}$ because these two vectors have different directions: $\vec{D}_{AB} \neq \vec{D}_{BA}$. In **Figure 2.3**, vector $\vec{D}_{BA}$ would be represented by a vector with an origin at point *B* and an end at point *A*, indicating vector $\vec{D}_{BA}$ points to the southwest, which is exactly 180° opposite to the direction of vector $\vec{D}_{AB}$. We say that vector $\vec{D}_{BA}$ is **antiparallel** to vector $\vec{D}_{AB}$ and write $\vec{D}_{AB} = -\vec{D}_{BA}$, where the minus sign indicates the antiparallel direction.

Two vectors that have identical directions are said to be **parallel vectors**—meaning, they are *parallel* to each other. Two parallel vectors $\vec{A}$ and $\vec{B}$ are equal, denoted by $\vec{A} = \vec{B}$, if and only if they have equal magnitudes $\left| \vec{A} \right| = \left| \vec{B} \right|$.

Two vectors with directions perpendicular to each other are said to be **orthogonal vectors**. These relations between vectors are illustrated in **Figure 2.5**.

Figure 2.5 Various relations between two vectors $\vec{A}$ and $\vec{B}$. (a) $\vec{A} \neq \vec{B}$ because $A \neq B$. (b) $\vec{A} \neq \vec{B}$ because they are not parallel and $A \neq B$. (c) $\vec{A} \neq -\vec{A}$ because they have different directions (even though $\left|\vec{A}\right| = \left|-\vec{A}\right| = A$). (d) $\vec{A} = \vec{B}$ because they are parallel *and* have identical magnitudes $A = B$. (e) $\vec{A} \neq \vec{B}$ because they have different directions (are not parallel); here, their directions differ by $90°$ —meaning, they are orthogonal.

 2.1 Check Your Understanding Two motorboats named *Alice* and *Bob* are moving on a lake. Given the information about their velocity vectors in each of the following situations, indicate whether their velocity vectors are equal or otherwise. (a) *Alice* moves north at 6 knots and *Bob* moves west at 6 knots. (b) *Alice* moves west at 6 knots and *Bob* moves west at 3 knots. (c) *Alice* moves northeast at 6 knots and *Bob* moves south at 3 knots. (d) *Alice* moves northeast at 6 knots and *Bob* moves southwest at 6 knots. (e) *Alice* moves northeast at 2 knots and *Bob* moves closer to the shore northeast at 2 knots.

Algebra of Vectors in One Dimension

Vectors can be multiplied by scalars, added to other vectors, or subtracted from other vectors. We can illustrate these vector concepts using an example of the fishing trip seen in **Figure 2.6**.

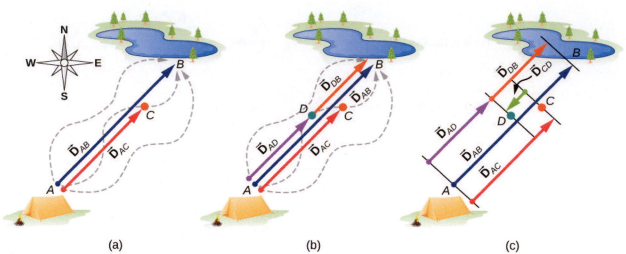

Figure 2.6 Displacement vectors for a fishing trip. (a) Stopping to rest at point *C* while walking from camp (point *A*) to the pond (point *B*). (b) Going back for the dropped tackle box (point *D*). (c) Finishing up at the fishing pond.

Suppose your friend departs from point *A* (the campsite) and walks in the direction to point *B* (the fishing pond), but, along the way, stops to rest at some point *C* located three-quarters of the distance between *A* and *B*, beginning from point *A* (**Figure 2.6**(a)). What is his displacement vector $\vec{D}_{AC}$ when he reaches point *C*? We know that if he walks all the way to *B*, his displacement vector relative to *A* is $\vec{D}_{AB}$, which has magnitude $D_{AB} = 6\,\text{km}$ and a direction of northeast. If he walks only a 0.75 fraction of the total distance, maintaining the northeasterly direction, at point *C* he must be $0.75 D_{AB} = 4.5\,\text{km}$ away from the campsite at *A*. So, his displacement vector at the rest point *C* has magnitude $D_{AC} = 4.5\,\text{km} = 0.75 D_{AB}$ and is parallel to the displacement vector $\vec{D}_{AB}$. All of this can be stated succinctly in the form of the following **vector equation**:

$$\vec{D}_{AC} = 0.75\,\vec{D}_{AB}.$$

In a vector equation, both sides of the equation are vectors. The previous equation is an example of a vector multiplied by a positive scalar (number) $\alpha = 0.75$. The result, $\vec{D}_{AC}$, of such a multiplication is a new vector with a direction parallel to the direction of the original vector $\vec{D}_{AB}$.

In general, when a vector $\vec{A}$ is multiplied by a *positive* scalar α, the result is a new vector $\vec{B}$ that is *parallel* to $\vec{A}$:

$$\vec{B} = \alpha\,\vec{A}. \tag{2.1}$$

The magnitude $\left|\vec{B}\right|$ of this new vector is obtained by multiplying the magnitude $\left|\vec{A}\right|$ of the original vector, as expressed by the **scalar equation**:

$$B = |\alpha|A. \tag{2.2}$$

In a scalar equation, both sides of the equation are numbers. **Equation 2.2** is a scalar equation because the magnitudes of vectors are scalar quantities (and positive numbers). If the scalar α is *negative* in the vector equation **Equation 2.1**, then the magnitude $\left|\vec{B}\right|$ of the new vector is still given by **Equation 2.2**, but the direction of the new vector $\vec{B}$ is *antiparallel* to the direction of $\vec{A}$. These principles are illustrated in **Figure 2.7**(a) by two examples where the length of vector $\vec{A}$ is 1.5 units. When $\alpha = 2$, the new vector $\vec{B} = 2\vec{A}$ has length $B = 2A = 3.0$ units (twice as long as the original vector) and is parallel to the original vector. When $\alpha = -2$, the new vector $\vec{C} = -2\vec{A}$ has length $C = |-2|A = 3.0$ units (twice as long as the original vector) and is antiparallel to the original vector.

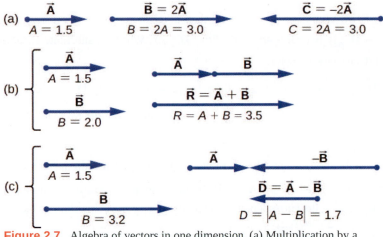

Figure 2.7 Algebra of vectors in one dimension. (a) Multiplication by a scalar. (b) Addition of two vectors ($\vec{R}$ is called the *resultant* of vectors $\vec{A}$ and $\vec{B}$). (c) Subtraction of two vectors ($\vec{D}$ is the difference of vectors $\vec{A}$ and $\vec{B}$).

Now suppose your fishing buddy departs from point *A* (the campsite), walking in the direction to point *B* (the fishing hole), but he realizes he lost his tackle box when he stopped to rest at point *C* (located three-quarters of the distance between *A* and *B*, beginning from point *A*). So, he turns back and retraces his steps in the direction toward the campsite and finds the box lying on the path at some point *D* only 1.2 km away from point *C* (see **Figure 2.6**(b)). What is his displacement vector $\vec{D}_{AD}$ when he finds the box at point *D*? What is his displacement vector $\vec{D}_{DB}$ from point *D* to the hole? We have already established that at rest point *C* his displacement vector is $\vec{D}_{AC} = 0.75\vec{D}_{AB}$. Starting at point *C*, he walks southwest (toward the campsite), which means his new displacement vector $\vec{D}_{CD}$ from point *C* to point *D* is antiparallel to $\vec{D}_{AB}$. Its magnitude $\left|\vec{D}_{CD}\right|$ is $D_{CD} = 1.2\,\text{km} = 0.2D_{AB}$, so his second displacement vector is $\vec{D}_{CD} = -0.2\vec{D}_{AB}$. His total displacement $\vec{D}_{AD}$ relative to the campsite is the **vector sum** of the two displacement vectors: vector $\vec{D}_{AC}$ (from the campsite to the rest point) and vector $\vec{D}_{CD}$ (from the rest point to the point where he finds his box):

$$\vec{D}_{AD} = \vec{D}_{AC} + \vec{D}_{CD}. \tag{2.3}$$

The vector sum of two (or more) vectors is called the **resultant vector** or, for short, the *resultant*. When the vectors on the right-hand-side of **Equation 2.3** are known, we can find the resultant $\vec{D}_{AD}$ as follows:

$$\vec{D}_{AD} = \vec{D}_{AC} + \vec{D}_{CD} = 0.75\vec{D}_{AB} - 0.2\vec{D}_{AB} = (0.75 - 0.2)\vec{D}_{AB} = 0.55\vec{D}_{AB}. \tag{2.4}$$

When your friend finally reaches the pond at *B*, his displacement vector $\vec{D}_{AB}$ from point *A* is the vector sum of his displacement vector $\vec{\mathbf{D}}_{AD}$ from point A to point D and his displacement vector $\vec{\mathbf{D}}_{DB}$ from point D to the fishing hole: $\vec{\mathbf{D}}_{AB} = \vec{\mathbf{D}}_{AD} + \vec{\mathbf{D}}_{DB}$ (see **Figure 2.6**(c)). This means his displacement vector $\vec{\mathbf{D}}_{DB}$ is the **difference of two vectors**:

$$\vec{\mathbf{D}}_{DB} = \vec{\mathbf{D}}_{AB} - \vec{\mathbf{D}}_{AD} = \vec{\mathbf{D}}_{AB} + (- \vec{\mathbf{D}}_{AD}). \tag{2.5}$$

Notice that a difference of two vectors is nothing more than a vector sum of two vectors because the second term in **Equation 2.5** is vector $- \vec{\mathbf{D}}_{AD}$ (which is antiparallel to $\vec{\mathbf{D}}_{AD}$). When we substitute **Equation 2.4** into **Equation 2.5**, we obtain the second displacement vector:

$$\vec{\mathbf{D}}_{DB} = \vec{\mathbf{D}}_{AB} - \vec{\mathbf{D}}_{AD} = \vec{\mathbf{D}}_{AB} - 0.55 \vec{\mathbf{D}}_{AB} = (1.0 - 0.55) \vec{\mathbf{D}}_{AB} = 0.45 \vec{\mathbf{D}}_{AB}. \tag{2.6}$$

This result means your friend walked $D_{DB} = 0.45 D_{AB} = 0.45(6.0 \text{ km}) = 2.7 \text{ km}$ from the point where he finds his tackle box to the fishing hole.

When vectors $\vec{\mathbf{A}}$ and $\vec{\mathbf{B}}$ lie along a line (that is, in one dimension), such as in the camping example, their resultant $\vec{\mathbf{R}} = \vec{\mathbf{A}} + \vec{\mathbf{B}}$ and their difference $\vec{\mathbf{D}} = \vec{\mathbf{A}} - \vec{\mathbf{B}}$ both lie along the same direction. We can illustrate the addition or subtraction of vectors by drawing the corresponding vectors to scale in one dimension, as shown in **Figure 2.7**.

To illustrate the resultant when $\vec{\mathbf{A}}$ and $\vec{\mathbf{B}}$ are two parallel vectors, we draw them along one line by placing the origin of one vector at the end of the other vector in head-to-tail fashion (see **Figure 2.7**(b)). The magnitude of this resultant is the sum of their magnitudes: $R = A + B$. The direction of the resultant is parallel to both vectors. When vector $\vec{\mathbf{A}}$ is antiparallel to vector $\vec{\mathbf{B}}$, we draw them along one line in either head-to-head fashion (**Figure 2.7**(c)) or tail-to-tail fashion. The magnitude of the vector difference, then, is the *absolute value* $D = |A - B|$ of the difference of their magnitudes. The direction of the difference vector $\vec{\mathbf{D}}$ is parallel to the direction of the longer vector.

In general, in one dimension—as well as in higher dimensions, such as in a plane or in space—we can add any number of vectors and we can do so in any order because the addition of vectors is **commutative**,

$$\vec{\mathbf{A}} + \vec{\mathbf{B}} = \vec{\mathbf{B}} + \vec{\mathbf{A}}, \tag{2.7}$$

and **associative**,

$$(\vec{\mathbf{A}} + \vec{\mathbf{B}}) + \vec{\mathbf{C}} = \vec{\mathbf{A}} + (\vec{\mathbf{B}} + \vec{\mathbf{C}}). \tag{2.8}$$

Moreover, multiplication by a scalar is **distributive**:

$$\alpha_1 \vec{\mathbf{A}} + \alpha_2 \vec{\mathbf{A}} = (\alpha_1 + \alpha_2) \vec{\mathbf{A}}. \tag{2.9}$$

We used the distributive property in **Equation 2.4** and **Equation 2.6**.

When adding many vectors in one dimension, it is convenient to use the concept of a **unit vector**. A unit vector, which is denoted by a letter symbol with a hat, such as $\hat{\mathbf{u}}$, has a magnitude of one and does not have any physical unit so that $\left| \hat{\mathbf{u}} \right| \equiv u = 1$. The only role of a unit vector is to specify direction. For example, instead of saying vector $\vec{\mathbf{D}}_{AB}$ has a magnitude of 6.0 km and a direction of northeast, we can introduce a unit vector $\hat{\mathbf{u}}$ that points to the northeast and say succinctly that $\vec{\mathbf{D}}_{AB} = (6.0\,\text{km})\hat{\mathbf{u}}$. Then the southwesterly direction is simply given by the unit vector $-\hat{\mathbf{u}}$. In this way, the displacement of 6.0 km in the southwesterly direction is expressed by the vector

$$\vec{\mathbf{D}}_{BA} = (-6.0\,\text{km})\hat{\mathbf{u}}.$$

Example 2.1

A Ladybug Walker

A long measuring stick rests against a wall in a physics laboratory with its 200-cm end at the floor. A ladybug lands on the 100-cm mark and crawls randomly along the stick. It first walks 15 cm toward the floor, then it walks 56 cm toward the wall, then it walks 3 cm toward the floor again. Then, after a brief stop, it continues for 25 cm toward the floor and then, again, it crawls up 19 cm toward the wall before coming to a complete rest (**Figure 2.8**). Find the vector of its total displacement and its final resting position on the stick.

Strategy

If we choose the direction along the stick toward the floor as the direction of unit vector $\hat{\mathbf{u}}$, then the direction toward the floor is $+\hat{\mathbf{u}}$ and the direction toward the wall is $-\hat{\mathbf{u}}$. The ladybug makes a total of five displacements:

$$\vec{\mathbf{D}}_1 = (15\,\text{cm})(+\hat{\mathbf{u}}),$$
$$\vec{\mathbf{D}}_2 = (56\,\text{cm})(-\hat{\mathbf{u}}),$$
$$\vec{\mathbf{D}}_3 = (3\,\text{cm})(+\hat{\mathbf{u}}),$$
$$\vec{\mathbf{D}}_4 = (25\,\text{cm})(+\hat{\mathbf{u}}),\ \text{and}$$
$$\vec{\mathbf{D}}_5 = (19\,\text{cm})(-\hat{\mathbf{u}}).$$

The total displacement $\vec{\mathbf{D}}$ is the resultant of all its displacement vectors.

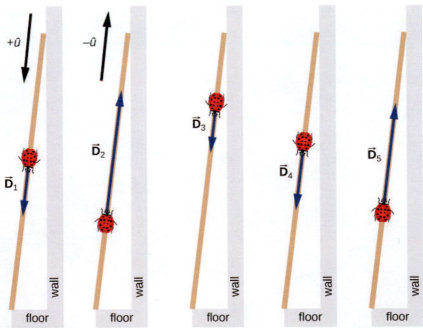

Figure 2.8 Five displacements of the ladybug. Note that in this schematic drawing, magnitudes of displacements are not drawn to scale. (credit "ladybug": modification of work by "Persian Poet Gal"/Wikimedia Commons)

Solution

The resultant of all the displacement vectors is

$$\vec{D} = \vec{D}_1 + \vec{D}_2 + \vec{D}_3 + \vec{D}_4 + \vec{D}_5$$
$$= (15 \text{ cm})(+\hat{u}) + (56 \text{ cm})(-\hat{u}) + (3 \text{ cm})(+\hat{u}) + (25 \text{ cm})(+\hat{u}) + (19 \text{ cm})(-\hat{u})$$
$$= (15 - 56 + 3 + 25 - 19)\text{cm}\,\hat{u}$$
$$= -32 \text{ cm}\,\hat{u}.$$

In this calculation, we use the distributive law given by **Equation 2.9**. The result reads that the total displacement vector points away from the 100-cm mark (initial landing site) toward the end of the meter stick that touches the wall. The end that touches the wall is marked 0 cm, so the final position of the ladybug is at the (100 – 32)cm = 68-cm mark.

 2.2 Check Your Understanding A cave diver enters a long underwater tunnel. When her displacement with respect to the entry point is 20 m, she accidentally drops her camera, but she doesn't notice it missing until she is some 6 m farther into the tunnel. She swims back 10 m but cannot find the camera, so she decides to end the dive. How far from the entry point is she? Taking the positive direction out of the tunnel, what is her displacement vector relative to the entry point?

Algebra of Vectors in Two Dimensions

When vectors lie in a plane—that is, when they are in two dimensions—they can be multiplied by scalars, added to other vectors, or subtracted from other vectors in accordance with the general laws expressed by **Equation 2.1**, **Equation 2.2**, **Equation 2.7**, and **Equation 2.8**. However, the addition rule for two vectors in a plane becomes more complicated than the rule for vector addition in one dimension. We have to use the laws of geometry to construct resultant vectors, followed by trigonometry to find vector magnitudes and directions. This geometric approach is commonly used in navigation (**Figure 2.9**). In this section, we need to have at hand two rulers, a triangle, a protractor, a pencil, and an eraser for drawing vectors to scale by geometric constructions.

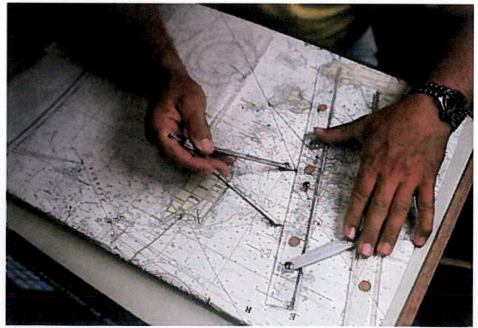

Figure 2.9 In navigation, the laws of geometry are used to draw resultant displacements on nautical maps.

For a geometric construction of the sum of two vectors in a plane, we follow **the parallelogram rule**. Suppose two vectors $\vec{A}$ and $\vec{B}$ are at the arbitrary positions shown in **Figure 2.10**. Translate either one of them in parallel to the beginning of the other vector, so that after the translation, both vectors have their origins at the same point. Now, at the end of vector $\vec{A}$ we draw a line parallel to vector $\vec{B}$ and at the end of vector $\vec{B}$ we draw a line parallel to vector $\vec{A}$ (the dashed lines in **Figure 2.10**). In this way, we obtain a parallelogram. From the origin of the two vectors we draw a diagonal that is the resultant $\vec{R}$ of the two vectors: $\vec{R} = \vec{A} + \vec{B}$ (**Figure 2.10**(a)). The other diagonal of this parallelogram is the vector difference of the two vectors $\vec{D} = \vec{A} - \vec{B}$, as shown in **Figure 2.10**(b). Notice that the end of the difference vector is placed at the end of vector $\vec{A}$.

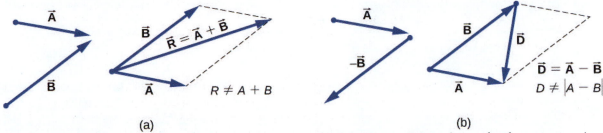

Figure 2.10 The parallelogram rule for the addition of two vectors. Make the parallel translation of each vector to a point where their origins (marked by the dot) coincide and construct a parallelogram with two sides on the vectors and the other two sides (indicated by dashed lines) parallel to the vectors. (a) Draw the resultant vector $\vec{R}$ along the diagonal of the parallelogram from the common point to the opposite corner. Length R of the resultant vector is *not* equal to the sum of the magnitudes of the two vectors. (b) Draw the difference vector $\vec{D} = \vec{A} - \vec{B}$ along the diagonal connecting the ends of the vectors. Place the origin of vector $\vec{D}$ at the end of vector $\vec{B}$ and the end (arrowhead) of vector $\vec{D}$ at the end of vector $\vec{A}$. Length D of the difference vector is *not* equal to the difference of magnitudes of the two vectors.

It follows from the parallelogram rule that neither the magnitude of the resultant vector nor the magnitude of the difference vector can be expressed as a simple sum or difference of magnitudes A and B, because the length of a diagonal cannot be expressed as a simple sum of side lengths. When using a geometric construction to find magnitudes $\left|\vec{R}\right|$ and $\left|\vec{D}\right|$, we have to use trigonometry laws for triangles, which may lead to complicated algebra. There are two ways to circumvent this algebraic complexity. One way is to use the method of components, which we examine in the next section. The other way is to draw the vectors to scale, as is done in navigation, and read approximate vector lengths and angles (directions) from the graphs. In this section we examine the second approach.

If we need to add three or more vectors, we repeat the parallelogram rule for the pairs of vectors until we find the resultant of all of the resultants. For three vectors, for example, we first find the resultant of vector 1 and vector 2, and then we find the resultant of this resultant and vector 3. The order in which we select the pairs of vectors does not matter because the operation of vector addition is commutative and associative (see **Equation 2.7** and **Equation 2.8**). Before we state a general rule that follows from repetitive applications of the parallelogram rule, let's look at the following example.

Suppose you plan a vacation trip in Florida. Departing from Tallahassee, the state capital, you plan to visit your uncle Joe in Jacksonville, see your cousin Vinny in Daytona Beach, stop for a little fun in Orlando, see a circus performance in Tampa, and visit the University of Florida in Gainesville. Your route may be represented by five displacement vectors $\vec{A}$, $\vec{B}$, $\vec{C}$, $\vec{D}$, and $\vec{E}$, which are indicated by the red vectors in **Figure 2.11**. What is your total displacement when you reach Gainesville? The total displacement is the vector sum of all five displacement vectors, which may be found by using the parallelogram rule four times. Alternatively, recall that the displacement vector has its beginning at the initial position (Tallahassee) and its end at the final position (Gainesville), so the total displacement vector can be drawn directly as an arrow connecting Tallahassee with Gainesville (see the green vector in **Figure 2.11**). When we use the parallelogram rule four times, the resultant $\vec{R}$ we obtain is exactly this green vector connecting Tallahassee with Gainesville: $\vec{R} = \vec{A} + \vec{B} + \vec{C} + \vec{D} + \vec{E}$.

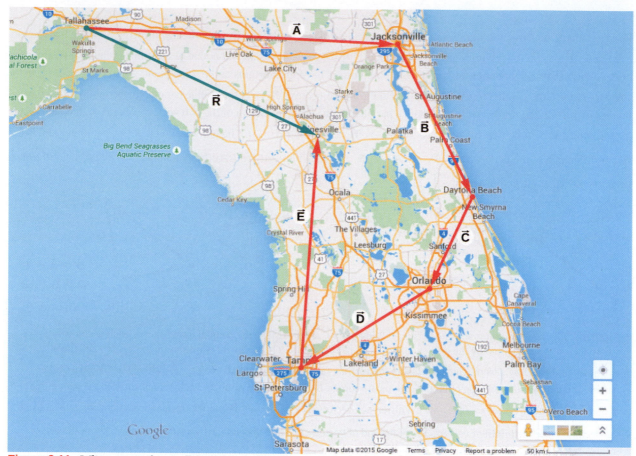

Figure 2.11 When we use the parallelogram rule four times, we obtain the resultant vector
$$\vec{R} = \vec{A} + \vec{B} + \vec{C} + \vec{D} + \vec{E}$$, which is the green vector connecting Tallahassee with Gainesville.

Drawing the resultant vector of many vectors can be generalized by using the following **tail-to-head geometric construction**. Suppose we want to draw the resultant vector $\vec{R}$ of four vectors $\vec{A}$, $\vec{B}$, $\vec{C}$, and $\vec{D}$ (**Figure 2.12**(a)). We select any one of the vectors as the first vector and make a parallel translation of a second vector to a position where the origin ("tail") of the second vector coincides with the end ("head") of the first vector. Then, we select a third vector and make a parallel translation of the third vector to a position where the origin of the third vector coincides with the end of the second vector. We repeat this procedure until all the vectors are in a head-to-tail arrangement like the one shown in **Figure 2.12**. We draw the resultant vector $\vec{R}$ by connecting the origin ("tail") of the first vector with the end ("head") of the last vector. The end of the resultant vector is at the end of the last vector. Because the addition of vectors is associative and commutative, we obtain the same resultant vector regardless of which vector we choose to be first, second, third, or fourth in this construction.

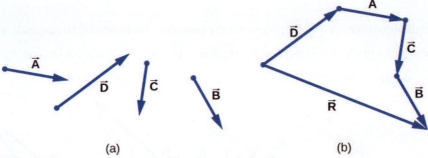

Figure 2.12 Tail-to-head method for drawing the resultant vector $\vec{R} = \vec{A} + \vec{B} + \vec{C} + \vec{D}$. (a) Four vectors of different magnitudes and directions. (b) Vectors in (a) are translated to new positions where the origin ("tail") of one vector is at the end ("head") of another vector. The resultant vector is drawn from the origin ("tail") of the first vector to the end ("head") of the last vector in this arrangement.

Example 2.2

Geometric Construction of the Resultant

The three displacement vectors $\vec{A}$, $\vec{B}$, and $\vec{C}$ in **Figure 2.13** are specified by their magnitudes $A = 10.0$, $B = 7.0$, and $C = 8.0$, respectively, and by their respective direction angles with the horizontal direction $\alpha = 35°$, $\beta = -110°$, and $\gamma = 30°$. The physical units of the magnitudes are centimeters. Choose a convenient scale and use a ruler and a protractor to find the following vector sums: (a) $\vec{R} = \vec{A} + \vec{B}$, (b) $\vec{D} = \vec{A} - \vec{B}$, and (c) $\vec{S} = \vec{A} - 3\vec{B} + \vec{C}$.

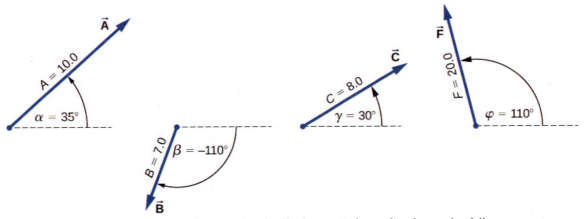

Figure 2.13 Vectors used in **Example 2.2** and in the Check Your Understanding feature that follows.

Strategy

In geometric construction, to find a vector means to find its magnitude and its direction angle with the horizontal direction. The strategy is to draw to scale the vectors that appear on the right-hand side of the equation and construct the resultant vector. Then, use a ruler and a protractor to read the magnitude of the resultant and the direction angle. For parts (a) and (b) we use the parallelogram rule. For (c) we use the tail-to-head method.

Solution

For parts (a) and (b), we attach the origin of vector $\vec{B}$ to the origin of vector $\vec{A}$, as shown in **Figure 2.14**, and construct a parallelogram. The shorter diagonal of this parallelogram is the sum $\vec{A} + \vec{B}$. The longer of the diagonals is the difference $\vec{A} - \vec{B}$. We use a ruler to measure the lengths of the diagonals, and a protractor to measure the angles with the horizontal. For the resultant $\vec{R}$, we obtain $R = 5.8$ cm and $\theta_R \approx 0°$. For the difference $\vec{D}$, we obtain $D = 16.2$ cm and $\theta_D = 49.3°$, which are shown in **Figure 2.14**.

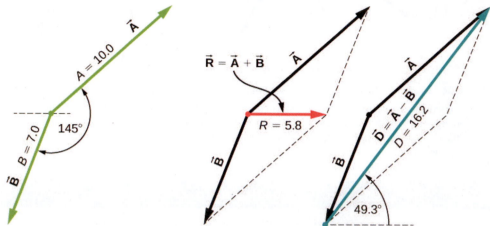

Figure 2.14 Using the parallelogram rule to solve (a) (finding the resultant, red) and (b) (finding the difference, blue).

For (c), we can start with vector $-3\vec{B}$ and draw the remaining vectors tail-to-head as shown in **Figure 2.15**. In vector addition, the order in which we draw the vectors is unimportant, but drawing the vectors to scale is very important. Next, we draw vector $\vec{S}$ from the origin of the first vector to the end of the last vector and place the arrowhead at the end of $\vec{S}$. We use a ruler to measure the length of $\vec{S}$, and find that its magnitude is $S = 36.9$ cm. We use a protractor and find that its direction angle is $\theta_S = 52.9°$. This solution is shown in **Figure 2.15**.

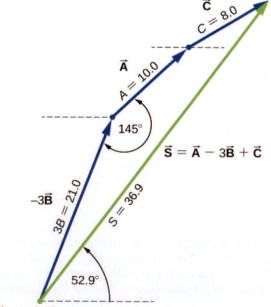

Figure 2.15 Using the tail-to-head method to solve (c) (finding vector $\vec{S}$, green).

2.3 **Check Your Understanding** Using the three displacement vectors $\vec{A}$, $\vec{B}$, and $\vec{F}$ in **Figure 2.13**, choose a convenient scale, and use a ruler and a protractor to find vector $\vec{G}$ given by the vector equation $\vec{G} = \vec{A} + 2\vec{B} - \vec{F}$.

 Observe the addition of vectors in a plane by visiting this **vector calculator (https://openstaxcollege.org/l/ 21compveccalc)** and this **Phet simulation (https://openstaxcollege.org/l/21phetvecaddsim)** .

2.2 | Coordinate Systems and Components of a Vector

Learning Objectives

By the end of this section, you will be able to:

* Describe vectors in two and three dimensions in terms of their components, using unit vectors along the axes.
* Distinguish between the vector components of a vector and the scalar components of a vector.
* Explain how the magnitude of a vector is defined in terms of the components of a vector.
* Identify the direction angle of a vector in a plane.
* Explain the connection between polar coordinates and Cartesian coordinates in a plane.

Vectors are usually described in terms of their components in a coordinate system. Even in everyday life we naturally invoke the concept of orthogonal projections in a rectangular coordinate system. For example, if you ask someone for directions to a particular location, you will more likely be told to go 40 km east and 30 km north than 50 km in the direction $37°$ north of east.

In a rectangular (Cartesian) xy-coordinate system in a plane, a point in a plane is described by a pair of coordinates (x, y). In a similar fashion, a vector $\vec{A}$ in a plane is described by a pair of its *vector* coordinates. The x-coordinate of vector $\vec{A}$ is called its x-component and the y-coordinate of vector $\vec{A}$ is called its y-component. The vector x-component is a vector denoted by $\vec{A}_x$. The vector y-component is a vector denoted by $\vec{A}_y$. In the Cartesian system, the x and y **vector components** of a vector are the orthogonal projections of this vector onto the x- and y-axes, respectively. In this way, following the parallelogram rule for vector addition, each vector on a Cartesian plane can be expressed as the vector sum of its vector components:

$$\vec{A} = \vec{A}_x + \vec{A}_y.$$

(2.10)

As illustrated in **Figure 2.16**, vector $\vec{A}$ is the diagonal of the rectangle where the x-component $\vec{A}_x$ is the side parallel to the x-axis and the y-component $\vec{A}_y$ is the side parallel to the y-axis. Vector component $\vec{A}_x$ is orthogonal to vector component $\vec{A}_y$.

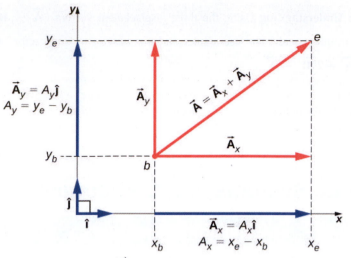

Figure 2.16 Vector $\vec{A}$ in a plane in the Cartesian coordinate system is the vector sum of its vector x- and y-components. The x-vector component $\vec{A}_x$ is the orthogonal projection of vector $\vec{A}$ onto the x-axis. The y-vector component $\vec{A}_y$ is the orthogonal projection of vector $\vec{A}$ onto the y-axis. The numbers A_x and A_y that multiply the unit vectors are the scalar components of the vector.

It is customary to denote the positive direction on the x-axis by the unit vector $\hat{i}$ and the positive direction on the y-axis by the unit vector $\hat{j}$. **Unit vectors of the axes**, $\hat{i}$ and $\hat{j}$, define two orthogonal directions in the plane. As shown in **Figure 2.16**, the x- and y- components of a vector can now be written in terms of the unit vectors of the axes:

$$\begin{cases} \vec{A}_x = A_x \, \hat{i} \\ \vec{A}_y = A_y \, \hat{j}. \end{cases} \tag{2.11}$$

The vectors $\vec{A}_x$ and $\vec{A}_y$ defined by **Equation 2.11** are the *vector components* of vector $\vec{A}$. The numbers A_x and A_y that define the vector components in **Equation 2.11** are the **scalar components** of vector $\vec{A}$. Combining **Equation 2.10** with **Equation 2.11**, we obtain **the component form of a vector**:

$$\vec{A} = A_x \, \hat{i} + A_y \, \hat{j}. \tag{2.12}$$

If we know the coordinates $b(x_b, y_b)$ of the origin point of a vector (where b stands for "beginning") and the coordinates $e(x_e, y_e)$ of the end point of a vector (where e stands for "end"), we can obtain the scalar components of a vector simply by subtracting the origin point coordinates from the end point coordinates:

$$\begin{cases} A_x = x_e - x_b \\ A_y = y_e - y_b. \end{cases} \tag{2.13}$$

Example 2.3

Displacement of a Mouse Pointer

A mouse pointer on the display monitor of a computer at its initial position is at point (6.0 cm, 1.6 cm) with respect to the lower left-side corner. If you move the pointer to an icon located at point (2.0 cm, 4.5 cm), what is the displacement vector of the pointer?

Strategy

The origin of the *xy*-coordinate system is the lower left-side corner of the computer monitor. Therefore, the unit vector $\hat{i}$ on the *x*-axis points horizontally to the right and the unit vector $\hat{j}$ on the *y*-axis points vertically upward. The origin of the displacement vector is located at point $b(6.0, 1.6)$ and the end of the displacement vector is located at point $e(2.0, 4.5)$. Substitute the coordinates of these points into **Equation 2.13** to find the scalar components D_x and D_y of the displacement vector $\vec{D}$. Finally, substitute the coordinates into **Equation 2.12** to write the displacement vector in the vector component form.

Solution

We identify $x_b = 6.0$, $x_e = 2.0$, $y_b = 1.6$, and $y_e = 4.5$, where the physical unit is 1 cm. The scalar *x*- and *y*-components of the displacement vector are

$$D_x = x_e - x_b = (2.0 - 6.0)\text{cm} = -4.0\,\text{cm},$$
$$D_y = y_e - y_b = (4.5 - 1.6)\text{cm} = +2.9\,\text{cm}.$$

The vector component form of the displacement vector is

$$\vec{D} = D_x\,\hat{i} + D_y\,\hat{j} = (-4.0\,\text{cm})\,\hat{i} + (2.9\,\text{cm})\,\hat{j} = (-4.0\,\hat{i} + 2.9\,\hat{j})\text{cm}. \qquad (2.14)$$

This solution is shown in **Figure 2.17**.

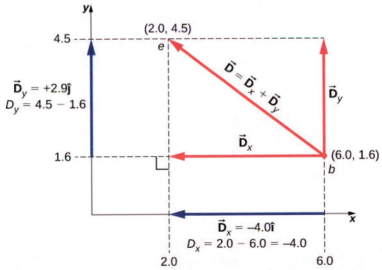

Figure 2.17 The graph of the displacement vector. The vector points from the origin point at *b* to the end point at *e*.

Significance

Notice that the physical unit—here, 1 cm—can be placed either with each component immediately before the unit vector or globally for both components, as in **Equation 2.14**. Often, the latter way is more convenient because it is simpler.

The vector x-component $\vec{D}_x = -4.0\,\hat{i} = 4.0(-\hat{i})$ of the displacement vector has the magnitude $\left|\vec{D}_x\right| = \left|-4.0\right|\left|\hat{i}\right| = 4.0$ because the magnitude of the unit vector is $\left|\hat{i}\right| = 1$. Notice, too, that the direction of the x-component is $-\hat{i}$, which is antiparallel to the direction of the $+x$-axis; hence, the x-component vector $\vec{D}_x$ points to the left, as shown in **Figure 2.17**. The scalar x-component of vector $\vec{D}$ is $D_x = -4.0$.

Similarly, the vector y-component $\vec{D}_y = +2.9\,\hat{j}$ of the displacement vector has magnitude $\left|\vec{D}_y\right| = \left|2.9\right|\left|\hat{j}\right| = 2.9$ because the magnitude of the unit vector is $\left|\hat{j}\right| = 1$. The direction of the y-component is $+\hat{j}$, which is parallel to the direction of the $+y$-axis. Therefore, the y-component vector $\vec{D}_y$ points up, as seen in **Figure 2.17**. The scalar y-component of vector $\vec{D}$ is $D_y = +2.9$. The displacement vector $\vec{D}$ is the resultant of its two *vector* components.

The vector component form of the displacement vector **Equation 2.14** tells us that the mouse pointer has been moved on the monitor 4.0 cm to the left and 2.9 cm upward from its initial position.

 2.4 Check Your Understanding A blue fly lands on a sheet of graph paper at a point located 10.0 cm to the right of its left edge and 8.0 cm above its bottom edge and walks slowly to a point located 5.0 cm from the left edge and 5.0 cm from the bottom edge. Choose the rectangular coordinate system with the origin at the lower left-side corner of the paper and find the displacement vector of the fly. Illustrate your solution by graphing.

When we know the scalar components A_x and A_y of a vector $\vec{A}$, we can find its magnitude A and its direction angle θ_A. The **direction angle**—or direction, for short—is the angle the vector forms with the positive direction on the x-axis. The angle θ_A is measured in the *counterclockwise direction* from the $+x$-axis to the vector (**Figure 2.18**). Because the lengths A, A_x, and A_y form a right triangle, they are related by the Pythagorean theorem:

$$A^2 = A_x^2 + A_y^2 \Leftrightarrow A = \sqrt{A_x^2 + A_y^2}. \tag{2.15}$$

This equation works even if the scalar components of a vector are negative. The direction angle θ_A of a vector is defined via the tangent function of angle θ_A in the triangle shown in **Figure 2.18**:

$$\tan\theta = \frac{A_y}{A_x} \tag{2.16}$$

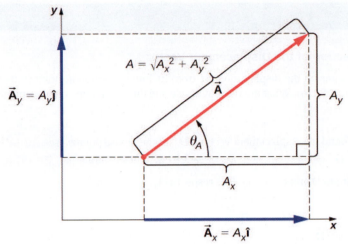

Figure 2.18 When the vector lies either in the first quadrant or in the fourth quadrant, where component A_x is positive (Figure 2.19), the direction angle θ_A in Equation 2.16) is identical to the angle θ
.

When the vector lies either in the first quadrant or in the fourth quadrant, where component A_x is positive (**Figure 2.19**), the angle θ in **Equation 2.16** is identical to the direction angle θ_A. For vectors in the fourth quadrant, angle θ is negative, which means that for these vectors, direction angle θ_A is measured *clockwise* from the positive x-axis. Similarly, for vectors in the second quadrant, angle θ is negative. When the vector lies in either the second or third quadrant, where component A_x is negative, the direction angle is $\theta_A = \theta + 180°$ (**Figure 2.19**).

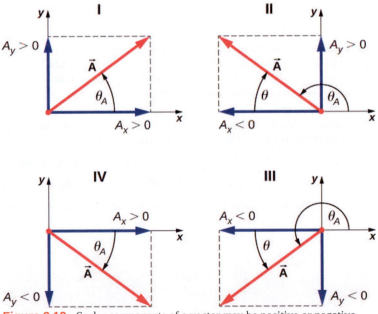

Figure 2.19 Scalar components of a vector may be positive or negative. Vectors in the first quadrant (I) have both scalar components positive and vectors in the third quadrant have both scalar components negative. For vectors in quadrants II and III, the direction angle of a vector is $\theta_A = \theta + 180°$.

Example 2.4

Magnitude and Direction of the Displacement Vector

You move a mouse pointer on the display monitor from its initial position at point (6.0 cm, 1.6 cm) to an icon located at point (2.0 cm, 4.5 cm). What are the magnitude and direction of the displacement vector of the pointer?

Strategy

In **Example 2.3**, we found the displacement vector $\vec{D}$ of the mouse pointer (see **Equation 2.14**). We identify its scalar components $D_x = -4.0$ cm and $D_y = +2.9$ cm and substitute into **Equation 2.15** and **Equation 2.16** to find the magnitude D and direction θ_D, respectively.

Solution

The magnitude of vector $\vec{D}$ is

$$D = \sqrt{D_x^2 + D_y^2} = \sqrt{(-4.0\,\text{cm})^2 + (2.9\,\text{cm})^2} = \sqrt{(4.0)^2 + (2.9)^2}\,\text{cm} = 4.9\,\text{cm}.$$

The direction angle is

$$\tan\theta = \frac{D_y}{D_x} = \frac{+2.9\,\text{cm}}{-4.0\,\text{cm}} = -0.725 \quad \Rightarrow \quad \theta = \tan^{-1}(-0.725) = -35.9°.$$

Vector $\vec{D}$ lies in the second quadrant, so its direction angle is

$$\theta_D = \theta + 180° = -35.9° + 180° = 144.1°.$$

 2.5 **Check Your Understanding** If the displacement vector of a blue fly walking on a sheet of graph paper is $\vec{D} = (-5.00\,\hat{i} - 3.00\,\hat{j})$cm, find its magnitude and direction.

In many applications, the magnitudes and directions of vector quantities are known and we need to find the resultant of many vectors. For example, imagine 400 cars moving on the Golden Gate Bridge in San Francisco in a strong wind. Each car gives the bridge a different push in various directions and we would like to know how big the resultant push can possibly be. We have already gained some experience with the geometric construction of vector sums, so we know the task of finding the resultant by drawing the vectors and measuring their lengths and angles may become intractable pretty quickly, leading to huge errors. Worries like this do not appear when we use analytical methods. The very first step in an analytical approach is to find vector components when the direction and magnitude of a vector are known.

Let us return to the right triangle in **Figure 2.18**. The quotient of the adjacent side A_x to the hypotenuse A is the cosine function of direction angle θ_A, $A_x/A = \cos\theta_A$, and the quotient of the opposite side A_y to the hypotenuse A is the sine function of θ_A, $A_y/A = \sin\theta_A$. When magnitude A and direction θ_A are known, we can solve these relations for the scalar components:

$$\begin{cases} A_x = A\cos\theta_A \\ A_y = A\sin\theta_A \end{cases} \tag{2.17}$$

When calculating vector components with **Equation 2.17**, care must be taken with the angle. The direction angle θ_A of a vector is the angle measured *counterclockwise* from the positive direction on the *x*-axis to the vector. The clockwise measurement gives a negative angle.

Example 2.5

Components of Displacement Vectors

A rescue party for a missing child follows a search dog named Trooper. Trooper wanders a lot and makes many trial sniffs along many different paths. Trooper eventually finds the child and the story has a happy ending, but his displacements on various legs seem to be truly convoluted. On one of the legs he walks 200.0 m southeast, then he runs north some 300.0 m. On the third leg, he examines the scents carefully for 50.0 m in the direction $30°$ west of north. On the fourth leg, Trooper goes directly south for 80.0 m, picks up a fresh scent and turns $23°$ west of south for 150.0 m. Find the scalar components of Trooper's displacement vectors and his displacement vectors in vector component form for each leg.

Strategy

Let's adopt a rectangular coordinate system with the positive x-axis in the direction of geographic east, with the positive y-direction pointed to geographic north. Explicitly, the unit vector $\hat{i}$ of the x-axis points east and the unit vector $\hat{j}$ of the y-axis points north. Trooper makes five legs, so there are five displacement vectors. We start by identifying their magnitudes and direction angles, then we use **Equation 2.17** to find the scalar components of the displacements and **Equation 2.12** for the displacement vectors.

Solution

On the first leg, the displacement magnitude is $L_1 = 200.0\,\text{m}$ and the direction is southeast. For direction angle θ_1 we can take either $45°$ measured clockwise from the east direction or $45° + 270°$ measured counterclockwise from the east direction. With the first choice, $\theta_1 = -45°$. With the second choice, $\theta_1 = +315°$. We can use either one of these two angles. The components are

$$L_{1x} = L_1 \cos \theta_1 = (200.0\,\text{m}) \cos 315° = 141.4\,\text{m},$$
$$L_{1y} = L_1 \sin \theta_1 = (200.0\,\text{m}) \sin 315° = -141.4\,\text{m}.$$

The displacement vector of the first leg is

$$\vec{L}_1 = L_{1x}\,\hat{i} + L_{1y}\,\hat{j} = (141.4\,\hat{i} - 141.4\,\hat{j})\,\text{m}.$$

On the second leg of Trooper's wanderings, the magnitude of the displacement is $L_2 = 300.0\,\text{m}$ and the direction is north. The direction angle is $\theta_2 = +90°$. We obtain the following results:

$$L_{2x} = L_2 \cos \theta_2 = (300.0\,\text{m}) \cos 90° = 0.0,$$
$$L_{2y} = L_2 \sin \theta_2 = (300.0\,\text{m}) \sin 90° = 300.0\,\text{m},$$
$$\vec{L}_2 = L_{2x}\,\hat{i} + L_{2y}\,\hat{j} = (300.0\,\text{m})\,\hat{j}.$$

On the third leg, the displacement magnitude is $L_3 = 50.0\,\text{m}$ and the direction is $30°$ west of north. The direction angle measured counterclockwise from the eastern direction is $\theta_3 = 30° + 90° = +120°$. This gives the following answers:

$$L_{3x} = L_3 \cos \theta_3 = (50.0\,\text{m}) \cos 120° = -25.0\,\text{m},$$
$$L_{3y} = L_3 \sin \theta_3 = (50.0\,\text{m}) \sin 120° = +43.3\,\text{m},$$
$$\vec{L}_3 = L_{3x}\,\hat{i} + L_{3y}\,\hat{j} = (-25.0\,\hat{i} + 43.3\,\hat{j})\,\text{m}.$$

On the fourth leg of the excursion, the displacement magnitude is $L_4 = 80.0\,\text{m}$ and the direction is south. The direction angle can be taken as either $\theta_4 = -90°$ or $\theta_4 = +270°$. We obtain

$$L_{4x} = L_4 \cos \theta_4 = (80.0 \text{ m}) \cos(-90°) = 0,$$
$$L_{4y} = L_4 \sin \theta_4 = (80.0 \text{ m}) \sin(-90°) = -80.0 \text{ m},$$
$$\overrightarrow{\mathbf{L}}_4 = L_{4x} \hat{\mathbf{i}} + L_{4y} \hat{\mathbf{j}} = (-80.0 \text{ m}) \hat{\mathbf{j}}.$$

On the last leg, the magnitude is $L_5 = 150.0$ m and the angle is $\theta_5 = -23° + 270° = +247°$ (23° west of south), which gives

$$L_{5x} = L_5 \cos \theta_5 = (150.0 \text{ m}) \cos 247° = -58.6 \text{ m},$$
$$L_{5y} = L_5 \sin \theta_5 = (150.0 \text{ m}) \sin 247° = -138.1 \text{ m},$$
$$\overrightarrow{\mathbf{L}}_5 = L_{5x} \hat{\mathbf{i}} + L_{5y} \hat{\mathbf{j}} = (-58.6 \hat{\mathbf{i}} - 138.1 \hat{\mathbf{j}}) \text{m}.$$

2.6   Check Your Understanding If Trooper runs 20 m west before taking a rest, what is his displacement vector?

Polar Coordinates

To describe locations of points or vectors in a plane, we need two orthogonal directions. In the Cartesian coordinate system these directions are given by unit vectors $\hat{\mathbf{i}}$ and $\hat{\mathbf{j}}$ along the x-axis and the y-axis, respectively. The Cartesian coordinate system is very convenient to use in describing displacements and velocities of objects and the forces acting on them. However, it becomes cumbersome when we need to describe the rotation of objects. When describing rotation, we usually work in the **polar coordinate system**.

In the polar coordinate system, the location of point P in a plane is given by two **polar coordinates** (**Figure 2.20**). The first polar coordinate is the **radial coordinate** r, which is the distance of point P from the origin. The second polar coordinate is an angle φ that the radial vector makes with some chosen direction, usually the positive x-direction. In polar coordinates, angles are measured in radians, or rads. The radial vector is attached at the origin and points away from the origin to point P. This radial direction is described by a unit radial vector $\hat{\mathbf{r}}$. The second unit vector $\hat{\mathbf{t}}$ is a vector orthogonal to the radial direction $\hat{\mathbf{r}}$. The positive $+\hat{\mathbf{t}}$ direction indicates how the angle φ changes in the counterclockwise direction. In this way, a point P that has coordinates (x, y) in the rectangular system can be described equivalently in the polar coordinate system by the two polar coordinates (r, φ). **Equation 2.17** is valid for any vector, so we can use it to express the x- and y-coordinates of vector $\overrightarrow{\mathbf{r}}$. In this way, we obtain the connection between the polar coordinates and rectangular coordinates of point P:

$$\begin{cases} x = r \cos \varphi \\ y = r \sin \varphi \end{cases}. \qquad (2.18)$$

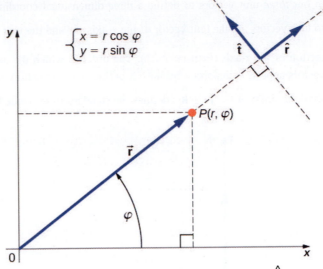

Figure 2.20 Using polar coordinates, the unit vector $\hat{\mathbf{r}}$ defines the positive direction along the radius r (radial direction) and, orthogonal to it, the unit vector $\hat{\mathbf{t}}$ defines the positive direction of rotation by the angle φ.

Example 2.6

Polar Coordinates

A treasure hunter finds one silver coin at a location 20.0 m away from a dry well in the direction $20°$ north of east and finds one gold coin at a location 10.0 m away from the well in the direction $20°$ north of west. What are the polar and rectangular coordinates of these findings with respect to the well?

Strategy

The well marks the origin of the coordinate system and east is the +x-direction. We identify radial distances from the locations to the origin, which are $r_S = 20.0 \, \text{m}$ (for the silver coin) and $r_G = 10.0 \, \text{m}$ (for the gold coin). To find the angular coordinates, we convert $20°$ to radians: $20° = \pi 20/180 = \pi/9$. We use **Equation 2.18** to find the x- and y-coordinates of the coins.

Solution

The angular coordinate of the silver coin is $\varphi_S = \pi/9$, whereas the angular coordinate of the gold coin is $\varphi_G = \pi - \pi/9 = 8\pi/9$. Hence, the polar coordinates of the silver coin are $(r_S, \varphi_S) = (20.0 \, \text{m}, \pi/9)$ and those of the gold coin are $(r_G, \varphi_G) = (10.0 \, \text{m}, 8\pi/9)$. We substitute these coordinates into **Equation 2.18** to obtain rectangular coordinates. For the gold coin, the coordinates are

$$\begin{cases} x_G = r_G \cos \varphi_G = (10.0 \, \text{m}) \cos 8\pi/9 = -9.4 \, \text{m} \\ y_G = r_G \sin \varphi_G = (10.0 \, \text{m}) \sin 8\pi/9 = 3.4 \, \text{m} \end{cases} \Rightarrow (x_G, y_G) = (-9.4 \, \text{m}, 3.4 \, \text{m}).$$

For the silver coin, the coordinates are

$$\begin{cases} x_S = r_S \cos \varphi_S = (20.0 \, \text{m}) \cos \pi/9 = 18.9 \, \text{m} \\ y_S = r_S \sin \varphi_S = (20.0 \, \text{m}) \sin \pi/9 = 6.8 \, \text{m} \end{cases} \Rightarrow (x_S, y_S) = (18.9 \, \text{m}, 6.8 \, \text{m}).$$

Vectors in Three Dimensions

To specify the location of a point in space, we need three coordinates (x, y, z), where coordinates x and y specify locations in a plane, and coordinate z gives a vertical position above or below the plane. Three-dimensional space has three orthogonal directions, so we need not two but *three* unit vectors to define a three-dimensional coordinate system. In the Cartesian coordinate system, the first two unit vectors are the unit vector of the *x*-axis $\hat{\mathbf{i}}$ and the unit vector of the *y*-axis $\hat{\mathbf{j}}$. The third unit vector $\hat{\mathbf{k}}$ is the direction of the *z*-axis (**Figure 2.21**). The order in which the axes are labeled, which is the order in which the three unit vectors appear, is important because it defines the orientation of the coordinate system. The order *x-y-z*, which is equivalent to the order $\hat{\mathbf{i}}$ - $\hat{\mathbf{j}}$ - $\hat{\mathbf{k}}$, defines the standard right-handed coordinate system (positive orientation).

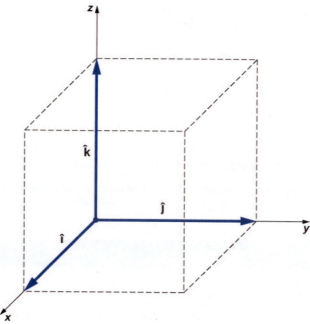

Figure 2.21 Three unit vectors define a Cartesian system in three-dimensional space. The order in which these unit vectors appear defines the orientation of the coordinate system. The order shown here defines the right-handed orientation.

In three-dimensional space, vector $\vec{\mathbf{A}}$ has three vector components: the *x*-component $\vec{\mathbf{A}}_x = A_x \hat{\mathbf{i}}$, which is the part of vector $\vec{\mathbf{A}}$ along the *x*-axis; the *y*-component $\vec{\mathbf{A}}_y = A_y \hat{\mathbf{j}}$, which is the part of $\vec{\mathbf{A}}$ along the *y*-axis; and the *z*-component $\vec{\mathbf{A}}_z = A_z \hat{\mathbf{k}}$, which is the part of the vector along the *z*-axis. A vector in three-dimensional space is the vector sum of its three vector components (**Figure 2.22**):

$$\vec{\mathbf{A}} = A_x \hat{\mathbf{i}} + A_y \hat{\mathbf{j}} + A_z \hat{\mathbf{k}}. \qquad (2.19)$$

If we know the coordinates of its origin $b(x_b, y_b, z_b)$ and of its end $e(x_e, y_e, z_e)$, its scalar components are obtained by taking their differences: A_x and A_y are given by **Equation 2.13** and the *z*-component is given by

$$A_z = z_e - z_b. \qquad (2.20)$$

Magnitude *A* is obtained by generalizing **Equation 2.15** to three dimensions:

$$A = \sqrt{A_x^2 + A_y^2 + A_z^2}. \qquad \textbf{(2.21)}$$

This expression for the vector magnitude comes from applying the Pythagorean theorem twice. As seen in **Figure 2.22**, the diagonal in the *xy*-plane has length $\sqrt{A_x^2 + A_y^2}$ and its square adds to the square A_z^2 to give A^2. Note that when the *z*-component is zero, the vector lies entirely in the *xy*-plane and its description is reduced to two dimensions.

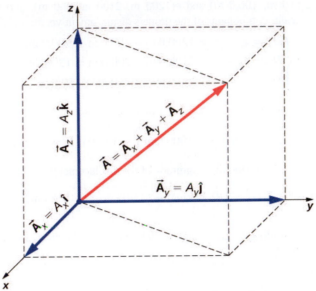

Figure 2.22 A vector in three-dimensional space is the vector sum of its three vector components.

Example 2.7

Takeoff of a Drone

During a takeoff of IAI Heron (**Figure 2.23**), its position with respect to a control tower is 100 m above the ground, 300 m to the east, and 200 m to the north. One minute later, its position is 250 m above the ground, 1200 m to the east, and 2100 m to the north. What is the drone's displacement vector with respect to the control tower? What is the magnitude of its displacement vector?

Figure 2.23 The drone IAI Heron in flight. (credit: SSgt Reynaldo Ramon, USAF)

Strategy

We take the origin of the Cartesian coordinate system as the control tower. The direction of the +*x*-axis is given by unit vector $\hat{\mathbf{i}}$ to the east, the direction of the +y-axis is given by unit vector $\hat{\mathbf{j}}$ to the north, and the direction of the +z-axis is given by unit vector $\hat{\mathbf{k}}$, which points up from the ground. The drone's first position is the origin (or, equivalently, the beginning) of the displacement vector and its second position is the end of the displacement vector.

Solution

We identify $b(300.0 \text{ m}, 200.0 \text{ m}, 100.0 \text{ m})$ and $e(1200 \text{ m}, 2100 \text{ m}, 250 \text{ m})$, and use **Equation 2.13** and **Equation 2.20** to find the scalar components of the drone's displacement vector:

$$\begin{cases} D_x = x_e - x_b = 1200.0 \text{ m} - 300.0 \text{ m} = 900.0 \text{ m}, \\ D_y = y_e - y_b = 2100.0 \text{ m} - 200.0 \text{ m} = 1900.0 \text{ m}, \\ D_z = z_e - z_b = 250.0 \text{ m} - 100.0 \text{ m} = 150.0 \text{ m}. \end{cases}$$

We substitute these components into **Equation 2.19** to find the displacement vector:

$$\vec{\mathbf{D}} = D_x \hat{\mathbf{i}} + D_y \hat{\mathbf{j}} + D_z \hat{\mathbf{k}} = 900.0 \text{ m} \, \hat{\mathbf{i}} + 1900.0 \text{ m} \, \hat{\mathbf{j}} + 150.0 \text{ m} \, \hat{\mathbf{k}} = (0.90 \, \hat{\mathbf{i}} + 1.90 \, \hat{\mathbf{j}} + 0.15 \, \hat{\mathbf{k}}) \text{ km}.$$

We substitute into **Equation 2.21** to find the magnitude of the displacement:

$$D = \sqrt{D_x^2 + D_y^2 + D_z^2} = \sqrt{(0.90 \text{ km})^2 + (1.90 \text{ km})^2 + (0.15 \text{ km})^2} = 2.11 \text{ km}.$$

 2.7 Check Your Understanding If the average velocity vector of the drone in the displacement in **Example 2.7** is $\vec{\mathbf{u}} = (15.0 \, \hat{\mathbf{i}} + 31.7 \, \hat{\mathbf{j}} + 2.5 \, \hat{\mathbf{k}}) \text{m/s}$, what is the magnitude of the drone's velocity vector?

2.3 | Algebra of Vectors

Learning Objectives

By the end of this section, you will be able to:

- Apply analytical methods of vector algebra to find resultant vectors and to solve vector equations for unknown vectors.
- Interpret physical situations in terms of vector expressions.

Vectors can be added together and multiplied by scalars. Vector addition is associative (**Equation 2.8**) and commutative (**Equation 2.7**), and vector multiplication by a sum of scalars is distributive (**Equation 2.9**). Also, scalar multiplication by a sum of vectors is distributive:

$$\alpha(\vec{\mathbf{A}} + \vec{\mathbf{B}}) = \alpha \vec{\mathbf{A}} + \alpha \vec{\mathbf{B}}. \tag{2.22}$$

In this equation, α is any number (a scalar). For example, a vector antiparallel to vector $\vec{\mathbf{A}} = A_x \hat{\mathbf{i}} + A_y \hat{\mathbf{j}} + A_z \hat{\mathbf{k}}$ can be expressed simply by multiplying $\vec{\mathbf{A}}$ by the scalar $\alpha = -1$:

$$-\vec{\mathbf{A}} = -A_x \hat{\mathbf{i}} - A_y \hat{\mathbf{j}} - A_z \hat{\mathbf{k}}. \tag{2.23}$$

Example 2.8

Direction of Motion

In a Cartesian coordinate system where $\hat{\mathbf{i}}$ denotes geographic east, $\hat{\mathbf{j}}$ denotes geographic north, and $\hat{\mathbf{k}}$ denotes altitude above sea level, a military convoy advances its position through unknown territory with velocity $\vec{\mathbf{v}} = (4.0\,\hat{\mathbf{i}} + 3.0\,\hat{\mathbf{j}} + 0.1\,\hat{\mathbf{k}})\text{km/h}$. If the convoy had to retreat, in what geographic direction would it be moving?

Solution

The velocity vector has the third component $\vec{\mathbf{v}}_z = (+0.1\text{km/h})\hat{\mathbf{k}}$, which says the convoy is climbing at a rate of 100 m/h through mountainous terrain. At the same time, its velocity is 4.0 km/h to the east and 3.0 km/h to the north, so it moves on the ground in direction $\tan^{-1}(3/4) \approx 37°$ north of east. If the convoy had to retreat, its new velocity vector $\vec{\mathbf{u}}$ would have to be antiparallel to $\vec{\mathbf{v}}$ and be in the form $\vec{\mathbf{u}} = -\alpha\,\vec{\mathbf{v}}$, where α is a positive number. Thus, the velocity of the retreat would be $\vec{\mathbf{u}} = \alpha(-4.0\,\hat{\mathbf{i}} - 3.0\,\hat{\mathbf{j}} - 0.1\,\hat{\mathbf{k}})\text{km/h}$. The negative sign of the third component indicates the convoy would be descending. The direction angle of the retreat velocity is $\tan^{-1}(-3\alpha/-4\alpha) \approx 37°$ south of west. Therefore, the convoy would be moving on the ground in direction $37°$ south of west while descending on its way back.

The generalization of the number zero to vector algebra is called the **null vector**, denoted by $\vec{\mathbf{0}}$. All components of the null vector are zero, $\vec{\mathbf{0}} = 0\,\hat{\mathbf{i}} + 0\,\hat{\mathbf{j}} + 0\,\hat{\mathbf{k}}$, so the null vector has no length and no direction.

Two vectors $\vec{\mathbf{A}}$ and $\vec{\mathbf{B}}$ are **equal vectors** if and only if their difference is the null vector:

$$\vec{\mathbf{0}} = \vec{\mathbf{A}} - \vec{\mathbf{B}} = (A_x\,\hat{\mathbf{i}} + A_y\,\hat{\mathbf{j}} + A_z\,\hat{\mathbf{k}}) - (B_x\,\hat{\mathbf{i}} + B_y\,\hat{\mathbf{j}} + B_z\,\hat{\mathbf{k}}) = (A_x - B_x)\,\hat{\mathbf{i}} + (A_y - B_y)\,\hat{\mathbf{j}} + (A_z - B_z)\,\hat{\mathbf{k}}.$$

This vector equation means we must have simultaneously $A_x - B_x = 0$, $A_y - B_y = 0$, and $A_z - B_z = 0$. Hence, we can write $\vec{\mathbf{A}} = \vec{\mathbf{B}}$ if and only if the corresponding components of vectors $\vec{\mathbf{A}}$ and $\vec{\mathbf{B}}$ are equal:

$$\vec{\mathbf{A}} = \vec{\mathbf{B}} \Leftrightarrow \begin{cases} A_x = B_x \\ A_y = B_y. \\ A_z = B_z \end{cases} \qquad (2.24)$$

Two vectors are equal when their corresponding scalar components are equal.

Resolving vectors into their scalar components (i.e., finding their scalar components) and expressing them analytically in vector component form (given by **Equation 2.19**) allows us to use vector algebra to find sums or differences of many vectors *analytically* (i.e., without using graphical methods). For example, to find the resultant of two vectors $\vec{\mathbf{A}}$ and $\vec{\mathbf{B}}$, we simply add them component by component, as follows:

$$\vec{\mathbf{R}} = \vec{\mathbf{A}} + \vec{\mathbf{B}} = (A_x\,\hat{\mathbf{i}} + A_y\,\hat{\mathbf{j}} + A_z\,\hat{\mathbf{k}}) + (B_x\,\hat{\mathbf{i}} + B_y\,\hat{\mathbf{j}} + B_z\,\hat{\mathbf{k}}) = (A_x + B_x)\,\hat{\mathbf{i}} + (A_y + B_y)\,\hat{\mathbf{j}} + (A_z + B_z)\,\hat{\mathbf{k}}.$$

In this way, using **Equation 2.24**, scalar components of the resultant vector $\vec{\mathbf{R}} = R_x\,\hat{\mathbf{i}} + R_y\,\hat{\mathbf{j}} + R_z\,\hat{\mathbf{k}}$ are the sums of corresponding scalar components of vectors $\vec{\mathbf{A}}$ and $\vec{\mathbf{B}}$:

$$\begin{cases} R_x = A_x + B_x, \\ R_y = A_y + B_y, \\ R_z = A_z + B_z. \end{cases}$$

Analytical methods can be used to find components of a resultant of many vectors. For example, if we are to sum up N vectors $\vec{F}_1$, $\vec{F}_2$, $\vec{F}_3$, ..., $\vec{F}_N$, where each vector is $\vec{F}_k = F_{kx}\,\hat{i} + F_{ky}\,\hat{j} + F_{kz}\,\hat{k}$, the resultant vector $\vec{F}_R$ is

$$\vec{F}_R = \vec{F}_1 + \vec{F}_2 + \vec{F}_3 + ... + \vec{F}_N = \sum_{k=1}^{N} \vec{F}_k = \sum_{k=1}^{N} \left(F_{kx}\,\hat{i} + F_{ky}\,\hat{j} + F_{kz}\,\hat{k} \right)$$

$$= \left(\sum_{k=1}^{N} F_{kx} \right)\hat{i} + \left(\sum_{k=1}^{N} F_{ky} \right)\hat{j} + \left(\sum_{k=1}^{N} F_{kz} \right)\hat{k}.$$

Therefore, scalar components of the resultant vector are

$$\begin{cases} F_{Rx} = \displaystyle\sum_{k=1}^{N} F_{kx} = F_{1x} + F_{2x} + ... + F_{Nx} \\[2mm] F_{Ry} = \displaystyle\sum_{k=1}^{N} F_{ky} = F_{1y} + F_{2y} + ... + F_{Ny} \\[2mm] F_{Rz} = \displaystyle\sum_{k=1}^{N} F_{kz} = F_{1z} + F_{2z} + ... + F_{Nz}. \end{cases}$$

(2.25)

Having found the scalar components, we can write the resultant in vector component form:

$$\vec{F}_R = F_{Rx}\,\hat{i} + F_{Ry}\,\hat{j} + F_{Rz}\,\hat{k}.$$

Analytical methods for finding the resultant and, in general, for solving vector equations are very important in physics because many physical quantities are vectors. For example, we use this method in kinematics to find resultant displacement vectors and resultant velocity vectors, in mechanics to find resultant force vectors and the resultants of many derived vector quantities, and in electricity and magnetism to find resultant electric or magnetic vector fields.

Example 2.9

Analytical Computation of a Resultant

Three displacement vectors $\vec{A}$, $\vec{B}$, and $\vec{C}$ in a plane (**Figure 2.13**) are specified by their magnitudes $A = 10.0$, $B = 7.0$, and $C = 8.0$, respectively, and by their respective direction angles with the horizontal direction $\alpha = 35°$, $\beta = -110°$, and $\gamma = 30°$. The physical units of the magnitudes are centimeters. Resolve the vectors to their scalar components and find the following vector sums: (a) $\vec{R} = \vec{A} + \vec{B} + \vec{C}$, (b) $\vec{D} = \vec{A} - \vec{B}$, and (c) $\vec{S} = \vec{A} - 3\vec{B} + \vec{C}$.

Strategy

First, we use **Equation 2.17** to find the scalar components of each vector and then we express each vector in its vector component form given by **Equation 2.12**. Then, we use analytical methods of vector algebra to find the resultants.

Solution

We resolve the given vectors to their scalar components:

$$\begin{cases} A_x = A \cos \alpha = (10.0 \text{ cm}) \cos 35° = 8.19 \text{ cm} \\ A_y = A \sin \alpha = (10.0 \text{ cm}) \sin 35° = 5.73 \text{ cm} \end{cases}$$

$$\begin{cases} B_x = B \cos \beta = (7.0 \text{ cm}) \cos (-110°) = -2.39 \text{ cm} \\ B_y = B \sin \beta = (7.0 \text{ cm}) \sin (-110°) = -6.58 \text{ cm} \end{cases}$$

$$\begin{cases} C_x = C \cos \gamma = (8.0 \text{ cm}) \cos 30° = 6.93 \text{ cm} \\ C_y = C \sin \gamma = (8.0 \text{ cm}) \sin 30° = 4.00 \text{ cm} \end{cases}$$

For (a) we may substitute directly into **Equation 2.25** to find the scalar components of the resultant:

$$\begin{cases} R_x = A_x + B_x + C_x = 8.19 \text{ cm} - 2.39 \text{ cm} + 6.93 \text{ cm} = 12.73 \text{ cm} \\ R_y = A_y + B_y + C_y = 5.73 \text{ cm} - 6.58 \text{ cm} + 4.00 \text{ cm} = 3.15 \text{ cm} \end{cases}$$

Therefore, the resultant vector is $\vec{R} = R_x \hat{i} + R_y \hat{j} = (12.7 \hat{i} + 3.1 \hat{j})$cm.

For (b), we may want to write the vector difference as

$$\vec{D} = \vec{A} - \vec{B} = (A_x \hat{i} + A_y \hat{j}) - (B_x \hat{i} + B_y \hat{j}) = (A_x - B_x) \hat{i} + (A_y - B_y) \hat{j}.$$

Then, the scalar components of the vector difference are

$$\begin{cases} D_x = A_x - B_x = 8.19 \text{ cm} - (-2.39 \text{ cm}) = 10.58 \text{ cm} \\ D_y = A_y - B_y = 5.73 \text{ cm} - (-6.58 \text{ cm}) = 12.31 \text{ cm} \end{cases}$$

Hence, the difference vector is $\vec{D} = D_x \hat{i} + D_y \hat{j} = (10.6 \hat{i} + 12.3 \hat{j})$cm.

For (c), we can write vector $\vec{S}$ in the following explicit form:

$$\vec{S} = \vec{A} - 3\vec{B} + \vec{C} = (A_x \hat{i} + A_y \hat{j}) - 3(B_x \hat{i} + B_y \hat{j}) + (C_x \hat{i} + C_y \hat{j})$$

$$= (A_x - 3B_x + C_x) \hat{i} + (A_y - 3B_y + C_y) \hat{j}.$$

Then, the scalar components of $\vec{S}$ are

$$\begin{cases} S_x = A_x - 3B_x + C_x = 8.19 \text{ cm} - 3(-2.39 \text{ cm}) + 6.93 \text{ cm} = 22.29 \text{ cm} \\ S_y = A_y - 3B_y + C_y = 5.73 \text{ cm} - 3(-6.58 \text{ cm}) + 4.00 \text{ cm} = 29.47 \text{ cm} \end{cases}$$

The vector is $\vec{S} = S_x \hat{i} + S_y \hat{j} = (22.3 \hat{i} + 29.5 \hat{j})$cm.

Significance

Having found the vector components, we can illustrate the vectors by graphing or we can compute magnitudes and direction angles, as shown in **Figure 2.24**. Results for the magnitudes in (b) and (c) can be compared with results for the same problems obtained with the graphical method, shown in **Figure 2.14** and **Figure 2.15**. Notice that the analytical method produces exact results and its accuracy is not limited by the resolution of a ruler or a protractor, as it was with the graphical method used in **Example 2.2** for finding this same resultant.

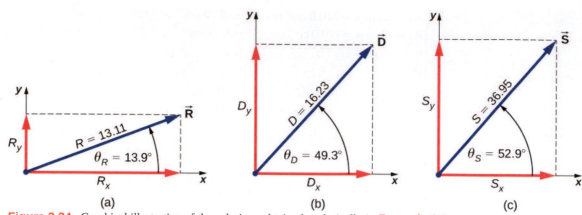

Figure 2.24 Graphical illustration of the solutions obtained analytically in **Example 2.9**.

 2.8 **Check Your Understanding** Three displacement vectors $\vec{A}$, $\vec{B}$, and $\vec{F}$ (**Figure 2.13**) are specified by their magnitudes $A = 10.00$, $B = 7.00$, and $F = 20.00$, respectively, and by their respective direction angles with the horizontal direction $\alpha = 35°$, $\beta = -110°$, and $\varphi = 110°$. The physical units of the magnitudes are centimeters. Use the analytical method to find vector $\vec{G} = \vec{A} + 2\vec{B} - \vec{F}$. Verify that $G = 28.15$ cm and that $\theta_G = -68.65°$.

Example 2.10

The Tug-of-War Game

Four dogs named Astro, Balto, Clifford, and Dug play a tug-of-war game with a toy (**Figure 2.25**). Astro pulls on the toy in direction $\alpha = 55°$ south of east, Balto pulls in direction $\beta = 60°$ east of north, and Clifford pulls in direction $\gamma = 55°$ west of north. Astro pulls strongly with 160.0 units of force (N), which we abbreviate as $A = 160.0$ N. Balto pulls even stronger than Astro with a force of magnitude $B = 200.0$ N, and Clifford pulls with a force of magnitude $C = 140.0$ N. When Dug pulls on the toy in such a way that his force balances out the resultant of the other three forces, the toy does not move in any direction. With how big a force and in what direction must Dug pull on the toy for this to happen?

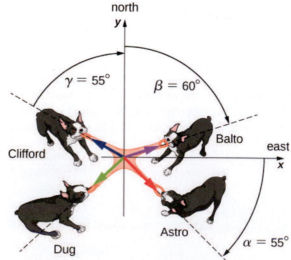

Figure 2.25 Four dogs play a tug-of-war game with a toy.

Strategy

We assume that east is the direction of the positive *x*-axis and north is the direction of the positive *y*-axis. As in **Example 2.9**, we have to resolve the three given forces— $\vec{A}$ (the pull from Astro), $\vec{B}$ (the pull from Balto), and $\vec{C}$ (the pull from Clifford)—into their scalar components and then find the scalar components of the resultant vector $\vec{R} = \vec{A} + \vec{B} + \vec{C}$. When the pulling force $\vec{D}$ from Dug balances out this resultant, the sum of $\vec{D}$ and $\vec{R}$ must give the null vector $\vec{D} + \vec{R} = \vec{0}$. This means that $\vec{D} = -\vec{R}$, so the pull from Dug must be antiparallel to $\vec{R}$.

Solution

The direction angles are $\theta_A = -\alpha = -55°$, $\theta_B = 90° - \beta = 30°$, and $\theta_C = 90° + \gamma = 145°$, and substituting them into **Equation 2.17** gives the scalar components of the three given forces:

$$\begin{cases} A_x = A\cos\theta_A = (160.0\,\text{N})\cos(-55°) = +91.8\,\text{N} \\ A_y = A\sin\theta_A = (160.0\,\text{N})\sin(-55°) = -131.1\,\text{N} \end{cases}$$
$$\begin{cases} B_x = B\cos\theta_B = (200.0\,\text{N})\cos 30° = +173.2\,\text{N} \\ B_y = B\sin\theta_B = (200.0\,\text{N})\sin 30° = +100.0\,\text{N} \end{cases}$$
$$\begin{cases} C_x = C\cos\theta_C = (140.0\,\text{N})\cos 145° = -114.7\,\text{N} \\ C_y = C\sin\theta_C = (140.0\,\text{N})\sin 145° = +80.3\,\text{N} \end{cases}$$

Now we compute scalar components of the resultant vector $\vec{R} = \vec{A} + \vec{B} + \vec{C}$:

$$\begin{cases} R_x = A_x + B_x + C_x = +91.8\,\text{N} + 173.2\,\text{N} - 114.7\,\text{N} = +150.3\,\text{N} \\ R_y = A_y + B_y + C_y = -131.1\,\text{N} + 100.0\,\text{N} + 80.3\,\text{N} = +49.2\,\text{N} \end{cases}$$

The antiparallel vector to the resultant $\vec{R}$ is

$$\vec{D} = -\vec{R} = -R_x\,\hat{i} - R_y\,\hat{j} = (-150.3\,\hat{i} - 49.2\,\hat{j})\,\text{N}.$$

The magnitude of Dug's pulling force is

$$D = \sqrt{D_x^2 + D_y^2} = \sqrt{(-150.3)^2 + (-49.2)^2}\,\text{N} = 158.1\,\text{N}.$$

The direction of Dug's pulling force is

$$\theta = \tan^{-1}\left(\frac{D_y}{D_x}\right) = \tan^{-1}\left(\frac{-49.2\,\text{N}}{-150.3\,\text{N}}\right) = \tan^{-1}\left(\frac{49.2}{150.3}\right) = 18.1°.$$

Dong pulls in the direction 18.1° south of west because both components are negative, which means the pull vector lies in the third quadrant (**Figure 2.19**).

 2.9 Check Your Understanding Suppose that Balto in **Example 2.10** leaves the game to attend to more important matters, but Astro, Clifford, and Dug continue playing. Astro and Clifford's pull on the toy does not change, but Dug runs around and bites on the toy in a different place. With how big a force and in what direction must Dug pull on the toy now to balance out the combined pulls from Clifford and Astro? Illustrate this situation by drawing a vector diagram indicating all forces involved.

Example 2.11

Vector Algebra

Find the magnitude of the vector $\vec{C}$ that satisfies the equation $2\vec{A} - 6\vec{B} + 3\vec{C} = 2\hat{j}$, where $\vec{A} = \hat{i} - 2\hat{k}$ and $\vec{B} = -\hat{j} + \hat{k}/2$.

Strategy

We first solve the given equation for the unknown vector $\vec{C}$. Then we substitute $\vec{A}$ and $\vec{B}$; group the terms along each of the three directions $\hat{i}$, $\hat{j}$, and $\hat{k}$; and identify the scalar components C_x, C_y, and C_z. Finally, we substitute into **Equation 2.21** to find magnitude C.

Solution

$$2\vec{A} - 6\vec{B} + 3\vec{C} = 2\hat{j}$$

$$3\vec{C} = 2\hat{j} - 2\vec{A} + 6\vec{B}$$

$$\vec{C} = \tfrac{2}{3}\hat{j} - \tfrac{2}{3}\vec{A} + 2\vec{B}$$

$$= \tfrac{2}{3}\hat{j} - \tfrac{2}{3}(\hat{i} - 2\hat{k}) + 2\left(-\hat{j} + \tfrac{\hat{k}}{2}\right) = \tfrac{2}{3}\hat{j} - \tfrac{2}{3}\hat{i} + \tfrac{4}{3}\hat{k} - 2\hat{j} + \hat{k}$$

$$= -\tfrac{2}{3}\hat{i} + \left(\tfrac{2}{3} - 2\right)\hat{j} + \left(\tfrac{4}{3} + 1\right)\hat{k}$$

$$= -\tfrac{2}{3}\hat{i} - \tfrac{4}{3}\hat{j} + \tfrac{7}{3}\hat{k}.$$

The components are $C_x = -2/3$, $C_y = -4/3$, and $C_z = 7/3$, and substituting into **Equation 2.21** gives

$$C = \sqrt{C_x^2 + C_y^2 + C_z^2} = \sqrt{(-2/3)^2 + (-4/3)^2 + (7/3)^2} = \sqrt{23}/3.$$

Example 2.12

Displacement of a Skier

Starting at a ski lodge, a cross-country skier goes 5.0 km north, then 3.0 km west, and finally 4.0 km southwest before taking a rest. Find his total displacement vector relative to the lodge when he is at the rest point. How far and in what direction must he ski from the rest point to return directly to the lodge?

Strategy

We assume a rectangular coordinate system with the origin at the ski lodge and with the unit vector $\hat{i}$ pointing east and the unit vector $\hat{j}$ pointing north. There are three displacements: $\vec{D}_1$, $\vec{D}_2$, and $\vec{D}_3$. We identify their magnitudes as $D_1 = 5.0\,\text{km}$, $D_2 = 3.0\,\text{km}$, and $D_3 = 4.0\,\text{km}$. We identify their directions are the angles $\theta_1 = 90°$, $\theta_2 = 180°$, and $\theta_3 = 180° + 45° = 225°$. We resolve each displacement vector to its scalar components and substitute the components into **Equation 2.25** to obtain the scalar components of the resultant displacement $\vec{D}$ from the lodge to the rest point. On the way back from the rest point to the lodge, the displacement is $\vec{B} = -\vec{D}$. Finally, we find the magnitude and direction of $\vec{B}$.

Solution

Scalar components of the displacement vectors are

$$\begin{cases} D_{1x} = D_1 \cos\theta_1 = (5.0\,\text{km})\cos 90° = 0 \\ D_{1y} = D_1 \sin\theta_1 = (5.0\,\text{km})\sin 90° = 5.0\,\text{km} \end{cases}$$

$$\begin{cases} D_{2x} = D_2 \cos\theta_2 = (3.0\,\text{km})\cos 180° = -3.0\,\text{km} \\ D_{2y} = D_2 \sin\theta_2 = (3.0\,\text{km})\sin 180° = 0 \end{cases}$$

$$\begin{cases} D_{3x} = D_3 \cos\theta_3 = (4.0\,\text{km})\cos 225° = -2.8\,\text{km} \\ D_{3y} = D_3 \sin\theta_3 = (4.0\,\text{km})\sin 225° = -2.8\,\text{km} \end{cases}$$

Scalar components of the net displacement vector are

$$\begin{cases} D_x = D_{1x} + D_{2x} + D_{3x} = (0 - 3.0 - 2.8)\text{km} = -5.8\,\text{km} \\ D_y = D_{1y} + D_{2y} + D_{3y} = (5.0 + 0 - 2.8)\text{km} = +2.2\,\text{km} \end{cases}$$

Hence, the skier's net displacement vector is $\vec{D} = D_x\,\hat{i} + D_y\,\hat{j} = (-5.8\,\hat{i} + 2.2\,\hat{j})\text{km}$. On the way back to the lodge, his displacement is $\vec{B} = -\vec{D} = -(-5.8\,\hat{i} + 2.2\,\hat{j})\text{km} = (5.8\,\hat{i} - 2.2\,\hat{j})\text{km}$. Its magnitude is $B = \sqrt{B_x^2 + B_y^2} = \sqrt{(5.8)^2 + (-2.2)^2}\,\text{km} = 6.2\,\text{km}$ and its direction angle is $\theta = \tan^{-1}(-2.2/5.8) = -20.8°$. Therefore, to return to the lodge, he must go 6.2 km in a direction about $21°$ south of east.

Significance

Notice that no figure is needed to solve this problem by the analytical method. Figures are required when using a graphical method; however, we can check if our solution makes sense by sketching it, which is a useful final step in solving any vector problem.

Example 2.13

Displacement of a Jogger

A jogger runs up a flight of 200 identical steps to the top of a hill and then runs along the top of the hill 50.0 m before he stops at a drinking fountain (**Figure 2.26**). His displacement vector from point A at the bottom of the steps to point B at the fountain is $\vec{D}_{AB} = (-90.0\,\hat{i} + 30.0\,\hat{j})\text{m}$. What is the height and width of each step in the flight? What is the actual distance the jogger covers? If he makes a loop and returns to point A, what is his net displacement vector?

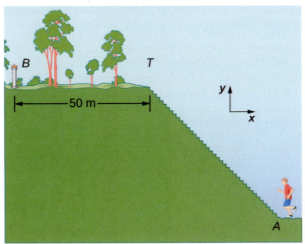

Figure 2.26 A jogger runs up a flight of steps.

Strategy

The displacement vector $\vec{D}_{AB}$ is the vector sum of the jogger's displacement vector $\vec{D}_{AT}$ along the stairs (from point A at the bottom of the stairs to point T at the top of the stairs) and his displacement vector $\vec{D}_{TB}$ on the top of the hill (from point A at the top of the stairs to the fountain at point T). We must find the horizontal and the vertical components of $\vec{D}_{AT}$. If each step has width w and height h, the horizontal component of $\vec{D}_{AT}$ must have a length of $200w$ and the vertical component must have a length of $200h$. The actual distance the jogger covers is the sum of the distance he runs up the stairs and the distance of 50.0 m that he runs along the top of the hill.

Solution

In the coordinate system indicated in **Figure 2.26**, the jogger's displacement vector on the top of the hill is $\vec{D}_{TB} = (-50.0 \text{ m})\,\hat{i}$. His net displacement vector is

$$\vec{D}_{AB} = \vec{D}_{AT} + \vec{D}_{TB}.$$

Therefore, his displacement vector $\vec{D}_{TB}$ along the stairs is

$$\vec{D}_{AT} = \vec{D}_{AB} - \vec{D}_{TB} = (-90.0\,\hat{i} + 30.0\,\hat{j})\text{m} - (-50.0 \text{ m})\,\hat{i} = [(-90.0 + 50.0)\,\hat{i} + 30.0\,\hat{j}\,)]\text{m}$$
$$= (-40.0\,\hat{i} + 30.0\,\hat{j}\,)\text{m}.$$

Its scalar components are $D_{ATx} = -40.0 \text{ m}$ and $D_{ATy} = 30.0 \text{ m}$. Therefore, we must have

$$200w = |-40.0|\text{m and } 200h = 30.0 \text{ m}.$$

Hence, the step width is $w = 40.0 \text{ m}/200 = 0.2 \text{ m} = 20$ cm, and the step height is $h = 30.0 \text{ m}/200 = 0.15 \text{ m} = 15$ cm. The distance that the jogger covers along the stairs is

$$D_{AT} = \sqrt{D_{ATx}^2 + D_{ATy}^2} = \sqrt{(-40.0)^2 + (30.0)^2}\,\text{m} = 50.0 \text{ m}.$$

Thus, the actual distance he runs is $D_{AT} + D_{TB} = 50.0 \text{ m} + 50.0 \text{ m} = 100.0 \text{ m}$. When he makes a loop and comes back from the fountain to his initial position at point A, the total distance he covers is twice this distance, or 200.0 m. However, his net displacement vector is zero, because when his final position is the same as his initial position, the scalar components of his net displacement vector are zero (**Equation 2.13**).

In many physical situations, we often need to know the direction of a vector. For example, we may want to know the direction of a magnetic field vector at some point or the direction of motion of an object. We have already said direction is given by a unit vector, which is a dimensionless entity—that is, it has no physical units associated with it. When the vector in question lies along one of the axes in a Cartesian system of coordinates, the answer is simple, because then its unit vector of direction is either parallel or antiparallel to the direction of the unit vector of an axis. For example, the direction of vector $\vec{d} = -5 \text{ m}\,\hat{i}$ is unit vector $\hat{d} = -\hat{i}$. The general rule of finding the unit vector $\hat{V}$ of direction for any vector $\vec{V}$ is to divide it by its magnitude V:

$$\hat{V} = \frac{\vec{V}}{V}. \tag{2.26}$$

We see from this expression that the unit vector of direction is indeed dimensionless because the numerator and the denominator in **Equation 2.26** have the same physical unit. In this way, **Equation 2.26** allows us to express the unit vector of direction in terms of unit vectors of the axes. The following example illustrates this principle.

Example 2.14

The Unit Vector of Direction

If the velocity vector of the military convoy in **Example 2.8** is $\vec{v} = (4.000\,\hat{i} + 3.000\,\hat{j} + 0.100\,\hat{k})$km/h, what is the unit vector of its direction of motion?

Strategy

The unit vector of the convoy's direction of motion is the unit vector $\hat{v}$ that is parallel to the velocity vector. The unit vector is obtained by dividing a vector by its magnitude, in accordance with **Equation 2.26**.

Solution

The magnitude of the vector $\vec{v}$ is

$$v = \sqrt{v_x^2 + v_y^2 + v_z^2} = \sqrt{4.000^2 + 3.000^2 + 0.100^2}\,\text{km/h} = 5.001\,\text{km/h}.$$

To obtain the unit vector $\hat{v}$, divide $\vec{v}$ by its magnitude:

$$\hat{v} = \frac{\vec{v}}{v} = \frac{(4.000\,\hat{i} + 3.000\,\hat{j} + 0.100\,\hat{k})\text{km/h}}{5.001\,\text{km/h}}$$

$$= \frac{(4.000\,\hat{i} + 3.000\,\hat{j} + 0.100\,\hat{k})}{5.001}$$

$$= \frac{4.000}{5.001}\hat{i} + \frac{3.000}{5.001}\hat{j} + \frac{0.100}{5.001}\hat{k}$$

$$= (79.98\,\hat{i} + 59.99\,\hat{j} + 2.00\,\hat{k}) \times 10^{-2}.$$

Significance

Note that when using the analytical method with a calculator, it is advisable to carry out your calculations to at least three decimal places and then round off the final answer to the required number of significant figures, which is the way we performed calculations in this example. If you round off your partial answer too early, you risk your final answer having a huge numerical error, and it may be far off from the exact answer or from a value measured in an experiment.

 2.10 **Check Your Understanding** Verify that vector $\hat{v}$ obtained in **Example 2.14** is indeed a unit vector by computing its magnitude. If the convoy in **Example 2.8** was moving across a desert flatland—that is, if the third component of its velocity was zero—what is the unit vector of its direction of motion? Which geographic direction does it represent?

2.4 | Products of Vectors

Learning Objectives

By the end of this section, you will be able to:

- Explain the difference between the scalar product and the vector product of two vectors.
- Determine the scalar product of two vectors.
- Determine the vector product of two vectors.
- Describe how the products of vectors are used in physics.

A vector can be multiplied by another vector but may not be divided by another vector. There are two kinds of products of vectors used broadly in physics and engineering. One kind of multiplication is a *scalar multiplication of two vectors*.

Taking a scalar product of two vectors results in a number (a scalar), as its name indicates. Scalar products are used to define work and energy relations. For example, the work that a force (a vector) performs on an object while causing its displacement (a vector) is defined as a scalar product of the force vector with the displacement vector. A quite different kind of multiplication is a *vector multiplication of vectors*. Taking a vector product of two vectors returns as a result a vector, as its name suggests. Vector products are used to define other derived vector quantities. For example, in describing rotations, a vector quantity called *torque* is defined as a vector product of an applied force (a vector) and its distance from pivot to force (a vector). It is important to distinguish between these two kinds of vector multiplications because the scalar product is a scalar quantity and a vector product is a vector quantity.

The Scalar Product of Two Vectors (the Dot Product)

Scalar multiplication of two vectors yields a scalar product.

Scalar Product (Dot Product)

The **scalar product** $\vec{A} \cdot \vec{B}$ of two vectors $\vec{A}$ and $\vec{B}$ is a number defined by the equation

$$\vec{A} \cdot \vec{B} = AB \cos \varphi, \tag{2.27}$$

where φ is the angle between the vectors (shown in **Figure 2.27**). The scalar product is also called the **dot product** because of the dot notation that indicates it.

In the definition of the dot product, the direction of angle φ does not matter, and φ can be measured from either of the two vectors to the other because $\cos \varphi = \cos(-\varphi) = \cos(2\pi - \varphi)$. The dot product is a negative number when $90° < \varphi \leq 180°$ and is a positive number when $0° \leq \varphi < 90°$. Moreover, the dot product of two parallel vectors is $\vec{A} \cdot \vec{B} = AB \cos 0° = AB$, and the dot product of two antiparallel vectors is $\vec{A} \cdot \vec{B} = AB \cos 180° = -AB$. The scalar product of two *orthogonal vectors* vanishes: $\vec{A} \cdot \vec{B} = AB \cos 90° = 0$. The scalar product of a vector with itself is the square of its magnitude:

$$\vec{A}^2 \equiv \vec{A} \cdot \vec{A} = AA \cos 0° = A^2. \tag{2.28}$$

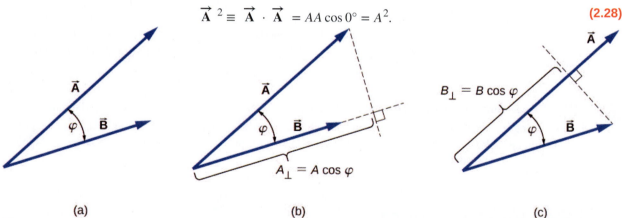

(a)   (b)   (c)

Figure 2.27   The scalar product of two vectors. (a) The angle between the two vectors. (b) The orthogonal projection $A_\perp$ of vector $\vec{A}$ onto the direction of vector $\vec{B}$. (c) The orthogonal projection $B_\perp$ of vector $\vec{B}$ onto the direction of vector $\vec{A}$.

Example 2.15

The Scalar Product

For the vectors shown in **Figure 2.13**, find the scalar product $\vec{A} \cdot \vec{F}$.

Strategy

From **Figure 2.13**, the magnitudes of vectors $\vec{A}$ and $\vec{F}$ are $A = 10.0$ and $F = 20.0$. Angle θ, between them, is the difference: $\theta = \varphi - \alpha = 110° - 35° = 75°$. Substituting these values into **Equation 2.27** gives the scalar product.

Solution

A straightforward calculation gives us

$$\vec{A} \cdot \vec{F} = AF \cos\theta = (10.0)(20.0)\cos 75° = 51.76.$$

 2.11 **Check Your Understanding** For the vectors given in **Figure 2.13**, find the scalar products $\vec{A} \cdot \vec{B}$ and $\vec{F} \cdot \vec{C}$.

In the Cartesian coordinate system, scalar products of the unit vector of an axis with other unit vectors of axes always vanish because these unit vectors are orthogonal:

$$\hat{i} \cdot \hat{j} = \left|\hat{i}\right|\left|\hat{j}\right| \cos 90° = (1)(1)(0) = 0, \qquad (2.29)$$

$$\hat{i} \cdot \hat{k} = \left|\hat{i}\right|\left|\hat{k}\right| \cos 90° = (1)(1)(0) = 0,$$

$$\hat{k} \cdot \hat{j} = \left|\hat{k}\right|\left|\hat{j}\right| \cos 90° = (1)(1)(0) = 0.$$

In these equations, we use the fact that the magnitudes of all unit vectors are one: $\left|\hat{i}\right| = \left|\hat{j}\right| = \left|\hat{k}\right| = 1$. For unit vectors of the axes, **Equation 2.28** gives the following identities:

$$\hat{i} \cdot \hat{i} = i^2 = \hat{j} \cdot \hat{j} = j^2 = \hat{k} \cdot \hat{k} = k^2 = 1. \qquad (2.30)$$

The scalar product $\vec{A} \cdot \vec{B}$ can also be interpreted as either the product of B with the orthogonal projection $A_\perp$ of vector $\vec{A}$ onto the direction of vector $\vec{B}$ (**Figure 2.27**(b)) or the product of A with the orthogonal projection $B_\perp$ of vector $\vec{B}$ onto the direction of vector $\vec{A}$ (**Figure 2.27**(c)):

$$\begin{aligned}\vec{A} \cdot \vec{B} &= AB \cos\varphi \\ &= B(A \cos\varphi) = BA_\perp \\ &= A(B \cos\varphi) = AB_\perp .\end{aligned}$$

For example, in the rectangular coordinate system in a plane, the scalar x-component of a vector is its dot product with the unit vector $\hat{i}$, and the scalar y-component of a vector is its dot product with the unit vector $\hat{j}$:

$$\begin{cases}\vec{A} \cdot \hat{i} = \left|\vec{A}\right|\left|\hat{i}\right| \cos\theta_A = A \cos\theta_A = A_x \\ \vec{A} \cdot \hat{j} = \left|\vec{A}\right|\left|\hat{j}\right| \cos(90° - \theta_A) = A \sin\theta_A = A_y\end{cases}$$

Scalar multiplication of vectors is commutative,

$$\vec{A} \cdot \vec{B} = \vec{B} \cdot \vec{A}, \qquad (2.31)$$

and obeys the distributive law:

$$\vec{A} \cdot (\vec{B} + \vec{C}) = \vec{A} \cdot \vec{B} + \vec{A} \cdot \vec{C}. \qquad (2.32)$$

We can use the commutative and distributive laws to derive various relations for vectors, such as expressing the dot product of two vectors in terms of their scalar components.

 2.12

Check Your Understanding For vector $\vec{A} = A_x\,\hat{i} + A_y\,\hat{j} + A_z\,\hat{k}$ in a rectangular coordinate system, use **Equation 2.29** through **Equation 2.32** to show that $\vec{A} \cdot \hat{i} = A_x$ $\vec{A} \cdot \hat{j} = A_y$ and $\vec{A} \cdot \hat{k} = A_z$.

When the vectors in **Equation 2.27** are given in their vector component forms,

$$\vec{A} = A_x\,\hat{i} + A_y\,\hat{j} + A_z\,\hat{k} \text{ and } \vec{B} = B_x\,\hat{i} + B_y\,\hat{j} + B_z\,\hat{k},$$

we can compute their scalar product as follows:

$$\begin{aligned}
\vec{A} \cdot \vec{B} &= (A_x\,\hat{i} + A_y\,\hat{j} + A_z\,\hat{k}) \cdot (B_x\,\hat{i} + B_y\,\hat{j} + B_z\,\hat{k}) \\
&= A_x B_x\,\hat{i} \cdot \hat{i} + A_x B_y\,\hat{i} \cdot \hat{j} + A_x B_z\,\hat{i} \cdot \hat{k} \\
&\quad + A_y B_x\,\hat{j} \cdot \hat{i} + A_y B_y\,\hat{j} \cdot \hat{j} + A_y B_z\,\hat{j} \cdot \hat{k} \\
&\quad + A_z B_x\,\hat{k} \cdot \hat{i} + A_z B_y\,\hat{k} \cdot \hat{j} + A_z B_z\,\hat{k} \cdot \hat{k}.
\end{aligned}$$

Since scalar products of two different unit vectors of axes give zero, and scalar products of unit vectors with themselves give one (see **Equation 2.29** and **Equation 2.30**), there are only three nonzero terms in this expression. Thus, the scalar product simplifies to

$$\vec{A} \cdot \vec{B} = A_x B_x + A_y B_y + A_z B_z. \qquad (2.33)$$

We can use **Equation 2.33** for the scalar product in terms of scalar components of vectors to find the angle between two vectors. When we divide **Equation 2.27** by AB, we obtain the equation for $\cos \varphi$, into which we substitute **Equation 2.33**:

$$\cos \varphi = \frac{\vec{A} \cdot \vec{B}}{AB} = \frac{A_x B_x + A_y B_y + A_z B_z}{AB}. \qquad (2.34)$$

Angle φ between vectors $\vec{A}$ and $\vec{B}$ is obtained by taking the inverse cosine of the expression in **Equation 2.34**.

Example 2.16

Angle between Two Forces

Three dogs are pulling on a stick in different directions, as shown in **Figure 2.28**. The first dog pulls with force $\vec{F}_1 = (10.0\,\hat{i} - 20.4\,\hat{j} + 2.0\,\hat{k})$N , the second dog pulls with force $\vec{F}_2 = (-15.0\,\hat{i} - 6.2\,\hat{k})$N , and the third dog pulls with force $\vec{F}_3 = (5.0\,\hat{i} + 12.5\,\hat{j})$N . What is the angle between forces $\vec{F}_1$ and $\vec{F}_2$?

Figure 2.28 Three dogs are playing with a stick.

Strategy

The components of force vector $\vec{F}_1$ are $F_{1x} = 10.0\,\text{N}$, $F_{1y} = -20.4\,\text{N}$, and $F_{1z} = 2.0\,\text{N}$, whereas those of force vector $\vec{F}_2$ are $F_{2x} = -15.0\,\text{N}$, $F_{2y} = 0.0\,\text{N}$, and $F_{2z} = -6.2\,\text{N}$. Computing the scalar product of these vectors and their magnitudes, and substituting into **Equation 2.34** gives the angle of interest.

Solution

The magnitudes of forces $\vec{F}_1$ and $\vec{F}_2$ are

$$F_1 = \sqrt{F_{1x}^2 + F_{1y}^2 + F_{1z}^2} = \sqrt{10.0^2 + 20.4^2 + 2.0^2}\,\text{N} = 22.8\,\text{N}$$

and

$$F_2 = \sqrt{F_{2x}^2 + F_{2y}^2 + F_{2z}^2} = \sqrt{15.0^2 + 6.2^2}\,\text{N} = 16.2\,\text{N}.$$

Substituting the scalar components into **Equation 2.33** yields the scalar product

$$\begin{aligned}
\vec{F}_1 \cdot \vec{F}_2 &= F_{1x}F_{2x} + F_{1y}F_{2y} + F_{1z}F_{2z} \\
&= (10.0\,\text{N})(-15.0\,\text{N}) + (-20.4\,\text{N})(0.0\,\text{N}) + (2.0\,\text{N})(-6.2\,\text{N}) \\
&= -162.4\,\text{N}^2.
\end{aligned}$$

Finally, substituting everything into **Equation 2.34** gives the angle

$$\cos\varphi = \frac{\vec{F}_1 \cdot \vec{F}_2}{F_1 F_2} = \frac{-162.4\,\text{N}^2}{(22.8\,\text{N})(16.2\,\text{N})} = -0.439 \Rightarrow \varphi = \cos^{-1}(-0.439) = 116.0°.$$

Significance

Notice that when vectors are given in terms of the unit vectors of axes, we can find the angle between them without knowing the specifics about the geographic directions the unit vectors represent. Here, for example, the +x-direction might be to the east and the +y-direction might be to the north. But, the angle between the forces in the problem is the same if the +x-direction is to the west and the +y-direction is to the south.

2.13 **Check Your Understanding** Find the angle between forces $\vec{F}_1$ and $\vec{F}_3$ in **Example 2.16**.

Example 2.17

The Work of a Force

When force $\vec{F}$ pulls on an object and when it causes its displacement $\vec{D}$, we say the force performs work. The amount of work the force does is the scalar product $\vec{F} \cdot \vec{D}$. If the stick in **Example 2.16** moves momentarily and gets displaced by vector $\vec{D} = (-7.9\,\hat{j} - 4.2\,\hat{k})$ cm, how much work is done by the third dog in **Example 2.16**?

Strategy

We compute the scalar product of displacement vector $\vec{D}$ with force vector $\vec{F}_3 = (5.0\,\hat{i} + 12.5\,\hat{j})$N, which is the pull from the third dog. Let's use W_3 to denote the work done by force $\vec{F}_3$ on displacement $\vec{D}$.

Solution

Calculating the work is a straightforward application of the dot product:

$$\begin{aligned}
W_3 &= \vec{F}_3 \cdot \vec{D} = F_{3x}D_x + F_{3y}D_y + F_{3z}D_z \\
&= (5.0\,\text{N})(0.0\,\text{cm}) + (12.5\,\text{N})(-7.9\,\text{cm}) + (0.0\,\text{N})(-4.2\,\text{cm}) \\
&= -98.7\,\text{N}\cdot\text{cm}.
\end{aligned}$$

Significance

The SI unit of work is called the joule (J), where $1\,\text{J} = 1\,\text{N}\cdot\text{m}$. The unit $\text{cm}\cdot\text{N}$ can be written as $10^{-2}\,\text{m}\cdot\text{N} = 10^{-2}\,\text{J}$, so the answer can be expressed as $W_3 = -0.9875\,\text{J} \approx -1.0\,\text{J}$.

 2.14 **Check Your Understanding** How much work is done by the first dog and by the second dog in **Example 2.16** on the displacement in **Example 2.17**?

The Vector Product of Two Vectors (the Cross Product)

Vector multiplication of two vectors yields a vector product.

Vector Product (Cross Product)

The **vector product** of two vectors $\vec{A}$ and $\vec{B}$ is denoted by $\vec{A} \times \vec{B}$ and is often referred to as a **cross product**. The vector product is a vector that has its direction perpendicular to both vectors $\vec{A}$ and $\vec{B}$. In other words, vector $\vec{A} \times \vec{B}$ is perpendicular to the plane that contains vectors $\vec{A}$ and $\vec{B}$, as shown in **Figure 2.29**. The magnitude of the vector product is defined as

$$\left| \vec{A} \times \vec{B} \right| = AB \sin\varphi, \tag{2.35}$$

where angle φ, between the two vectors, is measured from vector $\vec{A}$ (first vector in the product) to vector $\vec{B}$ (second vector in the product), as indicated in **Figure 2.29**, and is between $0°$ and $180°$.

According to **Equation 2.35**, the vector product vanishes for pairs of vectors that are either parallel $(\varphi = 0°)$ or antiparallel $(\varphi = 180°)$ because $\sin 0° = \sin 180° = 0$.

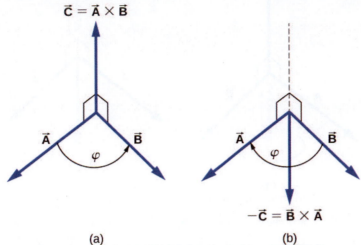

(a)   (b)

Figure 2.29 The vector product of two vectors is drawn in three-dimensional space. (a) The vector product $\vec{A} \times \vec{B}$ is a vector perpendicular to the plane that contains vectors $\vec{A}$ and $\vec{B}$. Small squares drawn in perspective mark right angles between $\vec{A}$ and $\vec{C}$, and between $\vec{B}$ and $\vec{C}$ so that if $\vec{A}$ and $\vec{B}$ lie on the floor, vector $\vec{C}$ points vertically upward to the ceiling. (b) The vector product $\vec{B} \times \vec{A}$ is a vector antiparallel to vector $\vec{A} \times \vec{B}$.

On the line perpendicular to the plane that contains vectors $\vec{A}$ and $\vec{B}$ there are two alternative directions—either up or down, as shown in **Figure 2.29**—and the direction of the vector product may be either one of them. In the standard right-handed orientation, where the angle between vectors is measured counterclockwise from the first vector, vector $\vec{A} \times \vec{B}$ points *upward*, as seen in **Figure 2.29**(a). If we reverse the order of multiplication, so that now $\vec{B}$ comes first in the product, then vector $\vec{B} \times \vec{A}$ must point *downward*, as seen in **Figure 2.29**(b). This means that vectors $\vec{A} \times \vec{B}$ and $\vec{B} \times \vec{A}$ are *antiparallel* to each other and that vector multiplication is *not* commutative but *anticommutative*. The **anticommutative property** means the vector product reverses the sign when the order of multiplication is reversed:

$$\vec{A} \times \vec{B} = -\vec{B} \times \vec{A}. \tag{2.36}$$

The **corkscrew right-hand rule** is a common mnemonic used to determine the direction of the vector product. As shown in **Figure 2.30**, a corkscrew is placed in a direction perpendicular to the plane that contains vectors $\vec{A}$ and $\vec{B}$, and its handle is turned in the direction from the first to the second vector in the product. The direction of the cross product is given by the progression of the corkscrew.

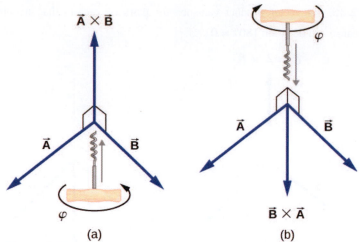

Figure 2.30 The corkscrew right-hand rule can be used to determine the direction of the cross product $\vec{A} \times \vec{B}$. Place a corkscrew in the direction perpendicular to the plane that contains vectors $\vec{A}$ and $\vec{B}$, and turn it in the direction from the first to the second vector in the product. The direction of the cross product is given by the progression of the corkscrew. (a) Upward movement means the cross-product vector points up. (b) Downward movement means the cross-product vector points downward.

Example 2.18

The Torque of a Force

The mechanical advantage that a familiar tool called a *wrench* provides (**Figure 2.31**) depends on magnitude F of the applied force, on its direction with respect to the wrench handle, and on how far from the nut this force is applied. The distance R from the nut to the point where force vector $\vec{F}$ is attached is represented by the radial vector $\vec{R}$. The physical vector quantity that makes the nut turn is called *torque* (denoted by $\vec{\tau}$), and it is the vector product of the distance between the pivot to force with the force: $\vec{\tau} = \vec{R} \times \vec{F}$.

To loosen a rusty nut, a 20.00-N force is applied to the wrench handle at angle $\varphi = 40°$ and at a distance of 0.25 m from the nut, as shown in **Figure 2.31**(a). Find the magnitude and direction of the torque applied to the nut. What would the magnitude and direction of the torque be if the force were applied at angle $\varphi = 45°$, as shown in **Figure 2.31**(b)? For what value of angle φ does the torque have the largest magnitude?

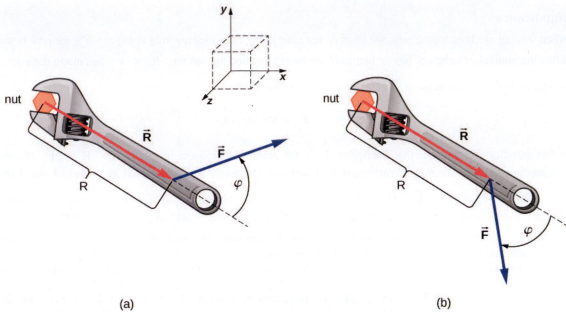

Figure 2.31 A wrench provides grip and mechanical advantage in applying torque to turn a nut. (a) Turn counterclockwise to loosen the nut. (b) Turn clockwise to tighten the nut.

Strategy

We adopt the frame of reference shown in **Figure 2.31**, where vectors $\vec{R}$ and $\vec{F}$ lie in the xy-plane and the origin is at the position of the nut. The radial direction along vector $\vec{R}$ (pointing away from the origin) is the reference direction for measuring the angle φ because $\vec{R}$ is the first vector in the vector product $\vec{\tau} = \vec{R} \times \vec{F}$. Vector $\vec{\tau}$ must lie along the z-axis because this is the axis that is perpendicular to the xy-plane, where both $\vec{R}$ and $\vec{F}$ lie. To compute the magnitude τ, we use **Equation 2.35**. To find the direction of $\vec{\tau}$, we use the corkscrew right-hand rule (**Figure 2.30**).

Solution

For the situation in (a), the corkscrew rule gives the direction of $\vec{R} \times \vec{F}$ in the positive direction of the z-axis. Physically, it means the torque vector $\vec{\tau}$ points out of the page, perpendicular to the wrench handle. We identify $F = 20.00$ N and $R = 0.25$ m, and compute the magnitude using **Equation 2.35**:

$$\tau = \left| \vec{R} \times \vec{F} \right| = RF \sin \varphi = (0.25 \text{ m})(20.00 \text{ N}) \sin 40° = 3.21 \text{ N} \cdot \text{m}.$$

For the situation in (b), the corkscrew rule gives the direction of $\vec{R} \times \vec{F}$ in the negative direction of the z-axis. Physically, it means the vector $\vec{\tau}$ points into the page, perpendicular to the wrench handle. The magnitude of this torque is

$$\tau = \left| \vec{R} \times \vec{F} \right| = RF \sin \varphi = (0.25 \text{ m})(20.00 \text{ N}) \sin 45° = 3.53 \text{ N} \cdot \text{m}.$$

The torque has the largest value when $\sin \varphi = 1$, which happens when $\varphi = 90°$. Physically, it means the wrench is most effective—giving us the best mechanical advantage—when we apply the force perpendicular to the wrench handle. For the situation in this example, this best-torque value is $\tau_{\text{best}} = RF = (0.25 \text{ m})(20.00 \text{ N}) = 5.00 \text{ N} \cdot \text{m}$.

Significance

When solving mechanics problems, we often do not need to use the corkscrew rule at all, as we'll see now in the following equivalent solution. Notice that once we have identified that vector $\overrightarrow{\mathbf{R}} \times \overrightarrow{\mathbf{F}}$ lies along the z-axis, we can write this vector in terms of the unit vector $\hat{\mathbf{k}}$ of the z-axis:

$$\overrightarrow{\mathbf{R}} \times \overrightarrow{\mathbf{F}} = RF \sin \varphi \, \hat{\mathbf{k}}.$$

In this equation, the number that multiplies $\hat{\mathbf{k}}$ is the scalar z-component of the vector $\overrightarrow{\mathbf{R}} \times \overrightarrow{\mathbf{F}}$. In the computation of this component, care must be taken that the angle φ is measured *counterclockwise* from $\overrightarrow{\mathbf{R}}$ (first vector) to $\overrightarrow{\mathbf{F}}$ (second vector). Following this principle for the angles, we obtain $RF \sin(+40°) = +3.2 \, \text{N} \cdot \text{m}$ for the situation in (a), and we obtain $RF \sin(-45°) = -3.5 \, \text{N} \cdot \text{m}$ for the situation in (b). In the latter case, the angle is negative because the graph in **Figure 2.31** indicates the angle is measured clockwise; but, the same result is obtained when this angle is measured counterclockwise because $+(360° - 45°) = +315°$ and $\sin(+315°) = \sin(-45°)$. In this way, we obtain the solution without reference to the corkscrew rule. For the situation in (a), the solution is $\overrightarrow{\mathbf{R}} \times \overrightarrow{\mathbf{F}} = +3.2 \, \text{N} \cdot \text{m} \, \hat{\mathbf{k}}$; for the situation in (b), the solution is $\overrightarrow{\mathbf{R}} \times \overrightarrow{\mathbf{F}} = -3.5 \, \text{N} \cdot \text{m} \, \hat{\mathbf{k}}$.

 2.15 **Check Your Understanding** For the vectors given in **Figure 2.13**, find the vector products $\overrightarrow{\mathbf{A}} \times \overrightarrow{\mathbf{B}}$ and $\overrightarrow{\mathbf{C}} \times \overrightarrow{\mathbf{F}}$.

Similar to the dot product (**Equation 2.32**), the cross product has the following distributive property:

$$\overrightarrow{\mathbf{A}} \times (\overrightarrow{\mathbf{B}} + \overrightarrow{\mathbf{C}}) = \overrightarrow{\mathbf{A}} \times \overrightarrow{\mathbf{B}} + \overrightarrow{\mathbf{A}} \times \overrightarrow{\mathbf{C}}. \tag{2.37}$$

The distributive property is applied frequently when vectors are expressed in their component forms, in terms of unit vectors of Cartesian axes.

When we apply the definition of the cross product, **Equation 2.35**, to unit vectors $\hat{\mathbf{i}}$, $\hat{\mathbf{j}}$, and $\hat{\mathbf{k}}$ that define the positive x-, y-, and z-directions in space, we find that

$$\hat{\mathbf{i}} \times \hat{\mathbf{i}} = \hat{\mathbf{j}} \times \hat{\mathbf{j}} = \hat{\mathbf{k}} \times \hat{\mathbf{k}} = 0. \tag{2.38}$$

All other cross products of these three unit vectors must be vectors of unit magnitudes because $\hat{\mathbf{i}}$, $\hat{\mathbf{j}}$, and $\hat{\mathbf{k}}$ are orthogonal. For example, for the pair $\hat{\mathbf{i}}$ and $\hat{\mathbf{j}}$, the magnitude is $\left| \hat{\mathbf{i}} \times \hat{\mathbf{j}} \right| = ij \sin 90° = (1)(1)(1) = 1$. The direction of the vector product $\hat{\mathbf{i}} \times \hat{\mathbf{j}}$ must be orthogonal to the xy-plane, which means it must be along the z-axis. The only unit vectors along the z-axis are $-\hat{\mathbf{k}}$ or $+\hat{\mathbf{k}}$. By the corkscrew rule, the direction of vector $\hat{\mathbf{i}} \times \hat{\mathbf{j}}$ must be parallel to the positive z-axis. Therefore, the result of the multiplication $\hat{\mathbf{i}} \times \hat{\mathbf{j}}$ is identical to $+\hat{\mathbf{k}}$. We can repeat similar reasoning for the remaining pairs of unit vectors. The results of these multiplications are

$$\begin{cases} \hat{i} \times \hat{j} = + \hat{k}, \\ \hat{j} \times \hat{k} = + \hat{i}, \\ \hat{k} \times \hat{i} = + \hat{j}. \end{cases} \tag{2.39}$$

Notice that in **Equation 2.39**, the three unit vectors $\hat{i}$, $\hat{j}$, and $\hat{k}$ appear in the *cyclic order* shown in a diagram in **Figure 2.32**(a). The cyclic order means that in the product formula, $\hat{i}$ follows $\hat{k}$ and comes before $\hat{j}$, or $\hat{k}$ follows $\hat{j}$ and comes before $\hat{i}$, or $\hat{j}$ follows $\hat{i}$ and comes before $\hat{k}$. The cross product of two different unit vectors is always a third unit vector. When two unit vectors in the cross product appear in the cyclic order, the result of such a multiplication is the remaining unit vector, as illustrated in **Figure 2.32**(b). When unit vectors in the cross product appear in a different order, the result is a unit vector that is antiparallel to the remaining unit vector (i.e., the result is with the minus sign, as shown by the examples in **Figure 2.32**(c) and **Figure 2.32**(d). In practice, when the task is to find cross products of vectors that are given in vector component form, this rule for the cross-multiplication of unit vectors is very useful.

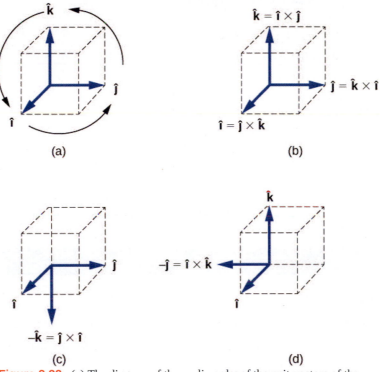

Figure 2.32 (a) The diagram of the cyclic order of the unit vectors of the axes. (b) The only cross products where the unit vectors appear in the cyclic order. These products have the positive sign. (c, d) Two examples of cross products where the unit vectors do not appear in the cyclic order. These products have the negative sign.

Suppose we want to find the cross product $\vec{A} \times \vec{B}$ for vectors $\vec{A} = A_x \hat{i} + A_y \hat{j} + A_z \hat{k}$ and $\vec{B} = B_x \hat{i} + B_y \hat{j} + B_z \hat{k}$. We can use the distributive property (**Equation 2.37**), the anticommutative property (**Equation 2.36**), and the results in **Equation 2.38** and **Equation 2.39** for unit vectors to perform the following algebra:

$$
\begin{aligned}
\vec{\mathbf{A}} \times \vec{\mathbf{B}} &= (A_x \hat{\mathbf{i}} + A_y \hat{\mathbf{j}} + A_z \hat{\mathbf{k}}) \times (B_x \hat{\mathbf{i}} + B_y \hat{\mathbf{j}} + B_z \hat{\mathbf{k}}) \\
&= A_x \hat{\mathbf{i}} \times (B_x \hat{\mathbf{i}} + B_y \hat{\mathbf{j}} + B_z \hat{\mathbf{k}}) + A_y \hat{\mathbf{j}} \times (B_x \hat{\mathbf{i}} + B_y \hat{\mathbf{j}} + B_z \hat{\mathbf{k}}) + A_z \hat{\mathbf{k}} \times (B_x \hat{\mathbf{i}} + B_y \hat{\mathbf{j}} + B_z \hat{\mathbf{k}}) \\
&= \quad A_x B_x \hat{\mathbf{i}} \times \hat{\mathbf{i}} + A_x B_y \hat{\mathbf{i}} \times \hat{\mathbf{j}} + A_x B_z \hat{\mathbf{i}} \times \hat{\mathbf{k}} \\
&\quad + A_y B_x \hat{\mathbf{j}} \times \hat{\mathbf{i}} + A_y B_y \hat{\mathbf{j}} \times \hat{\mathbf{j}} + A_y B_z \hat{\mathbf{j}} \times \hat{\mathbf{k}} \\
&\quad + A_z B_x \hat{\mathbf{k}} \times \hat{\mathbf{i}} + A_z B_y \hat{\mathbf{k}} \times \hat{\mathbf{j}} + A_z B_z \hat{\mathbf{k}} \times \hat{\mathbf{k}} \\
&= \quad A_x B_x(0) + A_x B_y(+\hat{\mathbf{k}}) + A_x B_z(-\hat{\mathbf{j}}) \\
&\quad + A_y B_x(-\hat{\mathbf{k}}) + A_y B_y(0) + A_y B_z(+\hat{\mathbf{i}}) \\
&\quad + A_z B_x(+\hat{\mathbf{j}}) + A_z B_y(-\hat{\mathbf{i}}) + A_z B_z(0).
\end{aligned}
$$

When performing algebraic operations involving the cross product, be very careful about keeping the correct order of multiplication because the cross product is anticommutative. The last two steps that we still have to do to complete our task are, first, grouping the terms that contain a common unit vector and, second, factoring. In this way we obtain the following very useful expression for the computation of the cross product:

$$
\vec{\mathbf{C}} = \vec{\mathbf{A}} \times \vec{\mathbf{B}} = (A_y B_z - A_z B_y)\hat{\mathbf{i}} + (A_z B_x - A_x B_z)\hat{\mathbf{j}} + (A_x B_y - A_y B_x)\hat{\mathbf{k}}. \tag{2.40}
$$

In this expression, the scalar components of the cross-product vector are

$$
\begin{cases}
C_x = A_y B_z - A_z B_y, \\
C_y = A_z B_x - A_x B_z, \\
C_z = A_x B_y - A_y B_x.
\end{cases} \tag{2.41}
$$

When finding the cross product, in practice, we can use either **Equation 2.35** or **Equation 2.40**, depending on which one of them seems to be less complex computationally. They both lead to the same final result. One way to make sure if the final result is correct is to use them both.

Example 2.19

A Particle in a Magnetic Field

When moving in a magnetic field, some particles may experience a magnetic force. Without going into details—a detailed study of magnetic phenomena comes in later chapters—let's acknowledge that the magnetic field $\vec{\mathbf{B}}$ is a vector, the magnetic force $\vec{\mathbf{F}}$ is a vector, and the velocity $\vec{\mathbf{u}}$ of the particle is a vector. The magnetic force vector is proportional to the vector product of the velocity vector with the magnetic field vector, which we express as $\vec{\mathbf{F}} = \zeta \vec{\mathbf{u}} \times \vec{\mathbf{B}}$. In this equation, a constant ζ takes care of the consistency in physical units, so we can omit physical units on vectors $\vec{\mathbf{u}}$ and $\vec{\mathbf{B}}$. In this example, let's assume the constant ζ is positive.

A particle moving in space with velocity vector $\vec{\mathbf{u}} = -5.0\hat{\mathbf{i}} - 2.0\hat{\mathbf{j}} + 3.5\hat{\mathbf{k}}$ enters a region with a magnetic field and experiences a magnetic force. Find the magnetic force $\vec{\mathbf{F}}$ on this particle at the entry point to the region where the magnetic field vector is (a) $\vec{\mathbf{B}} = 7.2\hat{\mathbf{i}} - \hat{\mathbf{j}} - 2.4\hat{\mathbf{k}}$ and (b) $\vec{\mathbf{B}} = 4.5\hat{\mathbf{k}}$. In each case, find magnitude F of the magnetic force and angle θ the force vector $\vec{\mathbf{F}}$ makes with the given magnetic field vector $\vec{\mathbf{B}}$.

Strategy

First, we want to find the vector product $\vec{u} \times \vec{B}$, because then we can determine the magnetic force using $\vec{F} = \zeta \vec{u} \times \vec{B}$. Magnitude F can be found either by using components, $F = \sqrt{F_x^2 + F_y^2 + F_z^2}$, or by computing the magnitude $\left| \vec{u} \times \vec{B} \right|$ directly using **Equation 2.35**. In the latter approach, we would have to find the angle between vectors $\vec{u}$ and $\vec{B}$. When we have $\vec{F}$, the general method for finding the direction angle θ involves the computation of the scalar product $\vec{F} \cdot \vec{B}$ and substitution into **Equation 2.34**. To compute the vector product we can either use **Equation 2.40** or compute the product directly, whichever way is simpler.

Solution

The components of the velocity vector are $u_x = -5.0$, $u_y = -2.0$, and $u_z = 3.5$.

(a) The components of the magnetic field vector are $B_x = 7.2$, $B_y = -1.0$, and $B_z = -2.4$. Substituting them into **Equation 2.41** gives the scalar components of vector $\vec{F} = \zeta \vec{u} \times \vec{B}$:

$$\begin{cases} F_x = \zeta(u_y B_z - u_z B_y) = \zeta[(-2.0)(-2.4) - (3.5)(-1.0)] = 8.3\zeta \\ F_y = \zeta(u_z B_x - u_x B_z) = \zeta[(3.5)(7.2) - (-5.0)(-2.4)] = 13.2\zeta \\ F_z = \zeta(u_x B_y - u_y B_x) = \zeta[(-5.0)(-1.0) - (-2.0)(7.2)] = 19.4\zeta \end{cases}.$$

Thus, the magnetic force is $\vec{F} = \zeta(8.3\,\hat{i} + 13.2\,\hat{j} + 19.4\,\hat{k})$ and its magnitude is

$$F = \sqrt{F_x^2 + F_y^2 + F_z^2} = \zeta\sqrt{(8.3)^2 + (13.2)^2 + (19.4)^2} = 24.9\zeta.$$

To compute angle θ, we may need to find the magnitude of the magnetic field vector,

$$B = \sqrt{B_x^2 + B_y^2 + B_z^2} = \sqrt{(7.2)^2 + (-1.0)^2 + (-2.4)^2} = 7.6,$$

and the scalar product $\vec{F} \cdot \vec{B}$:

$$\vec{F} \cdot \vec{B} = F_x B_x + F_y B_y + F_z B_z = (8.3\zeta)(7.2) + (13.2\zeta)(-1.0) + (19.4\zeta)(-2.4) = 0.$$

Now, substituting into **Equation 2.34** gives angle θ:

$$\cos\theta = \frac{\vec{F} \cdot \vec{B}}{FB} = \frac{0}{(18.2\zeta)(7.6)} = 0 \Rightarrow \theta = 90°.$$

Hence, the magnetic force vector is perpendicular to the magnetic field vector. (We could have saved some time if we had computed the scalar product earlier.)

(b) Because vector $\vec{B} = 4.5\,\hat{k}$ has only one component, we can perform the algebra quickly and find the vector product directly:

$$\vec{F} = \zeta \vec{u} \times \vec{B} = \zeta(-5.0\,\hat{i} - 2.0\,\hat{j} + 3.5\,\hat{k}) \times (4.5\,\hat{k})$$

$$= \zeta[(-5.0)(4.5)\,\hat{i} \times \hat{k} + (-2.0)(4.5)\,\hat{j} \times \hat{k} + (3.5)(4.5)\,\hat{k} \times \hat{k}]$$

$$= \zeta[-22.5(-\hat{j}) - 9.0(+\hat{i}) + 0] = \zeta(-9.0\,\hat{i} + 22.5\,\hat{j}).$$

The magnitude of the magnetic force is

$$F = \sqrt{F_x^2 + F_y^2 + F_z^2} = \zeta\sqrt{(-9.0)^2 + (22.5)^2 + (0.0)^2} = 24.2\zeta.$$

Because the scalar product is

$$\overrightarrow{\mathbf{F}} \cdot \overrightarrow{\mathbf{B}} = F_x B_x + F_y B_y + F_z B_z = (-9.0\zeta)(0) + (22.5\zeta)(0) + (0)(4.5) = 0,$$

the magnetic force vector $\overrightarrow{\mathbf{F}}$ is perpendicular to the magnetic field vector $\overrightarrow{\mathbf{B}}$.

Significance

Even without actually computing the scalar product, we can predict that the magnetic force vector must always be perpendicular to the magnetic field vector because of the way this vector is constructed. Namely, the magnetic force vector is the vector product $\overrightarrow{\mathbf{F}} = \zeta \overrightarrow{\mathbf{u}} \times \overrightarrow{\mathbf{B}}$ and, by the definition of the vector product (see **Figure 2.29**), vector $\overrightarrow{\mathbf{F}}$ must be perpendicular to both vectors $\overrightarrow{\mathbf{u}}$ and $\overrightarrow{\mathbf{B}}$.

 2.16 Check Your Understanding Given two vectors $\overrightarrow{\mathbf{A}} = -\hat{\mathbf{i}} + \hat{\mathbf{j}}$ and $\overrightarrow{\mathbf{B}} = 3\hat{\mathbf{i}} - \hat{\mathbf{j}}$, find (a) $\overrightarrow{\mathbf{A}} \times \overrightarrow{\mathbf{B}}$, (b) $\left| \overrightarrow{\mathbf{A}} \times \overrightarrow{\mathbf{B}} \right|$, (c) the angle between $\overrightarrow{\mathbf{A}}$ and $\overrightarrow{\mathbf{B}}$, and (d) the angle between $\overrightarrow{\mathbf{A}} \times \overrightarrow{\mathbf{B}}$ and vector $\overrightarrow{\mathbf{C}} = \hat{\mathbf{i}} + \hat{\mathbf{k}}$.

In conclusion to this section, we want to stress that "dot product" and "cross product" are entirely different mathematical objects that have different meanings. The dot product is a scalar; the cross product is a vector. Later chapters use the terms *dot product* and *scalar product* interchangeably. Similarly, the terms *cross product* and *vector product* are used interchangeably.

CHAPTER 2 REVIEW

KEY TERMS

anticommutative property change in the order of operation introduces the minus sign

antiparallel vectors two vectors with directions that differ by $180°$

associative terms can be grouped in any fashion

commutative operations can be performed in any order

component form of a vector a vector written as the vector sum of its components in terms of unit vectors

corkscrew right-hand rule a rule used to determine the direction of the vector product

cross product the result of the vector multiplication of vectors is a vector called a cross product; also called a vector product

difference of two vectors vector sum of the first vector with the vector antiparallel to the second

direction angle in a plane, an angle between the positive direction of the x-axis and the vector, measured counterclockwise from the axis to the vector

displacement change in position

distributive multiplication can be distributed over terms in summation

dot product the result of the scalar multiplication of two vectors is a scalar called a dot product; also called a scalar product

equal vectors two vectors are equal if and only if all their corresponding components are equal; alternately, two parallel vectors of equal magnitudes

magnitude length of a vector

null vector a vector with all its components equal to zero

orthogonal vectors two vectors with directions that differ by exactly $90°$, synonymous with perpendicular vectors

parallel vectors two vectors with exactly the same direction angles

parallelogram rule geometric construction of the vector sum in a plane

polar coordinate system an orthogonal coordinate system where location in a plane is given by polar coordinates

polar coordinates a radial coordinate and an angle

radial coordinate distance to the origin in a polar coordinate system

resultant vector vector sum of two (or more) vectors

scalar a number, synonymous with a scalar quantity in physics

scalar component a number that multiplies a unit vector in a vector component of a vector

scalar equation equation in which the left-hand and right-hand sides are numbers

scalar product the result of the scalar multiplication of two vectors is a scalar called a scalar product; also called a dot product

scalar quantity quantity that can be specified completely by a single number with an appropriate physical unit

tail-to-head geometric construction geometric construction for drawing the resultant vector of many vectors

unit vector vector of a unit magnitude that specifies direction; has no physical unit

unit vectors of the axes unit vectors that define orthogonal directions in a plane or in space

vector mathematical object with magnitude and direction

vector components orthogonal components of a vector; a vector is the vector sum of its vector components.

vector equation   equation in which the left-hand and right-hand sides are vectors

vector product   the result of the vector multiplication of vectors is a vector called a vector product; also called a cross product

vector quantity   physical quantity described by a mathematical vector—that is, by specifying both its magnitude and its direction; synonymous with a vector in physics

vector sum   resultant of the combination of two (or more) vectors

KEY EQUATIONS

Multiplication by a scalar (vector equation)	$\vec{B} = \alpha \vec{A}$
Multiplication by a scalar (scalar equation for magnitudes)	$B = |\alpha|A$
Resultant of two vectors	$\vec{D}_{AD} = \vec{D}_{AC} + \vec{D}_{CD}$
Commutative law	$\vec{A} + \vec{B} = \vec{B} + \vec{A}$
Associative law	$(\vec{A} + \vec{B}) + \vec{C} = \vec{A} + (\vec{B} + \vec{C})$
Distributive law	$\alpha_1 \vec{A} + \alpha_2 \vec{A} = (\alpha_1 + \alpha_2) \vec{A}$
The component form of a vector in two dimensions	$\vec{A} = A_x \hat{i} + A_y \hat{j}$
Scalar components of a vector in two dimensions	$\begin{cases} A_x = x_e - x_b \\ A_y = y_e - y_b \end{cases}$
Magnitude of a vector in a plane	$A = \sqrt{A_x^2 + A_y^2}$
The direction angle of a vector in a plane	$\theta_A = \tan^{-1}\left(\dfrac{A_y}{A_x}\right)$
Scalar components of a vector in a plane	$\begin{cases} A_x = A \cos \theta_A \\ A_y = A \sin \theta_A \end{cases}$
Polar coordinates in a plane	$\begin{cases} x = r \cos \varphi \\ y = r \sin \varphi \end{cases}$
The component form of a vector in three dimensions	$\vec{A} = A_x \hat{i} + A_y \hat{j} + A_z \hat{k}$
The scalar z-component of a vector in three dimensions	$A_z = z_e - z_b$
Magnitude of a vector in three dimensions	$A = \sqrt{A_x^2 + A_y^2 + A_z^2}$
Distributive property	$\alpha(\vec{A} + \vec{B}) = \alpha \vec{A} + \alpha \vec{B}$
Antiparallel vector to $\vec{A}$	$-\vec{A} = -A_x \hat{i} - A_y \hat{j} - A_z \hat{k}$

Equal vectors	$\vec{\mathbf{A}} = \vec{\mathbf{B}} \Leftrightarrow \begin{cases} A_x = B_x \\ A_y = B_y \\ A_z = B_z \end{cases}$
Components of the resultant of N vectors	$\begin{cases} F_{Rx} = \sum\limits_{k=1}^{N} F_{kx} = F_{1x} + F_{2x} + \ldots + F_{Nx} \\ F_{Ry} = \sum\limits_{k=1}^{N} F_{ky} = F_{1y} + F_{2y} + \ldots + F_{Ny} \\ F_{Rz} = \sum\limits_{k=1}^{N} F_{kz} = F_{1z} + F_{2z} + \ldots + F_{Nz} \end{cases}$
General unit vector	$\hat{\mathbf{V}} = \dfrac{\vec{\mathbf{V}}}{V}$
Definition of the scalar product	$\vec{\mathbf{A}} \cdot \vec{\mathbf{B}} = AB \cos \varphi$
Commutative property of the scalar product	$\vec{\mathbf{A}} \cdot \vec{\mathbf{B}} = \vec{\mathbf{B}} \cdot \vec{\mathbf{A}}$
Distributive property of the scalar product	$\vec{\mathbf{A}} \cdot (\vec{\mathbf{B}} + \vec{\mathbf{C}}) = \vec{\mathbf{A}} \cdot \vec{\mathbf{B}} + \vec{\mathbf{A}} \cdot \vec{\mathbf{C}}$
Scalar product in terms of scalar components of vectors	$\vec{\mathbf{A}} \cdot \vec{\mathbf{B}} = A_x B_x + A_y B_y + A_z B_z$
Cosine of the angle between two vectors	$\cos \varphi = \dfrac{\vec{\mathbf{A}} \cdot \vec{\mathbf{B}}}{AB}$
Dot products of unit vectors	$\hat{\mathbf{i}} \cdot \hat{\mathbf{j}} = \hat{\mathbf{j}} \cdot \hat{\mathbf{k}} = \hat{\mathbf{k}} \cdot \hat{\mathbf{i}} = 0$
Magnitude of the vector product (definition)	$\left| \vec{\mathbf{A}} \times \vec{\mathbf{B}} \right| = AB \sin \varphi$
Anticommutative property of the vector product	$\vec{\mathbf{A}} \times \vec{\mathbf{B}} = -\vec{\mathbf{B}} \times \vec{\mathbf{A}}$
Distributive property of the vector product	$\vec{\mathbf{A}} \times (\vec{\mathbf{B}} + \vec{\mathbf{C}}) = \vec{\mathbf{A}} \times \vec{\mathbf{B}} + \vec{\mathbf{A}} \times \vec{\mathbf{C}}$
Cross products of unit vectors	$\begin{cases} \hat{\mathbf{i}} \times \hat{\mathbf{j}} = +\hat{\mathbf{k}}, \\ \hat{\mathbf{j}} \times \hat{\mathbf{k}} = +\hat{\mathbf{i}}, \\ \hat{\mathbf{k}} \times \hat{\mathbf{i}} = +\hat{\mathbf{j}}. \end{cases}$
The cross product in terms of scalar components of vectors	$\vec{\mathbf{A}} \times \vec{\mathbf{B}} = (A_y B_z - A_z B_y)\hat{\mathbf{i}} + (A_z B_x - A_x B_z)\hat{\mathbf{j}} + (A_x B_y - A_y B_x)\hat{\mathbf{k}}$

SUMMARY

2.1 Scalars and Vectors

- A vector quantity is any quantity that has magnitude and direction, such as displacement or velocity. Vector quantities are represented by mathematical objects called vectors.

- Geometrically, vectors are represented by arrows, with the end marked by an arrowhead. The length of the vector is its magnitude, which is a positive scalar. On a plane, the direction of a vector is given by the angle the vector makes with a reference direction, often an angle with the horizontal. The direction angle of a vector is a scalar.

- Two vectors are equal if and only if they have the same magnitudes and directions. Parallel vectors have the same direction angles but may have different magnitudes. Antiparallel vectors have direction angles that differ by $180°$. Orthogonal vectors have direction angles that differ by $90°$.

- When a vector is multiplied by a scalar, the result is another vector of a different length than the length of the original vector. Multiplication by a positive scalar does not change the original direction; only the magnitude is affected. Multiplication by a negative scalar reverses the original direction. The resulting vector is antiparallel to the original vector. Multiplication by a scalar is distributive. Vectors can be divided by nonzero scalars but cannot be divided by vectors.

- Two or more vectors can be added to form another vector. The vector sum is called the resultant vector. We can add vectors to vectors or scalars to scalars, but we cannot add scalars to vectors. Vector addition is commutative and associative.

- To construct a resultant vector of two vectors in a plane geometrically, we use the parallelogram rule. To construct a resultant vector of many vectors in a plane geometrically, we use the tail-to-head method.

2.2 Coordinate Systems and Components of a Vector

- Vectors are described in terms of their components in a coordinate system. In two dimensions (in a plane), vectors have two components. In three dimensions (in space), vectors have three components.

- A vector component of a vector is its part in an axis direction. The vector component is the product of the unit vector of an axis with its scalar component along this axis. A vector is the resultant of its vector components.

- Scalar components of a vector are differences of coordinates, where coordinates of the origin are subtracted from end point coordinates of a vector. In a rectangular system, the magnitude of a vector is the square root of the sum of the squares of its components.

- In a plane, the direction of a vector is given by an angle the vector has with the positive *x*-axis. This direction angle is measured counterclockwise. The scalar *x*-component of a vector can be expressed as the product of its magnitude with the cosine of its direction angle, and the scalar *y*-component can be expressed as the product of its magnitude with the sine of its direction angle.

- In a plane, there are two equivalent coordinate systems. The Cartesian coordinate system is defined by unit vectors $\hat{\mathbf{i}}$ and $\hat{\mathbf{j}}$ along the *x*-axis and the *y*-axis, respectively. The polar coordinate system is defined by the radial unit vector $\hat{\mathbf{r}}$, which gives the direction from the origin, and a unit vector $\hat{\mathbf{t}}$, which is perpendicular (orthogonal) to the radial direction.

2.3 Algebra of Vectors

- Analytical methods of vector algebra allow us to find resultants of sums or differences of vectors without having to draw them. Analytical methods of vector addition are exact, contrary to graphical methods, which are approximate.

- Analytical methods of vector algebra are used routinely in mechanics, electricity, and magnetism. They are important mathematical tools of physics.

2.4 Products of Vectors

- There are two kinds of multiplication for vectors. One kind of multiplication is the scalar product, also known as the dot product. The other kind of multiplication is the vector product, also known as the cross product. The scalar product of vectors is a number (scalar). The vector product of vectors is a vector.

- Both kinds of multiplication have the distributive property, but only the scalar product has the commutative property. The vector product has the anticommutative property, which means that when we change the order in which two vectors are multiplied, the result acquires a minus sign.

- The scalar product of two vectors is obtained by multiplying their magnitudes with the cosine of the angle between them. The scalar product of orthogonal vectors vanishes; the scalar product of antiparallel vectors is negative.

- The vector product of two vectors is a vector perpendicular to both of them. Its magnitude is obtained by multiplying their magnitudes by the sine of the angle between them. The direction of the vector product can be determined by the corkscrew right-hand rule. The vector product of two either parallel or antiparallel vectors vanishes. The magnitude of the vector product is largest for orthogonal vectors.

- The scalar product of vectors is used to find angles between vectors and in the definitions of derived scalar physical quantities such as work or energy.

- The cross product of vectors is used in definitions of derived vector physical quantities such as torque or magnetic force, and in describing rotations.

CONCEPTUAL QUESTIONS

2.1 Scalars and Vectors

1. A weather forecast states the temperature is predicted to be $-5\,°C$ the following day. Is this temperature a vector or a scalar quantity? Explain.

2. Which of the following is a vector: a person's height, the altitude on Mt. Everest, the velocity of a fly, the age of Earth, the boiling point of water, the cost of a book, Earth's population, or the acceleration of gravity?

3. Give a specific example of a vector, stating its magnitude, units, and direction.

4. What do vectors and scalars have in common? How do they differ?

5. Suppose you add two vectors $\vec{A}$ and $\vec{B}$. What relative direction between them produces the resultant with the greatest magnitude? What is the maximum magnitude? What relative direction between them produces the resultant with the smallest magnitude? What is the minimum magnitude?

6. Is it possible to add a scalar quantity to a vector quantity?

7. Is it possible for two vectors of different magnitudes to add to zero? Is it possible for three vectors of different magnitudes to add to zero? Explain.

8. Does the odometer in an automobile indicate a scalar or a vector quantity?

9. When a 10,000-m runner competing on a 400-m track crosses the finish line, what is the runner's net displacement? Can this displacement be zero? Explain.

10. A vector has zero magnitude. Is it necessary to specify its direction? Explain.

11. Can a magnitude of a vector be negative?

12. Can the magnitude of a particle's displacement be greater that the distance traveled?

13. If two vectors are equal, what can you say about their components? What can you say about their magnitudes? What can you say about their directions?

14. If three vectors sum up to zero, what geometric condition do they satisfy?

2.2 Coordinate Systems and Components of a Vector

15. Give an example of a nonzero vector that has a component of zero.

16. Explain why a vector cannot have a component greater than its own magnitude.

17. If two vectors are equal, what can you say about their components?

18. If vectors $\vec{A}$ and $\vec{B}$ are orthogonal, what is the component of $\vec{B}$ along the direction of $\vec{A}$? What is the component of $\vec{A}$ along the direction of $\vec{B}$?

19. If one of the two components of a vector is not zero, can the magnitude of the other vector component of this vector be zero?

20. If two vectors have the same magnitude, do their components have to be the same?

2.4 Products of Vectors

21. What is wrong with the following expressions? How can you correct them? (a) $C = \vec{A}\,\vec{B}$, (b) $\vec{C} = \vec{A}\cdot\vec{B}$, (c) $C = \vec{A}\times\vec{B}$, (d) $C = A\vec{B}$, (e) $C + 2\vec{A} = B$, (f) $\vec{C} = A\times\vec{B}$, (g) $\vec{A}\cdot\vec{B} = \vec{A}\times\vec{B}$, (h) $\vec{C} = 2\vec{A}\cdot\vec{B}$, (i)

$C = \vec{A} / \vec{B}$, and (j) $C = \vec{A} /B$.

22. If the cross product of two vectors vanishes, what can you say about their directions?

23. If the dot product of two vectors vanishes, what can you say about their directions?

24. What is the dot product of a vector with the cross product that this vector has with another vector?

PROBLEMS

2.1 Scalars and Vectors

25. A scuba diver makes a slow descent into the depths of the ocean. His vertical position with respect to a boat on the surface changes several times. He makes the first stop 9.0 m from the boat but has a problem with equalizing the pressure, so he ascends 3.0 m and then continues descending for another 12.0 m to the second stop. From there, he ascends 4 m and then descends for 18.0 m, ascends again for 7 m and descends again for 24.0 m, where he makes a stop, waiting for his buddy. Assuming the positive direction up to the surface, express his net vertical displacement vector in terms of the unit vector. What is his distance to the boat?

26. In a tug-of-war game on one campus, 15 students pull on a rope at both ends in an effort to displace the central knot to one side or the other. Two students pull with force 196 N each to the right, four students pull with force 98 N each to the left, five students pull with force 62 N each to the left, three students pull with force 150 N each to the right, and one student pulls with force 250 N to the left. Assuming the positive direction to the right, express the net pull on the knot in terms of the unit vector. How big is the net pull on the knot? In what direction?

27. Suppose you walk 18.0 m straight west and then 25.0 m straight north. How far are you from your starting point and what is the compass direction of a line connecting your starting point to your final position? Use a graphical method.

28. For the vectors given in the following figure, use a graphical method to find the following resultants: (a) $\vec{A} + \vec{B}$, (b) $\vec{C} + \vec{B}$, (c) $\vec{D} + \vec{F}$, (d) $\vec{A} - \vec{B}$, (e) $\vec{D} - \vec{F}$, (f) $\vec{A} + 2\vec{F}$, (g) $\vec{C} - 2\vec{D} + 3\vec{F}$; and (h) $\vec{A} - 4\vec{D} + 2\vec{F}$.

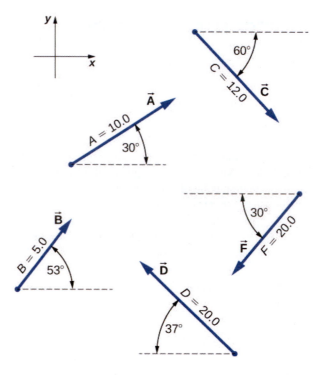

29. A delivery man starts at the post office, drives 40 km north, then 20 km west, then 60 km northeast, and finally 50 km north to stop for lunch. Use a graphical method to find his net displacement vector.

30. An adventurous dog strays from home, runs three blocks east, two blocks north, one block east, one block north, and two blocks west. Assuming that each block is about 100 m, how far from home and in what direction is the dog? Use a graphical method.

31. In an attempt to escape a desert island, a castaway builds a raft and sets out to sea. The wind shifts a great deal during the day and he is blown along the following directions: 2.50 km and $45.0°$ north of west, then 4.70 km and $60.0°$ south of east, then 1.30 km and $25.0°$ south of west, then 5.10 km straight east, then 1.70 km and $5.00°$ east of north, then 7.20 km and $55.0°$ south of west, and finally 2.80 km and $10.0°$ north of east. Use a graphical method to find the castaway's final position relative to the island.

32. A small plane flies 40.0 km in a direction $60°$ north of east and then flies 30.0 km in a direction $15°$ north of east. Use a graphical method to find the total distance the plane covers from the starting point and the direction of the path to the final position.

33. A trapper walks a 5.0-km straight-line distance from his cabin to the lake, as shown in the following figure. Use a graphical method (the parallelogram rule) to determine the trapper's displacement directly to the east and displacement directly to the north that sum up to his resultant displacement vector. If the trapper walked only in directions east and north, zigzagging his way to the lake, how many kilometers would he have to walk to get to the lake?

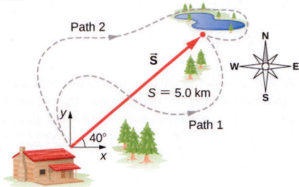

34. A surveyor measures the distance across a river that flows straight north by the following method. Starting directly across from a tree on the opposite bank, the surveyor walks 100 m along the river to establish a baseline. She then sights across to the tree and reads that the angle from the baseline to the tree is $35°$. How wide is the river?

35. A pedestrian walks 6.0 km east and then 13.0 km north. Use a graphical method to find the pedestrian's resultant displacement and geographic direction.

36. The magnitudes of two displacement vectors are $A = 20$ m and $B = 6$ m. What are the largest and the smallest values of the magnitude of the resultant $\vec{R} = \vec{A} + \vec{B}$?

2.2 Coordinate Systems and Components of a Vector

37. Assuming the $+x$-axis is horizontal and points to the right, resolve the vectors given in the following figure to their scalar components and express them in vector component form.

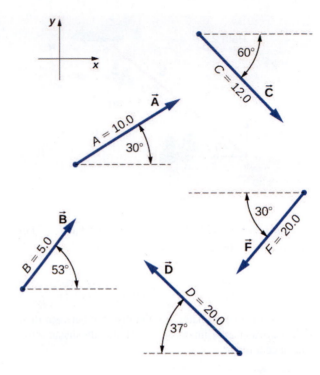

38. Suppose you walk 18.0 m straight west and then 25.0 m straight north. How far are you from your starting point? What is your displacement vector? What is the direction of your displacement? Assume the $+x$-axis is horizontal to the right.

39. You drive 7.50 km in a straight line in a direction $15°$ east of north. (a) Find the distances you would have to drive straight east and then straight north to arrive at the same point. (b) Show that you still arrive at the same point if the east and north legs are reversed in order. Assume the $+x$-axis is to the east.

40. A sledge is being pulled by two horses on a flat terrain. The net force on the sledge can be expressed in the Cartesian coordinate system as vector $\vec{F} = (-2980.0\,\hat{i} + 8200.0\,\hat{j})\text{N}$, where $\hat{i}$ and $\hat{j}$ denote directions to the east and north, respectively. Find the magnitude and direction of the pull.

41. A trapper walks a 5.0-km straight-line distance from her cabin to the lake, as shown in the following figure. Determine the east and north components of her displacement vector. How many more kilometers would she have to walk if she walked along the component displacements? What is her displacement vector?

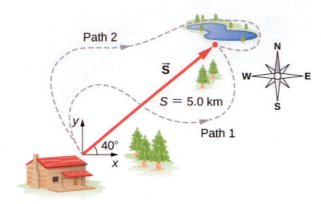

42. The polar coordinates of a point are $4\pi/3$ and 5.50 m. What are its Cartesian coordinates?

43. Two points in a plane have polar coordinates $P_1(2.500\text{ m}, \pi/6)$ and $P_2(3.800\text{ m}, 2\pi/3)$. Determine their Cartesian coordinates and the distance between them in the Cartesian coordinate system. Round the distance to a nearest centimeter.

44. A chameleon is resting quietly on a lanai screen, waiting for an insect to come by. Assume the origin of a Cartesian coordinate system at the lower left-hand corner of the screen and the horizontal direction to the right as the +*x*-direction. If its coordinates are (2.000 m, 1.000 m), (a) how far is it from the corner of the screen? (b) What is its location in polar coordinates?

45. Two points in the Cartesian plane are A(2.00 m, −4.00 m) and B(−3.00 m, 3.00 m). Find the distance between them and their polar coordinates.

46. A fly enters through an open window and zooms around the room. In a Cartesian coordinate system with three axes along three edges of the room, the fly changes its position from point b(4.0 m, 1.5 m, 2.5 m) to point e(1.0 m, 4.5 m, 0.5 m). Find the scalar components of the fly's displacement vector and express its displacement vector in vector component form. What is its magnitude?

2.3 Algebra of Vectors

47. For vectors $\overrightarrow{\mathbf{B}} = -\hat{\mathbf{i}} - 4\hat{\mathbf{j}}$ and $\overrightarrow{\mathbf{A}} = -3\hat{\mathbf{i}} - 2\hat{\mathbf{j}}$, calculate (a) $\overrightarrow{\mathbf{A}} + \overrightarrow{\mathbf{B}}$ and its magnitude and direction angle, and (b) $\overrightarrow{\mathbf{A}} - \overrightarrow{\mathbf{B}}$ and its magnitude and direction angle.

48. A particle undergoes three consecutive displacements given by vectors $\overrightarrow{\mathbf{D}}_1 = (3.0\hat{\mathbf{i}} - 4.0\hat{\mathbf{j}} - 2.0\hat{\mathbf{k}})\text{mm}$,

$\overrightarrow{\mathbf{D}}_2 = (1.0\hat{\mathbf{i}} - 7.0\hat{\mathbf{j}} + 4.0\hat{\mathbf{k}})\text{mm}$, and $\overrightarrow{\mathbf{D}}_3 = (-7.0\hat{\mathbf{i}} + 4.0\hat{\mathbf{j}} + 1.0\hat{\mathbf{k}})\text{mm}$. (a) Find the resultant displacement vector of the particle. (b) What is the magnitude of the resultant displacement? (c) If all displacements were along one line, how far would the particle travel?

49. Given two displacement vectors $\overrightarrow{\mathbf{A}} = (3.00\hat{\mathbf{i}} - 4.00\hat{\mathbf{j}} + 4.00\hat{\mathbf{k}})\text{m}$ and $\overrightarrow{\mathbf{B}} = (2.00\hat{\mathbf{i}} + 3.00\hat{\mathbf{j}} - 7.00\hat{\mathbf{k}})\text{m}$, find the displacements and their magnitudes for (a) $\overrightarrow{\mathbf{C}} = \overrightarrow{\mathbf{A}} + \overrightarrow{\mathbf{B}}$ and (b) $\overrightarrow{\mathbf{D}} = 2\overrightarrow{\mathbf{A}} - \overrightarrow{\mathbf{B}}$.

50. A small plane flies 40.0 km in a direction 60° north of east and then flies 30.0 km in a direction 15° north of east. Use the analytical method to find the total distance the plane covers from the starting point, and the geographic direction of its displacement vector. What is its displacement vector?

51. In an attempt to escape a desert island, a castaway builds a raft and sets out to sea. The wind shifts a great deal during the day, and she is blown along the following straight lines: 2.50 km and 45.0° north of west, then 4.70 km and 60.0° south of east, then 1.30 km and 25.0° south of west, then 5.10 km due east, then 1.70 km and 5.00° east of north, then 7.20 km and 55.0° south of west, and finally 2.80 km and 10.0° north of east. Use the analytical method to find the resultant vector of all her displacement vectors. What is its magnitude and direction?

52. Assuming the +*x*-axis is horizontal to the right for the vectors given in the following figure, use the analytical method to find the following resultants: (a) $\overrightarrow{\mathbf{A}} + \overrightarrow{\mathbf{B}}$, (b) $\overrightarrow{\mathbf{C}} + \overrightarrow{\mathbf{B}}$, (c) $\overrightarrow{\mathbf{D}} + \overrightarrow{\mathbf{F}}$, (d) $\overrightarrow{\mathbf{A}} - \overrightarrow{\mathbf{B}}$, (e) $\overrightarrow{\mathbf{D}} - \overrightarrow{\mathbf{F}}$, (f) $\overrightarrow{\mathbf{A}} + 2\overrightarrow{\mathbf{F}}$, (g) $\overrightarrow{\mathbf{C}} - 2\overrightarrow{\mathbf{D}} + 3\overrightarrow{\mathbf{F}}$, and (h) $\overrightarrow{\mathbf{A}} - 4\overrightarrow{\mathbf{D}} + 2\overrightarrow{\mathbf{F}}$.

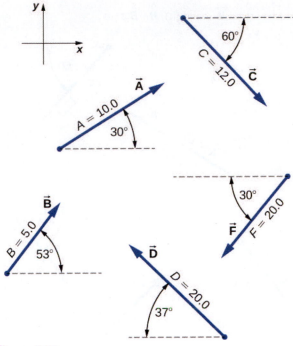

Figure 2.33

53. Given the vectors in the preceding figure, find vector $\vec{R}$ that solves equations (a) $\vec{D} + \vec{R} = \vec{F}$ and (b) $\vec{C} - 2\vec{D} + 5\vec{R} = 3\vec{F}$. Assume the +x-axis is horizontal to the right.

54. A delivery man starts at the post office, drives 40 km north, then 20 km west, then 60 km northeast, and finally 50 km north to stop for lunch. Use the analytical method to determine the following: (a) Find his net displacement vector. (b) How far is the restaurant from the post office? (c) If he returns directly from the restaurant to the post office, what is his displacement vector on the return trip? (d) What is his compass heading on the return trip? Assume the +x-axis is to the east.

55. An adventurous dog strays from home, runs three blocks east, two blocks north, and one block east, one block north, and two blocks west. Assuming that each block is about a 100 yd, use the analytical method to find the dog's net displacement vector, its magnitude, and its direction. Assume the +x-axis is to the east. How would your answer be affected if each block was about 100 m?

56. If $\vec{D} = (6.00\,\hat{i} - 8.00\,\hat{j})m$, $\vec{B} = (-8.00\,\hat{i} + 3.00\,\hat{j})m$, and $\vec{A} = (26.0\,\hat{i} + 19.0\,\hat{j})m$, find the unknown constants a and b such that $a\vec{D} + b\vec{B} + \vec{A} = \vec{0}$.

57. Given the displacement vector $\vec{D} = (3\,\hat{i} - 4\,\hat{j})m$, find the displacement vector $\vec{R}$ so that $\vec{D} + \vec{R} = -4D\,\hat{j}$.

58. Find the unit vector of direction for the following vector quantities: (a) Force $\vec{F} = (3.0\,\hat{i} - 2.0\,\hat{j})N$, (b) displacement $\vec{D} = (-3.0\,\hat{i} - 4.0\,\hat{j})m$, and (c) velocity $\vec{v} = (-5.00\,\hat{i} + 4.00\,\hat{j})m/s$.

59. At one point in space, the direction of the electric field vector is given in the Cartesian system by the unit vector $\hat{E} = 1/\sqrt{5}\,\hat{i} - 2/\sqrt{5}\,\hat{j}$. If the magnitude of the electric field vector is $E = 400.0$ V/m, what are the scalar components E_x, E_y, and E_z of the electric field vector $\vec{E}$ at this point? What is the direction angle θ_E of the electric field vector at this point?

60. A barge is pulled by the two tugboats shown in the following figure. One tugboat pulls on the barge with a force of magnitude 4000 units of force at 15° above the line AB (see the figure and the other tugboat pulls on the barge with a force of magnitude 5000 units of force at 12° below the line AB. Resolve the pulling forces to their scalar components and find the components of the resultant force pulling on the barge. What is the magnitude of the resultant pull? What is its direction relative to the line AB?

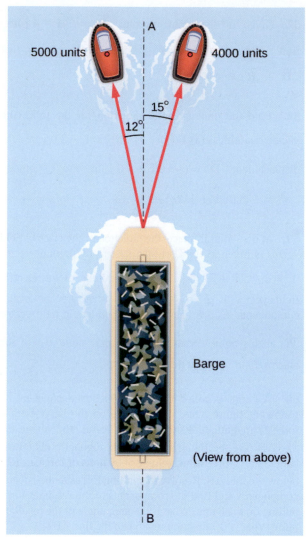

Figure 2.34

61. In the control tower at a regional airport, an air traffic controller monitors two aircraft as their positions change with respect to the control tower. One plane is a cargo carrier Boeing 747 and the other plane is a Douglas DC-3. The Boeing is at an altitude of 2500 m, climbing at $10°$ above the horizontal, and moving $30°$ north of west. The DC-3 is at an altitude of 3000 m, climbing at $5°$ above the horizontal, and cruising directly west. (a) Find the position vectors of the planes relative to the control tower. (b) What is the distance between the planes at the moment the air traffic controller makes a note about their positions?

2.4 Products of Vectors

62. Assuming the +x-axis is horizontal to the right for the vectors in the following figure, find the following scalar products: (a) $\vec{A} \cdot \vec{C}$, (b) $\vec{A} \cdot \vec{F}$, (c) $\vec{D} \cdot \vec{C}$, (d) $\vec{A} \cdot (\vec{F} + 2\vec{C})$, (e) $\hat{i} \cdot \vec{B}$, (f) $\hat{j} \cdot \vec{B}$, (g)

$(3\hat{i} - \hat{j}) \cdot \vec{B}$, and (h) $\vec{B} \cdot \vec{B}$.

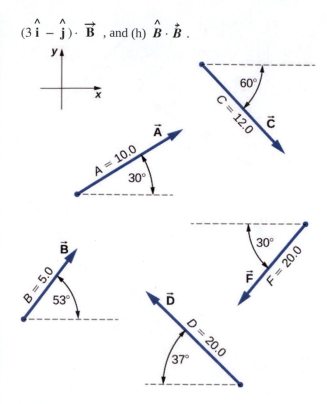

63. Assuming the +x-axis is horizontal to the right for the vectors in the preceding figure, find (a) the component of vector $\vec{A}$ along vector $\vec{C}$, (b) the component of vector $\vec{C}$ along vector $\vec{A}$, (c) the component of vector $\hat{i}$ along vector $\vec{F}$, and (d) the component of vector $\vec{F}$ along vector $\hat{i}$.

64. Find the angle between vectors for (a) $\vec{D} = (-3.0\hat{i} - 4.0\hat{j})$m and $\vec{A} = (-3.0\hat{i} + 4.0\hat{j})$m and (b) $\vec{D} = (2.0\hat{i} - 4.0\hat{j} + \hat{k})$m and $\vec{B} = (-2.0\hat{i} + 3.0\hat{j} + 2.0\hat{k})$m.

65. Find the angles that vector $\vec{D} = (2.0\hat{i} - 4.0\hat{j} + \hat{k})$m makes with the x-, y-, and z-axes.

66. Show that the force vector $\vec{D} = (2.0\hat{i} - 4.0\hat{j} + \hat{k})$N is orthogonal to the force vector $\vec{G} = (3.0\hat{i} + 4.0\hat{j} + 10.0\hat{k})$N.

67. Assuming the +x-axis is horizontal to the right for the vectors in the previous figure, find the following vector products: (a) $\vec{A} \times \vec{C}$, (b) $\vec{A} \times \vec{F}$, (c) $\vec{D} \times \vec{C}$, (d) $\vec{A} \times (\vec{F} + 2\vec{C})$, (e) $\hat{i} \times \vec{B}$, (f) $\hat{j} \times \vec{B}$, (g) $(3\hat{i} - \hat{j}) \times \vec{B}$, and (h) $\vec{B} \times \vec{B}$.

68. Find the cross product $\vec{A} \times \vec{C}$ for (a) $\vec{A} = 2.0\hat{i} - 4.0\hat{j} + \hat{k}$ and $\vec{C} = 3.0\hat{i} + 4.0\hat{j} + 10.0\hat{k}$, (b) $\vec{A} = 3.0\hat{i} + 4.0\hat{j} + 10.0\hat{k}$ and $\vec{C} = 2.0\hat{i} - 4.0\hat{j} + \hat{k}$, (c) $\vec{A} = -3.0\hat{i} - 4.0\hat{j}$ and $\vec{C} = -3.0\hat{i} + 4.0\hat{j}$, and (d) $\vec{C} = -2.0\hat{i} + 3.0\hat{j} + 2.0\hat{k}$ and $\vec{A} = -9.0\hat{j}$.

69. For the vectors in the earlier figure, find (a) $(\vec{A} \times \vec{F}) \cdot \vec{D}$, (b) $(\vec{A} \times \vec{F}) \cdot (\vec{D} \times \vec{B})$, and (c) $(\vec{A} \cdot \vec{F})(\vec{D} \times \vec{B})$.

70. (a) If $\vec{A} \times \vec{F} = \vec{B} \times \vec{F}$, can we conclude $\vec{A} = \vec{B}$? (b) If $\vec{A} \cdot \vec{F} = \vec{B} \cdot \vec{F}$, can we conclude $\vec{A} = \vec{B}$? (c) If $F\vec{A} = \vec{B}F$, can we conclude $\vec{A} = \vec{B}$? Why or why not?

ADDITIONAL PROBLEMS

71. You fly 32.0 km in a straight line in still air in the direction 35.0° south of west. (a) Find the distances you would have to fly due south and then due west to arrive at the same point. (b) Find the distances you would have to fly first in a direction 45.0° south of west and then in a direction 45.0° west of north. Note these are the components of the displacement along a different set of axes—namely, the one rotated by 45° with respect to the axes in (a).

72. Rectangular coordinates of a point are given by $(2, y)$ and its polar coordinates are given by $(r, \pi/6)$. Find y and r.

73. If the polar coordinates of a point are (r, φ) and its rectangular coordinates are (x, y), determine the polar coordinates of the following points: (a) $(-x, y)$, (b) $(-2x, -2y)$, and (c) $(3x, -3y)$.

74. Vectors $\vec{A}$ and $\vec{B}$ have identical magnitudes of 5.0 units. Find the angle between them if $\vec{A} + \vec{B} = 5\sqrt{2}\hat{j}$.

75. Starting at the island of Moi in an unknown archipelago, a fishing boat makes a round trip with two stops at the islands of Noi and Poi. It sails from Moi for 4.76 nautical miles (nmi) in a direction 37° north of east to Noi. From Noi, it sails 69° west of north to Poi. On its return leg from Poi, it sails 28° east of south. What distance does the boat sail between Noi and Poi? What distance does it sail between Moi and Poi? Express your answer both in nautical miles and in kilometers. Note: 1 nmi = 1852 m.

76. An air traffic controller notices two signals from two planes on the radar monitor. One plane is at altitude 800 m and in a 19.2-km horizontal distance to the tower in a direction 25° south of west. The second plane is at altitude 1100 m and its horizontal distance is 17.6 km and 20° south of west. What is the distance between these planes?

77. Show that when $\vec{A} + \vec{B} = \vec{C}$, then $C^2 = A^2 + B^2 + 2AB \cos\varphi$, where φ is the angle between vectors $\vec{A}$ and $\vec{B}$.

78. Four force vectors each have the same magnitude f. What is the largest magnitude the resultant force vector may have when these forces are added? What is the smallest magnitude of the resultant? Make a graph of both situations.

79. A skater glides along a circular path of radius 5.00 m in clockwise direction. When he coasts around one-half of the circle, starting from the west point, find (a) the magnitude of his displacement vector and (b) how far he actually skated. (c) What is the magnitude of his displacement vector when he skates all the way around the circle and comes back to the west point?

80. A stubborn dog is being walked on a leash by its owner. At one point, the dog encounters an interesting scent at some spot on the ground and wants to explore it in detail, but the owner gets impatient and pulls on the leash with force $\vec{F} = (98.0\hat{i} + 132.0\hat{j} + 32.0\hat{k})$N along the leash. (a) What is the magnitude of the pulling force? (b)

What angle does the leash make with the vertical?

81. If the velocity vector of a polar bear is $\vec{u} = (-18.0 \, \hat{i} - 13.0 \, \hat{j})$km/h, how fast and in what geographic direction is it heading? Here, $\hat{i}$ and $\hat{j}$ are directions to geographic east and north, respectively.

82. Find the scalar components of three-dimensional vectors $\vec{G}$ and $\vec{H}$ in the following figure and write the vectors in vector component form in terms of the unit vectors of the axes.

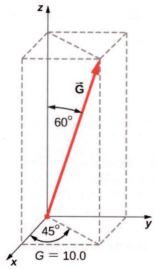

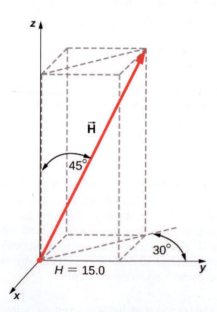

83. A diver explores a shallow reef off the coast of Belize. She initially swims 90.0 m north, makes a turn to the east and continues for 200.0 m, then follows a big grouper for 80.0 m in the direction $30°$ north of east. In the meantime, a local current displaces her by 150.0 m south. Assuming the current is no longer present, in what direction and how far should she now swim to come back to the point where she started?

84. A force vector $\vec{A}$ has x- and y-components, respectively, of -8.80 units of force and 15.00 units of force. The x- and y-components of force vector $\vec{B}$ are, respectively, 13.20 units of force and -6.60 units of force. Find the components of force vector $\vec{C}$ that satisfies the vector equation $\vec{A} - \vec{B} + 3\vec{C} = 0$.

85. Vectors $\vec{A}$ and $\vec{B}$ are two orthogonal vectors in the xy-plane and they have identical magnitudes. If $\vec{A} = 3.0 \, \hat{i} + 4.0 \, \hat{j}$, find $\vec{B}$.

86. For the three-dimensional vectors in the following figure, find (a) $\vec{G} \times \vec{H}$, (b) $\left| \vec{G} \times \vec{H} \right|$, and (c) $\vec{G} \cdot \vec{H}$.

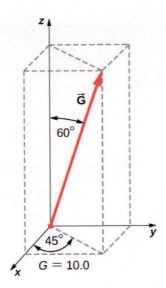

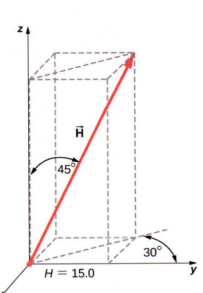

87. Show that $(\vec{\mathbf{B}} \times \vec{\mathbf{C}}) \cdot \vec{\mathbf{A}}$ is the volume of the parallelepiped, with edges formed by the three vectors in the following figure.

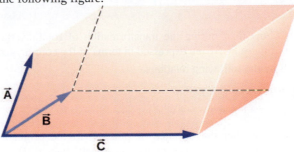

CHALLENGE PROBLEMS

88. Vector $\vec{\mathbf{B}}$ is 5.0 cm long and vector $\vec{\mathbf{A}}$ is 4.0 cm long. Find the angle between these two vectors when $\left| \vec{\mathbf{A}} + \vec{\mathbf{B}} \right| = 3.0 \, \text{cm}$ and $\left| \vec{\mathbf{A}} - \vec{\mathbf{B}} \right| = 3.0 \, \text{cm}$.

89. What is the component of the force vector $\vec{\mathbf{G}} = (3.0\,\hat{\mathbf{i}} + 4.0\,\hat{\mathbf{j}} + 10.0\,\hat{\mathbf{k}})\text{N}$ along the force vector $\vec{\mathbf{H}} = (1.0\,\hat{\mathbf{i}} + 4.0\,\hat{\mathbf{j}})\text{N}$?

90. The following figure shows a triangle formed by the three vectors $\vec{\mathbf{A}}$, $\vec{\mathbf{B}}$, and $\vec{\mathbf{C}}$. If vector $\vec{\mathbf{C}}'$ is drawn between the midpoints of vectors $\vec{\mathbf{A}}$ and $\vec{\mathbf{B}}$, show that $\vec{\mathbf{C}}' = \vec{\mathbf{C}}/2$.

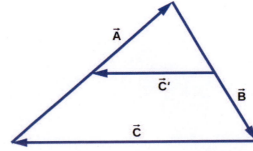

91. Distances between points in a plane do not change when a coordinate system is rotated. In other words, the magnitude of a vector is *invariant* under rotations of the coordinate system. Suppose a coordinate system S is rotated about its origin by angle φ to become a new coordinate system S', as shown in the following figure. A point in a plane has coordinates (x, y) in S and coordinates (x', y') in S'.

(a) Show that, during the transformation of rotation, the coordinates in S' are expressed in terms of the coordinates in S by the following relations:

$$\begin{cases} x' = x\cos\varphi + y\sin\varphi \\ y' = -x\sin\varphi + y\cos\varphi \end{cases}.$$

(b) Show that the distance of point P to the origin is invariant under rotations of the coordinate system. Here, you have to show that

$$\sqrt{x^2 + y^2} = \sqrt{x'^2 + y'^2}.$$

(c) Show that the distance between points P and Q is invariant under rotations of the coordinate system. Here, you have to show that

$$\sqrt{(x_P - x_Q)^2 + (y_P - y_Q)^2} = \sqrt{(x'_P - x'_Q)^2 + (y'_P - y'_Q)^2}.$$

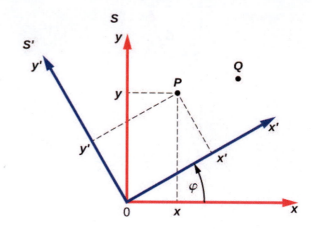

3 | MOTION ALONG A STRAIGHT LINE

Figure 3.1 A JR Central L0 series five-car maglev (magnetic levitation) train undergoing a test run on the Yamanashi Test Track. The maglev train's motion can be described using kinematics, the subject of this chapter. (credit: modification of work by "Maryland GovPics"/Flickr)

Chapter Outline

3.1 Position, Displacement, and Average Velocity

3.2 Instantaneous Velocity and Speed

3.3 Average and Instantaneous Acceleration

3.4 Motion with Constant Acceleration

3.5 Free Fall

3.6 Finding Velocity and Displacement from Acceleration

Introduction

Our universe is full of objects in motion. From the stars, planets, and galaxies; to the motion of people and animals; down to the microscopic scale of atoms and molecules—everything in our universe is in motion. We can describe motion using the two disciplines of kinematics and dynamics. We study dynamics, which is concerned with the causes of motion, in **Newton's Laws of Motion**; but, there is much to be learned about motion without referring to what causes it, and this is the study of kinematics. Kinematics involves describing motion through properties such as position, time, velocity, and acceleration.

A full treatment of **kinematics** considers motion in two and three dimensions. For now, we discuss motion in one dimension, which provides us with the tools necessary to study multidimensional motion. A good example of an object undergoing one-dimensional motion is the maglev (magnetic levitation) train depicted at the beginning of this chapter. As it travels, say, from Tokyo to Kyoto, it is at different positions along the track at various times in its journey, and therefore has displacements, or changes in position. It also has a variety of velocities along its path and it undergoes accelerations (changes in velocity). With the skills learned in this chapter we can calculate these quantities and average velocity. All these quantities can be described using kinematics, without knowing the train's mass or the forces involved.

3.1 | Position, Displacement, and Average Velocity

Learning Objectives
By the end of this section, you will be able to: • Define position, displacement, and distance traveled.   • Calculate the total displacement given the position as a function of time.   • Determine the total distance traveled.   • Calculate the average velocity given the displacement and elapsed time.

When you're in motion, the basic questions to ask are: Where are you? Where are you going? How fast are you getting there? The answers to these questions require that you specify your position, your displacement, and your average velocity—the terms we define in this section.

Position

To describe the motion of an object, you must first be able to describe its **position** (x): *where it is at any particular time*. More precisely, we need to specify its position relative to a convenient frame of reference. A frame of reference is an arbitrary set of axes from which the position and motion of an object are described. Earth is often used as a frame of reference, and we often describe the position of an object as it relates to stationary objects on Earth. For example, a rocket launch could be described in terms of the position of the rocket with respect to Earth as a whole, whereas a cyclist's position could be described in terms of where she is in relation to the buildings she passes **Figure 3.2**. In other cases, we use reference frames that are not stationary but are in motion relative to Earth. To describe the position of a person in an airplane, for example, we use the airplane, not Earth, as the reference frame. To describe the position of an object undergoing one-dimensional motion, we often use the variable x. Later in the chapter, during the discussion of free fall, we use the variable y.

Figure 3.2 These cyclists in Vietnam can be described by their position relative to buildings or a canal. Their motion can be described by their change in position, or displacement, in a frame of reference. (credit: modification of work by Suzan Black)

Displacement

If an object moves relative to a frame of reference—for example, if a professor moves to the right relative to a whiteboard **Figure 3.3**—then the object's position changes. This change in position is called **displacement**. The word *displacement* implies that an object has moved, or has been displaced. Although position is the numerical value of x along a straight line where an object might be located, displacement gives the *change* in position along this line. Since displacement indicates direction, it is a vector and can be either positive or negative, depending on the choice of positive direction. Also, an analysis of motion can have many displacements embedded in it. If right is positive and an object moves 2 m to the right, then 4 m to the left, the individual displacements are 2 m and -4 m, respectively.

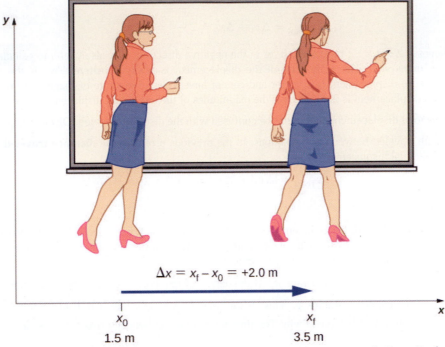

Figure 3.3 A professor paces left and right while lecturing. Her position relative to Earth is given by x. The +2.0-m displacement of the professor relative to Earth is represented by an arrow pointing to the right.

Displacement

Displacement Δx is the change in position of an object:

$$\Delta x = x_f - x_0, \tag{3.1}$$

where Δx is displacement, x_f is the final position, and x_0 is the initial position.

We use the uppercase Greek letter delta (Δ) to mean "change in" whatever quantity follows it; thus, Δx means *change in position* (final position less initial position). We always solve for displacement by subtracting initial position x_0 from final position x_f. Note that the SI unit for displacement is the meter, but sometimes we use kilometers or other units of length.

Keep in mind that when units other than meters are used in a problem, you may need to convert them to meters to complete the calculation (see **Appendix B**).

Objects in motion can also have a series of displacements. In the previous example of the pacing professor, the individual displacements are 2 m and −4 m, giving a total displacement of −2 m. We define **total displacement** Δx_{Total}, as *the sum of the individual displacements*, and express this mathematically with the equation

$$\Delta x_{Total} = \sum \Delta x_i, \tag{3.2}$$

where Δx_i are the individual displacements. In the earlier example,

$$\Delta x_1 = x_1 - x_0 = 2 - 0 = 2 \text{ m}.$$

Similarly,

$$\Delta x_2 = x_2 - x_1 = -2 - (2) = -4 \text{ m}.$$

Thus,

$$\Delta x_{\text{Total}} = \Delta x_1 + \Delta x_2 = 2 - 4 = -2 \text{ m}.$$

The total displacement is $2 - 4 = -2$ m to the left, or in the negative direction. It is also useful to calculate the magnitude of the displacement, or its size. The magnitude of the displacement is always positive. This is the absolute value of the displacement, because displacement is a vector and cannot have a negative value of magnitude. In our example, the magnitude of the total displacement is 2 m, whereas the magnitudes of the individual displacements are 2 m and 4 m.

The magnitude of the total displacement should not be confused with the distance traveled. Distance traveled x_{Total}, is the total length of the path traveled between two positions. In the previous problem, the **distance traveled** is the sum of the magnitudes of the individual displacements:

$$x_{\text{Total}} = |\Delta x_1| + |\Delta x_2| = 2 + 4 = 6 \text{ m}.$$

Average Velocity

To calculate the other physical quantities in kinematics we must introduce the time variable. The time variable allows us not only to state where the object is (its position) during its motion, but also how fast it is moving. How fast an object is moving is given by the rate at which the position changes with time.

For each position x_i, we assign a particular time t_i. If the details of the motion at each instant are not important, the rate is usually expressed as the **average velocity** $\bar{v}$. This vector quantity is simply the total displacement between two points divided by the time taken to travel between them. The time taken to travel between two points is called the **elapsed time** Δt.

Average Velocity

If x_1 and x_2 are the positions of an object at times t_1 and t_2, respectively, then

$$\text{Average velocity} = \bar{v} = \frac{\text{Displacement between two points}}{\text{Elapsed time between two points}} \tag{3.3}$$

$$\bar{v} = \frac{\Delta x}{\Delta t} = \frac{x_2 - x_1}{t_2 - t_1}.$$

It is important to note that the average velocity is a vector and can be negative, depending on positions x_1 and x_2.

Example 3.1

Delivering Flyers

Jill sets out from her home to deliver flyers for her yard sale, traveling due east along her street lined with houses. At 0.5 km and 9 minutes later she runs out of flyers and has to retrace her steps back to her house to get more. This takes an additional 9 minutes. After picking up more flyers, she sets out again on the same path, continuing where she left off, and ends up 1.0 km from her house. This third leg of her trip takes 15 minutes. At this point she turns back toward her house, heading west. After 1.75 km and 25 minutes she stops to rest.

    a. What is Jill's total displacement to the point where she stops to rest?

    b. What is the magnitude of the final displacement?

    c. What is the average velocity during her entire trip?

    d. What is the total distance traveled?

    e. Make a graph of position versus time.

A sketch of Jill's movements is shown in **Figure 3.4**.

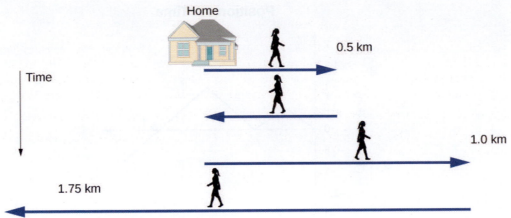

Figure 3.4 Timeline of Jill's movements.

Strategy

The problem contains data on the various legs of Jill's trip, so it would be useful to make a table of the physical quantities. We are given position and time in the wording of the problem so we can calculate the displacements and the elapsed time. We take east to be the positive direction. From this information we can find the total displacement and average velocity. Jill's home is the starting point x_0. The following table gives Jill's time and position in the first two columns, and the displacements are calculated in the third column.

Time t_i (min)	Position x_i (km)	Displacement Δx_i (km)
$t_0 = 0$	$x_0 = 0$	$\Delta x_0 = 0$
$t_1 = 9$	$x_1 = 0.5$	$\Delta x_1 = x_1 - x_0 = 0.5$
$t_2 = 18$	$x_2 = 0$	$\Delta x_2 = x_2 - x_1 = -0.5$
$t_3 = 33$	$x_3 = 1.0$	$\Delta x_3 = x_3 - x_2 = 1.0$
$t_4 = 58$	$x_4 = -0.75$	$\Delta x_4 = x_4 - x_3 = -1.75$

Solution

a. From the above table, the total displacement is

$$\sum \Delta x_i = 0.5 - 0.5 + 1.0 - 1.75 \text{ km} = -0.75 \text{ km}.$$

b. The magnitude of the total displacement is $|-0.75| \text{ km} = 0.75 \text{ km}$.

c. Average velocity $= \dfrac{\text{Total displacement}}{\text{Elapsed time}} = \bar{v} = \dfrac{-0.75 \text{ km}}{58 \text{ min}} = -0.013 \text{ km/min}$

d. The total distance traveled (sum of magnitudes of individual displacements) is

$$x_{\text{Total}} = \sum |\Delta x_i| = 0.5 + 0.5 + 1.0 + 1.75 \text{ km} = 3.75 \text{ km}.$$

e. We can graph Jill's position versus time as a useful aid to see the motion; the graph is shown in **Figure 3.5**.

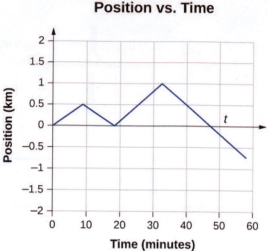

Figure 3.5 This graph depicts Jill's position versus time. The average velocity is the slope of a line connecting the initial and final points.

Significance

Jill's total displacement is −0.75 km, which means at the end of her trip she ends up 0.75 km due west of her home. The average velocity means if someone was to walk due west at 0.013 km/min starting at the same time Jill left her home, they both would arrive at the final stopping point at the same time. Note that if Jill were to end her trip at her house, her total displacement would be zero, as well as her average velocity. The total distance traveled during the 58 minutes of elapsed time for her trip is 3.75 km.

 3.1 Check Your Understanding A cyclist rides 3 km west and then turns around and rides 2 km east. (a) What is his displacement? (b) What is the distance traveled? (c) What is the magnitude of his displacement?

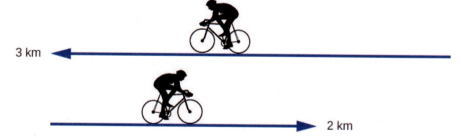

3.2 | Instantaneous Velocity and Speed

Learning Objectives

By the end of this section, you will be able to:

- Explain the difference between average velocity and instantaneous velocity.
- Describe the difference between velocity and speed.
- Calculate the instantaneous velocity given the mathematical equation for the velocity.
- Calculate the speed given the instantaneous velocity.

We have now seen how to calculate the average velocity between two positions. However, since objects in the real world move continuously through space and time, we would like to find the velocity of an object at any single point. We can find the velocity of the object anywhere along its path by using some fundamental principles of calculus. This section gives us better insight into the physics of motion and will be useful in later chapters.

Instantaneous Velocity

The quantity that tells us how fast an object is moving anywhere along its path is the **instantaneous velocity**, usually called simply *velocity*. It is the average velocity between two points on the path in the limit that the time (and therefore the displacement) between the two points approaches zero. To illustrate this idea mathematically, we need to express position x as a continuous function of t denoted by $x(t)$. The expression for the average velocity between two points using this notation is $\bar{v} = \dfrac{x(t_2) - x(t_1)}{t_2 - t_1}$. To find the instantaneous velocity at any position, we let $t_1 = t$ and $t_2 = t + \Delta t$. After inserting these expressions into the equation for the average velocity and taking the limit as $\Delta t \to 0$, we find the expression for the instantaneous velocity:

$$v(t) = \lim_{\Delta t \to 0} \frac{x(t + \Delta t) - x(t)}{\Delta t} = \frac{dx(t)}{dt}.$$

Instantaneous Velocity

The instantaneous velocity of an object is the limit of the average velocity as the elapsed time approaches zero, or the derivative of x with respect to t:

$$v(t) = \frac{d}{dt} x(t). \tag{3.4}$$

Like average velocity, instantaneous velocity is a vector with dimension of length per time. The instantaneous velocity at a specific time point t_0 is the rate of change of the position function, which is the slope of the position function $x(t)$ at t_0. **Figure 3.6** shows how the average velocity $\bar{v} = \dfrac{\Delta x}{\Delta t}$ between two times approaches the instantaneous velocity at t_0.

The instantaneous velocity is shown at time t_0, which happens to be at the maximum of the position function. The slope of the position graph is zero at this point, and thus the instantaneous velocity is zero. At other times, t_1, t_2, and so on, the instantaneous velocity is not zero because the slope of the position graph would be positive or negative. If the position function had a minimum, the slope of the position graph would also be zero, giving an instantaneous velocity of zero there as well. Thus, the zeros of the velocity function give the minimum and maximum of the position function.

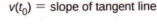

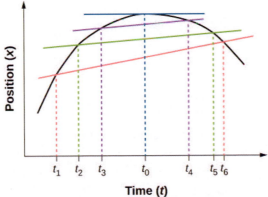

Figure 3.6 In a graph of position versus time, the instantaneous velocity is the slope of the tangent line at a given point. The average velocities $\bar{v} = \dfrac{\Delta x}{\Delta t} = \dfrac{x_f - x_i}{t_f - t_i}$ between times $\Delta t = t_6 - t_1$, $\Delta t = t_5 - t_2$, and $\Delta t = t_4 - t_3$ are shown. When $\Delta t \to 0$, the average velocity approaches the instantaneous velocity at $t = t_0$.

Example 3.2

Finding Velocity from a Position-Versus-Time Graph

Given the position-versus-time graph of **Figure 3.7**, find the velocity-versus-time graph.

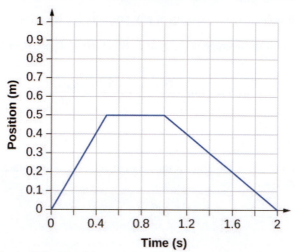

Figure 3.7 The object starts out in the positive direction, stops for a short time, and then reverses direction, heading back toward the origin. Notice that the object comes to rest instantaneously, which would require an infinite force. Thus, the graph is an approximation of motion in the real world. (The concept of force is discussed in **Newton's Laws of Motion**.)

Strategy

The graph contains three straight lines during three time intervals. We find the velocity during each time interval by taking the slope of the line using the grid.

Solution

Time interval 0 s to 0.5 s: $\bar{v} = \dfrac{\Delta x}{\Delta t} = \dfrac{0.5 \text{ m} - 0.0 \text{ m}}{0.5 \text{ s} - 0.0 \text{ s}} = 1.0 \text{ m/s}$

Time interval 0.5 s to 1.0 s: $\bar{v} = \dfrac{\Delta x}{\Delta t} = \dfrac{0.5 \text{ m} - 0.5 \text{ m}}{1.0 \text{ s} - 0.5 \text{ s}} = 0.0 \text{ m/s}$

Time interval 1.0 s to 2.0 s: $\bar{v} = \dfrac{\Delta x}{\Delta t} = \dfrac{0.0 \text{ m} - 0.5 \text{ m}}{2.0 \text{ s} - 1.0 \text{ s}} = -0.5 \text{ m/s}$

The graph of these values of velocity versus time is shown in **Figure 3.8**.

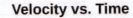

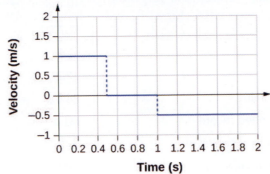

Figure 3.8 The velocity is positive for the first part of the trip, zero when the object is stopped, and negative when the object reverses direction.

Significance

During the time interval between 0 s and 0.5 s, the object's position is moving away from the origin and the position-versus-time curve has a positive slope. At any point along the curve during this time interval, we can find the instantaneous velocity by taking its slope, which is +1 m/s, as shown in **Figure 3.8**. In the subsequent time interval, between 0.5 s and 1.0 s, the position doesn't change and we see the slope is zero. From 1.0 s to 2.0 s, the object is moving back toward the origin and the slope is −0.5 m/s. The object has reversed direction and has a negative velocity.

Speed

In everyday language, most people use the terms *speed* and *velocity* interchangeably. In physics, however, they do not have the same meaning and are distinct concepts. One major difference is that speed has no direction; that is, speed is a scalar.

We can calculate the **average speed** by finding the total distance traveled divided by the elapsed time:

$$\text{Average speed} = \bar{s} = \frac{\text{Total distance}}{\text{Elapsed time}}. \tag{3.5}$$

Average speed is not necessarily the same as the magnitude of the average velocity, which is found by dividing the magnitude of the total displacement by the elapsed time. For example, if a trip starts and ends at the same location, the total displacement is zero, and therefore the average velocity is zero. The average speed, however, is not zero, because the total distance traveled is greater than zero. If we take a road trip of 300 km and need to be at our destination at a certain time, then we would be interested in our average speed.

However, we can calculate the **instantaneous speed** from the magnitude of the instantaneous velocity:

$$\text{Instantaneous speed} = |v(t)|. \tag{3.6}$$

If a particle is moving along the *x*-axis at +7.0 m/s and another particle is moving along the same axis at −7.0 m/s, they have different velocities, but both have the same speed of 7.0 m/s. Some typical speeds are shown in the following table.

Speed	m/s	mi/h
Continental drift	10^{-7}	2×10^{-7}
Brisk walk	1.7	3.9
Cyclist	4.4	10
Sprint runner	12.2	27
Rural speed limit	24.6	56
Official land speed record	341.1	763
Speed of sound at sea level	343	768
Space shuttle on reentry	7800	17,500
Escape velocity of Earth*	11,200	25,000
Orbital speed of Earth around the Sun	29,783	66,623
Speed of light in a vacuum	299,792,458	670,616,629

Table 3.1 Speeds of Various Objects *Escape velocity is the velocity at which an object must be launched so that it overcomes Earth's gravity and is not pulled back toward Earth.

Calculating Instantaneous Velocity

When calculating instantaneous velocity, we need to specify the explicit form of the position function $x(t)$. For the moment, let's use polynomials $x(t) = At^n$, because they are easily differentiated using the power rule of calculus:

$$\frac{dx(t)}{dt} = nAt^{n-1}. \tag{3.7}$$

The following example illustrates the use of **Equation 3.7**.

Example 3.3

Instantaneous Velocity Versus Average Velocity

The position of a particle is given by $x(t) = 3.0t + 0.5t^3$ m.

a. Using **Equation 3.4** and **Equation 3.7**, find the instantaneous velocity at $t = 2.0$ s.

b. Calculate the average velocity between 1.0 s and 3.0 s.

Strategy

Equation 3.4 gives the instantaneous velocity of the particle as the derivative of the position function. Looking at the form of the position function given, we see that it is a polynomial in t. Therefore, we can use **Equation 3.7**, the power rule from calculus, to find the solution. We use **Equation 3.6** to calculate the average velocity of the particle.

Solution

a. $v(t) = \dfrac{dx(t)}{dt} = 3.0 + 1.5t^2$ m/s.

Substituting $t = 2.0$ s into this equation gives $v(2.0 \text{ s}) = [3.0 + 1.5(2.0)^2]$ m/s $= 9.0$ m/s.

b. To determine the average velocity of the particle between 1.0 s and 3.0 s, we calculate the values of $x(1.0$ s) and $x(3.0$ s):

$$x(1.0 \text{ s}) = \left[(3.0)(1.0) + 0.5(1.0)^3\right] \text{m} = 3.5 \text{ m}$$

$$x(3.0 \text{ s}) = \left[(3.0)(3.0) + 0.5(3.0)^3\right] \text{m} = 22.5 \text{ m}.$$

Then the average velocity is

$$\bar{v} = \frac{x(3.0 \text{ s}) - x(1.0 \text{ s})}{t(3.0 \text{ s}) - t(1.0 \text{ s})} = \frac{22.5 - 3.5 \text{ m}}{3.0 - 1.0 \text{ s}} = 9.5 \text{ m/s}.$$

Significance

In the limit that the time interval used to calculate $\bar{v}$ goes to zero, the value obtained for $\bar{v}$ converges to the value of v.

Example 3.4

Instantaneous Velocity Versus Speed

Consider the motion of a particle in which the position is $x(t) = 3.0t - 3t^2$ m .

a. What is the instantaneous velocity at $t = 0.25$ s, $t = 0.50$ s, and $t = 1.0$ s?

b. What is the speed of the particle at these times?

Strategy

The instantaneous velocity is the derivative of the position function and the speed is the magnitude of the instantaneous velocity. We use **Equation 3.4** and **Equation 3.7** to solve for instantaneous velocity.

Solution

a. $v(t) = \dfrac{dx(t)}{dt} = 3.0 - 6.0t$ m/s $v(0.25 \text{ s}) = 1.50$ m/s, $v(0.5 \text{ s}) = 0$ m/s, $v(1.0 \text{ s}) = -3.0$ m/s

b. Speed $= |v(t)| = 1.50$ m/s, 0.0 m/s, and 3.0 m/s

Significance

The velocity of the particle gives us direction information, indicating the particle is moving to the left (west) or right (east). The speed gives the magnitude of the velocity. By graphing the position, velocity, and speed as functions of time, we can understand these concepts visually **Figure 3.9**. In (a), the graph shows the particle moving in the positive direction until $t = 0.5$ s, when it reverses direction. The reversal of direction can also be seen in (b) at 0.5 s where the velocity is zero and then turns negative. At 1.0 s it is back at the origin where it started. The particle's velocity at 1.0 s in (b) is negative, because it is traveling in the negative direction. But in (c), however, its speed is positive and remains positive throughout the travel time. We can also interpret velocity as the slope of the position-versus-time graph. The slope of $x(t)$ is decreasing toward zero, becoming zero at 0.5 s and increasingly negative thereafter. This analysis of comparing the graphs of position, velocity, and speed helps catch errors in calculations. The graphs must be consistent with each other and help interpret the calculations.

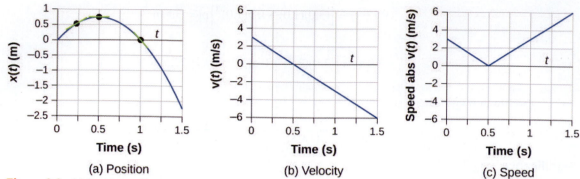

Figure 3.9 (a) Position: $x(t)$ versus time. (b) Velocity: $v(t)$ versus time. The slope of the position graph is the velocity. A rough comparison of the slopes of the tangent lines in (a) at 0.25 s, 0.5 s, and 1.0 s with the values for velocity at the corresponding times indicates they are the same values. (c) Speed: $|v(t)|$ versus time. Speed is always a positive number.

 3.2   **Check Your Understanding** The position of an object as a function of time is $x(t) = -3t^2$ m . (a) What is the velocity of the object as a function of time? (b) Is the velocity ever positive? (c) What are the velocity and speed at $t = 1.0$ s?

3.3 | Average and Instantaneous Acceleration

Learning Objectives

By the end of this section, you will be able to:

- Calculate the average acceleration between two points in time.
- Calculate the instantaneous acceleration given the functional form of velocity.
- Explain the vector nature of instantaneous acceleration and velocity.
- Explain the difference between average acceleration and instantaneous acceleration.
- Find instantaneous acceleration at a specified time on a graph of velocity versus time.

The importance of understanding acceleration spans our day-to-day experience, as well as the vast reaches of outer space and the tiny world of subatomic physics. In everyday conversation, to *accelerate* means to speed up; applying the brake pedal causes a vehicle to slow down. We are familiar with the acceleration of our car, for example. The greater the acceleration, the greater the change in velocity over a given time. Acceleration is widely seen in experimental physics. In linear particle accelerator experiments, for example, subatomic particles are accelerated to very high velocities in collision experiments, which tell us information about the structure of the subatomic world as well as the origin of the universe. In space, cosmic rays are subatomic particles that have been accelerated to very high energies in supernovas (exploding massive stars) and active galactic nuclei. It is important to understand the processes that accelerate cosmic rays because these rays contain highly penetrating radiation that can damage electronics flown on spacecraft, for example.

Average Acceleration

The formal definition of acceleration is consistent with these notions just described, but is more inclusive.

Average Acceleration

Average acceleration is the rate at which velocity changes:

$$\bar{a} = \frac{\Delta v}{\Delta t} = \frac{v_{\text{f}} - v_0}{t_{\text{f}} - t_0},$$ **(3.8)**

where $\bar{a}$ is **average acceleration**, v is velocity, and t is time. (The bar over the a means *average* acceleration.)

Because acceleration is velocity in meters per second divided by time in seconds, the SI units for acceleration are often abbreviated m/s^2—that is, meters per second squared or meters per second per second. This literally means by how many meters per second the velocity changes every second. Recall that velocity is a vector—it has both magnitude and direction—which means that a change in velocity can be a change in magnitude (or speed), but it can also be a change in direction. For example, if a runner traveling at 10 km/h due east slows to a stop, reverses direction, and continues her run at 10 km/h due west, her velocity has changed as a result of the change in direction, although the *magnitude* of the velocity is the same in both directions. Thus, acceleration occurs when velocity changes in magnitude (an increase or decrease in speed) or in direction, or both.

Acceleration as a Vector

Acceleration is a vector in the same direction as the *change* in velocity, Δv. Since velocity is a vector, it can change in magnitude or in direction, or both. Acceleration is, therefore, a change in speed or direction, or both.

Keep in mind that although acceleration is in the direction of the change in velocity, it is not always in the direction of motion. When an object slows down, its acceleration is opposite to the direction of its motion. Although this is commonly referred to as *deceleration* **Figure 3.10**, we say the train is accelerating in a direction opposite to its direction of motion.

Figure 3.10   A subway train in Sao Paulo, Brazil, decelerates as it comes into a station. It is accelerating in a direction opposite to its direction of motion. (credit: modification of work by Yusuke Kawasaki)

The term *deceleration* can cause confusion in our analysis because it is not a vector and it does not point to a specific direction with respect to a coordinate system, so we do not use it. Acceleration is a vector, so we must choose the appropriate sign for it in our chosen coordinate system. In the case of the train in **Figure 3.10**, acceleration is *in the negative direction in the chosen coordinate system*, so we say the train is undergoing negative acceleration.

If an object in motion has a velocity in the positive direction with respect to a chosen origin and it acquires a constant negative acceleration, the object eventually comes to a rest and reverses direction. If we wait long enough, the object passes through the origin going in the opposite direction. This is illustrated in **Figure 3.11**.

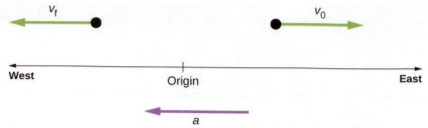

Figure 3.11 An object in motion with a velocity vector toward the east under negative acceleration comes to a rest and reverses direction. It passes the origin going in the opposite direction after a long enough time.

Example 3.5

Calculating Average Acceleration: A Racehorse Leaves the Gate

A racehorse coming out of the gate accelerates from rest to a velocity of 15.0 m/s due west in 1.80 s. What is its average acceleration?

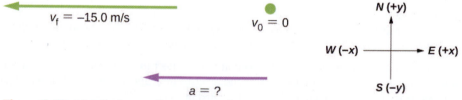

Figure 3.12 Racehorses accelerating out of the gate. (credit: modification of work by Jon Sullivan)

Strategy

First we draw a sketch and assign a coordinate system to the problem **Figure 3.13**. This is a simple problem, but it always helps to visualize it. Notice that we assign east as positive and west as negative. Thus, in this case, we have negative velocity.

Figure 3.13 Identify the coordinate system, the given information, and what you want to determine.

We can solve this problem by identifying Δv and Δt from the given information, and then calculating the average acceleration directly from the equation $\bar{a} = \dfrac{\Delta v}{\Delta t} = \dfrac{v_f - v_0}{t_f - t_0}$.

Solution

First, identify the knowns: $v_0 = 0$, $v_f = -15.0$ m/s (the negative sign indicates direction toward the west), $\Delta t = 1.80$ s.

Second, find the change in velocity. Since the horse is going from zero to −15.0 m/s, its change in velocity equals its final velocity:

$$\Delta v = v_f - v_0 = v_f = -15.0 \text{ m/s}.$$

Last, substitute the known values (Δv and Δt) and solve for the unknown $\bar{a}$:

$$\bar{a} = \frac{\Delta v}{\Delta t} = \frac{-15.0 \text{ m/s}}{1.80 \text{ s}} = -8.33 \text{m/s}^2.$$

Significance

The negative sign for acceleration indicates that acceleration is toward the west. An acceleration of 8.33 m/s² due west means the horse increases its velocity by 8.33 m/s due west each second; that is, 8.33 meters per second per second, which we write as 8.33 m/s². This is truly an average acceleration, because the ride is not smooth. We see later that an acceleration of this magnitude would require the rider to hang on with a force nearly equal to his weight.

3.3 **Check Your Understanding** Protons in a linear accelerator are accelerated from rest to 2.0×10^7 m/s in 10^{-4} s. What is the average acceleration of the protons?

Instantaneous Acceleration

Instantaneous acceleration a, or *acceleration at a specific instant in time*, is obtained using the same process discussed for instantaneous velocity. That is, we calculate the average acceleration between two points in time separated by Δt and let Δt approach zero. The result is the derivative of the velocity function $v(t)$, which is **instantaneous acceleration** and is expressed mathematically as

$$a(t) = \frac{d}{dt}v(t). \qquad (3.9)$$

Thus, similar to velocity being the derivative of the position function, instantaneous acceleration is the derivative of the velocity function. We can show this graphically in the same way as instantaneous velocity. In **Figure 3.14**, instantaneous acceleration at time t_0 is the slope of the tangent line to the velocity-versus-time graph at time t_0. We see that average acceleration $\bar{a} = \frac{\Delta v}{\Delta t}$ approaches instantaneous acceleration as Δt approaches zero. Also in part (a) of the figure, we see that velocity has a maximum when its slope is zero. This time corresponds to the zero of the acceleration function. In part (b), instantaneous acceleration at the minimum velocity is shown, which is also zero, since the slope of the curve is zero there, too. Thus, for a given velocity function, the zeros of the acceleration function give either the minimum or the maximum velocity.

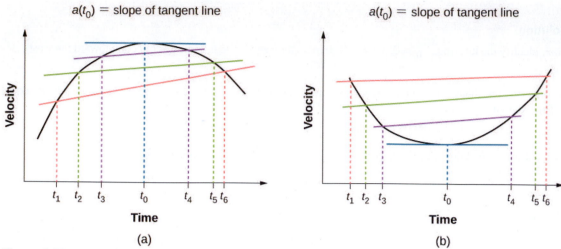

Figure 3.14 In a graph of velocity versus time, instantaneous acceleration is the slope of the tangent line. (a) Shown is average acceleration $\bar{a} = \dfrac{\Delta v}{\Delta t} = \dfrac{v_f - v_i}{t_f - t_i}$ between times $\Delta t = t_6 - t_1$, $\Delta t = t_5 - t_2$, and $\Delta t = t_4 - t_3$. When $\Delta t \to 0$, the average acceleration approaches instantaneous acceleration at time t_0. In view (a), instantaneous acceleration is shown for the point on the velocity curve at maximum velocity. At this point, instantaneous acceleration is the slope of the tangent line, which is zero. At any other time, the slope of the tangent line—and thus instantaneous acceleration—would not be zero. (b) Same as (a) but shown for instantaneous acceleration at minimum velocity.

To illustrate this concept, let's look at two examples. First, a simple example is shown using **Figure 3.9**(b), the velocity-versus-time graph of **Example 3.3**, to find acceleration graphically. This graph is depicted in **Figure 3.15**(a), which is a straight line. The corresponding graph of acceleration versus time is found from the slope of velocity and is shown in **Figure 3.15**(b). In this example, the velocity function is a straight line with a constant slope, thus acceleration is a constant. In the next example, the velocity function has a more complicated functional dependence on time.

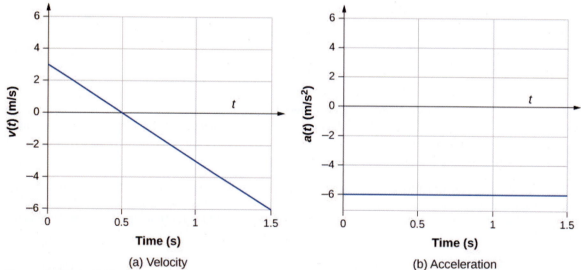

Figure 3.15 (a, b) The velocity-versus-time graph is linear and has a negative constant slope (a) that is equal to acceleration, shown in (b).

If we know the functional form of velocity, $v(t)$, we can calculate instantaneous acceleration $a(t)$ at any time point in the motion using **Equation 3.9**.

Example 3.6

Calculating Instantaneous Acceleration

A particle is in motion and is accelerating. The functional form of the velocity is $v(t) = 20t - 5t^2$ m/s.

- a. Find the functional form of the acceleration.
- b. Find the instantaneous velocity at $t = 1, 2, 3,$ and 5 s.
- c. Find the instantaneous acceleration at $t = 1, 2, 3,$ and 5 s.
- d. Interpret the results of (c) in terms of the directions of the acceleration and velocity vectors.

Strategy

We find the functional form of acceleration by taking the derivative of the velocity function. Then, we calculate the values of instantaneous velocity and acceleration from the given functions for each. For part (d), we need to compare the directions of velocity and acceleration at each time.

Solution

a. $a(t) = \dfrac{dv(t)}{dt} = 20 - 10t$ m/s^2

b. $v(1 \text{ s}) = 15$ m/s, $v(2 \text{ s}) = 20$ m/s, $v(3 \text{ s}) = 15$ m/s, $v(5 \text{ s}) = -25$ m/s

c. $a(1 \text{ s}) = 10$ m/s^2, $a(2 \text{ s}) = 0$ m/s^2, $a(3 \text{ s}) = -10$ m/s^2, $a(5 \text{ s}) = -30$ m/s^2

d. At $t = 1$ s, velocity $v(1 \text{ s}) = 15$ m/s is positive and acceleration is positive, so both velocity and acceleration are in the same direction. The particle is moving faster.

At $t = 2$ s, velocity has increased to $v(2 \text{ s}) = 20$ m/s, where it is maximum, which corresponds to the time when the acceleration is zero. We see that the maximum velocity occurs when the slope of the velocity function is zero, which is just the zero of the acceleration function.

At $t = 3$ s, velocity is $v(3 \text{ s}) = 15$ m/s and acceleration is negative. The particle has reduced its velocity and the acceleration vector is negative. The particle is slowing down.

At $t = 5$ s, velocity is $v(5 \text{ s}) = -25$ m/s and acceleration is increasingly negative. Between the times $t = 3$ s and $t = 5$ s the particle has decreased its velocity to zero and then become negative, thus reversing its direction. The particle is now speeding up again, but in the opposite direction.

We can see these results graphically in **Figure 3.16**.

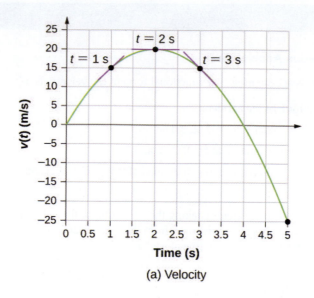

(a) Velocity

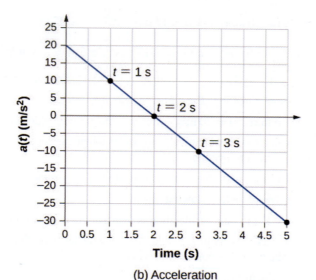

(b) Acceleration

Figure 3.16 (a) Velocity versus time. Tangent lines are indicated at times 1, 2, and 3 s. The slopes of the tangent lines are the accelerations. At $t = 3$ s, velocity is positive. At $t = 5$ s, velocity is negative, indicating the particle has reversed direction. (b) Acceleration versus time. Comparing the values of accelerations given by the black dots with the corresponding slopes of the tangent lines (slopes of lines through black dots) in (a), we see they are identical.

Significance

By doing both a numerical and graphical analysis of velocity and acceleration of the particle, we can learn much about its motion. The numerical analysis complements the graphical analysis in giving a total view of the motion. The zero of the acceleration function corresponds to the maximum of the velocity in this example. Also in this example, when acceleration is positive and in the same direction as velocity, velocity increases. As acceleration tends toward zero, eventually becoming negative, the velocity reaches a maximum, after which it starts decreasing. If we wait long enough, velocity also becomes negative, indicating a reversal of direction. A real-world example of this type of motion is a car with a velocity that is increasing to a maximum, after which it starts slowing down, comes to a stop, then reverses direction.

3.4 **Check Your Understanding** An airplane lands on a runway traveling east. Describe its acceleration.

Getting a Feel for Acceleration

You are probably used to experiencing acceleration when you step into an elevator, or step on the gas pedal in your car. However, acceleration is happening to many other objects in our universe with which we don't have direct contact. **Table 3.2** presents the acceleration of various objects. We can see the magnitudes of the accelerations extend over many orders of magnitude.

Acceleration	Value (m/s^2)
High-speed train	0.25
Elevator	2
Cheetah	5
Object in a free fall without air resistance near the surface of Earth	9.8
Space shuttle maximum during launch	29
Parachutist peak during normal opening of parachute	59
F16 aircraft pulling out of a dive	79
Explosive seat ejection from aircraft	147
Sprint missile	982
Fastest rocket sled peak acceleration	1540
Jumping flea	3200
Baseball struck by a bat	30,000
Closing jaws of a trap-jaw ant	1,000,000
Proton in the large Hadron collider	1.9×10^9

Table 3.2 Typical Values of Acceleration (credit: Wikipedia: Orders of Magnitude (acceleration))

In this table, we see that typical accelerations vary widely with different objects and have nothing to do with object size or how massive it is. Acceleration can also vary widely with time during the motion of an object. A drag racer has a large acceleration just after its start, but then it tapers off as the vehicle reaches a constant velocity. Its average acceleration can be quite different from its instantaneous acceleration at a particular time during its motion. **Figure 3.17** compares graphically average acceleration with instantaneous acceleration for two very different motions.

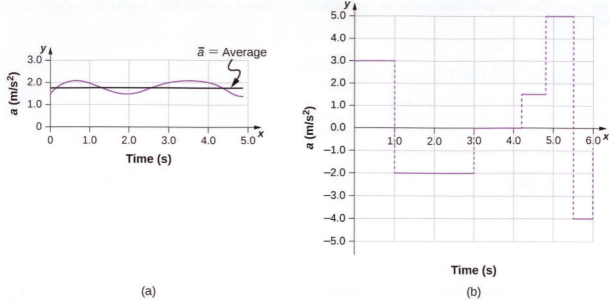

Figure 3.17 Graphs of instantaneous acceleration versus time for two different one-dimensional motions. (a) Acceleration varies only slightly and is always in the same direction, since it is positive. The average over the interval is nearly the same as the acceleration at any given time. (b) Acceleration varies greatly, perhaps representing a package on a post office conveyor belt that is accelerated forward and backward as it bumps along. It is necessary to consider small time intervals (such as from 0–1.0 s) with constant or nearly constant acceleration in such a situation.

 Learn about position, velocity, and acceleration graphs. Move the little man back and forth with a mouse and plot his motion. Set the position, velocity, or acceleration and let the simulation move the man for you. Visit **this link (https://openstaxcollege.org/l/21movmansimul)** to use the moving man simulation.

3.4 | Motion with Constant Acceleration

Learning Objectives

By the end of this section, you will be able to:

- Identify which equations of motion are to be used to solve for unknowns.
- Use appropriate equations of motion to solve a two-body pursuit problem.

You might guess that the greater the acceleration of, say, a car moving away from a stop sign, the greater the car's displacement in a given time. But, we have not developed a specific equation that relates acceleration and displacement. In this section, we look at some convenient equations for kinematic relationships, starting from the definitions of displacement, velocity, and acceleration. We first investigate a single object in motion, called single-body motion. Then we investigate the motion of two objects, called **two-body pursuit problems**.

Notation

First, let us make some simplifications in notation. Taking the initial time to be zero, as if time is measured with a stopwatch, is a great simplification. Since elapsed time is $\Delta t = t_f - t_0$, taking $t_0 = 0$ means that $\Delta t = t_f$, the final time on the stopwatch. When initial time is taken to be zero, we use the subscript 0 to denote initial values of position and velocity. That is, x_0 *is the initial position* and v_0 *is the initial velocity*. We put no subscripts on the final values. That is, *t is the final time, x is the final position*, and *v is the final velocity*. This gives a simpler expression for elapsed time, $\Delta t = t$. It also simplifies the expression for x displacement, which is now $\Delta x = x - x_0$. Also, it simplifies the expression for change in velocity, which is now $\Delta v = v - v_0$. To summarize, using the simplified notation, with the initial time taken to be zero,

$$\Delta t = t$$
$$\Delta x = x - x_0$$
$$\Delta v = v - v_0,$$

where the subscript 0 denotes an initial value and the absence of a subscript denotes a final value in whatever motion is under consideration.

We now make the important assumption that *acceleration is constant*. This assumption allows us to avoid using calculus to find instantaneous acceleration. Since acceleration is constant, the average and instantaneous accelerations are equal—that is,

$$\bar{a} = a = \text{constant.}$$

Thus, we can use the symbol a for acceleration at all times. Assuming acceleration to be constant does not seriously limit the situations we can study nor does it degrade the accuracy of our treatment. For one thing, acceleration *is* constant in a great number of situations. Furthermore, in many other situations we can describe motion accurately by assuming a constant acceleration equal to the average acceleration for that motion. Lastly, for motion during which acceleration changes drastically, such as a car accelerating to top speed and then braking to a stop, motion can be considered in separate parts, each of which has its own constant acceleration.

Displacement and Position from Velocity

To get our first two equations, we start with the definition of average velocity:

$$\bar{v} = \frac{\Delta x}{\Delta t}.$$

Substituting the simplified notation for Δx and Δt yields

$$\bar{v} = \frac{x - x_0}{t}.$$

Solving for x gives us

$$x = x_0 + \bar{v}t, \tag{3.10}$$

where the average velocity is

$$\bar{v} = \frac{v_0 + v}{2}. \tag{3.11}$$

The equation $\bar{v} = \frac{v_0 + v}{2}$ reflects the fact that when acceleration is constant, $\bar{v}$ is just the simple average of the initial and final velocities. **Figure 3.18** illustrates this concept graphically. In part (a) of the figure, acceleration is constant, with velocity increasing at a constant rate. The average velocity during the 1-h interval from 40 km/h to 80 km/h is 60 km/h:

$$\bar{v} = \frac{v_0 + v}{2} = \frac{40 \text{ km/h} + 80 \text{ km/h}}{2} = 60 \text{ km/h.}$$

In part (b), acceleration is not constant. During the 1-h interval, velocity is closer to 80 km/h than 40 km/h. Thus, the average velocity is greater than in part (a).

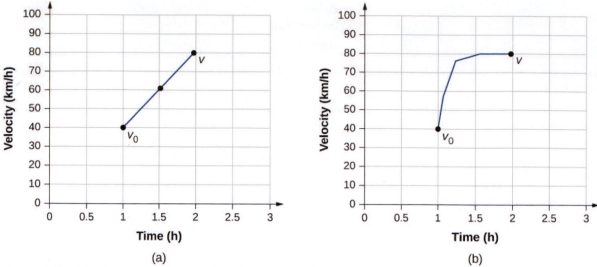

Figure 3.18 (a) Velocity-versus-time graph with constant acceleration showing the initial and final velocities v_0 and v.
The average velocity is $\frac{1}{2}(v_0 + v) = 60 \text{ km/h}$. (b) Velocity-versus-time graph with an acceleration that changes with time.
The average velocity is not given by $\frac{1}{2}(v_0 + v)$, but is greater than 60 km/h.

Solving for Final Velocity from Acceleration and Time

We can derive another useful equation by manipulating the definition of acceleration:

$$a = \frac{\Delta v}{\Delta t}.$$

Substituting the simplified notation for Δv and Δt gives us

$$a = \frac{v - v_0}{t} \quad \text{(constant } a\text{)}.$$

Solving for v yields

$$v = v_0 + at \quad \text{(constant } a\text{).} \tag{3.12}$$

Example 3.7

Calculating Final Velocity

An airplane lands with an initial velocity of 70.0 m/s and then decelerates at 1.50 m/s² for 40.0 s. What is its final velocity?

Strategy

First, we identify the knowns: $v_0 = 70 \text{ m/s}$, $a = -1.50 \text{ m/s}^2$, $t = 40 \text{ s}$.

Second, we identify the unknown; in this case, it is final velocity v_f.

Last, we determine which equation to use. To do this we figure out which kinematic equation gives the unknown in terms of the knowns. We calculate the final velocity using **Equation 3.12**, $v = v_0 + at$.

Solution

Substitute the known values and solve:

$$v = v_0 + at = 70.0 \text{ m/s} + \left(-1.50 \text{ m/s}^2\right)(40.0 \text{ s}) = 10.0 \text{ m/s}.$$

Figure 3.19 is a sketch that shows the acceleration and velocity vectors.

Figure 3.19 The airplane lands with an initial velocity of 70.0 m/s and slows to a final velocity of 10.0 m/s before heading for the terminal. Note the acceleration is negative because its direction is opposite to its velocity, which is positive.

Significance

The final velocity is much less than the initial velocity, as desired when slowing down, but is still positive (see figure). With jet engines, reverse thrust can be maintained long enough to stop the plane and start moving it backward, which is indicated by a negative final velocity, but is not the case here.

In addition to being useful in problem solving, the equation $v = v_0 + at$ gives us insight into the relationships among velocity, acceleration, and time. We can see, for example, that

- Final velocity depends on how large the acceleration is and how long it lasts
- If the acceleration is zero, then the final velocity equals the initial velocity ($v = v_0$), as expected (in other words, velocity is constant)
- If a is negative, then the final velocity is less than the initial velocity

All these observations fit our intuition. Note that it is always useful to examine basic equations in light of our intuition and experience to check that they do indeed describe nature accurately.

Solving for Final Position with Constant Acceleration

We can combine the previous equations to find a third equation that allows us to calculate the final position of an object experiencing constant acceleration. We start with

$$v = v_0 + at.$$

Adding v_0 to each side of this equation and dividing by 2 gives

$$\frac{v_0 + v}{2} = v_0 + \frac{1}{2}at.$$

Since $\frac{v_0 + v}{2} = \bar{v}$ for constant acceleration, we have

$$\bar{v} = v_0 + \frac{1}{2}at.$$

Now we substitute this expression for $\bar{v}$ into the equation for displacement, $x = x_0 + \bar{v}t$, yielding

$$x = x_0 + v_0 t + \frac{1}{2}at^2 \quad \text{(constant } a\text{).} \tag{3.13}$$

Example 3.8

Calculating Displacement of an Accelerating Object

Dragsters can achieve an average acceleration of 26.0 m/s^2. Suppose a dragster accelerates from rest at this rate for 5.56 s **Figure 3.20**. How far does it travel in this time?

Figure 3.20 U.S. Army Top Fuel pilot Tony "The Sarge" Schumacher begins a race with a controlled burnout. (credit: Lt. Col. William Thurmond. Photo Courtesy of U.S. Army.)

Strategy

First, let's draw a sketch **Figure 3.21**. We are asked to find displacement, which is x if we take x_0 to be zero. (Think about x_0 as the starting line of a race. It can be anywhere, but we call it zero and measure all other positions relative to it.) We can use the equation $x = x_0 + v_0 t + \frac{1}{2}at^2$ when we identify v_0, a, and t from the statement of the problem.

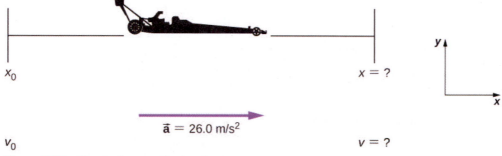

Figure 3.21 Sketch of an accelerating dragster.

Solution

First, we need to identify the knowns. Starting from rest means that $v_0 = 0$, a is given as 26.0 m/s^2 and t is given as 5.56 s.

Second, we substitute the known values into the equation to solve for the unknown:

$$x = x_0 + v_0 t + \frac{1}{2}at^2.$$

Since the initial position and velocity are both zero, this equation simplifies to

$$x = \frac{1}{2}at^2.$$

Substituting the identified values of a and t gives

$$x = \frac{1}{2}(26.0 \text{ m/s}^2)(5.56 \text{ s})^2 = 402 \text{ m}.$$

Significance

If we convert 402 m to miles, we find that the distance covered is very close to one-quarter of a mile, the standard distance for drag racing. So, our answer is reasonable. This is an impressive displacement to cover in only 5.56 s, but top-notch dragsters can do a quarter mile in even less time than this. If the dragster were given an initial velocity, this would add another term to the distance equation. If the same acceleration and time are used in the equation, the distance covered would be much greater.

What else can we learn by examining the equation $x = x_0 + v_0 t + \frac{1}{2}at^2$? We can see the following relationships:

- Displacement depends on the square of the elapsed time when acceleration is not zero. In **Example 3.8**, the dragster covers only one-fourth of the total distance in the first half of the elapsed time.

- If acceleration is zero, then initial velocity equals average velocity $(v_0 = \bar{v})$, and $x = x_0 + v_0 t + \frac{1}{2}at^2$ becomes $x = x_0 + v_0 t$.

Solving for Final Velocity from Distance and Acceleration

A fourth useful equation can be obtained from another algebraic manipulation of previous equations. If we solve $v = v_0 + at$ for t, we get

$$t = \frac{v - v_0}{a}.$$

Substituting this and $\bar{v} = \frac{v_0 + v}{2}$ into $x = x_0 + \bar{v}t$, we get

$$v^2 = v_0^2 + 2a(x - x_0) \quad \text{(constant } a\text{)}. \qquad (3.14)$$

Example 3.9

Calculating Final Velocity

Calculate the final velocity of the dragster in **Example 3.8** without using information about time.

Strategy

The equation $v^2 = v_0^2 + 2a(x - x_0)$ is ideally suited to this task because it relates velocities, acceleration, and displacement, and no time information is required.

Solution

First, we identify the known values. We know that $v_0 = 0$, since the dragster starts from rest. We also know that $x - x_0 = 402$ m (this was the answer in **Example 3.8**). The average acceleration was given by $a = 26.0 \text{ m/s}^2$.

Second, we substitute the knowns into the equation $v^2 = v_0^2 + 2a(x - x_0)$ and solve for v:

$$v^2 = 0 + 2(26.0 \text{ m/s}^2)(402 \text{ m}).$$

Thus,

$$v^2 = 2.09 \times 10^4 \text{ m}^2/\text{s}^2$$

$$v = \sqrt{2.09 \times 10^4 \text{ m}^2/\text{s}^2} = 145 \text{ m/s}.$$

Significance

A velocity of 145 m/s is about 522 km/h, or about 324 mi/h, but even this breakneck speed is short of the record for the quarter mile. Also, note that a square root has two values; we took the positive value to indicate a velocity in the same direction as the acceleration.

An examination of the equation $v^2 = v_0^2 + 2a(x - x_0)$ can produce additional insights into the general relationships among physical quantities:

- The final velocity depends on how large the acceleration is and the distance over which it acts.

- For a fixed acceleration, a car that is going twice as fast doesn't simply stop in twice the distance. It takes much farther to stop. (This is why we have reduced speed zones near schools.)

Putting Equations Together

In the following examples, we continue to explore one-dimensional motion, but in situations requiring slightly more algebraic manipulation. The examples also give insight into problem-solving techniques. The note that follows is provided for easy reference to the equations needed. Be aware that these equations are not independent. In many situations we have two unknowns and need two equations from the set to solve for the unknowns. We need as many equations as there are unknowns to solve a given situation.

Summary of Kinematic Equations (constant *a*)

$$x = x_0 + \bar{v}t$$

$$\bar{v} = \frac{v_0 + v}{2}$$

$$v = v_0 + at$$

$$x = x_0 + v_0 t + \frac{1}{2}at^2$$

$$v^2 = v_0^2 + 2a(x - x_0)$$

Before we get into the examples, let's look at some of the equations more closely to see the behavior of acceleration at extreme values. Rearranging **Equation 3.12**, we have

$$a = \frac{v - v_0}{t}.$$

From this we see that, for a finite time, if the difference between the initial and final velocities is small, the acceleration is small, approaching zero in the limit that the initial and final velocities are equal. On the contrary, in the limit $t \rightarrow 0$ for a finite difference between the initial and final velocities, acceleration becomes infinite.

Similarly, rearranging **Equation 3.14**, we can express acceleration in terms of velocities and displacement:

$$a = \frac{v^2 - v_0^2}{2(x - x_0)}.$$

Thus, for a finite difference between the initial and final velocities acceleration becomes infinite in the limit the displacement approaches zero. Acceleration approaches zero in the limit the difference in initial and final velocities approaches zero for a finite displacement.

Example 3.10

How Far Does a Car Go?

On dry concrete, a car can decelerate at a rate of 7.00 m/s^2, whereas on wet concrete it can decelerate at only 5.00 m/s^2. Find the distances necessary to stop a car moving at 30.0 m/s (about 110 km/h) on (a) dry concrete and (b) wet concrete. (c) Repeat both calculations and find the displacement from the point where the driver sees a traffic light turn red, taking into account his reaction time of 0.500 s to get his foot on the brake.

Strategy

First, we need to draw a sketch **Figure 3.22**. To determine which equations are best to use, we need to list all the known values and identify exactly what we need to solve for.

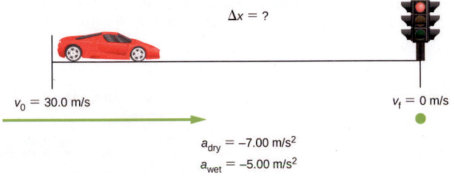

Figure 3.22   Sample sketch to visualize deceleration and stopping distance of a car.

Solution

a. First, we need to identify the knowns and what we want to solve for. We know that $v_0 = 30.0$ m/s, $v = 0$, and $a = -7.00$ m/s^2 (a is negative because it is in a direction opposite to velocity). We take x_0 to be zero. We are looking for displacement Δx, or $x - x_0$.

Second, we identify the equation that will help us solve the problem. The best equation to use is

$$v^2 = v_0^2 + 2a(x - x_0).$$

This equation is best because it includes only one unknown, x. We know the values of all the other variables in this equation. (Other equations would allow us to solve for x, but they require us to know the stopping time, t, which we do not know. We could use them, but it would entail additional calculations.) Third, we rearrange the equation to solve for x:

$$x - x_0 = \frac{v^2 - v_0^2}{2a}$$

and substitute the known values:

$$x - 0 = \frac{0^2 - (30.0 \text{ m/s})^2}{2(-7.00 \text{m/s}^2)}.$$

Thus,

$$x = 64.3 \text{ m on dry concrete.}$$

b. This part can be solved in exactly the same manner as (a). The only difference is that the acceleration is -5.00 m/s^2. The result is

$$x_{\text{wet}} = 90.0 \text{ m on wet concrete.}$$

c. When the driver reacts, the stopping distance is the same as it is in (a) and (b) for dry and wet concrete. So, to answer this question, we need to calculate how far the car travels during the reaction time, and then add that to the stopping time. It is reasonable to assume the velocity remains constant during the driver's reaction time.

To do this, we, again, identify the knowns and what we want to solve for. We know that $\bar{v} = 30.0 \text{ m/s}$, $t_{\text{reaction}} = 0.500 \text{ s}$, and $a_{\text{reaction}} = 0$. We take $x_{0\text{-reaction}}$ to be zero. We are looking for x_{reaction}.

Second, as before, we identify the best equation to use. In this case, $x = x_0 + \bar{v}t$ works well because the only unknown value is x, which is what we want to solve for.

Third, we substitute the knowns to solve the equation:

$$x = 0 + (30.0 \text{ m/s})(0.500 \text{ s}) = 15.0 \text{ m}.$$

This means the car travels 15.0 m while the driver reacts, making the total displacements in the two cases of dry and wet concrete 15.0 m greater than if he reacted instantly.

Last, we then add the displacement during the reaction time to the displacement when braking (**Figure 3.23**),

$$x_{\text{braking}} + x_{\text{reaction}} = x_{\text{total}},$$

and find (a) to be 64.3 m + 15.0 m = 79.3 m when dry and (b) to be 90.0 m + 15.0 m = 105 m when wet.

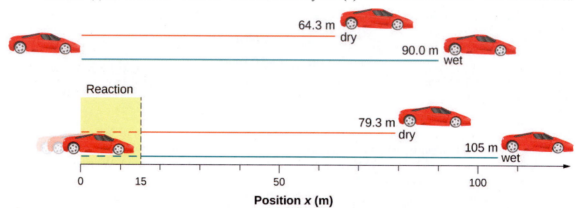

Figure 3.23 The distance necessary to stop a car varies greatly, depending on road conditions and driver reaction time. Shown here are the braking distances for dry and wet pavement, as calculated in this example, for a car traveling initially at 30.0 m/s. Also shown are the total distances traveled from the point when the driver first sees a light turn red, assuming a 0.500-s reaction time.

Significance

The displacements found in this example seem reasonable for stopping a fast-moving car. It should take longer to stop a car on wet pavement than dry. It is interesting that reaction time adds significantly to the displacements, but more important is the general approach to solving problems. We identify the knowns and the quantities to be determined, then find an appropriate equation. If there is more than one unknown, we need as many independent equations as there are unknowns to solve. There is often more than one way to solve a problem. The various parts of this example can, in fact, be solved by other methods, but the solutions presented here are the shortest.

Example 3.11

Calculating Time

Suppose a car merges into freeway traffic on a 200-m-long ramp. If its initial velocity is 10.0 m/s and it accelerates at 2.00 m/s², how long does it take the car to travel the 200 m up the ramp? (Such information might be useful to a traffic engineer.)

Strategy

First, we draw a sketch **Figure 3.24**. We are asked to solve for time t. As before, we identify the known quantities to choose a convenient physical relationship (that is, an equation with one unknown, t.)

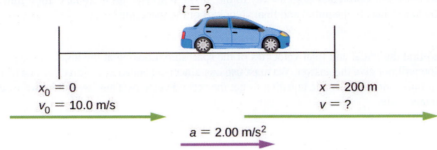

$t = ?$

$x_0 = 0$
$v_0 = 10.0 \text{ m/s}$

$x = 200 \text{ m}$
$v = ?$

$a = 2.00 \text{ m/s}^2$

Figure 3.24   Sketch of a car accelerating on a freeway ramp.

Solution

Again, we identify the knowns and what we want to solve for. We know that $x_0 = 0$,

$v_0 = 10$ m/s, $a = 2.00$ m/s^2, and $x = 200$ m.

We need to solve for t. The equation $x = x_0 + v_0 t + \frac{1}{2}at^2$ works best because the only unknown in the equation

is the variable t, for which we need to solve. From this insight we see that when we input the knowns into the equation, we end up with a quadratic equation.

We need to rearrange the equation to solve for t, then substituting the knowns into the equation:

$$200 \text{ m} = 0 \text{ m} + (10.0 \text{ m/s})t + \frac{1}{2}\left(2.00 \text{ m/s}^2\right)t^2.$$

We then simplify the equation. The units of meters cancel because they are in each term. We can get the units of seconds to cancel by taking $t = t$ s, where t is the magnitude of time and s is the unit. Doing so leaves

$$200 = 10t + t^2.$$

We then use the quadratic formula to solve for t,

$$t^2 + 10t - 200 = 0$$

$$t = \frac{-b \pm \sqrt{b^2 - 4ac}}{2a},$$

which yields two solutions: $t = 10.0$ and $t = -20.0$. A negative value for time is unreasonable, since it would mean the event happened 20 s before the motion began. We can discard that solution. Thus,

$$t = 10.0 \text{ s}.$$

Significance

Whenever an equation contains an unknown squared, there are two solutions. In some problems both solutions are meaningful; in others, only one solution is reasonable. The 10.0-s answer seems reasonable for a typical freeway on-ramp.

 3.5   Check Your Understanding A manned rocket accelerates at a rate of 20 m/s^2 during launch. How long does it take the rocket to reach a velocity of 400 m/s?

Example 3.12

Acceleration of a Spaceship

A spaceship has left Earth's orbit and is on its way to the Moon. It accelerates at 20 m/s^2 for 2 min and covers a distance of 1000 km. What are the initial and final velocities of the spaceship?

Strategy

We are asked to find the initial and final velocities of the spaceship. Looking at the kinematic equations, we see that one equation will not give the answer. We must use one kinematic equation to solve for one of the velocities and substitute it into another kinematic equation to get the second velocity. Thus, we solve two of the kinematic equations simultaneously.

Solution

First we solve for v_0 using $x = x_0 + v_0 t + \frac{1}{2}at^2$:

$$x - x_0 = v_0 t + \frac{1}{2}at^2$$

$$1.0 \times 10^6 \text{ m} = v_0(120.0 \text{ s}) + \frac{1}{2}(20.0 \text{ m/s}^2)(120.0 \text{ s})^2$$

$$v_0 = 7133.3 \text{ m/s}.$$

Then we substitute v_0 into $v = v_0 + at$ to solve for the final velocity:

$$v = v_0 + at = 7133.3 \text{ m/s} + (20.0 \text{ m/s}^2)(120.0 \text{ s}) = 9533.3 \text{ m/s}.$$

Significance

There are six variables in displacement, time, velocity, and acceleration that describe motion in one dimension. The initial conditions of a given problem can be many combinations of these variables. Because of this diversity, solutions may not be as easy as simple substitutions into one of the equations. This example illustrates that solutions to kinematics may require solving two simultaneous kinematic equations.

With the basics of kinematics established, we can go on to many other interesting examples and applications. In the process of developing kinematics, we have also glimpsed a general approach to problem solving that produces both correct answers and insights into physical relationships. The next level of complexity in our kinematics problems involves the motion of two interrelated bodies, called *two-body pursuit problems*.

Two-Body Pursuit Problems

Up until this point we have looked at examples of motion involving a single body. Even for the problem with two cars and the stopping distances on wet and dry roads, we divided this problem into two separate problems to find the answers. In a **two-body pursuit problem**, the motions of the objects are coupled—meaning, the unknown we seek depends on the motion of both objects. To solve these problems we write the equations of motion for each object and then solve them simultaneously to find the unknown. This is illustrated in **Figure 3.25**.

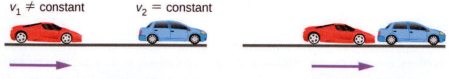

Figure 3.25   A two-body pursuit scenario where car 2 has a constant velocity and car 1 is behind with a constant acceleration. Car 1 catches up with car 2 at a later time.

The time and distance required for car 1 to catch car 2 depends on the initial distance car 1 is from car 2 as well as the velocities of both cars and the acceleration of car 1. The kinematic equations describing the motion of both cars must be solved to find these unknowns.

Consider the following example.

Example 3.13

Cheetah Catching a Gazelle

A cheetah waits in hiding behind a bush. The cheetah spots a gazelle running past at 10 m/s. At the instant the gazelle passes the cheetah, the cheetah accelerates from rest at 4 m/s^2 to catch the gazelle. (a) How long does it take the cheetah to catch the gazelle? (b) What is the displacement of the gazelle and cheetah?

Strategy

We use the set of equations for constant acceleration to solve this problem. Since there are two objects in motion, we have separate equations of motion describing each animal. But what links the equations is a common parameter that has the same value for each animal. If we look at the problem closely, it is clear the common parameter to each animal is their position x at a later time t. Since they both start at $x_0 = 0$, their displacements are the same at a later time t, when the cheetah catches up with the gazelle. If we pick the equation of motion that solves for the displacement for each animal, we can then set the equations equal to each other and solve for the unknown, which is time.

Solution

a. Equation for the gazelle: The gazelle has a constant velocity, which is its average velocity, since it is not accelerating. Therefore, we use **Equation 3.10** with $x_0 = 0$:

$$x = x_0 + \bar{v}t = \bar{v}t.$$

Equation for the cheetah: The cheetah is accelerating from rest, so we use **Equation 3.13** with $x_0 = 0$ and $v_0 = 0$:

$$x = x_0 + v_0 t + \frac{1}{2}at^2 = \frac{1}{2}at^2.$$

Now we have an equation of motion for each animal with a common parameter, which can be eliminated to find the solution. In this case, we solve for t:

$$x = \bar{v}t = \frac{1}{2}at^2$$

$$t = \frac{2\bar{v}}{a}.$$

The gazelle has a constant velocity of 10 m/s, which is its average velocity. The acceleration of the cheetah is 4 m/s^2. Evaluating t, the time for the cheetah to reach the gazelle, we have

$$t = \frac{2\bar{v}}{a} = \frac{2(10)}{4} = 5 \text{ s}.$$

b. To get the displacement, we use either the equation of motion for the cheetah or the gazelle, since they should both give the same answer.
Displacement of the cheetah:

$$x = \frac{1}{2}at^2 = \frac{1}{2}(4)(5)^2 = 50 \text{ m}.$$

Displacement of the gazelle:

$$x = \bar{v}t = 10(5) = 50 \text{ m}.$$

We see that both displacements are equal, as expected.

Significance

It is important to analyze the motion of each object and to use the appropriate kinematic equations to describe the individual motion. It is also important to have a good visual perspective of the two-body pursuit problem to see the common parameter that links the motion of both objects.

 3.6 **Check Your Understanding** A bicycle has a constant velocity of 10 m/s. A person starts from rest and runs to catch up to the bicycle in 30 s. What is the acceleration of the person?

3.5 | Free Fall

Learning Objectives

By the end of this section, you will be able to:

- Use the kinematic equations with the variables y and g to analyze free-fall motion.
- Describe how the values of the position, velocity, and acceleration change during a free fall.
- Solve for the position, velocity, and acceleration as functions of time when an object is in a free fall.

An interesting application of **Equation 3.4** through **Equation 3.14** is called *free fall*, which describes the motion of an object falling in a gravitational field, such as near the surface of Earth or other celestial objects of planetary size. Let's assume the body is falling in a straight line perpendicular to the surface, so its motion is one-dimensional. For example, we can estimate the depth of a vertical mine shaft by dropping a rock into it and listening for the rock to hit the bottom. But "falling," in the context of free fall, does not necessarily imply the body is moving from a greater height to a lesser height. If a ball is thrown upward, the equations of free fall apply equally to its ascent as well as its descent.

Gravity

The most remarkable and unexpected fact about falling objects is that if air resistance and friction are negligible, then in a given location all objects fall toward the center of Earth with the *same constant acceleration, independent of their mass*. This experimentally determined fact is unexpected because we are so accustomed to the effects of air resistance and friction that we expect light objects to fall slower than heavy ones. Until Galileo Galilei (1564–1642) proved otherwise, people believed that a heavier object has a greater acceleration in a free fall. We now know this is not the case. In the absence of air resistance, heavy objects arrive at the ground at the same time as lighter objects when dropped from the same height **Figure 3.26**.

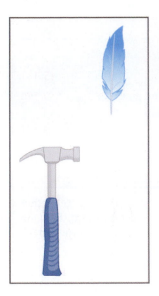

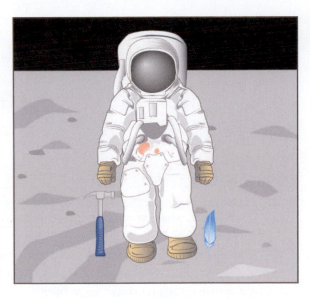

In air   In a vacuum   In a vacuum (the hard way)

Figure 3.26 A hammer and a feather fall with the same constant acceleration if air resistance is negligible. This is a general characteristic of gravity not unique to Earth, as astronaut David R. Scott demonstrated in 1971 on the Moon, where the acceleration from gravity is only 1.67 m/s^2 and there is no atmosphere.

In the real world, air resistance can cause a lighter object to fall slower than a heavier object of the same size. A tennis ball reaches the ground after a baseball dropped at the same time. (It might be difficult to observe the difference if the height is not large.) Air resistance opposes the motion of an object through the air, and friction between objects—such as between clothes and a laundry chute or between a stone and a pool into which it is dropped—also opposes motion between them.

For the ideal situations of these first few chapters, an object *falling without air resistance or friction* is defined to be in **free fall**. The force of gravity causes objects to fall toward the center of Earth. The acceleration of free-falling objects is therefore called **acceleration due to gravity**. Acceleration due to gravity is constant, which means we can apply the kinematic equations to any falling object where air resistance and friction are negligible. This opens to us a broad class of interesting situations.

Acceleration due to gravity is so important that its magnitude is given its own symbol, g. It is constant at any given location on Earth and has the average value

$$g = 9.81 \text{ m/s}^2 \quad (\text{or } 32.2 \text{ ft/s}^2).$$

Although g varies from 9.78 m/s^2 to 9.83 m/s^2, depending on latitude, altitude, underlying geological formations, and local topography, let's use an average value of 9.8 m/s^2 rounded to two significant figures in this text unless specified otherwise. Neglecting these effects on the value of g as a result of position on Earth's surface, as well as effects resulting from Earth's rotation, we take the direction of acceleration due to gravity to be downward (toward the center of Earth). In fact, its direction *defines* what we call vertical. Note that whether acceleration a in the kinematic equations has the value $+g$ or $-g$ depends on how we define our coordinate system. If we define the upward direction as positive, then $a = -g = -9.8 \text{ m/s}^2$, and if we define the downward direction as positive, then $a = g = 9.8 \text{ m/s}^2$.

One-Dimensional Motion Involving Gravity

The best way to see the basic features of motion involving gravity is to start with the simplest situations and then progress toward more complex ones. So, we start by considering straight up-and-down motion with no air resistance or friction. These assumptions mean the velocity (if there is any) is vertical. If an object is dropped, we know the initial velocity is zero when in free fall. When the object has left contact with whatever held or threw it, the object is in free fall. When the object is thrown, it has the same initial speed in free fall as it did before it was released. When the object comes in contact with the ground or any other object, it is no longer in free fall and its acceleration of g is no longer valid. Under these circumstances, the motion is one-dimensional and has constant acceleration of magnitude g. We represent vertical displacement with the symbol y.

Kinematic Equations for Objects in Free Fall

We assume here that acceleration equals $-g$ (with the positive direction upward).

$$v = v_0 - gt \tag{3.15}$$

$$y = y_0 + v_0 t - \frac{1}{2}gt^2 \tag{3.16}$$

$$v^2 = v_0^2 - 2g(y - y_0) \tag{3.17}$$

Problem-Solving Strategy: Free Fall

1. Decide on the sign of the acceleration of gravity. In **Equation 3.15** through **Equation 3.17**, acceleration g is negative, which says the positive direction is upward and the negative direction is downward. In some problems, it may be useful to have acceleration g as positive, indicating the positive direction is downward.

2. Draw a sketch of the problem. This helps visualize the physics involved.

3. Record the knowns and unknowns from the problem description. This helps devise a strategy for selecting the appropriate equations to solve the problem.

4. Decide which of **Equation 3.15** through **Equation 3.17** are to be used to solve for the unknowns.

Example 3.14

Free Fall of a Ball

Figure 3.27 shows the positions of a ball, at 1-s intervals, with an initial velocity of 4.9 m/s downward, that is thrown from the top of a 98-m-high building. (a) How much time elapses before the ball reaches the ground? (b) What is the velocity when it arrives at the ground?

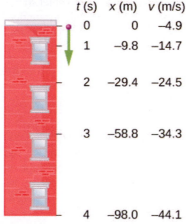

t (s)	x (m)	v (m/s)
0	0	−4.9
1	−9.8	−14.7
2	−29.4	−24.5
3	−58.8	−34.3
4	−98.0	−44.1

Figure 3.27 The positions and velocities at 1-s intervals of a ball thrown downward from a tall building at 4.9 m/s.

Strategy

Choose the origin at the top of the building with the positive direction upward and the negative direction downward. To find the time when the position is −98 m, we use **Equation 3.16**, with $y_0 = 0$, $v_0 = -4.9$ m/s, and $g = 9.8$ m/s^2.

Solution

a. Substitute the given values into the equation:

$$y = y_0 + v_0 t - \frac{1}{2}gt^2$$
$$-98.0 \text{ m} = 0 - (4.9 \text{ m/s})t - \frac{1}{2}(9.8 \text{ m/s}^2)t^2.$$

This simplifies to

$$t^2 + t - 20 = 0.$$

This is a quadratic equation with roots $t = -5.0$ s and $t = 4.0$ s. The positive root is the one we are interested in, since time $t = 0$ is the time when the ball is released at the top of the building. (The time $t = -5.0$ s represents the fact that a ball thrown upward from the ground would have been in the air for 5.0 s when it passed by the top of the building moving downward at 4.9 m/s.)

b. Using **Equation 3.15**, we have

$$v = v_0 - gt = -4.9 \text{ m/s} - (9.8 \text{ m/s}^2)(4.0 \text{ s}) = -44.1 \text{ m/s}.$$

Significance

For situations when two roots are obtained from a quadratic equation in the time variable, we must look at the physical significance of both roots to determine which is correct. Since $t = 0$ corresponds to the time when the ball was released, the negative root would correspond to a time before the ball was released, which is not physically meaningful. When the ball hits the ground, its velocity is not immediately zero, but as soon as the ball interacts with the ground, its acceleration is not g and it accelerates with a different value over a short time to zero velocity. This problem shows how important it is to establish the correct coordinate system and to keep the signs of g in the kinematic equations consistent.

Example 3.15

Vertical Motion of a Baseball

A batter hits a baseball straight upward at home plate and the ball is caught 5.0 s after it is struck **Figure 3.28**. (a) What is the initial velocity of the ball? (b) What is the maximum height the ball reaches? (c) How long does it take to reach the maximum height? (d) What is the acceleration at the top of its path? (e) What is the velocity of the ball when it is caught? Assume the ball is hit and caught at the same location.

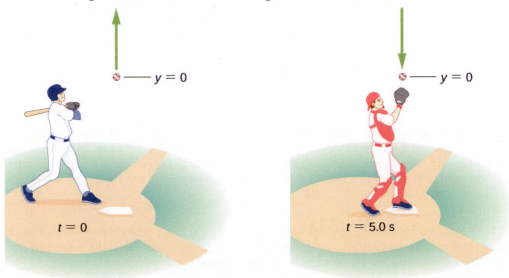

Figure 3.28   A baseball hit straight up is caught by the catcher 5.0 s later.

Strategy

Choose a coordinate system with a positive y-axis that is straight up and with an origin that is at the spot where the ball is hit and caught.

Solution

a. **Equation 3.16** gives

$$y = y_0 + v_0 t - \frac{1}{2}gt^2$$

$$0 = 0 + v_0(5.0 \text{ s}) - \frac{1}{2}\left(9.8 \text{ m/s}^2\right)(5.0 \text{ s})^2,$$

which gives $v_0 = 24.5$ m/s.

b. At the maximum height, $v = 0$. With $v_0 = 24.5$ m/s, **Equation 3.17** gives

$$v^2 = v_0^2 - 2g(y - y_0)$$

$$0 = (24.5 \text{ m/s})^2 - 2(9.8 \text{ m/s}^2)(y - 0)$$

or

$$y = 30.6 \text{ m.}$$

c. To find the time when $v = 0$, we use **Equation 3.15**:

$$v = v_0 - gt$$

$$0 = 24.5 \text{ m/s} - (9.8 \text{ m/s}^2)t.$$

This gives $t = 2.5$ s. Since the ball rises for 2.5 s, the time to fall is 2.5 s.

d. The acceleration is 9.8 m/s^2 everywhere, even when the velocity is zero at the top of the path. Although the velocity is zero at the top, it is changing at the rate of 9.8 m/s^2 downward.

e. The velocity at $t = 5.0$s can be determined with **Equation 3.15**:

$$\begin{aligned} v &= v_0 - gt \\ &= 24.5 \text{ m/s} - 9.8 \text{ m/s}^2(5.0 \text{ s}) \\ &= -24.5 \text{ m/s.} \end{aligned}$$

Significance

The ball returns with the speed it had when it left. This is a general property of free fall for any initial velocity. We used a single equation to go from throw to catch, and did not have to break the motion into two segments, upward and downward. We are used to thinking that the effect of gravity is to create free fall downward toward Earth. It is important to understand, as illustrated in this example, that objects moving upward away from Earth are also in a state of free fall.

 3.7 Check Your Understanding A chunk of ice breaks off a glacier and falls 30.0 m before it hits the water. Assuming it falls freely (there is no air resistance), how long does it take to hit the water? Which quantity increases faster, the speed of the ice chunk or its distance traveled?

Example 3.16

Rocket Booster

A small rocket with a booster blasts off and heads straight upward. When at a height of 5.0 km and velocity of 200.0 m/s, it releases its booster. (a) What is the maximum height the booster attains? (b) What is the velocity of the booster at a height of 6.0 km? Neglect air resistance.

Figure 3.29 A rocket releases its booster at a given height and velocity. How high and how fast does the booster go?

Strategy

We need to select the coordinate system for the acceleration of gravity, which we take as negative downward. We are given the initial velocity of the booster and its height. We consider the point of release as the origin. We know the velocity is zero at the maximum position within the acceleration interval; thus, the velocity of the booster is zero at its maximum height, so we can use this information as well. From these observations, we use **Equation 3.17**, which gives us the maximum height of the booster. We also use **Equation 3.17** to give the velocity at 6.0 km. The initial velocity of the booster is 200.0 m/s.

Solution

a. From **Equation 3.17**, $v^2 = v_0^2 - 2g(y - y_0)$. With $v = 0$ and $y_0 = 0$, we can solve for y:

$$y = \frac{v_0^2}{2g} = \frac{(2.0 \times 10^2 \, \text{m/s})^2}{2(9.8 \, \text{m/s}^2)} = 2040.8 \, \text{m}.$$

This solution gives the maximum height of the booster in our coordinate system, which has its origin at the point of release, so the maximum height of the booster is roughly 7.0 km.

b. An altitude of 6.0 km corresponds to $y = 1.0 \times 10^3$ m in the coordinate system we are using. The other initial conditions are $y_0 = 0$, and $v_0 = 200.0$ m/s .

We have, from **Equation 3.17**,

$$v^2 = (200.0 \text{ m/s})^2 - 2(9.8 \text{ m/s}^2)(1.0 \times 10^3 \text{ m}) \Rightarrow v = \pm 142.8 \text{ m/s}.$$

Significance

We have both a positive and negative solution in (b). Since our coordinate system has the positive direction upward, the +142.8 m/s corresponds to a positive upward velocity at 6000 m during the upward leg of the trajectory of the booster. The value $v = -142.8$ m/s corresponds to the velocity at 6000 m on the downward leg. This example is also important in that an object is given an initial velocity at the origin of our coordinate system, but the origin is at an altitude above the surface of Earth, which must be taken into account when forming the solution.

 Visit **this site (https://openstaxcollege.org/l/21equatgraph)** to learn about graphing polynomials. The shape of the curve changes as the constants are adjusted. View the curves for the individual terms (for example, $y = bx$) to see how they add to generate the polynomial curve.

3.6 | Finding Velocity and Displacement from Acceleration

Learning Objectives

By the end of this section, you will be able to:

- Derive the kinematic equations for constant acceleration using integral calculus.
- Use the integral formulation of the kinematic equations in analyzing motion.
- Find the functional form of velocity versus time given the acceleration function.
- Find the functional form of position versus time given the velocity function.

This section assumes you have enough background in calculus to be familiar with integration. In **Instantaneous Velocity and Speed** and **Average and Instantaneous Acceleration** we introduced the kinematic functions of velocity and acceleration using the derivative. By taking the derivative of the position function we found the velocity function, and likewise by taking the derivative of the velocity function we found the acceleration function. Using integral calculus, we can work backward and calculate the velocity function from the acceleration function, and the position function from the velocity function.

Kinematic Equations from Integral Calculus

Let's begin with a particle with an acceleration $a(t)$ which is a known function of time. Since the time derivative of the velocity function is acceleration,

$$\frac{d}{dt}v(t) = a(t),$$

we can take the indefinite integral of both sides, finding

$$\int \frac{d}{dt}v(t)dt = \int a(t)dt + C_1,$$

where C_1 is a constant of integration. Since $\int \frac{d}{dt}v(t)dt = v(t)$, the velocity is given by

$$v(t) = \int a(t)dt + C_1. \tag{3.18}$$

Similarly, the time derivative of the position function is the velocity function,

$$\frac{d}{dt}x(t) = v(t).$$

Thus, we can use the same mathematical manipulations we just used and find

$$x(t) = \int v(t)dt + C_2,$$

(3.19)

where C_2 is a second constant of integration.

We can derive the kinematic equations for a constant acceleration using these integrals. With $a(t) = a$ a constant, and doing the integration in **Equation 3.18**, we find

$$v(t) = \int a\,dt + C_1 = at + C_1.$$

If the initial velocity is $v(0) = v_0$, then

$$v_0 = 0 + C_1.$$

Then, $C_1 = v_0$ and

$$v(t) = v_0 + at,$$

which is **Equation 3.12**. Substituting this expression into **Equation 3.19** gives

$$x(t) = \int (v_0 + at)dt + C_2.$$

Doing the integration, we find

$$x(t) = v_0 t + \frac{1}{2}at^2 + C_2.$$

If $x(0) = x_0$, we have

$$x_0 = 0 + 0 + C_2;$$

so, $C_2 = x_0$. Substituting back into the equation for $x(t)$, we finally have

$$x(t) = x_0 + v_0 t + \frac{1}{2}at^2,$$

which is **Equation 3.13**.

Example 3.17

Motion of a Motorboat

A motorboat is traveling at a constant velocity of 5.0 m/s when it starts to decelerate to arrive at the dock. Its acceleration is $a(t) = -\frac{1}{4}t$ m/s^3. (a) What is the velocity function of the motorboat? (b) At what time does the velocity reach zero? (c) What is the position function of the motorboat? (d) What is the displacement of the motorboat from the time it begins to decelerate to when the velocity is zero? (e) Graph the velocity and position functions.

Strategy

(a) To get the velocity function we must integrate and use initial conditions to find the constant of integration. (b) We set the velocity function equal to zero and solve for t. (c) Similarly, we must integrate to find the position function and use initial conditions to find the constant of integration. (d) Since the initial position is taken to be zero, we only have to evaluate the position function at $t = 0$.

Solution

We take $t = 0$ to be the time when the boat starts to decelerate.

a. From the functional form of the acceleration we can solve **Equation 3.18** to get $v(t)$:

$$v(t) = \int a(t)dt + C_1 = \int -\tfrac{1}{4}t\,dt + C_1 = -\tfrac{1}{8}t^2 + C_1.$$

At $t = 0$ we have $v(0) = 5.0$ m/s $= 0 + C_1$, so $C_1 = 5.0$ m/s or $v(t) = 5.0$ m/s $-\tfrac{1}{8}t^2$.

b. $v(t) = 0 = 5.0$ m/s $-\tfrac{1}{8}t^2$ m/s^3 $\Rightarrow t = 6.3$ s

c. Solve **Equation 3.19**:

$$x(t) = \int v(t)dt + C_2 = \int (5.0 - \tfrac{1}{8}t^2)dt + C_2 = 5.0t\text{m/s} - \tfrac{1}{24}t^3 \text{ m/s}^3 + C_2.$$

At $t = 0$, we set $x(0) = 0 = x_0$, since we are only interested in the displacement from when the boat starts to decelerate. We have

$$x(0) = 0 = C_2.$$

Therefore, the equation for the position is

$$x(t) = 5.0t - \tfrac{1}{24}t^3.$$

d. Since the initial position is taken to be zero, we only have to evaluate the position function at the time when the velocity is zero. This occurs at $t = 6.3$ s. Therefore, the displacement is

$$x(6.3) = 5.0(6.3 \text{ s}) - \tfrac{1}{24}(6.3 \text{ s}) = 21.1 \text{ m}.$$

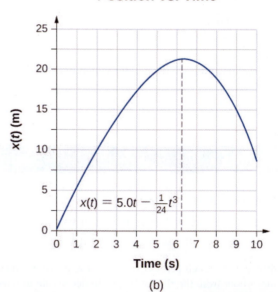

Figure 3.30 (a) Velocity of the motorboat as a function of time. The motorboat decreases its velocity to zero in 6.3 s. At times greater than this, velocity becomes negative—meaning, the boat is reversing direction. (b) Position of the motorboat as a function of time. At $t = 6.3$ s, the velocity is zero and the boat has stopped. At times greater than this, the velocity becomes negative—meaning, if the boat continues to move with the same acceleration, it reverses direction and heads back toward where it originated.

Significance

The acceleration function is linear in time so the integration involves simple polynomials. In **Figure 3.30**, we see that if we extend the solution beyond the point when the velocity is zero, the velocity becomes negative and the boat reverses direction. This tells us that solutions can give us information outside our immediate interest and we should be careful when interpreting them.

 3.8   **Check Your Understanding** A particle starts from rest and has an acceleration function $5 - 10t$ m/s^2. (a) What is the velocity function? (b) What is the position function? (c) When is the velocity zero?

CHAPTER 3 REVIEW

KEY TERMS

acceleration due to gravity acceleration of an object as a result of gravity

average acceleration the rate of change in velocity; the change in velocity over time

average speed the total distance traveled divided by elapsed time

average velocity the displacement divided by the time over which displacement occurs under constant acceleration

displacement the change in position of an object

distance traveled the total length of the path traveled between two positions

elapsed time the difference between the ending time and the beginning time

free fall the state of movement that results from gravitational force only

instantaneous acceleration acceleration at a specific point in time

instantaneous speed the absolute value of the instantaneous velocity

instantaneous velocity the velocity at a specific instant or time point

kinematics the description of motion through properties such as position, time, velocity, and acceleration

position the location of an object at a particular time

total displacement the sum of individual displacements over a given time period

two-body pursuit problem a kinematics problem in which the unknowns are calculated by solving the kinematic equations simultaneously for two moving objects

KEY EQUATIONS

Displacement	$\Delta x = x_\mathrm{f} - x_\mathrm{i}$
Total displacement	$\Delta x_\mathrm{Total} = \sum \Delta x_\mathrm{i}$
Average velocity (for constant acceleration)	$\bar{v} = \dfrac{\Delta x}{\Delta t} = \dfrac{x_2 - x_1}{t_2 - t_1}$
Instantaneous velocity	$v(t) = \dfrac{dx(t)}{dt}$
Average speed	$\text{Average speed} = \bar{s} = \dfrac{\text{Total distance}}{\text{Elapsed time}}$
Instantaneous speed	$\text{Instantaneous speed} = |v(t)|$
Average acceleration	$\bar{a} = \dfrac{\Delta v}{\Delta t} = \dfrac{v_f - v_0}{t_f - t_0}$
Instantaneous acceleration	$a(t) = \dfrac{dv(t)}{dt}$
Position from average velocity	$x = x_0 + \bar{v}t$
Average velocity	$\bar{v} = \dfrac{v_0 + v}{2}$
Velocity from acceleration	$v = v_0 + at \ \ (\text{constant } a)$

Position from velocity and acceleration	$x = x_0 + v_0 t + \frac{1}{2}at^2$ (constant a)
Velocity from distance	$v^2 = v_0^2 + 2a(x - x_0)$ (constant a)
Velocity of free fall	$v = v_0 - gt$ (positive upward)
Height of free fall	$y = y_0 + v_0 t - \frac{1}{2}gt^2$
Velocity of free fall from height	$v^2 = v_0^2 - 2g(y - y_0)$
Velocity from acceleration	$v(t) = \int a(t)dt + C_1$
Position from velocity	$x(t) = \int v(t)dt + C_2$

SUMMARY

3.1 Position, Displacement, and Average Velocity

- Kinematics is the description of motion without considering its causes. In this chapter, it is limited to motion along a straight line, called one-dimensional motion.

- Displacement is the change in position of an object. The SI unit for displacement is the meter. Displacement has direction as well as magnitude.

- Distance traveled is the total length of the path traveled between two positions.

- Time is measured in terms of change. The time between two position points x_1 and x_2 is $\Delta t = t_2 - t_1$. Elapsed time for an event is $\Delta t = t_f - t_0$, where t_f is the final time and t_0 is the initial time. The initial time is often taken to be zero.

- Average velocity $\bar{v}$ is defined as displacement divided by elapsed time. If x_1, t_1 and x_2, t_2 are two position time points, the average velocity between these points is

$$\bar{v} = \frac{\Delta x}{\Delta t} = \frac{x_2 - x_1}{t_2 - t_1}.$$

3.2 Instantaneous Velocity and Speed

- Instantaneous velocity is a continuous function of time and gives the velocity at any point in time during a particle's motion. We can calculate the instantaneous velocity at a specific time by taking the derivative of the position function, which gives us the functional form of instantaneous velocity $v(t)$.

- Instantaneous velocity is a vector and can be negative.

- Instantaneous speed is found by taking the absolute value of instantaneous velocity, and it is always positive.

- Average speed is total distance traveled divided by elapsed time.

- The slope of a position-versus-time graph at a specific time gives instantaneous velocity at that time.

3.3 Average and Instantaneous Acceleration

- Acceleration is the rate at which velocity changes. Acceleration is a vector; it has both a magnitude and direction. The SI unit for acceleration is meters per second squared.

- Acceleration can be caused by a change in the magnitude or the direction of the velocity, or both.

- Instantaneous acceleration $a(t)$ is a continuous function of time and gives the acceleration at any specific time during the motion. It is calculated from the derivative of the velocity function. Instantaneous acceleration is the slope of the velocity-versus-time graph.

- Negative acceleration (sometimes called deceleration) is acceleration in the negative direction in the chosen coordinate system.

3.4 Motion with Constant Acceleration

- When analyzing one-dimensional motion with constant acceleration, identify the known quantities and choose the appropriate equations to solve for the unknowns. Either one or two of the kinematic equations are needed to solve for the unknowns, depending on the known and unknown quantities.

- Two-body pursuit problems always require two equations to be solved simultaneously for the unknowns.

3.5 Free Fall

- An object in free fall experiences constant acceleration if air resistance is negligible.

- On Earth, all free-falling objects have an acceleration g due to gravity, which averages $g = 9.81 \text{ m/s}^2$.

- For objects in free fall, the upward direction is normally taken as positive for displacement, velocity, and acceleration.

3.6 Finding Velocity and Displacement from Acceleration

- Integral calculus gives us a more complete formulation of kinematics.

- If acceleration $a(t)$ is known, we can use integral calculus to derive expressions for velocity $v(t)$ and position $x(t)$.

- If acceleration is constant, the integral equations reduce to **Equation 3.12** and **Equation 3.13** for motion with constant acceleration.

CONCEPTUAL QUESTIONS

3.1 Position, Displacement, and Average Velocity

1. Give an example in which there are clear distinctions among distance traveled, displacement, and magnitude of displacement. Identify each quantity in your example specifically.

2. Under what circumstances does distance traveled equal magnitude of displacement? What is the only case in which magnitude of displacement and displacement are exactly the same?

3. Bacteria move back and forth using their flagella (structures that look like little tails). Speeds of up to 50 μm/s (50×10^{-6} m/s) have been observed. The total distance traveled by a bacterium is large for its size, whereas its displacement is small. Why is this?

4. Give an example of a device used to measure time and identify what change in that device indicates a change in time.

5. Does a car's odometer measure distance traveled or displacement?

6. During a given time interval the average velocity of an object is zero. What can you say conclude about its displacement over the time interval?

3.2 Instantaneous Velocity and Speed

7. There is a distinction between average speed and the magnitude of average velocity. Give an example that illustrates the difference between these two quantities.

8. Does the speedometer of a car measure speed or velocity?

9. If you divide the total distance traveled on a car trip (as determined by the odometer) by the elapsed time of the trip, are you calculating average speed or magnitude of average velocity? Under what circumstances are these two quantities the same?

10. How are instantaneous velocity and instantaneous speed related to one another? How do they differ?

3.3 Average and Instantaneous Acceleration

11. Is it possible for speed to be constant while acceleration is not zero?

12. Is it possible for velocity to be constant while acceleration is not zero? Explain.

13. Give an example in which velocity is zero yet acceleration is not.

14. If a subway train is moving to the left (has a negative velocity) and then comes to a stop, what is the direction of its acceleration? Is the acceleration positive or negative?

15. Plus and minus signs are used in one-dimensional motion to indicate direction. What is the sign of an acceleration that reduces the magnitude of a negative velocity? Of a positive velocity?

3.4 Motion with Constant Acceleration

16. When analyzing the motion of a single object, what is the required number of known physical variables that are needed to solve for the unknown quantities using the kinematic equations?

17. State two scenarios of the kinematics of single object where three known quantities require two kinematic equations to solve for the unknowns.

3.5 Free Fall

18. What is the acceleration of a rock thrown straight upward on the way up? At the top of its flight? On the way down? Assume there is no air resistance.

19. An object that is thrown straight up falls back to Earth. This is one-dimensional motion. (a) When is its velocity zero? (b) Does its velocity change direction? (c) Does the acceleration have the same sign on the way up as on the way down?

20. Suppose you throw a rock nearly straight up at a coconut in a palm tree and the rock just misses the coconut on the way up but hits the coconut on the way down. Neglecting air resistance and the slight horizontal variation in motion to account for the hit and miss of the coconut, how does the speed of the rock when it hits the coconut on the way down compare with what it would have been if it had hit the coconut on the way up? Is it more likely to dislodge the coconut on the way up or down? Explain.

21. The severity of a fall depends on your speed when you strike the ground. All factors but the acceleration from gravity being the same, how many times higher could a safe fall on the Moon than on Earth (gravitational acceleration on the Moon is about one-sixth that of the Earth)?

22. How many times higher could an astronaut jump on the Moon than on Earth if her takeoff speed is the same in both locations (gravitational acceleration on the Moon is about on-sixth of that on Earth)?

3.6 Finding Velocity and Displacement from Acceleration

23. When given the acceleration function, what additional information is needed to find the velocity function and position function?

PROBLEMS

3.1 Position, Displacement, and Average Velocity

24. Consider a coordinate system in which the positive x axis is directed upward vertically. What are the positions of a particle (a) 5.0 m directly above the origin and (b) 2.0 m below the origin?

25. A car is 2.0 km west of a traffic light at $t = 0$ and 5.0 km east of the light at $t = 6.0$ min. Assume the origin of the coordinate system is the light and the positive x direction is eastward. (a) What are the car's position vectors at these two times? (b) What is the car's displacement between 0 min and 6.0 min?

26. The Shanghai maglev train connects Longyang Road to Pudong International Airport, a distance of 30 km. The journey takes 8 minutes on average. What is the maglev train's average velocity?

27. The position of a particle moving along the x-axis is given by $x(t) = 4.0 - 2.0t$ m. (a) At what time does the particle cross the origin? (b) What is the displacement of the particle between $t = 3.0$ s and $t = 6.0$ s?

28. A cyclist rides 8.0 km east for 20 minutes, then he turns and heads west for 8 minutes and 3.2 km. Finally, he rides east for 16 km, which takes 40 minutes. (a) What is the final displacement of the cyclist? (b) What is his average velocity?

29. On February 15, 2013, a superbolide meteor (brighter than the Sun) entered Earth's atmosphere over Chelyabinsk, Russia, and exploded at an altitude of 23.5 km. Eyewitnesses could feel the intense heat from the fireball, and the blast wave from the explosion blew out windows in buildings. The blast wave took approximately 2 minutes 30 seconds to reach ground level. (a) What was the average velocity of the blast wave? b) Compare this with the speed of sound, which is 343 m/s at sea level.

3.2 Instantaneous Velocity and Speed

30. A woodchuck runs 20 m to the right in 5 s, then turns and runs 10 m to the left in 3 s. (a) What is the average velocity of the woodchuck? (b) What is its average speed?

31. Sketch the velocity-versus-time graph from the following position-versus-time graph.

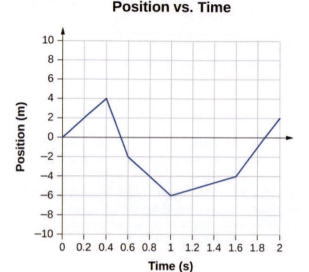

32. Sketch the velocity-versus-time graph from the following position-versus-time graph.

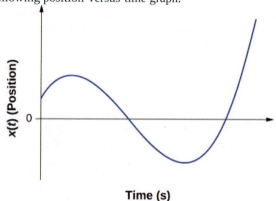

33. Given the following velocity-versus-time graph, sketch the position-versus-time graph.

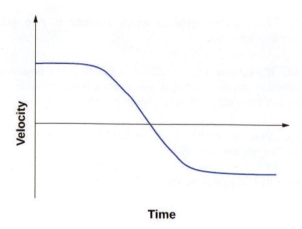

34. An object has a position function $x(t) = 5t$ m. (a) What is the velocity as a function of time? (b) Graph the position function and the velocity function.

35. A particle moves along the x-axis according to $x(t) = 10t - 2t^2$ m. (a) What is the instantaneous velocity at $t = 2$ s and $t = 3$ s? (b) What is the instantaneous speed at these times? (c) What is the average velocity between $t = 2$ s and $t = 3$ s?

36. Unreasonable results. A particle moves along the x-axis according to $x(t) = 3t^3 + 5t$. At what time is the velocity of the particle equal to zero? Is this reasonable?

3.3 Average and Instantaneous Acceleration

37. A cheetah can accelerate from rest to a speed of 30.0 m/s in 7.00 s. What is its acceleration?

38. Dr. John Paul Stapp was a U.S. Air Force officer who studied the effects of extreme acceleration on the human body. On December 10, 1954, Stapp rode a rocket sled, accelerating from rest to a top speed of 282 m/s (1015 km/h) in 5.00 s and was brought jarringly back to rest in only 1.40 s. Calculate his (a) acceleration in his direction of motion and (b) acceleration opposite to his direction of motion. Express each in multiples of g (9.80 m/s^2) by taking its ratio to the acceleration of gravity.

39. Sketch the acceleration-versus-time graph from the following velocity-versus-time graph.

Velocity vs. Time

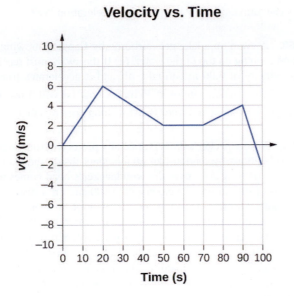

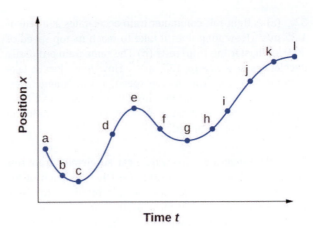

40. A commuter backs her car out of her garage with an acceleration of 1.40 m/s². (a) How long does it take her to reach a speed of 2.00 m/s? (b) If she then brakes to a stop in 0.800 s, what is her acceleration?

41. Assume an intercontinental ballistic missile goes from rest to a suborbital speed of 6.50 km/s in 60.0 s (the actual speed and time are classified). What is its average acceleration in meters per second and in multiples of g (9.80 m/s²)?

42. An airplane, starting from rest, moves down the runway at constant acceleration for 18 s and then takes off at a speed of 60 m/s. What is the average acceleration of the plane?

3.4 Motion with Constant Acceleration

43. A particle moves in a straight line at a constant velocity of 30 m/s. What is its displacement between t = 0 and t = 5.0 s?

44. A particle moves in a straight line with an initial velocity of 0 m/s and a constant acceleration of 30 m/s². If $t = 0$ at $x = 0$, what is the particle's position at $t = 5$ s?

45. A particle moves in a straight line with an initial velocity of 30 m/s and constant acceleration 30 m/s². (a) What is its displacement at $t = 5$ s? (b) What is its velocity at this same time?

46. (a) Sketch a graph of velocity versus time corresponding to the graph of displacement versus time given in the following figure. (b) Identify the time or times (t_a, t_b, t_c, etc.) at which the instantaneous velocity has the greatest positive value. (c) At which times is it zero? (d) At which times is it negative?

47. (a) Sketch a graph of acceleration versus time corresponding to the graph of velocity versus time given in the following figure. (b) Identify the time or times (t_a, t_b, t_c, etc.) at which the acceleration has the greatest positive value. (c) At which times is it zero? (d) At which times is it negative?

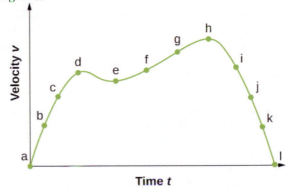

48. A particle has a constant acceleration of 6.0 m/s². (a) If its initial velocity is 2.0 m/s, at what time is its displacement 5.0 m? (b) What is its velocity at that time?

49. At t = 10 s, a particle is moving from left to right with a speed of 5.0 m/s. At t = 20 s, the particle is moving right to left with a speed of 8.0 m/s. Assuming the particle's acceleration is constant, determine (a) its acceleration, (b) its initial velocity, and (c) the instant when its velocity is zero.

50. A well-thrown ball is caught in a well-padded mitt. If the acceleration of the ball is 2.10×10^4 m/s², and 1.85 ms (1 ms = 10^{-3} s) elapses from the time the ball first touches the mitt until it stops, what is the initial velocity of the ball?

51. A bullet in a gun is accelerated from the firing chamber to the end of the barrel at an average rate of 6.20×10^5 m/s² for 8.10×10^{-4} s . What is its muzzle velocity (that is, its final velocity)?

52. (a) A light-rail commuter train accelerates at a rate of 1.35 m/s^2. How long does it take to reach its top speed of 80.0 km/h, starting from rest? (b) The same train ordinarily decelerates at a rate of 1.65 m/s^2. How long does it take to come to a stop from its top speed? (c) In emergencies, the train can decelerate more rapidly, coming to rest from 80.0 km/h in 8.30 s. What is its emergency acceleration in meters per second squared?

53. While entering a freeway, a car accelerates from rest at a rate of 2.40 m/s^2 for 12.0 s. (a) Draw a sketch of the situation. (b) List the knowns in this problem. (c) How far does the car travel in those 12.0 s? To solve this part, first identify the unknown, then indicate how you chose the appropriate equation to solve for it. After choosing the equation, show your steps in solving for the unknown, check your units, and discuss whether the answer is reasonable. (d) What is the car's final velocity? Solve for this unknown in the same manner as in (c), showing all steps explicitly.

54. Unreasonable results At the end of a race, a runner decelerates from a velocity of 9.00 m/s at a rate of 2.00 m/s^2. (a) How far does she travel in the next 5.00 s? (b) What is her final velocity? (c) Evaluate the result. Does it make sense?

55. Blood is accelerated from rest to 30.0 cm/s in a distance of 1.80 cm by the left ventricle of the heart. (a) Make a sketch of the situation. (b) List the knowns in this problem. (c) How long does the acceleration take? To solve this part, first identify the unknown, then discuss how you chose the appropriate equation to solve for it. After choosing the equation, show your steps in solving for the unknown, checking your units. (d) Is the answer reasonable when compared with the time for a heartbeat?

56. During a slap shot, a hockey player accelerates the puck from a velocity of 8.00 m/s to 40.0 m/s in the same direction. If this shot takes $3.33 \times 10^{-2} \text{ s}$, what is the distance over which the puck accelerates?

57. A powerful motorcycle can accelerate from rest to 26.8 m/s (100 km/h) in only 3.90 s. (a) What is its average acceleration? (b) How far does it travel in that time?

58. Freight trains can produce only relatively small accelerations. (a) What is the final velocity of a freight train that accelerates at a rate of 0.0500 m/s^2 for 8.00 min, starting with an initial velocity of 4.00 m/s? (b) If the train can slow down at a rate of 0.550 m/s^2, how long will it take to come to a stop from this velocity? (c) How far will it travel in each case?

59. A fireworks shell is accelerated from rest to a velocity of 65.0 m/s over a distance of 0.250 m. (a) Calculate the acceleration. (b) How long did the acceleration last?

60. A swan on a lake gets airborne by flapping its wings and running on top of the water. (a) If the swan must reach a velocity of 6.00 m/s to take off and it accelerates from rest at an average rate of 0.35 m/s^2, how far will it travel before becoming airborne? (b) How long does this take?

61. A woodpecker's brain is specially protected from large accelerations by tendon-like attachments inside the skull. While pecking on a tree, the woodpecker's head comes to a stop from an initial velocity of 0.600 m/s in a distance of only 2.00 mm. (a) Find the acceleration in meters per second squared and in multiples of g, where $g = 9.80 \text{ m/} s^2$. (b) Calculate the stopping time. (c) The tendons cradling the brain stretch, making its stopping distance 4.50 mm (greater than the head and, hence, less acceleration of the brain). What is the brain's acceleration, expressed in multiples of g?

62. An unwary football player collides with a padded goalpost while running at a velocity of 7.50 m/s and comes to a full stop after compressing the padding and his body 0.350 m. (a) What is his acceleration? (b) How long does the collision last?

63. A care package is dropped out of a cargo plane and lands in the forest. If we assume the care package speed on impact is 54 m/s (123 mph), then what is its acceleration? Assume the trees and snow stops it over a distance of 3.0 m.

64. An express train passes through a station. It enters with an initial velocity of 22.0 m/s and decelerates at a rate of 0.150 m/s^2 as it goes through. The station is 210.0 m long. (a) How fast is it going when the nose leaves the station? (b) How long is the nose of the train in the station? (c) If the train is 130 m long, what is the velocity of the end of the train as it leaves? (d) When does the end of the train leave the station?

65. Unreasonable results Dragsters can actually reach a top speed of 145.0 m/s in only 4.45 s. (a) Calculate the average acceleration for such a dragster. (b) Find the final velocity of this dragster starting from rest and accelerating at the rate found in (a) for 402.0 m (a quarter mile) without using any information on time. (c) Why is the final velocity greater than that used to find the average acceleration? (*Hint*: Consider whether the assumption of constant acceleration is valid for a dragster. If not, discuss whether the acceleration would be greater at the beginning or end of the run and what effect that would have on the final velocity.)

3.5 Free Fall

66. Calculate the displacement and velocity at times of (a) 0.500 s, (b) 1.00 s, (c) 1.50 s, and (d) 2.00 s for a ball thrown straight up with an initial velocity of 15.0 m/s. Take the point of release to be $y_0 = 0$.

67. Calculate the displacement and velocity at times of (a) 0.500 s, (b) 1.00 s, (c) 1.50 s, (d) 2.00 s, and (e) 2.50 s for a rock thrown straight down with an initial velocity of 14.0 m/s from the Verrazano Narrows Bridge in New York City. The roadway of this bridge is 70.0 m above the water.

68. A basketball referee tosses the ball straight up for the starting tip-off. At what velocity must a basketball player leave the ground to rise 1.25 m above the floor in an attempt to get the ball?

69. A rescue helicopter is hovering over a person whose boat has sunk. One of the rescuers throws a life preserver straight down to the victim with an initial velocity of 1.40 m/s and observes that it takes 1.8 s to reach the water. (a) List the knowns in this problem. (b) How high above the water was the preserver released? Note that the downdraft of the helicopter reduces the effects of air resistance on the falling life preserver, so that an acceleration equal to that of gravity is reasonable.

70. Unreasonable results A dolphin in an aquatic show jumps straight up out of the water at a velocity of 15.0 m/s. (a) List the knowns in this problem. (b) How high does his body rise above the water? To solve this part, first note that the final velocity is now a known, and identify its value. Then, identify the unknown and discuss how you chose the appropriate equation to solve for it. After choosing the equation, show your steps in solving for the unknown, checking units, and discuss whether the answer is reasonable. (c) How long a time is the dolphin in the air? Neglect any effects resulting from his size or orientation.

71. A diver bounces straight up from a diving board, avoiding the diving board on the way down, and falls feet first into a pool. She starts with a velocity of 4.00 m/s and her takeoff point is 1.80 m above the pool. (a) What is her highest point above the board? (b) How long a time are her feet in the air? (c) What is her velocity when her feet hit the water?

72. (a) Calculate the height of a cliff if it takes 2.35 s for a rock to hit the ground when it is thrown straight up from the cliff with an initial velocity of 8.00 m/s. (b) How long a time would it take to reach the ground if it is thrown straight down with the same speed?

73. A very strong, but inept, shot putter puts the shot straight up vertically with an initial velocity of 11.0 m/s.

How long a time does he have to get out of the way if the shot was released at a height of 2.20 m and he is 1.80 m tall?

74. You throw a ball straight up with an initial velocity of 15.0 m/s. It passes a tree branch on the way up at a height of 7.0 m. How much additional time elapses before the ball passes the tree branch on the way back down?

75. A kangaroo can jump over an object 2.50 m high. (a) Considering just its vertical motion, calculate its vertical speed when it leaves the ground. (b) How long a time is it in the air?

76. Standing at the base of one of the cliffs of Mt. Arapiles in Victoria, Australia, a hiker hears a rock break loose from a height of 105.0 m. He can't see the rock right away, but then does, 1.50 s later. (a) How far above the hiker is the rock when he can hear it? (b) How much time does he have to move before the rock hits his head?

77. There is a 250-m-high cliff at Half Dome in Yosemite National Park in California. Suppose a boulder breaks loose from the top of this cliff. (a) How fast will it be going when it strikes the ground? (b) Assuming a reaction time of 0.300 s, how long a time will a tourist at the bottom have to get out of the way after hearing the sound of the rock breaking loose (neglecting the height of the tourist, which would become negligible anyway if hit)? The speed of sound is 335.0 m/s on this day.

3.6 Finding Velocity and Displacement from Acceleration

78. The acceleration of a particle varies with time according to the equation $a(t) = pt^2 - qt^3$. Initially, the velocity and position are zero. (a) What is the velocity as a function of time? (b) What is the position as a function of time?

79. Between $t = 0$ and $t = t_0$, a rocket moves straight upward with an acceleration given by $a(t) = A - Bt^{1/2}$, where A and B are constants. (a) If x is in meters and t is in seconds, what are the units of A and B? (b) If the rocket starts from rest, how does the velocity vary between $t = 0$ and $t = t_0$? (c) If its initial position is zero, what is the rocket's position as a function of time during this same time interval?

80. The velocity of a particle moving along the x-axis varies with time according to $v(t) = A + Bt^{-1}$, where $A = 2$ m/s, $B = 0.25$ m, and $1.0\,\text{s} \le t \le 8.0\,\text{s}$. Determine the acceleration and position of the particle at $t = 2.0$ s and $t = 5.0$ s. Assume that $x(t = 1\,\text{s}) = 0$.

81. A particle at rest leaves the origin with its velocity increasing with time according to $v(t) = 3.2t$ m/s. At 5.0 s, the particle's velocity starts decreasing according to [16.0 − 1.5(t − 5.0)] m/s. This decrease continues until $t = 11.0$ s, after which the particle's velocity remains constant at 7.0 m/s. (a) What is the acceleration of the particle as a function of time? (b) What is the position of the particle at $t = 2.0$ s, $t = 7.0$ s, and $t = 12.0$ s?

ADDITIONAL PROBLEMS

82. Professional baseball player Nolan Ryan could pitch a baseball at approximately 160.0 km/h. At that average velocity, how long did it take a ball thrown by Ryan to reach home plate, which is 18.4 m from the pitcher's mound? Compare this with the average reaction time of a human to a visual stimulus, which is 0.25 s.

83. An airplane leaves Chicago and makes the 3000-km trip to Los Angeles in 5.0 h. A second plane leaves Chicago one-half hour later and arrives in Los Angeles at the same time. Compare the average velocities of the two planes. Ignore the curvature of Earth and the difference in altitude between the two cities.

84. Unreasonable Results A cyclist rides 16.0 km east, then 8.0 km west, then 8.0 km east, then 32.0 km west, and finally 11.2 km east. If his average velocity is 24 km/h, how long did it take him to complete the trip? Is this a reasonable time?

85. An object has an acceleration of $+1.2\,\text{cm/s}^2$. At $t = 4.0\,\text{s}$, its velocity is $-3.4\,\text{cm/s}$. Determine the object's velocities at $t = 1.0\,\text{s}$ and $t = 6.0\,\text{s}$.

86. A particle moves along the x-axis according to the equation $x(t) = 2.0 - 4.0t^2$ m. What are the velocity and acceleration at $t = 2.0$ s and $t = 5.0$ s?

87. A particle moving at constant acceleration has velocities of $2.0\,\text{m/s}$ at $t = 2.0$ s and $-7.6\,\text{m/s}$ at $t = 5.2$ s. What is the acceleration of the particle?

88. A train is moving up a steep grade at constant velocity (see following figure) when its caboose breaks loose and starts rolling freely along the track. After 5.0 s, the caboose is 30 m behind the train. What is the acceleration of the caboose?

89. An electron is moving in a straight line with a velocity of 4.0×10^5 m/s. It enters a region 5.0 cm long where it undergoes an acceleration of 6.0×10^{12} m/s^2 along the same straight line. (a) What is the electron's velocity when it emerges from this region? b) How long does the electron take to cross the region?

90. An ambulance driver is rushing a patient to the hospital. While traveling at 72 km/h, she notices the traffic light at the upcoming intersections has turned amber. To reach the intersection before the light turns red, she must travel 50 m in 2.0 s. (a) What minimum acceleration must the ambulance have to reach the intersection before the light turns red? (b) What is the speed of the ambulance when it reaches the intersection?

91. A motorcycle that is slowing down uniformly covers 2.0 successive km in 80 s and 120 s, respectively. Calculate (a) the acceleration of the motorcycle and (b) its velocity at the beginning and end of the 2-km trip.

92. A cyclist travels from point A to point B in 10 min. During the first 2.0 min of her trip, she maintains a uniform acceleration of $0.090\,\text{m/s}^2$. She then travels at constant velocity for the next 5.0 min. Next, she decelerates at a constant rate so that she comes to a rest at point B 3.0 min later. (a) Sketch the velocity-versus-time graph for the trip. (b) What is the acceleration during the last 3 min? (c) How far does the cyclist travel?

93. Two trains are moving at 30 m/s in opposite directions on the same track. The engineers see simultaneously that they are on a collision course and apply the brakes when they are 1000 m apart. Assuming both trains have the same acceleration, what must this acceleration be if the trains are to stop just short of colliding?

94. A 10.0-m-long truck moving with a constant velocity of 97.0 km/h passes a 3.0-m-long car moving with a constant velocity of 80.0 km/h. How much time elapses between the moment the front of the truck is even with the back of the car and the moment the back of the truck is even with the front of the car?

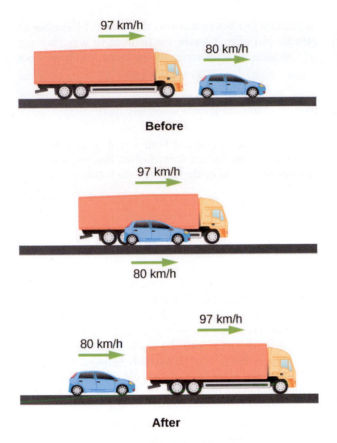

97 km/h

80 km/h

Before

97 km/h

80 km/h

80 km/h

97 km/h

After

95. A police car waits in hiding slightly off the highway. A speeding car is spotted by the police car doing 40 m/s. At the instant the speeding car passes the police car, the police car accelerates from rest at 4 m/s^2 to catch the speeding car. How long does it take the police car to catch the speeding car?

96. Pablo is running in a half marathon at a velocity of 3 m/s. Another runner, Jacob, is 50 meters behind Pablo with the same velocity. Jacob begins to accelerate at 0.05 m/s^2. (a) How long does it take Jacob to catch Pablo? (b) What is the distance covered by Jacob? (c) What is the final velocity of Jacob?

97. Unreasonable results A runner approaches the finish line and is 75 m away; her average speed at this position is 8 m/s. She decelerates at this point at 0.5 m/s^2. How long does it take her to cross the finish line from 75 m away? Is this reasonable?

98. An airplane accelerates at 5.0 m/s^2 for 30.0 s. During this time, it covers a distance of 10.0 km. What are the initial and final velocities of the airplane?

99. Compare the distance traveled of an object that undergoes a change in velocity that is twice its initial velocity with an object that changes its velocity by four times its initial velocity over the same time period. The accelerations of both objects are constant.

100. An object is moving east with a constant velocity and is at position x_0 at time $t_0 = 0$. (a) With what acceleration must the object have for its total displacement to be zero at a later time t? (b) What is the physical interpretation of the solution in the case for $t \to \infty$?

101. A ball is thrown straight up. It passes a 2.00-m-high window 7.50 m off the ground on its path up and takes 1.30 s to go past the window. What was the ball's initial velocity?

102. A coin is dropped from a hot-air balloon that is 300 m above the ground and rising at 10.0 m/s upward. For the coin, find (a) the maximum height reached, (b) its position and velocity 4.00 s after being released, and (c) the time before it hits the ground.

103. A soft tennis ball is dropped onto a hard floor from a height of 1.50 m and rebounds to a height of 1.10 m. (a) Calculate its velocity just before it strikes the floor. (b) Calculate its velocity just after it leaves the floor on its way back up. (c) Calculate its acceleration during contact with the floor if that contact lasts 3.50 ms $(3.50 \times 10^{-3}$ s$)$ (d) How much did the ball compress during its collision with the floor, assuming the floor is absolutely rigid?

104. Unreasonable results. A raindrop falls from a cloud 100 m above the ground. Neglect air resistance. What is the speed of the raindrop when it hits the ground? Is this a reasonable number?

105. Compare the time in the air of a basketball player who jumps 1.0 m vertically off the floor with that of a player who jumps 0.3 m vertically.

106. Suppose that a person takes 0.5 s to react and move his hand to catch an object he has dropped. (a) How far does the object fall on Earth, where $g = 9.8$ m/s^2? (b) How far does the object fall on the Moon, where the acceleration due to gravity is 1/6 of that on Earth?

107. A hot-air balloon rises from ground level at a constant velocity of 3.0 m/s. One minute after liftoff, a sandbag is dropped accidentally from the balloon. Calculate (a) the time it takes for the sandbag to reach the ground and (b) the velocity of the sandbag when it hits the ground.

108. (a) A world record was set for the men's 100-m dash in the 2008 Olympic Games in Beijing by Usain Bolt of Jamaica. Bolt "coasted" across the finish line with a time of 9.69 s. If we assume that Bolt accelerated for 3.00 s to reach his maximum speed, and maintained that speed for the rest of the race, calculate his maximum speed and his acceleration. (b) During the same Olympics, Bolt also set the world record in the 200-m dash with a time of 19.30 s. Using the same assumptions as for the 100-m dash, what was his maximum speed for this race?

109. An object is dropped from a height of 75.0 m above ground level. (a) Determine the distance traveled during the first second. (b) Determine the final velocity at which the object hits the ground. (c) Determine the distance traveled during the last second of motion before hitting the ground.

110. A steel ball is dropped onto a hard floor from a height of 1.50 m and rebounds to a height of 1.45 m. (a) Calculate

its velocity just before it strikes the floor. (b) Calculate its velocity just after it leaves the floor on its way back up. (c) Calculate its acceleration during contact with the floor if that contact lasts 0.0800 ms $(8.00 \times 10^{-5}$ s) (d) How much did the ball compress during its collision with the floor, assuming the floor is absolutely rigid?

111. An object is dropped from a roof of a building of height h. During the last second of its descent, it drops a distance $h/3$. Calculate the height of the building.

CHALLENGE PROBLEMS

112. In a 100-m race, the winner is timed at 11.2 s. The second-place finisher's time is 11.6 s. How far is the second-place finisher behind the winner when she crosses the finish line? Assume the velocity of each runner is constant throughout the race.

113. The position of a particle moving along the x-axis varies with time according to $x(t) = 5.0t^2 - 4.0t^3$ m. Find (a) the velocity and acceleration of the particle as functions of time, (b) the velocity and acceleration at $t = 2.0$ s, (c) the time at which the position is a maximum, (d) the time at which the velocity is zero, and (e) the maximum position.

114. A cyclist sprints at the end of a race to clinch a victory. She has an initial velocity of 11.5 m/s and accelerates at a rate of 0.500 m/s² for 7.00 s. (a) What is her final velocity? (b) The cyclist continues at this velocity to

the finish line. If she is 300 m from the finish line when she starts to accelerate, how much time did she save? (c) The second-place winner was 5.00 m ahead when the winner started to accelerate, but he was unable to accelerate, and traveled at 11.8 m/s until the finish line. What was the difference in finish time in seconds between the winner and runner-up? How far back was the runner-up when the winner crossed the finish line?

115. In 1967, New Zealander Burt Munro set the world record for an Indian motorcycle, on the Bonneville Salt Flats in Utah, of 295.38 km/h. The one-way course was 8.00 km long. Acceleration rates are often described by the time it takes to reach 96.0 km/h from rest. If this time was 4.00 s and Burt accelerated at this rate until he reached his maximum speed, how long did it take Burt to complete the course?

4 | MOTION IN TWO AND THREE DIMENSIONS

Figure 4.1 The Red Arrows is the aerobatics display team of Britain's Royal Air Force. Based in Lincolnshire, England, they perform precision flying shows at high speeds, which requires accurate measurement of position, velocity, and acceleration in three dimensions. (credit: modification of work by Phil Long)

Chapter Outline

4.1 Displacement and Velocity Vectors

4.2 Acceleration Vector

4.3 Projectile Motion

4.4 Uniform Circular Motion

4.5 Relative Motion in One and Two Dimensions

Introduction

To give a complete description of kinematics, we must explore motion in two and three dimensions. After all, most objects in our universe do not move in straight lines; rather, they follow curved paths. From kicked footballs to the flight paths of birds to the orbital motions of celestial bodies and down to the flow of blood plasma in your veins, most motion follows curved trajectories.

Fortunately, the treatment of motion in one dimension in the previous chapter has given us a foundation on which to build, as the concepts of position, displacement, velocity, and acceleration defined in one dimension can be expanded to two and three dimensions. Consider the Red Arrows, also known as the Royal Air Force Aerobatic team of the United Kingdom. Each jet follows a unique curved trajectory in three-dimensional airspace, as well as has a unique velocity and acceleration. Thus, to describe the motion of any of the jets accurately, we must assign to each jet a unique position vector in three dimensions as well as a unique velocity and acceleration vector. We can apply the same basic equations for displacement, velocity, and acceleration we derived in **Motion Along a Straight Line** to describe the motion of the jets in two and three dimensions, but with some modifications—in particular, the inclusion of vectors.

In this chapter we also explore two special types of motion in two dimensions: projectile motion and circular motion. Last, we conclude with a discussion of relative motion. In the chapter-opening picture, each jet has a relative motion with respect to any other jet in the group or to the people observing the air show on the ground.

4.1 | Displacement and Velocity Vectors

Learning Objectives

By the end of this section, you will be able to:

- Calculate position vectors in a multidimensional displacement problem.
- Solve for the displacement in two or three dimensions.
- Calculate the velocity vector given the position vector as a function of time.
- Calculate the average velocity in multiple dimensions.

Displacement and velocity in two or three dimensions are straightforward extensions of the one-dimensional definitions. However, now they are vector quantities, so calculations with them have to follow the rules of vector algebra, not scalar algebra.

Displacement Vector

To describe motion in two and three dimensions, we must first establish a coordinate system and a convention for the axes. We generally use the coordinates x, y, and z to locate a particle at point $P(x, y, z)$ in three dimensions. If the particle is moving, the variables x, y, and z are functions of time (t):

$$x = x(t) \quad y = y(t) \quad z = z(t). \tag{4.1}$$

The **position vector** from the origin of the coordinate system to point P is $\vec{r}(t)$. In unit vector notation, introduced in **Coordinate Systems and Components of a Vector**, $\vec{r}(t)$ is

$$\vec{r}(t) = x(t)\hat{i} + y(t)\hat{j} + z(t)\hat{k}. \tag{4.2}$$

Figure 4.2 shows the coordinate system and the vector to point P, where a particle could be located at a particular time t. Note the orientation of the x, y, and z axes. This orientation is called a right-handed coordinate system (**Coordinate Systems and Components of a Vector**) and it is used throughout the chapter.

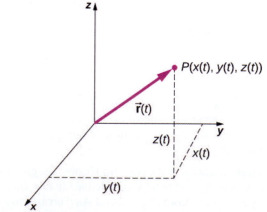

Figure 4.2 A three-dimensional coordinate system with a particle at position $P(x(t), y(t), z(t))$.

With our definition of the position of a particle in three-dimensional space, we can formulate the three-dimensional displacement. **Figure 4.3** shows a particle at time t_1 located at P_1 with position vector $\vec{r}(t_1)$. At a later time t_2, the particle is located at P_2 with position vector $\vec{r}(t_2)$. The **displacement vector** $\Delta\vec{r}$ is found by subtracting $\vec{r}(t_1)$ from $\vec{r}(t_2)$:

$$\Delta\vec{r} = \vec{r}(t_2) - \vec{r}(t_1). \tag{4.3}$$

Vector addition is discussed in **Vectors**. Note that this is the same operation we did in one dimension, but now the vectors are in three-dimensional space.

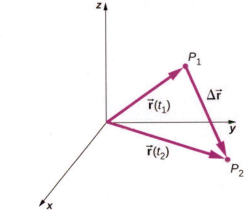

Figure 4.3 The displacement $\Delta\vec{r} = \vec{r}(t_2) - \vec{r}(t_1)$ is the vector from P_1 to P_2.

The following examples illustrate the concept of displacement in multiple dimensions.

Example 4.1

Polar Orbiting Satellite

A satellite is in a circular polar orbit around Earth at an altitude of 400 km—meaning, it passes directly overhead at the North and South Poles. What is the magnitude and direction of the displacement vector from when it is directly over the North Pole to when it is at $-45°$ latitude?

Strategy

We make a picture of the problem to visualize the solution graphically. This will aid in our understanding of the displacement. We then use unit vectors to solve for the displacement.

Solution

Figure 4.4 shows the surface of Earth and a circle that represents the orbit of the satellite. Although satellites are moving in three-dimensional space, they follow trajectories of ellipses, which can be graphed in two dimensions. The position vectors are drawn from the center of Earth, which we take to be the origin of the coordinate system, with the y-axis as north and the x-axis as east. The vector between them is the displacement of the satellite. We take the radius of Earth as 6370 km, so the length of each position vector is 6770 km.

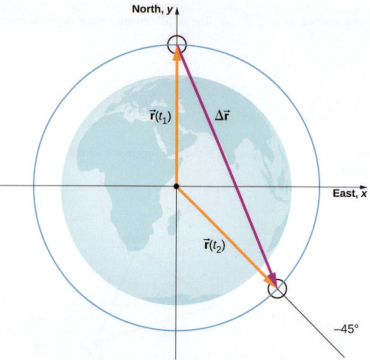

Figure 4.4 Two position vectors are drawn from the center of Earth, which is the origin of the coordinate system, with the *y*-axis as north and the *x*-axis as east. The vector between them is the displacement of the satellite.

In unit vector notation, the position vectors are

$$\vec{r}(t_1) = 6770. \text{ km } \hat{j}$$

$$\vec{r}(t_2) = 6770. \text{ km }(\cos 45°) \hat{i} + 6770. \text{ km }(\sin(-45°)) \hat{j}.$$

Evaluating the sine and cosine, we have

$$\vec{r}(t_1) = 6770. \hat{j}$$

$$\vec{r}(t_2) = 4787 \hat{i} - 4787 \hat{j}.$$

Now we can find $\Delta \vec{r}$, the displacement of the satellite:

$$\Delta \vec{r} = \vec{r}(t_2) - \vec{r}(t_1) = 4787 \hat{i} - 11{,}557 \hat{j}.$$

The magnitude of the displacement is $|\Delta \vec{r}| = \sqrt{(4787)^2 + (-11{,}557)^2} = 12{,}509$ km. The angle the displacement makes with the *x*-axis is $\theta = \tan^{-1}\left(\dfrac{-11{,}557}{4787}\right) = -67.5°$.

Significance

Plotting the displacement gives information and meaning to the unit vector solution to the problem. When plotting the displacement, we need to include its components as well as its magnitude and the angle it makes with a chosen axis—in this case, the *x*-axis (**Figure 4.5**).

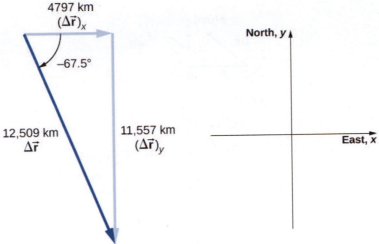

Figure 4.5 Displacement vector with components, angle, and magnitude.

Note that the satellite took a curved path along its circular orbit to get from its initial position to its final position in this example. It also could have traveled 4787 km east, then 11,557 km south to arrive at the same location. Both of these paths are longer than the length of the displacement vector. In fact, the displacement vector gives the shortest path between two points in one, two, or three dimensions.

Many applications in physics can have a series of displacements, as discussed in the previous chapter. The total displacement is the sum of the individual displacements, only this time, we need to be careful, because we are adding vectors. We illustrate this concept with an example of Brownian motion.

Example 4.2

Brownian Motion

Brownian motion is a chaotic random motion of particles suspended in a fluid, resulting from collisions with the molecules of the fluid. This motion is three-dimensional. The displacements in numerical order of a particle undergoing Brownian motion could look like the following, in micrometers (**Figure 4.6**):

$$\Delta \vec{r}_1 = 2.0\hat{i} + \hat{j} + 3.0\hat{k}$$

$$\Delta \vec{r}_2 = -\hat{i} + 3.0\hat{k}$$

$$\Delta \vec{r}_3 = 4.0\hat{i} - 2.0\hat{j} + \hat{k}$$

$$\Delta \vec{r}_4 = -3.0\hat{i} + \hat{j} + 2.0\hat{k}.$$

What is the total displacement of the particle from the origin?

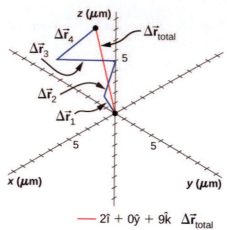

Figure 4.6 Trajectory of a particle undergoing random displacements of Brownian motion. The total displacement is shown in red.

Solution

We form the sum of the displacements and add them as vectors:

$$\Delta \vec{r}_{\text{Total}} = \sum \Delta \vec{r}_i = \Delta \vec{r}_1 + \Delta \vec{r}_2 + \Delta \vec{r}_3 + \Delta \vec{r}_4$$

$$= (2.0 - 1.0 + 4.0 - 3.0)\,\hat{i} + (1.0 + 0 - 2.0 + 1.0)\,\hat{j} + (3.0 + 3.0 + 1.0 + 2.0)\,\hat{k}$$

$$= 2.0\,\hat{i} + 0\,\hat{j} + 9.0\,\hat{k}\,\mu m.$$

To complete the solution, we express the displacement as a magnitude and direction,

$$\left|\Delta \vec{r}_{\text{Total}}\right| = \sqrt{2.0^2 + 0^2 + 9.0^2} = 9.2\ \mu m, \quad \theta = \tan^{-1}\left(\frac{9}{2}\right) = 77°,$$

with respect to the x-axis in the xz-plane.

Significance

From the figure we can see the magnitude of the total displacement is less than the sum of the magnitudes of the individual displacements.

Velocity Vector

In the previous chapter we found the instantaneous velocity by calculating the derivative of the position function with respect to time. We can do the same operation in two and three dimensions, but we use vectors. The instantaneous **velocity vector** is now

$$\vec{v}(t) = \lim_{\Delta t \to 0} \frac{\vec{r}(t + \Delta t) - \vec{r}(t)}{\Delta t} = \frac{d\vec{r}}{dt}. \tag{4.4}$$

Let's look at the relative orientation of the position vector and velocity vector graphically. In **Figure 4.7** we show the vectors $\vec{r}(t)$ and $\vec{r}(t + \Delta t)$, which give the position of a particle moving along a path represented by the gray line. As Δt goes to zero, the velocity vector, given by **Equation 4.4**, becomes tangent to the path of the particle at time t.

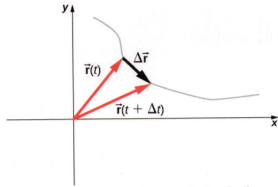

Figure 4.7 A particle moves along a path given by the gray line. In the limit as Δt approaches zero, the velocity vector becomes tangent to the path of the particle.

Equation 4.4 can also be written in terms of the components of $\overrightarrow{\mathbf{v}}(t)$. Since

$$\overrightarrow{\mathbf{r}}(t) = x(t)\,\hat{\mathbf{i}} + y(t)\,\hat{\mathbf{j}} + z(t)\,\hat{\mathbf{k}},$$

we can write

$$\overrightarrow{\mathbf{v}}(t) = v_x(t)\,\hat{\mathbf{i}} + v_y(t)\,\hat{\mathbf{j}} + v_z(t)\,\hat{\mathbf{k}} \tag{4.5}$$

where

$$v_x(t) = \frac{dx(t)}{dt}, \quad v_y(t) = \frac{dy(t)}{dt}, \quad v_z(t) = \frac{dz(t)}{dt}. \tag{4.6}$$

If only the average velocity is of concern, we have the vector equivalent of the one-dimensional average velocity for two and three dimensions:

$$\overrightarrow{\mathbf{v}}_{avg} = \frac{\overrightarrow{\mathbf{r}}(t_2) - \overrightarrow{\mathbf{r}}(t_1)}{t_2 - t_1}. \tag{4.7}$$

Example 4.3

Calculating the Velocity Vector

The position function of a particle is $\overrightarrow{\mathbf{r}}(t) = 2.0t^2\,\hat{\mathbf{i}} + (2.0 + 3.0t)\,\hat{\mathbf{j}} + 5.0t\,\hat{\mathbf{k}}$ m. (a) What is the instantaneous velocity and speed at $t = 2.0$ s? (b) What is the average velocity between 1.0 s and 3.0 s?

Solution

Using **Equation 4.5** and **Equation 4.6**, and taking the derivative of the position function with respect to time, we find

(a) $v(t) = \dfrac{d\mathbf{r}(t)}{dt} = 4.0t\,\hat{\mathbf{i}} + 3.0\,\hat{\mathbf{j}} + 5.0\,\hat{\mathbf{k}}\,\text{m/s}$

$\vec{\mathbf{v}}\,(2.0s) = 8.0\,\hat{\mathbf{i}} + 3.0\,\hat{\mathbf{j}} + 5.0\,\hat{\mathbf{k}}\,\text{m/s}$

Speed $\left|\vec{\mathbf{v}}\,(2.0\text{ s})\right| = \sqrt{8^2 + 3^2 + 5^2} = 9.9\text{ m/s}.$

(b) From **Equation 4.7**,

$$\vec{\mathbf{v}}_{\text{avg}} = \frac{\vec{\mathbf{r}}\,(t_2) - \vec{\mathbf{r}}\,(t_1)}{t_2 - t_1} = \frac{\vec{\mathbf{r}}\,(3.0\text{ s}) - \vec{\mathbf{r}}\,(1.0\text{ s})}{3.0\text{ s} - 1.0\text{ s}} = \frac{(18\,\hat{\mathbf{i}} + 11\,\hat{\mathbf{j}} + 15\,\hat{\mathbf{k}})\,\text{m} - (2\,\hat{\mathbf{i}} + 5\,\hat{\mathbf{j}} + 5\,\hat{\mathbf{k}})\,\text{m}}{2.0\text{ s}}$$

$$= \frac{(16\,\hat{\mathbf{i}} + 6\,\hat{\mathbf{j}} + 10\,\hat{\mathbf{k}})\,\text{m}}{2.0\text{ s}} = 8.0\,\hat{\mathbf{i}} + 3.0\,\hat{\mathbf{j}} + 5.0\,\hat{\mathbf{k}}\,\text{m/s}.$$

Significance

We see the average velocity is the same as the instantaneous velocity at $t = 2.0$ s, as a result of the velocity function being linear. This need not be the case in general. In fact, most of the time, instantaneous and average velocities are not the same.

 4.1 **Check Your Understanding** The position function of a particle is $\vec{\mathbf{r}}\,(t) = 3.0t^3\,\hat{\mathbf{i}} + 4.0\,\hat{\mathbf{j}}$. (a) What is the instantaneous velocity at $t = 3$ s? (b) Is the average velocity between 2 s and 4 s equal to the instantaneous velocity at $t = 3$ s?

The Independence of Perpendicular Motions

When we look at the three-dimensional equations for position and velocity written in unit vector notation, **Equation 4.2** and **Equation 4.5**, we see the components of these equations are separate and unique functions of time that do not depend on one another. Motion along the x direction has no part of its motion along the y and z directions, and similarly for the other two coordinate axes. Thus, the motion of an object in two or three dimensions can be divided into separate, independent motions along the perpendicular axes of the coordinate system in which the motion takes place.

To illustrate this concept with respect to displacement, consider a woman walking from point A to point B in a city with square blocks. The woman taking the path from A to B may walk east for so many blocks and then north (two perpendicular directions) for another set of blocks to arrive at B. How far she walks east is affected only by her motion eastward. Similarly, how far she walks north is affected only by her motion northward.

Independence of Motion

In the kinematic description of motion, we are able to treat the horizontal and vertical components of motion separately. In many cases, motion in the horizontal direction does not affect motion in the vertical direction, and vice versa.

An example illustrating the independence of vertical and horizontal motions is given by two baseballs. One baseball is dropped from rest. At the same instant, another is thrown horizontally from the same height and it follows a curved path. A stroboscope captures the positions of the balls at fixed time intervals as they fall (**Figure 4.8**).

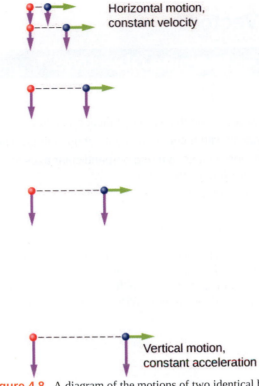

Figure 4.8 A diagram of the motions of two identical balls: one falls from rest and the other has an initial horizontal velocity. Each subsequent position is an equal time interval. Arrows represent the horizontal and vertical velocities at each position. The ball on the right has an initial horizontal velocity whereas the ball on the left has no horizontal velocity. Despite the difference in horizontal velocities, the vertical velocities and positions are identical for both balls, which shows the vertical and horizontal motions are independent.

It is remarkable that for each flash of the strobe, the vertical positions of the two balls are the same. This similarity implies vertical motion is independent of whether the ball is moving horizontally. (Assuming no air resistance, the vertical motion of a falling object is influenced by gravity only, not by any horizontal forces.) Careful examination of the ball thrown horizontally shows it travels the same horizontal distance between flashes. This is because there are no additional forces on the ball in the horizontal direction after it is thrown. This result means horizontal velocity is constant and is affected neither by vertical motion nor by gravity (which is vertical). Note this case is true for ideal conditions only. In the real world, air resistance affects the speed of the balls in both directions.

The two-dimensional curved path of the horizontally thrown ball is composed of two independent one-dimensional motions (horizontal and vertical). The key to analyzing such motion, called *projectile motion*, is to resolve it into motions along perpendicular directions. Resolving two-dimensional motion into perpendicular components is possible because the components are independent.

4.2 | Acceleration Vector

Learning Objectives

By the end of this section, you will be able to:

- Calculate the acceleration vector given the velocity function in unit vector notation.
- Describe the motion of a particle with a constant acceleration in three dimensions.
- Use the one-dimensional motion equations along perpendicular axes to solve a problem in two or three dimensions with a constant acceleration.
- Express the acceleration in unit vector notation.

Instantaneous Acceleration

In addition to obtaining the displacement and velocity vectors of an object in motion, we often want to know its **acceleration vector** at any point in time along its trajectory. This acceleration vector is the instantaneous acceleration and it can be obtained from the derivative with respect to time of the velocity function, as we have seen in a previous chapter. The only difference in two or three dimensions is that these are now vector quantities. Taking the derivative with respect to time $\vec{v}(t)$, we find

$$\vec{a}(t) = \lim_{t \to 0} \frac{\vec{v}(t + \Delta t) - \vec{v}(t)}{\Delta t} = \frac{d\vec{v}(t)}{dt}. \tag{4.8}$$

The acceleration in terms of components is

$$\vec{a}(t) = \frac{dv_x(t)}{dt}\hat{i} + \frac{dv_y(t)}{dt}\hat{j} + \frac{dv_z(t)}{dt}\hat{k}. \tag{4.9}$$

Also, since the velocity is the derivative of the position function, we can write the acceleration in terms of the second derivative of the position function:

$$\vec{a}(t) = \frac{d^2x(t)}{dt^2}\hat{i} + \frac{d^2y(t)}{dt^2}\hat{j} + \frac{d^2z(t)}{dt^2}\hat{k}. \tag{4.10}$$

Example 4.4

Finding an Acceleration Vector

A particle has a velocity of $\vec{v}(t) = 5.0t\,\hat{i} + t^2\,\hat{j} - 2.0t^3\,\hat{k}$ m/s. (a) What is the acceleration function? (b) What is the acceleration vector at $t = 2.0$ s? Find its magnitude and direction.

Solution

(a) We take the first derivative with respect to time of the velocity function to find the acceleration. The derivative is taken component by component:

$$\vec{a}(t) = 5.0\,\hat{i} + 2.0t\,\hat{j} - 6.0t^2\,\hat{k}\ \text{m/s}^2.$$

(b) Evaluating $\vec{a}\,(2.0\,\text{s}) = 5.0\,\hat{i} + 4.0\,\hat{j} - 24.0\,\hat{k}\,\text{m/s}^2$ gives us the direction in unit vector notation. The magnitude of the acceleration is $|\vec{a}\,(2.0\,\text{s})| = \sqrt{5.0^2 + 4.0^2 + (-24.0)^2} = 24.8\,\text{m/s}^2$.

Significance

In this example we find that acceleration has a time dependence and is changing throughout the motion. Let's consider a different velocity function for the particle.

Example 4.5

Finding a Particle Acceleration

A particle has a position function $\vec{r}\,(t) = (10t - t^2)\,\hat{i} + 5t\,\hat{j} + 5t\,\hat{k}\,\text{m}$. (a) What is the velocity? (b) What is the acceleration? (c) Describe the motion from $t = 0$ s.

Strategy

We can gain some insight into the problem by looking at the position function. It is linear in y and z, so we know the acceleration in these directions is zero when we take the second derivative. Also, note that the position in the x direction is zero for $t = 0$ s and $t = 10$ s.

Solution

(a) Taking the derivative with respect to time of the position function, we find

$$\vec{v}\,(t) = (10 - 2t)\,\hat{i} + 5\,\hat{j} + 5\,\hat{k}\,\text{m/s}.$$

The velocity function is linear in time in the x direction and is constant in the y and z directions.

(b) Taking the derivative of the velocity function, we find

$$\vec{a}\,(t) = -2\,\hat{i}\,\text{m/s}^2.$$

The acceleration vector is a constant in the negative x-direction.

(c) The trajectory of the particle can be seen in **Figure 4.9**. Let's look in the y and z directions first. The particle's position increases steadily as a function of time with a constant velocity in these directions. In the x direction, however, the particle follows a path in positive x until $t = 5$ s, when it reverses direction. We know this from looking at the velocity function, which becomes zero at this time and negative thereafter. We also know this because the acceleration is negative and constant—meaning, the particle is decelerating, or accelerating in the negative direction. The particle's position reaches 25 m, where it then reverses direction and begins to accelerate in the negative x direction. The position reaches zero at $t = 10$ s.

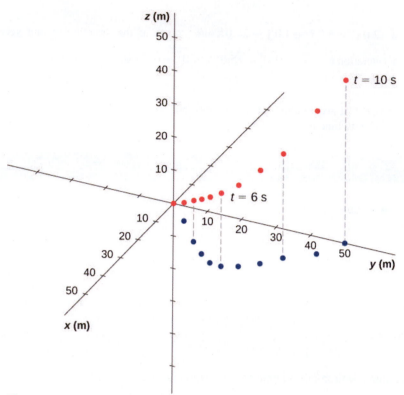

Figure 4.9 The particle starts at point $(x, y, z) = (0, 0, 0)$ with position vector $\vec{r} = 0$. The projection of the trajectory onto the xy-plane is shown. The values of y and z increase linearly as a function of time, whereas x has a turning point at $t = 5$ s and 25 m, when it reverses direction. At this point, the x component of the velocity becomes negative. At $t = 10$ s, the particle is back to 0 m in the x direction.

Significance

By graphing the trajectory of the particle, we can better understand its motion, given by the numerical results of the kinematic equations.

 4.2 Check Your Understanding Suppose the acceleration function has the form $\vec{a}(t) = a\,\hat{i} + b\,\hat{j} + c\,\hat{k}$ m/s^2, where a, b, and c are constants. What can be said about the functional form of the velocity function?

Constant Acceleration

Multidimensional motion with constant acceleration can be treated the same way as shown in the previous chapter for one-dimensional motion. Earlier we showed that three-dimensional motion is equivalent to three one-dimensional motions, each along an axis perpendicular to the others. To develop the relevant equations in each direction, let's consider the two-dimensional problem of a particle moving in the xy plane with constant acceleration, ignoring the z-component for the moment. The acceleration vector is

$$\vec{a} = a_{0x}\,\hat{i} + a_{0y}\,\hat{j}.$$

Each component of the motion has a separate set of equations similar to **Equation 3.10–Equation 3.14** of the previous chapter on one-dimensional motion. We show only the equations for position and velocity in the x- and y-directions. A similar set of kinematic equations could be written for motion in the z-direction:

$$x(t) = x_0 + (v_x)_{avg} t \tag{4.11}$$

$$v_x(t) = v_{0x} + a_x t \tag{4.12}$$

$$x(t) = x_0 + v_{0x} t + \frac{1}{2} a_x t^2 \tag{4.13}$$

$$v_x^2(t) = v_{0x}^2 + 2a_x(x - x_0) \tag{4.14}$$

$$y(t) = y_0 + (v_y)_{avg} t \tag{4.15}$$

$$v_y(t) = v_{0y} + a_y t \tag{4.16}$$

$$y(t) = y_0 + v_{0y} t + \frac{1}{2} a_y t^2 \tag{4.17}$$

$$v_y^2(t) = v_{0y}^2 + 2a_y(y - y_0). \tag{4.18}$$

Here the subscript 0 denotes the initial position or velocity. **Equation 4.11** to **Equation 4.18** can be substituted into **Equation 4.2** and **Equation 4.5** without the z-component to obtain the position vector and velocity vector as a function of time in two dimensions:

$$\vec{r}(t) = x(t)\,\hat{i} + y(t)\,\hat{j} \text{ and } \vec{v}(t) = v_x(t)\,\hat{i} + v_y(t)\,\hat{j}.$$

The following example illustrates a practical use of the kinematic equations in two dimensions.

Example 4.6

A Skier

Figure 4.10 shows a skier moving with an acceleration of 2.1 m/s^2 down a slope of $15°$ at $t = 0$. With the origin of the coordinate system at the front of the lodge, her initial position and velocity are

$$\vec{r}(0) = (75.0\,\hat{i} - 50.0\,\hat{j})\,\text{m}$$

and

$$\vec{v}(0) = (4.1\,\hat{i} - 1.1\,\hat{j})\,\text{m/s}.$$

(a) What are the x- and y-components of the skier's position and velocity as functions of time? (b) What are her position and velocity at $t = 10.0$ s?

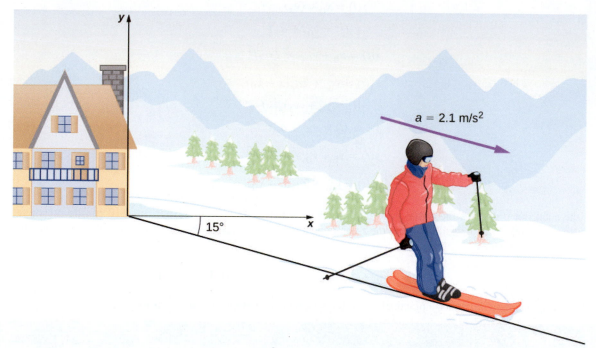

Figure 4.10 A skier has an acceleration of 2.1 m/s^2 down a slope of $15°$. The origin of the coordinate system is at the ski lodge.

Strategy

Since we are evaluating the components of the motion equations in the x and y directions, we need to find the components of the acceleration and put them into the kinematic equations. The components of the acceleration are found by referring to the coordinate system in **Figure 4.10**. Then, by inserting the components of the initial position and velocity into the motion equations, we can solve for her position and velocity at a later time t.

Solution

(a) The origin of the coordinate system is at the top of the hill with y-axis vertically upward and the x-axis horizontal. By looking at the trajectory of the skier, the x-component of the acceleration is positive and the y-component is negative. Since the angle is $15°$ down the slope, we find

$$a_x = (2.1 \text{ m/s}^2) \cos(15°) = 2.0 \text{ m/s}^2$$

$$a_y = (-2.1 \text{ m/s}^2) \sin 15° = -0.54 \text{ m/s}^2.$$

Inserting the initial position and velocity into **Equation 4.12** and **Equation 4.13** for x, we have

$$x(t) = 75.0 \text{ m} + (4.1 \text{ m/s})t + \frac{1}{2}(2.0 \text{ m/s}^2)t^2$$

$$v_x(t) = 4.1 \text{ m/s} + (2.0 \text{ m/s}^2)t.$$

For y, we have

$$y(t) = -50.0 \text{ m} + (-1.1 \text{ m/s})t + \frac{1}{2}(-0.54 \text{ m/s}^2)t^2$$

$$v_y(t) = -1.1 \text{ m/s} + (-0.54 \text{ m/s}^2)t.$$

(b) Now that we have the equations of motion for x and y as functions of time, we can evaluate them at $t = 10.0$ s:

$$x(10.0 \text{ s}) = 75.0 \text{ m} + (4.1 \text{ m/s}^2)(10.0 \text{ s}) + \frac{1}{2}(2.0 \text{ m/s}^2)(10.0 \text{ s})^2 = 216.0 \text{ m}$$

$$v_x(10.0 \text{ s}) = 4.1 \text{ m/s} + (2.0 \text{ m/s}^2)(10.0 \text{ s}) = 24.1 \text{m/s}$$

$$y(10.0 \text{ s}) = -50.0 \text{ m} + (-1.1 \text{ m/s})(10.0 \text{ s}) + \frac{1}{2}(-0.54 \text{ m/s}^2)(10.0 \text{ s})^2 = -88.0 \text{ m}$$

$$v_y(10.0 \text{ s}) = -1.1 \text{ m/s} + (-0.54 \text{ m/s}^2)(10.0 \text{ s}) = -6.5 \text{ m/s}.$$

The position and velocity at $t = 10.0$ s are, finally,

$$\vec{r}(10.0 \text{ s}) = (216.0 \, \hat{i} - 88.0 \, \hat{j}) \text{ m}$$

$$\vec{v}(10.0 \text{ s}) = (24.1 \, \hat{i} - 6.5 \, \hat{j}) \text{m/s}.$$

The magnitude of the velocity of the skier at 10.0 s is 25 m/s, which is 60 mi/h.

Significance

It is useful to know that, given the initial conditions of position, velocity, and acceleration of an object, we can find the position, velocity, and acceleration at any later time.

With **Equation 4.8** through **Equation 4.10** we have completed the set of expressions for the position, velocity, and acceleration of an object moving in two or three dimensions. If the trajectories of the objects look something like the "Red Arrows" in the opening picture for the chapter, then the expressions for the position, velocity, and acceleration can be quite complicated. In the sections to follow we examine two special cases of motion in two and three dimensions by looking at projectile motion and circular motion.

 At this **University of Colorado Boulder website (https://openstaxcollege.org/l/21phetmotladyb)** , you can explore the position velocity and acceleration of a ladybug with an interactive simulation that allows you to change these parameters.

4.3 | Projectile Motion

Learning Objectives

By the end of this section, you will be able to:

- Use one-dimensional motion in perpendicular directions to analyze projectile motion.
- Calculate the range, time of flight, and maximum height of a projectile that is launched and impacts a flat, horizontal surface.
- Find the time of flight and impact velocity of a projectile that lands at a different height from that of launch.
- Calculate the trajectory of a projectile.

Projectile motion is the motion of an object thrown or projected into the air, subject only to acceleration as a result of gravity. The applications of projectile motion in physics and engineering are numerous. Some examples include meteors as they enter Earth's atmosphere, fireworks, and the motion of any ball in sports. Such objects are called *projectiles* and their path is called a **trajectory**. The motion of falling objects as discussed in **Motion Along a Straight Line** is a simple one-dimensional type of projectile motion in which there is no horizontal movement. In this section, we consider two-dimensional projectile motion, and our treatment neglects the effects of air resistance.

The most important fact to remember here is that *motions along perpendicular axes are independent* and thus can be analyzed separately. We discussed this fact in **Displacement and Velocity Vectors**, where we saw that vertical and horizontal motions are independent. The key to analyzing two-dimensional projectile motion is to break it into two motions: one along the horizontal axis and the other along the vertical. (This choice of axes is the most sensible because acceleration resulting from gravity is vertical; thus, there is no acceleration along the horizontal axis when air resistance is negligible.) As is customary, we call the horizontal axis the *x*-axis and the vertical axis the *y*-axis. It is not required that we use this choice of axes; it is simply convenient in the case of gravitational acceleration. In other cases we may choose a different set of axes. **Figure 4.11** illustrates the notation for displacement, where we define $\vec{s}$ to be the total displacement, and $\vec{x}$ and $\vec{y}$ are its component vectors along the horizontal and vertical axes, respectively. The magnitudes of these vectors are *s, x,* and *y.*

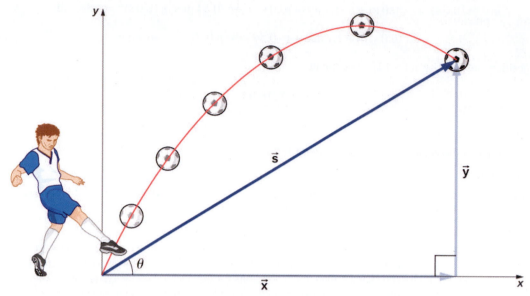

Figure 4.11 The total displacement *s* of a soccer ball at a point along its path. The vector $\vec{s}$ has components $\vec{x}$ and $\vec{y}$ along the horizontal and vertical axes. Its magnitude is *s* and it makes an angle θ with the horizontal.

To describe projectile motion completely, we must include velocity and acceleration, as well as displacement. We must find their components along the *x*- and *y*-axes. Let's assume all forces except gravity (such as air resistance and friction, for example) are negligible. Defining the positive direction to be upward, the components of acceleration are then very simple:

$$a_y = -g = -9.8 \text{ m/s}^2 \ (-32 \text{ ft/s}^2).$$

Because gravity is vertical, $a_x = 0.$ If $a_x = 0,$ this means the initial velocity in the *x* direction is equal to the final velocity in the *x* direction, or $v_x = v_{0x}.$ With these conditions on acceleration and velocity, we can write the kinematic **Equation 4.11** through **Equation 4.18** for motion in a uniform gravitational field, including the rest of the kinematic equations for a constant acceleration from **Motion with Constant Acceleration**. The kinematic equations for motion in a uniform gravitational field become kinematic equations with $a_y = -g, \ a_x = 0$:

Horizontal Motion

$$v_{0x} = v_x, \ x = x_0 + v_x t \tag{4.19}$$

Vertical Motion

$$y = y_0 + \frac{1}{2}(v_{0y} + v_y)t \tag{4.20}$$

$$v_y = v_{0y} - gt \tag{4.21}$$

$$y = y_0 + v_{0y}t - \frac{1}{2}gt^2 \tag{4.22}$$

$$v_y^2 = v_{0y}^2 - 2g(y - y_0) \tag{4.23}$$

Using this set of equations, we can analyze projectile motion, keeping in mind some important points.

Problem-Solving Strategy: Projectile Motion

1. Resolve the motion into horizontal and vertical components along the *x*- and *y*-axes. The magnitudes of the components of displacement $\vec{s}$ along these axes are *x* and *y.* The magnitudes of the components of velocity

$\vec{v}$ are $v_x = v\cos\theta$ and $v_y = v\sin\theta$, where v is the magnitude of the velocity and θ is its direction relative to the horizontal, as shown in **Figure 4.12**.

2. Treat the motion as two independent one-dimensional motions: one horizontal and the other vertical. Use the kinematic equations for horizontal and vertical motion presented earlier.

3. Solve for the unknowns in the two separate motions: one horizontal and one vertical. Note that the only common variable between the motions is time t. The problem-solving procedures here are the same as those for one-dimensional kinematics and are illustrated in the following solved examples.

4. Recombine quantities in the horizontal and vertical directions to find the total displacement $\vec{s}$ and velocity $\vec{v}$. Solve for the magnitude and direction of the displacement and velocity using

$$s = \sqrt{x^2 + y^2}, \quad \theta = \tan^{-1}(y/x), \quad v = \sqrt{v_x^2 + v_y^2},$$

where θ is the direction of the displacement $\vec{s}$.

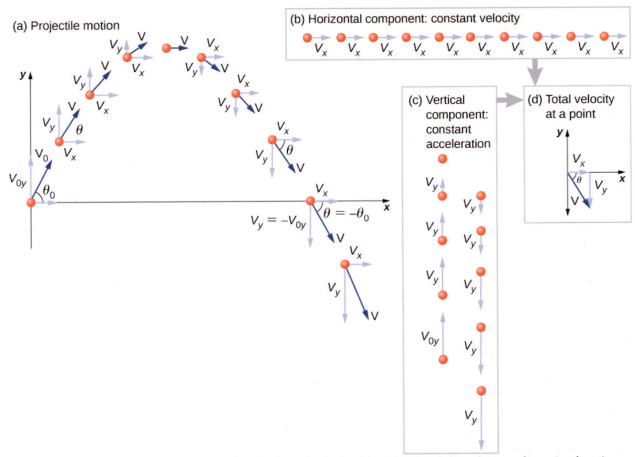

Figure 4.12 (a) We analyze two-dimensional projectile motion by breaking it into two independent one-dimensional motions along the vertical and horizontal axes. (b) The horizontal motion is simple, because $a_x = 0$ and v_x is a constant. (c) The velocity in the vertical direction begins to decrease as the object rises. At its highest point, the vertical velocity is zero. As the object falls toward Earth again, the vertical velocity increases again in magnitude but points in the opposite direction to the initial vertical velocity. (d) The x and y motions are recombined to give the total velocity at any given point on the trajectory.

Example 4.7

A Fireworks Projectile Explodes High and Away

During a fireworks display, a shell is shot into the air with an initial speed of 70.0 m/s at an angle of $75.0°$ above the horizontal, as illustrated in **Figure 4.13**. The fuse is timed to ignite the shell just as it reaches its highest point above the ground. (a) Calculate the height at which the shell explodes. (b) How much time passes between the launch of the shell and the explosion? (c) What is the horizontal displacement of the shell when it explodes? (d) What is the total displacement from the point of launch to the highest point?

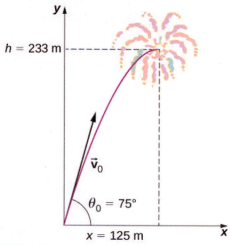

Figure 4.13 The trajectory of a fireworks shell. The fuse is set to explode the shell at the highest point in its trajectory, which is found to be at a height of 233 m and 125 m away horizontally.

Strategy

The motion can be broken into horizontal and vertical motions in which $a_x = 0$ and $a_y = -g$. We can then define x_0 and y_0 to be zero and solve for the desired quantities.

Solution

(a) By "height" we mean the altitude or vertical position y above the starting point. The highest point in any trajectory, called the *apex*, is reached when $v_y = 0$. Since we know the initial and final velocities, as well as the initial position, we use the following equation to find y:

$$v_y^2 = v_{0y}^2 - 2g(y - y_0).$$

Because y_0 and v_y are both zero, the equation simplifies to

$$0 = v_{0y}^2 - 2gy.$$

Solving for y gives

$$y = \frac{v_{0y}^2}{2g}.$$

Now we must find v_{0y}, the component of the initial velocity in the y direction. It is given by $v_{0y} = v_0 \sin\theta_0$, where v_0 is the initial velocity of 70.0 m/s and $\theta_0 = 75°$ is the initial angle. Thus,

$$v_{0y} = v_0 \sin\theta = (70.0 \text{ m/s})\sin 75° = 67.6 \text{ m/s}$$

and y is

$$y = \frac{(67.6 \text{ m/s})^2}{2(9.80 \text{ m/s}^2)}.$$

Thus, we have

$$y = 233 \text{ m.}$$

Note that because up is positive, the initial vertical velocity is positive, as is the maximum height, but the acceleration resulting from gravity is negative. Note also that the maximum height depends only on the vertical component of the initial velocity, so that any projectile with a 67.6-m/s initial vertical component of velocity reaches a maximum height of 233 m (neglecting air resistance). The numbers in this example are reasonable for large fireworks displays, the shells of which do reach such heights before exploding. In practice, air resistance is not completely negligible, so the initial velocity would have to be somewhat larger than that given to reach the same height.

(b) As in many physics problems, there is more than one way to solve for the time the projectile reaches its highest point. In this case, the easiest method is to use $v_y = v_{0y} - gt$. Because $v_y = 0$ at the apex, this equation reduces to simply

$$0 = v_{0y} - gt$$

or

$$t = \frac{v_{0y}}{g} = \frac{67.6 \text{ m/s}}{9.80 \text{ m/s}^2} = 6.90\text{s.}$$

This time is also reasonable for large fireworks. If you are able to see the launch of fireworks, notice that several seconds pass before the shell explodes. Another way of finding the time is by using $y = y_0 + \frac{1}{2}(v_{0y} + v_y)t$. This is left for you as an exercise to complete.

(c) Because air resistance is negligible, $a_x = 0$ and the horizontal velocity is constant, as discussed earlier. The horizontal displacement is the horizontal velocity multiplied by time as given by $x = x_0 + v_x t$, where x_0 is equal to zero. Thus,

$$x = v_x t,$$

where v_x is the x-component of the velocity, which is given by

$$v_x = v_0 \cos\theta = (70.0 \text{ m/s})\cos 75° = 18.1 \text{ m/s.}$$

Time t for both motions is the same, so x is

$$x = (18.1 \text{ m/s})6.90 \text{ s} = 125 \text{ m.}$$

Horizontal motion is a constant velocity in the absence of air resistance. The horizontal displacement found here could be useful in keeping the fireworks fragments from falling on spectators. When the shell explodes, air resistance has a major effect, and many fragments land directly below.

(d) The horizontal and vertical components of the displacement were just calculated, so all that is needed here is to find the magnitude and direction of the displacement at the highest point:

$$\vec{s} = 125\,\hat{i} + 233\,\hat{j}$$

$$|\vec{s}| = \sqrt{125^2 + 233^2} = 264 \text{ m}$$

$$\theta = \tan^{-1}\left(\frac{233}{125}\right) = 61.8°.$$

Note that the angle for the displacement vector is less than the initial angle of launch. To see why this is, review **Figure 4.11**, which shows the curvature of the trajectory toward the ground level.

When solving **Example 4.7**(a), the expression we found for y is valid for any projectile motion when air resistance is negligible. Call the maximum height $y = h$. Then,

$$h = \frac{v_{0y}^2}{2g}.$$

This equation defines the *maximum height of a projectile above its launch position* and it depends only on the vertical component of the initial velocity.

 4.3   **Check Your Understanding** A rock is thrown horizontally off a cliff 100.0 m high with a velocity of 15.0 m/s. (a) Define the origin of the coordinate system. (b) Which equation describes the horizontal motion? (c) Which equations describe the vertical motion? (d) What is the rock's velocity at the point of impact?

Example 4.8

Calculating Projectile Motion: Tennis Player

A tennis player wins a match at Arthur Ashe stadium and hits a ball into the stands at 30 m/s and at an angle 45° above the horizontal (**Figure 4.14**). On its way down, the ball is caught by a spectator 10 m above the point where the ball was hit. (a) Calculate the time it takes the tennis ball to reach the spectator. (b) What are the magnitude and direction of the ball's velocity at impact?

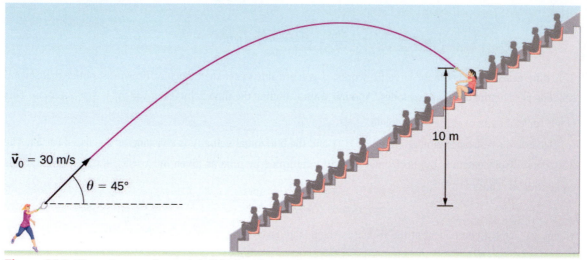

Figure 4.14   The trajectory of a tennis ball hit into the stands.

Strategy

Again, resolving this two-dimensional motion into two independent one-dimensional motions allows us to solve for the desired quantities. The time a projectile is in the air is governed by its vertical motion alone. Thus, we solve for t first. While the ball is rising and falling vertically, the horizontal motion continues at a constant velocity. This example asks for the final velocity. Thus, we recombine the vertical and horizontal results to obtain $\vec{v}$ at final time t, determined in the first part of the example.

Solution

(a) While the ball is in the air, it rises and then falls to a final position 10.0 m higher than its starting altitude. We can find the time for this by using **Equation 4.22**:

$$y = y_0 + v_{0y}t - \frac{1}{2}gt^2.$$

If we take the initial position y_0 to be zero, then the final position is $y = 10$ m. The initial vertical velocity is the vertical component of the initial velocity:

$$v_{0y} = v_0 \sin\theta_0 = (30.0 \text{ m/s})\sin 45° = 21.2 \text{ m/s}.$$

Substituting into **Equation 4.22** for y gives us

$$10.0 \text{ m} = (21.2 \text{ m/s})t - (4.90 \text{ m/s}^2)t^2.$$

Rearranging terms gives a quadratic equation in t:

$$(4.90 \text{ m/s}^2)t^2 - (21.2 \text{ m/s})t + 10.0 \text{ m} = 0.$$

Use of the quadratic formula yields $t = 3.79$ s and $t = 0.54$ s. Since the ball is at a height of 10 m at two times during its trajectory—once on the way up and once on the way down—we take the longer solution for the time it takes the ball to reach the spectator:

$$t = 3.79 \text{ s}.$$

The time for projectile motion is determined completely by the vertical motion. Thus, any projectile that has an initial vertical velocity of 21.2 m/s and lands 10.0 m below its starting altitude spends 3.79 s in the air.

(b) We can find the final horizontal and vertical velocities v_x and v_y with the use of the result from (a). Then, we can combine them to find the magnitude of the total velocity vector $\vec{v}$ and the angle θ it makes with the horizontal. Since v_x is constant, we can solve for it at any horizontal location. We choose the starting point because we know both the initial velocity and the initial angle. Therefore,

$$v_x = v_0 \cos\theta_0 = (30 \text{ m/s})\cos 45° = 21.2 \text{ m/s}.$$

The final vertical velocity is given by **Equation 4.21**:

$$v_y = v_{0y} - gt.$$

Since v_{0y} was found in part (a) to be 21.2 m/s, we have

$$v_y = 21.2 \text{ m/s} - 9.8 \text{ m/s}^2(3.79 \text{ s}) = -15.9 \text{ m/s}.$$

The magnitude of the final velocity $\vec{v}$ is

$$v = \sqrt{v_x^2 + v_y^2} = \sqrt{(21.2 \text{ m/s})^2 + (-15.9 \text{ m/s})^2} = 26.5 \text{ m/s}.$$

The direction θ_v is found using the inverse tangent:

$$\theta_v = \tan^{-1}\left(\frac{v_y}{v_x}\right) = \tan^{-1}\left(\frac{21.2}{-15.9}\right) = -53.1°.$$

Significance

(a) As mentioned earlier, the time for projectile motion is determined completely by the vertical motion. Thus, any projectile that has an initial vertical velocity of 21.2 m/s and lands 10.0 m below its starting altitude spends 3.79 s in the air. (b) The negative angle means the velocity is 53.1° below the horizontal at the point of impact. This result is consistent with the fact that the ball is impacting at a point on the other side of the apex of the trajectory and therefore has a negative y component of the velocity. The magnitude of the velocity is less than the magnitude of the initial velocity we expect since it is impacting 10.0 m above the launch elevation.

Time of Flight, Trajectory, and Range

Of interest are the time of flight, trajectory, and range for a projectile launched on a flat horizontal surface and impacting on the same surface. In this case, kinematic equations give useful expressions for these quantities, which are derived in the following sections.

Time of flight

We can solve for the time of flight of a projectile that is both launched and impacts on a flat horizontal surface by performing some manipulations of the kinematic equations. We note the position and displacement in y must be zero at launch and at impact on an even surface. Thus, we set the displacement in y equal to zero and find

$$y - y_0 = v_{0y}t - \frac{1}{2}gt^2 = (v_0 \sin\theta_0)t - \frac{1}{2}gt^2 = 0.$$

Factoring, we have

$$t\left(v_0 \sin\theta_0 - \frac{gt}{2}\right) = 0.$$

Solving for t gives us

$$T_{\text{tof}} = \frac{2(v_0 \sin\theta_0)}{g}. \qquad (4.24)$$

This is the **time of flight** for a projectile both launched and impacting on a flat horizontal surface. **Equation 4.24** does not apply when the projectile lands at a different elevation than it was launched, as we saw in **Example 4.8** of the tennis player hitting the ball into the stands. The other solution, $t = 0$, corresponds to the time at launch. The time of flight is linearly proportional to the initial velocity in the y direction and inversely proportional to g. Thus, on the Moon, where gravity is one-sixth that of Earth, a projectile launched with the same velocity as on Earth would be airborne six times as long.

Trajectory

The trajectory of a projectile can be found by eliminating the time variable t from the kinematic equations for arbitrary t and solving for $y(x)$. We take $x_0 = y_0 = 0$ so the projectile is launched from the origin. The kinematic equation for x gives

$$x = v_{0x}t \Rightarrow t = \frac{x}{v_{0x}} = \frac{x}{v_0 \cos\theta_0}.$$

Substituting the expression for t into the equation for the position $y = (v_0 \sin\theta_0)t - \frac{1}{2}gt^2$ gives

$$y = (v_0 \sin\theta_0)\left(\frac{x}{v_0 \cos\theta_0}\right) - \frac{1}{2}g\left(\frac{x}{v_0 \cos\theta_0}\right)^2.$$

Rearranging terms, we have

$$y = (\tan\theta_0)x - \left[\frac{g}{2(v_0 \cos\theta_0)^2}\right]x^2. \qquad (4.25)$$

This trajectory equation is of the form $y = ax + bx^2,$ which is an equation of a parabola with coefficients

$$a = \tan\theta_0, \quad b = -\frac{g}{2(v_0 \cos\theta_0)^2}.$$

Range

From the trajectory equation we can also find the **range**, or the horizontal distance traveled by the projectile. Factoring **Equation 4.25**, we have

$$y = x\left[\tan\theta_0 - \frac{g}{2(v_0 \cos\theta_0)^2}x\right].$$

The position y is zero for both the launch point and the impact point, since we are again considering only a flat horizontal surface. Setting $y = 0$ in this equation gives solutions $x = 0$, corresponding to the launch point, and

$$x = \frac{2v_0^2 \sin\theta_0 \cos\theta_0}{g},$$

corresponding to the impact point. Using the trigonometric identity $2\sin\theta\cos\theta = \sin2\theta$ and setting $x = R$ for range, we find

$$R = \frac{v_0^2 \sin 2\theta_0}{g}.$$

(4.26)

Note particularly that **Equation 4.26** is valid only for launch and impact on a horizontal surface. We see the range is directly proportional to the square of the initial speed v_0 and $\sin 2\theta_0$, and it is inversely proportional to the acceleration of gravity. Thus, on the Moon, the range would be six times greater than on Earth for the same initial velocity. Furthermore, we see from the factor $\sin 2\theta_0$ that the range is maximum at $45°$. These results are shown in **Figure 4.15**. In (a) we see that the greater the initial velocity, the greater the range. In (b), we see that the range is maximum at $45°$. This is true only for conditions neglecting air resistance. If air resistance is considered, the maximum angle is somewhat smaller. It is interesting that the same range is found for two initial launch angles that sum to $90°$. The projectile launched with the smaller angle has a lower apex than the higher angle, but they both have the same range.

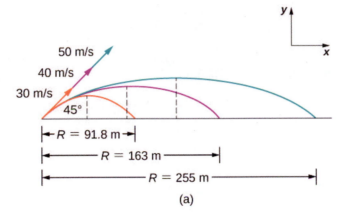

(a)

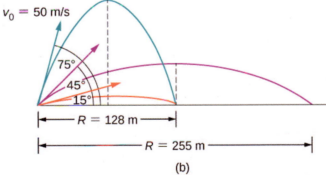

(b)

Figure 4.15 Trajectories of projectiles on level ground. (a) The greater the initial speed v_0, the greater the range for a given initial angle. (b) The effect of initial angle θ_0 on the range of a projectile with a given initial speed. Note that the range is the same for initial angles of $15°$ and $75°$, although the maximum heights of those paths are different.

Example 4.9

Comparing Golf Shots

A golfer finds himself in two different situations on different holes. On the second hole he is 120 m from the green and wants to hit the ball 90 m and let it run onto the green. He angles the shot low to the ground at $30°$ to the horizontal to let the ball roll after impact. On the fourth hole he is 90 m from the green and wants to let the ball drop with a minimum amount of rolling after impact. Here, he angles the shot at $70°$ to the horizontal to minimize rolling after impact. Both shots are hit and impacted on a level surface.

(a) What is the initial speed of the ball at the second hole?

(b) What is the initial speed of the ball at the fourth hole?

(c) Write the trajectory equation for both cases.

(d) Graph the trajectories.

Strategy

We see that the range equation has the initial speed and angle, so we can solve for the initial speed for both (a) and (b). When we have the initial speed, we can use this value to write the trajectory equation.

Solution

(a) $R = \dfrac{v_0^2 \sin 2\theta_0}{g} \Rightarrow v_0 = \sqrt{\dfrac{Rg}{\sin 2\theta_0}} = \sqrt{\dfrac{90.0 \text{ m}(9.8 \text{ m/s}^2)}{\sin(2(70°))}} = 37.0 \text{ m/s}$

(b) $R = \dfrac{v_0^2 \sin 2\theta_0}{g} \Rightarrow v_0 = \sqrt{\dfrac{Rg}{\sin 2\theta_0}} = \sqrt{\dfrac{90.0 \text{ m}(9.8 \text{ m/s}^2)}{\sin(2(30°))}} = 31.9 \text{ m/s}$

(c)

$$y = x\left[\tan\theta_0 - \dfrac{g}{2(v_0 \cos\theta_0)^2}x\right]$$

Second hole: $y = x\left[\tan 70° - \dfrac{9.8 \text{ m/s}^2}{2[(37.0 \text{ m/s})(\cos 70°)]^2}x\right] = 2.75x - 0.0306x^2$

Fourth hole: $y = x\left[\tan 30° - \dfrac{9.8 \text{ m/s}^2}{2[(31.9 \text{ m/s})(\cos 30°)]^2}x\right] = 0.58x - 0.0064x^2$

(d) Using a graphing utility, we can compare the two trajectories, which are shown in **Figure 4.16**.

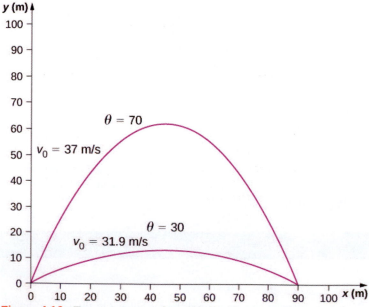

Figure 4.16 Two trajectories of a golf ball with a range of 90 m. The impact points of both are at the same level as the launch point.

Significance

The initial speed for the shot at $70°$ is greater than the initial speed of the shot at $30°$. Note from **Figure 4.16** that two projectiles launched at the same speed but at different angles have the same range if the launch angles add to $90°$. The launch angles in this example add to give a number greater than $90°$. Thus, the shot at $70°$ has to have a greater launch speed to reach 90 m, otherwise it would land at a shorter distance.

 4.4   Check Your Understanding If the two golf shots in **Example 4.9** were launched at the same speed, which shot would have the greatest range?

When we speak of the range of a projectile on level ground, we assume R is very small compared with the circumference of Earth. If, however, the range is large, Earth curves away below the projectile and the acceleration resulting from gravity changes direction along the path. The range is larger than predicted by the range equation given earlier because the projectile has farther to fall than it would on level ground, as shown in **Figure 4.17**, which is based on a drawing in Newton's *Principia*. If the initial speed is great enough, the projectile goes into orbit. Earth's surface drops 5 m every 8000 m. In 1 s an object falls 5 m without air resistance. Thus, if an object is given a horizontal velocity of 8000 m/s (or 18,000 mi/hr) near Earth's surface, it will go into orbit around the planet because the surface continuously falls away from the object. This is roughly the speed of the Space Shuttle in a low Earth orbit when it was operational, or any satellite in a low Earth orbit. These and other aspects of orbital motion, such as Earth's rotation, are covered in greater depth in **Gravitation**.

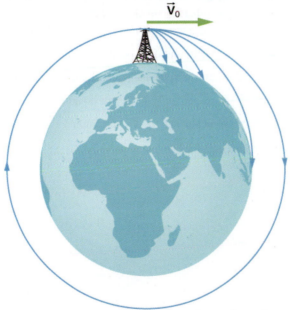

Figure 4.17   Projectile to satellite. In each case shown here, a projectile is launched from a very high tower to avoid air resistance. With increasing initial speed, the range increases and becomes longer than it would be on level ground because Earth curves away beneath its path. With a speed of 8000 m/s, orbit is achieved.

At **PhET Explorations: Projectile Motion (https://openstaxcollege.org/l/21phetpromot)** , learn about projectile motion in terms of the launch angle and initial velocity.

4.4 | Uniform Circular Motion

Uniform circular motion is a specific type of motion in which an object travels in a circle with a constant speed. For example, any point on a propeller spinning at a constant rate is executing uniform circular motion. Other examples are the second, minute, and hour hands of a watch. It is remarkable that points on these rotating objects are actually accelerating, although the rotation rate is a constant. To see this, we must analyze the motion in terms of vectors.

Centripetal Acceleration

In one-dimensional kinematics, objects with a constant speed have zero acceleration. However, in two- and three-dimensional kinematics, even if the speed is a constant, a particle can have acceleration if it moves along a curved trajectory such as a circle. In this case the velocity vector is changing, or $d\vec{v}/dt \neq 0$. This is shown in **Figure 4.18**. As the particle moves counterclockwise in time Δt on the circular path, its position vector moves from $\vec{r}(t)$ to $\vec{r}(t+\Delta t)$. The velocity vector has constant magnitude and is tangent to the path as it changes from $\vec{v}(t)$ to $\vec{v}(t+\Delta t)$, changing its direction only. Since the velocity vector $\vec{v}(t)$ is perpendicular to the position vector $\vec{r}(t)$, the triangles formed by the position vectors and $\Delta\vec{r}$, and the velocity vectors and $\Delta\vec{v}$ are similar. Furthermore, since $\left|\vec{r}(t)\right| = \left|\vec{r}(t+\Delta t)\right|$ and $\left|\vec{v}(t)\right| = \left|\vec{v}(t+\Delta t)\right|$, the two triangles are isosceles. From these facts we can make the assertion

$$\frac{\Delta v}{v} = \frac{\Delta r}{r} \text{ or } \Delta v = \frac{v}{r}\Delta r.$$

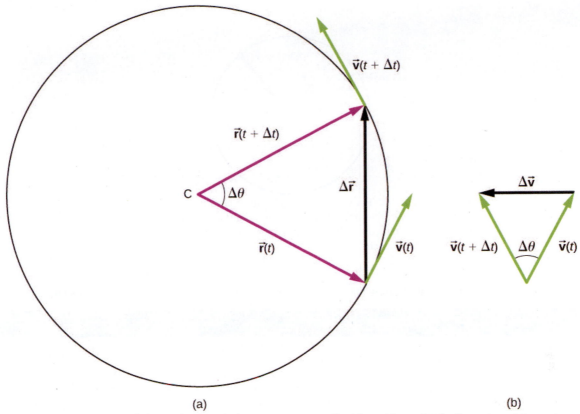

(a)   (b)

Figure 4.18 (a) A particle is moving in a circle at a constant speed, with position and velocity vectors at times t and $t + \Delta t$. (b) Velocity vectors forming a triangle. The two triangles in the figure are similar. The vector $\Delta \vec{v}$ points toward the center of the circle in the limit $\Delta t \to 0$.

We can find the magnitude of the acceleration from

$$a = \lim_{\Delta t \to 0}\left(\frac{\Delta v}{\Delta t}\right) = \frac{v}{r}\left(\lim_{\Delta t \to 0}\frac{\Delta r}{\Delta t}\right) = \frac{v^2}{r}.$$

The direction of the acceleration can also be found by noting that as Δt and therefore $\Delta \theta$ approach zero, the vector $\Delta \vec{v}$ approaches a direction perpendicular to $\vec{v}$. In the limit $\Delta t \to 0$, $\Delta \vec{v}$ is perpendicular to $\vec{v}$. Since $\vec{v}$ is tangent to the circle, the acceleration $d\vec{v}/dt$ points toward the center of the circle. Summarizing, a particle moving in a circle at a constant speed has an acceleration with magnitude

$$a_C = \frac{v^2}{r}. \tag{4.27}$$

The direction of the acceleration vector is toward the center of the circle (**Figure 4.19**). This is a radial acceleration and is called the **centripetal acceleration**, which is why we give it the subscript c. The word *centripetal* comes from the Latin words *centrum* (meaning "center") and *petere* (meaning "to seek"), and thus takes the meaning "center seeking."

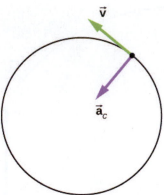

Figure 4.19 The centripetal acceleration vector points toward the center of the circular path of motion and is an acceleration in the radial direction. The velocity vector is also shown and is tangent to the circle.

Let's investigate some examples that illustrate the relative magnitudes of the velocity, radius, and centripetal acceleration.

Example 4.10

Creating an Acceleration of 1 *g*

A jet is flying at 134.1 m/s along a straight line and makes a turn along a circular path level with the ground. What does the radius of the circle have to be to produce a centripetal acceleration of 1 *g* on the pilot and jet toward the center of the circular trajectory?

Strategy

Given the speed of the jet, we can solve for the radius of the circle in the expression for the centripetal acceleration.

Solution

Set the centripetal acceleration equal to the acceleration of gravity: $9.8 \text{ m/s}^2 = v^2/r$.

Solving for the radius, we find

$$r = \frac{(134.1 \text{ m/s})^2}{9.8 \text{ m/s}^2} = 1835 \text{ m} = 1.835 \text{ km}.$$

Significance

To create a greater acceleration than *g* on the pilot, the jet would either have to decrease the radius of its circular trajectory or increase its speed on its existing trajectory or both.

 4.5 **Check Your Understanding** A flywheel has a radius of 20.0 cm. What is the speed of a point on the edge of the flywheel if it experiences a centripetal acceleration of 900.0 cm/s^2?

Centripetal acceleration can have a wide range of values, depending on the speed and radius of curvature of the circular path. Typical centripetal accelerations are given in the following table.

Object	Centripetal Acceleration (m/s^2 or factors of *g*)
Earth around the Sun	5.93×10^{-3}

Table 4.1 Typical Centripetal Accelerations

Object	Centripetal Acceleration (m/s² or factors of g)
Moon around the Earth	2.73×10^{-3}
Satellite in geosynchronous orbit	0.233
Outer edge of a CD when playing	5.78
Jet in a barrel roll	(2–3 g)
Roller coaster	(5 g)
Electron orbiting a proton in a simple Bohr model of the atom	9.0×10^{22}

Table 4.1 Typical Centripetal Accelerations

Equations of Motion for Uniform Circular Motion

A particle executing circular motion can be described by its position vector $\vec{r}(t)$. **Figure 4.20** shows a particle executing circular motion in a counterclockwise direction. As the particle moves on the circle, its position vector sweeps out the angle θ with the x-axis. Vector $\vec{r}(t)$ making an angle θ with the x-axis is shown with its components along the x- and y-axes. The magnitude of the position vector is $A = |\vec{r}(t)|$ and is also the radius of the circle, so that in terms of its components,

$$\vec{r}(t) = A\cos\omega t\ \hat{i} + A\sin\omega t\ \hat{j}. \tag{4.28}$$

Here, ω is a constant called the **angular frequency** of the particle. The angular frequency has units of radians (rad) per second and is simply the number of radians of angular measure through which the particle passes per second. The angle θ that the position vector has at any particular time is ωt.

If T is the period of motion, or the time to complete one revolution (2π rad), then

$$\omega = \frac{2\pi}{T}.$$

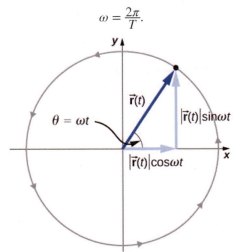

Figure 4.20 The position vector for a particle in circular motion with its components along the x- and y-axes. The particle moves counterclockwise. Angle θ is the angular frequency ω in radians per second multiplied by t.

Velocity and acceleration can be obtained from the position function by differentiation:

$$\vec{v}(t) = \frac{d\vec{r}(t)}{dt} = -A\omega\sin\omega t\,\hat{i} + A\omega\cos\omega t\,\hat{j}. \qquad (4.29)$$

It can be shown from **Figure 4.20** that the velocity vector is tangential to the circle at the location of the particle, with magnitude $A\omega$. Similarly, the acceleration vector is found by differentiating the velocity:

$$\vec{a}(t) = \frac{d\vec{v}(t)}{dt} = -A\omega^2\cos\omega t\,\hat{i} - A\omega^2\sin\omega t\,\hat{j}. \qquad (4.30)$$

From this equation we see that the acceleration vector has magnitude $A\omega^2$ and is directed opposite the position vector, toward the origin, because $\vec{a}(t) = -\omega^2\vec{r}(t)$.

Example 4.11

Circular Motion of a Proton

A proton has speed 5×10^6 m/s and is moving in a circle in the xy plane of radius $r = 0.175$ m. What is its position in the xy plane at time $t = 2.0\times10^{-7}$ s $= 200$ ns? At $t = 0$, the position of the proton is 0.175 m $\hat{i}$ and it circles counterclockwise. Sketch the trajectory.

Solution

From the given data, the proton has period and angular frequency:

$$T = \frac{2\pi r}{v} = \frac{2\pi(0.175\text{ m})}{5.0\times10^6\text{ m/s}} = 2.20\times10^{-7}\text{ s}$$

$$\omega = \frac{2\pi}{T} = \frac{2\pi}{2.20\times10^{-7}\text{ s}} = 2.856\times10^7\text{ rad/s}.$$

The position of the particle at $t = 2.0\times10^{-7}$ s with $A = 0.175$ m is

$$\vec{r}(2.0\times10^{-7}\text{ s}) = A\cos\omega(2.0\times10^{-7}\text{ s})\,\hat{i} + A\sin\omega(2.0\times10^{-7}\text{ s})\,\hat{j}\text{ m}$$

$$= 0.175\cos[(2.856\times10^7\text{ rad/s})(2.0\times10^{-7}\text{ s})]\,\hat{i}$$

$$+ 0.175\sin[(2.856\times10^7\text{ rad/s})(2.0\times10^{-7}\text{ s})]\,\hat{j}\text{ m}$$

$$= 0.175\cos(5.712\text{ rad})\,\hat{i} + 0.175\sin(5.712\text{ rad})\,\hat{j} = 0.147\,\hat{i} - 0.095\,\hat{j}\text{ m}.$$

From this result we see that the proton is located slightly below the x-axis. This is shown in **Figure 4.21**.

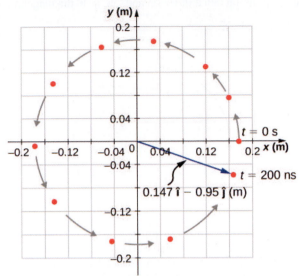

Figure 4.21 Position vector of the proton at $t = 2.0 \times 10^{-7}\,\text{s} = 200\,\text{ns}$. The trajectory of the proton is shown. The angle through which the proton travels along the circle is 5.712 rad, which a little less than one complete revolution.

Significance

We picked the initial position of the particle to be on the *x*-axis. This was completely arbitrary. If a different starting position were given, we would have a different final position at *t* = 200 ns.

Nonuniform Circular Motion

Circular motion does not have to be at a constant speed. A particle can travel in a circle and speed up or slow down, showing an acceleration in the direction of the motion.

In uniform circular motion, the particle executing circular motion has a constant speed and the circle is at a fixed radius. If the speed of the particle is changing as well, then we introduce an additional acceleration in the direction tangential to the circle. Such accelerations occur at a point on a top that is changing its spin rate, or any accelerating rotor. In **Displacement and Velocity Vectors** we showed that centripetal acceleration is the time rate of change of the direction of the velocity vector. If the speed of the particle is changing, then it has a **tangential acceleration** that is the time rate of change of the magnitude of the velocity:

$$a_{\text{T}} = \frac{d\left| \vec{v} \right|}{dt}. \qquad (4.31)$$

The direction of tangential acceleration is tangent to the circle whereas the direction of centripetal acceleration is radially inward toward the center of the circle. Thus, a particle in circular motion with a tangential acceleration has a **total acceleration** that is the vector sum of the centripetal and tangential accelerations:

$$\vec{a} = \vec{a}_{\text{C}} + \vec{a}_{\text{T}}. \qquad (4.32)$$

The acceleration vectors are shown in **Figure 4.22**. Note that the two acceleration vectors $\vec{a}_C$ and $\vec{a}_T$ are perpendicular to each other, with $\vec{a}_C$ in the radial direction and $\vec{a}_T$ in the tangential direction. The total acceleration $\vec{a}$ points at an angle between $\vec{a}_C$ and $\vec{a}_T$.

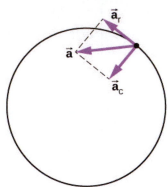

Figure 4.22 The centripetal acceleration points toward the center of the circle. The tangential acceleration is tangential to the circle at the particle's position. The total acceleration is the vector sum of the tangential and centripetal accelerations, which are perpendicular.

Example 4.12

Total Acceleration during Circular Motion

A particle moves in a circle of radius $r = 2.0$ m. During the time interval from $t = 1.5$ s to $t = 4.0$ s its speed varies with time according to

$$v(t) = c_1 - \frac{c_2}{t^2}, \quad c_1 = 4.0 \text{ m/s}, \ c_2 = 6.0 \text{ m} \cdot \text{s}.$$

What is the total acceleration of the particle at $t = 2.0$ s?

Strategy

We are given the speed of the particle and the radius of the circle, so we can calculate centripetal acceleration easily. The direction of the centripetal acceleration is toward the center of the circle. We find the magnitude of the tangential acceleration by taking the derivative with respect to time of $|v(t)|$ using **Equation 4.31** and evaluating it at $t = 2.0$ s. We use this and the magnitude of the centripetal acceleration to find the total acceleration.

Solution

Centripetal acceleration is

$$v(2.0\text{s}) = \left(4.0 - \frac{6.0}{(2.0)^2}\right)\text{m/s} = 2.5 \text{ m/s}$$

$$a_C = \frac{v^2}{r} = \frac{(2.5 \text{ m/s})^2}{2.0 \text{ m}} = 3.1 \text{ m/s}^2$$

directed toward the center of the circle. Tangential acceleration is

$$a_T = \left|\frac{d\vec{v}}{dt}\right| = \frac{2c_2}{t^3} = \frac{12.0}{(2.0)^3}\text{m/s}^2 = 1.5 \text{ m/s}^2.$$

Total acceleration is

$$|\vec{a}| = \sqrt{3.1^2 + 1.5^2}\text{m/s}^2 = 3.44 \text{ m/s}^2$$

and $\theta = \tan^{-1}\frac{3.1}{1.5} = 64°$ from the tangent to the circle. See **Figure 4.23**.

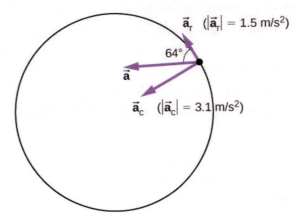

Figure 4.23 The tangential and centripetal acceleration vectors. The net acceleration $\vec{a}$ is the vector sum of the two accelerations.

Significance

The directions of centripetal and tangential accelerations can be described more conveniently in terms of a polar coordinate system, with unit vectors in the radial and tangential directions. This coordinate system, which is used for motion along curved paths, is discussed in detail later in the book.

4.5 | Relative Motion in One and Two Dimensions

Learning Objectives

By the end of this section, you will be able to:

- Explain the concept of reference frames.
- Write the position and velocity vector equations for relative motion.
- Draw the position and velocity vectors for relative motion.
- Analyze one-dimensional and two-dimensional relative motion problems using the position and velocity vector equations.

Motion does not happen in isolation. If you're riding in a train moving at 10 m/s east, this velocity is measured relative to the ground on which you're traveling. However, if another train passes you at 15 m/s east, your velocity relative to this other train is different from your velocity relative to the ground. Your velocity relative to the other train is 5 m/s west. To explore this idea further, we first need to establish some terminology.

Reference Frames

To discuss relative motion in one or more dimensions, we first introduce the concept of **reference frames**. When we say an object has a certain velocity, we must state it has a velocity with respect to a given reference frame. In most examples we have examined so far, this reference frame has been Earth. If you say a person is sitting in a train moving at 10 m/s east, then you imply the person on the train is moving relative to the surface of Earth at this velocity, and Earth is the reference frame. We can expand our view of the motion of the person on the train and say Earth is spinning in its orbit around the Sun, in which case the motion becomes more complicated. In this case, the solar system is the reference frame. In summary, all discussion of relative motion must define the reference frames involved. We now develop a method to refer to reference frames in relative motion.

Relative Motion in One Dimension

We introduce relative motion in one dimension first, because the velocity vectors simplify to having only two possible directions. Take the example of the person sitting in a train moving east. If we choose east as the positive direction and Earth as the reference frame, then we can write the velocity of the train with respect to the Earth as $\vec{v}_{TE} = 10 \text{ m/s } \hat{i}$ east, where the subscripts TE refer to train and Earth. Let's now say the person gets up out of her seat and walks toward the back of the train at 2 m/s. This tells us she has a velocity relative to the reference frame of the train. Since the person is walking west, in the negative direction, we write her velocity with respect to the train as $\vec{v}_{PT} = -2 \text{ m/s } \hat{i}$. We can add the two velocity vectors to find the velocity of the person with respect to Earth. This **relative velocity** is written as

$$\vec{v}_{PE} = \vec{v}_{PT} + \vec{v}_{TE}. \tag{4.33}$$

Note the ordering of the subscripts for the various reference frames in **Equation 4.33**. The subscripts for the coupling reference frame, which is the train, appear consecutively in the right-hand side of the equation. **Figure 4.24** shows the correct order of subscripts when forming the vector equation.

$$\vec{v}_{PE} = \vec{v}_{PT} + \vec{v}_{TE}$$

Figure 4.24 When constructing the vector equation, the subscripts for the coupling reference frame appear consecutively on the inside. The subscripts on the left-hand side of the equation are the same as the two outside subscripts on the right-hand side of the equation.

Adding the vectors, we find $\vec{v}_{PE} = 8 \text{ m/s } \hat{i}$, so the person is moving 8 m/s east with respect to Earth. Graphically, this is shown in **Figure 4.25**.

Figure 4.25 Velocity vectors of the train with respect to Earth, person with respect to the train, and person with respect to Earth.

Relative Velocity in Two Dimensions

We can now apply these concepts to describing motion in two dimensions. Consider a particle P and reference frames S and S', as shown in **Figure 4.26**. The position of the origin of S' as measured in S is $\vec{r}_{S'S}$, the position of P as measured in S' is $\vec{r}_{PS'}$, and the position of P as measured in S is $\vec{r}_{PS}$.

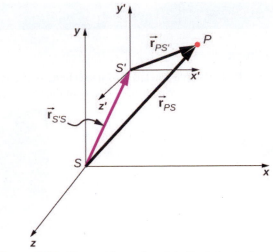

Figure 4.26 The positions of particle P relative to frames S and S' are $\vec{r}_{PS}$ and $\vec{r}_{PS'}$, respectively.

From **Figure 4.26** we see that

$$\vec{r}_{PS} = \vec{r}_{PS'} + \vec{r}_{S'S}. \qquad (4.34)$$

The relative velocities are the time derivatives of the position vectors. Therefore,

$$\vec{v}_{PS} = \vec{v}_{PS'} + \vec{v}_{S'S}. \qquad (4.35)$$

The velocity of a particle relative to S is equal to its velocity relative to S′ plus the velocity of S′ relative to S.

We can extend **Equation 4.35** to any number of reference frames. For particle P with velocities $\vec{v}_{PA}$, $\vec{v}_{PB}$, and $\vec{v}_{PC}$ in frames A, B, and C,

$$\vec{v}_{PC} = \vec{v}_{PA} + \vec{v}_{AB} + \vec{v}_{BC}. \qquad (4.36)$$

We can also see how the accelerations are related as observed in two reference frames by differentiating **Equation 4.35**:

$$\vec{a}_{PS} = \vec{a}_{PS'} + \vec{a}_{S'S}. \qquad (4.37)$$

We see that if the velocity of S' relative to S is a constant, then $\vec{a}_{S'S} = 0$ and

$$\vec{a}_{PS} = \vec{a}_{PS'}. \qquad (4.38)$$

This says the acceleration of a particle is the same as measured by two observers moving at a constant velocity relative to each other.

Example 4.13

Motion of a Car Relative to a Truck

A truck is traveling south at a speed of 70 km/h toward an intersection. A car is traveling east toward the intersection at a speed of 80 km/h (**Figure 4.27**). What is the velocity of the car relative to the truck?

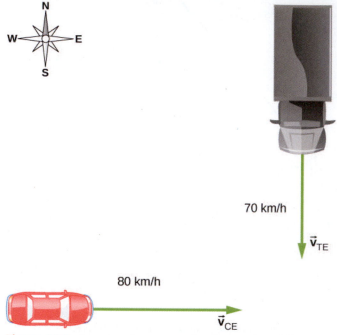

Figure 4.27   A car travels east toward an intersection while a truck travels south toward the same intersection.

Strategy

First, we must establish the reference frame common to both vehicles, which is Earth. Then, we write the velocities of each with respect to the reference frame of Earth, which enables us to form a vector equation that links the car, the truck, and Earth to solve for the velocity of the car with respect to the truck.

Solution

The velocity of the car with respect to Earth is $\vec{v}_{CE} = 80 \text{ km/h } \hat{i}$. The velocity of the truck with respect to Earth is $\vec{v}_{TE} = -70 \text{ km/h } \hat{j}$. Using the velocity addition rule, the relative motion equation we are seeking is

$$\vec{v}_{CT} = \vec{v}_{CE} + \vec{v}_{ET}.$$

Here, $\vec{v}_{CT}$ is the velocity of the car with respect to the truck, and Earth is the connecting reference frame. Since we have the velocity of the truck with respect to Earth, the negative of this vector is the velocity of Earth with respect to the truck: $\vec{v}_{ET} = -\vec{v}_{TE}$. The vector diagram of this equation is shown in **Figure 4.28**.

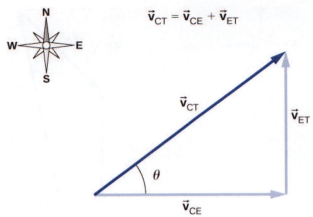

Figure 4.28 Vector diagram of the vector equation
$\vec{v}_{CT} = \vec{v}_{CE} + \vec{v}_{ET}$.

We can now solve for the velocity of the car with respect to the truck:

$$\left|\vec{v}_{CT}\right| = \sqrt{(80.0 \text{ km/h})^2 + (70.0 \text{ km/h})^2} = 106. \text{ km/h}$$

and

$$\theta = \tan^{-1}\left(\frac{70.0}{80.0}\right) = 41.2° \text{ north of east.}$$

Significance

Drawing a vector diagram showing the velocity vectors can help in understanding the relative velocity of the two objects.

4.6 Check Your Understanding A boat heads north in still water at 4.5 m/s directly across a river that is running east at 3.0 m/s. What is the velocity of the boat with respect to Earth?

Example 4.14

Flying a Plane in a Wind

A pilot must fly his plane due north to reach his destination. The plane can fly at 300 km/h in still air. A wind is blowing out of the northeast at 90 km/h. (a) What is the speed of the plane relative to the ground? (b) In what direction must the pilot head her plane to fly due north?

Strategy

The pilot must point her plane somewhat east of north to compensate for the wind velocity. We need to construct a vector equation that contains the velocity of the plane with respect to the ground, the velocity of the plane with respect to the air, and the velocity of the air with respect to the ground. Since these last two quantities are known, we can solve for the velocity of the plane with respect to the ground. We can graph the vectors and use this diagram to evaluate the magnitude of the plane's velocity with respect to the ground. The diagram will also tell us the angle the plane's velocity makes with north with respect to the air, which is the direction the pilot must head her plane.

Solution

The vector equation is $\vec{v}_{PG} = \vec{v}_{PA} + \vec{v}_{AG}$, where P = plane, A = air, and G = ground. From the geometry in **Figure 4.29**, we can solve easily for the magnitude of the velocity of the plane with respect to the ground and the angle of the plane's heading, θ.

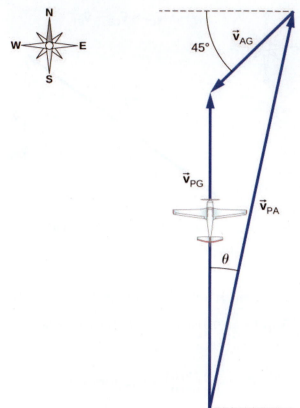

Figure 4.29 Vector diagram for **Equation 4.34** showing the vectors $\vec{v}_{PA}$, $\vec{v}_{AG}$, and $\vec{v}_{PG}$.

(a) Known quantities:

$$\left| \vec{v}_{PA} \right| = 300 \text{ km/h}$$

$$\left| \vec{v}_{AG} \right| = 90 \text{ km/h}$$

Substituting into the equation of motion, we obtain $\left| \vec{v}_{PG} \right| = 230 \text{ km/h}$.

(b) The angle $\theta = \tan^{-1} \frac{63.64}{300} = 12°$ east of north.

CHAPTER 4 REVIEW

KEY TERMS

acceleration vector instantaneous acceleration found by taking the derivative of the velocity function with respect to time in unit vector notation

angular frequency ω, rate of change of an angle with which an object that is moving on a circular path

centripetal acceleration component of acceleration of an object moving in a circle that is directed radially inward toward the center of the circle

displacement vector vector from the initial position to a final position on a trajectory of a particle

position vector vector from the origin of a chosen coordinate system to the position of a particle in two- or three-dimensional space

projectile motion motion of an object subject only to the acceleration of gravity

range maximum horizontal distance a projectile travels

reference frame coordinate system in which the position, velocity, and acceleration of an object at rest or moving is measured

relative velocity velocity of an object as observed from a particular reference frame, or the velocity of one reference frame with respect to another reference frame

tangential acceleration magnitude of which is the time rate of change of speed. Its direction is tangent to the circle.

time of flight elapsed time a projectile is in the air

total acceleration vector sum of centripetal and tangential accelerations

trajectory path of a projectile through the air

velocity vector vector that gives the instantaneous speed and direction of a particle; tangent to the trajectory

KEY EQUATIONS

Position vector	$\vec{r}(t) = x(t)\,\hat{i} + y(t)\,\hat{j} + z(t)\,\hat{k}$
Displacement vector	$\Delta\vec{r} = \vec{r}(t_2) - \vec{r}(t_1)$
Velocity vector	$\vec{v}(t) = \lim\limits_{\Delta t \to 0} \dfrac{\vec{r}(t + \Delta t) - \vec{r}(t)}{\Delta t} = \dfrac{d\vec{r}}{dt}$
Velocity in terms of components	$\vec{v}(t) = v_x(t)\,\hat{i} + v_y(t)\,\hat{j} + v_z(t)\,\hat{k}$
Velocity components	$v_x(t) = \dfrac{dx(t)}{dt} \quad v_y(t) = \dfrac{dy(t)}{dt} \quad v_z(t) = \dfrac{dz(t)}{dt}$
Average velocity	$\vec{v}_{avg} = \dfrac{\vec{r}(t_2) - \vec{r}(t_1)}{t_2 - t_1}$
Instantaneous acceleration	$\vec{a}(t) = \lim\limits_{t \to 0} \dfrac{\vec{v}(t + \Delta t) - \vec{v}(t)}{\Delta t} = \dfrac{d\vec{v}(t)}{dt}$
Instantaneous acceleration, component form	$\vec{a}(t) = \dfrac{dv_x(t)}{dt}\,\hat{i} + \dfrac{dv_y(t)}{dt}\,\hat{j} + \dfrac{dv_z(t)}{dt}\,\hat{k}$
Instantaneous acceleration as second derivatives of position	$\vec{a}(t) = \dfrac{d^2 x(t)}{dt^2}\,\hat{i} + \dfrac{d^2 y(t)}{dt^2}\,\hat{j} + \dfrac{d^2 z(t)}{dt^2}\,\hat{k}$

Time of flight	$T_{\text{tof}} = \dfrac{2(v_0 \sin\theta_0)}{g}$
Trajectory	$y = (\tan\theta_0)x - \left[\dfrac{g}{2(v_0 \cos\theta_0)^2}\right]x^2$
Range	$R = \dfrac{v_0^2 \sin 2\theta_0}{g}$
Centripetal acceleration	$a_C = \dfrac{v^2}{r}$
Position vector, uniform circular motion	$\overrightarrow{\mathbf{r}}(t) = A\cos\omega t\ \hat{\mathbf{i}} + A\sin\omega t\ \hat{\mathbf{j}}$
Velocity vector, uniform circular motion	$\overrightarrow{\mathbf{v}}(t) = \dfrac{d\overrightarrow{\mathbf{r}}(t)}{dt} = -A\omega \sin\omega t\ \hat{\mathbf{i}} + A\omega \cos\omega t\ \hat{\mathbf{j}}$
Acceleration vector, uniform circular motion	$\overrightarrow{\mathbf{a}}(t) = \dfrac{d\overrightarrow{\mathbf{v}}(t)}{dt} = -A\omega^2 \cos\omega t\ \hat{\mathbf{i}} - A\omega^2 \sin\omega t\ \hat{\mathbf{j}}$
Tangential acceleration	$a_T = \dfrac{d\left|\overrightarrow{\mathbf{v}}\right|}{dt}$
Total acceleration	$\overrightarrow{\mathbf{a}} = \overrightarrow{\mathbf{a}}_C + \overrightarrow{\mathbf{a}}_T$
Position vector in frame S is the position vector in frame S' plus the vector from the origin of S to the origin of S'	$\overrightarrow{\mathbf{r}}_{PS} = \overrightarrow{\mathbf{r}}_{PS'} + \overrightarrow{\mathbf{r}}_{S'S}$
Relative velocity equation connecting two reference frames	$\overrightarrow{\mathbf{v}}_{PS} = \overrightarrow{\mathbf{v}}_{PS'} + \overrightarrow{\mathbf{v}}_{S'S}$
Relative velocity equation connecting more than two reference frames	$\overrightarrow{\mathbf{v}}_{PC} = \overrightarrow{\mathbf{v}}_{PA} + \overrightarrow{\mathbf{v}}_{AB} + \overrightarrow{\mathbf{v}}_{BC}$
Relative acceleration equation	$\overrightarrow{\mathbf{a}}_{PS} = \overrightarrow{\mathbf{a}}_{PS'} + \overrightarrow{\mathbf{a}}_{S'S}$

SUMMARY

4.1 Displacement and Velocity Vectors

- The position function $\overrightarrow{\mathbf{r}}(t)$ gives the position as a function of time of a particle moving in two or three dimensions. Graphically, it is a vector from the origin of a chosen coordinate system to the point where the particle is located at a specific time.

- The displacement vector $\Delta\overrightarrow{\mathbf{r}}$ gives the shortest distance between any two points on the trajectory of a particle in two or three dimensions.

- Instantaneous velocity gives the speed and direction of a particle at a specific time on its trajectory in two or three dimensions, and is a vector in two and three dimensions.

- The velocity vector is tangent to the trajectory of the particle.

- Displacement $\overrightarrow{\mathbf{r}}(t)$ can be written as a vector sum of the one-dimensional displacements $\overrightarrow{x}(t)$, $\overrightarrow{y}(t)$, $\overrightarrow{z}(t)$ along the x, y, and z directions.

- Velocity $\overrightarrow{\mathbf{v}}(t)$ can be written as a vector sum of the one-dimensional velocities $v_x(t)$, $v_y(t)$, $v_z(t)$ along the x, y, and z directions.

- Motion in any given direction is independent of motion in a perpendicular direction.

4.2 Acceleration Vector

- In two and three dimensions, the acceleration vector can have an arbitrary direction and does not necessarily point along a given component of the velocity.

- The instantaneous acceleration is produced by a change in velocity taken over a very short (infinitesimal) time period. Instantaneous acceleration is a vector in two or three dimensions. It is found by taking the derivative of the velocity function with respect to time.

- In three dimensions, acceleration $\vec{a}(t)$ can be written as a vector sum of the one-dimensional accelerations $a_x(t)$, $a_y(t)$, and $a_z(t)$ along the x-, y-, and z-axes.

- The kinematic equations for constant acceleration can be written as the vector sum of the constant acceleration equations in the x, y, and z directions.

4.3 Projectile Motion

- Projectile motion is the motion of an object subject only to the acceleration of gravity, where the acceleration is constant, as near the surface of Earth.

- To solve projectile motion problems, we analyze the motion of the projectile in the horizontal and vertical directions using the one-dimensional kinematic equations for x and y.

- The time of flight of a projectile launched with initial vertical velocity v_{0y} on an even surface is given by

$$T_{tof} = \frac{2(v_0 \sin\theta)}{g}.$$

This equation is valid only when the projectile lands at the same elevation from which it was launched.

- The maximum horizontal distance traveled by a projectile is called the range. Again, the equation for range is valid only when the projectile lands at the same elevation from which it was launched.

4.4 Uniform Circular Motion

- Uniform circular motion is motion in a circle at constant speed.

- Centripetal acceleration $\vec{a}_C$ is the acceleration a particle must have to follow a circular path. Centripetal acceleration always points toward the center of rotation and has magnitude $a_C = v^2/r$.

- Nonuniform circular motion occurs when there is tangential acceleration of an object executing circular motion such that the speed of the object is changing. This acceleration is called tangential acceleration $\vec{a}_T$. The magnitude of tangential acceleration is the time rate of change of the magnitude of the velocity. The tangential acceleration vector is tangential to the circle, whereas the centripetal acceleration vector points radially inward toward the center of the circle. The total acceleration is the vector sum of tangential and centripetal accelerations.

- An object executing uniform circular motion can be described with equations of motion. The position vector of the object is $\vec{r}(t) = A\cos\omega t\,\hat{\mathbf{i}} + A\sin\omega t\,\hat{\mathbf{j}}$, where A is the magnitude $\left|\vec{r}(t)\right|$, which is also the radius of the circle, and ω is the angular frequency.

4.5 Relative Motion in One and Two Dimensions

- When analyzing motion of an object, the reference frame in terms of position, velocity, and acceleration needs to be specified.

- Relative velocity is the velocity of an object as observed from a particular reference frame, and it varies with the choice of reference frame.

- If S and S' are two reference frames moving relative to each other at a constant velocity, then the velocity of an object relative to S is equal to its velocity relative to S' plus the velocity of S' relative to S.

- If two reference frames are moving relative to each other at a constant velocity, then the accelerations of an object as observed in both reference frames are equal.

CONCEPTUAL QUESTIONS

4.1 Displacement and Velocity Vectors

1. What form does the trajectory of a particle have if the distance from any point A to point B is equal to the magnitude of the displacement from A to B?

2. Give an example of a trajectory in two or three dimensions caused by independent perpendicular motions.

3. If the instantaneous velocity is zero, what can be said about the slope of the position function?

4.2 Acceleration Vector

4. If the position function of a particle is a linear function of time, what can be said about its acceleration?

5. If an object has a constant x-component of the velocity and suddenly experiences an acceleration in the y direction, does the x-component of its velocity change?

6. If an object has a constant x-component of velocity and suddenly experiences an acceleration at an angle of $70°$ in the x direction, does the x-component of velocity change?

4.3 Projectile Motion

7. Answer the following questions for projectile motion on level ground assuming negligible air resistance, with the initial angle being neither $0°$ nor $90°$: (a) Is the velocity ever zero? (b) When is the velocity a minimum? A maximum? (c) Can the velocity ever be the same as the initial velocity at a time other than at $t = 0$? (d) Can the speed ever be the same as the initial speed at a time other than at $t = 0$?

8. Answer the following questions for projectile motion on level ground assuming negligible air resistance, with the initial angle being neither $0°$ nor $90°$: (a) Is the acceleration ever zero? (b) Is the vector **v** ever parallel or antiparallel to the vector **a**? (c) Is the vector **v** ever perpendicular to the vector **a**? If so, where is this located?

9. A dime is placed at the edge of a table so it hangs over slightly. A quarter is slid horizontally on the table surface perpendicular to the edge and hits the dime head on. Which coin hits the ground first?

4.4 Uniform Circular Motion

10. Can centripetal acceleration change the speed of a particle undergoing circular motion?

11. Can tangential acceleration change the speed of a particle undergoing circular motion?

4.5 Relative Motion in One and Two Dimensions

12. What frame or frames of reference do you use instinctively when driving a car? When flying in a commercial jet?

13. A basketball player dribbling down the court usually keeps his eyes fixed on the players around him. He is moving fast. Why doesn't he need to keep his eyes on the ball?

14. If someone is riding in the back of a pickup truck and throws a softball straight backward, is it possible for the ball to fall straight down as viewed by a person standing at the side of the road? Under what condition would this occur? How would the motion of the ball appear to the person who threw it?

15. The hat of a jogger running at constant velocity falls off the back of his head. Draw a sketch showing the path of the hat in the jogger's frame of reference. Draw its path as viewed by a stationary observer. Neglect air resistance.

16. A clod of dirt falls from the bed of a moving truck. It strikes the ground directly below the end of the truck. (a) What is the direction of its velocity relative to the truck just before it hits? (b) Is this the same as the direction of its velocity relative to ground just before it hits? Explain your answers.

PROBLEMS

4.1 Displacement and Velocity Vectors

17. The coordinates of a particle in a rectangular coordinate system are (1.0, –4.0, 6.0). What is the position vector of the particle?

18. The position of a particle changes from $\vec{r}_1 = (2.0\,\hat{i} + 3.0\,\hat{j})$cm to $\vec{r}_2 = (-4.0\,\hat{i} + 3.0\,\hat{j})$ cm. What is the particle's displacement?

19. The 18th hole at Pebble Beach Golf Course is a dogleg to the left of length 496.0 m. The fairway off the tee is taken to be the x direction. A golfer hits his tee shot a distance of 300.0 m, corresponding to a displacement $\Delta\vec{r}_1 = 300.0\,\text{m}\,\hat{i}$, and hits his second shot 189.0 m with a displacement $\Delta\vec{r}_2 = 172.0\,\text{m}\,\hat{i} + 80.3\,\text{m}\,\hat{j}$. What is the final displacement of the golf ball from the tee?

20. A bird flies straight northeast a distance of 95.0 km for 3.0 h. With the x-axis due east and the y-axis due north, what is the displacement in unit vector notation for the bird? What is the average velocity for the trip?

21. A cyclist rides 5.0 km due east, then 10.0 km $20°$ west of north. From this point she rides 8.0 km due west. What is the final displacement from where the cyclist started?

22. New York Rangers defenseman Daniel Girardi stands at the goal and passes a hockey puck 20 m and $45°$ from straight down the ice to left wing Chris Kreider waiting at the blue line. Kreider waits for Girardi to reach the blue line and passes the puck directly across the ice to him 10 m away. What is the final displacement of the puck? See the following figure.

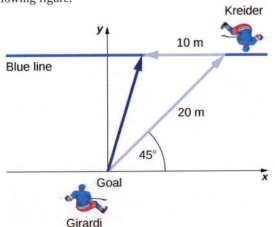

23. The position of a particle is $\vec{r}(t) = 4.0t^2\,\hat{i} - 3.0\,\hat{j} + 2.0t^3\,\hat{k}$ m. (a) What is the velocity of the particle at 0 s and at 1.0 s? (b) What is the average velocity between 0 s and 1.0 s?

24. Clay Matthews, a linebacker for the Green Bay Packers, can reach a speed of 10.0 m/s. At the start of a play, Matthews runs downfield at $45°$ with respect to the 50-yard line and covers 8.0 m in 1 s. He then runs straight down the field at $90°$ with respect to the 50-yard line for 12 m, with an elapsed time of 1.2 s. (a) What is Matthews' final displacement from the start of the play? (b) What is his average velocity?

25. The F-35B Lighting II is a short-takeoff and vertical landing fighter jet. If it does a vertical takeoff to 20.00-m height above the ground and then follows a flight path angled at $30°$ with respect to the ground for 20.00 km, what is the final displacement?

4.2 Acceleration Vector

26. The position of a particle is $\vec{r}(t) = (3.0t^2\,\hat{i} + 5.0\,\hat{j} - 6.0t\,\hat{k})$ m. (a) Determine its velocity and acceleration as functions of time. (b) What are its velocity and acceleration at time $t = 0$?

27. A particle's acceleration is $(4.0\,\hat{i} + 3.0\,\hat{j})$m/s^2. At $t = 0$, its position and velocity are zero. (a) What are the particle's position and velocity as functions of time? (b) Find the equation of the path of the particle. Draw the x- and y-axes and sketch the trajectory of the particle.

28. A boat leaves the dock at $t = 0$ and heads out into a lake with an acceleration of 2.0 m/s$^2\,\hat{i}$. A strong wind is pushing the boat, giving it an additional velocity of 2.0 m/s $\hat{i} + 1.0$ m/s $\hat{j}$. (a) What is the velocity of the boat at $t = 10$ s? (b) What is the position of the boat at $t = 10$s? Draw a sketch of the boat's trajectory and position at $t = 10$ s, showing the x- and y-axes.

29. The position of a particle for $t > 0$ is given by $\vec{r}(t) = (3.0t^2\,\hat{i} - 7.0t^3\,\hat{j} - 5.0t^{-2}\,\hat{k})$ m. (a) What is the velocity as a function of time? (b) What is the acceleration as a function of time? (c) What is the particle's velocity at $t = 2.0$ s? (d) What is its speed at $t = 1.0$ s and $t = 3.0$ s? (e) What is the average velocity between $t = 1.0$ s and $t = 2.0$ s?

30. The acceleration of a particle is a constant. At $t = 0$ the velocity of the particle is $(10\,\hat{\mathbf{i}} + 20\,\hat{\mathbf{j}})$ m/s. At $t = 4$ s the velocity is $10\,\hat{\mathbf{j}}$ m/s. (a) What is the particle's acceleration? (b) How do the position and velocity vary with time? Assume the particle is initially at the origin.

31. A particle has a position function $\vec{\mathbf{r}}(t) = \cos(1.0t)\,\hat{\mathbf{i}} + \sin(1.0t)\,\hat{\mathbf{j}} + t\,\hat{\mathbf{k}},$ where the arguments of the cosine and sine functions are in radians. (a) What is the velocity vector? (b) What is the acceleration vector?

32. A Lockheed Martin F-35 II Lighting jet takes off from an aircraft carrier with a runway length of 90 m and a takeoff speed 70 m/s at the end of the runway. Jets are catapulted into airspace from the deck of an aircraft carrier with two sources of propulsion: the jet propulsion and the catapult. At the point of leaving the deck of the aircraft carrier, the F-35's acceleration decreases to a constant acceleration of $5.0\,\text{m/s}^2$ at $30°$ with respect to the horizontal. (a) What is the initial acceleration of the F-35 on the deck of the aircraft carrier to make it airborne? (b) Write the position and velocity of the F-35 in unit vector notation from the point it leaves the deck of the aircraft carrier. (c) At what altitude is the fighter 5.0 s after it leaves the deck of the aircraft carrier? (d) What is its velocity and speed at this time? (e) How far has it traveled horizontally?

4.3 Projectile Motion

33. A bullet is shot horizontally from shoulder height (1.5 m) with an initial speed 200 m/s. (a) How much time elapses before the bullet hits the ground? (b) How far does the bullet travel horizontally?

34. A marble rolls off a tabletop 1.0 m high and hits the floor at a point 3.0 m away from the table's edge in the horizontal direction. (a) How long is the marble in the air? (b) What is the speed of the marble when it leaves the table's edge? (c) What is its speed when it hits the floor?

35. A dart is thrown horizontally at a speed of 10 m/s at the bull's-eye of a dartboard 2.4 m away, as in the following figure. (a) How far below the intended target does the dart hit? (b) What does your answer tell you about how proficient dart players throw their darts?

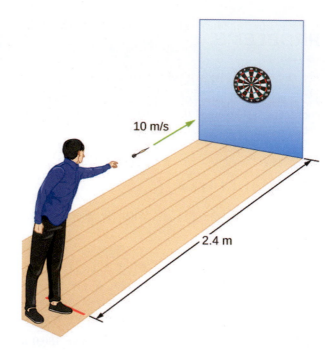

36. An airplane flying horizontally with a speed of 500 km/h at a height of 800 m drops a crate of supplies (see the following figure). If the parachute fails to open, how far in front of the release point does the crate hit the ground?

37. Suppose the airplane in the preceding problem fires a projectile horizontally in its direction of motion at a speed of 300 m/s relative to the plane. (a) How far in front of the release point does the projectile hit the ground? (b) What is its speed when it hits the ground?

38. A fastball pitcher can throw a baseball at a speed of 40 m/s (90 mi/h). (a) Assuming the pitcher can release the ball 16.7 m from home plate so the ball is moving horizontally, how long does it take the ball to reach home plate? (b) How far does the ball drop between the pitcher's hand and home plate?

39. A projectile is launched at an angle of $30°$ and lands 20 s later at the same height as it was launched. (a) What is the initial speed of the projectile? (b) What is the maximum altitude? (c) What is the range? (d) Calculate the displacement from the point of launch to the position on its trajectory at 15 s.

40. A basketball player shoots toward a basket 6.1 m away and 3.0 m above the floor. If the ball is released 1.8 m above the floor at an angle of $60°$ above the horizontal, what must the initial speed be if it were to go through the basket?

41. At a particular instant, a hot air balloon is 100 m in the air and descending at a constant speed of 2.0 m/s. At this exact instant, a girl throws a ball horizontally, relative to herself, with an initial speed of 20 m/s. When she lands, where will she find the ball? Ignore air resistance.

42. A man on a motorcycle traveling at a uniform speed of 10 m/s throws an empty can straight upward relative to himself with an initial speed of 3.0 m/s. Find the equation of the trajectory as seen by a police officer on the side of the road. Assume the initial position of the can is the point where it is thrown. Ignore air resistance.

43. An athlete can jump a distance of 8.0 m in the broad jump. What is the maximum distance the athlete can jump on the Moon, where the gravitational acceleration is one-sixth that of Earth?

44. The maximum horizontal distance a boy can throw a ball is 50 m. Assume he can throw with the same initial speed at all angles. How high does he throw the ball when he throws it straight upward?

45. A rock is thrown off a cliff at an angle of $53°$ with respect to the horizontal. The cliff is 100 m high. The initial speed of the rock is 30 m/s. (a) How high above the edge of the cliff does the rock rise? (b) How far has it moved horizontally when it is at maximum altitude? (c) How long after the release does it hit the ground? (d) What is the range of the rock? (e) What are the horizontal and vertical positions of the rock relative to the edge of the cliff at $t = 2.0$ s, $t = 4.0$ s, and $t = 6.0$ s?

46. Trying to escape his pursuers, a secret agent skis off a slope inclined at $30°$ below the horizontal at 60 km/h. To survive and land on the snow 100 m below, he must clear a gorge 60 m wide. Does he make it? Ignore air resistance.

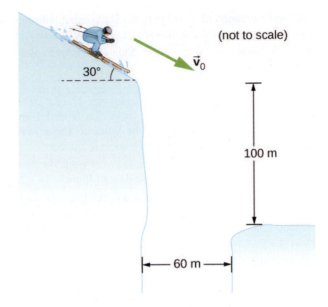

47. A golfer on a fairway is 70 m away from the green, which sits below the level of the fairway by 20 m. If the golfer hits the ball at an angle of $40°$ with an initial speed of 20 m/s, how close to the green does she come?

48. A projectile is shot at a hill, the base of which is 300 m away. The projectile is shot at $60°$ above the horizontal with an initial speed of 75 m/s. The hill can be approximated by a plane sloped at $20°$ to the horizontal. Relative to the coordinate system shown in the following figure, the equation of this straight line is $y = (\tan 20°)x - 109$. Where on the hill does the projectile land?

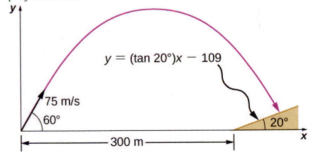

49. An astronaut on Mars kicks a soccer ball at an angle of $45°$ with an initial velocity of 15 m/s. If the acceleration of gravity on Mars is 3.7m/s^2, (a) what is the range of the soccer kick on a flat surface? (b) What would be the range of the same kick on the Moon, where gravity is one-sixth that of Earth?

50. Mike Powell holds the record for the long jump of 8.95 m, established in 1991. If he left the ground at an angle of $15°$, what was his initial speed?

51. MIT's robot cheetah can jump over obstacles 46 cm

high and has speed of 12.0 km/h. (a) If the robot launches itself at an angle of $60°$ at this speed, what is its maximum height? (b) What would the launch angle have to be to reach a height of 46 cm?

52. Mt. Asama, Japan, is an active volcano. In 2009, an eruption threw solid volcanic rocks that landed 1 km horizontally from the crater. If the volcanic rocks were launched at an angle of $40°$ with respect to the horizontal and landed 900 m below the crater, (a) what would be their initial velocity and (b) what is their time of flight?

53. Drew Brees of the New Orleans Saints can throw a football 23.0 m/s (50 mph). If he angles the throw at $10°$ from the horizontal, what distance does it go if it is to be caught at the same elevation as it was thrown?

54. The Lunar Roving Vehicle used in NASA's late *Apollo* missions reached an unofficial lunar land speed of 5.0 m/s by astronaut Eugene Cernan. If the rover was moving at this speed on a flat lunar surface and hit a small bump that projected it off the surface at an angle of $20°$, how long would it be "airborne" on the Moon?

55. A soccer goal is 2.44 m high. A player kicks the ball at a distance 10 m from the goal at an angle of $25°$. The ball hits the crossbar at the top of the goal. What is the initial speed of the soccer ball?

56. Olympus Mons on Mars is the largest volcano in the solar system, at a height of 25 km and with a radius of 312 km. If you are standing on the summit, with what initial velocity would you have to fire a projectile from a cannon horizontally to clear the volcano and land on the surface of Mars? Note that Mars has an acceleration of gravity of 3.7 m/s^2.

57. In 1999, Robbie Knievel was the first to jump the Grand Canyon on a motorcycle. At a narrow part of the canyon (69.0 m wide) and traveling 35.8 m/s off the takeoff ramp, he reached the other side. What was his launch angle?

58. You throw a baseball at an initial speed of 15.0 m/s at an angle of $30°$ with respect to the horizontal. What would the ball's initial speed have to be at $30°$ on a planet that has twice the acceleration of gravity as Earth to achieve the same range? Consider launch and impact on a horizontal surface.

59. Aaron Rodgers throws a football at 20.0 m/s to his wide receiver, who runs straight down the field at 9.4 m/s for 20.0 m. If Aaron throws the football when the wide receiver has reached 10.0 m, what angle does Aaron have to launch the ball so the receiver catches it at the 20.0 m

mark?

4.4 Uniform Circular Motion

60. A flywheel is rotating at 30 rev/s. What is the total angle, in radians, through which a point on the flywheel rotates in 40 s?

61. A particle travels in a circle of radius 10 m at a constant speed of 20 m/s. What is the magnitude of the acceleration?

62. Cam Newton of the Carolina Panthers throws a perfect football spiral at 8.0 rev/s. The radius of a pro football is 8.5 cm at the middle of the short side. What is the centripetal acceleration of the laces on the football?

63. A fairground ride spins its occupants inside a flying saucer-shaped container. If the horizontal circular path the riders follow has an 8.00-m radius, at how many revolutions per minute are the riders subjected to a centripetal acceleration equal to that of gravity?

64. A runner taking part in the 200-m dash must run around the end of a track that has a circular arc with a radius of curvature of 30.0 m. The runner starts the race at a constant speed. If she completes the 200-m dash in 23.2 s and runs at constant speed throughout the race, what is her centripetal acceleration as she runs the curved portion of the track?

65. What is the acceleration of Venus toward the Sun, assuming a circular orbit?

66. An experimental jet rocket travels around Earth along its equator just above its surface. At what speed must the jet travel if the magnitude of its acceleration is g?

67. A fan is rotating at a constant 360.0 rev/min. What is the magnitude of the acceleration of a point on one of its blades 10.0 cm from the axis of rotation?

68. A point located on the second hand of a large clock has a radial acceleration of 0.1cm/s^2. How far is the point from the axis of rotation of the second hand?

4.5 Relative Motion in One and Two Dimensions

69. The coordinate axes of the reference frame S' remain parallel to those of S, as S' moves away from S at a constant velocity $\vec{\mathbf{v}}_{S'} = (4.0\,\hat{\mathbf{i}} + 3.0\,\hat{\mathbf{j}} + 5.0\,\hat{\mathbf{k}})$ m/s. (a) If at time $t = 0$ the origins coincide, what is the position of the origin O' in the S frame as a function of time? (b) How is particle position for $\vec{r}(t)$ and $\vec{r}'(t)$, as measured in S and S', respectively, related? (c) What is the relationship between particle velocities $\vec{v}(t)$ and $\vec{v}'(t)$? (d) How are accelerations $\vec{a}(t)$ and $\vec{a}'(t)$ related?

70. The coordinate axes of the reference frame S' remain parallel to those of S, as S' moves away from S at a constant velocity $\vec{v}_{S'S} = (1.0\,\hat{i} + 2.0\,\hat{j} + 3.0\,\hat{k})t$ m/s . (a) If at time $t = 0$ the origins coincide, what is the position of origin O' in the S frame as a function of time? (b) How is particle position for $\vec{r}(t)$ and $\vec{r}'(t)$, as measured in S and S', respectively, related? (c) What is the relationship between particle velocities $\vec{v}(t)$ and $\vec{v}'(t)$? (d) How are accelerations $\vec{a}(t)$ and $\vec{a}'(t)$ related?

71. The velocity of a particle in reference frame A is $(2.0\,\hat{i} + 3.0\,\hat{j})$ m/s. The velocity of reference frame A with respect to reference frame B is $4.0\,\hat{k}$ m/s, and the velocity of reference frame B with respect to C is $2.0\,\hat{j}$ m/s. What is the velocity of the particle in reference frame C?

72. Raindrops fall vertically at 4.5 m/s relative to the earth. What does an observer in a car moving at 22.0 m/s in a straight line measure as the velocity of the raindrops?

73. A seagull can fly at a velocity of 9.00 m/s in still air. (a) If it takes the bird 20.0 min to travel 6.00 km straight

into an oncoming wind, what is the velocity of the wind? (b) If the bird turns around and flies with the wind, how long will it take the bird to return 6.00 km?

74. A ship sets sail from Rotterdam, heading due north at 7.00 m/s relative to the water. The local ocean current is 1.50 m/s in a direction $40.0°$ north of east. What is the velocity of the ship relative to Earth?

75. A boat can be rowed at 8.0 km/h in still water. (a) How much time is required to row 1.5 km downstream in a river moving 3.0 km/h relative to the shore? (b) How much time is required for the return trip? (c) In what direction must the boat be aimed to row straight across the river? (d) Suppose the river is 0.8 km wide. What is the velocity of the boat with respect to Earth and how much time is required to get to the opposite shore? (e) Suppose, instead, the boat is aimed straight across the river. How much time is required to get across and how far downstream is the boat when it reaches the opposite shore?

76. A small plane flies at 200 km/h in still air. If the wind blows directly out of the west at 50 km/h, (a) in what direction must the pilot head her plane to move directly north across land and (b) how long does it take her to reach a point 300 km directly north of her starting point?

77. A cyclist traveling southeast along a road at 15 km/h feels a wind blowing from the southwest at 25 km/h. To a stationary observer, what are the speed and direction of the wind?

78. A river is moving east at 4 m/s. A boat starts from the dock heading $30°$ north of west at 7 m/s. If the river is 1800 m wide, (a) what is the velocity of the boat with respect to Earth and (b) how long does it take the boat to cross the river?

ADDITIONAL PROBLEMS

79. A Formula One race car is traveling at 89.0 m/s along a straight track enters a turn on the race track with radius of curvature of 200.0 m. What centripetal acceleration must the car have to stay on the track?

80. A particle travels in a circular orbit of radius 10 m. Its speed is changing at a rate of $15.0\,\text{m/s}^2$ at an instant when its speed is 40.0 m/s. What is the magnitude of the acceleration of the particle?

81. The driver of a car moving at 90.0 km/h presses down on the brake as the car enters a circular curve of radius 150.0 m. If the speed of the car is decreasing at a rate

of 9.0 km/h each second, what is the magnitude of the acceleration of the car at the instant its speed is 60.0 km/h?

82. A race car entering the curved part of the track at the Daytona 500 drops its speed from 85.0 m/s to 80.0 m/s in 2.0 s. If the radius of the curved part of the track is 316.0 m, calculate the total acceleration of the race car at the beginning and ending of reduction of speed.

83. An elephant is located on Earth's surface at a latitude λ. Calculate the centripetal acceleration of the elephant resulting from the rotation of Earth around its polar axis. Express your answer in terms of λ, the radius R_E of

Earth, and time T for one rotation of Earth. Compare your answer with g for $\lambda = 40°$.

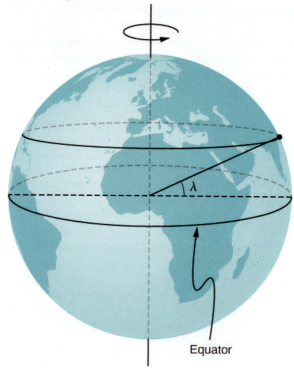

Equator

84. A proton in a synchrotron is moving in a circle of radius 1 km and increasing its speed by $v(t) = c_1 + c_2 t^2$, where $c_1 = 2.0 \times 10^5$ m/s, $c_2 = 10^5$ m/s³. (a) What is the proton's total acceleration at $t = 5.0$ s? (b) At what time does the expression for the velocity become unphysical?

85. A propeller blade at rest starts to rotate from $t = 0$ s to $t = 5.0$ s with a tangential acceleration of the tip of the blade at 3.00 m/s². The tip of the blade is 1.5 m from the axis of rotation. At $t = 5.0$ s, what is the total acceleration of the tip of the blade?

86. A particle is executing circular motion with a constant angular frequency of $\omega = 4.00$ rad/s. If time $t = 0$ corresponds to the position of the particle being located at $y = 0$ m and $x = 5$ m, (a) what is the position of the particle at $t = 10$ s? (b) What is its velocity at this time? (c) What is its acceleration?

87. A particle's centripetal acceleration is $a_C = 4.0$ m/s² at $t = 0$ s. It is executing uniform circular motion about an axis at a distance of 5.0 m. What is its velocity at $t = 10$ s?

88. A rod 3.0 m in length is rotating at 2.0 rev/s about an axis at one end. Compare the centripetal accelerations at radii of (a) 1.0 m, (b) 2.0 m, and (c) 3.0 m.

89. A particle located initially at $(1.5 \hat{j} + 4.0 \hat{k})$m undergoes a displacement of $(2.5 \hat{i} + 3.2 \hat{j} - 1.2 \hat{k})$ m. What is the final position of the particle?

90. The position of a particle is given by $\vec{r}(t) = (50 \text{ m/s})t \hat{i} - (4.9 \text{ m/s}^2)t^2 \hat{j}$. (a) What are the particle's velocity and acceleration as functions of time? (b) What are the initial conditions to produce the motion?

91. A spaceship is traveling at a constant velocity of $\vec{v}(t) = 250.0 \hat{i}$ m/s when its rockets fire, giving it an acceleration of $\vec{a}(t) = (3.0 \hat{i} + 4.0 \hat{k})$m/s². What is its velocity 5 s after the rockets fire?

92. A crossbow is aimed horizontally at a target 40 m away. The arrow hits 30 cm below the spot at which it was aimed. What is the initial velocity of the arrow?

93. A long jumper can jump a distance of 8.0 m when he takes off at an angle of $45°$ with respect to the horizontal. Assuming he can jump with the same initial speed at all angles, how much distance does he lose by taking off at $30°$?

94. On planet Arcon, the maximum horizontal range of a projectile launched at 10 m/s is 20 m. What is the acceleration of gravity on this planet?

95. A mountain biker encounters a jump on a race course that sends him into the air at $60°$ to the horizontal. If he lands at a horizontal distance of 45.0 m and 20 m below his launch point, what is his initial speed?

96. Which has the greater centripetal acceleration, a car with a speed of 15.0 m/s along a circular track of radius 100.0 m or a car with a speed of 12.0 m/s along a circular track of radius 75.0 m?

97. A geosynchronous satellite orbits Earth at a distance of 42,250.0 km and has a period of 1 day. What is the centripetal acceleration of the satellite?

98. Two speedboats are traveling at the same speed relative to the water in opposite directions in a moving river. An observer on the riverbank sees the boats moving at 4.0 m/s and 5.0 m/s. (a) What is the speed of the boats relative to the river? (b) How fast is the river moving relative to the shore?

CHALLENGE PROBLEMS

99. World's Longest Par 3. The tee of the world's longest par 3 sits atop South Africa's Hanglip Mountain at 400.0 m above the green and can only be reached by helicopter. The horizontal distance to the green is 359.0 m. Neglect air resistance and answer the following questions. (a) If a golfer launches a shot that is $40°$ with respect to the horizontal, what initial velocity must she give the ball? (b) What is the time to reach the green?

100. When a field goal kicker kicks a football as hard as he can at $45°$ to the horizontal, the ball just clears the 3-m-high crossbar of the goalposts 45.7 m away. (a) What is the maximum speed the kicker can impart to the football? (b) In addition to clearing the crossbar, the football must be high enough in the air early during its flight to clear the reach of the onrushing defensive lineman. If the lineman is 4.6 m away and has a vertical reach of 2.5 m, can he block the 45.7-m field goal attempt? (c) What if the lineman is 1.0 m away?

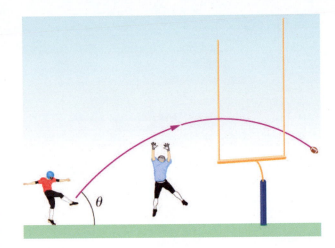

101. A truck is traveling east at 80 km/h. At an intersection 32 km ahead, a car is traveling north at 50 km/h. (a) How long after this moment will the vehicles be closest to each other? (b) How far apart will they be at that point?

5 | NEWTON'S LAWS OF MOTION

Figure 5.1 The Golden Gate Bridge, one of the greatest works of modern engineering, was the longest suspension bridge in the world in the year it opened, 1937. It is still among the 10 longest suspension bridges as of this writing. In designing and building a bridge, what physics must we consider? What forces act on the bridge? What forces keep the bridge from falling? How do the towers, cables, and ground interact to maintain stability?

Chapter Outline

5.1 Forces

5.2 Newton's First Law

5.3 Newton's Second Law

5.4 Mass and Weight

5.5 Newton's Third Law

5.6 Common Forces

5.7 Drawing Free-Body Diagrams

Introduction

When you drive across a bridge, you expect it to remain stable. You also expect to speed up or slow your car in response to traffic changes. In both cases, you deal with forces. The forces on the bridge are in equilibrium, so it stays in place. In contrast, the force produced by your car engine causes a change in motion. Isaac Newton discovered the laws of motion that describe these situations.

Forces affect every moment of your life. Your body is held to Earth by force and held together by the forces of charged particles. When you open a door, walk down a street, lift your fork, or touch a baby's face, you are applying forces. Zooming in deeper, your body's atoms are held together by electrical forces, and the core of the atom, called the nucleus, is held together by the strongest force we know—strong nuclear force.

5.1 | Forces

The study of motion is called *kinematics*, but kinematics only describes the way objects move—their velocity and their acceleration. **Dynamics** is the study of how forces affect the motion of objects and systems. It considers the causes of motion of objects and systems of interest, where a system is anything being analyzed. The foundation of dynamics are the laws of motion stated by Isaac Newton (1642–1727). These laws provide an example of the breadth and simplicity of principles under which nature functions. They are also universal laws in that they apply to situations on Earth and in space.

Newton's laws of motion were just one part of the monumental work that has made him legendary (**Figure 5.2**). The development of Newton's laws marks the transition from the Renaissance to the modern era. Not until the advent of modern physics was it discovered that Newton's laws produce a good description of motion only when the objects are moving at speeds much less than the speed of light and when those objects are larger than the size of most molecules (about 10^{-9} m in diameter). These constraints define the realm of Newtonian mechanics. At the beginning of the twentieth century, Albert Einstein (1879–1955) developed the theory of relativity and, along with many other scientists, quantum mechanics. Quantum mechanics does not have the constraints present in Newtonian physics. All of the situations we consider in this chapter, and all those preceding the introduction of relativity in **Relativity (http://cnx.org/content/m58555/latest/)** , are in the realm of Newtonian physics.

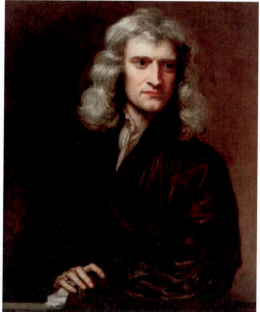

Figure 5.2　Isaac Newton (1642–1727) published his amazing work, *Philosophiae Naturalis Principia Mathematica*, in 1687. It proposed scientific laws that still apply today to describe the motion of objects (the laws of motion). Newton also discovered the law of gravity, invented calculus, and made great contributions to the theories of light and color.

Working Definition of Force

Dynamics is the study of the forces that cause objects and systems to move. To understand this, we need a working definition of force. An intuitive definition of **force**—that is, a push or a pull—is a good place to start. We know that a push or a pull has both magnitude and direction (therefore, it is a vector quantity), so we can define force as the push or pull on an object with a specific magnitude and direction. Force can be represented by vectors or expressed as a multiple of a standard force.

The push or pull on an object can vary considerably in either magnitude or direction. For example, a cannon exerts a strong force on a cannonball that is launched into the air. In contrast, Earth exerts only a tiny downward pull on a flea. Our everyday experiences also give us a good idea of how multiple forces add. If two people push in different directions on a third person, as illustrated in **Figure 5.3**, we might expect the total force to be in the direction shown. Since force is a vector, it adds just like other vectors. Forces, like other vectors, are represented by arrows and can be added using the familiar head-to-tail method or trigonometric methods. These ideas were developed in **Vectors**.

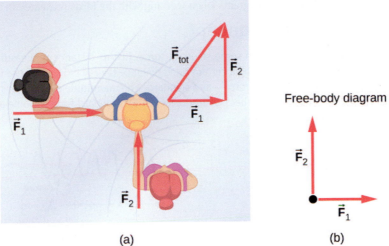

Figure 5.3 (a) An overhead view of two ice skaters pushing on a third skater. Forces are vectors and add like other vectors, so the total force on the third skater is in the direction shown. (b) A free-body diagram representing the forces acting on the third skater.

Figure 5.3(b) is our first example of a **free-body diagram**, which is a sketch showing all external forces acting on an object or system. The object or system is represented by a single isolated point (or free body), and only those forces acting *on* it that originate outside of the object or system—that is, **external forces**—are shown. (These forces are the only ones shown because only external forces acting on the free body affect its motion. We can ignore any internal forces within the body.) The forces are represented by vectors extending outward from the free body.

Free-body diagrams are useful in analyzing forces acting on an object or system, and are employed extensively in the study and application of Newton's laws of motion. You will see them throughout this text and in all your studies of physics. The following steps briefly explain how a free-body diagram is created; we examine this strategy in more detail in **Drawing Free-Body Diagrams**.

Problem-Solving Strategy: Drawing Free-Body Diagrams

1. Draw the object under consideration. If you are treating the object as a particle, represent the object as a point. Place this point at the origin of an *xy*-coordinate system.

2. Include all forces that act on the object, representing these forces as vectors. However, do not include the net force on the object or the forces that the object exerts on its environment.

3. Resolve all force vectors into *x*- and *y*-components.

4. Draw a separate free-body diagram for each object in the problem.

We illustrate this strategy with two examples of free-body diagrams (**Figure 5.4**). The terms used in this figure are explained in more detail later in the chapter.

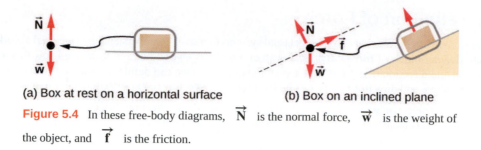

(a) Box at rest on a horizontal surface   (b) Box on an inclined plane

Figure 5.4 In these free-body diagrams, $\vec{N}$ is the normal force, $\vec{w}$ is the weight of the object, and $\vec{f}$ is the friction.

The steps given here are sufficient to guide you in this important problem-solving strategy. The final section of this chapter explains in more detail how to draw free-body diagrams when working with the ideas presented in this chapter.

Development of the Force Concept

A quantitative definition of force can be based on some standard force, just as distance is measured in units relative to a standard length. One possibility is to stretch a spring a certain fixed distance (**Figure 5.5**) and use the force it exerts to pull itself back to its relaxed shape—called a *restoring force*—as a standard. The magnitude of all other forces can be considered as multiples of this standard unit of force. Many other possibilities exist for standard forces. Some alternative definitions of force will be given later in this chapter.

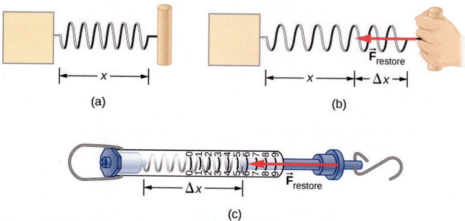

Figure 5.5 The force exerted by a stretched spring can be used as a standard unit of force. (a) This spring has a length x when undistorted. (b) When stretched a distance Δx, the spring exerts a restoring force $\vec{F}_{restore}$, which is reproducible. (c) A spring scale is one device that uses a spring to measure force. The force $\vec{F}_{restore}$ is exerted on whatever is attached to the hook. Here, this force has a magnitude of six units of the force standard being employed.

Let's analyze force more deeply. Suppose a physics student sits at a table, working diligently on his homework (**Figure 5.6**). What external forces act on him? Can we determine the origin of these forces?

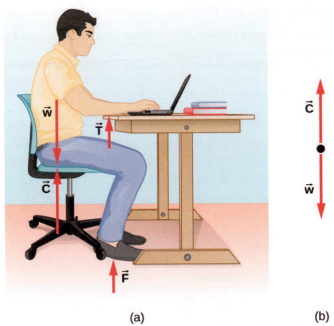

Figure 5.6 (a) The forces acting on the student are due to the chair, the table, the floor, and Earth's gravitational attraction. (b) In solving a problem involving the student, we may want to consider only the forces acting along the line running through his torso. A free-body diagram for this situation is shown.

In most situations, forces are grouped into two categories: *contact forces* and *field forces*. As you might guess, contact forces are due to direct physical contact between objects. For example, the student in **Figure 5.6** experiences the contact forces $\vec{C}$, $\vec{F}$, and $\vec{T}$, which are exerted by the chair on his posterior, the floor on his feet, and the table on his forearms, respectively. Field forces, however, act without the necessity of physical contact between objects. They depend on the presence of a "field" in the region of space surrounding the body under consideration. Since the student is in Earth's gravitational field, he feels a gravitational force $\vec{w}$; in other words, he has weight.

You can think of a field as a property of space that is detectable by the forces it exerts. Scientists think there are only four fundamental force fields in nature. These are the gravitational, electromagnetic, strong nuclear, and weak fields (we consider these four forces in nature later in this text). As noted for $\vec{w}$ in **Figure 5.6**, the gravitational field is responsible for the weight of a body. The forces of the electromagnetic field include those of static electricity and magnetism; they are also responsible for the attraction among atoms in bulk matter. Both the strong nuclear and the weak force fields are effective only over distances roughly equal to a length of scale no larger than an atomic nucleus (10^{-15} m). Their range is so small that neither field has influence in the macroscopic world of Newtonian mechanics.

Contact forces are fundamentally electromagnetic. While the elbow of the student in **Figure 5.6** is in contact with the tabletop, the atomic charges in his skin interact electromagnetically with the charges in the surface of the table. The net (total) result is the force $\vec{T}$. Similarly, when adhesive tape sticks to a piece of paper, the atoms of the tape are intermingled with those of the paper to cause a net electromagnetic force between the two objects. However, in the context of Newtonian mechanics, the electromagnetic origin of contact forces is not an important concern.

Vector Notation for Force

As previously discussed, force is a vector; it has both magnitude and direction. The SI unit of force is called the **newton** (abbreviated N), and 1 N is the force needed to accelerate an object with a mass of 1 kg at a rate of 1 m/s^2: $1 \text{ N} = 1 \text{ kg} \cdot \text{m/s}^2$. An easy way to remember the size of a newton is to imagine holding a small apple; it has a weight of about 1 N.

We can thus describe a two-dimensional force in the form $\vec{F} = a\,\hat{i} + b\,\hat{j}$ (the unit vectors $\hat{i}$ and $\hat{j}$ indicate the direction of these forces along the *x*-axis and the *y*-axis, respectively) and a three-dimensional force in the form $\vec{\mathbf{F}} = a\,\hat{\mathbf{i}} + b\,\hat{\mathbf{j}} + c\,\hat{\mathbf{k}}$. In **Figure 5.3**, let's suppose that ice skater 1, on the left side of the figure, pushes horizontally with a force of 30.0 N to the right; we represent this as $\vec{\mathbf{F}}_1 = 30.0\,\hat{\mathbf{i}}$ N. Similarly, if ice skater 2 pushes with a force of 40.0 N in the positive vertical direction shown, we would write $\vec{\mathbf{F}}_2 = 40.0\,\hat{\mathbf{j}}$ N. The resultant of the two forces causes a mass to accelerate—in this case, the third ice skater. This resultant is called the **net external force** $\vec{\mathbf{F}}_{net}$ and is found by taking the vector sum of all external forces acting on an object or system (thus, we can also represent net external force as $\sum \vec{\mathbf{F}}$):

$$\vec{\mathbf{F}}_{net} = \sum \vec{\mathbf{F}} = \vec{\mathbf{F}}_1 + \vec{\mathbf{F}}_2 + \cdots \qquad (5.1)$$

This equation can be extended to any number of forces.

In this example, we have $\vec{\mathbf{F}}_{net} = \sum \vec{\mathbf{F}} = \vec{\mathbf{F}}_1 + \vec{\mathbf{F}}_2 = 30.0\,\hat{\mathbf{i}} + 40.0\,\hat{\mathbf{j}}$ N. The hypotenuse of the triangle shown in **Figure 5.3** is the resultant force, or net force. It is a vector. To find its magnitude (the size of the vector, without regard to direction), we use the rule given in **Vectors**, taking the square root of the sum of the squares of the components:

$$F_{net} = \sqrt{(30.0\,\text{N})^2 + (40.0\,\text{N})^2} = 50.0\,\text{N}.$$

The direction is given by

$$\theta = \tan^{-1}\left(\frac{F_2}{F_1}\right) = \tan^{-1}\left(\frac{40.0}{30.0}\right) = 53.1°,$$

measured from the positive *x*-axis, as shown in the free-body diagram in **Figure 5.3**(b).

Let's suppose the ice skaters now push the third ice skater with $\vec{\mathbf{F}}_1 = 3.0\,\hat{\mathbf{i}} + 8.0\,\hat{\mathbf{j}}$ N and $\vec{\mathbf{F}}_2 = 5.0\,\hat{\mathbf{i}} + 4.0\,\hat{\mathbf{j}}$ N. What is the resultant of these two forces? We must recognize that force is a vector; therefore, we must add using the rules for vector addition:

$$\vec{\mathbf{F}}_{net} = \vec{\mathbf{F}}_1 + \vec{\mathbf{F}}_2 = \left(3.0\,\hat{\mathbf{i}} + 8.0\,\hat{\mathbf{j}}\right) + \left(5.0\,\hat{\mathbf{i}} + 4.0\,\hat{\mathbf{j}}\right) = 8.0\,\hat{\mathbf{i}} + 12\,\hat{\mathbf{j}}\ \text{N}$$

 5.1 **Check Your Understanding** Find the magnitude and direction of the net force in the ice skater example just given.

 View this **interactive simulation (https://openstaxcollege.org/l/21addvectors)** to learn how to add vectors. Drag vectors onto a graph, change their length and angle, and sum them together. The magnitude, angle, and components of each vector can be displayed in several formats.

5.2 | Newton's First Law

Learning Objectives

By the end of the section, you will be able to:

- Describe Newton's first law of motion
- Recognize friction as an external force
- Define inertia
- Identify inertial reference frames
- Calculate equilibrium for a system

Experience suggests that an object at rest remains at rest if left alone and that an object in motion tends to slow down and stop unless some effort is made to keep it moving. However, **Newton's first law** gives a deeper explanation of this observation.

Newton's First Law of Motion

A body at rest remains at rest or, if in motion, remains in motion at constant velocity unless acted on by a net external force.

Note the repeated use of the verb "remains." We can think of this law as preserving the status quo of motion. Also note the expression "constant velocity;" this means that the object maintains a path along a straight line, since neither the magnitude nor the direction of the velocity vector changes. We can use **Figure 5.7** to consider the two parts of Newton's first law.

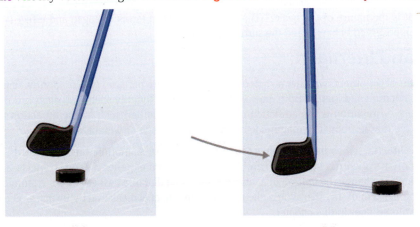

(a)   (b)

Figure 5.7 (a) A hockey puck is shown at rest; it remains at rest until an outside force such as a hockey stick changes its state of rest; (b) a hockey puck is shown in motion; it continues in motion in a straight line until an outside force causes it to change its state of motion. Although it is slick, an ice surface provides some friction that slows the puck.

Rather than contradicting our experience, Newton's first law says that there must be a cause for any change in velocity (a change in either magnitude or direction) to occur. This cause is a net external force, which we defined earlier in the chapter. An object sliding across a table or floor slows down due to the net force of friction acting on the object. If friction disappears, will the object still slow down?

The idea of cause and effect is crucial in accurately describing what happens in various situations. For example, consider what happens to an object sliding along a rough horizontal surface. The object quickly grinds to a halt. If we spray the surface with talcum powder to make the surface smoother, the object slides farther. If we make the surface even smoother by rubbing lubricating oil on it, the object slides farther yet. Extrapolating to a frictionless surface and ignoring air resistance, we can imagine the object sliding in a straight line indefinitely. Friction is thus the cause of slowing (consistent with Newton's first law). The object would not slow down if friction were eliminated.

Consider an air hockey table (**Figure 5.8**). When the air is turned off, the puck slides only a short distance before friction slows it to a stop. However, when the air is turned on, it creates a nearly frictionless surface, and the puck glides long distances without slowing down. Additionally, if we know enough about the friction, we can accurately predict how quickly the object slows down.

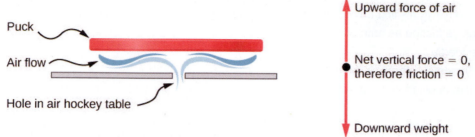

Figure 5.8   An air hockey table is useful in illustrating Newton's laws. When the air is off, friction quickly slows the puck; but when the air is on, it minimizes contact between the puck and the hockey table, and the puck glides far down the table.

Newton's first law is general and can be applied to anything from an object sliding on a table to a satellite in orbit to blood pumped from the heart. Experiments have verified that any change in velocity (speed or direction) must be caused by an external force. The idea of *generally applicable or universal laws* is important—it is a basic feature of all laws of physics. Identifying these laws is like recognizing patterns in nature from which further patterns can be discovered. The genius of Galileo, who first developed the idea for the first law of motion, and Newton, who clarified it, was to ask the fundamental question: "What is the cause?" Thinking in terms of cause and effect is fundamentally different from the typical ancient Greek approach, when questions such as "Why does a tiger have stripes?" would have been answered in Aristotelian fashion, such as "That is the nature of the beast." The ability to think in terms of cause and effect is the ability to make a connection between an observed behavior and the surrounding world.

Gravitation and Inertia

Regardless of the scale of an object, whether a molecule or a subatomic particle, two properties remain valid and thus of interest to physics: gravitation and inertia. Both are connected to mass. Roughly speaking, *mass* is a measure of the amount of matter in something. *Gravitation* is the attraction of one mass to another, such as the attraction between yourself and Earth that holds your feet to the floor. The magnitude of this attraction is your weight, and it is a force.

Mass is also related to **inertia**, the ability of an object to resist changes in its motion—in other words, to resist acceleration. Newton's first law is often called the **law of inertia**. As we know from experience, some objects have more inertia than others. It is more difficult to change the motion of a large boulder than that of a basketball, for example, because the boulder has more mass than the basketball. In other words, the inertia of an object is measured by its mass. The relationship between mass and weight is explored later in this chapter.

Inertial Reference Frames

Earlier, we stated Newton's first law as "A body at rest remains at rest or, if in motion, remains in motion at constant velocity unless acted on by a net external force." It can also be stated as "Every body remains in its state of uniform motion in a straight line unless it is compelled to change that state by forces acting on it." To Newton, "uniform motion in a straight line" meant constant velocity, which includes the case of zero velocity, or rest. Therefore, the first law says that the velocity of an object remains constant if the net force on it is zero.

Newton's first law is usually considered to be a statement about reference frames. It provides a method for identifying a special type of reference frame: the **inertial reference frame**. In principle, we can make the net force on a body zero. If its velocity relative to a given frame is constant, then that frame is said to be inertial. So by definition, an inertial reference frame is a reference frame in which Newton's first law is valid. Newton's first law applies to objects with constant velocity. From this fact, we can infer the following statement.

Are inertial frames common in nature? It turns out that well within experimental error, a reference frame at rest relative to the most distant, or "fixed," stars is inertial. All frames moving uniformly with respect to this fixed-star frame are also inertial. For example, a nonrotating reference frame attached to the Sun is, for all practical purposes, inertial, because its velocity relative to the fixed stars does not vary by more than one part in 10^{10}. Earth accelerates relative to the fixed stars because it rotates on its axis and revolves around the Sun; hence, a reference frame attached to its surface is not inertial. For most problems, however, such a frame serves as a sufficiently accurate approximation to an inertial frame, because the acceleration of a point on Earth's surface relative to the fixed stars is rather small ($< 3.4 \times 10^{-2}$ m/s^2). Thus, unless indicated otherwise, we consider reference frames fixed on Earth to be inertial.

Finally, no particular inertial frame is more special than any other. As far as the laws of nature are concerned, all inertial frames are equivalent. In analyzing a problem, we choose one inertial frame over another simply on the basis of convenience.

Newton's First Law and Equilibrium

Newton's first law tells us about the equilibrium of a system, which is the state in which the forces on the system are balanced. Returning to **Forces** and the ice skaters in **Figure 5.3**, we know that the forces $\vec{F}_1$ and $\vec{F}_2$ combine to form a resultant force, or the net external force: $\vec{F}_R = \vec{F}_{net} = \vec{F}_1 + \vec{F}_2$. To create equilibrium, we require a balancing force that will produce a net force of zero. This force must be equal in magnitude but opposite in direction to $\vec{F}_R$, which means the vector must be $-\vec{F}_R$. Referring to the ice skaters, for which we found $\vec{F}_R$ to be $30.0 \hat{i} + 40.0 \hat{j}$ N, we can determine the balancing force by simply finding $-\vec{F}_R = -30.0 \hat{i} - 40.0 \hat{j}$ N. See the free-body diagram in **Figure 5.3**(b).

We can give Newton's first law in vector form:

$$\vec{v} = \text{constant when } \vec{F}_{net} = \vec{0} \text{ N.} \tag{5.2}$$

This equation says that a net force of zero implies that the velocity $\vec{v}$ of the object is constant. (The word "constant" can indicate zero velocity.)

Newton's first law is deceptively simple. If a car is at rest, the only forces acting on the car are weight and the contact force of the pavement pushing up on the car (**Figure 5.9**). It is easy to understand that a nonzero net force is required to change the state of motion of the car. As a car moves with constant velocity, the friction force propels the car forward and opposes the drag force against it.

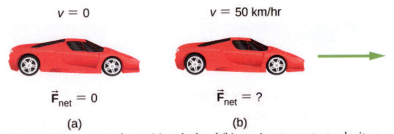

Figure 5.9 A car is shown (a) parked and (b) moving at constant velocity. How do Newton's laws apply to the parked car? What does the knowledge that the car is moving at constant velocity tell us about the net horizontal force on the car?

Example 5.1

When Does Newton's First Law Apply to Your Car?

Newton's laws can be applied to all physical processes involving force and motion, including something as mundane as driving a car.

(a) Your car is parked outside your house. Does Newton's first law apply in this situation? Why or why not?

(b) Your car moves at constant velocity down the street. Does Newton's first law apply in this situation? Why or why not?

Strategy

In (a), we are considering the first part of Newton's first law, dealing with a body at rest; in (b), we look at the second part of Newton's first law for a body in motion.

Solution

a. When your car is parked, all forces on the car must be balanced; the vector sum is 0 N. Thus, the net force is zero, and Newton's first law applies. The acceleration of the car is zero, and in this case, the velocity is also zero.

b. When your car is moving at constant velocity down the street, the net force must also be zero according to Newton's first law. The car's frictional force between the road and tires opposes the drag force on the car with the same magnitude, producing a net force of zero. The body continues in its state of constant velocity until the net force becomes nonzero. Realize that *a net force of zero means that an object is either at rest or moving with constant velocity, that is, it is not accelerating.* What do you suppose happens when the car accelerates? We explore this idea in the next section.

Significance

As this example shows, there are two kinds of equilibrium. In (a), the car is at rest; we say it is in *static equilibrium*. In (b), the forces on the car are balanced, but the car is moving; we say that it is in *dynamic equilibrium*. (We examine this idea in more detail in **Static Equilibrium and Elasticity**.) Again, it is possible for two (or more) forces to act on an object yet for the object to move. In addition, a net force of zero cannot produce acceleration.

 5.2   Check Your Understanding A skydiver opens his parachute, and shortly thereafter, he is moving at constant velocity. (a) What forces are acting on him? (b) Which force is bigger?

 Engage this **simulation (https://openstaxcollege.org/l/21forcemotion)** to predict, qualitatively, how an external force will affect the speed and direction of an object's motion. Explain the effects with the help of a free-body diagram. Use free-body diagrams to draw position, velocity, acceleration, and force graphs, and vice versa. Explain how the graphs relate to one another. Given a scenario or a graph, sketch all four graphs.

5.3 | Newton's Second Law

Learning Objectives

By the end of the section, you will be able to:

- Distinguish between external and internal forces
- Describe Newton's second law of motion
- Explain the dependence of acceleration on net force and mass

Newton's second law is closely related to his first law. It mathematically gives the cause-and-effect relationship between force and changes in motion. Newton's second law is quantitative and is used extensively to calculate what happens in situations involving a force. Before we can write down Newton's second law as a simple equation that gives the exact relationship of force, mass, and acceleration, we need to sharpen some ideas we mentioned earlier.

Force and Acceleration

First, what do we mean by a change in motion? The answer is that a change in motion is equivalent to a change in velocity. A change in velocity means, by definition, that there is acceleration. Newton's first law says that a net external force causes a change in motion; thus, we see that a *net external force causes nonzero acceleration*.

We defined external force in **Forces** as force acting on an object or system that originates outside of the object or system. Let's consider this concept further. An intuitive notion of *external* is correct—it is outside the system of interest. For example, in **Figure 5.10**(a), the system of interest is the car plus the person within it. The two forces exerted by the two students are external forces. In contrast, an internal force acts between elements of the system. Thus, the force the person in the car exerts to hang on to the steering wheel is an internal force between elements of the system of interest. Only external forces affect the motion of a system, according to Newton's first law. (The internal forces cancel each other out, as explained in the next section.) Therefore, we must define the boundaries of the system before we can determine which forces are external. Sometimes, the system is obvious, whereas at other times, identifying the boundaries of a system is more subtle. The concept of a system is fundamental to many areas of physics, as is the correct application of Newton's laws. This concept is revisited many times in the study of physics.

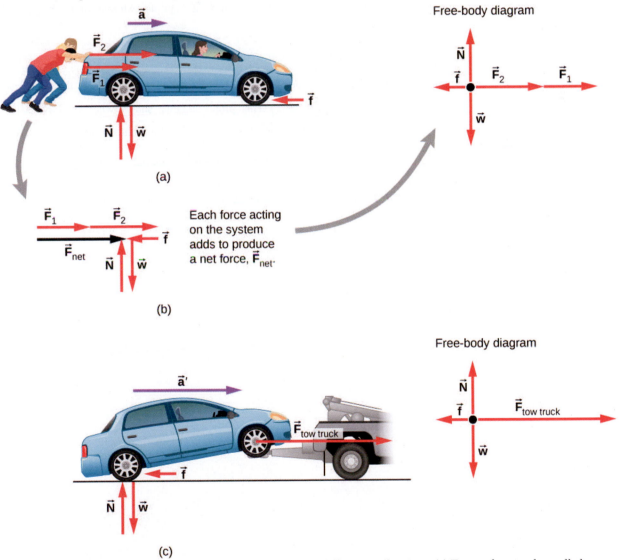

Figure 5.10 Different forces exerted on the same mass produce different accelerations. (a) Two students push a stalled car. All external forces acting on the car are shown. (b) The forces acting on the car are transferred to a coordinate plane (free-body diagram) for simpler analysis. (c) The tow truck can produce greater external force on the same mass, and thus greater acceleration.

From this example, you can see that different forces exerted on the same mass produce different accelerations. In **Figure 5.10**(a), the two students push a car with a driver in it. Arrows representing all external forces are shown. The system of interest is the car and its driver. The weight $\overrightarrow{\mathbf{w}}$ of the system and the support of the ground $\overrightarrow{\mathbf{N}}$ are also shown for completeness and are assumed to cancel (because there was no vertical motion and no imbalance of forces in the vertical direction to create a change in motion). The vector $\overrightarrow{\mathbf{f}}$ represents the friction acting on the car, and it acts to the left, opposing the motion of the car. (We discuss friction in more detail in the next chapter.) In **Figure 5.10**(b), all external forces acting on the system add together to produce the net force $\overrightarrow{\mathbf{F}}_{net}$. The free-body diagram shows all of the forces acting on the system of interest. The dot represents the center of mass of the system. Each force vector extends from this dot. Because there are two forces acting to the right, the vectors are shown collinearly. Finally, in **Figure 5.10**(c), a larger net external force produces a larger acceleration ($\overrightarrow{\mathbf{a}'} > \overrightarrow{\mathbf{a}}$) when the tow truck pulls the car.

It seems reasonable that acceleration would be directly proportional to and in the same direction as the net external force acting on a system. This assumption has been verified experimentally and is illustrated in **Figure 5.10**. To obtain an equation for Newton's second law, we first write the relationship of acceleration $\overrightarrow{\mathbf{a}}$ and net external force $\overrightarrow{\mathbf{F}}_{net}$ as the proportionality

$$\overrightarrow{\mathbf{a}} \propto \overrightarrow{\mathbf{F}}_{net}$$

where the symbol $\propto$ means "proportional to." (Recall from **Forces** that the net external force is the vector sum of all external forces and is sometimes indicated as $\sum \overrightarrow{\mathbf{F}}$.) This proportionality shows what we have said in words—acceleration is directly proportional to net external force. Once the system of interest is chosen, identify the external forces and ignore the internal ones. It is a tremendous simplification to disregard the numerous internal forces acting between objects within the system, such as muscular forces within the students' bodies, let alone the myriad forces between the atoms in the objects. Still, this simplification helps us solve some complex problems.

It also seems reasonable that acceleration should be inversely proportional to the mass of the system. In other words, the larger the mass (the inertia), the smaller the acceleration produced by a given force. As illustrated in **Figure 5.11**, the same net external force applied to a basketball produces a much smaller acceleration when it is applied to an SUV. The proportionality is written as

$$a \propto \frac{1}{m},$$

where m is the mass of the system and a is the magnitude of the acceleration. Experiments have shown that acceleration is exactly inversely proportional to mass, just as it is directly proportional to net external force.

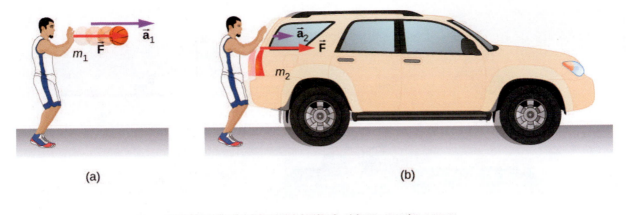

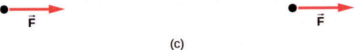

The free-body diagrams for both objects are the same.

(c)

Figure 5.11 The same force exerted on systems of different masses produces different accelerations. (a) A basketball player pushes on a basketball to make a pass. (Ignore the effect of gravity on the ball.) (b) The same player exerts an identical force on a stalled SUV and produces far less acceleration. (c) The free-body diagrams are identical, permitting direct comparison of the two situations. A series of patterns for free-body diagrams will emerge as you do more problems and learn how to draw them in **Drawing Free-Body Diagrams**.

It has been found that the acceleration of an object depends only on the net external force and the mass of the object. Combining the two proportionalities just given yields **Newton's second law**.

Newton's Second Law of Motion

The acceleration of a system is directly proportional to and in the same direction as the net external force acting on the system and is inversely proportion to its mass. In equation form, Newton's second law is

$$\vec{a} = \frac{\vec{F}_{net}}{m},$$

where $\vec{a}$ is the acceleration, $\vec{F}_{net}$ is the net force, and m is the mass. This is often written in the more familiar form

$$\vec{F}_{net} = \sum \vec{F} = m\vec{a}, \tag{5.3}$$

but the first equation gives more insight into what Newton's second law means. When only the magnitude of force and acceleration are considered, this equation can be written in the simpler scalar form:

$$F_{net} = ma. \tag{5.4}$$

The law is a cause-and-effect relationship among three quantities that is not simply based on their definitions. The validity of the second law is based on experimental verification. The free-body diagram, which you will learn to draw in **Drawing Free-Body Diagrams**, is the basis for writing Newton's second law.

Example 5.2

What Acceleration Can a Person Produce When Pushing a Lawn Mower?

Suppose that the net external force (push minus friction) exerted on a lawn mower is 51 N (about 11 lb.) parallel to the ground (**Figure 5.12**). The mass of the mower is 24 kg. What is its acceleration?

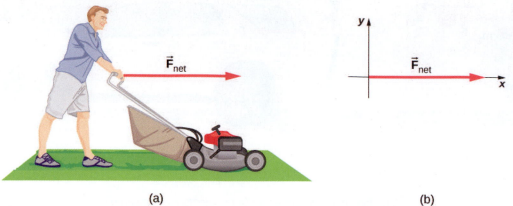

Figure 5.12 (a) The net force on a lawn mower is 51 N to the right. At what rate does the lawn mower accelerate to the right? (b) The free-body diagram for this problem is shown.

Strategy

This problem involves only motion in the horizontal direction; we are also given the net force, indicated by the single vector, but we can suppress the vector nature and concentrate on applying Newton's second law. Since F_{net} and m are given, the acceleration can be calculated directly from Newton's second law as $F_{net} = ma$.

Solution

The magnitude of the acceleration a is $a = F_{net}/m$. Entering known values gives

$$a = \frac{51\ \text{N}}{24\ \text{kg}}.$$

Substituting the unit of kilograms times meters per square second for newtons yields

$$a = \frac{51\ \text{kg} \cdot \text{m/s}^2}{24\ \text{kg}} = 2.1\ \text{m/s}^2.$$

Significance

The direction of the acceleration is the same direction as that of the net force, which is parallel to the ground. This is a result of the vector relationship expressed in Newton's second law, that is, the vector representing net force is the scalar multiple of the acceleration vector. There is no information given in this example about the individual external forces acting on the system, but we can say something about their relative magnitudes. For example, the force exerted by the person pushing the mower must be greater than the friction opposing the motion (since we know the mower moved forward), and the vertical forces must cancel because no acceleration occurs in the vertical direction (the mower is moving only horizontally). The acceleration found is small enough to be reasonable for a person pushing a mower. Such an effort would not last too long, because the person's top speed would soon be reached.

 5.3 Check Your Understanding At the time of its launch, the HMS *Titanic* was the most massive mobile object ever built, with a mass of 6.0×10^7 kg. If a force of 6 MN $(6 \times 10^6$ N$)$ was applied to the ship, what acceleration would it experience?

In the preceding example, we dealt with net force only for simplicity. However, several forces act on the lawn mower. The weight $\vec{\mathbf{w}}$ (discussed in detail in **Mass and Weight**) pulls down on the mower, toward the center of Earth; this produces a contact force on the ground. The ground must exert an upward force on the lawn mower, known as the normal force $\vec{\mathbf{N}}$, which we define in **Common Forces**. These forces are balanced and therefore do not produce vertical acceleration. In the next example, we show both of these forces. As you continue to solve problems using Newton's second law, be sure to show multiple forces.

Example 5.3

Which Force Is Bigger?

(a) The car shown in **Figure 5.13** is moving at a constant speed. Which force is bigger, $\overrightarrow{\mathbf{F}}_{\text{friction}}$ or $\overrightarrow{\mathbf{F}}_{\text{drag}}$? Explain.

(b) The same car is now accelerating to the right. Which force is bigger, $\overrightarrow{\mathbf{F}}_{\text{friction}}$ or $\overrightarrow{\mathbf{F}}_{\text{drag}}$? Explain.

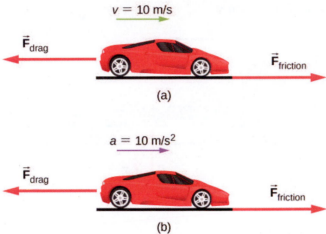

Figure 5.13 A car is shown (a) moving at constant speed and (b) accelerating. How do the forces acting on the car compare in each case? (a) What does the knowledge that the car is moving at constant velocity tell us about the net horizontal force on the car compared to the friction force? (b) What does the knowledge that the car is accelerating tell us about the horizontal force on the car compared to the friction force?

Strategy

We must consider Newton's first and second laws to analyze the situation. We need to decide which law applies; this, in turn, will tell us about the relationship between the forces.

Solution

a. The forces are equal. According to Newton's first law, if the net force is zero, the velocity is constant.

b. In this case, $\overrightarrow{\mathbf{F}}_{\text{friction}}$ must be larger than $\overrightarrow{\mathbf{F}}_{\text{drag}}$. According to Newton's second law, a net force is required to cause acceleration.

Significance

These questions may seem trivial, but they are commonly answered incorrectly. For a car or any other object to move, it must be accelerated from rest to the desired speed; this requires that the friction force be greater than the drag force. Once the car is moving at constant velocity, the net force must be zero; otherwise, the car will accelerate (gain speed). To solve problems involving Newton's laws, we must understand whether to apply Newton's first law (where $\sum \overrightarrow{\mathbf{F}} = \overrightarrow{\mathbf{0}}$) or Newton's second law (where $\sum \overrightarrow{\mathbf{F}}$ is not zero). This will be apparent as you see more examples and attempt to solve problems on your own.

Example 5.4

What Rocket Thrust Accelerates This Sled?

Before manned space flights, rocket sleds were used to test aircraft, missile equipment, and physiological effects on human subjects at high speeds. They consisted of a platform that was mounted on one or two rails and propelled by several rockets.

Calculate the magnitude of force exerted by each rocket, called its thrust T, for the four-rocket propulsion system shown in **Figure 5.14**. The sled's initial acceleration is 49 m/s^2, the mass of the system is 2100 kg, and the force of friction opposing the motion is 650 N.

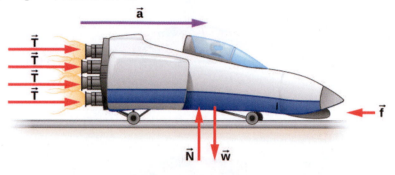

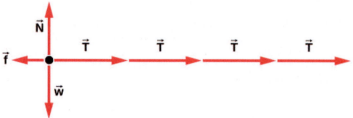

Free-body diagram

Figure 5.14 A sled experiences a rocket thrust that accelerates it to the right. Each rocket creates an identical thrust T. The system here is the sled, its rockets, and its rider, so none of the forces between these objects are considered. The arrow representing friction ($\vec{f}$) is drawn larger than scale.

Strategy

Although forces are acting both vertically and horizontally, we assume the vertical forces cancel because there is no vertical acceleration. This leaves us with only horizontal forces and a simpler one-dimensional problem. Directions are indicated with plus or minus signs, with right taken as the positive direction. See the free-body diagram in **Figure 5.14**.

Solution

Since acceleration, mass, and the force of friction are given, we start with Newton's second law and look for ways to find the thrust of the engines. We have defined the direction of the force and acceleration as acting "to the right," so we need to consider only the magnitudes of these quantities in the calculations. Hence we begin with

$$F_{\text{net}} = ma$$

where F_{net} is the net force along the horizontal direction. We can see from the figure that the engine thrusts add, whereas friction opposes the thrust. In equation form, the net external force is

$$F_{\text{net}} = 4T - f.$$

Substituting this into Newton's second law gives us

$$F_{\text{net}} = ma = 4T - f.$$

Using a little algebra, we solve for the total thrust $4T$:

$$4T = ma + f.$$

Substituting known values yields

$$4T = ma + f = (2100 \text{ kg})(49 \text{ m/s}^2) + 650 \text{ N}.$$

Therefore, the total thrust is

$$4T = 1.0 \times 10^5 \text{ N},$$

and the individual thrusts are

$$T = \frac{1.0 \times 10^5 \text{ N}}{4} = 2.5 \times 10^4 \text{ N}.$$

Significance

The numbers are quite large, so the result might surprise you. Experiments such as this were performed in the early 1960s to test the limits of human endurance, and the setup was designed to protect human subjects in jet fighter emergency ejections. Speeds of 1000 km/h were obtained, with accelerations of 45 g's. (Recall that g, acceleration due to gravity, is 9.80 m/s^2. When we say that acceleration is 45 g's, it is $45 \times 9.8 \text{ m/s}^2$, which is approximately 440 m/s^2.) Although living subjects are not used anymore, land speeds of 10,000 km/h have been obtained with a rocket sled.

In this example, as in the preceding one, the system of interest is obvious. We see in later examples that choosing the system of interest is crucial—and the choice is not always obvious.

Newton's second law is more than a definition; it is a relationship among acceleration, force, and mass. It can help us make predictions. Each of those physical quantities can be defined independently, so the second law tells us something basic and universal about nature.

 5.4   Check Your Understanding A 550-kg sports car collides with a 2200-kg truck, and during the collision, the net force on each vehicle is the force exerted by the other. If the magnitude of the truck's acceleration is 10 m/s^2, what is the magnitude of the sports car's acceleration?

Component Form of Newton's Second Law

We have developed Newton's second law and presented it as a vector equation in **Equation 5.3**. This vector equation can be written as three component equations:

$$\sum \vec{F}_x = m\vec{a}_x, \quad \sum \vec{F}_y = m\vec{a}_y, \quad \text{and} \quad \sum \vec{F}_z = m\vec{a}_z. \tag{5.5}$$

The second law is a description of how a body responds mechanically to its environment. The influence of the environment is the net force $\vec{F}_{\text{net}}$, the body's response is the acceleration $\vec{a}$, and the strength of the response is inversely proportional to the mass m. The larger the mass of an object, the smaller its response (its acceleration) to the influence of the environment (a given net force). Therefore, a body's mass is a measure of its inertia, as we explained in **Newton's First Law**.

Example 5.5

Force on a Soccer Ball

A 0.400-kg soccer ball is kicked across the field by a player; it undergoes acceleration given by $\vec{a} = 3.00\,\hat{i} + 7.00\,\hat{j}$ m/s^2. Find (a) the resultant force acting on the ball and (b) the magnitude and direction of the resultant force.

Strategy

The vectors in $\hat{i}$ and $\hat{j}$ format, which indicate force direction along the x-axis and the y-axis, respectively, are involved, so we apply Newton's second law in vector form.

Solution

a. We apply Newton's second law:

$$\vec{F}_{net} = m\,\vec{a} = (0.400\,\text{kg})\left(3.00\,\hat{i} + 7.00\,\hat{j}\ \text{m/s}^2\right) = 1.20\,\hat{i} + 2.80\,\hat{j}\ \text{N}.$$

b. Magnitude and direction are found using the components of $\vec{F}_{net}$:

$$F_{net} = \sqrt{(1.20\,\text{N})^2 + (2.80\,\text{N})^2} = 3.05\,\text{N} \text{ and } \theta = \tan^{-1}\left(\frac{2.80}{1.20}\right) = 66.8°.$$

Significance

We must remember that Newton's second law is a vector equation. In (a), we are multiplying a vector by a scalar to determine the net force in vector form. While the vector form gives a compact representation of the force vector, it does not tell us how "big" it is, or where it goes, in intuitive terms. In (b), we are determining the actual size (magnitude) of this force and the direction in which it travels.

Example 5.6

Mass of a Car

Find the mass of a car if a net force of $-600.0\,\hat{j}$ N produces an acceleration of $-0.2\,\hat{j}$ m/s^2.

Strategy

Vector division is not defined, so $m = \vec{F}_{net}/\vec{a}$ cannot be performed. However, mass m is a scalar, so we can use the scalar form of Newton's second law, $m = F_{net}/a$.

Solution

We use $m = F_{net}/a$ and substitute the magnitudes of the two vectors: $F_{net} = 600.0\,\text{N}$ and $a = 0.2\,\text{m/s}^2$. Therefore,

$$m = \frac{F_{net}}{a} = \frac{600.0\,\text{N}}{0.2\,\text{m/s}^2} = 3000\,\text{kg}.$$

Significance

Force and acceleration were given in the $\hat{i}$ and $\hat{j}$ format, but the answer, mass m, is a scalar and thus is not given in $\hat{i}$ and $\hat{j}$ form.

Example 5.7

Several Forces on a Particle

A particle of mass $m = 4.0 \, \text{kg}$ is acted upon by four forces of magnitudes. $F_1 = 10.0 \, \text{N}$, $F_2 = 40.0 \, \text{N}$, $F_3 = 5.0 \, \text{N}$, and $F_4 = 2.0 \, \text{N}$, with the directions as shown in the free-body diagram in **Figure 5.15**. What is the acceleration of the particle?

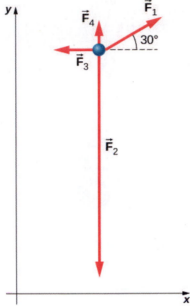

Figure 5.15 Four forces in the *xy*-plane are applied to a 4.0-kg particle.

Strategy

Because this is a two-dimensional problem, we must use a free-body diagram. First, $\vec{F}_1$ must be resolved into *x*- and *y*-components. We can then apply the second law in each direction.

Solution

We draw a free-body diagram as shown in **Figure 5.15**. Now we apply Newton's second law. We consider all vectors resolved into *x*- and *y*-components:

$$\sum F_x = ma_x \qquad\qquad \sum F_y = ma_y$$

$$F_{1x} - F_{3x} = ma_x \qquad\qquad F_{1y} + F_{4y} - F_{2y} = ma_y$$

$$F_1 \cos 30° - F_{3x} = ma_x \qquad\qquad F_1 \sin 30° + F_{4y} - F_{2y} = ma_y$$

$$(10.0 \, \text{N})(\cos 30°) - 5.0 \, \text{N} = (4.0 \, \text{kg})a_x \qquad (10.0 \, \text{N})(\sin 30°) + 2.0 \, \text{N} - 40.0 \, \text{N} = (4.0 \, \text{kg})a_y$$

$$a_x = 0.92 \, \text{m/s}^2. \qquad\qquad a_y = -8.3 \, \text{m/s}^2.$$

Thus, the net acceleration is

$$\vec{a} = \left(0.92 \, \hat{i} - 8.3 \, \hat{j}\right) \text{m/s}^2,$$

which is a vector of magnitude $8.4 \, \text{m/s}^2$ directed at $276°$ to the positive *x*-axis.

Significance

Numerous examples in everyday life can be found that involve three or more forces acting on a single object,

such as cables running from the Golden Gate Bridge or a football player being tackled by three defenders. We can
see that the solution of this example is just an extension of what we have already done.

 5.5   Check Your Understanding A car has forces acting on it, as shown below. The mass of the car is 1000.0
kg. The road is slick, so friction can be ignored. (a) What is the net force on the car? (b) What is the acceleration
of the car?

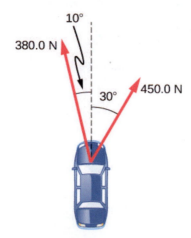

Newton's Second Law and Momentum

Newton actually stated his second law in terms of momentum: "The instantaneous rate at which a body's momentum
changes is equal to the net force acting on the body." ("Instantaneous rate" implies that the derivative is involved.) This can
be given by the vector equation

$$\vec{F}_{net} = \frac{d\vec{p}}{dt}. \tag{5.6}$$

This means that Newton's second law addresses the central question of motion: What causes a change in motion of an
object? Momentum was described by Newton as "quantity of motion," a way of combining both the velocity of an object
and its mass. We devote **Linear Momentum and Collisions** to the study of momentum.

For now, it is sufficient to define *momentum* $\vec{p}$ as the product of the mass of the object m and its velocity $\vec{v}$:

$$\vec{p} = m\vec{v} . \tag{5.7}$$

Since velocity is a vector, so is momentum.

It is easy to visualize momentum. A train moving at 10 m/s has more momentum than one that moves at 2 m/s. In everyday
life, we speak of one sports team as "having momentum" when they score points against the opposing team.

If we substitute **Equation 5.7** into **Equation 5.6**, we obtain

$$\vec{F}_{net} = \frac{d\vec{p}}{dt} = \frac{d(m\vec{v})}{dt}.$$

When m is constant, we have

$$\vec{F}_{net} = m\frac{d(\vec{v})}{dt} = m\vec{a} .$$

Thus, we see that the momentum form of Newton's second law reduces to the form given earlier in this section.

Explore the **forces at work (https://openstaxcollege.org/l/21forcesatwork)** when **pulling a cart (https://openstaxcollege.org/l/21pullacart)** or pushing a refrigerator, crate, or person. Create an **applied force (https://openstaxcollege.org/l/21forcemotion)** and see how it makes objects move. Put **an object on a ramp (https://openstaxcollege.org/l/21ramp)** and see how it affects its motion.

5.4 | Mass and Weight

Learning Objectives

By the end of the section, you will be able to:

- Explain the difference between mass and weight
- Explain why falling objects on Earth are never truly in free fall
- Describe the concept of weightlessness

Mass and weight are often used interchangeably in everyday conversation. For example, our medical records often show our weight in kilograms but never in the correct units of newtons. In physics, however, there is an important distinction. Weight is the pull of Earth on an object. It depends on the distance from the center of Earth. Unlike weight, mass does not vary with location. The mass of an object is the same on Earth, in orbit, or on the surface of the Moon.

Units of Force

The equation $F_{net} = ma$ is used to define net force in terms of mass, length, and time. As explained earlier, the SI unit of force is the newton. Since $F_{net} = ma$,

$$1\,\text{N} = 1\,\text{kg} \cdot \text{m/s}^2.$$

Although almost the entire world uses the newton for the unit of force, in the United States, the most familiar unit of force is the pound (lb), where 1 N = 0.225 lb. Thus, a 225-lb person weighs 1000 N.

Weight and Gravitational Force

When an object is dropped, it accelerates toward the center of Earth. Newton's second law says that a net force on an object is responsible for its acceleration. If air resistance is negligible, the net force on a falling object is the gravitational force, commonly called its **weight** $\vec{w}$, or its force due to gravity acting on an object of mass m. Weight can be denoted as a vector because it has a direction; *down* is, by definition, the direction of gravity, and hence, weight is a downward force. The magnitude of weight is denoted as w. Galileo was instrumental in showing that, in the absence of air resistance, all objects fall with the same acceleration g. Using Galileo's result and Newton's second law, we can derive an equation for weight.

Consider an object with mass m falling toward Earth. It experiences only the downward force of gravity, which is the weight $\vec{w}$. Newton's second law says that the magnitude of the net external force on an object is $\vec{F}_{net} = m\vec{a}$. We know that the acceleration of an object due to gravity is $\vec{g}$, or $\vec{a} = \vec{g}$. Substituting these into Newton's second law gives us the following equations.

Weight

The gravitational force on a mass is its weight. We can write this in vector form, where $\vec{w}$ is weight and m is mass, as

$$\vec{w} = m\vec{g}.$$ (5.8)

In scalar form, we can write

$$w = mg.$$ (5.9)

Since $g = 9.80 \text{ m/s}^2$ on Earth, the weight of a 1.00-kg object on Earth is 9.80 N:

$$w = mg = (1.00 \text{ kg})(9.80 \text{ m/s}^2) = 9.80 \text{ N}.$$

When the net external force on an object is its weight, we say that it is in **free fall**, that is, the only force acting on the object is gravity. However, when objects on Earth fall downward, they are never truly in free fall because there is always some upward resistance force from the air acting on the object.

Acceleration due to gravity g varies slightly over the surface of Earth, so the weight of an object depends on its location and is not an intrinsic property of the object. Weight varies dramatically if we leave Earth's surface. On the Moon, for example, acceleration due to gravity is only 1.67 m/s^2. A 1.0-kg mass thus has a weight of 9.8 N on Earth and only about 1.7 N on the Moon.

The broadest definition of weight in this sense is that the weight of an object is the gravitational force on it from the nearest large body, such as Earth, the Moon, or the Sun. This is the most common and useful definition of weight in physics. It differs dramatically, however, from the definition of weight used by NASA and the popular media in relation to space travel and exploration. When they speak of "weightlessness" and "microgravity," they are referring to the phenomenon we call "free fall" in physics. We use the preceding definition of weight, force $\vec{w}$ due to gravity acting on an object of mass m, and we make careful distinctions between free fall and actual weightlessness.

Be aware that weight and mass are different physical quantities, although they are closely related. Mass is an intrinsic property of an object: It is a quantity of matter. The quantity or amount of matter of an object is determined by the numbers of atoms and molecules of various types it contains. Because these numbers do not vary, in Newtonian physics, mass does not vary; therefore, its response to an applied force does not vary. In contrast, weight is the gravitational force acting on an object, so it does vary depending on gravity. For example, a person closer to the center of Earth, at a low elevation such as New Orleans, weighs slightly more than a person who is located in the higher elevation of Denver, even though they may have the same mass.

It is tempting to equate mass to weight, because most of our examples take place on Earth, where the weight of an object varies only a little with the location of the object. In addition, it is difficult to count and identify all of the atoms and molecules in an object, so mass is rarely determined in this manner. If we consider situations in which $\vec{g}$ is a constant on Earth, we see that weight $\vec{w}$ is directly proportional to mass m, since $\vec{w} = m\vec{g}$, that is, the more massive an object is, the more it weighs. Operationally, the masses of objects are determined by comparison with the standard kilogram, as we discussed in **Units and Measurement**. But by comparing an object on Earth with one on the Moon, we can easily see a variation in weight but not in mass. For instance, on Earth, a 5.0-kg object weighs 49 N; on the Moon, where g is 1.67 m/s^2, the object weighs 8.4 N. However, the mass of the object is still 5.0 kg on the Moon.

Example 5.8

Clearing a Field

A farmer is lifting some moderately heavy rocks from a field to plant crops. He lifts a stone that weighs 40.0 lb. (about 180 N). What force does he apply if the stone accelerates at a rate of 1.5 m/s^2?

Strategy

We were given the weight of the stone, which we use in finding the net force on the stone. However, we also need to know its mass to apply Newton's second law, so we must apply the equation for weight, $w = mg$, to determine the mass.

Solution

No forces act in the horizontal direction, so we can concentrate on vertical forces, as shown in the following free-body diagram. We label the acceleration to the side; technically, it is not part of the free-body diagram, but it helps to remind us that the object accelerates upward (so the net force is upward).

$$F = ?$$

$$\uparrow a = 1.5 \text{ m/s}^2$$

$$w = 180 \text{ N}$$

$$
\begin{aligned}
w &= mg \\
m &= \frac{w}{g} = \frac{180 \text{ N}}{9.8 \text{ m/s}^2} = 18 \text{ kg} \\
\sum F &= ma \\
F - w &= ma \\
F - 180 \text{ N} &= (18 \text{ kg})(1.5 \text{ m/s}^2) \\
F - 180 \text{ N} &= 27 \text{ N} \\
F &= 207 \text{ N} = 210 \text{ N to two significant figures}
\end{aligned}
$$

Significance

To apply Newton's second law as the primary equation in solving a problem, we sometimes have to rely on other equations, such as the one for weight or one of the kinematic equations, to complete the solution.

5.6 **Check Your Understanding** For **Example 5.8**, find the acceleration when the farmer's applied force is 230.0 N.

Can you avoid the boulder field and land safely just before your fuel runs out, as Neil Armstrong did in 1969? This **version of the classic video game (https://openstaxcollege.org/l/21lunarlander)** accurately simulates the real motion of the lunar lander, with the correct mass, thrust, fuel consumption rate, and lunar gravity. The real lunar lander is hard to control.

Use this **interactive simulation (https://openstaxcollege.org/l/21gravityorbits)** to move the Sun, Earth, Moon, and space station to see the effects on their gravitational forces and orbital paths. Visualize the sizes and distances between different heavenly bodies, and turn off gravity to see what would happen without it.

5.5 | Newton's Third Law

Learning Objectives

By the end of the section, you will be able to:

- State Newton's third law of motion
- Identify the action and reaction forces in different situations
- Apply Newton's third law to define systems and solve problems of motion

We have thus far considered force as a push or a pull; however, if you think about it, you realize that no push or pull ever occurs by itself. When you push on a wall, the wall pushes back on you. This brings us to **Newton's third law**.

Newton's Third Law of Motion

Whenever one body exerts a force on a second body, the first body experiences a force that is equal in magnitude and opposite in direction to the force that it exerts. Mathematically, if a body A exerts a force $\vec{F}$ on body B, then B

simultaneously exerts a force $-\overrightarrow{\mathbf{F}}$ on A, or in vector equation form,

$$\overrightarrow{\mathbf{F}}_{AB} = -\overrightarrow{\mathbf{F}}_{BA}. \tag{5.10}$$

Newton's third law represents a certain symmetry in nature: Forces always occur in pairs, and one body cannot exert a force on another without experiencing a force itself. We sometimes refer to this law loosely as "action-reaction," where the force exerted is the action and the force experienced as a consequence is the reaction. Newton's third law has practical uses in analyzing the origin of forces and understanding which forces are external to a system.

We can readily see Newton's third law at work by taking a look at how people move about. Consider a swimmer pushing off the side of a pool (Figure 5.16). She pushes against the wall of the pool with her feet and accelerates in the direction opposite that of her push. The wall has exerted an equal and opposite force on the swimmer. You might think that two equal and opposite forces would cancel, but they do not *because they act on different systems*. In this case, there are two systems that we could investigate: the swimmer and the wall. If we select the swimmer to be the system of interest, as in the figure, then $F_{\text{wall on feet}}$ is an external force on this system and affects its motion. The swimmer moves in the direction of this force. In contrast, the force $F_{\text{feet on wall}}$ acts on the wall, not on our system of interest. Thus, $F_{\text{feet on wall}}$ does not directly affect the motion of the system and does not cancel $F_{\text{wall on feet}}$. The swimmer pushes in the direction opposite that in which she wishes to move. The reaction to her push is thus in the desired direction. In a free-body diagram, such as the one shown in Figure 5.16, we never include both forces of an action-reaction pair; in this case, we only use $F_{\text{wall on feet}}$, not $F_{\text{feet on wall}}$.

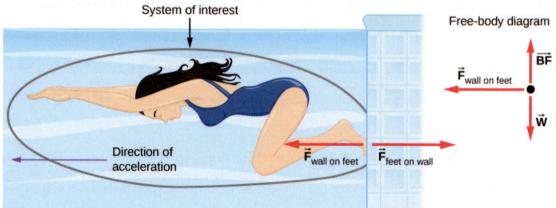

Figure 5.16 When the swimmer exerts a force on the wall, she accelerates in the opposite direction; in other words, the net external force on her is in the direction opposite of $F_{\text{feet on wall}}$. This opposition occurs because, in accordance with Newton's third law, the wall exerts a force $F_{\text{wall on feet}}$ on the swimmer that is equal in magnitude but in the direction opposite to the one she exerts on it. The line around the swimmer indicates the system of interest. Thus, the free-body diagram shows only $F_{\text{wall on feet}}$, w (the gravitational force), and BF, which is the buoyant force of the water supporting the swimmer's weight. The vertical forces w and BF cancel because there is no vertical acceleration.

Other examples of Newton's third law are easy to find:

- As a professor paces in front of a whiteboard, he exerts a force backward on the floor. The floor exerts a reaction force forward on the professor that causes him to accelerate forward.

- A car accelerates forward because the ground pushes forward on the drive wheels, in reaction to the drive wheels pushing backward on the ground. You can see evidence of the wheels pushing backward when tires spin on a gravel road and throw the rocks backward.

- Rockets move forward by expelling gas backward at high velocity. This means the rocket exerts a large backward force on the gas in the rocket combustion chamber; therefore, the gas exerts a large reaction force forward on the rocket. This reaction force, which pushes a body forward in response to a backward force, is called **thrust**. It is a common misconception that rockets propel themselves by pushing on the ground or on the air behind them. They actually work better in a vacuum, where they can more readily expel the exhaust gases.

- Helicopters create lift by pushing air down, thereby experiencing an upward reaction force.

- Birds and airplanes also fly by exerting force on the air in a direction opposite that of whatever force they need. For example, the wings of a bird force air downward and backward to get lift and move forward.

- An octopus propels itself in the water by ejecting water through a funnel from its body, similar to a jet ski.

- When a person pulls down on a vertical rope, the rope pulls up on the person (**Figure 5.17**).

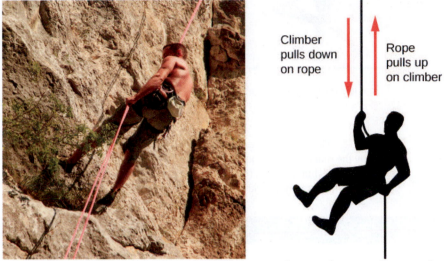

Figure 5.17 When the mountain climber pulls down on the rope, the rope pulls up on the mountain climber. (credit left: modification of work by Cristian Bortes)

There are two important features of Newton's third law. First, the forces exerted (the action and reaction) are always equal in magnitude but opposite in direction. Second, these forces are acting on different bodies or systems: *A*'s force acts on *B* and *B*'s force acts on *A*. In other words, the two forces are distinct forces that do not act on the same body. Thus, they do not cancel each other.

For the situation shown in **Figure 5.6**, the third law indicates that because the chair is pushing upward on the boy with force $\vec{C}$, he is pushing downward on the chair with force $-\vec{C}$. Similarly, he is pushing downward with forces $-\vec{F}$ and $-\vec{T}$ on the floor and table, respectively. Finally, since Earth pulls downward on the boy with force $\vec{w}$, he pulls upward on Earth with force $-\vec{w}$. If that student were to angrily pound the table in frustration, he would quickly learn the painful lesson (avoidable by studying Newton's laws) that the table hits back just as hard.

A person who is walking or running applies Newton's third law instinctively. For example, the runner in **Figure 5.18** pushes backward on the ground so that it pushes him forward.

Figure 5.18 The runner experiences Newton's third law. (a) A force is exerted by the runner on the ground. (b) The reaction force of the ground on the runner pushes him forward. (credit "runner": modification of work by "Greenwich Photography"/Flickr)

Example 5.9

Forces on a Stationary Object

The package in **Figure 5.19** is sitting on a scale. The forces on the package are $\vec{S}$, which is due to the scale, and $-\vec{w}$, which is due to Earth's gravitational field. The reaction forces that the package exerts are $-\vec{S}$ on the scale and $\vec{w}$ on Earth. Because the package is not accelerating, application of the second law yields

$$\vec{S} - \vec{w} = m\vec{a} = \vec{0},$$

so

$$\vec{S} = \vec{w}.$$

Thus, the scale reading gives the magnitude of the package's weight. However, the scale does not measure the weight of the package; it measures the force $-\vec{S}$ on its surface. If the system is accelerating, $\vec{S}$ and $-\vec{w}$ would not be equal, as explained in **Applications of Newton's Laws**.

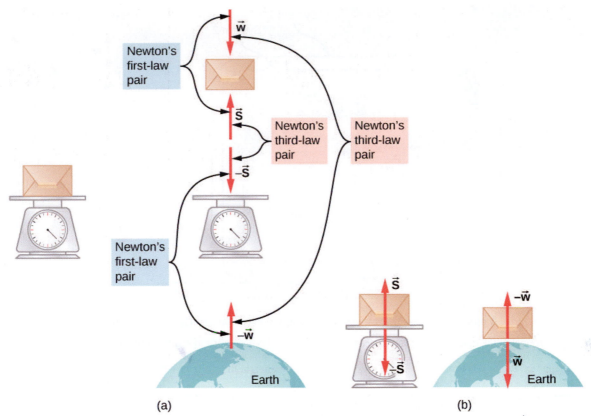

Figure 5.19 (a) The forces on a package sitting on a scale, along with their reaction forces. The force $\vec{w}$ is the weight of the package (the force due to Earth's gravity) and $\vec{S}$ is the force of the scale on the package. (b) Isolation of the package-scale system and the package-Earth system makes the action and reaction pairs clear.

Example 5.10

Getting Up to Speed: Choosing the Correct System

A physics professor pushes a cart of demonstration equipment to a lecture hall (**Figure 5.20**). Her mass is 65.0 kg, the cart's mass is 12.0 kg, and the equipment's mass is 7.0 kg. Calculate the acceleration produced when the professor exerts a backward force of 150 N on the floor. All forces opposing the motion, such as friction on the cart's wheels and air resistance, total 24.0 N.

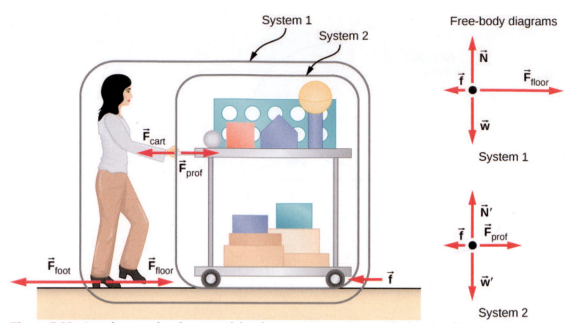

Figure 5.20 A professor pushes the cart with her demonstration equipment. The lengths of the arrows are proportional to the magnitudes of the forces (except for $\vec{f}$, because it is too small to drawn to scale). System 1 is appropriate for this example, because it asks for the acceleration of the entire group of objects. Only $\vec{F}_{floor}$ and $\vec{f}$ are external forces acting on System 1 along the line of motion. All other forces either cancel or act on the outside world. System 2 is chosen for the next example so that $\vec{F}_{prof}$ is an external force and enters into Newton's second law. The free-body diagrams, which serve as the basis for Newton's second law, vary with the system chosen.

Strategy

Since they accelerate as a unit, we define the system to be the professor, cart, and equipment. This is System 1 in **Figure 5.20**. The professor pushes backward with a force F_{foot} of 150 N. According to Newton's third law, the floor exerts a forward reaction force F_{floor} of 150 N on System 1. Because all motion is horizontal, we can assume there is no net force in the vertical direction. Therefore, the problem is one-dimensional along the horizontal direction. As noted, friction f opposes the motion and is thus in the opposite direction of F_{floor}. We do not include the forces F_{prof} or F_{cart} because these are internal forces, and we do not include F_{foot} because it acts on the floor, not on the system. There are no other significant forces acting on System 1. If the net external force can be found from all this information, we can use Newton's second law to find the acceleration as requested. See the free-body diagram in the figure.

Solution

Newton's second law is given by

$$a = \frac{F_{net}}{m}.$$

The net external force on System 1 is deduced from **Figure 5.20** and the preceding discussion to be

$$F_{net} = F_{floor} - f = 150\,\text{N} - 24.0\,\text{N} = 126\,\text{N}.$$

The mass of System 1 is

$$m = (65.0 + 12.0 + 7.0)\,\text{kg} = 84\,\text{kg}.$$

These values of F_{net} and m produce an acceleration of

$$a = \frac{F_{net}}{m} = \frac{126\,N}{84\,kg} = 1.5\,m/s^2.$$

Significance

None of the forces between components of System 1, such as between the professor's hands and the cart, contribute to the net external force because they are internal to System 1. Another way to look at this is that forces between components of a system cancel because they are equal in magnitude and opposite in direction. For example, the force exerted by the professor on the cart results in an equal and opposite force back on the professor. In this case, both forces act on the same system and therefore cancel. Thus, internal forces (between components of a system) cancel. Choosing System 1 was crucial to solving this problem.

Example 5.11

Force on the Cart: Choosing a New System

Calculate the force the professor exerts on the cart in **Figure 5.20**, using data from the previous example if needed.

Strategy

If we define the system of interest as the cart plus the equipment (System 2 in **Figure 5.20**), then the net external force on System 2 is the force the professor exerts on the cart minus friction. The force she exerts on the cart, F_{prof}, is an external force acting on System 2. F_{prof} was internal to System 1, but it is external to System 2 and thus enters Newton's second law for this system.

Solution

Newton's second law can be used to find F_{prof}. We start with

$$a = \frac{F_{net}}{m}.$$

The magnitude of the net external force on System 2 is

$$F_{net} = F_{prof} - f.$$

We solve for F_{prof}, the desired quantity:

$$F_{prof} = F_{net} + f.$$

The value of f is given, so we must calculate net F_{net}. That can be done because both the acceleration and the mass of System 2 are known. Using Newton's second law, we see that

$$F_{net} = ma,$$

where the mass of System 2 is 19.0 kg ($m = 12.0\,kg + 7.0\,kg$) and its acceleration was found to be $a = 1.5\,m/s^2$ in the previous example. Thus,

$$F_{net} = ma = (19.0\,kg)(1.5\,m/s^2) = 29\,N.$$

Now we can find the desired force:

$$F_{prof} = F_{net} + f = 29\,N + 24.0\,N = 53\,N.$$

Significance

This force is significantly less than the 150-N force the professor exerted backward on the floor. Not all of that 150-N force is transmitted to the cart; some of it accelerates the professor. The choice of a system is an important analytical step both in solving problems and in thoroughly understanding the physics of the situation (which are not necessarily the same things).

 5.7 Check Your Understanding Two blocks are at rest and in contact on a frictionless surface as shown below, with $m_1 = 2.0\,\text{kg}$, $m_2 = 6.0\,\text{kg}$, and applied force 24 N. (a) Find the acceleration of the system of blocks. (b) Suppose that the blocks are later separated. What force will give the second block, with the mass of 6.0 kg, the same acceleration as the system of blocks?

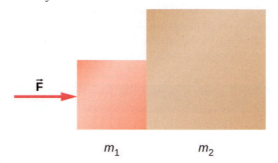

 View this **video (https://openstaxcollege.org/l/21actionreact)** to watch examples of action and reaction.

 View this **video (https://openstaxcollege.org/l/21NewtonsLaws)** to watch examples of Newton's laws and internal and external forces.

5.6 | Common Forces

Learning Objectives

By the end of the section, you will be able to:

- Define normal and tension forces
- Distinguish between real and fictitious forces
- Apply Newton's laws of motion to solve problems involving a variety of forces

Forces are given many names, such as push, pull, thrust, and weight. Traditionally, forces have been grouped into several categories and given names relating to their source, how they are transmitted, or their effects. Several of these categories are discussed in this section, together with some interesting applications. Further examples of forces are discussed later in this text.

A Catalog of Forces: Normal, Tension, and Other Examples of Forces

A catalog of forces will be useful for reference as we solve various problems involving force and motion. These forces include normal force, tension, friction, and spring force.

Normal force

Weight (also called the force of gravity) is a pervasive force that acts at all times and must be counteracted to keep an object from falling. You must support the weight of a heavy object by pushing up on it when you hold it stationary, as illustrated in **Figure 5.21**(a). But how do inanimate objects like a table support the weight of a mass placed on them, such as shown in **Figure 5.21**(b)? When the bag of dog food is placed on the table, the table sags slightly under the load. This would be noticeable if the load were placed on a card table, but even a sturdy oak table deforms when a force is applied to it. Unless an object is deformed beyond its limit, it will exert a restoring force much like a deformed spring (or a trampoline or diving board). The greater the deformation, the greater the restoring force. Thus, when the load is placed on the table, the table sags until the restoring force becomes as large as the weight of the load. At this point, the net external force on the load is zero. That is the situation when the load is stationary on the table. The table sags quickly and the sag is slight, so we do not notice it. But it is similar to the sagging of a trampoline when you climb onto it.

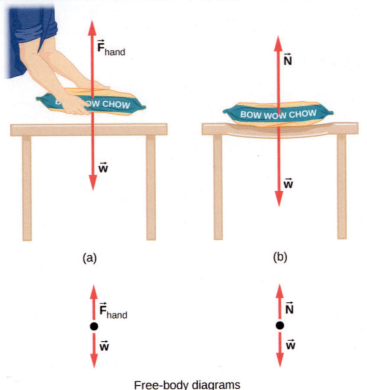

Free-body diagrams

Figure 5.21 (a) The person holding the bag of dog food must supply an upward force $\vec{F}_{hand}$ equal in magnitude and opposite in direction to the weight of the food $\vec{w}$ so that it doesn't drop to the ground. (b) The card table sags when the dog food is placed on it, much like a stiff trampoline. Elastic restoring forces in the table grow as it sags until they supply a force $\vec{N}$ equal in magnitude and opposite in direction to the weight of the load.

We must conclude that whatever supports a load, be it animate or not, must supply an upward force equal to the weight of the load, as we assumed in a few of the previous examples. If the force supporting the weight of an object, or a load, is perpendicular to the surface of contact between the load and its support, this force is defined as a **normal force** and here is given by the symbol $\vec{N}$. (This is not the newton unit for force, or N.) The word *normal* means perpendicular to a surface. This means that the normal force experienced by an object resting on a horizontal surface can be expressed in vector form as follows:

$$\vec{N} = -m\vec{g} .$$

(5.11)

In scalar form, this becomes

$$N = mg.$$

(5.12)

The normal force can be less than the object's weight if the object is on an incline.

Example 5.12

Weight on an Incline

Consider the skier on the slope in **Figure 5.22**. Her mass including equipment is 60.0 kg. (a) What is her acceleration if friction is negligible? (b) What is her acceleration if friction is 45.0 N?

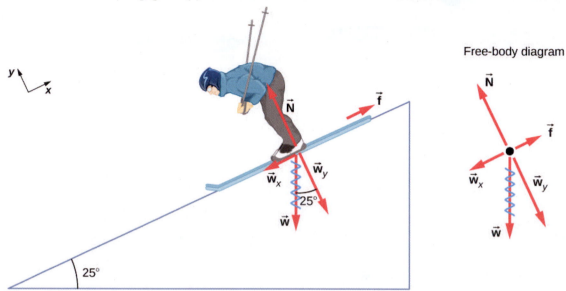

Figure 5.22 Since the acceleration is parallel to the slope and acting down the slope, it is most convenient to project all forces onto a coordinate system where one axis is parallel to the slope and the other is perpendicular to it (axes shown to the left of the skier). $\vec{\mathbf{N}}$ is perpendicular to the slope and $\vec{\mathbf{f}}$ is parallel to the slope, but $\vec{\mathbf{w}}$ has components along both axes, namely, w_y and w_x. Here, $\vec{\mathbf{w}}$ has a squiggly line to show that it has been replaced by these components. The force $\vec{\mathbf{N}}$ is equal in magnitude to w_y, so there is no acceleration perpendicular to the slope, but f is less than w_x, so there is a downslope acceleration (along the axis parallel to the slope).

Strategy

This is a two-dimensional problem, since not all forces on the skier (the system of interest) are parallel. The approach we have used in two-dimensional kinematics also works well here. Choose a convenient coordinate system and project the vectors onto its axes, creating two one-dimensional problems to solve. The most convenient coordinate system for motion on an incline is one that has one coordinate parallel to the slope and one perpendicular to the slope. (Motions along mutually perpendicular axes are independent.) We use x and y for the parallel and perpendicular directions, respectively. This choice of axes simplifies this type of problem, because there is no motion perpendicular to the slope and the acceleration is downslope. Regarding the forces, friction is drawn in opposition to motion (friction always opposes forward motion) and is always parallel to the slope, w_x is drawn parallel to the slope and downslope (it causes the motion of the skier down the slope), and w_y is drawn as the component of weight perpendicular to the slope. Then, we can consider the separate problems of forces parallel to the slope and forces perpendicular to the slope.

Solution

The magnitude of the component of weight parallel to the slope is

$$w_x = w \sin 25° = mg \sin 25°,$$

and the magnitude of the component of the weight perpendicular to the slope is

$$w_y = w \cos 25° = mg \cos 25°.$$

a. Neglect friction. Since the acceleration is parallel to the slope, we need only consider forces parallel to the slope. (Forces perpendicular to the slope add to zero, since there is no acceleration in that direction.) The forces parallel to the slope are the component of the skier's weight parallel to slope w_x and friction f. Using Newton's second law, with subscripts to denote quantities parallel to the slope,

$$a_x = \frac{F_{\text{net }x}}{m}$$

where $F_{\text{net }x} = w_x - mg \sin 25°$, assuming no friction for this part. Therefore,

$$a_x = \frac{F_{\text{net }x}}{m} = \frac{mg \sin 25°}{m} = g \sin 25°$$

$$\left(9.80 \text{ m/s}^2\right)(0.4226) = 4.14 \text{ m/s}^2$$

is the acceleration.

b. Include friction. We have a given value for friction, and we know its direction is parallel to the slope and it opposes motion between surfaces in contact. So the net external force is

$$F_{\text{net }x} = w_x - f.$$

Substituting this into Newton's second law, $a_x = F_{\text{net }x}/m$, gives

$$a_x = \frac{F_{\text{net }x}}{m} = \frac{w_x - f}{m} = \frac{mg \sin 25° - f}{m}.$$

We substitute known values to obtain

$$a_x = \frac{(60.0 \text{ kg})\left(9.80 \text{ m/s}^2\right)(0.4226) - 45.0 \text{ N}}{60.0 \text{ kg}}.$$

This gives us

$$a_x = 3.39 \text{ m/s}^2,$$

which is the acceleration parallel to the incline when there is 45.0 N of opposing friction.

Significance

Since friction always opposes motion between surfaces, the acceleration is smaller when there is friction than when there is none. It is a general result that if friction on an incline is negligible, then the acceleration down the incline is $a = g \sin \theta$, regardless of mass. As discussed previously, all objects fall with the same acceleration in the absence of air resistance. Similarly, all objects, regardless of mass, slide down a frictionless incline with the same acceleration (if the angle is the same).

When an object rests on an incline that makes an angle θ with the horizontal, the force of gravity acting on the object is divided into two components: a force acting perpendicular to the plane, w_y, and a force acting parallel to the plane, w_x (**Figure 5.23**). The normal force $\vec{\mathbf{N}}$ is typically equal in magnitude and opposite in direction to the perpendicular component of the weight w_y. The force acting parallel to the plane, w_x, causes the object to accelerate down the incline.

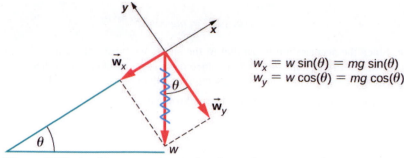

Figure 5.23 An object rests on an incline that makes an angle θ with the horizontal.

Be careful when resolving the weight of the object into components. If the incline is at an angle θ to the horizontal, then the magnitudes of the weight components are

$$w_x = w \sin \theta = mg \sin \theta$$

and

$$w_y = w \cos \theta = mg \cos \theta.$$

We use the second equation to write the normal force experienced by an object resting on an inclined plane:

$$N = mg \cos \theta. \tag{5.13}$$

Instead of memorizing these equations, it is helpful to be able to determine them from reason. To do this, we draw the right angle formed by the three weight vectors. The angle θ of the incline is the same as the angle formed between w and w_y.

Knowing this property, we can use trigonometry to determine the magnitude of the weight components:

$$\cos \theta = \frac{w_y}{w}, \quad w_y = w \cos \theta = mg \sin \theta$$

$$\sin \theta = \frac{w_x}{w}, \quad w_x = w \sin \theta = mg \sin \theta.$$

 5.8 **Check Your Understanding** A force of 1150 N acts parallel to a ramp to push a 250-kg gun safe into a moving van. The ramp is frictionless and inclined at $17°$. (a) What is the acceleration of the safe up the ramp? (b) If we consider friction in this problem, with a friction force of 120 N, what is the acceleration of the safe?

Tension

A **tension** is a force along the length of a medium; in particular, it is a pulling force that acts along a stretched flexible connector, such as a rope or cable. The word "tension" comes from a Latin word meaning "to stretch." Not coincidentally, the flexible cords that carry muscle forces to other parts of the body are called *tendons*.

Any flexible connector, such as a string, rope, chain, wire, or cable, can only exert a pull parallel to its length; thus, a force carried by a flexible connector is a tension with a direction parallel to the connector. Tension is a pull in a connector. Consider the phrase: "You can't push a rope." Instead, tension force pulls outward along the two ends of a rope.

Consider a person holding a mass on a rope, as shown in **Figure 5.24**. If the 5.00-kg mass in the figure is stationary, then its acceleration is zero and the net force is zero. The only external forces acting on the mass are its weight and the tension supplied by the rope. Thus,

$$F_{\text{net}} = T - w = 0,$$

where T and w are the magnitudes of the tension and weight, respectively, and their signs indicate direction, with up being positive. As we proved using Newton's second law, the tension equals the weight of the supported mass:

$$T = w = mg. \qquad\qquad (5.14)$$

Thus, for a 5.00-kg mass (neglecting the mass of the rope), we see that

$$T = mg = (5.00 \,\text{kg})(9.80 \,\text{m/s}^2) = 49.0 \,\text{N}.$$

If we cut the rope and insert a spring, the spring would extend a length corresponding to a force of 49.0 N, providing a direct observation and measure of the tension force in the rope.

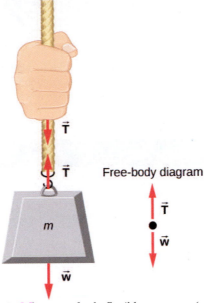

Free-body diagram

Figure 5.24 When a perfectly flexible connector (one requiring no force to bend it) such as this rope transmits a force $\vec{T}$, that force must be parallel to the length of the rope, as shown. By Newton's third law, the rope pulls with equal force but in opposite directions on the hand and the supported mass (neglecting the weight of the rope). The rope is the medium that carries the equal and opposite forces between the two objects. The tension anywhere in the rope between the hand and the mass is equal. Once you have determined the tension in one location, you have determined the tension at all locations along the rope.

Flexible connectors are often used to transmit forces around corners, such as in a hospital traction system, a tendon, or a bicycle brake cable. If there is no friction, the tension transmission is undiminished; only its direction changes, and it is always parallel to the flexible connector, as shown in **Figure 5.25**.

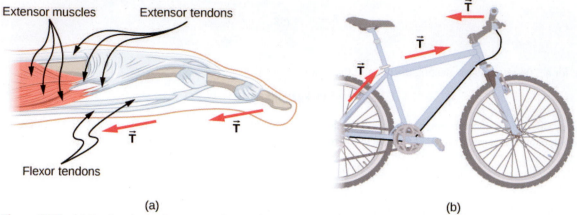

Extensor muscles **Extensor tendons**

Flexor tendons

(a)

(b)

Figure 5.25 (a) Tendons in the finger carry force T from the muscles to other parts of the finger, usually changing the force's direction but not its magnitude (the tendons are relatively friction free). (b) The brake cable on a bicycle carries the tension T from the brake lever on the handlebars to the brake mechanism. Again, the direction but not the magnitude of T is changed.

Example 5.13

What Is the Tension in a Tightrope?

Calculate the tension in the wire supporting the 70.0-kg tightrope walker shown in **Figure 5.26**.

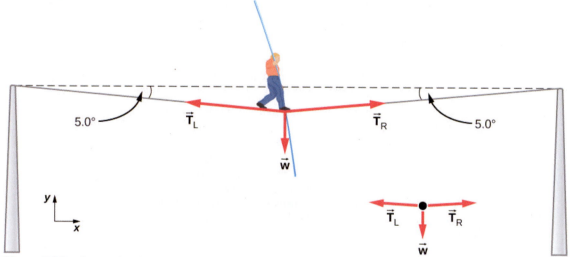

Figure 5.26 The weight of a tightrope walker causes a wire to sag by $5.0°$. The system of interest is the point in the wire at which the tightrope walker is standing.

Strategy

As you can see in **Figure 5.26**, the wire is bent under the person's weight. Thus, the tension on either side of the person has an upward component that can support his weight. As usual, forces are vectors represented pictorially by arrows that have the same direction as the forces and lengths proportional to their magnitudes. The system is the tightrope walker, and the only external forces acting on him are his weight $\vec{w}$ and the two tensions $\vec{T}_L$ (left tension) and $\vec{T}_R$ (right tension). It is reasonable to neglect the weight of the wire. The net external force is zero, because the system is static. We can use trigonometry to find the tensions. One conclusion is possible at the outset—we can see from **Figure 5.26**(b) that the magnitudes of the tensions T_L and T_R must be equal. We know this because there is no horizontal acceleration in the rope and the only forces acting to the left and right are

T_L and T_R. Thus, the magnitude of those horizontal components of the forces must be equal so that they cancel each other out.

Whenever we have two-dimensional vector problems in which no two vectors are parallel, the easiest method of solution is to pick a convenient coordinate system and project the vectors onto its axes. In this case, the best coordinate system has one horizontal axis (x) and one vertical axis (y).

Solution

First, we need to resolve the tension vectors into their horizontal and vertical components. It helps to look at a new free-body diagram showing all horizontal and vertical components of each force acting on the system (**Figure 5.27**).

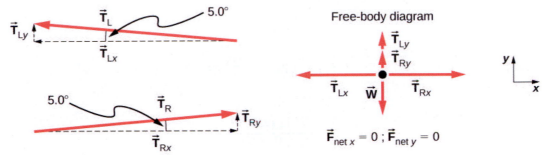

Figure 5.27 When the vectors are projected onto vertical and horizontal axes, their components along these axes must add to zero, since the tightrope walker is stationary. The small angle results in T being much greater than w.

Consider the horizontal components of the forces (denoted with a subscript x):

$$F_{net\,x} = T_{Rx} - T_{Lx}.$$

The net external horizontal force $F_{net\,x} = 0$, since the person is stationary. Thus,

$$\begin{aligned} F_{net\,x} &= 0 = T_{Rx} - T_{Lx} \\ T_{Lx} &= T_{Rx}. \end{aligned}$$

Now observe **Figure 5.27**. You can use trigonometry to determine the magnitude of T_L and T_R:

$$\cos 5.0° = \frac{T_{Lx}}{T_L}, \quad T_{Lx} = T_L \cos 5.0°$$

$$\cos 5.0° = \frac{T_{Rx}}{T_R}, \quad T_{Rx} = T_R \cos 5.0°.$$

Equating T_{Lx} and T_{Rx}:

$$T_L \cos 5.0° = T_R \cos 5.0°.$$

Thus,

$$T_L = T_R = T,$$

as predicted. Now, considering the vertical components (denoted by a subscript y), we can solve for T. Again, since the person is stationary, Newton's second law implies that $F_{net\,y} = 0$. Thus, as illustrated in the free-body diagram,

$$F_{net\,y} = T_{Ly} + T_{Ry} - w = 0.$$

We can use trigonometry to determine the relationships among T_{Ly}, T_{Ry}, and T. As we determined from the analysis in the horizontal direction, $T_L = T_R = T$:

$$\sin 5.0° \ = \ \frac{T_{Ly}}{T_L}, \quad T_{Ly} = T_L \sin 5.0° = T \sin 5.0°$$

$$\sin 5.0° \ = \ \frac{T_{Ry}}{T_R}, \quad T_{Ry} = T_R \sin 5.0° = T \sin 5.0°.$$

Now we can substitute the vales for T_{Ly} and T_{Ry}, into the net force equation in the vertical direction:

$$
\begin{aligned}
F_{net\ y} \ &= \ T_{Ly} + T_{Ry} - w = 0 \\
F_{net\ y} \ &= \ T \sin 5.0° + T \sin 5.0° - w = 0 \\
2T \sin 5.0° - w \ &= \ 0 \\
2T \sin 5.0° \ &= \ w
\end{aligned}
$$

and

$$T = \frac{w}{2 \sin 5.0°} = \frac{mg}{2 \sin 5.0°},$$

so

$$T = \frac{(70.0\ \text{kg})(9.80\ \text{m/s}^2)}{2(0.0872)},$$

and the tension is

$$T = 3930\ \text{N}.$$

Significance

The vertical tension in the wire acts as a force that supports the weight of the tightrope walker. The tension is almost six times the 686-N weight of the tightrope walker. Since the wire is nearly horizontal, the vertical component of its tension is only a fraction of the tension in the wire. The large horizontal components are in opposite directions and cancel, so most of the tension in the wire is not used to support the weight of the tightrope walker.

If we wish to create a large tension, all we have to do is exert a force perpendicular to a taut flexible connector, as illustrated in **Figure 5.26**. As we saw in **Example 5.13**, the weight of the tightrope walker acts as a force perpendicular to the rope. We saw that the tension in the rope is related to the weight of the tightrope walker in the following way:

$$T = \frac{w}{2 \sin \theta}.$$

We can extend this expression to describe the tension T created when a perpendicular force $(F_\perp)$ is exerted at the middle of a flexible connector:

$$T = \frac{F_\perp}{2 \sin \theta}.$$

The angle between the horizontal and the bent connector is represented by θ. In this case, T becomes large as θ approaches zero. Even the relatively small weight of any flexible connector will cause it to sag, since an infinite tension would result if it were horizontal (i.e., $\theta = 0$ and $\sin \theta = 0$). For example, **Figure 5.28** shows a situation where we wish to pull a car out of the mud when no tow truck is available. Each time the car moves forward, the chain is tightened to keep it as straight as possible. The tension in the chain is given by $T = \frac{F_\perp}{2 \sin \theta}$, and since θ is small, T is large. This situation is analogous to the tightrope walker, except that the tensions shown here are those transmitted to the car and the tree rather than those acting at the point where $F_\perp$ is applied.

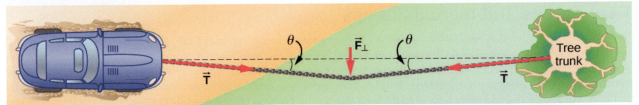

Figure 5.28 We can create a large tension in the chain—and potentially a big mess—by pushing on it perpendicular to its length, as shown.

5.9 **Check Your Understanding** One end of a 3.0-m rope is tied to a tree; the other end is tied to a car stuck in the mud. The motorist pulls sideways on the midpoint of the rope, displacing it a distance of 0.25 m. If he exerts a force of 200.0 N under these conditions, determine the force exerted on the car.

In **Applications of Newton's Laws**, we extend the discussion on tension in a cable to include cases in which the angles shown are not equal.

Friction

Friction is a resistive force opposing motion or its tendency. Imagine an object at rest on a horizontal surface. The net force acting on the object must be zero, leading to equality of the weight and the normal force, which act in opposite directions. If the surface is tilted, the normal force balances the component of the weight perpendicular to the surface. If the object does not slide downward, the component of the weight parallel to the inclined plane is balanced by friction. Friction is discussed in greater detail in the next chapter.

Spring force

A spring is a special medium with a specific atomic structure that has the ability to restore its shape, if deformed. To restore its shape, a spring exerts a restoring force that is proportional to and in the opposite direction in which it is stretched or compressed. This is the statement of a law known as Hooke's law, which has the mathematical form

$$\vec{\mathbf{F}} = -k\,\vec{\mathbf{x}}.$$

The constant of proportionality k is a measure of the spring's stiffness. The line of action of this force is parallel to the spring axis, and the sense of the force is in the opposite direction of the displacement vector (**Figure 5.29**). The displacement must be measured from the relaxed position; $x = 0$ when the spring is relaxed.

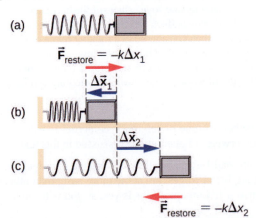

Figure 5.29 A spring exerts its force proportional to a displacement, whether it is compressed or stretched. (a) The spring is in a relaxed position and exerts no force on the block. (b) The spring is compressed by displacement $\Delta\vec{\mathbf{x}}_1$ of the object and exerts restoring force $-k\Delta\vec{\mathbf{x}}_1$. (c) The spring is stretched by displacement $\Delta\vec{\mathbf{x}}_2$ of the object and exerts restoring force $-k\Delta\vec{\mathbf{x}}_2$.

Real Forces and Inertial Frames

There is another distinction among forces: Some forces are real, whereas others are not. *Real forces* have some physical origin, such as a gravitational pull. In contrast, *fictitious forces* arise simply because an observer is in an accelerating or noninertial frame of reference, such as one that rotates (like a merry-go-round) or undergoes linear acceleration (like a car slowing down). For example, if a satellite is heading due north above Earth's Northern Hemisphere, then to an observer on Earth, it will appear to experience a force to the west that has no physical origin. Instead, Earth is rotating toward the east and moves east under the satellite. In Earth's frame, this looks like a westward force on the satellite, or it can be interpreted as a violation of Newton's first law (the law of inertia). We can identify a fictitious force by asking the question, "What is the reaction force?" If we cannot name the reaction force, then the force we are considering is fictitious. In the example of the satellite, the reaction force would have to be an eastward force on Earth. Recall that an inertial frame of reference is one in which all forces are real and, equivalently, one in which Newton's laws have the simple forms given in this chapter.

Earth's rotation is slow enough that Earth is nearly an inertial frame. You ordinarily must perform precise experiments to observe fictitious forces and the slight departures from Newton's laws, such as the effect just described. On a large scale, such as for the rotation of weather systems and ocean currents, the effects can be easily observed (**Figure 5.30**).

Figure 5.30 Hurricane Fran is shown heading toward the southeastern coast of the United States in September 1996. Notice the characteristic "eye" shape of the hurricane. This is a result of the Coriolis effect, which is the deflection of objects (in this case, air) when considered in a rotating frame of reference, like the spin of Earth. This hurricane shows a counter-clockwise rotation because it is a low pressure storm. (credit "runner": modification of work by "Greenwich Photography"/Flickr)

The crucial factor in determining whether a frame of reference is inertial is whether it accelerates or rotates relative to a known inertial frame. Unless stated otherwise, all phenomena discussed in this text are in inertial frames.

The forces discussed in this section are real forces, but they are not the only real forces. Lift and thrust, for example, are more specialized real forces. In the long list of forces, are some more basic than others? Are some different manifestations of the same underlying force? The answer to both questions is yes, as you will see in the treatment of modern physics later in the text.

 Explore forces and motion in this **interactive simulation (https://openstaxcollege.org/l/21ramp)** as you push household objects up and down a ramp. Lower and raise the ramp to see how the angle of inclination affects the parallel forces. Graphs show forces, energy, and work.

 Stretch and compress springs in this **activity (https://openstaxcollege.org/l/21hookeslaw)** to explore the relationships among force, spring constant, and displacement. Investigate what happens when two springs are connected in series and in parallel.

5.7 | Drawing Free-Body Diagrams

Learning Objectives

By the end of the section, you will be able to:

- Explain the rules for drawing a free-body diagram
- Construct free-body diagrams for different situations

The first step in describing and analyzing most phenomena in physics involves the careful drawing of a free-body diagram. Free-body diagrams have been used in examples throughout this chapter. Remember that a free-body diagram must only include the external forces acting on the body of interest. Once we have drawn an accurate free-body diagram, we can apply Newton's first law if the body is in equilibrium (balanced forces; that is, $F_{net} = 0$) or Newton's second law if the body is accelerating (unbalanced force; that is, $F_{net} \neq 0$).

In **Forces**, we gave a brief problem-solving strategy to help you understand free-body diagrams. Here, we add some details to the strategy that will help you in constructing these diagrams.

Problem-Solving Strategy: Constructing Free-Body Diagrams

Observe the following rules when constructing a free-body diagram:

1. Draw the object under consideration; it does not have to be artistic. At first, you may want to draw a circle around the object of interest to be sure you focus on labeling the forces acting on the object. If you are treating the object as a particle (no size or shape and no rotation), represent the object as a point. We often place this point at the origin of an *xy*-coordinate system.

2. Include all forces that act on the object, representing these forces as vectors. Consider the types of forces described in **Common Forces**—normal force, friction, tension, and spring force—as well as weight and applied force. Do not include the net force on the object. With the exception of gravity, all of the forces we have discussed require direct contact with the object. However, forces that the object exerts on its environment must not be included. We never include both forces of an action-reaction pair.

3. Convert the free-body diagram into a more detailed diagram showing the *x*- and *y*-components of a given force (this is often helpful when solving a problem using Newton's first or second law). In this case, place a squiggly line through the original vector to show that it is no longer in play—it has been replaced by its *x*- and *y*-components.

4. If there are two or more objects, or bodies, in the problem, draw a separate free-body diagram for each object.

Note: If there is acceleration, we do not directly include it in the free-body diagram; however, it may help to indicate acceleration outside the free-body diagram. You can label it in a different color to indicate that it is separate from the free-body diagram.

Let's apply the problem-solving strategy in drawing a free-body diagram for a sled. In **Figure 5.31**(a), a sled is pulled by force **P** at an angle of $30°$. In part (b), we show a free-body diagram for this situation, as described by steps 1 and 2 of the problem-solving strategy. In part (c), we show all forces in terms of their *x*- and *y*-components, in keeping with step 3.

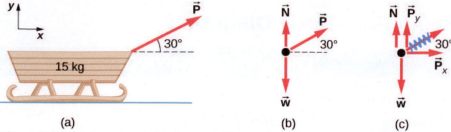

Figure 5.31 (a) A moving sled is shown as (b) a free-body diagram and (c) a free-body diagram with force components.

Example 5.14

Two Blocks on an Inclined Plane

Construct the free-body diagram for object A and object B in **Figure 5.32**.

Strategy

We follow the four steps listed in the problem-solving strategy.

Solution

We start by creating a diagram for the first object of interest. In **Figure 5.32**(a), object A is isolated (circled) and represented by a dot.

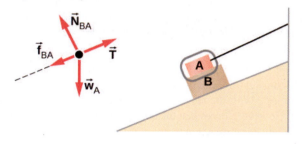

$\vec{w}_A$ = weight of block A

$\vec{T}$ = tension

$\vec{N}_{BA}$ = normal force exerted by B on A

$\vec{f}_{BA}$ = friction force exerted by B on A

(a)

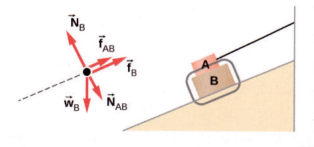

$\vec{w}_B$ = weight of block B

$\vec{N}_{AB}$ = normal force exerted by A on B

$\vec{N}_B$ = normal force exerted by the incline plane on B

$\vec{f}_{AB}$ = friction force exerted by A on B

$\vec{f}_B$ = friction force exerted by the incline plane on B

(b)

Figure 5.32 (a) The free-body diagram for isolated object A. (b) The free-body diagram for isolated object B. Comparing the two drawings, we see that friction acts in the opposite direction in the two figures. Because object A experiences a force that tends to pull it to the right, friction must act to the left. Because object B experiences a component of its weight that pulls it to the left, down the incline, the friction force must oppose it and act up the ramp. Friction always acts opposite the intended direction of motion.

We now include any force that acts on the body. Here, no applied force is present. The weight of the object acts as a force pointing vertically downward, and the presence of the cord indicates a force of tension pointing away from the object. Object A has one interface and hence experiences a normal force, directed away from the interface. The source of this force is object B, and this normal force is labeled accordingly. Since object B has a tendency to slide down, object A has a tendency to slide up with respect to the interface, so the friction f_{BA} is directed downward parallel to the inclined plane.

As noted in step 4 of the problem-solving strategy, we then construct the free-body diagram in **Figure 5.32**(b) using the same approach. Object B experiences two normal forces and two friction forces due to the presence of two contact surfaces. The interface with the inclined plane exerts external forces of N_B and f_B, and the interface with object B exerts the normal force N_{AB} and friction f_{AB}; N_{AB} is directed away from object B, and f_{AB} is opposing the tendency of the relative motion of object B with respect to object A.

Significance

The object under consideration in each part of this problem was circled in gray. When you are first learning how to draw free-body diagrams, you will find it helpful to circle the object before deciding what forces are acting on that particular object. This focuses your attention, preventing you from considering forces that are not acting on the body.

Example 5.15

Two Blocks in Contact

A force is applied to two blocks in contact, as shown.

Strategy

Draw a free-body diagram for each block. Be sure to consider Newton's third law at the interface where the two blocks touch.

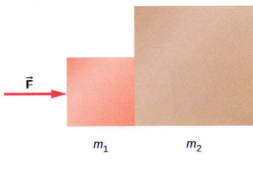

Solution

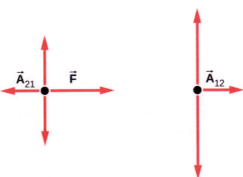

Significance

$\vec{A}_{21}$ is the action force of block 2 on block 1. $\vec{A}_{12}$ is the reaction force of block 1 on block 2. We use these free-body diagrams in **Applications of Newton's Laws**.

Example 5.16

Block on the Table (Coupled Blocks)

A block rests on the table, as shown. A light rope is attached to it and runs over a pulley. The other end of the rope is attached to a second block. The two blocks are said to be coupled. Block m_2 exerts a force due to its weight, which causes the system (two blocks and a string) to accelerate.

Strategy

We assume that the string has no mass so that we do not have to consider it as a separate object. Draw a free-body diagram for each block.

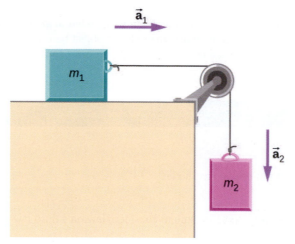

Solution

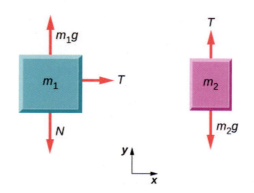

Significance

Each block accelerates (notice the labels shown for $\vec{a}_1$ and $\vec{a}_2$); however, assuming the string remains taut, they accelerate at the same rate. Thus, we have $\vec{a}_1 = \vec{a}_2$. If we were to continue solving the problem, we could simply call the acceleration $\vec{a}$. Also, we use two free-body diagrams because we are usually finding tension T, which may require us to use a system of two equations in this type of problem. The tension is the same on both m_1 and m_2.

5.10 **Check Your Understanding** (a) Draw the free-body diagram for the situation shown. (b) Redraw it showing components; use *x*-axes parallel to the two ramps.

 View this **simulation (https://openstaxcollege.org/l/21forcemotion)** to predict, qualitatively, how an external force will affect the speed and direction of an object's motion. Explain the effects with the help of a free-body diagram. Use free-body diagrams to draw position, velocity, acceleration, and force graphs, and vice versa. Explain how the graphs relate to one another. Given a scenario or a graph, sketch all four graphs.

CHAPTER 5 REVIEW

KEY TERMS

dynamics study of how forces affect the motion of objects and systems

external force force acting on an object or system that originates outside of the object or system

force push or pull on an object with a specific magnitude and direction; can be represented by vectors or expressed as a multiple of a standard force

free fall situation in which the only force acting on an object is gravity

free-body diagram sketch showing all external forces acting on an object or system; the system is represented by a single isolated point, and the forces are represented by vectors extending outward from that point

Hooke's law in a spring, a restoring force proportional to and in the opposite direction of the imposed displacement

inertia ability of an object to resist changes in its motion

inertial reference frame reference frame moving at constant velocity relative to an inertial frame is also inertial; a reference frame accelerating relative to an inertial frame is not inertial

law of inertia see Newton's first law of motion

net external force vector sum of all external forces acting on an object or system; causes a mass to accelerate

newton SI unit of force; 1 N is the force needed to accelerate an object with a mass of 1 kg at a rate of 1 m/s^2

Newton's first law of motion body at rest remains at rest or, if in motion, remains in motion at constant velocity unless acted on by a net external force; also known as the law of inertia

Newton's second law of motion acceleration of a system is directly proportional to and in the same direction as the net external force acting on the system and is inversely proportional to its mass

Newton's third law of motion whenever one body exerts a force on a second body, the first body experiences a force that is equal in magnitude and opposite in direction to the force that it exerts

normal force force supporting the weight of an object, or a load, that is perpendicular to the surface of contact between the load and its support; the surface applies this force to an object to support the weight of the object

tension pulling force that acts along a stretched flexible connector, such as a rope or cable

thrust reaction force that pushes a body forward in response to a backward force

weight force $\vec{\mathbf{w}}$ due to gravity acting on an object of mass m

KEY EQUATIONS

Net external force	$\vec{\mathbf{F}}_{net} = \sum \vec{\mathbf{F}} = \vec{\mathbf{F}}_1 + \vec{\mathbf{F}}_2 + \cdots$
Newton's first law	$\vec{\mathbf{v}} = \text{constant when } \vec{\mathbf{F}}_{net} = \vec{\mathbf{0}} \text{ N}$
Newton's second law, vector form	$\vec{\mathbf{F}}_{net} = \sum \vec{\mathbf{F}} = m\vec{\mathbf{a}}$
Newton's second law, scalar form	$F_{net} = ma$
Newton's second law, component form	$\sum \vec{\mathbf{F}}_x = m\vec{\mathbf{a}}_x, \sum \vec{\mathbf{F}}_y = m\vec{\mathbf{a}}_y, \text{ and } \sum \vec{\mathbf{F}}_z = m\vec{\mathbf{a}}_z.$
Newton's second law, momentum form	$\vec{\mathbf{F}}_{net} = \dfrac{d\vec{\mathbf{p}}}{dt}$
Definition of weight, vector form	$\vec{\mathbf{w}} = m\vec{\mathbf{g}}$

Definition of weight, scalar form	$w = mg$
Newton's third law	$\vec{\mathbf{F}}_{AB} = -\vec{\mathbf{F}}_{BA}$
Normal force on an object resting on a horizontal surface, vector form	$\vec{\mathbf{N}} = -m\vec{\mathbf{g}}$
Normal force on an object resting on a horizontal surface, scalar form	$N = mg$
Normal force on an object resting on an inclined plane, scalar form	$N = mg\cos\theta$
Tension in a cable supporting an object of mass m at rest, scalar form	$T = w = mg$

SUMMARY

5.1 Forces

- Dynamics is the study of how forces affect the motion of objects, whereas kinematics simply describes the way objects move.

- Force is a push or pull that can be defined in terms of various standards, and it is a vector that has both magnitude and direction.

- External forces are any outside forces that act on a body. A free-body diagram is a drawing of all external forces acting on a body.

- The SI unit of force is the newton (N).

5.2 Newton's First Law

- According to Newton's first law, there must be a cause for any change in velocity (a change in either magnitude or direction) to occur. This law is also known as the law of inertia.

- Friction is an external force that causes an object to slow down.

- Inertia is the tendency of an object to remain at rest or remain in motion. Inertia is related to an object's mass.

- If an object's velocity relative to a given frame is constant, then the frame is inertial. This means that for an inertial reference frame, Newton's first law is valid.

- Equilibrium is achieved when the forces on a system are balanced.

- A net force of zero means that an object is either at rest or moving with constant velocity; that is, it is not accelerating.

5.3 Newton's Second Law

- An external force acts on a system from outside the system, as opposed to internal forces, which act between components within the system.

- Newton's second law of motion says that the net external force on an object with a certain mass is directly proportional to and in the same direction as the acceleration of the object.

- Newton's second law can also describe net force as the instantaneous rate of change of momentum. Thus, a net external force causes nonzero acceleration.

5.4 Mass and Weight

- Mass is the quantity of matter in a substance.

- The weight of an object is the net force on a falling object, or its gravitational force. The object experiences acceleration due to gravity.

- Some upward resistance force from the air acts on all falling objects on Earth, so they can never truly be in free fall.

- Careful distinctions must be made between free fall and weightlessness using the definition of weight as force due to gravity acting on an object of a certain mass.

5.5 Newton's Third Law

- Newton's third law of motion represents a basic symmetry in nature, with an experienced force equal in magnitude and opposite in direction to an exerted force.

- Two equal and opposite forces do not cancel because they act on different systems.

- Action-reaction pairs include a swimmer pushing off a wall, helicopters creating lift by pushing air down, and an octopus propelling itself forward by ejecting water from its body. Rockets, airplanes, and cars are pushed forward by a thrust reaction force.

- Choosing a system is an important analytical step in understanding the physics of a problem and solving it.

5.6 Common Forces

- When an object rests on a surface, the surface applies a force to the object that supports the weight of the object. This supporting force acts perpendicular to and away from the surface. It is called a normal force.

- When an object rests on a nonaccelerating horizontal surface, the magnitude of the normal force is equal to the weight of the object.

- When an object rests on an inclined plane that makes an angle θ with the horizontal surface, the weight of the object can be resolved into components that act perpendicular and parallel to the surface of the plane.

- The pulling force that acts along a stretched flexible connector, such as a rope or cable, is called tension. When a rope supports the weight of an object at rest, the tension in the rope is equal to the weight of the object. If the object is accelerating, tension is greater than weight, and if it is decelerating, tension is less than weight.

- The force of friction is a force experienced by a moving object (or an object that has a tendency to move) parallel to the interface opposing the motion (or its tendency).

- The force developed in a spring obeys Hooke's law, according to which its magnitude is proportional to the displacement and has a sense in the opposite direction of the displacement.

- Real forces have a physical origin, whereas fictitious forces occur because the observer is in an accelerating or noninertial frame of reference.

5.7 Drawing Free-Body Diagrams

- To draw a free-body diagram, we draw the object of interest, draw all forces acting on that object, and resolve all force vectors into x- and y-components. We must draw a separate free-body diagram for each object in the problem.

- A free-body diagram is a useful means of describing and analyzing all the forces that act on a body to determine equilibrium according to Newton's first law or acceleration according to Newton's second law.

CONCEPTUAL QUESTIONS

5.1 Forces

1. What properties do forces have that allow us to classify them as vectors?

5.2 Newton's First Law

2. Taking a frame attached to Earth as inertial, which of the following objects cannot have inertial frames attached to them, and which are inertial reference frames?

(a) A car moving at constant velocity

(b) A car that is accelerating

(c) An elevator in free fall

(d) A space capsule orbiting Earth

(e) An elevator descending uniformly

3. A woman was transporting an open box of cupcakes to a school party. The car in front of her stopped suddenly; she applied her brakes immediately. She was wearing her seat belt and suffered no physical harm (just a great deal of embarrassment), but the cupcakes flew into the dashboard and became "smushcakes." Explain what happened.

5.3 Newton's Second Law

4. Why can we neglect forces such as those holding a body together when we apply Newton's second law?

5. A rock is thrown straight up. At the top of the trajectory, the velocity is momentarily zero. Does this imply that the force acting on the object is zero? Explain your answer.

5.4 Mass and Weight

6. What is the relationship between weight and mass? Which is an intrinsic, unchanging property of a body?

7. How much does a 70-kg astronaut weight in space, far from any celestial body? What is her mass at this location?

8. Which of the following statements is accurate?

(a) Mass and weight are the same thing expressed in different units.

(b) If an object has no weight, it must have no mass.

(c) If the weight of an object varies, so must the mass.

(d) Mass and inertia are different concepts.

(e) Weight is always proportional to mass.

9. When you stand on Earth, your feet push against it with a force equal to your weight. Why doesn't Earth accelerate away from you?

10. How would you give the value of $\vec{g}$ in vector form?

5.5 Newton's Third Law

11. Identify the action and reaction forces in the following situations: (a) Earth attracts the Moon, (b) a boy kicks a football, (c) a rocket accelerates upward, (d) a car accelerates forward, (e) a high jumper leaps, and (f) a bullet is shot from a gun.

12. Suppose that you are holding a cup of coffee in your hand. Identify all forces on the cup and the reaction to each force.

13. (a) Why does an ordinary rifle recoil (kick backward) when fired? (b) The barrel of a recoilless rifle is open at both ends. Describe how Newton's third law applies when one is fired. (c) Can you safely stand close behind one when it is fired?

5.6 Common Forces

14. A table is placed on a rug. Then a book is placed on the table. What does the floor exert a normal force on?

15. A particle is moving to the right. (a) Can the force on it be acting to the left? If yes, what would happen? (b) Can that force be acting downward? If yes, why?

5.7 Drawing Free-Body Diagrams

16. In completing the solution for a problem involving forces, what do we do after constructing the free-body diagram? That is, what do we apply?

17. If a book is located on a table, how many forces should be shown in a free-body diagram of the book? Describe them.

18. If the book in the previous question is in free fall, how many forces should be shown in a free-body diagram of the book? Describe them.

PROBLEMS

5.1 Forces

19. Two ropes are attached to a tree, and forces of $\vec{F}_1 = 2.0\,\hat{i} + 4.0\,\hat{j}$ N and $\vec{F}_2 = 3.0\,\hat{i} + 6.0\,\hat{j}$ N are applied. The forces are coplanar (in the same plane). (a) What is the resultant (net force) of these two force vectors? (b) Find the magnitude and direction of this net force.

20. A telephone pole has three cables pulling as shown from above, with $\vec{F}_1 = \left(300.0\,\hat{i} + 500.0\,\hat{j}\right)$, $\vec{F}_2 = -200.0\,\hat{i}$, and $\vec{F}_3 = -800.0\,\hat{j}$. (a) Find the net force on the telephone pole in component form. (b) Find the magnitude and direction of this net force.

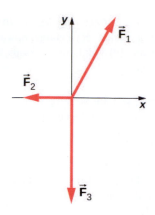

21. Two teenagers are pulling on ropes attached to a tree. The angle between the ropes is $30.0°$. David pulls with a force of 400.0 N and Stephanie pulls with a force of 300.0 N. (a) Find the component form of the net force. (b) Find the magnitude of the resultant (net) force on the tree and the angle it makes with David's rope.

5.2 Newton's First Law

22. Two forces of $\vec{F}_1 = \frac{75.0}{\sqrt{2}}\left(\hat{i} - \hat{j}\right)\text{N}$ and $\vec{F}_2 = \frac{150.0}{\sqrt{2}}\left(\hat{i} - \hat{j}\right)\text{N}$ act on an object. Find the third force $\vec{F}_3$ that is needed to balance the first two forces.

23. While sliding a couch across a floor, Andrea and Jennifer exert forces $\vec{F}_A$ and $\vec{F}_J$ on the couch. Andrea's force is due north with a magnitude of 130.0 N and Jennifer's force is $32°$ east of north with a magnitude of 180.0 N. (a) Find the net force in component form. (b) Find the magnitude and direction of the net force. (c) If Andrea and Jennifer's housemates, David and Stephanie, disagree with the move and want to prevent its relocation, with what combined force $\vec{F}_{DS}$ should they push so that the couch does not move?

5.3 Newton's Second Law

24. Andrea, a 63.0-kg sprinter, starts a race with an acceleration of $4.200\ \text{m/s}^2$. What is the net external force on her?

25. If the sprinter from the previous problem accelerates at that rate for 20.00 m and then maintains that velocity for the remainder of a 100.00-m dash, what will her time be for the race?

26. A cleaner pushes a 4.50-kg laundry cart in such a way that the net external force on it is 60.0 N. Calculate the magnitude of his cart's acceleration.

27. Astronauts in orbit are apparently weightless. This means that a clever method of measuring the mass of astronauts is needed to monitor their mass gains or losses, and adjust their diet. One way to do this is to exert a known force on an astronaut and measure the acceleration produced. Suppose a net external force of 50.0 N is exerted, and an astronaut's acceleration is measured to be $0.893\ \text{m/s}^2$. (a) Calculate her mass. (b) By exerting a force on the astronaut, the vehicle in which she orbits experiences an equal and opposite force. Use this knowledge to find an equation for the acceleration of the system (astronaut and spaceship) that would be measured by a nearby observer. (c) Discuss how this would affect the measurement of the astronaut's acceleration. Propose a method by which recoil of the vehicle is avoided.

28. In **Figure 5.12**, the net external force on the 24-kg mower is given as 51 N. If the force of friction opposing the motion is 24 N, what force F (in newtons) is the person exerting on the mower? Suppose the mower is moving at 1.5 m/s when the force F is removed. How far will the mower go before stopping?

29. The rocket sled shown below decelerates at a rate of $196\ \text{m/s}^2$. What force is necessary to produce this deceleration? Assume that the rockets are off. The mass of the system is 2.10×10^3 kg.

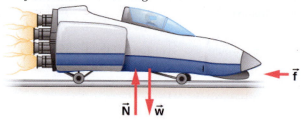

30. If the rocket sled shown in the previous problem starts with only one rocket burning, what is the magnitude of this acceleration? Assume that the mass of the system is 2.10×10^3 kg, the thrust T is 2.40×10^4 N, and the force of friction opposing the motion is 650.0 N. (b) Why is the acceleration not one-fourth of what it is with all rockets burning?

31. What is the deceleration of the rocket sled if it comes to rest in 1.10 s from a speed of 1000.0 km/h? (Such deceleration caused one test subject to black out and have temporary blindness.)

32. Suppose two children push horizontally, but in exactly opposite directions, on a third child in a wagon. The first child exerts a force of 75.0 N, the second exerts a force of 90.0 N, friction is 12.0 N, and the mass of the third child plus wagon is 23.0 kg. (a) What is the system of interest if the acceleration of the child in the wagon is to be calculated? (See the free-body diagram.) (b) Calculate the acceleration. (c) What would the acceleration be if friction were 15.0 N?

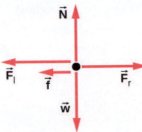

33. A powerful motorcycle can produce an acceleration of 3.50 m/s^2 while traveling at 90.0 km/h. At that speed, the forces resisting motion, including friction and air resistance, total 400.0 N. (Air resistance is analogous to air friction. It always opposes the motion of an object.) What is the magnitude of the force that motorcycle exerts backward on the ground to produce its acceleration if the mass of the motorcycle with rider is 245 kg?

34. A car with a mass of 1000.0 kg accelerates from 0 to 90.0 km/h in 10.0 s. (a) What is its acceleration? (b) What is the net force on the car?

35. The driver in the previous problem applies the brakes when the car is moving at 90.0 km/h, and the car comes to rest after traveling 40.0 m. What is the net force on the car during its deceleration?

36. An 80.0-kg passenger in an SUV traveling at 1.00×10^3 km/h is wearing a seat belt. The driver slams on the brakes and the SUV stops in 45.0 m. Find the force of the seat belt on the passenger.

37. A particle of mass 2.0 kg is acted on by a single force $\overrightarrow{\textbf{F}}_1 = 18 \hat{\textbf{i}}$ N. (a) What is the particle's acceleration? (b) If the particle starts at rest, how far does it travel in the first 5.0 s?

38. Suppose that the particle of the previous problem also experiences forces $\overrightarrow{\textbf{F}}_2 = -15 \hat{\textbf{i}}$ N and $\overrightarrow{\textbf{F}}_3 = 6.0 \hat{\textbf{j}}$ N. What is its acceleration in this case?

39. Find the acceleration of the body of mass 5.0 kg shown below.

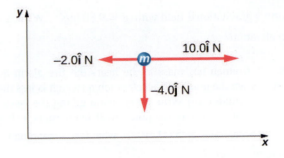

40. In the following figure, the horizontal surface on which this block slides is frictionless. If the two forces acting on it each have magnitude $F = 30.0$ N and $M = 10.0$ kg, what is the magnitude of the resulting acceleration of the block?

5.4 Mass and Weight

41. The weight of an astronaut plus his space suit on the Moon is only 250 N. (a) How much does the suited astronaut weigh on Earth? (b) What is the mass on the Moon? On Earth?

42. Suppose the mass of a fully loaded module in which astronauts take off from the Moon is 1.00×10^4 kg. The thrust of its engines is 3.00×10^4 N. (a) Calculate the module's magnitude of acceleration in a vertical takeoff from the Moon. (b) Could it lift off from Earth? If not, why not? If it could, calculate the magnitude of its acceleration.

43. A rocket sled accelerates at a rate of 49.0 m/s^2. Its passenger has a mass of 75.0 kg. (a) Calculate the horizontal component of the force the seat exerts against his body. Compare this with his weight using a ratio. (b) Calculate the direction and magnitude of the total force the seat exerts against his body.

44. Repeat the previous problem for a situation in which the rocket sled decelerates at a rate of 201 m/s^2. In this problem, the forces are exerted by the seat and the seat belt.

45. A body of mass 2.00 kg is pushed straight upward by a 25.0 N vertical force. What is its acceleration?

46. A car weighing 12,500 N starts from rest and accelerates to 83.0 km/h in 5.00 s. The friction force is 1350 N. Find the applied force produced by the engine.

47. A body with a mass of 10.0 kg is assumed to be in

Earth's gravitational field with $g = 9.80 \text{ m/s}^2$. What is its acceleration?

48. A fireman has mass m; he hears the fire alarm and slides down the pole with acceleration a (which is less than g in magnitude). (a) Write an equation giving the vertical force he must apply to the pole. (b) If his mass is 90.0 kg and he accelerates at 5.00 m/s^2, what is the magnitude of his applied force?

49. A baseball catcher is performing a stunt for a television commercial. He will catch a baseball (mass 145 g) dropped from a height of 60.0 m above his glove. His glove stops the ball in 0.0100 s. What is the force exerted by his glove on the ball?

50. When the Moon is directly overhead at sunset, the force by Earth on the Moon, F_{EM}, is essentially at $90°$ to the force by the Sun on the Moon, F_{SM}, as shown below. Given that $F_{EM} = 1.98 \times 10^{20} \text{ N}$ and $F_{SM} = 4.36 \times 10^{20} \text{ N}$, all other forces on the Moon are negligible, and the mass of the Moon is $7.35 \times 10^{22} \text{ kg}$, determine the magnitude of the Moon's acceleration.

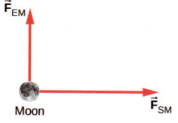

5.5 Newton's Third Law

51. (a) What net external force is exerted on a 1100.0-kg artillery shell fired from a battleship if the shell is accelerated at $2.40 \times 10^4 \text{ m/s}^2$? (b) What is the magnitude of the force exerted on the ship by the artillery shell, and why?

52. A brave but inadequate rugby player is being pushed backward by an opposing player who is exerting a force of 800.0 N on him. The mass of the losing player plus equipment is 90.0 kg, and he is accelerating backward at 1.20 m/s^2. (a) What is the force of friction between the losing player's feet and the grass? (b) What force does the winning player exert on the ground to move forward if his mass plus equipment is 110.0 kg?

53. A history book is lying on top of a physics book on a desk, as shown below; a free-body diagram is also shown. The history and physics books weigh 14 N and 18 N, respectively. Identify each force on each book with a

double subscript notation (for instance, the contact force of the history book pressing against physics book can be described as $\vec{F}_{HP}$), and determine the value of each of these forces, explaining the process used.

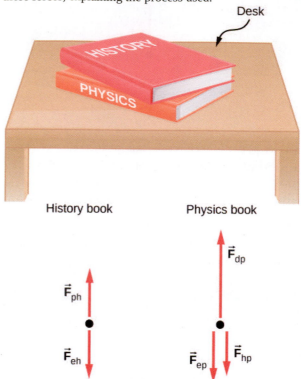

54. A truck collides with a car, and during the collision, the net force on each vehicle is essentially the force exerted by the other. Suppose the mass of the car is 550 kg, the mass of the truck is 2200 kg, and the magnitude of the truck's acceleration is 10 m/s^2. Find the magnitude of the car's acceleration.

5.6 Common Forces

55. A leg is suspended in a traction system, as shown below. (a) Which pulley in the figure is used to calculate the force exerted on the foot? (b) What is the tension in the rope? Here $\vec{T}$ is the tension, $\vec{w}_{leg}$ is the weight of the leg, and $\vec{w}$ is the weight of the load that provides the tension.

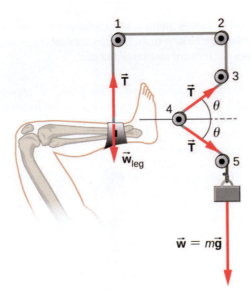

$$\vec{w} = m\vec{g}$$

56. Suppose the shinbone in the preceding image was a femur in a traction setup for a broken bone, with pulleys and rope available. How might we be able to increase the force along the femur using the same weight?

57. Two teams of nine members each engage in tug-of-war. Each of the first team's members has an average mass of 68 kg and exerts an average force of 1350 N horizontally. Each of the second team's members has an average mass of 73 kg and exerts an average force of 1365 N horizontally. (a) What is magnitude of the acceleration of the two teams, and which team wins? (b) What is the tension in the section of rope between the teams?

58. What force does a trampoline have to apply to Jennifer, a 45.0-kg gymnast, to accelerate her straight up at $7.50 \, \text{m/s}^2$? The answer is independent of the velocity of the gymnast—she can be moving up or down or can be instantly stationary.

59. (a) Calculate the tension in a vertical strand of spider web if a spider of mass $2.00 \times 10^{-5} \, \text{kg}$ hangs motionless on it. (b) Calculate the tension in a horizontal strand of spider web if the same spider sits motionless in the middle of it much like the tightrope walker in **Figure 5.26**. The strand sags at an angle of $12°$ below the horizontal. Compare this with the tension in the vertical strand (find their ratio).

60. Suppose Kevin, a 60.0-kg gymnast, climbs a rope. (a) What is the tension in the rope if he climbs at a constant speed? (b) What is the tension in the rope if he accelerates upward at a rate of $1.50 \, \text{m/s}^2$?

61. Show that, as explained in the text, a force $F_\perp$ exerted on a flexible medium at its center and perpendicular to its length (such as on the tightrope wire in **Figure 5.26**)

gives rise to a tension of magnitude $T = F_\perp / 2 \sin(\theta)$.

62. Consider **Figure 5.28**. The driver attempts to get the car out of the mud by exerting a perpendicular force of 610.0 N, and the distance she pushes in the middle of the rope is 1.00 m while she stands 6.00 m away from the car on the left and 6.00 m away from the tree on the right. What is the tension T in the rope, and how do you find the answer?

63. A bird has a mass of 26 g and perches in the middle of a stretched telephone line. (a) Show that the tension in the line can be calculated using the equation $T = \dfrac{mg}{2 \sin \theta}$. Determine the tension when (b) $\theta = 5°$ and (c) $\theta = 0.5°$. Assume that each half of the line is straight.

64. One end of a 30-m rope is tied to a tree; the other end is tied to a car stuck in the mud. The motorist pulls sideways on the midpoint of the rope, displacing it a distance of 2 m. If he exerts a force of 80 N under these conditions, determine the force exerted on the car.

65. Consider the baby being weighed in the following figure. (a) What is the mass of the infant and basket if a scale reading of 55 N is observed? (b) What is tension T_1 in the cord attaching the baby to the scale? (c) What is tension T_2 in the cord attaching the scale to the ceiling, if the scale has a mass of 0.500 kg? (d) Sketch the situation, indicating the system of interest used to solve each part. The masses of the cords are negligible.

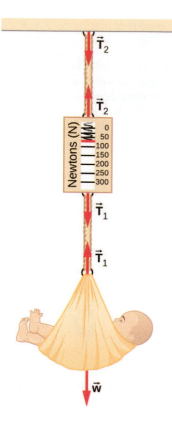

66. What force must be applied to a 100.0-kg crate on a frictionless plane inclined at $30°$ to cause an acceleration of 2.0 m/s^2 up the plane?

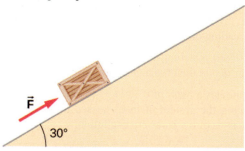

67. A 2.0-kg block is on a perfectly smooth ramp that makes an angle of $30°$ with the horizontal. (a) What is the block's acceleration down the ramp and the force of the ramp on the block? (b) What force applied upward along and parallel to the ramp would allow the block to move with constant velocity?

5.7 Drawing Free-Body Diagrams

68. A ball of mass m hangs at rest, suspended by a string. (a) Sketch all forces. (b) Draw the free-body diagram for the ball.

69. A car moves along a horizontal road. Draw a free-body diagram; be sure to include the friction of the road that opposes the forward motion of the car.

70. A runner pushes against the track, as shown. (a) Provide a free-body diagram showing all the forces on the runner. (*Hint:* Place all forces at the center of his body, and include his weight.) (b) Give a revised diagram showing the *xy*-component form.

(credit: modification of work by "Greenwich Photography"/Flickr)

71. The traffic light hangs from the cables as shown. Draw a free-body diagram on a coordinate plane for this situation.

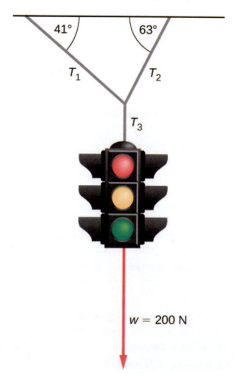

ADDITIONAL PROBLEMS

72. Two small forces, $\vec{F}_1 = -2.40\,\hat{i} - 6.10t\,\hat{j}$ N and $\vec{F}_2 = 8.50\,\hat{i} - 9.70\,\hat{j}$ N, are exerted on a rogue asteroid by a pair of space tractors. (a) Find the net force. (b) What are the magnitude and direction of the net force? (c) If the mass of the asteroid is 125 kg, what acceleration does it experience (in vector form)? (d) What are the magnitude and direction of the acceleration?

73. Two forces of 25 and 45 N act on an object. Their directions differ by $70°$. The resulting acceleration has magnitude of $10.0\ \text{m/s}^2$. What is the mass of the body?

74. A force of 1600 N acts parallel to a ramp to push a 300-kg piano into a moving van. The ramp is inclined at $20°$. (a) What is the acceleration of the piano up the ramp? (b) What is the velocity of the piano when it reaches the top if the ramp is 4.0 m long and the piano starts from rest?

75. Draw a free-body diagram of a diver who has entered the water, moved downward, and is acted on by an upward force due to the water which balances the weight (that is, the diver is suspended).

76. For a swimmer who has just jumped off a diving board, assume air resistance is negligible. The swimmer has a mass of 80.0 kg and jumps off a board 10.0 m above the water. Three seconds after entering the water, her downward motion is stopped. What average upward force did the water exert on her?

77. (a) Find an equation to determine the magnitude of the net force required to stop a car of mass m, given that the initial speed of the car is v_0 and the stopping distance is x.

(b) Find the magnitude of the net force if the mass of the car is 1050 kg, the initial speed is 40.0 km/h, and the stopping distance is 25.0 m.

78. A sailboat has a mass of 1.50×10^3 kg and is acted on by a force of 2.00×10^3 N toward the east, while the wind acts behind the sails with a force of 3.00×10^3 N in a direction $45°$ north of east. Find the magnitude and direction of the resulting acceleration.

79. Find the acceleration of the body of mass 10.0 kg shown below.

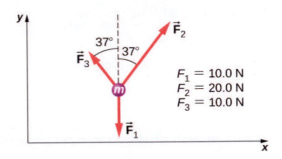

$F_1 = 10.0$ N
$F_2 = 20.0$ N
$F_3 = 10.0$ N

80. A body of mass 2.0 kg is moving along the x-axis with a speed of 3.0 m/s at the instant represented below. (a) What is the acceleration of the body? (b) What is the body's velocity 10.0 s later? (c) What is its displacement after 10.0 s?

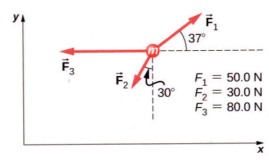

$F_1 = 50.0$ N
$F_2 = 30.0$ N
$F_3 = 80.0$ N

81. Force $\vec{F}_B$ has twice the magnitude of force $\vec{F}_A$. Find the direction in which the particle accelerates in this figure.

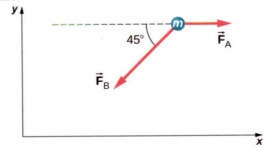

82. Shown below is a body of mass 1.0 kg under the influence of the forces $\vec{F}_A$, $\vec{F}_B$, and $m\vec{g}$. If the body accelerates to the left at $20\ \text{m/s}^2$, what are $\vec{F}_A$ and $\vec{F}_B$?

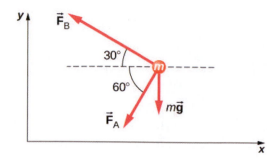

83. A force acts on a car of mass m so that the speed v of the car increases with position x as $v = kx^2$, where k is constant and all quantities are in SI units. Find the force acting on the car as a function of position.

84. A 7.0-N force parallel to an incline is applied to a 1.0-kg crate. The ramp is tilted at $20°$ and is frictionless. (a) What is the acceleration of the crate? (b) If all other conditions are the same but the ramp has a friction force of 1.9 N, what is the acceleration?

85. Two boxes, A and B, are at rest. Box A is on level ground, while box B rests on an inclined plane tilted at angle θ with the horizontal. (a) Write expressions for the normal force acting on each block. (b) Compare the two forces; that is, tell which one is larger or whether they are equal in magnitude. (c) If the angle of incline is $10°$, which force is greater?

86. A mass of 250.0 g is suspended from a spring hanging vertically. The spring stretches 6.00 cm. How much will the spring stretch if the suspended mass is 530.0 g?

87. As shown below, two identical springs, each with the spring constant 20 N/m, support a 15.0-N weight. (a) What is the tension in spring A? (b) What is the amount of stretch of spring A from the rest position?

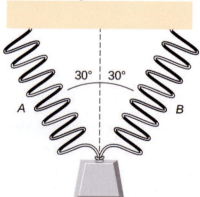

88. Shown below is a 30.0-kg block resting on a frictionless ramp inclined at $60°$ to the horizontal. The block is held by a spring that is stretched 5.0 cm. What is the force constant of the spring?

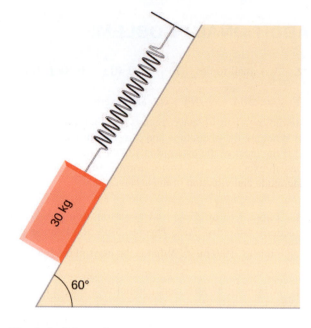

89. In building a house, carpenters use nails from a large box. The box is suspended from a spring twice during the day to measure the usage of nails. At the beginning of the day, the spring stretches 50 cm. At the end of the day, the spring stretches 30 cm. What fraction or percentage of the nails have been used?

90. A force is applied to a block to move it up a $30°$ incline. The incline is frictionless. If $F = 65.0\,\text{N}$ and $M = 5.00\,\text{kg}$, what is the magnitude of the acceleration of the block?

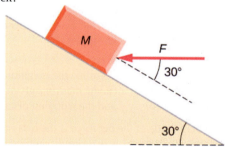

91. Two forces are applied to a 5.0-kg object, and it accelerates at a rate of $2.0\,\text{m/s}^2$ in the positive y-direction. If one of the forces acts in the positive x-direction with magnitude 12.0 N, find the magnitude of the other force.

92. The block on the right shown below has more mass than the block on the left ($m_2 > m_1$). Draw free-body diagrams for each block.

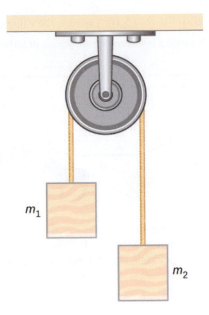

CHALLENGE PROBLEMS

93. If two tugboats pull on a disabled vessel, as shown here in an overhead view, the disabled vessel will be pulled along the direction indicated by the result of the exerted forces. (a) Draw a free-body diagram for the vessel. Assume no friction or drag forces affect the vessel. (b) Did you include all forces in the overhead view in your free-body diagram? Why or why not?

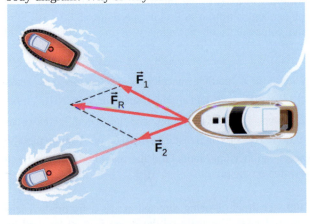

94. A 10.0-kg object is initially moving east at 15.0 m/s. Then a force acts on it for 2.00 s, after which it moves northwest, also at 15.0 m/s. What are the magnitude and direction of the average force that acted on the object over the 2.00-s interval?

95. On June 25, 1983, shot-putter Udo Beyer of East Germany threw the 7.26-kg shot 22.22 m, which at that time was a world record. (a) If the shot was released at a height of 2.20 m with a projection angle of $45.0°$, what was its initial velocity? (b) If while in Beyer's hand the shot was accelerated uniformly over a distance of 1.20 m, what was the net force on it?

96. A body of mass m moves in a horizontal direction such that at time t its position is given by $x(t) = at^4 + bt^3 + ct$, where a, b, and c are constants. (a) What is the acceleration of the body? (b) What is the time-dependent force acting on the body?

97. A body of mass m has initial velocity v_0 in the positive x-direction. It is acted on by a constant force F for time t until the velocity becomes zero; the force continues to act on the body until its velocity becomes $-v_0$ in the same amount of time. Write an expression for the total distance the body travels in terms of the variables indicated.

98. The velocities of a 3.0-kg object at $t = 6.0$ s and $t = 8.0$ s are $(3.0\,\hat{i} - 6.0\,\hat{j} + 4.0\,\hat{k})$ m/s and $(-2.0\,\hat{i} + 4.0\,\hat{k})$ m/s, respectively. If the object is moving at constant acceleration, what is the force acting on it?

99. A 120-kg astronaut is riding in a rocket sled that is sliding along an inclined plane. The sled has a horizontal component of acceleration of $5.0\,\text{m/s}^2$ and a downward component of $3.8\,\text{m/s}^2$. Calculate the magnitude of the force on the rider by the sled. (*Hint:* Remember that gravitational acceleration must be considered.)

100. Two forces are acting on a 5.0-kg object that moves with acceleration $2.0\,\text{m/s}^2$ in the positive y-direction. If one of the forces acts in the positive x-direction and has magnitude of 12 N, what is the magnitude of the other force?

101. Suppose that you are viewing a soccer game from a helicopter above the playing field. Two soccer players simultaneously kick a stationary soccer ball on the flat field; the soccer ball has mass 0.420 kg. The first player kicks with force 162 N at 9.0° north of west. At the same instant, the second player kicks with force 215 N at 15° east of south. Find the acceleration of the ball in $\hat{\textbf{i}}$ and $\hat{\textbf{j}}$ form.

102. A 10.0-kg mass hangs from a spring that has the spring constant 535 N/m. Find the position of the end of the spring away from its rest position. (Use $g = 9.80 \text{ m/s}^2$.)

103. A 0.0502-kg pair of fuzzy dice is attached to the rearview mirror of a car by a short string. The car accelerates at constant rate, and the dice hang at an angle of 3.20° from the vertical because of the car's acceleration. What is the magnitude of the acceleration of the car?

104. At a circus, a donkey pulls on a sled carrying a small clown with a force given by $2.48 \hat{\textbf{i}} + 4.33 \hat{\textbf{j}}$ N. A horse pulls on the same sled, aiding the hapless donkey, with a force of $6.56 \hat{\textbf{i}} + 5.33 \hat{\textbf{j}}$ N. The mass of the sled is 575 kg. Using $\hat{\textbf{i}}$ and $\hat{\textbf{j}}$ form for the answer to each problem, find (a) the net force on the sled when the two animals act together, (b) the acceleration of the sled, and (c) the velocity after 6.50 s.

105. Hanging from the ceiling over a baby bed, well out of baby's reach, is a string with plastic shapes, as shown here. The string is taut (there is no slack), as shown by the straight segments. Each plastic shape has the same mass m, and they are equally spaced by a distance d, as shown. The angles labeled θ describe the angle formed by the end of the string and the ceiling at each end. The center length of sting is horizontal. The remaining two segments each form an angle with the horizontal, labeled ϕ. Let T_1 be the tension in the leftmost section of the string, T_2 be the tension in the section adjacent to it, and T_3 be the tension in the horizontal segment. (a) Find an equation for the tension in each section of the string in terms of the variables m, g, and θ. (b) Find the angle ϕ in terms of

the angle θ. (c) If $\theta = 5.10°$, what is the value of ϕ? (d) Find the distance x between the endpoints in terms of d and θ.

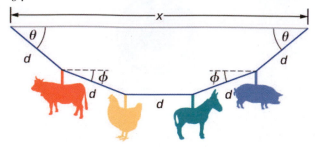

106. A bullet shot from a rifle has mass of 10.0 g and travels to the right at 350 m/s. It strikes a target, a large bag of sand, penetrating it a distance of 34.0 cm. Find the magnitude and direction of the retarding force that slows and stops the bullet.

107. An object is acted on by three simultaneous forces: $\overrightarrow{\textbf{F}}_1 = \left(-3.00 \hat{\textbf{i}} + 2.00 \hat{\textbf{j}}\right)$ N, $\overrightarrow{\textbf{F}}_2 = \left(6.00 \hat{\textbf{i}} - 4.00 \hat{\textbf{j}}\right)$ N, and $\overrightarrow{\textbf{F}}_3 = \left(2.00 \hat{\textbf{i}} + 5.00 \hat{\textbf{j}}\right)$ N. The object experiences acceleration of 4.23 m/s^2. (a) Find the acceleration vector in terms of m. (b) Find the mass of the object. (c) If the object begins from rest, find its speed after 5.00 s. (d) Find the components of the velocity of the object after 5.00 s.

108. In a particle accelerator, a proton has mass 1.67×10^{-27} kg and an initial speed of 2.00×10^5 m/s. It moves in a straight line, and its speed increases to 9.00×10^5 m/s in a distance of 10.0 cm. Assume that the acceleration is constant. Find the magnitude of the force exerted on the proton.

109. A drone is being directed across a frictionless ice-covered lake. The mass of the drone is 1.50 kg, and its velocity is $3.00 \hat{\textbf{i}}$ m/s. After 10.0 s, the velocity is $9.00 \hat{\textbf{i}} + 4.00 \hat{\textbf{j}}$ m/s. If a constant force in the horizontal direction is causing this change in motion, find (a) the components of the force and (b) the magnitude of the force.

6 | APPLICATIONS OF NEWTON'S LAWS

Figure 6.1 Stock cars racing in the Grand National Divisional race at Iowa Speedway in May, 2015. Cars often reach speeds of 200 mph (320 km/h). (credit: modification of work by Erik Schneider/U.S. Navy)

Chapter Outline

6.1 Solving Problems with Newton's Laws

6.2 Friction

6.3 Centripetal Force

6.4 Drag Force and Terminal Speed

Introduction

Car racing has grown in popularity in recent years. As each car moves in a curved path around the turn, its wheels also spin rapidly. The wheels complete many revolutions while the car makes only part of one (a circular arc). How can we describe the velocities, accelerations, and forces involved? What force keeps a racecar from spinning out, hitting the wall bordering the track? What provides this force? Why is the track banked? We answer all of these questions in this chapter as we expand our consideration of Newton's laws of motion.

6.1 | Solving Problems with Newton's Laws

Learning Objectives

By the end of the section, you will be able to:

- Apply problem-solving techniques to solve for quantities in more complex systems of forces
- Use concepts from kinematics to solve problems using Newton's laws of motion
- Solve more complex equilibrium problems
- Solve more complex acceleration problems
- Apply calculus to more advanced dynamics problems

Success in problem solving is necessary to understand and apply physical principles. We developed a pattern of analyzing and setting up the solutions to problems involving Newton's laws in **Newton's Laws of Motion**; in this chapter, we continue to discuss these strategies and apply a step-by-step process.

Problem-Solving Strategies

We follow here the basics of problem solving presented earlier in this text, but we emphasize specific strategies that are useful in applying Newton's laws of motion. Once you identify the physical principles involved in the problem and determine that they include Newton's laws of motion, you can apply these steps to find a solution. These techniques also reinforce concepts that are useful in many other areas of physics. Many problem-solving strategies are stated outright in the worked examples, so the following techniques should reinforce skills you have already begun to develop.

Problem-Solving Strategy: Applying Newton's Laws of Motion

1. Identify the physical principles involved by listing the givens and the quantities to be calculated.

2. Sketch the situation, using arrows to represent all forces.

3. Determine the system of interest. The result is a free-body diagram that is essential to solving the problem.

4. Apply Newton's second law to solve the problem. If necessary, apply appropriate kinematic equations from the chapter on motion along a straight line.

5. Check the solution to see whether it is reasonable.

Let's apply this problem-solving strategy to the challenge of lifting a grand piano into a second-story apartment. Once we have determined that Newton's laws of motion are involved (if the problem involves forces), it is particularly important to draw a careful sketch of the situation. Such a sketch is shown in **Figure 6.2**(a). Then, as in **Figure 6.2**(b), we can represent all forces with arrows. Whenever sufficient information exists, it is best to label these arrows carefully and make the length and direction of each correspond to the represented force.

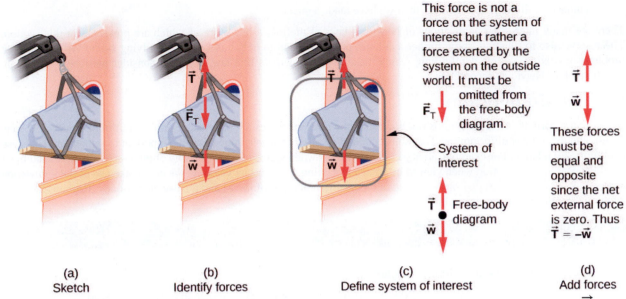

This force is not a force on the system of interest but rather a force exerted by the system on the outside world. It must be omitted from the free-body diagram.

System of interest

Free-body diagram

These forces must be equal and opposite since the net external force is zero. Thus $\vec{T} = -\vec{w}$

(a)	(b)	(c)	(d)
Sketch	Identify forces	Define system of interest	Add forces

Figure 6.2 (a) A grand piano is being lifted to a second-story apartment. (b) Arrows are used to represent all forces: $\vec{T}$ is the tension in the rope above the piano, $\vec{F}_T$ is the force that the piano exerts on the rope, and $\vec{w}$ is the weight of the piano. All other forces, such as the nudge of a breeze, are assumed to be negligible. (c) Suppose we are given the piano's mass and asked to find the tension in the rope. We then define the system of interest as shown and draw a free-body diagram. Now $\vec{F}_T$ is no longer shown, because it is not a force acting on the system of interest; rather, $\vec{F}_T$ acts on the outside world. (d) Showing only the arrows, the head-to-tail method of addition is used. It is apparent that if the piano is stationary, $\vec{T} = -\vec{w}$.

As with most problems, we next need to identify what needs to be determined and what is known or can be inferred from the problem as stated, that is, make a list of knowns and unknowns. It is particularly crucial to identify the system of interest, since Newton's second law involves only external forces. We can then determine which forces are external and which are internal, a necessary step to employ Newton's second law. (See **Figure 6.2**(c).) Newton's third law may be used to identify whether forces are exerted between components of a system (internal) or between the system and something outside (external). As illustrated in **Newton's Laws of Motion**, the system of interest depends on the question we need to answer. Only forces are shown in free-body diagrams, not acceleration or velocity. We have drawn several free-body diagrams in previous worked examples. **Figure 6.2**(c) shows a free-body diagram for the system of interest. Note that no internal forces are shown in a free-body diagram.

Once a free-body diagram is drawn, we apply Newton's second law. This is done in **Figure 6.2**(d) for a particular situation. In general, once external forces are clearly identified in free-body diagrams, it should be a straightforward task to put them into equation form and solve for the unknown, as done in all previous examples. If the problem is one-dimensional—that is, if all forces are parallel—then the forces can be handled algebraically. If the problem is two-dimensional, then it must be broken down into a pair of one-dimensional problems. We do this by projecting the force vectors onto a set of axes chosen for convenience. As seen in previous examples, the choice of axes can simplify the problem. For example, when an incline is involved, a set of axes with one axis parallel to the incline and one perpendicular to it is most convenient. It is almost always convenient to make one axis parallel to the direction of motion, if this is known. Generally, just write Newton's second law in components along the different directions. Then, you have the following equations:

$$\sum F_x = ma_x, \quad \sum F_y = ma_y.$$

(If, for example, the system is accelerating horizontally, then you can then set $a_y = 0$.) We need this information to determine unknown forces acting on a system.

As always, we must check the solution. In some cases, it is easy to tell whether the solution is reasonable. For example, it is reasonable to find that friction causes an object to slide down an incline more slowly than when no friction exists. In practice, intuition develops gradually through problem solving; with experience, it becomes progressively easier to judge whether an answer is reasonable. Another way to check a solution is to check the units. If we are solving for force and end up with units of millimeters per second, then we have made a mistake.

There are many interesting applications of Newton's laws of motion, a few more of which are presented in this section. These serve also to illustrate some further subtleties of physics and to help build problem-solving skills. We look first at problems involving particle equilibrium, which make use of Newton's first law, and then consider particle acceleration, which involves Newton's second law.

Particle Equilibrium

Recall that a particle in equilibrium is one for which the external forces are balanced. Static equilibrium involves objects at rest, and dynamic equilibrium involves objects in motion without acceleration, but it is important to remember that these conditions are relative. For example, an object may be at rest when viewed from our frame of reference, but the same object would appear to be in motion when viewed by someone moving at a constant velocity. We now make use of the knowledge attained in **Newton's Laws of Motion**, regarding the different types of forces and the use of free-body diagrams, to solve additional problems in particle equilibrium.

Example 6.1

Different Tensions at Different Angles

Consider the traffic light (mass of 15.0 kg) suspended from two wires as shown in **Figure 6.3**. Find the tension in each wire, neglecting the masses of the wires.

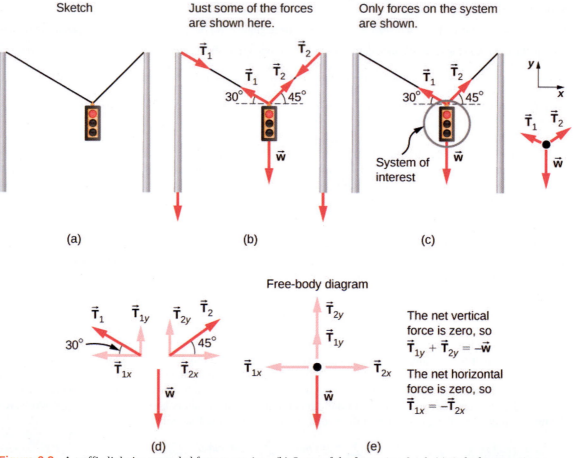

Figure 6.3 A traffic light is suspended from two wires. (b) Some of the forces involved. (c) Only forces acting on the system are shown here. The free-body diagram for the traffic light is also shown. (d) The forces projected onto vertical (*y*) and horizontal (*x*) axes. The horizontal components of the tensions must cancel, and the sum of the vertical components of the tensions must equal the weight of the traffic light. (e) The free-body diagram shows the vertical and horizontal forces acting on the traffic light.

Strategy

The system of interest is the traffic light, and its free-body diagram is shown in **Figure 6.3**(c). The three forces involved are not parallel, and so they must be projected onto a coordinate system. The most convenient coordinate system has one axis vertical and one horizontal, and the vector projections on it are shown in **Figure 6.3**(d). There are two unknowns in this problem (T_1 and T_2), so two equations are needed to find them. These two equations come from applying Newton's second law along the vertical and horizontal axes, noting that the net external force is zero along each axis because acceleration is zero.

Solution

First consider the horizontal or *x*-axis:

$$F_{\text{net } x} = T_{2x} + T_{1x} = 0.$$

Thus, as you might expect,

$$|T_{1x}| = |T_{2x}|.$$

This gives us the following relationship:

$$T_1 \cos 30° = T_2 \cos 45°.$$

Thus,

$$T_2 = 1.225T_1.$$

Note that T_1 and T_2 are not equal in this case because the angles on either side are not equal. It is reasonable that T_2 ends up being greater than T_1 because it is exerted more vertically than T_1.

Now consider the force components along the vertical or *y*-axis:

$$F_{\text{net } y} = T_{1y} + T_{2y} - w = 0.$$

This implies

$$T_{1y} + T_{2y} = w.$$

Substituting the expressions for the vertical components gives

$$T_1 \sin 30° + T_2 \sin 45° = w.$$

There are two unknowns in this equation, but substituting the expression for T_2 in terms of T_1 reduces this to one equation with one unknown:

$$T_1(0.500) + (1.225T_1)(0.707) = w = mg,$$

which yields

$$1.366T_1 = (15.0 \, \text{kg})(9.80 \, \text{m/s}^2).$$

Solving this last equation gives the magnitude of T_1 to be

$$T_1 = 108 \, \text{N}.$$

Finally, we find the magnitude of T_2 by using the relationship between them, $T_2 = 1.225T_1$, found above. Thus we obtain

$$T_2 = 132 \, \text{N}.$$

Significance

Both tensions would be larger if both wires were more horizontal, and they will be equal if and only if the angles on either side are the same (as they were in the earlier example of a tightrope walker in **Newton's Laws of**

Motion.

Particle Acceleration

We have given a variety of examples of particles in equilibrium. We now turn our attention to particle acceleration problems, which are the result of a nonzero net force. Refer again to the steps given at the beginning of this section, and notice how they are applied to the following examples.

Example 6.2

Drag Force on a Barge

Two tugboats push on a barge at different angles (**Figure 6.4**). The first tugboat exerts a force of 2.7×10^5 N in the x-direction, and the second tugboat exerts a force of 3.6×10^5 N in the y-direction. The mass of the barge is 5.0×10^6 kg and its acceleration is observed to be 7.5×10^{-2} m/s^2 in the direction shown. What is the drag force of the water on the barge resisting the motion? (*Note:* Drag force is a frictional force exerted by fluids, such as air or water. The drag force opposes the motion of the object. Since the barge is flat bottomed, we can assume that the drag force is in the direction opposite of motion of the barge.)

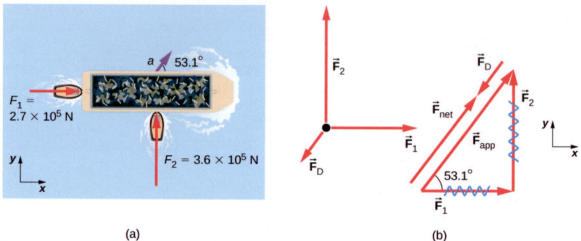

(a)   (b)

Figure 6.4 (a) A view from above of two tugboats pushing on a barge. (b) The free-body diagram for the ship contains only forces acting in the plane of the water. It omits the two vertical forces—the weight of the barge and the buoyant force of the water supporting it cancel and are not shown. Note that $\overrightarrow{\textbf{F}}_{\text{app}}$ is the total applied force of the tugboats.

Strategy

The directions and magnitudes of acceleration and the applied forces are given in **Figure 6.4**(a). We define the total force of the tugboats on the barge as $\overrightarrow{\textbf{F}}_{\text{app}}$ so that

$$\overrightarrow{\textbf{F}}_{\text{app}} = \overrightarrow{\textbf{F}}_1 + \overrightarrow{\textbf{F}}_2.$$

The drag of the water $\overrightarrow{\textbf{F}}_{\text{D}}$ is in the direction opposite to the direction of motion of the boat; this force thus works against $\overrightarrow{\textbf{F}}_{\text{app}}$, as shown in the free-body diagram in **Figure 6.4**(b). The system of interest here is the barge, since the forces on it are given as well as its acceleration. Because the applied forces are perpendicular, the x- and y-axes are in the same direction as $\overrightarrow{\textbf{F}}_1$ and $\overrightarrow{\textbf{F}}_2$. The problem quickly becomes a one-dimensional problem along the direction of $\overrightarrow{\mathbf{F}}_{app}$, since friction is in the direction opposite to $\overrightarrow{\mathbf{F}}_{app}$. Our strategy is to find the magnitude and direction of the net applied force $\overrightarrow{\mathbf{F}}_{app}$ and then apply Newton's second law to solve for the drag force $\overrightarrow{\mathbf{F}}_D$.

Solution

Since F_x and F_y are perpendicular, we can find the magnitude and direction of $\overrightarrow{\mathbf{F}}_{app}$ directly. First, the resultant magnitude is given by the Pythagorean theorem:

$$F_{app} = \sqrt{F_1^2 + F_2^2} = \sqrt{(2.7 \times 10^5\,\text{N})^2 + (3.6 \times 10^5\ \text{N})^2} = 4.5 \times 10^5\ \text{N}.$$

The angle is given by

$$\theta = \tan^{-1}\left(\frac{F_2}{F_1}\right) = \tan^{-1}\left(\frac{3.6 \times 10^5\ \text{N}}{2.7 \times 10^5\ \text{N}}\right) = 53.1°.$$

From Newton's first law, we know this is the same direction as the acceleration. We also know that $\overrightarrow{\mathbf{F}}_D$ is in the opposite direction of $\overrightarrow{\mathbf{F}}_{app}$, since it acts to slow down the acceleration. Therefore, the net external force is in the same direction as $\overrightarrow{\mathbf{F}}_{app}$, but its magnitude is slightly less than $\overrightarrow{\mathbf{F}}_{app}$. The problem is now one-dimensional. From the free-body diagram, we can see that

$$F_{net} = F_{app} - F_D.$$

However, Newton's second law states that

$$F_{net} = ma.$$

Thus,

$$F_{app} - F_D = ma.$$

This can be solved for the magnitude of the drag force of the water F_D in terms of known quantities:

$$F_D = F_{app} - ma.$$

Substituting known values gives

$$F_D = \left(4.5 \times 10^5\ \text{N}\right) - \left(5.0 \times 10^6\ \text{kg}\right)\left(7.5 \times 10^{-2}\ \text{m/s}^2\right) = 7.5 \times 10^4\ \text{N}.$$

The direction of $\overrightarrow{\mathbf{F}}_D$ has already been determined to be in the direction opposite to $\overrightarrow{\mathbf{F}}_{app}$, or at an angle of $53°$ south of west.

Significance

The numbers used in this example are reasonable for a moderately large barge. It is certainly difficult to obtain larger accelerations with tugboats, and small speeds are desirable to avoid running the barge into the docks. Drag is relatively small for a well-designed hull at low speeds, consistent with the answer to this example, where F_D is less than 1/600th of the weight of the ship.

In **Newton's Laws of Motion**, we discussed the normal force, which is a contact force that acts normal to the surface so that an object does not have an acceleration perpendicular to the surface. The bathroom scale is an excellent example of a normal force acting on a body. It provides a quantitative reading of how much it must push upward to support the weight of an object. But can you predict what you would see on the dial of a bathroom scale if you stood on it during an elevator ride? Will you see a value greater than your weight when the elevator starts up? What about when the elevator moves upward at a constant speed? Take a guess before reading the next example.

Example 6.3

What Does the Bathroom Scale Read in an Elevator?

Figure 6.5 shows a 75.0-kg man (weight of about 165 lb.) standing on a bathroom scale in an elevator. Calculate the scale reading: (a) if the elevator accelerates upward at a rate of $1.20 \, \text{m/s}^2$, and (b) if the elevator moves upward at a constant speed of 1 m/s.

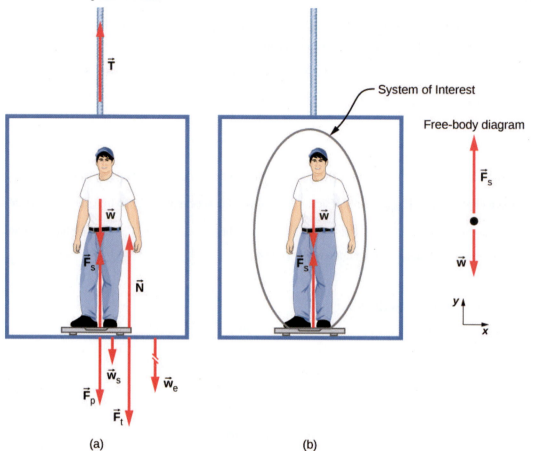

(a)   (b)

Figure 6.5 (a) The various forces acting when a person stands on a bathroom scale in an elevator. The arrows are approximately correct for when the elevator is accelerating upward—broken arrows represent forces too large to be drawn to scale. $\vec{T}$ is the tension in the supporting cable, $\vec{w}$ is the weight of the person, $\vec{w}_s$ is the weight of the scale, $\vec{w}_e$ is the weight of the elevator, $\vec{F}_s$ is the force of the scale on the person, $\vec{F}_p$ is the force of the person on the scale, $\vec{F}_t$ is the force of the scale on the floor of the elevator, and $\vec{N}$ is the force of the floor upward on the scale. (b) The free-body diagram shows only the external forces acting *on* the designated system of interest—the person—and is the diagram we use for the solution of the problem.

Strategy

If the scale at rest is accurate, its reading equals $\vec{F}_p$, the magnitude of the force the person exerts downward on it. **Figure 6.5**(a) shows the numerous forces acting on the elevator, scale, and person. It makes this one-dimensional problem look much more formidable than if the person is chosen to be the system of interest and a free-body diagram is drawn, as in **Figure 6.5**(b). Analysis of the free-body diagram using Newton's laws can produce answers to both **Figure 6.5**(a) and (b) of this example, as well as some other questions that might arise.

The only forces acting on the person are his weight $\overrightarrow{\mathbf{w}}$ and the upward force of the scale $\overrightarrow{\mathbf{F}}_s$. According to Newton's third law, $\overrightarrow{\mathbf{F}}_p$ and $\overrightarrow{\mathbf{F}}_s$ are equal in magnitude and opposite in direction, so that we need to find F_s in order to find what the scale reads. We can do this, as usual, by applying Newton's second law,

$$\overrightarrow{\mathbf{F}}_{net} = m\,\overrightarrow{\mathbf{a}}.$$

From the free-body diagram, we see that $\overrightarrow{\mathbf{F}}_{net} = \overrightarrow{\mathbf{F}}_s - \overrightarrow{\mathbf{w}}$, so we have

$$F_s - w = ma.$$

Solving for F_s gives us an equation with only one unknown:

$$F_s = ma + w,$$

or, because $w = mg$, simply

$$F_s = ma + mg.$$

No assumptions were made about the acceleration, so this solution should be valid for a variety of accelerations in addition to those in this situation. (*Note:* We are considering the case when the elevator is accelerating upward. If the elevator is accelerating downward, Newton's second law becomes $F_s - w = -ma$.)

Solution

a. We have $a = 1.20\,\text{m/s}^2$, so that

$$F_s = (75.0\,\text{kg})(9.80\,\text{m/s}^2) + (75.0\,\text{kg})(1.20\,\text{m/s}^2)$$

yielding

$$F_s = 825\,\text{N}.$$

b. Now, what happens when the elevator reaches a constant upward velocity? Will the scale still read more than his weight? For any constant velocity—up, down, or stationary—acceleration is zero because $a = \dfrac{\Delta v}{\Delta t}$ and $\Delta v = 0$. Thus,

$$F_s = ma + mg = 0 + mg$$

or

$$F_s = (75.0\,\text{kg})(9.80\,\text{m/s}^2),$$

which gives

$$F_s = 735\,\text{N}.$$

Significance

The scale reading in **Figure 6.5**(a) is about 185 lb. What would the scale have read if he were stationary? Since his acceleration would be zero, the force of the scale would be equal to his weight:

$$F_{net} = ma = 0 = F_s - w$$
$$F_s = w = mg$$
$$F_s = (75.0\,\text{kg})(9.80\,\text{m/s}^2) = 735\,\text{N}.$$

Thus, the scale reading in the elevator is greater than his 735-N (165-lb.) weight. This means that the scale is pushing up on the person with a force greater than his weight, as it must in order to accelerate him upward. Clearly, the greater the acceleration of the elevator, the greater the scale reading, consistent with what you feel in rapidly accelerating versus slowly accelerating elevators. In **Figure 6.5**(b), the scale reading is 735 N, which equals the person's weight. This is the case whenever the elevator has a constant velocity—moving up, moving down, or stationary.

 6.1   Check Your Understanding Now calculate the scale reading when the elevator accelerates downward at a rate of 1.20 m/s^2.

The solution to the previous example also applies to an elevator accelerating downward, as mentioned. When an elevator accelerates downward, a is negative, and the scale reading is *less* than the weight of the person. If a constant downward velocity is reached, the scale reading again becomes equal to the person's weight. If the elevator is in free fall and accelerating downward at g, then the scale reading is zero and the person appears to be weightless.

Example 6.4

Two Attached Blocks

Figure 6.6 shows a block of mass m_1 on a frictionless, horizontal surface. It is pulled by a light string that passes over a frictionless and massless pulley. The other end of the string is connected to a block of mass m_2. Find the acceleration of the blocks and the tension in the string in terms of m_1, m_2, and g.

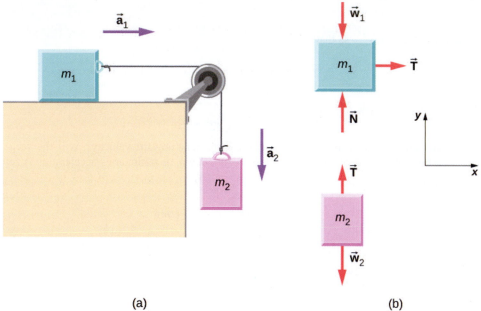

(a)   (b)

Figure 6.6   (a) Block 1 is connected by a light string to block 2. (b) The free-body diagrams of the blocks.

Strategy

We draw a free-body diagram for each mass separately, as shown in **Figure 6.6**. Then we analyze each one to find the required unknowns. The forces on block 1 are the gravitational force, the contact force of the surface, and the tension in the string. Block 2 is subjected to the gravitational force and the string tension. Newton's second law applies to each, so we write two vector equations:

For block 1: $\vec{T} + \vec{w}_1 + \vec{N} = m_1 \vec{a}_1$

For block 2: $\vec{T} + \vec{w}_2 = m_2 \vec{a}_2$.

Notice that $\vec{\mathbf{T}}$ is the same for both blocks. Since the string and the pulley have negligible mass, and since there is no friction in the pulley, the tension is the same throughout the string. We can now write component equations for each block. All forces are either horizontal or vertical, so we can use the same horizontal/vertical coordinate system for both objects

Solution

The component equations follow from the vector equations above. We see that block 1 has the vertical forces balanced, so we ignore them and write an equation relating the x-components. There are no horizontal forces on block 2, so only the y-equation is written. We obtain these results:

$$
\begin{array}{cc}
\textbf{Block 1} & \textbf{Block 2} \\
\sum F_x = ma_x & \sum F_y = ma_y \\
T_x = m_1 a_{1x} & T_y - m_2 g = m_2 a_{2y}.
\end{array}
$$

When block 1 moves to the right, block 2 travels an equal distance downward; thus, $a_{1x} = -a_{2y}$. Writing the common acceleration of the blocks as $a = a_{1x} = -a_{2y}$, we now have

$$T = m_1 a$$

and

$$T - m_2 g = -m_2 a.$$

From these two equations, we can express a and T in terms of the masses m_1 and m_2, and g :

$$a = \frac{m_2}{m_1 + m_2} g$$

and

$$T = \frac{m_1 m_2}{m_1 + m_2} g.$$

Significance

Notice that the tension in the string is *less* than the weight of the block hanging from the end of it. A common error in problems like this is to set $T = m_2 g$. You can see from the free-body diagram of block 2 that cannot be correct if the block is accelerating.

 6.2   Check Your Understanding Calculate the acceleration of the system, and the tension in the string, when the masses are $m_1 = 5.00\,\text{kg}$ and $m_2 = 3.00\,\text{kg}$.

Example 6.5

Atwood Machine

A classic problem in physics, similar to the one we just solved, is that of the Atwood machine, which consists of a rope running over a pulley, with two objects of different mass attached. It is particularly useful in understanding the connection between force and motion. In **Figure 6.7**, $m_1 = 2.00\,\text{kg}$ and $m_2 = 4.00\,\text{kg}$. Consider the pulley to be frictionless. (a) If m_2 is released, what will its acceleration be? (b) What is the tension in the string?

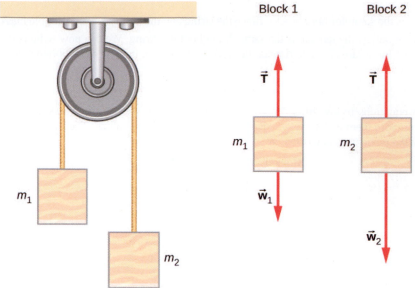

Figure 6.7   An Atwood machine and free-body diagrams for each of the two blocks.

Strategy

We draw a free-body diagram for each mass separately, as shown in the figure. Then we analyze each diagram to find the required unknowns. This may involve the solution of simultaneous equations. It is also important to note the similarity with the previous example. As block 2 accelerates with acceleration a_2 in the downward direction, block 1 accelerates upward with acceleration a_1. Thus, $a = a_1 = -a_2$.

Solution

a. We have

$$\text{For } m_1, \ \sum F_y = T - m_1 g = m_1 a. \qquad \text{For } m_2, \ \sum F_y = T - m_2 g = -m_2 a.$$

(The negative sign in front of $m_2 a$ indicates that m_2 accelerates downward; both blocks accelerate at the same rate, but in opposite directions.) Solve the two equations simultaneously (subtract them) and the result is

$$(m_2 - m_1)g = (m_1 + m_2)a.$$

Solving for a:

$$a = \frac{m_2 - m_1}{m_1 + m_2}g = \frac{4 \text{ kg} - 2 \text{ kg}}{4 \text{ kg} + 2 \text{ kg}}(9.8 \text{ m/s}^2) = 3.27 \text{ m/s}^2.$$

b. Observing the first block, we see that

$$T - m_1 g = m_1 a$$
$$T = m_1(g + a) = (2 \text{ kg})(9.8 \text{ m/s}^2 + 3.27 \text{ m/s}^2) = 26.1 \text{ N}.$$

Significance

The result for the acceleration given in the solution can be interpreted as the ratio of the unbalanced force on the system, $(m_2 - m_1)g$, to the total mass of the system, $m_1 + m_2$. We can also use the Atwood machine to measure local gravitational field strength.

6.3 **Check Your Understanding** Determine a general formula in terms of m_1, m_2 and g for calculating the tension in the string for the Atwood machine shown above.

Newton's Laws of Motion and Kinematics

Physics is most interesting and most powerful when applied to general situations that involve more than a narrow set of physical principles. Newton's laws of motion can also be integrated with other concepts that have been discussed previously in this text to solve problems of motion. For example, forces produce accelerations, a topic of kinematics, and hence the relevance of earlier chapters.

When approaching problems that involve various types of forces, acceleration, velocity, and/or position, listing the givens and the quantities to be calculated will allow you to identify the principles involved. Then, you can refer to the chapters that deal with a particular topic and solve the problem using strategies outlined in the text. The following worked example illustrates how the problem-solving strategy given earlier in this chapter, as well as strategies presented in other chapters, is applied to an integrated concept problem.

Example 6.6

What Force Must a Soccer Player Exert to Reach Top Speed?

A soccer player starts at rest and accelerates forward, reaching a velocity of 8.00 m/s in 2.50 s. (a) What is her average acceleration? (b) What average force does the ground exert forward on the runner so that she achieves this acceleration? The player's mass is 70.0 kg, and air resistance is negligible.

Strategy

To find the answers to this problem, we use the problem-solving strategy given earlier in this chapter. The solutions to each part of the example illustrate how to apply specific problem-solving steps. In this case, we do not need to use all of the steps. We simply identify the physical principles, and thus the knowns and unknowns; apply Newton's second law; and check to see whether the answer is reasonable.

Solution

a. We are given the initial and final velocities (zero and 8.00 m/s forward); thus, the change in velocity is $\Delta v = 8.00 \text{ m/s}$. We are given the elapsed time, so $\Delta t = 2.50 \text{ s}$. The unknown is acceleration, which can be found from its definition:

$$a = \frac{\Delta v}{\Delta t}.$$

Substituting the known values yields

$$a = \frac{8.00 \text{ m/s}}{2.50 \text{ s}} = 3.20 \text{ m/s}^2.$$

b. Here we are asked to find the average force the ground exerts on the runner to produce this acceleration. (Remember that we are dealing with the force or forces acting on the object of interest.) This is the reaction force to that exerted by the player backward against the ground, by Newton's third law. Neglecting air resistance, this would be equal in magnitude to the net external force on the player, since this force causes her acceleration. Since we now know the player's acceleration and are given her mass, we can use Newton's second law to find the force exerted. That is,

$$F_{\text{net}} = ma.$$

Substituting the known values of m and a gives

$$F_{\text{net}} = (70.0 \text{ kg})(3.20 \text{ m/s}^2) = 224 \text{ N}.$$

This is a reasonable result: The acceleration is attainable for an athlete in good condition. The force is about 50 pounds, a reasonable average force.

Significance

This example illustrates how to apply problem-solving strategies to situations that include topics from different chapters. The first step is to identify the physical principles, the knowns, and the unknowns involved in the problem. The second step is to solve for the unknown, in this case using Newton's second law. Finally, we check our answer to ensure it is reasonable. These techniques for integrated concept problems will be useful in applications of physics outside of a physics course, such as in your profession, in other science disciplines, and in everyday life.

 6.4   Check Your Understanding The soccer player stops after completing the play described above, but now notices that the ball is in position to be stolen. If she now experiences a force of 126 N to attempt to steal the ball, which is 2.00 m away from her, how long will it take her to get to the ball?

Example 6.7

What Force Acts on a Model Helicopter?

A 1.50-kg model helicopter has a velocity of $5.00 \, \hat{j}$ m/s at $t = 0$. It is accelerated at a constant rate for two seconds (2.00 s) after which it has a velocity of $\left(6.00 \, \hat{i} + 12.00 \, \hat{j}\right)$m/s. What is the magnitude of the resultant force acting on the helicopter during this time interval?

Strategy

We can easily set up a coordinate system in which the x-axis ($\hat{i}$ direction) is horizontal, and the y-axis ($\hat{j}$ direction) is vertical. We know that $\Delta t = 2.00 s$ and $(6.00 \, \hat{i} + 12.00 \, \hat{j}$ m/s) $- (5.00 \, \hat{j}$ m/s). From this, we can calculate the acceleration by the definition; we can then apply Newton's second law.

Solution

We have

$$a = \frac{\Delta v}{\Delta t} = \frac{(6.00 \, \hat{i} + 12.00 \, \hat{j} \, \text{m/s}) - (5.00 \, \hat{j} \, \text{m/s})}{2.00 \, \text{s}} = 3.00 \, \hat{i} + 3.50 \, \hat{j} \, \text{m/s}^2$$

$$\sum \vec{F} = m \vec{a} = (1.50 \, \text{kg})(3.00 \, \hat{i} + 3.50 \, \hat{j} \, \text{m/s}^2) = 4.50 \, \hat{i} + 5.25 \, \hat{j} \, \text{N}.$$

The magnitude of the force is now easily found:

$$F = \sqrt{(4.50 \, \text{N})^2 + (5.25 \, \text{N})^2} = 6.91 \, \text{N}.$$

Significance

The original problem was stated in terms of $\hat{i} - \hat{j}$ vector components, so we used vector methods. Compare this example with the previous example.

 6.5   Check Your Understanding Find the direction of the resultant for the 1.50-kg model helicopter.

Example 6.8

Baggage Tractor

Figure 6.8(a) shows a baggage tractor pulling luggage carts from an airplane. The tractor has mass 650.0 kg, while cart A has mass 250.0 kg and cart B has mass 150.0 kg. The driving force acting for a brief period of time accelerates the system from rest and acts for 3.00 s. (a) If this driving force is given by $F = (820.0t)\,\text{N}$, find the speed after 3.00 seconds. (b) What is the horizontal force acting on the connecting cable between the tractor and cart A at this instant?

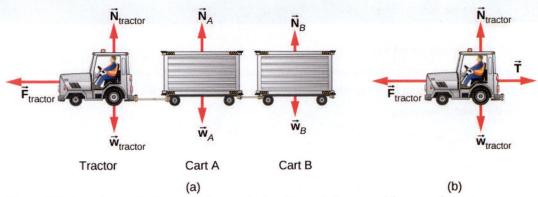

Figure 6.8 (a) A free-body diagram is shown, which indicates all the external forces on the system consisting of the tractor and baggage carts for carrying airline luggage. (b) A free-body diagram of the tractor only is shown isolated in order to calculate the tension in the cable to the carts.

Strategy

A free-body diagram shows the driving force of the tractor, which gives the system its acceleration. We only need to consider motion in the horizontal direction. The vertical forces balance each other and it is not necessary to consider them. For part b, we make use of a free-body diagram of the tractor alone to determine the force between it and cart A. This exposes the coupling force $\vec{T}$, which is our objective.

Solution

a. $\sum F_x = m_{\text{system}} a_x$ and $\sum F_x = 820.0t$, so

$$
\begin{aligned}
820.0t &= (650.0 + 250.0 + 150.0)a \\
a &= 0.7809t.
\end{aligned}
$$

Since acceleration is a function of time, we can determine the velocity of the tractor by using $a = \dfrac{dv}{dt}$ with the initial condition that $v_0 = 0$ at $t = 0$. We integrate from $t = 0$ to $t = 3$:

$$
dv = adt, \quad \int_0^3 dv = \int_0^{3.00} adt = \int_0^{3.00} 0.7809t\,dt, \quad v = 0.3905t^2 \Big|_0^{3.00} = 3.51\ \text{m/s}.
$$

b. Refer to the free-body diagram in **Figure 6.8**(b).

$$
\begin{aligned}
\sum F_x &= m_{\text{tractor}} a_x \\
820.0t - T &= m_{\text{tractor}}(0.7805)t \\
(820.0)(3.00) - T &= (650.0)(0.7805)(3.00) \\
T &= 938\ \text{N}.
\end{aligned}
$$

Significance

Since the force varies with time, we must use calculus to solve this problem. Notice how the total mass of the system was important in solving **Figure 6.8**(a), whereas only the mass of the truck (since it supplied the force) was of use in **Figure 6.8**(b).

Recall that $v = \dfrac{ds}{dt}$ and $a = \dfrac{dv}{dt}$. If acceleration is a function of time, we can use the calculus forms developed in **Motion Along a Straight Line**, as shown in this example. However, sometimes acceleration is a function of displacement. In this case, we can derive an important result from these calculus relations. Solving for dt in each, we have $dt = \frac{ds}{v}$ and $dt = \frac{dv}{a}$. Now, equating these expressions, we have $\frac{ds}{v} = \frac{dv}{a}$. We can rearrange this to obtain $a\,ds = v\,dv$.

Example 6.9

Motion of a Projectile Fired Vertically

A 10.0-kg mortar shell is fired vertically upward from the ground, with an initial velocity of 50.0 m/s (see **Figure 6.9**). Determine the maximum height it will travel if atmospheric resistance is measured as $F_D = (0.0100v^2)$ N, where v is the speed at any instant.

(a)   (b)

Figure 6.9 (a) The mortar fires a shell straight up; we consider the friction force provided by the air. (b) A free-body diagram is shown which indicates all the forces on the mortar shell. (credit a: modification of work by OS541/DoD; The appearance of U.S. Department of Defense (DoD) visual information does not imply or constitute DoD endorsement.)

Strategy

The known force on the mortar shell can be related to its acceleration using the equations of motion. Kinematics can then be used to relate the mortar shell's acceleration to its position.

Solution

Initially, $y_0 = 0$ and $v_0 = 50.0$ m/s. At the maximum height $y = h$, $v = 0$. The free-body diagram shows F_D to act downward, because it slows the upward motion of the mortar shell. Thus, we can write

$$\sum F_y = ma_y$$

$$-F_D - w = ma_y$$
$$-0.0100v^2 - 98.0 = 10.0a$$
$$a = -0.00100v^2 - 9.80.$$

The acceleration depends on v and is therefore variable. Since $a = f(v)$, we can relate a to v using the rearrangement described above,

$$a\,ds = v\,dv.$$

We replace ds with dy because we are dealing with the vertical direction,

$$ady = vdv, \quad (-0.00100v^2 - 9.80)dy = vdv.$$

We now separate the variables (v's and dv's on one side; dy on the other):

$$\int_0^h dy = \int_{50.0}^0 \frac{vdv}{(-0.00100v^2 - 9.80)}$$

$$\int_0^h dy = -\int_{50.0}^0 \frac{vdv}{(0.00100v^2 + 9.80)} = (-5 \times 10^3)\ln(0.00100v^2 + 9.80)\Big|_{50.0}^0.$$

Thus, $h = 114\text{ m}$.

Significance

Notice the need to apply calculus since the force is not constant, which also means that acceleration is not constant. To make matters worse, the force depends on v (not t), and so we must use the trick explained prior to the example. The answer for the height indicates a lower elevation if there were air resistance. We will deal with the effects of air resistance and other drag forces in greater detail in **Drag Force and Terminal Speed**.

 6.6 Check Your Understanding If atmospheric resistance is neglected, find the maximum height for the mortar shell. Is calculus required for this solution?

 Explore the forces at work in this **simulation (https://openstaxcollege.org/l/21forcesatwork)** when you try to push a filing cabinet. Create an applied force and see the resulting frictional force and total force acting on the cabinet. Charts show the forces, position, velocity, and acceleration vs. time. View a free-body diagram of all the forces (including gravitational and normal forces).

6.2 | Friction

Learning Objectives

By the end of the section, you will be able to:

- Describe the general characteristics of friction
- List the various types of friction
- Calculate the magnitude of static and kinetic friction, and use these in problems involving Newton's laws of motion

When a body is in motion, it has resistance because the body interacts with its surroundings. This resistance is a force of friction. Friction opposes relative motion between systems in contact but also allows us to move, a concept that becomes obvious if you try to walk on ice. Friction is a common yet complex force, and its behavior still not completely understood.

Still, it is possible to understand the circumstances in which it behaves.

Static and Kinetic Friction

The basic definition of **friction** is relatively simple to state.

Friction

Friction is a force that opposes relative motion between systems in contact.

There are several forms of friction. One of the simpler characteristics of sliding friction is that it is parallel to the contact surfaces between systems and is always in a direction that opposes motion or attempted motion of the systems relative to each other. If two systems are in contact and moving relative to one another, then the friction between them is called kinetic friction. For example, friction slows a hockey puck sliding on ice. When objects are stationary, static friction can act between them; the static friction is usually greater than the kinetic friction between two objects.

Static and Kinetic Friction

If two systems are in contact and stationary relative to one another, then the friction between them is called **static friction**. If two systems are in contact and moving relative to one another, then the friction between them is called **kinetic friction**.

Imagine, for example, trying to slide a heavy crate across a concrete floor—you might push very hard on the crate and not move it at all. This means that the static friction responds to what you do—it increases to be equal to and in the opposite direction of your push. If you finally push hard enough, the crate seems to slip suddenly and starts to move. Now static friction gives way to kinetic friction. Once in motion, it is easier to keep it in motion than it was to get it started, indicating that the kinetic frictional force is less than the static frictional force. If you add mass to the crate, say by placing a box on top of it, you need to push even harder to get it started and also to keep it moving. Furthermore, if you oiled the concrete you would find it easier to get the crate started and keep it going (as you might expect).

Figure 6.10 is a crude pictorial representation of how friction occurs at the interface between two objects. Close-up inspection of these surfaces shows them to be rough. Thus, when you push to get an object moving (in this case, a crate), you must raise the object until it can skip along with just the tips of the surface hitting, breaking off the points, or both. A considerable force can be resisted by friction with no apparent motion. The harder the surfaces are pushed together (such as if another box is placed on the crate), the more force is needed to move them. Part of the friction is due to adhesive forces between the surface molecules of the two objects, which explains the dependence of friction on the nature of the substances. For example, rubber-soled shoes slip less than those with leather soles. Adhesion varies with substances in contact and is a complicated aspect of surface physics. Once an object is moving, there are fewer points of contact (fewer molecules adhering), so less force is required to keep the object moving. At small but nonzero speeds, friction is nearly independent of speed.

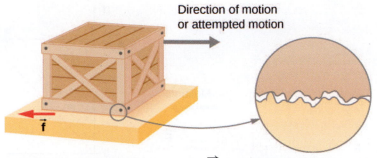

Figure 6.10 Frictional forces, such as $\vec{f}$, always oppose motion or attempted motion between objects in contact. Friction arises in part because of the roughness of the surfaces in contact, as seen in the expanded view. For the object to move, it must rise to where the peaks of the top surface can skip along the bottom surface. Thus, a force is required just to set the object in motion. Some of the peaks will be broken off, also requiring a force to maintain motion. Much of the friction is actually due to attractive forces between molecules making up the two objects, so that even perfectly smooth surfaces are not friction-free. (In fact, perfectly smooth, clean surfaces of similar materials would adhere, forming a bond called a "cold weld.")

The magnitude of the frictional force has two forms: one for static situations (static friction), the other for situations involving motion (kinetic friction). What follows is an approximate empirical (experimentally determined) model only. These equations for static and kinetic friction are not vector equations.

Magnitude of Static Friction

The magnitude of static friction f_s is

$$f_s \leq \mu_s N, \tag{6.1}$$

where μ_s is the coefficient of static friction and N is the magnitude of the normal force.

The symbol $\leq$ means *less than or equal to*, implying that static friction can have a maximum value of $\mu_s N$. Static friction is a responsive force that increases to be equal and opposite to whatever force is exerted, up to its maximum limit. Once the applied force exceeds $f_s(\text{max})$, the object moves. Thus,

$$f_s(\text{max}) = \mu_s N.$$

Magnitude of Kinetic Friction

The magnitude of kinetic friction f_k is given by

$$f_k = \mu_k N, \tag{6.2}$$

where μ_k is the coefficient of kinetic friction.

A system in which $f_k = \mu_k N$ is described as a system in which *friction behaves simply*. The transition from static friction to kinetic friction is illustrated in **Figure 6.11**.

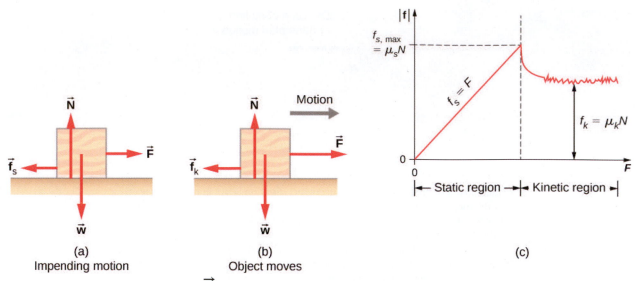

(a)
Impending motion

(b)
Object moves

(c)

Figure 6.11 (a) The force of friction $\vec{f}$ between the block and the rough surface opposes the direction of the applied force $\vec{F}$. The magnitude of the static friction balances that of the applied force. This is shown in the left side of the graph in (c). (b) At some point, the magnitude of the applied force is greater than the force of kinetic friction, and the block moves to the right. This is shown in the right side of the graph. (c) The graph of the frictional force versus the applied force; note that $f_s(\text{max}) > f_k$. This means that $\mu_s > \mu_k$.

As you can see in **Table 6.1**, the coefficients of kinetic friction are less than their static counterparts. The approximate values of μ are stated to only one or two digits to indicate the approximate description of friction given by the preceding two equations.

System	Static Friction μ_s	Kinetic Friction μ_k
Rubber on dry concrete	1.0	0.7
Rubber on wet concrete	0.5-0.7	0.3-0.5
Wood on wood	0.5	0.3
Waxed wood on wet snow	0.14	0.1
Metal on wood	0.5	0.3
Steel on steel (dry)	0.6	0.3
Steel on steel (oiled)	0.05	0.03
Teflon on steel	0.04	0.04
Bone lubricated by synovial fluid	0.016	0.015
Shoes on wood	0.9	0.7
Shoes on ice	0.1	0.05
Ice on ice	0.1	0.03
Steel on ice	0.4	0.02

Table 6.1 Approximate Coefficients of Static and Kinetic Friction

Equation 6.1 and **Equation 6.2** include the dependence of friction on materials and the normal force. The direction of friction is always opposite that of motion, parallel to the surface between objects, and perpendicular to the normal force. For example, if the crate you try to push (with a force parallel to the floor) has a mass of 100 kg, then the normal force is equal to its weight,

$$w = mg = (100\ \text{kg})(9.80\ \text{m/s}^2) = 980\ \text{N},$$

perpendicular to the floor. If the coefficient of static friction is 0.45, you would have to exert a force parallel to the floor greater than

$$f_s(\text{max}) = \mu_s N = (0.45)(980\ \text{N}) = 440\ \text{N}$$

to move the crate. Once there is motion, friction is less and the coefficient of kinetic friction might be 0.30, so that a force of only

$$f_k = \mu_k N = (0.30)(980\ \text{N}) = 290\ \text{N}$$

keeps it moving at a constant speed. If the floor is lubricated, both coefficients are considerably less than they would be without lubrication. Coefficient of friction is a unitless quantity with a magnitude usually between 0 and 1.0. The actual value depends on the two surfaces that are in contact.

Many people have experienced the slipperiness of walking on ice. However, many parts of the body, especially the joints, have much smaller coefficients of friction—often three or four times less than ice. A joint is formed by the ends of two bones, which are connected by thick tissues. The knee joint is formed by the lower leg bone (the tibia) and the thighbone (the femur). The hip is a ball (at the end of the femur) and socket (part of the pelvis) joint. The ends of the bones in the joint are covered by cartilage, which provides a smooth, almost-glassy surface. The joints also produce a fluid (synovial fluid) that reduces friction and wear. A damaged or arthritic joint can be replaced by an artificial joint (**Figure 6.12**). These replacements can be made of metals (stainless steel or titanium) or plastic (polyethylene), also with very small coefficients of friction.

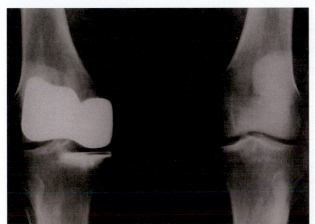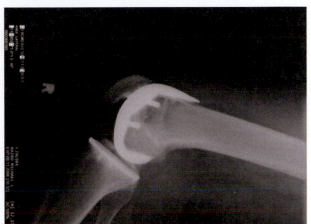

Figure 6.12 Artificial knee replacement is a procedure that has been performed for more than 20 years. These post-operative X-rays show a right knee joint replacement. (credit: modification of work by Mike Baird)

Natural lubricants include saliva produced in our mouths to aid in the swallowing process, and the slippery mucus found between organs in the body, allowing them to move freely past each other during heartbeats, during breathing, and when a person moves. Hospitals and doctor's clinics commonly use artificial lubricants, such as gels, to reduce friction.

The equations given for static and kinetic friction are empirical laws that describe the behavior of the forces of friction. While these formulas are very useful for practical purposes, they do not have the status of mathematical statements that represent general principles (e.g., Newton's second law). In fact, there are cases for which these equations are not even good approximations. For instance, neither formula is accurate for lubricated surfaces or for two surfaces siding across each other at high speeds. Unless specified, we will not be concerned with these exceptions.

Example 6.10

Static and Kinetic Friction

A 20.0-kg crate is at rest on a floor as shown in **Figure 6.13**. The coefficient of static friction between the crate and floor is 0.700 and the coefficient of kinetic friction is 0.600. A horizontal force $\overrightarrow{\mathbf{P}}$ is applied to the crate.

Find the force of friction if (a) $\overrightarrow{\mathbf{P}} = 20.0\ \text{N}$, (b) $\overrightarrow{\mathbf{P}} = 30.0\ \text{N}$, (c) $\overrightarrow{\mathbf{P}} = 120.0\ \text{N}$, and (d) $\overrightarrow{\mathbf{P}} = 180.0\ \text{N}$.

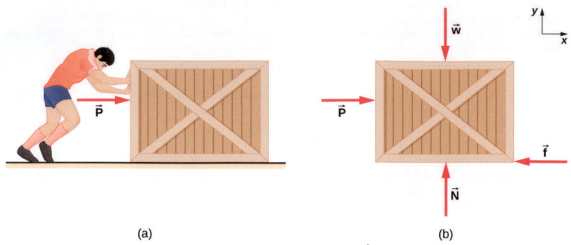

(a)   (b)

Figure 6.13 (a) A crate on a horizontal surface is pushed with a force $\vec{P}$. (b) The forces on the crate. Here, $\vec{f}$ may represent either the static or the kinetic frictional force.

Strategy

The free-body diagram of the crate is shown in **Figure 6.13**(b). We apply Newton's second law in the horizontal and vertical directions, including the friction force in opposition to the direction of motion of the box.

Solution

Newton's second law GIVES

$$\sum F_x = ma_x \qquad \sum F_y = ma_y$$
$$P - f = ma_x \qquad N - w = 0.$$

Here we are using the symbol f to represent the frictional force since we have not yet determined whether the crate is subject to station friction or kinetic friction. We do this whenever we are unsure what type of friction is acting. Now the weight of the crate is

$$w = (20.0\,\text{kg})(9.80\,\text{m/s}^2) = 196\,\text{N},$$

which is also equal to N. The maximum force of static friction is therefore $(0.700)(196\,\text{N}) = 137\,\text{N}$. As long as $\vec{P}$ is less than 137 N, the force of static friction keeps the crate stationary and $f_s = \vec{P}$. Thus, (a) $f_s = 20.0\,\text{N}$, (b) $f_s = 30.0\,\text{N}$, and (c) $f_s = 120.0\,\text{N}$.

(d) If $\vec{P} = 180.0\,\text{N}$, the applied force is greater than the maximum force of static friction (137 N), so the crate can no longer remain at rest. Once the crate is in motion, kinetic friction acts. Then

$$f_k = \mu_k N = (0.600)(196\,\text{N}) = 118\,\text{N},$$

and the acceleration is

$$a_x = \frac{\vec{P} - f_k}{m} = \frac{180.0\,\text{N} - 118\,\text{N}}{20.0\,\text{kg}} = 3.10\,\text{m/s}^2.$$

Significance

This example illustrates how we consider friction in a dynamics problem. Notice that static friction has a value that matches the applied force, until we reach the maximum value of static friction. Also, no motion can occur until the applied force equals the force of static friction, but the force of kinetic friction will then become smaller.

6.7 **Check Your Understanding** A block of mass 1.0 kg rests on a horizontal surface. The frictional coefficients for the block and surface are $\mu_s = 0.50$ and $\mu_k = 0.40$. (a) What is the minimum horizontal force required to move the block? (b) What is the block's acceleration when this force is applied?

Friction and the Inclined Plane

One situation where friction plays an obvious role is that of an object on a slope. It might be a crate being pushed up a ramp to a loading dock or a skateboarder coasting down a mountain, but the basic physics is the same. We usually generalize the sloping surface and call it an inclined plane but then pretend that the surface is flat. Let's look at an example of analyzing motion on an inclined plane with friction.

Example 6.11

Downhill Skier

A skier with a mass of 62 kg is sliding down a snowy slope at a constant velocity. Find the coefficient of kinetic friction for the skier if friction is known to be 45.0 N.

Strategy

The magnitude of kinetic friction is given as 45.0 N. Kinetic friction is related to the normal force N by $f_k = \mu_k N$; thus, we can find the coefficient of kinetic friction if we can find the normal force on the skier. The normal force is always perpendicular to the surface, and since there is no motion perpendicular to the surface, the normal force should equal the component of the skier's weight perpendicular to the slope. (See **Figure 6.14**, which repeats a figure from the chapter on Newton's laws of motion.)

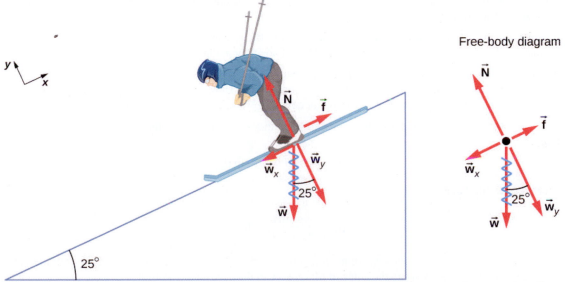

Figure 6.14 The motion of the skier and friction are parallel to the slope, so it is most convenient to project all forces onto a coordinate system where one axis is parallel to the slope and the other is perpendicular (axes shown to left of skier). The normal force $\vec{N}$ is perpendicular to the slope, and friction $\vec{f}$ is parallel to the slope, but the skier's weight $\vec{w}$ has components along both axes, namely $\vec{w}_y$ and $\vec{w}_x$. The normal force $\vec{N}$ is equal in magnitude to $\vec{w}_y$, so there is no motion perpendicular to the slope. However, $\vec{f}$ is equal to $\vec{w}_x$ in magnitude, so there is a constant velocity down the slope (along the x-axis).

We have

$$N = w_y = w \cos 25° = mg \cos 25°.$$

Substituting this into our expression for kinetic friction, we obtain

$$f_k = \mu_k\, mg \cos 25°,$$

which can now be solved for the coefficient of kinetic friction μ_k.

Solution

Solving for μ_k gives

$$\mu_k = \frac{f_k}{N} = \frac{f_k}{w \cos 25°} = \frac{f_k}{mg \cos 25°}.$$

Substituting known values on the right-hand side of the equation,

$$\mu_k = \frac{45.0\,\text{N}}{(62\,\text{kg})(9.80\,\text{m/s}^2)(0.906)} = 0.082.$$

Significance

This result is a little smaller than the coefficient listed in **Table 6.1** for waxed wood on snow, but it is still reasonable since values of the coefficients of friction can vary greatly. In situations like this, where an object of mass m slides down a slope that makes an angle θ with the horizontal, friction is given by $f_k = \mu_k\, mg \cos \theta$.

All objects slide down a slope with constant acceleration under these circumstances.

We have discussed that when an object rests on a horizontal surface, the normal force supporting it is equal in magnitude to its weight. Furthermore, simple friction is always proportional to the normal force. When an object is not on a horizontal surface, as with the inclined plane, we must find the force acting on the object that is directed perpendicular to the surface; it is a component of the weight.

We now derive a useful relationship for calculating coefficient of friction on an inclined plane. Notice that the result applies only for situations in which the object slides at constant speed down the ramp.

An object slides down an inclined plane at a constant velocity if the net force on the object is zero. We can use this fact to measure the coefficient of kinetic friction between two objects. As shown in **Example 6.10**, the kinetic friction on a slope is $f_k = \mu_k\, mg \cos \theta$. The component of the weight down the slope is equal to $mg \sin \theta$ (see the free-body diagram in **Figure 6.14**). These forces act in opposite directions, so when they have equal magnitude, the acceleration is zero. Writing these out,

$$\mu_k\, mg \cos \theta = mg \sin \theta.$$

Solving for μ_k, we find that

$$\mu_k = \frac{mg \sin \theta}{mg \cos \theta} = \tan \theta.$$

Put a coin on a book and tilt it until the coin slides at a constant velocity down the book. You might need to tap the book lightly to get the coin to move. Measure the angle of tilt relative to the horizontal and find μ_k. Note that the coin does not start to slide at all until an angle greater than θ is attained, since the coefficient of static friction is larger than the coefficient of kinetic friction. Think about how this may affect the value for μ_k and its uncertainty.

Atomic-Scale Explanations of Friction

The simpler aspects of friction dealt with so far are its macroscopic (large-scale) characteristics. Great strides have been made in the atomic-scale explanation of friction during the past several decades. Researchers are finding that the atomic nature of friction seems to have several fundamental characteristics. These characteristics not only explain some of the simpler aspects of friction—they also hold the potential for the development of nearly friction-free environments that could save hundreds of billions of dollars in energy which is currently being converted (unnecessarily) into heat.

Figure 6.15 illustrates one macroscopic characteristic of friction that is explained by microscopic (small-scale) research. We have noted that friction is proportional to the normal force, but not to the amount of area in contact, a somewhat counterintuitive notion. When two rough surfaces are in contact, the actual contact area is a tiny fraction of the total area because only high spots touch. When a greater normal force is exerted, the actual contact area increases, and we find that the friction is proportional to this area.

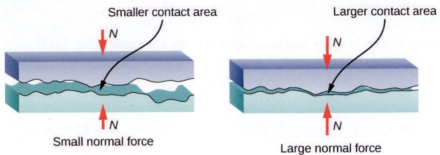

Figure 6.15 Two rough surfaces in contact have a much smaller area of actual contact than their total area. When the normal force is larger as a result of a larger applied force, the area of actual contact increases, as does friction.

However, the atomic-scale view promises to explain far more than the simpler features of friction. The mechanism for how heat is generated is now being determined. In other words, why do surfaces get warmer when rubbed? Essentially, atoms are linked with one another to form lattices. When surfaces rub, the surface atoms adhere and cause atomic lattices to vibrate—essentially creating sound waves that penetrate the material. The sound waves diminish with distance, and their energy is converted into heat. Chemical reactions that are related to frictional wear can also occur between atoms and molecules on the surfaces. **Figure 6.16** shows how the tip of a probe drawn across another material is deformed by atomic-scale friction. The force needed to drag the tip can be measured and is found to be related to shear stress, which is discussed in **Static Equilibrium and Elasticity**. The variation in shear stress is remarkable (more than a factor of 10^{12}) and difficult to predict theoretically, but shear stress is yielding a fundamental understanding of a large-scale phenomenon known since ancient times—friction.

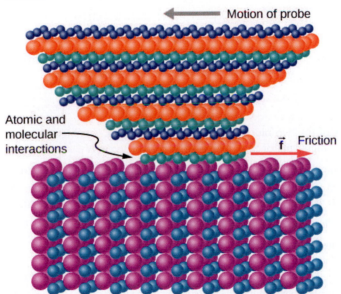

Figure 6.16 The tip of a probe is deformed sideways by frictional force as the probe is dragged across a surface. Measurements of how the force varies for different materials are yielding fundamental insights into the atomic nature of friction.

Describe a **model for friction (https://openstaxcollege.org/l/21friction)** on a molecular level. Describe matter in terms of molecular motion. The description should include diagrams to support the description; how the temperature affects the image; what are the differences and similarities between solid, liquid, and gas particle motion; and how the size and speed of gas molecules relate to everyday objects.

Example 6.12

Sliding Blocks

The two blocks of **Figure 6.17** are attached to each other by a massless string that is wrapped around a frictionless pulley. When the bottom 4.00-kg block is pulled to the left by the constant force $\vec{P}$, the top 2.00-kg block slides across it to the right. Find the magnitude of the force necessary to move the blocks at constant speed. Assume that the coefficient of kinetic friction between all surfaces is 0.400.

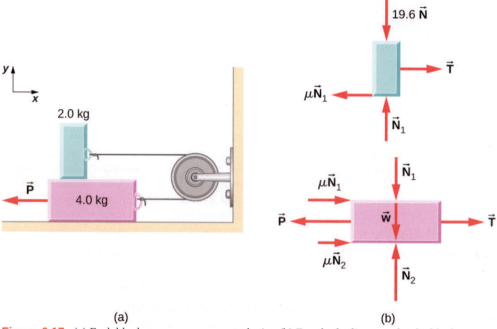

Figure 6.17 (a) Each block moves at constant velocity. (b) Free-body diagrams for the blocks.

Strategy

We analyze the motions of the two blocks separately. The top block is subjected to a contact force exerted by the bottom block. The components of this force are the normal force N_1 and the frictional force $-0.400N_1$.

Other forces on the top block are the tension $T\mathbf{i}$ in the string and the weight of the top block itself, 19.6 N. The bottom block is subjected to contact forces due to the top block and due to the floor. The first contact force has components $-N_1$ and $0.400N_1$, which are simply reaction forces to the contact forces that the bottom block exerts on the top block. The components of the contact force of the floor are N_2 and $0.400N_2$. Other forces on this block are $-P$, the tension $T\mathbf{i}$, and the weight -39.2 N.

Solution

Since the top block is moving horizontally to the right at constant velocity, its acceleration is zero in both the horizontal and the vertical directions. From Newton's second law,

$$\sum F_x = m_1 a_x \qquad \sum F_y = m_1 a_y$$
$$T - 0.400N_1 = 0 \qquad N_1 - 19.6\,\text{N} = 0.$$

Solving for the two unknowns, we obtain $N_1 = 19.6\,\text{N}$ and $T = 0.40N_1 = 7.84\,\text{N}$. The bottom block is also not accelerating, so the application of Newton's second law to this block gives

$$\sum F_x = m_2 a_x \qquad \sum F_y = m_2 a_y$$
$$T - P + 0.400\,N_1 + 0.400\,N_2 = 0 \qquad N_2 - 39.2\,\text{N} - N_1 = 0.$$

The values of N_1 and T were found with the first set of equations. When these values are substituted into the second set of equations, we can determine N_2 and P. They are

$$N_2 = 58.8\,\text{N} \quad\text{and}\quad P = 39.2\,\text{N}.$$

Significance

Understanding what direction in which to draw the friction force is often troublesome. Notice that each friction force labeled in **Figure 6.17** acts in the direction opposite the motion of its corresponding block.

Example 6.13

A Crate on an Accelerating Truck

A 50.0-kg crate rests on the bed of a truck as shown in **Figure 6.18**. The coefficients of friction between the surfaces are $\mu_k = 0.300$ and $\mu_s = 0.400$. Find the frictional force on the crate when the truck is accelerating forward relative to the ground at (a) 2.00 m/s^2, and (b) 5.00 m/s^2.

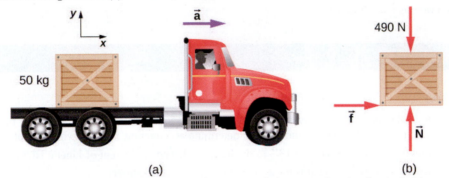

(a) (b)

Figure 6.18 (a) A crate rests on the bed of the truck that is accelerating forward. (b) The free-body diagram of the crate.

Strategy

The forces on the crate are its weight and the normal and frictional forces due to contact with the truck bed. We start by *assuming* that the crate is not slipping. In this case, the static frictional force f_s acts on the crate. Furthermore, the accelerations of the crate and the truck are equal.

Solution

a. Application of Newton's second law to the crate, using the reference frame attached to the ground, yields

$$\sum F_x = ma_x \qquad\qquad \sum F_y = ma_y$$
$$f_s = (50.0\,\text{kg})(2.00\,\text{m/s}^2) \qquad N - 4.90\times10^2\,\text{N} = (50.0\,\text{kg})(0)$$
$$= 1.00\times10^2\,\text{N} \qquad\qquad\qquad N = 4.90\times10^2\,\text{N}.$$

We can now check the validity of our no-slip assumption. The maximum value of the force of static friction is

$$\mu_s N = (0.400)(4.90\times10^2\,\text{N}) = 196\,\text{N},$$

whereas the *actual* force of static friction that acts when the truck accelerates forward at 2.00 m/s^2 is only 1.00×10^2 N. Thus, the assumption of no slipping is valid.

b. If the crate is to move with the truck when it accelerates at 5.0 m/s^2, the force of static friction must be

$$f_s = ma_x = (50.0 \text{ kg})(5.00 \text{ m/s}^2) = 250 \text{ N}.$$

Since this exceeds the maximum of 196 N, the crate must slip. The frictional force is therefore kinetic and is

$$f_k = \mu_k N = (0.300)(4.90 \times 10^2 \text{ N}) = 147 \text{ N}.$$

The horizontal acceleration of the crate relative to the ground is now found from

$$\sum F_x = ma_x$$
$$147 \text{ N} = (50.0 \text{ kg})a_x,$$
$$\text{so } a_x = 2.94 \text{ m/s}^2.$$

Significance

Relative to the ground, the truck is accelerating forward at 5.0 m/s^2 and the crate is accelerating forward at 2.94 m/s^2. Hence the crate is sliding backward relative to the bed of the truck with an acceleration $2.94 \text{ m/s}^2 - 5.00 \text{ m/s}^2 = -2.06 \text{ m/s}^2$.

Example 6.14

Snowboarding

Earlier, we analyzed the situation of a downhill skier moving at constant velocity to determine the coefficient of kinetic friction. Now let's do a similar analysis to determine acceleration. The snowboarder of **Figure 6.19** glides down a slope that is inclined at $\theta = 13^0$ to the horizontal. The coefficient of kinetic friction between the board and the snow is $\mu_k = 0.20.$ What is the acceleration of the snowboarder?

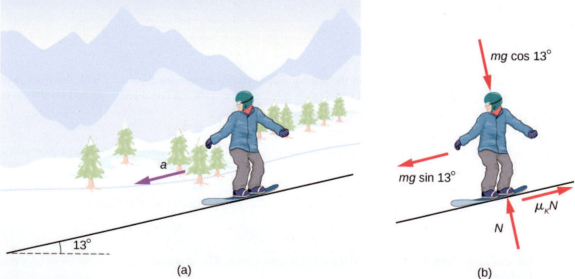

Figure 6.19 (a) A snowboarder glides down a slope inclined at 13° to the horizontal. (b) The free-body diagram of the snowboarder.

Strategy

The forces acting on the snowboarder are her weight and the contact force of the slope, which has a component normal to the incline and a component along the incline (force of kinetic friction). Because she moves along the slope, the most convenient reference frame for analyzing her motion is one with the x-axis along and the y-axis perpendicular to the incline. In this frame, both the normal and the frictional forces lie along coordinate axes, the components of the weight are $mg \sin \theta$ along the slope and $mg \cos \theta$ at right angles into the slope, and the only acceleration is along the x-axis $(a_y = 0)$.

Solution

We can now apply Newton's second law to the snowboarder:

$$\sum F_x = ma_x \qquad \sum F_y = ma_y$$
$$mg \sin \theta - \mu_k N = ma_x \qquad N - mg \cos \theta = m(0).$$

From the second equation, $N = mg \cos \theta$. Upon substituting this into the first equation, we find

$$a_x = g(\sin \theta - \mu_k \cos \theta)$$
$$= g(\sin 13° - 0.20 \cos 13°) = 0.29 \text{ m/s}^2.$$

Significance

Notice from this equation that if θ is small enough or μ_k is large enough, a_x is negative, that is, the snowboarder slows down.

 6.8 Check Your Understanding The snowboarder is now moving down a hill with incline $10.0°$. What is the skier's acceleration?

6.3 | Centripetal Force

Learning Objectives

By the end of the section, you will be able to:

- Explain the equation for centripetal acceleration
- Apply Newton's second law to develop the equation for centripetal force
- Use circular motion concepts in solving problems involving Newton's laws of motion

In **Motion in Two and Three Dimensions**, we examined the basic concepts of circular motion. An object undergoing circular motion, like one of the race cars shown at the beginning of this chapter, must be accelerating because it is changing the direction of its velocity. We proved that this centrally directed acceleration, called centripetal acceleration, is given by the formula

$$a_c = \frac{v^2}{r}$$

where v is the velocity of the object, directed along a tangent line to the curve at any instant. If we know the angular velocity ω, then we can use

$$a_c = r\omega^2.$$

Angular velocity gives the rate at which the object is turning through the curve, in units of rad/s. This acceleration acts along the radius of the curved path and is thus also referred to as a radial acceleration.

An acceleration must be produced by a force. Any force or combination of forces can cause a centripetal or radial acceleration. Just a few examples are the tension in the rope on a tether ball, the force of Earth's gravity on the Moon, friction between roller skates and a rink floor, a banked roadway's force on a car, and forces on the tube of a spinning centrifuge. Any net force causing uniform circular motion is called a **centripetal force**. The direction of a centripetal force is toward the center of curvature, the same as the direction of centripetal acceleration. According to Newton's second law of motion, net force is mass times acceleration: $F_{net} = ma$. For uniform circular motion, the acceleration is the centripetal acceleration: $a = a_c$. Thus, the magnitude of centripetal force F_c is

$$F_c = ma_c.$$

By substituting the expressions for centripetal acceleration a_c $(a_c = \frac{v^2}{r}; a_c = r\omega^2)$, we get two expressions for the centripetal force F_c in terms of mass, velocity, angular velocity, and radius of curvature:

$$F_c = m\frac{v^2}{r}; \quad F_c = mr\omega^2. \tag{6.3}$$

You may use whichever expression for centripetal force is more convenient. Centripetal force $\vec{F}_c$ is always perpendicular to the path and points to the center of curvature, because $\vec{a}_c$ is perpendicular to the velocity and points to the center of curvature. Note that if you solve the first expression for r, you get

$$r = \frac{mv^2}{F_c}.$$

This implies that for a given mass and velocity, a large centripetal force causes a small radius of curvature—that is, a tight curve, as in **Figure 6.20**.

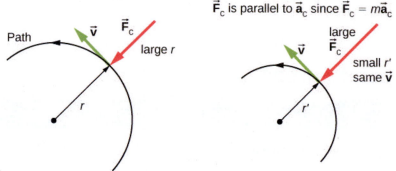

Figure 6.20 The frictional force supplies the centripetal force and is numerically equal to it. Centripetal force is perpendicular to velocity and causes uniform circular motion. The larger the F_c, the smaller the radius of curvature r and the sharper the curve. The second curve has the same v, but a larger F_c produces a smaller r'.

Example 6.15

What Coefficient of Friction Do Cars Need on a Flat Curve?

(a) Calculate the centripetal force exerted on a 900.0-kg car that negotiates a 500.0-m radius curve at 25.00 m/s.
(b) Assuming an unbanked curve, find the minimum static coefficient of friction between the tires and the road, static friction being the reason that keeps the car from slipping (**Figure 6.21**).

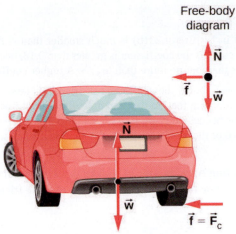

Figure 6.21 This car on level ground is moving away and turning to the left. The centripetal force causing the car to turn in a circular path is due to friction between the tires and the road. A minimum coefficient of friction is needed, or the car will move in a larger-radius curve and leave the roadway.

Strategy

a. We know that $F_c = \frac{mv^2}{r}$. Thus,

$$F_c = \frac{mv^2}{r} = \frac{(900.0 \text{ kg})(25.00 \text{ m/s})^2}{(500.0 \text{ m})} = 1125 \text{ N}.$$

b. **Figure 6.21** shows the forces acting on the car on an unbanked (level ground) curve. Friction is to the left, keeping the car from slipping, and because it is the only horizontal force acting on the car, the friction is the centripetal force in this case. We know that the maximum static friction (at which the tires roll but do not slip) is $\mu_s N$, where μ_s is the static coefficient of friction and N is the normal force. The normal force equals the car's weight on level ground, so $N = mg$. Thus the centripetal force in this situation is

$$F_c \equiv f = \mu_s N = \mu_s mg.$$

Now we have a relationship between centripetal force and the coefficient of friction. Using the equation

$$F_c = m\frac{v^2}{r},$$

we obtain

$$m\frac{v^2}{r} = \mu_s mg.$$

We solve this for μ_s, noting that mass cancels, and obtain

$$\mu_s = \frac{v^2}{rg}.$$

Substituting the knowns,

$$\mu_s = \frac{(25.00 \text{ m/s})^2}{(500.0 \text{ m})(9.80 \text{ m/s}^2)} = 0.13.$$

(Because coefficients of friction are approximate, the answer is given to only two digits.)

Significance

The coefficient of friction found in **Figure 6.21**(b) is much smaller than is typically found between tires and roads. The car still negotiates the curve if the coefficient is greater than 0.13, because static friction is a responsive force, able to assume a value less than but no more than $\mu_s N$. A higher coefficient would also allow the car to negotiate the curve at a higher speed, but if the coefficient of friction is less, the safe speed would be less than 25 m/s. Note that mass cancels, implying that, in this example, it does not matter how heavily loaded the car is to negotiate the turn. Mass cancels because friction is assumed proportional to the normal force, which in turn is proportional to mass. If the surface of the road were banked, the normal force would be less, as discussed next.

 6.9 Check Your Understanding A car moving at 96.8 km/h travels around a circular curve of radius 182.9 m on a flat country road. What must be the minimum coefficient of static friction to keep the car from slipping?

Banked Curves

Let us now consider **banked curves**, where the slope of the road helps you negotiate the curve (**Figure 6.22**). The greater the angle θ, the faster you can take the curve. Race tracks for bikes as well as cars, for example, often have steeply banked curves. In an "ideally banked curve," the angle θ is such that you can negotiate the curve at a certain speed without the aid of friction between the tires and the road. We will derive an expression for θ for an ideally banked curve and consider an example related to it.

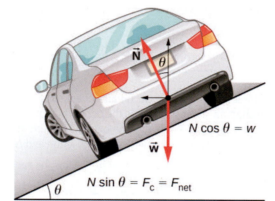

Figure 6.22 The car on this banked curve is moving away and turning to the left.

For **ideal banking**, the net external force equals the horizontal centripetal force in the absence of friction. The components of the normal force N in the horizontal and vertical directions must equal the centripetal force and the weight of the car, respectively. In cases in which forces are not parallel, it is most convenient to consider components along perpendicular axes—in this case, the vertical and horizontal directions.

Figure 6.22 shows a free-body diagram for a car on a frictionless banked curve. If the angle θ is ideal for the speed and radius, then the net external force equals the necessary centripetal force. The only two external forces acting on the car are its weight $\vec{w}$ and the normal force of the road $\vec{N}$. (A frictionless surface can only exert a force perpendicular to the surface—that is, a normal force.) These two forces must add to give a net external force that is horizontal toward the center of curvature and has magnitude mv^2/r. Because this is the crucial force and it is horizontal, we use a coordinate system with vertical and horizontal axes. Only the normal force has a horizontal component, so this must equal the centripetal force, that is,

$$N \sin \theta = \frac{mv^2}{r}.$$

Because the car does not leave the surface of the road, the net vertical force must be zero, meaning that the vertical components of the two external forces must be equal in magnitude and opposite in direction. From **Figure 6.22**, we see that the vertical component of the normal force is $N \cos \theta$, and the only other vertical force is the car's weight. These must be equal in magnitude; thus,

$$N \cos \theta = mg.$$

Now we can combine these two equations to eliminate N and get an expression for θ, as desired. Solving the second equation for $N = mg/(\cos\theta)$ and substituting this into the first yields

$$mg \frac{\sin \theta}{\cos \theta} = \frac{mv^2}{r}$$

$$mg \tan \theta = \frac{mv^2}{r}$$

$$\tan \theta = \frac{v^2}{rg}.$$

Taking the inverse tangent gives

$$\theta = \tan^{-1}\left(\frac{v^2}{rg}\right). \tag{6.4}$$

This expression can be understood by considering how θ depends on v and r. A large θ is obtained for a large v and a small r. That is, roads must be steeply banked for high speeds and sharp curves. Friction helps, because it allows you to take the curve at greater or lower speed than if the curve were frictionless. Note that θ does not depend on the mass of the vehicle.

Example 6.16

What Is the Ideal Speed to Take a Steeply Banked Tight Curve?

Curves on some test tracks and race courses, such as Daytona International Speedway in Florida, are very steeply banked. This banking, with the aid of tire friction and very stable car configurations, allows the curves to be taken at very high speed. To illustrate, calculate the speed at which a 100.0-m radius curve banked at $31.0°$ should be driven if the road were frictionless.

Strategy

We first note that all terms in the expression for the ideal angle of a banked curve except for speed are known; thus, we need only rearrange it so that speed appears on the left-hand side and then substitute known quantities.

Solution

Starting with

$$\tan \theta = \frac{v^2}{rg},$$

we get

$$v = \sqrt{rg \tan \theta}.$$

Noting that $\tan 31.0° = 0.609$, we obtain

$$v = \sqrt{(100.0 \text{ m})(9.80 \text{ m/s}^2)(0.609)} = 24.4 \text{ m/s}.$$

Significance

This is just about 165 km/h, consistent with a very steeply banked and rather sharp curve. Tire friction enables a vehicle to take the curve at significantly higher speeds.

Airplanes also make turns by banking. The lift force, due to the force of the air on the wing, acts at right angles to the wing. When the airplane banks, the pilot is obtaining greater lift than necessary for level flight. The vertical component of lift balances the airplane's weight, and the horizontal component accelerates the plane. The banking angle shown in **Figure 6.23** is given by θ. We analyze the forces in the same way we treat the case of the car rounding a banked curve.

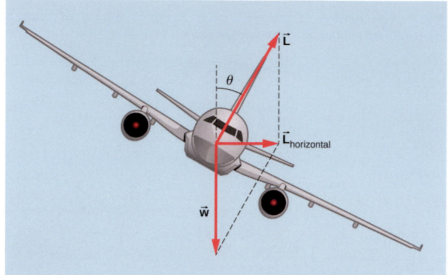

Figure 6.23 In a banked turn, the horizontal component of lift is unbalanced and accelerates the plane. The normal component of lift balances the plane's weight. The banking angle is given by θ. Compare the vector diagram with that shown in **Figure 6.22**.

 Join the **ladybug (https://openstaxcollege.org/l/21ladybug)** in an exploration of rotational motion. Rotate the merry-go-round to change its angle or choose a constant angular velocity or angular acceleration. Explore how circular motion relates to the bug's *xy*-position, velocity, and acceleration using vectors or graphs.

 A circular motion requires a force, the so-called centripetal force, which is directed to the axis of rotation. This simplified **model of a carousel (https://openstaxcollege.org/l/21carousel)** demonstrates this force.

Inertial Forces and Noninertial (Accelerated) Frames: The Coriolis Force

What do taking off in a jet airplane, turning a corner in a car, riding a merry-go-round, and the circular motion of a tropical cyclone have in common? Each exhibits inertial forces—forces that merely seem to arise from motion, because the observer's frame of reference is accelerating or rotating. When taking off in a jet, most people would agree it feels as if you are being pushed back into the seat as the airplane accelerates down the runway. Yet a physicist would say that *you* tend to remain stationary while the *seat* pushes forward on you. An even more common experience occurs when you make a tight curve in your car—say, to the right (**Figure 6.24**). You feel as if you are thrown (that is, *forced*) toward the left relative to the car. Again, a physicist would say that *you* are going in a straight line (recall Newton's first law) but the *car* moves to the right, not that you are experiencing a force from the left.

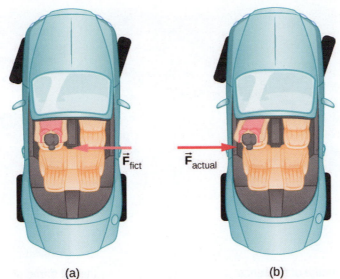

(a)   (b)

Figure 6.24   (a) The car driver feels herself forced to the left relative to the car when she makes a right turn. This is an inertial force arising from the use of the car as a frame of reference. (b) In Earth's frame of reference, the driver moves in a straight line, obeying Newton's first law, and the car moves to the right. There is no force to the left on the driver relative to Earth. Instead, there is a force to the right on the car to make it turn.

We can reconcile these points of view by examining the frames of reference used. Let us concentrate on people in a car. Passengers instinctively use the car as a frame of reference, whereas a physicist might use Earth. The physicist might make this choice because Earth is nearly an inertial frame of reference, in which all forces have an identifiable physical origin. In such a frame of reference, Newton's laws of motion take the form given in **Newton's Laws of Motion**. The car is a **noninertial frame of reference** because it is accelerated to the side. The force to the left sensed by car passengers is an **inertial force** having no physical origin (it is due purely to the inertia of the passenger, not to some physical cause such as tension, friction, or gravitation). The car, as well as the driver, is actually accelerating to the right. This inertial force is said to be an inertial force because it does not have a physical origin, such as gravity.

A physicist will choose whatever reference frame is most convenient for the situation being analyzed. There is no problem to a physicist in including inertial forces and Newton's second law, as usual, if that is more convenient, for example, on a merry-go-round or on a rotating planet. Noninertial (accelerated) frames of reference are used when it is useful to do so. Different frames of reference must be considered in discussing the motion of an astronaut in a spacecraft traveling at speeds near the speed of light, as you will appreciate in the study of the special theory of relativity.

Let us now take a mental ride on a merry-go-round—specifically, a rapidly rotating playground merry-go-round (**Figure 6.25**). You take the merry-go-round to be your frame of reference because you rotate together. When rotating in that noninertial frame of reference, you feel an inertial force that tends to throw you off; this is often referred to as a *centrifugal force* (not to be confused with centripetal force). Centrifugal force is a commonly used term, but it does not actually exist. You must hang on tightly to counteract your inertia (which people often refer to as centrifugal force). In Earth's frame of reference, there is no force trying to throw you off; we emphasize that centrifugal force is a fiction. You must hang on to make yourself go in a circle because otherwise you would go in a straight line, right off the merry-go-round, in keeping with Newton's first law. But the force you exert acts toward the center of the circle.

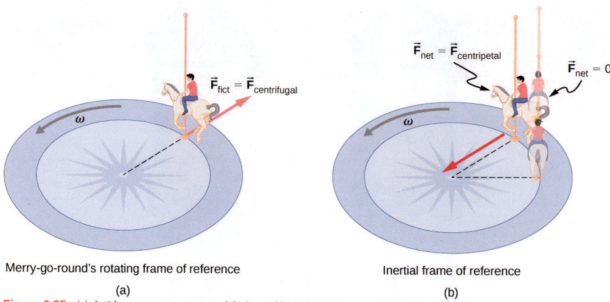

Merry-go-round's rotating frame of reference

(a)

Inertial frame of reference

(b)

Figure 6.25 (a) A rider on a merry-go-round feels as if he is being thrown off. This inertial force is sometimes mistakenly called the centrifugal force in an effort to explain the rider's motion in the rotating frame of reference. (b) In an inertial frame of reference and according to Newton's laws, it is his inertia that carries him off (the unshaded rider has $F_{net} = 0$ and heads in a straight line). A force, $F_{centripetal}$, is needed to cause a circular path.

This inertial effect, carrying you away from the center of rotation if there is no centripetal force to cause circular motion, is put to good use in centrifuges (**Figure 6.26**). A centrifuge spins a sample very rapidly, as mentioned earlier in this chapter. Viewed from the rotating frame of reference, the inertial force throws particles outward, hastening their sedimentation. The greater the angular velocity, the greater the centrifugal force. But what really happens is that the inertia of the particles carries them along a line tangent to the circle while the test tube is forced in a circular path by a centripetal force.

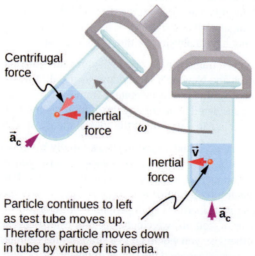

Figure 6.26 Centrifuges use inertia to perform their task. Particles in the fluid sediment settle out because their inertia carries them away from the center of rotation. The large angular velocity of the centrifuge quickens the sedimentation. Ultimately, the particles come into contact with the test tube walls, which then supply the centripetal force needed to make them move in a circle of constant radius.

Let us now consider what happens if something moves in a rotating frame of reference. For example, what if you slide a ball directly away from the center of the merry-go-round, as shown in **Figure 6.27**? The ball follows a straight path relative to

Earth (assuming negligible friction) and a path curved to the right on the merry-go-round's surface. A person standing next to the merry-go-round sees the ball moving straight and the merry-go-round rotating underneath it. In the merry-go-round's frame of reference, we explain the apparent curve to the right by using an inertial force, called the **Coriolis force**, which causes the ball to curve to the right. The Coriolis force can be used by anyone in that frame of reference to explain why objects follow curved paths and allows us to apply Newton's laws in noninertial frames of reference.

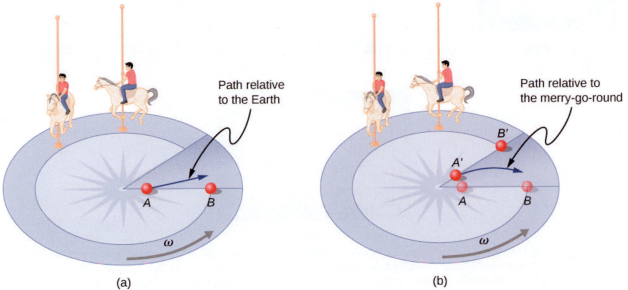

Figure 6.27 Looking down on the counterclockwise rotation of a merry-go-round, we see that a ball slid straight toward the edge follows a path curved to the right. The person slides the ball toward point *B*, starting at point *A*. Both points rotate to the shaded positions (*A'* and *B'*) shown in the time that the ball follows the curved path in the rotating frame and a straight path in Earth's frame.

Up until now, we have considered Earth to be an inertial frame of reference with little or no worry about effects due to its rotation. Yet such effects *do* exist—in the rotation of weather systems, for example. Most consequences of Earth's rotation can be qualitatively understood by analogy with the merry-go-round. Viewed from above the North Pole, Earth rotates counterclockwise, as does the merry-go-round in **Figure 6.27**. As on the merry-go-round, any motion in Earth's Northern Hemisphere experiences a Coriolis force to the right. Just the opposite occurs in the Southern Hemisphere; there, the force is to the left. Because Earth's angular velocity is small, the Coriolis force is usually negligible, but for large-scale motions, such as wind patterns, it has substantial effects.

The Coriolis force causes hurricanes in the Northern Hemisphere to rotate in the counterclockwise direction, whereas tropical cyclones in the Southern Hemisphere rotate in the clockwise direction. (The terms hurricane, typhoon, and tropical storm are regionally specific names for cyclones, which are storm systems characterized by low pressure centers, strong winds, and heavy rains.) **Figure 6.28** helps show how these rotations take place. Air flows toward any region of low pressure, and tropical cyclones contain particularly low pressures. Thus winds flow toward the center of a tropical cyclone or a low-pressure weather system at the surface. In the Northern Hemisphere, these inward winds are deflected to the right, as shown in the figure, producing a counterclockwise circulation at the surface for low-pressure zones of any type. Low pressure at the surface is associated with rising air, which also produces cooling and cloud formation, making low-pressure patterns quite visible from space. Conversely, wind circulation around high-pressure zones is clockwise in the Southern Hemisphere but is less visible because high pressure is associated with sinking air, producing clear skies.

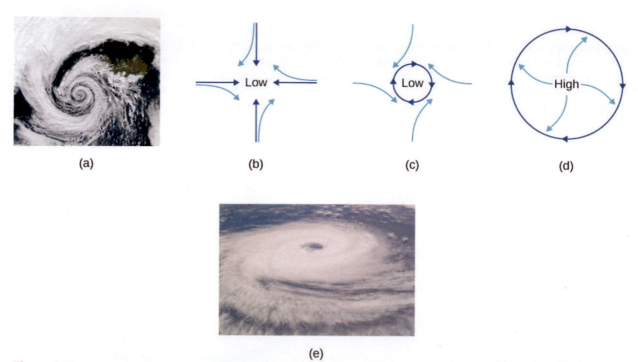

Figure 6.28 (a) The counterclockwise rotation of this Northern Hemisphere hurricane is a major consequence of the Coriolis force. (b) Without the Coriolis force, air would flow straight into a low-pressure zone, such as that found in tropical cyclones. (c) The Coriolis force deflects the winds to the right, producing a counterclockwise rotation. (d) Wind flowing away from a high-pressure zone is also deflected to the right, producing a clockwise rotation. (e) The opposite direction of rotation is produced by the Coriolis force in the Southern Hemisphere, leading to tropical cyclones. (credit a and credit e: modifications of work by NASA)

The rotation of tropical cyclones and the path of a ball on a merry-go-round can just as well be explained by inertia and the rotation of the system underneath. When noninertial frames are used, inertial forces, such as the Coriolis force, must be invented to explain the curved path. There is no identifiable physical source for these inertial forces. In an inertial frame, inertia explains the path, and no force is found to be without an identifiable source. Either view allows us to describe nature, but a view in an inertial frame is the simplest in the sense that all forces have origins and explanations.

6.4 | Drag Force and Terminal Speed

Learning Objectives

By the end of the section, you will be able to:

- Express the drag force mathematically
- Describe applications of the drag force
- Define terminal velocity
- Determine an object's terminal velocity given its mass

Another interesting force in everyday life is the force of drag on an object when it is moving in a fluid (either a gas or a liquid). You feel the drag force when you move your hand through water. You might also feel it if you move your hand during a strong wind. The faster you move your hand, the harder it is to move. You feel a smaller drag force when you tilt your hand so only the side goes through the air—you have decreased the area of your hand that faces the direction of motion.

Drag Forces

Like friction, the **drag force** always opposes the motion of an object. Unlike simple friction, the drag force is proportional to some function of the velocity of the object in that fluid. This functionality is complicated and depends upon the shape of the object, its size, its velocity, and the fluid it is in. For most large objects such as cyclists, cars, and baseballs not moving too slowly, the magnitude of the drag force F_D is proportional to the square of the speed of the object. We can write this relationship mathematically as $F_D \propto v^2$. When taking into account other factors, this relationship becomes

$$F_D = \tfrac{1}{2} C \rho A v^2, \tag{6.5}$$

where C is the drag coefficient, A is the area of the object facing the fluid, and ρ is the density of the fluid. (Recall that density is mass per unit volume.) This equation can also be written in a more generalized fashion as $F_D = b v^2$, where b is a constant equivalent to $0.5 C \rho A$. We have set the exponent n for these equations as 2 because when an object is moving at high velocity through air, the magnitude of the drag force is proportional to the square of the speed. As we shall see in **Fluid Mechanics**, for small particles moving at low speeds in a fluid, the exponent n is equal to 1.

Drag Force

Drag force F_D is proportional to the square of the speed of the object. Mathematically,

$$F_D = \tfrac{1}{2} C \rho A v^2,$$

where C is the drag coefficient, A is the area of the object facing the fluid, and ρ is the density of the fluid.

Athletes as well as car designers seek to reduce the drag force to lower their race times (**Figure 6.29**). Aerodynamic shaping of an automobile can reduce the drag force and thus increase a car's gas mileage.

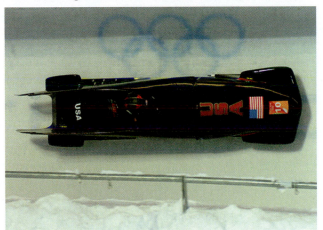

Figure 6.29 From racing cars to bobsled racers, aerodynamic shaping is crucial to achieving top speeds. Bobsleds are designed for speed and are shaped like a bullet with tapered fins. (credit: "U.S. Army"/Wikimedia Commons)

The value of the drag coefficient C is determined empirically, usually with the use of a wind tunnel (**Figure 6.30**).

Figure 6.30 NASA researchers test a model plane in a wind tunnel. (credit: NASA/Ames)

The drag coefficient can depend upon velocity, but we assume that it is a constant here. **Table 6.2** lists some typical drag coefficients for a variety of objects. Notice that the drag coefficient is a dimensionless quantity. At highway speeds, over 50% of the power of a car is used to overcome air drag. The most fuel-efficient cruising speed is about 70–80 km/h (about 45–50 mi/h). For this reason, during the 1970s oil crisis in the United States, maximum speeds on highways were set at about 90 km/h (55 mi/h).

Object	C
Airfoil	0.05
Toyota Camry	0.28
Ford Focus	0.32
Honda Civic	0.36
Ferrari Testarossa	0.37
Dodge Ram Pickup	0.43
Sphere	0.45
Hummer H2 SUV	0.64
Skydiver (feet first)	0.70
Bicycle	0.90
Skydiver (horizontal)	1.0
Circular flat plate	1.12

Table 6.2 Typical Values of Drag Coefficient C

Substantial research is under way in the sporting world to minimize drag. The dimples on golf balls are being redesigned, as are the clothes that athletes wear. Bicycle racers and some swimmers and runners wear full bodysuits. Australian Cathy Freeman wore a full body suit in the 2000 Sydney Olympics and won a gold medal in the 400-m race. Many swimmers in the 2008 Beijing Olympics wore (Speedo) body suits; it might have made a difference in breaking many world records (**Figure 6.31**). Most elite swimmers (and cyclists) shave their body hair. Such innovations can have the effect of slicing away milliseconds in a race, sometimes making the difference between a gold and a silver medal. One consequence is that careful and precise guidelines must be continuously developed to maintain the integrity of the sport.

Figure 6.31 Body suits, such as this LZR Racer Suit, have been credited with aiding in many world records after their release in 2008. Smoother "skin" and more compression forces on a swimmer's body provide at least 10% less drag. (credit: NASA/Kathy Barnstorff)

Terminal Velocity

Some interesting situations connected to Newton's second law occur when considering the effects of drag forces upon a moving object. For instance, consider a skydiver falling through air under the influence of gravity. The two forces acting on him are the force of gravity and the drag force (ignoring the small buoyant force). The downward force of gravity remains constant regardless of the velocity at which the person is moving. However, as the person's velocity increases, the magnitude of the drag force increases until the magnitude of the drag force is equal to the gravitational force, thus producing a net force of zero. A zero net force means that there is no acceleration, as shown by Newton's second law. At this point, the person's velocity remains constant and we say that the person has reached his **terminal velocity** (v_T). Since F_D is proportional to the speed squared, a heavier skydiver must go faster for F_D to equal his weight. Let's see how this works out more quantitatively.

At the terminal velocity,

$$F_{net} = mg - F_D = ma = 0.$$

Thus,

$$mg = F_D.$$

Using the equation for drag force, we have

$$mg = \frac{1}{2}C\rho A v_T^2.$$

Solving for the velocity, we obtain

$$v_T = \sqrt{\frac{2mg}{\rho CA}}.$$

Assume the density of air is $\rho = 1.21 \, \text{kg/m}^3$. A 75-kg skydiver descending head first has a cross-sectional area of approximately $A = 0.18 \, \text{m}^2$ and a drag coefficient of approximately $C = 0.70$. We find that

$$v_T = \sqrt{\frac{2(75 \, \text{kg})(9.80 \, \text{m/s}^2)}{(1.21 \, \text{kg/m}^3)(0.70)(0.18 \, \text{m}^2)}} = 98 \, \text{m/s} = 350 \, \text{km/h}.$$

This means a skydiver with a mass of 75 kg achieves a terminal velocity of about 350 km/h while traveling in a pike (head first) position, minimizing the area and his drag. In a spread-eagle position, that terminal velocity may decrease to about 200 km/h as the area increases. This terminal velocity becomes much smaller after the parachute opens.

Example 6.17

Terminal Velocity of a Skydiver

Find the terminal velocity of an 85-kg skydiver falling in a spread-eagle position.

Strategy

At terminal velocity, $F_{net} = 0$. Thus, the drag force on the skydiver must equal the force of gravity (the person's weight). Using the equation of drag force, we find $mg = \frac{1}{2}\rho C A v^2$.

Solution

The terminal velocity v_T can be written as

$$v_T = \sqrt{\frac{2mg}{\rho C A}} = \sqrt{\frac{2(85 \, \text{kg})(9.80 \, \text{m/s}^2)}{(1.21 \, \text{kg/m}^3)(1.0)(0.70 \, \text{m}^2)}} = 44 \, \text{m/s}.$$

Significance

This result is consistent with the value for v_T mentioned earlier. The 75-kg skydiver going feet first had a terminal velocity of $v_T = 98 \, \text{m/s}$. He weighed less but had a smaller frontal area and so a smaller drag due to the air.

 6.10 Check Your Understanding Find the terminal velocity of a 50-kg skydiver falling in spread-eagle fashion.

The size of the object that is falling through air presents another interesting application of air drag. If you fall from a 5-m-high branch of a tree, you will likely get hurt—possibly fracturing a bone. However, a small squirrel does this all the time, without getting hurt. You do not reach a terminal velocity in such a short distance, but the squirrel does.

The following interesting quote on animal size and terminal velocity is from a 1928 essay by a British biologist, J. B. S. Haldane, titled "On Being the Right Size."

"To the mouse and any smaller animal, [gravity] presents practically no dangers. You can drop a mouse down a thousand-yard mine shaft; and, on arriving at the bottom, it gets a slight shock and walks away, provided that the ground is fairly soft. A rat is killed, a man is broken, and a horse splashes. For the resistance presented to movement by the air is proportional to the surface of the moving object. Divide an animal's length, breadth, and height each by ten; its weight is reduced to a thousandth, but its surface only to a hundredth. So the resistance to falling in the case of the small animal is relatively ten times greater than the driving force."

The above quadratic dependence of air drag upon velocity does not hold if the object is very small, is going very slow, or is in a denser medium than air. Then we find that the drag force is proportional just to the velocity. This relationship is given by Stokes' law.

Stokes' Law

For a spherical object falling in a medium, the drag force is

$$F_s = 6\pi r \eta v, \tag{6.6}$$

where r is the radius of the object, η is the viscosity of the fluid, and v is the object's velocity.

Good examples of Stokes' law are provided by microorganisms, pollen, and dust particles. Because each of these objects is so small, we find that many of these objects travel unaided only at a constant (terminal) velocity. Terminal velocities for bacteria (size about $1 \, \mu m$) can be about $2 \, \mu m/s$. To move at a greater speed, many bacteria swim using flagella (organelles shaped like little tails) that are powered by little motors embedded in the cell.

Sediment in a lake can move at a greater terminal velocity (about $5 \, \mu m/s$), so it can take days for it to reach the bottom of the lake after being deposited on the surface.

If we compare animals living on land with those in water, you can see how drag has influenced evolution. Fish, dolphins, and even massive whales are streamlined in shape to reduce drag forces. Birds are streamlined and migratory species that fly large distances often have particular features such as long necks. Flocks of birds fly in the shape of a spearhead as the flock forms a streamlined pattern (**Figure 6.32**). In humans, one important example of streamlining is the shape of sperm, which need to be efficient in their use of energy.

Figure 6.32 Geese fly in a V formation during their long migratory travels. This shape reduces drag and energy consumption for individual birds, and also allows them a better way to communicate. (credit: modification of work by "Julo"/Wikimedia Commons)

 In lecture demonstrations, we do **measurements of the drag force (https://openstax.org/l/21dragforce)** on different objects. The objects are placed in a uniform airstream created by a fan. Calculate the Reynolds number and the drag coefficient.

The Calculus of Velocity-Dependent Frictional Forces

When a body slides across a surface, the frictional force on it is approximately constant and given by $\mu_k N$. Unfortunately, the frictional force on a body moving through a liquid or a gas does not behave so simply. This drag force is generally a complicated function of the body's velocity. However, for a body moving in a straight line at moderate speeds through a liquid such as water, the frictional force can often be approximated by

$$f_R = -bv,$$

where b is a constant whose value depends on the dimensions and shape of the body and the properties of the liquid, and v is the velocity of the body. Two situations for which the frictional force can be represented by this equation are a motorboat moving through water and a small object falling slowly through a liquid.

Let's consider the object falling through a liquid. The free-body diagram of this object with the positive direction downward is shown in **Figure 6.33**. Newton's second law in the vertical direction gives the differential equation

$$mg - bv = m\frac{dv}{dt},$$

where we have written the acceleration as dv/dt. As v increases, the frictional force $-bv$ increases until it matches mg. At this point, there is no acceleration and the velocity remains constant at the terminal velocity v_T. From the previous equation,

$$mg - bv_T = 0,$$

so

$$v_T = \frac{mg}{b}.$$

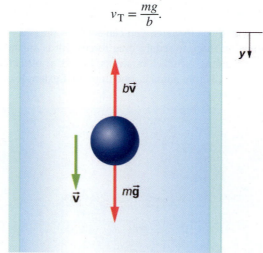

Figure 6.33 Free-body diagram of an object falling through a resistive medium.

We can find the object's velocity by integrating the differential equation for v. First, we rearrange terms in this equation to obtain

$$\frac{dv}{g - (b/m)v} = dt.$$

Assuming that $v = 0$ at $t = 0$, integration of this equation yields

$$\int_0^v \frac{dv'}{g - (b/m)v'} = \int_0^t dt',$$

or

$$-\frac{m}{b}\ln\left(g - \frac{b}{m}v'\right)\Big|_0^v = t'\big|_0^t,$$

where v' and t' are dummy variables of integration. With the limits given, we find

$$-\frac{m}{b}[\ln\left(g - \frac{b}{m}v\right) - \ln g] = t.$$

Since $\ln A - \ln B = \ln(A/B)$, and $\ln(A/B) = x$ implies $e^x = A/B$, we obtain

$$\frac{g - (bv/m)}{g} = e^{-bt/m},$$

and

$$v = \frac{mg}{b}(1 - e^{-bt/m}).$$

Notice that as $t \to \infty$, $v \to mg/b = v_T$, which is the terminal velocity.

The position at any time may be found by integrating the equation for v. With $v = dy/dt$,

$$dy = \frac{mg}{b}(1 - e^{-bt/m})dt.$$

Assuming $y = 0$ when $t = 0$,

$$\int_0^y dy' = \frac{mg}{b} \int_0^t (1 - e^{-bt'/m})dt',$$

which integrates to

$$y = \frac{mg}{b}t + \frac{m^2 g}{b^2}(e^{-bt/m} - 1).$$

Example 6.18

Effect of the Resistive Force on a Motorboat

A motorboat is moving across a lake at a speed v_0 when its motor suddenly freezes up and stops. The boat then slows down under the frictional force $f_R = -bv$. (a) What are the velocity and position of the boat as functions of time? (b) If the boat slows down from 4.0 to 1.0 m/s in 10 s, how far does it travel before stopping?

Solution

a. With the motor stopped, the only horizontal force on the boat is $f_R = -bv$, so from Newton's second law,

$$m\frac{dv}{dt} = -bv,$$

which we can write as

$$\frac{dv}{v} = -\frac{b}{m}dt.$$

Integrating this equation between the time zero when the velocity is v_0 and the time t when the velocity is v, we have

$$\int_0^v \frac{dv'}{v'} = -\frac{b}{m} \int_0^t dt'.$$

Thus,

$$\ln\frac{v}{v_0} = -\frac{b}{m}t,$$

which, since $\ln A = x$ implies $e^x = A$, we can write this as

$$v = v_0 e^{-bt/m}.$$

Now from the definition of velocity,

$$\frac{dx}{dt} = v_0 e^{-bt/m},$$

so we have

$$dx = v_0 e^{-bt/m} dt.$$

With the initial position zero, we have

$$\int_0^x dx' = v_0 \int_0^t e^{-bt'/m} dt',$$

and

$$x = -\frac{mv_0}{b} e^{-bt'/m} \Big|_0^t = \frac{mv_0}{b}(1 - e^{-bt/m}).$$

As time increases, $e^{-bt/m} \to 0$, and the position of the boat approaches a limiting value

$$x_{\text{max}} = \frac{mv_0}{b}.$$

Although this tells us that the boat takes an infinite amount of time to reach x_{max}, the boat effectively stops after a reasonable time. For example, at $t = 10m/b$, we have

$$v = v_0 e^{-10} \simeq 4.5 \times 10^{-5} v_0,$$

whereas we also have

$$x = x_{\text{max}}(1 - e^{-10}) \simeq 0.99995 x_{\text{max}}.$$

Therefore, the boat's velocity and position have essentially reached their final values.

b. With $v_0 = 4.0$ m/s and $v = 1.0$ m/s, we have 1.0 m/s $= (4.0$ m/s$)e^{-(b/m)(10\,\text{s})}$, so

$$\ln 0.25 = -\ln 4.0 = -\frac{b}{m}(10\,\text{s}),$$

and

$$\frac{b}{m} = \frac{1}{10}\ln 4.0\ \text{s}^{-1} = 0.14\ \text{s}^{-1}.$$

Now the boat's limiting position is

$$x_{\text{max}} = \frac{mv_0}{b} = \frac{4.0\ \text{m/s}}{0.14\ \text{s}^{-1}} = 29\ \text{m}.$$

Significance

In the both of the previous examples, we found "limiting" values. The terminal velocity is the same as the limiting velocity, which is the velocity of the falling object after a (relatively) long time has passed. Similarly, the limiting distance of the boat is the distance the boat will travel after a long amount of time has passed. Due to the properties of exponential decay, the time involved to reach either of these values is actually not too long (certainly not an infinite amount of time!) but they are quickly found by taking the limit to infinity.

 6.11 Check Your Understanding Suppose the resistive force of the air on a skydiver can be approximated by $f = -bv^2$. If the terminal velocity of a 100-kg skydiver is 60 m/s, what is the value of b?

CHAPTER 6 REVIEW

KEY TERMS

banked curve curve in a road that is sloping in a manner that helps a vehicle negotiate the curve

centripetal force any net force causing uniform circular motion

Coriolis force inertial force causing the apparent deflection of moving objects when viewed in a rotating frame of reference

drag force force that always opposes the motion of an object in a fluid; unlike simple friction, the drag force is proportional to some function of the velocity of the object in that fluid

friction force that opposes relative motion or attempts at motion between systems in contact

ideal banking sloping of a curve in a road, where the angle of the slope allows the vehicle to negotiate the curve at a certain speed without the aid of friction between the tires and the road; the net external force on the vehicle equals the horizontal centripetal force in the absence of friction

inertial force force that has no physical origin

kinetic friction force that opposes the motion of two systems that are in contact and moving relative to each other

noninertial frame of reference accelerated frame of reference

static friction force that opposes the motion of two systems that are in contact and are not moving relative to each other

terminal velocity constant velocity achieved by a falling object, which occurs when the weight of the object is balanced by the upward drag force

KEY EQUATIONS

Magnitude of static friction	$f_s \leq \mu_s N$
Magnitude of kinetic friction	$f_k = \mu_k N$
Centripetal force	$F_c = m\frac{v^2}{r}$ or $F_c = mr\omega^2$
Ideal angle of a banked curve	$\tan \theta = \frac{v^2}{rg}$
Drag force	$F_D = \frac{1}{2}C\rho A v^2$
Stokes' law	$F_s = 6\pi r \eta v$

SUMMARY

6.1 Solving Problems with Newton's Laws

- Newton's laws of motion can be applied in numerous situations to solve motion problems.

- Some problems contain multiple force vectors acting in different directions on an object. Be sure to draw diagrams, resolve all force vectors into horizontal and vertical components, and draw a free-body diagram. Always analyze the direction in which an object accelerates so that you can determine whether $F_{net} = ma$ or $F_{net} = 0$.

- The normal force on an object is not always equal in magnitude to the weight of the object. If an object is accelerating vertically, the normal force is less than or greater than the weight of the object. Also, if the object is on an inclined plane, the normal force is always less than the full weight of the object.

- Some problems contain several physical quantities, such as forces, acceleration, velocity, or position. You can apply concepts from kinematics and dynamics to solve these problems.

6.2 Friction

- Friction is a contact force that opposes the motion or attempted motion between two systems. Simple friction is proportional to the normal force N supporting the two systems.

- The magnitude of static friction force between two materials stationary relative to each other is determined using the coefficient of static friction, which depends on both materials.

- The kinetic friction force between two materials moving relative to each other is determined using the coefficient of kinetic friction, which also depends on both materials and is always less than the coefficient of static friction.

6.3 Centripetal Force

- Centripetal force $\vec{F}_c$ is a "center-seeking" force that always points toward the center of rotation. It is perpendicular to linear velocity and has the magnitude

$$F_c = ma_c.$$

- Rotating and accelerated frames of reference are noninertial. Inertial forces, such as the Coriolis force, are needed to explain motion in such frames.

6.4 Drag Force and Terminal Speed

- Drag forces acting on an object moving in a fluid oppose the motion. For larger objects (such as a baseball) moving at a velocity in air, the drag force is determined using the drag coefficient (typical values are given in **Table 6.2**), the area of the object facing the fluid, and the fluid density.

- For small objects (such as a bacterium) moving in a denser medium (such as water), the drag force is given by Stokes' law.

CONCEPTUAL QUESTIONS

6.1 Solving Problems with Newton's Laws

1. To simulate the apparent weightlessness of space orbit, astronauts are trained in the hold of a cargo aircraft that is accelerating downward at g. Why do they appear to be weightless, as measured by standing on a bathroom scale, in this accelerated frame of reference? Is there any difference between their apparent weightlessness in orbit and in the aircraft?

6.2 Friction

2. The glue on a piece of tape can exert forces. Can these forces be a type of simple friction? Explain, considering especially that tape can stick to vertical walls and even to ceilings.

3. When you learn to drive, you discover that you need to let up slightly on the brake pedal as you come to a stop or the car will stop with a jerk. Explain this in terms of the relationship between static and kinetic friction.

4. When you push a piece of chalk across a chalkboard, it sometimes screeches because it rapidly alternates between slipping and sticking to the board. Describe this process in more detail, in particular, explaining how it is related to the fact that kinetic friction is less than static friction.

(The same slip-grab process occurs when tires screech on pavement.)

5. A physics major is cooking breakfast when she notices that the frictional force between her steel spatula and Teflon frying pan is only 0.200 N. Knowing the coefficient of kinetic friction between the two materials, she quickly calculates the normal force. What is it?

6.3 Centripetal Force

6. If you wish to reduce the stress (which is related to centripetal force) on high-speed tires, would you use large- or small-diameter tires? Explain.

7. Define centripetal force. Can any type of force (for example, tension, gravitational force, friction, and so on) be a centripetal force? Can any combination of forces be a centripetal force?

8. If centripetal force is directed toward the center, why do you feel that you are 'thrown' away from the center as a car goes around a curve? Explain.

9. Race car drivers routinely cut corners, as shown below (Path 2). Explain how this allows the curve to be taken at the greatest speed.

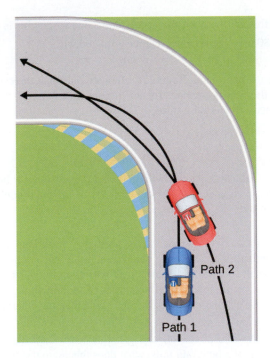

10. Many amusement parks have rides that make vertical loops like the one shown below. For safety, the cars are attached to the rails in such a way that they cannot fall off. If the car goes over the top at just the right speed, gravity alone will supply the centripetal force. What other force acts and what is its direction if:

(a) The car goes over the top at faster than this speed?

(b) The car goes over the top at slower than this speed?

11. What causes water to be removed from clothes in a spin-dryer?

12. As a skater forms a circle, what force is responsible for making his turn? Use a free-body diagram in your answer.

13. Suppose a child is riding on a merry-go-round at a distance about halfway between its center and edge. She has a lunch box resting on wax paper, so that there is very little friction between it and the merry-go-round. Which path shown below will the lunch box take when she lets go? The lunch box leaves a trail in the dust on the merry-go-round. Is that trail straight, curved to the left, or curved to the right? Explain your answer.

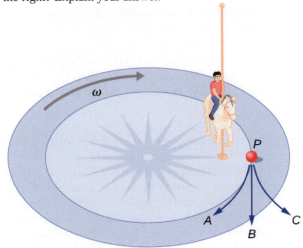

Merry-go-round's rotating
frame of reference

14. Do you feel yourself thrown to either side when you negotiate a curve that is ideally banked for your car's speed? What is the direction of the force exerted on you by the car seat?

15. Suppose a mass is moving in a circular path on a frictionless table as shown below. In Earth's frame of reference, there is no centrifugal force pulling the mass away from the center of rotation, yet there is a force stretching the string attaching the mass to the nail. Using concepts related to centripetal force and Newton's third law, explain what force stretches the string, identifying its physical origin.

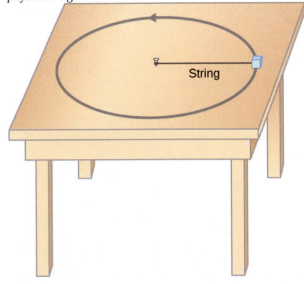

16. When a toilet is flushed or a sink is drained, the water (and other material) begins to rotate about the drain on the way down. Assuming no initial rotation and a flow initially directly straight toward the drain, explain what causes the rotation and which direction it has in the Northern Hemisphere. (Note that this is a small effect and in most toilets the rotation is caused by directional water jets.) Would the direction of rotation reverse if water were forced up the drain?

17. A car rounds a curve and encounters a patch of ice with a very low coefficient of kinetic fiction. The car slides off the road. Describe the path of the car as it leaves the road.

18. In one amusement park ride, riders enter a large vertical barrel and stand against the wall on its horizontal floor. The barrel is spun up and the floor drops away. Riders feel as if they are pinned to the wall by a force something like the gravitational force. This is an inertial force sensed and used by the riders to explain events in the rotating frame of reference of the barrel. Explain in an inertial frame of reference (Earth is nearly one) what pins the riders to the wall, and identify all forces acting on them.

19. Two friends are having a conversation. Anna says a satellite in orbit is in free fall because the satellite keeps falling toward Earth. Tom says a satellite in orbit is not in free fall because the acceleration due to gravity is not 9.80 m/s^2. Who do you agree with and why?

20. A nonrotating frame of reference placed at the center of the Sun is very nearly an inertial one. Why is it not exactly an inertial frame?

6.4 Drag Force and Terminal Speed

21. Athletes such as swimmers and bicyclists wear body suits in competition. Formulate a list of pros and cons of such suits.

22. Two expressions were used for the drag force experienced by a moving object in a liquid. One depended upon the speed, while the other was proportional to the square of the speed. In which types of motion would each of these expressions be more applicable than the other one?

23. As cars travel, oil and gasoline leaks onto the road surface. If a light rain falls, what does this do to the control of the car? Does a heavy rain make any difference?

24. Why can a squirrel jump from a tree branch to the ground and run away undamaged, while a human could break a bone in such a fall?

PROBLEMS

6.1 Solving Problems with Newton's Laws

25. A 30.0-kg girl in a swing is pushed to one side and held at rest by a horizontal force $\vec{F}$ so that the swing ropes are $30.0°$ with respect to the vertical. (a) Calculate the tension in each of the two ropes supporting the swing under these conditions. (b) Calculate the magnitude of $\vec{F}$.

26. Find the tension in each of the three cables supporting the traffic light if it weighs 2.00×10^2 N.

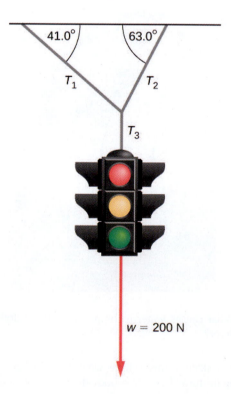

27. Three forces act on an object, considered to be a

particle, which moves with constant velocity $v = (3\hat{i} - 2\hat{j})$ m/s. Two of the forces are

$$\vec{F}_1 = (3\hat{i} + 5\hat{j} - 6\hat{k})\,N$$ and

$$\vec{F}_2 = (4\hat{i} - 7\hat{j} + 2\hat{k})\,N.$$ Find the third force.

28. A flea jumps by exerting a force of 1.20×10^{-5} N straight down on the ground. A breeze blowing on the flea parallel to the ground exerts a force of 0.500×10^{-6} N on the flea while the flea is still in contact with the ground. Find the direction and magnitude of the acceleration of the flea if its mass is 6.00×10^{-7} kg. Do not neglect the gravitational force.

29. Two muscles in the back of the leg pull upward on the Achilles tendon, as shown below. (These muscles are called the medial and lateral heads of the gastrocnemius muscle.) Find the magnitude and direction of the total force on the Achilles tendon. What type of movement could be caused by this force?

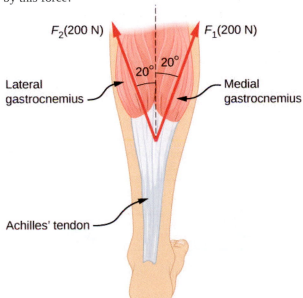

30. After a mishap, a 76.0-kg circus performer clings to a trapeze, which is being pulled to the side by another circus artist, as shown here. Calculate the tension in the two ropes if the person is momentarily motionless. Include a free-body diagram in your solution.

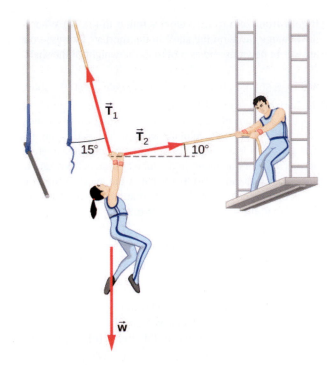

31. A 35.0-kg dolphin decelerates from 12.0 to 7.50 m/s in 2.30 s to join another dolphin in play. What average force was exerted to slow the first dolphin if it was moving horizontally? (The gravitational force is balanced by the buoyant force of the water.)

32. When starting a foot race, a 70.0-kg sprinter exerts an average force of 650 N backward on the ground for 0.800 s. (a) What is his final speed? (b) How far does he travel?

33. A large rocket has a mass of 2.00×10^6 kg at takeoff, and its engines produce a thrust of 3.50×10^7 N. (a) Find its initial acceleration if it takes off vertically. (b) How long does it take to reach a velocity of 120 km/h straight up, assuming constant mass and thrust?

34. A basketball player jumps straight up for a ball. To do this, he lowers his body 0.300 m and then accelerates through this distance by forcefully straightening his legs. This player leaves the floor with a vertical velocity sufficient to carry him 0.900 m above the floor. (a) Calculate his velocity when he leaves the floor. (b) Calculate his acceleration while he is straightening his legs. He goes from zero to the velocity found in (a) in a distance of 0.300 m. (c) Calculate the force he exerts on the floor to do this, given that his mass is 110.0 kg.

35. A 2.50-kg fireworks shell is fired straight up from a mortar and reaches a height of 110.0 m. (a) Neglecting air resistance (a poor assumption, but we will make it for this example), calculate the shell's velocity when it leaves the mortar. (b) The mortar itself is a tube 0.450 m long. Calculate the average acceleration of the shell in the tube as it goes from zero to the velocity found in (a). (c) What is the average force on the shell in the mortar? Express your answer in newtons and as a ratio to the weight of the shell.

36. A 0.500-kg potato is fired at an angle of $80.0°$ above the horizontal from a PVC pipe used as a "potato gun" and reaches a height of 110.0 m. (a) Neglecting air resistance, calculate the potato's velocity when it leaves the gun. (b) The gun itself is a tube 0.450 m long. Calculate the average acceleration of the potato in the tube as it goes from zero to the velocity found in (a). (c) What is the average force on the potato in the gun? Express your answer in newtons and as a ratio to the weight of the potato.

37. An elevator filled with passengers has a mass of 1.70×10^3 kg. (a) The elevator accelerates upward from rest at a rate of 1.20 m/s^2 for 1.50 s. Calculate the tension in the cable supporting the elevator. (b) The elevator continues upward at constant velocity for 8.50 s. What is the tension in the cable during this time? (c) The elevator decelerates at a rate of 0.600 m/s^2 for 3.00 s. What is the tension in the cable during deceleration? (d) How high has the elevator moved above its original starting point, and what is its final velocity?

38. A 20.0-g ball hangs from the roof of a freight car by a string. When the freight car begins to move, the string makes an angle of $35.0°$ with the vertical. (a) What is the acceleration of the freight car? (b) What is the tension in the string?

39. A student's backpack, full of textbooks, is hung from a spring scale attached to the ceiling of an elevator. When the elevator is accelerating downward at 3.8 m/s^2, the scale reads 60 N. (a) What is the mass of the backpack? (b) What does the scale read if the elevator moves upward while speeding up at a rate 3.8 m/s^2? (c) What does the scale read if the elevator moves upward at constant velocity? (d) If the elevator had no brakes and the cable supporting it were to break loose so that the elevator could fall freely, what would the spring scale read?

40. A service elevator takes a load of garbage, mass 10.0 kg, from a floor of a skyscraper under construction, down to ground level, accelerating downward at a rate of 1.2 m/s^2. Find the magnitude of the force the garbage exerts on the floor of the service elevator?

41. A roller coaster car starts from rest at the top of a track 30.0 m long and inclined at $20.0°$ to the horizontal. Assume that friction can be ignored. (a) What is the acceleration of the car? (b) How much time elapses before it reaches the bottom of the track?

42. The device shown below is the Atwood's machine considered in **Example 6.5**. Assuming that the masses of the string and the frictionless pulley are negligible, (a) find an equation for the acceleration of the two blocks; (b) find an equation for the tension in the string; and (c) find both the acceleration and tension when block 1 has mass 2.00 kg and block 2 has mass 4.00 kg.

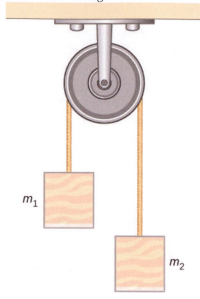

43. Two blocks are connected by a massless rope as shown below. The mass of the block on the table is 4.0 kg and the hanging mass is 1.0 kg. The table and the pulley are frictionless. (a) Find the acceleration of the system. (b) Find the tension in the rope. (c) Find the speed with which the hanging mass hits the floor if it starts from rest and is initially located 1.0 m from the floor.

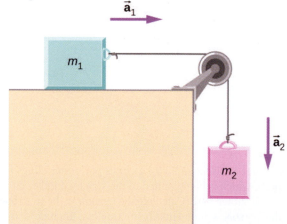

44. Shown below are two carts connected by a cord that passes over a small frictionless pulley. Each cart rolls freely with negligible friction. Calculate the acceleration of the carts and the tension in the cord.

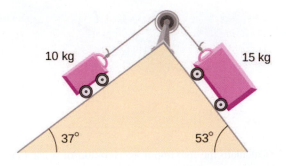

45. A 2.00 kg block (mass 1) and a 4.00 kg block (mass 2) are connected by a light string as shown; the inclination of the ramp is $40.0°$. Friction is negligible. What is (a) the acceleration of each block and (b) the tension in the string?

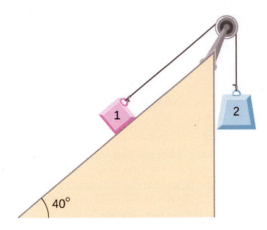

6.2 Friction

46. (a) When rebuilding his car's engine, a physics major must exert 3.00×10^2 N of force to insert a dry steel piston into a steel cylinder. What is the normal force between the piston and cylinder? (b) What force would he have to exert if the steel parts were oiled?

47. (a) What is the maximum frictional force in the knee joint of a person who supports 66.0 kg of her mass on that knee? (b) During strenuous exercise, it is possible to exert forces to the joints that are easily 10 times greater than the weight being supported. What is the maximum force of friction under such conditions? The frictional forces in joints are relatively small in all circumstances except when the joints deteriorate, such as from injury or arthritis. Increased frictional forces can cause further damage and pain.

48. Suppose you have a 120-kg wooden crate resting on a wood floor, with coefficient of static friction 0.500 between these wood surfaces. (a) What maximum force can you exert horizontally on the crate without moving it? (b) If you continue to exert this force once the crate starts to slip, what will its acceleration then be? The coefficient of sliding friction is known to be 0.300 for this situation.

49. (a) If half of the weight of a small 1.00×10^3-kg utility truck is supported by its two drive wheels, what is the maximum acceleration it can achieve on dry concrete? (b) Will a metal cabinet lying on the wooden bed of the truck slip if it accelerates at this rate? (c) Solve both problems assuming the truck has four-wheel drive.

50. A team of eight dogs pulls a sled with waxed wood runners on wet snow (mush!). The dogs have average masses of 19.0 kg, and the loaded sled with its rider has a mass of 210 kg. (a) Calculate the acceleration of the dogs starting from rest if each dog exerts an average force of 185 N backward on the snow. (b) Calculate the force in the coupling between the dogs and the sled.

51. Consider the 65.0-kg ice skater being pushed by two others shown below. (a) Find the direction and magnitude of $\mathbf{F}_{tot}$, the total force exerted on her by the others, given that the magnitudes F_1 and F_2 are 26.4 N and 18.6 N, respectively. (b) What is her initial acceleration if she is initially stationary and wearing steel-bladed skates that point in the direction of $\mathbf{F}_{tot}$? (c) What is her acceleration assuming she is already moving in the direction of $\mathbf{F}_{tot}$?

(Remember that friction always acts in the direction opposite that of motion or attempted motion between surfaces in contact.)

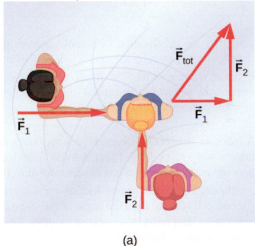

(a)

Free-body diagram

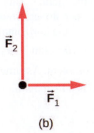

(b)

52. Show that the acceleration of any object down a

frictionless incline that makes an angle θ with the horizontal is $a = g\sin\theta$. (Note that this acceleration is independent of mass.)

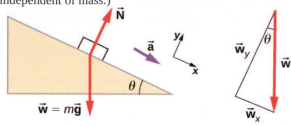

$$\vec{w} = m\vec{g}$$

53. Show that the acceleration of any object down an incline where friction behaves simply (that is, where $f_k = \mu_k N$) is $a = g(\sin\theta - \mu_k \cos\theta)$. Note that the acceleration is independent of mass and reduces to the expression found in the previous problem when friction becomes negligibly small ($\mu_k = 0$).

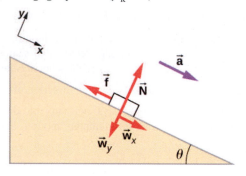

54. Calculate the deceleration of a snow boarder going up a $5.00°$ slope, assuming the coefficient of friction for waxed wood on wet snow. The result of the preceding problem may be useful, but be careful to consider the fact that the snow boarder is going uphill.

55. A machine at a post office sends packages out a chute and down a ramp to be loaded into delivery vehicles. (a) Calculate the acceleration of a box heading down a $10.0°$ slope, assuming the coefficient of friction for a parcel on waxed wood is 0.100. (b) Find the angle of the slope down which this box could move at a constant velocity. You can neglect air resistance in both parts.

56. If an object is to rest on an incline without slipping, then friction must equal the component of the weight of the object parallel to the incline. This requires greater and greater friction for steeper slopes. Show that the maximum angle of an incline above the horizontal for which an object will not slide down is $\theta = \tan^{-1}\mu_s$. You may use the result of the previous problem. Assume that $a = 0$ and that static friction has reached its maximum value.

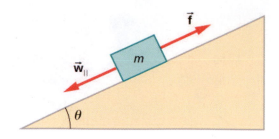

57. Calculate the maximum acceleration of a car that is heading down a $6.00°$ slope (one that makes an angle of $6.00°$ with the horizontal) under the following road conditions. You may assume that the weight of the car is evenly distributed on all four tires and that the coefficient of static friction is involved—that is, the tires are not allowed to slip during the deceleration. (Ignore rolling.) Calculate for a car: (a) On dry concrete. (b) On wet concrete. (c) On ice, assuming that $\mu_s = 0.100$, the same as for shoes on ice.

58. Calculate the maximum acceleration of a car that is heading up a $4.00°$ slope (one that makes an angle of $4.00°$ with the horizontal) under the following road conditions. Assume that only half the weight of the car is supported by the two drive wheels and that the coefficient of static friction is involved—that is, the tires are not allowed to slip during the acceleration. (Ignore rolling.) (a) On dry concrete. (b) On wet concrete. (c) On ice, assuming that $\mu_s = 0.100$, the same as for shoes on ice.

59. Repeat the preceding problem for a car with four-wheel drive.

60. A freight train consists of two 8.00×10^5-kg engines and 45 cars with average masses of 5.50×10^5 kg. (a) What force must each engine exert backward on the track to accelerate the train at a rate of $5.00 \times 10^{-2}\,\text{m/s}^2$ if the force of friction is $7.50 \times 10^5\,\text{N}$, assuming the engines exert identical forces? This is not a large frictional force for such a massive system. Rolling friction for trains is small, and consequently, trains are very energy-efficient transportation systems. (b) What is the force in the coupling between the 37th and 38th cars (this is the force each exerts on the other), assuming all cars have the same mass and that friction is evenly distributed among all of the cars and engines?

61. Consider the 52.0-kg mountain climber shown below. (a) Find the tension in the rope and the force that the mountain climber must exert with her feet on the vertical rock face to remain stationary. Assume that the force is exerted parallel to her legs. Also, assume negligible force exerted by her arms. (b) What is the minimum coefficient of friction between her shoes and the cliff?

62. A contestant in a winter sporting event pushes a 45.0-kg block of ice across a frozen lake as shown below. (a) Calculate the minimum force F he must exert to get the block moving. (b) What is its acceleration once it starts to move, if that force is maintained?

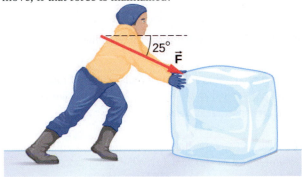

63. The contestant now pulls the block of ice with a rope over his shoulder at the same angle above the horizontal as shown below. Calculate the minimum force F he must exert to get the block moving. (b) What is its acceleration once it starts to move, if that force is maintained?

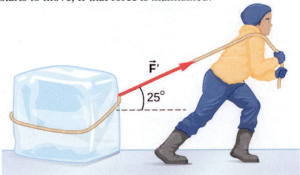

64. At a post office, a parcel that is a 20.0-kg box slides down a ramp inclined at $30.0°$ with the horizontal. The coefficient of kinetic friction between the box and plane is 0.0300. (a) Find the acceleration of the box. (b) Find the velocity of the box as it reaches the end of the plane, if the length of the plane is 2 m and the box starts at rest.

6.3 Centripetal Force

65. (a) A 22.0-kg child is riding a playground merry-go-round that is rotating at 40.0 rev/min. What centripetal force is exerted if he is 1.25 m from its center? (b) What centripetal force is exerted if the merry-go-round rotates at 3.00 rev/min and he is 8.00 m from its center? (c) Compare each force with his weight.

66. Calculate the centripetal force on the end of a 100-m (radius) wind turbine blade that is rotating at 0.5 rev/s. Assume the mass is 4 kg.

67. What is the ideal banking angle for a gentle turn of 1.20-km radius on a highway with a 105 km/h speed limit (about 65 mi/h), assuming everyone travels at the limit?

68. What is the ideal speed to take a 100.0-m-radius curve banked at a $20.0°$ angle?

69. (a) What is the radius of a bobsled turn banked at $75.0°$ and taken at 30.0 m/s, assuming it is ideally banked? (b) Calculate the centripetal acceleration. (c) Does this acceleration seem large to you?

70. Part of riding a bicycle involves leaning at the correct angle when making a turn, as seen below. To be stable, the force exerted by the ground must be on a line going through the center of gravity. The force on the bicycle wheel can be resolved into two perpendicular components—friction parallel to the road (this must supply the centripetal force) and the vertical normal force (which must equal the system's weight). (a) Show that θ (as defined as shown) is related to the speed v and radius of curvature r of the turn in the same way as for an ideally banked roadway—that is, $\theta = \tan^{-1}(v^2/rg)$. (b) Calculate θ for a 12.0-m/s turn of radius 30.0 m (as in a race).

Free-body diagram

$\vec{F}$ = sum of $\vec{N}$ and $\vec{F}_c$
$N = w$

71. If a car takes a banked curve at less than the ideal speed, friction is needed to keep it from sliding toward the inside of the curve (a problem on icy mountain roads). (a)

Calculate the ideal speed to take a 100.0 m radius curve banked at $15.0°$. (b) What is the minimum coefficient of friction needed for a frightened driver to take the same curve at 20.0 km/h?

72. Modern roller coasters have vertical loops like the one shown here. The radius of curvature is smaller at the top than on the sides so that the downward centripetal acceleration at the top will be greater than the acceleration due to gravity, keeping the passengers pressed firmly into their seats. (a) What is the speed of the roller coaster at the top of the loop if the radius of curvature there is 15.0 m and the downward acceleration of the car is 1.50 g? (b) How high above the top of the loop must the roller coaster start from rest, assuming negligible friction? (c) If it actually starts 5.00 m higher than your answer to (b), how much energy did it lose to friction? Its mass is 1.50×10^3 kg.

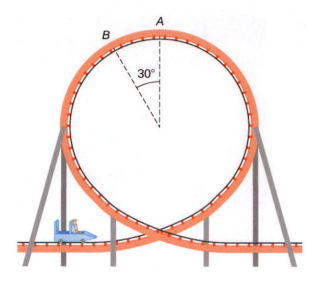

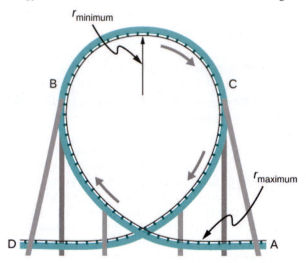

73. A child of mass 40.0 kg is in a roller coaster car that travels in a loop of radius 7.00 m. At point A the speed of the car is 10.0 m/s, and at point B, the speed is 10.5 m/s. Assume the child is not holding on and does not wear a seat belt. (a) What is the force of the car seat on the child at point A? (b) What is the force of the car seat on the child at point B? (c) What minimum speed is required to keep the child in his seat at point A?

74. In the simple Bohr model of the ground state of the hydrogen atom, the electron travels in a circular orbit around a fixed proton. The radius of the orbit is 5.28×10^{-11} m, and the speed of the electron is 2.18×10^6 m/s. The mass of an electron is 9.11×10^{-31} kg. What is the force on the electron?

75. Railroad tracks follow a circular curve of radius 500.0 m and are banked at an angle of $5.0°$. For trains of what speed are these tracks designed?

76. The CERN particle accelerator is circular with a circumference of 7.0 km. (a) What is the acceleration of the protons ($m = 1.67 \times 10^{-27}$ kg) that move around the accelerator at 5% of the speed of light? (The speed of light is $v = 3.00 \times 10^8$ m/s.) (b) What is the force on the protons?

77. A car rounds an unbanked curve of radius 65 m. If the coefficient of static friction between the road and car is 0.70, what is the maximum speed at which the car can traverse the curve without slipping?

78. A banked highway is designed for traffic moving at 90.0 km/h. The radius of the curve is 310 m. What is the angle of banking of the highway?

6.4 Drag Force and Terminal Speed

79. The terminal velocity of a person falling in air depends upon the weight and the area of the person facing the fluid. Find the terminal velocity (in meters per second and kilometers per hour) of an 80.0-kg skydiver falling in a pike (headfirst) position with a surface area of 0.140 m^2.

80. A 60.0-kg and a 90.0-kg skydiver jump from an airplane at an altitude of 6.00×10^3 m, both falling in the pike position. Make some assumption on their frontal areas and calculate their terminal velocities. How long will it take for each skydiver to reach the ground (assuming the time to reach terminal velocity is small)? Assume all values are accurate to three significant digits.

81. A 560-g squirrel with a surface area of $930\,\text{cm}^2$ falls from a 5.0-m tree to the ground. Estimate its terminal velocity. (Use a drag coefficient for a horizontal skydiver.) What will be the velocity of a 56-kg person hitting the ground, assuming no drag contribution in such a short distance?

82. To maintain a constant speed, the force provided by a car's engine must equal the drag force plus the force of friction of the road (the rolling resistance). (a) What are the drag forces at 70 km/h and 100 km/h for a Toyota Camry? (Drag area is $0.70\,\text{m}^2$) (b) What is the drag force at 70 km/h and 100 km/h for a Hummer H2? (Drag area is $2.44\,\text{m}^2$) Assume all values are accurate to three significant digits.

83. By what factor does the drag force on a car increase as it goes from 65 to 110 km/h?

84. Calculate the velocity a spherical rain drop would achieve falling from 5.00 km (a) in the absence of air drag (b) with air drag. Take the size across of the drop to be 4 mm, the density to be 1.00×10^3 kg/m^3, and the surface area to be πr^2.

85. Using Stokes' law, verify that the units for viscosity are kilograms per meter per second.

86. Find the terminal velocity of a spherical bacterium (diameter $2.00\,\mu\text{m}$) falling in water. You will first need to note that the drag force is equal to the weight at terminal velocity. Take the density of the bacterium to be 1.10×10^3 kg/m^3.

87. Stokes' law describes sedimentation of particles in liquids and can be used to measure viscosity. Particles in liquids achieve terminal velocity quickly. One can measure the time it takes for a particle to fall a certain distance and then use Stokes' law to calculate the viscosity of the liquid. Suppose a steel ball bearing (density 7.8×10^3 kg/m^3, diameter 3.0 mm) is dropped in a container of motor oil. It takes 12 s to fall a distance of 0.60 m. Calculate the viscosity of the oil.

88. Suppose that the resistive force of the air on a skydiver

can be approximated by $f = -bv^2$. If the terminal velocity of a 50.0-kg skydiver is 60.0 m/s, what is the value of b?

89. A small diamond of mass 10.0 g drops from a swimmer's earring and falls through the water, reaching a terminal velocity of 2.0 m/s. (a) Assuming the frictional force on the diamond obeys $f = -bv$, what is b? (b) How far does the diamond fall before it reaches 90 percent of its terminal speed?

90. (a) What is the final velocity of a car originally traveling at 50.0 km/h that decelerates at a rate of $0.400\,\text{m/s}^2$ for 50.0 s? Assume a coefficient of friction of 1.0. (b) What is unreasonable about the result? (c) Which premise is unreasonable, or which premises are inconsistent?

91. A 75.0-kg woman stands on a bathroom scale in an elevator that accelerates from rest to 30.0 m/s in 2.00 s. (a) Calculate the scale reading in newtons and compare it with her weight. (The scale exerts an upward force on her equal to its reading.) (b) What is unreasonable about the result? (c) Which premise is unreasonable, or which premises are inconsistent?

92. (a) Calculate the minimum coefficient of friction needed for a car to negotiate an unbanked 50.0 m radius curve at 30.0 m/s. (b) What is unreasonable about the result? (c) Which premises are unreasonable or inconsistent?

93. As shown below, if $M = 5.50\,\text{kg}$, what is the tension in string 1?

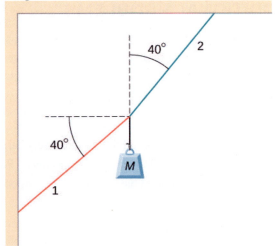

94. As shown below, if $F = 60.0\,\text{N}$ and $M = 4.00\,\text{kg}$, what is the magnitude of the acceleration of the suspended object? All surfaces are frictionless.

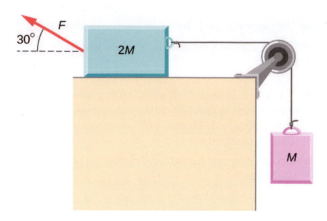

95. As shown below, if $M = 6.0\,\text{kg}$, what is the tension in the connecting string? The pulley and all surfaces are frictionless.

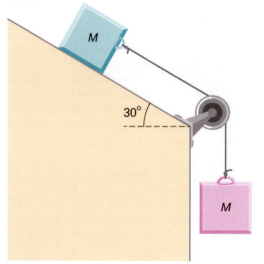

96. A small space probe is released from a spaceship. The space probe has mass 20.0 kg and contains 90.0 kg of fuel. It starts from rest in deep space, from the origin of a coordinate system based on the spaceship, and burns fuel at the rate of 3.00 kg/s. The engine provides a constant thrust of 120.0 N. (a) Write an expression for the mass of the space probe as a function of time, between 0 and 30 seconds, assuming that the engine ignites fuel beginning at $t = 0$. (b) What is the velocity after 15.0 s? (c) What is the position of the space probe after 15.0 s, with initial position at the origin? (d) Write an expression for the position as a function of time, for $t > 30.0\,\text{s}$.

97. A half-full recycling bin has mass 3.0 kg and is pushed up a $40.0°$ incline with constant speed under the action of a 26-N force acting up and parallel to the incline. The incline has friction. What magnitude force must act up and parallel to the incline for the bin to move down the incline at constant velocity?

98. A child has mass 6.0 kg and slides down a $35°$ incline with constant speed under the action of a 34-N force acting up and parallel to the incline. What is the coefficient of kinetic friction between the child and the surface of the incline?

ADDITIONAL PROBLEMS

99. The two barges shown here are coupled by a cable of negligible mass. The mass of the front barge is $2.00 \times 10^3\,\text{kg}$ and the mass of the rear barge is $3.00 \times 10^3\,\text{kg}$. A tugboat pulls the front barge with a horizontal force of magnitude $20.0 \times 10^3\,\text{N}$, and the frictional forces of the water on the front and rear barges are $8.00 \times 10^3\,\text{N}$ and $10.0 \times 10^3\,\text{N}$, respectively. Find the horizontal acceleration of the barges and the tension in the connecting cable.

100. If the order of the barges of the preceding exercise is reversed so that the tugboat pulls the 3.00×10^3-kg barge with a force of $20.0 \times 10^3\,\text{N}$, what are the acceleration of the barges and the tension in the coupling cable?

101. An object with mass m moves along the x-axis. Its position at any time is given by $x(t) = pt^3 + qt^2$ where p and q are constants. Find the net force on this object for any time t.

102. A helicopter with mass 2.35×10^4 kg has a position given by $\overrightarrow{\mathbf{r}}(t) = (0.020\, t^3)\,\hat{\mathbf{i}} + (2.2t)\,\hat{\mathbf{j}} - (0.060\, t^2)\,\hat{\mathbf{k}}$. Find the net force on the helicopter at $t = 3.0$ s.

103. Located at the origin, an electric car of mass m is at rest and in equilibrium. A time dependent force of $\overrightarrow{\mathbf{F}}(t)$ is applied at time $t = 0$, and its components are $F_x(t) = p + nt$ and $F_y(t) = qt$ where p, q, and n are constants. Find the position $\overrightarrow{\mathbf{r}}(t)$ and velocity $\overrightarrow{\mathbf{v}}(t)$ as functions of time t.

104. A particle of mass m is located at the origin. It is at rest and in equilibrium. A time-dependent force of $\overrightarrow{\mathbf{F}}(t)$ is applied at time $t = 0$, and its components are $F_x(t) = pt$ and $F_y(t) = n + qt$ where p, q, and n are constants. Find the position $\overrightarrow{\mathbf{r}}(t)$ and velocity $\overrightarrow{\mathbf{v}}(t)$ as functions of time t.

105. A 2.0-kg object has a velocity of $4.0\,\hat{\mathbf{i}}$ m/s at $t = 0$. A constant resultant force of $(2.0\,\hat{\mathbf{i}} + 4.0\,\hat{\mathbf{j}})\,\text{N}$ then acts on the object for 3.0 s. What is the magnitude of the object's velocity at the end of the 3.0-s interval?

106. A 1.5-kg mass has an acceleration of $(4.0\,\hat{\mathbf{i}} - 3.0\,\hat{\mathbf{j}})\,\text{m/s}^2$. Only two forces act on the mass. If one of the forces is $(2.0\,\hat{\mathbf{i}} - 1.4\,\hat{\mathbf{j}})\,\text{N}$, what is the magnitude of the other force?

107. A box is dropped onto a conveyor belt moving at 3.4 m/s. If the coefficient of friction between the box and the belt is 0.27, how long will it take before the box moves without slipping?

108. Shown below is a 10.0-kg block being pushed by a horizontal force $\overrightarrow{\mathbf{F}}$ of magnitude 200.0 N. The coefficient of kinetic friction between the two surfaces is 0.50. Find the acceleration of the block.

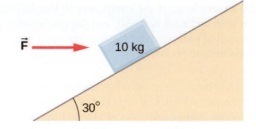

109. As shown below, the mass of block 1 is $m_1 = 4.0$ kg, while the mass of block 2 is $m_2 = 8.0$ kg. The coefficient of friction between m_1 and the inclined surface is $\mu_k = 0.40$. What is the acceleration of the system?

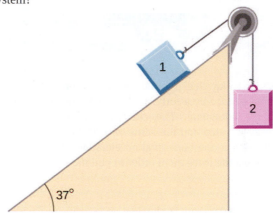

110. A student is attempting to move a 30-kg mini-fridge into her dorm room. During a moment of inattention, the mini-fridge slides down a 35 degree incline at constant speed when she applies a force of 25 N acting up and parallel to the incline. What is the coefficient of kinetic friction between the fridge and the surface of the incline?

111. A crate of mass 100.0 kg rests on a rough surface inclined at an angle of $37.0°$ with the horizontal. A massless rope to which a force can be applied parallel to the surface is attached to the crate and leads to the top of the incline. In its present state, the crate is just ready to slip and start to move down the plane. The coefficient of friction is 80% of that for the static case. (a) What is the coefficient of static friction? (b) What is the maximum force that can be applied upward along the plane on the rope and not move the block? (c) With a slightly greater applied force, the block will slide up the plane. Once it begins to move, what is its acceleration and what reduced force is necessary to keep it moving upward at constant speed? (d) If the block is given a slight nudge to get it started down the plane, what will be its acceleration in that direction? (e) Once the block begins to slide downward, what upward force on the rope is required to keep the block from accelerating downward?

112. A car is moving at high speed along a highway when the driver makes an emergency braking. The wheels become locked (stop rolling), and the resulting skid marks are 32.0 meters long. If the coefficient of kinetic friction between tires and road is 0.550, and the acceleration was constant during braking, how fast was the car going when the wheels became locked?

113. A crate having mass 50.0 kg falls horizontally off the back of the flatbed truck, which is traveling at 100 km/h. Find the value of the coefficient of kinetic friction between the road and crate if the crate slides 50 m on the road in coming to rest. The initial speed of the crate is the same as the truck, 100 km/h.

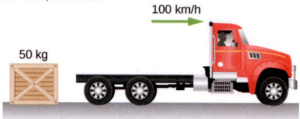
100 km/h

50 kg

114. A 15-kg sled is pulled across a horizontal, snow-covered surface by a force applied to a rope at 30 degrees with the horizontal. The coefficient of kinetic friction between the sled and the snow is 0.20. (a) If the force is 33 N, what is the horizontal acceleration of the sled? (b) What must the force be in order to pull the sled at constant velocity?

115. A 30.0-g ball at the end of a string is swung in a vertical circle with a radius of 25.0 cm. The rotational velocity is 200.0 cm/s. Find the tension in the string: (a) at the top of the circle, (b) at the bottom of the circle, and (c) at a distance of 12.5 cm from the center of the circle ($r = 12.5$ cm).

116. A particle of mass 0.50 kg starts moves through a circular path in the xy-plane with a position given by $\vec{r}(t) = (4.0\cos 3t)\,\hat{i} + (4.0\sin 3t)\,\hat{j}$ where r is in meters and t is in seconds. (a) Find the velocity and acceleration vectors as functions of time. (b) Show that the acceleration vector always points toward the center of the circle (and thus represents centripetal acceleration). (c) Find the centripetal force vector as a function of time.

117. A stunt cyclist rides on the interior of a cylinder 12 m in radius. The coefficient of static friction between the tires and the wall is 0.68. Find the value of the minimum speed for the cyclist to perform the stunt.

118. When a body of mass 0.25 kg is attached to a vertical massless spring, it is extended 5.0 cm from its unstretched length of 4.0 cm. The body and spring are placed on a horizontal frictionless surface and rotated about the held end of the spring at 2.0 rev/s. How far is the spring stretched?

119. A piece of bacon starts to slide down the pan when one side of a pan is raised up 5.0 cm. If the length of the pan from pivot to the raising point is 23.5 cm, what is the coefficient of static friction between the pan and the bacon?

120. A plumb bob hangs from the roof of a railroad car. The car rounds a circular track of radius 300.0 m at a speed of 90.0 km/h. At what angle relative to the vertical does the plumb bob hang?

121. An airplane flies at 120.0 m/s and banks at a $30°$ angle. If its mass is 2.50×10^3 kg, (a) what is the magnitude of the lift force? (b) what is the radius of the turn?

122. The position of a particle is given by $\vec{r}(t) = A\left(\cos \omega t\,\hat{i} + \sin \omega t\,\hat{j}\right)$, where ω is a constant. (a) Show that the particle moves in a circle of radius A. (b) Calculate $d\vec{r}/dt$ and then show that the speed of the particle is a constant $A\omega$. (c) Determine $d^2\vec{r}/dt^2$ and show that a is given by $a_c = r\omega^2$. (d) Calculate the centripetal force on the particle. [*Hint:* For (b) and (c), you will need to use $(d/dt)(\cos \omega t) = -\omega \sin \omega t$ and $(d/dt)(\sin \omega t) = \omega \cos \omega t$.

123. Two blocks connected by a string are pulled across a horizontal surface by a force applied to one of the blocks, as shown below. The coefficient of kinetic friction between the blocks and the surface is 0.25. If each block has an acceleration of $2.0\,\text{m/s}^2$ to the right, what is the magnitude F of the applied force?

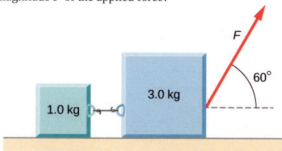

124. As shown below, the coefficient of kinetic friction between the surface and the larger block is 0.20, and the coefficient of kinetic friction between the surface and the smaller block is 0.30. If $F = 10\,\text{N}$ and $M = 1.0\,\text{kg}$, what is the tension in the connecting string?

125. In the figure, the coefficient of kinetic friction between the surface and the blocks is μ_k. If $M = 1.0\,\text{kg}$, find an expression for the magnitude of the acceleration of either block (in terms of F, μ_k, and g).

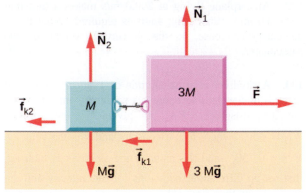

126. Two blocks are stacked as shown below, and rest on a frictionless surface. There is friction between the two blocks (coefficient of friction μ). An external force is applied to the top block at an angle θ with the horizontal. What is the maximum force F that can be applied for the two blocks to move together?

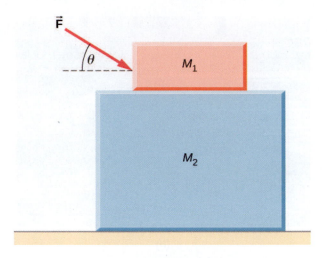

127. A box rests on the (horizontal) back of a truck. The coefficient of static friction between the box and the surface on which it rests is 0.24. What maximum distance can the truck travel (starting from rest and moving horizontally with constant acceleration) in 3.0 s without having the box slide?

128. A double-incline plane is shown below. The coefficient of friction on the left surface is 0.30, and on the right surface 0.16. Calculate the acceleration of the system.

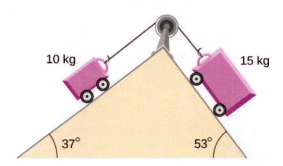

CHALLENGE PROBLEMS

129. In a later chapter, you will find that the weight of a particle varies with altitude such that $w = \dfrac{mgr_0{}^2}{r^2}$ where r_0 is the radius of Earth and r is the distance from Earth's center. If the particle is fired vertically with velocity v_0 from Earth's surface, determine its velocity as a function of position r. (*Hint:* use $a\,dr = v\,dv$, the rearrangement mentioned in the text.)

130. A large centrifuge, like the one shown below, is used to expose aspiring astronauts to accelerations similar to those experienced in rocket launches and atmospheric reentries. (a) At what angular velocity is the centripetal acceleration $10g$ if the rider is 15.0 m from the center of rotation? (b) The rider's cage hangs on a pivot at the end of the arm, allowing it to swing outward during rotation as shown in the bottom accompanying figure. At what angle θ below the horizontal will the cage hang when the centripetal acceleration is $10g$? (*Hint:* The arm supplies centripetal force and supports the weight of the cage. Draw a free-body diagram of the forces to see what the angle θ should be.)

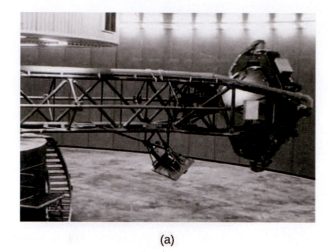

(a)

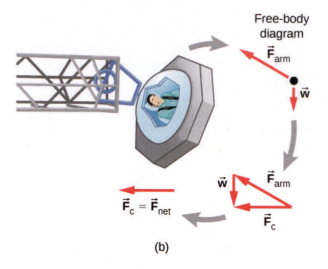

(b)

131. A car of mass 1000.0 kg is traveling along a level road at 100.0 km/h when its brakes are applied. Calculate the stopping distance if the coefficient of kinetic friction of the tires is 0.500. Neglect air resistance. (*Hint:* since the distance traveled is of interest rather than the time, x is the

desired independent variable and not t. Use the Chain Rule to change the variable: $\frac{dv}{dt} = \frac{dv}{dx}\frac{dx}{dt} = v\frac{dv}{dx}$.)

132. An airplane flying at 200.0 m/s makes a turn that takes 4.0 min. What bank angle is required? What is the percentage increase in the perceived weight of the passengers?

133. A skydiver is at an altitude of 1520 m. After 10.0 seconds of free fall, he opens his parachute and finds that the air resistance, F_D, is given by the formula $F_D = -bv$, where b is a constant and v is the velocity. If $b = 0.750$, and the mass of the skydiver is 82.0 kg, first set up differential equations for the velocity and the position, and then find: (a) the speed of the skydiver when the parachute opens, (b) the distance fallen before the parachute opens, (c) the terminal velocity after the parachute opens (find the limiting velocity), and (d) the time the skydiver is in the air after the parachute opens.

134. In a television commercial, a small, spherical bead of mass 4.00 g is released from rest at $t = 0$ in a bottle of liquid shampoo. The terminal speed is observed to be 2.00 cm/s. Find (a) the value of the constant b in the equation $v = \frac{mg}{b}(1 - e^{-bt/m})$, and (b) the value of the resistive force when the bead reaches terminal speed.

135. A boater and motor boat are at rest on a lake. Together, they have mass 200.0 kg. If the thrust of the motor is a constant force of 40.0 N in the direction of motion, and if the resistive force of the water is numerically equivalent to 2 times the speed v of the boat, set up and solve the differential equation to find: (a) the velocity of the boat at time t; (b) the limiting velocity (the velocity after a long time has passed).

7 | WORK AND KINETIC ENERGY

Figure 7.1 A sprinter exerts her maximum power with the greatest force in the short time her foot is in contact with the ground. This adds to her kinetic energy, preventing her from slowing down during the race. Pushing back hard on the track generates a reaction force that propels the sprinter forward to win at the finish. (credit: modification of work by Marie-Lan Nguyen)

Chapter Outline

7.1 Work

7.2 Kinetic Energy

7.3 Work-Energy Theorem

7.4 Power

Introduction

In this chapter, we discuss some basic physical concepts involved in every physical motion in the universe, going beyond the concepts of force and change in motion, which we discussed in **Motion in Two and Three Dimensions** and **Newton's Laws of Motion**. These concepts are work, kinetic energy, and power. We explain how these quantities are related to one another, which will lead us to a fundamental relationship called the work-energy theorem. In the next chapter, we generalize this idea to the broader principle of conservation of energy.

The application of Newton's laws usually requires solving differential equations that relate the forces acting on an object to the accelerations they produce. Often, an analytic solution is intractable or impossible, requiring lengthy numerical solutions or simulations to get approximate results. In such situations, more general relations, like the work-energy theorem (or the conservation of energy), can still provide useful answers to many questions and require a more modest amount of mathematical calculation. In particular, you will see how the work-energy theorem is useful in relating the speeds of a particle, at different points along its trajectory, to the forces acting on it, even when the trajectory is otherwise too complicated to deal with. Thus, some aspects of motion can be addressed with fewer equations and without vector decompositions.

7.1 | Work

In physics, **work** is done on an object when energy is transferred to the object. In other words, work is done when a force acts on something that undergoes a displacement from one position to another. Forces can vary as a function of position, and displacements can be along various paths between two points. We first define the increment of work dW done by a force $\vec{F}$ acting through an infinitesimal displacement $d\vec{r}$ as the dot product of these two vectors:

$$dW = \vec{F} \cdot d\vec{r} = \left| \vec{F} \right| \left| d\vec{r} \right| \cos\theta. \tag{7.1}$$

Then, we can add up the contributions for infinitesimal displacements, along a path between two positions, to get the total work.

Work Done by a Force

The work done by a force is the integral of the force with respect to displacement along the path of the displacement:

$$W_{AB} = \int_{\text{path } AB} \vec{F} \cdot d\vec{r}. \tag{7.2}$$

The vectors involved in the definition of the work done by a force acting on a particle are illustrated in **Figure 7.2**.

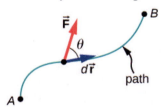

Figure 7.2 Vectors used to define work. The force acting on a particle and its infinitesimal displacement are shown at one point along the path between A and B. The infinitesimal work is the dot product of these two vectors; the total work is the integral of the dot product along the path.

We choose to express the dot product in terms of the magnitudes of the vectors and the cosine of the angle between them, because the meaning of the dot product for work can be put into words more directly in terms of magnitudes and angles. We could equally well have expressed the dot product in terms of the various components introduced in **Vectors**. In two dimensions, these were the x- and y-components in Cartesian coordinates, or the r- and φ-components in polar coordinates; in three dimensions, it was just x-, y-, and z-components. Which choice is more convenient depends on the situation. In words, you can express **Equation 7.1** for the work done by a force acting over a displacement as a product of one component acting parallel to the other component. From the properties of vectors, it doesn't matter if you take the component of the force parallel to the displacement or the component of the displacement parallel to the force—you get the same result either way.

Recall that the magnitude of a force times the cosine of the angle the force makes with a given direction is the component of the force in the given direction. The components of a vector can be positive, negative, or zero, depending on whether the angle between the vector and the component-direction is between $0°$ and $90°$ or $90°$ and $180°$, or is equal to $90°$. As a result, the work done by a force can be positive, negative, or zero, depending on whether the force is generally in the direction of the displacement, generally opposite to the displacement, or perpendicular to the displacement. The maximum work is done by a given force when it is along the direction of the displacement ($\cos\theta = \pm 1$), and zero work is done when the force is perpendicular to the displacement ($\cos\theta = 0$).

The units of work are units of force multiplied by units of length, which in the SI system is newtons times meters, $N \cdot m$. This combination is called a joule, for historical reasons that we will mention later, and is abbreviated as J. In the English system, still used in the United States, the unit of force is the pound (lb) and the unit of distance is the foot (ft), so the unit of work is the foot-pound $(ft \cdot lb)$.

Work Done by Constant Forces and Contact Forces

The simplest work to evaluate is that done by a force that is constant in magnitude and direction. In this case, we can factor out the force; the remaining integral is just the total displacement, which only depends on the end points A and B, but not on the path between them:

$$W_{AB} = \vec{F} \cdot \int_A^B d\vec{r} = \vec{F} \cdot (\vec{r}_B - \vec{r}_A) = |\vec{F}||\vec{r}_B - \vec{r}_A|\cos\theta \text{ (constant force).}$$

We can also see this by writing out **Equation 7.2** in Cartesian coordinates and using the fact that the components of the force are constant:

$$W_{AB} = \int_{path\ AB} \vec{F} \cdot d\vec{r} = \int_{path\ AB} (F_x dx + F_y dy + F_z dz) = F_x \int_A^B dx + F_y \int_A^B dy + F_z \int_A^B dz$$

$$= F_x(x_B - x_A) + F_y(y_B - y_A) + F_z(z_B - z_A) = \vec{F} \cdot (\vec{r}_B - \vec{r}_A).$$

Figure 7.3(a) shows a person exerting a constant force $\vec{F}$ along the handle of a lawn mower, which makes an angle θ with the horizontal. The horizontal displacement of the lawn mower, over which the force acts, is $\vec{d}$. The work done on the lawn mower is $W = \vec{F} \cdot \vec{d} = Fd\cos\theta$, which the figure also illustrates as the horizontal component of the force times the magnitude of the displacement.

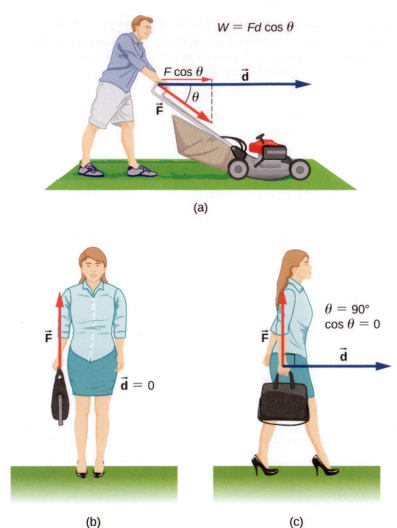

Figure 7.3 Work done by a constant force. (a) A person pushes a lawn mower with a constant force. The component of the force parallel to the displacement is the work done, as shown in the equation in the figure. (b) A person holds a briefcase. No work is done because the displacement is zero. (c) The person in (b) walks horizontally while holding the briefcase. No work is done because $\cos \theta$ is zero.

Figure 7.3(b) shows a person holding a briefcase. The person must exert an upward force, equal in magnitude to the weight of the briefcase, but this force does no work, because the displacement over which it acts is zero.

In **Figure 7.3**(c), where the person in (b) is walking horizontally with constant speed, the work done by the person on the briefcase is still zero, but now because the angle between the force exerted and the displacement is $90°$ ($\overrightarrow{\mathbf{F}}$ perpendicular to $\overrightarrow{\mathbf{d}}$) and $\cos 90° = 0$.

Example 7.1

Calculating the Work You Do to Push a Lawn Mower

How much work is done on the lawn mower by the person in **Figure 7.3**(a) if he exerts a constant force of 75.0 N at an angle $35°$ below the horizontal and pushes the mower 25.0 m on level ground?

Strategy

We can solve this problem by substituting the given values into the definition of work done on an object by a constant force, stated in the equation $W = Fd \cos \theta$. The force, angle, and displacement are given, so that only the work W is unknown.

Solution

The equation for the work is

$$W = Fd \cos \theta.$$

Substituting the known values gives

$$W = (75.0 \, \text{N})(25.0 \, \text{m})\cos(35.0°) = 1.54 \times 10^3 \, \text{J}.$$

Significance

Even though one and a half kilojoules may seem like a lot of work, we will see in **Potential Energy and Conservation of Energy** that it's only about as much work as you could do by burning one sixth of a gram of fat.

When you mow the grass, other forces act on the lawn mower besides the force you exert—namely, the contact force of the ground and the gravitational force of Earth. Let's consider the work done by these forces in general. For an object moving on a surface, the displacement $d\vec{r}$ is tangent to the surface. The part of the contact force on the object that is perpendicular to the surface is the normal force $\vec{N}$. Since the cosine of the angle between the normal and the tangent to a surface is zero, we have

$$dW_{\text{N}} = \vec{N} \cdot d\vec{r} = \vec{0}.$$

The normal force never does work under these circumstances. (Note that if the displacement $d\vec{r}$ did have a relative component perpendicular to the surface, the object would either leave the surface or break through it, and there would no longer be any normal contact force. However, if the object is more than a particle, and has an internal structure, the normal contact force can do work on it, for example, by displacing it or deforming its shape. This will be mentioned in the next chapter.)

The part of the contact force on the object that is parallel to the surface is friction, $\vec{f}$. For this object sliding along the surface, kinetic friction $\vec{f}_k$ is opposite to $d\vec{r}$, relative to the surface, so the work done by kinetic friction is negative. If the magnitude of $\vec{f}_k$ is constant (as it would be if all the other forces on the object were constant), then the work done by friction is

$$W_{\text{fr}} = \int_A^B \vec{f}_k \cdot d\vec{r} = -f_k \int_A^B |dr| = -f_k |l_{AB}|, \qquad (7.3)$$

where $|l_{AB}|$ is the path length on the surface. The force of static friction does no work in the reference frame between two surfaces because there is never displacement between the surfaces. As an external force, static friction can do work. Static friction can keep someone from sliding off a sled when the sled is moving and perform positive work on the person. If you're driving your car at the speed limit on a straight, level stretch of highway, the negative work done by air resistance is balanced by the positive work done by the static friction of the road on the drive wheels. You can pull the rug out from under an object in such a way that it slides backward relative to the rug, but forward relative to the floor. In this case, kinetic friction exerted by the rug on the object could be in the same direction as the displacement of the object, relative to the floor, and do positive work. The bottom line is that you need to analyze each particular case to determine the work done by the forces, whether positive, negative or zero.

Example 7.2

Moving a Couch

You decide to move your couch to a new position on your horizontal living room floor. The normal force on the couch is 1 kN and the coefficient of friction is 0.6. (a) You first push the couch 3 m parallel to a wall and then 1 m perpendicular to the wall (*A* to *B* in **Figure 7.4**). How much work is done by the frictional force? (b) You don't like the new position, so you move the couch straight back to its original position (*B* to *A* in **Figure 7.4**). What was the total work done against friction moving the couch away from its original position and back again?

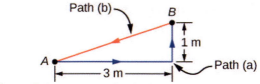

Figure 7.4   Top view of paths for moving a couch.

Strategy

The magnitude of the force of kinetic friction on the couch is constant, equal to the coefficient of friction times the normal force, $f_K = \mu_K N$. Therefore, the work done by it is $W_{\text{fr}} = -f_K d$, where *d* is the path length traversed.

The segments of the paths are the sides of a right triangle, so the path lengths are easily calculated. In part (b), you can use the fact that the work done against a force is the negative of the work done by the force.

Solution

a.   The work done by friction is

$$W = -(0.6)(1 \text{ kN})(3 \text{ m} + 1 \text{ m}) = -2.4 \text{ kJ}.$$

b.   The length of the path along the hypotenuse is $\sqrt{10}$ m, so the total work done against friction is

$$W = (0.6)(1 \text{ kN})(3 \text{ m} + 1 \text{ m} + \sqrt{10} \text{ m}) = 4.3 \text{ kJ}.$$

Significance

The total path over which the work of friction was evaluated began and ended at the same point (it was a closed path), so that the total displacement of the couch was zero. However, the total work was not zero. The reason is that forces like friction are classified as nonconservative forces, or dissipative forces, as we discuss in the next chapter.

   7.1   **Check Your Understanding** Can kinetic friction ever be a constant force for all paths?

The other force on the lawn mower mentioned above was Earth's gravitational force, or the weight of the mower. Near the surface of Earth, the gravitational force on an object of mass *m* has a constant magnitude, *mg*, and constant direction, vertically down. Therefore, the work done by gravity on an object is the dot product of its weight and its displacement. In many cases, it is convenient to express the dot product for gravitational work in terms of the *x*-, *y*-, and *z*-components of the vectors. A typical coordinate system has the *x*-axis horizontal and the *y*-axis vertically up. Then the gravitational force is $-mg\,\hat{\mathbf{j}}$, so the work done by gravity, over any path from *A* to *B*, is

$$W_{\text{grav, } AB} = -mg\,\hat{\mathbf{j}} \cdot (\vec{\mathbf{r}}_B - \vec{\mathbf{r}}_A) = -mg(y_B - y_A). \tag{7.4}$$

The work done by a constant force of gravity on an object depends only on the object's weight and the difference in height through which the object is displaced. Gravity does negative work on an object that moves upward ($y_B > y_A$), or, in other words, you must do positive work against gravity to lift an object upward. Alternately, gravity does positive work on an object that moves downward ($y_B < y_A$), or you do negative work against gravity to "lift" an object downward, controlling its descent so it doesn't drop to the ground. ("Lift" is used as opposed to "drop".)

Example 7.3

Shelving a Book

You lift an oversized library book, weighing 20 N, 1 m vertically down from a shelf, and carry it 3 m horizontally to a table (**Figure 7.5**). How much work does gravity do on the book? (b) When you're finished, you move the book in a straight line back to its original place on the shelf. What was the total work done against gravity, moving the book away from its original position on the shelf and back again?

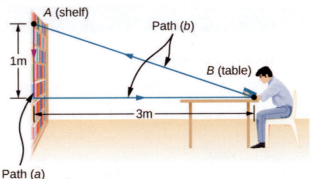

Figure 7.5 Side view of the paths for moving a book to and from a shelf.

Strategy

We have just seen that the work done by a constant force of gravity depends only on the weight of the object moved and the difference in height for the path taken, $W_{AB} = -mg(y_B - y_A)$. We can evaluate the difference in height to answer (a) and (b).

Solution

a. Since the book starts on the shelf and is lifted down $y_B - y_A = -1$ m, we have

$$W = -(20\,\text{N})(-1\,\text{m}) = 20\,\text{J}.$$

b. There is zero difference in height for any path that begins and ends at the same place on the shelf, so $W = 0$.

Significance

Gravity does positive work (20 J) when the book moves down from the shelf. The gravitational force between two objects is an attractive force, which does positive work when the objects get closer together. Gravity does zero work (0 J) when the book moves horizontally from the shelf to the table and negative work (−20 J) when the book moves from the table back to the shelf. The total work done by gravity is zero $[20\,\text{J} + 0\,\text{J} + (-20\,\text{J}) = 0]$.

Unlike friction or other dissipative forces, described in **Example 7.2**, the total work done against gravity, over any closed path, is zero. Positive work is done against gravity on the upward parts of a closed path, but an equal amount of negative work is done against gravity on the downward parts. In other words, work done *against* gravity, lifting an object *up*, is "given back" when the object comes back down. Forces like gravity (those that do zero work over any closed path) are classified as conservative forces and play an important role in physics.

 7.2 Check Your Understanding Can Earth's gravity ever be a constant force for all paths?

Work Done by Forces that Vary

In general, forces may vary in magnitude and direction at points in space, and paths between two points may be curved. The infinitesimal work done by a variable force can be expressed in terms of the components of the force and the displacement along the path,

$$dW = F_x dx + F_y dy + F_z dz.$$

Here, the components of the force are functions of position along the path, and the displacements depend on the equations of the path. (Although we chose to illustrate dW in Cartesian coordinates, other coordinates are better suited to some situations.) **Equation 7.2** defines the total work as a line integral, or the limit of a sum of infinitesimal amounts of work. The physical concept of work is straightforward: you calculate the work for tiny displacements and add them up. Sometimes the mathematics can seem complicated, but the following example demonstrates how cleanly they can operate.

Example 7.4

Work Done by a Variable Force over a Curved Path

An object moves along a parabolic path $y = (0.5\,\text{m}^{-1})x^2$ from the origin $A = (0, 0)$ to the point $B = (2\,\text{m}, 2\,\text{m})$ under the action of a force $\vec{F} = (5\,\text{N/m})y\,\hat{i} + (10\,\text{N/m})x\,\hat{j}$ (**Figure 7.6**). Calculate the work done.

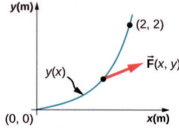

Figure 7.6 The parabolic path of a particle acted on by a given force.

Strategy

The components of the force are given functions of x and y. We can use the equation of the path to express y and dy in terms of x and dx; namely,

$$y = (0.5\,\text{m}^{-1})x^2 \text{ and } dy = 2(0.5\,\text{m}^{-1})xdx.$$

Then, the integral for the work is just a definite integral of a function of x.

Solution

The infinitesimal element of work is

$$
\begin{aligned}
dW &= F_x dx + F_y dy = (5\,\text{N/m})ydx + (10\,\text{N/m})xdy \\
&= (5\,\text{N/m})(0.5\,\text{m}^{-1})x^2 dx + (10\,\text{N/m})2(0.5\,\text{m}^{-1})x^2 dx = (12.5\,\text{N/m}^2)x^2 dx.
\end{aligned}
$$

The integral of x^2 is $x^3/3$, so

$$W = \int_0^{2\,\text{m}} (12.5\,\text{N/m}^2)x^2 dx = (12.5\,\text{N/m}^2)\frac{x^3}{3}\Big|_0^{2\,\text{m}} = (12.5\,\text{N/m}^2)\left(\frac{8}{3}\right) = 33.3\,\text{J}.$$

Significance

This integral was not hard to do. You can follow the same steps, as in this example, to calculate line integrals representing work for more complicated forces and paths. In this example, everything was given in terms of x- and y-components, which are easiest to use in evaluating the work in this case. In other situations, magnitudes and angles might be easier.

 7.3   Check Your Understanding Find the work done by the same force in **Example 7.4** over a cubic path, $y = (0.25 \text{ m}^{-2})x^3$, between the same points $A = (0, 0)$ and $B = (2 \text{ m}, 2 \text{ m})$.

You saw in **Example 7.4** that to evaluate a line integral, you could reduce it to an integral over a single variable or parameter. Usually, there are several ways to do this, which may be more or less convenient, depending on the particular case. In **Example 7.4**, we reduced the line integral to an integral over x, but we could equally well have chosen to reduce everything to a function of y. We didn't do that because the functions in y involve the square root and fractional exponents, which may be less familiar, but for illustrative purposes, we do this now. Solving for x and dx, in terms of y, along the parabolic path, we get

$$x = \sqrt{y/(0.5 \text{ m}^{-1})} = \sqrt{(2 \text{ m})y} \text{ and } dx = \sqrt{(2 \text{ m})} \times \tfrac{1}{2}dy/\sqrt{y} = dy/\sqrt{(2 \text{ m}^{-1})y}.$$

The components of the force, in terms of y, are

$$F_x = (5 \text{ N/m})y \text{ and } F_y = (10 \text{ N/m})x = (10 \text{ N/m})\sqrt{(2 \text{ m})y},$$

so the infinitesimal work element becomes

$$dW = F_x dx + F_y dy = \frac{(5 \text{ N/m})y\, dy}{\sqrt{(2 \text{ m}^{-1})y}} + (10 \text{ N/m})\sqrt{(2 \text{ m})y}\, dy$$

$$= (5 \text{ N} \cdot \text{m}^{-1/2})\left(\frac{1}{\sqrt{2}} + 2\sqrt{2}\right)\sqrt{y}\, dy = (17.7 \text{ N} \cdot \text{m}^{-1/2})y^{1/2}\, dy.$$

The integral of $y^{1/2}$ is $\tfrac{2}{3}y^{3/2}$, so the work done from A to B is

$$W = \int_0^{2 \text{ m}} (17.7 \text{ N} \cdot \text{m}^{-1/2})y^{1/2}\, dy = (17.7 \text{ N} \cdot \text{m}^{-1/2})\tfrac{2}{3}(2 \text{ m})^{3/2} = 33.3 \text{ J}.$$

As expected, this is exactly the same result as before.

One very important and widely applicable variable force is the force exerted by a perfectly elastic spring, which satisfies Hooke's law $\overrightarrow{\mathbf{F}} = -k\Delta\overrightarrow{\mathbf{x}}$, where k is the spring constant, and $\Delta\overrightarrow{\mathbf{x}} = \overrightarrow{\mathbf{x}} - \overrightarrow{\mathbf{x}}_{eq}$ is the displacement from the spring's unstretched (equilibrium) position (**Newton's Laws of Motion**). Note that the unstretched position is only the same as the equilibrium position if no other forces are acting (or, if they are, they cancel one another). Forces between molecules, or in any system undergoing small displacements from a stable equilibrium, behave approximately like a spring force.

To calculate the work done by a spring force, we can choose the x-axis along the length of the spring, in the direction of increasing length, as in **Figure 7.7**, with the origin at the equilibrium position $x_{eq} = 0$. (Then positive x corresponds to a stretch and negative x to a compression.) With this choice of coordinates, the spring force has only an x-component, $F_x = -kx$, and the work done when x changes from x_A to x_B is

$$W_{\text{spring, } AB} = \int_A^B F_x dx = -k\int_A^B x\, dx = -k\frac{x^2}{2}\Big|_A^B = -\tfrac{1}{2}k(x_B^2 - x_A^2). \qquad (7.5)$$

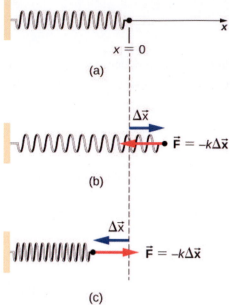

Figure 7.7 (a) The spring exerts no force at its equilibrium position. The spring exerts a force in the opposite direction to (b) an extension or stretch, and (c) a compression.

Notice that W_{AB} depends only on the starting and ending points, A and B, and is independent of the actual path between them, as long as it starts at A and ends at B. That is, the actual path could involve going back and forth before ending.

Another interesting thing to notice about **Equation 7.5** is that, for this one-dimensional case, you can readily see the correspondence between the work done by a force and the area under the curve of the force versus its displacement. Recall that, in general, a one-dimensional integral is the limit of the sum of infinitesimals, $f(x)dx$, representing the area of strips, as shown in **Figure 7.8**. In **Equation 7.5**, since $F = -kx$ is a straight line with slope $-k$, when plotted versus x, the "area" under the line is just an algebraic combination of triangular "areas," where "areas" above the x-axis are positive and those below are negative, as shown in **Figure 7.9**. The magnitude of one of these "areas" is just one-half the triangle's base, along the x-axis, times the triangle's height, along the force axis. (There are quotation marks around "area" because this base-height product has the units of work, rather than square meters.)

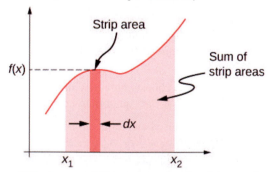

Figure 7.8 A curve of $f(x)$ versus x showing the area of an infinitesimal strip, $f(x)dx$, and the sum of such areas, which is the integral of $f(x)$ from x_1 to x_2.

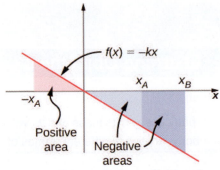

Figure 7.9 Curve of the spring force $f(x) = -kx$ versus x, showing areas under the line, between x_A and x_B, for both positive and negative values of x_A. When x_A is negative, the total area under the curve for the integral in **Equation 7.5** is the sum of positive and negative triangular areas. When x_A is positive, the total area under the curve is the difference between two negative triangles.

Example 7.5

Work Done by a Spring Force

A perfectly elastic spring requires 0.54 J of work to stretch 6 cm from its equilibrium position, as in **Figure 7.7**(b). (a) What is its spring constant k? (b) How much work is required to stretch it an additional 6 cm?

Strategy

Work "required" means work done against the spring force, which is the negative of the work in **Equation 7.5**, that is

$$W = \frac{1}{2}k(x_B^2 - x_A^2).$$

For part (a), $x_A = 0$ and $x_B = 6 \text{cm}$; for part (b), $x_B = 6 \text{cm}$ and $x_B = 12 \text{cm}$. In part (a), the work is given and you can solve for the spring constant; in part (b), you can use the value of k, from part (a), to solve for the work.

Solution

a. $W = 0.54 \text{ J} = \frac{1}{2}k[(6 \text{ cm})^2 - 0]$, so $k = 3 \text{ N/cm}$.

b. $W = \frac{1}{2}(3 \text{ N/cm})[(12 \text{ cm})^2 - (6 \text{ cm})^2] = 1.62 \text{ J}$.

Significance

Since the work done by a spring force is independent of the path, you only needed to calculate the difference in the quantity $\frac{1}{2}kx^2$ at the end points. Notice that the work required to stretch the spring from 0 to 12 cm is four times that required to stretch it from 0 to 6 cm, because that work depends on the square of the amount of stretch from equilibrium, $\frac{1}{2}kx^2$. In this circumstance, the work to stretch the spring from 0 to 12 cm is also equal to the work for a composite path from 0 to 6 cm followed by an additional stretch from 6 cm to 12 cm. Therefore, $4W(0 \text{ cm to } 6 \text{ cm}) = W(0 \text{ cm to } 6 \text{ cm}) + W(6 \text{ cm to } 12 \text{ cm})$, or $W(6 \text{ cm to } 12 \text{ cm}) = 3W(0 \text{ cm to } 6 \text{ cm})$, as we found above.

 7.4 Check Your Understanding The spring in **Example 7.5** is compressed 6 cm from its equilibrium length. (a) Does the spring force do positive or negative work and (b) what is the magnitude?

7.2 | Kinetic Energy

Learning Objectives

By the end of this section, you will be able to:

- Calculate the kinetic energy of a particle given its mass and its velocity or momentum
- Evaluate the kinetic energy of a body, relative to different frames of reference

It's plausible to suppose that the greater the velocity of a body, the greater effect it could have on other bodies. This does not depend on the direction of the velocity, only its magnitude. At the end of the seventeenth century, a quantity was introduced into mechanics to explain collisions between two perfectly elastic bodies, in which one body makes a head-on collision with an identical body at rest. The first body stops, and the second body moves off with the initial velocity of the first body. (If you have ever played billiards or croquet, or seen a model of Newton's Cradle, you have observed this type of collision.) The idea behind this quantity was related to the forces acting on a body and was referred to as "the energy of motion." Later on, during the eighteenth century, the name **kinetic energy** was given to energy of motion.

With this history in mind, we can now state the classical definition of kinetic energy. Note that when we say "classical," we mean non-relativistic, that is, at speeds much less that the speed of light. At speeds comparable to the speed of light, the special theory of relativity requires a different expression for the kinetic energy of a particle, as discussed in **Relativity (http://cnx.org/content/m58555/latest/)** .

Since objects (or systems) of interest vary in complexity, we first define the kinetic energy of a particle with mass m.

Kinetic Energy

The kinetic energy of a particle is one-half the product of the particle's mass m and the square of its speed v:

$$K = \frac{1}{2}mv^2. \tag{7.6}$$

We then extend this definition to any system of particles by adding up the kinetic energies of all the constituent particles:

$$K = \sum \frac{1}{2}mv^2. \tag{7.7}$$

Note that just as we can express Newton's second law in terms of either the rate of change of momentum or mass times the rate of change of velocity, so the kinetic energy of a particle can be expressed in terms of its mass and momentum ($\vec{\mathbf{p}} = m\vec{\mathbf{v}}$), instead of its mass and velocity. Since $v = p/m$, we see that

$$K = \frac{1}{2}m\left(\frac{p}{m}\right)^2 = \frac{p^2}{2m}$$

also expresses the kinetic energy of a single particle. Sometimes, this expression is more convenient to use than **Equation 7.6**.

The units of kinetic energy are mass times the square of speed, or $\text{kg} \cdot \text{m}^2/\text{s}^2$. But the units of force are mass times acceleration, $\text{kg} \cdot \text{m/s}^2$, so the units of kinetic energy are also the units of force times distance, which are the units of work, or joules. You will see in the next section that work and kinetic energy have the same units, because they are different forms of the same, more general, physical property.

Example 7.6

Kinetic Energy of an Object

(a) What is the kinetic energy of an 80-kg athlete, running at 10 m/s? (b) The Chicxulub crater in Yucatan, one of the largest existing impact craters on Earth, is thought to have been created by an asteroid, traveling at 22 km/s and releasing 4.2×10^{23} J of kinetic energy upon impact. What was its mass? (c) In nuclear reactors, thermal neutrons, traveling at about 2.2 km/s, play an important role. What is the kinetic energy of such a particle?

Strategy

To answer these questions, you can use the definition of kinetic energy in **Equation 7.6**. You also have to look up the mass of a neutron.

Solution

Don't forget to convert km into m to do these calculations, although, to save space, we omitted showing these conversions.

a. $K = \frac{1}{2}(80 \text{ kg})(10 \text{ m/s})^2 = 4.0 \text{ kJ}.$

b. $m = 2K/v^2 = 2(4.2 \times 10^{23} \text{ J})/(22 \text{ km/s})^2 = 1.7 \times 10^{15} \text{ kg}.$

c. $K = \frac{1}{2}(1.68 \times 10^{-27} \text{ kg})(2.2 \text{ km/s})^2 = 4.1 \times 10^{-21} \text{ J}.$

Significance

In this example, we used the way mass and speed are related to kinetic energy, and we encountered a very wide range of values for the kinetic energies. Different units are commonly used for such very large and very small values. The energy of the impactor in part (b) can be compared to the explosive yield of TNT and nuclear explosions, 1 megaton $= 4.18 \times 10^{15}$ J. The Chicxulub asteroid's kinetic energy was about a hundred million megatons. At the other extreme, the energy of subatomic particle is expressed in electron-volts, 1 eV $= 1.6 \times 10^{-19}$ J. The thermal neutron in part (c) has a kinetic energy of about one fortieth of an electron-volt.

 7.5   Check Your Understanding (a) A car and a truck are each moving with the same kinetic energy. Assume that the truck has more mass than the car. Which has the greater speed? (b) A car and a truck are each moving with the same speed. Which has the greater kinetic energy?

Because velocity is a relative quantity, you can see that the value of kinetic energy must depend on your frame of reference. You can generally choose a frame of reference that is suited to the purpose of your analysis and that simplifies your calculations. One such frame of reference is the one in which the observations of the system are made (likely an external frame). Another choice is a frame that is attached to, or moves with, the system (likely an internal frame). The equations for relative motion, discussed in **Motion in Two and Three Dimensions**, provide a link to calculating the kinetic energy of an object with respect to different frames of reference.

Example 7.7

Kinetic Energy Relative to Different Frames

A 75.0-kg person walks down the central aisle of a subway car at a speed of 1.50 m/s relative to the car, whereas the train is moving at 15.0 m/s relative to the tracks. (a) What is the person's kinetic energy relative to the car? (b) What is the person's kinetic energy relative to the tracks? (c) What is the person's kinetic energy relative to a frame moving with the person?

Strategy

Since speeds are given, we can use $\frac{1}{2}mv^2$ to calculate the person's kinetic energy. However, in part (a), the person's speed is relative to the subway car (as given); in part (b), it is relative to the tracks; and in part (c), it is zero. If we denote the car frame by C, the track frame by T, and the person by P, the relative velocities in part (b) are related by $\vec{v}_{PT} = \vec{v}_{PC} + \vec{v}_{CT}$. We can assume that the central aisle and the tracks lie along the same line, but the direction the person is walking relative to the car isn't specified, so we will give an answer for each possibility, $v_{PT} = v_{CT} \pm v_{PC}$, as shown in **Figure 7.10**.

Figure 7.10 The possible motions of a person walking in a train are (a) toward the front of the car and (b) toward the back of the car.

Solution

a. $K = \frac{1}{2}(75.0 \text{ kg})(1.50 \text{ m/s})^2 = 84.4 \text{ J}.$

b. $v_{PT} = (15.0 \pm 1.50) \text{ m/s}.$ Therefore, the two possible values for kinetic energy relative to the car are

$$K = \frac{1}{2}(75.0 \text{ kg})(13.5 \text{ m/s})^2 = 6.83 \text{ kJ}$$

and

$$K = \frac{1}{2}(75.0 \text{ kg})(16.5 \text{ m/s})^2 = 10.2 \text{ kJ}.$$

c. In a frame where $v_P = 0$, $K = 0$ as well.

Significance

You can see that the kinetic energy of an object can have very different values, depending on the frame of reference. However, the kinetic energy of an object can never be negative, since it is the product of the mass and the square of the speed, both of which are always positive or zero.

 7.6 Check Your Understanding You are rowing a boat parallel to the banks of a river. Your kinetic energy relative to the banks is less than your kinetic energy relative to the water. Are you rowing with or against the current?

The kinetic energy of a particle is a single quantity, but the kinetic energy of a system of particles can sometimes be divided into various types, depending on the system and its motion. For example, if all the particles in a system have the same velocity, the system is undergoing translational motion and has translational kinetic energy. If an object is rotating, it could have rotational kinetic energy, or if it's vibrating, it could have vibrational kinetic energy. The kinetic energy of a system, relative to an internal frame of reference, may be called internal kinetic energy. The kinetic energy associated with random molecular motion may be called thermal energy. These names will be used in later chapters of the book, when appropriate. Regardless of the name, every kind of kinetic energy is the same physical quantity, representing energy associated with motion.

Example 7.8

Special Names for Kinetic Energy

(a) A player lobs a mid-court pass with a 624-g basketball, which covers 15 m in 2 s. What is the basketball's horizontal translational kinetic energy while in flight? (b) An average molecule of air, in the basketball in part (a), has a mass of 29 u, and an average speed of 500 m/s, relative to the basketball. There are about 3×10^{23} molecules inside it, moving in random directions, when the ball is properly inflated. What is the average translational kinetic energy of the random motion of all the molecules inside, relative to the basketball? (c) How fast would the basketball have to travel relative to the court, as in part (a), so as to have a kinetic energy equal to the amount in part (b)?

Strategy

In part (a), first find the horizontal speed of the basketball and then use the definition of kinetic energy in terms of mass and speed, $K = \frac{1}{2}mv^2$. Then in part (b), convert unified units to kilograms and then use $K = \frac{1}{2}mv^2$ to get the average translational kinetic energy of one molecule, relative to the basketball. Then multiply by the number of molecules to get the total result. Finally, in part (c), we can substitute the amount of kinetic energy in part (b), and the mass of the basketball in part (a), into the definition $K = \frac{1}{2}mv^2$, and solve for v.

Solution

a. The horizontal speed is (15 m)/(2 s), so the horizontal kinetic energy of the basketball is

$$\frac{1}{2}(0.624\,\text{kg})(7.5\,\text{m/s})^2 = 17.6\,\text{J}.$$

b. The average translational kinetic energy of a molecule is

$$\frac{1}{2}(29\,\text{u})(1.66 \times 10^{-27}\,\text{kg/u})(500\,\text{m/s})^2 = 6.02 \times 10^{-21}\,\text{J},$$

and the total kinetic energy of all the molecules is

$$(3 \times 10^{23})(6.02 \times 10^{-21}\,\text{J}) = 1.80\,\text{kJ}.$$

c. $v = \sqrt{2(1.8\,\text{kJ})/(0.624\,\text{kg})} = 76.0\,\text{m/s}.$

Significance

In part (a), this kind of kinetic energy can be called the horizontal kinetic energy of an object (the basketball), relative to its surroundings (the court). If the basketball were spinning, all parts of it would have not just the average speed, but it would also have rotational kinetic energy. Part (b) reminds us that this kind of kinetic energy can be called internal or thermal kinetic energy. Notice that this energy is about a hundred times the energy in part (a). How to make use of thermal energy will be the subject of the chapters on thermodynamics. In part (c), since the energy in part (b) is about 100 times that in part (a), the speed should be about 10 times as big, which it is (76 compared to 7.5 m/s).

7.3 | Work-Energy Theorem

Learning Objectives

By the end of this section, you will be able to:

- Apply the work-energy theorem to find information about the motion of a particle, given the forces acting on it
- Use the work-energy theorem to find information about the forces acting on a particle, given information about its motion

We have discussed how to find the work done on a particle by the forces that act on it, but how is that work manifested in the motion of the particle? According to Newton's second law of motion, the sum of all the forces acting on a particle, or the net force, determines the rate of change in the momentum of the particle, or its motion. Therefore, we should consider the work done by all the forces acting on a particle, or the **net work**, to see what effect it has on the particle's motion.

Let's start by looking at the net work done on a particle as it moves over an infinitesimal displacement, which is the dot product of the net force and the displacement: $dW_{\text{net}} = \vec{\mathbf{F}}_{\text{net}} \cdot d\vec{\mathbf{r}}$. Newton's second law tells us that $\vec{\mathbf{F}}_{\text{net}} = m(d\vec{\mathbf{v}}/dt)$, so $dW_{\text{net}} = m(d\vec{\mathbf{v}}/dt) \cdot d\vec{\mathbf{r}}$. For the mathematical functions describing the motion of a physical particle, we can rearrange the differentials dt, etc., as algebraic quantities in this expression, that is,

$$dW_{\text{net}} = m\left(\frac{d\vec{\mathbf{v}}}{dt}\right) \cdot d\vec{\mathbf{r}} = md\vec{\mathbf{v}} \cdot \left(\frac{d\vec{\mathbf{r}}}{dt}\right) = m\vec{\mathbf{v}} \cdot d\vec{\mathbf{v}},$$

where we substituted the velocity for the time derivative of the displacement and used the commutative property of the dot product [**Equation 2.30**]. Since derivatives and integrals of scalars are probably more familiar to you at this point, we express the dot product in terms of Cartesian coordinates before we integrate between any two points A and B on the particle's trajectory. This gives us the net work done on the particle:

$$W_{\text{net, } AB} = \int_A^B (mv_x\, dv_x + mv_y\, dv_y + mv_z\, dv_z) \tag{7.8}$$

$$= \tfrac{1}{2}m \left| v_x^2 + v_y^2 + v_z^2 \right|_A^B = \left| \tfrac{1}{2}mv^2 \right|_A^B = K_B - K_A.$$

In the middle step, we used the fact that the square of the velocity is the sum of the squares of its Cartesian components, and in the last step, we used the definition of the particle's kinetic energy. This important result is called the **work-energy theorem** (**Figure 7.11**).

Work-Energy Theorem

The net work done on a particle equals the change in the particle's kinetic energy:

$$W_{\text{net}} = K_B - K_A. \tag{7.9}$$

Figure 7.11 Horse pulls are common events at state fairs. The work done by the horses pulling on the load results in a change in kinetic energy of the load, ultimately going faster. (credit: modification of work by "Jassen"/ Flickr)

According to this theorem, when an object slows down, its final kinetic energy is less than its initial kinetic energy, the change in its kinetic energy is negative, and so is the net work done on it. If an object speeds up, the net work done on it is positive. When calculating the net work, you must include all the forces that act on an object. If you leave out any forces that act on an object, or if you include any forces that don't act on it, you will get a wrong result.

The importance of the work-energy theorem, and the further generalizations to which it leads, is that it makes some types of calculations much simpler to accomplish than they would be by trying to solve Newton's second law. For example, in **Newton's Laws of Motion**, we found the speed of an object sliding down a frictionless plane by solving Newton's second law for the acceleration and using kinematic equations for constant acceleration, obtaining

$$v_{\text{f}}^2 = v_{\text{i}}^2 + 2g(s_{\text{f}} - s_{\text{i}})\sin\theta,$$

where s is the displacement down the plane.

We can also get this result from the work-energy theorem in **Equation 7.1**. Since only two forces are acting on the object-gravity and the normal force-and the normal force doesn't do any work, the net work is just the work done by gravity.

The work dW is the dot product of the force of gravity or $\overrightarrow{\mathbf{F}} = -mg\hat{j}$ and the displacement $\overrightarrow{dr} = dx\,\hat{i} + dy\,\hat{j}$. After taking the dot product and integrating from an initial position y_i to a final position y_f, one finds the net work as

$$W_{\text{net}} = W_{\text{grav}} = -mg(y_f - y_i),$$

where y is positive up. The work-energy theorem says that this equals the change in kinetic energy:

$$-mg(y_f - y_i) = \frac{1}{2}m(v_f^2 - v_i^2).$$

Using a right triangle, we can see that $(y_f - y_i) = (s_f - s_i)\sin\theta$, so the result for the final speed is the same.

What is gained by using the work-energy theorem? The answer is that for a frictionless plane surface, not much. However, Newton's second law is easy to solve only for this particular case, whereas the work-energy theorem gives the final speed for any shaped frictionless surface. For an arbitrary curved surface, the normal force is not constant, and Newton's second law may be difficult or impossible to solve analytically. Constant or not, for motion along a surface, the normal force never does any work, because it's perpendicular to the displacement. A calculation using the work-energy theorem avoids this difficulty and applies to more general situations.

Problem-Solving Strategy: Work-Energy Theorem

1. Draw a free-body diagram for each force on the object.

2. Determine whether or not each force does work over the displacement in the diagram. Be sure to keep any positive or negative signs in the work done.

3. Add up the total amount of work done by each force.

4. Set this total work equal to the change in kinetic energy and solve for any unknown parameter.

5. Check your answers. If the object is traveling at a constant speed or zero acceleration, the total work done should be zero and match the change in kinetic energy. If the total work is positive, the object must have sped up or increased kinetic energy. If the total work is negative, the object must have slowed down or decreased kinetic energy.

Example 7.9

Loop-the-Loop

The frictionless track for a toy car includes a loop-the-loop of radius R. How high, measured from the bottom of the loop, must the car be placed to start from rest on the approaching section of track and go all the way around the loop?

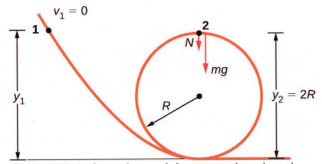

Figure 7.12 A frictionless track for a toy car has a loop-the-loop in it. How high must the car start so that it can go around the loop without falling off?

Strategy

The free-body diagram at the final position of the object is drawn in **Figure 7.12**. The gravitational work is the only work done over the displacement that is not zero. Since the weight points in the same direction as the net vertical displacement, the total work done by the gravitational force is positive. From the work-energy theorem, the starting height determines the speed of the car at the top of the loop,

$$-mg(y_2 - y_1) = \frac{1}{2}mv_2{}^2,$$

where the notation is shown in the accompanying figure. At the top of the loop, the normal force and gravity are both down and the acceleration is centripetal, so

$$a_{\text{top}} = \frac{F}{m} = \frac{N + mg}{m} = \frac{v_2^2}{R}.$$

The condition for maintaining contact with the track is that there must be some normal force, however slight; that is, $N > 0$. Substituting for v_2^2 and N, we can find the condition for y_1.

Solution

Implement the steps in the strategy to arrive at the desired result:

$$N = \frac{-mg + mv_2^2}{R} = \frac{-mg + 2mg(y_1 - 2R)}{R} > 0 \quad \text{or} \quad y_1 > \frac{5R}{2}.$$

Significance

On the surface of the loop, the normal component of gravity and the normal contact force must provide the centripetal acceleration of the car going around the loop. The tangential component of gravity slows down or speeds up the car. A child would find out how high to start the car by trial and error, but now that you know the work-energy theorem, you can predict the minimum height (as well as other more useful results) from physical principles. By using the work-energy theorem, you did not have to solve a differential equation to determine the height.

 7.7 **Check Your Understanding** Suppose the radius of the loop-the-loop in **Example 7.9** is 15 cm and the toy car starts from rest at a height of 45 cm above the bottom. What is its speed at the top of the loop?

 Visit Carleton College's site to see a **video (https://openstaxcollege.org/l/21carollvidrol)** of a looping rollercoaster.

In situations where the motion of an object is known, but the values of one or more of the forces acting on it are not known, you may be able to use the work-energy theorem to get some information about the forces. Work depends on the force and the distance over which it acts, so the information is provided via their product.

Example 7.10

Determining a Stopping Force

A bullet from a 0.22LR-caliber cartridge has a mass of 40 grains (2.60 g) and a muzzle velocity of 1100 ft./s (335 m/s). It can penetrate eight 1-inch pine boards, each with thickness 0.75 inches. What is the average stopping force exerted by the wood, as shown in **Figure 7.13**?

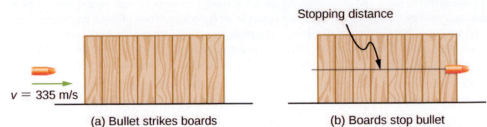

Figure 7.13 The boards exert a force to stop the bullet. As a result, the boards do work and the bullet loses kinetic energy.

Strategy

We can assume that under the general conditions stated, the bullet loses all its kinetic energy penetrating the boards, so the work-energy theorem says its initial kinetic energy is equal to the average stopping force times the distance penetrated. The change in the bullet's kinetic energy and the net work done stopping it are both negative, so when you write out the work-energy theorem, with the net work equal to the average force times the stopping distance, that's what you get. The total thickness of eight 1-inch pine boards that the bullet penetrates is $8 \times \frac{3}{4}$ in. $= 6$ in. $= 15.2$ cm.

Solution

Applying the work-energy theorem, we get

$$W_{\text{net}} = -F_{\text{ave}} \, \Delta s_{\text{stop}} = -K_{\text{initial}},$$

so

$$F_{\text{ave}} = \frac{\frac{1}{2}mv^2}{\Delta s_{\text{stop}}} = \frac{\frac{1}{2}(2.6 \times 10^{-3}\,\text{kg})(335\,\text{m/s})^2}{0.152\,\text{m}} = 960\,\text{N}.$$

Significance

We could have used Newton's second law and kinematics in this example, but the work-energy theorem also supplies an answer to less simple situations. The penetration of a bullet, fired vertically upward into a block of wood, is discussed in one section of Asif Shakur's recent article ["Bullet-Block Science Video Puzzle." *The Physics Teacher* (January 2015) 53(1): 15-16]. If the bullet is fired dead center into the block, it loses all its kinetic energy and penetrates slightly farther than if fired off-center. The reason is that if the bullet hits off-center, it has a little kinetic energy after it stops penetrating, because the block rotates. The work-energy theorem implies that a smaller change in kinetic energy results in a smaller penetration. You will understand more of the physics in this interesting article after you finish reading **Angular Momentum**.

 Learn more about work and energy in this **PhET simulation (https://openstaxcollege.org/l/ 21PhETSimRamp)** called "the ramp." Try changing the force pushing the box and the frictional force along the incline. The work and energy plots can be examined to note the total work done and change in kinetic energy of the box.

7.4 | Power

Learning Objectives
By the end of this section, you will be able to: • Relate the work done during a time interval to the power delivered • Find the power expended by a force acting on a moving body

The concept of work involves force and displacement; the work-energy theorem relates the net work done on a body to the difference in its kinetic energy, calculated between two points on its trajectory. None of these quantities or relations involves time explicitly, yet we know that the time available to accomplish a particular amount of work is frequently just as important to us as the amount itself. In the chapter-opening figure, several sprinters may have achieved the same velocity at the finish, and therefore did the same amount of work, but the winner of the race did it in the least amount of time.

We express the relation between work done and the time interval involved in doing it, by introducing the concept of power. Since work can vary as a function of time, we first define **average power** as the work done during a time interval, divided by the interval,

$$P_{\text{ave}} = \frac{\Delta W}{\Delta t}.$$ (7.10)

Then, we can define the **instantaneous power** (frequently referred to as just plain **power**).

Power

Power is defined as the rate of doing work, or the limit of the average power for time intervals approaching zero,

$$P = \frac{dW}{dt}.$$ (7.11)

If the power is constant over a time interval, the average power for that interval equals the instantaneous power, and the work done by the agent supplying the power is $W = P\Delta t$. If the power during an interval varies with time, then the work done is the time integral of the power,

$$W = \int P\,dt.$$

The work-energy theorem relates how work can be transformed into kinetic energy. Since there are other forms of energy as well, as we discuss in the next chapter, we can also define power as the rate of transfer of energy. Work and energy are measured in units of joules, so power is measured in units of joules per second, which has been given the SI name watts, abbreviation W: $1\,\text{J/s} = 1\,\text{W}$. Another common unit for expressing the power capability of everyday devices is horsepower: $1\,\text{hp} = 746\,\text{W}$.

Example 7.11

Pull-Up Power

An 80-kg army trainee does 10 pull-ups in 10 s (**Figure 7.14**). How much average power do the trainee's muscles supply moving his body? (*Hint:* Make reasonable estimates for any quantities needed.)

Figure 7.14 What is the power expended in doing ten pull-ups in ten seconds?

Strategy

The work done against gravity, going up or down a distance Δy, is $mg\Delta y$. (If you lift and lower yourself at constant speed, the force you exert cancels gravity over the whole pull-up cycle.) Thus, the work done by the trainee's muscles (moving, but not accelerating, his body) for a complete repetition (up and down) is $2mg\Delta y$. Let's assume that $\Delta y = 2\,\text{ft} \approx 60\,\text{cm}$. Also, assume that the arms comprise 10% of the body mass and are not included in the moving mass. With these assumptions, we can calculate the work done for 10 pull-ups and divide by 10 s to get the average power.

Solution

The result we get, applying our assumptions, is

$$P_{\text{ave}} = \frac{10 \times 2(0.9 \times 80\,\text{kg})(9.8\,\text{m/s}^2)(0.6\,\text{m})}{10\,\text{s}} = 850\,\text{W}.$$

Significance

This is typical for power expenditure in strenuous exercise; in everyday units, it's somewhat more than one horsepower $(1\,\text{hp} = 746\,\text{W})$.

 7.8 Check Your Understanding Estimate the power expended by a weightlifter raising a 150-kg barbell 2 m in 3 s.

The power involved in moving a body can also be expressed in terms of the forces acting on it. If a force $\vec{F}$ acts on a body that is displaced $d\vec{r}$ in a time dt, the power expended by the force is

$$P = \frac{dW}{dt} = \frac{\vec{F} \cdot d\vec{r}}{dt} = \vec{F} \cdot \left(\frac{d\vec{r}}{dt} \right) = \vec{F} \cdot \vec{v}, \tag{7.12}$$

where $\vec{v}$ is the velocity of the body. The fact that the limits implied by the derivatives exist, for the motion of a real body, justifies the rearrangement of the infinitesimals.

Example 7.12

Automotive Power Driving Uphill

How much power must an automobile engine expend to move a 1200-kg car up a 15% grade at 90 km/h (**Figure 7.15**)? Assume that 25% of this power is dissipated overcoming air resistance and friction.

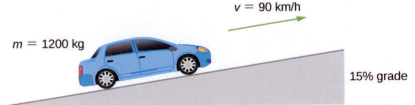

Figure 7.15 We want to calculate the power needed to move a car up a hill at constant speed.

Strategy

At constant velocity, there is no change in kinetic energy, so the net work done to move the car is zero. Therefore the power supplied by the engine to move the car equals the power expended against gravity and air resistance. By assumption, 75% of the power is supplied against gravity, which equals $m \vec{\textbf{g}} \cdot \vec{\textbf{v}} = mgv \sin \theta$, where θ is the angle of the incline. A 15% grade means $\tan \theta = 0.15$. This reasoning allows us to solve for the power required.

Solution

Carrying out the suggested steps, we find

$$0.75 \, P = mgv \sin(\tan^{-1} 0.15),$$

or

$$P = \frac{(1200 \times 9.8 \text{ N})(90 \text{ m}/3.6 \text{ s})\sin(8.53°)}{0.75} = 58 \text{ kW},$$

or about 78 hp. (You should supply the steps used to convert units.)

Significance

This is a reasonable amount of power for the engine of a small to mid-size car to supply $(1 \text{ hp} = 0.746 \text{ kW})$.

Note that this is only the power expended to move the car. Much of the engine's power goes elsewhere, for example, into waste heat. That's why cars need radiators. Any remaining power could be used for acceleration, or to operate the car's accessories.

CHAPTER 7 REVIEW

KEY TERMS

average power work done in a time interval divided by the time interval

kinetic energy energy of motion, one-half an object's mass times the square of its speed

net work work done by all the forces acting on an object

power (or instantaneous power) rate of doing work

work done when a force acts on something that undergoes a displacement from one position to another

work done by a force integral, from the initial position to the final position, of the dot product of the force and the infinitesimal displacement along the path over which the force acts

work-energy theorem net work done on a particle is equal to the change in its kinetic energy

KEY EQUATIONS

Work done by a force over an infinitesimal displacement	$dW = \vec{\mathbf{F}} \cdot d\vec{\mathbf{r}} = \left| \vec{\mathbf{F}} \right| \left| d\vec{\mathbf{r}} \right| \cos\theta$
Work done by a force acting along a path from A to B	$W_{AB} = \int_{pathAB} \vec{\mathbf{F}} \cdot d\vec{\mathbf{r}}$
Work done by a constant force of kinetic friction	$W_{\mathrm{fr}} = -f_k \left| l_{AB} \right|$
Work done going from A to B by Earth's gravity, near its surface	$W_{\mathrm{grav},AB} = -mg(y_B - y_A)$
Work done going from A to B by one-dimensional spring force	$W_{\mathrm{spring},AB} = -\left(\frac{1}{2}k\right)\left(x_B^2 - x_A^2\right)$
Kinetic energy of a non-relativistic particle	$K = \frac{1}{2}mv^2 = \frac{p^2}{2m}$
Work-energy theorem	$W_{\mathrm{net}} = K_B - K_A$
Power as rate of doing work	$P = \dfrac{dW}{dt}$
Power as the dot product of force and velocity	$P = \vec{\mathbf{F}} \cdot \vec{\mathbf{v}}$

SUMMARY

7.1 Work

- The infinitesimal increment of work done by a force, acting over an infinitesimal displacement, is the dot product of the force and the displacement.
- The work done by a force, acting over a finite path, is the integral of the infinitesimal increments of work done along the path.
- The work done *against* a force is the negative of the work done *by* the force.
- The work done by a normal or frictional contact force must be determined in each particular case.
- The work done by the force of gravity, on an object near the surface of Earth, depends only on the weight of the object and the difference in height through which it moved.
- The work done by a spring force, acting from an initial position to a final position, depends only on the spring constant and the squares of those positions.

7.2 Kinetic Energy

- The kinetic energy of a particle is the product of one-half its mass and the square of its speed, for non-relativistic speeds.

- The kinetic energy of a system is the sum of the kinetic energies of all the particles in the system.

- Kinetic energy is relative to a frame of reference, is always positive, and is sometimes given special names for different types of motion.

7.3 Work-Energy Theorem

- Because the net force on a particle is equal to its mass times the derivative of its velocity, the integral for the net work done on the particle is equal to the change in the particle's kinetic energy. This is the work-energy theorem.

- You can use the work-energy theorem to find certain properties of a system, without having to solve the differential equation for Newton's second law.

7.4 Power

- Power is the rate of doing work; that is, the derivative of work with respect to time.

- Alternatively, the work done, during a time interval, is the integral of the power supplied over the time interval.

- The power delivered by a force, acting on a moving particle, is the dot product of the force and the particle's velocity.

CONCEPTUAL QUESTIONS

7.1 Work

1. Give an example of something we think of as work in everyday circumstances that is not work in the scientific sense. Is energy transferred or changed in form in your example? If so, explain how this is accomplished without doing work.

2. Give an example of a situation in which there is a force and a displacement, but the force does no work. Explain why it does no work.

3. Describe a situation in which a force is exerted for a long time but does no work. Explain.

4. A body moves in a circle at constant speed. Does the centripetal force that accelerates the body do any work? Explain.

5. Suppose you throw a ball upward and catch it when it returns at the same height. How much work does the gravitational force do on the ball over its entire trip?

6. Why is it more difficult to do sit-ups while on a slant board than on a horizontal surface? (See below.)

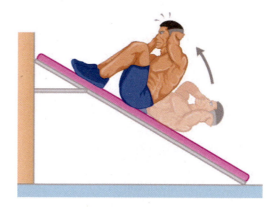

7. As a young man, Tarzan climbed up a vine to reach his tree house. As he got older, he decided to build and use a staircase instead. Since the work of the gravitational force *mg* is path independent, what did the King of the Apes gain in using stairs?

7.2 Kinetic Energy

8. A particle of m has a velocity of $v_x \hat{i} + v_y \hat{j} + v_z \hat{k}$. Is its kinetic energy given by $m(v_x^2 \hat{i} + v_y^2 \hat{j} + v_z^2 \hat{k})/2$? If not, what is the correct expression?

9. One particle has mass m and a second particle has mass $2m$. The second particle is moving with speed v and the first with speed $2v$. How do their kinetic energies compare?

10. A person drops a pebble of mass m_1 from a height h, and it hits the floor with kinetic energy K. The person drops another pebble of mass m_2 from a height of $2h$, and it hits the floor with the same kinetic energy K. How do the masses of the pebbles compare?

7.3 Work-Energy Theorem

11. The person shown below does work on the lawn mower. Under what conditions would the mower gain energy from the person pushing the mower? Under what conditions would it lose energy?

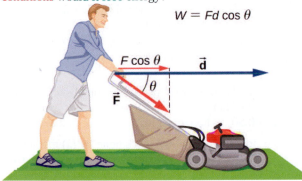

$$W = Fd \cos \theta$$

12. Work done on a system puts energy into it. Work done by a system removes energy from it. Give an example for each statement.

13. Two marbles of masses m and $2m$ are dropped from a height h. Compare their kinetic energies when they reach the ground.

14. Compare the work required to accelerate a car of mass 2000 kg from 30.0 to 40.0 km/h with that required for an acceleration from 50.0 to 60.0 km/h.

15. Suppose you are jogging at constant velocity. Are you doing any work on the environment and vice versa?

16. Two forces act to double the speed of a particle, initially moving with kinetic energy of 1 J. One of the forces does 4 J of work. How much work does the other force do?

7.4 Power

17. Most electrical appliances are rated in watts. Does this rating depend on how long the appliance is on? (When off, it is a zero-watt device.) Explain in terms of the definition of power.

18. Explain, in terms of the definition of power, why energy consumption is sometimes listed in kilowatt-hours rather than joules. What is the relationship between these two energy units?

19. A spark of static electricity, such as that you might receive from a doorknob on a cold dry day, may carry a few hundred watts of power. Explain why you are not injured by such a spark.

20. Does the work done in lifting an object depend on how fast it is lifted? Does the power expended depend on how fast it is lifted?

21. Can the power expended by a force be negative?

22. How can a 50-W light bulb use more energy than a 1000-W oven?

PROBLEMS

7.1 Work

23. How much work does a supermarket checkout attendant do on a can of soup he pushes 0.600 m horizontally with a force of 5.00 N?

24. A 75.0-kg person climbs stairs, gaining 2.50 m in height. Find the work done to accomplish this task.

25. (a) Calculate the work done on a 1500-kg elevator car by its cable to lift it 40.0 m at constant speed, assuming friction averages 100 N. (b) What is the work done on the lift by the gravitational force in this process? (c) What is the total work done on the lift?

26. Suppose a car travels 108 km at a speed of 30.0 m/s, and uses 2.0 gal of gasoline. Only 30% of the gasoline goes into useful work by the force that keeps the car moving at constant speed despite friction. (The energy content of gasoline is about 140 MJ/gal.) (a) What is the magnitude of the force exerted to keep the car moving at constant speed?

(b) If the required force is directly proportional to speed, how many gallons will be used to drive 108 km at a speed of 28.0 m/s?

27. Calculate the work done by an 85.0-kg man who pushes a crate 4.00 m up along a ramp that makes an angle of 20.0° with the horizontal (see below). He exerts a force of 500 N on the crate parallel to the ramp and moves at a constant speed. Be certain to include the work he does on the crate and on his body to get up the ramp.

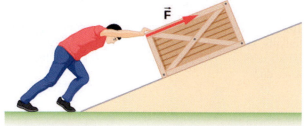

28. How much work is done by the boy pulling his sister 30.0 m in a wagon as shown below? Assume no friction acts on the wagon.

29. A shopper pushes a grocery cart 20.0 m at constant speed on level ground, against a 35.0 N frictional force. He pushes in a direction 25.0° below the horizontal. (a) What is the work done on the cart by friction? (b) What is the work done on the cart by the gravitational force? (c) What is the work done on the cart by the shopper? (d) Find the force the shopper exerts, using energy considerations. (e) What is the total work done on the cart?

30. Suppose the ski patrol lowers a rescue sled and victim, having a total mass of 90.0 kg, down a 60.0° slope at constant speed, as shown below. The coefficient of friction between the sled and the snow is 0.100. (a) How much work is done by friction as the sled moves 30.0 m along the hill? (b) How much work is done by the rope on the sled in this distance? (c) What is the work done by the gravitational force on the sled? (d) What is the total work done?

31. A constant 20-N force pushes a small ball in the direction of the force over a distance of 5.0 m. What is the work done by the force?

32. A toy cart is pulled a distance of 6.0 m in a straight line across the floor. The force pulling the cart has a magnitude of 20 N and is directed at 37° above the horizontal. What is the work done by this force?

33. A 5.0-kg box rests on a horizontal surface. The coefficient of kinetic friction between the box and surface is $\mu_K = 0.50$. A horizontal force pulls the box at constant velocity for 10 cm. Find the work done by (a) the applied horizontal force, (b) the frictional force, and (c) the net force.

34. A sled plus passenger with total mass 50 kg is pulled 20 m across the snow ($\mu_k = 0.20$) at constant velocity by a force directed 25° above the horizontal. Calculate (a) the work of the applied force, (b) the work of friction, and (c) the total work.

35. Suppose that the sled plus passenger of the preceding problem is pushed 20 m across the snow at constant velocity by a force directed 30° below the horizontal. Calculate (a) the work of the applied force, (b) the work of friction, and (c) the total work.

36. How much work does the force $F(x) = (-2.0/x)\,\text{N}$ do on a particle as it moves from $x = 2.0\,\text{m}$ to $x = 5.0\,\text{m}$?

37. How much work is done against the gravitational force on a 5.0-kg briefcase when it is carried from the ground floor to the roof of the Empire State Building, a vertical climb of 380 m?

38. It takes 500 J of work to compress a spring 10 cm. What is the force constant of the spring?

39. A bungee cord is essentially a very long rubber band that can stretch up to four times its unstretched length. However, its spring constant varies over its stretch [see Menz, P.G. "The Physics of Bungee Jumping." *The Physics Teacher* (November 1993) 31: 483-487]. Take the length of the cord to be along the x-direction and define the stretch x as the length of the cord l minus its un-stretched length l_0; that is, $x = l - l_0$ (see below). Suppose a particular bungee cord has a spring constant, for $0 \leq x \leq 4.88 \text{ m}$, of $k_1 = 204 \text{ N/m}$ and for $4.88 \text{ m} \leq x$, of $k_2 = 111 \text{ N/m}$. (Recall that the spring constant is the slope of the force $F(x)$ versus its stretch x.) (a) What is the tension in the cord when the stretch is 16.7 m (the maximum desired for a given jump)? (b) How much work must be done against the elastic force of the bungee cord to stretch it 16.7 m?

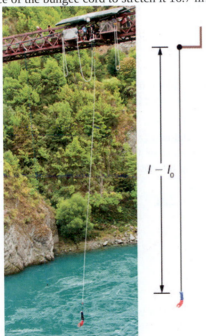

Figure 7.16 (credit: modification of work by Graeme Churchard)

40. A bungee cord exerts a nonlinear elastic force of magnitude $F(x) = k_1 x + k_2 x^3$, where x is the distance the cord is stretched, $k_1 = 204 \text{ N/m}$ and $k_2 = -0.233 \text{ N/m}^3$. How much work must be done on the cord to stretch it 16.7 m?

41. Engineers desire to model the magnitude of the elastic force of a bungee cord using the equation

$$F(x) = a\left[\frac{x + 9 \text{ m}}{9 \text{ m}} - \left(\frac{9 \text{ m}}{x + 9 \text{ m}}\right)^2\right],$$

where x is the stretch of the cord along its length and a is a constant. If it takes 22.0 kJ of work to stretch the cord by 16.7 m, determine the value of the constant a.

42. A particle moving in the xy-plane is subject to a force

$$\vec{F}(x, y) = (50 \text{ N} \cdot \text{m}^2)\frac{(x\hat{\mathbf{i}} + y\hat{\mathbf{j}})}{(x^2 + y^2)^{3/2}},$$

where x and y are in meters. Calculate the work done on the particle by this force, as it moves in a straight line from the point (3 m, 4 m) to the point (8 m, 6 m).

43. A particle moves along a curved path $y(x) = (10 \text{ m})\{1 + \cos[(0.1 \text{ m}^{-1})x]\}$, from $x = 0$ to $x = 10\pi \text{ m}$, subject to a tangential force of variable magnitude $F(x) = (10 \text{ N})\sin[(0.1 \text{ m}^{-1})x]$. How much work does the force do? (*Hint:* Consult a table of integrals or use a numerical integration program.)

7.2 Kinetic Energy

44. Compare the kinetic energy of a 20,000-kg truck moving at 110 km/h with that of an 80.0-kg astronaut in orbit moving at 27,500 km/h.

45. (a) How fast must a 3000-kg elephant move to have the same kinetic energy as a 65.0-kg sprinter running at 10.0 m/s? (b) Discuss how the larger energies needed for the movement of larger animals would relate to metabolic rates.

46. Estimate the kinetic energy of a 90,000-ton aircraft carrier moving at a speed of at 30 knots. You will need to look up the definition of a nautical mile to use in converting the unit for speed, where 1 knot equals 1 nautical mile per hour.

47. Calculate the kinetic energies of (a) a 2000.0-kg automobile moving at 100.0 km/h; (b) an 80.-kg runner sprinting at 10. m/s; and (c) a 9.1×10^{-31}-kg electron moving at 2.0×10^7 m/s.

48. A 5.0-kg body has three times the kinetic energy of an 8.0-kg body. Calculate the ratio of the speeds of these bodies.

49. An 8.0-g bullet has a speed of 800 m/s. (a) What is its kinetic energy? (b) What is its kinetic energy if the speed is halved?

7.3 Work-Energy Theorem

50. (a) Calculate the force needed to bring a 950-kg car to rest from a speed of 90.0 km/h in a distance of 120 m (a fairly typical distance for a non-panic stop). (b) Suppose instead the car hits a concrete abutment at full speed and is brought to a stop in 2.00 m. Calculate the force exerted on the car and compare it with the force found in part (a).

51. A car's bumper is designed to withstand a 4.0-km/h (1.1-m/s) collision with an immovable object without damage to the body of the car. The bumper cushions the shock by absorbing the force over a distance. Calculate the magnitude of the average force on a bumper that collapses 0.200 m while bringing a 900-kg car to rest from an initial speed of 1.1 m/s.

52. Boxing gloves are padded to lessen the force of a blow. (a) Calculate the force exerted by a boxing glove on an opponent's face, if the glove and face compress 7.50 cm during a blow in which the 7.00-kg arm and glove are brought to rest from an initial speed of 10.0 m/s. (b) Calculate the force exerted by an identical blow in the gory old days when no gloves were used, and the knuckles and face would compress only 2.00 cm. Assume the change in mass by removing the glove is negligible. (c) Discuss the magnitude of the force with glove on. Does it seem high enough to cause damage even though it is lower than the force with no glove?

53. Using energy considerations, calculate the average force a 60.0-kg sprinter exerts backward on the track to accelerate from 2.00 to 8.00 m/s in a distance of 25.0 m, if he encounters a headwind that exerts an average force of 30.0 N against him.

54. A 5.0-kg box has an acceleration of 2.0 m/s^2 when it is pulled by a horizontal force across a surface with $\mu_K = 0.50$. Find the work done over a distance of 10 cm by (a) the horizontal force, (b) the frictional force, and (c) the net force. (d) What is the change in kinetic energy of the box?

55. A constant 10-N horizontal force is applied to a 20-kg cart at rest on a level floor. If friction is negligible, what is the speed of the cart when it has been pushed 8.0 m?

56. In the preceding problem, the 10-N force is applied at an angle of $45°$ below the horizontal. What is the speed of the cart when it has been pushed 8.0 m?

57. Compare the work required to stop a 100-kg crate sliding at 1.0 m/s and an 8.0-g bullet traveling at 500 m/s.

58. A wagon with its passenger sits at the top of a hill. The wagon is given a slight push and rolls 100 m down a

$10°$ incline to the bottom of the hill. What is the wagon's speed when it reaches the end of the incline. Assume that the retarding force of friction is negligible.

59. An 8.0-g bullet with a speed of 800 m/s is shot into a wooden block and penetrates 20 cm before stopping. What is the average force of the wood on the bullet? Assume the block does not move.

60. A 2.0-kg block starts with a speed of 10 m/s at the bottom of a plane inclined at $37°$ to the horizontal. The coefficient of sliding friction between the block and plane is $\mu_k = 0.30$. (a) Use the work-energy principle to determine how far the block slides along the plane before momentarily coming to rest. (b) After stopping, the block slides back down the plane. What is its speed when it reaches the bottom? (*Hint:* For the round trip, only the force of friction does work on the block.)

61. When a 3.0-kg block is pushed against a massless spring of force constant constant 4.5×10^3 N/m, the spring is compressed 8.0 cm. The block is released, and it slides 2.0 m (from the point at which it is released) across a horizontal surface before friction stops it. What is the coefficient of kinetic friction between the block and the surface?

62. A small block of mass 200 g starts at rest at A, slides to B where its speed is $v_B = 8.0$ m/s, then slides along the horizontal surface a distance 10 m before coming to rest at C. (See below.) (a) What is the work of friction along the curved surface? (b) What is the coefficient of kinetic friction along the horizontal surface?

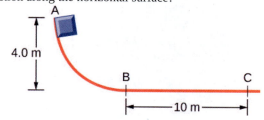

63. A small object is placed at the top of an incline that is essentially frictionless. The object slides down the incline onto a rough horizontal surface, where it stops in 5.0 s after traveling 60 m. (a) What is the speed of the object at the bottom of the incline and its acceleration along the horizontal surface? (b) What is the height of the incline?

64. When released, a 100-g block slides down the path shown below, reaching the bottom with a speed of 4.0 m/s. How much work does the force of friction do?

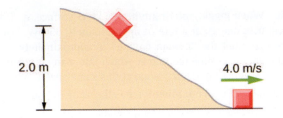

65. A 0.22LR-caliber bullet like that mentioned in **Example 7.10** is fired into a door made of a single thickness of 1-inch pine boards. How fast would the bullet be traveling after it penetrated through the door?

66. A sled starts from rest at the top of a snow-covered incline that makes a $22°$ angle with the horizontal. After sliding 75 m down the slope, its speed is 14 m/s. Use the work-energy theorem to calculate the coefficient of kinetic friction between the runners of the sled and the snowy surface.

7.4 Power

67. A person in good physical condition can put out 100 W of useful power for several hours at a stretch, perhaps by pedaling a mechanism that drives an electric generator. Neglecting any problems of generator efficiency and practical considerations such as resting time: (a) How many people would it take to run a 4.00-kW electric clothes dryer? (b) How many people would it take to replace a large electric power plant that generates 800 MW?

68. What is the cost of operating a 3.00-W electric clock for a year if the cost of electricity is $0.0900 per $kW \cdot h$?

69. A large household air conditioner may consume 15.0 kW of power. What is the cost of operating this air conditioner 3.00 h per day for 30.0 d if the cost of electricity is $0.110 per $kW \cdot h$?

70. (a) What is the average power consumption in watts of an appliance that uses 5.00 $kW \cdot h$ of energy per day? (b) How many joules of energy does this appliance consume in a year?

71. (a) What is the average useful power output of a person who does 6.00×10^6 J of useful work in 8.00 h? (b) Working at this rate, how long will it take this person to lift 2000 kg of bricks 1.50 m to a platform? (Work done to lift his body can be omitted because it is not considered useful output here.)

72. A 500-kg dragster accelerates from rest to a final speed of 110 m/s in 400 m (about a quarter of a mile) and encounters an average frictional force of 1200 N. What is its average power output in watts and horsepower if this takes 7.30 s?

73. (a) How long will it take an 850-kg car with a useful power output of 40.0 hp (1 hp equals 746 W) to reach a speed of 15.0 m/s, neglecting friction? (b) How long will this acceleration take if the car also climbs a 3.00-m high hill in the process?

74. (a) Find the useful power output of an elevator motor that lifts a 2500-kg load a height of 35.0 m in 12.0 s, if it also increases the speed from rest to 4.00 m/s. Note that the total mass of the counterbalanced system is 10,000 kg—so that only 2500 kg is raised in height, but the full 10,000 kg is accelerated. (b) What does it cost, if electricity is $0.0900 per $kW \cdot h$?

75. (a) How long would it take a 1.50×10^5-kg airplane with engines that produce 100 MW of power to reach a speed of 250 m/s and an altitude of 12.0 km if air resistance were negligible? (b) If it actually takes 900 s, what is the power? (c) Given this power, what is the average force of air resistance if the airplane takes 1200 s? (*Hint:* You must find the distance the plane travels in 1200 s assuming constant acceleration.)

76. Calculate the power output needed for a 950-kg car to climb a $2.00°$ slope at a constant 30.0 m/s while encountering wind resistance and friction totaling 600 N.

77. A man of mass 80 kg runs up a flight of stairs 20 m high in 10 s. (a) how much power is used to lift the man? (b) If the man's body is 25% efficient, how much power does he expend?

78. The man of the preceding problem consumes approximately 1.05×10^7 J (2500 food calories) of energy per day in maintaining a constant weight. What is the average power he produces over a day? Compare this with his power production when he runs up the stairs.

79. An electron in a television tube is accelerated uniformly from rest to a speed of 8.4×10^7 m/s over a distance of 2.5 cm. What is the power delivered to the electron at the instant that its displacement is 1.0 cm?

80. Coal is lifted out of a mine a vertical distance of 50 m by an engine that supplies 500 W to a conveyer belt. How much coal per minute can be brought to the surface? Ignore the effects of friction.

81. A girl pulls her 15-kg wagon along a flat sidewalk by applying a 10-N force at $37°$ to the horizontal. Assume that friction is negligible and that the wagon starts from rest. (a) How much work does the girl do on the wagon in the first 2.0 s? (b) How much instantaneous power does she exert at $t = 2.0$ s?

82. A typical automobile engine has an efficiency of 25%. Suppose that the engine of a 1000-kg automobile has a maximum power output of 140 hp. What is the maximum grade that the automobile can climb at 50 km/h if the frictional retarding force on it is 300 N?

83. When jogging at 13 km/h on a level surface, a 70-kg man uses energy at a rate of approximately 850 W. Using the facts that the "human engine" is approximately 25% efficient, determine the rate at which this man uses energy when jogging up a $5.0°$ slope at this same speed. Assume that the frictional retarding force is the same in both cases.

ADDITIONAL PROBLEMS

84. A cart is pulled a distance D on a flat, horizontal surface by a constant force F that acts at an angle θ with the horizontal direction. The other forces on the object during this time are gravity (F_w), normal forces (F_{N1}) and (F_{N2}), and rolling frictions F_{r1} and F_{r2}, as shown below. What is the work done by each force?

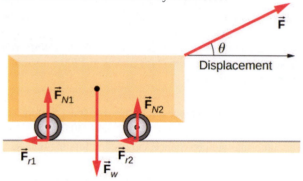

85. Consider a particle on which several forces act, one of which is known to be constant in time: $\overrightarrow{\mathbf{F}}_1 = (3\,\text{N})\,\hat{\mathbf{i}} + (4\,\text{N})\,\hat{\mathbf{j}}$. As a result, the particle moves along the x-axis from $x = 0$ to $x = 5$ m in some time interval. What is the work done by $\overrightarrow{\mathbf{F}}_1$?

86. Consider a particle on which several forces act, one of which is known to be constant in time: $\overrightarrow{\mathbf{F}}_1 = (3\,\text{N})\,\hat{\mathbf{i}} + (4\,\text{N})\,\hat{\mathbf{j}}$. As a result, the particle moves first along the x-axis from $x = 0$ to $x = 5$ m and then parallel to the y-axis from $y = 0$ to $y = 6$ m. What is the work done by $\overrightarrow{\mathbf{F}}_1$?

87. Consider a particle on which several forces act, one of which is known to be constant in time: $\overrightarrow{\mathbf{F}}_1 = (3\,\text{N})\,\hat{\mathbf{i}} + (4\,\text{N})\,\hat{\mathbf{j}}$. As a result, the particle moves along a straight path from a Cartesian coordinate of (0 m, 0 m) to (5 m, 6 m). What is the work done by $\overrightarrow{\mathbf{F}}_1$?

88. Consider a particle on which a force acts that depends on the position of the particle. This force is given by

$\overrightarrow{\mathbf{F}}_1 = (2y)\,\hat{\mathbf{i}} + (3x)\,\hat{\mathbf{j}}$. Find the work done by this force when the particle moves from the origin to a point 5 meters to the right on the x-axis.

89. A boy pulls a 5-kg cart with a 20-N force at an angle of $30°$ above the horizontal for a length of time. Over this time frame, the cart moves a distance of 12 m on the horizontal floor. (a) Find the work done on the cart by the boy. (b) What will be the work done by the boy if he pulled with the same force horizontally instead of at an angle of $30°$ above the horizontal over the same distance?

90. A crate of mass 200 kg is to be brought from a site on the ground floor to a third floor apartment. The workers know that they can either use the elevator first, then slide it along the third floor to the apartment, or first slide the crate to another location marked C below, and then take the elevator to the third floor and slide it on the third floor a shorter distance. The trouble is that the third floor is very rough compared to the ground floor. Given that the coefficient of kinetic friction between the crate and the ground floor is 0.100 and between the crate and the third floor surface is 0.300, find the work needed by the workers for each path shown from A to E. Assume that the force the workers need to do is just enough to slide the crate at constant velocity (zero acceleration). *Note:* The work by the elevator against the force of gravity is not done by the workers.

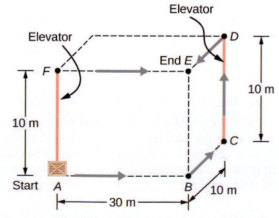

91. A hockey puck of mass 0.17 kg is shot across a rough floor with the roughness different at different places, which can be described by a position-dependent coefficient of

kinetic friction. For a puck moving along the x-axis, the coefficient of kinetic friction is the following function of x, where x is in m: $\mu(x) = 0.1 + 0.05x$. Find the work done by the kinetic frictional force on the hockey puck when it has moved (a) from $x = 0$ to $x = 2\,\mathrm{m}$, and (b) from $x = 2\,\mathrm{m}$ to $x = 4\,\mathrm{m}$.

92. A horizontal force of 20 N is required to keep a 5.0 kg box traveling at a constant speed up a frictionless incline for a vertical height change of 3.0 m. (a) What is the work done by gravity during this change in height? (b) What is the work done by the normal force? (c) What is the work done by the horizontal force?

93. A 7.0-kg box slides along a horizontal frictionless floor at 1.7 m/s and collides with a relatively massless spring that compresses 23 cm before the box comes to a stop. (a) How much kinetic energy does the box have before it collides with the spring? (b) Calculate the work done by the spring. (c) Determine the spring constant of the spring.

94. You are driving your car on a straight road with a coefficient of friction between the tires and the road of 0.55. A large piece of debris falls in front of your view and you immediate slam on the brakes, leaving a skid mark of 30.5 m (100-feet) long before coming to a stop. A policeman sees your car stopped on the road, looks at the skid mark, and gives you a ticket for traveling over the 13.4 m/s (30 mph) speed limit. Should you fight the speeding ticket in court?

95. A crate is being pushed across a rough floor surface. If no force is applied on the crate, the crate will slow down and come to a stop. If the crate of mass 50 kg moving at speed 8 m/s comes to rest in 10 seconds, what is the rate at which the frictional force on the crate takes energy away from the crate?

96. Suppose a horizontal force of 20 N is required to maintain a speed of 8 m/s of a 50 kg crate. (a) What is the power of this force? (b) Note that the acceleration of the crate is zero despite the fact that 20 N force acts on the crate horizontally. What happens to the energy given to the crate as a result of the work done by this 20 N force?

97. Grains from a hopper falls at a rate of 10 kg/s vertically onto a conveyor belt that is moving horizontally at a constant speed of 2 m/s. (a) What force is needed to keep the conveyor belt moving at the constant velocity? (b) What is the minimum power of the motor driving the conveyor belt?

98. A cyclist in a race must climb a $5°$ hill at a speed of 8 m/s. If the mass of the bike and the biker together is 80 kg, what must be the power output of the biker to achieve the goal?

CHALLENGE PROBLEMS

99. Shown below is a 40-kg crate that is pushed at constant velocity a distance 8.0 m along a $30°$ incline by the horizontal force $\vec{\mathbf{F}}$. The coefficient of kinetic friction between the crate and the incline is $\mu_k = 0.40$. Calculate the work done by (a) the applied force, (b) the frictional force, (c) the gravitational force, and (d) the net force.

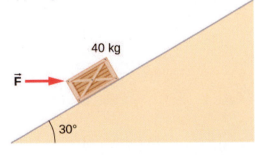

100. The surface of the preceding problem is modified so that the coefficient of kinetic friction is decreased. The same horizontal force is applied to the crate, and after being pushed 8.0 m, its speed is 5.0 m/s. How much work is now done by the force of friction? Assume that the crate starts at rest.

101. The force $F(x)$ varies with position, as shown below. Find the work done by this force on a particle as it moves from $x = 1.0\,\mathrm{m}$ to $x = 5.0\,\mathrm{m}$.

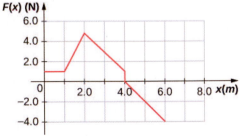

102. Find the work done by the same force in **Example 7.4**, between the same points, $A = (0,\,0)$ and $B = (2\,\mathrm{m},\,2\,\mathrm{m})$, over a circular arc of radius 2 m, centered at $(0,\,2\,\mathrm{m})$. Evaluate the path integral using Cartesian coordinates. (*Hint:* You will probably need to consult a table of integrals.)

103. Answer the preceding problem using polar coordinates.

104. Find the work done by the same force in **Example 7.4**, between the same points, $A = (0, 0)$ and $B = (2\,\text{m}, 2\,\text{m})$, over a circular arc of radius 2 m, centered at $(2\,\text{m}, 0)$. Evaluate the path integral using Cartesian coordinates. (*Hint:* You will probably need to consult a table of integrals.)

105. Answer the preceding problem using polar coordinates.

106. Constant power P is delivered to a car of mass m by its engine. Show that if air resistance can be ignored, the distance covered in a time t by the car, starting from rest, is given by $s = (8P/9m)^{1/2} t^{3/2}$.

107. Suppose that the air resistance a car encounters is independent of its speed. When the car travels at 15 m/ s, its engine delivers 20 hp to its wheels. (a) What is the power delivered to the wheels when the car travels at 30 m/ s? (b) How much energy does the car use in covering 10 km at 15 m/s? At 30 m/s? Assume that the engine is 25% efficient. (c) Answer the same questions if the force of air resistance is proportional to the speed of the automobile. (d) What do these results, plus your experience with gasoline consumption, tell you about air resistance?

108. Consider a linear spring, as in **Figure 7.7**(a), with mass M uniformly distributed along its length. The left end of the spring is fixed, but the right end, at the equilibrium position $x = 0$, is moving with speed v in the x-direction.

What is the total kinetic energy of the spring? (*Hint:* First express the kinetic energy of an infinitesimal element of the spring dm in terms of the total mass, equilibrium length, speed of the right-hand end, and position along the spring; then integrate.)

8 | POTENTIAL ENERGY AND CONSERVATION OF ENERGY

Figure 8.1 Shown here is part of a Ball Machine sculpture by George Rhoads. A ball in this contraption is lifted, rolls, falls, bounces, and collides with various objects, but throughout its travels, its kinetic energy changes in definite, predictable amounts, which depend on its position and the objects with which it interacts. (credit: modification of work by Roland Tanglao)

Chapter Outline

8.1 Potential Energy of a System

8.2 Conservative and Non-Conservative Forces

8.3 Conservation of Energy

8.4 Potential Energy Diagrams and Stability

8.5 Sources of Energy

Introduction

In George Rhoads' rolling ball sculpture, the principle of conservation of energy governs the changes in the ball's kinetic energy and relates them to changes and transfers for other types of energy associated with the ball's interactions. In this chapter, we introduce the important concept of potential energy. This will enable us to formulate the law of conservation of mechanical energy and to apply it to simple systems, making solving problems easier. In the final section on sources of energy, we will consider energy transfers and the general law of conservation of energy. Throughout this book, the law of conservation of energy will be applied in increasingly more detail, as you encounter more complex and varied systems, and other forms of energy.

8.1 | Potential Energy of a System

In **Work**, we saw that the work done on an object by the constant gravitational force, near the surface of Earth, over any displacement is a function only of the difference in the positions of the end-points of the displacement. This property allows us to define a different kind of energy for the system than its kinetic energy, which is called **potential energy**. We consider various properties and types of potential energy in the following subsections.

Potential Energy Basics

In **Motion in Two and Three Dimensions**, we analyzed the motion of a projectile, like kicking a football in **Figure 8.2**. For this example, let's ignore friction and air resistance. As the football rises, the work done by the gravitational force on the football is negative, because the ball's displacement is positive vertically and the force due to gravity is negative vertically. We also noted that the ball slowed down until it reached its highest point in the motion, thereby decreasing the ball's kinetic energy. This loss in kinetic energy translates to a gain in gravitational potential energy of the football-Earth system.

As the football falls toward Earth, the work done on the football is now positive, because the displacement and the gravitational force both point vertically downward. The ball also speeds up, which indicates an increase in kinetic energy. Therefore, energy is converted from gravitational potential energy back into kinetic energy.

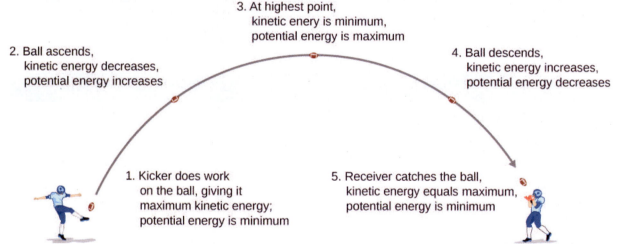

Figure 8.2 As a football starts its descent toward the wide receiver, gravitational potential energy is converted back into kinetic energy.

Based on this scenario, we can define the difference of potential energy from point A to point B as the negative of the work done:

$$\Delta U_{AB} = U_B - U_A = -W_{AB}.$$ (8.1)

This formula explicitly states a **potential energy difference**, not just an absolute potential energy. Therefore, we need to define potential energy at a given position in such a way as to state standard values of potential energy on their own, rather than potential energy differences. We do this by rewriting the potential energy function in terms of an arbitrary constant,

$$\Delta U = U(\overrightarrow{\mathbf{r}}) - U(\overrightarrow{\mathbf{r}}_0). \tag{8.2}$$

The choice of the potential energy at a starting location of $\overrightarrow{\mathbf{r}}_0$ is made out of convenience in the given problem. Most importantly, whatever choice is made should be stated and kept consistent throughout the given problem. There are some well-accepted choices of initial potential energy. For example, the lowest height in a problem is usually defined as zero potential energy, or if an object is in space, the farthest point away from the system is often defined as zero potential energy. Then, the potential energy, with respect to zero at $\overrightarrow{\mathbf{r}}_0$, is just $U(\overrightarrow{r})$.

As long as there is no friction or air resistance, the change in kinetic energy of the football equals the change in gravitational potential energy of the football. This can be generalized to any potential energy:

$$\Delta K_{AB} = \Delta U_{AB}. \tag{8.3}$$

Let's look at a specific example, choosing zero potential energy for gravitational potential energy at convenient points.

Example 8.1

Basic Properties of Potential Energy

A particle moves along the x-axis under the action of a force given by $F = -ax^2$, where $a = 3 \text{ N/m}^2$. (a) What is the difference in its potential energy as it moves from $x_A = 1 \text{ m}$ to $x_B = 2 \text{ m}$? (b) What is the particle's potential energy at $x = 1 \text{ m}$ with respect to a given 0.5 J of potential energy at $x = 0$?

Strategy

(a) The difference in potential energy is the negative of the work done, as defined by **Equation 8.1**. The work is defined in the previous chapter as the dot product of the force with the distance. Since the particle is moving forward in the x-direction, the dot product simplifies to a multiplication ($\hat{\mathbf{i}} \cdot \hat{\mathbf{i}} = 1$). To find the total work done, we need to integrate the function between the given limits. After integration, we can state the work or the potential energy. (b) The potential energy function, with respect to zero at $x = 0$, is the indefinite integral encountered in part (a), with the constant of integration determined from **Equation 8.3**. Then, we substitute the x-value into the function of potential energy to calculate the potential energy at $x = 1 \text{ m}$.

Solution

a. The work done by the given force as the particle moves from coordinate x to $x + dx$ in one dimension is

$$dW = \overrightarrow{\mathbf{F}} \cdot d\overrightarrow{\mathbf{r}} = Fdx = -ax^2 dx.$$

Substituting this expression into **Equation 8.1**, we obtain

$$\Delta U = -W = \int_{x_1}^{x_2} ax^2 dx = \frac{1}{3}(3 \text{ N/m}^2)x^2 \Big|_{1 \text{ m}}^{2 \text{ m}} = 7 \text{ J}.$$

b. The indefinite integral for the potential energy function in part (a) is

$$U(x) = \frac{1}{3}ax^3 + \text{const.},$$

and we want the constant to be determined by

$$U(0) = 0.5 \text{ J}.$$

Thus, the potential energy with respect to zero at $x = 0$ is just

$$U(x) = \frac{1}{3}ax^3 + 0.5 \text{ J.}$$

Therefore, the potential energy at $x = 1$ m is

$$U(1 \text{ m}) = \frac{1}{3}\left(3 \text{ N/m}^2\right)(1 \text{ m})^3 + 0.5 \text{ J} = 1.5 \text{ J.}$$

Significance

In this one-dimensional example, any function we can integrate, independent of path, is conservative. Notice how we applied the definition of potential energy difference to determine the potential energy function with respect to zero at a chosen point. Also notice that the potential energy, as determined in part (b), at $x = 1$ m is $U(1 \text{ m}) = 1$ J and at $x = 2$ m is $U(2 \text{ m}) = 8$ J; their difference is the result in part (a).

 8.1   Check Your Understanding In **Example 8.1**, what are the potential energies of the particle at $x = 1$ m and $x = 2$ m with respect to zero at $x = 1.5$ m ? Verify that the difference of potential energy is still 7 J.

Systems of Several Particles

In general, a system of interest could consist of several particles. The difference in the potential energy of the system is the negative of the work done by gravitational or elastic forces, which, as we will see in the next section, are conservative forces. The potential energy difference depends only on the initial and final positions of the particles, and on some parameters that characterize the interaction (like mass for gravity or the spring constant for a Hooke's law force).

It is important to remember that potential energy is a property of the interactions between objects in a chosen system, and not just a property of each object. This is especially true for electric forces, although in the examples of potential energy we consider below, parts of the system are either so big (like Earth, compared to an object on its surface) or so small (like a massless spring), that the changes those parts undergo are negligible if included in the system.

Types of Potential Energy

For each type of interaction present in a system, you can label a corresponding type of potential energy. The total potential energy of the system is the sum of the potential energies of all the types. (This follows from the additive property of the dot product in the expression for the work done.) Let's look at some specific examples of types of potential energy discussed in **Work**. First, we consider each of these forces when acting separately, and then when both act together.

Gravitational potential energy near Earth's surface

The system of interest consists of our planet, Earth, and one or more particles near its surface (or bodies small enough to be considered as particles, compared to Earth). The gravitational force on each particle (or body) is just its weight mg near the surface of Earth, acting vertically down. According to Newton's third law, each particle exerts a force on Earth of equal magnitude but in the opposite direction. Newton's second law tells us that the magnitude of the acceleration produced by each of these forces on Earth is mg divided by Earth's mass. Since the ratio of the mass of any ordinary object to the mass of Earth is vanishingly small, the motion of Earth can be completely neglected. Therefore, we consider this system to be a group of single-particle systems, subject to the uniform gravitational force of Earth.

In **Work**, the work done on a body by Earth's uniform gravitational force, near its surface, depended on the mass of the body, the acceleration due to gravity, and the difference in height the body traversed, as given by **Equation 7.4**. By definition, this work is the negative of the difference in the gravitational potential energy, so that difference is

$$\Delta U_{\text{grav}} = -W_{\text{grav, } AB} = mg(y_B - y_A). \tag{8.4}$$

You can see from this that the gravitational potential energy function, near Earth's surface, is

$$U(y) = mgy + \text{const.} \tag{8.5}$$

You can choose the value of the constant, as described in the discussion of **Equation 8.2**; however, for solving most problems, the most convenient constant to choose is zero for when $y = 0,$ which is the lowest vertical position in the problem.

Example 8.2

Gravitational Potential Energy of a Hiker

The summit of Great Blue Hill in Milton, MA, is 147 m above its base and has an elevation above sea level of 195 m (**Figure 8.3**). (Its Native American name, *Massachusett*, was adopted by settlers for naming the Bay Colony and state near its location.) A 75-kg hiker ascends from the base to the summit. What is the gravitational potential energy of the hiker-Earth system with respect to zero gravitational potential energy at base height, when the hiker is (a) at the base of the hill, (b) at the summit, and (c) at sea level, afterward?

Figure 8.3 Sketch of the profile of Great Blue Hill, Milton, MA. The altitudes of the three levels are indicated.

Strategy

First, we need to pick an origin for the *y*-axis and then determine the value of the constant that makes the potential energy zero at the height of the base. Then, we can determine the potential energies from **Equation 8.5**, based on the relationship between the zero potential energy height and the height at which the hiker is located.

Solution

a. Let's choose the origin for the *y*-axis at base height, where we also want the zero of potential energy to be. This choice makes the constant equal to zero and

$$U(\text{base}) = U(0) = 0.$$

b. At the summit, $y = 147 \, \text{m}$, so

$$U(\text{summit}) = U(147 \, \text{m}) = mgh = (75 \times 9.8 \, \text{N})(147 \, \text{m}) = 108 \, \text{kJ}.$$

c. At sea level, $y = (147 - 195)\text{m} = -48 \, \text{m}$, so

$$U(\text{sea-level}) = (75 \times 9.8 \, \text{N})(-48 \, \text{m}) = -35.3 \, \text{kJ}.$$

Significance

Besides illustrating the use of **Equation 8.4** and **Equation 8.5**, the values of gravitational potential energy we found are reasonable. The gravitational potential energy is higher at the summit than at the base, and lower at sea level than at the base. Gravity does work on you on your way up, too! It does negative work and not quite as much (in magnitude), as your muscles do. But it certainly does work. Similarly, your muscles do work on your way down, as negative work. The numerical values of the potential energies depend on the choice of zero of potential energy, but the physically meaningful differences of potential energy do not. [Note that since **Equation 8.2** is a difference, the numerical values do not depend on the origin of coordinates.]

 8.2 Check Your Understanding What are the values of the gravitational potential energy of the hiker at the base, summit, and sea level, with respect to a sea-level zero of potential energy?

Elastic potential energy

In **Work**, we saw that the work done by a perfectly elastic spring, in one dimension, depends only on the spring constant and the squares of the displacements from the unstretched position, as given in **Equation 7.5**. This work involves only the properties of a Hooke's law interaction and not the properties of real springs and whatever objects are attached to them. Therefore, we can define the difference of elastic potential energy for a spring force as the negative of the work done by the spring force in this equation, before we consider systems that embody this type of force. Thus,

$$\Delta U = -W_{AB} = \frac{1}{2}k(x_B^2 - x_A^2), \tag{8.6}$$

where the object travels from point A to point B. The potential energy function corresponding to this difference is

$$U(x) = \frac{1}{2}kx^2 + \text{const.} \tag{8.7}$$

If the spring force is the only force acting, it is simplest to take the zero of potential energy at $x = 0$, when the spring is at its unstretched length. Then, the constant is **Equation 8.7** is zero. (Other choices may be more convenient if other forces are acting.)

Example 8.3

Spring Potential Energy

A system contains a perfectly elastic spring, with an unstretched length of 20 cm and a spring constant of 4 N/cm. (a) How much elastic potential energy does the spring contribute when its length is 23 cm? (b) How much more potential energy does it contribute if its length increases to 26 cm?

Strategy

When the spring is at its unstretched length, it contributes nothing to the potential energy of the system, so we can use **Equation 8.7** with the constant equal to zero. The value of x is the length minus the unstretched length. When the spring is expanded, the spring's displacement or difference between its relaxed length and stretched length should be used for the x-value in calculating the potential energy of the spring.

Solution

a. The displacement of the spring is $x = 23\text{ cm} - 20\text{ cm} = 3\text{ cm}$, so the contributed potential energy is

$$U = \frac{1}{2}kx^2 = \frac{1}{2}(4\text{ N/cm})(3\text{ cm})^2 = 0.18\text{ J}.$$

b. When the spring's displacement is $x = 26\text{ cm} - 20\text{ cm} = 6\text{ cm}$, the potential energy is

$$U = \frac{1}{2}kx^2 = \frac{1}{2}(4\text{ N/cm})(6\text{ cm})^2 = 0.72\text{ J},\text{ which is a 0.54-J increase over the amount in part (a).}$$

Significance

Calculating the elastic potential energy and potential energy differences from **Equation 8.7** involves solving for the potential energies based on the given lengths of the spring. Since U depends on x^2, the potential energy for a compression (negative x) is the same as for an extension of equal magnitude.

 8.3   Check Your Understanding When the length of the spring in **Example 8.3** changes from an initial value of 22.0 cm to a final value, the elastic potential energy it contributes changes by -0.0800 J. Find the final length.

Gravitational and elastic potential energy

A simple system embodying both gravitational and elastic types of potential energy is a one-dimensional, vertical mass-spring system. This consists of a massive particle (or block), hung from one end of a perfectly elastic, massless spring, the other end of which is fixed, as illustrated in **Figure 8.4**.

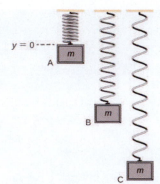

Figure 8.4 A vertical mass-spring system, with the y-axis pointing downward. The mass is initially at an unstretched spring length, point A. Then it is released, expanding past point B to point C, where it comes to a stop.

First, let's consider the potential energy of the system. We need to define the constant in the potential energy function of **Equation 8.5**. Often, the ground is a suitable choice for when the gravitational potential energy is zero; however, in this case, the highest point or when $y = 0$ is a convenient location for zero gravitational potential energy. Note that this choice is arbitrary, and the problem can be solved correctly even if another choice is picked.

We must also define the elastic potential energy of the system and the corresponding constant, as detailed in **Equation 8.7**. This is where the spring is unstretched, or at the $y = 0$ position.

If we consider that the total energy of the system is conserved, then the energy at point A equals point C. The block is placed just on the spring so its initial kinetic energy is zero. By the setup of the problem discussed previously, both the gravitational potential energy and elastic potential energy are equal to zero. Therefore, the initial energy of the system is zero. When the block arrives at point C, its kinetic energy is zero. However, it now has both gravitational potential energy and elastic potential energy. Therefore, we can solve for the distance y that the block travels before coming to a stop:

$$K_A \;+\; U_A = K_C + U_C$$

$$0 \;=\; 0 + mgy_C + \left(\tfrac{1}{2}ky_C\right)^2$$

$$y_C \;=\; \frac{-2mg}{k}$$

Figure 8.5 A bungee jumper transforms gravitational potential energy at the start of the jump into elastic potential energy at the bottom of the jump.

Example 8.4

Potential Energy of a Vertical Mass-Spring System

A block weighing $1.2\,\text{N}$ is hung from a spring with a spring constant of $6.0\,\text{N/m}$, as shown in **Figure 8.4**. (a) What is the maximum expansion of the spring, as seen at point C? (b) What is the total potential energy at point B, halfway between A and C? (c) What is the speed of the block at point B?

Strategy

In part (a) we calculate the distance y_C as discussed in the previous text. Then in part (b), we use half of the y value to calculate the potential energy at point B using equations **Equation 8.4** and **Equation 8.6**. This energy must be equal to the kinetic energy, **Equation 7.6**, at point B since the initial energy of the system is zero. By calculating the kinetic energy at point B, we can now calculate the speed of the block at point B.

Solution

a. Since the total energy of the system is zero at point A as discussed previously, the maximum expansion of the spring is calculated to be:

$$y_C = \frac{-2mg}{k}$$

$$y_C = \frac{-2(1.2\,\text{N})}{(6.0\,\text{N/m})} = -0.40\,\text{m}$$

b. The position of y_B is half of the position at y_C or $-0.20\,\text{m}$. The total potential energy at point B would therefore be:

$$U_B = mgy_B + \left(\tfrac{1}{2}ky_B\right)^2$$

$$U_B = (1.2\,\text{N})(-0.20\,\text{m}) + \left(\tfrac{1}{6}(6\text{N/m})(-0.20\,\text{m})\right)^2$$

$$U_B = -0.12\,\text{J}$$

c. The mass of the block is the weight divided by gravity.

$$m = \frac{F_w}{g} = \frac{1.2 \text{ N}}{9.8 \text{m/s}^2} = 0.12 \text{kg}$$

The kinetic energy at point B therefore is 0.12 J because the total energy is zero. Therefore, the speed of the block at point B is equal to

$$K = \tfrac{1}{2}mv$$

$$v = \sqrt{\frac{2K}{m}} = \sqrt{\frac{2(0.12 \text{ J })}{(0.12 \text{kg})}} = 1.4 \text{m/s}$$

Significance

Even though the potential energy due to gravity is relative to a chosen zero location, the solutions to this problem would be the same if the zero energy points were chosen at different locations.

 8.4 Check Your Understanding Suppose the mass in **Equation 8.6** is doubled while keeping the all other conditions the same. Would the maximum expansion of the spring increase, decrease, or remain the same? Would the speed at point B be larger, smaller, or the same compared to the original mass?

 View this **simulation (https://openstaxcollege.org/l/21conenerskat)** to learn about conservation of energy with a skater! Build tracks, ramps and jumps for the skater and view the kinetic energy, potential energy and friction as he moves. You can also take the skater to different planets or even space!

A sample chart of a variety of energies is shown in **Table 8.1** to give you an idea about typical energy values associated with certain events. Some of these are calculated using kinetic energy, whereas others are calculated by using quantities found in a form of potential energy that may not have been discussed at this point.

Object/phenomenon	Energy in joules
Big Bang	10^{68}
Annual world energy use	4.0×10^{20}
Large fusion bomb (9 megaton)	3.8×10^{16}
Hiroshima-size fission bomb (10 kiloton)	4.2×10^{13}
1 barrel crude oil	5.9×10^{9}
1 ton TNT	4.2×10^{9}
1 gallon of gasoline	1.2×10^{8}
Daily adult food intake (recommended)	1.2×10^{7}
1000-kg car at 90 km/h	3.1×10^{5}
Tennis ball at 100 km/h	22
Mosquito $\left(10^{-2} \text{ g at } 0.5 \text{ m/s}\right)$	1.3×10^{-6}
Single electron in a TV tube beam	4.0×10^{-15}
Energy to break one DNA strand	10^{-19}

Table 8.1 Energy of Various Objects and Phenomena

8.2 | Conservative and Non-Conservative Forces

Learning Objectives

By the end of this section, you will be able to:

- Characterize a conservative force in several different ways
- Specify mathematical conditions that must be satisfied by a conservative force and its components
- Relate the conservative force between particles of a system to the potential energy of the system
- Calculate the components of a conservative force in various cases

In **Potential Energy and Conservation of Energy**, any transition between kinetic and potential energy conserved the total energy of the system. This was path independent, meaning that we can start and stop at any two points in the problem, and the total energy of the system—kinetic plus potential—at these points are equal to each other. This is characteristic of a **conservative force**. We dealt with conservative forces in the preceding section, such as the gravitational force and spring force. When comparing the motion of the football in **Figure 8.2**, the total energy of the system never changes, even though the gravitational potential energy of the football increases, as the ball rises relative to ground and falls back to the initial gravitational potential energy when the football player catches the ball. **Non-conservative forces** are dissipative forces such as friction or air resistance. These forces take energy away from the system as the system progresses, energy that you can't get back. These forces are path dependent; therefore it matters where the object starts and stops.

Conservative Force

The work done by a conservative force is independent of the path; in other words, the work done by a conservative force is the same for any path connecting two points:

$$W_{AB, \text{path-1}} = \int_{AB, \text{path-1}} \vec{F}_{\text{cons}} \cdot d\vec{r} = W_{AB, \text{path-2}} = \int_{AB, \text{path-2}} \vec{F}_{\text{cons}} \cdot d\vec{r}. \tag{8.8}$$

The work done by a non-conservative force depends on the path taken.

Equivalently, a force is conservative if the work it does around any closed path is zero:

$$W_{\text{closed path}} = \oint \vec{F}_{\text{cons}} \cdot d\vec{r} = 0. \tag{8.9}$$

[In **Equation 8.9**, we use the notation of a circle in the middle of the integral sign for a line integral over a closed path, a notation found in most physics and engineering texts.] **Equation 8.8** and **Equation 8.9** are equivalent because any closed path is the sum of two paths: the first going from A to B, and the second going from B to A. The work done going along a path from B to A is the negative of the work done going along the same path from A to B, where A and B are any two points on the closed path:

$$0 = \int \vec{F}_{\text{cons}} \cdot d\vec{r} = \int_{AB, \text{path-1}} \vec{F}_{\text{cons}} \cdot d\vec{r} + \int_{BA, \text{path-2}} \vec{F}_{\text{cons}} \cdot d\vec{r}$$

$$= \int_{AB, \text{path-1}} \vec{F}_{\text{cons}} \cdot d\vec{r} - \int_{AB, \text{path-2}} \vec{F}_{\text{cons}} \cdot d\vec{r} = 0.$$

You might ask how we go about proving whether or not a force is conservative, since the definitions involve any and all paths from A to B, or any and all closed paths, but to do the integral for the work, you have to choose a particular path. One answer is that the work done is independent of path if the infinitesimal work $\vec{F} \cdot d\vec{r}$ is an **exact differential**, the way the infinitesimal net work was equal to the exact differential of the kinetic energy, $dW_{\text{net}} = m\vec{v} \cdot d\vec{v} = d\frac{1}{2}mv^2$,

when we derived the work-energy theorem in **Work-Energy Theorem**. There are mathematical conditions that you can use to test whether the infinitesimal work done by a force is an exact differential, and the force is conservative.

These conditions only involve differentiation and are thus relatively easy to apply. In two dimensions, the condition for $\overrightarrow{\mathbf{F}} \cdot d\overrightarrow{\mathbf{r}} = F_x dx + F_y dy$ to be an exact differential is

$$\frac{dF_x}{dy} = \frac{dF_y}{dx}. \tag{8.10}$$

You may recall that the work done by the force in **Example 7.4** depended on the path. For that force,

$$F_x = (5 \text{ N/m})y \text{ and } F_y = (10 \text{ N/m})x.$$

Therefore,

$$(dF_x/dy) = 5 \text{ N/m} \neq (dF_y/dx) = 10 \text{ N/m},$$

which indicates it is a non-conservative force. Can you see what you could change to make it a conservative force?

Figure 8.6 A grinding wheel applies a non-conservative force, because the work done depends on how many rotations the wheel makes, so it is path-dependent. (credit: modification of work by Grantez Stephens, U.S. Navy)

Example 8.5

Conservative or Not?

Which of the following two-dimensional forces are conservative and which are not? Assume a and b are constants with appropriate units:

(a) $axy^3\,\hat{\mathbf{i}} + ayx^3\,\hat{\mathbf{j}}$, (b) $a\left[(y^2/x)\hat{\mathbf{i}} + 2y\ln(x/b)\hat{\mathbf{j}}\right]$, (c) $\dfrac{ax\,\hat{\mathbf{i}} + ay\,\hat{\mathbf{j}}}{x^2 + y^2}$

Strategy

Apply the condition stated in **Equation 8.10**, namely, using the derivatives of the components of each force indicated. If the derivative of the y-component of the force with respect to x is equal to the derivative of the x-component of the force with respect to y, the force is a conservative force, which means the path taken for potential energy or work calculations always yields the same results.

Solution

a. $\dfrac{dF_x}{dy} = \dfrac{d(axy^3)}{dy} = 3axy^2$ and $\dfrac{dF_y}{dx} = \dfrac{d(ayx^3)}{dx} = 3ayx^2$, so this force is non-conservative.

b. $\dfrac{dF_x}{dy} = \dfrac{d\left(ay^2/x\right)}{dy} = \dfrac{2ay}{x}$ and $\dfrac{dF_y}{dx} = \dfrac{d(2ay\ln(x/b))}{dx} = \dfrac{2ay}{x}$, so this force is conservative.

c. $\dfrac{dF_x}{dy} = \dfrac{d\left(ax/\left(x^2+y^2\right)\right)}{dy} = -\dfrac{ax(2y)}{\left(x^2+y^2\right)^2} = \dfrac{dF_y}{dx} = \dfrac{d\left(ay/\left(x^2+y^2\right)\right)}{dx}$, again conservative.

Significance

The conditions in **Equation 8.10** are derivatives as functions of a single variable; in three dimensions, similar conditions exist that involve more derivatives.

 8.5 **Check Your Understanding** A two-dimensional, conservative force is zero on the *x*- and *y*-axes, and satisfies the condition $(dF_x/dy) = (dF_y/dx) = \left(4\ \text{N/m}^3\right)xy$. What is the magnitude of the force at the point $x = y = 1$ m?

Before leaving this section, we note that non-conservative forces do not have potential energy associated with them because the energy is lost to the system and can't be turned into useful work later. So there is always a conservative force associated with every potential energy. We have seen that potential energy is defined in relation to the work done by conservative forces. That relation, **Equation 8.1**, involved an integral for the work; starting with the force and displacement, you integrated to get the work and the change in potential energy. However, integration is the inverse operation of differentiation; you could equally well have started with the potential energy and taken its derivative, with respect to displacement, to get the force. The infinitesimal increment of potential energy is the dot product of the force and the infinitesimal displacement,

$$dU = -\overrightarrow{\mathbf{F}} \cdot d\overrightarrow{\mathbf{l}} = -F_l dl.$$

Here, we chose to represent the displacement in an arbitrary direction by $d\overrightarrow{\mathbf{l}}$, so as not to be restricted to any particular coordinate direction. We also expressed the dot product in terms of the magnitude of the infinitesimal displacement and the component of the force in its direction. Both these quantities are scalars, so you can divide by dl to get

$$F_l = -\dfrac{dU}{dl}. \qquad (8.11)$$

This equation gives the relation between force and the potential energy associated with it. In words, the component of a conservative force, in a particular direction, equals the negative of the derivative of the corresponding potential energy, with respect to a displacement in that direction. For one-dimensional motion, say along the *x*-axis, **Equation 8.11** give the entire vector force, $\overline{\mathbf{F}} = F_x\,\hat{\mathbf{i}} = -\dfrac{\partial U}{\partial x}\hat{\mathbf{i}}$.

In two dimensions,

$$\overline{\mathbf{F}} = F_x\,\hat{\mathbf{i}} + F_y\,\hat{\mathbf{j}} = -\left(\dfrac{\partial U}{\partial x}\right)\hat{\mathbf{i}} - \left(\dfrac{\partial U}{\partial y}\right)\hat{\mathbf{j}}.$$

From this equation, you can see why **Equation 8.11** is the condition for the work to be an exact differential, in terms of the derivatives of the components of the force. In general, a partial derivative notation is used. If a function has many variables in it, the derivative is taken only of the variable the partial derivative specifies. The other variables are held constant. In three dimensions, you add another term for the *z*-component, and the result is that the force is the negative of the gradient of the potential energy. However, we won't be looking at three-dimensional examples just yet.

Example 8.6

Force due to a Quartic Potential Energy

The potential energy for a particle undergoing one-dimensional motion along the x-axis is

$$U(x) = \tfrac{1}{4}cx^4,$$

where $c = 8 \text{ N/m}^3$. Its total energy at $x = 0$ is 2 J, and it is not subject to any non-conservative forces. Find (a) the positions where its kinetic energy is zero and (b) the forces at those positions.

Strategy

(a) We can find the positions where $K = 0$, so the potential energy equals the total energy of the given system.

(b) Using **Equation 8.11**, we can find the force evaluated at the positions found from the previous part, since the mechanical energy is conserved.

Solution

a. The total energy of the system of 2 J equals the quartic elastic energy as given in the problem,

$$2 \text{ J} = \tfrac{1}{4}\left(8 \text{ N/m}^3\right){x_{\text{f}}}^4.$$

Solving for x_{f} results in $x_{\text{f}} = \pm 1 \text{ m}$.

b. From **Equation 8.11**,

$$F_x = -dU/dx = -cx^3.$$

Thus, evaluating the force at $\pm 1 \text{ m}$, we get

$$\vec{\textbf{F}} = -(8 \text{ N/m}^3)(\pm 1 \text{ m})^3\, \hat{\textbf{i}} = \pm 8 \text{ N}\, \hat{\textbf{i}}.$$

At both positions, the magnitude of the forces is 8 N and the directions are toward the origin, since this is the potential energy for a restoring force.

Significance

Finding the force from the potential energy is mathematically easier than finding the potential energy from the force, because differentiating a function is generally easier than integrating one.

 8.6 Check Your Understanding Find the forces on the particle in **Example 8.6** when its kinetic energy is 1.0 J at $x = 0$.

8.3 | Conservation of Energy

Learning Objectives

By the end of this section, you will be able to:

- Formulate the principle of conservation of mechanical energy, with or without the presence of non-conservative forces
- Use the conservation of mechanical energy to calculate various properties of simple systems

In this section, we elaborate and extend the result we derived in **Potential Energy of a System**, where we re-wrote the work-energy theorem in terms of the change in the kinetic and potential energies of a particle. This will lead us to a discussion of the important principle of the conservation of mechanical energy. As you continue to examine other topics in physics, in later chapters of this book, you will see how this conservation law is generalized to encompass other types of energy and energy transfers. The last section of this chapter provides a preview.

The terms 'conserved quantity' and 'conservation law' have specific, scientific meanings in physics, which are different from the everyday meanings associated with the use of these words. (The same comment is also true about the scientific and everyday uses of the word 'work.') In everyday usage, you could conserve water by not using it, or by using less of it, or by re-using it. Water is composed of molecules consisting of two atoms of hydrogen and one of oxygen. Bring these atoms together to form a molecule and you create water; dissociate the atoms in such a molecule and you destroy water. However, in scientific usage, a **conserved quantity** for a system stays constant, changes by a definite amount that is transferred to other systems, and/or is converted into other forms of that quantity. A conserved quantity, in the scientific sense, can be transformed, but not strictly created or destroyed. Thus, there is no physical law of conservation of water.

Systems with a Single Particle or Object

We first consider a system with a single particle or object. Returning to our development of **Equation 8.2**, recall that we first separated all the forces acting on a particle into conservative and non-conservative types, and wrote the work done by each type of force as a separate term in the work-energy theorem. We then replaced the work done by the conservative forces by the change in the potential energy of the particle, combining it with the change in the particle's kinetic energy to get **Equation 8.2**. Now, we write this equation without the middle step and define the sum of the kinetic and potential energies, $K + U = E$; to be the **mechanical energy** of the particle.

Conservation of Energy

The mechanical energy E of a particle stays constant unless forces outside the system or non-conservative forces do work on it, in which case, the change in the mechanical energy is equal to the work done by the non-conservative forces:

$$W_{\text{nc, }AB} = \Delta(K + U)_{AB} = \Delta E_{AB}. \tag{8.12}$$

This statement expresses the concept of **energy conservation** for a classical particle as long as there is no non-conservative work. Recall that a classical particle is just a point mass, is nonrelativistic, and obeys Newton's laws of motion. In **Relativity (http://cnx.org/content/m58555/latest/)** , we will see that conservation of energy still applies to a non-classical particle, but for that to happen, we have to make a slight adjustment to the definition of energy.

It is sometimes convenient to separate the case where the work done by non-conservative forces is zero, either because no such forces are assumed present, or, like the normal force, they do zero work when the motion is parallel to the surface. Then

$$0 = W_{\text{nc, }AB} = \Delta(K + U)_{AB} = \Delta E_{AB}. \tag{8.13}$$

In this case, the conservation of mechanical energy can be expressed as follows: The mechanical energy of a particle does not change if all the non-conservative forces that may act on it do no work. Understanding the concept of energy conservation is the important thing, not the particular equation you use to express it.

Problem-Solving Strategy: Conservation of Energy

1. Identify the body or bodies to be studied (the system). Often, in applications of the principle of mechanical energy conservation, we study more than one body at the same time.

2. Identify all forces acting on the body or bodies.

3. Determine whether each force that does work is conservative. If a non-conservative force (e.g., friction) is doing work, then mechanical energy is not conserved. The system must then be analyzed with non-conservative work, **Equation 8.13**.

4. For every force that does work, choose a reference point and determine the potential energy function for the force. The reference points for the various potential energies do not have to be at the same location.

5. Apply the principle of mechanical energy conservation by setting the sum of the kinetic energies and potential energies equal at every point of interest.

Example 8.7

Simple Pendulum

A particle of mass m is hung from the ceiling by a massless string of length 1.0 m, as shown in **Figure 8.7**. The particle is released from rest, when the angle between the string and the downward vertical direction is $30°$. What is its speed when it reaches the lowest point of its arc?

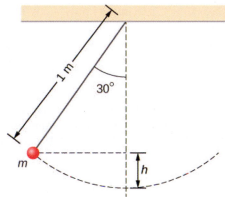

Figure 8.7 A particle hung from a string constitutes a simple pendulum. It is shown when released from rest, along with some distances used in analyzing the motion.

Strategy

Using our problem-solving strategy, the first step is to define that we are interested in the particle-Earth system. Second, only the gravitational force is acting on the particle, which is conservative (step 3). We neglect air resistance in the problem, and no work is done by the string tension, which is perpendicular to the arc of the motion. Therefore, the mechanical energy of the system is conserved, as represented by **Equation 8.13**, $0 = \Delta(K + U)$. Because the particle starts from rest, the increase in the kinetic energy is just the kinetic energy at the lowest point. This increase in kinetic energy equals the decrease in the gravitational potential energy, which we can calculate from the geometry. In step 4, we choose a reference point for zero gravitational potential energy to be at the lowest vertical point the particle achieves, which is mid-swing. Lastly, in step 5, we set the sum of energies at the highest point (initial) of the swing to the lowest point (final) of the swing to ultimately solve for the final speed.

Solution

We are neglecting non-conservative forces, so we write the energy conservation formula relating the particle at the highest point (initial) and the lowest point in the swing (final) as

$$K_i + U_i = K_f + U_f.$$

Since the particle is released from rest, the initial kinetic energy is zero. At the lowest point, we define the gravitational potential energy to be zero. Therefore our conservation of energy formula reduces to

$$0 + mgh = \frac{1}{2}mv^2 + 0$$
$$v = \sqrt{2gh}.$$

The vertical height of the particle is not given directly in the problem. This can be solved for by using trigonometry and two givens: the length of the pendulum and the angle through which the particle is vertically pulled up. Looking at the diagram, the vertical dashed line is the length of the pendulum string. The vertical height is labeled h. The other partial length of the vertical string can be calculated with trigonometry. That piece is solved for by

$$\cos\theta = x/L, \ x = L\cos\theta.$$

Therefore, by looking at the two parts of the string, we can solve for the height h,

$$
\begin{aligned}
x + h &= L \\
L\cos\theta + h &= L \\
h &= L - L\cos\theta = L(1 - \cos\theta).
\end{aligned}
$$

We substitute this height into the previous expression solved for speed to calculate our result:

$$v = \sqrt{2gL(1 - \cos\theta)} = \sqrt{2(9.8 \text{ m/s}^2)(1 \text{ m})(1 - \cos 30°)} = 1.62 \text{ m/s}.$$

Significance

We found the speed directly from the conservation of mechanical energy, without having to solve the differential equation for the motion of a pendulum (see **Oscillations**). We can approach this problem in terms of bar graphs of total energy. Initially, the particle has all potential energy, being at the highest point, and no kinetic energy. When the particle crosses the lowest point at the bottom of the swing, the energy moves from the potential energy column to the kinetic energy column. Therefore, we can imagine a progression of this transfer as the particle moves between its highest point, lowest point of the swing, and back to the highest point (**Figure 8.8**). As the particle travels from the lowest point in the swing to the highest point on the far right hand side of the diagram, the energy bars go in reverse order from (c) to (b) to (a).

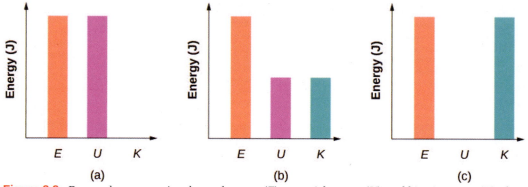

Figure 8.8 Bar graphs representing the total energy (E), potential energy (U), and kinetic energy (K) of the particle in different positions. (a) The total energy of the system equals the potential energy and the kinetic energy is zero, which is found at the highest point the particle reaches. (b) The particle is midway between the highest and lowest point, so the kinetic energy plus potential energy bar graphs equal the total energy. (c) The particle is at the lowest point of the swing, so the kinetic energy bar graph is the highest and equal to the total energy of the system.

 8.7 **Check Your Understanding** How high above the bottom of its arc is the particle in the simple pendulum above, when its speed is 0.81 m/s?

Example 8.8

Air Resistance on a Falling Object

A helicopter is hovering at an altitude of 1 km when a panel from its underside breaks loose and plummets to the ground (**Figure 8.9**). The mass of the panel is 15 kg, and it hits the ground with a speed of 45 m/s. How much mechanical energy was dissipated by air resistance during the panel's descent?

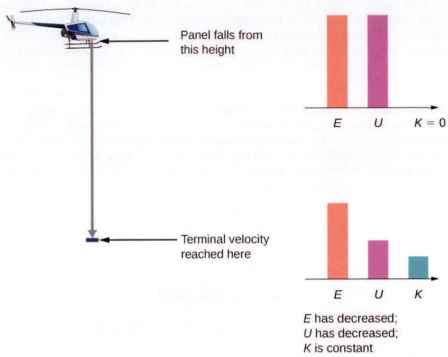

Figure 8.9 A helicopter loses a panel that falls until it reaches terminal velocity of 45 m/s. How much did air resistance contribute to the dissipation of energy in this problem?

Strategy

Step 1: Here only one body is being investigated.

Step 2: Gravitational force is acting on the panel, as well as air resistance, which is stated in the problem.

Step 3: Gravitational force is conservative; however, the non-conservative force of air resistance does negative work on the falling panel, so we can use the conservation of mechanical energy, in the form expressed by **Equation 8.12**, to find the energy dissipated. This energy is the magnitude of the work:

$$\Delta E_{diss} = \left| W_{nc,if} \right| = \left| \Delta(K+U)_{if} \right|.$$

Step 4: The initial kinetic energy, at $y_i = 1\,\text{km}$, is zero. We set the gravitational potential energy to zero at ground level out of convenience.

Step 5: The non-conservative work is set equal to the energies to solve for the work dissipated by air resistance.

Solution

The mechanical energy dissipated by air resistance is the algebraic sum of the gain in the kinetic energy and loss in potential energy. Therefore the calculation of this energy is

$$\begin{aligned} \Delta E_{diss} &= \left| K_f - K_i + U_f - U_i \right| \\ &= \left| \tfrac{1}{2}(15\,\text{kg})(45\,\text{m/s})^2 - 0 + 0 - (15\,\text{kg})(9.8\,\text{m/s}^2)(1000\,\text{m}) \right| = 130\,\text{kJ}. \end{aligned}$$

Significance

Most of the initial mechanical energy of the panel (U_i), 147 kJ, was lost to air resistance. Notice that we were able to calculate the energy dissipated without knowing what the force of air resistance was, only that it was dissipative.

8.8 Check Your Understanding You probably recall that, neglecting air resistance, if you throw a projectile straight up, the time it takes to reach its maximum height equals the time it takes to fall from the maximum height back to the starting height. Suppose you cannot neglect air resistance, as in **Example 8.8**. Is the time the projectile takes to go up (a) greater than, (b) less than, or (c) equal to the time it takes to come back down? Explain.

In these examples, we were able to use conservation of energy to calculate the speed of a particle just at particular points in its motion. But the method of analyzing particle motion, starting from energy conservation, is more powerful than that. More advanced treatments of the theory of mechanics allow you to calculate the full time dependence of a particle's motion, for a given potential energy. In fact, it is often the case that a better model for particle motion is provided by the form of its kinetic and potential energies, rather than an equation for force acting on it. (This is especially true for the quantum mechanical description of particles like electrons or atoms.)

We can illustrate some of the simplest features of this energy-based approach by considering a particle in one-dimensional motion, with potential energy $U(x)$ and no non-conservative interactions present. **Equation 8.12** and the definition of velocity require

$$K = \tfrac{1}{2}mv^2 = E - U(x)$$

$$v = \frac{dx}{dt} = \sqrt{\frac{2(E - U(x))}{m}}.$$

Separate the variables x and t and integrate, from an initial time $t = 0$ to an arbitrary time, to get

$$t = \int_0^t dt = \int_{x_0}^x \frac{dt}{\sqrt{2[E - U(x)]/m}}. \tag{8.14}$$

If you can do the integral in **Equation 8.14**, then you can solve for x as a function of t.

Example 8.9

Constant Acceleration

Use the potential energy $U(x) = -E(x/x_0)$, for $E > 0$, in **Equation 8.14** to find the position x of a particle as a function of time t.

Strategy

Since we know how the potential energy changes as a function of x, we can substitute for $U(x)$ in **Equation 8.14**, integrate, and then solve for x. This results in an expression of x as a function of time with constants of energy E, mass m, and the initial position x_0.

Solution

Following the first two suggested steps in the above strategy,

$$t = \int_{x_0}^x \frac{dx}{\sqrt{(2E/mx_0)(x_0 - x)}} = \frac{1}{\sqrt{(2E/mx_0)}} \left| -2\sqrt{(x_0 - x)} \right|_{x_0}^x = -\frac{2\sqrt{(x_0 - x)}}{\sqrt{(2E/mx_0)}}.$$

Solving for the position, we obtain $x(t) = x_0 - \tfrac{1}{2}(E/mx_0)t^2$.

Significance

The position as a function of time, for this potential, represents one-dimensional motion with constant acceleration, $a = (E/mx_0)$, starting at rest from position x_0. This is not so surprising, since this is a potential energy for a constant force, $F = -dU/dx = E/x_0$, and $a = F/m$.

8.9 Check Your Understanding What potential energy $U(x)$ can you substitute in **Equation 8.13** that will result in motion with constant velocity of 2 m/s for a particle of mass 1 kg and mechanical energy 1 J?

We will look at another more physically appropriate example of the use of **Equation 8.13** after we have explored some further implications that can be drawn from the functional form of a particle's potential energy.

Systems with Several Particles or Objects

Systems generally consist of more than one particle or object. However, the conservation of mechanical energy, in one of the forms in **Equation 8.12** or **Equation 8.13**, is a fundamental law of physics and applies to any system. You just have to include the kinetic and potential energies of all the particles, and the work done by all the non-conservative forces acting on them. Until you learn more about the dynamics of systems composed of many particles, in **Linear Momentum and Collisions**, **Fixed-Axis Rotation**, and **Angular Momentum**, it is better to postpone discussing the application of energy conservation to then.

8.4 | Potential Energy Diagrams and Stability

Learning Objectives
By the end of this section, you will be able to: • Create and interpret graphs of potential energy • Explain the connection between stability and potential energy

Often, you can get a good deal of useful information about the dynamical behavior of a mechanical system just by interpreting a graph of its potential energy as a function of position, called a **potential energy diagram**. This is most easily accomplished for a one-dimensional system, whose potential energy can be plotted in one two-dimensional graph—for example, $U(x)$ versus x—on a piece of paper or a computer program. For systems whose motion is in more than one dimension, the motion needs to be studied in three-dimensional space. We will simplify our procedure for one-dimensional motion only.

First, let's look at an object, freely falling vertically, near the surface of Earth, in the absence of air resistance. The mechanical energy of the object is conserved, $E = K + U$, and the potential energy, with respect to zero at ground level, is $U(y) = mgy$, which is a straight line through the origin with slope mg. In the graph shown in **Figure 8.10**, the x-axis is the height above the ground y and the y-axis is the object's energy.

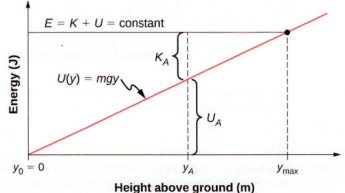

Figure 8.10 The potential energy graph for an object in vertical free fall, with various quantities indicated.

The line at energy E represents the constant mechanical energy of the object, whereas the kinetic and potential energies, K_A and U_A, are indicated at a particular height y_A. You can see how the total energy is divided between kinetic and potential energy as the object's height changes. Since kinetic energy can never be negative, there is a maximum potential energy and a maximum height, which an object with the given total energy cannot exceed:

$$K = E - U \geq 0,$$
$$U \leq E.$$

If we use the gravitational potential energy reference point of zero at $y_0,$ we can rewrite the gravitational potential energy U as mgy. Solving for y results in

$$y \leq E/mg = y_{max}.$$

We note in this expression that the quantity of the total energy divided by the weight (mg) is located at the maximum height of the particle, or y_{max}. At the maximum height, the kinetic energy and the speed are zero, so if the object were initially traveling upward, its velocity would go through zero there, and y_{max} would be a turning point in the motion. At ground level, $y_0 = 0$, the potential energy is zero, and the kinetic energy and the speed are maximum:

$$U_0 \;=\; 0 = E - K_0,$$
$$E \;=\; K_0 = \tfrac{1}{2}mv_0{}^2,$$
$$v_0 \;=\; \pm\sqrt{2E/m}.$$

The maximum speed $\pm v_0$ gives the initial velocity necessary to reach y_{max}, the maximum height, and $-v_0$ represents the final velocity, after falling from y_{max}. You can read all this information, and more, from the potential energy diagram we have shown.

Consider a mass-spring system on a frictionless, stationary, horizontal surface, so that gravity and the normal contact force do no work and can be ignored (**Figure 8.11**). This is like a one-dimensional system, whose mechanical energy E is a constant and whose potential energy, with respect to zero energy at zero displacement from the spring's unstretched length, $x = 0,$ is $U(x) = \tfrac{1}{2}kx^2$.

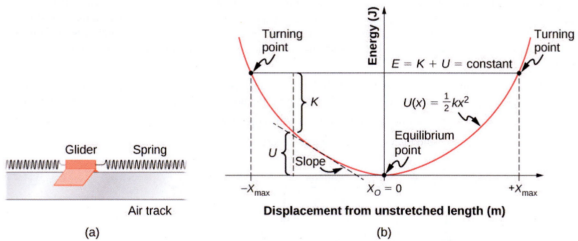

Figure 8.11 (a) A glider between springs on an air track is an example of a horizontal mass-spring system. (b) The potential energy diagram for this system, with various quantities indicated.

You can read off the same type of information from the potential energy diagram in this case, as in the case for the body in vertical free fall, but since the spring potential energy describes a variable force, you can learn more from this graph. As for the object in vertical free fall, you can deduce the physically allowable range of motion and the maximum values of distance and speed, from the limits on the kinetic energy, $0 \leq K \leq E$. Therefore, $K = 0$ and $U = E$ at a **turning point**, of which there are two for the elastic spring potential energy,

$$x_{max} = \pm\sqrt{2E/k}.$$

The glider's motion is confined to the region between the turning points, $-x_{max} \leq x \leq x_{max}$. This is true for any (positive) value of E because the potential energy is unbounded with respect to x. For this reason, as well as the shape of the potential energy curve, $U(x)$ is called an infinite potential well. At the bottom of the potential well, $x = 0$, $U = 0$ and the kinetic energy is a maximum, $K = E$, so $v_{\text{max}} = \pm\sqrt{2E/m}$.

However, from the slope of this potential energy curve, you can also deduce information about the force on the glider and its acceleration. We saw earlier that the negative of the slope of the potential energy is the spring force, which in this case is also the net force, and thus is proportional to the acceleration. When $x = 0$, the slope, the force, and the acceleration are all zero, so this is an **equilibrium point**. The negative of the slope, on either side of the equilibrium point, gives a force pointing back to the equilibrium point, $F = \pm kx$, so the equilibrium is termed stable and the force is called a restoring force. This implies that $U(x)$ has a relative minimum there. If the force on either side of an equilibrium point has a direction opposite from that direction of position change, the equilibrium is termed unstable, and this implies that $U(x)$ has a relative maximum there.

Example 8.10

Quartic and Quadratic Potential Energy Diagram

The potential energy for a particle undergoing one-dimensional motion along the x-axis is $U(x) = 2(x^4 - x^2)$, where U is in joules and x is in meters. The particle is not subject to any non-conservative forces and its mechanical energy is constant at $E = -0.25 \text{ J}$. (a) Is the motion of the particle confined to any regions on the x-axis, and if so, what are they? (b) Are there any equilibrium points, and if so, where are they and are they stable or unstable?

Strategy

First, we need to graph the potential energy as a function of x. The function is zero at the origin, becomes negative as x increases in the positive or negative directions (x^2 is larger than x^4 for $x < 1$), and then becomes positive at sufficiently large $|x|$. Your graph should look like a double potential well, with the zeros determined by solving the equation $U(x) = 0$, and the extremes determined by examining the first and second derivatives of $U(x)$, as shown in **Figure 8.12**.

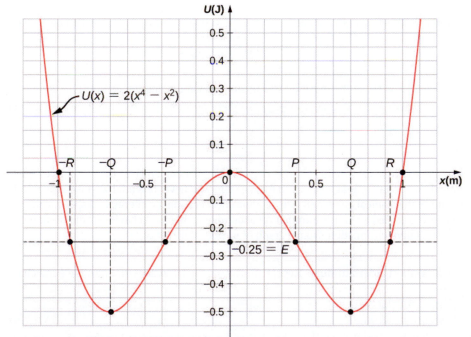

Figure 8.12 The potential energy graph for a one-dimensional, quartic and quadratic potential energy, with various quantities indicated.

You can find the values of (a) the allowed regions along the x-axis, for the given value of the mechanical energy, from the condition that the kinetic energy can't be negative, and (b) the equilibrium points and their stability from the properties of the force (stable for a relative minimum and unstable for a relative maximum of potential energy).

You can just eyeball the graph to reach qualitative answers to the questions in this example. That, after all, is the value of potential energy diagrams. You can see that there are two allowed regions for the motion $(E > U)$ and three equilibrium points (slope $dU/dx = 0$), of which the central one is unstable $\left(d^2 U/dx^2 < 0\right)$, and the other two are stable $\left(d^2 U/dx^2 > 0\right)$.

Solution

a. To find the allowed regions for x, we use the condition

$$K = E - U = -\tfrac{1}{4} - 2\left(x^4 - x^2\right) \geq 0.$$

If we complete the square in x^2, this condition simplifies to $2\left(x^2 - \tfrac{1}{2}\right)^2 \leq \tfrac{1}{4}$, which we can solve to obtain

$$\tfrac{1}{2} - \sqrt{\tfrac{1}{8}} \leq x^2 \leq \tfrac{1}{2} + \sqrt{\tfrac{1}{8}}.$$

This represents two allowed regions, $x_P \leq x \leq x_R$ and $-x_R \leq x \leq -x_P$, where $x_P = 0.38$ and $x_R = 0.92$ (in meters).

b. To find the equilibrium points, we solve the equation

$$dU/dx = 8x^3 - 4x = 0$$

and find $x = 0$ and $x = \pm x_Q$, where $x_Q = 1/\sqrt{2} = 0.707$ (meters). The second derivative

$$d^2 U/dx^2 = 24x^2 - 4$$

is negative at $x = 0$, so that position is a relative maximum and the equilibrium there is unstable. The second derivative is positive at $x = \pm x_Q$, so these positions are relative minima and represent stable equilibria.

Significance

The particle in this example can oscillate in the allowed region about either of the two stable equilibrium points we found, but it does not have enough energy to escape from whichever potential well it happens to initially be in. The conservation of mechanical energy and the relations between kinetic energy and speed, and potential energy and force, enable you to deduce much information about the qualitative behavior of the motion of a particle, as well as some quantitative information, from a graph of its potential energy.

 8.10 **Check Your Understanding** Repeat **Example 8.10** when the particle's mechanical energy is +0.25 J.

Before ending this section, let's practice applying the method based on the potential energy of a particle to find its position as a function of time, for the one-dimensional, mass-spring system considered earlier in this section.

Example 8.11

Sinusoidal Oscillations

Find $x(t)$ for a particle moving with a constant mechanical energy $E > 0$ and a potential energy $U(x) = \frac{1}{2}kx^2$, when the particle starts from rest at time $t = 0$.

Strategy

We follow the same steps as we did in **Example 8.9**. Substitute the potential energy U into **Equation 8.14** and factor out the constants, like m or k. Integrate the function and solve the resulting expression for position, which is now a function of time.

Solution

Substitute the potential energy in **Equation 8.14** and integrate using an integral solver found on a web search:

$$t = \int_{x_0}^{x} \frac{dx}{\sqrt{(k/m)\left[(2E/k) - x^2\right]}} = \sqrt{\frac{m}{k}}\left[\sin^{-1}\left(\frac{x}{\sqrt{2E/k}}\right) - \sin^{-1}\left(\frac{x_0}{\sqrt{2E/k}}\right)\right].$$

From the initial conditions at $t = 0$, the initial kinetic energy is zero and the initial potential energy is $\frac{1}{2}kx_0^2 = E$, from which you can see that $x_0/\sqrt{(2E/k)} = \pm 1$ and $\sin^{-1}(\pm) = \pm 90^0$. Now you can solve for x:

$$x(t) = \sqrt{(2E/k)} \sin\left[(\sqrt{k/m})t \pm 90^0\right] = \pm\sqrt{(2E/k)} \cos\left[(\sqrt{k/m})t\right].$$

Significance

A few paragraphs earlier, we referred to this mass-spring system as an example of a harmonic oscillator. Here, we anticipate that a harmonic oscillator executes sinusoidal oscillations with a maximum displacement of $\sqrt{(2E/k)}$ (called the amplitude) and a rate of oscillation of $(1/2\pi)\sqrt{k/m}$ (called the frequency). Further discussions about oscillations can be found in **Oscillations**.

 8.11 **Check Your Understanding** Find $x(t)$ for the mass-spring system in **Example 8.11** if the particle starts from $x_0 = 0$ at $t = 0$. What is the particle's initial velocity?

8.5 | Sources of Energy

Learning Objectives

By the end of this section, you will be able to:

- Describe energy transformations and conversions in general terms
- Explain what it means for an energy source to be renewable or nonrenewable

In this chapter, we have studied energy. We learned that energy can take different forms and can be transferred from one form to another. You will find that energy is discussed in many everyday, as well as scientific, contexts, because it is involved in all physical processes. It will also become apparent that many situations are best understood, or most easily conceptualized, by considering energy. So far, no experimental results have contradicted the conservation of energy. In fact, whenever measurements have appeared to conflict with energy conservation, new forms of energy have been discovered or recognized in accordance with this principle.

What are some other forms of energy? Many of these are covered in later chapters (also see **Figure 8.13**), but let's detail a few here:

- Atoms and molecules inside all objects are in random motion. The internal kinetic energy from these random motions is called *thermal energy*, because it is related to the temperature of the object. Note that thermal energy can also be transferred from one place to another, not transformed or converted, by the familiar processes of conduction, convection, and radiation. In this case, the energy is known as *heat energy*.

- *Electrical energy* is a common form that is converted to many other forms and does work in a wide range of practical situations.

- Fuels, such as gasoline and food, have *chemical energy*, which is potential energy arising from their molecular structure. Chemical energy can be converted into thermal energy by reactions like oxidation. Chemical reactions can also produce electrical energy, such as in batteries. Electrical energy can, in turn, produce thermal energy and light, such as in an electric heater or a light bulb.

- Light is just one kind of electromagnetic radiation, or *radiant energy*, which also includes radio, infrared, ultraviolet, X-rays, and gamma rays. All bodies with thermal energy can radiate energy in electromagnetic waves.

- *Nuclear energy* comes from reactions and processes that convert measurable amounts of mass into energy. Nuclear energy is transformed into radiant energy in the Sun, into thermal energy in the boilers of nuclear power plants, and then into electrical energy in the generators of power plants. These and all other forms of energy can be transformed into one another and, to a certain degree, can be converted into mechanical work.

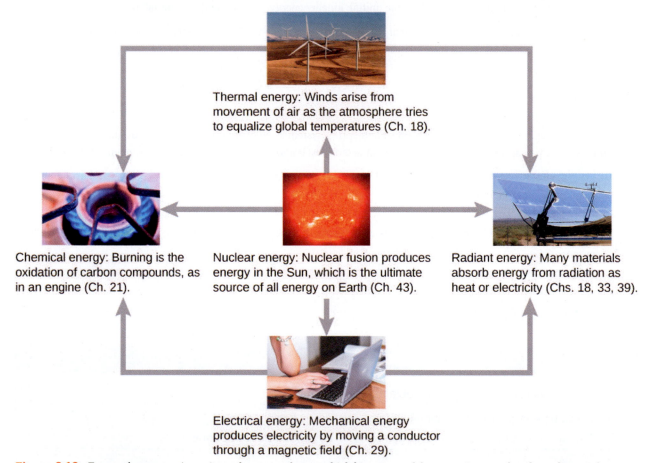

Figure 8.13 Energy that we use in society takes many forms, which be converted from one into another depending on the process involved. We will study many of these forms of energy in later chapters in this text. (credit "sun": modification of work by EIT - SOHO Consortium, ESA, NASA credit "solar panels": "modification of work by "kjkolb"/Wikimedia Commons; credit "gas burner": modification of work by Steven Depolo)

The transformation of energy from one form into another happens all the time. The chemical energy in food is converted into thermal energy through metabolism; light energy is converted into chemical energy through photosynthesis. Another example of energy conversion occurs in a solar cell. Sunlight impinging on a solar cell produces electricity, which can be used to run electric motors or heat water. In an example encompassing many steps, the chemical energy contained in coal is converted into thermal energy as it burns in a furnace, to transform water into steam, in a boiler. Some of the thermal energy in the steam is then converted into mechanical energy as it expands and spins a turbine, which is connected to a generator to produce electrical energy. In these examples, not all of the initial energy is converted into the forms mentioned, because some energy is always transferred to the environment.

Energy is an important element at all levels of society. We live in a very interdependent world, and access to adequate and reliable energy resources is crucial for economic growth and for maintaining the quality of our lives. The principal energy resources used in the world are shown in **Figure 8.14**. The figure distinguishes between two major types of energy sources: **renewable** and **non-renewable**, and further divides each type into a few more specific kinds. Renewable sources are energy sources that are replenished through naturally occurring, ongoing processes, on a time scale that is much shorter than the anticipated lifetime of the civilization using the source. Non-renewable sources are depleted once some of the energy they contain is extracted and converted into other kinds of energy. The natural processes by which non-renewable sources are formed typically take place over geological time scales.

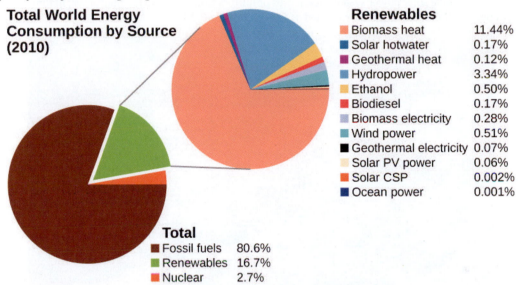

Figure 8.14 World energy consumption by source; the percentage of renewables is increasing, accounting for 19% in 2012.

Our most important non-renewable energy sources are fossil fuels, such as coal, petroleum, and natural gas. These account for about 81% of the world's energy consumption, as shown in the figure. Burning fossil fuels creates chemical reactions that transform potential energy, in the molecular structures of the reactants, into thermal energy and products. This thermal energy can be used to heat buildings or to operate steam-driven machinery. Internal combustion and jet engines convert some of the energy of rapidly expanding gases, released from burning gasoline, into mechanical work. Electrical power generation is mostly derived from transferring energy in expanding steam, via turbines, into mechanical work, which rotates coils of wire in magnetic fields to generate electricity. Nuclear energy is the other non-renewable source shown in **Figure 8.14** and supplies about 3% of the world's consumption. Nuclear reactions release energy by transforming potential energy, in the structure of nuclei, into thermal energy, analogous to energy release in chemical reactions. The thermal energy obtained from nuclear reactions can be transferred and converted into other forms in the same ways that energy from fossil fuels are used.

An unfortunate byproduct of relying on energy produced from the combustion of fossil fuels is the release of carbon dioxide into the atmosphere and its contribution to global warming. Nuclear energy poses environmental problems as well, including the safety and disposal of nuclear waste. Besides these important consequences, reserves of non-renewable sources of energy are limited and, given the rapidly growing rate of world energy consumption, may not last for more than a few hundred years. Considerable effort is going on to develop and expand the use of renewable sources of energy, involving a significant percentage of the world's physicists and engineers.

Four of the renewable energy sources listed in **Figure 8.14**—those using material from plants as fuel (biomass heat, ethanol, biodiesel, and biomass electricity)—involve the same types of energy transformations and conversions as just discussed for fossil and nuclear fuels. The other major types of renewable energy sources are hydropower, wind power, geothermal power, and solar power.

Hydropower is produced by converting the gravitational potential energy of falling or flowing water into kinetic energy and then into work to run electric generators or machinery. Converting the mechanical energy in ocean surface waves and tides is in development. Wind power also converts kinetic energy into work, which can be used directly to generate electricity, operate mills, and propel sailboats.

The interior of Earth has a great deal of thermal energy, part of which is left over from its original formation (gravitational potential energy converted into thermal energy) and part of which is released from radioactive minerals (a form of natural nuclear energy). It will take a very long time for this geothermal energy to escape into space, so people generally regard it as a renewable source, when actually, it's just inexhaustible on human time scales.

The source of solar power is energy carried by the electromagnetic waves radiated by the Sun. Most of this energy is carried by visible light and infrared (heat) radiation. When suitable materials absorb electromagnetic waves, radiant energy is converted into thermal energy, which can be used to heat water, or when concentrated, to make steam and generate electricity (**Figure 8.15**). However, in another important physical process, known as the photoelectric effect, energetic radiation impinging on certain materials is directly converted into electricity. Materials that do this are called photovoltaics (PV in **Figure 8.14**). Some solar power systems use lenses or mirrors to concentrate the Sun's rays, before converting their energy through photovoltaics, and these are qualified as CSP in **Figure 8.14**.

Figure 8.15   Solar cell arrays found in a sunny area converting the solar energy into stored electrical energy. (credit: modification of work by Sarah Swenty, U.S. Fish and Wildlife Service)

As we finish this chapter on energy and work, it is relevant to draw some distinctions between two sometimes misunderstood terms in the area of energy use. As we mentioned earlier, the "law of conservation of energy" is a very useful principle in analyzing physical processes. It cannot be proven from basic principles but is a very good bookkeeping device, and no exceptions have ever been found. It states that the total amount of energy in an isolated system always remains constant. Related to this principle, but remarkably different from it, is the important philosophy of energy conservation. This concept has to do with seeking to decrease the amount of energy used by an individual or group through reducing activities (e.g., turning down thermostats, diving fewer kilometers) and/or increasing conversion efficiencies in the performance of a particular task, such as developing and using more efficient room heaters, cars that have greater miles-per-gallon ratings, energy-efficient compact fluorescent lights, etc.

Since energy in an isolated system is not destroyed, created, or generated, you might wonder why we need to be concerned about our energy resources, since energy is a conserved quantity. The problem is that the final result of most energy transformations is waste heat, that is, work that has been "degraded" in the energy transformation. We will discuss this idea in more detail in the chapters on thermodynamics.

CHAPTER 8 REVIEW

KEY TERMS

conservative force force that does work independent of path

conserved quantity one that cannot be created or destroyed, but may be transformed between different forms of itself

energy conservation total energy of an isolated system is constant

equilibrium point position where the assumed conservative, net force on a particle, given by the slope of its potential energy curve, is zero

exact differential is the total differential of a function and requires the use of partial derivatives if the function involves more than one dimension

mechanical energy sum of the kinetic and potential energies

non-conservative force force that does work that depends on path

non-renewable energy source that is not renewable, but is depleted by human consumption

potential energy function of position, energy possessed by an object relative to the system considered

potential energy diagram graph of a particle's potential energy as a function of position

potential energy difference negative of the work done acting between two points in space

renewable energy source that is replenished by natural processes, over human time scales

turning point position where the velocity of a particle, in one-dimensional motion, changes sign

KEY EQUATIONS

Difference of potential energy	$\Delta U_{AB} = U_B - U_A = -W_{AB}$
Potential energy with respect to zero of potential energy at $\vec{r}_0$	$\Delta U = U(\vec{r}) - U(\vec{r}_0)$
Gravitational potential energy near Earth's surface	$U(y) = mgy + \text{const.}$
Potential energy for an ideal spring	$U(x) = \frac{1}{2}kx^2 + \text{const.}$
Work done by conservative force over a closed path	$W_{\text{closed path}} = \oint \vec{F}_{\text{cons}} \cdot d\vec{r} = 0$
Condition for conservative force in two dimensions	$\left(\dfrac{dF_x}{dy}\right) = \left(\dfrac{dF_y}{dx}\right)$
Conservative force is the negative derivative of potential energy	$F_l = -\dfrac{dU}{dl}$
Conservation of energy with no non-conservative forces	$0 = W_{nc, AB} = \Delta(K + U)_{AB} = \Delta E_{AB}.$

SUMMARY

8.1 Potential Energy of a System

- For a single-particle system, the difference of potential energy is the opposite of the work done by the forces acting on the particle as it moves from one position to another.

- Since only differences of potential energy are physically meaningful, the zero of the potential energy function can be chosen at a convenient location.

- The potential energies for Earth's constant gravity, near its surface, and for a Hooke's law force are linear and quadratic functions of position, respectively.

8.2 Conservative and Non-Conservative Forces

- A conservative force is one for which the work done is independent of path. Equivalently, a force is conservative if the work done over any closed path is zero.

- A non-conservative force is one for which the work done depends on the path.

- For a conservative force, the infinitesimal work is an exact differential. This implies conditions on the derivatives of the force's components.

- The component of a conservative force, in a particular direction, equals the negative of the derivative of the potential energy for that force, with respect to a displacement in that direction.

8.3 Conservation of Energy

- A conserved quantity is a physical property that stays constant regardless of the path taken.

- A form of the work-energy theorem says that the change in the mechanical energy of a particle equals the work done on it by non-conservative forces.

- If non-conservative forces do no work and there are no external forces, the mechanical energy of a particle stays constant. This is a statement of the conservation of mechanical energy and there is no change in the total mechanical energy.

- For one-dimensional particle motion, in which the mechanical energy is constant and the potential energy is known, the particle's position, as a function of time, can be found by evaluating an integral that is derived from the conservation of mechanical energy.

8.4 Potential Energy Diagrams and Stability

- Interpreting a one-dimensional potential energy diagram allows you to obtain qualitative, and some quantitative, information about the motion of a particle.

- At a turning point, the potential energy equals the mechanical energy and the kinetic energy is zero, indicating that the direction of the velocity reverses there.

- The negative of the slope of the potential energy curve, for a particle, equals the one-dimensional component of the conservative force on the particle. At an equilibrium point, the slope is zero and is a stable (unstable) equilibrium for a potential energy minimum (maximum).

8.5 Sources of Energy

- Energy can be transferred from one system to another and transformed or converted from one type into another. Some of the basic types of energy are kinetic, potential, thermal, and electromagnetic.

- Renewable energy sources are those that are replenished by ongoing natural processes, over human time scales. Examples are wind, water, geothermal, and solar power.

- Non-renewable energy sources are those that are depleted by consumption, over human time scales. Examples are fossil fuel and nuclear power.

CONCEPTUAL QUESTIONS

8.1 Potential Energy of a System

1. The kinetic energy of a system must always be positive or zero. Explain whether this is true for the potential energy of a system.

2. The force exerted by a diving board is conservative, provided the internal friction is negligible. Assuming friction is negligible, describe changes in the potential energy of a diving board as a swimmer drives from it, starting just before the swimmer steps on the board until just after his feet leave it.

3. Describe the gravitational potential energy transfers and transformations for a javelin, starting from the point at which an athlete picks up the javelin and ending when the javelin is stuck into the ground after being thrown.

4. A couple of soccer balls of equal mass are kicked off the ground at the same speed but at different angles. Soccer ball A is kicked off at an angle slightly above the horizontal, whereas ball B is kicked slightly below the vertical. How do each of the following compare for ball A and ball B? (a) The initial kinetic energy and (b) the change in gravitational potential energy from the ground to the highest point? If the energy in part (a) differs from part (b), explain why there is a difference between the two energies.

5. What is the dominant factor that affects the speed of an object that started from rest down a frictionless incline if the only work done on the object is from gravitational forces?

6. Two people observe a leaf falling from a tree. One person is standing on a ladder and the other is on the ground. If each person were to compare the energy of the leaf observed, would each person find the following to be the same or different for the leaf, from the point where it falls off the tree to when it hits the ground: (a) the kinetic energy of the leaf; (b) the change in gravitational potential energy; (c) the final gravitational potential energy?

8.2 Conservative and Non-Conservative Forces

7. What is the physical meaning of a non-conservative force?

8. A bottle rocket is shot straight up in the air with a speed 30 m/s . If the air resistance is ignored, the bottle would go up to a height of approximately 46 m . However, the rocket goes up to only 35 m before returning to the ground. What happened? Explain, giving only a qualitative response.

9. An external force acts on a particle during a trip from one point to another and back to that same point. This particle is only effected by conservative forces. Does this particle's kinetic energy and potential energy change as a result of this trip?

8.3 Conservation of Energy

10. When a body slides down an inclined plane, does the work of friction depend on the body's initial speed? Answer the same question for a body sliding down a curved surface.

11. Consider the following scenario. A car for which friction is *not* negligible accelerates from rest down a hill, running out of gasoline after a short distance (see below). The driver lets the car coast farther down the hill, then up

and over a small crest. He then coasts down that hill into a gas station, where he brakes to a stop and fills the tank with gasoline. Identify the forms of energy the car has, and how they are changed and transferred in this series of events.

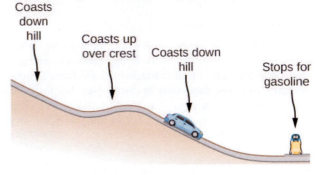

12. A dropped ball bounces to one-half its original height. Discuss the energy transformations that take place.

13. "$E = K + U$ constant is a special case of the work-energy theorem." Discuss this statement.

14. In a common physics demonstration, a bowling ball is suspended from the ceiling by a rope.

The professor pulls the ball away from its equilibrium position and holds it adjacent to his nose, as shown below. He releases the ball so that it swings directly away from him. Does he get struck by the ball on its return swing? What is he trying to show in this demonstration?

15. A child jumps up and down on a bed, reaching a higher height after each bounce. Explain how the child can increase his maximum gravitational potential energy with each bounce.

16. Can a non-conservative force increase the mechanical energy of the system?

17. Neglecting air resistance, how much would I have to raise the vertical height if I wanted to double the impact speed of a falling object?

18. A box is dropped onto a spring at its equilibrium position. The spring compresses with the box attached and comes to rest. Since the spring is in the vertical position, does the change in the gravitational potential energy of the box while the spring is compressing need to be considered in this problem?

PROBLEMS

8.1 Potential Energy of a System

19. Using values from **Table 8.1**, how many DNA molecules could be broken by the energy carried by a single electron in the beam of an old-fashioned TV tube? (These electrons were not dangerous in themselves, but they did create dangerous X-rays. Later-model tube TVs had shielding that absorbed X-rays before they escaped and exposed viewers.)

20. If the energy in fusion bombs were used to supply the energy needs of the world, how many of the 9-megaton variety would be needed for a year's supply of energy (using data from **Equation 8.7**)?

21. A camera weighing 10 N falls from a small drone hovering 20 m overhead and enters free fall. What is the gravitational potential energy change of the camera from the drone to the ground if you take a reference point of (a) the ground being zero gravitational potential energy? (b) The drone being zero gravitational potential energy? What is the gravitational potential energy of the camera (c) before it falls from the drone and (d) after the camera lands on the ground if the reference point of zero gravitational potential energy is taken to be a second person looking out of a building 30 m from the ground?

22. Someone drops a $50-g$ pebble off of a docked cruise ship, 70.0 m from the water line. A person on a dock 3.0 m from the water line holds out a net to catch the pebble. (a) How much work is done on the pebble by gravity during the drop? (b) What is the change in the gravitational potential energy during the drop? If the gravitational potential energy is zero at the water line, what is the gravitational potential energy (c) when the pebble is dropped? (d) When it reaches the net? What if the gravitational potential energy was 30.0 Joules at water level? (e) Find the answers to the same questions in (c) and (d).

23. A cat's crinkle ball toy of mass 15 g is thrown straight up with an initial speed of 3 m/s. Assume in this problem that air drag is negligible. (a) What is the kinetic energy of the ball as it leaves the hand? (b) How much work is done by the gravitational force during the ball's rise to its peak? (c) What is the change in the gravitational potential energy of the ball during the rise to its peak? (d) If the gravitational potential energy is taken to be zero at the point where it leaves your hand, what is the gravitational potential energy when it reaches the maximum height? (e) What if the gravitational potential energy is taken to be zero at the maximum height the ball reaches, what would the gravitational potential energy be when it leaves the hand? (f) What is the maximum height the ball reaches?

8.2 Conservative and Non-Conservative Forces

24. A force $F(x) = (3.0/x)\,\text{N}$ acts on a particle as it moves along the positive x-axis. (a) How much work does the force do on the particle as it moves from $x = 2.0$ m to $x = 5.0$ m? (b) Picking a convenient reference point of the potential energy to be zero at $x = \infty$, find the potential energy for this force.

25. A force $F(x) = \left(-5.0x^2 + 7.0x\right)\text{N}$ acts on a particle. (a) How much work does the force do on the particle as it moves from $x = 2.0$ m to $x = 5.0$ m? (b) Picking a convenient reference point of the potential energy to be zero at $x = \infty$, find the potential energy for this force.

26. Find the force corresponding to the potential energy $U(x) = -a/x + b/x^2$.

27. The potential energy function for either one of the two atoms in a diatomic molecule is often approximated by $U(x) = -a/x^{12} - b/x^6$ where x is the distance between the atoms. (a) At what distance of seperation does the potential energy have a local minimum (not at $x = \infty$)? (b) What is the force on an atom at this separation? (c) How does the force vary with the separation distance?

28. A particle of mass 2.0 kg moves under the influence of the force $F(x) = (3/\sqrt{x})\,\text{N}$. If its speed at $x = 2.0$ m is $v = 6.0$ m/s, what is its speed at $x = 7.0$ m?

29. A particle of mass 2.0 kg moves under the influence of the force $F(x) = \left(-5x^2 + 7x\right)\text{N}$. If its speed at $x = -4.0$ m is $v = 20.0$ m/s, what is its speed at $x = 4.0$ m?

30. A crate on rollers is being pushed without frictional loss of energy across the floor of a freight car (see the following figure). The car is moving to the right with a constant speed v_0. If the crate starts at rest relative to the freight car, then from the work-energy theorem, $Fd = mv^2/2$, where d, the distance the crate moves, and

v, the speed of the crate, are both measured relative to the freight car. (a) To an observer at rest beside the tracks, what distance d' is the crate pushed when it moves the distance d in the car? (b) What are the crate's initial and final speeds v_0' and v' as measured by the observer beside the tracks? (c) Show that $Fd' = m(v')^2/2 - m(v'_0)^2/2$ and, consequently, that work is equal to the change in kinetic energy in both reference systems.

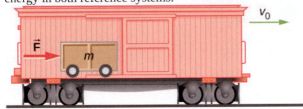

8.3 Conservation of Energy

31. A boy throws a ball of mass 0.25 kg straight upward with an initial speed of 20 m/s When the ball returns to the boy, its speed is 17 m/s How much much work does air resistance do on the ball during its flight?

32. A mouse of mass 200 g falls 100 m down a vertical mine shaft and lands at the bottom with a speed of 8.0 m/s. During its fall, how much work is done on the mouse by air resistance?

33. Using energy considerations and assuming negligible air resistance, show that a rock thrown from a bridge 20.0 m above water with an initial speed of 15.0 m/s strikes the water with a speed of 24.8 m/s independent of the direction thrown. (*Hint:* show that $K_i + U_i = K_f + U_f$)

34. A 1.0-kg ball at the end of a 2.0-m string swings in a vertical plane. At its lowest point the ball is moving with a speed of 10 m/s. (a) What is its speed at the top of its path? (b) What is the tension in the string when the ball is at the bottom and at the top of its path?

35. Ignoring details associated with friction, extra forces exerted by arm and leg muscles, and other factors, we can consider a pole vault as the conversion of an athlete's running kinetic energy to gravitational potential energy. If an athlete is to lift his body 4.8 m during a vault, what speed must he have when he plants his pole?

36. Tarzan grabs a vine hanging vertically from a tall tree when he is running at 9.0 m/s. (a) How high can he swing upward? (b) Does the length of the vine affect this height?

37. Assume that the force of a bow on an arrow behaves like the spring force. In aiming the arrow, an archer pulls the bow back 50 cm and holds it in position with a force of 150 N. If the mass of the arrow is 50 g and the "spring" is massless, what is the speed of the arrow immediately after it leaves the bow?

38. A $100 - \text{kg}$ man is skiing across level ground at a speed of 8.0 m/s when he comes to the small slope 1.8 m higher than ground level shown in the following figure. (a) If the skier coasts up the hill, what is his speed when he reaches the top plateau? Assume friction between the snow and skis is negligible. (b) What is his speed when he reaches the upper level if an $80 - \text{N}$ frictional force acts on the skis?

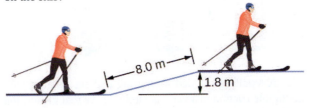

39. A sled of mass 70 kg starts from rest and slides down a $10°$ incline 80 m long. It then travels for 20 m horizontally before starting back up an $8°$ incline. It travels 80 m along this incline before coming to rest. What is the magnitude of the net work done on the sled by friction?

40. A girl on a skateboard (total mass of 40 kg) is moving at a speed of 10 m/s at the bottom of a long ramp. The ramp is inclined at $20°$ with respect to the horizontal. If she travels 14.2 m upward along the ramp before stopping, what is the net frictional force on her?

41. A baseball of mass 0.25 kg is hit at home plate with a speed of 40 m/s. When it lands in a seat in the left-field bleachers a horizontal distance 120 m from home plate, it is moving at 30 m/s. If the ball lands 20 m above the spot where it was hit, how much work is done on it by air resistance?

42. A small block of mass m slides without friction around the loop-the-loop apparatus shown below. (a) If the block starts from rest at A, what is its speed at B? (b) What is the force of the track on the block at B?

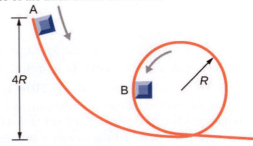

43. The massless spring of a spring gun has a force constant $k = 12 \text{ N/cm}$. When the gun is aimed vertically, a 15-g projectile is shot to a height of 5.0 m above the end of the expanded spring. (See below.) How much was the spring compressed initially?

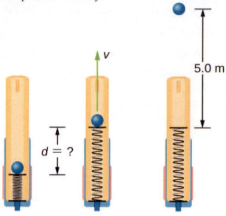

44. A small ball is tied to a string and set rotating with negligible friction in a vertical circle. Prove that the tension in the string at the bottom of the circle exceeds that at the top of the circle by eight times the weight of the ball. Assume the ball's speed is zero as it sails over the top of the circle and there is no additional energy added to the ball during rotation.

8.4 Potential Energy Diagrams and Stability

45. A mysterious constant force of 10 N acts horizontally on everything. The direction of the force is found to be always pointed toward a wall in a big hall. Find the potential energy of a particle due to this force when it is at a distance x from the wall, assuming the potential energy at the wall to be zero.

46. A single force $F(x) = -4.0x$ (in newtons) acts on a 1.0-kg body. When $x = 3.5$ m, the speed of the body is 4.0 m/s. What is its speed at $x = 2.0$ m?

47. A particle of mass 4.0 kg is constrained to move along the x-axis under a single force $F(x) = -cx^3$, where $c = 8.0$ N/m^3. The particle's speed at A, where $x_A = 1.0$ m, is 6.0 m/s. What is its speed at B, where $x_B = -2.0$ m?

48. The force on a particle of mass 2.0 kg varies with position according to $F(x) = -3.0x^2$ (x in meters, $F(x)$ in newtons). The particle's velocity at $x = 2.0$ m is 5.0 m/s. Calculate the mechanical energy of the particle using (a) the origin as the reference point and (b) $x = 4.0$ m as the reference point. (c) Find the particle's velocity at $x = 1.0$ m. Do this part of the problem for each reference point.

49. A 4.0-kg particle moving along the x-axis is acted upon by the force whose functional form appears below. The velocity of the particle at $x = 0$ is $v = 6.0$ m/s. Find the particle's speed at $x =$ (a)2.0 m, (b)4.0 m, (c)10.0 m, (d) Does the particle turn around at some point and head back toward the origin? (e) Repeat part (d) if $v = 2.0$ m/s at $x = 0$.

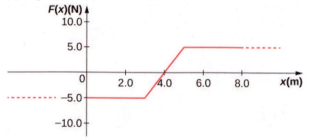

50. A particle of mass 0.50 kg moves along the x-axis with a potential energy whose dependence on x is shown below. (a) What is the force on the particle at $x = 2.0$, 5.0, 8.0, and 12 m? (b) If the total mechanical energy E of the particle is −6.0 J, what are the minimum and maximum positions of the particle? (c) What are these positions if $E = 2.0$ J? (d) If $E = 16$ J, what are the speeds of the particle at the positions listed in part (a)?

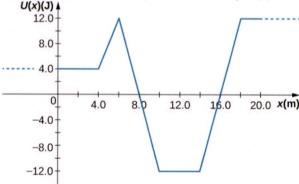

51. (a) Sketch a graph of the potential energy function $U(x) = kx^2/2 + Ae^{-\alpha x^2}$, where k, A, and α are constants. (b) What is the force corresponding to this potential energy? (c) Suppose a particle of mass m moving with this potential energy has a velocity v_a when its position is $x = a$. Show that the particle does not pass through the origin unless $A \le \dfrac{mv_a^2 + ka^2}{2\left(1 - e^{-\alpha a^2}\right)}$.

8.5 Sources of Energy

52. In the cartoon movie *Pocahontas* *(https://openstaxcollege.org/l/21pocahontclip)*, Pocahontas runs to the edge of a cliff and jumps off, showcasing the fun side of her personality. (a) If she is running at 3.0 m/s before jumping off the cliff and she hits the water at the bottom of the cliff at 20.0 m/s, how high is the cliff? Assume negligible air drag in this cartoon. (b) If she jumped off the same cliff from a standstill, how fast would she be falling right before she hit the water?

53. In the reality television show **"Amazing Race" (https://openstaxcollege.org/l/21amazraceclip)** , a contestant is firing 12-kg watermelons from a slingshot to hit targets down the field. The slingshot is pulled back 1.5 m and the watermelon is considered to be at ground level. The launch point is 0.3 m from the ground and the targets are 10 m horizontally away. Calculate the spring constant of the slingshot.

54. In the **Back to the Future movies (https://openstaxcollege.org/l/21bactofutclip)** , a DeLorean car of mass 1230 kg travels at 88 miles per hour to venture back to the future. (a) What is the kinetic energy of the DeLorean? (b) What spring constant would be needed to stop this DeLorean in a distance of 0.1m?

55. In the **Hunger Games movie (https://openstaxcollege.org/l/21HungGamesclip)** , Katniss Everdeen fires a 0.0200-kg arrow from ground level to pierce an apple up on a stage. The spring constant of the bow is 330 N/m and she pulls the arrow back a distance of 0.55 m. The apple on the stage is 5.00 m higher than the launching point of the arrow. At what speed does the arrow (a) leave the bow? (b) strike the apple?

56. In a **"Top Fail" video (https://openstaxcollege.org/l/21topfailvideo)** , two women run at each other and collide by hitting exercise balls together. If each woman has a mass of 50 kg, which includes the exercise ball, and one woman runs to the right at 2.0 m/s and the other is running toward her at 1.0 m/s, (a) how much total kinetic energy is there in the system? (b) If energy is conserved after the collision and each exercise ball has a mass of 2.0 kg, how fast would the balls fly off toward the camera?

57. In a **Coyote/Road Runner cartoon clip (https://openstaxcollege.org/l/21coyroadcarcl)** , a spring expands quickly and sends the coyote into a rock. If the spring extended 5 m and sent the coyote of mass 20 kg to a speed of 15 m/s, (a) what is the spring constant of this spring? (b) If the coyote were sent vertically into the air with the energy given to him by the spring, how high could he go if there were no non-conservative forces?

58. In an iconic movie scene, **Forrest Gump (https://openstaxcollege.org/l/21ForrGumpvid)** runs around the country. If he is running at a constant speed of 3 m/s, would it take him more or less energy to run uphill or downhill and why?

59. In the movie *Monty Python and the Holy Grail*

(https://openstaxcollege.org/l/21monpytmovcl) a cow is catapulted from the top of a castle wall over to the people down below. The gravitational potential energy is set to zero at ground level. The cow is launched from a spring of spring constant 1.1×10^4 N/m that is expanded 0.5 m from equilibrium. If the castle is 9.1 m tall and the mass of the cow is 110 kg, (a) what is the gravitational potential energy of the cow at the top of the castle? (b) What is the elastic spring energy of the cow before the catapult is released? (c) What is the speed of the cow right before it lands on the ground?

60. A 60.0-kg skier with an initial speed of 12.0 m/s coasts up a 2.50-m high rise as shown. Find her final speed at the top, given that the coefficient of friction between her skis and the snow is 0.80.

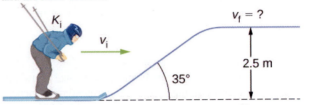

61. (a) How high a hill can a car coast up (engines disengaged) if work done by friction is negligible and its initial speed is 110 km/h? (b) If, in actuality, a 750-kg car with an initial speed of 110 km/h is observed to coast up a hill to a height 22.0 m above its starting point, how much thermal energy was generated by friction? (c) What is the average force of friction if the hill has a slope of $2.5°$ above the horizontal?

62. A 5.00×10^5 -kg subway train is brought to a stop from a speed of 0.500 m/s in 0.400 m by a large spring bumper at the end of its track. What is the spring constant k of the spring?

63. A pogo stick has a spring with a spring constant of 2.5×10^4 N/m, which can be compressed 12.0 cm. To what maximum height from the uncompressed spring can a child jump on the stick using only the energy in the spring, if the child and stick have a total mass of 40 kg?

64. A block of mass 500 g is attached to a spring of spring constant 80 N/m (see the following figure). The other end of the spring is attached to a support while the mass rests on a rough surface with a coefficient of friction of 0.20 that is inclined at angle of $30°$. The block is pushed along the surface till the spring compresses by 10 cm and is then released from rest. (a) How much potential energy was stored in the block-spring-support system when the block was just released? (b) Determine the speed of the block when it crosses the point when the spring is neither compressed nor stretched. (c) Determine the position of the block where it just comes to rest on its way up the incline.

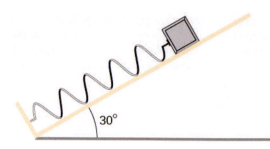

65. A block of mass 200 g is attached at the end of a massless spring of spring constant 50 N/m. The other end of the spring is attached to the ceiling and the mass is released at a height considered to be where the gravitational potential energy is zero. (a) What is the net potential energy of the block at the instant the block is at the lowest point? (b) What is the net potential energy of the block at the midpoint of its descent? (c) What is the speed of the block at the midpoint of its descent?

66. A T-shirt cannon launches a shirt at 5.00 m/s from a platform height of 3.00 m from ground level. How fast will the shirt be traveling if it is caught by someone whose hands are (a) 1.00 m from ground level? (b) 4.00 m from ground level? Neglect air drag.

67. A child (32 kg) jumps up and down on a trampoline.

ADDITIONAL PROBLEMS

69. A massless spring with force constant $k = 200$ N/m hangs from the ceiling. A 2.0-kg block is attached to the free end of the spring and released. If the block falls 17 cm before starting back upwards, how much work is done by friction during its descent?

70. A particle of mass 2.0 kg moves under the influence of the force $F(x) = (-5x^2 + 7x)$ N. Suppose a frictional force also acts on the particle. If the particle's speed when it starts at $x = -4.0$ m is 0.0 m/s and when it arrives at $x = 4.0$ m is 9.0 m/s, how much work is done on it by the frictional force between $x = -4.0$ m and $x = 4.0$ m?

71. Block 2 shown below slides along a frictionless table as block 1 falls. Both blocks are attached by a frictionless pulley. Find the speed of the blocks after they have each moved 2.0 m. Assume that they start at rest and that the pulley has negligible mass. Use $m_1 = 2.0$ kg and $m_2 = 4.0$ kg.

The trampoline exerts a spring restoring force on the child with a constant of 5000 N/m. At the highest point of the bounce, the child is 1.0 m above the level surface of the trampoline. What is the compression distance of the trampoline? Neglect the bending of the legs or any transfer of energy of the child into the trampoline while jumping.

68. Shown below is a box of mass m_1 that sits on a frictionless incline at an angle above the horizontal $\theta = 30°$. This box is connected by a relatively massless string, over a frictionless pulley, and finally connected to a box at rest over the ledge, labeled m_2. If m_1 and m_2 are a height h above the ground and $m_2 >> m_1$: (a) What is the initial gravitational potential energy of the system? (b) What is the final kinetic energy of the system?

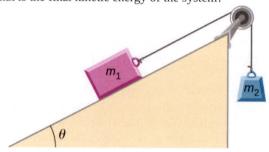

72. A body of mass m and negligible size starts from rest and slides down the surface of a frictionless solid sphere of radius R. (See below.) Prove that the body leaves the sphere when $\theta = \cos^{-1}(2/3)$.

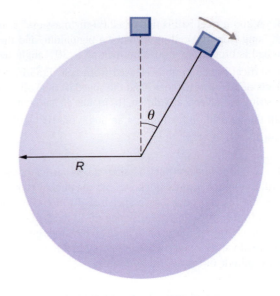

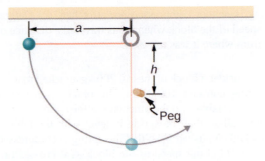

73. A mysterious force acts on all particles along a particular line and always points towards a particular point P on the line. The magnitude of the force on a particle increases as the cube of the distance from that point; that is $F \propto r^3$, if the distance from P to the position of the particle is r. Let b be the proportionality constant, and write the magnitude of the force as $F = br^3$. Find the potential energy of a particle subjected to this force when the particle is at a distance D from P, assuming the potential energy to be zero when the particle is at P.

74. An object of mass 10 kg is released at point A, slides to the bottom of the $30°$ incline, then collides with a horizontal massless spring, compressing it a maximum distance of 0.75 m. (See below.) The spring constant is 500 M/m, the height of the incline is 2.0 m, and the horizontal surface is frictionless. (a) What is the speed of the object at the bottom of the incline? (b) What is the work of friction on the object while it is on the incline? (c) The spring recoils and sends the object back toward the incline. What is the speed of the object when it reaches the base of the incline? (d) What vertical distance does it move back up the incline?

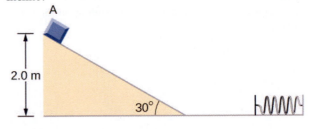

75. Shown below is a small ball of mass m attached to a string of length a. A small peg is located a distance h below the point where the string is supported. If the ball is released when the string is horizontal, show that h must be greater than $3a/5$ if the ball is to swing completely around the peg.

76. A block leaves a frictionless inclined surface horizontally after dropping off by a height h. Find the horizontal distance D where it will land on the floor, in terms of h, H, and g.

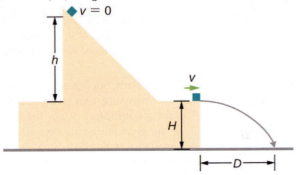

77. A block of mass m, after sliding down a frictionless incline, strikes another block of mass M that is attached to a spring of spring constant k (see below). The blocks stick together upon impact and travel together. (a) Find the compression of the spring in terms of m, M, h, g, and k when the combination comes to rest. Hint: The speed of the combined blocks $m + M$ (v_2) is based on the speed of block m just prior to the collision with the block M (v_1) based on the equation $v_2 = (m / m) + M (v_1)$. This will be discussed further in the chapter on Linear Momentum and Collisions. (b) The loss of kinetic energy as a result of the bonding of the two masses upon impact is stored in the so-called binding energy of the two masses. Calculate the binding energy.

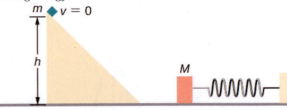

78. A block of mass 300 g is attached to a spring of spring constant 100 N/m. The other end of the spring is attached to a support while the block rests on a smooth horizontal table and can slide freely without any friction. The block is pushed horizontally till the spring compresses by 12 cm, and then the block is released from rest. (a) How much potential energy was stored in the block-spring support system when the block was just released? (b) Determine the speed of the block when it crosses the point when the spring is neither compressed nor stretched. (c) Determine the speed of the block when it has traveled a distance of 20 cm from where it was released.

79. Consider a block of mass 0.200 kg attached to a spring of spring constant 100 N/m. The block is placed on a frictionless table, and the other end of the spring is attached to the wall so that the spring is level with the table. The block is then pushed in so that the spring is compressed by 10.0 cm. Find the speed of the block as it crosses (a) the point when the spring is not stretched, (b) 5.00 cm to the left of point in (a), and (c) 5.00 cm to the right of point in (a).

80. A skier starts from rest and slides downhill. What will be the speed of the skier if he drops by 20 meters in vertical height? Ignore any air resistance (which will, in reality, be quite a lot), and any friction between the skis and the snow.

81. Repeat the preceding problem, but this time, suppose that the work done by air resistance cannot be ignored. Let the work done by the air resistance when the skier goes from A to B along the given hilly path be −2000 J. The work done by air resistance is negative since the air resistance acts in the opposite direction to the displacement. Supposing the mass of the skier is 50 kg, what is the speed of the skier at point B?

82. Two bodies are interacting by a conservative force. Show that the mechanical energy of an isolated system consisting of two bodies interacting with a conservative force is conserved. (*Hint*: Start by using Newton's third law and the definition of work to find the work done on each body by the conservative force.)

83. In an amusement park, a car rolls in a track as shown below. Find the speed of the car at A, B, and C. Note that the work done by the rolling friction is zero since the displacement of the point at which the rolling friction acts on the tires is momentarily at rest and therefore has a zero displacement.

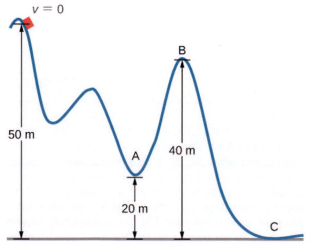

84. A 200-g steel ball is tied to a 2.00-m "massless" string and hung from the ceiling to make a pendulum, and then, the ball is brought to a position making a $30°$ angle with the vertical direction and released from rest. Ignoring the effects of the air resistance, find the speed of the ball when the string (a) is vertically down, (b) makes an angle of $20°$ with the vertical and (c) makes an angle of $10°$ with the vertical.

85. A 300 g hockey puck is shot across an ice-covered pond. Before the hockey puck was hit, the puck was at rest. After the hit, the puck has a speed of 40 m/s. The puck comes to rest after going a distance of 30 m. (a) Describe how the energy of the puck changes over time, giving the numerical values of any work or energy involved. (b) Find the magnitude of the net friction force.

86. A projectile of mass 2 kg is fired with a speed of 20 m/s at an angle of $30°$ with respect to the horizontal. (a) Calculate the initial total energy of the projectile given that the reference point of zero gravitational potential energy at the launch position. (b) Calculate the kinetic energy at the highest vertical position of the projectile. (c) Calculate the gravitational potential energy at the highest vertical position. (d) Calculate the maximum height that the projectile reaches. Compare this result by solving the same problem using your knowledge of projectile motion.

87. An artillery shell is fired at a target 200 m above the ground. When the shell is 100 m in the air, it has a speed of 100 m/s. What is its speed when it hits its target? Neglect air friction.

88. How much energy is lost to a dissipative drag force if a 60-kg person falls at a constant speed for 15 meters?

89. A box slides on a frictionless surface with a total energy of 50 J. It hits a spring and compresses the spring a distance of 25 cm from equilibrium. If the same box with the same initial energy slides on a rough surface, it only compresses the spring a distance of 15 cm, how much energy must have been lost by sliding on the rough surface?

9 | LINEAR MOMENTUM AND COLLISIONS

Figure 9.1 The concepts of impulse, momentum, and center of mass are crucial for a major-league baseball player to successfully get a hit. If he misjudges these quantities, he might break his bat instead. (credit: modification of work by "Cathy T"/Flickr)

Chapter Outline

Introduction

The concepts of work, energy, and the work-energy theorem are valuable for two primary reasons: First, they are powerful computational tools, making it much easier to analyze complex physical systems than is possible using Newton's laws directly (for example, systems with nonconstant forces); and second, the observation that the total energy of a closed system is conserved means that the system can only evolve in ways that are consistent with energy conservation. In other words, a system cannot evolve randomly; it can only change in ways that conserve energy.

In this chapter, we develop and define another conserved quantity, called *linear momentum*, and another relationship (the *impulse-momentum theorem*), which will put an additional constraint on how a system evolves in time. Conservation of momentum is useful for understanding collisions, such as that shown in the above image. It is just as powerful, just as important, and just as useful as conservation of energy and the work-energy theorem.

9.1 | Linear Momentum

Our study of kinetic energy showed that a complete understanding of an object's motion must include both its mass and its velocity ($K = (1/2)mv^2$). However, as powerful as this concept is, it does not include any information about the direction of the moving object's velocity vector. We'll now define a physical quantity that includes direction.

Like kinetic energy, this quantity includes both mass and velocity; like kinetic energy, it is a way of characterizing the "quantity of motion" of an object. It is given the name **momentum** (from the Latin word *movimentum*, meaning "movement"), and it is represented by the symbol p.

Momentum

The momentum p of an object is the product of its mass and its velocity:

$$\vec{\mathbf{p}} = m\vec{\mathbf{v}}.$$

(9.1)

Figure 9.2 The velocity and momentum vectors for the ball are in the same direction. The mass of the ball is about 0.5 kg, so the momentum vector is about half the length of the velocity vector because momentum is velocity time mass. (credit: modification of work by Ben Sutherland)

As shown in **Figure 9.2**, momentum is a vector quantity (since velocity is). This is one of the things that makes momentum useful and not a duplication of kinetic energy. It is perhaps most useful when determining whether an object's motion is difficult to change (**Figure 9.3**) or easy to change (**Figure 9.4**).

Figure 9.3 This supertanker transports a huge mass of oil; as a consequence, it takes a long time for a force to change its (comparatively small) velocity. (credit: modification of work by "the_tahoe_guy"/Flickr)

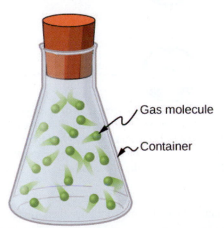

Figure 9.4 Gas molecules can have very large velocities, but these velocities change nearly instantaneously when they collide with the container walls or with each other. This is primarily because their masses are so tiny.

Unlike kinetic energy, momentum depends equally on an object's mass and velocity. For example, as you will learn when you study thermodynamics, the average speed of an air molecule at room temperature is approximately 500 m/s, with an average molecular mass of 6×10^{-25} kg ; its momentum is thus

$$p_{\text{molecule}} = \left(6 \times 10^{-25} \text{ kg}\right)\left(500 \frac{\text{m}}{\text{s}}\right) = 3 \times 10^{-22} \frac{\text{kg} \cdot \text{m}}{\text{s}}.$$

For comparison, a typical automobile might have a speed of only 15 m/s, but a mass of 1400 kg, giving it a momentum of

$$p_{\text{car}} = (1400 \text{ kg})\left(15 \frac{\text{m}}{\text{s}}\right) = 21,000 \frac{\text{kg} \cdot \text{m}}{\text{s}}.$$

These momenta are different by 27 orders of magnitude, or a factor of a billion billion billion!

9.2 | Impulse and Collisions

We have defined momentum to be the product of mass and velocity. Therefore, if an object's velocity should change (due to the application of a force on the object), then necessarily, its momentum changes as well. This indicates a connection between momentum and force. The purpose of this section is to explore and describe that connection.

Suppose you apply a force on a free object for some amount of time. Clearly, the larger the force, the larger the object's change of momentum will be. Alternatively, the more time you spend applying this force, again the larger the change of momentum will be, as depicted in **Figure 9.5**. The amount by which the object's motion changes is therefore proportional to the magnitude of the force, and also to the time interval over which the force is applied.

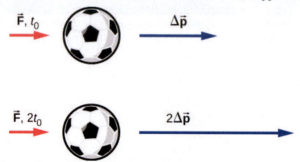

Figure 9.5 The change in momentum of an object is proportional to the length of time during which the force is applied. If a force is exerted on the lower ball for twice as long as on the upper ball, then the change in the momentum of the lower ball is twice that of the upper ball.

Mathematically, if a quantity is proportional to two (or more) things, then it is proportional to the product of those things. The product of a force and a time interval (over which that force acts) is called **impulse**, and is given the symbol $\vec{\mathbf{J}}$.

Impulse

Let $\vec{\mathbf{F}}(t)$ be the force applied to an object over some differential time interval dt (**Figure 9.6**). The resulting impulse on the object is defined as

$$d\vec{\mathbf{J}} \equiv \vec{\mathbf{F}}(t)dt. \tag{9.2}$$

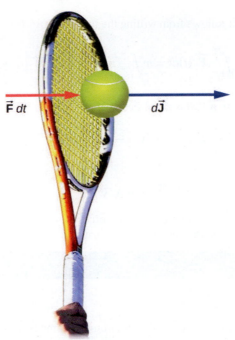

Figure 9.6 A force applied by a tennis racquet to a tennis ball over a time interval generates an impulse acting on the ball.

The total impulse over the interval $t_f - t_i$ is

$$\vec{J} = \int_{t_i}^{t_f} d\vec{J} \quad \text{or} \quad \vec{J} \equiv \int_{t_i}^{t_f} \vec{F}(t)dt. \tag{9.3}$$

Equation 9.2 and **Equation 9.3** together say that when a force is applied for an infinitesimal time interval dt, it causes an infinitesimal impulse $d\vec{J}$, and the total impulse given to the object is defined to be the sum (integral) of all these infinitesimal impulses.

To calculate the impulse using **Equation 9.3**, we need to know the force function $F(t)$, which we often don't. However, a result from calculus is useful here: Recall that the average value of a function over some interval is calculated by

$$f(x)_{ave} = \frac{1}{\Delta x}\int_{x_i}^{x_f} f(x)dx$$

where $\Delta x = x_f - x_i$. Applying this to the time-dependent force function, we obtain

$$\vec{F}_{ave} = \frac{1}{\Delta t}\int_{t_i}^{t_f} \vec{F}(t)dt. \tag{9.4}$$

Therefore, from **Equation 9.3**,

$$\vec{J} = \vec{F}_{ave}\Delta t. \tag{9.5}$$

The idea here is that you can calculate the impulse on the object even if you don't know the details of the force as a function of time; you only need the average force. In fact, though, the process is usually reversed: You determine the impulse (by measurement or calculation) and then calculate the average force that caused that impulse.

To calculate the impulse, a useful result follows from writing the force in **Equation 9.3** as $\overrightarrow{\mathbf{F}}(t) = m\overrightarrow{\mathbf{a}}(t)$:

$$\overrightarrow{\mathbf{J}} = \int_{t_i}^{t_f} \overrightarrow{\mathbf{F}}(t)dt = m\int_{t_i}^{t_f}\overrightarrow{\mathbf{a}}(t)dt = m\left[\overrightarrow{\mathbf{v}}(t_f) - \overrightarrow{\mathbf{v}}_i\right].$$

For a constant force $\overrightarrow{\mathbf{F}}_{ave} = \overrightarrow{\mathbf{F}} = m\overrightarrow{\mathbf{a}}$, this simplifies to

$$\overrightarrow{\mathbf{J}} = m\overrightarrow{\mathbf{a}}\Delta t = m\overrightarrow{\mathbf{v}}_f - m\overrightarrow{\mathbf{v}}_i = m(\overrightarrow{\mathbf{v}}_f - \overrightarrow{\mathbf{v}}_i).$$

That is,

$$\overrightarrow{\mathbf{J}} = m\Delta\overrightarrow{\mathbf{v}}. \tag{9.6}$$

Note that the integral form, **Equation 9.3**, applies to constant forces as well; in that case, since the force is independent of time, it comes out of the integral, which can then be trivially evaluated.

Example 9.1

The Arizona Meteor Crater

Approximately 50,000 years ago, a large (radius of 25 m) iron-nickel meteorite collided with Earth at an estimated speed of 1.28×10^4 m/s in what is now the northern Arizona desert, in the United States. The impact produced a crater that is still visible today (**Figure 9.7**); it is approximately 1200 m (three-quarters of a mile) in diameter, 170 m deep, and has a rim that rises 45 m above the surrounding desert plain. Iron-nickel meteorites typically have a density of $\rho = 7970 \text{ kg/m}^3$. Use impulse considerations to estimate the average force and the maximum force that the meteor applied to Earth during the impact.

Figure 9.7 The Arizona Meteor Crater in Flagstaff, Arizona (often referred to as the Barringer Crater after the person who first suggested its origin and whose family owns the land). (credit: modification of work by "Shane.torgerson"/Wikimedia Commons)

Strategy

It is conceptually easier to reverse the question and calculate the force that Earth applied on the meteor in order to stop it. Therefore, we'll calculate the force on the meteor and then use Newton's third law to argue that the force from the meteor on Earth was equal in magnitude and opposite in direction.

Using the given data about the meteor, and making reasonable guesses about the shape of the meteor and impact time, we first calculate the impulse using **Equation 9.6**. We then use the relationship between force and impulse **Equation 9.5** to estimate the average force during impact. Next, we choose a reasonable force function for the impact event, calculate the average value of that function **Equation 9.4**, and set the resulting expression equal to the calculated average force. This enables us to solve for the maximum force.

Solution

Define upward to be the +y-direction. For simplicity, assume the meteor is traveling vertically downward prior to impact. In that case, its initial velocity is $\vec{\mathbf{v}}_i = -v_i\,\hat{\mathbf{j}}$, and the force Earth exerts on the meteor points upward, $\vec{\mathbf{F}}(t) = +F(t)\,\hat{\mathbf{j}}$. The situation at $t = 0$ is depicted below.

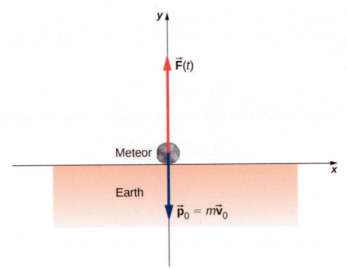

The average force during the impact is related to the impulse by

$$\vec{\mathbf{F}}_{\text{ave}} = \frac{\vec{\mathbf{J}}}{\Delta t}.$$

From **Equation 9.6**, $\vec{\mathbf{J}} = m\Delta\vec{\mathbf{v}}$, so we have

$$\vec{\mathbf{F}}_{\text{ave}} = \frac{m\Delta\vec{\mathbf{v}}}{\Delta t}.$$

The mass is equal to the product of the meteor's density and its volume:

$$m = \rho V.$$

If we assume (guess) that the meteor was roughly spherical, we have

$$V = \frac{4}{3}\pi R^3.$$

Thus we obtain

$$\vec{\mathbf{F}}_{\text{ave}} = \frac{\rho V\Delta\vec{\mathbf{v}}}{\Delta t} = \frac{\rho\left(\frac{4}{3}\pi R^3\right)\left(\vec{\mathbf{v}}_f - \vec{\mathbf{v}}_i\right)}{\Delta t}.$$

The problem says the velocity at impact was -1.28×10^4 m/s $\hat{\mathbf{j}}$ (the final velocity is zero); also, we guess that the primary impact lasted about $t_{\text{max}} = 2\,\text{s}$. Substituting these values gives

$$\vec{\mathbf{F}}_{ave} = \frac{\left(7970\,\frac{kg}{m^3}\right)\left[\frac{4}{3}\,\pi\,(25\ m)^3\right]\left[0\,\frac{m}{s} - \left(-1.28\times10^4\,\frac{m}{s}\,\hat{\mathbf{j}}\right)\right]}{2\ s}.$$

$$= +\left(3.33\times10^{12}\ N\right)\hat{\mathbf{j}}$$

This is the average force applied during the collision. Notice that this force vector points in the same direction as the change of velocity vector $\Delta\vec{\mathbf{v}}$.

Next, we calculate the maximum force. The impulse is related to the force function by

$$\vec{\mathbf{J}} = \int_{t_i}^{t\,max}\vec{\mathbf{F}}\,(t)dt.$$

We need to make a reasonable choice for the force as a function of time. We define $t = 0$ to be the moment the meteor first touches the ground. Then we assume the force is a maximum at impact, and rapidly drops to zero. A function that does this is

$$F(t) = F_{max}\,e^{-t^2/\left(2\tau^2\right)}.$$

(The parameter τ represents how rapidly the force decreases to zero.) The average force is

$$F_{ave} = \frac{1}{\Delta t}\int_0^{t\,max}F_{max}e^{-t^2/\left(2\tau^2\right)}\,dt$$

where $\Delta t = t_{max} - 0\ s$. Since we already have a numeric value for F_{ave}, we can use the result of the integral to obtain F_{max}.

Choosing $\tau = \frac{1}{e}t_{max}$ (this is a common choice, as you will see in later chapters), and guessing that $t_{max} = 2\ s$, this integral evaluates to

$$F_{avg} = 0.458\,F_{max}.$$

Thus, the maximum force has a magnitude of

$$0.458F_{max} = 3.33\times10^{12}\ N$$
$$F_{max} = 7.27\times10^{12}\ N.$$

The complete force function, including the direction, is

$$\vec{\mathbf{F}}\,(t) = \left(7.27\times10^{12}\ N\right)e^{-t^2/\left(8s^2\right)}\,\hat{\mathbf{y}}.$$

This is the force Earth applied to the meteor; by Newton's third law, the force the meteor applied to Earth is

$$\vec{\mathbf{F}}\,(t) = -\left(7.27\times10^{12}\ N\right)e^{-t^2/\left(8s^2\right)}\,\hat{\mathbf{y}}$$

which is the answer to the original question.

Significance

The graph of this function contains important information. Let's graph (the magnitude of) both this function and the average force together (**Figure 9.8**).

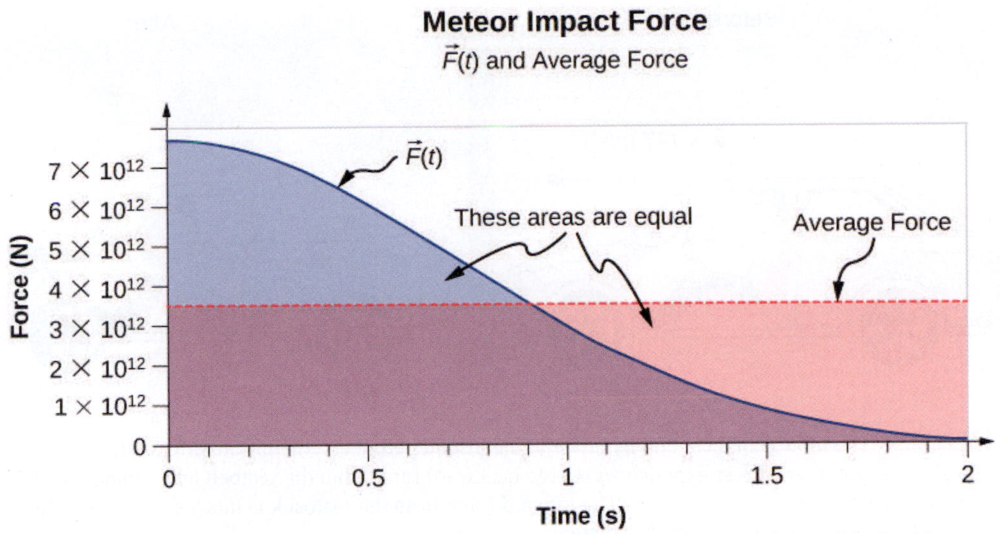

Meteor Impact Force
$\vec{F}(t)$ and Average Force

Figure 9.8 A graph of the average force (in red) and the force as a function of time (blue) of the meteor impact. The areas under the curves are equal to each other, and are numerically equal to the applied impulse.

Notice that the area under each plot has been filled in. For the plot of the (constant) force F_{ave}, the area is a rectangle, corresponding to $F_{ave} \Delta t = J$. As for the plot of $F(t)$, recall from calculus that the area under the plot of a function is numerically equal to the integral of that function, over the specified interval; so here, that is $\int_{0}^{t_{max}} F(t)dt = J$. Thus, the areas are equal, and both represent the impulse that the meteor applied to Earth during the two-second impact. The average force on Earth sounds like a huge force, and it is. Nevertheless, Earth barely noticed it. The acceleration Earth obtained was just

$$\vec{a} = \frac{-\vec{F}_{ave}}{M_{Earth}} = \frac{-\left(3.33 \times 10^{12} \text{ N}\right)\hat{j}}{5.97 \times 10^{24} \text{ kg}} = -\left(5.6 \times 10^{-13} \frac{\text{m}}{\text{s}^2}\right)\hat{j}$$

which is completely immeasurable. That said, the impact created seismic waves that nowadays could be detected by modern monitoring equipment.

Example 9.2

The Benefits of Impulse

A car traveling at 27 m/s collides with a building. The collision with the building causes the car to come to a stop in approximately 1 second. The driver, who weighs 860 N, is protected by a combination of a variable-tension seatbelt and an airbag (**Figure 9.9**). (In effect, the driver collides with the seatbelt and airbag and *not* with the building.) The airbag and seatbelt slow his velocity, such that he comes to a stop in approximately 2.5 s.

   a.   What average force does the driver experience during the collision?

   b.   Without the seatbelt and airbag, his collision time (with the steering wheel) would have been approximately 0.20 s. What force would he experience in this case?

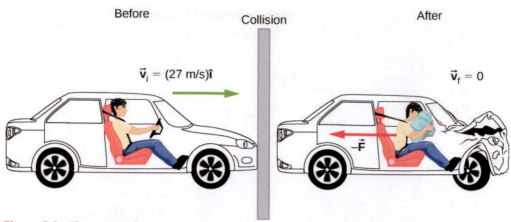

Figure 9.9 The motion of a car and its driver at the instant before and the instant after colliding with the wall. The restrained driver experiences a large backward force from the seatbelt and airbag, which causes his velocity to decrease to zero. (The forward force from the seatback is much smaller than the backward force, so we neglect it in the solution.)

Strategy

We are given the driver's weight, his initial and final velocities, and the time of collision; we are asked to calculate a force. Impulse seems the right way to tackle this; we can combine **Equation 9.5** and **Equation 9.6**.

Solution

a. Define the $+x$-direction to be the direction the car is initially moving. We know

$$\vec{J} = \vec{F}\,\Delta t$$

and

$$\vec{J} = m\Delta\vec{v}.$$

Since J is equal to both those things, they must be equal to each other:

$$\vec{F}\,\Delta t = m\Delta\vec{v}.$$

We need to convert this weight to the equivalent mass, expressed in SI units:

$$\frac{860\ \text{N}}{9.8\ \text{m/s}^2} = 87.8\ \text{kg}.$$

Remembering that $\Delta\vec{v} = \vec{v}_f - \vec{v}_i$, and noting that the final velocity is zero, we solve for the force:

$$\vec{F} = m\frac{0 - v_i\,\hat{i}}{\Delta t} = (87.8\ \text{kg})\left(\frac{-(27\ \text{m/s})\,\hat{i}}{2.5\ \text{s}}\right) = -(948\ \text{N})\,\hat{i}.$$

The negative sign implies that the force slows him down. For perspective, this is about 1.1 times his own weight.

b. Same calculation, just the different time interval:

$$\vec{F} = (87.8\ \text{kg})\left(\frac{-(27\ \text{m/s})\,\hat{i}}{0.20\ \text{s}}\right) = -(11{,}853\ \text{N})\,\hat{i}$$

which is about 14 times his own weight. Big difference!

Significance

You see that the value of an airbag is how greatly it reduces the force on the vehicle occupants. For this reason, they have been required on all passenger vehicles in the United States since 1991, and have been commonplace throughout Europe and Asia since the mid-1990s. The change of momentum in a crash is the same, with or without an airbag; the force, however, is vastly different.

Effect of Impulse

Since an impulse is a force acting for some amount of time, it causes an object's motion to change. Recall **Equation 9.6**:

$$\vec{J} = m\Delta\vec{v} .$$

Because $m\vec{v}$ is the momentum of a system, $m\Delta\vec{v}$ is the *change* of momentum $\Delta\vec{p}$. This gives us the following relation, called the **impulse-momentum theorem** (or relation).

Impulse-Momentum Theorem

An impulse applied to a system changes the system's momentum, and that change of momentum is exactly equal to the impulse that was applied:

$$\vec{J} = \Delta\vec{p} .$$

(9.7)

The impulse-momentum theorem is depicted graphically in **Figure 9.10**.

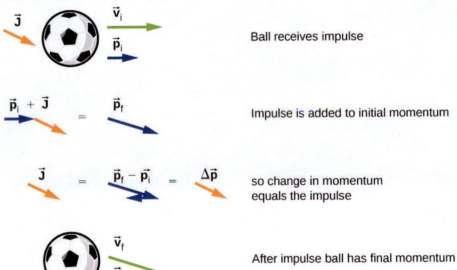

Figure 9.10 Illustration of impulse-momentum theorem. (a) A ball with initial velocity $\vec{v}_0$ and momentum $\vec{p}_0$ receives an impulse $\vec{J}$. (b) This impulse is added vectorially to the initial momentum. (c) Thus, the impulse equals the change in momentum, $\vec{J} = \Delta\vec{p}$. (d) After the impulse, the ball moves off with its new momentum $\vec{p}_f$.

There are two crucial concepts in the impulse-momentum theorem:

1. Impulse is a vector quantity; an impulse of, say, $-(10\,\text{N}\cdot\text{s})\,\hat{i}$ is very different from an impulse of $+(10\,\text{N}\cdot\text{s})\,\hat{i}$; they cause completely opposite changes of momentum.

2.   An impulse does not cause momentum; rather, it causes a *change* in the momentum of an object. Thus, you must subtract the final momentum from the initial momentum, and—since momentum is also a vector quantity—you must take careful account of the signs of the momentum vectors.

The most common questions asked in relation to impulse are to calculate the applied force, or the change of velocity that occurs as a result of applying an impulse. The general approach is the same.

Problem-Solving Strategy: Impulse-Momentum Theorem

1.   Express the impulse as force times the relevant time interval.

2.   Express the impulse as the change of momentum, usually $m\Delta v$.

3.   Equate these and solve for the desired quantity.

Example 9.3

Moving the *Enterprise*

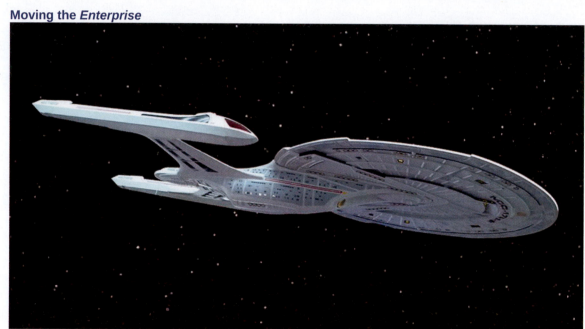

Figure 9.11   The fictional starship *Enterprise* from the Star Trek adventures operated on so-called "impulse engines" that combined matter with antimatter to produce energy.

"Mister Sulu, take us out; ahead one-quarter impulse." With this command, Captain Kirk of the starship *Enterprise* (**Figure 9.11**) has his ship start from rest to a final speed of $v_f = 1/4\!\left(3.0 \times 10^8 \text{ m/s}\right)$. Assuming this maneuver is completed in 60 s, what average force did the impulse engines apply to the ship?

Strategy

We are asked for a force; we know the initial and final speeds (and hence the change in speed), and we know the time interval over which this all happened. In particular, we know the amount of time that the force acted. This suggests using the impulse-momentum relation. To use that, though, we need the mass of the *Enterprise*. An internet search gives a best estimate of the mass of the *Enterprise* (in the 2009 movie) as $2 \times 10^9 \text{ kg}$.

Solution

Because this problem involves only one direction (i.e., the direction of the force applied by the engines), we only need the scalar form of the impulse-momentum theorem **Equation 9.7**, which is

$$\Delta p = J$$

with

$$\Delta p = m\Delta v$$

and

$$J = F\Delta t.$$

Equating these expressions gives

$$F\Delta t = m\Delta v.$$

Solving for the magnitude of the force and inserting the given values leads to

$$F = \frac{m\Delta v}{\Delta t} = \frac{\left(2 \times 10^9 \text{ kg}\right)\left(7.5 \times 10^7 \text{ m/s}\right)}{60 \text{ s}} = 2.5 \times 10^{15} \text{ N}.$$

Significance

This is an unimaginably huge force. It goes almost without saying that such a force would kill everyone on board instantly, as well as destroying every piece of equipment. Fortunately, the *Enterprise* has "inertial dampeners." It is left as an exercise for the reader's imagination to determine how these work.

 9.1 Check Your Understanding The U.S. Air Force uses "10gs" (an acceleration equal to $10 \times 9.8 \text{ m/s}^2$) as the maximum acceleration a human can withstand (but only for several seconds) and survive. How much time must the *Enterprise* spend accelerating if the humans on board are to experience an average of at most 10gs of acceleration? (Assume the inertial dampeners are offline.)

Example 9.4

The iPhone Drop

Apple released its iPhone 6 Plus in November 2014. According to many reports, it was originally supposed to have a screen made from sapphire, but that was changed at the last minute for a hardened glass screen. Reportedly, this was because the sapphire screen cracked when the phone was dropped. What force did the iPhone 6 Plus experience as a result of being dropped?

Strategy

The force the phone experiences is due to the impulse applied to it by the floor when the phone collides with the floor. Our strategy then is to use the impulse-momentum relationship. We calculate the impulse, estimate the impact time, and use this to calculate the force.

We need to make a couple of reasonable estimates, as well as find technical data on the phone itself. First, let's suppose that the phone is most often dropped from about chest height on an average-height person. Second, assume that it is dropped from rest, that is, with an initial vertical velocity of zero. Finally, we assume that the phone bounces very little—the height of its bounce is assumed to be negligible.

Solution

Define upward to be the +y-direction. A typical height is approximately $h = 1.5 \text{ m}$ and, as stated, $\overrightarrow{\textbf{v}}_i = (0 \text{ m/s})\hat{\textbf{i}}$. The average force on the phone is related to the impulse the floor applies on it during the collision:

$$\overrightarrow{\textbf{F}}_{ave} = \frac{\overrightarrow{\textbf{J}}}{\Delta t}.$$

The impulse $\overrightarrow{\textbf{J}}$ equals the change in momentum,

$$\vec{J} = \Delta \vec{p}$$

so

$$\vec{F}_{ave} = \frac{\Delta \vec{p}}{\Delta t}.$$

Next, the change of momentum is

$$\Delta \vec{p} = m\Delta \vec{v}.$$

We need to be careful with the velocities here; this is the change of velocity due to the collision with the floor. But the phone also has an initial drop velocity [$\vec{v}_i = (0 \text{ m/s}) \hat{j}$], so we label our velocities. Let:

- $\vec{v}_i =$ the initial velocity with which the phone was dropped (zero, in this example)

- $\vec{v}_1 =$ the velocity the phone had the instant just before it hit the floor

- $\vec{v}_2 =$ the final velocity of the phone as a result of hitting the floor

Figure 9.12 shows the velocities at each of these points in the phone's trajectory.

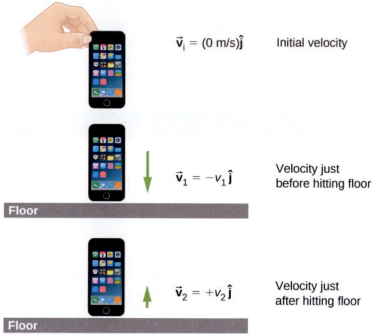

$\vec{v}_i = (0 \text{ m/s})\hat{j}$ Initial velocity

$\vec{v}_1 = -v_1\hat{j}$ Velocity just before hitting floor

Floor

$\vec{v}_2 = +v_2\hat{j}$ Velocity just after hitting floor

Floor

Figure 9.12 (a) The initial velocity of the phone is zero, just after the person drops it. (b) Just before the phone hits the floor, its velocity is $\vec{v}_1$, which is unknown at the moment, except for its direction, which is downward $(-\hat{j})$. (c) After bouncing off the floor, the phone has a velocity $\vec{v}_2$, which is also unknown, except for its direction, which is upward $(+\hat{j})$.

With these definitions, the change of momentum of the phone during the collision with the floor is

$$m\Delta \vec{v} = m\left(\vec{v}_2 - \vec{v}_1\right).$$

Since we assume the phone doesn't bounce at all when it hits the floor (or at least, the bounce height is negligible), then $\vec{\mathbf{v}}_2$ is zero, so

$$m\Delta\vec{\mathbf{v}} = m\left[0 - \left(-v_1\,\hat{\mathbf{j}}\right)\right]$$

$$m\Delta\vec{\mathbf{v}} = +mv_1\,\hat{\mathbf{j}}.$$

We can get the speed of the phone just before it hits the floor using either kinematics or conservation of energy. We'll use conservation of energy here; you should re-do this part of the problem using kinematics and prove that you get the same answer.

First, define the zero of potential energy to be located at the floor. Conservation of energy then gives us:

$$E_{\mathrm{i}} = E_1$$

$$K_{\mathrm{i}} + U_{\mathrm{i}} = K_1 + U_1$$

$$\tfrac{1}{2}mv_{\mathrm{i}}^2 + mgh_{\mathrm{drop}} = \tfrac{1}{2}mv_1^2 + mgh_{\mathrm{floor}}.$$

Defining $h_{\mathrm{floor}} = 0$ and using $\vec{\mathbf{v}}_{\mathrm{i}} = (0\ \mathrm{m/s})\,\hat{\mathbf{j}}$ gives

$$\tfrac{1}{2}mv_1^2 = mgh_{\mathrm{drop}}$$

$$v_1 = \pm\sqrt{2gh_{\mathrm{drop}}}.$$

Because v_1 is a vector magnitude, it must be positive. Thus, $m\Delta v = mv_1 = m\sqrt{2gh_{\mathrm{drop}}}$. Inserting this result into the expression for force gives

$$\vec{\mathbf{F}} = \frac{\Delta\vec{\mathbf{p}}}{\Delta t}$$

$$= \frac{m\Delta\vec{\mathbf{v}}}{\Delta t}$$

$$= \frac{+mv_1\,\hat{\mathbf{j}}}{\Delta t}$$

$$= \frac{m\sqrt{2gh}}{\Delta t}\,\hat{\mathbf{j}}.$$

Finally, we need to estimate the collision time. One common way to estimate a collision time is to calculate how long the object would take to travel its own length. The phone is moving at 5.4 m/s just before it hits the floor, and it is 0.14 m long, giving an estimated collision time of 0.026 s. Inserting the given numbers, we obtain

$$\vec{\mathbf{F}} = \frac{(0.172\ \mathrm{kg})\sqrt{2(9.8\ \mathrm{m/s}^2)(1.5\ \mathrm{m})}}{0.026\ \mathrm{s}}\,\hat{\mathbf{j}} = (36\ \mathrm{N})\,\hat{\mathbf{j}}.$$

Significance

The iPhone itself weighs just $(0.172\ \mathrm{kg})(9.81\ \mathrm{m/s}^2) = 1.68\ \mathrm{N}$; the force the floor applies to it is therefore over 20 times its weight.

  9.2 Check Your Understanding What if we had assumed the phone *did* bounce on impact? Would this have increased the force on the iPhone, decreased it, or made no difference?

Momentum and Force

In **Example 9.3**, we obtained an important relationship:

$$\overrightarrow{\mathbf{F}}_{\text{ave}} = \frac{\Delta \overrightarrow{\mathbf{p}}}{\Delta t}. \tag{9.8}$$

In words, the average force applied to an object is equal to the change of the momentum that the force causes, divided by the time interval over which this change of momentum occurs. This relationship is very useful in situations where the collision time Δt is small, but measureable; typical values would be 1/10th of a second, or even one thousandth of a second. Car crashes, punting a football, or collisions of subatomic particles would meet this criterion.

For a *continuously* changing momentum—due to a continuously changing force—this becomes a powerful conceptual tool. In the limit $\Delta t \to dt$, **Equation 9.2** becomes

$$\overrightarrow{\mathbf{F}} = \frac{d \overrightarrow{\mathbf{p}}}{dt}. \tag{9.9}$$

This says that the rate of change of the system's momentum (implying that momentum is a function of time) is exactly equal to the net applied force (also, in general, a function of time). This is, in fact, Newton's second law, written in terms of momentum rather than acceleration. This is the relationship Newton himself presented in his *Principia Mathematica* (although he called it "quantity of motion" rather than "momentum").

If the mass of the system remains constant, **Equation 9.3** reduces to the more familiar form of Newton's second law. We can see this by substituting the definition of momentum:

$$\overrightarrow{\mathbf{F}} = \frac{d(m \overrightarrow{\mathbf{v}})}{dt} = m\frac{d \overrightarrow{\mathbf{v}}}{dt} = m \overrightarrow{\mathbf{a}}.$$

The assumption of constant mass allowed us to pull m out of the derivative. If the mass is not constant, we cannot use this form of the second law, but instead must start from **Equation 9.3**. Thus, one advantage to expressing force in terms of changing momentum is that it allows for the mass of the system to change, as well as the velocity; this is a concept we'll explore when we study the motion of rockets.

Newton's Second Law of Motion in Terms of Momentum

The net external force on a system is equal to the rate of change of the momentum of that system caused by the force:

$$\overrightarrow{\mathbf{F}} = \frac{d \overrightarrow{\mathbf{p}}}{dt}.$$

Although **Equation 9.3** allows for changing mass, as we will see in **Rocket Propulsion**, the relationship between momentum and force remains useful when the mass of the system is constant, as in the following example.

Example 9.5

Calculating Force: Venus Williams' Tennis Serve

During the 2007 French Open, Venus Williams hit the fastest recorded serve in a premier women's match, reaching a speed of 58 m/s (209 km/h). What is the average force exerted on the 0.057-kg tennis ball by Venus Williams' racquet? Assume that the ball's speed just after impact is 58 m/s, as shown in **Figure 9.13**, that the initial horizontal component of the velocity before impact is negligible, and that the ball remained in contact with the racquet for 5.0 ms.

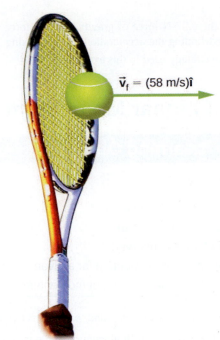

Figure 9.13 The final velocity of the tennis ball is $\vec{v}_f = (58 \text{ m/s}) \hat{i}$.

Strategy

This problem involves only one dimension because the ball starts from having no horizontal velocity component before impact. Newton's second law stated in terms of momentum is then written as

$$\vec{F} = \frac{d\vec{p}}{dt}.$$

As noted above, when mass is constant, the change in momentum is given by

$$\Delta p = m\Delta v = m(v_f - v_i)$$

where we have used scalars because this problem involves only one dimension. In this example, the velocity just after impact and the time interval are given; thus, once Δp is calculated, we can use $F = \frac{\Delta p}{\Delta t}$ to find the force.

Solution

To determine the change in momentum, insert the values for the initial and final velocities into the equation above:

$$\begin{aligned}
\Delta p &= m(v_f - v_i) \\
&= (0.057 \text{ kg})(58 \text{ m/s} - 0 \text{ m/s}) \\
&= 3.3 \frac{\text{kg} \cdot \text{m}}{\text{s}}.
\end{aligned}$$

Now the magnitude of the net external force can be determined by using

$$F = \frac{\Delta p}{\Delta t} = \frac{3.3 \frac{\text{kg} \cdot \text{m}}{\text{s}}}{5.0 \times 10^{-3} \text{ s}} = 6.6 \times 10^2 \text{ N}.$$

where we have retained only two significant figures in the final step.

Significance

This quantity was the average force exerted by Venus Williams' racquet on the tennis ball during its brief impact (note that the ball also experienced the 0.57-N force of gravity, but that force was not due to the racquet). This problem could also be solved by first finding the acceleration and then using $F = ma$, but one additional step would be required compared with the strategy used in this example.

9.3 | Conservation of Linear Momentum

Learning Objectives

By the end of this section, you will be able to:

*   Explain the meaning of "conservation of momentum"
*   Correctly identify if a system is, or is not, closed
*   Define a system whose momentum is conserved
*   Mathematically express conservation of momentum for a given system
*   Calculate an unknown quantity using conservation of momentum

Recall Newton's third law: When two objects of masses m_1 and m_2 interact (meaning that they apply forces on each other), the force that object 2 applies to object 1 is equal in magnitude and opposite in direction to the force that object 1 applies on object 2. Let:

* $\vec{\mathbf{F}}_{21} = $ the force on m_1 from m_2

* $\vec{\mathbf{F}}_{12} = $ the force on m_2 from m_1

Then, in symbols, Newton's third law says

$$\vec{\mathbf{F}}_{21} = -\vec{\mathbf{F}}_{12}$$
$$m_1 \vec{\mathbf{a}}_1 = -m_2 \vec{\mathbf{a}}_2.$$

(9.10)

(Recall that these two forces do not cancel because they are applied to different objects. F_{21} causes m_1 to accelerate, and F_{12} causes m_2 to accelerate.)

Although the magnitudes of the forces on the objects are the same, the accelerations are not, simply because the masses (in general) are different. Therefore, the changes in velocity of each object are different:

$$\frac{d\vec{\mathbf{v}}_1}{dt} \neq \frac{d\vec{\mathbf{v}}_2}{dt}.$$

However, the products of the mass and the change of velocity *are* equal (in magnitude):

$$m_1 \frac{d\vec{\mathbf{v}}_1}{dt} = -m_2 \frac{d\vec{\mathbf{v}}_2}{dt}.$$

(9.11)

It's a good idea, at this point, to make sure you're clear on the physical meaning of the derivatives in **Equation 9.3**. Because of the interaction, each object ends up getting its velocity changed, by an amount dv. Furthermore, the interaction occurs over a time interval dt, which means that the change of velocities also occurs over dt. This time interval is the same for each object.

Let's assume, for the moment, that the masses of the objects do not change during the interaction. (We'll relax this restriction later.) In that case, we can pull the masses inside the derivatives:

$$\frac{d}{dt}(m_1 \vec{\mathbf{v}}_1) = -\frac{d}{dt}(m_2 \vec{\mathbf{v}}_2)$$

(9.12)

and thus

$$\frac{d\vec{p}_1}{dt} = -\frac{d\vec{p}_2}{dt}.$$ (9.13)

This says that *the rate at which momentum changes is the same for both objects*. The masses are different, and the changes of velocity are different, but the rate of change of the product of m and $\vec{v}$ are the same.

Physically, this means that during the interaction of the two objects (m_1 and m_2), both objects have their momentum changed; but those changes are identical in magnitude, though opposite in sign. For example, the momentum of object 1 might increase, which means that the momentum of object 2 decreases by exactly the same amount.

In light of this, let's re-write **Equation 9.12** in a more suggestive form:

$$\frac{d\vec{p}_1}{dt} + \frac{d\vec{p}_2}{dt} = 0.$$ (9.14)

This says that during the interaction, although object 1's momentum changes, and object 2's momentum also changes, these two changes cancel each other out, so that the total change of momentum of the two objects together is zero.

Since the total combined momentum of the two objects together never changes, then we could write

$$\frac{d}{dt}\left(\vec{p}_1 + \vec{p}_2\right) = 0$$ (9.15)

from which it follows that

$$\vec{p}_1 + \vec{p}_2 = \text{constant}.$$ (9.16)

As shown in **Figure 9.14**, the total momentum of the system before and after the collision remains the same.

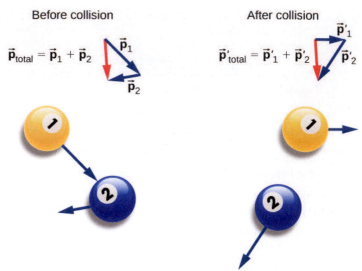

Figure 9.14 Before the collision, the two billiard balls travel with momenta $\vec{p}_1$ and $\vec{p}_3$. The total momentum of the system is the sum of these, as shown by the red vector labeled $\vec{p}_{total}$ on the left. After the collision, the two billiard balls travel with different momenta $\vec{p}'_1$ and $\vec{p}'_3$. The total momentum, however, has not changed, as shown by the red vector arrow $\vec{p}'_{total}$ on the right.

Generalizing this result to N objects, we obtain

$$\vec{p}_1 + \vec{p}_2 + \vec{p}_3 + \cdots + \vec{p}_N = \text{constant} \qquad \text{(9.17)}$$

$$\sum_{j=1}^{N} \vec{p}_j = \text{constant}.$$

Equation 9.17 is the definition of the total (or net) momentum of a system of N interacting objects, along with the statement that the total momentum of a system of objects is constant in time—or better, is conserved.

Conservation Laws

If the value of a physical quantity is constant in time, we say that the quantity is conserved.

Requirements for Momentum Conservation

There is a complication, however. A system must meet two requirements for its momentum to be conserved:

1. *The mass of the system must remain constant during the interaction.*
    As the objects interact (apply forces on each other), they may *transfer* mass from one to another; but any mass one object gains is balanced by the loss of that mass from another. The total mass of the system of objects, therefore, remains unchanged as time passes:
    $$\left[\frac{dm}{dt}\right]_{\text{system}} = 0.$$

2. *The net external force on the system must be zero.*
    As the objects collide, or explode, and move around, they exert forces on each other. However, all of these forces are internal to the system, and thus each of these internal forces is balanced by another internal force that is equal in

magnitude and opposite in sign. As a result, the change in momentum caused by each internal force is cancelled by another momentum change that is equal in magnitude and opposite in direction. Therefore, internal forces cannot change the total momentum of a system because the changes sum to zero. However, if there is some external force that acts on all of the objects (gravity, for example, or friction), then this force changes the momentum of the system as a whole; that is to say, the momentum of the system is changed by the external force. Thus, for the momentum of the system to be conserved, we must have

$$\vec{F}_{ext} = \vec{0}\,.$$

A system of objects that meets these two requirements is said to be a **closed system** (also called an isolated system). Thus, the more compact way to express this is shown below.

Law of Conservation of Momentum

The total momentum of a closed system is conserved:

$$\sum_{j=1}^{N} \vec{p}_j = \text{constant.}$$

This statement is called the **Law of Conservation of Momentum**. Along with the conservation of energy, it is one of the foundations upon which all of physics stands. All our experimental evidence supports this statement: from the motions of galactic clusters to the quarks that make up the proton and the neutron, and at every scale in between. *In a closed system, the total momentum never changes.*

Note that there absolutely *can* be external forces acting on the system; but for the system's momentum to remain constant, these external forces have to cancel, so that the *net* external force is zero. Billiard balls on a table all have a weight force acting on them, but the weights are balanced (canceled) by the normal forces, so there is no *net* force.

The Meaning of 'System'

A **system** (mechanical) is the collection of objects in whose motion (kinematics and dynamics) you are interested. If you are analyzing the bounce of a ball on the ground, you are probably only interested in the motion of the ball, and not of Earth; thus, the ball is your system. If you are analyzing a car crash, the two cars together compose your system (**Figure 9.15**).

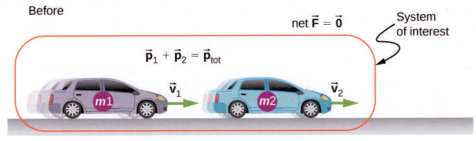

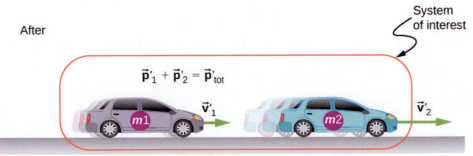

Figure 9.15 The two cars together form the system that is to be analyzed. It is important to remember that the contents (the mass) of the system do not change before, during, or after the objects in the system interact.

Example 9.6

Colliding Carts

Two carts in a physics lab roll on a level track, with negligible friction. These carts have small magnets at their ends, so that when they collide, they stick together (**Figure 9.16**). The first cart has a mass of 675 grams and is rolling at 0.75 m/s to the right; the second has a mass of 500 grams and is rolling at 1.33 m/s, also to the right. After the collision, what is the velocity of the two joined carts?

Figure 9.16 Two lab carts collide and stick together after the collision.

Strategy

We have a collision. We're given masses and initial velocities; we're asked for the final velocity. This all suggests using conservation of momentum as a method of solution. However, we can only use it if we have a closed system. So we need to be sure that the system we choose has no net external force on it, and that its mass is not changed by the collision.

Defining the system to be the two carts meets the requirements for a closed system: The combined mass of the two carts certainly doesn't change, and while the carts definitely exert forces on each other, those forces are internal to the system, so they do not change the momentum of the system as a whole. In the vertical direction, the weights of the carts are canceled by the normal forces on the carts from the track.

Solution

Conservation of momentum is

$$\vec{p}_f = \vec{p}_i.$$

Define the direction of their initial velocity vectors to be the +x-direction. The initial momentum is then

$$\vec{p}_i = m_1 v_1 \hat{i} + m_2 v_2 \hat{i}.$$

The final momentum of the now-linked carts is

$$\vec{p}_f = (m_1 + m_2) \vec{v}_f.$$

Equating:

$$(m_1 + m_2) \vec{v}_f = m_1 v_1 \hat{i} + m_2 v_2 \hat{i}$$

$$\vec{v}_f = \left(\frac{m_1 v_1 + m_2 v_2}{m_1 + m_2}\right)\hat{i}.$$

Substituting the given numbers:

$$\vec{v}_f = \left[\frac{(0.675 \text{ kg})(0.75 \text{ m/s}) + (0.5 \text{ kg})(1.33 \text{ m/s})}{1.175 \text{ kg}} \right] \hat{i}$$

$$= (0.997 \text{ m/s}) \hat{i}.$$

Significance

The principles that apply here to two laboratory carts apply identically to all objects of whatever type or size. Even for photons, the concepts of momentum and conservation of momentum are still crucially important even at that scale. (Since they are massless, the momentum of a photon is defined very differently from the momentum of ordinary objects. You will learn about this when you study quantum physics.)

 9.3 Check Your Understanding Suppose the second, smaller cart had been initially moving to the left. What would the sign of the final velocity have been in this case?

Example 9.7

A Bouncing Superball

A superball of mass 0.25 kg is dropped from rest from a height of $h = 1.50 \text{ m}$ above the floor. It bounces with no loss of energy and returns to its initial height (**Figure 9.17**).

   a. What is the superball's change of momentum during its bounce on the floor?

   b. What was Earth's change of momentum due to the ball colliding with the floor?

   c. What was Earth's change of velocity as a result of this collision?

(This example shows that you have to be careful about defining your system.)

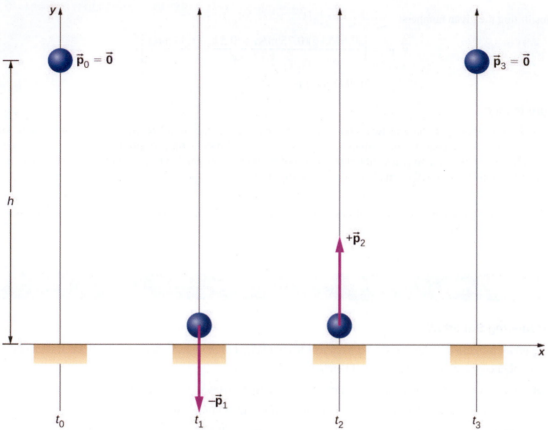

Figure 9.17 A superball is dropped to the floor (t_0), hits the floor (t_1), bounces (t_2), and returns to its initial height (t_3).

Strategy

Since we are asked only about the ball's change of momentum, we define our system to be the ball. But this is clearly not a closed system; gravity applies a downward force on the ball while it is falling, and the normal force from the floor applies a force during the bounce. Thus, we cannot use conservation of momentum as a strategy. Instead, we simply determine the ball's momentum just before it collides with the floor and just after, and calculate the difference. We have the ball's mass, so we need its velocities.

Solution

   a. Since this is a one-dimensional problem, we use the scalar form of the equations. Let:

      ◦ $p_0 =$ the magnitude of the ball's momentum at time t_0, the moment it was released; since it was dropped from rest, this is zero.

      ◦ $p_1 =$ the magnitude of the ball's momentum at time t_1, the instant just before it hits the floor.

      ◦ $p_2 =$ the magnitude of the ball's momentum at time t_2, just after it loses contact with the floor after the bounce.

     The ball's change of momentum is

$$\Delta \vec{p} = \vec{p}_2 - \vec{p}_1$$

$$= p_2 \,\hat{\mathbf{j}} - \left(-p_1 \,\hat{\mathbf{j}}\right)$$

$$= (p_2 + p_1)\,\hat{\mathbf{j}}.$$

Its velocity just before it hits the floor can be determined from either conservation of energy or kinematics. We use kinematics here; you should re-solve it using conservation of energy and confirm you get the same result.

We want the velocity just before it hits the ground (at time t_1). We know its initial velocity $v_0 = 0$ (at time t_0), the height it falls, and its acceleration; we don't know the fall time. We could calculate that, but instead we use

$$\vec{v}_1 = -\hat{\mathbf{j}}\sqrt{2gy} = -5.4 \text{ m/s}\,\hat{\mathbf{j}}.$$

Thus the ball has a momentum of

$$\vec{p}_1 = -(0.25 \text{ kg})\left(-5.4 \text{ m/s}\,\hat{\mathbf{j}}\right)$$

$$= -(1.4 \text{ kg} \cdot \text{m/s})\,\hat{\mathbf{j}}.$$

We don't have an easy way to calculate the momentum after the bounce. Instead, we reason from the symmetry of the situation.

Before the bounce, the ball starts with zero velocity and falls 1.50 m under the influence of gravity, achieving some amount of momentum just before it hits the ground. On the return trip (after the bounce), it starts with some amount of momentum, rises the same 1.50 m it fell, and ends with zero velocity. Thus, the motion after the bounce was the mirror image of the motion before the bounce. From this symmetry, it must be true that the ball's momentum after the bounce must be equal and opposite to its momentum before the bounce. (This is a subtle but crucial argument; make sure you understand it before you go on.) Therefore,

$$\vec{p}_2 = -\vec{p}_1 = +(1.4 \text{ kg} \cdot \text{m/s})\,\hat{\mathbf{j}}.$$

Thus, the ball's change of velocity during the bounce is

$$\Delta \vec{p} = \vec{p}_2 - \vec{p}_1$$

$$= (1.4 \text{ kg} \cdot \text{m/s})\,\hat{\mathbf{j}} - (-1.4 \text{ kg} \cdot \text{m/s})\,\hat{\mathbf{j}}$$

$$= +(2.8 \text{ kg} \cdot \text{m/s})\,\hat{\mathbf{j}}.$$

b. What was Earth's change of momentum due to the ball colliding with the floor?

Your instinctive response may well have been either "zero; the Earth is just too massive for that tiny ball to have affected it" or possibly, "more than zero, but utterly negligible." But no—if we re-define our system to be the Superball + Earth, then this system is closed (neglecting the gravitational pulls of the Sun, the Moon, and the other planets in the solar system), and therefore the total change of momentum of this new system must be zero. Therefore, Earth's change of momentum is exactly the same magnitude:

$$\Delta \vec{p}_{\text{Earth}} = -2.8 \text{ kg} \cdot \text{m/s}\,\hat{\mathbf{j}}.$$

c. What was Earth's change of velocity as a result of this collision?

This is where your instinctive feeling is probably correct:

$$\Delta \vec{v}_{Earth} = \frac{\Delta \vec{p}_{Earth}}{M_{Earth}}$$

$$= -\frac{2.8 \text{ kg} \cdot \text{m/s}}{5.97 \times 10^{24} \text{ kg}} \hat{j}$$

$$= -\left(4.7 \times 10^{-25} \text{ m/s}\right) \hat{j}.$$

This change of Earth's velocity *is* utterly negligible.

Significance

It is important to realize that the answer to part (c) is not a velocity; it is a change of velocity, which is a very different thing. Nevertheless, to give you a feel for just how small that change of velocity is, suppose you were moving with a velocity of 4.7×10^{-25} m/s. At this speed, it would take you about 7 million years to travel a distance equal to the diameter of a hydrogen atom.

 9.4 Check Your Understanding Would the ball's change of momentum have been larger, smaller, or the same, if it had collided with the floor and stopped (without bouncing)?

Would the ball's change of momentum have been larger, smaller, or the same, if it had collided with the floor and stopped (without bouncing)?

Example 9.8

Ice Hockey 1

Two hockey pucks of identical mass are on a flat, horizontal ice hockey rink. The red puck is motionless; the blue puck is moving at 2.5 m/s to the left (**Figure 9.18**). It collides with the motionless red puck. The pucks have a mass of 15 g. After the collision, the red puck is moving at 2.5 m/s, to the left. What is the final velocity of the blue puck?

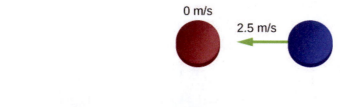

Figure 9.18 Two identical hockey pucks colliding. The top diagram shows the pucks the instant before the collision, and the bottom diagram show the pucks the instant after the collision. The net external force is zero.

Strategy

We're told that we have two colliding objects, we're told the masses and initial velocities, and one final velocity; we're asked for both final velocities. Conservation of momentum seems like a good strategy. Define the system to be the two pucks; there's no friction, so we have a closed system.

Before you look at the solution, what do you think the answer will be?

The blue puck final velocity will be:

- zero

- 2.5 m/s to the left

- 2.5 m/s to the right

- 1.25 m/s to the left

- 1.25 m/s to the right

- something else

Solution

Define the $+x$-direction to point to the right. Conservation of momentum then reads

$$\vec{p}_f = \vec{p}_i$$

$$mv_{r_f}\hat{i} + mv_{b_f}\hat{i} = mv_{r_i}\hat{i} - mv_{b_i}\hat{i}.$$

Before the collision, the momentum of the system is entirely and only in the blue puck. Thus,

$$mv_{r_f}\hat{i} + mv_{b_f}\hat{i} = -mv_{b_i}\hat{i}$$

$$v_{r_f}\hat{i} + v_{b_f}\hat{i} = -v_{b_i}\hat{i}.$$

(Remember that the masses of the pucks are equal.) Substituting numbers:

$$-(2.5 \text{ m/s})\hat{i} + \vec{v}_{b_f} = -(2.5 \text{ m/s})\hat{i}$$

$$\vec{v}_{b_f} = 0.$$

Significance

Evidently, the two pucks simply exchanged momentum. The blue puck transferred all of its momentum to the red puck. In fact, this is what happens in similar collision where $m_1 = m_2$.

 9.5 Check Your Understanding Even if there were some friction on the ice, it is still possible to use conservation of momentum to solve this problem, but you would need to impose an additional condition on the problem. What is that additional condition?

Example 9.9

Landing of *Philae*

On November 12, 2014, the European Space Agency successfully landed a probe named *Philae* on Comet 67P/Churyumov/Gerasimenko (**Figure 9.19**). During the landing, however, the probe actually landed three times, because it bounced twice. Let's calculate how much the comet's speed changed as a result of the first bounce.

Figure 9.19 An artist's rendering of *Philae* landing on a comet. (credit: modification of work by "DLR German Aerospace Center"/Flickr)

Let's define upward to be the $+y$-direction, perpendicular to the surface of the comet, and $y = 0$ to be at the surface of the comet. Here's what we know:

- The mass of Comet 67P: $M_c = 1.0 \times 10^{13}$ kg

- The acceleration due to the comet's gravity: $\vec{a} = -\left(5.0 \times 10^{-3} \text{ m/s}^2\right)\hat{j}$

- *Philae's* mass: $M_p = 96$ kg

- Initial touchdown speed: $\vec{v}_1 = -(1.0 \text{ m/s})\,\hat{j}$

- Initial upward speed due to first bounce: $\vec{v}_2 = (0.38 \text{ m/s})\,\hat{j}$

- Landing impact time: $\Delta t = 1.3$ s

Strategy

We're asked for how much the comet's speed changed, but we don't know much about the comet, beyond its mass and the acceleration its gravity causes. However, we *are* told that the *Philae* lander collides with (lands on) the comet, and bounces off of it. A collision suggests momentum as a strategy for solving this problem.

If we define a system that consists of both *Philae* and Comet 67/P, then there is no net external force on this system, and thus the momentum of this system is conserved. (We'll neglect the gravitational force of the sun.) Thus, if we calculate the change of momentum of the lander, we automatically have the change of momentum of the comet. Also, the comet's change of velocity is directly related to its change of momentum as a result of the lander "colliding" with it.

Solution

Let $\vec{p}_1$ be *Philae's* momentum at the moment just before touchdown, and $\vec{p}_2$ be its momentum just after the first bounce. Then its momentum just before landing was

$$\vec{p}_1 = M_p \vec{v}_1 = (96 \text{ kg})\left(-1.0 \text{ m/s}\,\hat{j}\right) = -(96 \text{ kg} \cdot \text{m/s})\,\hat{j}$$

and just after was

$$\vec{p}_2 = M_p \vec{v}_2 = (96 \text{ kg})\left(+0.38 \text{ m/s } \hat{j}\right) = (36.5 \text{ kg} \cdot \text{m/s}) \hat{j}.$$

Therefore, the lander's change of momentum during the first bounce is

$$\Delta \vec{p} = \vec{p}_2 - \vec{p}_1$$

$$= (36.5 \text{ kg} \cdot \text{m/s}) \hat{j} - \left(-96.0 \text{ kg} \cdot \text{m/s } \hat{j}\right) = (133 \text{ kg} \cdot \text{m/s}) \hat{j}$$

Notice how important it is to include the negative sign of the initial momentum.

Now for the comet. Since momentum of the system must be conserved, the *comet's* momentum changed by exactly the negative of this:

$$\Delta \vec{p}_c = -\Delta \vec{p} = -(133 \text{ kg} \cdot \text{m/s}) \hat{j}.$$

Therefore, its change of velocity is

$$\Delta \vec{v}_c = \frac{\Delta \vec{p}_c}{M_c} = \frac{-(133 \text{ kg} \cdot \text{m/s}) \hat{j}}{1.0 \times 10^{13} \text{ kg}} = -\left(1.33 \times 10^{-11} \text{ m/s}\right) \hat{j}.$$

Significance

This is a very small change in velocity, about a thousandth of a billionth of a meter per second. Crucially, however, it is *not* zero.

 9.6   Check Your Understanding The changes of momentum for *Philae* and for Comet 67/P were equal (in magnitude). Were the impulses experienced by *Philae* and the comet equal? How about the forces? How about the changes of kinetic energies?

9.4 | Types of Collisions

Learning Objectives

By the end of this section, you will be able to:

- Identify the type of collision
- Correctly label a collision as elastic or inelastic
- Use kinetic energy along with momentum and impulse to analyze a collision

Although momentum is conserved in all interactions, not all interactions (collisions or explosions) are the same. The possibilities include:

- A single object can explode into multiple objects (one-to-many).
- Multiple objects can collide and stick together, forming a single object (many-to-one).
- Multiple objects can collide and bounce off of each other, remaining as multiple objects (many-to-many). If they do bounce off each other, then they may recoil at the same speeds with which they approached each other before the collision, or they may move off more slowly.

It's useful, therefore, to categorize different types of interactions, according to how the interacting objects move before and after the interaction.

One-to-Many

The first possibility is that a single object may break apart into two or more pieces. An example of this is a firecracker, or a bow and arrow, or a rocket rising through the air toward space. These can be difficult to analyze if the number of fragments after the collision is more than about three or four; but nevertheless, the total momentum of the system before and after the explosion is identical.

Note that if the object is initially motionless, then the system (which is just the object) has no momentum and no kinetic energy. After the explosion, the net momentum of all the pieces of the object must sum to zero (since the momentum of this closed system cannot change). However, the system *will* have a great deal of kinetic energy after the explosion, although it had none before. Thus, we see that, although the momentum of the system is conserved in an explosion, the kinetic energy of the system most definitely is not; it increases. This interaction—one object becoming many, with an increase of kinetic energy of the system—is called an **explosion**.

Where does the energy come from? Does conservation of energy still hold? Yes; some form of potential energy is converted to kinetic energy. In the case of gunpowder burning and pushing out a bullet, chemical potential energy is converted to kinetic energy of the bullet, and of the recoiling gun. For a bow and arrow, it is elastic potential energy in the bowstring.

Many-to-One

The second possibility is the reverse: that two or more objects collide with each other and stick together, thus (after the collision) forming one single composite object. The total mass of this composite object is the sum of the masses of the original objects, and the new single object moves with a velocity dictated by the conservation of momentum. However, it turns out again that, although the total momentum of the system of objects remains constant, the kinetic energy doesn't; but this time, the kinetic energy decreases. This type of collision is called **inelastic**.

In the extreme case, multiple objects collide, stick together, and remain motionless after the collision. Since the objects are all motionless after the collision, the final kinetic energy is also zero; the loss of kinetic energy is a maximum. Such a collision is said to be **perfectly inelastic**.

Many-to-Many

The extreme case on the other end is if two or more objects approach each other, collide, and bounce off each other, moving away from each other at the same relative speed at which they approached each other. In this case, the total kinetic energy of the system is conserved. Such an interaction is called **elastic**.

In any interaction of a closed system of objects, the total momentum of the system is conserved ($\vec{\mathbf{p}}_f = \vec{\mathbf{p}}_i$) but the kinetic energy may not be:

- If $0 < K_f < K_i$, the collision is inelastic.

- If $K_f = 0$, the collision is perfectly inelastic.

- If $K_f = K_i$, the collision is elastic.

- If $K_f > K_i$, the interaction is an explosion.

The point of all this is that, in analyzing a collision or explosion, you can use both momentum and kinetic energy.

Problem-Solving Strategy: Collisions

A closed system always conserves momentum; it might also conserve kinetic energy, but very often it doesn't. Energy-momentum problems confined to a plane (as ours are) usually have two unknowns. Generally, this approach works well:

1. Define a closed system.

2. Write down the expression for conservation of momentum.

3. If kinetic energy is conserved, write down the expression for conservation of kinetic energy; if not, write down the expression for the change of kinetic energy.

4. You now have two equations in two unknowns, which you solve by standard methods.

Example 9.10

Formation of a Deuteron

A proton (mass 1.67×10^{-27} kg) collides with a neutron (with essentially the same mass as the proton) to form a particle called a *deuteron*. What is the velocity of the deuteron if it is formed from a proton moving with velocity 7.0×10^6 m/s to the left and a neutron moving with velocity 4.0×10^6 m/s to the right?

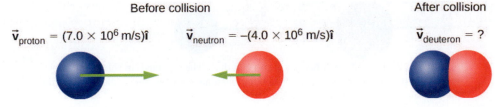

Strategy

Define the system to be the two particles. This is a collision, so we should first identify what kind. Since we are told the two particles form a single particle after the collision, this means that the collision is perfectly inelastic. Thus, kinetic energy is not conserved, but momentum is. Thus, we use conservation of energy to determine the final velocity of the system.

Solution

Treat the two particles as having identical masses M. Use the subscripts p, n, and d for proton, neutron, and deuteron, respectively. This is a one-dimensional problem, so we have

$$Mv_p - Mv_n = 2Mv_d.$$

The masses divide out:

$$v_p - v_n = 2v_d$$
$$7.0 \times 10^6 \text{ m/s} - 4.0 \times 10^6 \text{ m/s} = 2v_d$$
$$v_d = 1.5 \times 10^6 \text{ m/s}.$$

The velocity is thus $\vec{v}_d = \left(1.5 \times 10^6 \text{ m/s}\right)\hat{i}$.

Significance

This is essentially how particle colliders like the Large Hadron Collider work: They accelerate particles up to very high speeds (large momenta), but in opposite directions. This maximizes the creation of so-called "daughter particles."

Example 9.11

Ice Hockey 2

(This is a variation of an earlier example.)

Two ice hockey pucks of different masses are on a flat, horizontal hockey rink. The red puck has a mass of 15 grams, and is motionless; the blue puck has a mass of 12 grams, and is moving at 2.5 m/s to the left. It collides with the motionless red puck (**Figure 9.20**). If the collision is perfectly elastic, what are the final velocities of the two pucks?

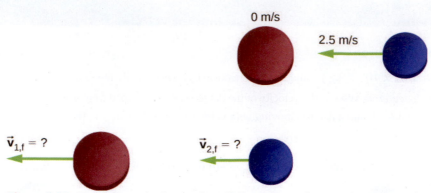

Figure 9.20 Two different hockey pucks colliding. The top diagram shows the pucks the instant before the collision, and the bottom diagram show the pucks the instant after the collision. The net external force is zero.

Strategy

We're told that we have two colliding objects, and we're told their masses and initial velocities, and one final velocity; we're asked for both final velocities. Conservation of momentum seems like a good strategy; define the system to be the two pucks. There is no friction, so we have a closed system. We have two unknowns (the two final velocities), but only one equation. The comment about the collision being perfectly elastic is the clue; it suggests that kinetic energy is also conserved in this collision. That gives us our second equation.

The initial momentum and initial kinetic energy of the system resides entirely and only in the second puck (the blue one); the collision transfers some of this momentum and energy to the first puck.

Solution

Conservation of momentum, in this case, reads

$$p_i = p_f$$
$$m_2 v_{2,i} = m_1 v_{1,f} + m_2 v_{2,f}.$$

Conservation of kinetic energy reads

$$K_i = K_f$$
$$\tfrac{1}{2} m_2 v_{2,i}^2 = \tfrac{1}{2} m_1 v_{1,f}^2 + \tfrac{1}{2} m_2 v_{2,f}^2.$$

There are our two equations in two unknowns. The algebra is tedious but not terribly difficult; you definitely should work it through. The solution is

$$v_{1,f} = \frac{(m_1 - m_2) v_{1,i} + 2 m_2 v_{2,i}}{m_1 + m_2}$$
$$v_{2f} = \frac{(m_2 - m_1) v_{2,i} + 2 m_1 v_{1,i}}{m_1 + m_2}.$$

Substituting the given numbers, we obtain

$$v_{1,f} = 2.22 \tfrac{m}{s}$$
$$v_{2,f} = -0.28 \tfrac{m}{s}.$$

Significance

Notice that after the collision, the blue puck is moving to the right; its direction of motion was reversed. The red puck is now moving to the left.

9.7 **Check Your Understanding** There is a second solution to the system of equations solved in this example (because the energy equation is quadratic): $v_{1,f} = -2.5$ m/s, $v_{2,f} = 0$. This solution is unacceptable on physical grounds; what's wrong with it?

Example 9.12

Thor vs. Iron Man

The 2012 movie "The Avengers" has a scene where Iron Man and Thor fight. At the beginning of the fight, Thor throws his hammer at Iron Man, hitting him and throwing him slightly up into the air and against a small tree, which breaks. From the video, Iron Man is standing still when the hammer hits him. The distance between Thor and Iron Man is approximately 10 m, and the hammer takes about 1 s to reach Iron Man after Thor releases it. The tree is about 2 m behind Iron Man, which he hits in about 0.75 s. Also from the video, Iron Man's trajectory to the tree is very close to horizontal. Assuming Iron Man's total mass is 200 kg:

a.   Estimate the mass of Thor's hammer

b.   Estimate how much kinetic energy was lost in this collision

Strategy

After the collision, Thor's hammer is in contact with Iron Man for the entire time, so this is a perfectly inelastic collision. Thus, with the correct choice of a closed system, we expect momentum is conserved, but not kinetic energy. We use the given numbers to estimate the initial momentum, the initial kinetic energy, and the final kinetic energy. Because this is a one-dimensional problem, we can go directly to the scalar form of the equations.

Solution

a.   First, we posit conservation of momentum. For that, we need a closed system. The choice here is the system (hammer + Iron Man), from the time of collision to the moment just before Iron Man and the hammer hit the tree. Let:

- $M_H = $ mass of the hammer

- $M_I = $ mass of Iron Man

- $v_H = $ velocity of the hammer before hitting Iron Man

- $v = $ combined velocity of Iron Man + hammer after the collision

Again, Iron Man's initial velocity was zero. Conservation of momentum here reads:

$$M_H v_H = (M_H + M_I)v.$$

We are asked to find the mass of the hammer, so we have

$$
\begin{aligned}
M_H v_H &= M_H v + M_I v \\
M_H (v_H - v) &= M_I v \\
M_H &= \frac{M_I v}{v_H - v} \\
&= \frac{(200 \text{ kg})\left(\frac{2 \text{ m}}{0.75 \text{ s}}\right)}{10 \frac{\text{m}}{\text{s}} - \left(\frac{2 \text{ m}}{0.75 \text{ s}}\right)} \\
&= 73 \text{ kg}.
\end{aligned}
$$

Considering the uncertainties in our estimates, this should be expressed with just one significant figure; thus, $M_H = 7 \times 10^1$ kg.

b.   The initial kinetic energy of the system, like the initial momentum, is all in the hammer:

$$
\begin{aligned}
K_i &= \tfrac{1}{2}M_H v_H^2 \\
&= \tfrac{1}{2}(70\ \text{kg})(10\ \text{m/s})^2 \\
&= 3500\ \text{J}.
\end{aligned}
$$

After the collision,

$$
\begin{aligned}
K_f &= \tfrac{1}{2}(M_H + M_I)v^2 \\
&= \tfrac{1}{2}(70\ \text{kg} + 200\ \text{kg})(2.67\ \text{m/s})^2 \\
&= 960\ \text{J}.
\end{aligned}
$$

Thus, there was a loss of $3500\ \text{J} - 960\ \text{J} = 2540\ \text{J}$.

Significance

From other scenes in the movie, Thor apparently can control the hammer's velocity with his mind. It is possible, therefore, that he mentally causes the hammer to maintain its initial velocity of 10 m/s while Iron Man is being driven backward toward the tree. If so, this would represent an external force on our system, so it would not be closed. Thor's mental control of his hammer is beyond the scope of this book, however.

Example 9.13

Analyzing a Car Crash

At a stoplight, a large truck (3000 kg) collides with a motionless small car (1200 kg). The truck comes to an instantaneous stop; the car slides straight ahead, coming to a stop after sliding 10 meters. The measured coefficient of friction between the car's tires and the road was 0.62. How fast was the truck moving at the moment of impact?

Strategy

At first it may seem we don't have enough information to solve this problem. Although we know the initial speed of the car, we don't know the speed of the truck (indeed, that's what we're asked to find), so we don't know the initial momentum of the system. Similarly, we know the final speed of the truck, but not the speed of the car immediately after impact. The fact that the car eventually slid to a speed of zero doesn't help with the final momentum, since an external friction force caused that. Nor can we calculate an impulse, since we don't know the collision time, or the amount of time the car slid before stopping. A useful strategy is to impose a restriction on the analysis.

Suppose we define a system consisting of just the truck and the car. The momentum of this system isn't conserved, because of the friction between the car and the road. But if we *could* find the speed of the car the instant after impact—before friction had any measurable effect on the car—then we could consider the momentum of the system to be conserved, with that restriction.

Can we find the final speed of the car? Yes; we invoke the work-kinetic energy theorem.

Solution

First, define some variables. Let:

- M_c and M_T be the masses of the car and truck, respectively

- $v_{T,i}$ and $v_{T,f}$ be the velocities of the truck before and after the collision, respectively

- $v_{c,i}$ and $v_{c,f}$ Z be the velocities of the car before and after the collision, respectively

- K_i and K_f be the kinetic energies of the car immediately after the collision, and after the car has stopped sliding (so $K_f = 0$).

- d be the distance the car slides after the collision before eventually coming to a stop.

Since we actually want the initial speed of the truck, and since the truck is not part of the work-energy calculation, let's start with conservation of momentum. For the car + truck system, conservation of momentum reads

$$p_i = p_f$$
$$M_c v_{c,i} + M_T v_{T,i} = M_c v_{c,f} + M_T v_{T,f}.$$

Since the car's initial velocity was zero, as was the truck's final velocity, this simplifies to

$$v_{T,i} = \frac{M_c}{M_T} v_{c,f}.$$

So now we need the car's speed immediately after impact. Recall that

$$W = \Delta K$$

where

$$\begin{aligned} \Delta K &= K_f - K_i \\ &= 0 - \frac{1}{2} M_c v_{c,f}^2. \end{aligned}$$

Also,

$$W = \vec{\mathbf{F}} \cdot \vec{\mathbf{d}} = Fd\cos\theta.$$

The work is done over the distance the car slides, which we've called d. Equating:

$$Fd\cos\theta = -\frac{1}{2} M_c v_{c,f}^2.$$

Friction is the force on the car that does the work to stop the sliding. With a level road, the friction force is

$$F = \mu_k M_c g.$$

Since the angle between the directions of the friction force vector and the displacement d is $180°$, and $\cos(180°) = -1$, we have

$$-(\mu_k M_c g)d = -\frac{1}{2} M_c v_{c,f}^2$$

(Notice that the car's mass divides out; evidently the mass of the car doesn't matter.)

Solving for the car's speed immediately after the collision gives

$$v_{c,f} = \sqrt{2\mu_k gd}.$$

Substituting the given numbers:

$$\begin{aligned} v_{c,f} &= \sqrt{2(0.62)\left(9.81 \frac{\text{m}}{\text{s}^2}\right)(10\,\text{m})}. \\ &= 11.0\,\text{m/s}. \end{aligned}$$

Now we can calculate the initial speed of the truck:

$$v_{T,i} = \left(\frac{1200\,\text{kg}}{3000\,\text{kg}}\right)\left(11.0 \frac{\text{m}}{\text{s}}\right) = 4.4\,\text{m/s}.$$

Significance

This is an example of the type of analysis done by investigators of major car accidents. A great deal of legal and financial consequences depend on an accurate analysis and calculation of momentum and energy.

9.8 Check Your Understanding Suppose there had been no friction (the collision happened on ice); that would make μ_k zero, and thus $v_{c,f} = \sqrt{2\mu_k g d} = 0$, which is obviously wrong. What is the mistake in this conclusion?

Subatomic Collisions and Momentum

Conservation of momentum is crucial to our understanding of atomic and subatomic particles because much of what we know about these particles comes from collision experiments.

At the beginning of the twentieth century, there was considerable interest in, and debate about, the structure of the atom. It was known that atoms contain two types of electrically charged particles: negatively charged electrons and positively charged protons. (The existence of an electrically neutral particle was suspected, but would not be confirmed until 1932.) The question was, how were these particles arranged in the atom? Were they distributed uniformly throughout the volume of the atom (as J.J. Thomson proposed), or arranged at the corners of regular polygons (which was Gilbert Lewis' model), or rings of negative charge that surround the positively charged nucleus—rather like the planetary rings surrounding Saturn (as suggested by Hantaro Nagaoka), or something else?

The New Zealand physicist Ernest Rutherford (along with the German physicist Hans Geiger and the British physicist Ernest Marsden) performed the crucial experiment in 1909. They bombarded a thin sheet of gold foil with a beam of high-energy (that is, high-speed) alpha-particles (the nucleus of a helium atom). The alpha-particles collided with the gold atoms, and their subsequent velocities were detected and analyzed, using conservation of momentum and conservation of energy.

If the charges of the gold atoms were distributed uniformly (per Thomson), then the alpha-particles should collide with them and nearly all would be deflected through many angles, all small; the Nagaoka model would produce a similar result. If the atoms were arranged as regular polygons (Lewis), the alpha-particles would deflect at a relatively small number of angles.

What *actually* happened is that nearly *none* of the alpha-particles were deflected. Those that were, were deflected at large angles, some close to $180°$ —those alpha-particles reversed direction completely (**Figure 9.21**). None of the existing atomic models could explain this. Eventually, Rutherford developed a model of the atom that was much closer to what we now have—again, using conservation of momentum and energy as his starting point.

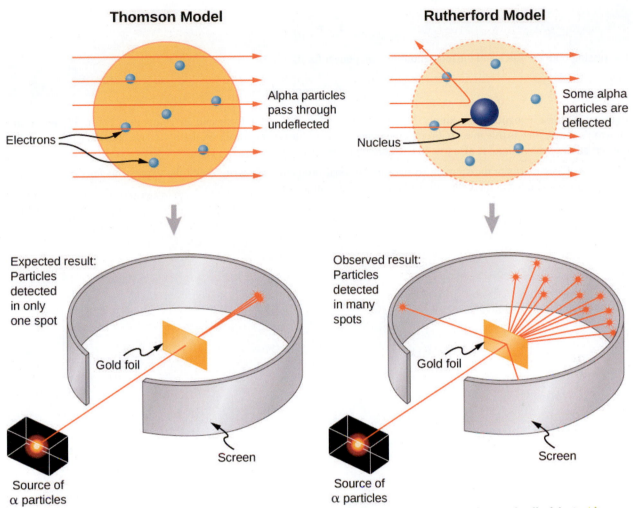

Figure 9.21 The Thomson and Rutherford models of the atom. The Thomson model predicted that nearly all of the incident alpha-particles would be scattered and at small angles. Rutherford and Geiger found that nearly none of the alpha particles were scattered, but those few that were deflected did so through very large angles. The results of Rutherford's experiments were inconsistent with the Thomson model. Rutherford used conservation of momentum and energy to develop a new, and better model of the atom—the nuclear model.

9.5 | Collisions in Multiple Dimensions

Learning Objectives

By the end of this section, you will be able to:

- Express momentum as a two-dimensional vector
- Write equations for momentum conservation in component form
- Calculate momentum in two dimensions, as a vector quantity

It is far more common for collisions to occur in two dimensions; that is, the angle between the initial velocity vectors is neither zero nor $180°$. Let's see what complications arise from this.

The first idea we need is that momentum is a vector; like all vectors, it can be expressed as a sum of perpendicular components (usually, though not always, an x-component and a y-component, and a z-component if necessary). Thus, when we write down the statement of conservation of momentum for a problem, our momentum vectors can be, and usually will be, expressed in component form.

The second idea we need comes from the fact that momentum is related to force:

$$\vec{F} = \frac{d\vec{p}}{dt}.$$

Expressing both the force and the momentum in component form,

$$F_x = \frac{dp_x}{dt}, \quad F_y = \frac{dp_y}{dt}, \quad F_z = \frac{dp_z}{dt}.$$

Remember, these equations are simply Newton's second law, in vector form and in component form. We know that Newton's second law is true in each direction, independently of the others. It follows therefore (via Newton's third law) that conservation of momentum is also true in each direction independently.

These two ideas motivate the solution to two-dimensional problems: We write down the expression for conservation of momentum twice: once in the x-direction and once in the y-direction.

$$
\begin{aligned}
p_{f,x} &= p_{1,i,x} + p_{2,i,x} \\
p_{f,y} &= p_{1,i,y} + p_{2,i,y}
\end{aligned}
\tag{9.18}
$$

This procedure is shown graphically in **Figure 9.22**.

Break initial momentum
into x and y components

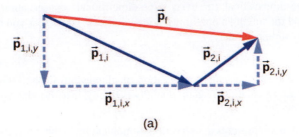

(a)

Add x and y components
to obtain x and y components
of final momentum

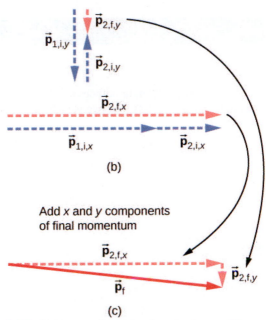

(b)

Add x and y components
of final momentum

(c)

Figure 9.22 (a) For two-dimensional momentum problems, break the initial momentum vectors into their x- and y-components. (b) Add the x- and y-components together separately. This gives you the x- and y-components of the final momentum, which are shown as red dashed vectors. (c) Adding these components together gives the final momentum.

We solve each of these two component equations independently to obtain the x- and y-components of the desired velocity vector:

$$v_{f,\,x} = \frac{m_1 v_{1,\text{i},\,x} + m_2 v_{2,\text{i},\,x}}{m}$$

$$v_{f,\,y} = \frac{m_1 v_{1,\text{i},\,y} + m_2 v_{2,\text{i},\,y}}{m}.$$

(Here, m represents the total mass of the system.) Finally, combine these components using the Pythagorean theorem,

$$v_f = \left| \vec{v}_f \right| = \sqrt{v_{f,\,x}^2 + v_{f,\,y}^2}.$$

Problem-Solving Strategy: Conservation of Momentum in Two Dimensions

The method for solving a two-dimensional (or even three-dimensional) conservation of momentum problem is generally the same as the method for solving a one-dimensional problem, except that you have to conserve momentum in both (or all three) dimensions simultaneously:

1. Identify a closed system.

2. Write down the equation that represents conservation of momentum in the *x*-direction, and solve it for the desired quantity. If you are calculating a vector quantity (velocity, usually), this will give you the *x*-component of the vector.

3. Write down the equation that represents conservation of momentum in the *y*-direction, and solve. This will give you the *y*-component of your vector quantity.

4. Assuming you are calculating a vector quantity, use the Pythagorean theorem to calculate its magnitude, using the results of steps 3 and 4.

Example 9.14

Traffic Collision

A small car of mass 1200 kg traveling east at 60 km/hr collides at an intersection with a truck of mass 3000 kg that is traveling due north at 40 km/hr (**Figure 9.23**). The two vehicles are locked together. What is the velocity of the combined wreckage?

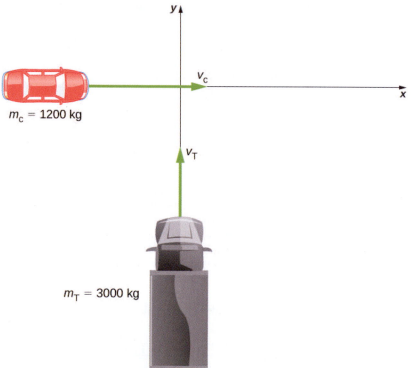

Figure 9.23 A large truck moving north is about to collide with a small car moving east. The final momentum vector has both *x*- and *y*-components.

Strategy

First off, we need a closed system. The natural system to choose is the (car + truck), but this system is not closed; friction from the road acts on both vehicles. We avoid this problem by restricting the question to finding the velocity at the instant just after the collision, so that friction has not yet had any effect on the system. With that restriction, momentum is conserved for this system.

Since there are two directions involved, we do conservation of momentum twice: once in the x-direction and once in the y-direction.

Solution

Before the collision the total momentum is

$$\vec{\mathbf{p}} = m_c \vec{\mathbf{v}}_c + m_T \vec{\mathbf{v}}_T.$$

After the collision, the wreckage has momentum

$$\vec{\mathbf{p}} = (m_c + m_T) \vec{\mathbf{v}}_w.$$

Since the system is closed, momentum must be conserved, so we have

$$m_c \vec{\mathbf{v}}_c + m_T \vec{\mathbf{v}}_T = (m_c + m_T) \vec{\mathbf{v}}_w.$$

We have to be careful; the two initial momenta are not parallel. We must add vectorially (**Figure 9.24**).

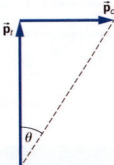

Figure 9.24 Graphical addition of momentum vectors. Notice that, although the car's velocity is larger than the truck's, its momentum is smaller.

If we define the $+x$-direction to point east and the $+y$-direction to point north, as in the figure, then (conveniently),

$$\vec{\mathbf{p}}_c = p_c \hat{\mathbf{i}} = m_c v_c \hat{\mathbf{i}}$$
$$\vec{\mathbf{p}}_T = p_T \hat{\mathbf{j}} = m_T v_T \hat{\mathbf{j}}.$$

Therefore, in the x-direction:

$$m_c v_c = (m_c + m_T) v_{w,\,x}$$
$$v_{w,\,x} = \left(\frac{m_c}{m_c + m_T}\right) v_c$$

and in the y-direction:

$$m_T v_T = (m_c + m_T) v_{w,\,y}$$
$$v_{w,\,y} = \left(\frac{m_T}{m_c + m_T}\right) v_T.$$

Applying the Pythagorean theorem gives

$$|\vec{\mathbf{v}}_w| = \sqrt{\left[\left(\frac{m_c}{m_c + m_t}\right)v_c\right]^2 + \left[\left(\frac{m_t}{m_c + m_t}\right)v_t\right]^2}$$

$$= \sqrt{\left[\left(\frac{1200\text{ kg}}{4200\text{ kg}}\right)\left(16.67\,\tfrac{\text{m}}{\text{s}}\right)\right]^2 + \left[\left(\frac{3000\text{ kg}}{4200\text{ kg}}\right)\left(11.1\,\tfrac{\text{m}}{\text{s}}\right)\right]^2}$$

$$= \sqrt{\left(4.76\,\tfrac{\text{m}}{\text{s}}\right)^2 + \left(7.93\,\tfrac{\text{m}}{\text{s}}\right)^2}$$

$$= 9.25\,\tfrac{\text{m}}{\text{s}} \approx 33.3\,\tfrac{\text{km}}{\text{hr}}.$$

As for its direction, using the angle shown in the figure,

$$\theta = \tan^{-1}\left(\frac{v_{w,\,x}}{v_{w,\,y}}\right) = \tan^{-1}\left(\frac{7.93\text{ m/s}}{4.76\text{ m/s}}\right) = 59°.$$

This angle is east of north, or $31°$ counterclockwise from the $+x$-direction.

Significance

As a practical matter, accident investigators usually work in the "opposite direction"; they measure the distance of skid marks on the road (which gives the stopping distance) and use the work-energy theorem along with conservation of momentum to determine the speeds and directions of the cars prior to the collision. We saw that analysis in an earlier section.

9.9 Check Your Understanding Suppose the initial velocities were *not* at right angles to each other. How would this change both the physical result and the mathematical analysis of the collision?

Example 9.15

Exploding Scuba Tank

A common scuba tank is an aluminum cylinder that weighs 31.7 pounds empty (**Figure 9.25**). When full of compressed air, the internal pressure is between 2500 and 3000 psi (pounds per square inch). Suppose such a tank, which had been sitting motionless, suddenly explodes into three pieces. The first piece, weighing 10 pounds, shoots off horizontally at 235 miles per hour; the second piece (7 pounds) shoots off at 172 miles per hour, also in the horizontal plane, but at a $19°$ angle to the first piece. What is the mass and initial velocity of the third piece? (Do all work, and express your final answer, in SI units.)

Figure 9.25 A scuba tank explodes into three pieces.

Strategy

To use conservation of momentum, we need a closed system. If we define the system to be the scuba tank, this is not a closed system, since gravity is an external force. However, the problem asks for the just the initial velocity of the third piece, so we can neglect the effect of gravity and consider the tank by itself as a closed system. Notice that, for this system, the initial momentum vector is zero.

We choose a coordinate system where all the motion happens in the *xy*-plane. We then write down the equations for conservation of momentum in each direction, thus obtaining the *x*- and *y*-components of the momentum of the third piece, from which we obtain its magnitude (via the Pythagorean theorem) and its direction. Finally, dividing this momentum by the mass of the third piece gives us the velocity.

Solution

First, let's get all the conversions to SI units out of the way:

$$31.7 \text{ lb} \times \frac{1 \text{ kg}}{2.2 \text{ lb}} \rightarrow 14.4 \text{ kg}$$

$$10 \text{ lb} \rightarrow 4.5 \text{ kg}$$
$$235 \frac{\text{miles}}{\text{hour}} \times \frac{1 \text{ hour}}{3600 \text{ s}} \times \frac{1609 \text{ m}}{\text{mile}} = 105 \frac{\text{m}}{\text{s}}$$

$$7 \text{ lb} \rightarrow 3.2 \text{ kg}$$
$$172 \frac{\text{mile}}{\text{hour}} = 77 \frac{\text{m}}{\text{s}}$$
$$m_3 = 14.4 \text{ kg} - (4.5 \text{ kg} + 3.2 \text{ kg}) = 6.7 \text{ kg}.$$

Now apply conservation of momentum in each direction.

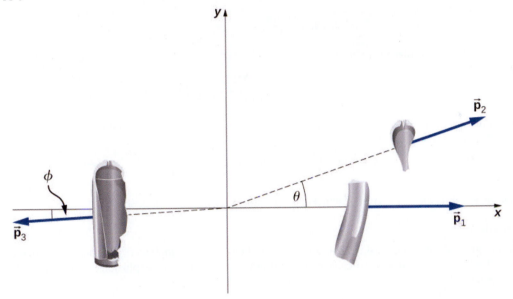

x-direction:

$$
\begin{aligned}
p_{f,x} &= p_{0,x} \\
p_{1,x} + p_{2,x} + p_{3,x} &= 0 \\
m_1 v_{1,x} + m_2 v_{2,x} + p_{3,x} &= 0 \\
p_{3,x} &= -m_1 v_{1,x} - m_2 v_{2,x}
\end{aligned}
$$

y-direction:

$$
\begin{aligned}
p_{f,y} &= p_{0,y} \\
p_{1,y} + p_{2,y} + p_{3,y} &= 0 \\
m_1 v_{1,y} + m_2 v_{2,y} + p_{3,y} &= 0 \\
p_{3,y} &= -m_1 v_{1,y} - m_2 v_{2,y}
\end{aligned}
$$

From our chosen coordinate system, we write the *x*-components as

$$p_{3,x} = -m_1 v_1 - m_2 v_2 \cos\theta$$
$$= -(14.5 \text{ kg})(105 \tfrac{\text{m}}{\text{s}}) - (4.5 \text{ kg})(77 \tfrac{\text{m}}{\text{s}})\cos(19°)$$
$$= -1850 \tfrac{\text{kg} \cdot \text{m}}{\text{s}}.$$

For the y-direction, we have

$$p_{3y} = 0 - m_2 v_2 \sin\theta$$
$$= -(4.5 \text{ kg})(77 \tfrac{\text{m}}{\text{s}})\sin(19°)$$
$$= -113 \tfrac{\text{kg} \cdot \text{m}}{\text{s}}.$$

This gives the magnitude of p_3:

$$p_3 = \sqrt{p_{3,x}^2 + p_{3,y}^2}$$
$$= \sqrt{\left(-1850 \tfrac{\text{kg} \cdot \text{m}}{\text{s}}\right)^2 + \left(-113 \tfrac{\text{kg} \cdot \text{m}}{\text{s}}\right)}$$
$$= 1854 \tfrac{\text{kg} \cdot \text{m}}{\text{s}}.$$

The velocity of the third piece is therefore

$$v_3 = \frac{p_3}{m_3} = \frac{1854 \tfrac{\text{kg} \cdot \text{m}}{\text{s}}}{6.7 \text{ kg}} = 277 \tfrac{\text{m}}{\text{s}}.$$

The direction of its velocity vector is the same as the direction of its momentum vector:

$$\phi = \tan^{-1}\left(\frac{p_{3,y}}{p_{3,x}}\right) = \tan^{-1}\left(\frac{113 \tfrac{\text{kg} \cdot \text{m}}{\text{s}}}{1850 \tfrac{\text{kg} \cdot \text{m}}{\text{s}}}\right) = 3.5°.$$

Because ϕ is below the $-x$-axis, the actual angle is $183.5°$ from the $+x$-direction.

Significance

The enormous velocities here are typical; an exploding tank of any compressed gas can easily punch through the wall of a house and cause significant injury, or death. Fortunately, such explosions are extremely rare, on a percentage basis.

 9.10 Check Your Understanding Notice that the mass of the air in the tank was neglected in the analysis and solution. How would the solution method changed if the air was included? How large a difference do you think it would make in the final answer?

9.6 | Center of Mass

Learning Objectives

By the end of this section, you will be able to:

- Explain the meaning and usefulness of the concept of center of mass
- Calculate the center of mass of a given system
- Apply the center of mass concept in two and three dimensions
- Calculate the velocity and acceleration of the center of mass

We have been avoiding an important issue up to now: When we say that an object moves (more correctly, accelerates) in a way that obeys Newton's second law, we have been ignoring the fact that all objects are actually made of many constituent particles. A car has an engine, steering wheel, seats, passengers; a football is leather and rubber surrounding air; a brick is made of atoms. There are many different types of particles, and they are generally not distributed uniformly in the object. How do we include these facts into our calculations?

Then too, an extended object might change shape as it moves, such as a water balloon or a cat falling (**Figure 9.26**). This implies that the constituent particles are applying internal forces on each other, in addition to the external force that is acting on the object as a whole. We want to be able to handle this, as well.

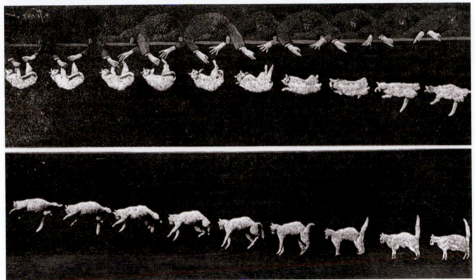

Figure 9.26 As the cat falls, its body performs complicated motions so it can land on its feet, but one point in the system moves with the simple uniform acceleration of gravity.

The problem before us, then, is to determine what part of an extended object is obeying Newton's second law when an external force is applied and to determine how the motion of the object as a whole is affected by both the internal and external forces.

Be warned: To treat this new situation correctly, we must be rigorous and completely general. We won't make any assumptions about the nature of the object, or of its constituent particles, or either the internal or external forces. Thus, the arguments will be complex.

Internal and External Forces

Suppose we have an extended object of mass M, made of N interacting particles. Let's label their masses as m_j, where $j = 1, 2, 3, \ldots, N$. Note that

$$M = \sum_{j=1}^{N} m_j.$$

(9.19)

If we apply some net **external force** $\vec{F}_{ext}$ on the object, every particle experiences some "share" or some fraction of that external force. Let:

$$\vec{f}_j^{\,ext} = \text{the fraction of the external force that the } j\text{th particle experiences.}$$

Notice that these fractions of the total force are not necessarily equal; indeed, they virtually never are. (They *can* be, but they usually aren't.) In general, therefore,

$$\vec{f}_1^{\,ext} \neq \vec{f}_2^{\,ext} \neq \cdots \neq \vec{f}_N^{\,ext}.$$

Next, we assume that each of the particles making up our object can interact (apply forces on) every other particle of the object. We won't try to guess what kind of forces they are; but since these forces are the result of particles of the object acting on other particles of the same object, we refer to them as **internal forces** $\vec{f}_j^{\,int}$; thus:

$\vec{\mathbf{f}}_j^{\,\text{int}} = $ the net internal force that the jth particle experiences from all the other particles that make up the object.

Now, the *net* force, internal plus external, on the jth particle is the vector sum of these:

$$\vec{\mathbf{f}}_j = \vec{\mathbf{f}}_j^{\,\text{int}} + \vec{\mathbf{f}}_j^{\,\text{ext}}. \tag{9.20}$$

where again, this is for all N particles; $j = 1, 2, 3, \ldots, N$.

As a result of this fractional force, the momentum of each particle gets changed:

$$\vec{\mathbf{f}}_j = \frac{d\vec{\mathbf{p}}_j}{dt} \tag{9.21}$$

$$\vec{\mathbf{f}}_j^{\,\text{int}} + \vec{\mathbf{f}}_j^{\,\text{ext}} = \frac{d\vec{\mathbf{p}}_j}{dt}.$$

The net force $\vec{\mathbf{F}}$ on the *object* is the vector sum of these forces:

$$\vec{\mathbf{F}}_{\text{net}} = \sum_{j=1}^{N}\left(\vec{\mathbf{f}}_j^{\,\text{int}} + \vec{\mathbf{f}}_j^{\,\text{ext}}\right) \tag{9.22}$$

$$= \sum_{j=1}^{N}\vec{\mathbf{f}}_j^{\,\text{int}} + \sum_{j=1}^{N}\vec{\mathbf{f}}_j^{\,\text{ext}}.$$

This net force changes the momentum of the object as a whole, and the net change of momentum of the object must be the vector sum of all the individual changes of momentum of all of the particles:

$$\vec{\mathbf{F}}_{\text{net}} = \sum_{j=1}^{N}\frac{d\vec{\mathbf{p}}_j}{dt}. \tag{9.23}$$

Combining **Equation 9.22** and **Equation 9.23** gives

$$\sum_{j=1}^{N}\vec{\mathbf{f}}_j^{\,\text{int}} + \sum_{j=1}^{N}\vec{\mathbf{f}}_j^{\,\text{ext}} = \sum_{j=1}^{N}\frac{d\vec{\mathbf{p}}_j}{dt}. \tag{9.24}$$

Let's now think about these summations. First consider the internal forces term; remember that each $\vec{\mathbf{f}}_j^{\,\text{int}}$ is the force on the jth particle from the other particles in the object. But by Newton's third law, for every one of these forces, there must be another force that has the same magnitude, but the opposite sign (points in the opposite direction). These forces do not cancel; however, that's not what we're doing in the summation. Rather, we're simply *mathematically adding up* all the internal force vectors. That is, in general, the internal forces for any individual part of the object won't cancel, but when all the internal forces are added up, the internal forces must cancel in pairs. It follows, therefore, that the sum of all the internal forces must be zero:

$$\sum_{j=1}^{N}\vec{\mathbf{f}}_j^{\,\text{int}} = 0.$$

(This argument is subtle, but crucial; take plenty of time to completely understand it.)

For the external forces, this summation is simply the total external force that was applied to the whole object:

$$\sum_{j=1}^{N}\vec{\mathbf{f}}_j^{\,\text{ext}} = \vec{\mathbf{F}}_{\text{ext}}.$$

As a result,

$$\vec{\mathbf{F}}_{\text{ext}} = \sum_{j=1}^{N}\frac{d\vec{\mathbf{p}}_j}{dt}. \tag{9.25}$$

This is an important result. **Equation 9.25** tells us that the total change of momentum of the entire object (all N particles) is due only to the external forces; the internal forces do not change the momentum of the object as a whole. This is why you can't lift yourself in the air by standing in a basket and pulling up on the handles: For the system of you + basket, your upward pulling force is an internal force.

Force and Momentum

Remember that our actual goal is to determine the equation of motion for the entire object (the entire system of particles). To that end, let's define:

$\vec{p}_{CM} = $ the total momentum of the system of N particles (the reason for the subscript will become clear shortly)

Then we have

$$\vec{p}_{CM} \equiv \sum_{j=1}^{N} \vec{p}_j,$$

and therefore **Equation 9.25** can be written simply as

$$\vec{F} = \frac{d\vec{p}_{CM}}{dt}. \tag{9.26}$$

Since this change of momentum is caused by only the net external force, we have dropped the "ext" subscript.

This is Newton's second law, but now for the entire extended object. If this feels a bit anticlimactic, remember what is hiding inside it: $\vec{p}_{CM}$ is the vector sum of the momentum of (in principle) hundreds of thousands of billions of billions of particles (6.02×10^{23}), all caused by one simple net external force—a force that you can calculate.

Center of Mass

Our next task is to determine what part of the extended object, if any, is obeying **Equation 9.26**.

It's tempting to take the next step; does the following equation mean anything?

$$\vec{F} = M\vec{a} \tag{9.27}$$

If it *does* mean something (acceleration of what, exactly?), then we could write

$$M\vec{a} = \frac{d\vec{p}_{CM}}{dt}$$

and thus

$$M\vec{a} = \sum_{j=1}^{N} \frac{d\vec{p}_j}{dt} = \frac{d}{dt}\sum_{j=1}^{N} \vec{p}_j.$$

which follows because the derivative of a sum is equal to the sum of the derivatives.

Now, $\vec{p}_j$ is the momentum of the jth particle. Defining the positions of the constituent particles (relative to some coordinate system) as $\vec{r}_j = (x_j, y_j, z_j)$, we thus have

$$\vec{p}_j = m_j\vec{v}_j = m_j\frac{d\vec{r}_j}{dt}.$$

Substituting back, we obtain

$$M\overrightarrow{\mathbf{a}} = \frac{d}{dt}\sum_{j=1}^{N}m_j\frac{d\overrightarrow{\mathbf{r}}_j}{dt}$$

$$= \frac{d^2}{dt^2}\sum_{j=1}^{N}m_j\overrightarrow{\mathbf{r}}_j.$$

Dividing both sides by M (the total mass of the extended object) gives us

$$\overrightarrow{\mathbf{a}} = \frac{d^2}{dt^2}\left(\frac{1}{M}\sum_{j=1}^{N}m_j\overrightarrow{\mathbf{r}}_j\right). \tag{9.28}$$

Thus, the point in the object that traces out the trajectory dictated by the applied force in **Equation 9.27** is inside the parentheses in **Equation 9.28**.

Looking at this calculation, notice that (inside the parentheses) we are calculating the product of each particle's mass with its position, adding all N of these up, and dividing this sum by the total mass of particles we summed. This is reminiscent of an average; inspired by this, we'll (loosely) interpret it to be the weighted average position of the mass of the extended object. It's actually called the **center of mass** of the object. Notice that the position of the center of mass has units of meters; that suggests a definition:

$$\overrightarrow{\mathbf{r}}_{\text{CM}} \equiv \frac{1}{M}\sum_{j=1}^{N}m_j\overrightarrow{\mathbf{r}}_j. \tag{9.29}$$

So, the point that obeys **Equation 9.26** (and therefore **Equation 9.27** as well) is the center of mass of the object, which is located at the position vector $\overrightarrow{\mathbf{r}}_{\text{CM}}$.

It may surprise you to learn that there does not have to be any actual mass at the center of mass of an object. For example, a hollow steel sphere with a vacuum inside it is spherically symmetrical (meaning its mass is uniformly distributed about the center of the sphere); all of the sphere's mass is out on its surface, with no mass inside. But it can be shown that the center of mass of the sphere is at its geometric center, which seems reasonable. Thus, there is no mass at the position of the center of mass of the sphere. (Another example is a doughnut.) The procedure to find the center of mass is illustrated in **Figure 9.27**.

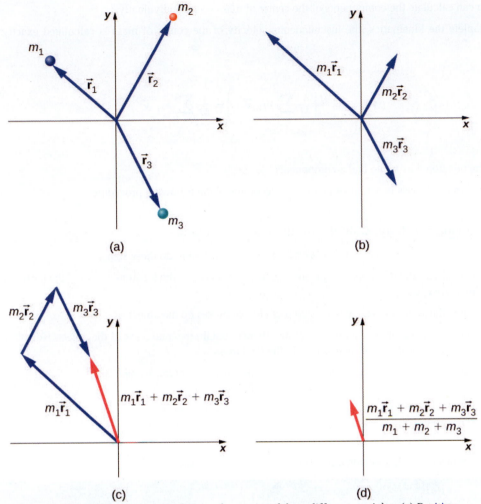

Figure 9.27 Finding the center of mass of a system of three different particles. (a) Position vectors are created for each object. (b) The position vectors are multiplied by the mass of the corresponding object. (c) The scaled vectors from part (b) are added together. (d) The final vector is divided by the total mass. This vector points to the center of mass of the system. Note that no mass is actually present at the center of mass of this system.

Since $\vec{r}_j = x_j \hat{i} + y_j \hat{j} + z_j \hat{k}$, it follows that:

$$r_{CM,x} = \frac{1}{M} \sum_{j=1}^{N} m_j x_j \tag{9.30}$$

$$r_{CM,y} = \frac{1}{M} \sum_{j=1}^{N} m_j y_j \tag{9.31}$$

$$r_{CM,z} = \frac{1}{M} \sum_{j=1}^{N} m_j z_j \tag{9.32}$$

and thus

$$\vec{r}_{CM} = r_{CM,x} \hat{i} + r_{CM,y} \hat{j} + r_{CM,z} \hat{k}$$

$$r_{CM} = |\vec{r}_{CM}| = \left(r_{CM,x}^2 + r_{CM,y}^2 + r_{CM,z}^2 \right)^{1/2}.$$

Therefore, you can calculate the components of the center of mass vector individually.

Finally, to complete the kinematics, the instantaneous velocity of the center of mass is calculated exactly as you might suspect:

$$\vec{v}_{CM} = \frac{d}{dt}\left(\frac{1}{M}\sum_{j=1}^{N} m_j \vec{r}_j\right) = \frac{1}{M}\sum_{j=1}^{N} m_j \vec{v}_j \qquad (9.33)$$

and this, like the position, has x-, y-, and z-components.

To calculate the center of mass in actual situations, we recommend the following procedure:

Problem-Solving Strategy: Calculating the Center of Mass

The center of mass of an object is a position vector. Thus, to calculate it, do these steps:

1. Define your coordinate system. Typically, the origin is placed at the location of one of the particles. This is not required, however.

2. Determine the x, y, z-coordinates of each particle that makes up the object.

3. Determine the mass of each particle, and sum them to obtain the total mass of the object. Note that the mass of the object at the origin *must* be included in the total mass.

4. Calculate the x-, y-, and z-components of the center of mass vector, using **Equation 9.30**, **Equation 9.31**, and **Equation 9.32**.

5. If required, use the Pythagorean theorem to determine its magnitude.

Here are two examples that will give you a feel for what the center of mass is.

Example 9.16

Center of Mass of the Earth-Moon System

Using data from text appendix, determine how far the center of mass of the Earth-moon system is from the center of Earth. Compare this distance to the radius of Earth, and comment on the result. Ignore the other objects in the solar system.

Strategy

We get the masses and separation distance of the Earth and moon, impose a coordinate system, and use **Equation 9.29** with just $N = 2$ objects. We use a subscript "e" to refer to Earth, and subscript "m" to refer to the moon.

Solution

Define the origin of the coordinate system as the center of Earth. Then, with just two objects, **Equation 9.29** becomes

$$R = \frac{m_e r_e + m_m r_m}{m_e + m_m}.$$

From **Appendix D**,

$$m_e = 5.97 \times 10^{24} \text{ kg}$$

$$m_m = 7.36 \times 10^{22} \text{ kg}$$

$$r_m = 3.82 \times 10^{8} \text{ m}.$$

We defined the center of Earth as the origin, so $r_e = 0$ m. Inserting these into the equation for R gives

$$R = \frac{(5.97 \times 10^{24} \text{ kg})(0 \text{ m}) + (7.36 \times 10^{22} \text{ kg})(3.82 \times 10^8 \text{ m})}{5.97 \times 10^{24} \text{ kg} + 7.36 \times 10^{22} \text{ kg}}$$

$$= 4.64 \times 10^6 \text{ m}.$$

Significance

The radius of Earth is 6.37×10^6 m, so the center of mass of the Earth-moon system is $(6.37 - 4.64)$ $\times 10^6$ m $= 1.73 \times 10^6$ m $= 1730$ km (roughly 1080 miles) *below* the surface of Earth. The location of the center of mass is shown (not to scale).

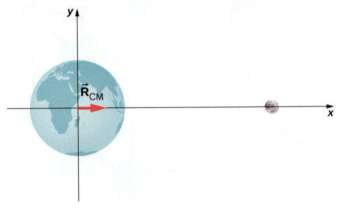

9.11 **Check Your Understanding** Suppose we included the sun in the system. Approximately where would the center of mass of the Earth-moon-sun system be located? (Feel free to actually calculate it.)

Example 9.17

Center of Mass of a Salt Crystal

Figure 9.28 shows a single crystal of sodium chloride—ordinary table salt. The sodium and chloride ions form a single unit, NaCl. When multiple NaCl units group together, they form a cubic lattice. The smallest possible cube (called the *unit cell*) consists of four sodium ions and four chloride ions, alternating. The length of one edge of this cube (i.e., the bond length) is 2.36×10^{-10} m. Find the location of the center of mass of the unit cell. Specify it either by its coordinates $(r_{CM,x}, r_{CM,y}, r_{CM,z})$, or by r_{CM} and two angles.

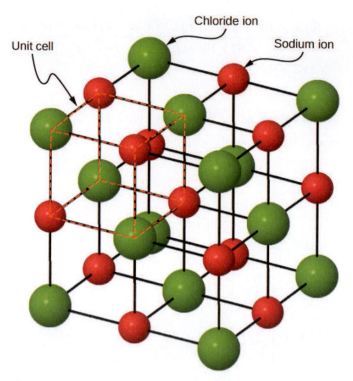

Figure 9.28 A drawing of a sodium chloride (NaCl) crystal.

Strategy

We can look up all the ion masses. If we impose a coordinate system on the unit cell, this will give us the positions of the ions. We can then apply **Equation 9.30**, **Equation 9.31**, and **Equation 9.32** (along with the Pythagorean theorem).

Solution

Define the origin to be at the location of the chloride ion at the bottom left of the unit cell. **Figure 9.29** shows the coordinate system.

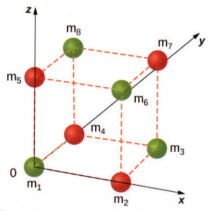

Figure 9.29 A single unit cell of a NaCl crystal.

There are eight ions in this crystal, so $N = 8$:

$$\vec{r}_{CM} = \frac{1}{M} \sum_{j=1}^{8} m_j \vec{r}_j.$$

The mass of each of the chloride ions is

$$35.453\text{u} \times \frac{1.660 \times 10^{-27} \text{ kg}}{\text{u}} = 5.885 \times 10^{-26} \text{ kg}$$

so we have

$$m_1 = m_3 = m_6 = m_8 = 5.885 \times 10^{-26} \text{ kg}.$$

For the sodium ions,

$$m_2 = m_4 = m_5 = m_7 = 3.816 \times 10^{-26} \text{ kg}.$$

The total mass of the unit cell is therefore

$$M = (4)(5.885 \times 10^{-26} \text{ kg}) + (4)(3.816 \times 10^{-26} \text{ kg}) = 3.880 \times 10^{-25} \text{ kg}.$$

From the geometry, the locations are

$$\vec{r}_1 = 0$$

$$\vec{r}_2 = (2.36 \times 10^{-10} \text{ m})\hat{i}$$

$$\vec{r}_3 = r_{3x}\hat{i} + r_{3y}\hat{j} = (2.36 \times 10^{-10} \text{ m})\hat{i} + (2.36 \times 10^{-10} \text{ m})\hat{j}$$

$$\vec{r}_4 = (2.36 \times 10^{-10} \text{ m})\hat{j}$$

$$\vec{r}_5 = (2.36 \times 10^{-10} \text{ m})\vec{k}$$

$$\vec{r}_6 = r_{6x}\hat{i} + r_{6z}\hat{k} = (2.36 \times 10^{-10} \text{ m})\hat{i} + (2.36 \times 10^{-10} \text{ m})\hat{k}$$

$$\vec{r}_7 = r_{7x}\hat{i} + r_{7y}\hat{j} + r_{7z}\hat{k} = (2.36 \times 10^{-10} \text{ m})\hat{i} + (2.36 \times 10^{-10} \text{ m})\hat{j} + (2.36 \times 10^{-10} \text{ m})\hat{k}$$

$$\vec{r}_8 = r_{8y}\hat{j} + r_{8z}\hat{k} = (2.36 \times 10^{-10} \text{ m})\hat{j} + (2.36 \times 10^{-10} \text{ m})\hat{k}.$$

Substituting:

$$|\vec{r}_{CM,x}| = \sqrt{r_{CM,x}^2 + r_{CM,y}^2 + r_{CM,z}^2}$$

$$= \frac{1}{M}\sum_{j=1}^{8} m_j(r_x)_j$$

$$= \frac{1}{M}(m_1 r_{1x} + m_2 r_{2x} + m_3 r_{3x} + m_4 r_{4x} + m_5 r_{5x} + m_6 r_{6x} + m_7 r_{7x} + m_8 r_{8x})$$

$$= \frac{1}{3.8804 \times 10^{-25} \text{ kg}}[(5.885 \times 10^{-26} \text{ kg})(0 \text{ m}) + (3.816 \times 10^{-26} \text{ kg})(2.36 \times 10^{-10} \text{ m})$$

$$+ (5.885 \times 10^{-26} \text{ kg})(2.36 \times 10^{-10} \text{ m})$$

$$+ (3.816 \times 10^{-26} \text{ kg})(2.36 \times 10^{-10} \text{ m}) + 0 + 0$$

$$+ (3.816 \times 10^{-26} \text{ kg})(2.36 \times 10^{-10} \text{ m}) + 0]$$

$$= 1.18 \times 10^{-10} \text{ m}.$$

Similar calculations give $r_{CM,y} = r_{CM,z} = 1.18 \times 10^{-10}$ m (you could argue that this must be true, by symmetry, but it's a good idea to check).

Significance

Although this is a great exercise to determine the center of mass given a Chloride ion at the origin, in fact the origin could be chosen at any location. Therefore, there is no meaningful application of the center of mass of a unit cell beyond as an exercise.

9.12   Check Your Understanding Suppose you have a macroscopic salt crystal (that is, a crystal that is large enough to be visible with your unaided eye). It is made up of a *huge* number of unit cells. Is the center of mass of this crystal necessarily at the geometric center of the crystal?

Two crucial concepts come out of these examples:

1. As with all problems, you must define your coordinate system and origin. For center-of-mass calculations, it often makes sense to choose your origin to be located at one of the masses of your system. That choice automatically defines its distance in **Equation 9.29** to be zero. However, you must still include the mass of the object at your origin in your calculation of M, the total mass **Equation 9.19**. In the Earth-moon system example, this means including the mass of Earth. If you hadn't, you'd have ended up with the center of mass of the system being at the center of the moon, which is clearly wrong.

2. In the second example (the salt crystal), notice that there is no mass at all at the location of the center of mass. This is an example of what we stated above, that there does not have to be any actual mass at the center of mass of an object.

Center of Mass of Continuous Objects

If the object in question has its mass distributed uniformly in space, rather than as a collection of discrete particles, then $m_j \rightarrow dm$, and the summation becomes an integral:

$$\vec{\mathbf{r}}_{CM} = \frac{1}{M} \int \vec{\mathbf{r}} \, dm. \tag{9.34}$$

In this context, r is a characteristic dimension of the object (the radius of a sphere, the length of a long rod). To generate an integrand that can actually be calculated, you need to express the differential mass element dm as a function of the mass density of the continuous object, and the dimension r. An example will clarify this.

Example 9.18

CM of a Uniform Thin Hoop

Find the center of mass of a uniform thin hoop (or ring) of mass M and radius r.

Strategy

First, the hoop's symmetry suggests the center of mass should be at its geometric center. If we define our coordinate system such that the origin is located at the center of the hoop, the integral should evaluate to zero.

We replace dm with an expression involving the density of the hoop and the radius of the hoop. We then have an expression we can actually integrate. Since the hoop is described as "thin," we treat it as a one-dimensional object, neglecting the thickness of the hoop. Therefore, its density is expressed as the number of kilograms of material per meter. Such a density is called a **linear mass density**, and is given the symbol λ; this is the Greek letter "lambda," which is the equivalent of the English letter "l" (for "linear").

Since the hoop is described as uniform, this means that the linear mass density λ is constant. Thus, to get our expression for the differential mass element dm, we multiply λ by a differential length of the hoop, substitute, and integrate (with appropriate limits for the definite integral).

Solution

First, define our coordinate system and the relevant variables (**Figure 9.30**).

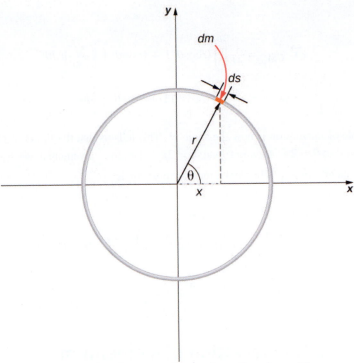

Figure 9.30 Finding the center of mass of a uniform hoop. We express the coordinates of a differential piece of the hoop, and then integrate around the hoop.

The center of mass is calculated with **Equation 9.34**:

$$\overrightarrow{\mathbf{r}}_{CM} = \frac{1}{M}\int_a^b \overrightarrow{\mathbf{r}}\ dm.$$

We have to determine the limits of integration a and b. Expressing $\overrightarrow{\mathbf{r}}$ in component form gives us

$$\overrightarrow{\mathbf{r}}_{CM} = \frac{1}{M}\int_a^b \left[(r\cos\theta)\ \hat{\mathbf{i}} + (r\sin\theta)\ \hat{\mathbf{j}}\right]dm.$$

In the diagram, we highlighted a piece of the hoop that is of differential length ds; it therefore has a differential mass $dm = \lambda ds$. Substituting:

$$\overrightarrow{\mathbf{r}}_{CM} = \frac{1}{M}\int_a^b \left[(r\cos\theta)\ \hat{\mathbf{i}} + (r\sin\theta)\ \hat{\mathbf{j}}\right]\lambda ds.$$

However, the arc length ds subtends a differential angle $d\theta$, so we have

$$ds = rd\theta$$

and thus

$$\overrightarrow{\mathbf{r}}_{CM} = \frac{1}{M}\int_a^b \left[(r\cos\theta)\ \hat{\mathbf{i}} + (r\sin\theta)\ \hat{\mathbf{j}}\right]\lambda rd\theta.$$

One more step: Since λ is the linear mass density, it is computed by dividing the total mass by the length of the hoop:

$$\lambda = \frac{M}{2\pi r}$$

giving us

$$\vec{\mathbf{r}}_{CM} = \frac{1}{M}\int_a^b\left[(r\cos\theta)\,\hat{\mathbf{i}} + (r\sin\theta)\,\hat{\mathbf{j}}\right]\left(\frac{M}{2\pi r}\right)r\,d\theta$$

$$= \frac{1}{2\pi}\int_a^b\left[(r\cos\theta)\,\hat{\mathbf{i}} + (r\sin\theta)\,\hat{\mathbf{j}}\right]d\theta.$$

Notice that the variable of integration is now the angle θ. This tells us that the limits of integration (around the circular hoop) are $\theta = 0$ to $\theta = 2\pi$, so $a = 0$ and $b = 2\pi$. Also, for convenience, we separate the integral into the x- and y-components of $\vec{\mathbf{r}}_{CM}$. The final integral expression is

$$\vec{\mathbf{r}}_{CM} = r_{CM,x}\,\hat{\mathbf{i}} + r_{CM,y}\,\hat{\mathbf{j}}$$

$$= \left[\frac{1}{2\pi}\int_0^{2\pi}(r\cos\theta)d\theta\right]\hat{\mathbf{i}} + \left[\frac{1}{2\pi}\int_0^{2\pi}(r\sin\theta)d\theta\right]\hat{\mathbf{j}}$$

$$= 0\,\hat{\mathbf{i}} + 0\,\hat{\mathbf{j}} = \vec{\mathbf{0}}$$

as expected.

Center of Mass and Conservation of Momentum

How does all this connect to conservation of momentum?

Suppose you have N objects with masses $m_1, m_2, m_3, ...m_N$ and initial velocities $\vec{\mathbf{v}}_1, \vec{\mathbf{v}}_2, \vec{\mathbf{v}}_3, ..., \vec{\mathbf{v}}_N$. The center of mass of the objects is

$$\vec{\mathbf{r}}_{CM} = \frac{1}{M}\sum_{j=1}^{N}m_j\,\vec{\mathbf{r}}_j.$$

Its velocity is

$$\vec{\mathbf{v}}_{CM} = \frac{d\,\vec{\mathbf{r}}_{CM}}{dt} = \frac{1}{M}\sum_{j=1}^{N}m_j\frac{d\,\vec{\mathbf{r}}_j}{dt} \tag{9.35}$$

and thus the initial momentum of the center of mass is

$$\left[M\frac{d\,\vec{\mathbf{r}}_{CM}}{dt}\right]_i = \sum_{j=1}^{N}m_j\frac{d\,\vec{\mathbf{r}}_{j,\,i}}{dt}$$

$$M\,\vec{\mathbf{v}}_{CM,i} = \sum_{j=1}^{N}m_j\,\vec{\mathbf{v}}_{j,\,i}.$$

After these masses move and interact with each other, the momentum of the center of mass is

$$M\,\vec{\mathbf{v}}_{CM,f} = \sum_{j=1}^{N}m_j\,\vec{\mathbf{v}}_{j,\,f}.$$

But conservation of momentum tells us that the right-hand side of both equations must be equal, which says

$$M\,\vec{\mathbf{v}}_{CM,f} = M\,\vec{\mathbf{v}}_{CM,i}. \tag{9.36}$$

This result implies that conservation of momentum is expressed in terms of the center of mass of the system. Notice that as an object moves through space with no net external force acting on it, an individual particle of the object may accelerate in various directions, with various magnitudes, depending on the net internal force acting on that object at any time. (Remember, it is only the vector sum of all the internal forces that vanishes, not the internal force on a single particle.) Thus, such a particle's momentum will not be constant—but the momentum of the entire extended object will be, in accord with **Equation 9.36**.

Equation 9.36 implies another important result: Since M represents the mass of the entire system of particles, it is necessarily constant. (If it isn't, we don't have a closed system, so we can't expect the system's momentum to be conserved.) As a result, **Equation 9.36** implies that, for a closed system,

$$\vec{v}_{CM,f} = \vec{v}_{CM,i}. \tag{9.37}$$

That is to say, *in the absence of an external force, the velocity of the center of mass never changes*.

You might be tempted to shrug and say, "Well yes, that's just Newton's first law," but remember that Newton's first law discusses the constant velocity of a particle, whereas **Equation 9.37** applies to the center of mass of a (possibly vast) collection of interacting particles, and that there may not be any particle at the center of mass at all! So, this really is a remarkable result.

Example 9.19

Fireworks Display

When a fireworks rocket explodes, thousands of glowing fragments fly outward in all directions, and fall to Earth in an elegant and beautiful display (**Figure 9.31**). Describe what happens, in terms of conservation of momentum and center of mass.

Figure 9.31 These exploding fireworks are a vivid example of conservation of momentum and the motion of the center of mass.

The picture shows radial symmetry about the central points of the explosions; this suggests the idea of center of mass. We can also see the parabolic motion of the glowing particles; this brings to mind projectile motion ideas.

Solution

Initially, the fireworks rocket is launched and flies more or less straight upward; this is the cause of the more-or-less-straight, white trail going high into the sky below the explosion in the upper-right of the picture (the yellow explosion). This trail is not parabolic because the explosive shell, during its launch phase, is actually a rocket; the impulse applied to it by the ejection of the burning fuel applies a force on the shell during the rise-time interval. (This is a phenomenon we will study in the next section.) The shell has multiple forces on it; thus, it is not in free-fall prior to the explosion.

At the instant of the explosion, the thousands of glowing fragments fly outward in a radially symmetrical pattern. The symmetry of the explosion is the result of all the internal forces summing to zero $\left(\sum_j \vec{f}_j^{\,int} = 0 \right)$; for every internal force, there is another that is equal in magnitude and opposite in direction.

However, as we learned above, these internal forces cannot change the momentum of the center of mass of the (now exploded) shell. Since the rocket force has now vanished, the center of mass of the shell is now a projectile (the only force on it is gravity), so its trajectory does become parabolic. The two red explosions on the left show the path of their centers of mass at a slightly longer time after explosion compared to the yellow explosion on the upper right.

In fact, if you look carefully at all three explosions, you can see that the glowing trails are not truly radially symmetric; rather, they are somewhat denser on one side than the other. Specifically, the yellow explosion and the lower middle explosion are slightly denser on their right sides, and the upper-left explosion is denser on its left side. This is because of the momentum of their centers of mass; the differing trail densities are due to the momentum each piece of the shell had at the moment of its explosion. The fragment for the explosion on the upper left of the picture had a momentum that pointed upward and to the left; the middle fragment's momentum pointed upward and slightly to the right; and the right-side explosion clearly upward and to the right (as evidenced by the white rocket exhaust trail visible below the yellow explosion).

Finally, each fragment is a projectile on its own, thus tracing out thousands of glowing parabolas.

Significance

In the discussion above, we said, "…the center of mass of the shell is now a projectile (the only force on it is gravity)…." This is not quite accurate, for there may not be any mass at all at the center of mass; in which case, there could not be a force acting on it. This is actually just verbal shorthand for describing the fact that the gravitational forces on all the particles act so that the center of mass changes position exactly as if all the mass of the shell were always located at the position of the center of mass.

 9.13 Check Your Understanding How would the firework display change in deep space, far away from any source of gravity?

You may sometimes hear someone describe an explosion by saying something like, "the fragments of the exploded object always move in a way that makes sure that the center of mass continues to move on its original trajectory." This makes it sound as if the process is somewhat magical: how can it be that, in *every* explosion, it *always* works out that the fragments move in just the right way so that the center of mass' motion is unchanged? Phrased this way, it would be hard to believe no explosion ever does anything differently.

The explanation of this apparently astonishing coincidence is: We defined the center of mass precisely so this is exactly what we would get. Recall that first we defined the momentum of the system:

$$\vec{p}_{CM} = \sum_{j=1}^{N} \frac{d\vec{p}_j}{dt}.$$

We then concluded that the net external force on the system (if any) changed this momentum:

$$\vec{F} = \frac{d\vec{p}_{CM}}{dt}$$

and then—and here's the point—we defined an acceleration that would obey Newton's second law. That is, we demanded that we should be able to write

$$\vec{a} = \frac{\vec{F}}{M}$$

which requires that

$$\vec{a} = \frac{d^2}{dt^2}\left(\frac{1}{M}\sum_{j=1}^{N} m_j \vec{r}_j\right).$$

where the quantity inside the parentheses is the center of mass of our system. So, it's not astonishing that the center of mass obeys Newton's second law; we defined it so that it would.

9.7 | Rocket Propulsion

Now we deal with the case where the mass of an object is changing. We analyze the motion of a rocket, which changes its velocity (and hence its momentum) by ejecting burned fuel gases, thus causing it to accelerate in the opposite direction of the velocity of the ejected fuel (see **Figure 9.32**). Specifically: A fully fueled rocket ship in deep space has a total mass m_0 (this mass includes the initial mass of the fuel). At some moment in time, the rocket has a velocity $\vec{v}$ and mass m; this mass is a combination of the mass of the empty rocket and the mass of the remaining unburned fuel it contains. (We refer to m as the "instantaneous mass" and $\vec{v}$ as the "instantaneous velocity.") The rocket accelerates by burning the fuel it carries and ejecting the burned exhaust gases. If the burn rate of the fuel is constant, and the velocity at which the exhaust is ejected is also constant, what is the change of velocity of the rocket as a result of burning all of its fuel?

Figure 9.32 The space shuttle had a number of reusable parts. Solid fuel boosters on either side were recovered and refueled after each flight, and the entire orbiter returned to Earth for use in subsequent flights. The large liquid fuel tank was expended. The space shuttle was a complex assemblage of technologies, employing both solid and liquid fuel, and pioneering ceramic tiles as reentry heat shields. As a result, it permitted multiple launches as opposed to single-use rockets. (credit: modification of work by NASA)

Physical Analysis

Here's a description of what happens, so that you get a feel for the physics involved.

- As the rocket engines operate, they are continuously ejecting burned fuel gases, which have both mass and velocity, and therefore some momentum. By conservation of momentum, the rocket's momentum changes by this same amount (with the opposite sign). We will assume the burned fuel is being ejected at a constant rate, which means the rate of change of the rocket's momentum is also constant. By **Equation 9.9**, this represents a constant force on the rocket.

- However, as time goes on, the mass of the rocket (which includes the mass of the remaining fuel) continuously decreases. Thus, even though the force on the rocket is constant, the resulting acceleration is not; it is continuously increasing.

- So, the total change of the rocket's velocity will depend on the amount of mass of fuel that is burned, and that dependence is not linear.

The problem has the mass and velocity of the rocket changing; also, the total mass of ejected gases is changing. If we define our system to be the rocket + fuel, then this is a closed system (since the rocket is in deep space, there are no external forces acting on this system); as a result, momentum is conserved for this system. Thus, we can apply conservation of momentum to answer the question (**Figure 9.33**).

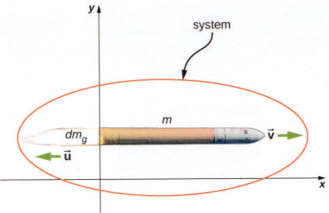

Figure 9.33 The rocket accelerates to the right due to the expulsion of some of its fuel mass to the left. Conservation of momentum enables us to determine the resulting change of velocity. The mass m is the instantaneous total mass of the rocket (i.e., mass of rocket body plus mass of fuel at that point in time). (credit: modification of work by NASA/Bill Ingalls)

At the same moment that the total instantaneous rocket mass is m (i.e., m is the mass of the rocket body plus the mass of the fuel at that point in time), we define the rocket's instantaneous velocity to be $\vec{v} = v\,\hat{i}$ (in the +x-direction); this velocity is measured relative to an inertial reference system (the Earth, for example). Thus, the initial momentum of the system is

$$\vec{p}_i = mv\,\hat{i}.$$

The rocket's engines are burning fuel at a constant rate and ejecting the exhaust gases in the −x-direction. During an infinitesimal time interval dt, the engines eject a (positive) infinitesimal mass of gas dm_g at velocity $\vec{u} = -u\,\hat{i}$; note that although the rocket velocity $v\,\hat{i}$ is measured with respect to Earth, the exhaust gas velocity is measured with respect to the (moving) rocket. Measured with respect to the Earth, therefore, the exhaust gas has velocity $(v - u)\,\hat{i}$.

As a consequence of the ejection of the fuel gas, the rocket's mass decreases by dm_g , and its velocity increases by $dv\,\hat{i}$. Therefore, including both the change for the rocket and the change for the exhaust gas, the final momentum of the system is

$$\vec{p}_f = \vec{p}_{rocket} + \vec{p}_{gas}$$
$$= (m - dm_g)(v + dv)\,\hat{i} + dm_g(v - u)\,\hat{i}.$$

Since all vectors are in the x-direction, we drop the vector notation. Applying conservation of momentum, we obtain

$$p_i = p_f$$
$$mv = (m - dm_g)(v + dv) + dm_g(v - u)$$
$$mv = mv + m\,dv - dm_g v - dm_g\,dv + dm_g v - dm_g u$$
$$m\,dv = dm_g\,dv + dm_g v.$$

Now, dm_g and dv are each very small; thus, their product $dm_g\,dv$ is very, very small, much smaller than the other two terms in this expression. We neglect this term, therefore, and obtain:

$$m\,dv = dm_g u.$$

Our next step is to remember that, since dm_g represents an increase in the mass of ejected gases, it must also represent a decrease of mass of the rocket:

$$dm_g = -dm.$$

Replacing this, we have

$$mdv = -dmu$$

or

$$dv = -u\frac{dm}{m}.$$

Integrating from the initial mass m_i to the final mass m of the rocket gives us the result we are after:

$$\int_{v_i}^{v} dv = -u\int_{m_i}^{m}\frac{1}{m}dm$$
$$v - v_i = u\ln\left(\frac{m_i}{m}\right)$$

and thus our final answer is

$$\Delta v = u\ln\left(\frac{m_i}{m}\right). \tag{9.38}$$

This result is called the **rocket equation**. It was originally derived by the Soviet physicist Konstantin Tsiolkovsky in 1897. It gives us the change of velocity that the rocket obtains from burning a mass of fuel that decreases the total rocket mass from m_0 down to m. As expected, the relationship between Δv and the change of mass of the rocket is nonlinear.

Problem-Solving Strategy: Rocket Propulsion

In rocket problems, the most common questions are finding the change of velocity due to burning some amount of fuel for some amount of time; or to determine the acceleration that results from burning fuel.

1. To determine the change of velocity, use the rocket equation **Equation 9.38**.

2. To determine the acceleration, determine the force by using the impulse-momentum theorem, using the rocket equation to determine the change of velocity.

Example 9.20

Thrust on a Spacecraft

A spacecraft is moving in gravity-free space along a straight path when its pilot decides to accelerate forward. He turns on the thrusters, and burned fuel is ejected at a constant rate of 2.0×10^2 kg/s , at a speed (relative to the rocket) of 2.5×10^2 m/s . The initial mass of the spacecraft and its unburned fuel is 2.0×10^4 kg , and the thrusters are on for 30 s.

a. What is the thrust (the force applied to the rocket by the ejected fuel) on the spacecraft?

b. What is the spacecraft's acceleration as a function of time?

c. What are the spacecraft's accelerations at $t = 0, 15, 30,$ and 35 s?

Strategy

a. The force on the spacecraft is equal to the rate of change of the momentum of the fuel.

b. Knowing the force from part (a), we can use Newton's second law to calculate the consequent acceleration. The key here is that, although the force applied to the spacecraft is constant (the fuel is being ejected at a constant rate), the mass of the spacecraft isn't; thus, the acceleration caused by the force won't be constant. We expect to get a function $a(t)$, therefore.

c. We'll use the function we obtain in part (b), and just substitute the numbers given. Important: We expect that the acceleration will get larger as time goes on, since the mass being accelerated is continuously decreasing (fuel is being ejected from the rocket).

Solution

a. The momentum of the ejected fuel gas is

$$p = m_g v.$$

The ejection velocity $v = 2.5 \times 10^2 \, \text{m/s}$ is constant, and therefore the force is

$$F = \frac{dp}{dt} = v \frac{dm_g}{dt} = -v \frac{dm}{dt}.$$

Now, $\dfrac{dm_g}{dt}$ is the rate of change of the mass of the fuel; the problem states that this is $2.0 \times 10^2 \, \text{kg/s}$.

Substituting, we get

$$
\begin{aligned}
F &= v \frac{dm_g}{dt} \\
&= \left(2.5 \times 10^2 \, \tfrac{\text{m}}{\text{s}}\right)\!\left(2.0 \times 10^2 \, \tfrac{\text{kg}}{\text{s}}\right) \\
&= 5 \times 10^4 \, \text{N}.
\end{aligned}
$$

b. Above, we defined m to be the combined mass of the empty rocket plus however much unburned fuel it contained: $m = m_R + m_g$. From Newton's second law,

$$a = \frac{F}{m} = \frac{F}{m_R + m_g}.$$

The force is constant and the empty rocket mass m_R is constant, but the fuel mass m_g is decreasing at a uniform rate; specifically:

$$m_g = m_g(t) = m_{g0} - \left(\frac{dm_g}{dt}\right)t.$$

This gives us

$$a(t) = \frac{F}{m_{g_i} - \left(\frac{dm_g}{dt}\right)t} = \frac{F}{M - \left(\frac{dm_g}{dt}\right)t}.$$

Notice that, as expected, the acceleration is a function of time. Substituting the given numbers:

$$a(t) = \frac{5 \times 10^4 \, \text{N}}{2.0 \times 10^4 \, \text{kg} - \left(2.0 \times 10^2 \, \tfrac{\text{kg}}{\text{s}}\right)t}.$$

c. At $t = 0 \, \text{s}$:

$$a(0 \, \text{s}) = \frac{5 \times 10^4 \, \text{N}}{2.0 \times 10^4 \, \text{kg} - \left(2.0 \times 10^2 \, \tfrac{\text{kg}}{\text{s}}\right)(0 \, \text{s})} = 2.5 \frac{\text{m}}{\text{s}^2}.$$

At $t = 15 \, \text{s}$, $a(15 \, \text{s}) = 2.9 \, \text{m/s}^2$.

At $t = 30 \, \text{s}$, $a(30 \, \text{s}) = 3.6 \, \text{m/s}^2$.

Acceleration is increasing, as we expected.

Significance

Notice that the acceleration is not constant; as a result, any dynamical quantities must be calculated either using integrals, or (more easily) conservation of total energy.

9.14   **Check Your Understanding** What is the physical difference (or relationship) between $\frac{dm}{dt}$ and $\frac{dm_g}{dt}$

in this example?

Rocket in a Gravitational Field

Let's now analyze the velocity change of the rocket during the launch phase, from the surface of Earth. To keep the math manageable, we'll restrict our attention to distances for which the acceleration caused by gravity can be treated as a constant g.

The analysis is similar, except that now there is an external force of $\overrightarrow{\mathbf{F}} = -mg\,\hat{\mathbf{j}}$ acting on our system. This force applies

an impulse $d\overrightarrow{\mathbf{J}} = \overrightarrow{\mathbf{F}}\,dt = -mgdt\,\hat{\mathbf{j}}$, which is equal to the change of momentum. This gives us

$$d\overrightarrow{\mathbf{p}} = d\overrightarrow{\mathbf{J}}$$

$$\overrightarrow{\mathbf{p}}_f - \overrightarrow{\mathbf{p}}_i = -mgdt\,\hat{\mathbf{j}}$$

$$\left[(m - dm_g)(v + dv) + dm_g(v - u) - mv\right]\hat{\mathbf{j}} = -mgdt\,\hat{\mathbf{j}}$$

and so

$$mdv - dm_g u = -mgdt$$

where we have again neglected the term $dm_g\,dv$ and dropped the vector notation. Next we replace dm_g with $-dm$:

$$mdv + dmu = -mgdt$$

$$mdv = -dmu - mgdt.$$

Dividing through by m gives

$$dv = -u\frac{dm}{m} - gdt$$

and integrating, we have

$$\Delta v = u\ln\left(\frac{m_i}{m}\right) - g\Delta t. \qquad (9.39)$$

Unsurprisingly, the rocket's velocity is affected by the (constant) acceleration of gravity.

Remember that Δt is the burn time of the fuel. Now, in the absence of gravity, **Equation 9.38** implies that it makes no difference how much time it takes to burn the entire mass of fuel; the change of velocity does not depend on Δt. However, in the presence of gravity, it matters a lot. The $-g\,\Delta t$ term in **Equation 9.39** tells us that the *longer* the burn time is, the *smaller* the rocket's change of velocity will be. This is the reason that the launch of a rocket is so spectacular at the first moment of liftoff: It's essential to burn the fuel as quickly as possible, to get as large a Δv as possible.

CHAPTER 9 REVIEW

KEY TERMS

center of mass weighted average position of the mass

closed system system for which the mass is constant and the net external force on the system is zero

elastic collision that conserves kinetic energy

explosion single object breaks up into multiple objects; kinetic energy is not conserved in explosions

external force force applied to an extended object that changes the momentum of the extended object as a whole

impulse effect of applying a force on a system for a time interval; this time interval is usually small, but does not have to be

impulse-momentum theorem change of momentum of a system is equal to the impulse applied to the system

inelastic collision that does not conserve kinetic energy

internal force force that the simple particles that make up an extended object exert on each other. Internal forces can be attractive or repulsive

Law of Conservation of Momentum total momentum of a closed system cannot change

linear mass density λ, expressed as the number of kilograms of material per meter

momentum measure of the quantity of motion that an object has; it takes into account both how fast the object is moving, and its mass; specifically, it is the product of mass and velocity; it is a vector quantity

perfectly inelastic collision after which all objects are motionless, the final kinetic energy is zero, and the loss of kinetic energy is a maximum

rocket equation derived by the Soviet physicist Konstantin Tsiolkovsky in 1897, it gives us the change of velocity that the rocket obtains from burning a mass of fuel that decreases the total rocket mass from m_i down to m

system object or collection of objects whose motion is currently under investigation; however, your system is defined at the start of the problem, you must keep that definition for the entire problem

KEY EQUATIONS

Definition of momentum	$\vec{p} = m\vec{v}$
Impulse	$\vec{J} \equiv \int_{t_i}^{t_f} \vec{F}(t)dt$ or $\vec{J} = \vec{F}_{ave}\Delta t$
Impulse-momentum theorem	$\vec{J} = \Delta\vec{p}$
Average force from momentum	$\vec{F} = \dfrac{\Delta\vec{p}}{\Delta t}$
Instantaneous force from momentum (Newton's second law)	$\vec{F}(t) = \dfrac{d\vec{p}}{dt}$
Conservation of momentum	$\dfrac{d\vec{p}_1}{dt} + \dfrac{d\vec{p}_2}{dt} = 0$ or $\vec{p}_1 + \vec{p}_2 = \text{constant}$
Generalized conservation of momentum	$\displaystyle\sum_{j=1}^{N} \vec{p}_j = \text{constant}$

Conservation of momentum in two dimensions	$p_{f, x} = p_{1,i, x} + p_{2,i, x}$ $p_{f, y} = p_{1,i, y} + p_{2,i, y}$
External forces	$\vec{\mathbf{F}}_{ext} = \sum_{j=1}^{N} \dfrac{d\vec{\mathbf{p}}_j}{dt}$
Newton's second law for an extended object	$\vec{\mathbf{F}} = \dfrac{d\vec{\mathbf{p}}_{CM}}{dt}$
Acceleration of the center of mass	$\vec{\mathbf{a}}_{CM} = \dfrac{d^2}{dt^2}\left(\dfrac{1}{M}\sum_{j=1}^{N} m_j \vec{\mathbf{r}}_j\right) = \dfrac{1}{M}\sum_{j=1}^{N} m_j \vec{\mathbf{a}}_j$
Position of the center of mass for a system of particles	$\vec{\mathbf{r}}_{CM} \equiv \dfrac{1}{M}\sum_{j=1}^{N} m_j \vec{\mathbf{r}}_j$
Velocity of the center of mass	$\vec{\mathbf{v}}_{CM} = \dfrac{d}{dt}\left(\dfrac{1}{M}\sum_{j=1}^{N} m_j \vec{\mathbf{r}}_j\right) = \dfrac{1}{M}\sum_{j=1}^{N} m_j \vec{\mathbf{v}}_j$
Position of the center of mass of a continuous object	$\vec{\mathbf{r}}_{CM} \equiv \dfrac{1}{M}\int \vec{\mathbf{r}}\, dm$
Rocket equation	$\Delta v = u \ln\left(\dfrac{m_i}{m}\right)$

SUMMARY

9.1 Linear Momentum

- The motion of an object depends on its mass as well as its velocity. Momentum is a concept that describes this. It is a useful and powerful concept, both computationally and theoretically. The SI unit for momentum is kg · m/s.

9.2 Impulse and Collisions

- When a force is applied on an object for some amount of time, the object experiences an impulse.
- This impulse is equal to the object's change of momentum.
- Newton's second law in terms of momentum states that the net force applied to a system equals the rate of change of the momentum that the force causes.

9.3 Conservation of Linear Momentum

- The law of conservation of momentum says that the momentum of a closed system is constant in time (conserved).
- A closed (or isolated) system is defined to be one for which the mass remains constant, and the net external force is zero.
- The total momentum of a system is conserved *only* when the system is closed.

9.4 Types of Collisions

- An elastic collision is one that conserves kinetic energy.
- An inelastic collision does not conserve kinetic energy.
- Momentum is conserved regardless of whether or not kinetic energy is conserved.
- Analysis of kinetic energy changes and conservation of momentum together allow the final velocities to be calculated in terms of initial velocities and masses in one-dimensional, two-body collisions.

9.5 Collisions in Multiple Dimensions

- The approach to two-dimensional collisions is to choose a convenient coordinate system and break the motion into components along perpendicular axes.

- Momentum is conserved in both directions simultaneously and independently.

- The Pythagorean theorem gives the magnitude of the momentum vector using the *x*- and *y*-components, calculated using conservation of momentum in each direction.

9.6 Center of Mass

- An extended object (made up of many objects) has a defined position vector called the center of mass.

- The center of mass can be thought of, loosely, as the average location of the total mass of the object.

- The center of mass of an object traces out the trajectory dictated by Newton's second law, due to the net external force.

- The internal forces within an extended object cannot alter the momentum of the extended object as a whole.

9.7 Rocket Propulsion

- A rocket is an example of conservation of momentum where the mass of the system is not constant, since the rocket ejects fuel to provide thrust.

- The rocket equation gives us the change of velocity that the rocket obtains from burning a mass of fuel that decreases the total rocket mass.

CONCEPTUAL QUESTIONS

9.1 Linear Momentum

1. An object that has a small mass and an object that has a large mass have the same momentum. Which object has the largest kinetic energy?

2. An object that has a small mass and an object that has a large mass have the same kinetic energy. Which mass has the largest momentum?

9.2 Impulse and Collisions

3. Is it possible for a small force to produce a larger impulse on a given object than a large force? Explain.

4. Why is a 10-m fall onto concrete far more dangerous than a 10-m fall onto water?

5. What external force is responsible for changing the momentum of a car moving along a horizontal road?

6. A piece of putty and a tennis ball with the same mass are thrown against a wall with the same velocity. Which object experience a greater impulse from the wall or are the impulses equal? Explain.

9.3 Conservation of Linear Momentum

7. Under what circumstances is momentum conserved?

8. Can momentum be conserved for a system if there are external forces acting on the system? If so, under what conditions? If not, why not?

9. Explain in terms of momentum and Newton's laws how a car's air resistance is due in part to the fact that it pushes air in its direction of motion.

10. Can objects in a system have momentum while the momentum of the system is zero? Explain your answer.

11. A sprinter accelerates out of the starting blocks. Can you consider him as a closed system? Explain.

12. A rocket in deep space (zero gravity) accelerates by firing hot gas out of its thrusters. Does the rocket constitute a closed system? Explain.

9.4 Types of Collisions

13. Two objects of equal mass are moving with equal and opposite velocities when they collide. Can all the kinetic energy be lost in the collision?

14. Describe a system for which momentum is conserved but mechanical energy is not. Now the reverse: Describe a system for which kinetic energy is conserved but momentum is not.

9.5 Collisions in Multiple Dimensions

15. Momentum for a system can be conserved in one direction while not being conserved in another. What is the angle between the directions? Give an example.

9.6 Center of Mass

16. Suppose a fireworks shell explodes, breaking into

three large pieces for which air resistance is negligible. How does the explosion affect the motion of the center of mass? How would it be affected if the pieces experienced significantly more air resistance than the intact shell?

9.7 Rocket Propulsion

17. It is possible for the velocity of a rocket to be greater than the exhaust velocity of the gases it ejects. When that is the case, the gas velocity and gas momentum are in the same direction as that of the rocket. How is the rocket still able to obtain thrust by ejecting the gases?

PROBLEMS

9.1 Linear Momentum

18. An elephant and a hunter are having a confrontation.

$m_E = 2000.0$ kg $m_{hunter} = 90.0$ kg

$m_{dart} = 0.0400$ kg

$\vec{v}_{dart} = (600 \text{ m/s})(-\hat{i})$

$\vec{v}_E = (7.50 \text{ m/s})\hat{i}$ $\vec{v}_{hunter} = (7.40 \text{ m/s})\hat{i}$

a. Calculate the momentum of the 2000.0-kg elephant charging the hunter at a speed of 7.50 m/s.
b. Calculate the ratio of the elephant's momentum to the momentum of a 0.0400-kg tranquilizer dart fired at a speed of 600 m/s.
c. What is the momentum of the 90.0-kg hunter running at 7.40 m/s after missing the elephant?

19. A skater of mass 40 kg is carrying a box of mass 5 kg. The skater has a speed of 5 m/s with respect to the floor and is gliding without any friction on a smooth surface.
a. Find the momentum of the box with respect to the floor.
b. Find the momentum of the box with respect to the floor after she puts the box down on the frictionless skating surface.

20. A car of mass 2000 kg is moving with a constant velocity of 10 m/s due east. What is the momentum of the car?

21. The mass of Earth is 5.97×10^{24} kg and its orbital radius is an average of 1.50×10^{11} m. Calculate the magnitude of its linear momentum at the location in the

diagram.

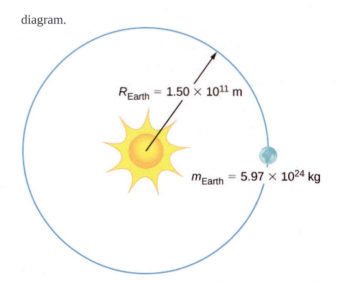

22. If a rainstorm drops 1 cm of rain over an area of 10 km^2 in the period of 1 hour, what is the momentum of the rain that falls in one second? Assume the terminal velocity of a raindrop is 10 m/s.

23. What is the average momentum of an avalanche that moves a 40-cm-thick layer of snow over an area of 100 m by 500 m over a distance of 1 km down a hill in 5.5 s? Assume a density of 350 kg/m^3 for the snow.

24. What is the average momentum of a 70.0-kg sprinter who runs the 100-m dash in 9.65 s?

9.2 Impulse and Collisions

25. A 75.0-kg person is riding in a car moving at 20.0 m/s when the car runs into a bridge abutment (see the following figure).

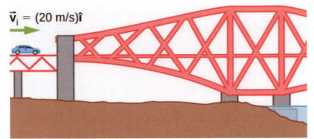

$\vec{\mathbf{v}}_i = (20 \text{ m/s})\hat{\mathbf{i}}$

    a. Calculate the average force on the person if he is stopped by a padded dashboard that compresses an average of 1.00 cm.

    b. Calculate the average force on the person if he is stopped by an air bag that compresses an average of 15.0 cm.

26. One hazard of space travel is debris left by previous missions. There are several thousand objects orbiting Earth that are large enough to be detected by radar, but there are far greater numbers of very small objects, such as flakes of paint. Calculate the force exerted by a 0.100-mg chip of paint that strikes a spacecraft window at a relative speed of 4.00×10^3 m/s , given the collision lasts 6.00×10^{-8} s .

27. A cruise ship with a mass of 1.00×10^7 kg strikes a pier at a speed of 0.750 m/s. It comes to rest after traveling 6.00 m, damaging the ship, the pier, and the tugboat captain's finances. Calculate the average force exerted on the pier using the concept of impulse. (*Hint*: First calculate the time it took to bring the ship to rest, assuming a constant force.)

$\vec{\mathbf{v}}_i = (0.750 \text{ m/s})\hat{\mathbf{i}}$

28. Calculate the final speed of a 110-kg rugby player who is initially running at 8.00 m/s but collides head-on with a padded goalpost and experiences a backward force of 1.76×10^4 N for 5.50×10^{-2} s .

29. Water from a fire hose is directed horizontally against a wall at a rate of 50.0 kg/s and a speed of 42.0 m/s. Calculate the force exerted on the wall, assuming the water's horizontal momentum is reduced to zero.

30. A 0.450-kg hammer is moving horizontally at 7.00 m/s when it strikes a nail and comes to rest after driving the nail 1.00 cm into a board. Assume constant acceleration of the hammer-nail pair.

    a. Calculate the duration of the impact.

    b. What was the average force exerted on the nail?

31. What is the momentum (as a function of time) of a 5.0-kg particle moving with a velocity $\vec{\mathbf{v}}(t) = \left(2.0\,\hat{\mathbf{i}} + 4.0t\,\hat{\mathbf{j}}\right)$ m/s ? What is the net force acting on this particle?

32. The *x*-component of a force on a 46-g golf ball by a 7-iron versus time is plotted in the following figure:

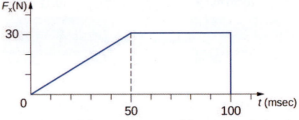

    a. Find the *x*-component of the impulse during the intervals

        i. [0, 50 ms], and

       ii. [50 ms, 100 ms]

    b. Find the change in the *x*-component of the momentum during the intervals

      iii. [0, 50 ms], and

      iv. [50 ms, 100 ms]

33. A hockey puck of mass 150 g is sliding due east on a frictionless table with a speed of 10 m/s. Suddenly, a constant force of magnitude 5 N and direction due north is applied to the puck for 1.5 s. Find the north and east components of the momentum at the end of the 1.5-s interval.

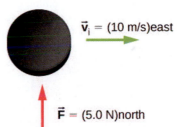

$\vec{\mathbf{v}}_i = (10 \text{ m/s})\text{east}$

$\vec{\mathbf{F}} = (5.0 \text{ N})\text{north}$

34. A ball of mass 250 g is thrown with an initial velocity of 25 m/s at an angle of $30°$ with the horizontal direction. Ignore air resistance. What is the momentum of the ball after 0.2 s? (Do this problem by finding the components of the momentum first, and then constructing the magnitude and direction of the momentum vector from the components.)

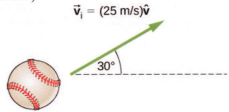

$\vec{\mathbf{v}}_i = (25 \text{ m/s})\hat{\mathbf{v}}$

$30°$

9.3 Conservation of Linear Momentum

35. Train cars are coupled together by being bumped into one another. Suppose two loaded train cars are moving toward one another, the first having a mass of 1.50×10^5 kg and a velocity of $(0.30 \text{ m/s}) \hat{\mathbf{i}}$, and the second having a mass of 1.10×10^5 kg and a velocity of $-(0.12 \text{ m/s}) \hat{\mathbf{i}}$. What is their final velocity?

36. Two identical pucks collide elastically on an air hockey table. Puck 1 was originally at rest; puck 2 has an incoming speed of 6.00 m/s and scatters at an angle of $30°$ with respect to its incoming direction. What is the velocity (magnitude and direction) of puck 1 after the collision?

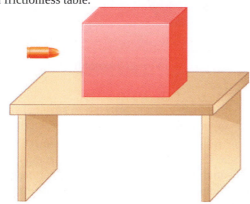

37. The figure below shows a bullet of mass 200 g traveling horizontally towards the east with speed 400 m/s, which strikes a block of mass 1.5 kg that is initially at rest on a frictionless table.

After striking the block, the bullet is embedded in the block and the block and the bullet move together as one unit.

a. What is the magnitude and direction of the velocity of the block/bullet combination immediately after the impact?
b. What is the magnitude and direction of the impulse by the block on the bullet?
c. What is the magnitude and direction of the impulse from the bullet on the block?
d. If it took 3 ms for the bullet to change the speed from 400 m/s to the final speed after impact, what is the average force between the block and the bullet during this time?

38. A 20-kg child is coasting at 3.3 m/s over flat ground in a 4.0-kg wagon. The child drops a 1.0-kg ball out the back of the wagon. What is the final speed of the child and wagon?

39. A 4.5 kg puffer fish expands to 40% of its mass by taking in water. When the puffer fish is threatened, it releases the water toward the threat to move quickly forward. What is the ratio of the speed of the puffer fish forward to the speed of the expelled water backwards?

40. Explain why a cannon recoils when it fires a shell.

41. Two figure skaters are coasting in the same direction, with the leading skater moving at 5.5 m/s and the trailing skating moving at 6.2 m/s. When the trailing skater catches up with the leading skater, he picks her up without applying any horizontal forces on his skates. If the trailing skater is 50% heavier than the 50-kg leading skater, what is their speed after he picks her up?

42. A 2000-kg railway freight car coasts at 4.4 m/s underneath a grain terminal, which dumps grain directly down into the freight car. If the speed of the loaded freight car must not go below 3.0 m/s, what is the maximum mass of grain that it can accept?

9.4 Types of Collisions

43. A 5.50-kg bowling ball moving at 9.00 m/s collides with a 0.850-kg bowling pin, which is scattered at an angle to the initial direction of the bowling ball and with a speed of 15.0 m/s.

a. Calculate the final velocity (magnitude and direction) of the bowling ball.
b. Is the collision elastic?

44. Ernest Rutherford (the first New Zealander to be awarded the Nobel Prize in Chemistry) demonstrated that nuclei were very small and dense by scattering helium-4 nuclei from gold-197 nuclei. The energy of the incoming helium nucleus was 8.00×10^{-13} J , and the masses of the helium and gold nuclei were 6.68×10^{-27} kg and

3.29×10^{-25} kg, respectively (note that their mass ratio is 4 to 197).

a. If a helium nucleus scatters to an angle of $120°$ during an elastic collision with a gold nucleus, calculate the helium nucleus's final speed and the final velocity (magnitude and direction) of the gold nucleus.

b. What is the final kinetic energy of the helium nucleus?

45. A 90.0-kg ice hockey player hits a 0.150-kg puck, giving the puck a velocity of 45.0 m/s. If both are initially at rest and if the ice is frictionless, how far does the player recoil in the time it takes the puck to reach the goal 15.0 m away?

46. A 100-g firecracker is launched vertically into the air and explodes into two pieces at the peak of its trajectory. If a 72-g piece is projected horizontally to the left at 20 m/s, what is the speed and direction of the other piece?

47. In an elastic collision, a 400-kg bumper car collides directly from behind with a second, identical bumper car that is traveling in the same direction. The initial speed of the leading bumper car is 5.60 m/s and that of the trailing car is 6.00 m/s. Assuming that the mass of the drivers is much, much less than that of the bumper cars, what are their final speeds?

48. Repeat the preceding problem if the mass of the leading bumper car is 30.0% greater than that of the trailing bumper car.

49. An alpha particle (^{4}He) undergoes an elastic collision with a stationary uranium nucleus (^{235}U). What percent of the kinetic energy of the alpha particle is transferred to the uranium nucleus? Assume the collision is one-dimensional.

50. You are standing on a very slippery icy surface and throw a 1-kg football horizontally at a speed of 6.7 m/s. What is your velocity when you release the football? Assume your mass is 65 kg.

51. A 35-kg child rides a relatively massless sled down a hill and then coasts along the flat section at the bottom, where a second 35-kg child jumps on the sled as it passes by her. If the speed of the sled is 3.5 m/s before the second child jumps on, what is its speed after she jumps on?

52. A boy sleds down a hill and onto a frictionless ice-covered lake at 10.0 m/s. In the middle of the lake is a 1000-kg boulder. When the sled crashes into the boulder, he is propelled backwards from the boulder. The collision is an elastic collision. If the boy's mass is 40.0 kg and the sled's mass is 2.50 kg, what is the speed of the sled and the boulder after the collision?

9.5 Collisions in Multiple Dimensions

53. A 0.90-kg falcon is diving at 28.0 m/s at a downward angle of $35°$. It catches a 0.325-kg dove from behind in midair. What is their combined velocity after impact if the pigeon's initial velocity was 7.00 m/s directed horizontally? Note that $\hat{v}_{1,i}$ is a unit vector pointing in the direction in which the falcon is initially flying.

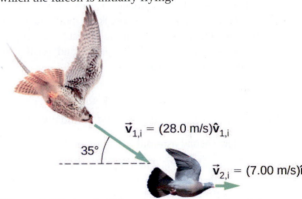

Figure 9.34 (credit "falcon": modification of work by "USFWS Mountain-Prairie"/Flickr; credit "pigeon": modification of work by Jacob Spinks)

54. A billiard ball, labeled 1, moving horizontally strikes another billiard ball, labeled 2, at rest. Before impact, ball 1 was moving at a speed of 3.00 m/s, and after impact it is moving at 0.50 m/s at 50° from the original direction. If the two balls have equal masses of 300 g, what is the velocity of the ball 2 after the impact?

55. A projectile of mass 2.0 kg is fired in the air at an angle of $40.0°$ to the horizon at a speed of 50.0 m/s. At the highest point in its flight, the projectile breaks into three parts of mass 1.0 kg, 0.7 kg, and 0.3 kg. The 1.0-kg part falls straight down after breakup with an initial speed of 10.0 m/s, the 0.7-kg part moves in the original forward direction, and the 0.3-kg part goes straight up.

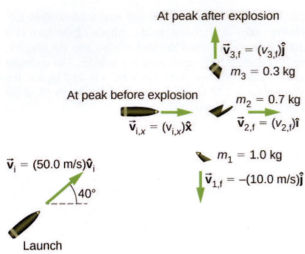

At peak after explosion

$\vec{v}_{3,f} = (v_{3,f})\hat{j}$

$m_3 = 0.3$ kg

At peak before explosion $m_2 = 0.7$ kg

$\vec{v}_{i,x} = (v_{i,x})\hat{x}$ $\vec{v}_{2,f} = (v_{2,f})\hat{i}$

$m_1 = 1.0$ kg

$\vec{v}_i = (50.0$ m/s$)\hat{v}_i$ $\vec{v}_{1,f} = -(10.0$ m/s$)\hat{j}$

40°

Launch

a. Find the speeds of the 0.3-kg and 0.7-kg pieces immediately after the break-up.
b. How high from the break-up point does the 0.3-kg piece go before coming to rest?
c. Where does the 0.7-kg piece land relative to where it was fired from?

56. Two asteroids collide and stick together. The first asteroid has mass of 15×10^3 kg and is initially moving at 770 m/s. The second asteroid has mass of 20×10^3 kg and is moving at 1020 m/s. Their initial velocities made an angle of 20° with respect to each other. What is the final speed and direction with respect to the velocity of the first asteroid?

57. A 200-kg rocket in deep space moves with a velocity of $(121$ m/s$)\hat{i} + (38.0$ m/s$)\hat{j}$. Suddenly, it explodes into three pieces, with the first (78 kg) moving at $-(321$ m/s$)\hat{i} + (228$ m/s$)\hat{j}$ and the second (56 kg) moving at $(16.0$ m/s$)\hat{i} - (88.0$ m/s$)\hat{j}$. Find the velocity of the third piece.

58. A proton traveling at 3.0×10^6 m/s scatters elastically from an initially stationary alpha particle and is deflected at an angle of 85° with respect to its initial velocity. Given that the alpha particle has four times the mass of the proton, what percent of its initial kinetic energy does the proton retain after the collision?

59. Three 70-kg deer are standing on a flat 200-kg rock that is on an ice-covered pond. A gunshot goes off and the dear scatter, with deer A running at $(15$ m/s$)\hat{i} + (5.0$ m/s$)\hat{j}$, deer B running at $(-12$ m/s$)\hat{i} + (8.0$ m/s$)\hat{j}$, and deer C running at

$(1.2$ m/s$)\hat{i} - (18.0$ m/s$)\hat{j}$. What is the velocity of the rock on which they were standing?

60. A family is skating. The father (75 kg) skates at 8.2 m/s and collides and sticks to the mother (50 kg), who was initially moving at 3.3 m/s and at 45° with respect to the father's velocity. The pair then collides with their daughter (30 kg), who was stationary, and the three slide off together. What is their final velocity?

61. An oxygen atom (mass 16 u) moving at 733 m/s at 15.0° with respect to the $\hat{i}$ direction collides and sticks to an oxygen molecule (mass 32 u) moving at 528 m/s at 128° with respect to the $\hat{i}$ direction. The two stick together to form ozone. What is the final velocity of the ozone molecule?

62. Two cars of the same mass approach an extremely icy four-way perpendicular intersection. Car A travels northward at 30 m/s and car B is travelling eastward. They collide and stick together, traveling at 28° north of east. What was the initial velocity of car B?

9.6 Center of Mass

63. Three point masses are placed at the corners of a triangle as shown in the figure below.

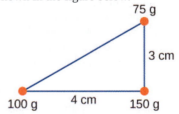

Find the center of mass of the three-mass system.

64. Two particles of masses m_1 and m_2 separated by a horizontal distance D are released from the same height h at the same time. Find the vertical position of the center of mass of these two particles at a time before the two particles strike the ground. Assume no air resistance.

65. Two particles of masses m_1 and m_2 separated by a horizontal distance D are let go from the same height h at different times. Particle 1 starts at $t = 0$, and particle 2 is let go at $t = T$. Find the vertical position of the center of mass at a time before the first particle strikes the ground. Assume no air resistance.

66. Two particles of masses m_1 and m_2 move uniformly in different circles of radii R_1 and R_2 about origin in the x, y-plane. The x- and y-coordinates of the center of mass

and that of particle 1 are given as follows (where length is in meters and t in seconds):
$x_1(t) = 4\cos(2t)$, $y_1(t) = 4\sin(2t)$

and:
$x_{CM}(t) = 3\cos(2t)$, $y_{CM}(t) = 3\sin(2t)$.

    a. Find the radius of the circle in which particle 1 moves.
    b. Find the x- and y-coordinates of particle 2 and the radius of the circle this particle moves.

67. Two particles of masses m_1 and m_2 move uniformly in different circles of radii R_1 and R_2 about the origin in the x, y-plane. The coordinates of the two particles in meters are given as follows ($z = 0$ for both). Here t is in seconds:

$$x_1(t) = 4\cos(2t)$$
$$y_1(t) = 4\sin(2t)$$
$$x_2(t) = 2\cos\left(3t - \frac{\pi}{2}\right)$$
$$y_2(t) = 2\sin\left(3t - \frac{\pi}{2}\right)$$

    a. Find the radii of the circles of motion of both particles.
    b. Find the x- and y-coordinates of the center of mass.
    c. Decide if the center of mass moves in a circle by plotting its trajectory.

68. Find the center of mass of a one-meter long rod, made of 50 cm of iron (density $8\ \frac{\text{g}}{\text{cm}^3}$) and 50 cm of aluminum (density $2.7\ \frac{\text{g}}{\text{cm}^3}$).

69. Find the center of mass of a rod of length L whose mass density changes from one end to the other quadratically. That is, if the rod is laid out along the x-axis with one end at the origin and the other end at $x = L$, the density is given by $\rho(x) = \rho_0 + (\rho_1 - \rho_0)\left(\frac{x}{L}\right)^2$, where ρ_0 and ρ_1 are constant values.

70. Find the center of mass of a rectangular block of length a and width b that has a nonuniform density such that when the rectangle is placed in the x,y-plane with one corner at the origin and the block placed in the first quadrant with the two edges along the x- and y-axes, the density is given by $\rho(x, y) = \rho_0 x$, where ρ_0 is a constant.

71. Find the center of mass of a rectangular material of length a and width b made up of a material of nonuniform

density. The density is such that when the rectangle is placed in the xy-plane, the density is given by $\rho(x, y) = \rho_0 xy$.

72. A cube of side a is cut out of another cube of side b as shown in the figure below.

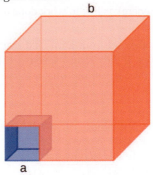

Find the location of the center of mass of the structure. (*Hint:* Think of the missing part as a negative mass overlapping a positive mass.)

73. Find the center of mass of a cone of uniform density that has a radius R at the base, height h, and mass M. Let the origin be at the center of the base of the cone and have $+z$ going through the cone vertex.

74. Find the center of mass of a thin wire of mass m and length L bent in a semicircular shape. Let the origin be at the center of the semicircle and have the wire arc from the $+x$ axis, cross the $+y$ axis, and terminate at the $-x$ axis.

75. Find the center of mass of a uniform thin semicircular plate of radius R. Let the origin be at the center of the semicircle, the plate arc from the $+x$ axis to the $-x$ axis, and the z axis be perpendicular to the plate.

76. Find the center of mass of a sphere of mass M and radius R and a cylinder of mass m, radius r, and height h arranged as shown below.

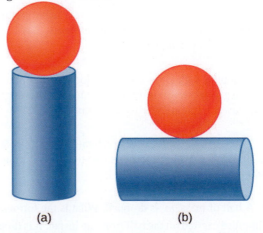

(a)   (b)

Express your answers in a coordinate system that has the origin at the center of the cylinder.

9.7 Rocket Propulsion

77. (a) A 5.00-kg squid initially at rest ejects 0.250 kg of fluid with a velocity of 10.0 m/s. What is the recoil velocity of the squid if the ejection is done in 0.100 s and there is a 5.00-N frictional force opposing the squid's movement?

(b) How much energy is lost to work done against friction?

78. A rocket takes off from Earth and reaches a speed of 100 m/s in 10.0 s. If the exhaust speed is 1500 m/s and the mass of fuel burned is 100 kg, what was the initial mass of the rocket?

79. Repeat the preceding problem but for a rocket that takes off from a space station, where there is no gravity other than the negligible gravity due to the space station.

80. How much fuel would be needed for a 1000-kg rocket (this is its mass with no fuel) to take off from Earth and reach 1000 m/s in 30 s? The exhaust speed is 1000 m/s.

81. What exhaust speed is required to accelerate a rocket in deep space from 800 m/s to 1000 m/s in 5.0 s if the total rocket mass is 1200 kg and the rocket only has 50 kg of fuel left?

82. Unreasonable Results Squids have been reported to jump from the ocean and travel 30.0 m (measured horizontally) before re-entering the water.

(a) Calculate the initial speed of the squid if it leaves the water at an angle of 20.0°, assuming negligible lift from the air and negligible air resistance.

(b) The squid propels itself by squirting water. What fraction of its mass would it have to eject in order to achieve the speed found in the previous part? The water is ejected at 12.0 m/s; gravitational force and friction are neglected.

(c) What is unreasonable about the results?

(d) Which premise is unreasonable, or which premises are inconsistent?

ADDITIONAL PROBLEMS

83. Two 70-kg canoers paddle in a single, 50-kg canoe. Their paddling moves the canoe at 1.2 m/s with respect to the water, and the river they're in flows at 4 m/s with respect to the land. What is their momentum with respect to the land?

84. Which has a larger magnitude of momentum: a 3000-kg elephant moving at 40 km/h or a 60-kg cheetah moving at 112 km/h?

85. A driver applies the brakes and reduces the speed of her car by 20%, without changing the direction in which the car is moving. By how much does the car's momentum change?

86. You friend claims that momentum is mass multiplied by velocity, so things with more mass have more momentum. Do you agree? Explain.

87. Dropping a glass on a cement floor is more likely to break the glass than if it is dropped from the same height on a grass lawn. Explain in terms of the impulse.

88. Your 1500-kg sports car accelerates from 0 to 30 m/s in 10 s. What average force is exerted on it during this acceleration?

89. A ball of mass m is dropped. What is the formula for the impulse exerted on the ball from the instant it is dropped to an arbitrary time τ later? Ignore air resistance.

90. Repeat the preceding problem, but including a drag force due to air of $f_{\text{drag}} = -b \overrightarrow{v}$.

91. A 5.0-g egg falls from a 90-cm-high counter onto the floor and breaks. What impulse is exerted by the floor on the egg?

92. A car crashes into a large tree that does not move. The car goes from 30 m/s to 0 in 1.3 m. (a) What impulse is applied to the driver by the seatbelt, assuming he follows the same motion as the car? (b) What is the average force applied to the driver by the seatbelt?

93. Two hockey players approach each other head on, each traveling at the same speed v_i. They collide and get tangled together, falling down and moving off at a speed $v_i/5$. What is the ratio of their masses?

94. You are coasting on your 10-kg bicycle at 15 m/s and a 5.0-g bug splatters on your helmet. The bug was initially moving at 2.0 m/s in the same direction as you. If your mass is 60 kg, (a) what is the initial momentum of you plus your bicycle? (b) What is the initial momentum of the bug? (c) What is your change in velocity due to the collision with the bug? (d) What would the change in velocity have been if the bug were traveling in the opposite direction?

95. A load of gravel is dumped straight down into a 30 000-kg freight car coasting at 2.2 m/s on a straight section of a railroad. If the freight car's speed after receiving the gravel is 1.5 m/s, what mass of gravel did it receive?

96. Two carts on a straight track collide head on. The first cart was moving at 3.6 m/s in the positive x direction and the second was moving at 2.4 m/s in the opposite direction. After the collision, the second car continues moving in its initial direction of motion at 0.24 m/s. If the mass of the second car is 5.0 times that of the first, what is the mass and final velocity of the first car?

97. A 100-kg astronaut finds himself separated from his spaceship by 10 m and moving away from the spaceship at 0.1 m/s. To get back to the spaceship, he throws a 10-kg tool bag away from the spaceship at 5.0 m/s. How long will he take to return to the spaceship?

98. Derive the equations giving the final speeds for two objects that collide elastically, with the mass of the objects being m_1 and m_2 and the initial speeds being $v_{1,i}$ and $v_{2,i} = 0$ (i.e., second object is initially stationary).

99. Repeat the preceding problem for the case when the initial speed of the second object is nonzero.

100. A child sleds down a hill and collides at 5.6 m/s into a stationary sled that is identical to his. The child is launched forward at the same speed, leaving behind the two sleds that lock together and slide forward more slowly. What is the speed of the two sleds after this collision?

101. For the preceding problem, find the final speed of each sled for the case of an elastic collision.

102. A 90-kg football player jumps vertically into the air to catch a 0.50-kg football that is thrown essentially horizontally at him at 17 m/s. What is his horizontal speed after catching the ball?

103. Three skydivers are plummeting earthward. They are initially holding onto each other, but then push apart. Two skydivers of mass 70 and 80 kg gain horizontal velocities of 1.2 m/s north and 1.4 m/s southeast, respectively. What is the horizontal velocity of the third skydiver, whose mass is 55 kg?

104. Two billiard balls are at rest and touching each other on a pool table. The cue ball travels at 3.8 m/s along the line of symmetry between these balls and strikes them simultaneously. If the collision is elastic, what is the velocity of the three balls after the collision?

105. A billiard ball traveling at $(2.2 \text{ m/s}) \hat{i} - (0.4 \text{ m/s}) \hat{j}$ collides with a wall that is aligned in the $\hat{j}$ direction. Assuming the collision is elastic, what is the final velocity of the ball?

106. Two identical billiard balls collide. The first one is initially traveling at $(2.2 \text{ m/s}) \hat{i} - (0.4 \text{ m/s}) \hat{j}$ and the second one at $-(1.4 \text{ m/s}) \hat{i} + (2.4 \text{ m/s}) \hat{j}$. Suppose they collide when the center of ball 1 is at the origin and the center of ball 2 is at the point $(2R, 0)$ where R is the radius of the balls. What is the final velocity of each ball?

107. Repeat the preceding problem if the balls collide when the center of ball 1 is at the origin and the center of ball 2 is at the point $(0, 2R)$.

108. Repeat the preceding problem if the balls collide when the center of ball 1 is at the origin and the center of ball 2 is at the point $(\sqrt{3}R/2, R/2)$

109. Where is the center of mass of a semicircular wire of radius R that is centered on the origin, begins and ends on the x axis, and lies in the x,y plane?

110. Where is the center of mass of a slice of pizza that was cut into eight equal slices? Assume the origin is at the apex of the slice and measure angles with respect to an edge of the slice. The radius of the pizza is R.

111. If 1% of the Earth's mass were transferred to the Moon, how far would the center of mass of the Earth-Moon-population system move? The mass of the Earth is $5.97 \times 10^{24} \text{ kg}$ and that of the Moon is $7.34 \times 10^{22} \text{ kg}$. The radius of the Moon's orbit is about $3.84 \times 10^5 \text{ m}$.

112. You friend wonders how a rocket continues to climb into the sky once it is sufficiently high above the surface of Earth so that its expelled gasses no longer push on the surface. How do you respond?

113. To increase the acceleration of a rocket, should you throw rocks out of the front window of the rocket or out of the back window?

CHALLENGE PROBLEMS

114. A 65-kg person jumps from the first floor window of a burning building and lands almost vertically on the ground with a horizontal velocity of 3 m/s and vertical velocity of −9 m/s. Upon impact with the ground he is brought to rest in a short time. The force experienced by his feet depends on whether he keeps his knees stiff or bends them. Find the force on his feet in each case.

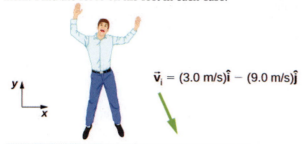

$$\vec{v}_i = (3.0 \text{ m/s})\hat{i} - (9.0 \text{ m/s})\hat{j}$$

    a. First find the impulse on the person from the impact on the ground. Calculate both its magnitude and direction.
    b. Find the average force on the feet if the person keeps his leg stiff and straight and his center of mass drops by only 1 cm vertically and 1 cm horizontally during the impact.
    c. Find the average force on the feet if the person bends his legs throughout the impact so that his center of mass drops by 50 cm vertically and 5 cm horizontally during the impact.
    d. Compare the results of part (b) and (c), and draw conclusions about which way is better.

You will need to find the time the impact lasts by making reasonable assumptions about the deceleration. Although the force is not constant during the impact, working with constant average force for this problem is acceptable.

115. Two projectiles of mass m_1 and m_2 are fired at the same speed but in opposite directions from two launch sites separated by a distance D. They both reach the same spot in their highest point and strike there. As a result of the impact they stick together and move as a single body afterwards. Find the place they will land.

116. Two identical objects (such as billiard balls) have a one-dimensional collision in which one is initially motionless. After the collision, the moving object is stationary and the other moves with the same speed as the other originally had. Show that both momentum and kinetic energy are conserved.

117. A ramp of mass M is at rest on a horizontal surface. A small cart of mass m is placed at the top of the ramp and released.

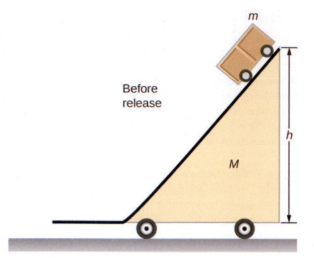

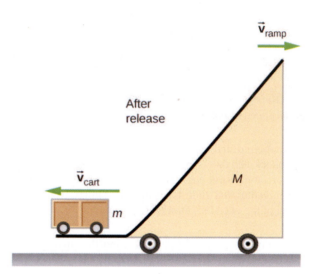

What are the velocities of the ramp and the cart relative to the ground at the instant the cart leaves the ramp?

118. Find the center of mass of the structure given in the figure below. Assume a uniform thickness of 20 cm, and a uniform density of 1 g/cm^3.

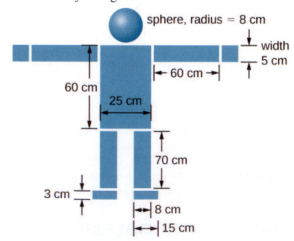

10 | FIXED-AXIS ROTATION

Figure 10.1 Brazos wind farm in west Texas. As of 2012, wind farms in the US had a power output of 60 gigawatts, enough capacity to power 15 million homes for a year. (credit: modification of work by U.S. Department of Energy)

Chapter Outline

10.1 Rotational Variables

10.2 Rotation with Constant Angular Acceleration

10.3 Relating Angular and Translational Quantities

10.4 Moment of Inertia and Rotational Kinetic Energy

10.5 Calculating Moments of Inertia

10.6 Torque

10.7 Newton's Second Law for Rotation

10.8 Work and Power for Rotational Motion

Introduction

In previous chapters, we described motion (kinematics) and how to change motion (dynamics), and we defined important concepts such as energy for objects that can be considered as point masses. Point masses, by definition, have no shape and so can only undergo translational motion. However, we know from everyday life that rotational motion is also very important and that many objects that move have both translation and rotation. The wind turbines in our chapter opening image are a prime example of how rotational motion impacts our daily lives, as the market for clean energy sources continues to grow.

We begin to address rotational motion in this chapter, starting with fixed-axis rotation. Fixed-axis rotation describes the rotation around a fixed axis of a rigid body; that is, an object that does not deform as it moves. We will show how to apply all the ideas we've developed up to this point about translational motion to an object rotating around a fixed axis. In the next chapter, we extend these ideas to more complex rotational motion, including objects that both rotate and translate, and objects that do not have a fixed rotational axis.

10.1 | Rotational Variables

So far in this text, we have mainly studied translational motion, including the variables that describe it: displacement, velocity, and acceleration. Now we expand our description of motion to rotation—specifically, rotational motion about a fixed axis. We will find that rotational motion is described by a set of related variables similar to those we used in translational motion.

Angular Velocity

Uniform circular motion (discussed previously in **Motion in Two and Three Dimensions**) is motion in a circle at constant speed. Although this is the simplest case of rotational motion, it is very useful for many situations, and we use it here to introduce rotational variables.

In **Figure 10.2**, we show a particle moving in a circle. The coordinate system is fixed and serves as a frame of reference to define the particle's position. Its position vector from the origin of the circle to the particle sweeps out the angle θ, which increases in the counterclockwise direction as the particle moves along its circular path. The angle θ is called the **angular position** of the particle. As the particle moves in its circular path, it also traces an arc length s.

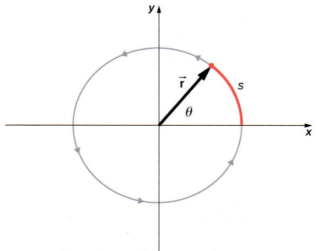

Figure 10.2 A particle follows a circular path. As it moves counterclockwise, it sweeps out a positive angle θ with respect to the x-axis and traces out an arc length s.

The angle is related to the radius of the circle and the arc length by

$$\theta = \frac{s}{r}.$$

(10.1)

The angle θ, the angular position of the particle along its path, has units of radians (rad). There are 2π radians in $360°$. Note that the radian measure is a ratio of length measurements, and therefore is a dimensionless quantity. As the particle moves along its circular path, its angular position changes and it undergoes angular displacements $\Delta\theta$.

We can assign vectors to the quantities in **Equation 10.1**. The angle $\vec{\theta}$ is a vector out of the page in **Figure 10.2**. The angular position vector $\vec{r}$ and the arc length $\vec{s}$ both lie in the plane of the page. These three vectors are related to each other by

$$\vec{s} = \vec{\theta} \times \vec{r}. \tag{10.2}$$

That is, the arc length is the cross product of the angle vector and the position vector, as shown in **Figure 10.3**.

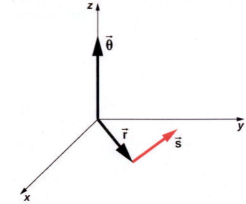

Figure 10.3 The angle vector points along the z-axis and the position vector and arc length vector both lie in the xy-plane. We see that $\vec{s} = \vec{\theta} \times \vec{r}$. All three vectors are perpendicular to each other.

The magnitude of the **angular velocity**, denoted by ω, is the time rate of change of the angle θ as the particle moves in its circular path. The **instantaneous angular velocity** is defined as the limit in which $\Delta t \to 0$ in the average angular velocity $\bar{\omega} = \frac{\Delta\theta}{\Delta t}$:

$$\omega = \lim_{\Delta t \to 0} \frac{\Delta\theta}{\Delta t} = \frac{d\theta}{dt}, \tag{10.3}$$

where θ is the angle of rotation (**Figure 10.2**). The units of angular velocity are radians per second (rad/s). Angular velocity can also be referred to as the rotation rate in radians per second. In many situations, we are given the rotation rate in revolutions/s or cycles/s. To find the angular velocity, we must multiply revolutions/s by 2π, since there are 2π radians in one complete revolution. Since the direction of a positive angle in a circle is counterclockwise, we take counterclockwise rotations as being positive and clockwise rotations as negative.

We can see how angular velocity is related to the tangential speed of the particle by differentiating **Equation 10.1** with respect to time. We rewrite **Equation 10.1** as

$$s = r\theta.$$

Taking the derivative with respect to time and noting that the radius r is a constant, we have

$$\frac{ds}{dt} = \frac{d}{dt}(r\theta) = \theta\frac{dr}{dt} + r\frac{d\theta}{dt} = r\frac{d\theta}{dt}$$

where $\theta\frac{dr}{dt} = 0$. Here $\frac{ds}{dt}$ is just the tangential speed v_t of the particle in **Figure 10.2**. Thus, by using **Equation 10.3**, we arrive at

$$v_t = r\omega. \tag{10.4}$$

That is, the tangential speed of the particle is its angular velocity times the radius of the circle. From **Equation 10.4**, we see that the tangential speed of the particle increases with its distance from the axis of rotation for a constant angular velocity. This effect is shown in **Figure 10.4**. Two particles are placed at different radii on a rotating disk with a constant angular velocity. As the disk rotates, the tangential speed increases linearly with the radius from the axis of rotation. In **Figure 10.4**, we see that $v_1 = r_1\omega_1$ and $v_2 = r_2\omega_2$. But the disk has a constant angular velocity, so $\omega_1 = \omega_2$. This means $\frac{v_1}{r_1} = \frac{v_2}{r_2}$ or $v_2 = \left(\frac{r_2}{r_1}\right)v_1$. Thus, since $r_2 > r_1$, $v_2 > v_1$.

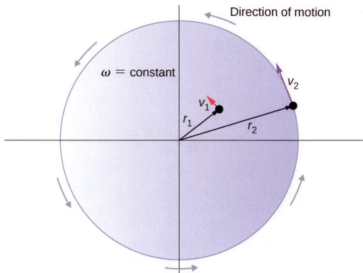

Figure 10.4 Two particles on a rotating disk have different tangential speeds, depending on their distance to the axis of rotation.

Up until now, we have discussed the magnitude of the angular velocity $\omega = d\theta/dt$, which is a scalar quantity—the change in angular position with respect to time. The vector $\vec{\omega}$ is the vector associated with the angular velocity and points along the axis of rotation. This is useful because when a rigid body is rotating, we want to know both the axis of rotation and the direction that the body is rotating about the axis, clockwise or counterclockwise. The angular velocity $\vec{\omega}$ gives us this information. The angular velocity $\vec{\omega}$ has a direction determined by what is called the right-hand rule. The right-hand rule is such that if the fingers of your right hand wrap counterclockwise from the *x*-axis (the direction in which θ increases) toward the *y*-axis, your thumb points in the direction of the positive *z*-axis (**Figure 10.5**). An angular velocity $\vec{\omega}$ that points along the positive *z*-axis therefore corresponds to a counterclockwise rotation, whereas an angular velocity $\vec{\omega}$ that points along the negative *z*-axis corresponds to a clockwise rotation.

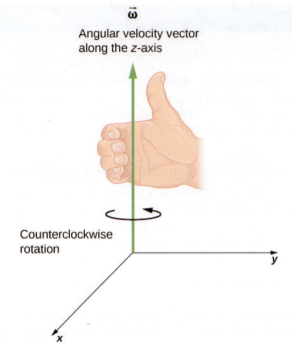

Figure 10.5 For counterclockwise rotation in the coordinate system shown, the angular velocity points in the positive z-direction by the right-hand-rule.

Similar to **Equation 10.2**, one can state a cross product relation to the vector of the tangential velocity as stated in **Equation 10.4**. Therefore, we have

$$\vec{v} = \vec{\omega} \times \vec{r} \, . \tag{10.5}$$

That is, the tangential velocity is the cross product of the angular velocity and the position vector, as shown in **Figure 10.6**. From part (a) of this figure, we see that with the angular velocity in the positive z-direction, the rotation in the xy-plane is counterclockwise. In part (b), the angular velocity is in the negative z-direction, giving a clockwise rotation in the xy-plane.

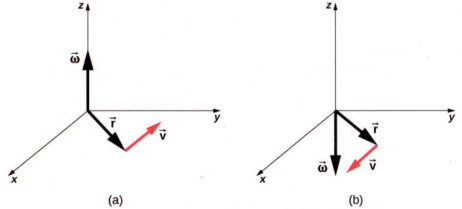

Figure 10.6 The vectors shown are the angular velocity, position, and tangential velocity. (a) The angular velocity points in the positive z-direction, giving a counterclockwise rotation in the xy-plane. (b) The angular velocity points in the negative z-direction, giving a clockwise rotation.

Example 10.1

Rotation of a Flywheel

A flywheel rotates such that it sweeps out an angle at the rate of $\theta = \omega t = (45.0 \text{ rad/s})t$ radians. The wheel rotates counterclockwise when viewed in the plane of the page. (a) What is the angular velocity of the flywheel? (b) What direction is the angular velocity? (c) How many radians does the flywheel rotate through in 30 s? (d) What is the tangential speed of a point on the flywheel 10 cm from the axis of rotation?

Strategy

The functional form of the angular position of the flywheel is given in the problem as $\theta(t) = \omega t$, so by taking the derivative with respect to time, we can find the angular velocity. We use the right-hand rule to find the angular velocity. To find the angular displacement of the flywheel during 30 s, we seek the angular displacement $\Delta\theta$, where the change in angular position is between 0 and 30 s. To find the tangential speed of a point at a distance from the axis of rotation, we multiply its distance times the angular velocity of the flywheel.

Solution

a. $\omega = \dfrac{d\theta}{dt} = 45 \text{ rad/s}$. We see that the angular velocity is a constant.

b. By the right-hand rule, we curl the fingers in the direction of rotation, which is counterclockwise in the plane of the page, and the thumb points in the direction of the angular velocity, which is out of the page.

c. $\Delta\theta = \theta(30 \text{ s}) - \theta(0 \text{ s}) = 45.0(30.0 \text{ s}) - 45.0(0 \text{ s}) = 1350.0 \text{ rad}$.

d. $v_t = r\omega = (0.1 \text{ m})(45.0 \text{ rad/s}) = 4.5 \text{ m/s}$.

Significance

In 30 s, the flywheel has rotated through quite a number of revolutions, about 215 if we divide the angular displacement by 2π. A massive flywheel can be used to store energy in this way, if the losses due to friction are minimal. Recent research has considered superconducting bearings on which the flywheel rests, with zero energy loss due to friction.

Angular Acceleration

We have just discussed angular velocity for uniform circular motion, but not all motion is uniform. Envision an ice skater spinning with his arms outstretched—when he pulls his arms inward, his angular velocity increases. Or think about a computer's hard disk slowing to a halt as the angular velocity decreases. We will explore these situations later, but we can already see a need to define an **angular acceleration** for describing situations where ω changes. The faster the change in ω, the greater the angular acceleration. We define the **instantaneous angular acceleration** α as the derivative of angular velocity with respect to time:

$$\alpha = \lim_{\Delta t \to 0} \frac{\Delta\omega}{\Delta t} = \frac{d\omega}{dt} = \frac{d^2\theta}{dt^2}, \tag{10.6}$$

where we have taken the limit of the average angular acceleration, $\bar{\alpha} = \frac{\Delta\omega}{\Delta t}$ as $\Delta t \to 0$.

The units of angular acceleration are (rad/s)/s, or rad/s^2.

In the same way as we defined the vector associated with angular velocity $\vec{\omega}$, we can define $\vec{\alpha}$, the vector associated with angular acceleration (**Figure 10.7**). If the angular velocity is along the positive z-axis, as in **Figure 10.5**, and $\frac{d\omega}{dt}$ is positive, then the angular acceleration $\vec{\alpha}$ is positive and points along the $+z$- axis. Similarly, if the angular velocity $\vec{\omega}$ is along the positive z-axis and $\frac{d\omega}{dt}$ is negative, then the angular acceleration is negative and points along the $+z$-

axis.

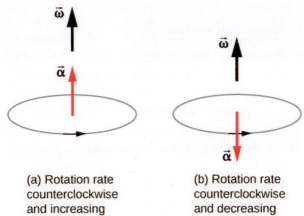

(a) Rotation rate counterclockwise and increasing

(b) Rotation rate counterclockwise and decreasing

Figure 10.7 The rotation is counterclockwise in both (a) and (b) with the angular velocity in the same direction. (a) The angular acceleration is in the same direction as the angular velocity, which increases the rotation rate. (b) The angular acceleration is in the opposite direction to the angular velocity, which decreases the rotation rate.

We can express the tangential acceleration vector as a cross product of the angular acceleration and the position vector. This expression can be found by taking the time derivative of $\vec{v} = \vec{\omega} \times \vec{r}$ and is left as an exercise:

$$\vec{a} = \vec{\alpha} \times \vec{r}.$$
(10.7)

The vector relationships for the angular acceleration and tangential acceleration are shown in **Figure 10.8**.

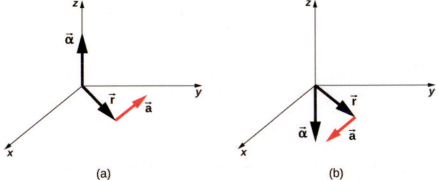

(a)   (b)

Figure 10.8 (a) The angular acceleration is the positive z-direction and produces a tangential acceleration in a counterclockwise sense. (b) The angular acceleration is in the negative z-direction and produces a tangential acceleration in the clockwise sense.

We can relate the tangential acceleration of a point on a rotating body at a distance from the axis of rotation in the same way that we related the tangential speed to the angular velocity. If we differentiate **Equation 10.4** with respect to time, noting that the radius r is constant, we obtain

$$a_{\text{t}} = r\alpha.$$
(10.8)

Thus, the tangential acceleration a_t is the radius times the angular acceleration. **Equation 10.4** and **Equation 10.8** are important for the discussion of rolling motion (see **Angular Momentum**).

Let's apply these ideas to the analysis of a few simple fixed-axis rotation scenarios. Before doing so, we present a problem-solving strategy that can be applied to rotational kinematics: the description of rotational motion.

Problem-Solving Strategy: Rotational Kinematics

1. Examine the situation to determine that rotational kinematics (rotational motion) is involved.

2. Identify exactly what needs to be determined in the problem (identify the unknowns). A sketch of the situation is useful.

3. Make a complete list of what is given or can be inferred from the problem as stated (identify the knowns).

4. Solve the appropriate equation or equations for the quantity to be determined (the unknown). It can be useful to think in terms of a translational analog, because by now you are familiar with the equations of translational motion.

5. Substitute the known values along with their units into the appropriate equation and obtain numerical solutions complete with units. Be sure to use units of radians for angles.

6. Check your answer to see if it is reasonable: Does your answer make sense?

Now let's apply this problem-solving strategy to a few specific examples.

Example 10.2

A Spinning Bicycle Wheel

A bicycle mechanic mounts a bicycle on the repair stand and starts the rear wheel spinning from rest to a final angular velocity of 250 rpm in 5.00 s. (a) Calculate the average angular acceleration in rad/s^2. (b) If she now hits the brakes, causing an angular acceleration of -87.3 rad/s^2, how long does it take the wheel to stop?

Strategy

The average angular acceleration can be found directly from its definition $\bar{\alpha} = \frac{\Delta\omega}{\Delta t}$ because the final angular velocity and time are given. We see that $\Delta\omega = \omega_{\text{final}} - \omega_{\text{initial}} = 250 \text{ rev/min}$ and Δt is 5.00 s. For part (b), we know the angular acceleration and the initial angular velocity. We can find the stopping time by using the definition of average angular acceleration and solving for Δt, yielding

$$\Delta t = \frac{\Delta\omega}{\alpha}.$$

Solution

a. Entering known information into the definition of angular acceleration, we get

$$\bar{\alpha} = \frac{\Delta\omega}{\Delta t} = \frac{250 \text{ rpm}}{5.00 \text{ s}}.$$

Because $\Delta\omega$ is in revolutions per minute (rpm) and we want the standard units of rad/s^2 for angular acceleration, we need to convert from rpm to rad/s:

$$\Delta\omega = 250\frac{\text{rev}}{\text{min}} \cdot \frac{2\pi \text{ rad}}{\text{rev}} \cdot \frac{1 \text{ min}}{60 \text{ s}} = 26.2\frac{\text{rad}}{\text{s}}.$$

Entering this quantity into the expression for α, we get

$$\alpha = \frac{\Delta\omega}{\Delta t} = \frac{26.2 \text{ rad/s}}{5.00 \text{ s}} = 5.24 \text{ rad/s}^2.$$

b. Here the angular velocity decreases from 26.2 rad/s (250 rpm) to zero, so that $\Delta\omega$ is -26.2 rad/s, and α is given to be -87.3 rad/s^2. Thus,

$$\Delta t = \frac{-26.2 \text{ rad/s}}{-87.3 \text{ rad/s}^2} = 0.300 \text{ s.}$$

Significance

Note that the angular acceleration as the mechanic spins the wheel is small and positive; it takes 5 s to produce an appreciable angular velocity. When she hits the brake, the angular acceleration is large and negative. The angular velocity quickly goes to zero.

 10.1 Check Your Understanding The fan blades on a turbofan jet engine (shown below) accelerate from rest up to a rotation rate of 40.0 rev/s in 20 s. The increase in angular velocity of the fan is constant in time. (The GE90-110B1 turbofan engine mounted on a Boeing 777, as shown, is currently the largest turbofan engine in the world, capable of thrusts of 330–510 kN.)

(a) What is the average angular acceleration?

(b) What is the instantaneous angular acceleration at any time during the first 20 s?

Figure 10.9 (credit: "Bubinator"/ Wikimedia Commons)

Example 10.3

Wind Turbine

A wind turbine (**Figure 10.10**) in a wind farm is being shut down for maintenance. It takes 30 s for the turbine to go from its operating angular velocity to a complete stop in which the angular velocity function is $\omega(t) = [(ts^{-1} - 30.0)^2 / 100.0] \text{rad/s}$. If the turbine is rotating counterclockwise looking into the page, (a) what are the directions of the angular velocity and acceleration vectors? (b) What is the average angular acceleration? (c) What is the instantaneous angular acceleration at $t = 0.0, 15.0, 30.0 \text{ s}$?

Figure 10.10 A wind turbine that is rotating counterclockwise, as seen head on.

Strategy

a. We are given the rotational sense of the turbine, which is counterclockwise in the plane of the page. Using the right hand rule (**Figure 10.5**), we can establish the directions of the angular velocity and acceleration vectors.

b. We calculate the initial and final angular velocities to get the average angular acceleration. We establish the sign of the angular acceleration from the results in (a).

c. We are given the functional form of the angular velocity, so we can find the functional form of the angular acceleration function by taking its derivative with respect to time.

Solution

a. Since the turbine is rotating counterclockwise, angular velocity $\vec{\omega}$ points out of the page. But since the angular velocity is decreasing, the angular acceleration $\vec{\alpha}$ points into the page, in the opposite sense to the angular velocity.

b. The initial angular velocity of the turbine, setting $t = 0$, is $\omega = 9.0 \, \text{rad/s}$. The final angular velocity is zero, so the average angular acceleration is

$$\bar{\alpha} = \frac{\Delta\omega}{\Delta t} = \frac{\omega - \omega_0}{t - t_0} = \frac{0 - 9.0 \, \text{rad/s}}{30.0 - 0 \, \text{s}} = -0.3 \, \text{rad/s}^2.$$

c. Taking the derivative of the angular velocity with respect to time gives $\alpha = \frac{d\omega}{dt} = (t - 30.0)/50.0 \, \text{rad/s}^2$

$$\alpha(0.0 \, \text{s}) = -0.6 \, \text{rad/s}^2, \; \alpha(15.0 \, \text{s}) = -0.3 \, \text{rad/s}^2, \; \text{and} \; \alpha(30.0 \, \text{s}) = 0 \, \text{rad/s}.$$

Significance

We found from the calculations in (a) and (b) that the angular acceleration α and the average angular acceleration $\bar{\alpha}$ are negative. The turbine has an angular acceleration in the opposite sense to its angular velocity.

We now have a basic vocabulary for discussing fixed-axis rotational kinematics and relationships between rotational variables. We discuss more definitions and connections in the next section.

10.2 | Rotation with Constant Angular Acceleration

Learning Objectives

By the end of this section, you will be able to:

- Derive the kinematic equations for rotational motion with constant angular acceleration
- Select from the kinematic equations for rotational motion with constant angular acceleration the appropriate equations to solve for unknowns in the analysis of systems undergoing fixed-axis rotation
- Use solutions found with the kinematic equations to verify the graphical analysis of fixed-axis rotation with constant angular acceleration

In the preceding section, we defined the rotational variables of angular displacement, angular velocity, and angular acceleration. In this section, we work with these definitions to derive relationships among these variables and use these relationships to analyze rotational motion for a rigid body about a fixed axis under a constant angular acceleration. This analysis forms the basis for rotational kinematics. If the angular acceleration is constant, the equations of rotational kinematics simplify, similar to the equations of linear kinematics discussed in **Motion along a Straight Line** and **Motion in Two and Three Dimensions**. We can then use this simplified set of equations to describe many applications in physics and engineering where the angular acceleration of the system is constant. Rotational kinematics is also a prerequisite to the discussion of rotational dynamics later in this chapter.

Kinematics of Rotational Motion

Using our intuition, we can begin to see how the rotational quantities θ, ω, α, and t are related to one another. For example, we saw in the preceding section that if a flywheel has an angular acceleration in the same direction as its angular velocity vector, its angular velocity increases with time and its angular displacement also increases. On the contrary, if the angular acceleration is opposite to the angular velocity vector, its angular velocity decreases with time. We can describe these physical situations and many others with a consistent set of rotational kinematic equations under a constant angular acceleration. The method to investigate rotational motion in this way is called **kinematics of rotational motion**.

To begin, we note that if the system is rotating under a constant acceleration, then the average angular velocity follows a simple relation because the angular velocity is increasing linearly with time. The average angular velocity is just half the sum of the initial and final values:

$$\bar{\omega} = \frac{\omega_0 + \omega_f}{2}. \tag{10.9}$$

From the definition of the average angular velocity, we can find an equation that relates the angular position, average angular velocity, and time:

$$\bar{\omega} = \frac{\Delta\theta}{\Delta t}.$$

Solving for θ, we have

$$\theta_f = \theta_0 + \bar{\omega}\, t, \tag{10.10}$$

where we have set $t_0 = 0$. This equation can be very useful if we know the average angular velocity of the system. Then we could find the angular displacement over a given time period. Next, we find an equation relating ω, α, and t. To determine this equation, we start with the definition of angular acceleration:

$$\alpha = \frac{d\omega}{dt}.$$

We rearrange this to get $\alpha dt = d\omega$ and then we integrate both sides of this equation from initial values to final values, that is, from t_0 to t and ω_0 to ω_f. In uniform rotational motion, the angular acceleration is constant so it can be pulled out of the integral, yielding two definite integrals:

$$\alpha \int_{t_0}^{t} dt' = \int_{\omega_0}^{\omega_f} d\omega.$$

Setting $t_0 = 0$, we have

$$\alpha t = \omega_f - \omega_0.$$

We rearrange this to obtain

$$\omega_f = \omega_0 + \alpha t, \tag{10.11}$$

where ω_0 is the initial angular velocity. **Equation 10.11** is the rotational counterpart to the linear kinematics equation $v_f = v_0 + at$. With **Equation 10.11**, we can find the angular velocity of an object at any specified time t given the initial angular velocity and the angular acceleration.

Let's now do a similar treatment starting with the equation $\omega = \dfrac{d\theta}{dt}$. We rearrange it to obtain $\omega dt = d\theta$ and integrate both sides from initial to final values again, noting that the angular acceleration is constant and does not have a time dependence. However, this time, the angular velocity is not constant (in general), so we substitute in what we derived above:

$$\int_{t_0}^{t_f} (\omega_0 + \alpha t') dt' = \int_{\theta_0}^{\theta_f} d\theta;$$

$$\int_{t_0}^{t} \omega_0 \, dt + \int_{t_0}^{t} \alpha t \, dt = \int_{\theta_0}^{\theta_f} d\theta = \left[\omega_0 t' + \alpha \left(\frac{(t')^2}{2} \right) \right]_{t_0}^{t} = \omega_0 t + \alpha \left(\frac{t^2}{2} \right) = \theta_f - \theta_0,$$

where we have set $t_0 = 0$. Now we rearrange to obtain

$$\theta_f = \theta_0 + \omega_0 t + \frac{1}{2} \alpha t^2. \tag{10.12}$$

Equation 10.12 is the rotational counterpart to the linear kinematics equation found in **Motion Along a Straight Line** for position as a function of time. This equation gives us the angular position of a rotating rigid body at any time t given the initial conditions (initial angular position and initial angular velocity) and the angular acceleration.

We can find an equation that is independent of time by solving for t in **Equation 10.11** and substituting into **Equation 10.12**. **Equation 10.12** becomes

$$\begin{aligned}
\theta_f &= \theta_0 + \omega_0\left(\frac{\omega_f - \omega_0}{\alpha}\right) + \frac{1}{2}\alpha\left(\frac{\omega_f - \omega_0}{\alpha}\right)^2 \\
&= \theta_0 + \frac{\omega_0\omega_f}{\alpha} - \frac{\omega_0^2}{\alpha} + \frac{1}{2}\frac{\omega_f^2}{\alpha} - \frac{\omega_0\omega_f}{\alpha} + \frac{1}{2}\frac{\omega_0^2}{\alpha} \\
&= \theta_0 + \frac{1}{2}\frac{\omega_f^2}{\alpha} - \frac{1}{2}\frac{\omega_0^2}{\alpha}, \\
\theta_f - \theta_0 &= \frac{\omega_f^2 - \omega_0^2}{2\alpha}
\end{aligned}$$

or

$$\omega_f^2 = \omega_0^2 + 2\alpha(\Delta\theta). \qquad (10.13)$$

Equation 10.10 through **Equation 10.13** describe fixed-axis rotation for constant acceleration and are summarized in **Table 10.1**.

Angular displacement from average angular velocity	$\theta_f = \theta_0 + \bar{\omega}\,t$
Angular velocity from angular acceleration	$\omega_f = \omega_0 + \alpha t$
Angular displacement from angular velocity and angular acceleration	$\theta_f = \theta_0 + \omega_0 t + \frac{1}{2}\alpha t^2$
Angular velocity from angular displacement and angular acceleration	$\omega_f^2 = \omega_0{}^2 + 2\alpha(\Delta\theta)$

Table 10.1 Kinematic Equations

Applying the Equations for Rotational Motion

Now we can apply the key kinematic relations for rotational motion to some simple examples to get a feel for how the equations can be applied to everyday situations.

Example 10.4

Calculating the Acceleration of a Fishing Reel

A deep-sea fisherman hooks a big fish that swims away from the boat, pulling the fishing line from his fishing reel. The whole system is initially at rest, and the fishing line unwinds from the reel at a radius of 4.50 cm from its axis of rotation. The reel is given an angular acceleration of $110\ \text{rad/s}^2$ for 2.00 s (**Figure 10.11**).

(a) What is the final angular velocity of the reel after 2 s?

(b) How many revolutions does the reel make?

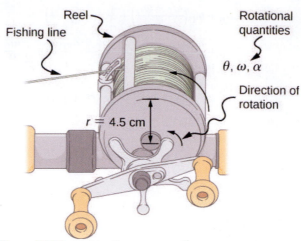

Figure 10.11 Fishing line coming off a rotating reel moves linearly.

Strategy

Identify the knowns and compare with the kinematic equations for constant acceleration. Look for the appropriate equation that can be solved for the unknown, using the knowns given in the problem description.

Solution

a. We are given α and t and want to determine ω. The most straightforward equation to use is $\omega_f = \omega_0 + \alpha t$, since all terms are known besides the unknown variable we are looking for. We are given that $\omega_0 = 0$ (it starts from rest), so

$$\omega_f = 0 + (110 \text{ rad/s}^2)(2.00 \text{ s}) = 220 \text{ rad/s}.$$

b. We are asked to find the number of revolutions. Because $1 \text{ rev} = 2\pi \text{ rad}$, we can find the number of revolutions by finding θ in radians. We are given α and t, and we know ω_0 is zero, so we can obtain θ by using

$$\theta_f = \theta_i + \omega_i t + \frac{1}{2}\alpha t^2$$
$$= 0 + 0 + (0.500)\left(110 \text{ rad/s}^2\right)(2.00 \text{ s})^2 = 220 \text{ rad}.$$

Converting radians to revolutions gives

$$\text{Number of rev} = (220 \text{ rad})\frac{1 \text{ rev}}{2\pi \text{ rad}} = 35.0 \text{ rev}.$$

Significance

This example illustrates that relationships among rotational quantities are highly analogous to those among linear quantities. The answers to the questions are realistic. After unwinding for two seconds, the reel is found to spin at 220 rad/s, which is 2100 rpm. (No wonder reels sometimes make high-pitched sounds.)

In the preceding example, we considered a fishing reel with a positive angular acceleration. Now let us consider what happens with a negative angular acceleration.

Example 10.5

Calculating the Duration When the Fishing Reel Slows Down and Stops

Now the fisherman applies a brake to the spinning reel, achieving an angular acceleration of -300 rad/s^2. How long does it take the reel to come to a stop?

Strategy

We are asked to find the time t for the reel to come to a stop. The initial and final conditions are different from those in the previous problem, which involved the same fishing reel. Now we see that the initial angular velocity is $\omega_0 = 220$ rad/s and the final angular velocity ω is zero. The angular acceleration is given as $\alpha = -300$ rad/s^2. Examining the available equations, we see all quantities but t are known in $\omega_f = \omega_0 + \alpha t$, making it easiest to use this equation.

Solution

The equation states

$$\omega_f = \omega_0 + \alpha t.$$

We solve the equation algebraically for t and then substitute the known values as usual, yielding

$$t = \frac{\omega_f - \omega_0}{\alpha} = \frac{0 - 220.0 \text{ rad/s}}{-300.0 \text{ rad/s}^2} = 0.733 \text{ s}.$$

Significance

Note that care must be taken with the signs that indicate the directions of various quantities. Also, note that the time to stop the reel is fairly small because the acceleration is rather large. Fishing lines sometimes snap because of the accelerations involved, and fishermen often let the fish swim for a while before applying brakes on the reel. A tired fish is slower, requiring a smaller acceleration.

 10.2 Check Your Understanding A centrifuge used in DNA extraction spins at a maximum rate of 7000 rpm, producing a "g-force" on the sample that is 6000 times the force of gravity. If the centrifuge takes 10 seconds to come to rest from the maximum spin rate: (a) What is the angular acceleration of the centrifuge? (b) What is the angular displacement of the centrifuge during this time?

Example 10.6

Angular Acceleration of a Propeller

Figure 10.12 shows a graph of the angular velocity of a propeller on an aircraft as a function of time. Its angular velocity starts at 30 rad/s and drops linearly to 0 rad/s over the course of 5 seconds. (a) Find the angular acceleration of the object and verify the result using the kinematic equations. (b) Find the angle through which the propeller rotates during these 5 seconds and verify your result using the kinematic equations.

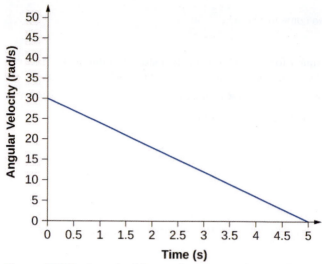

Figure 10.12   A graph of the angular velocity of a propeller versus time.

Strategy

a.   Since the angular velocity varies linearly with time, we know that the angular acceleration is constant and does not depend on the time variable. The angular acceleration is the slope of the angular velocity vs. time graph, $\alpha = \frac{d\omega}{dt}$. To calculate the slope, we read directly from **Figure 10.12**, and see that $\omega_0 = 30$ rad/s

at $t = 0$ s and $\omega_f = 0$ rad/s at $t = 5$ s . Then, we can verify the result using $\omega = \omega_0 + \alpha t$.

b.   We use the equation $\omega = \frac{d\theta}{dt}$; since the time derivative of the angle is the angular velocity, we can find

the angular displacement by integrating the angular velocity, which from the figure means taking the area under the angular velocity graph. In other words:

$$\int_{\theta_0}^{\theta_f} d\theta = \theta_f - \theta_0 = \int_{t_0}^{t_f} \omega(t) dt.$$

Then we use the kinematic equations for constant acceleration to verify the result.

Solution

a.   Calculating the slope, we get

$$\alpha = \frac{\omega - \omega_0}{t - t_0} = \frac{(0 - 30.0)\ \text{rad/s}}{(5.0 - 0)\ \text{s}} = -6.0\ \text{rad/s}^2.$$

We see that this is exactly **Equation 10.11** with a little rearranging of terms.

b.   We can find the area under the curve by calculating the area of the right triangle, as shown in **Figure 10.13**.

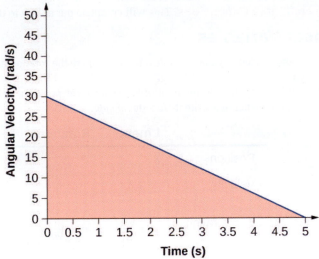

Figure 10.13 The area under the curve is the area of the right triangle.

$$\Delta\theta = \text{area(triangle)};$$
$$\Delta\theta = \frac{1}{2}(30 \text{ rad/s})(5 \text{ s}) = 75 \text{ rad}.$$

We verify the solution using **Equation 10.12**:

$$\theta_f = \theta_0 + \omega_0 t + \frac{1}{2}\alpha t^2.$$

Setting $\theta_0 = 0$, we have

$$\theta_0 = (30.0 \text{ rad/s})(5.0 \text{ s}) + \frac{1}{2}(-6.0 \text{ rad/s}^2)(5.0 \text{ rad/s})^2 = 150.0 - 75.0 = 75.0 \text{ rad}.$$

This verifies the solution found from finding the area under the curve.

Significance

We see from part (b) that there are alternative approaches to analyzing fixed-axis rotation with constant acceleration. We started with a graphical approach and verified the solution using the rotational kinematic equations. Since $\alpha = \frac{d\omega}{dt}$, we could do the same graphical analysis on an angular acceleration-vs.-time curve.

The area under an α-vs.-t curve gives us the change in angular velocity. Since the angular acceleration is constant in this section, this is a straightforward exercise.

10.3 | Relating Angular and Translational Quantities

Learning Objectives

By the end of this section, you will be able to:

- Given the linear kinematic equation, write the corresponding rotational kinematic equation
- Calculate the linear distances, velocities, and accelerations of points on a rotating system given the angular velocities and accelerations

In this section, we relate each of the rotational variables to the translational variables defined in **Motion Along a Straight Line** and **Motion in Two and Three Dimensions**. This will complete our ability to describe rigid-body rotations.

Angular vs. Linear Variables

In **Rotational Variables**, we introduced angular variables. If we compare the rotational definitions with the definitions of linear kinematic variables from **Motion Along a Straight Line** and **Motion in Two and Three Dimensions**, we find that there is a mapping of the linear variables to the rotational ones. Linear position, velocity, and acceleration have their rotational counterparts, as we can see when we write them side by side:

	Linear	**Rotational**
Position	x	θ
Velocity	$v = \dfrac{dx}{dt}$	$\omega = \dfrac{d\theta}{dt}$
Acceleration	$a = \dfrac{dv}{dt}$	$\alpha = \dfrac{d\omega}{dt}$

Let's compare the linear and rotational variables individually. The linear variable of position has physical units of meters, whereas the angular position variable has dimensionless units of radians, as can be seen from the definition of $\theta = \frac{s}{r}$, which is the ratio of two lengths. The linear velocity has units of m/s, and its counterpart, the angular velocity, has units of rad/s. In **Rotational Variables**, we saw in the case of circular motion that the linear tangential speed of a particle at a radius r from the axis of rotation is related to the angular velocity by the relation $v_t = r\omega$. This could also apply to points on a rigid body rotating about a fixed axis. Here, we consider only circular motion. In circular motion, both uniform and nonuniform, there exists a centripetal acceleration (**Motion in Two and Three Dimensions**). The centripetal acceleration vector points inward from the particle executing circular motion toward the axis of rotation. The derivation of the magnitude of the centripetal acceleration is given in **Motion in Two and Three Dimensions**. From that derivation, the magnitude of the centripetal acceleration was found to be

$$a_c = \frac{v_t^2}{r},$$ (10.14)

where r is the radius of the circle.

Thus, in uniform circular motion when the angular velocity is constant and the angular acceleration is zero, we have a linear acceleration—that is, centripetal acceleration—since the tangential speed in **Equation 10.14** is a constant. If nonuniform circular motion is present, the rotating system has an angular acceleration, and we have both a linear centripetal acceleration that is changing (because v_t is changing) as well as a linear tangential acceleration. These relationships are shown in **Figure 10.14**, where we show the centripetal and tangential accelerations for uniform and nonuniform circular motion.

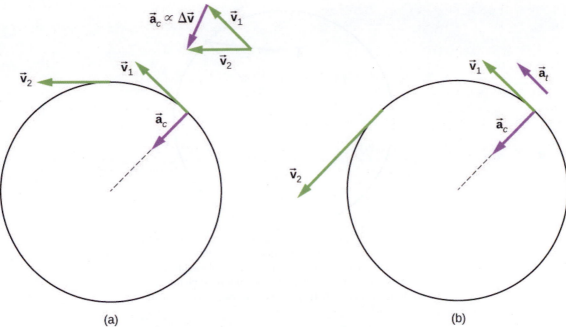

Figure 10.14 (a) Uniform circular motion: The centripetal acceleration a_c has its vector inward toward the axis of rotation. There is no tangential acceleration. (b) Nonuniform circular motion: An angular acceleration produces an inward centripetal acceleration that is changing in magnitude, plus a tangential acceleration a_t.

The centripetal acceleration is due to the change in the direction of tangential velocity, whereas the tangential acceleration is due to any change in the magnitude of the tangential velocity. The tangential and centripetal acceleration vectors $\vec{a}_t$ and $\vec{a}_c$ are always perpendicular to each other, as seen in **Figure 10.14**. To complete this description, we can assign a **total linear acceleration** vector to a point on a rotating rigid body or a particle executing circular motion at a radius r from a fixed axis. The total linear acceleration vector $\vec{a}$ is the vector sum of the centripetal and tangential accelerations,

$$\vec{a} = \vec{a}_c + \vec{a}_t. \tag{10.15}$$

The total linear acceleration vector in the case of nonuniform circular motion points at an angle between the centripetal and tangential acceleration vectors, as shown in **Figure 10.15**. Since $\vec{a}_c \perp \vec{a}_t$, the magnitude of the total linear acceleration is

$$\left| \vec{a} \right| = \sqrt{a_c^2 + a_t^2}.$$

Note that if the angular acceleration is zero, the total linear acceleration is equal to the centripetal acceleration.

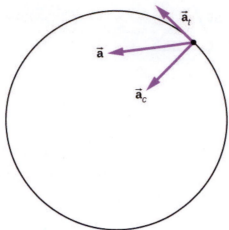

Figure 10.15 A particle is executing circular motion and has an angular acceleration. The total linear acceleration of the particle is the vector sum of the centripetal acceleration and tangential acceleration vectors. The total linear acceleration vector is at an angle in between the centripetal and tangential accelerations.

Relationships between Rotational and Translational Motion

We can look at two relationships between rotational and translational motion.

1. Generally speaking, the linear kinematic equations have their rotational counterparts. **Table 10.2** lists the four linear kinematic equations and the corresponding rotational counterpart. The two sets of equations look similar to each other, but describe two different physical situations, that is, rotation and translation.

Rotational	Translational
$\theta_f = \theta_0 + \bar{\omega} t$	$x = x_0 + \bar{v} t$
$\omega_f = \omega_0 + \alpha t$	$v_f = v_0 + at$
$\theta_f = \theta_0 + \omega_0 t + \frac{1}{2}\alpha t^2$	$x_f = x_0 + v_0 t + \frac{1}{2}at^2$
$\omega_f^2 = \omega_0^2 + 2\alpha(\Delta\theta)$	$v_f^2 = v_0^2 + 2a(\Delta x)$

Table 10.2 Rotational and Translational Kinematic Equations

2. The second correspondence has to do with relating linear and rotational variables in the special case of circular motion. This is shown in **Table 10.3**, where in the third column, we have listed the connecting equation that relates the linear variable to the rotational variable. The rotational variables of angular velocity and acceleration have subscripts that indicate their definition in circular motion.

Rotational	Translational	Relationship ($r = $ radius)
θ	s	$\theta = \frac{s}{r}$
ω	v_t	$\omega = \frac{v_t}{r}$
α	a_t	$\alpha = \frac{a_t}{r}$

Table 10.3 Rotational and Translational Quantities: Circular Motion

Rotational	Translational	Relationship (r = radius)
	a_c	$a_c = \dfrac{v_t^2}{r}$

Table 10.3 Rotational and Translational Quantities: Circular Motion

Example 10.7

Linear Acceleration of a Centrifuge

A centrifuge has a radius of 20 cm and accelerates from a maximum rotation rate of 10,000 rpm to rest in 30 seconds under a constant angular acceleration. It is rotating counterclockwise. What is the magnitude of the total acceleration of a point at the tip of the centrifuge at $t = 29.0$s? What is the direction of the total acceleration vector?

Strategy

With the information given, we can calculate the angular acceleration, which then will allow us to find the tangential acceleration. We can find the centripetal acceleration at $t = 0$ by calculating the tangential speed at this time. With the magnitudes of the accelerations, we can calculate the total linear acceleration. From the description of the rotation in the problem, we can sketch the direction of the total acceleration vector.

Solution

The angular acceleration is

$$\alpha = \frac{\omega - \omega_0}{t} = \frac{0 - (1.0 \times 10^4)2\pi/60.0 \, \text{s(rad/s)}}{30.0 \, \text{s}} = -34.9 \, \text{rad/s}^2.$$

Therefore, the tangential acceleration is

$$a_t = r\alpha = 0.2 \, \text{m}(-34.9 \, \text{rad/s}^2) = -7.0 \, \text{m/s}^2.$$

The angular velocity at $t = 29.0$ s is

$$\omega = \omega_0 + \alpha t = 1.0 \times 10^4 \left(\frac{2\pi}{60.0 \, \text{s}}\right) + \left(-34.9 \, \text{rad/s}^2\right)(29.0 \, \text{s})$$

$$= 1047.2 \, \text{rad/s} - 1012.71 = 35.1 \, \text{rad/s}.$$

Thus, the tangential speed at $t = 29.0$ s is

$$v_t = r\omega = 0.2 \, \text{m}(35.1 \, \text{rad/s}) = 7.0 \, \text{m/s}.$$

We can now calculate the centripetal acceleration at $t = 29.0$ s :

$$a_c = \frac{v^2}{r} = \frac{(7.0 \, \text{m/s})^2}{0.2 \, \text{m}} = 245.0 \, \text{m/s}^2.$$

Since the two acceleration vectors are perpendicular to each other, the magnitude of the total linear acceleration is

$$\left|\vec{a}\right| = \sqrt{a_c^2 + a_t^2} = \sqrt{(245.0)^2 + (-7.0)^2} = 245.1 \, \text{m/s}^2.$$

Since the centrifuge has a negative angular acceleration, it is slowing down. The total acceleration vector is as shown in **Figure 10.16**. The angle with respect to the centripetal acceleration vector is

$$\theta = \tan^{-1}\frac{-7.0}{245.0} = -1.6°.$$

The negative sign means that the total acceleration vector is angled toward the clockwise direction.

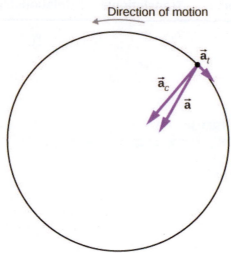

Direction of motion

Figure 10.16 The centripetal, tangential, and total acceleration vectors. The centrifuge is slowing down, so the tangential acceleration is clockwise, opposite the direction of rotation (counterclockwise).

Significance

From **Figure 10.16**, we see that the tangential acceleration vector is opposite the direction of rotation. The magnitude of the tangential acceleration is much smaller than the centripetal acceleration, so the total linear acceleration vector will make a very small angle with respect to the centripetal acceleration vector.

  10.3 Check Your Understanding A boy jumps on a merry-go-round with a radius of 5 m that is at rest. It starts accelerating at a constant rate up to an angular velocity of 5 rad/s in 20 seconds. What is the distance travelled by the boy?

  Check out this **PhET simulation (https://openstaxcollege.org/l/21rotatingdisk)** to change the parameters of a rotating disk (the initial angle, angular velocity, and angular acceleration), and place bugs at different radial distances from the axis. The simulation then lets you explore how circular motion relates to the bugs' *xy*-position, velocity, and acceleration using vectors or graphs.

10.4 | Moment of Inertia and Rotational Kinetic Energy

Learning Objectives

By the end of this section, you will be able to:

- Describe the differences between rotational and translational kinetic energy
- Define the physical concept of moment of inertia in terms of the mass distribution from the rotational axis
- Explain how the moment of inertia of rigid bodies affects their rotational kinetic energy
- Use conservation of mechanical energy to analyze systems undergoing both rotation and translation
- Calculate the angular velocity of a rotating system when there are energy losses due to nonconservative forces

So far in this chapter, we have been working with rotational kinematics: the description of motion for a rotating rigid body with a fixed axis of rotation. In this section, we define two new quantities that are helpful for analyzing properties of rotating objects: moment of inertia and rotational kinetic energy. With these properties defined, we will have two important tools we need for analyzing rotational dynamics.

Rotational Kinetic Energy

Any moving object has kinetic energy. We know how to calculate this for a body undergoing translational motion, but how about for a rigid body undergoing rotation? This might seem complicated because each point on the rigid body has a different velocity. However, we can make use of angular velocity—which is the same for the entire rigid body—to express the kinetic energy for a rotating object. **Figure 10.17** shows an example of a very energetic rotating body: an electric grindstone propelled by a motor. Sparks are flying, and noise and vibration are generated as the grindstone does its work. This system has considerable energy, some of it in the form of heat, light, sound, and vibration. However, most of this energy is in the form of **rotational kinetic energy**.

Figure 10.17 The rotational kinetic energy of the grindstone is converted to heat, light, sound, and vibration. (credit: Zachary David Bell, US Navy)

Energy in rotational motion is not a new form of energy; rather, it is the energy associated with rotational motion, the same as kinetic energy in translational motion. However, because kinetic energy is given by $K = \frac{1}{2}mv^2$, and velocity is a quantity that is different for every point on a rotating body about an axis, it makes sense to find a way to write kinetic energy in terms of the variable ω, which is the same for all points on a rigid rotating body. For a single particle rotating around a fixed axis, this is straightforward to calculate. We can relate the angular velocity to the magnitude of the translational velocity using the relation $v_t = \omega r$, where r is the distance of the particle from the axis of rotation and v_t is its tangential speed. Substituting into the equation for kinetic energy, we find

$$K = \tfrac{1}{2}mv_t^2 = \tfrac{1}{2}m(\omega r)^2 = \tfrac{1}{2}(mr^2)\omega^2.$$

In the case of a rigid rotating body, we can divide up any body into a large number of smaller masses, each with a mass m_j and distance to the axis of rotation r_j, such that the total mass of the body is equal to the sum of the individual masses: $M = \sum_j m_j$. Each smaller mass has tangential speed v_j, where we have dropped the subscript t for the moment. The total kinetic energy of the rigid rotating body is

$$K = \sum_j \tfrac{1}{2}m_j v_j^2 = \sum_j \tfrac{1}{2}m_j (r_j \omega_j)^2$$

and since $\omega_j = \omega$ for all masses,

$$K = \frac{1}{2}\left(\sum_j m_j r_j^2\right)\omega^2. \tag{10.16}$$

The units of **Equation 10.16** are joules (J). The equation in this form is complete, but awkward; we need to find a way to generalize it.

Moment of Inertia

If we compare **Equation 10.16** to the way we wrote kinetic energy in **Work and Kinetic Energy**, $\left(\frac{1}{2}mv^2\right)$, this suggests we have a new rotational variable to add to our list of our relations between rotational and translational variables. The quantity $\sum_j m_j r_j^2$ is the counterpart for mass in the equation for rotational kinetic energy. This is an important new term for rotational motion. This quantity is called the **moment of inertia** I, with units of $\text{kg} \cdot \text{m}^2$:

$$I = \sum_j m_j r_j^2. \tag{10.17}$$

For now, we leave the expression in summation form, representing the moment of inertia of a system of point particles rotating about a fixed axis. We note that the moment of inertia of a single point particle about a fixed axis is simply mr^2, with r being the distance from the point particle to the axis of rotation. In the next section, we explore the integral form of this equation, which can be used to calculate the moment of inertia of some regular-shaped rigid bodies.

The moment of inertia is the quantitative measure of rotational inertia, just as in translational motion, and mass is the quantitative measure of linear inertia—that is, the more massive an object is, the more inertia it has, and the greater is its resistance to change in linear velocity. Similarly, the greater the moment of inertia of a rigid body or system of particles, the greater is its resistance to change in angular velocity about a fixed axis of rotation. It is interesting to see how the moment of inertia varies with r, the distance to the axis of rotation of the mass particles in **Equation 10.17**. Rigid bodies and systems of particles with more mass concentrated at a greater distance from the axis of rotation have greater moments of inertia than bodies and systems of the same mass, but concentrated near the axis of rotation. In this way, we can see that a hollow cylinder has more rotational inertia than a solid cylinder of the same mass when rotating about an axis through the center. Substituting **Equation 10.17** into **Equation 10.16**, the expression for the kinetic energy of a rotating rigid body becomes

$$K = \frac{1}{2}I\omega^2. \tag{10.18}$$

We see from this equation that the kinetic energy of a rotating rigid body is directly proportional to the moment of inertia and the square of the angular velocity. This is exploited in flywheel energy-storage devices, which are designed to store large amounts of rotational kinetic energy. Many carmakers are now testing flywheel energy storage devices in their automobiles, such as the flywheel, or kinetic energy recovery system, shown in **Figure 10.18**.

Figure 10.18 A KERS (kinetic energy recovery system) flywheel used in cars. (credit: "cmonville"/Flickr)

The rotational and translational quantities for kinetic energy and inertia are summarized in **Table 10.4**. The relationship column is not included because a constant doesn't exist by which we could multiply the rotational quantity to get the translational quantity, as can be done for the variables in **Table 10.3**.

Rotational	Translational
$I = \sum_j m_j r_j^2$	m
$K = \frac{1}{2}I\omega^2$	$K = \frac{1}{2}mv^2$

Table 10.4 Rotational and Translational Kinetic Energies and Inertia

Example 10.8

Moment of Inertia of a System of Particles

Six small washers are spaced 10 cm apart on a rod of negligible mass and 0.5 m in length. The mass of each washer is 20 g. The rod rotates about an axis located at 25 cm, as shown in **Figure 10.19**. (a) What is the moment of inertia of the system? (b) If the two washers closest to the axis are removed, what is the moment of inertia of the remaining four washers? (c) If the system with six washers rotates at 5 rev/s, what is its rotational kinetic energy?

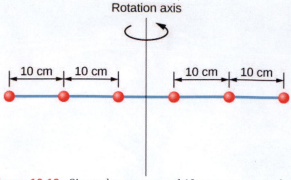

Figure 10.19 Six washers are spaced 10 cm apart on a rod of
negligible mass and rotating about a vertical axis.

Strategy

a. We use the definition for moment of inertia for a system of particles and perform the summation to
    evaluate this quantity. The masses are all the same so we can pull that quantity in front of the summation
    symbol.

b. We do a similar calculation.

c. We insert the result from (a) into the expression for rotational kinetic energy.

Solution

a. $I = \sum_j m_j r_j^2 = (0.02 \text{ kg})(2 \times (0.25 \text{ m})^2 + 2 \times (0.15 \text{ m})^2 + 2 \times (0.05 \text{ m})^2) = 0.0035 \text{ kg} \cdot \text{m}^2$.

b. $I = \sum_j m_j r_j^2 = (0.02 \text{ kg})(2 \times (0.25 \text{ m})^2 + 2 \times (0.15 \text{ m})^2) = 0.0034 \text{ kg} \cdot \text{m}^2$.

c. $K = \frac{1}{2}I\omega^2 = \frac{1}{2}(0.0035 \text{ kg} \cdot \text{m}^2)(5.0 \times 2\pi \text{ rad/s})^2 = 1.73 \text{ J}$.

Significance

We can see the individual contributions to the moment of inertia. The masses close to the axis of rotation have a
very small contribution. When we removed them, it had a very small effect on the moment of inertia.

In the next section, we generalize the summation equation for point particles and develop a method to calculate moments of
inertia for rigid bodies. For now, though, **Figure 10.20** gives values of rotational inertia for common object shapes around
specified axes.

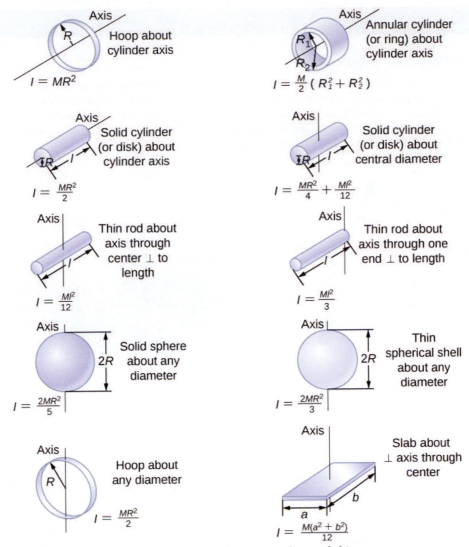

Figure 10.20 Values of rotational inertia for common shapes of objects.

Applying Rotational Kinetic Energy

Now let's apply the ideas of rotational kinetic energy and the moment of inertia table to get a feeling for the energy associated with a few rotating objects. The following examples will also help get you comfortable using these equations. First, let's look at a general problem-solving strategy for rotational energy.

> ### Problem-Solving Strategy: Rotational Energy
>
> 1. Determine that energy or work is involved in the rotation.
>
> 2. Determine the system of interest. A sketch usually helps.
>
> 3. Analyze the situation to determine the types of work and energy involved.
>
> 4. If there are no losses of energy due to friction and other nonconservative forces, mechanical energy is conserved, that is, $K_i + U_i = K_f + U_f$.
>
> 5. If nonconservative forces are present, mechanical energy is not conserved, and other forms of energy, such as heat and light, may enter or leave the system. Determine what they are and calculate them as necessary.
>
> 6. Eliminate terms wherever possible to simplify the algebra.
>
> 7. Evaluate the numerical solution to see if it makes sense in the physical situation presented in the wording of the problem.

Example 10.9

Calculating Helicopter Energies

A typical small rescue helicopter has four blades: Each is 4.00 m long and has a mass of 50.0 kg (**Figure 10.21**). The blades can be approximated as thin rods that rotate about one end of an axis perpendicular to their length. The helicopter has a total loaded mass of 1000 kg. (a) Calculate the rotational kinetic energy in the blades when they rotate at 300 rpm. (b) Calculate the translational kinetic energy of the helicopter when it flies at 20.0 m/s, and compare it with the rotational energy in the blades.

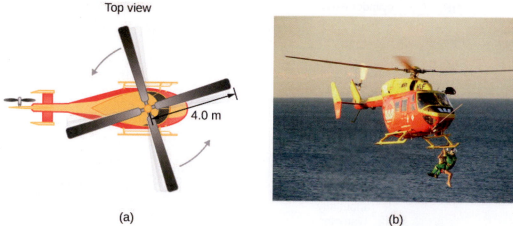

Figure 10.21 (a) Sketch of a four-blade helicopter. (b) A water rescue operation featuring a helicopter from the Auckland Westpac Rescue Helicopter Service. (credit b: modification of work by "111 Emergency"/Flickr)

Strategy

Rotational and translational kinetic energies can be calculated from their definitions. The wording of the problem gives all the necessary constants to evaluate the expressions for the rotational and translational kinetic energies.

Solution

a. The rotational kinetic energy is

$$K = \frac{1}{2}I\omega^2.$$

We must convert the angular velocity to radians per second and calculate the moment of inertia before we can find K. The angular velocity ω is

$$\omega = \frac{300 \text{ rev}}{1.00 \text{ min}} \frac{2\pi \text{ rad}}{1 \text{ rev}} \frac{1.00 \text{ min}}{60.0 \text{ s}} = 31.4 \frac{\text{rad}}{\text{s}}.$$

The moment of inertia of one blade is that of a thin rod rotated about its end, listed in **Figure 10.20**. The total I is four times this moment of inertia because there are four blades. Thus,

$$I = 4\frac{Ml^2}{3} = 4 \times \frac{(50.0 \text{ kg})(4.00 \text{ m})^2}{3} = 1067.0 \text{ kg} \cdot \text{m}^2.$$

Entering ω and I into the expression for rotational kinetic energy gives

$$K = 0.5(1067 \text{ kg} \cdot \text{m}^2)(31.4 \text{ rad/s})^2 = 5.26 \times 10^5 \text{ J}.$$

b. Entering the given values into the equation for translational kinetic energy, we obtain

$$K = \frac{1}{2}mv^2 = (0.5)(1000.0 \text{ kg})(20.0 \text{ m/s})^2 = 2.00 \times 10^5 \text{ J}.$$

To compare kinetic energies, we take the ratio of translational kinetic energy to rotational kinetic energy. This ratio is

$$\frac{2.00 \times 10^5 \text{ J}}{5.26 \times 10^5 \text{ J}} = 0.380.$$

Significance

The ratio of translational energy to rotational kinetic energy is only 0.380. This ratio tells us that most of the kinetic energy of the helicopter is in its spinning blades.

Example 10.10

Energy in a Boomerang

A person hurls a boomerang into the air with a velocity of 30.0 m/s at an angle of $40.0°$ with respect to the horizontal (**Figure 10.22**). It has a mass of 1.0 kg and is rotating at 10.0 rev/s. The moment of inertia of the boomerang is given as $I = \frac{1}{12}mL^2$ where $L = 0.7 \text{ m}$. (a) What is the total energy of the boomerang when it leaves the hand? (b) How high does the boomerang go from the elevation of the hand, neglecting air resistance?

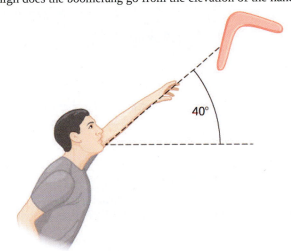

Figure 10.22 A boomerang is hurled into the air at an initial angle of $40°$.

Strategy

We use the definitions of rotational and linear kinetic energy to find the total energy of the system. The problem states to neglect air resistance, so we don't have to worry about energy loss. In part (b), we use conservation of mechanical energy to find the maximum height of the boomerang.

Solution

a. Moment of inertia: $I = \frac{1}{12}mL^2 = \frac{1}{12}(1.0 \text{ kg})(0.7\text{m})^2 = 0.041 \text{ kg} \cdot \text{m}^2$.

Angular velocity: $\omega = (10.0 \text{ rev/s})(2\pi) = 62.83 \text{ rad/s}$.

The rotational kinetic energy is therefore

$$K_R = \frac{1}{2}(0.041 \text{ kg} \cdot \text{m}^2)(62.83 \text{ rad/s})^2 = 80.93 \text{ J}.$$

The translational kinetic energy is

$$K_T = \frac{1}{2}mv^2 = \frac{1}{2}(1.0 \text{ kg})(30.0 \text{ m/s})^2 = 450.0 \text{ J}.$$

Thus, the total energy in the boomerang is

$$K_{\text{Total}} = K_{\text{R}} + K_{\text{T}} = 80.93 + 450.0 = 530.93 \text{ J}.$$

b. We use conservation of mechanical energy. Since the boomerang is launched at an angle, we need to write the total energies of the system in terms of its linear kinetic energies using the velocity in the x- and y-directions. The total energy when the boomerang leaves the hand is

$$E_{\text{Before}} = \tfrac{1}{2}mv_x^2 + \tfrac{1}{2}mv_y^2 + \tfrac{1}{2}I\omega^2.$$

The total energy at maximum height is

$$E_{\text{Final}} = \tfrac{1}{2}mv_x^2 + \tfrac{1}{2}I\omega^2 + mgh.$$

By conservation of mechanical energy, $E_{\text{Before}} = E_{\text{Final}}$ so we have, after canceling like terms,

$$\tfrac{1}{2}mv_y^2 = mgh.$$

Since $v_y = 30.0 \text{ m/s}(\sin 40°) = 19.28 \text{ m/s}$, we find

$$h = \frac{(19.28 \text{ m/s})^2}{2(9.8 \text{ m/s}^2)} = 18.97 \text{ m}.$$

Significance

In part (b), the solution demonstrates how energy conservation is an alternative method to solve a problem that normally would be solved using kinematics. In the absence of air resistance, the rotational kinetic energy was not a factor in the solution for the maximum height.

 10.4 **Check Your Understanding** A nuclear submarine propeller has a moment of inertia of $800.0 \text{ kg} \cdot \text{m}^2$.

If the submerged propeller has a rotation rate of 4.0 rev/s when the engine is cut, what is the rotation rate of the propeller after 5.0 s when water resistance has taken 50,000 J out of the system?

10.5 | Calculating Moments of Inertia

Learning Objectives

By the end of this section, you will be able to:

- Calculate the moment of inertia for uniformly shaped, rigid bodies
- Apply the parallel axis theorem to find the moment of inertia about any axis parallel to one already known
- Calculate the moment of inertia for compound objects

In the preceding section, we defined the moment of inertia but did not show how to calculate it. In this section, we show how to calculate the moment of inertia for several standard types of objects, as well as how to use known moments of inertia to find the moment of inertia for a shifted axis or for a compound object. This section is very useful for seeing how to apply a general equation to complex objects (a skill that is critical for more advanced physics and engineering courses).

Moment of Inertia

We defined the moment of inertia I of an object to be $I = \sum_i m_i r_i^2$ for all the point masses that make up the object.

Because r is the distance to the axis of rotation from each piece of mass that makes up the object, the moment of inertia for any object depends on the chosen axis. To see this, let's take a simple example of two masses at the end of a massless (negligibly small mass) rod (**Figure 10.23**) and calculate the moment of inertia about two different axes. In this case, the summation over the masses is simple because the two masses at the end of the barbell can be approximated as point masses, and the sum therefore has only two terms.

In the case with the axis in the center of the barbell, each of the two masses m is a distance R away from the axis, giving a moment of inertia of

$$I_1 = mR^2 + mR^2 = 2mR^2.$$

In the case with the axis at the end of the barbell—passing through one of the masses—the moment of inertia is

$$I_2 = m(0)^2 + m(2R)^2 = 4mR^2.$$

From this result, we can conclude that it is twice as hard to rotate the barbell about the end than about its center.

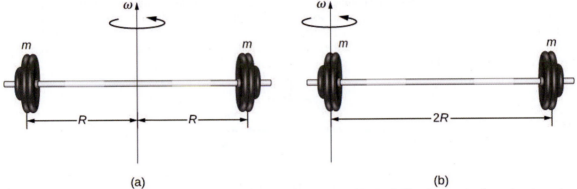

(a)   (b)

Figure 10.23 (a) A barbell with an axis of rotation through its center; (b) a barbell with an axis of rotation through one end.

In this example, we had two point masses and the sum was simple to calculate. However, to deal with objects that are not point-like, we need to think carefully about each of the terms in the equation. The equation asks us to sum over each 'piece of mass' a certain distance from the axis of rotation. But what exactly does each 'piece of mass' mean? Recall that in our derivation of this equation, each piece of mass had the same magnitude of velocity, which means the whole piece had to have a single distance r to the axis of rotation. However, this is not possible unless we take an infinitesimally small piece of mass dm, as shown in **Figure 10.24**.

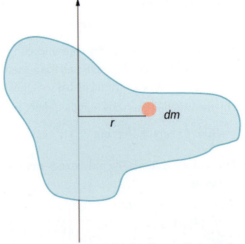

Figure 10.24 Using an infinitesimally small piece of mass to calculate the contribution to the total moment of inertia.

The need to use an infinitesimally small piece of mass dm suggests that we can write the moment of inertia by evaluating an integral over infinitesimal masses rather than doing a discrete sum over finite masses:

$$I = \sum_i m_i r_i^2 \quad \text{becomes} \quad I = \int r^2 \, dm. \tag{10.19}$$

This, in fact, is the form we need to generalize the equation for complex shapes. It is best to work out specific examples in detail to get a feel for how to calculate the moment of inertia for specific shapes. This is the focus of most of the rest of this section.

A uniform thin rod with an axis through the center

Consider a uniform (density and shape) thin rod of mass M and length L as shown in **Figure 10.25**. We want a thin rod so that we can assume the cross-sectional area of the rod is small and the rod can be thought of as a string of masses along a one-dimensional straight line. In this example, the axis of rotation is perpendicular to the rod and passes through the midpoint for simplicity. Our task is to calculate the moment of inertia about this axis. We orient the axes so that the z-axis is the axis of rotation and the x-axis passes through the length of the rod, as shown in the figure. This is a convenient choice because we can then integrate along the x-axis.

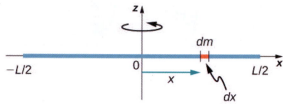

Figure 10.25 Calculation of the moment of inertia I for a uniform thin rod about an axis through the center of the rod.

We define dm to be a small element of mass making up the rod. The moment of inertia integral is an integral over the mass distribution. However, we know how to integrate over space, not over mass. We therefore need to find a way to relate mass to spatial variables. We do this using the **linear mass density** λ of the object, which is the mass per unit length. Since the mass density of this object is uniform, we can write

$$\lambda = \frac{m}{l} \quad \text{or} \quad m = \lambda l.$$

If we take the differential of each side of this equation, we find

$$dm = d(\lambda l) = \lambda(dl)$$

since λ is constant. We chose to orient the rod along the x-axis for convenience—this is where that choice becomes very helpful. Note that a piece of the rod dl lies completely along the x-axis and has a length dx; in fact, $dl = dx$ in this situation. We can therefore write $dm = \lambda(dx)$, giving us an integration variable that we know how to deal with. The distance of each piece of mass dm from the axis is given by the variable x, as shown in the figure. Putting this all together, we obtain

$$I = \int r^2 \, dm = \int x^2 \, dm = \int x^2 \lambda dx.$$

The last step is to be careful about our limits of integration. The rod extends from $x = -L/2$ to $x = L/2$, since the axis is in the middle of the rod at $x = 0$. This gives us

$$I = \int_{-L/2}^{L/2} x^2 \lambda dx = \lambda \frac{x^3}{3} \Bigg|_{-L/2}^{L/2} = \lambda \left(\frac{1}{3}\right)\left[\left(\frac{L}{2}\right)^3 - \left(\frac{-L}{2}\right)^3\right]$$

$$= \lambda\left(\frac{1}{3}\right)\frac{L^3}{8}(2) = \frac{M}{L}\left(\frac{1}{3}\right)\frac{L^3}{8}(2) = \frac{1}{12}ML^2.$$

Next, we calculate the moment of inertia for the same uniform thin rod but with a different axis choice so we can compare the results. We would expect the moment of inertia to be smaller about an axis through the center of mass than the endpoint axis, just as it was for the barbell example at the start of this section. This happens because more mass is distributed farther from the axis of rotation.

A uniform thin rod with axis at the end

Now consider the same uniform thin rod of mass M and length L, but this time we move the axis of rotation to the end of the rod. We wish to find the moment of inertia about this new axis (**Figure 10.26**). The quantity dm is again defined to be a small element of mass making up the rod. Just as before, we obtain

$$I = \int r^2\, dm = \int x^2\, dm = \int x^2 \lambda dx.$$

However, this time we have different limits of integration. The rod extends from $x = 0$ to $x = L$, since the axis is at the end of the rod at $x = 0$. Therefore we find

$$I = \int_0^L x^2 \lambda dx = \lambda \frac{x^3}{3}\Big|_0^L = \lambda\left(\frac{1}{3}\right)[(L)^3 - (0)^3]$$

$$= \lambda\left(\frac{1}{3}\right)L^3 = \frac{M}{L}\left(\frac{1}{3}\right)L^3 = \frac{1}{3}ML^2.$$

Figure 10.26 Calculation of the moment of inertia I for a uniform thin rod about an axis through the end of the rod.

Note the rotational inertia of the rod about its endpoint is larger than the rotational inertia about its center (consistent with the barbell example) by a factor of four.

The Parallel-Axis Theorem

The similarity between the process of finding the moment of inertia of a rod about an axis through its middle and about an axis through its end is striking, and suggests that there might be a simpler method for determining the moment of inertia for a rod about any axis parallel to the axis through the center of mass. Such an axis is called a **parallel axis**. There is a theorem for this, called the **parallel-axis theorem**, which we state here but do not derive in this text.

Parallel-Axis Theorem

Let m be the mass of an object and let d be the distance from an axis through the object's center of mass to a new axis. Then we have

$$I_{\text{parallel-axis}} = I_{\text{center of mass}} + md^2. \tag{10.20}$$

Let's apply this to the rod examples solved above:

$$I_{\text{end}} = I_{\text{center of mass}} + md^2 = \frac{1}{12}mL^2 + m\left(\frac{L}{2}\right)^2 = \left(\frac{1}{12} + \frac{1}{4}\right)mL^2 = \frac{1}{3}mL^2.$$

This result agrees with our more lengthy calculation from above. This is a useful equation that we apply in some of the examples and problems.

10.5   Check Your Understanding What is the moment of inertia of a cylinder of radius R and mass m about an axis through a point on the surface, as shown below?

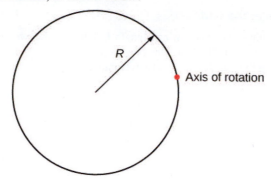

A uniform thin disk about an axis through the center

Integrating to find the moment of inertia of a two-dimensional object is a little bit trickier, but one shape is commonly done at this level of study—a uniform thin disk about an axis through its center (**Figure 10.27**).

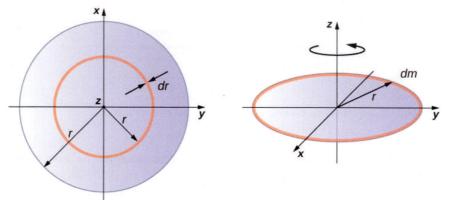

Figure 10.27   Calculating the moment of inertia for a thin disk about an axis through its center.

Since the disk is thin, we can take the mass as distributed entirely in the *xy*-plane. We again start with the relationship for the **surface mass density**, which is the mass per unit surface area. Since it is uniform, the surface mass density σ is constant:

$$\sigma = \frac{m}{A} \quad \text{or} \quad \sigma A = m, \text{ so } dm = \sigma(dA).$$

Now we use a simplification for the area. The area can be thought of as made up of a series of thin rings, where each ring is a mass increment dm of radius r equidistant from the axis, as shown in part (b) of the figure. The infinitesimal area of each ring dA is therefore given by the length of each ring ($2\pi r$) times the infinitesimmal width of each ring dr:

$$A = \pi r^2, \, dA = d(\pi r^2) = \pi dr^2 = 2\pi r dr.$$

The full area of the disk is then made up from adding all the thin rings with a radius range from 0 to R. This radius range then becomes our limits of integration for dr, that is, we integrate from $r = 0$ to $r = R$. Putting this all together, we have

$$I = \int_0^R r^2 \sigma(2\pi r)dr = 2\pi\sigma \int_0^R r^3 dr = 2\pi\sigma \frac{r^4}{4}\Big|_0^R = 2\pi\sigma\left(\frac{R^4}{4} - 0\right)$$

$$= 2\pi\frac{m}{A}\left(\frac{R^4}{4}\right) = 2\pi\frac{m}{\pi R^2}\left(\frac{R^4}{4}\right) = \frac{1}{2}mR^2.$$

Note that this agrees with the value given in **Figure 10.20**.

Calculating the moment of inertia for compound objects

Now consider a compound object such as that in **Figure 10.28**, which depicts a thin disk at the end of a thin rod. This cannot be easily integrated to find the moment of inertia because it is not a uniformly shaped object. However, if we go back to the initial definition of moment of inertia as a summation, we can reason that a compound object's moment of inertia can be found from the sum of each part of the object:

$$I_{\text{total}} = \sum_i I_i. \qquad\qquad (10.21)$$

It is important to note that the moments of inertia of the objects in **Equation 10.21** are *about a common axis*. In the case of this object, that would be a rod of length L rotating about its end, and a thin disk of radius R rotating about an axis shifted off of the center by a distance $L + R$, where R is the radius of the disk. Let's define the mass of the rod to be m_r and the mass of the disk to be m_d.

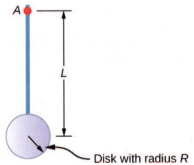

Figure 10.28 Compound object consisting of a disk at the end of a rod. The axis of rotation is located at *A*.

The moment of inertia of the rod is simply $\frac{1}{3}m_r L^2$, but we have to use the parallel-axis theorem to find the moment of inertia of the disk about the axis shown. The moment of inertia of the disk about its center is $\frac{1}{2}m_d R^2$ and we apply the parallel-axis theorem $I_{\text{parallel-axis}} = I_{\text{center of mass}} + md^2$ to find

$$I_{\text{parallel-axis}} = \frac{1}{2}m_d R^2 + m_d (L + R)^2.$$

Adding the moment of inertia of the rod plus the moment of inertia of the disk with a shifted axis of rotation, we find the moment of inertia for the compound object to be

$$I_{\text{total}} = \frac{1}{3}m_r L^2 + \frac{1}{2}m_d R^2 + m_d (L + R)^2.$$

Applying moment of inertia calculations to solve problems

Now let's examine some practical applications of moment of inertia calculations.

Example 10.11

Person on a Merry-Go-Round

A 25-kg child stands at a distance $r = 1.0 \text{ m}$ from the axis of a rotating merry-go-round (**Figure 10.29**). The merry-go-round can be approximated as a uniform solid disk with a mass of 500 kg and a radius of 2.0 m. Find the moment of inertia of this system.

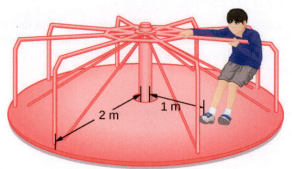

Figure 10.29 Calculating the moment of inertia for a child on a merry-go-round.

Strategy

This problem involves the calculation of a moment of inertia. We are given the mass and distance to the axis of rotation of the child as well as the mass and radius of the merry-go-round. Since the mass and size of the child are much smaller than the merry-go-round, we can approximate the child as a point mass. The notation we use is $m_c = 25$ kg, $r_c = 1.0$ m, $m_m = 500$ kg, $r_m = 2.0$ m .

Our goal is to find $I_{total} = \sum_i I_i$.

Solution

For the child, $I_c = m_c r^2$, and for the merry-go-round, $I_m = \frac{1}{2} m_m r^2$. Therefore

$$I_{total} = 25(1)^2 + \frac{1}{2}(500)(2)^2 = 25 + 1000 = 1025 \text{ kg} \cdot \text{m}^2.$$

Significance

The value should be close to the moment of inertia of the merry-go-round by itself because it has much more mass distributed away from the axis than the child does.

Example 10.12

Rod and Solid Sphere

Find the moment of inertia of the rod and solid sphere combination about the two axes as shown below. The rod has length 0.5 m and mass 2.0 kg. The radius of the sphere is 20.0 cm and has mass 1.0 kg.

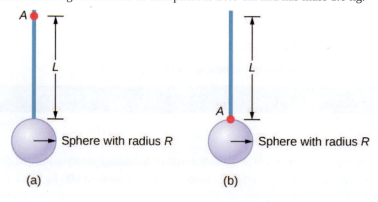

Strategy

Since we have a compound object in both cases, we can use the parallel-axis theorem to find the moment of inertia about each axis. In (a), the center of mass of the sphere is located at a distance $L + R$ from the axis of rotation. In (b), the center of mass of the sphere is located a distance R from the axis of rotation. In both cases, the moment of inertia of the rod is about an axis at one end. Refer to **Table 10.4** for the moments of inertia for the individual objects.

a. $I_{\text{total}} = \sum_i I_i = I_{\text{Rod}} + I_{\text{Sphere}};$

$I_{\text{Sphere}} = I_{\text{center of mass}} + m_{\text{Sphere}}(L + R)^2 = \frac{2}{5}m_{\text{Sphere}}R^2 + m_{\text{Sphere}}(L + R)^2;$

$I_{\text{total}} = I_{\text{Rod}} + I_{\text{Sphere}} = \frac{1}{3}m_{\text{Rod}}L^2 + \frac{2}{5}m_{\text{Sphere}}R^2 + m_{\text{Sphere}}(L + R)^2;$

$I_{\text{total}} = \frac{1}{3}(2.0\,\text{kg})(0.5\,\text{m})^2 + \frac{2}{5}(1.0\,\text{kg})(0.2\,\text{m})^2 + (1.0\,\text{kg})(0.5\,\text{m} + 0.2\,\text{m})^2;$

$I_{\text{total}} = (0.167 + 0.016 + 0.490)\,\text{kg}\cdot\text{m}^2 = 0.673\,\text{kg}\cdot\text{m}^2.$

b. $I_{\text{Sphere}} = \frac{2}{5}m_{\text{Sphere}}R^2 + m_{\text{Sphere}}R^2;$

$I_{\text{total}} = I_{\text{Rod}} + I_{\text{Sphere}} = \frac{1}{3}m_{\text{Rod}}L^2 + \frac{2}{5}m_{\text{Sphere}}R^2 + m_{\text{Sphere}}R^2;$

$I_{\text{total}} = \frac{1}{3}(2.0\,\text{kg})(0.5\,\text{m})^2 + \frac{2}{5}(1.0\,\text{kg})(0.2\,\text{m})^2 + (1.0\,\text{kg})(0.2\,\text{m})^2;$

$I_{\text{total}} = (0.167 + 0.016 + 0.04)\,\text{kg}\cdot\text{m}^2 = 0.223\,\text{kg}\cdot\text{m}^2.$

Significance

Using the parallel-axis theorem eases the computation of the moment of inertia of compound objects. We see that the moment of inertia is greater in (a) than (b). This is because the axis of rotation is closer to the center of mass of the system in (b). The simple analogy is that of a rod. The moment of inertia about one end is $\frac{1}{3}mL^2$, but the moment of inertia through the center of mass along its length is $\frac{1}{12}mL^2$.

Example 10.13

Angular Velocity of a Pendulum

A pendulum in the shape of a rod (**Figure 10.30**) is released from rest at an angle of $30°$. It has a length 30 cm and mass 300 g. What is its angular velocity at its lowest point?

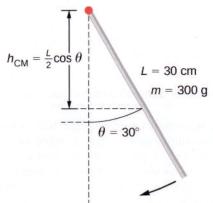

$h_{\text{CM}} = \frac{L}{2}\cos\theta$

$L = 30$ cm
$m = 300$ g

$\theta = 30°$

Figure 10.30 A pendulum in the form of a rod is released from rest at an angle of $30°$.

Strategy

Use conservation of energy to solve the problem. At the point of release, the pendulum has gravitational potential energy, which is determined from the height of the center of mass above its lowest point in the swing. At the bottom of the swing, all of the gravitational potential energy is converted into rotational kinetic energy.

Solution

The change in potential energy is equal to the change in rotational kinetic energy, $\Delta U + \Delta K = 0$.

At the top of the swing: $U = mgh_{cm} = mg\frac{L}{2}(\cos\theta)$. At the bottom of the swing, $U = mg\frac{L}{2}$.

At the top of the swing, the rotational kinetic energy is $K = 0$. At the bottom of the swing, $K = \frac{1}{2}I\omega^2$.

Therefore:

$$\Delta U + \Delta K = 0 \Rightarrow (mg\frac{L}{2}(1 - \cos\theta) - 0) + (0 - \frac{1}{2}I\omega^2) = 0$$

or

$$\frac{1}{2}I\omega^2 = mg\frac{L}{2}(1 - \cos\theta).$$

Solving for ω, we have

$$\omega = \sqrt{mg\frac{L}{I}(1 - \cos\theta)} = \sqrt{mg\frac{L}{1/3mL^2}(1 - \cos\theta)} = \sqrt{g\frac{3}{L}(1 - \cos\theta)}.$$

Inserting numerical values, we have

$$\omega = \sqrt{9.8 \text{ m/s}^2\frac{3}{0.3 \text{ m}}(1 - \cos 30)} = 3.6 \text{ rad/s}.$$

Significance

Note that the angular velocity of the pendulum does not depend on its mass.

10.6 | Torque

Learning Objectives

By the end of this section, you will be able to:

- Describe how the magnitude of a torque depends on the magnitude of the lever arm and the angle the force vector makes with the lever arm
- Determine the sign (positive or negative) of a torque using the right-hand rule
- Calculate individual torques about a common axis and sum them to find the net torque

An important quantity for describing the dynamics of a rotating rigid body is torque. We see the application of torque in many ways in our world. We all have an intuition about torque, as when we use a large wrench to unscrew a stubborn bolt. Torque is at work in unseen ways, as when we press on the accelerator in a car, causing the engine to put additional torque on the drive train. Or every time we move our bodies from a standing position, we apply a torque to our limbs. In this section, we define torque and make an argument for the equation for calculating torque for a rigid body with fixed-axis rotation.

Defining Torque

So far we have defined many variables that are rotational equivalents to their translational counterparts. Let's consider what the counterpart to force must be. Since forces change the translational motion of objects, the rotational counterpart must be related to changing the rotational motion of an object about an axis. We call this rotational counterpart **torque**.

In everyday life, we rotate objects about an axis all the time, so intuitively we already know much about torque. Consider, for example, how we rotate a door to open it. First, we know that a door opens slowly if we push too close to its hinges; it is more efficient to rotate a door open if we push far from the hinges. Second, we know that we should push perpendicular to the plane of the door; if we push parallel to the plane of the door, we are not able to rotate it. Third, the larger the force, the more effective it is in opening the door; the harder you push, the more rapidly the door opens. The first point implies that the farther the force is applied from the axis of rotation, the greater the angular acceleration; the second implies that the effectiveness depends on the angle at which the force is applied; the third implies that the magnitude of the force must also be part of the equation. Note that for rotation in a plane, torque has two possible directions. Torque is either clockwise or counterclockwise relative to the chosen pivot point. **Figure 10.31** shows counterclockwise rotations.

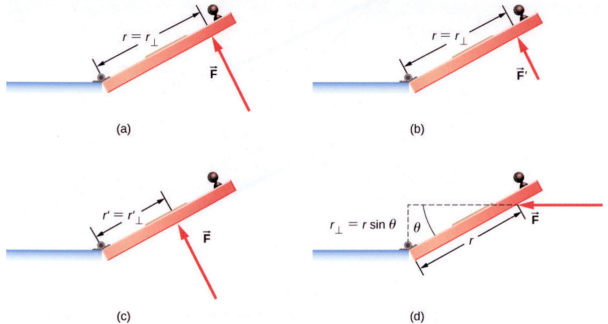

Figure 10.31 Torque is the turning or twisting effectiveness of a force, illustrated here for door rotation on its hinges (as viewed from overhead). Torque has both magnitude and direction. (a) A counterclockwise torque is produced by a force $\vec{F}$ acting at a distance r from the hinges (the pivot point). (b) A smaller counterclockwise torque is produced when a smaller force $\vec{F}'$ acts at the same distance r from the hinges. (c) The same force as in (a) produces a smaller counterclockwise torque when applied at a smaller distance from the hinges. (d) A smaller counterclockwise torque is produced by the same magnitude force as (a) acting at the same distance as (a) but at an angle θ that is less than $90°$.

Now let's consider how to define torques in the general three-dimensional case.

Torque

When a force $\vec{F}$ is applied to a point P whose position is $\vec{r}$ relative to O (**Figure 10.32**), the torque $\vec{\tau}$ around O is

$$\vec{\tau} = \vec{r} \times \vec{F}.$$

(10.22)

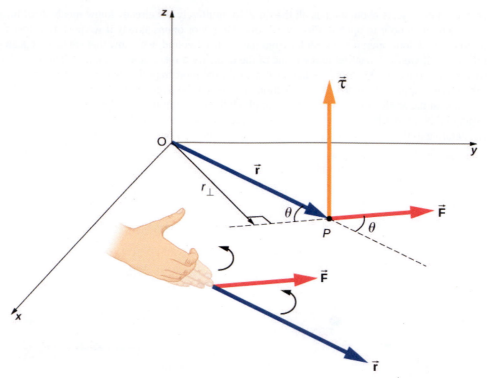

Figure 10.32 The torque is perpendicular to the plane defined by $\vec{r}$ and $\vec{F}$ and its direction is determined by the right-hand rule.

From the definition of the cross product, the torque $\vec{\tau}$ is perpendicular to the plane containing $\vec{r}$ and $\vec{F}$ and has magnitude

$$\left|\vec{\tau}\right| = \left|\vec{r} \times \vec{F}\right| = rF\sin\theta,$$

where θ is the angle between the vectors $\vec{r}$ and $\vec{F}$. The SI unit of torque is newtons times meters, usually written as $\text{N}\cdot\text{m}$. The quantity $r_{\perp} = r\sin\theta$ is the perpendicular distance from O to the line determined by the vector $\vec{F}$ and is called the **lever arm**. Note that the greater the lever arm, the greater the magnitude of the torque. In terms of the lever arm, the magnitude of the torque is

$$\left|\vec{\tau}\right| = r_{\perp}\, F. \qquad\qquad (10.23)$$

The cross product $\vec{r} \times \vec{F}$ also tells us the sign of the torque. In **Figure 10.32**, the cross product $\vec{r} \times \vec{F}$ is along the positive z-axis, which by convention is a positive torque. If $\vec{r} \times \vec{F}$ is along the negative z-axis, this produces a negative torque.

If we consider a disk that is free to rotate about an axis through the center, as shown in **Figure 10.33**, we can see how the angle between the radius $\vec{r}$ and the force $\vec{F}$ affects the magnitude of the torque. If the angle is zero, the torque is zero; if the angle is $90°$, the torque is maximum. The torque in **Figure 10.33** is positive because the direction of the torque by the right-hand rule is out of the page along the positive z-axis. The disk rotates counterclockwise due to the torque, in the same direction as a positive angular acceleration.

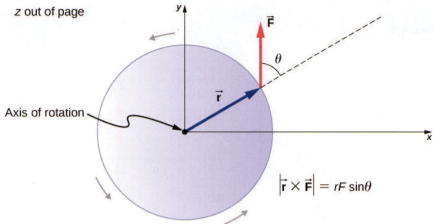

Figure 10.33 A disk is free to rotate about its axis through the center. The magnitude of the torque on the disk is $rF\sin\theta$. When $\theta = 0°$, the torque is zero and the disk does not rotate. When $\theta = 90°$, the torque is maximum and the disk rotates with maximum angular acceleration.

Any number of torques can be calculated about a given axis. The individual torques add to produce a net torque about the axis. When the appropriate sign (positive or negative) is assigned to the magnitudes of individual torques about a specified axis, the net torque about the axis is the sum of the individual torques:

$$\vec{\tau}_{net} = \sum_i |\vec{\tau}_i|. \tag{10.24}$$

Calculating Net Torque for Rigid Bodies on a Fixed Axis

In the following examples, we calculate the torque both abstractly and as applied to a rigid body.

We first introduce a problem-solving strategy.

Problem-Solving Strategy: Finding Net Torque

1. Choose a coordinate system with the pivot point or axis of rotation as the origin of the selected coordinate system.

2. Determine the angle between the lever arm $\vec{r}$ and the force vector.

3. Take the cross product of $\vec{r}$ and $\vec{F}$ to determine if the torque is positive or negative about the pivot point or axis.

4. Evaluate the magnitude of the torque using $r_\perp F$.

5. Assign the appropriate sign, positive or negative, to the magnitude.

6. Sum the torques to find the net torque.

Example 10.14

Calculating Torque

Four forces are shown in **Figure 10.34** at particular locations and orientations with respect to a given xy-coordinate system. Find the torque due to each force about the origin, then use your results to find the net torque about the origin.

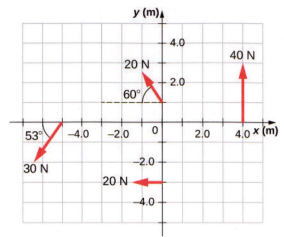

Figure 10.34 Four forces producing torques.

Strategy

This problem requires calculating torque. All known quantities—forces with directions and lever arms—are given in the figure. The goal is to find each individual torque and the net torque by summing the individual torques. Be careful to assign the correct sign to each torque by using the cross product of $\vec{r}$ and the force vector $\vec{F}$.

Solution

Use $\left| \vec{\tau} \right| = r_\perp F = rF\sin\theta$ to find the magnitude and $\vec{\tau} = \vec{r} \times \vec{F}$ to determine the sign of the torque.

The torque from force 40 N in the first quadrant is given by $(4)(40)\sin 90° = 160\,\text{N}\cdot\text{m}$.

The cross product of $\vec{r}$ and $\vec{F}$ is out of the page, positive.

The torque from force 20 N in the third quadrant is given by $-(3)(20)\sin 90° = -60\,\text{N}\cdot\text{m}$.

The cross product of $\vec{r}$ and $\vec{F}$ is into the page, so it is negative.

The torque from force 30 N in the third quadrant is given by $(5)(30)\sin 53° = 120\,\text{N}\cdot\text{m}$.

The cross product of $\vec{r}$ and $\vec{F}$ is out of the page, positive.

The torque from force 20 N in the second quadrant is given by $(1)(20)\sin 30° = 10\,\text{N}\cdot\text{m}$.

The cross product of $\vec{r}$ and $\vec{F}$ is out of the page.

The net torque is therefore $\tau_{\text{net}} = \sum_i \left| \tau_i \right| = 160 - 60 + 120 + 10 = 230\,\text{N}\cdot\text{m}$.

Significance

Note that each force that acts in the counterclockwise direction has a positive torque, whereas each force that acts in the clockwise direction has a negative torque. The torque is greater when the distance, force, or perpendicular components are greater.

Example 10.15

Calculating Torque on a rigid body

Figure 10.35 shows several forces acting at different locations and angles on a flywheel. We have $\left|\vec{F}_1\right| = 20 \text{ N}$, $\left|\vec{F}_2\right| = 30 \text{ N}$, $\left|\vec{F}_3\right| = 30 \text{ N}$, and $r = 0.5 \text{ m}$. Find the net torque on the flywheel about an axis through the center.

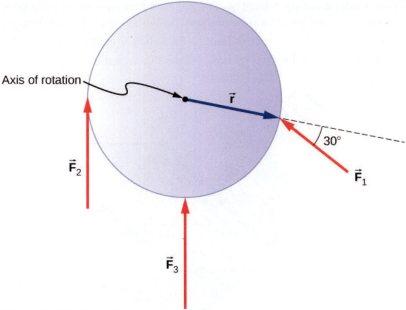

Figure 10.35 Three forces acting on a flywheel.

Strategy

We calculate each torque individually, using the cross product, and determine the sign of the torque. Then we sum the torques to find the net torque.

Solution

We start with $\vec{F}_1$. If we look at **Figure 10.35**, we see that $\vec{F}_1$ makes an angle of $90° + 60°$ with the radius vector $\vec{r}$. Taking the cross product, we see that it is out of the page and so is positive. We also see this from calculating its magnitude:

$$\left|\vec{\tau}_1\right| = rF_1 \sin 150° = 0.5 \text{ m}(20 \text{ N})(0.5) = 5.0 \text{ N} \cdot \text{m}.$$

Next we look at $\vec{F}_2$. The angle between $\vec{F}_2$ and $\vec{r}$ is $90°$ and the cross product is into the page so the torque is negative. Its value is

$$\left|\vec{\tau}_2\right| = -rF_2 \sin 90° = -0.5 \text{ m}(30 \text{ N}) = -15.0 \text{ N} \cdot \text{m}.$$

When we evaluate the torque due to $\vec{F}_3$, we see that the angle it makes with $\vec{r}$ is zero so $\vec{r} \times \vec{F}_3 = 0$. Therefore, $\vec{F}_3$ does not produce any torque on the flywheel.

We evaluate the sum of the torques:

$$\tau_{\text{net}} = \sum_i \left|\tau_i\right| = 5 - 15 = -10 \text{ N} \cdot \text{m}.$$

Significance

The axis of rotation is at the center of mass of the flywheel. Since the flywheel is on a fixed axis, it is not free to translate. If it were on a frictionless surface and not fixed in place, $\vec{\textbf{F}}_3$ would cause the flywheel to translate, as well as $\vec{\textbf{F}}_1$. Its motion would be a combination of translation and rotation.

10.6 **Check Your Understanding** A large ocean-going ship runs aground near the coastline, similar to the fate of the *Costa Concordia*, and lies at an angle as shown below. Salvage crews must apply a torque to right the ship in order to float the vessel for transport. A force of 5.0×10^5 N acting at point *A* must be applied to right the ship. What is the torque about the point of contact of the ship with the ground (**Figure 10.36**)?

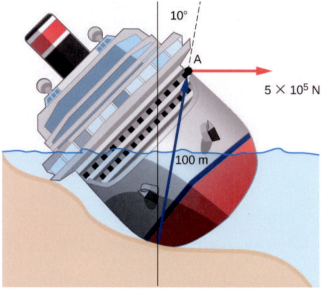

Figure 10.36 A ship runs aground and tilts, requiring torque to be applied to return the vessel to an upright position.

10.7 | Newton's Second Law for Rotation

Learning Objectives
By the end of this section, you will be able to:
• Calculate the torques on rotating systems about a fixed axis to find the angular acceleration
• Explain how changes in the moment of inertia of a rotating system affect angular acceleration with a fixed applied torque

In this section, we put together all the pieces learned so far in this chapter to analyze the dynamics of rotating rigid bodies. We have analyzed motion with kinematics and rotational kinetic energy but have not yet connected these ideas with force and/or torque. In this section, we introduce the rotational equivalent to Newton's second law of motion and apply it to rigid bodies with fixed-axis rotation.

Newton's Second Law for Rotation

We have thus far found many counterparts to the translational terms used throughout this text, most recently, torque, the rotational analog to force. This raises the question: Is there an analogous equation to Newton's second law, $\Sigma \vec{\textbf{F}} = m\vec{\textbf{a}}$, which involves torque and rotational motion? To investigate this, we start with Newton's second law for a single particle rotating around an axis and executing circular motion. Let's exert a force $\vec{F}$ on a point mass m that is at a distance r from a pivot point (**Figure 10.37**). The particle is constrained to move in a circular path with fixed radius and the force is tangent to the circle. We apply Newton's second law to determine the magnitude of the acceleration $a = F/m$ in the direction of $\vec{F}$. Recall that the magnitude of the tangential acceleration is proportional to the magnitude of the angular acceleration by $a = r\alpha$. Substituting this expression into Newton's second law, we obtain

$$F = mr\alpha.$$

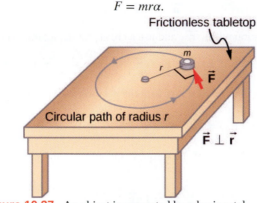

Figure 10.37 An object is supported by a horizontal frictionless table and is attached to a pivot point by a cord that supplies centripetal force. A force $\vec{F}$ is applied to the object perpendicular to the radius r, causing it to accelerate about the pivot point. The force is perpendicular to r.

Multiply both sides of this equation by r,

$$rF = mr^2\alpha.$$

Note that the left side of this equation is the torque about the axis of rotation, where r is the lever arm and F is the force, perpendicular to r. Recall that the moment of inertia for a point particle is $I = mr^2$. The torque applied perpendicularly to the point mass in **Figure 10.37** is therefore

$$\tau = I\alpha.$$

The torque on the particle is equal to the moment of inertia about the rotation axis times the angular acceleration. We can generalize this equation to a rigid body rotating about a fixed axis.

Newton's Second Law for Rotation

If more than one torque acts on a rigid body about a fixed axis, then the sum of the torques equals the moment of inertia times the angular acceleration:

$$\sum_i \tau_i = I\alpha. \qquad\qquad \textbf{(10.25)}$$

The term $I\alpha$ is a scalar quantity and can be positive or negative (counterclockwise or clockwise) depending upon the sign of the net torque. Remember the convention that counterclockwise angular acceleration is positive. Thus, if a rigid body is rotating clockwise and experiences a positive torque (counterclockwise), the angular acceleration is positive.

Equation 10.25 is **Newton's second law for rotation** and tells us how to relate torque, moment of inertia, and rotational kinematics. This is called the equation for **rotational dynamics**. With this equation, we can solve a whole class of problems involving force and rotation. It makes sense that the relationship for how much force it takes to rotate a body would include the moment of inertia, since that is the quantity that tells us how easy or hard it is to change the rotational motion of an object.

Deriving Newton's Second Law for Rotation in Vector Form

As before, when we found the angular acceleration, we may also find the torque vector. The second law $\Sigma \vec{F} = m \vec{a}$ tells us the relationship between net force and how to change the translational motion of an object. We have a vector rotational equivalent of this equation, which can be found by using **Equation 10.7** and **Figure 10.8**. **Equation 10.7** relates the angular acceleration to the position and tangential acceleration vectors:

$$\vec{a} = \vec{\alpha} \times \vec{r} .$$

We form the cross product of this equation with $\vec{r}$ and use a cross product identity (note that $\vec{r} \cdot \vec{\alpha} = 0$):

$$\vec{r} \times \vec{a} = \vec{r} \times (\vec{\alpha} \times \vec{r}) = \vec{\alpha}(\vec{r} \cdot \vec{r}) - \vec{r}(\vec{r} \cdot \vec{\alpha}) = \vec{\alpha}(\vec{r} \cdot \vec{r}) = \vec{\alpha} r^2 .$$

We now form the cross product of Newton's second law with the position vector $\vec{r}$,

$$\Sigma(\vec{r} \times \vec{F}) = \vec{r} \times (m \vec{a}) = m \vec{r} \times \vec{a} = mr^2 \vec{\alpha} .$$

Identifying the first term on the left as the sum of the torques, and mr^2 as the moment of inertia, we arrive at Newton's second law of rotation in vector form:

$$\Sigma \vec{\tau} = I \vec{\alpha} . \tag{10.26}$$

This equation is exactly **Equation 10.25** but with the torque and angular acceleration as vectors. An important point is that the torque vector is in the same direction as the angular acceleration.

Applying the Rotational Dynamics Equation

Before we apply the rotational dynamics equation to some everyday situations, let's review a general problem-solving strategy for use with this category of problems.

Problem-Solving Strategy: Rotational Dynamics

1. Examine the situation to determine that torque and mass are involved in the rotation. Draw a careful sketch of the situation.

2. Determine the system of interest.

3. Draw a free-body diagram. That is, draw and label all external forces acting on the system of interest.

4. Identify the pivot point. If the object is in equilibrium, it must be in equilibrium for all possible pivot points—chose the one that simplifies your work the most.

5. Apply $\displaystyle\sum_{i} \tau_i = I\alpha$, the rotational equivalent of Newton's second law, to solve the problem. Care must be taken to use the correct moment of inertia and to consider the torque about the point of rotation.

6. As always, check the solution to see if it is reasonable.

Example 10.16

Calculating the Effect of Mass Distribution on a Merry-Go-Round

Consider the father pushing a playground merry-go-round in **Figure 10.38**. He exerts a force of 250 N at the edge of the 50.0-kg merry-go-round, which has a 1.50-m radius. Calculate the angular acceleration produced (a) when no one is on the merry-go-round and (b) when an 18.0-kg child sits 1.25 m away from the center. Consider the merry-go-round itself to be a uniform disk with negligible friction.

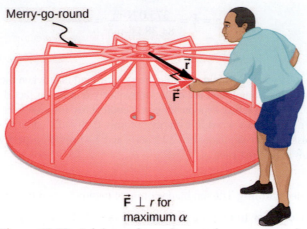

Merry-go-round

$\vec{F} \perp r$ for
maximum α

Figure 10.38 A father pushes a playground merry-go-round at its edge and perpendicular to its radius to achieve maximum torque.

Strategy

The net torque is given directly by the expression $\sum_i \tau_i = I\alpha$, To solve for α, we must first calculate the net torque τ (which is the same in both cases) and moment of inertia I (which is greater in the second case).

Solution

a.   The moment of inertia of a solid disk about this axis is given in **Figure 10.20** to be

$$\frac{1}{2}MR^2.$$

We have $M = 50.0\,\text{kg}$ and $R = 1.50\,\text{m}$, so

$$I = (0.500)(50.0\,\text{kg})(1.50\,\text{m})^2 = 56.25\,\text{kg-m}^2.$$

To find the net torque, we note that the applied force is perpendicular to the radius and friction is negligible, so that

$$\tau = rF\sin\theta = (1.50\,\text{m})(250.0\,\text{N}) = 375.0\,\text{N-m}.$$

Now, after we substitute the known values, we find the angular acceleration to be

$$\alpha = \frac{\tau}{I} = \frac{375.0\,\text{N-m}}{56.25\,\text{kg-m}^2} = 6.67\frac{\text{rad}}{\text{s}^2}.$$

b.   We expect the angular acceleration for the system to be less in this part because the moment of inertia is greater when the child is on the merry-go-round. To find the total moment of inertia I, we first find the child's moment of inertia I_c by approximating the child as a point mass at a distance of 1.25 m from the axis. Then

$$I_c = mR^2 = (18.0\,\text{kg})(1.25\,\text{m})^2 = 28.13\,\text{kg-m}^2.$$

The total moment of inertia is the sum of the moments of inertia of the merry-go-round and the child (about the same axis):

$$I = 28.13\,\text{kg-m}^2 + 56.25\,\text{kg-m}^2 = 84.38\,\text{kg-m}^2.$$

Substituting known values into the equation for α gives

$$\alpha = \frac{\tau}{I} = \frac{375.0\ \text{N-m}}{84.38\ \text{kg-m}^2} = 4.44 \frac{\text{rad}}{\text{s}^2}.$$

Significance

The angular acceleration is less when the child is on the merry-go-round than when the merry-go-round is empty, as expected. The angular accelerations found are quite large, partly due to the fact that friction was considered to be negligible. If, for example, the father kept pushing perpendicularly for 2.00 s, he would give the merry-go-round an angular velocity of 13.3 rad/s when it is empty but only 8.89 rad/s when the child is on it. In terms of revolutions per second, these angular velocities are 2.12 rev/s and 1.41 rev/s, respectively. The father would end up running at about 50 km/h in the first case.

 10.7 **Check Your Understanding** The fan blades on a jet engine have a moment of inertia $30.0\ \text{kg-m}^2$. In 10 s, they rotate counterclockwise from rest up to a rotation rate of 20 rev/s. (a) What torque must be applied to the blades to achieve this angular acceleration? (b) What is the torque required to bring the fan blades rotating at 20 rev/s to a rest in 20 s?

10.8 | Work and Power for Rotational Motion

Learning Objectives
By the end of this section, you will be able to: • Use the work-energy theorem to analyze rotation to find the work done on a system when it is rotated about a fixed axis for a finite angular displacement • Solve for the angular velocity of a rotating rigid body using the work-energy theorem • Find the power delivered to a rotating rigid body given the applied torque and angular velocity • Summarize the rotational variables and equations and relate them to their translational counterparts

Thus far in the chapter, we have extensively addressed kinematics and dynamics for rotating rigid bodies around a fixed axis. In this final section, we define work and power within the context of rotation about a fixed axis, which has applications to both physics and engineering. The discussion of work and power makes our treatment of rotational motion almost complete, with the exception of rolling motion and angular momentum, which are discussed in **Angular Momentum**. We begin this section with a treatment of the work-energy theorem for rotation.

Work for Rotational Motion

Now that we have determined how to calculate kinetic energy for rotating rigid bodies, we can proceed with a discussion of the work done on a rigid body rotating about a fixed axis. **Figure 10.39** shows a rigid body that has rotated through an angle $d\theta$ from A to B while under the influence of a force $\vec{\mathbf{F}}$. The external force $\vec{\mathbf{F}}$ is applied to point P, whose position is $\vec{\mathbf{r}}$, and the rigid body is constrained to rotate about a fixed axis that is perpendicular to the page and passes through O. The rotational axis is fixed, so the vector $\vec{\mathbf{r}}$ moves in a circle of radius r, and the vector $d\vec{\mathbf{s}}$ is perpendicular to $\vec{\mathbf{r}}$.

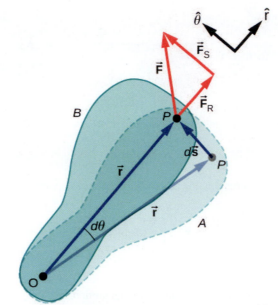

Figure 10.39 A rigid body rotates through an angle $d\theta$ from A to B by the action of an external force $\vec{F}$ applied to point P.

From **Equation 10.2**, we have

$$\vec{s} = \vec{\theta} \times \vec{r} .$$

Thus,

$$d\vec{s} = d(\vec{\theta} \times \vec{r}) = d\vec{\theta} \times \vec{r} + d\vec{r} \times \vec{\theta} = d\vec{\theta} \times \vec{r} .$$

Note that $d\vec{r}$ is zero because $\vec{r}$ is fixed on the rigid body from the origin O to point P. Using the definition of work, we obtain

$$W = \int \sum \vec{F} \cdot d\vec{s} = \int \sum \vec{F} \cdot (d\vec{\theta} \times \vec{r}) = \int d\vec{\theta} \cdot (\vec{r} \times \sum \vec{F})$$

where we used the identity $\vec{a} \cdot (\vec{b} \times \vec{c}) = \vec{b} \cdot (\vec{c} \times \vec{a})$. Noting that $(\vec{r} \times \sum \vec{F}) = \sum \vec{\tau}$, we arrive at the expression for the **rotational work** done on a rigid body:

$$W = \int \sum \vec{\tau} \cdot d\vec{\theta} . \tag{10.27}$$

The total work done on a rigid body is the sum of the torques integrated over the angle through which the body rotates. The incremental work is

$$dW = \left(\sum_i \tau_i \right) d\theta \tag{10.28}$$

where we have taken the dot product in **Equation 10.27**, leaving only torques along the axis of rotation. In a rigid body, all particles rotate through the same angle; thus the work of every external force is equal to the torque times the common incremental angle $d\theta$. The quantity $\left(\sum_i \tau_i \right)$ is the net torque on the body due to external forces.

Similarly, we found the kinetic energy of a rigid body rotating around a fixed axis by summing the kinetic energy of each particle that makes up the rigid body. Since the work-energy theorem $W_i = \Delta K_i$ is valid for each particle, it is valid for the sum of the particles and the entire body.

Work-Energy Theorem for Rotation

The work-energy theorem for a rigid body rotating around a fixed axis is

$$W_{AB} = K_B - K_A \tag{10.29}$$

where

$$K = \tfrac{1}{2}I\omega^2$$

and the rotational work done by a net force rotating a body from point A to point B is

$$W_{AB} = \int_{\theta_A}^{\theta_B} \left(\sum_i \tau_i\right) d\theta. \tag{10.30}$$

We give a strategy for using this equation when analyzing rotational motion.

Problem-Solving Strategy: Work-Energy Theorem for Rotational Motion

1. Identify the forces on the body and draw a free-body diagram. Calculate the torque for each force.
2. Calculate the work done during the body's rotation by every torque.
3. Apply the work-energy theorem by equating the net work done on the body to the change in rotational kinetic energy.

Let's look at two examples and use the work-energy theorem to analyze rotational motion.

Example 10.17

Rotational Work and Energy

A $12.0\,\text{N}\cdot\text{m}$ torque is applied to a flywheel that rotates about a fixed axis and has a moment of inertia of $30.0\,\text{kg}\cdot\text{m}^2$. If the flywheel is initially at rest, what is its angular velocity after it has turned through eight revolutions?

Strategy

We apply the work-energy theorem. We know from the problem description what the torque is and the angular displacement of the flywheel. Then we can solve for the final angular velocity.

Solution

The flywheel turns through eight revolutions, which is 16π radians. The work done by the torque, which is constant and therefore can come outside the integral in **Equation 10.30**, is

$$W_{AB} = \tau(\theta_B - \theta_A).$$

We apply the work-energy theorem:

$$W_{AB} = \tau(\theta_B - \theta_A) = \tfrac{1}{2}I\omega_B^2 - \tfrac{1}{2}I\omega_A^2.$$

With $\tau = 12.0\,\text{N}\cdot\text{m}$, $\theta_B - \theta_A = 16.0\pi\,\text{rad}$, $I = 30.0\,\text{kg}\cdot\text{m}^2$, and $\omega_A = 0$, we have

$$12.0\,\text{N-m}(16.0\pi\,\text{rad}) = \tfrac{1}{2}(30.0\,\text{kg}\cdot\text{m}^2)(\omega_B^2) - 0.$$

Therefore,

$$\omega_B = 6.3 \text{ rad/s.}$$

This is the angular velocity of the flywheel after eight revolutions.

Significance

The work-energy theorem provides an efficient way to analyze rotational motion, connecting torque with rotational kinetic energy.

Example 10.18

Rotational Work: A Pulley

A string wrapped around the pulley in **Figure 10.40** is pulled with a constant downward force $\overrightarrow{\mathbf{F}}$ of magnitude 50 N. The radius R and moment of inertia I of the pulley are 0.10 m and 2.5×10^{-3} kg-m^2, respectively. If the string does not slip, what is the angular velocity of the pulley after 1.0 m of string has unwound? Assume the pulley starts from rest.

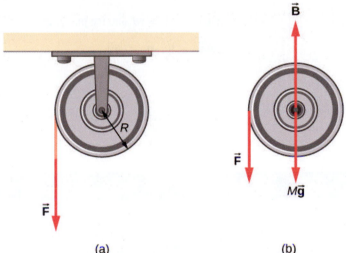

(a)   **(b)**

Figure 10.40 (a) A string is wrapped around a pulley of radius R. (b) The free-body diagram.

Strategy

Looking at the free-body diagram, we see that neither $\overrightarrow{\mathbf{B}}$, the force on the bearings of the pulley, nor $M\overrightarrow{\mathbf{g}}$, the weight of the pulley, exerts a torque around the rotational axis, and therefore does no work on the pulley. As the pulley rotates through an angle θ, $\overrightarrow{\mathbf{F}}$ acts through a distance d such that $d = R\theta$.

Solution

Since the torque due to $\overrightarrow{\mathbf{F}}$ has magnitude $\tau = RF$, we have

$$W = \tau\theta = (FR)\theta = Fd.$$

If the force on the string acts through a distance of 1.0 m, we have, from the work-energy theorem,

$$W_{AB} = K_B - K_A$$
$$Fd = \tfrac{1}{2}I\omega^2 - 0$$
$$(50.0\,\text{N})(1.0\,\text{m}) = \tfrac{1}{2}(2.5 \times 10^{-3}\,\text{kg-m}^2)\omega^2.$$

Solving for ω, we obtain

$$\omega = 200.0\,\text{rad/s}.$$

Power for Rotational Motion

Power always comes up in the discussion of applications in engineering and physics. Power for rotational motion is equally as important as power in linear motion and can be derived in a similar way as in linear motion when the force is a constant.

The linear power when the force is a constant is $P = \overrightarrow{\textbf{F}} \cdot \overrightarrow{\textbf{v}}$. If the net torque is constant over the angular displacement, **Equation 10.25** simplifies and the net torque can be taken out of the integral. In the following discussion, we assume the net torque is constant. We can apply the definition of power derived in **Power** to rotational motion. From **Work and Kinetic Energy**, the instantaneous power (or just power) is defined as the rate of doing work,

$$P = \frac{dW}{dt}.$$

If we have a constant net torque, **Equation 10.25** becomes $W = \tau\theta$ and the power is

$$P = \frac{dW}{dt} = \frac{d}{dt}(\tau\theta) = \tau\frac{d\theta}{dt}$$

or

$$P = \tau\omega. \tag{10.31}$$

Example 10.19

Torque on a Boat Propeller

A boat engine operating at 9.0×10^4 W is running at 300 rev/min. What is the torque on the propeller shaft?

Strategy

We are given the rotation rate in rev/min and the power consumption, so we can easily calculate the torque.

Solution

$$300.0\,\text{rev/min} = 31.4\,\text{rad/s};$$

$$\tau = \frac{P}{\omega} = \frac{9.0 \times 10^4\,\text{N} \cdot \text{m/s}}{31.4\,\text{rad/s}} = 2864.8\,\text{N} \cdot \text{m}.$$

Significance

It is important to note the radian is a dimensionless unit because its definition is the ratio of two lengths. It therefore does not appear in the solution.

 10.8 **Check Your Understanding** A constant torque of $500\,\text{kN} \cdot \text{m}$ is applied to a wind turbine to keep it rotating at 6 rad/s. What is the power required to keep the turbine rotating?

Rotational and Translational Relationships Summarized

The rotational quantities and their linear analog are summarized in three tables. **Table 10.5** summarizes the rotational variables for circular motion about a fixed axis with their linear analogs and the connecting equation, except for the centripetal acceleration, which stands by itself. **Table 10.6** summarizes the rotational and translational kinematic equations. **Table 10.7** summarizes the rotational dynamics equations with their linear analogs.

Rotational	Translational	Relationship
θ	x	$\theta = \frac{s}{r}$
ω	v_t	$\omega = \frac{v_t}{r}$
α	a_t	$\alpha = \frac{a_t}{r}$
	a_c	$a_c = \frac{v_t^2}{r}$

Table 10.5 Rotational and Translational Variables: Summary

Rotational	Translational
$\theta_f = \theta_0 + \bar{\omega} t$	$x = x_0 + \bar{v}\, t$
$\omega_f = \omega_0 + \alpha t$	$v_f = v_0 + at$
$\theta_f = \theta_0 + \omega_0 t + \frac{1}{2}\alpha t^2$	$x_f = x_0 + v_0 t + \frac{1}{2}at^2$
$\omega_f^2 = \omega^2{}_0 + 2\alpha(\Delta\theta)$	$v_f^2 = v^2{}_0 + 2a(\Delta x)$

Table 10.6 Rotational and Translational Kinematic Equations: Summary

Rotational	Translational
$I = \sum\limits_i m_i r_i^2$	m
$K = \frac{1}{2}I\omega^2$	$K = \frac{1}{2}mv^2$
$\sum\limits_i \tau_i = I\alpha$	$\sum\limits_i \vec{\mathbf{F}}_i = m\,\vec{\mathbf{a}}$
$W_{AB} = \int\limits_{\theta_A}^{\theta_B}\left(\sum\limits_i \tau_i\right)d\theta$	$W = \int \vec{\mathbf{F}} \cdot d\,\vec{\mathbf{s}}$
$P = \tau\omega$	$P = \vec{\mathbf{F}} \cdot \vec{\mathbf{v}}$

Table 10.7 Rotational and Translational Equations: Dynamics

CHAPTER 10 REVIEW

KEY TERMS

angular acceleration time rate of change of angular velocity

angular position angle a body has rotated through in a fixed coordinate system

angular velocity time rate of change of angular position

instantaneous angular acceleration derivative of angular velocity with respect to time

instantaneous angular velocity derivative of angular position with respect to time

kinematics of rotational motion describes the relationships among rotation angle, angular velocity, angular acceleration, and time

lever arm perpendicular distance from the line that the force vector lies on to a given axis

linear mass density the mass per unit length λ of a one dimensional object

moment of inertia rotational mass of rigid bodies that relates to how easy or hard it will be to change the angular velocity of the rotating rigid body

Newton's second law for rotation sum of the torques on a rotating system equals its moment of inertia times its angular acceleration

parallel axis axis of rotation that is parallel to an axis about which the moment of inertia of an object is known

parallel-axis theorem if the moment of inertia is known for a given axis, it can be found for any axis parallel to it

rotational dynamics analysis of rotational motion using the net torque and moment of inertia to find the angular acceleration

rotational kinetic energy kinetic energy due to the rotation of an object; this is part of its total kinetic energy

rotational work work done on a rigid body due to the sum of the torques integrated over the angle through with the body rotates

surface mass density mass per unit area σ of a two dimensional object

torque cross product of a force and a lever arm to a given axis

total linear acceleration vector sum of the centripetal acceleration vector and the tangential acceleration vector

work-energy theorem for rotation the total rotational work done on a rigid body is equal to the change in rotational kinetic energy of the body

KEY EQUATIONS

Angular position	$\theta = \frac{s}{r}$
Angular velocity	$\omega = \lim_{\Delta t \to 0} \frac{\Delta \theta}{\Delta t} = \frac{d\theta}{dt}$
Tangential speed	$v_{\text{t}} = r\omega$
Angular acceleration	$\alpha = \lim_{\Delta t \to 0} \frac{\Delta \omega}{\Delta t} = \frac{d\omega}{dt} = \frac{d^2\theta}{dt^2}$
Tangential acceleration	$a_{\text{t}} = r\alpha$
Average angular velocity	$\bar{\omega} = \frac{\omega_0 + \omega_{\text{f}}}{2}$
Angular displacement	$\theta_{\text{f}} = \theta_0 + \bar{\omega} t$

Angular velocity from constant angular acceleration	$\omega_{\mathrm{f}} = \omega_0 + \alpha t$
Angular velocity from displacement and constant angular acceleration	$\theta_{\mathrm{f}} = \theta_0 + \omega_0 t + \frac{1}{2}\alpha t^2$
Change in angular velocity	$\omega_{\mathrm{f}}^2 = \omega_0^2 + 2\alpha(\Delta\theta)$
Total acceleration	$\vec{\mathbf{a}} = \vec{\mathbf{a}}_c + \vec{\mathbf{a}}_t$
Rotational kinetic energy	$K = \frac{1}{2}\left(\sum_j m_j r_j^2\right)\omega^2$
Moment of inertia	$I = \sum_j m_j r_j^2$
Rotational kinetic energy in terms of the moment of inertia of a rigid body	$K = \frac{1}{2}I\omega^2$
Moment of inertia of a continuous object	$I = \int r^2\, dm$
Parallel-axis theorem	$I_{\text{parallel-axis}} = I_{\text{center of mass}} + md^2$
Moment of inertia of a compound object	$I_{\text{total}} = \sum_i I_i$
Torque vector	$\vec{\boldsymbol{\tau}} = \vec{\mathbf{r}} \times \vec{\mathbf{F}}$
Magnitude of torque	$\left|\vec{\boldsymbol{\tau}}\right| = r_\perp F$
Total torque	$\tau_{\text{net}} = \sum_i |\tau_i|$
Newton's second law for rotation	$\sum_i \tau_i = I\alpha$
Incremental work done by a torque	$dW = \left(\sum_i \tau_i\right)d\theta$
Work-energy theorem	$W_{AB} = K_B - K_A$
Rotational work done by net force	$W_{AB} = \int_{\theta_A}^{\theta_B}\left(\sum_i \tau_i\right)d\theta$
Rotational power	$P = \tau\omega$

SUMMARY

10.1 Rotational Variables

- The angular position θ of a rotating body is the angle the body has rotated through in a fixed coordinate system, which serves as a frame of reference.
- The angular velocity of a rotating body about a fixed axis is defined as $\omega(\text{rad/s})$, the rotational rate of the body in radians per second. The instantaneous angular velocity of a rotating body $\omega = \lim_{\Delta t \to 0} \frac{\Delta\omega}{\Delta t} = \frac{d\theta}{dt}$ is the derivative

with respect to time of the angular position θ, found by taking the limit $\Delta t \to 0$ in the average angular velocity $\bar{\omega} = \frac{\Delta \theta}{\Delta t}$. The angular velocity relates v_t to the tangential speed of a point on the rotating body through the relation $v_t = r\omega$, where r is the radius to the point and v_t is the tangential speed at the given point.

- The angular velocity $\vec{\omega}$ is found using the right-hand rule. If the fingers curl in the direction of rotation about a fixed axis, the thumb points in the direction of $\vec{\omega}$ (see **Figure 10.5**).

- If the system's angular velocity is not constant, then the system has an angular acceleration. The average angular acceleration over a given time interval is the change in angular velocity over this time interval, $\bar{\alpha} = \frac{\Delta \omega}{\Delta t}$. The instantaneous angular acceleration is the time derivative of angular velocity, $\alpha = \lim\limits_{\Delta t \to 0} \frac{\Delta \omega}{\Delta t} = \frac{d\omega}{dt}$. The angular acceleration $\vec{\alpha}$ is found by locating the angular velocity. If a rotation rate of a rotating body is decreasing, the angular acceleration is in the opposite direction to $\vec{\omega}$. If the rotation rate is increasing, the angular acceleration is in the same direction as $\vec{\omega}$.

- The tangential acceleration of a point at a radius from the axis of rotation is the angular acceleration times the radius to the point.

10.2 Rotation with Constant Angular Acceleration

- The kinematics of rotational motion describes the relationships among rotation angle (angular position), angular velocity, angular acceleration, and time.

- For a constant angular acceleration, the angular velocity varies linearly. Therefore, the average angular velocity is 1/2 the initial plus final angular velocity over a given time period:

$$\bar{\omega} = \frac{\omega_0 + \omega_f}{2}.$$

- We used a graphical analysis to find solutions to fixed-axis rotation with constant angular acceleration. From the relation $\omega = \frac{d\theta}{dt}$, we found that the area under an angular velocity-vs.-time curve gives the angular displacement,

$$\theta_f - \theta_0 = \Delta\theta = \int_{t_0}^{t} \omega(t)dt.$$ The results of the graphical analysis were verified using the kinematic equations for constant angular acceleration. Similarly, since $\alpha = \frac{d\omega}{dt}$, the area under an angular acceleration-vs.-time graph gives the change in angular velocity: $\omega_f - \omega_0 = \Delta\omega = \int_{t_0}^{t} \alpha(t)dt$.

10.3 Relating Angular and Translational Quantities

- The linear kinematic equations have their rotational counterparts such that there is a mapping $x \to \theta, \ v \to \omega, \ a \to \alpha$.

- A system undergoing uniform circular motion has a constant angular velocity, but points at a distance r from the rotation axis have a linear centripetal acceleration.

- A system undergoing nonuniform circular motion has an angular acceleration and therefore has both a linear centripetal and linear tangential acceleration at a point a distance r from the axis of rotation.

- The total linear acceleration is the vector sum of the centripetal acceleration vector and the tangential acceleration vector. Since the centripetal and tangential acceleration vectors are perpendicular to each other for circular motion, the magnitude of the total linear acceleration is $\left| \vec{a} \right| = \sqrt{a_c^2 + a_t^2}$.

10.4 Moment of Inertia and Rotational Kinetic Energy

- The rotational kinetic energy is the kinetic energy of rotation of a rotating rigid body or system of particles, and is given by $K = \frac{1}{2}I\omega^2$, where I is the moment of inertia, or "rotational mass" of the rigid body or system of particles.

- The moment of inertia for a system of point particles rotating about a fixed axis is $I = \sum_j m_j r_j^2$, where m_j is the mass of the point particle and r_j is the distance of the point particle to the rotation axis. Because of the r^2 term, the moment of inertia increases as the square of the distance to the fixed rotational axis. The moment of inertia is the rotational counterpart to the mass in linear motion.

- In systems that are both rotating and translating, conservation of mechanical energy can be used if there are no nonconservative forces at work. The total mechanical energy is then conserved and is the sum of the rotational and translational kinetic energies, and the gravitational potential energy.

10.5 Calculating Moments of Inertia

- Moments of inertia can be found by summing or integrating over every 'piece of mass' that makes up an object, multiplied by the square of the distance of each 'piece of mass' to the axis. In integral form the moment of inertia is $I = \int r^2 \, dm$.

- Moment of inertia is larger when an object's mass is farther from the axis of rotation.

- It is possible to find the moment of inertia of an object about a new axis of rotation once it is known for a parallel axis. This is called the parallel axis theorem given by $I_{\text{parallel-axis}} = I_{\text{center of mass}} + md^2$, where d is the distance from the initial axis to the parallel axis.

- Moment of inertia for a compound object is simply the sum of the moments of inertia for each individual object that makes up the compound object.

10.6 Torque

- The magnitude of a torque about a fixed axis is calculated by finding the lever arm to the point where the force is applied and using the relation $\left| \vec{\tau} \right| = r_\perp F$, where $r_\perp$ is the perpendicular distance from the axis to the line upon which the force vector lies.

- The sign of the torque is found using the right hand rule. If the page is the plane containing $\vec{r}$ and $\vec{F}$, then $\vec{r} \times \vec{F}$ is out of the page for positive torques and into the page for negative torques.

- The net torque can be found from summing the individual torques about a given axis.

10.7 Newton's Second Law for Rotation

- Newton's second law for rotation, $\sum_i \tau_i = I\alpha$, says that the sum of the torques on a rotating system about a fixed axis equals the product of the moment of inertia and the angular acceleration. This is the rotational analog to Newton's second law of linear motion.

- In the vector form of Newton's second law for rotation, the torque vector $\vec{\tau}$ is in the same direction as the angular acceleration $\vec{\alpha}$. If the angular acceleration of a rotating system is positive, the torque on the system is also positive, and if the angular acceleration is negative, the torque is negative.

10.8 Work and Power for Rotational Motion

- The incremental work dW in rotating a rigid body about a fixed axis is the sum of the torques about the axis times the incremental angle $d\theta$.

- The total work done to rotate a rigid body through an angle θ about a fixed axis is the sum of the torques integrated over the angular displacement. If the torque is a constant as a function of θ, then $W_{AB} = \tau(\theta_B - \theta_A)$.

- The work-energy theorem relates the rotational work done to the change in rotational kinetic energy: $W_{AB} = K_B - K_A$ where $K = \frac{1}{2}I\omega^2$.

- The power delivered to a system that is rotating about a fixed axis is the torque times the angular velocity, $P = \tau\omega$.

CONCEPTUAL QUESTIONS

10.1 Rotational Variables

1. A clock is mounted on the wall. As you look at it, what is the direction of the angular velocity vector of the second hand?

2. What is the value of the angular acceleration of the second hand of the clock on the wall?

3. A baseball bat is swung. Do all points on the bat have the same angular velocity? The same tangential speed?

4. The blades of a blender on a counter are rotating clockwise as you look into it from the top. If the blender is put to a greater speed what direction is the angular acceleration of the blades?

10.2 Rotation with Constant Angular Acceleration

5. If a rigid body has a constant angular acceleration, what is the functional form of the angular velocity in terms of the time variable?

6. If a rigid body has a constant angular acceleration, what is the functional form of the angular position?

7. If the angular acceleration of a rigid body is zero, what is the functional form of the angular velocity?

8. A massless tether with a masses tied to both ends rotates about a fixed axis through the center. Can the total acceleration of the tether/mass combination be zero if the angular velocity is constant?

10.3 Relating Angular and Translational Quantities

9. Explain why centripetal acceleration changes the direction of velocity in circular motion but not its magnitude.

10. In circular motion, a tangential acceleration can change the magnitude of the velocity but not its direction. Explain your answer.

11. Suppose a piece of food is on the edge of a rotating microwave oven plate. Does it experience nonzero tangential acceleration, centripetal acceleration, or both when: (a) the plate starts to spin faster? (b) The plate rotates at constant angular velocity? (c) The plate slows to a halt?

10.4 Moment of Inertia and Rotational Kinetic Energy

12. What if another planet the same size as Earth were put into orbit around the Sun along with Earth. Would the moment of inertia of the system increase, decrease, or stay the same?

13. A solid sphere is rotating about an axis through its center at a constant rotation rate. Another hollow sphere of the same mass and radius is rotating about its axis through the center at the same rotation rate. Which sphere has a greater rotational kinetic energy?

10.5 Calculating Moments of Inertia

14. If a child walks toward the center of a merry-go-round, does the moment of inertia increase or decrease?

15. A discus thrower rotates with a discus in his hand before letting it go. (a) How does his moment of inertia change after releasing the discus? (b) What would be a good approximation to use in calculating the moment of inertia of the discus thrower and discus?

16. Does increasing the number of blades on a propeller increase or decrease its moment of inertia, and why?

17. The moment of inertia of a long rod spun around an axis through one end perpendicular to its length is $mL^2/3$. Why is this moment of inertia greater than it would be if you spun a point mass m at the location of the center of mass of the rod (at $L/2$) (that would be $mL^2/4$)?

18. Why is the moment of inertia of a hoop that has a mass M and a radius R greater than the moment of inertia of a disk that has the same mass and radius?

10.6 Torque

19. What three factors affect the torque created by a force relative to a specific pivot point?

20. Give an example in which a small force exerts a large torque. Give another example in which a large force exerts a small torque.

21. When reducing the mass of a racing bike, the greatest benefit is realized from reducing the mass of the tires and wheel rims. Why does this allow a racer to achieve greater accelerations than would an identical reduction in the mass of the bicycle's frame?

22. Can a single force produce a zero torque?

23. Can a set of forces have a net torque that is zero and a net force that is not zero?

24. Can a set of forces have a net force that is zero and a net torque that is not zero?

25. In the expression $\overrightarrow{\mathbf{r}} \times \overrightarrow{\mathbf{F}}$ can $\left| \overrightarrow{\mathbf{r}} \right|$ ever be less than the lever arm? Can it be equal to the lever arm?

10.7 Newton's Second Law for Rotation

26. If you were to stop a spinning wheel with a constant force, where on the wheel would you apply the force to produce the maximum negative acceleration?

27. A rod is pivoted about one end. Two forces $\overrightarrow{\mathbf{F}}$ and $-\overrightarrow{\mathbf{F}}$ are applied to it. Under what circumstances will the rod not rotate?

PROBLEMS

10.1 Rotational Variables

28. Calculate the angular velocity of Earth.

29. A track star runs a 400-m race on a 400-m circular track in 45 s. What is his angular velocity assuming a constant speed?

30. A wheel rotates at a constant rate of 2.0×10^3 rev/min . (a) What is its angular velocity in radians per second? (b) Through what angle does it turn in 10 s? Express the solution in radians and degrees.

31. A particle moves 3.0 m along a circle of radius 1.5 m. (a) Through what angle does it rotate? (b) If the particle makes this trip in 1.0 s at a constant speed, what is its angular velocity? (c) What is its acceleration?

32. A compact disc rotates at 500 rev/min. If the diameter of the disc is 120 mm, (a) what is the tangential speed of a point at the edge of the disc? (b) At a point halfway to the center of the disc?

33. Unreasonable results. The propeller of an aircraft is spinning at 10 rev/s when the pilot shuts off the engine. The propeller reduces its angular velocity at a constant $2.0 \, \text{rad/s}^2$ for a time period of 40 s. What is the rotation rate of the propeller in 40 s? Is this a reasonable situation?

34. A gyroscope slows from an initial rate of 32.0 rad/s at a rate of $0.700 \, \text{rad/s}^2$. How long does it take to come to rest?

35. On takeoff, the propellers on a UAV (unmanned aerial vehicle) increase their angular velocity from rest at a rate of $\omega = (25.0t) \, \text{rad/s}$ for 3.0 s. (a) What is the instantaneous angular velocity of the propellers at $t = 2.0 \, \text{s}$? (b) What is the angular acceleration?

36. The angular position of a rod varies as $20.0t^2$ radians from time $t = 0$. The rod has two beads on it as shown in the following figure, one at 10 cm from the rotation axis and the other at 20 cm from the rotation axis. (a) What is the instantaneous angular velocity of the rod at $t = 5 \, \text{s}$? (b) What is the angular acceleration of the rod? (c) What are the tangential speeds of the beads at $t = 5 \, \text{s}$? (d) What are the tangential accelerations of the beads at $t = 5 \, \text{s}$? (e) What are the centripetal accelerations of the beads at $t = 5 \, \text{s}$?

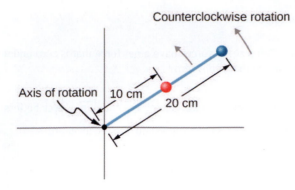

Counterclockwise rotation

Axis of rotation | 10 cm
20 cm

10.2 Rotation with Constant Angular Acceleration

37. A wheel has a constant angular acceleration of 5.0 rad/s^2. Starting from rest, it turns through 300 rad. (a) What is its final angular velocity? (b) How much time elapses while it turns through the 300 radians?

38. During a 6.0-s time interval, a flywheel with a constant angular acceleration turns through 500 radians that acquire an angular velocity of 100 rad/s. (a) What is the angular velocity at the beginning of the 6.0 s? (b) What is the angular acceleration of the flywheel?

39. The angular velocity of a rotating rigid body increases from 500 to 1500 rev/min in 120 s. (a) What is the angular acceleration of the body? (b) Through what angle does it turn in this 120 s?

40. A flywheel slows from 600 to 400 rev/min while rotating through 40 revolutions. (a) What is the angular acceleration of the flywheel? (b) How much time elapses during the 40 revolutions?

41. A wheel 1.0 m in diameter rotates with an angular acceleration of 4.0 rad/s^2. (a) If the wheel's initial angular velocity is 2.0 rad/s, what is its angular velocity after 10 s? (b) Through what angle does it rotate in the 10-s interval? (c) What are the tangential speed and acceleration of a point on the rim of the wheel at the end of the 10-s interval?

42. A vertical wheel with a diameter of 50 cm starts from rest and rotates with a constant angular acceleration of 5.0 rad/s^2 around a fixed axis through its center counterclockwise. (a) Where is the point that is initially at the bottom of the wheel at $t = 10 \text{ s}$? (b) What is the point's linear acceleration at this instant?

43. A circular disk of radius 10 cm has a constant angular acceleration of 1.0 rad/s^2; at $t = 0$ its angular velocity is 2.0 rad/s. (a) Determine the disk's angular velocity at $t = 5.0 \text{ s}$. (b) What is the angle it has rotated through

during this time? (c) What is the tangential acceleration of a point on the disk at $t = 5.0 \text{ s}$?

44. The angular velocity vs. time for a fan on a hovercraft is shown below. (a) What is the angle through which the fan blades rotate in the first 8 seconds? (b) Verify your result using the kinematic equations.

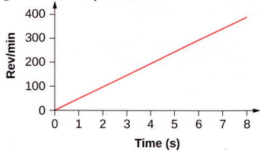

45. A rod of length 20 cm has two beads attached to its ends. The rod with beads starts rotating from rest. If the beads are to have a tangential speed of 20 m/s in 7 s, what is the angular acceleration of the rod to achieve this?

10.3 Relating Angular and Translational Quantities

46. At its peak, a tornado is 60.0 m in diameter and carries 500 km/h winds. What is its angular velocity in revolutions per second?

47. A man stands on a merry-go-round that is rotating at 2.5 rad/s. If the coefficient of static friction between the man's shoes and the merry-go-round is $\mu_S = 0.5$, how far from the axis of rotation can he stand without sliding?

48. An ultracentrifuge accelerates from rest to 100,000 rpm in 2.00 min. (a) What is the average angular acceleration in rad/s^2? (b) What is the tangential acceleration of a point 9.50 cm from the axis of rotation? (c) What is the centripetal acceleration in m/s^2 and multiples of g of this point at full rpm? (d) What is the total distance travelled by a point 9.5 cm from the axis of rotation of the ultracentrifuge?

49. A wind turbine is rotating counterclockwise at 0.5 rev/s and slows to a stop in 10 s. Its blades are 20 m in length. (a) What is the angular acceleration of the turbine? (b) What is the centripetal acceleration of the tip of the blades at $t = 0 \text{ s}$? (c) What is the magnitude and direction of the total linear acceleration of the tip of the blades at $t = 0 \text{ s}$?

50. What is (a) the angular speed and (b) the linear speed of a point on Earth's surface at latitude $30°$ N. Take the radius of the Earth to be 6309 km. (c) At what latitude would your linear speed be 10 m/s?

forearm?

51. A child with mass 40 kg sits on the edge of a merry-go-round at a distance of 3.0 m from its axis of rotation. The merry-go-round accelerates from rest up to 0.4 rev/s in 10 s. If the coefficient of static friction between the child and the surface of the merry-go-round is 0.6, does the child fall off before 5 s?

52. A bicycle wheel with radius 0.3 m rotates from rest to 3 rev/s in 5 s. What is the magnitude and direction of the total acceleration vector at the edge of the wheel at 1.0 s?

53. The angular velocity of a flywheel with radius 1.0 m varies according to $\omega(t) = 2.0t$. Plot $a_c(t)$ and $a_t(t)$ from $t = 0$ to 3.0 s for $r = 1.0\,\text{m}$. Analyze these results to explain when $a_c \gg a_t$ and when $a_c \ll a_t$ for a point on the flywheel at a radius of 1.0 m.

10.4 Moment of Inertia and Rotational Kinetic Energy

54. A system of point particles is shown in the following figure. Each particle has mass 0.3 kg and they all lie in the same plane. (a) What is the moment of inertia of the system about the given axis? (b) If the system rotates at 5 rev/s, what is its rotational kinetic energy?

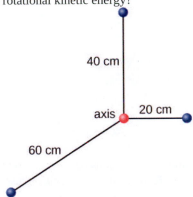

55. (a) Calculate the rotational kinetic energy of Earth on its axis. (b) What is the rotational kinetic energy of Earth in its orbit around the Sun?

56. Calculate the rotational kinetic energy of a 12-kg motorcycle wheel if its angular velocity is 120 rad/s and its inner radius is 0.280 m and outer radius 0.330 m.

57. A baseball pitcher throws the ball in a motion where there is rotation of the forearm about the elbow joint as well as other movements. If the linear velocity of the ball relative to the elbow joint is 20.0 m/s at a distance of 0.480 m from the joint and the moment of inertia of the forearm is $0.500\,\text{kg-m}^2$, what is the rotational kinetic energy of the

58. A diver goes into a somersault during a dive by tucking her limbs. If her rotational kinetic energy is 100 J and her moment of inertia in the tuck is $9.0\,\text{kg}\cdot\text{m}^2$, what is her rotational rate during the somersault?

59. An aircraft is coming in for a landing at 300 meters height when the propeller falls off. The aircraft is flying at 40.0 m/s horizontally. The propeller has a rotation rate of 20 rev/s, a moment of inertia of $70.0\,\text{kg-m}^2$, and a mass of 200 kg. Neglect air resistance. (a) With what translational velocity does the propeller hit the ground? (b) What is the rotation rate of the propeller at impact?

60. If air resistance is present in the preceding problem and reduces the propeller's rotational kinetic energy at impact by 30%, what is the propeller's rotation rate at impact?

61. A neutron star of mass 2×10^{30} kg and radius 10 km rotates with a period of 0.02 seconds. What is its rotational kinetic energy?

62. An electric sander consisting of a rotating disk of mass 0.7 kg and radius 10 cm rotates at 15 rev/sec. When applied to a rough wooden wall the rotation rate decreases by 20%. (a) What is the final rotational kinetic energy of the rotating disk? (b) How much has its rotational kinetic energy decreased?

63. A system consists of a disk of mass 2.0 kg and radius 50 cm upon which is mounted an annular cylinder of mass 1.0 kg with inner radius 20 cm and outer radius 30 cm (see below). The system rotates about an axis through the center of the disk and annular cylinder at 10 rev/s. (a) What is the moment of inertia of the system? (b) What is its rotational kinetic energy?

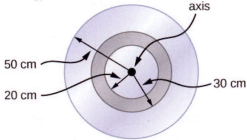

10.5 Calculating Moments of Inertia

64. While punting a football, a kicker rotates his leg about the hip joint. The moment of inertia of the leg is $3.75\,\text{kg-m}^2$ and its rotational kinetic energy is 175 J. (a) What is the angular velocity of the leg? (b) What is the velocity of tip of the punter's shoe if it is 1.05 m from the hip joint?

65. Using the parallel axis theorem, what is the moment of inertia of the rod of mass *m* about the axis shown below?

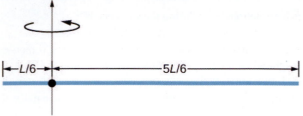

66. Find the moment of inertia of the rod in the previous problem by direct integration.

67. A uniform rod of mass 1.0 kg and length 2.0 m is free to rotate about one end (see the following figure). If the rod is released from rest at an angle of $60°$ with respect to the horizontal, what is the speed of the tip of the rod as it passes the horizontal position?

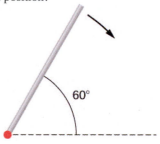

68. A pendulum consists of a rod of mass 2 kg and length 1 m with a solid sphere at one end with mass 0.3 kg and radius 20 cm (see the following figure). If the pendulum is released from rest at an angle of $30°$, what is the angular velocity at the lowest point?

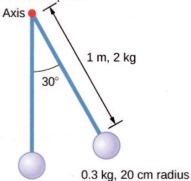

69. A solid sphere of radius 10 cm is allowed to rotate freely about an axis. The sphere is given a sharp blow so that its center of mass starts from the position shown in the following figure with speed 15 cm/s. What is the maximum angle that the diameter makes with the vertical?

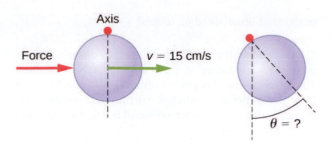

70. Calculate the moment of inertia by direct integration of a thin rod of mass *M* and length *L* about an axis through the rod at *L*/3, as shown below. Check your answer with the parallel-axis theorem.

10.6 Torque

71. Two flywheels of negligible mass and different radii are bonded together and rotate about a common axis (see below). The smaller flywheel of radius 30 cm has a cord that has a pulling force of 50 N on it. What pulling force needs to be applied to the cord connecting the larger flywheel of radius 50 cm such that the combination does not rotate?

F = ?

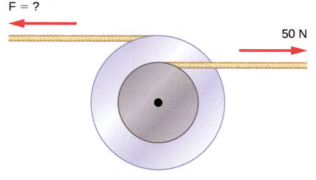

72. The cylindrical head bolts on a car are to be tightened with a torque of 62.0 N · m. If a mechanic uses a wrench of length 20 cm, what perpendicular force must he exert on the end of the wrench to tighten a bolt correctly?

73. (a) When opening a door, you push on it perpendicularly with a force of 55.0 N at a distance of 0.850 m from the hinges. What torque are you exerting relative to the hinges? (b) Does it matter if you push at the same height as the hinges? There is only one pair of hinges.

74. When tightening a bolt, you push perpendicularly on a wrench with a force of 165 N at a distance of 0.140 m from the center of the bolt. How much torque are you exerting in newton-meters (relative to the center of the bolt)?

75. What hanging mass must be placed on the cord to keep the pulley from rotating (see the following figure)? The mass on the frictionless plane is 5.0 kg. The inner radius of the pulley is 20 cm and the outer radius is 30 cm.

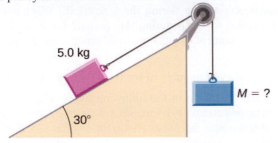

76. A simple pendulum consists of a massless tether 50 cm in length connected to a pivot and a small mass of 1.0 kg attached at the other end. What is the torque about the pivot when the pendulum makes an angle of $40°$ with respect to the vertical?

77. Calculate the torque about the z-axis that is out of the page at the origin in the following figure, given that $F_1 = 3\,N$, $F_2 = 2\,N$, $F_3 = 3\,N$, $F_4 = 1.8\,N$.

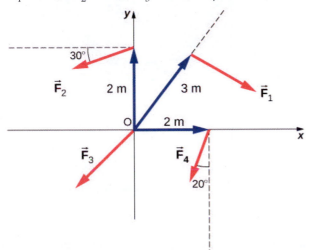

78. A seesaw has length 10.0 m and uniform mass 10.0 kg and is resting at an angle of $30°$ with respect to the ground (see the following figure). The pivot is located at 6.0 m. What magnitude of force needs to be applied perpendicular to the seesaw at the raised end so as to allow the seesaw to barely start to rotate?

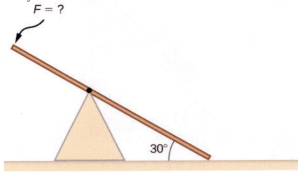

79. A pendulum consists of a rod of mass 1 kg and length

1 m connected to a pivot with a solid sphere attached at the other end with mass 0.5 kg and radius 30 cm. What is the torque about the pivot when the pendulum makes an angle of $30°$ with respect to the vertical?

80. A torque of $5.00 \times 10^3\,N\cdot m$ is required to raise a drawbridge (see the following figure). What is the tension necessary to produce this torque? Would it be easier to raise the drawbridge if the angle θ were larger or smaller?

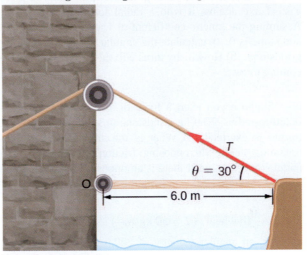

81. A horizontal beam of length 3 m and mass 2.0 kg has a mass of 1.0 kg and width 0.2 m sitting at the end of the beam (see the following figure). What is the torque of the system about the support at the wall?

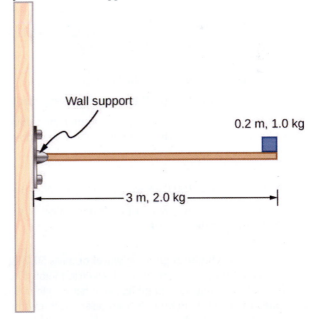

82. What force must be applied to end of a rod along the x-axis of length 2.0 m in order to produce a torque on the rod about the origin of $8.0\hat{\mathbf{k}}\,N\cdot m$?

83. What is the torque about the origin of the force

$(5.0\,\hat{\mathbf{i}} - 2.0\,\hat{\mathbf{j}} + 1.0\,\hat{\mathbf{k}})\,\text{N}$ if it is applied at the point whose position is: $\vec{\mathbf{r}} = \left(-2.0\,\hat{\mathbf{i}} + 4.0\,\hat{\mathbf{j}}\right)\text{m}$?

10.7 Newton's Second Law for Rotation

84. You have a grindstone (a disk) that is 90.0 kg, has a 0.340-m radius, and is turning at 90.0 rpm, and you press a steel axe against it with a radial force of 20.0 N. (a) Assuming the kinetic coefficient of friction between steel and stone is 0.20, calculate the angular acceleration of the grindstone. (b) How many turns will the stone make before coming to rest?

85. Suppose you exert a force of 180 N tangential to a 0.280-m-radius, 75.0-kg grindstone (a solid disk). (a)What torque is exerted? (b) What is the angular acceleration assuming negligible opposing friction? (c) What is the angular acceleration if there is an opposing frictional force of 20.0 N exerted 1.50 cm from the axis?

86. A flywheel ($I = 50\,\text{kg-m}^2$) starting from rest acquires an angular velocity of 200.0 rad/s while subject to a constant torque from a motor for 5 s. (a) What is the angular acceleration of the flywheel? (b) What is the magnitude of the torque?

87. A constant torque is applied to a rigid body whose moment of inertia is $4.0\,\text{kg-m}^2$ around the axis of rotation. If the wheel starts from rest and attains an angular velocity of 20.0 rad/s in 10.0 s, what is the applied torque?

88. A torque of 50.0 N-m is applied to a grinding wheel ($I = 20.0\,\text{kg-m}^2$) for 20 s. (a) If it starts from rest, what is the angular velocity of the grinding wheel after the torque is removed? (b) Through what angle does the wheel move while the torque is applied?

89. A flywheel ($I = 100.0\,\text{kg-m}^2$) rotating at 500.0 rev/min is brought to rest by friction in 2.0 min. What is the frictional torque on the flywheel?

90. A uniform cylindrical grinding wheel of mass 50.0 kg and diameter 1.0 m is turned on by an electric motor. The friction in the bearings is negligible. (a) What torque must be applied to the wheel to bring it from rest to 120 rev/min in 20 revolutions? (b) A tool whose coefficient of kinetic friction with the wheel is 0.60 is pressed perpendicularly against the wheel with a force of 40.0 N. What torque must be supplied by the motor to keep the wheel rotating at a constant angular velocity?

91. Suppose when Earth was created, it was not rotating.

However, after the application of a uniform torque after 6 days, it was rotating at 1 rev/day. (a) What was the angular acceleration during the 6 days? (b) What torque was applied to Earth during this period? (c) What force tangent to Earth at its equator would produce this torque?

92. A pulley of moment of inertia $2.0\,\text{kg-m}^2$ is mounted on a wall as shown in the following figure. Light strings are wrapped around two circumferences of the pulley and weights are attached. What are (a) the angular acceleration of the pulley and (b) the linear acceleration of the weights? Assume the following data: $r_1 = 50\,\text{cm}$, $r_2 = 20\,\text{cm}$, $m_1 = 1.0\,\text{kg}$, $m_2 = 2.0\,\text{kg}$

93. A block of mass 3 kg slides down an inclined plane at an angle of $45°$ with a massless tether attached to a pulley with mass 1 kg and radius 0.5 m at the top of the incline (see the following figure). The pulley can be approximated as a disk. The coefficient of kinetic friction on the plane is 0.4. What is the acceleration of the block?

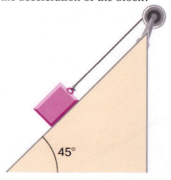

94. The cart shown below moves across the table top as the block falls. What is the acceleration of the cart? Neglect friction and assume the following data:

$m_1 = 2.0 \, \text{kg}, \, m_2 = 4.0 \, \text{kg}, \, I = 0.4 \, \text{kg-m}^2, \, r = 20 \, \text{cm}$

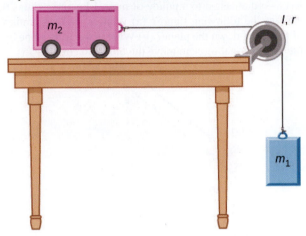

95. A uniform rod of mass and length is held vertically by two strings of negligible mass, as shown below. (a) Immediately after the string is cut, what is the linear acceleration of the free end of the stick? (b) Of the middle of the stick?

96. A thin stick of mass 0.2 kg and length $L = 0.5 \, \text{m}$ is attached to the rim of a metal disk of mass $M = 2.0 \, \text{kg}$ and radius $R = 0.3 \, \text{m}$. The stick is free to rotate around a horizontal axis through its other end (see the following figure). (a) If the combination is released with the stick horizontal, what is the speed of the center of the disk when the stick is vertical? (b) What is the acceleration of the center of the disk at the instant the stick is released? (c) At the instant the stick passes through the vertical?

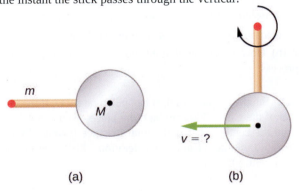

(a) (b)

10.8 Work and Power for Rotational Motion

97. A wind turbine rotates at 20 rev/min. If its power output is 2.0 MW, what is the torque produced on the turbine from the wind?

98. A clay cylinder of radius 20 cm on a potter's wheel spins at a constant rate of 10 rev/s. The potter applies a force of 10 N to the clay with his hands where the coefficient of friction is 0.1 between his hands and the clay. What is the power that the potter has to deliver to the wheel to keep it rotating at this constant rate?

99. A uniform cylindrical grindstone has a mass of 10 kg and a radius of 12 cm. (a) What is the rotational kinetic energy of the grindstone when it is rotating at $1.5 \times 10^3 \, \text{rev/min}$? (b) After the grindstone's motor is turned off, a knife blade is pressed against the outer edge of the grindstone with a perpendicular force of 5.0 N. The coefficient of kinetic friction between the grindstone and the blade is 0.80. Use the work energy theorem to determine how many turns the grindstone makes before it stops.

100. A uniform disk of mass 500 kg and radius 0.25 m is mounted on frictionless bearings so it can rotate freely around a vertical axis through its center (see the following figure). A cord is wrapped around the rim of the disk and pulled with a force of 10 N. (a) How much work has the force done at the instant the disk has completed three revolutions, starting from rest? (b) Determine the torque due to the force, then calculate the work done by this torque at the instant the disk has completed three revolutions? (c) What is the angular velocity at that instant? (d) What is the power output of the force at that instant?

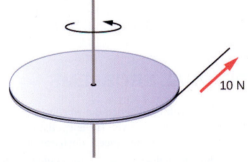

101. A propeller is accelerated from rest to an angular velocity of 1000 rev/min over a period of 6.0 seconds by a constant torque of $2.0 \times 10^3 \, \text{N} \cdot \text{m}$. (a) What is the moment of inertia of the propeller? (b) What power is being provided to the propeller 3.0 s after it starts rotating?

102. A sphere of mass 1.0 kg and radius 0.5 m is attached to the end of a massless rod of length 3.0 m. The rod rotates about an axis that is at the opposite end of the sphere (see below). The system rotates horizontally about the axis at a constant 400 rev/min. After rotating at this angular speed in a vacuum, air resistance is introduced and provides a force 0.15 N on the sphere opposite to the direction of motion. What is the power provided by air resistance to the system 100.0 s after air resistance is introduced?

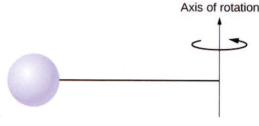

Axis of rotation

103. A uniform rod of length L and mass M is held vertically with one end resting on the floor as shown below. When the rod is released, it rotates around its lower end until it hits the floor. Assuming the lower end of the rod does not slip, what is the linear velocity of the upper end when it hits the floor?

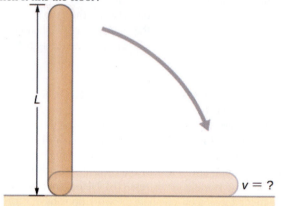

$v = ?$

104. An athlete in a gym applies a constant force of 50 N to the pedals of a bicycle to keep the rotation rate of the wheel at 10 rev/s. The length of the pedal arms is 30 cm. What is the power delivered to the bicycle by the athlete?

ADDITIONAL PROBLEMS

107. A cyclist is riding such that the wheels of the bicycle have a rotation rate of 3.0 rev/s. If the cyclist brakes such that the rotation rate of the wheels decrease at a rate of 0.3 rev/s^2, how long does it take for the cyclist to come to a complete stop?

108. Calculate the angular velocity of the orbital motion of Earth around the Sun.

109. A phonograph turntable rotating at 33 1/3 rev/min slows down and stops in 1.0 min. (a) What is the turntable's angular acceleration assuming it is constant? (b) How many revolutions does the turntable make while stopping?

105. A 2-kg block on a frictionless inclined plane at $40°$ has a cord attached to a pulley of mass 1 kg and radius 20 cm (see the following figure). (a) What is the acceleration of the block down the plane? (b) What is the work done by the gravitational force to move the block 50 cm?

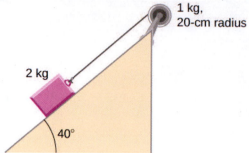

1 kg,
20-cm radius

2 kg

40°

106. Small bodies of mass m_1 and m_2 are attached to opposite ends of a thin rigid rod of length L and mass M. The rod is mounted so that it is free to rotate in a horizontal plane around a vertical axis (see below). What distance d from m_1 should the rotational axis be so that a minimum amount of work is required to set the rod rotating at an angular velocity ω?

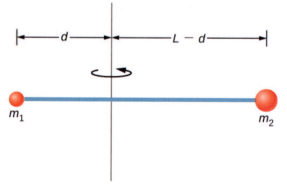

m_1

m_2

$\leftarrow d \rightarrow$ $\leftarrow L - d \rightarrow$

110. With the aid of a string, a gyroscope is accelerated from rest to 32 rad/s in 0.40 s under a constant angular acceleration. (a) What is its angular acceleration in rad/s^2? (b) How many revolutions does it go through in the process?

111. Suppose a piece of dust has fallen on a CD. If the spin rate of the CD is 500 rpm, and the piece of dust is 4.3 cm from the center, what is the total distance traveled by the dust in 3 minutes? (Ignore accelerations due to getting the CD rotating.)

112. A system of point particles is rotating about a fixed axis at 4 rev/s. The particles are fixed with respect to each other. The masses and distances to the axis of the point particles are $m_1 = 0.1$ kg, $r_1 = 0.2$ m, $m_2 = 0.05$ kg, $r_2 = 0.4$ m, $m_3 = 0.5$ kg, $r_3 = 0.01$ m. (a) What is the moment of inertia of the system? (b) What is the rotational kinetic energy of the system?

113. Calculate the moment of inertia of a skater given the following information. (a) The 60.0-kg skater is approximated as a cylinder that has a 0.110-m radius. (b) The skater with arms extended is approximated by a cylinder that is 52.5 kg, has a 0.110-m radius, and has two 0.900-m-long arms which are 3.75 kg each and extend straight out from the cylinder like rods rotated about their ends.

114. A stick of length 1.0 m and mass 6.0 kg is free to rotate about a horizontal axis through the center. Small bodies of masses 4.0 and 2.0 kg are attached to its two ends (see the following figure). The stick is released from the horizontal position. What is the angular velocity of the stick when it swings through the vertical?

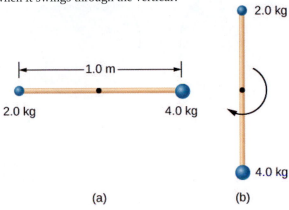

(a)   (b)

115. A pendulum consists of a rod of length 2 m and mass 3 kg with a solid sphere of mass 1 kg and radius 0.3 m attached at one end. The axis of rotation is as shown below. What is the angular velocity of the pendulum at its lowest point if it is released from rest at an angle of 30°?

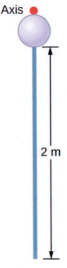

116. Calculate the torque of the 40-N force around the axis through O and perpendicular to the plane of the page as shown below.

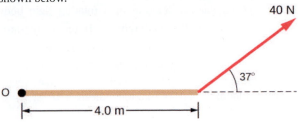

117. Two children push on opposite sides of a door during play. Both push horizontally and perpendicular to the door. One child pushes with a force of 17.5 N at a distance of 0.600 m from the hinges, and the second child pushes at a distance of 0.450 m. What force must the second child exert to keep the door from moving? Assume friction is negligible.

118. The force of $20\hat{j}$ N is applied at $\vec{r} = (4.0\hat{i} - 2.0\hat{j})$ m. What is the torque of this force about the origin?

119. An automobile engine can produce 200 N · m of torque. Calculate the angular acceleration produced if 95.0% of this torque is applied to the drive shaft, axle, and rear wheels of a car, given the following information. The car is suspended so that the wheels can turn freely. Each wheel acts like a 15.0-kg disk that has a 0.180-m radius. The walls of each tire act like a 2.00-kg annular ring that has inside radius of 0.180 m and outside radius of 0.320 m. The tread of each tire acts like a 10.0-kg hoop of radius 0.330 m. The 14.0-kg axle acts like a rod that has a 2.00-cm radius. The 30.0-kg drive shaft acts like a rod that has a 3.20-cm radius.

120. A grindstone with a mass of 50 kg and radius 0.8 m maintains a constant rotation rate of 4.0 rev/s by a motor while a knife is pressed against the edge with a force of 5.0 N. The coefficient of kinetic friction between the grindstone and the blade is 0.8. What is the power provided by the motor to keep the grindstone at the constant rotation rate?

CHALLENGE PROBLEMS

121. The angular acceleration of a rotating rigid body is given by $\alpha = (2.0 - 3.0t)\,\text{rad/s}^2$. If the body starts rotating from rest at $t = 0$, (a) what is the angular velocity? (b) Angular position? (c) What angle does it rotate through in 10 s? (d) Where does the vector perpendicular to the axis of rotation indicating $0°$ at $t = 0$ lie at $t = 10\,\text{s}$?

122. Earth's day has increased by 0.002 s in the last century. If this increase in Earth's period is constant, how long will it take for Earth to come to rest?

123. A disk of mass m, radius R, and area A has a surface mass density $\sigma = \frac{mr}{AR}$ (see the following figure). What is the moment of inertia of the disk about an axis through the center?

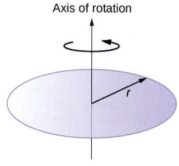

Axis of rotation

124. Zorch, an archenemy of Rotation Man, decides to slow Earth's rotation to once per 28.0 h by exerting an opposing force at and parallel to the equator. Rotation Man is not immediately concerned, because he knows Zorch can only exert a force of $4.00 \times 10^7\,\text{N}$ (a little greater than a Saturn V rocket's thrust). How long must Zorch push with this force to accomplish his goal? (This period gives Rotation Man time to devote to other villains.)

125. A cord is wrapped around the rim of a solid cylinder of radius 0.25 m, and a constant force of 40 N is exerted on the cord shown, as shown in the following figure. The cylinder is mounted on frictionless bearings, and its moment of inertia is $6.0\,\text{kg}\cdot\text{m}^2$. (a) Use the work energy theorem to calculate the angular velocity of the cylinder after 5.0 m of cord have been removed. (b) If the 40-N force is replaced by a 40-N weight, what is the angular velocity of the cylinder after 5.0 m of cord have unwound?

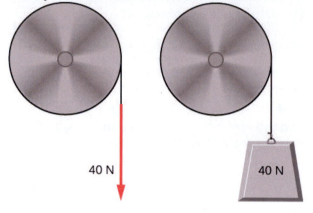

40 N   40 N

11 | ANGULAR MOMENTUM

Figure 11.1 A helicopter has its main lift blades rotating to keep the aircraft airborne. Due to conservation of angular momentum, the body of the helicopter would want to rotate in the opposite sense to the blades, if it were not for the small rotor on the tail of the aircraft, which provides thrust to stabilize it.

Chapter Outline

11.1 Rolling Motion

11.2 Angular Momentum

11.3 Conservation of Angular Momentum

11.4 Precession of a Gyroscope

Introduction

Angular momentum is the rotational counterpart of linear momentum. Any massive object that rotates about an axis carries angular momentum, including rotating flywheels, planets, stars, hurricanes, tornadoes, whirlpools, and so on. The helicopter shown in the chapter-opening picture can be used to illustrate the concept of angular momentum. The lift blades spin about a vertical axis through the main body and carry angular momentum. The body of the helicopter tends to rotate in the opposite sense in order to conserve angular momentum. The small rotors at the tail of the aircraft provide a counter thrust against the body to prevent this from happening, and the helicopter stabilizes itself. The concept of conservation of angular momentum is discussed later in this chapter. In the main part of this chapter, we explore the intricacies of angular momentum of rigid bodies such as a top, and also of point particles and systems of particles. But to be complete, we start with a discussion of rolling motion, which builds upon the concepts of the previous chapter.

11.1 | Rolling Motion

Rolling motion is that common combination of rotational and translational motion that we see everywhere, every day. Think about the different situations of wheels moving on a car along a highway, or wheels on a plane landing on a runway, or wheels on a robotic explorer on another planet. Understanding the forces and torques involved in **rolling motion** is a crucial factor in many different types of situations.

For analyzing rolling motion in this chapter, refer to **Figure 10.20** in **Fixed-Axis Rotation** to find moments of inertia of some common geometrical objects. You may also find it useful in other calculations involving rotation.

Rolling Motion without Slipping

People have observed rolling motion without slipping ever since the invention of the wheel. For example, we can look at the interaction of a car's tires and the surface of the road. If the driver depresses the accelerator to the floor, such that the tires spin without the car moving forward, there must be kinetic friction between the wheels and the surface of the road. If the driver depresses the accelerator slowly, causing the car to move forward, then the tires roll without slipping. It is surprising to most people that, in fact, the bottom of the wheel is at rest with respect to the ground, indicating there must be static friction between the tires and the road surface. In **Figure 11.2**, the bicycle is in motion with the rider staying upright. The tires have contact with the road surface, and, even though they are rolling, the bottoms of the tires deform slightly, do not slip, and are at rest with respect to the road surface for a measurable amount of time. There must be static friction between the tire and the road surface for this to be so.

(a)

(b)

Figure 11.2 (a) The bicycle moves forward, and its tires do not slip. The bottom of the slightly deformed tire is at rest with respect to the road surface for a measurable amount of time. (b) This image shows that the top of a rolling wheel appears blurred by its motion, but the bottom of the wheel is instantaneously at rest. (credit a: modification of work by Nelson Lourenço; credit b: modification of work by Colin Rose)

To analyze rolling without slipping, we first derive the linear variables of velocity and acceleration of the center of mass of the wheel in terms of the angular variables that describe the wheel's motion. The situation is shown in **Figure 11.3**.

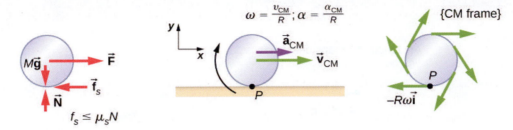

(a) Forces on the wheel   (b) Wheel rolls without slipping   (c) Point P has velocity vector in the negative direction with respect to the center of mass of the wheel

Figure 11.3 (a) A wheel is pulled across a horizontal surface by a force $\overrightarrow{\mathbf{F}}$. The force of static friction $\overrightarrow{\mathbf{f}}_S$, $\left|\overrightarrow{\mathbf{f}}_S\right| \leq \mu_S N$ is large enough to keep it from slipping. (b) The linear velocity and acceleration vectors of the center of mass and the relevant expressions for ω and α. Point P is at rest relative to the surface. (c) Relative to the center of mass (CM) frame, point P has linear velocity $-R\omega\,\hat{\mathbf{i}}$.

From **Figure 11.3**(a), we see the force vectors involved in preventing the wheel from slipping. In (b), point P that touches the surface is at rest relative to the surface. Relative to the center of mass, point P has velocity $-R\omega\,\hat{\mathbf{i}}$, where R is the radius of the wheel and ω is the wheel's angular velocity about its axis. Since the wheel is rolling, the velocity of P with respect to the surface is its velocity with respect to the center of mass plus the velocity of the center of mass with respect to the surface:

$$\overrightarrow{\mathbf{v}}_P = -R\omega\,\hat{\mathbf{i}} + v_{\text{CM}}\,\hat{\mathbf{i}}.$$

Since the velocity of P relative to the surface is zero, $v_P = 0$, this says that

$$v_{\text{CM}} = R\omega. \tag{11.1}$$

Thus, the velocity of the wheel's center of mass is its radius times the angular velocity about its axis. We show the correspondence of the linear variable on the left side of the equation with the angular variable on the right side of the equation. This is done below for the linear acceleration.

If we differentiate **Equation 11.1** on the left side of the equation, we obtain an expression for the linear acceleration of the center of mass. On the right side of the equation, R is a constant and since $\alpha = \dfrac{d\omega}{dt}$, we have

$$a_{\text{CM}} = R\alpha. \tag{11.2}$$

Furthermore, we can find the distance the wheel travels in terms of angular variables by referring to **Figure 11.4**. As the wheel rolls from point A to point B, its outer surface maps onto the ground by exactly the distance travelled, which is d_{CM}.

We see from **Figure 11.4** that the length of the outer surface that maps onto the ground is the arc length $R\theta$. Equating the two distances, we obtain

$$d_{CM} = R\theta. \tag{11.3}$$

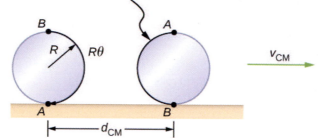

Arc length AB maps onto wheel's surface

Figure 11.4 As the wheel rolls on the surface, the arc length $R\theta$ from A to B maps onto the surface, corresponding to the distance d_{CM} that the center of mass has moved.

Example 11.1

Rolling Down an Inclined Plane

A solid cylinder rolls down an inclined plane without slipping, starting from rest. It has mass m and radius r. (a) What is its acceleration? (b) What condition must the coefficient of static friction μ_S satisfy so the cylinder does not slip?

Strategy

Draw a sketch and free-body diagram, and choose a coordinate system. We put x in the direction down the plane and y upward perpendicular to the plane. Identify the forces involved. These are the normal force, the force of gravity, and the force due to friction. Write down Newton's laws in the x- and y-directions, and Newton's law for rotation, and then solve for the acceleration and force due to friction.

Solution

a. The free-body diagram and sketch are shown in **Figure 11.5**, including the normal force, components of the weight, and the static friction force. There is barely enough friction to keep the cylinder rolling without slipping. Since there is no slipping, the magnitude of the friction force is less than or equal to $\mu_S N$. Writing down Newton's laws in the x- and y-directions, we have

$$\sum F_x = ma_x; \quad \sum F_y = ma_y.$$

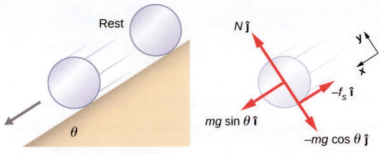

Figure 11.5 A solid cylinder rolls down an inclined plane without slipping from rest. The coordinate system has *x* in the direction down the inclined plane and *y* perpendicular to the plane. The free-body diagram is shown with the normal force, the static friction force, and the components of the weight $m\vec{g}$. Friction makes the cylinder roll down the plane rather than slip.

Substituting in from the free-body diagram,

$$mg\sin\theta - f_S = m(a_{CM})_x,$$
$$N - mg\cos\theta = 0,$$
$$f_S \leq \mu_S N,$$

we can then solve for the linear acceleration of the center of mass from these equations:

$$(a_{CM})_x = g(\sin\theta - \mu_S\cos\theta).$$

However, it is useful to express the linear acceleration in terms of the moment of inertia. For this, we write down Newton's second law for rotation,

$$\sum \tau_{CM} = I_{CM}\alpha.$$

The torques are calculated about the axis through the center of mass of the cylinder. The only nonzero torque is provided by the friction force. We have

$$f_S r = I_{CM}\alpha.$$

Finally, the linear acceleration is related to the angular acceleration by

$$(a_{CM})_x = r\alpha.$$

These equations can be used to solve for a_{CM}, α, and f_S in terms of the moment of inertia, where we have dropped the *x*-subscript. We write a_{CM} in terms of the vertical component of gravity and the friction force, and make the following substitutions.

$$a_{CM} = g\sin\theta - \frac{f_S}{m}$$

$$f_S = \frac{I_{CM}\alpha}{r} = \frac{I_{CM}a_{CM}}{r^2}$$

From this we obtain

$$a_{CM} = g \sin\theta - \frac{I_{CM} a_{CM}}{mr^2},$$

$$= \frac{mg \sin\theta}{m + (I_{CM}/r^2)}.$$

Note that this result is independent of the coefficient of static friction, μ_S.

Since we have a solid cylinder, from **Figure 10.20**, we have $I_{CM} = mr^2/2$ and

$$a_{CM} = \frac{mg \sin\theta}{m + (mr^2/2r^2)} = \frac{2}{3}g \sin\theta.$$

Therefore, we have

$$\alpha = \frac{a_{CM}}{r} = \frac{2}{3r}g \sin\theta.$$

b. Because slipping does not occur, $f_S \leq \mu_S N$. Solving for the friction force,

$$f_S = I_{CM}\frac{\alpha}{r} = I_{CM}\frac{(a_{CM})}{r^2} = \frac{I_{CM}}{r^2}\left(\frac{mg \sin\theta}{m + (I_{CM}/r^2)}\right) = \frac{mg I_{CM} \sin\theta}{mr^2 + I_{CM}}.$$

Substituting this expression into the condition for no slipping, and noting that $N = mg \cos\theta$, we have

$$\frac{mg I_{CM} \sin\theta}{mr^2 + I_{CM}} \leq \mu_S mg \cos\theta$$

or

$$\mu_S \geq \frac{\tan\theta}{1 + (mr^2/I_{CM})}.$$

For the solid cylinder, this becomes

$$\mu_S \geq \frac{\tan\theta}{1 + (2mr^2/mr^2)} = \frac{1}{3}\tan\theta.$$

Significance

a. The linear acceleration is linearly proportional to $\sin\theta$. Thus, the greater the angle of the incline, the greater the linear acceleration, as would be expected. The angular acceleration, however, is linearly proportional to $\sin\theta$ and inversely proportional to the radius of the cylinder. Thus, the larger the radius, the smaller the angular acceleration.

b. For no slipping to occur, the coefficient of static friction must be greater than or equal to $(1/3)\tan\theta$. Thus, the greater the angle of incline, the greater the coefficient of static friction must be to prevent the cylinder from slipping.

 11.1 Check Your Understanding A hollow cylinder is on an incline at an angle of $60°$. The coefficient of static friction on the surface is $\mu_S = 0.6$. (a) Does the cylinder roll without slipping? (b) Will a solid cylinder roll without slipping?

It is worthwhile to repeat the equation derived in this example for the acceleration of an object rolling without slipping:

$$a_{CM} = \frac{mg \sin \theta}{m + (I_{CM}/r^2)}. \tag{11.4}$$

This is a very useful equation for solving problems involving rolling without slipping. Note that the acceleration is less than that for an object sliding down a frictionless plane with no rotation. The acceleration will also be different for two rotating cylinders with different rotational inertias.

Rolling Motion with Slipping

In the case of rolling motion with slipping, we must use the coefficient of kinetic friction, which gives rise to the kinetic friction force since static friction is not present. The situation is shown in **Figure 11.6**. In the case of slipping, $v_{CM} - R\omega \neq 0$, because point P on the wheel is not at rest on the surface, and $v_P \neq 0$. Thus, $\omega \neq \frac{v_{CM}}{R}$, $\alpha \neq \frac{a_{CM}}{R}$.

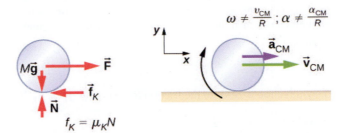

(a) Forces on wheel   (b) Wheel is rolling and slipping

Figure 11.6   (a) Kinetic friction arises between the wheel and the surface because the wheel is slipping. (b) The simple relationships between the linear and angular variables are no longer valid.

Example 11.2

Rolling Down an Inclined Plane with Slipping

A solid cylinder rolls down an inclined plane from rest and undergoes slipping (**Figure 11.7**). It has mass m and radius r. (a) What is its linear acceleration? (b) What is its angular acceleration about an axis through the center of mass?

Strategy

Draw a sketch and free-body diagram showing the forces involved. The free-body diagram is similar to the no-slipping case except for the friction force, which is kinetic instead of static. Use Newton's second law to solve for the acceleration in the x-direction. Use Newton's second law of rotation to solve for the angular acceleration.

Solution

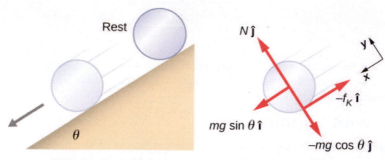

Rest

$N\,\hat{\jmath}$

$mg \sin\theta\,\hat{\imath}$

$-f_K\,\hat{\imath}$

$-mg \cos\theta\,\hat{\jmath}$

Wheel rolls and slips | Free-body diagram

Figure 11.7 A solid cylinder rolls down an inclined plane from rest and undergoes slipping. The coordinate system has x in the direction down the inclined plane and y upward perpendicular to the plane. The free-body diagram shows the normal force, kinetic friction force, and the components of the weight $m\,\vec{g}$.

The sum of the forces in the y-direction is zero, so the friction force is now $f_k = \mu_k N = \mu_k mg\cos\theta$.

Newton's second law in the x-direction becomes

$$\sum F_x = ma_x,$$
$$mg\sin\theta - \mu_k mg\cos\theta = m(a_{CM})_x,$$

or

$$(a_{CM})_x = g(\sin\theta - \mu_K\cos\theta).$$

The friction force provides the only torque about the axis through the center of mass, so Newton's second law of rotation becomes

$$\sum \tau_{CM} = I_{CM}\alpha,$$
$$f_k r = I_{CM}\alpha = \tfrac{1}{2}mr^2\alpha.$$

Solving for α, we have

$$\alpha = \frac{2f_k}{mr} = \frac{2\mu_k g\cos\theta}{r}.$$

Significance

We write the linear and angular accelerations in terms of the coefficient of kinetic friction. The linear acceleration is the same as that found for an object sliding down an inclined plane with kinetic friction. The angular acceleration about the axis of rotation is linearly proportional to the normal force, which depends on the cosine of the angle of inclination. As $\theta \to 90°$, this force goes to zero, and, thus, the angular acceleration goes to zero.

Conservation of Mechanical Energy in Rolling Motion

In the preceding chapter, we introduced rotational kinetic energy. Any rolling object carries rotational kinetic energy, as well as translational kinetic energy and potential energy if the system requires. Including the gravitational potential energy, the total mechanical energy of an object rolling is

$$E_T = \tfrac{1}{2}mv_{CM}^2 + \tfrac{1}{2}I_{CM}\omega^2 + mgh.$$

In the absence of any nonconservative forces that would take energy out of the system in the form of heat, the total energy of a rolling object without slipping is conserved and is constant throughout the motion. Examples where energy is not conserved are a rolling object that is slipping, production of heat as a result of kinetic friction, and a rolling object encountering air resistance.

You may ask why a rolling object that is not slipping conserves energy, since the static friction force is nonconservative. The answer can be found by referring back to **Figure 11.3**. Point P in contact with the surface is at rest with respect to the surface. Therefore, its infinitesimal displacement $d\vec{r}$ with respect to the surface is zero, and the incremental work done by the static friction force is zero. We can apply energy conservation to our study of rolling motion to bring out some interesting results.

Example 11.3

Curiosity Rover

The *Curiosity* rover, shown in **Figure 11.8**, was deployed on Mars on August 6, 2012. The wheels of the rover have a radius of 25 cm. Suppose astronauts arrive on Mars in the year 2050 and find the now-inoperative *Curiosity* on the side of a basin. While they are dismantling the rover, an astronaut accidentally loses a grip on one of the wheels, which rolls without slipping down into the bottom of the basin 25 meters below. If the wheel has a mass of 5 kg, what is its velocity at the bottom of the basin?

Figure 11.8 The NASA Mars Science Laboratory rover *Curiosity* during testing on June 3, 2011. The location is inside the Spacecraft Assembly Facility at NASA's Jet Propulsion Laboratory in Pasadena, California. (credit: NASA/JPL-Caltech)

Strategy

We use mechanical energy conservation to analyze the problem. At the top of the hill, the wheel is at rest and has only potential energy. At the bottom of the basin, the wheel has rotational and translational kinetic energy, which must be equal to the initial potential energy by energy conservation. Since the wheel is rolling without slipping, we use the relation $v_{CM} = r\omega$ to relate the translational variables to the rotational variables in the energy conservation equation. We then solve for the velocity. From **Figure 11.8**, we see that a hollow cylinder is a good approximation for the wheel, so we can use this moment of inertia to simplify the calculation.

Solution

Energy at the top of the basin equals energy at the bottom:

$$mgh = \frac{1}{2}mv_{CM}^2 + \frac{1}{2}I_{CM}\omega^2.$$

The known quantities are $I_{CM} = mr^2$, $r = 0.25$ m, and $h = 25.0$ m.

We rewrite the energy conservation equation eliminating ω by using $\omega = \frac{v_{CM}}{r}$. We have

$$mgh = \frac{1}{2}mv_{CM}^2 + \frac{1}{2}mr^2\frac{v_{CM}^2}{r^2}$$

or

$$gh = \frac{1}{2}v_{CM}^2 + \frac{1}{2}v_{CM}^2 \Rightarrow v_{CM} = \sqrt{gh}.$$

On Mars, the acceleration of gravity is 3.71 m/s^2, which gives the magnitude of the velocity at the bottom of the basin as

$$v_{CM} = \sqrt{(3.71 \text{ m/s}^2)25.0 \text{ m}} = 9.63 \text{ m/s}.$$

Significance

This is a fairly accurate result considering that Mars has very little atmosphere, and the loss of energy due to air resistance would be minimal. The result also assumes that the terrain is smooth, such that the wheel wouldn't encounter rocks and bumps along the way.

Also, in this example, the kinetic energy, or energy of motion, is equally shared between linear and rotational motion. If we look at the moments of inertia in **Figure 10.20**, we see that the hollow cylinder has the largest moment of inertia for a given radius and mass. If the wheels of the rover were solid and approximated by solid cylinders, for example, there would be more kinetic energy in linear motion than in rotational motion. This would give the wheel a larger linear velocity than the hollow cylinder approximation. Thus, the solid cylinder would reach the bottom of the basin faster than the hollow cylinder.

11.2 | Angular Momentum

Learning Objectives

By the end of this section, you will be able to:

- Describe the vector nature of angular momentum
- Find the total angular momentum and torque about a designated origin of a system of particles
- Calculate the angular momentum of a rigid body rotating about a fixed axis
- Calculate the torque on a rigid body rotating about a fixed axis
- Use conservation of angular momentum in the analysis of objects that change their rotation rate

Why does Earth keep on spinning? What started it spinning to begin with? Why doesn't Earth's gravitational attraction not bring the Moon crashing in toward Earth? And how does an ice skater manage to spin faster and faster simply by pulling her arms in? Why does she not have to exert a torque to spin faster?

Questions like these have answers based in angular momentum, the rotational analog to linear momentum. In this chapter, we first define and then explore angular momentum from a variety of viewpoints. First, however, we investigate the angular momentum of a single particle. This allows us to develop angular momentum for a system of particles and for a rigid body.

Angular Momentum of a Single Particle

Figure 11.9 shows a particle at a position $\vec{r}$ with linear momentum $\vec{p} = m\vec{v}$ with respect to the origin. Even if the particle is not rotating about the origin, we can still define an angular momentum in terms of the position vector and the linear momentum.

Angular Momentum of a Particle

The **angular momentum** $\vec{l}$ of a particle is defined as the cross-product of $\vec{r}$ and $\vec{p}$, and is perpendicular to the plane containing $\vec{r}$ and $\vec{p}$:

$$\vec{l} = \vec{r} \times \vec{p}. \tag{11.5}$$

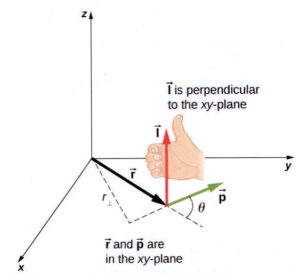

Figure 11.9 In three-dimensional space, the position vector $\vec{r}$ locates a particle in the *xy*-plane with linear momentum $\vec{p}$. The angular momentum with respect to the origin is $\vec{l} = \vec{r} \times \vec{p}$, which is in the *z*-direction. The direction of $\vec{l}$ is given by the right-hand rule, as shown.

The intent of choosing the direction of the angular momentum to be perpendicular to the plane containing $\vec{r}$ and $\vec{p}$ is similar to choosing the direction of torque to be perpendicular to the plane of $\vec{r}$ and $\vec{F}$, as discussed in **Fixed-Axis Rotation**. The magnitude of the angular momentum is found from the definition of the cross-product,

$$l = rp\sin\theta,$$

where θ is the angle between $\vec{r}$ and $\vec{p}$. The units of angular momentum are $\text{kg} \cdot \text{m}^2/\text{s}$.

As with the definition of torque, we can define a lever arm $r_\perp$ that is the perpendicular distance from the momentum vector $\vec{p}$ to the origin, $r_\perp = r\sin\theta$. With this definition, the magnitude of the angular momentum becomes

$$l = r_\perp p = r_\perp mv.$$

We see that if the direction of $\vec{p}$ is such that it passes through the origin, then $\theta = 0$, and the angular momentum is zero because the lever arm is zero. In this respect, the magnitude of the angular momentum depends on the choice of origin.

If we take the time derivative of the angular momentum, we arrive at an expression for the torque on the particle:

$$\frac{d\vec{l}}{dt} = \frac{d\vec{r}}{dt} \times \vec{p} + \vec{r} \times \frac{d\vec{p}}{dt} = \vec{v} \times m\vec{v} + \vec{r} \times \frac{d\vec{p}}{dt} = \vec{r} \times \frac{d\vec{p}}{dt}.$$

Here we have used the definition of $\vec{p}$ and the fact that a vector crossed into itself is zero. From Newton's second law,

$\frac{d\vec{p}}{dt} = \sum \vec{F}$, the net force acting on the particle, and the definition of the net torque, we can write

$$\frac{d\vec{l}}{dt} = \sum \vec{\tau} . \qquad\qquad (11.6)$$

Note the similarity with the linear result of Newton's second law, $\frac{d\vec{p}}{dt} = \sum \vec{F}$. The following problem-solving strategy can serve as a guideline for calculating the angular momentum of a particle.

Problem-Solving Strategy: Angular Momentum of a Particle

1. Choose a coordinate system about which the angular momentum is to be calculated.
2. Write down the radius vector to the point particle in unit vector notation.
3. Write the linear momentum vector of the particle in unit vector notation.
4. Take the cross product $\vec{l} = \vec{r} \times \vec{p}$ and use the right-hand rule to establish the direction of the angular momentum vector.
5. See if there is a time dependence in the expression of the angular momentum vector. If there is, then a torque exists about the origin, and use $\frac{d\vec{l}}{dt} = \sum \vec{\tau}$ to calculate the torque. If there is no time dependence in the expression for the angular momentum, then the net torque is zero.

Example 11.4

Angular Momentum and Torque on a Meteor

A meteor enters Earth's atmosphere (**Figure 11.10**) and is observed by someone on the ground before it burns up in the atmosphere. The vector $\vec{r} = 25\,\text{km}\,\hat{i} + 25\,\text{km}\,\hat{j}$ gives the position of the meteor with respect to the observer. At the instant the observer sees the meteor, it has linear momentum $\vec{p} = 15.0\,\text{kg}(-2.0\text{km/s}\,\hat{j})$, and it is accelerating at a constant $2.0\,\text{m/s}^2(-\hat{j})$ along its path, which for our purposes can be taken as a straight line. (a) What is the angular momentum of the meteor about the origin, which is at the location of the observer? (b) What is the torque on the meteor about the origin?

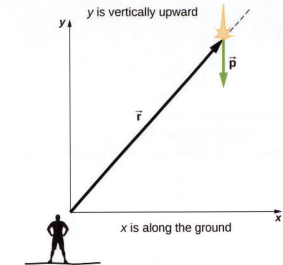

Figure 11.10 An observer on the ground sees a meteor at position $\vec{r}$ with linear momentum $\vec{p}$.

Strategy

We resolve the acceleration into x- and y-components and use the kinematic equations to express the velocity as a function of acceleration and time. We insert these expressions into the linear momentum and then calculate the angular momentum using the cross-product. Since the position and momentum vectors are in the xy-plane, we expect the angular momentum vector to be along the z-axis. To find the torque, we take the time derivative of the angular momentum.

Solution

The meteor is entering Earth's atmosphere at an angle of $90.0°$ below the horizontal, so the components of the acceleration in the x- and y-directions are

$$a_x = 0, \quad a_y = -2.0 \, \text{m/s}^2.$$

We write the velocities using the kinematic equations.

$$v_x = 0, \quad v_y = -2.0 \times 10^3 \, \text{m/s} - (2.0 \, \text{m/s}^2)t.$$

a. The angular momentum is

$$\vec{l} = \vec{r} \times \vec{p} = (25.0 \, \text{km} \, \hat{i} + 25.0 \, \text{km} \, \hat{j}) \times 15.0 \, \text{kg}(0 \, \hat{i} + v_y \, \hat{j})$$

$$= 15.0 \, \text{kg}[25.0 \, \text{km}(v_y) \hat{k}]$$

$$= 15.0 \, \text{kg}[2.50 \times 10^4 \, \text{m}(-2.0 \times 10^3 \, \text{m/s} - (2.0 \, \text{m/s}^2)t) \hat{k}].$$

At $t = 0$, the angular momentum of the meteor about the origin is

$$\vec{l}_0 = 15.0 \, \text{kg}[2.50 \times 10^4 \, \text{m}(-2.0 \times 10^3 \, \text{m/s}) \hat{k}] = 7.50 \times 10^8 \, \text{kg} \cdot \text{m}^2/\text{s}(-\hat{k}).$$

This is the instant that the observer sees the meteor.

b. To find the torque, we take the time derivative of the angular momentum. Taking the time derivative of $\vec{l}$ as a function of time, which is the second equation immediately above, we have

$$\frac{d\vec{l}}{dt} = -15.0 \, \text{kg}(2.50 \times 10^4 \, \text{m})(2.0 \, \text{m/s}^2) \hat{k}.$$

Then, since $\frac{d\vec{l}}{dt} = \sum \vec{\tau}$, we have

$$\sum \vec{\tau} = -7.5 \times 10^5 \, \text{N} \cdot \text{m} \hat{\mathbf{k}}.$$

The units of torque are given as newton-meters, not to be confused with joules. As a check, we note that the lever arm is the *x*-component of the vector $\vec{r}$ in **Figure 11.10** since it is perpendicular to the force acting on the meteor, which is along its path. By Newton's second law, this force is

$$\vec{F} = ma(-\hat{\mathbf{j}}) = 15.0 \, \text{kg}(2.0 \, \text{m/s}^2)(-\hat{\mathbf{j}}) = 30.0 \, \text{kg} \cdot \text{m/s}^2(-\hat{\mathbf{j}}).$$

The lever arm is

$$\vec{r}_\perp = 2.5 \times 10^4 \, \text{m} \, \hat{\mathbf{i}}.$$

Thus, the torque is

$$\sum \vec{\tau} = \vec{r}_\perp \times \vec{F} = (2.5 \times 10^4 \, \text{m} \, \hat{\mathbf{i}}) \times (-30.0 \, \text{kg} \cdot \text{m/s}^2 \hat{\mathbf{j}}),$$

$$= 7.5 \times 10^5 \, \text{N} \cdot \text{m}(-\hat{\mathbf{k}}).$$

Significance

Since the meteor is accelerating downward toward Earth, its radius and velocity vector are changing. Therefore, since $\vec{l} = \vec{r} \times \vec{p}$, the angular momentum is changing as a function of time. The torque on the meteor about the origin, however, is constant, because the lever arm $\vec{r}_\perp$ and the force on the meteor are constants.

This example is important in that it illustrates that the angular momentum depends on the choice of origin about which it is calculated. The methods used in this example are also important in developing angular momentum for a system of particles and for a rigid body.

 11.2 Check Your Understanding A proton spiraling around a magnetic field executes circular motion in the plane of the paper, as shown below. The circular path has a radius of 0.4 m and the proton has velocity 4.0×10^6 m/s . What is the angular momentum of the proton about the origin?

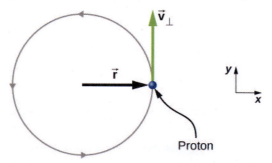

Angular Momentum of a System of Particles

The angular momentum of a system of particles is important in many scientific disciplines, one being astronomy. Consider a spiral galaxy, a rotating island of stars like our own Milky Way. The individual stars can be treated as point particles, each of which has its own angular momentum. The vector sum of the individual angular momenta give the total angular momentum of the galaxy. In this section, we develop the tools with which we can calculate the total angular momentum of a system of particles.

In the preceding section, we introduced the angular momentum of a single particle about a designated origin. The expression for this angular momentum is $\vec{l} = \vec{r} \times \vec{p}$, where the vector $\vec{r}$ is from the origin to the particle, and $\vec{p}$ is the particle's linear momentum. If we have a system of N particles, each with position vector from the origin given by $\vec{r}_i$ and each having momentum $\vec{p}_i$, then the total angular momentum of the system of particles about the origin is the vector sum of the individual angular momenta about the origin. That is,

$$\vec{L} = \vec{l}_1 + \vec{l}_2 + \cdots + \vec{l}_N. \tag{11.7}$$

Similarly, if particle i is subject to a net torque $\vec{\tau}_i$ about the origin, then we can find the net torque about the origin due to the system of particles by differentiating **Equation 11.7**:

$$\frac{d\vec{L}}{dt} = \sum_i \frac{d\vec{l}_i}{dt} = \sum_i \vec{\tau}_i.$$

The sum of the individual torques produces a net external torque on the system, which we designate $\sum \vec{\tau}$. Thus,

$$\frac{d\vec{L}}{dt} = \sum \vec{\tau}. \tag{11.8}$$

Equation 11.8 states that *the rate of change of the total angular momentum of a system is equal to the net external torque acting on the system when both quantities are measured with respect to a given origin.* **Equation 11.8** can be applied to any system that has net angular momentum, including rigid bodies, as discussed in the next section.

Example 11.5

Angular Momentum of Three Particles

Referring to **Figure 11.11**(a), determine the total angular momentum due to the three particles about the origin. (b) What is the rate of change of the angular momentum?

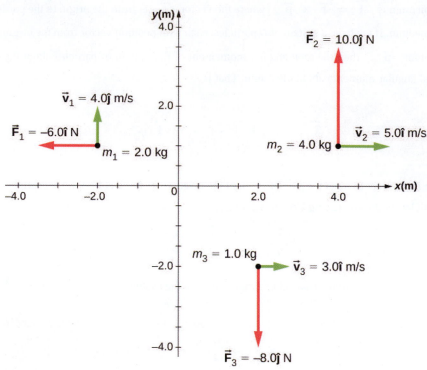

Figure 11.11 Three particles in the *xy*-plane with different position and momentum vectors.

Strategy

Write down the position and momentum vectors for the three particles. Calculate the individual angular momenta and add them as vectors to find the total angular momentum. Then do the same for the torques.

Solution

a. Particle 1: $\vec{r}_1 = -2.0\,\mathrm{m}\,\hat{i} + 1.0\,\mathrm{m}\,\hat{j}$, $\vec{p}_1 = 2.0\,\mathrm{kg}(4.0\,\mathrm{m/s}\,\hat{j}) = 8.0\,\mathrm{kg}\cdot\mathrm{m/s}\,\hat{j}$,

$$\vec{l}_1 = \vec{r}_1 \times \vec{p}_1 = -16.0\,\mathrm{kg}\cdot\mathrm{m}^2/\mathrm{s}\,\hat{k}.$$

Particle 2: $\vec{r}_2 = 4.0\,\mathrm{m}\,\hat{i} + 1.0\,\mathrm{m}\,\hat{j}$, $\vec{p}_2 = 4.0\,\mathrm{kg}(5.0\,\mathrm{m/s}\,\hat{i}) = 20.0\,\mathrm{kg}\cdot\mathrm{m/s}\,\hat{i}$,

$$\vec{l}_2 = \vec{r}_2 \times \vec{p}_2 = -20.0\,\mathrm{kg}\cdot\mathrm{m}^2/\mathrm{s}\,\hat{k}.$$

Particle 3: $\vec{r}_3 = 2.0\,\mathrm{m}\,\hat{i} - 2.0\,\mathrm{m}\,\hat{j}$, $\vec{p}_3 = 1.0\,\mathrm{kg}(3.0\,\mathrm{m/s}\,\hat{i}) = 3.0\,\mathrm{kg}\cdot\mathrm{m/s}\,\hat{i}$,

$$\vec{l}_3 = \vec{r}_3 \times \vec{p}_3 = 6.0\,\mathrm{kg}\cdot\mathrm{m}^2/\mathrm{s}\,\hat{k}.$$

We add the individual angular momenta to find the total about the origin:

$$\vec{l}_T = \vec{l}_1 + \vec{l}_2 + \vec{l}_3 = -30\,\mathrm{kg}\cdot\mathrm{m}^2/\mathrm{s}\,\hat{k}.$$

b. The individual forces and lever arms are

$$\vec{r}_{1\perp} = 1.0\,\text{m}\,\hat{j}, \quad \vec{F}_1 = -6.0\,\text{N}\,\hat{i}, \quad \vec{\tau}_1 = 6.0\,\text{N}\cdot\text{m}\,\hat{k}$$

$$\vec{r}_{2\perp} = 4.0\,\text{m}\,\hat{i}, \quad \vec{F}_2 = 10.0\,\text{N}\,\hat{j}, \quad \vec{\tau}_2 = 40.0\,\text{N}\cdot\text{m}\,\hat{k}$$

$$\vec{r}_{3\perp} = 2.0\,\text{m}\,\hat{i}, \quad \vec{F}_3 = -8.0\,\text{N}\,\hat{j}, \quad \vec{\tau}_3 = -16.0\,\text{N}\cdot\text{m}\,\hat{k}.$$

Therefore:

$$\sum_i \vec{\tau}_i = \vec{\tau}_1 + \vec{\tau}_2 + \vec{\tau}_3 = 30\,\text{N}\cdot\text{m}\,\hat{k}.$$

Significance

This example illustrates the superposition principle for angular momentum and torque of a system of particles. Care must be taken when evaluating the radius vectors $\vec{r}_i$ of the particles to calculate the angular momenta, and the lever arms, $\vec{r}_{i\perp}$ to calculate the torques, as they are completely different quantities.

Angular Momentum of a Rigid Body

We have investigated the angular momentum of a single particle, which we generalized to a system of particles. Now we can use the principles discussed in the previous section to develop the concept of the angular momentum of a rigid body. Celestial objects such as planets have angular momentum due to their spin and orbits around stars. In engineering, anything that rotates about an axis carries angular momentum, such as flywheels, propellers, and rotating parts in engines. Knowledge of the angular momenta of these objects is crucial to the design of the system in which they are a part.

To develop the angular momentum of a rigid body, we model a rigid body as being made up of small mass segments, Δm_i.

In **Figure 11.12**, a rigid body is constrained to rotate about the z-axis with angular velocity ω. All mass segments that make up the rigid body undergo circular motion about the z-axis with the same angular velocity. Part (a) of the figure shows mass segment Δm_i with position vector $\vec{r}_i$ from the origin and radius R_i to the z-axis. The magnitude of its tangential velocity is $v_i = R_i\omega$. Because the vectors $\vec{v}_i$ and $\vec{r}_i$ are perpendicular to each other, the magnitude of the angular momentum of this mass segment is

$$l_i = r_i(\Delta m v_i)\sin 90°.$$

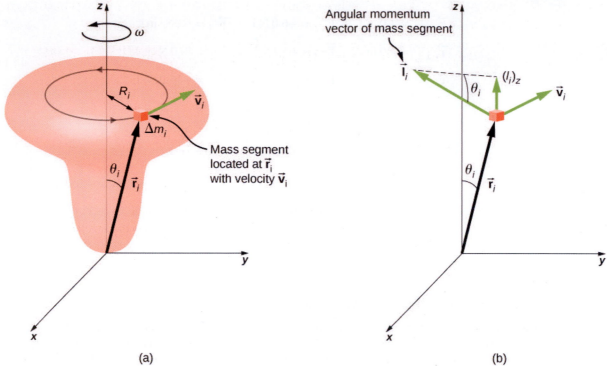

Figure 11.12 (a) A rigid body is constrained to rotate around the z-axis. The rigid body is symmetrical about the z-axis. A mass segment Δm_i is located at position $\vec{r}_i$, which makes angle θ_i with respect to the z-axis. The circular motion of an infinitesimal mass segment is shown. (b) $\vec{l}_i$ is the angular momentum of the mass segment and has a component along the z-axis ($\vec{l}_i)_z$.

Using the right-hand rule, the angular momentum vector points in the direction shown in part (b). The sum of the angular momenta of all the mass segments contains components both along and perpendicular to the axis of rotation. Every mass segment has a perpendicular component of the angular momentum that will be cancelled by the perpendicular component of an identical mass segment on the opposite side of the rigid body. Thus, the component along the axis of rotation is the only component that gives a nonzero value when summed over all the mass segments. From part (b), the component of $\vec{l}_i$ along the axis of rotation is

$$(l_i)_z = l_i \sin\theta_i = (r_i \Delta m_i v_i)\sin\theta_i,$$
$$= (r_i \sin\theta_i)(\Delta m_i v_i) = R_i \Delta m_i v_i.$$

The net angular momentum of the rigid body along the axis of rotation is

$$L = \sum_i (\vec{l}_i)_z = \sum_i R_i \Delta m_i v_i = \sum_i R_i \Delta m_i (R_i \omega) = \omega \sum_i \Delta m_i (R_i)^2.$$

The summation $\sum_i \Delta m_i (R_i)^2$ is simply the moment of inertia I of the rigid body about the axis of rotation. For a thin hoop rotating about an axis perpendicular to the plane of the hoop, all of the R_i's are equal to R so the summation reduces to $R^2 \sum_i \Delta m_i = mR^2$, which is the moment of inertia for a thin hoop found in **Figure 10.20**. Thus, the magnitude of the angular momentum along the axis of rotation of a rigid body rotating with angular velocity ω about the axis is

$$L = I\omega. \tag{11.9}$$

This equation is analogous to the magnitude of the linear momentum $p = mv$. The direction of the angular momentum vector is directed along the axis of rotation given by the right-hand rule.

Example 11.6

Angular Momentum of a Robot Arm

A robot arm on a Mars rover like *Curiosity* shown in **Figure 11.8** is 1.0 m long and has forceps at the free end to pick up rocks. The mass of the arm is 2.0 kg and the mass of the forceps is 1.0 kg. See **Figure 11.13**. The robot arm and forceps move from rest to $\omega = 0.1\pi$ rad/s in 0.1 s. It rotates down and picks up a Mars rock that has mass 1.5 kg. The axis of rotation is the point where the robot arm connects to the rover. (a) What is the angular momentum of the robot arm by itself about the axis of rotation after 0.1 s when the arm has stopped accelerating? (b) What is the angular momentum of the robot arm when it has the Mars rock in its forceps and is rotating upwards? (c) When the arm does not have a rock in the forceps, what is the torque about the point where the arm connects to the rover when it is accelerating from rest to its final angular velocity?

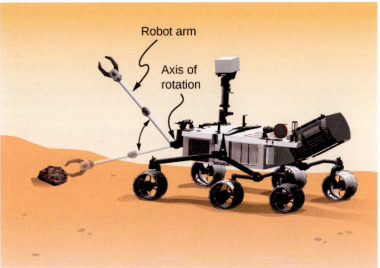

Figure 11.13 A robot arm on a Mars rover swings down and picks up a Mars rock. (credit: modification of work by NASA/JPL-Caltech)

Strategy

We use **Equation 11.9** to find angular momentum in the various configurations. When the arm is rotating downward, the right-hand rule gives the angular momentum vector directed out of the page, which we will call the positive z-direction. When the arm is rotating upward, the right-hand rule gives the direction of the angular momentum vector into the page or in the negative z-direction. The moment of inertia is the sum of the individual moments of inertia. The arm can be approximated with a solid rod, and the forceps and Mars rock can be approximated as point masses located at a distance of 1 m from the origin. For part (c), we use Newton's second law of motion for rotation to find the torque on the robot arm.

Solution

a. Writing down the individual moments of inertia, we have

Robot arm: $I_R = \frac{1}{3}m_R r^2 = \frac{1}{3}(2.00 \text{ kg})(1.00 \text{ m})^2 = \frac{2}{3} \text{kg} \cdot \text{m}^2$.

Forceps: $I_F = m_F r^2 = (1.0 \text{ kg})(1.0 \text{ m})^2 = 1.0 \text{ kg} \cdot \text{m}^2$.

Mars rock: $I_{MR} = m_{MR} r^2 = (1.5 \text{ kg})(1.0 \text{ m})^2 = 1.5 \text{ kg} \cdot \text{m}^2$.

Therefore, without the Mars rock, the total moment of inertia is

$$I_{Total} = I_R + I_F = 1.67 \text{ kg} \cdot \text{m}^2$$

and the magnitude of the angular momentum is

$$L = I\omega = 1.67 \, \text{kg} \cdot \text{m}^2 (0.1\pi \, \text{rad/s}) = 0.17\pi \, \text{kg} \cdot \text{m}^2/\text{s}.$$

The angular momentum vector is directed out of the page in the $\hat{\mathbf{k}}$ direction since the robot arm is rotating counterclockwise.

b. We must include the Mars rock in the calculation of the moment of inertia, so we have

$$I_{\text{Total}} = I_{\text{R}} + I_{\text{F}} + I_{\text{MR}} = 3.17 \, \text{kg} \cdot \text{m}^2$$

and

$$L = I\omega = 3.17 \, \text{kg} \cdot \text{m}^2 (0.1\pi \, \text{rad/s}) = 0.32\pi \, \text{kg} \cdot \text{m}^2/\text{s}.$$

Now the angular momentum vector is directed into the page in the $-\hat{\mathbf{k}}$ direction, by the right-hand rule, since the robot arm is now rotating clockwise.

c. We find the torque when the arm does not have the rock by taking the derivative of the angular momentum using **Equation 11.8** $\frac{d\vec{\mathbf{L}}}{dt} = \sum \vec{\boldsymbol{\tau}}$. But since $L = I\omega$, and understanding that the direction of the angular momentum and torque vectors are along the axis of rotation, we can suppress the vector notation and find

$$\frac{dL}{dt} = \frac{d(I\omega)}{dt} = I\frac{d\omega}{dt} = I\alpha = \sum \tau,$$

which is Newton's second law for rotation. Since $\alpha = \frac{0.1\pi \, \text{rad/s}}{0.1 \, \text{s}} = \pi \, \text{rad/s}^2$, we can calculate the net torque:

$$\sum \tau = I\alpha = 1.67 \, \text{kg} \cdot \text{m}^2 (\pi \, \text{rad/s}^2) = 1.67\pi \, \text{N} \cdot \text{m}.$$

Significance

The angular momentum in (a) is less than that of (b) due to the fact that the moment of inertia in (b) is greater than (a), while the angular velocity is the same.

 11.3 Check Your Understanding Which has greater angular momentum: a solid sphere of mass m rotating at a constant angular frequency ω_0 about the z-axis, or a solid cylinder of same mass and rotation rate about the z-axis?

 Visit the **University of Colorado's Interactive Simulation of Angular Momentum (https://openstaxcollege.org/l/21angmomintsim)** to learn more about angular momentum.

11.3 | Conservation of Angular Momentum

Learning Objectives

By the end of this section, you will be able to:

- Apply conservation of angular momentum to determine the angular velocity of a rotating system in which the moment of inertia is changing
- Explain how the rotational kinetic energy changes when a system undergoes changes in both moment of inertia and angular velocity

So far, we have looked at the angular momentum of systems consisting of point particles and rigid bodies. We have also analyzed the torques involved, using the expression that relates the external net torque to the change in angular momentum, **Equation 11.8**. Examples of systems that obey this equation include a freely spinning bicycle tire that slows over time due to torque arising from friction, or the slowing of Earth's rotation over millions of years due to frictional forces exerted on tidal deformations.

However, suppose there is no net external torque on the system, $\sum \vec{\tau} = 0$. In this case, **Equation 11.8** becomes the **law of conservation of angular momentum**.

> ## Law of Conservation of Angular Momentum
>
> The angular momentum of a system of particles around a point in a fixed inertial reference frame is conserved if there is no net external torque around that point:
>
> $$\frac{d\vec{L}}{dt} = 0 \tag{11.10}$$
>
> or
>
> $$\vec{L} = \vec{l}_1 + \vec{l}_2 + \cdots + \vec{l}_N = \text{constant.} \tag{11.11}$$

Note that the *total* angular momentum $\vec{L}$ is conserved. Any of the individual angular momenta can change as long as their sum remains constant. This law is analogous to linear momentum being conserved when the external force on a system is zero.

As an example of conservation of angular momentum, **Figure 11.14** shows an ice skater executing a spin. The net torque on her is very close to zero because there is relatively little friction between her skates and the ice. Also, the friction is exerted very close to the pivot point. Both $\left|\vec{F}\right|$ and $\left|\vec{r}\right|$ are small, so $\left|\vec{\tau}\right|$ is negligible. Consequently, she can spin for quite some time. She can also increase her rate of spin by pulling her arms and legs in. Why does pulling her arms and legs in increase her rate of spin? The answer is that her angular momentum is constant, so that

$$L' = L$$

or

$$I'\omega' = I\omega,$$

where the primed quantities refer to conditions after she has pulled in her arms and reduced her moment of inertia. Because I' is smaller, the angular velocity ω' must increase to keep the angular momentum constant.

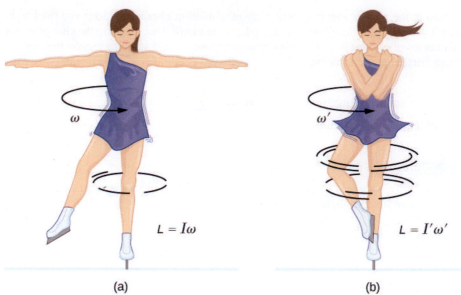

Figure 11.14 (a) An ice skater is spinning on the tip of her skate with her arms extended. Her angular momentum is conserved because the net torque on her is negligibly small. (b) Her rate of spin increases greatly when she pulls in her arms, decreasing her moment of inertia. The work she does to pull in her arms results in an increase in rotational kinetic energy.

It is interesting to see how the rotational kinetic energy of the skater changes when she pulls her arms in. Her initial rotational energy is

$$K_{\text{Rot}} = \frac{1}{2}I\omega^2,$$

whereas her final rotational energy is

$$K'_{\text{Rot}} = \frac{1}{2}I'(\omega')^2.$$

Since $I'\omega' = I\omega$, we can substitute for ω' and find

$$K'_{\text{Rot}} = \frac{1}{2}I'(\omega')^2 = \frac{1}{2}I'\left(\frac{I}{I'}\omega\right)^2 = \frac{1}{2}I\omega^2\left(\frac{I}{I'}\right) = K_{\text{Rot}}\left(\frac{I}{I'}\right).$$

Because her moment of inertia has decreased, $I' < I$, her final rotational kinetic energy has increased. The source of this additional rotational kinetic energy is the work required to pull her arms inward. Note that the skater's arms do not move in a perfect circle—they spiral inward. This work causes an increase in the rotational kinetic energy, while her angular momentum remains constant. Since she is in a frictionless environment, no energy escapes the system. Thus, if she were to extend her arms to their original positions, she would rotate at her original angular velocity and her kinetic energy would return to its original value.

The solar system is another example of how conservation of angular momentum works in our universe. Our solar system was born from a huge cloud of gas and dust that initially had rotational energy. Gravitational forces caused the cloud to contract, and the rotation rate increased as a result of conservation of angular momentum (**Figure 11.15**).

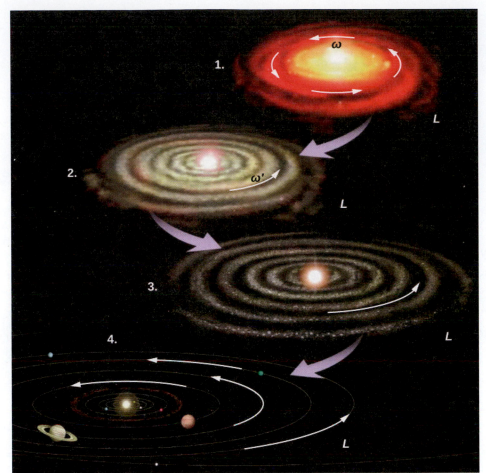

Figure 11.15 The solar system coalesced from a cloud of gas and dust that was originally rotating. The orbital motions and spins of the planets are in the same direction as the original spin and conserve the angular momentum of the parent cloud. (credit: modification of work by NASA)

We continue our discussion with an example that has applications to engineering.

Example 11.7

Coupled Flywheels

A flywheel rotates without friction at an angular velocity $\omega_0 = 600\ \text{rev/min}$ on a frictionless, vertical shaft of negligible rotational inertia. A second flywheel, which is at rest and has a moment of inertia three times that of the rotating flywheel, is dropped onto it (**Figure 11.16**). Because friction exists between the surfaces, the flywheels very quickly reach the same rotational velocity, after which they spin together. (a) Use the law of conservation of angular momentum to determine the angular velocity ω of the combination. (b) What fraction of the initial kinetic energy is lost in the coupling of the flywheels?

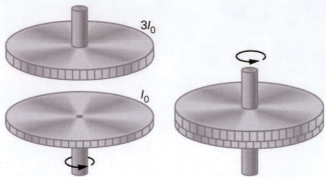

Figure 11.16 Two flywheels are coupled and rotate together.

Strategy

Part (a) is straightforward to solve for the angular velocity of the coupled system. We use the result of (a) to compare the initial and final kinetic energies of the system in part (b).

Solution

a. No external torques act on the system. The force due to friction produces an internal torque, which does not affect the angular momentum of the system. Therefore conservation of angular momentum gives

$$I_0 \omega_0 = (I_0 + 3I_0)\omega,$$
$$\omega = \frac{1}{4}\omega_0 = 150 \text{ rev/min} = 15.7 \text{ rad/s}.$$

b. Before contact, only one flywheel is rotating. The rotational kinetic energy of this flywheel is the initial rotational kinetic energy of the system, $\frac{1}{2}I_0 \omega_0^2$. The final kinetic energy is $\frac{1}{2}(4I_0)\omega^2 = \frac{1}{2}(4I_0)\left(\frac{\omega_0}{4}\right)^2 = \frac{1}{8}I_0 \omega_0^2$.

Therefore, the ratio of the final kinetic energy to the initial kinetic energy is

$$\frac{\frac{1}{8}I_0 \omega_0^2}{\frac{1}{2}I_0 \omega_0^2} = \frac{1}{4}.$$

Thus, 3/4 of the initial kinetic energy is lost to the coupling of the two flywheels.

Significance

Since the rotational inertia of the system increased, the angular velocity decreased, as expected from the law of conservation of angular momentum. In this example, we see that the final kinetic energy of the system has decreased, as energy is lost to the coupling of the flywheels. Compare this to the example of the skater in **Figure 11.14** doing work to bring her arms inward and adding rotational kinetic energy.

 11.4 Check Your Understanding A merry-go-round at a playground is rotating at 4.0 rev/min. Three children jump on and increase the moment of inertia of the merry-go-round/children rotating system by 25%. What is the new rotation rate?

Example 11.8

Dismount from a High Bar

An 80.0-kg gymnast dismounts from a high bar. He starts the dismount at full extension, then tucks to complete a number of revolutions before landing. His moment of inertia when fully extended can be approximated as a rod of length 1.8 m and when in the tuck a rod of half that length. If his rotation rate at full extension is 1.0 rev/s and he enters the tuck when his center of mass is at 3.0 m height moving horizontally to the floor, how many revolutions can he execute if he comes out of the tuck at 1.8 m height? See **Figure 11.17**.

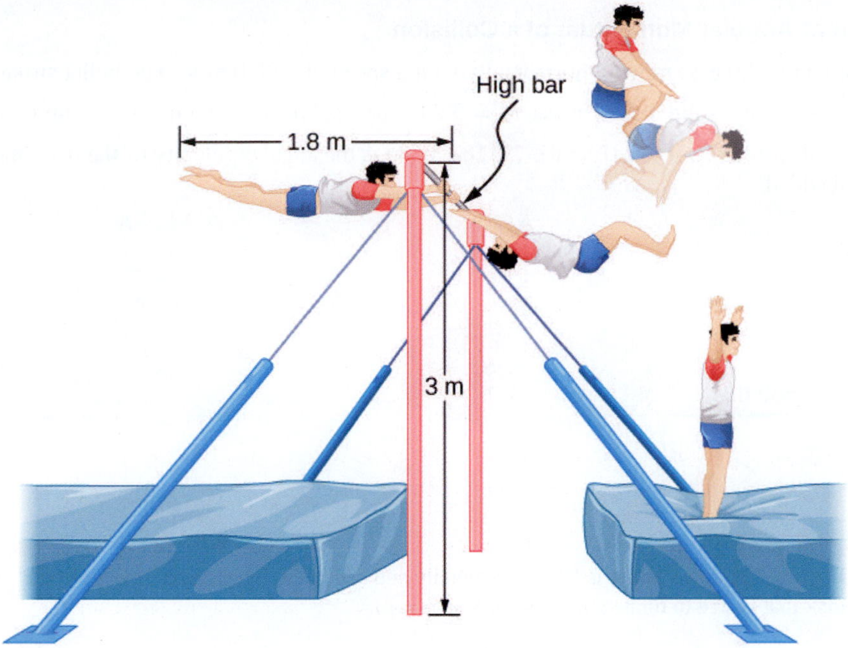

Figure 11.17 A gymnast dismounts from a high bar and executes a number of revolutions in the tucked position before landing upright.

Strategy

Using conservation of angular momentum, we can find his rotation rate when in the tuck. Using the equations of kinematics, we can find the time interval from a height of 3.0 m to 1.8 m. Since he is moving horizontally with respect to the ground, the equations of free fall simplify. This will allow the number of revolutions that can be executed to be calculated. Since we are using a ratio, we can keep the units as rev/s and don't need to convert to radians/s.

Solution

The moment of inertia at full extension is $I_0 = \frac{1}{12}mL^2 = \frac{1}{12}80.0 \, \text{kg}(1.8 \, \text{m})^2 = 21.6 \, \text{kg} \cdot \text{m}^2$.

The moment of inertia in the tuck is $I_f = \frac{1}{12}mL_f^2 = \frac{1}{12}80.0 \, \text{kg}(0.9 \, \text{m})^2 = 5.4 \, \text{kg} \cdot \text{m}^2$.

Conservation of angular momentum: $I_f \omega_f = I_0 \omega_0 \Rightarrow \omega_f = \frac{I_0 \omega_0}{I_f} = \frac{21.6 \, \text{kg} \cdot \text{m}^2(1.0 \, \text{rev/s})}{5.4 \, \text{kg} \cdot \text{m}^2} = 4.0 \, \text{rev/s}$.

Time interval in the tuck: $t = \sqrt{\frac{2h}{g}} = \sqrt{\frac{2(3.0 - 1.8)\text{m}}{9.8 \, \text{m/s}}} = 0.5 \, \text{s}$.

In 0.5 s, he will be able to execute two revolutions at 4.0 rev/s.

Significance

Note that the number of revolutions he can complete will depend on how long he is in the air. In the problem, he is exiting the high bar horizontally to the ground. He could also exit at an angle with respect to the ground, giving him more or less time in the air depending on the angle, positive or negative, with respect to the ground. Gymnasts must take this into account when they are executing their dismounts.

Example 11.9

Conservation of Angular Momentum of a Collision

A bullet of mass $m = 2.0\,\text{g}$ is moving horizontally with a speed of $500.0\,\text{m/s}$. The bullet strikes and becomes embedded in the edge of a solid disk of mass $M = 3.2\,\text{kg}$ and radius $R = 0.5\,\text{m}$. The cylinder is free to rotate around its axis and is initially at rest (**Figure 11.18**). What is the angular velocity of the disk immediately after the bullet is embedded?

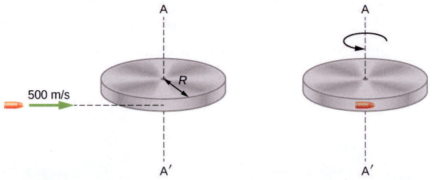

Figure 11.18 A bullet is fired horizontally and becomes embedded in the edge of a disk that is free to rotate about its vertical axis.

Strategy

For the system of the bullet and the cylinder, no external torque acts along the vertical axis through the center of the disk. Thus, the angular momentum along this axis is conserved. The initial angular momentum of the bullet is mvR, which is taken about the rotational axis of the disk the moment before the collision. The initial angular momentum of the cylinder is zero. Thus, the net angular momentum of the system is mvR. Since angular momentum is conserved, the initial angular momentum of the system is equal to the angular momentum of the bullet embedded in the disk immediately after impact.

Solution

The initial angular momentum of the system is

$$L_i = mvR.$$

The moment of inertia of the system with the bullet embedded in the disk is

$$I = mR^2 + \tfrac{1}{2}MR^2 = \left(m + \tfrac{M}{2}\right)R^2.$$

The final angular momentum of the system is

$$L_f = I\omega_f.$$

Thus, by conservation of angular momentum, $L_i = L_f$ and

$$mvR = \left(m + \tfrac{M}{2}\right)R^2\omega_f.$$

Solving for ω_f,

$$\omega_f = \frac{mvR}{(m + M/2)R^2} = \frac{(2.0 \times 10^{-3}\,\text{kg})(500.0\,\text{m/s})}{(2.0 \times 10^{-3}\,\text{kg} + 1.6\,\text{kg})(0.50\,\text{m})} = 1.2\,\text{rad/s}.$$

Significance

The system is composed of both a point particle and a rigid body. Care must be taken when formulating the angular momentum before and after the collision. Just before impact the angular momentum of the bullet is taken about the rotational axis of the disk.

11.4 | Precession of a Gyroscope

Figure 11.19 shows a gyroscope, defined as a spinning disk in which the axis of rotation is free to assume any orientation. When spinning, the orientation of the spin axis is unaffected by the orientation of the body that encloses it. The body or vehicle enclosing the gyroscope can be moved from place to place and the orientation of the spin axis will remain the same. This makes gyroscopes very useful in navigation, especially where magnetic compasses can't be used, such as in manned and unmanned spacecraft, intercontinental ballistic missiles, unmanned aerial vehicles, and satellites like the Hubble Space Telescope.

Figure 11.19 A gyroscope consists of a spinning disk about an axis that is free to assume any orientation.

We illustrate the **precession** of a gyroscope with an example of a top in the next two figures. If the top is placed on a flat surface near the surface of Earth at an angle to the vertical and is not spinning, it will fall over, due to the force of gravity producing a torque acting on its center of mass. This is shown in **Figure 11.20**(a). However, if the top is spinning on its axis, rather than topple over due to this torque, it precesses about the vertical, shown in part (b) of the figure. This is due to the torque on the center of mass, which provides the change in angular momentum.

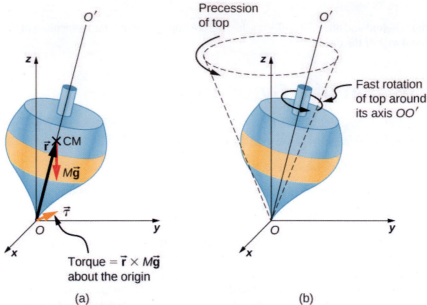

Figure 11.20 (a) If the top is not spinning, there is a torque $\vec{r} \times M\vec{g}$ about the origin, and the top falls over. (b) If the top is spinning about its axis OO', it doesn't fall over but precesses about the z-axis.

Figure 11.21 shows the forces acting on a spinning top. The torque produced is perpendicular to the angular momentum vector $\vec{L}$ according to $d\vec{L} = \vec{\tau}\, dt$, but not its magnitude. The top *precesses* around a vertical axis, since the torque is always horizontal and perpendicular to $\vec{L}$. If the top is *not* spinning, it acquires angular momentum in the direction of the torque, and it rotates around a horizontal axis, falling over just as we would expect.

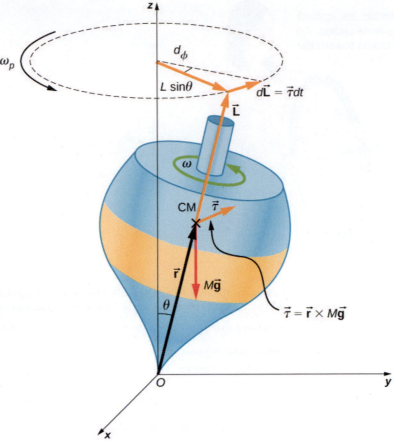

Figure 11.21 The force of gravity acting on the center of mass produces a torque $\vec{\tau}$ in the direction perpendicular to $\vec{L}$. The magnitude of $\vec{L}$ doesn't change but its direction does, and the top precesses about the z-axis.

We can experience this phenomenon first hand by holding a spinning bicycle wheel and trying to rotate it about an axis perpendicular to the spin axis. As shown in **Figure 11.22**, the person applies forces perpendicular to the spin axis in an attempt to rotate the wheel, but instead, the wheel axis starts to change direction to her left due to the applied torque.

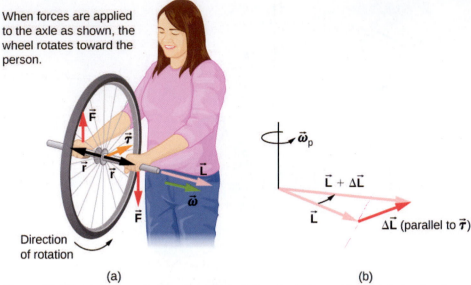

When forces are applied to the axle as shown, the wheel rotates toward the person.

Direction of rotation

(a)   (b)

Figure 11.22 (a) A person holding the spinning bike wheel lifts it with her right hand and pushes down with her left hand in an attempt to rotate the wheel. This action creates a torque directly toward her. This torque causes a change in angular momentum $\Delta \vec{L}$ in exactly the same direction. (b) A vector diagram depicting how $\Delta \vec{L}$ and $\vec{L}$ add, producing a new angular momentum pointing more toward the person. The wheel moves toward the person, perpendicular to the forces she exerts on it.

We all know how easy it is for a bicycle to tip over when sitting on it at rest. But when riding the bicycle at a good pace, tipping it over involves changing the angular momentum vector of the spinning wheels.

 View the video on **gyroscope precession (https://openstaxcollege.org/l/21gyrovideo)** for a complete demonstration of precession of the bicycle wheel.

Also, when a spinning disk is put in a box such as a Blu-Ray player, try to move it. It is easy to translate the box in a given direction but difficult to rotate it about an axis perpendicular to the axis of the spinning disk, since we are putting a torque on the box that will cause the angular momentum vector of the spinning disk to precess.

We can calculate the precession rate of the top in **Figure 11.21**. From **Figure 11.21**, we see that the magnitude of the torque is

$$\tau = rMg \sin \theta.$$

Thus,

$$dL = rMg \sin \theta \, dt.$$

The angle the top precesses through in time dt is

$$d\phi = \frac{dL}{L \sin \theta} = \frac{rMg \sin \theta}{L \sin \theta} dt = \frac{rMg}{L} dt.$$

The precession angular velocity is $\omega_P = \frac{d\phi}{dt}$ and from this equation we see that

$$\omega_P = \frac{rMg}{L}. \text{ or, since } L = I\omega,$$

$$\omega_P = \frac{rMg}{I\omega}. \tag{11.12}$$

In this derivation, we assumed that $\omega_P \ll \omega$, that is, that the precession angular velocity is much less than the angular velocity of the gyroscope disk. The precession angular velocity adds a small component to the angular momentum along the z-axis. This is seen in a slight bob up and down as the gyroscope processes, referred to as nutation.

Earth itself acts like a gigantic gyroscope. Its angular momentum is along its axis and currently points at Polaris, the North Star. But Earth is slowly precessing (once in about 26,000 years) due to the torque of the Sun and the Moon on its nonspherical shape.

Example 11.10

Period of Precession

A gyroscope spins with its tip on the ground and is spinning with negligible frictional resistance. The disk of the gyroscope has mass 0.3 kg and is spinning at 20 rev/s. Its center of mass is 5.0 cm from the pivot and the radius of the disk is 5.0 cm. What is the precessional period of the gyroscope?

Strategy

We use **Equation 11.12** to find the precessional angular velocity of the gyroscope. This allows us to find the period of precession.

Solution

The moment of inertia of the disk is

$$I = \tfrac{1}{2}mr^2 = \tfrac{1}{2}(0.30 \text{ kg})(0.05 \text{ m})^2 = 3.75 \times 10^{-4} \text{ kg} \cdot \text{m}^2.$$

The angular velocity of the disk is

$$20.0 \text{ rev/s} = 20.0(2\pi) \text{ rad/s} = 125.66 \text{ rad/s}.$$

We can now substitute in **Equation 11.12**. The precessional angular velocity is

$$\omega_P = \frac{rMg}{I\omega} = \frac{(0.05 \text{ m})(0.3 \text{ kg})(9.8 \text{ m/s}^2)}{(3.75 \times 10^{-4} \text{ kg} \cdot \text{m}^2)(125.66 \text{ rad/s})} = 3.12 \text{ rad/s}.$$

The precessional period of the gyroscope is

$$T_P = \frac{2\pi}{3.12 \text{ rad/s}} = 2.0 \text{ s}.$$

Significance

The precessional angular frequency of the gyroscope, 3.12 rad/s, or about 0.5 rev/s, is much less than the angular velocity 20 rev/s of the gyroscope disk. Therefore, we don't expect a large component of the angular momentum to arise due to precession, and **Equation 11.12** is a good approximation of the precessional angular velocity.

 11.5 Check Your Understanding A top has a precession frequency of 5.0 rad/s on Earth. What is its precession frequency on the Moon?

CHAPTER 11 REVIEW

KEY TERMS

angular momentum rotational analog of linear momentum, found by taking the product of moment of inertia and angular velocity

law of conservation of angular momentum angular momentum is conserved, that is, the initial angular momentum is equal to the final angular momentum when no external torque is applied to the system

precession circular motion of the pole of the axis of a spinning object around another axis due to a torque

rolling motion combination of rotational and translational motion with or without slipping

KEY EQUATIONS

Velocity of center of mass of rolling object	$v_{CM} = R\omega$
Acceleration of center of mass of rolling object	$a_{CM} = R\alpha$
Displacement of center of mass of rolling object	$d_{CM} = R\theta$
Acceleration of an object rolling without slipping	$a_{CM} = \dfrac{mg\sin\theta}{m + (I_{CM}/r^2)}$
Angular momentum	$\vec{l} = \vec{r} \times \vec{p}$
Derivative of angular momentum equals torque	$\dfrac{d\vec{l}}{dt} = \sum \vec{\tau}$
Angular momentum of a system of particles	$\vec{L} = \vec{l}_1 + \vec{l}_2 + \cdots + \vec{l}_N$
For a system of particles, derivative of angular momentum equals torque	$\dfrac{d\vec{L}}{dt} = \sum \vec{\tau}$
Angular momentum of a rotating rigid body	$L = I\omega$
Conservation of angular momentum	$\dfrac{d\vec{L}}{dt} = 0$
Conservation of angular momentum	$\vec{L} = \vec{l}_1 + \vec{l}_2 + \cdots + \vec{l}_N = \text{constant}$
Precessional angular velocity	$\omega_P = \dfrac{rMg}{I\omega}$

SUMMARY

11.1 Rolling Motion

- In rolling motion without slipping, a static friction force is present between the rolling object and the surface. The relations $v_{CM} = R\omega$, $a_{CM} = R\alpha$, and $d_{CM} = R\theta$ all apply, such that the linear velocity, acceleration, and distance of the center of mass are the angular variables multiplied by the radius of the object.

- In rolling motion with slipping, a kinetic friction force arises between the rolling object and the surface. In this case, $v_{CM} \neq R\omega$, $a_{CM} \neq R\alpha$, and $d_{CM} \neq R\theta$.

- Energy conservation can be used to analyze rolling motion. Energy is conserved in rolling motion without slipping. Energy is not conserved in rolling motion with slipping due to the heat generated by kinetic friction.

11.2 Angular Momentum

- The angular momentum $\vec{l} = \vec{r} \times \vec{p}$ of a single particle about a designated origin is the vector product of the position vector in the given coordinate system and the particle's linear momentum.

- The angular momentum $\vec{l} = \sum_i \vec{l}_i$ of a system of particles about a designated origin is the vector sum of the individual momenta of the particles that make up the system.

- The net torque on a system about a given origin is the time derivative of the angular momentum about that origin: $\frac{d\vec{L}}{dt} = \sum \vec{\tau}$.

- A rigid rotating body has angular momentum $L = I\omega$ directed along the axis of rotation. The time derivative of the angular momentum $\frac{dL}{dt} = \sum \tau$ gives the net torque on a rigid body and is directed along the axis of rotation.

11.3 Conservation of Angular Momentum

- In the absence of external torques, a system's total angular momentum is conserved. This is the rotational counterpart to linear momentum being conserved when the external force on a system is zero.

- For a rigid body that changes its angular momentum in the absence of a net external torque, conservation of angular momentum gives $I_f \omega_f = I_i \omega_i$. This equation says that the angular velocity is inversely proportional to the moment of inertia. Thus, if the moment of inertia decreases, the angular velocity must increase to conserve angular momentum.

- Systems containing both point particles and rigid bodies can be analyzed using conservation of angular momentum. The angular momentum of all bodies in the system must be taken about a common axis.

11.4 Precession of a Gyroscope

- When a gyroscope is set on a pivot near the surface of Earth, it precesses around a vertical axis, since the torque is always horizontal and perpendicular to $\vec{L}$. If the gyroscope is not spinning, it acquires angular momentum in the direction of the torque, and it rotates about a horizontal axis, falling over just as we would expect.

- The precessional angular velocity is given by $\omega_P = \frac{rMg}{I\omega}$, where r is the distance from the pivot to the center of mass of the gyroscope, I is the moment of inertia of the gyroscope's spinning disk, M is its mass, and ω is the angular frequency of the gyroscope disk.

CONCEPTUAL QUESTIONS

11.1 Rolling Motion

1. Can a round object released from rest at the top of a frictionless incline undergo rolling motion?

2. A cylindrical can of radius R is rolling across a horizontal surface without slipping. (a) After one complete revolution of the can, what is the distance that its center of mass has moved? (b) Would this distance be greater or smaller if slipping occurred?

3. A wheel is released from the top on an incline. Is the wheel most likely to slip if the incline is steep or gently sloped?

4. Which rolls down an inclined plane faster, a hollow cylinder or a solid sphere? Both have the same mass and radius.

5. A hollow sphere and a hollow cylinder of the same radius and mass roll up an incline without slipping and have the same initial center of mass velocity. Which object reaches a greater height before stopping?

11.2 Angular Momentum

6. Can you assign an angular momentum to a particle without first defining a reference point?

7. For a particle traveling in a straight line, are there any points about which the angular momentum is zero? Assume the line intersects the origin.

8. Under what conditions does a rigid body have angular momentum but not linear momentum?

9. If a particle is moving with respect to a chosen origin it has linear momentum. What conditions must exist for this particle's angular momentum to be zero about the chosen origin?

10. If you know the velocity of a particle, can you say anything about the particle's angular momentum?

11.3 Conservation of Angular Momentum

11. What is the purpose of the small propeller at the back of a helicopter that rotates in the plane perpendicular to the large propeller?

12. Suppose a child walks from the outer edge of a rotating merry-go-round to the inside. Does the angular velocity of the merry-go-round increase, decrease, or remain the same? Explain your answer. Assume the merry-go-round is spinning without friction.

13. As the rope of a tethered ball winds around a pole, what happens to the angular velocity of the ball?

14. Suppose the polar ice sheets broke free and floated toward Earth's equator without melting. What would happen to Earth's angular velocity?

15. Explain why stars spin faster when they collapse.

16. Competitive divers pull their limbs in and curl up their bodies when they do flips. Just before entering the water, they fully extend their limbs to enter straight down (see below). Explain the effect of both actions on their angular velocities. Also explain the effect on their angular momentum.

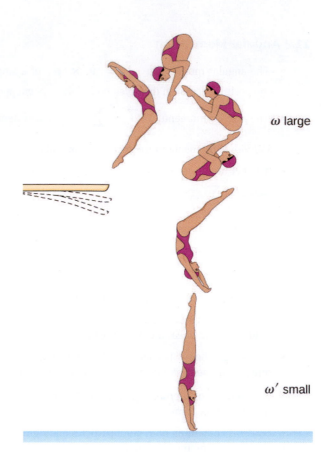

ω large

ω' small

11.4 Precession of a Gyroscope

17. Gyroscopes used in guidance systems to indicate directions in space must have an angular momentum that does not change in direction. When placed in the vehicle, they are put in a compartment that is separated from the main fuselage, such that changes in the orientation of the fuselage does not affect the orientation of the gyroscope. If the space vehicle is subjected to large forces and accelerations how can the direction of the gyroscopes angular momentum be constant at all times?

18. Earth precesses about its vertical axis with a period of 26,000 years. Discuss whether **Equation 11.12** can be used to calculate the precessional angular velocity of Earth.

PROBLEMS

11.1 Rolling Motion

19. What is the angular velocity of a 75.0-cm-diameter tire on an automobile traveling at 90.0 km/h?

20. A boy rides his bicycle 2.00 km. The wheels have radius 30.0 cm. What is the total angle the tires rotate

through during his trip?

21. If the boy on the bicycle in the preceding problem accelerates from rest to a speed of 10.0 m/s in 10.0 s, what is the angular acceleration of the tires?

22. Formula One race cars have 66-cm-diameter tires. If a Formula One averages a speed of 300 km/h during a race, what is the angular displacement in revolutions of the wheels if the race car maintains this speed for 1.5 hours?

23. A marble rolls down an incline at $30°$ from rest. (a) What is its acceleration? (b) How far does it go in 3.0 s?

24. Repeat the preceding problem replacing the marble with a solid cylinder. Explain the new result.

25. A rigid body with a cylindrical cross-section is released from the top of a $30°$ incline. It rolls 10.0 m to the bottom in 2.60 s. Find the moment of inertia of the body in terms of its mass m and radius r.

26. A yo-yo can be thought of a solid cylinder of mass m and radius r that has a light string wrapped around its circumference (see below). One end of the string is held fixed in space. If the cylinder falls as the string unwinds without slipping, what is the acceleration of the cylinder?

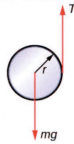

27. A solid cylinder of radius 10.0 cm rolls down an incline with slipping. The angle of the incline is $30°$. The coefficient of kinetic friction on the surface is 0.400. What is the angular acceleration of the solid cylinder? What is the linear acceleration?

28. A bowling ball rolls up a ramp 0.5 m high without slipping to storage. It has an initial velocity of its center of mass of 3.0 m/s. (a) What is its velocity at the top of the ramp? (b) If the ramp is 1 m high does it make it to the top?

29. A 40.0-kg solid cylinder is rolling across a horizontal surface at a speed of 6.0 m/s. How much work is required to stop it?

30. A 40.0-kg solid sphere is rolling across a horizontal surface with a speed of 6.0 m/s. How much work is required to stop it? Compare results with the preceding problem.

31. A solid cylinder rolls up an incline at an angle of $20°$. If it starts at the bottom with a speed of 10 m/s, how far up the incline does it travel?

32. A solid cylindrical wheel of mass M and radius R is pulled by a force $\vec{F}$ applied to the center of the wheel at $37°$ to the horizontal (see the following figure). If the

wheel is to roll without slipping, what is the maximum value of $\left|\vec{F}\right|$? The coefficients of static and kinetic friction are $\mu_S = 0.40$ and $\mu_k = 0.30$.

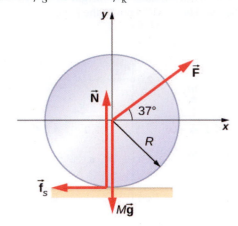

33. A hollow cylinder that is rolling without slipping is given a velocity of 5.0 m/s and rolls up an incline to a vertical height of 1.0 m. If a hollow sphere of the same mass and radius is given the same initial velocity, how high vertically does it roll up the incline?

11.2 Angular Momentum

34. A 0.2-kg particle is travelling along the line $y = 2.0 \text{ m}$ with a velocity 5.0 m/s . What is the angular momentum of the particle about the origin?

35. A bird flies overhead from where you stand at an altitude of 300.0 m and at a speed horizontal to the ground of 20.0 m/s. The bird has a mass of 2.0 kg. The radius vector to the bird makes an angle θ with respect to the ground. The radius vector to the bird and its momentum vector lie in the xy-plane. What is the bird's angular momentum about the point where you are standing?

36. A Formula One race car with mass 750.0 kg is speeding through a course in Monaco and enters a circular turn at 220.0 km/h in the counterclockwise direction about the origin of the circle. At another part of the course, the car enters a second circular turn at 180 km/h also in the counterclockwise direction. If the radius of curvature of the first turn is 130.0 m and that of the second is 100.0 m, compare the angular momenta of the race car in each turn taken about the origin of the circular turn.

37. A particle of mass 5.0 kg has position vector $\vec{r} = (2.0 \hat{i} - 3.0 \hat{j})$m at a particular instant of time when its velocity is $\vec{v} = (3.0 \hat{i})$m/s with respect to the origin. (a) What is the angular momentum of the particle? (b) If a force $\vec{F} = 5.0 \hat{j}$ N acts on the particle at this instant, what is the torque about the origin?

38. Use the right-hand rule to determine the directions of the angular momenta about the origin of the particles as shown below. The z-axis is out of the page.

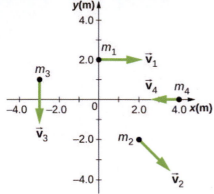

39. Suppose the particles in the preceding problem have masses $m_1 = 0.10$ kg, $m_2 = 0.20$ kg, $m_3 = 0.30$ kg, $m_4 = 0.40$ kg. The velocities of the particles are $v_1 = 2.0 \hat{i}$ m/s, $v_2 = (3.0 \hat{i} - 3.0 \hat{j})$m/s, $v_3 = -1.5 \hat{j}$ m/s, $v_4 = -4.0 \hat{i}$ m/s. (a) Calculate the angular momentum of each particle about the origin. (b) What is the total angular momentum of the four-particle system about the origin?

40. Two particles of equal mass travel with the same speed in opposite directions along parallel lines separated by a distance d. Show that the angular momentum of this two-particle system is the same no matter what point is used as the reference for calculating the angular momentum.

41. An airplane of mass 4.0×10^4 kg flies horizontally at an altitude of 10 km with a constant speed of 250 m/s relative to Earth. (a) What is the magnitude of the airplane's angular momentum relative to a ground observer directly below the plane? (b) Does the angular momentum change as the airplane flies along a constant altitude?

42. At a particular instant, a 1.0-kg particle's position is $\vec{r} = (2.0 \hat{i} - 4.0 \hat{j} + 6.0 \hat{k})$m, its velocity is $\vec{v} = (-1.0 \hat{i} + 4.0 \hat{j} + 1.0 \hat{k})$m/s, and the force on it is $\vec{F} = (10.0 \hat{i} + 15.0 \hat{j})$N. (a) What is the angular momentum of the particle about the origin? (b) What is the torque on the particle about the origin? (c) What is the time rate of change of the particle's angular momentum at this instant?

43. A particle of mass m is dropped at the point $(-d, 0)$

and falls vertically in Earth's gravitational field $-g \hat{j}$.

(a) What is the expression for the angular momentum of the particle around the z-axis, which points directly out of the page as shown below? (b) Calculate the torque on the particle around the z-axis. (c) Is the torque equal to the time rate of change of the angular momentum?

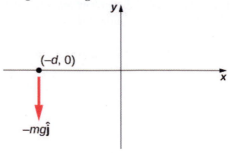

44. (a) Calculate the angular momentum of Earth in its orbit around the Sun. (b) Compare this angular momentum with the angular momentum of Earth about its axis.

45. A boulder of mass 20 kg and radius 20 cm rolls down a hill 15 m high from rest. What is its angular momentum when it is half way down the hill? (b) At the bottom?

46. A satellite is spinning at 6.0 rev/s. The satellite consists of a main body in the shape of a sphere of radius 2.0 m and mass 10,000 kg, and two antennas projecting out from the center of mass of the main body that can be approximated with rods of length 3.0 m each and mass 10 kg. The antenna's lie in the plane of rotation. What is the angular momentum of the satellite?

47. A propeller consists of two blades each 3.0 m in length and mass 120 kg each. The propeller can be approximated by a single rod rotating about its center of mass. The propeller starts from rest and rotates up to 1200 rpm in 30 seconds at a constant rate. (a) What is the angular momentum of the propeller at $t = 10$ s; $t = 20$ s? (b) What is the torque on the propeller?

48. A pulsar is a rapidly rotating neutron star. The Crab nebula pulsar in the constellation Taurus has a period of 33.5×10^{-3} s, radius 10.0 km, and mass 2.8×10^{30} kg. The pulsar's rotational period will increase over time due to the release of electromagnetic radiation, which doesn't change its radius but reduces its rotational energy. (a) What is the angular momentum of the pulsar? (b) Suppose the angular velocity decreases at a rate of 10^{-14} rad/s^2. What is the torque on the pulsar?

49. The blades of a wind turbine are 30 m in length and rotate at a maximum rotation rate of 20 rev/min. (a) If the blades are 6000 kg each and the rotor assembly has three blades, calculate the angular momentum of the turbine at this rotation rate. (b) What is the torque require to rotate the blades up to the maximum rotation rate in 5 minutes?

50. A roller coaster has mass 3000.0 kg and needs to make it safely through a vertical circular loop of radius 50.0 m. What is the minimum angular momentum of the coaster at the bottom of the loop to make it safely through? Neglect friction on the track. Take the coaster to be a point particle.

51. A mountain biker takes a jump in a race and goes airborne. The mountain bike is travelling at 10.0 m/s before it goes airborne. If the mass of the front wheel on the bike is 750 g and has radius 35 cm, what is the angular momentum of the spinning wheel in the air the moment the bike leaves the ground?

11.3 Conservation of Angular Momentum

52. A disk of mass 2.0 kg and radius 60 cm with a small mass of 0.05 kg attached at the edge is rotating at 2.0 rev/s. The small mass suddenly separates from the disk. What is the disk's final rotation rate?

53. The Sun's mass is 2.0×10^{30} kg, its radius is 7.0×10^5 km, and it has a rotational period of approximately 28 days. If the Sun should collapse into a white dwarf of radius 3.5×10^3 km, what would its period be if no mass were ejected and a sphere of uniform density can model the Sun both before and after?

54. A cylinder with rotational inertia $I_1 = 2.0 \, \text{kg} \cdot \text{m}^2$ rotates clockwise about a vertical axis through its center with angular speed $\omega_1 = 5.0 \, \text{rad/s}$. A second cylinder with rotational inertia $I_2 = 1.0 \, \text{kg} \cdot \text{m}^2$ rotates counterclockwise about the same axis with angular speed $\omega_2 = 8.0 \, \text{rad/s}$. If the cylinders couple so they have the same rotational axis what is the angular speed of the combination? What percentage of the original kinetic energy is lost to friction?

55. A diver off the high board imparts an initial rotation with his body fully extended before going into a tuck and executing three back somersaults before hitting the water. If his moment of inertia before the tuck is $16.9 \, \text{kg} \cdot \text{m}^2$ and after the tuck during the somersaults is $4.2 \, \text{kg} \cdot \text{m}^2$, what rotation rate must he impart to his body directly off the board and before the tuck if he takes 1.4 s to execute the somersaults before hitting the water?

56. An Earth satellite has its apogee at 2500 km above the surface of Earth and perigee at 500 km above the surface of Earth. At apogee its speed is 730 m/s. What is its speed at

perigee? Earth's radius is 6370 km (see below).

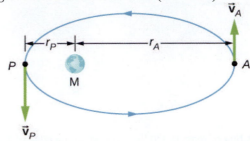

57. A Molniya orbit is a highly eccentric orbit of a communication satellite so as to provide continuous communications coverage for Scandinavian countries and adjacent Russia. The orbit is positioned so that these countries have the satellite in view for extended periods in time (see below). If a satellite in such an orbit has an apogee at 40,000.0 km as measured from the center of Earth and a velocity of 3.0 km/s, what would be its velocity at perigee measured at 200.0 km altitude?

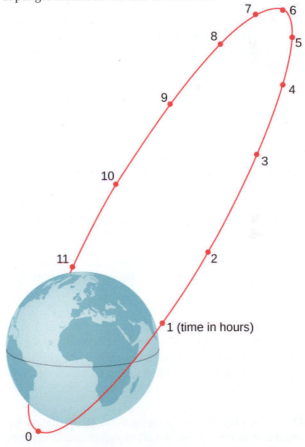

58. Shown below is a small particle of mass 20 g that is moving at a speed of 10.0 m/s when it collides and sticks to the edge of a uniform solid cylinder. The cylinder is free to rotate about its axis through its center and is perpendicular to the page. The cylinder has a mass of 0.5 kg and a radius of 10 cm, and is initially at rest. (a) What is the angular velocity of the system after the collision? (b) How much kinetic energy is lost in the collision?

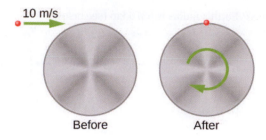

Before   After

59. A bug of mass 0.020 kg is at rest on the edge of a solid cylindrical disk $(M = 0.10 \, \text{kg}, R = 0.10 \, \text{m})$ rotating in a horizontal plane around the vertical axis through its center. The disk is rotating at 10.0 rad/s. The bug crawls to the center of the disk. (a) What is the new angular velocity of the disk? (b) What is the change in the kinetic energy of the system? (c) If the bug crawls back to the outer edge of the disk, what is the angular velocity of the disk then? (d) What is the new kinetic energy of the system? (e) What is the cause of the increase and decrease of kinetic energy?

60. A uniform rod of mass 200 g and length 100 cm is free to rotate in a horizontal plane around a fixed vertical axis through its center, perpendicular to its length. Two small beads, each of mass 20 g, are mounted in grooves along the rod. Initially, the two beads are held by catches on opposite sides of the rod's center, 10 cm from the axis of rotation. With the beads in this position, the rod is rotating with an angular velocity of 10.0 rad/s. When the catches are released, the beads slide outward along the rod. (a) What is the rod's angular velocity when the beads reach the ends of the rod? (b) What is the rod's angular velocity if the beads fly off the rod?

61. A merry-go-round has a radius of 2.0 m and a moment of inertia $300 \, \text{kg} \cdot \text{m}^2$. A boy of mass 50 kg runs tangent to the rim at a speed of 4.0 m/s and jumps on. If the merry-go-round is initially at rest, what is the angular velocity after the boy jumps on?

62. A playground merry-go-round has a mass of 120 kg and a radius of 1.80 m and it is rotating with an angular velocity of 0.500 rev/s. What is its angular velocity after a 22.0-kg child gets onto it by grabbing its outer edge? The child is initially at rest.

63. Three children are riding on the edge of a merry-go-round that is 100 kg, has a 1.60-m radius, and is spinning at 20.0 rpm. The children have masses of 22.0, 28.0, and 33.0 kg. If the child who has a mass of 28.0 kg moves to the center of the merry-go-round, what is the new angular velocity in rpm?

64. (a) Calculate the angular momentum of an ice skater spinning at 6.00 rev/s given his moment of inertia is $0.400 \, \text{kg} \cdot \text{m}^2$. (b) He reduces his rate of spin (his angular velocity) by extending his arms and increasing his moment of inertia. Find the value of his moment of inertia if his angular velocity decreases to 1.25 rev/s. (c) Suppose instead he keeps his arms in and allows friction of the ice to slow him to 3.00 rev/s. What average torque was exerted if this takes 15.0 s?

65. Twin skaters approach one another as shown below and lock hands. (a) Calculate their final angular velocity, given each had an initial speed of 2.50 m/s relative to the ice. Each has a mass of 70.0 kg, and each has a center of mass located 0.800 m from their locked hands. You may approximate their moments of inertia to be that of point masses at this radius. (b) Compare the initial kinetic energy and final kinetic energy.

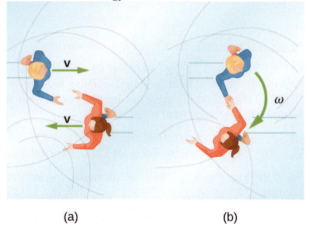

(a)   (b)

66. A baseball catcher extends his arm straight up to catch a fast ball with a speed of 40 m/s. The baseball is 0.145 kg and the catcher's arm length is 0.5 m and mass 4.0 kg. (a) What is the angular velocity of the arm immediately after catching the ball as measured from the arm socket? (b) What is the torque applied if the catcher stops the rotation of his arm 0.3 s after catching the ball?

67. In 2015, in Warsaw, Poland, Olivia Oliver of Nova Scotia broke the world record for being the fastest spinner on ice skates. She achieved a record 342 rev/min, beating the existing Guinness World Record by 34 rotations. If an ice skater extends her arms at that rotation rate, what would be her new rotation rate? Assume she can be approximated by a 45-kg rod that is 1.7 m tall with a radius of 15 cm in the record spin. With her arms stretched take the approximation of a rod of length 130 cm with 10% of her body mass aligned perpendicular to the spin axis. Neglect frictional forces.

68. A satellite in a geosynchronous circular orbit is 42,164.0 km from the center of Earth. A small asteroid collides with the satellite sending it into an elliptical orbit of apogee 45,000.0 km. What is the speed of the satellite at apogee? Assume its angular momentum is conserved.

69. A gymnast does cartwheels along the floor and then

launches herself into the air and executes several flips in a tuck while she is airborne. If her moment of inertia when executing the cartwheels is $13.5 \, \text{kg} \cdot \text{m}^2$ and her spin rate is 0.5 rev/s, how many revolutions does she do in the air if her moment of inertia in the tuck is $3.4 \, \text{kg} \cdot \text{m}^2$ and she has 2.0 s to do the flips in the air?

70. The centrifuge at NASA Ames Research Center has a radius of 8.8 m and can produce forces on its payload of 20 *g*s or 20 times the force of gravity on Earth. (a) What is the angular momentum of a 20-kg payload that experiences 10 *g*s in the centrifuge? (b) If the driver motor was turned off in (a) and the payload lost 10 kg, what would be its new spin rate, taking into account there are no frictional forces present?

71. A ride at a carnival has four spokes to which pods are attached that can hold two people. The spokes are each 15 m long and are attached to a central axis. Each spoke has mass 200.0 kg, and the pods each have mass 100.0 kg. If the ride spins at 0.2 rev/s with each pod containing two 50.0-kg children, what is the new spin rate if all the children jump off the ride?

72. An ice skater is preparing for a jump with turns and has his arms extended. His moment of inertia is $1.8 \, \text{kg} \cdot \text{m}^2$ while his arms are extended, and he is spinning at 0.5 rev/s. If he launches himself into the air at 9.0 m/s at an angle of $45°$ with respect to the ice, how many revolutions can he execute while airborne if his moment of inertia in the air is $0.5 \, \text{kg} \cdot \text{m}^2$?

73. A space station consists of a giant rotating hollow cylinder of mass 10^6 kg including people on the station and a radius of 100.00 m. It is rotating in space at 3.30 rev/min in order to produce artificial gravity. If 100 people of an average mass of 65.00 kg spacewalk to an awaiting spaceship, what is the new rotation rate when all the people are off the station?

74. Neptune has a mass of 1.0×10^{26} kg and is 4.5×10^9 km from the Sun with an orbital period of 165 years. Planetesimals in the outer primordial solar system 4.5 billion years ago coalesced into Neptune over hundreds of millions of years. If the primordial disk that evolved into our present day solar system had a radius of 10^{11}

km and if the matter that made up these planetesimals that later became Neptune was spread out evenly on the edges of it, what was the orbital period of the outer edges of the primordial disk?

11.4 Precession of a Gyroscope

75. A gyroscope has a 0.5-kg disk that spins at 40 rev/s. The center of mass of the disk is 10 cm from a pivot which is also the radius of the disk. What is the precession angular velocity?

76. The precession angular velocity of a gyroscope is 1.0 rad/s. If the mass of the rotating disk is 0.4 kg and its radius is 30 cm, as well as the distance from the center of mass to the pivot, what is the rotation rate in rev/s of the disk?

77. The axis of Earth makes a $23.5°$ angle with a direction perpendicular to the plane of Earth's orbit. As shown below, this axis precesses, making one complete rotation in 25,780 y.

(a) Calculate the change in angular momentum in half this time.

(b) What is the average torque producing this change in angular momentum?

(c) If this torque were created by a pair of forces acting at the most effective point on the equator, what would the magnitude of each force be?

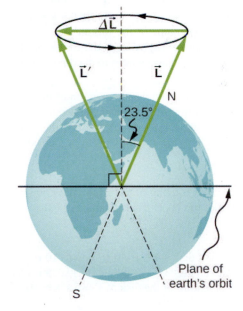

ADDITIONAL PROBLEMS

78. A marble is rolling across the floor at a speed of 7.0 m/s when it starts up a plane inclined at $30°$ to the horizontal.

(a) How far along the plane does the marble travel before coming to a rest? (b) How much time elapses while the marble moves up the plane?

79. Repeat the preceding problem replacing the marble with a hollow sphere. Explain the new results.

80. The mass of a hoop of radius 1.0 m is 6.0 kg. It rolls across a horizontal surface with a speed of 10.0 m/s. (a) How much work is required to stop the hoop? (b) If the hoop starts up a surface at $30°$ to the horizontal with a speed of 10.0 m/s, how far along the incline will it travel before stopping and rolling back down?

81. Repeat the preceding problem for a hollow sphere of the same radius and mass and initial speed. Explain the differences in the results.

82. A particle has mass 0.5 kg and is traveling along the line $x = 5.0$ m at 2.0 m/s in the positive y-direction. What is the particle's angular momentum about the origin?

83. A 4.0-kg particle moves in a circle of radius 2.0 m. The angular momentum of the particle varies in time according to $l = 5.0t^2$. (a) What is the torque on the particle about the center of the circle at $t = 3.4$ s? (b) What is the angular velocity of the particle at $t = 3.4$ s?

84. A proton is accelerated in a cyclotron to 5.0×10^6 m/s in 0.01 s. The proton follows a circular path. If the radius of the cyclotron is 0.5 km, (a) What is the angular momentum of the proton about the center at its maximum speed? (b) What is the torque on the proton about the center as it accelerates to maximum speed?

85. (a) What is the angular momentum of the Moon in its orbit around Earth? (b) How does this angular momentum compare with the angular momentum of the Moon on its axis? Remember that the Moon keeps one side toward Earth at all times.

86. A DVD is rotating at 500 rpm. What is the angular momentum of the DVD if has a radius of 6.0 cm and mass 20.0 g?

87. A potter's disk spins from rest up to 10 rev/s in 15 s. The disk has a mass 3.0 kg and radius 30.0 cm. What is the angular momentum of the disk at $t = 5$ s, $t = 10$ s?

88. Suppose you start an antique car by exerting a force of 300 N on its crank for 0.250 s. What is the angular

momentum given to the engine if the handle of the crank is 0.300 m from the pivot and the force is exerted to create maximum torque the entire time?

89. A solid cylinder of mass 2.0 kg and radius 20 cm is rotating counterclockwise around a vertical axis through its center at 600 rev/min. A second solid cylinder of the same mass and radius is rotating clockwise around the same vertical axis at 900 rev/min. If the cylinders couple so that they rotate about the same vertical axis, what is the angular velocity of the combination?

90. A boy stands at the center of a platform that is rotating without friction at 1.0 rev/s. The boy holds weights as far from his body as possible. At this position the total moment of inertia of the boy, platform, and weights is $5.0\ \text{kg} \cdot \text{m}^2$. The boy draws the weights in close to his body, thereby decreasing the total moment of inertia to $1.5\ \text{kg} \cdot \text{m}^2$. (a) What is the final angular velocity of the platform? (b) By how much does the rotational kinetic energy increase?

91. Eight children, each of mass 40 kg, climb on a small merry-go-round. They position themselves evenly on the outer edge and join hands. The merry-go-round has a radius of 4.0 m and a moment of inertia $1000.0\ \text{kg} \cdot \text{m}^2$. After the merry-go-round is given an angular velocity of 6.0 rev/min, the children walk inward and stop when they are 0.75 m from the axis of rotation. What is the new angular velocity of the merry-go-round? Assume there is negligible frictional torque on the structure.

92. A thin meter stick of mass 150 g rotates around an axis perpendicular to the stick's long axis at an angular velocity of 240 rev/min. What is the angular momentum of the stick if the rotation axis (a) passes through the center of the stick? (b) Passes through one end of the stick?

93. A satellite in the shape of a sphere of mass 20,000 kg and radius 5.0 m is spinning about an axis through its center of mass. It has a rotation rate of 8.0 rev/s. Two antennas deploy in the plane of rotation extending from the center of mass of the satellite. Each antenna can be approximated as a rod has mass 200.0 kg and length 7.0 m. What is the new rotation rate of the satellite?

94. A top has moment of inertia $3.2 \times 10^{-4}\ \text{kg} \cdot \text{m}^2$ and radius 4.0 cm from the center of mass to the pivot point. If it spins at 20.0 rev/s and is precessing, how many revolutions does it precess in 10.0 s?

CHALLENGE PROBLEMS

95. The truck shown below is initially at rest with solid cylindrical roll of paper sitting on its bed. If the truck moves forward with a uniform acceleration a, what distance s does it move before the paper rolls off its back end? (*Hint*: If the roll accelerates forward with a', then is accelerates backward relative to the truck with an acceleration $a - a'$. Also, $R\alpha = a - a'$.)

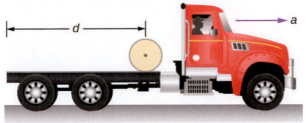

96. A bowling ball of radius 8.5 cm is tossed onto a bowling lane with speed 9.0 m/s. The direction of the toss is to the left, as viewed by the observer, so the bowling ball starts to rotate counterclockwise when in contact with the floor. The coefficient of kinetic friction on the lane is 0.3. (a) What is the time required for the ball to come to the point where it is not slipping? What is the distance d to the point where the ball is rolling without slipping?

97. A small ball of mass 0.50 kg is attached by a massless string to a vertical rod that is spinning as shown below. When the rod has an angular velocity of 6.0 rad/s, the string makes an angle of $30°$ with respect to the vertical. (a)

If the angular velocity is increased to 10.0 rad/s, what is the new angle of the string? (b) Calculate the initial and final angular momenta of the ball. (c) Can the rod spin fast enough so that the ball is horizontal?

98. A bug flying horizontally at 1.0 m/s collides and sticks to the end of a uniform stick hanging vertically. After the impact, the stick swings out to a maximum angle of $5.0°$ from the vertical before rotating back. If the mass of the stick is 10 times that of the bug, calculate the length of the stick.

12 | STATIC EQUILIBRIUM AND ELASTICITY

Figure 12.1 Two stilt walkers in standing position. All forces acting on each stilt walker balance out; neither changes its translational motion. In addition, all torques acting on each person balance out, and thus neither of them changes its rotational motion. The result is static equilibrium. (credit: modification of work by Stuart Redler)

Chapter Outline

12.1 Conditions for Static Equilibrium

12.2 Examples of Static Equilibrium

12.3 Stress, Strain, and Elastic Modulus

12.4 Elasticity and Plasticity

Introduction

In earlier chapters, you learned about forces and Newton's laws for translational motion. You then studied torques and the rotational motion of a body about a fixed axis of rotation. You also learned that static equilibrium means no motion at all and that dynamic equilibrium means motion without acceleration.

In this chapter, we combine the conditions for static translational equilibrium and static rotational equilibrium to describe situations typical for any kind of construction. What type of cable will support a suspension bridge? What type of foundation will support an office building? Will this prosthetic arm function correctly? These are examples of questions that contemporary engineers must be able to answer.

The elastic properties of materials are especially important in engineering applications, including bioengineering. For example, materials that can stretch or compress and then return to their original form or position make good shock absorbers. In this chapter, you will learn about some applications that combine equilibrium with elasticity to construct real structures that last.

12.1 | Conditions for Static Equilibrium

Learning Objectives
By the end of this section, you will be able to: • Identify the physical conditions of static equilibrium. • Draw a free-body diagram for a rigid body acted on by forces. • Explain how the conditions for equilibrium allow us to solve statics problems.

We say that a rigid body is in **equilibrium** when both its linear and angular acceleration are zero relative to an inertial frame of reference. This means that a body in equilibrium can be moving, but if so, its linear and angular velocities must be constant. We say that a rigid body is in **static equilibrium** when it is at rest *in our selected frame of reference*. Notice that the distinction between the state of rest and a state of uniform motion is artificial—that is, an object may be at rest in our selected frame of reference, yet to an observer moving at constant velocity relative to our frame, the same object appears to be in uniform motion with constant velocity. Because the motion is *relative*, what is in static equilibrium to us is in dynamic equilibrium to the moving observer, and vice versa. Since the laws of physics are identical for all inertial reference frames, in an inertial frame of reference, there is no distinction between static equilibrium and equilibrium.

According to Newton's second law of motion, the linear acceleration of a rigid body is caused by a net force acting on it, or

$$\sum_k \vec{F}_k = m\,\vec{a}_{CM}. \tag{12.1}$$

Here, the sum is of all external forces acting on the body, where m is its mass and $\vec{a}_{CM}$ is the linear acceleration of its center of mass (a concept we discussed in **Linear Momentum and Collisions** on linear momentum and collisions). In equilibrium, the linear acceleration is zero. If we set the acceleration to zero in **Equation 12.1**, we obtain the following equation:

First Equilibrium Condition

The first equilibrium condition for the static equilibrium of a rigid body expresses *translational* equilibrium:

$$\sum_k \vec{F}_k = \vec{0}. \tag{12.2}$$

The first equilibrium condition, **Equation 12.2**, is the equilibrium condition for forces, which we encountered when studying applications of Newton's laws.

This vector equation is equivalent to the following three scalar equations for the components of the net force:

$$\sum_k F_{kx} = 0, \quad \sum_k F_{ky} = 0, \quad \sum_k F_{kz} = 0. \tag{12.3}$$

Analogously to **Equation 12.1**, we can state that the rotational acceleration $\vec{\alpha}$ of a rigid body about a fixed axis of rotation is caused by the net torque acting on the body, or

$$\sum_k \vec{\tau}_k = I\,\vec{\alpha}. \tag{12.4}$$

Here I is the rotational inertia of the body in rotation about this axis and the summation is over *all* torques $\vec{\tau}_k$ of external forces in **Equation 12.2**. In equilibrium, the rotational acceleration is zero. By setting to zero the right-hand side of **Equation 12.4**, we obtain the second equilibrium condition:

Second Equilibrium Condition

The second equilibrium condition for the static equilibrium of a rigid body expresses *rotational* equilibrium:

$$\sum_k \vec{\tau}_k = \vec{0} \,. \qquad \qquad \text{(12.5)}$$

The second equilibrium condition, **Equation 12.5**, is the equilibrium condition for torques that we encountered when we studied rotational dynamics. It is worth noting that this equation for equilibrium is generally valid for rotational equilibrium about any axis of rotation (fixed or otherwise). Again, this vector equation is equivalent to three scalar equations for the vector components of the net torque:

$$\sum_k \tau_{kx} = 0, \qquad \sum_k \tau_{ky} = 0, \qquad \sum_k \tau_{kz} = 0. \qquad \qquad \text{(12.6)}$$

The second equilibrium condition means that in equilibrium, there is no net external torque to cause rotation about any axis.

The first and second equilibrium conditions are stated in a particular reference frame. The first condition involves only forces and is therefore independent of the origin of the reference frame. However, the second condition involves torque, which is defined as a cross product, $\vec{\tau}_k = \vec{r}_k \times \vec{F}_k$, where the position vector $\vec{r}_k$ with respect to the axis of rotation of the point where the force is applied enters the equation. Therefore, torque depends on the location of the axis in the reference frame. However, when rotational and translational equilibrium conditions hold simultaneously in one frame of reference, then they also hold in any other inertial frame of reference, so that the net torque about any axis of rotation is still zero. The explanation for this is fairly straightforward.

Suppose vector $\vec{R}$ is the position of the origin of a new inertial frame of reference S' in the old inertial frame of reference S. From our study of relative motion, we know that in the new frame of reference S', the position vector $\vec{r}'_k$ of the point where the force $\vec{F}_k$ is applied is related to $\vec{r}_k$ via the equation

$$\vec{r}'_k = \vec{r}_k - \vec{R} \,.$$

Now, we can sum all torques $\vec{\tau}'_k = \vec{r}'_k \times \vec{F}_k$ of all external forces in a new reference frame, S':

$$\sum_k \vec{\tau}'_k = \sum_k \vec{r}'_k \times \vec{F}_k = \sum_k (\vec{r}_k - \vec{R}) \times \vec{F}_k = \sum_k \vec{r}_k \times \vec{F}_k - \sum_k \vec{R} \times \vec{F}_k = \sum_k \vec{\tau}_k - \vec{R} \times \sum_k \vec{F}_k = \vec{0} \,.$$

In the final step in this chain of reasoning, we used the fact that in equilibrium in the old frame of reference, S, the first term vanishes because of **Equation 12.5** and the second term vanishes because of **Equation 12.2**. Hence, we see that the net torque in any inertial frame of reference S' is zero, provided that both conditions for equilibrium hold in an inertial frame of reference S.

The practical implication of this is that when applying equilibrium conditions for a rigid body, we are free to choose any point as the origin of the reference frame. Our choice of reference frame is dictated by the physical specifics of the problem we are solving. In one frame of reference, the mathematical form of the equilibrium conditions may be quite complicated, whereas in another frame, the same conditions may have a simpler mathematical form that is easy to solve. The origin of a selected frame of reference is called the pivot point.

In the most general case, equilibrium conditions are expressed by the six scalar equations (**Equation 12.3** and **Equation 12.6**). For planar equilibrium problems with rotation about a fixed axis, which we consider in this chapter, we can reduce the number of equations to three. The standard procedure is to adopt a frame of reference where the z-axis is the axis of rotation. With this choice of axis, the net torque has only a z-component, all forces that have non-zero torques lie in the xy-plane, and therefore contributions to the net torque come from only the x- and y-components of external forces. Thus, for planar problems with the axis of rotation perpendicular to the xy-plane, we have the following three equilibrium conditions for forces and torques:

$$F_{1x} + F_{2x} + \cdots + F_{Nx} = 0 \qquad \qquad \text{(12.7)}$$

$$F_{1y} + F_{2y} + \cdots + F_{Ny} = 0 \qquad \qquad \text{(12.8)}$$

$$\tau_1 + \tau_2 + \cdots + \tau_N = 0 \qquad \qquad \text{(12.9)}$$

where the summation is over all N external forces acting on the body and over their torques. In **Equation 12.9**, we simplified the notation by dropping the subscript z, but we understand here that the summation is over all contributions along the z-axis, which is the axis of rotation. In **Equation 12.9**, the z-component of torque $\vec{\tau}_k$ from the force $\vec{F}_k$ is

$$\tau_k = r_k F_k \sin \theta \qquad\qquad (12.10)$$

where r_k is the length of the lever arm of the force and F_k is the magnitude of the force (as you saw in **Fixed-Axis Rotation**). The angle θ is the angle between vectors $\vec{r}_k$ and $\vec{F}_k$, measuring *from vector* $\vec{r}_k$ *to vector* $\vec{F}_k$ in the *counterclockwise* direction (**Figure 12.2**). When using **Equation 12.10**, we often compute the magnitude of torque and assign its sense as either positive $(+)$ or negative $(-)$, depending on the direction of rotation caused by this torque alone. In **Equation 12.9**, net torque is the sum of terms, with each term computed from **Equation 12.10**, and each term must have the correct *sense*. Similarly, in **Equation 12.7**, we assign the $+$ sign to force components in the $+$ x-direction and the $-$ sign to components in the $-$ x-direction. The same rule must be consistently followed in **Equation 12.8**, when computing force components along the y-axis.

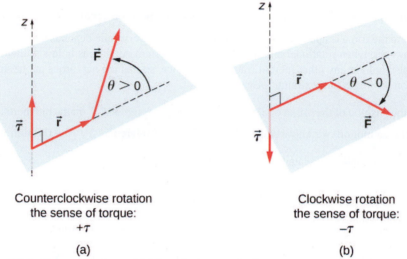

Counterclockwise rotation
the sense of torque:
$+\tau$

(a)

Clockwise rotation
the sense of torque:
$-\tau$

(b)

Figure 12.2 Torque of a force: (a) When the torque of a force causes counterclockwise rotation about the axis of rotation, we say that its *sense* is positive, which means the torque vector is parallel to the axis of rotation. (b) When torque of a force causes clockwise rotation about the axis, we say that its sense is negative, which means the torque vector is antiparallel to the axis of rotation.

 View this **demonstration (https://openstaxcollege.org/l/21rigsquare)** to see two forces act on a rigid square in two dimensions. At all times, the static equilibrium conditions given by **Equation 12.7** through **Equation 12.9** are satisfied. You can vary magnitudes of the forces and their lever arms and observe the effect these changes have on the square.

In many equilibrium situations, one of the forces acting on the body is its weight. In free-body diagrams, the weight vector is attached to the **center of gravity** of the body. For all practical purposes, the center of gravity is identical to the center of mass, as you learned in **Linear Momentum and Collisions** on linear momentum and collisions. Only in situations where a body has a large spatial extension so that the gravitational field is nonuniform throughout its volume, are the center of gravity and the center of mass located at different points. In practical situations, however, even objects as large as buildings or cruise ships are located in a uniform gravitational field on Earth's surface, where the acceleration due to gravity has a constant magnitude of $g = 9.8 \text{ m/s}^2$. In these situations, the center of gravity is identical to the center of mass. Therefore, throughout this chapter, we use the center of mass (CM) as the point where the weight vector is attached. Recall that the CM has a special physical meaning: When an external force is applied to a body at exactly its CM, the body as a whole undergoes translational motion and such a force does not cause rotation.

When the CM is located off the axis of rotation, a net **gravitational torque** occurs on an object. Gravitational torque is the torque caused by weight. This gravitational torque may rotate the object if there is no support present to balance it. The magnitude of the gravitational torque depends on how far away from the pivot the CM is located. For example, in the case of a tipping truck (**Figure 12.3**), the pivot is located on the line where the tires make contact with the road's surface. If the CM is located high above the road's surface, the gravitational torque may be large enough to turn the truck over. Passenger cars with a low-lying CM, close to the pavement, are more resistant to tipping over than are trucks.

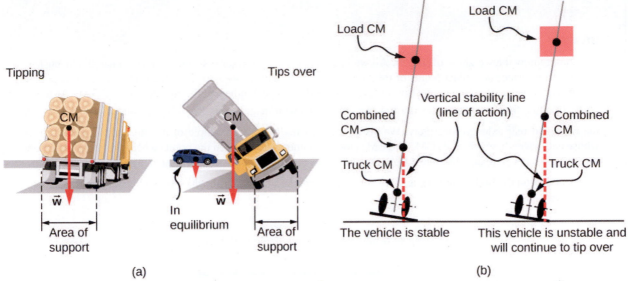

Figure 12.3 The distribution of mass affects the position of the center of mass (CM), where the weight vector $\vec{w}$ is attached. If the center of gravity is within the area of support, the truck returns to its initial position after tipping [see the left panel in (b)]. But if the center of gravity lies outside the area of support, the truck turns over [see the right panel in (b)]. Both vehicles in (b) are out of equilibrium. Notice that the car in (a) is in equilibrium: The low location of its center of gravity makes it hard to tip over.

 If you tilt a box so that one edge remains in contact with the table beneath it, then one edge of the base of support becomes a pivot. As long as the center of gravity of the box remains over the base of support, gravitational torque rotates the box back toward its original position of stable equilibrium. When the center of gravity moves outside of the base of support, gravitational torque rotates the box in the opposite direction, and the box rolls over. View this **demonstration (https://openstaxcollege.org/l/21unstable)** to experiment with stable and unstable positions of a box.

Example 12.1

Center of Gravity of a Car

A passenger car with a 2.5-m wheelbase has 52% of its weight on the front wheels on level ground, as illustrated in **Figure 12.4**. Where is the CM of this car located with respect to the rear axle?

Figure 12.4 The weight distribution between the axles of a car. Where is the center of gravity located? (credit "car": modification of work by Jane Whitney)

Strategy

We do not know the weight w of the car. All we know is that when the car rests on a level surface, $0.52w$ pushes down on the surface at contact points of the front wheels and $0.48w$ pushes down on the surface at contact points of the rear wheels. Also, the contact points are separated from each other by the distance $d = 2.5$ m. At these contact points, the car experiences normal reaction forces with magnitudes $F_F = 0.52w$ and $F_R = 0.48w$ on the front and rear axles, respectively. We also know that the car is an example of a rigid body in equilibrium whose entire weight w acts at its CM. The CM is located somewhere between the points where the normal reaction forces act, somewhere at a distance x from the point where F_R acts. Our task is to find x. Thus, we identify three forces acting on the body (the car), and we can draw a free-body diagram for the extended rigid body, as shown in **Figure 12.5**.

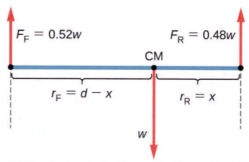

Figure 12.5 The free-body diagram for the car clearly indicates force vectors acting on the car and distances to the center of mass (CM). When CM is selected as the pivot point, these distances are lever arms of normal reaction forces. Notice that vector magnitudes and lever arms do not need to be drawn to scale, but all quantities of relevance must be clearly labeled.

We are almost ready to write down equilibrium conditions **Equation 12.7** through **Equation 12.9** for the car, but first we must decide on the reference frame. Suppose we choose the x-axis along the length of the car, the y-axis vertical, and the z-axis perpendicular to this xy-plane. With this choice we only need to write **Equation 12.7** and **Equation 12.9** because all the y-components are identically zero. Now we need to decide on the location of the pivot point. We can choose any point as the location of the axis of rotation (z-axis). Suppose we place the axis of rotation at CM, as indicated in the free-body diagram for the car. At this point, we are ready to write the equilibrium conditions for the car.

Solution

Each equilibrium condition contains only three terms because there are $N = 3$ forces acting on the car. The first equilibrium condition, **Equation 12.7**, reads

$$+F_F - w + F_R = 0. \tag{12.11}$$

This condition is trivially satisfied because when we substitute the data, **Equation 12.11** becomes $+0.52w - w + 0.48w = 0$. The second equilibrium condition, **Equation 12.9**, reads

$$\tau_F + \tau_w + \tau_R = 0 \tag{12.12}$$

where τ_F is the torque of force F_F, τ_w is the gravitational torque of force w, and τ_R is the torque of force F_R. When the pivot is located at CM, the gravitational torque is identically zero because the lever arm of the weight with respect to an axis that passes through CM is zero. The lines of action of both normal reaction forces are perpendicular to their lever arms, so in **Equation 12.10**, we have $|\sin \theta| = 1$ for both forces. From the free-body diagram, we read that torque τ_F causes clockwise rotation about the pivot at CM, so its sense is negative; and torque τ_R causes counterclockwise rotation about the pivot at CM, so its sense is positive. With this information, we write the second equilibrium condition as

$$-r_F F_F + r_R F_R = 0.$$

(12.13)

With the help of the free-body diagram, we identify the force magnitudes $F_R = 0.48w$ and $F_F = 0.52w$, and their corresponding lever arms $r_R = x$ and $r_F = d - x$. We can now write the second equilibrium condition, **Equation 12.13**, explicitly in terms of the unknown distance x:

$$-0.52(d - x)w + 0.48xw = 0.$$

(12.14)

Here the weight w cancels and we can solve the equation for the unknown position x of the CM. The answer is $x = 0.52d = 0.52(2.5\ \text{m}) = 1.3\ \text{m}.$

Solution

Choosing the pivot at the position of the front axle does not change the result. The free-body diagram for this pivot location is presented in **Figure 12.6**. For this choice of pivot point, the second equilibrium condition is

$$-r_w w + r_R F_R = 0.$$

(12.15)

When we substitute the quantities indicated in the diagram, we obtain

$$-(d - x)w + 0.48dw = 0.$$

(12.16)

The answer obtained by solving **Equation 12.13** is, again, $x = 0.52d = 1.3\ \text{m}.$

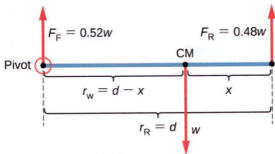

Figure 12.6 The equivalent free-body diagram for the car; the pivot is clearly indicated.

Significance

This example shows that when solving static equilibrium problems, we are free to choose the pivot location. For different choices of the pivot point we have different sets of equilibrium conditions to solve. However, all choices lead to the same solution to the problem.

 12.1 Check Your Understanding Solve **Example 12.1** by choosing the pivot at the location of the rear axle.

 12.2 Check Your Understanding Explain which one of the following situations satisfies both equilibrium conditions: (a) a tennis ball that does not spin as it travels in the air; (b) a pelican that is gliding in the air at a constant velocity at one altitude; or (c) a crankshaft in the engine of a parked car.

A special case of static equilibrium occurs when all external forces on an object act at or along the axis of rotation or when the spatial extension of the object can be disregarded. In such a case, the object can be effectively treated like a point mass. In this special case, we need not worry about the second equilibrium condition, **Equation 12.9**, because all torques are identically zero and the first equilibrium condition (for forces) is the only condition to be satisfied. The free-body diagram and problem-solving strategy for this special case were outlined in **Newton's Laws of Motion** and **Applications of Newton's Laws**. You will see a typical equilibrium situation involving only the first equilibrium condition in the next example.

 View this **demonstration (https://openstaxcollege.org/l/21pulleyknot)** to see three weights that are connected by strings over pulleys and tied together in a knot. You can experiment with the weights to see how they affect the equilibrium position of the knot and, at the same time, see the vector-diagram representation of the first equilibrium condition at work.

Example 12.2

A Breaking Tension

A small pan of mass 42.0 g is supported by two strings, as shown in **Figure 12.7**. The maximum tension that the string can support is 2.80 N. Mass is added gradually to the pan until one of the strings snaps. Which string is it? How much mass must be added for this to occur?

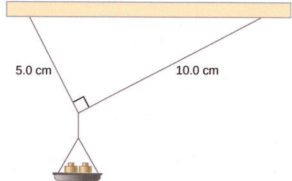

Figure 12.7 Mass is added gradually to the pan until one of the strings snaps.

Strategy

This mechanical system consisting of strings, masses, and the pan is in static equilibrium. Specifically, the knot that ties the strings to the pan is in static equilibrium. The knot can be treated as a point; therefore, we need only the first equilibrium condition. The three forces pulling at the knot are the tension $\vec{T}_1$ in the 5.0-cm string, the tension $\vec{T}_2$ in the 10.0-cm string, and the weight $\vec{w}$ of the pan holding the masses. We adopt a rectangular coordinate system with the y-axis pointing opposite to the direction of gravity and draw the free-body diagram for the knot (see **Figure 12.8**). To find the tension components, we must identify the direction angles α_1 and α_2 that the strings make with the horizontal direction that is the x-axis. As you can see in **Figure 12.7**, the strings make two sides of a right triangle. We can use the Pythagorean theorem to solve this triangle, shown in **Figure 12.8**, and find the sine and cosine of the angles α_1 and α_2. Then we can resolve the tensions into their rectangular components, substitute in the first condition for equilibrium (**Equation 12.7** and **Equation 12.8**), and solve for the tensions in the strings. The string with a greater tension will break first.

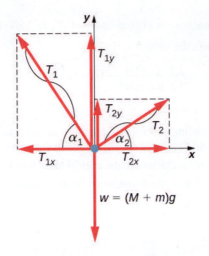

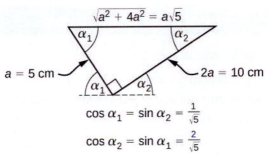

Figure 12.8 Free-body diagram for the knot in **Example 12.2**.

Solution

The weight w pulling on the knot is due to the mass M of the pan and mass m added to the pan, or $w = (M + m)g$. With the help of the free-body diagram in **Figure 12.8**, we can set up the equilibrium conditions for the knot:

$$\text{in the } x\text{-direction,} \quad -T_{1x} + T_{2x} = 0$$
$$\text{in the } y\text{-direction,} \quad +T_{1y} + T_{2y} - w = 0.$$

From the free-body diagram, the magnitudes of components in these equations are

$$T_{1x} = T_1 \cos \alpha_1 = T_1/\sqrt{5}, \quad T_{1y} = T_1 \sin \alpha_1 = 2T_1/\sqrt{5}$$
$$T_{2x} = T_2 \cos \alpha_2 = 2T_2/\sqrt{5}, \quad T_{2y} = T_2 \sin \alpha_2 = T_2/\sqrt{5}.$$

We substitute these components into the equilibrium conditions and simplify. We then obtain two equilibrium equations for the tensions:

$$\text{in } x\text{-direction,} \quad T_1 = 2T_2$$
$$\text{in } y\text{-direction,} \quad \frac{2T_1}{\sqrt{5}} + \frac{T_2}{\sqrt{5}} = (M + m)g.$$

The equilibrium equation for the x-direction tells us that the tension T_1 in the 5.0-cm string is twice the tension T_2 in the 10.0-cm string. Therefore, the shorter string will snap. When we use the first equation to eliminate T_2 from the second equation, we obtain the relation between the mass m on the pan and the tension T_1 in the shorter string:

$$2.5T_1/\sqrt{5} = (M + m)g.$$

The string breaks when the tension reaches the critical value of $T_1 = 2.80\,\text{N}$. The preceding equation can be solved for the critical mass m that breaks the string:

$$m = \frac{2.5}{\sqrt{5}} \frac{T_1}{g} - M = \frac{2.5}{\sqrt{5}} \frac{2.80\,\text{N}}{9.8\,\text{m/s}^2} - 0.042\,\text{kg} = 0.277\,\text{kg} = 277.0\,\text{g}.$$

Significance

Suppose that the mechanical system considered in this example is attached to a ceiling inside an elevator going up. As long as the elevator moves up at a constant speed, the result stays the same because the weight w does not change. If the elevator moves up with acceleration, the critical mass is smaller because the weight of $M + m$ becomes larger by an apparent weight due to the acceleration of the elevator. Still, in all cases the shorter string breaks first.

12.2 | Examples of Static Equilibrium

Learning Objectives

By the end of this section, you will be able to:

- Identify and analyze static equilibrium situations
- Set up a free-body diagram for an extended object in static equilibrium
- Set up and solve static equilibrium conditions for objects in equilibrium in various physical situations

All examples in this chapter are planar problems. Accordingly, we use equilibrium conditions in the component form of **Equation 12.7** to **Equation 12.9**. We introduced a problem-solving strategy in **Example 12.1** to illustrate the physical meaning of the equilibrium conditions. Now we generalize this strategy in a list of steps to follow when solving static equilibrium problems for extended rigid bodies. We proceed in five practical steps.

Problem-Solving Strategy: Static Equilibrium

1. Identify the object to be analyzed. For some systems in equilibrium, it may be necessary to consider more than one object. Identify all forces acting on the object. Identify the questions you need to answer. Identify the information given in the problem. In realistic problems, some key information may be implicit in the situation rather than provided explicitly.

2. Set up a free-body diagram for the object. (a) Choose the xy-reference frame for the problem. Draw a free-body diagram for the object, including only the forces that act on it. When suitable, represent the forces in terms of their components in the chosen reference frame. As you do this for each force, cross out the original force so that you do not erroneously include the same force twice in equations. Label all forces—you will need this for correct computations of net forces in the x- and y-directions. For an unknown force, the direction must be assigned arbitrarily; think of it as a 'working direction' or 'suspected direction.' The correct direction is determined by the sign that you obtain in the final solution. A plus sign $(+)$ means that the working direction is the actual direction. A minus sign $(-)$ means that the actual direction is opposite to the assumed working direction. (b) Choose the location of the rotation axis; in other words, choose the pivot point with respect to which you will compute torques of acting forces. On the free-body diagram, indicate the location of the pivot and the lever arms of acting forces—you will need this for correct computations of torques. In the selection of the pivot, keep in mind that the pivot can be placed anywhere you wish, but the guiding principle is that the best choice will simplify as much as possible the calculation of the net torque along the rotation axis.

3. Set up the equations of equilibrium for the object. (a) Use the free-body diagram to write a correct equilibrium condition **Equation 12.7** for force components in the x-direction. (b) Use the free-body diagram to write a correct equilibrium condition **Equation 12.11** for force components in the y-direction. (c) Use the free-body diagram to write a correct equilibrium condition **Equation 12.9** for torques along the axis of rotation. Use

Equation 12.10 to evaluate torque magnitudes and senses.

4. Simplify and solve the system of equations for equilibrium to obtain unknown quantities. At this point, your work involves algebra only. Keep in mind that the number of equations must be the same as the number of unknowns. If the number of unknowns is larger than the number of equations, the problem cannot be solved.

5. Evaluate the expressions for the unknown quantities that you obtained in your solution. Your final answers should have correct numerical values and correct physical units. If they do not, then use the previous steps to track back a mistake to its origin and correct it. Also, you may independently check for your numerical answers by shifting the pivot to a different location and solving the problem again, which is what we did in **Example 12.1**.

Note that setting up a free-body diagram for a rigid-body equilibrium problem is the most important component in the solution process. Without the correct setup and a correct diagram, you will not be able to write down correct conditions for equilibrium. Also note that a free-body diagram for an extended rigid body that may undergo rotational motion is different from a free-body diagram for a body that experiences only translational motion (as you saw in the chapters on Newton's laws of motion). In translational dynamics, a body is represented as its CM, where all forces on the body are attached and no torques appear. This does not hold true in rotational dynamics, where an extended rigid body cannot be represented by one point alone. The reason for this is that in analyzing rotation, we must identify torques acting on the body, and torque depends both on the acting force and on its lever arm. Here, the free-body diagram for an extended rigid body helps us identify external torques.

Example 12.3

The Torque Balance

Three masses are attached to a uniform meter stick, as shown in **Figure 12.9**. The mass of the meter stick is 150.0 g and the masses to the left of the fulcrum are $m_1 = 50.0$ g and $m_2 = 75.0$ g. Find the mass m_3 that

balances the system when it is attached at the right end of the stick, and the normal reaction force at the fulcrum when the system is balanced.

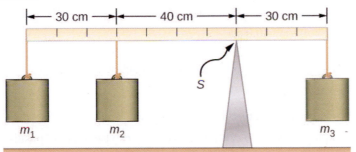

Figure 12.9 In a torque balance, a horizontal beam is supported at a fulcrum (indicated by S) and masses are attached to both sides of the fulcrum. The system is in static equilibrium when the beam does not rotate. It is balanced when the beam remains level.

Strategy

For the arrangement shown in the figure, we identify the following five forces acting on the meter stick:

$w_1 = m_1 g$ is the weight of mass m_1; $w_2 = m_2 g$ is the weight of mass m_2;

$w = mg$ is the weight of the entire meter stick; $w_3 = m_3 g$ is the weight of unknown mass m_3;

F_S is the normal reaction force at the support point S.

We choose a frame of reference where the direction of the y-axis is the direction of gravity, the direction of the x-axis is along the meter stick, and the axis of rotation (the z-axis) is perpendicular to the x-axis and passes through the support point S. In other words, we choose the pivot at the point where the meter stick touches the support. This is a natural choice for the pivot because this point does not move as the stick rotates. Now we are ready to set up the free-body diagram for the meter stick. We indicate the pivot and attach five vectors representing the five forces along the line representing the meter stick, locating the forces with respect to the pivot **Figure 12.10**. At this stage, we can identify the lever arms of the five forces given the information provided in the problem. For the three hanging masses, the problem is explicit about their locations along the stick, but the information about the location of the weight w is given implicitly. The key word here is "uniform." We know from our previous studies that the CM of a uniform stick is located at its midpoint, so this is where we attach the weight w, at the 50-cm mark.

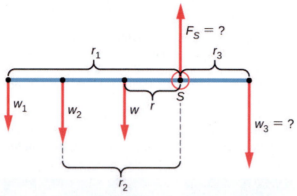

Figure 12.10 Free-body diagram for the meter stick. The pivot is chosen at the support point S.

Solution

With **Figure 12.9** and **Figure 12.10** for reference, we begin by finding the lever arms of the five forces acting on the stick:

$$
\begin{aligned}
r_1 &= 30.0\,\text{cm} + 40.0\,\text{cm} = 70.0\,\text{cm} \\
r_2 &= 40.0\,\text{cm} \\
r &= 50.0\,\text{cm} - 30.0\,\text{cm} = 20.0\,\text{cm} \\
r_S &= 0.0\,\text{cm (because } F_S \text{ is attached at the pivot)} \\
r_3 &= 30.0\,\text{cm}.
\end{aligned}
$$

Now we can find the five torques with respect to the chosen pivot:

$$
\begin{aligned}
\tau_1 &= +r_1 w_1 \sin 90° = +r_1 m_1 g \quad &\text{(counterclockwise rotation, positive sense)} \\
\tau_2 &= +r_2 w_2 \sin 90° = +r_2 m_2 g \quad &\text{(counterclockwise rotation, positive sense)} \\
\tau &= +rw \sin 90° = +rmg \quad &\text{(gravitational torque)} \\
\tau_S &= r_S F_S \sin \theta_S = 0 \quad &\text{(because } r_S = 0 \text{ cm)} \\
\tau_3 &= -r_3 w_3 \sin 90° = -r_3 m_3 g \quad &\text{(clockwise rotation, negative sense)}
\end{aligned}
$$

The second equilibrium condition (equation for the torques) for the meter stick is

$$\tau_1 + \tau_2 + \tau + \tau_S + \tau_3 = 0.$$

When substituting torque values into this equation, we can omit the torques giving zero contributions. In this way the second equilibrium condition is

$$+r_1 m_1 g + r_2 m_2 g + rmg - r_3 m_3 g = 0. \tag{12.17}$$

Selecting the $+y$-direction to be parallel to $\vec{F}_S$, the first equilibrium condition for the stick is

$$-w_1 - w_2 - w + F_S - w_3 = 0.$$

Substituting the forces, the first equilibrium condition becomes

$$-m_1 g - m_2 g - mg + F_S - m_3 g = 0. \tag{12.18}$$

We solve these equations simultaneously for the unknown values m_3 and F_S. In **Equation 12.17**, we cancel the g factor and rearrange the terms to obtain

$$r_3 m_3 = r_1 m_1 + r_2 m_2 + rm.$$

To obtain m_3 we divide both sides by r_3, so we have

$$
\begin{aligned}
m_3 &= \frac{r_1}{r_3} m_1 + \frac{r_2}{r_3} m_2 + \frac{r}{r_3} m \\
&= \frac{70}{30}(50.0\ \text{g}) + \frac{40}{30}(75.0\ \text{g}) + \frac{20}{30}(150.0\ \text{g}) = 316.0\tfrac{2}{3}\text{g} \simeq 317\ \text{g}.
\end{aligned}
\tag{12.19}
$$

To find the normal reaction force, we rearrange the terms in **Equation 12.18**, converting grams to kilograms:

$$
\begin{aligned}
F_S &= (m_1 + m_2 + m + m_3)g \\
&= (50.0 + 75.0 + 150.0 + 316.7) \times 10^{-3}\ \text{kg} \times 9.8\ \frac{\text{m}}{\text{s}^2} = 5.8\ \text{N}.
\end{aligned}
\tag{12.20}
$$

Significance

Notice that **Equation 12.17** is independent of the value of g. The torque balance may therefore be used to measure mass, since variations in g-values on Earth's surface do not affect these measurements. This is not the case for a spring balance because it measures the force.

 12.3 **Check Your Understanding** Repeat **Example 12.3** using the left end of the meter stick to calculate the torques; that is, by placing the pivot at the left end of the meter stick.

In the next example, we show how to use the first equilibrium condition (equation for forces) in the vector form given by **Equation 12.7** and **Equation 12.8**. We present this solution to illustrate the importance of a suitable choice of reference frame. Although all inertial reference frames are equivalent and numerical solutions obtained in one frame are the same as in any other, an unsuitable choice of reference frame can make the solution quite lengthy and convoluted, whereas a wise choice of reference frame makes the solution straightforward. We show this in the equivalent solution to the same problem. This particular example illustrates an application of static equilibrium to biomechanics.

Example 12.4

Forces in the Forearm

A weightlifter is holding a 50.0-lb weight (equivalent to 222.4 N) with his forearm, as shown in **Figure 12.11**. His forearm is positioned at $\beta = 60°$ with respect to his upper arm. The forearm is supported by a contraction of the biceps muscle, which causes a torque around the elbow. Assuming that the tension in the biceps acts along the vertical direction given by gravity, what tension must the muscle exert to hold the forearm at the position shown? What is the force on the elbow joint? Assume that the forearm's weight is negligible. Give your final answers in SI units.

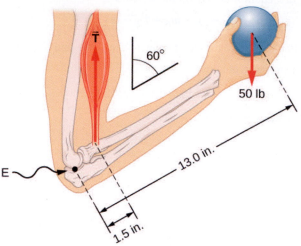

Figure 12.11 The forearm is rotated around the elbow (E) by a contraction of the biceps muscle, which causes tension $\overrightarrow{\mathbf{T}}_M$.

Strategy

We identify three forces acting on the forearm: the unknown force $\overrightarrow{\mathbf{F}}$ at the elbow; the unknown tension $\overrightarrow{\mathbf{T}}_M$ in the muscle; and the weight $\overrightarrow{\mathbf{w}}$ with magnitude $w = 50\,\text{lb}$. We adopt the frame of reference with the x-axis along the forearm and the pivot at the elbow. The vertical direction is the direction of the weight, which is the same as the direction of the upper arm. The x-axis makes an angle $\beta = 60°$ with the vertical. The y-axis is perpendicular to the x-axis. Now we set up the free-body diagram for the forearm. First, we draw the axes, the pivot, and the three vectors representing the three identified forces. Then we locate the angle β and represent each force by its x- and y-components, remembering to cross out the original force vector to avoid double counting. Finally, we label the forces and their lever arms. The free-body diagram for the forearm is shown in **Figure 12.12**. At this point, we are ready to set up equilibrium conditions for the forearm. Each force has x- and y-components; therefore, we have two equations for the first equilibrium condition, one equation for each component of the net force acting on the forearm.

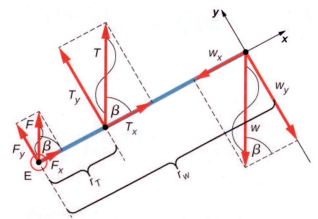

Figure 12.12 Free-body diagram for the forearm: The pivot is located at point E (elbow).

Notice that in our frame of reference, contributions to the second equilibrium condition (for torques) come only from the y-components of the forces because the x-components of the forces are all parallel to their lever arms, so that for any of them we have $\sin\theta = 0$ in **Equation 12.10**. For the y-components we have $\theta = \pm 90°$ in **Equation 12.10**. Also notice that the torque of the force at the elbow is zero because this force is attached at the pivot. So the contribution to the net torque comes only from the torques of T_y and of w_y.

Solution

We see from the free-body diagram that the x-component of the net force satisfies the equation

$$+F_x + T_x - w_x = 0 \tag{12.21}$$

and the y-component of the net force satisfies

$$+F_y + T_y - w_y = 0. \tag{12.22}$$

Equation 12.21 and **Equation 12.22** are two equations of the first equilibrium condition (for forces). Next, we read from the free-body diagram that the net torque along the axis of rotation is

$$+r_T T_y - r_w w_y = 0. \tag{12.23}$$

Equation 12.23 is the second equilibrium condition (for torques) for the forearm. The free-body diagram shows that the lever arms are $r_T = 1.5\ \text{in.}$ and $r_w = 13.0\ \text{in.}$ At this point, we do not need to convert inches into SI units, because as long as these units are consistent in **Equation 12.23**, they cancel out. Using the free-body diagram again, we find the magnitudes of the component forces:

$$
\begin{aligned}
F_x &= F\cos\beta = F\cos 60° = F/2 \\
T_x &= T\cos\beta = T\cos 60° = T/2 \\
w_x &= w\cos\beta = w\cos 60° = w/2 \\
F_y &= F\sin\beta = F\sin 60° = F\sqrt{3}/2 \\
T_y &= T\sin\beta = T\sin 60° = T\sqrt{3}/2 \\
w_y &= w\sin\beta = w\sin 60° = w\sqrt{3}/2.
\end{aligned}
$$

We substitute these magnitudes into **Equation 12.21**, **Equation 12.22**, and **Equation 12.23** to obtain, respectively,

$$
\begin{aligned}
F/2 + T/2 - w/2 &= 0 \\
F\sqrt{3}/2 + T\sqrt{3}/2 - w\sqrt{3}/2 &= 0 \\
r_T T\sqrt{3}/2 - r_w w\sqrt{3}/2 &= 0.
\end{aligned}
$$

When we simplify these equations, we see that we are left with only two independent equations for the two unknown force magnitudes, F and T, because **Equation 12.21** for the x-component is equivalent to **Equation 12.22** for the y-component. In this way, we obtain the first equilibrium condition for forces

$$F + T - w = 0 \tag{12.24}$$

and the second equilibrium condition for torques

$$r_T T - r_w w = 0. \tag{12.25}$$

The magnitude of tension in the muscle is obtained by solving **Equation 12.25**:

$$T = \frac{r_w}{r_T} w = \frac{13.0}{1.5}(50\ \text{lb}) = 433\tfrac{1}{3}\text{lb} \simeq 433.3\ \text{lb}.$$

The force at the elbow is obtained by solving **Equation 12.24**:

$$F = w - T = 50.0\ \text{lb} - 433.3\ \text{lb} = -383.3\ \text{lb}.$$

The negative sign in the equation tells us that the actual force at the elbow is antiparallel to the working direction adopted for drawing the free-body diagram. In the final answer, we convert the forces into SI units of force. The

answer is

$$F = 383.3 \text{ lb} = 383.3(4.448 \text{ N}) = 1705 \text{ N downward}$$
$$T = 433.3 \text{ lb} = 433.3(4.448 \text{ N}) = 1927 \text{ N upward}.$$

Significance

Two important issues here are worth noting. The first concerns conversion into SI units, which can be done at the very end of the solution as long as we keep consistency in units. The second important issue concerns the hinge joints such as the elbow. In the initial analysis of a problem, hinge joints should always be assumed to exert a force in an *arbitrary direction*, and then you must solve for all components of a hinge force independently. In this example, the elbow force happens to be vertical because the problem assumes the tension by the biceps to be vertical as well. Such a simplification, however, is not a general rule.

Solution

Suppose we adopt a reference frame with the direction of the y-axis along the 50-lb weight and the pivot placed at the elbow. In this frame, all three forces have only y-components, so we have only one equation for the first equilibrium condition (for forces). We draw the free-body diagram for the forearm as shown in **Figure 12.13**, indicating the pivot, the acting forces and their lever arms with respect to the pivot, and the angles θ_T and θ_w that the forces $\vec{T}_M$ and $\vec{w}$ (respectively) make with their lever arms. In the definition of torque given by **Equation 12.10**, the angle θ_T is the direction angle of the vector $\vec{T}_M$, counted *counterclockwise* from the radial direction of the lever arm that always points away from the pivot. By the same convention, the angle θ_w is measured *counterclockwise* from the radial direction of the lever arm to the vector $\vec{w}$. Done this way, the non-zero torques are most easily computed by directly substituting into **Equation 12.10** as follows:

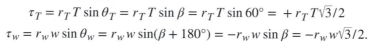

$$\tau_T = r_T T \sin\theta_T = r_T T \sin\beta = r_T T \sin 60° = + r_T T \sqrt{3}/2$$
$$\tau_w = r_w w \sin\theta_w = r_w w \sin(\beta + 180°) = -r_w w \sin\beta = -r_w w\sqrt{3}/2.$$

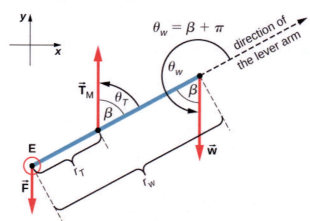

Figure 12.13 Free-body diagram for the forearm for the equivalent solution. The pivot is located at point E (elbow).

The second equilibrium condition, $\tau_T + \tau_w = 0$, can be now written as

$$r_T T\sqrt{3}/2 - r_w w\sqrt{3}/2 = 0. \qquad\qquad (12.26)$$

From the free-body diagram, the first equilibrium condition (for forces) is

$$-F + T - w = 0. \qquad\qquad (12.27)$$

Equation 12.26 is identical to **Equation 12.25** and gives the result $T = 433.3$ lb. **Equation 12.27** gives

$$F = T - w = 433.3 \text{ lb} - 50.0 \text{ lb} = 383.3 \text{ lb}.$$

We see that these answers are identical to our previous answers, but the second choice for the frame of reference leads to an equivalent solution that is simpler and quicker because it does not require that the forces be resolved into their rectangular components.

 12.4   Check Your Understanding Repeat **Example 12.4** assuming that the forearm is an object of uniform density that weighs 8.896 N.

Example 12.5

A Ladder Resting Against a Wall

A uniform ladder is $L = 5.0 \, \text{m}$ long and weighs 400.0 N. The ladder rests against a slippery vertical wall, as shown in **Figure 12.14**. The inclination angle between the ladder and the rough floor is $\beta = 53°$. Find the reaction forces from the floor and from the wall on the ladder and the coefficient of static friction μ_s at the interface of the ladder with the floor that prevents the ladder from slipping.

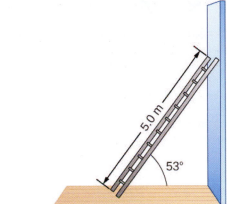

Figure 12.14 A 5.0-m-long ladder rests against a frictionless wall.

Strategy

We can identify four forces acting on the ladder. The first force is the normal reaction force N from the floor in the upward vertical direction. The second force is the static friction force $f = \mu_s N$ directed horizontally along the floor toward the wall—this force prevents the ladder from slipping. These two forces act on the ladder at its contact point with the floor. The third force is the weight w of the ladder, attached at its CM located midway between its ends. The fourth force is the normal reaction force F from the wall in the horizontal direction away from the wall, attached at the contact point with the wall. There are no other forces because the wall is slippery, which means there is no friction between the wall and the ladder. Based on this analysis, we adopt the frame of reference with the y-axis in the vertical direction (parallel to the wall) and the x-axis in the horizontal direction (parallel to the floor). In this frame, each force has either a horizontal component or a vertical component but not both, which simplifies the solution. We select the pivot at the contact point with the floor. In the free-body diagram for the ladder, we indicate the pivot, all four forces and their lever arms, and the angles between lever arms and the forces, as shown in **Figure 12.15**. With our choice of the pivot location, there is no torque either from the normal reaction force N or from the static friction f because they both act at the pivot.

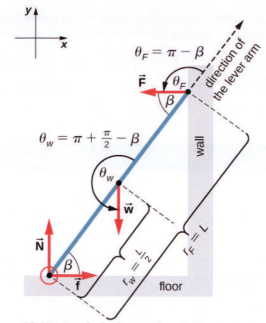

Figure 12.15 Free-body diagram for a ladder resting against a frictionless wall.

Solution

From the free-body diagram, the net force in the x-direction is

$$+f - F = 0 \tag{12.28}$$

the net force in the y-direction is

$$+N - w = 0 \tag{12.29}$$

and the net torque along the rotation axis at the pivot point is

$$\tau_w + \tau_F = 0. \tag{12.30}$$

where τ_w is the torque of the weight w and τ_F is the torque of the reaction F. From the free-body diagram, we identify that the lever arm of the reaction at the wall is $r_F = L = 5.0\,\text{m}$ and the lever arm of the weight is $r_w = L/2 = 2.5\,\text{m}$. With the help of the free-body diagram, we identify the angles to be used in **Equation 12.10** for torques: $\theta_F = 180° - \beta$ for the torque from the reaction force with the wall, and $\theta_w = 180° + (90° - \beta)$ for the torque due to the weight. Now we are ready to use **Equation 12.10** to compute torques:

$$\tau_w = r_w\,w\,\sin\theta_w = r_w\,w\,\sin(180° + 90° - \beta) = -\frac{L}{2}w\,\sin(90° - \beta) = -\frac{L}{2}w\cos\beta$$

$$\tau_F = r_F\,F\,\sin\theta_F = r_F\,F\,\sin(180° - \beta) = LF\sin\beta.$$

We substitute the torques into **Equation 12.30** and solve for F :

$$-\frac{L}{2}w\cos\beta + LF\sin\beta \;=\; 0 \tag{12.31}$$

$$F = \frac{w}{2}\cot\beta = \frac{400.0\,\text{N}}{2}\cot 53° \;=\; 150.7\,\text{N}$$

We obtain the normal reaction force with the floor by solving **Equation 12.29**: $N = w = 400.0\,\text{N}$. The magnitude of friction is obtained by solving **Equation 12.28**: $f = F = 150.7\,\text{N}$. The coefficient of static friction is $\mu_s = f/N = 150.7/400.0 = 0.377$.

The net force on the ladder at the contact point with the floor is the vector sum of the normal reaction from the floor and the static friction forces:

$$\vec{F}_{\text{floor}} = \vec{f} + \vec{N} = (150.7\,\text{N})(-\hat{i}) + (400.0\,\text{N})(+\hat{j}) = (-150.7\,\hat{i} + 400.0\,\hat{j})\,\text{N}.$$

Its magnitude is

$$F_{\text{floor}} = \sqrt{f^2 + N^2} = \sqrt{150.7^2 + 400.0^2}\,\text{N} = 427.4\,\text{N}$$

and its direction is

$$\varphi = \tan^{-1}(N/f) = \tan^{-1}(400.0/150.7) = 69.3°\ \text{above the floor.}$$

We should emphasize here two general observations of practical use. First, notice that when we choose a pivot point, there is no expectation that the system will actually pivot around the chosen point. The ladder in this example is not rotating at all but firmly stands on the floor; nonetheless, its contact point with the floor is a good choice for the pivot. Second, notice when we use **Equation 12.10** for the computation of individual torques, we do not need to resolve the forces into their normal and parallel components with respect to the direction of the lever arm, and we do not need to consider a sense of the torque. As long as the angle in **Equation 12.10** is correctly identified—with the help of a free-body diagram—as the angle measured counterclockwise from the direction of the lever arm to the direction of the force vector, **Equation 12.10** gives both the magnitude and the sense of the torque. This is because torque is the vector product of the lever-arm vector crossed with the force vector, and **Equation 12.10** expresses the rectangular component of this vector product along the axis of rotation.

Significance

This result is independent of the length of the ladder because L is cancelled in the second equilibrium condition, **Equation 12.31**. No matter how long or short the ladder is, as long as its weight is 400 N and the angle with the floor is $53°$, our results hold. But the ladder will slip if the net torque becomes negative in **Equation 12.31**. This happens for some angles when the coefficient of static friction is not great enough to prevent the ladder from slipping.

12.5 Check Your Understanding For the situation described in **Example 12.5**, determine the values of the coefficient μ_s of static friction for which the ladder starts slipping, given that β is the angle that the ladder makes with the floor.

Example 12.6

Forces on Door Hinges

A swinging door that weighs $w = 400.0\,\text{N}$ is supported by hinges A and B so that the door can swing about a vertical axis passing through the hinges **Figure 12.16**. The door has a width of $b = 1.00\,\text{m}$, and the door slab has a uniform mass density. The hinges are placed symmetrically at the door's edge in such a way that the door's weight is evenly distributed between them. The hinges are separated by distance $a = 2.00\,\text{m}$. Find the forces on the hinges when the door rests half-open.

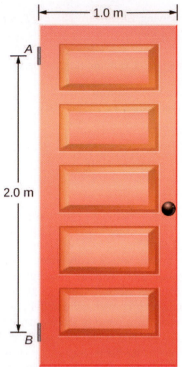

Figure 12.16 A 400-N swinging vertical door is supported by two hinges attached at points A and B.

Strategy

The forces that the door exerts on its hinges can be found by simply reversing the directions of the forces that the hinges exert on the door. Hence, our task is to find the forces from the hinges on the door. Three forces act on the door slab: an unknown force $\vec{A}$ from hinge A, an unknown force $\vec{B}$ from hinge B, and the known weight $\vec{w}$ attached at the center of mass of the door slab. The CM is located at the geometrical center of the door because the slab has a uniform mass density. We adopt a rectangular frame of reference with the y-axis along the direction of gravity and the x-axis in the plane of the slab, as shown in panel (a) of **Figure 12.17**, and resolve all forces into their rectangular components. In this way, we have four unknown component forces: two components of force $\vec{A}$ (A_x and A_y), and two components of force $\vec{B}$ (B_x and B_y). In the free-body diagram, we represent the two forces at the hinges by their vector components, whose assumed orientations are arbitrary. Because there are four unknowns (A_x, B_x, A_y, and B_y), we must set up four independent equations. One equation is the equilibrium condition for forces in the x-direction. The second equation is the equilibrium condition for forces in the y-direction. The third equation is the equilibrium condition for torques in rotation about a hinge. Because the weight is evenly distributed between the hinges, we have the fourth equation, $A_y = B_y$. To set up the equilibrium conditions, we draw a free-body diagram and choose the pivot point at the upper hinge, as shown in panel (b) of **Figure 12.17**. Finally, we solve the equations for the unknown force components and find the forces.

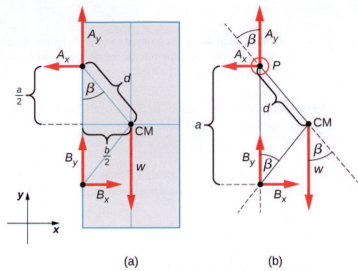

Figure 12.17 (a) Geometry and (b) free-body diagram for the door.

Solution

From the free-body diagram for the door we have the first equilibrium condition for forces:

in x-direction: $-A_x + B_x = 0 \implies A_x = B_x$

in y-direction: $+A_y + B_y - w = 0 \implies A_y = B_y = \dfrac{w}{2} = \dfrac{400.0\,\text{N}}{2} = 200.0\,\text{N}.$

We select the pivot at point P (upper hinge, per the free-body diagram) and write the second equilibrium condition for torques in rotation about point P:

$$\text{pivot at } P: \tau_w + \tau_{Bx} + \tau_{By} = 0. \tag{12.32}$$

We use the free-body diagram to find all the terms in this equation:

$$\tau_w = dw\sin(-\beta) = -dw\sin\beta = -dw\dfrac{b/2}{d} = -w\dfrac{b}{2}$$
$$\tau_{Bx} = aB_x\sin 90° = +aB_x$$
$$\tau_{By} = aB_y\sin 180° = 0.$$

In evaluating $\sin\beta$, we use the geometry of the triangle shown in part (a) of the figure. Now we substitute these torques into **Equation 12.32** and compute B_x:

$$\text{pivot at } P: -w\dfrac{b}{2} + aB_x = 0 \implies B_x = w\dfrac{b}{2a} = (400.0\,\text{N})\dfrac{1}{2 \cdot 2} = 100.0\,\text{N}.$$

Therefore the magnitudes of the horizontal component forces are $A_x = B_x = 100.0\,\text{N}.$ The forces on the door are

at the upper hinge: $\vec{F}_{A \text{ on door}} = -100.0\,\text{N}\,\hat{i} + 200.0\,\text{N}\,\hat{j}$

at the lower hinge: $\vec{F}_{B \text{ on door}} = +100.0\,\text{N}\,\hat{i} + 200.0\,\text{N}\,\hat{j}.$

The forces on the hinges are found from Newton's third law as

on the upper hinge: $\vec{F}_{\text{door on } A} = 100.0\,\text{N}\,\hat{i} - 200.0\,\text{N}\,\hat{j}$

on the lower hinge: $\vec{F}_{\text{door on } B} = -100.0\,\text{N}\,\hat{i} - 200.0\,\text{N}\,\hat{j}.$

Significance

Note that if the problem were formulated without the assumption of the weight being equally distributed between the two hinges, we wouldn't be able to solve it because the number of the unknowns would be greater than the number of equations expressing equilibrium conditions.

 12.6 **Check Your Understanding** Solve the problem in **Example 12.6** by taking the pivot position at the center of mass.

 12.7 **Check Your Understanding** A 50-kg person stands 1.5 m away from one end of a uniform 6.0-m-long scaffold of mass 70.0 kg. Find the tensions in the two vertical ropes supporting the scaffold.

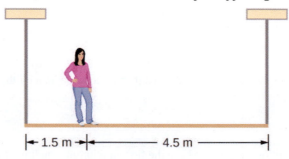

 12.8 **Check Your Understanding** A 400.0-N sign hangs from the end of a uniform strut. The strut is 4.0 m long and weighs 600.0 N. The strut is supported by a hinge at the wall and by a cable whose other end is tied to the wall at a point 3.0 m above the left end of the strut. Find the tension in the supporting cable and the force of the hinge on the strut.

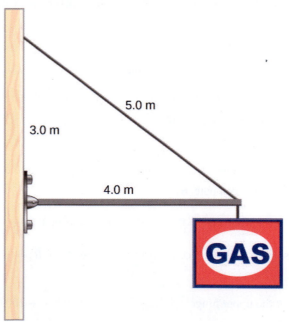

12.3 | Stress, Strain, and Elastic Modulus

Learning Objectives

By the end of this section, you will be able to:

* Explain the concepts of stress and strain in describing elastic deformations of materials
* Describe the types of elastic deformation of objects and materials

A model of a rigid body is an idealized example of an object that does not deform under the actions of external forces. It is very useful when analyzing mechanical systems—and many physical objects are indeed rigid to a great extent. The extent to which an object can be *perceived* as rigid depends on the physical properties of the material from which it is made. For example, a ping-pong ball made of plastic is brittle, and a tennis ball made of rubber is elastic when acted upon by squashing forces. However, under other circumstances, both a ping-pong ball and a tennis ball may bounce well as rigid bodies. Similarly, someone who designs prosthetic limbs may be able to approximate the mechanics of human limbs by modeling them as rigid bodies; however, the actual combination of bones and tissues is an elastic medium.

For the remainder of this chapter, we move from consideration of forces that affect the motion of an object to those that affect an object's shape. A change in shape due to the application of a force is known as a deformation. Even very small forces are known to cause some deformation. Deformation is experienced by objects or physical media under the action of external forces—for example, this may be squashing, squeezing, ripping, twisting, shearing, or pulling the objects apart. In the language of physics, two terms describe the forces on objects undergoing deformation: *stress* and *strain*.

Stress is a quantity that describes the magnitude of forces that cause deformation. Stress is generally defined as *force per unit area*. When forces pull on an object and cause its elongation, like the stretching of an elastic band, we call such stress a **tensile stress**. When forces cause a compression of an object, we call it a **compressive stress**. When an object is being squeezed from all sides, like a submarine in the depths of an ocean, we call this kind of stress a **bulk stress** (or **volume stress**). In other situations, the acting forces may be neither tensile nor compressive, and still produce a noticeable deformation. For example, suppose you hold a book tightly between the palms of your hands, then with one hand you press-and-pull on the front cover away from you, while with the other hand you press-and-pull on the back cover toward you. In such a case, when deforming forces act tangentially to the object's surface, we call them 'shear' forces and the stress they cause is called **shear stress**.

The SI unit of stress is the pascal (Pa). When one newton of force presses on a unit surface area of one meter squared, the resulting stress is one pascal:

$$\text{one pascal} = 1.0\,\text{Pa} = \frac{1.0\,\text{N}}{1.0\,\text{m}^2}.$$

In the British system of units, the unit of stress is 'psi,' which stands for 'pound per square inch' $\left(\text{lb/in}^2\right)$. Another unit that is often used for bulk stress is the atm (atmosphere). Conversion factors are

$$1\,\text{psi} = 6895\,\text{Pa} \quad \text{and} \quad 1\,\text{Pa} = 1.450 \times 10^{-4}\,\text{psi}$$

$$1\,\text{atm} = 1.013 \times 10^5\,\text{Pa} = 14.7\,\text{psi}.$$

An object or medium under stress becomes deformed. The quantity that describes this deformation is called **strain**. Strain is given as a fractional change in either length (under tensile stress) or volume (under bulk stress) or geometry (under shear stress). Therefore, strain is a dimensionless number. Strain under a tensile stress is called **tensile strain**, strain under bulk stress is called **bulk strain** (or **volume strain**), and that caused by shear stress is called **shear strain**.

The greater the stress, the greater the strain; however, the relation between strain and stress does not need to be linear. Only when stress is sufficiently low is the deformation it causes in direct proportion to the stress value. The proportionality constant in this relation is called the **elastic modulus**. In the linear limit of low stress values, the general relation between stress and strain is

stress = (elastic modulus) × strain. **(12.33)**

As we can see from dimensional analysis of this relation, the elastic modulus has the same physical unit as stress because strain is dimensionless.

We can also see from **Equation 12.33** that when an object is characterized by a large value of elastic modulus, the effect of stress is small. On the other hand, a small elastic modulus means that stress produces large strain and noticeable deformation. For example, a stress on a rubber band produces larger strain (deformation) than the same stress on a steel band of the same dimensions because the elastic modulus for rubber is two orders of magnitude smaller than the elastic modulus for steel.

The elastic modulus for tensile stress is called **Young's modulus**; that for the bulk stress is called the **bulk modulus**; and that for shear stress is called the **shear modulus**. Note that the relation between stress and strain is an *observed* relation, measured in the laboratory. Elastic moduli for various materials are measured under various physical conditions, such as varying temperature, and collected in engineering data tables for reference (**Table 12.1**). These tables are valuable references for industry and for anyone involved in engineering or construction. In the next section, we discuss strain-stress relations beyond the linear limit represented by **Equation 12.33**, in the full range of stress values up to a fracture point. In the remainder of this section, we study the linear limit expressed by **Equation 12.33**.

Material	Young's modulus $\times 10^{10}\,\text{Pa}$	Bulk modulus $\times 10^{10}\,\text{Pa}$	Shear modulus $\times 10^{10}\,\text{Pa}$
Aluminum	7.0	7.5	2.5
Bone (tension)	1.6	0.8	8.0
Bone (compression)	0.9		
Brass	9.0	6.0	3.5
Brick	1.5		
Concrete	2.0		
Copper	11.0	14.0	4.4
Crown glass	6.0	5.0	2.5
Granite	4.5	4.5	2.0
Hair (human)	1.0		
Hardwood	1.5		1.0
Iron	21.0	16.0	7.7
Lead	1.6	4.1	0.6
Marble	6.0	7.0	2.0
Nickel	21.0	17.0	7.8
Polystyrene	3.0		
Silk	6.0		
Spider thread	3.0		
Steel	20.0	16.0	7.5
Acetone		0.07	
Ethanol		0.09	
Glycerin		0.45	
Mercury		2.5	
Water		0.22	

Table 12.1 Approximate Elastic Moduli for Selected Materials

Tensile or Compressive Stress, Strain, and Young's Modulus

Tension or compression occurs when two antiparallel forces of equal magnitude act on an object along only one of its dimensions, in such a way that the object does not move. One way to envision such a situation is illustrated in **Figure 12.18**. A rod segment is either stretched or squeezed by a pair of forces acting along its length and perpendicular to its cross-section. The net effect of such forces is that the rod changes its length from the original length L_0 that it had before the forces appeared, to a new length L that it has under the action of the forces. This change in length $\Delta L = L - L_0$ may be either elongation (when L is larger than the original length L_0) or contraction (when L is smaller than the original length L_0). Tensile stress and strain occur when the forces are stretching an object, causing its elongation, and the length change ΔL is positive. Compressive stress and strain occur when the forces are contracting an object, causing its shortening, and the length change ΔL is negative.

In either of these situations, we define stress as the ratio of the deforming force $F_\perp$ to the cross-sectional area A of the object being deformed. The symbol $F_\perp$ that we reserve for the deforming force means that this force acts perpendicularly to the cross-section of the object. Forces that act parallel to the cross-section do not change the length of an object. The definition of the tensile stress is

$$\text{tensile stress} = \frac{F_\perp}{A}. \qquad (12.34)$$

Tensile strain is the measure of the deformation of an object under tensile stress and is defined as the fractional change of the object's length when the object experiences tensile stress

$$\text{tensile strain} = \frac{\Delta L}{L_0}. \qquad (12.35)$$

Compressive stress and strain are defined by the same formulas, **Equation 12.34** and **Equation 12.35**, respectively. The only difference from the tensile situation is that for compressive stress and strain, we take absolute values of the right-hand sides in **Equation 12.34** and **Equation 12.35**.

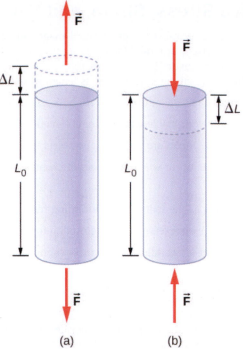

Figure 12.18 When an object is in either tension or compression, the net force on it is zero, but the object deforms by changing its original length L_0. (a) Tension: The rod is elongated by ΔL. (b) Compression: The rod is contracted by ΔL. In both cases, the deforming force acts along the length of the rod and perpendicular to its cross-section. In the linear range of low stress, the cross-sectional area of the rod does not change.

Young's modulus Y is the elastic modulus when deformation is caused by either tensile or compressive stress, and is defined by **Equation 12.33**. Dividing this equation by tensile strain, we obtain the expression for Young's modulus:

$$Y = \frac{\text{tensile stress}}{\text{tensile strain}} = \frac{F_\perp /A}{\Delta L/L_0} = \frac{F_\perp}{A}\frac{L_0}{\Delta L}. \qquad (12.36)$$

Example 12.7

Compressive Stress in a Pillar

A sculpture weighing 10,000 N rests on a horizontal surface at the top of a 6.0-m-tall vertical pillar **Figure 12.19**. The pillar's cross-sectional area is 0.20 m^2 and it is made of granite with a mass density of 2700 kg/m^3.

Find the compressive stress at the cross-section located 3.0 m below the top of the pillar and the value of the compressive strain of the top 3.0-m segment of the pillar.

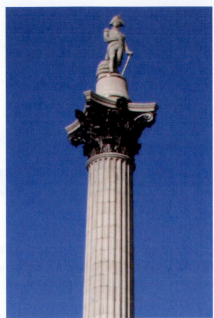

Figure 12.19 Nelson's Column in Trafalgar Square, London, England. (credit: modification of work by Cristian Bortes)

Strategy

First we find the weight of the 3.0-m-long top section of the pillar. The normal force that acts on the cross-section located 3.0 m down from the top is the sum of the pillar's weight and the sculpture's weight. Once we have the normal force, we use **Equation 12.34** to find the stress. To find the compressive strain, we find the value of Young's modulus for granite in **Table 12.1** and invert **Equation 12.36**.

Solution

The volume of the pillar segment with height $h = 3.0 \, \text{m}$ and cross-sectional area $A = 0.20 \, \text{m}^2$ is

$$V = Ah = (0.20 \, \text{m}^2)(3.0 \, \text{m}) = 0.60 \, \text{m}^3.$$

With the density of granite $\rho = 2.7 \times 10^3 \, \text{kg/m}^3$, the mass of the pillar segment is

$$m = \rho V = (2.7 \times 10^3 \, \text{kg/m}^3)(0.60 \, \text{m}^3) = 1.60 \times 10^3 \, \text{kg}.$$

The weight of the pillar segment is

$$w_p = mg = (1.60 \times 10^3 \, \text{kg})(9.80 \, \text{m/s}^2) = 1.568 \times 10^4 \, \text{N}.$$

The weight of the sculpture is $w_s = 1.0 \times 10^4 \, \text{N}$, so the normal force on the cross-sectional surface located 3.0 m below the sculpture is

$$F_\perp = w_p + w_s = (1.568 + 1.0) \times 10^4 \, \text{N} = 2.568 \times 10^4 \, \text{N}.$$

Therefore, the stress is

$$\text{stress} = \frac{F_\perp}{A} = \frac{2.568 \times 10^4 \, \text{N}}{0.20 \, \text{m}^2} = 1.284 \times 10^5 \, \text{Pa} = 128.4 \, \text{kPa}.$$

Young's modulus for granite is $Y = 4.5 \times 10^{10} \, \text{Pa} = 4.5 \times 10^7 \, \text{kPa}$. Therefore, the compressive strain at this position is

$$\text{strain} = \frac{\text{stress}}{Y} = \frac{128.4 \, \text{kPa}}{4.5 \times 10^7 \, \text{kPa}} = 2.85 \times 10^{-6}.$$

Significance

Notice that the normal force acting on the cross-sectional area of the pillar is not constant along its length, but varies from its smallest value at the top to its largest value at the bottom of the pillar. Thus, if the pillar has a uniform cross-sectional area along its length, the stress is largest at its base.

 12.9   Check Your Understanding Find the compressive stress and strain at the base of Nelson's column.

Example 12.8

Stretching a Rod

A 2.0-m-long steel rod has a cross-sectional area of $0.30\,\text{cm}^2$. The rod is a part of a vertical support that holds a heavy 550-kg platform that hangs attached to the rod's lower end. Ignoring the weight of the rod, what is the tensile stress in the rod and the elongation of the rod under the stress?

Strategy

First we compute the tensile stress in the rod under the weight of the platform in accordance with **Equation 12.34**. Then we invert **Equation 12.36** to find the rod's elongation, using $L_0 = 2.0\,\text{m}$. From **Table 12.1**, Young's modulus for steel is $Y = 2.0 \times 10^{11}\,\text{Pa}$.

Solution

Substituting numerical values into the equations gives us

$$\frac{F_\perp}{A} = \frac{(550\,\text{kg})(9.8\,\text{m/s}^2)}{3.0 \times 10^{-5}\,\text{m}^2} = 1.8 \times 10^8\,\text{Pa}$$

$$\Delta L = \frac{F_\perp}{A}\frac{L_0}{Y} = (1.8 \times 10^8\,\text{Pa})\frac{2.0\,\text{m}}{2.0 \times 10^{11}\,\text{Pa}} = 1.8 \times 10^{-3}\,\text{m} = 1.8\,\text{mm}.$$

Significance

Similarly as in the example with the column, the tensile stress in this example is not uniform along the length of the rod. Unlike in the previous example, however, if the weight of the rod is taken into consideration, the stress in the rod is largest at the top and smallest at the bottom of the rod where the equipment is attached.

 12.10   Check Your Understanding A 2.0-m-long wire stretches 1.0 mm when subjected to a load. What is the tensile strain in the wire?

Objects can often experience both compressive stress and tensile stress simultaneously **Figure 12.20**. One example is a long shelf loaded with heavy books that sags between the end supports under the weight of the books. The top surface of the shelf is in compressive stress and the bottom surface of the shelf is in tensile stress. Similarly, long and heavy beams sag under their own weight. In modern building construction, such bending strains can be almost eliminated with the use of I-beams **Figure 12.21**.

Figure 12.20 (a) An object bending downward experiences tensile stress (stretching) in the upper section and compressive stress (compressing) in the lower section. (b) Elite weightlifters often bend iron bars temporarily during lifting, as in the 2012 Olympics competition. (credit b: modification of work by Oleksandr Kocherzhenko)

Figure 12.21 Steel I-beams are used in construction to reduce bending strains. (credit: modification of work by "US Army Corps of Engineers Europe District"/Flickr)

A heavy box rests on a table supported by three columns. View this **demonstration (https://openstaxcollege.org/l/21movebox)** to move the box to see how the compression (or tension) in the columns is affected when the box changes its position.

Bulk Stress, Strain, and Modulus

When you dive into water, you feel a force pressing on every part of your body from all directions. What you are experiencing then is bulk stress, or in other words, **pressure**. Bulk stress always tends to decrease the volume enclosed by the surface of a submerged object. The forces of this "squeezing" are always perpendicular to the submerged surface **Figure 12.22**. The effect of these forces is to decrease the volume of the submerged object by an amount ΔV compared with the volume V_0 of the object in the absence of bulk stress. This kind of deformation is called bulk strain and is described by a change in volume relative to the original volume:

$$\text{bulk strain} = \frac{\Delta V}{V_0}. \tag{12.37}$$

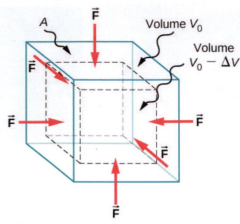

Figure 12.22 An object under increasing bulk stress always undergoes a decrease in its volume. Equal forces perpendicular to the surface act from all directions. The effect of these forces is to decrease the volume by the amount ΔV compared to the original volume, V_0.

The bulk strain results from the bulk stress, which is a force $F_\perp$ normal to a surface that presses on the unit surface area A of a submerged object. This kind of physical quantity, or pressure p, is defined as

$$\text{pressure} = p \equiv \frac{F_\perp}{A}. \tag{12.38}$$

We will study pressure in fluids in greater detail in **Fluid Mechanics**. An important characteristic of pressure is that it is a scalar quantity and does not have any particular direction; that is, pressure acts equally in all possible directions. When you submerge your hand in water, you sense the same amount of pressure acting on the top surface of your hand as on the bottom surface, or on the side surface, or on the surface of the skin between your fingers. What you are perceiving in this case is an increase in pressure Δp over what you are used to feeling when your hand is not submerged in water. What you feel when your hand is not submerged in the water is the **normal pressure** p_0 of one atmosphere, which serves as a reference point. The bulk stress is this increase in pressure, or Δp, over the normal level, p_0.

When the bulk stress increases, the bulk strain increases in response, in accordance with **Equation 12.33**. The proportionality constant in this relation is called the bulk modulus, B, or

$$B = \frac{\text{bulk stress}}{\text{bulk strain}} = -\frac{\Delta p}{\Delta V / V_0} = -\Delta p \frac{V_0}{\Delta V}. \tag{12.39}$$

The minus sign that appears in **Equation 12.39** is for consistency, to ensure that B is a positive quantity. Note that the minus sign $(-)$ is necessary because an increase Δp in pressure (a positive quantity) always causes a decrease ΔV in volume, and decrease in volume is a negative quantity. The reciprocal of the bulk modulus is called **compressibility** k, or

$$k = \frac{1}{B} = -\frac{\Delta V / V_0}{\Delta p}. \tag{12.40}$$

The term 'compressibility' is used in relation to fluids (gases and liquids). Compressibility describes the change in the volume of a fluid per unit increase in pressure. Fluids characterized by a large compressibility are relatively easy to compress. For example, the compressibility of water is 4.64×10^{-5}/atm and the compressibility of acetone is 1.45×10^{-4}/atm. This means that under a 1.0-atm increase in pressure, the relative decrease in volume is approximately three times as large for acetone as it is for water.

Example 12.9

Hydraulic Press

In a hydraulic press **Figure 12.23**, a 250-liter volume of oil is subjected to a 2300-psi pressure increase. If the compressibility of oil is 2.0×10^{-5} / atm, find the bulk strain and the absolute decrease in the volume of oil when the press is operating.

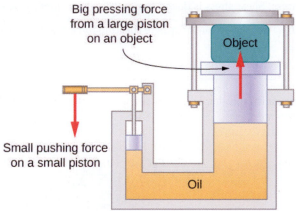

Figure 12.23 In a hydraulic press, when a small piston is displaced downward, the pressure in the oil is transmitted throughout the oil to the large piston, causing the large piston to move upward. A small force applied to a small piston causes a large pressing force, which the large piston exerts on an object that is either lifted or squeezed. The device acts as a mechanical lever.

Strategy

We must invert **Equation 12.40** to find the bulk strain. First, we convert the pressure increase from psi to atm, $\Delta p = 2300 \, \text{psi} = 2300/14.7 \, \text{atm} \approx 160 \, \text{atm}$, and identify $V_0 = 250 \, \text{L}$.

Solution

Substituting values into the equation, we have

$$\text{bulk strain} = \frac{\Delta V}{V_0} = \frac{\Delta p}{B} = k\Delta p = (2.0 \times 10^{-5}/\text{atm})(160 \, \text{atm}) = 0.0032$$

$$\text{answer:} \quad \Delta V = 0.0032 \, V_0 = 0.0032(250 \, \text{L}) = 0.78 \, \text{L}.$$

Significance

Notice that since the compressibility of water is 2.32 times larger than that of oil, if the working substance in the hydraulic press of this problem were changed to water, the bulk strain as well as the volume change would be 2.32 times larger.

 12.11 **Check Your Understanding** If the normal force acting on each face of a cubical 1.0-m^3 piece of steel is changed by $1.0 \times 10^7 \, \text{N}$, find the resulting change in the volume of the piece of steel.

Shear Stress, Strain, and Modulus

The concepts of shear stress and strain concern only solid objects or materials. Buildings and tectonic plates are examples of objects that may be subjected to shear stresses. In general, these concepts do not apply to fluids.

Shear deformation occurs when two antiparallel forces of equal magnitude are applied tangentially to opposite surfaces of a solid object, causing no deformation in the transverse direction to the line of force, as in the typical example of shear stress illustrated in **Figure 12.24**. Shear deformation is characterized by a gradual shift Δx of layers in the direction tangent to the acting forces. This gradation in Δx occurs in the transverse direction along some distance L_0. Shear strain is defined by the ratio of the largest displacement Δx to the transverse distance L_0

$$\text{shear strain} = \frac{\Delta x}{L_0}. \tag{12.41}$$

Shear strain is caused by shear stress. Shear stress is due to forces that act *parallel* to the surface. We use the symbol $F_{\parallel}$ for such forces. The magnitude $F_{\parallel}$ per surface area A where shearing force is applied is the measure of shear stress

$$\text{shear stress} = \frac{F_{\parallel}}{A}. \tag{12.42}$$

The shear modulus is the proportionality constant in **Equation 12.33** and is defined by the ratio of stress to strain. Shear modulus is commonly denoted by S:

$$S = \frac{\text{shear stress}}{\text{shear strain}} = \frac{F_{\parallel} / A}{\Delta x / L_0} = \frac{F_{\parallel}}{A} \frac{L_0}{\Delta x}. \tag{12.43}$$

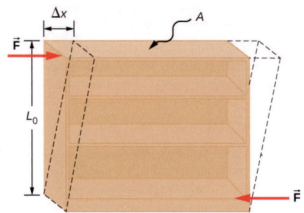

Figure 12.24 An object under shear stress: Two antiparallel forces of equal magnitude are applied tangentially to opposite parallel surfaces of the object. The dashed-line contour depicts the resulting deformation. There is no change in the direction transverse to the acting forces and the transverse length L_0 is

unaffected. Shear deformation is characterized by a gradual shift Δx of layers in the direction tangent to the forces.

Example 12.10

An Old Bookshelf

A cleaning person tries to move a heavy, old bookcase on a carpeted floor by pushing tangentially on the surface of the very top shelf. However, the only noticeable effect of this effort is similar to that seen in **Figure 12.24**, and it disappears when the person stops pushing. The bookcase is 180.0 cm tall and 90.0 cm wide with four 30.0-cm-deep shelves, all partially loaded with books. The total weight of the bookcase and books is 600.0 N. If the person gives the top shelf a 50.0-N push that displaces the top shelf horizontally by 15.0 cm relative to the motionless bottom shelf, find the shear modulus of the bookcase.

Strategy

The only pieces of relevant information are the physical dimensions of the bookcase, the value of the tangential force, and the displacement this force causes. We identify $F_\parallel = 50.0$ N, $\Delta x = 15.0$ cm, $L_0 = 180.0$ cm, and $A = (30.0 \text{ cm})(90.0 \text{ cm}) = 2700.0 \text{ cm}^2$, and we use **Equation 12.43** to compute the shear modulus.

Solution

Substituting numbers into the equations, we obtain for the shear modulus

$$S = \frac{F_\parallel}{A} \frac{L_0}{\Delta x} = \frac{50.0 \text{ N}}{2700.0 \text{ cm}^2} \frac{180.0 \text{ cm.}}{15.0 \text{ cm.}} = \frac{2}{9} \frac{\text{N}}{\text{cm}^2} = \frac{2}{9} \times 10^4 \frac{\text{N}}{\text{m}^2} = \frac{20}{9} \times 10^3 \text{ Pa} = 2.222 \text{ kPa.}$$

We can also find shear stress and strain, respectively:

$$\frac{F_\parallel}{A} = \frac{50.0 \text{ N}}{2700.0 \text{ cm}^2} = \frac{5}{27} \text{ kPa} = 185.2 \text{ Pa}$$

$$\frac{\Delta x}{L_0} = \frac{15.0 \text{ cm}}{180.0 \text{ cm}} = \frac{1}{12} = 0.083.$$

Significance

If the person in this example gave the shelf a healthy push, it might happen that the induced shear would collapse it to a pile of rubbish. Much the same shear mechanism is responsible for failures of earth-filled dams and levees; and, in general, for landslides.

 12.12 Check Your Understanding Explain why the concepts of Young's modulus and shear modulus do not apply to fluids.

12.4 | Elasticity and Plasticity

Learning Objectives
By the end of this section, you will be able to: • Explain the limit where a deformation of material is elastic   • Describe the range where materials show plastic behavior   • Analyze elasticity and plasticity on a stress-strain diagram

We referred to the proportionality constant between stress and strain as the elastic modulus. But why do we call it that? What does it mean for an object to be elastic and how do we describe its behavior?

Elasticity is the tendency of solid objects and materials to return to their original shape after the external forces (load) causing a deformation are removed. An object is **elastic** when it comes back to its original size and shape when the load is no longer present. Physical reasons for elastic behavior vary among materials and depend on the microscopic structure of the material. For example, the elasticity of polymers and rubbers is caused by stretching polymer chains under an applied force. In contrast, the elasticity of metals is caused by resizing and reshaping the crystalline cells of the lattices (which are the material structures of metals) under the action of externally applied forces.

The two parameters that determine the elasticity of a material are its *elastic modulus* and its *elastic limit*. A high elastic modulus is typical for materials that are hard to deform; in other words, materials that require a high load to achieve a significant strain. An example is a steel band. A low elastic modulus is typical for materials that are easily deformed under a load; for example, a rubber band. If the stress under a load becomes too high, then when the load is removed, the material no longer comes back to its original shape and size, but relaxes to a different shape and size: The material becomes permanently deformed. The **elastic limit** is the stress value beyond which the material no longer behaves elastically but becomes permanently deformed.

Our perception of an elastic material depends on both its elastic limit and its elastic modulus. For example, all rubbers are characterized by a low elastic modulus and a high elastic limit; hence, it is easy to stretch them and the stretch is noticeably large. Among materials with identical elastic limits, the most elastic is the one with the lowest elastic modulus.

When the load increases from zero, the resulting stress is in direct proportion to strain in the way given by **Equation 12.33**, but only when stress does not exceed some limiting value. For stress values within this linear limit, we can describe elastic behavior in analogy with Hooke's law for a spring. According to Hooke's law, the stretch value of a spring under an applied force is directly proportional to the magnitude of the force. Conversely, the response force from the spring to an applied stretch is directly proportional to the stretch. In the same way, the deformation of a material under a load is directly proportional to the load, and, conversely, the resulting stress is directly proportional to strain. The linearity limit (or the **proportionality limit**) is the largest stress value beyond which stress is no longer proportional to strain. Beyond the linearity limit, the relation between stress and strain is no longer linear. When stress becomes larger than the linearity limit but still within the elasticity limit, behavior is still elastic, but the relation between stress and strain becomes nonlinear.

For stresses beyond the elastic limit, a material exhibits **plastic behavior**. This means the material deforms irreversibly and does not return to its original shape and size, even when the load is removed. When stress is gradually increased beyond the elastic limit, the material undergoes plastic deformation. Rubber-like materials show an increase in stress with the increasing strain, which means they become more difficult to stretch and, eventually, they reach a fracture point where they break. Ductile materials such as metals show a gradual decrease in stress with the increasing strain, which means they become easier to deform as stress-strain values approach the breaking point. Microscopic mechanisms responsible for plasticity of materials are different for different materials.

We can graph the relationship between stress and strain on a **stress-strain diagram**. Each material has its own characteristic strain-stress curve. A typical stress-strain diagram for a ductile metal under a load is shown in **Figure 12.25**. In this figure, strain is a fractional elongation (not drawn to scale). When the load is gradually increased, the linear behavior (red line) that starts at the no-load point (the origin) ends at the linearity limit at point H. For further load increases beyond point H, the stress-strain relation is nonlinear but still elastic. In the figure, this nonlinear region is seen between points H and E. Ever larger loads take the stress to the elasticity limit E, where elastic behavior ends and plastic deformation begins. Beyond the elasticity limit, when the load is removed, for example at P, the material relaxes to a new shape and size along the green line. This is to say that the material becomes permanently deformed and does not come back to its initial shape and size when stress becomes zero.

The material undergoes plastic deformation for loads large enough to cause stress to go beyond the elasticity limit at E. The material continues to be plastically deformed until the stress reaches the fracture point (breaking point). Beyond the fracture point, we no longer have one sample of material, so the diagram ends at the fracture point. For the completeness of this qualitative description, it should be said that the linear, elastic, and plasticity limits denote a range of values rather than one sharp point.

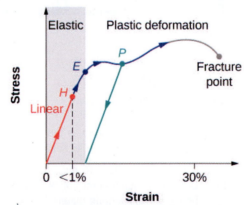

Figure 12.25 Typical stress-strain plot for a metal under a load: The graph ends at the fracture point. The arrows show the direction of changes under an ever-increasing load. Points H and E are the linearity and elasticity limits, respectively. Between points H and E, the behavior is nonlinear. The green line originating at P illustrates the metal's response when the load is removed. The permanent deformation has a strain value at the point where the green line intercepts the horizontal axis.

The value of stress at the fracture point is called breaking stress (or **ultimate stress**). Materials with similar elastic properties, such as two metals, may have very different breaking stresses. For example, ultimate stress for aluminum is

$2.2 \times 10^8 \, \text{Pa}$ and for steel it may be as high as $20.0 \times 10^8 \, \text{Pa}$, depending on the kind of steel. We can make a quick estimate, based on **Equation 12.34**, that for rods with a 1-in^2 cross-sectional area, the breaking load for an aluminum rod is 3.2×10^4 lb, and the breaking load for a steel rod is about nine times larger.

CHAPTER 12 REVIEW

KEY TERMS

breaking stress (ultimate stress) value of stress at the fracture point

bulk modulus elastic modulus for the bulk stress

bulk strain (or **volume strain**) strain under the bulk stress, given as fractional change in volume

bulk stress (or **volume stress**) stress caused by compressive forces, in all directions

center of gravity point where the weight vector is attached

compressibility reciprocal of the bulk modulus

compressive strain strain that occurs when forces are contracting an object, causing its shortening

compressive stress stress caused by compressive forces, only in one direction

elastic object that comes back to its original size and shape when the load is no longer present

elastic limit stress value beyond which material no longer behaves elastically and becomes permanently deformed

elastic modulus proportionality constant in linear relation between stress and strain, in SI pascals

equilibrium body is in equilibrium when its linear and angular accelerations are both zero relative to an inertial frame of reference

first equilibrium condition expresses translational equilibrium; all external forces acting on the body balance out and their vector sum is zero

gravitational torque torque on the body caused by its weight; it occurs when the center of gravity of the body is not located on the axis of rotation

linearity limit (proportionality limit) largest stress value beyond which stress is no longer proportional to strain

normal pressure pressure of one atmosphere, serves as a reference level for pressure

pascal (Pa) SI unit of stress, SI unit of pressure

plastic behavior material deforms irreversibly, does not go back to its original shape and size when load is removed and stress vanishes

pressure force pressing in normal direction on a surface per the surface area, the bulk stress in fluids

second equilibrium condition expresses rotational equilibrium; all torques due to external forces acting on the body balance out and their vector sum is zero

shear modulus elastic modulus for shear stress

shear strain strain caused by shear stress

shear stress stress caused by shearing forces

static equilibrium body is in static equilibrium when it is at rest in our selected inertial frame of reference

strain dimensionless quantity that gives the amount of deformation of an object or medium under stress

stress quantity that contains information about the magnitude of force causing deformation, defined as force per unit area

stress-strain diagram graph showing the relationship between stress and strain, characteristic of a material

tensile strain strain under tensile stress, given as fractional change in length, which occurs when forces are stretching an object, causing its elongation

tensile stress stress caused by tensile forces, only in one direction, which occurs when forces are stretching an object, causing its elongation

Young's modulus elastic modulus for tensile or compressive stress

KEY EQUATIONS

First Equilibrium Condition	$\sum_k \vec{\mathbf{F}}_k = \vec{\mathbf{0}}$
Second Equilibrium Condition	$\sum_k \vec{\tau}_k = \vec{\mathbf{0}}$
Linear relation between stress and strain	stress = (elastic modulus) × strain
Young's modulus	$Y = \dfrac{\text{tensile stress}}{\text{tensile strain}} = \dfrac{F_\perp}{A}\dfrac{L_0}{\Delta L}$
Bulk modulus	$B = \dfrac{\text{bulk stress}}{\text{bulk strain}} = -\Delta p\dfrac{V_0}{\Delta V}$
Shear modulus	$S = \dfrac{\text{shear stress}}{\text{shear strain}} = \dfrac{F_\parallel}{A}\dfrac{L_0}{\Delta x}$

SUMMARY

12.1 Conditions for Static Equilibrium

- A body is in equilibrium when it remains either in uniform motion (both translational and rotational) or at rest. When a body in a selected inertial frame of reference neither rotates nor moves in translational motion, we say the body is in static equilibrium in this frame of reference.

- Conditions for equilibrium require that the sum of all external forces acting on the body is zero (first condition of equilibrium), and the sum of all external torques from external forces is zero (second condition of equilibrium). These two conditions must be simultaneously satisfied in equilibrium. If one of them is not satisfied, the body is not in equilibrium.

- The free-body diagram for a body is a useful tool that allows us to count correctly all contributions from all external forces and torques acting on the body. Free-body diagrams for the equilibrium of an extended rigid body must indicate a pivot point and lever arms of acting forces with respect to the pivot.

12.2 Examples of Static Equilibrium

- A variety of engineering problems can be solved by applying equilibrium conditions for rigid bodies.

- In applications, identify all forces that act on a rigid body and note their lever arms in rotation about a chosen rotation axis. Construct a free-body diagram for the body. Net external forces and torques can be clearly identified from a correctly constructed free-body diagram. In this way, you can set up the first equilibrium condition for forces and the second equilibrium condition for torques.

- In setting up equilibrium conditions, we are free to adopt any inertial frame of reference and any position of the pivot point. All choices lead to one answer. However, some choices can make the process of finding the solution unduly complicated. We reach the same answer no matter what choices we make. The only way to master this skill is to practice.

12.3 Stress, Strain, and Elastic Modulus

- External forces on an object (or medium) cause its deformation, which is a change in its size and shape. The strength of the forces that cause deformation is expressed by stress, which in SI units is measured in the unit of pressure (pascal). The extent of deformation under stress is expressed by strain, which is dimensionless.

- For a small stress, the relation between stress and strain is linear. The elastic modulus is the proportionality constant in this linear relation.

- Tensile (or compressive) strain is the response of an object or medium to tensile (or compressive) stress. Here, the elastic modulus is called Young's modulus. Tensile (or compressive) stress causes elongation (or shortening) of the object or medium and is due to an external forces acting along only one direction perpendicular to the cross-section.

- Bulk strain is the response of an object or medium to bulk stress. Here, the elastic modulus is called the bulk modulus. Bulk stress causes a change in the volume of the object or medium and is caused by forces acting on the body from all directions, perpendicular to its surface. Compressibility of an object or medium is the reciprocal of its bulk modulus.

- Shear strain is the deformation of an object or medium under shear stress. The shear modulus is the elastic modulus in this case. Shear stress is caused by forces acting along the object's two parallel surfaces.

12.4 Elasticity and Plasticity

- An object or material is elastic if it comes back to its original shape and size when the stress vanishes. In elastic deformations with stress values lower than the proportionality limit, stress is proportional to strain. When stress goes beyond the proportionality limit, the deformation is still elastic but nonlinear up to the elasticity limit.

- An object or material has plastic behavior when stress is larger than the elastic limit. In the plastic region, the object or material does not come back to its original size or shape when stress vanishes but acquires a permanent deformation. Plastic behavior ends at the breaking point.

CONCEPTUAL QUESTIONS

12.1 Conditions for Static Equilibrium

1. What can you say about the velocity of a moving body that is in dynamic equilibrium?

2. Under what conditions can a rotating body be in equilibrium? Give an example.

3. What three factors affect the torque created by a force relative to a specific pivot point?

4. Mechanics sometimes put a length of pipe over the handle of a wrench when trying to remove a very tight bolt. How does this help?

For the next four problems, evaluate the statement as either true or false and explain your answer.

5. If there is only one external force (or torque) acting on an object, it cannot be in equilibrium.

6. If an object is in equilibrium there must be an even number of forces acting on it.

7. If an odd number of forces act on an object, the object cannot be in equilibrium.

8. A body moving in a circle with a constant speed is in rotational equilibrium.

9. What purpose is served by a long and flexible pole carried by wire-walkers?

12.2 Examples of Static Equilibrium

10. Is it possible to rest a ladder against a rough wall when

the floor is frictionless?

11. Show how a spring scale and a simple fulcrum can be used to weigh an object whose weight is larger than the maximum reading on the scale.

12. A painter climbs a ladder. Is the ladder more likely to slip when the painter is near the bottom or near the top?

12.3 Stress, Strain, and Elastic Modulus

Note: Unless stated otherwise, the weights of the wires, rods, and other elements are assumed to be negligible. Elastic moduli of selected materials are given in **Table 12.1**.

13. Why can a squirrel jump from a tree branch to the ground and run away undamaged, while a human could break a bone in such a fall?

14. When a glass bottle full of vinegar warms up, both the vinegar and the glass expand, but the vinegar expands significantly more with temperature than does the glass. The bottle will break if it is filled up to its very tight cap. Explain why and how a pocket of air above the vinegar prevents the bottle from breaking.

15. A thin wire strung between two nails in the wall is used to support a large picture. Is the wire likely to snap if it is strung tightly or if it is strung so that it sags considerably?

16. Review the relationship between stress and strain. Can you find any similarities between the two quantities?

17. What type of stress are you applying when you press

on the ends of a wooden rod? When you pull on its ends?

18. Can compressive stress be applied to a rubber band?

19. Can Young's modulus have a negative value? What about the bulk modulus?

20. If a hypothetical material has a negative bulk modulus, what happens when you squeeze a piece of it?

21. Discuss how you might measure the bulk modulus of a liquid.

12.4 Elasticity and Plasticity

Note: Unless stated otherwise, the weights of the wires, rods, and other elements are assumed to be negligible. Elastic moduli of selected materials are given in **Table 12.1**.

22. What is meant when a fishing line is designated as "a 10-lb test?"

23. Steel rods are commonly placed in concrete before it sets. What is the purpose of these rods?

PROBLEMS

12.1 Conditions for Static Equilibrium

24. When tightening a bolt, you push perpendicularly on a wrench with a force of 165 N at a distance of 0.140 m from the center of the bolt. How much torque are you exerting relative to the center of the bolt?

25. When opening a door, you push on it perpendicularly with a force of 55.0 N at a distance of 0.850 m from the hinges. What torque are you exerting relative to the hinges?

26. Find the magnitude of the tension in each supporting cable shown below. In each case, the weight of the suspended body is 100.0 N and the masses of the cables are negligible.

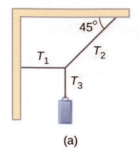

(a)

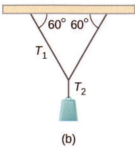

(b)

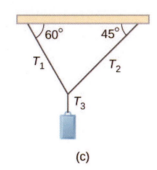

(c)

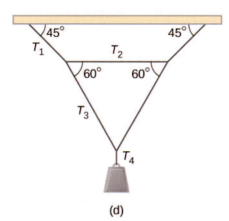

(d)

27. What force must be applied at point *P* to keep the structure shown in equilibrium? The weight of the structure is negligible.

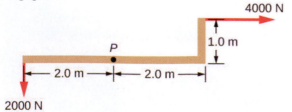

28. Is it possible to apply a force at *P* to keep in equilibrium the structure shown? The weight of the structure is negligible.

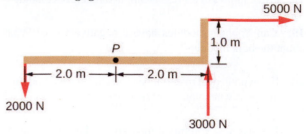

29. Two children push on opposite sides of a door during play. Both push horizontally and perpendicular to the door. One child pushes with a force of 17.5 N at a distance of 0.600 m from the hinges, and the second child pushes at a distance of 0.450 m. What force must the second child exert to keep the door from moving? Assume friction is negligible.

30. A small 1000-kg SUV has a wheel base of 3.0 m. If 60% if its weight rests on the front wheels, how far behind the front wheels is the wagon's center of mass?

31. The uniform seesaw is balanced at its center of mass, as seen below. The smaller boy on the right has a mass of 40.0 kg. What is the mass of his friend?

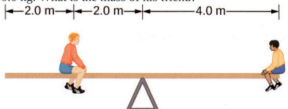

12.2 Examples of Static Equilibrium

32. A uniform plank rests on a level surface as shown below. The plank has a mass of 30 kg and is 6.0 m long. How much mass can be placed at its right end before it tips? (*Hint:* When the board is about to tip over, it makes contact with the surface only along the edge that becomes a momentary axis of rotation.)

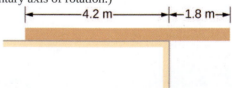

33. The uniform seesaw shown below is balanced on a fulcrum located 3.0 m from the left end. The smaller boy on the right has a mass of 40 kg and the bigger boy on the left has a mass 80 kg. What is the mass of the board?

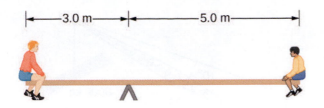

34. In order to get his car out of the mud, a man ties one end of a rope to the front bumper and the other end to a tree 15 m away, as shown below. He then pulls on the center of the rope with a force of 400 N, which causes its center to be displaced 0.30 m, as shown. What is the force of the rope on the car?

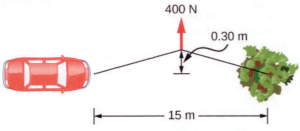

35. A uniform 40.0-kg scaffold of length 6.0 m is supported by two light cables, as shown below. An 80.0-kg painter stands 1.0 m from the left end of the scaffold, and his painting equipment is 1.5 m from the right end. If the tension in the left cable is twice that in the right cable, find the tensions in the cables and the mass of the equipment.

36. When the structure shown below is supported at point P, it is in equilibrium. Find the magnitude of force F and the force applied at P. The weight of the structure is negligible.

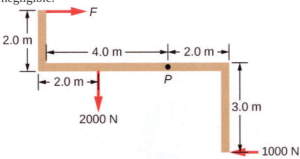

37. To get up on the roof, a person (mass 70.0 kg) places a 6.00-m aluminum ladder (mass 10.0 kg) against the house on a concrete pad with the base of the ladder 2.00 m from the house. The ladder rests against a plastic rain gutter, which we can assume to be frictionless. The center of mass of the ladder is 2.00 m from the bottom. The person is

standing 3.00 m from the bottom. Find the normal reaction and friction forces on the ladder at its base.

38. A uniform horizontal strut weighs 400.0 N. One end of the strut is attached to a hinged support at the wall, and the other end of the strut is attached to a sign that weighs 200.0 N. The strut is also supported by a cable attached between the end of the strut and the wall. Assuming that the entire weight of the sign is attached at the very end of the strut, find the tension in the cable and the force at the hinge of the strut.

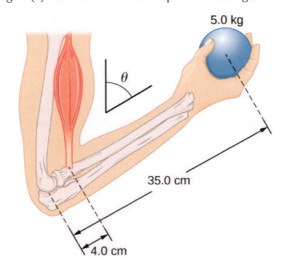

39. The forearm shown below is positioned at an angle θ with respect to the upper arm, and a 5.0-kg mass is held in the hand. The total mass of the forearm and hand is 3.0 kg, and their center of mass is 15.0 cm from the elbow. (a) What is the magnitude of the force that the biceps muscle exerts on the forearm for $\theta = 60°$? (b) What is the magnitude of the force on the elbow joint for the same angle? (c) How do these forces depend on the angle θ?

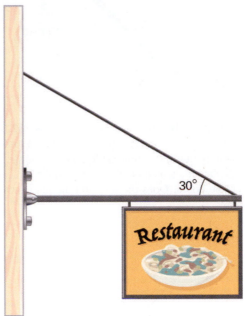

40. The uniform boom shown below weighs 3000 N. It is supported by the horizontal guy wire and by the hinged support at point *A*. What are the forces on the boom due to the wire and due to the support at *A*? Does the force at *A* act along the boom?

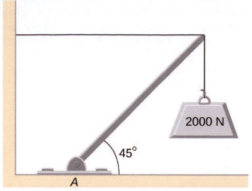

41. The uniform boom shown below weighs 700 N, and the object hanging from its right end weighs 400 N. The boom is supported by a light cable and by a hinge at the wall. Calculate the tension in the cable and the force on the hinge on the boom. Does the force on the hinge act along the boom?

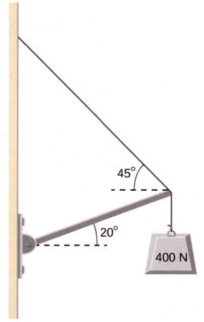

42. A 12.0-m boom, *AB*, of a crane lifting a 3000-kg load is shown below. The center of mass of the boom is at its geometric center, and the mass of the boom is 1000 kg. For the position shown, calculate tension *T* in the cable and the force at the axle *A*.

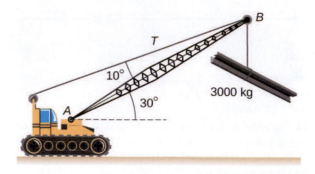

43. A uniform trapdoor shown below is 1.0 m by 1.5 m and weighs 300 N. It is supported by a single hinge (H), and by a light rope tied between the middle of the door and the floor. The door is held at the position shown, where its slab makes a 30° angle with the horizontal floor and the rope makes a 20° angle with the floor. Find the tension in the rope and the force at the hinge.

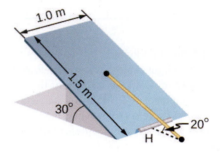

44. A 90-kg man walks on a sawhorse, as shown below. The sawhorse is 2.0 m long and 1.0 m high, and its mass is 25.0 kg. Calculate the normal reaction force on each leg at the contact point with the floor when the man is 0.5 m from the far end of the sawhorse. (*Hint:* At each end, find the total reaction force first. This reaction force is the vector sum of two reaction forces, each acting along one leg. The normal reaction force at the contact point with the floor is the normal (with respect to the floor) component of this force.)

12.3 Stress, Strain, and Elastic Modulus

45. The "lead" in pencils is a graphite composition with a Young's modulus of approximately $1.0 \times 10^9 \, \text{N/m}^2$. Calculate the change in length of the lead in an automatic pencil if you tap it straight into the pencil with a force of 4.0 N. The lead is 0.50 mm in diameter and 60 mm long.

46. TV broadcast antennas are the tallest artificial structures on Earth. In 1987, a 72.0-kg physicist placed himself and 400 kg of equipment at the top of a 610-m-high antenna to perform gravity experiments. By how much was the antenna compressed, if we consider it to be equivalent to a steel cylinder 0.150 m in radius?

47. By how much does a 65.0-kg mountain climber stretch her 0.800-cm diameter nylon rope when she hangs 35.0 m below a rock outcropping? (For nylon, $Y = 1.35 \times 10^9 \, \text{Pa}$.)

48. When water freezes, its volume increases by 9.05%. What force per unit area is water capable of exerting on a container when it freezes?

49. A farmer making grape juice fills a glass bottle to the brim and caps it tightly. The juice expands more than the glass when it warms up, in such a way that the volume increases by 0.2%. Calculate the force exerted by the juice per square centimeter if its bulk modulus is $1.8 \times 10^9 \, \text{N/m}^2$, assuming the bottle does not break.

50. A disk between vertebrae in the spine is subjected to a shearing force of 600.0 N. Find its shear deformation, using the shear modulus of $1.0 \times 10^9 \, \text{N/m}^2$. The disk is equivalent to a solid cylinder 0.700 cm high and 4.00 cm in diameter.

51. A vertebra is subjected to a shearing force of 500.0 N. Find the shear deformation, taking the vertebra to be a cylinder 3.00 cm high and 4.00 cm in diameter. How does your result compare with the result obtained in the preceding problem? Are spinal problems more common in disks than in vertebrae?

52. Calculate the force a piano tuner applies to stretch a steel piano wire by 8.00 mm, if the wire is originally 1.35 m long and its diameter is 0.850 mm.

53. A 20.0-m-tall hollow aluminum flagpole is equivalent in strength to a solid cylinder 4.00 cm in diameter. A strong wind bends the pole as much as a horizontal 900.0-N force on the top would do. How far to the side does the top of the pole flex?

54. A copper wire of diameter 1.0 cm stretches 1.0% when it is used to lift a load upward with an acceleration of $2.0 \, \text{m/s}^2$. What is the weight of the load?

55. As an oil well is drilled, each new section of drill pipe supports its own weight and the weight of the pipe and the drill bit beneath it. Calculate the stretch in a new 6.00-m-long steel pipe that supports a 100-kg drill bit and a 3.00-km length of pipe with a linear mass density of 20.0 kg/m. Treat the pipe as a solid cylinder with a 5.00-cm diameter.

56. A large uniform cylindrical steel rod of density $\rho = 7.8 \, \text{g/cm}^3$ is 2.0 m long and has a diameter of 5.0 cm. The rod is fastened to a concrete floor with its long axis vertical. What is the normal stress in the rod at the cross-section located at (a) 1.0 m from its lower end? (b) 1.5 m from the lower end?

57. A 90-kg mountain climber hangs from a nylon rope and stretches it by 25.0 cm. If the rope was originally 30.0 m long and its diameter is 1.0 cm, what is Young's modulus for the nylon?

58. A suspender rod of a suspension bridge is 25.0 m long. If the rod is made of steel, what must its diameter be so that it does not stretch more than 1.0 cm when a 2.5×10^4-kg truck passes by it? Assume that the rod supports all of the weight of the truck.

59. A copper wire is 1.0 m long and its diameter is 1.0 mm. If the wire hangs vertically, how much weight must be added to its free end in order to stretch it 3.0 mm?

60. A 100-N weight is attached to a free end of a metallic wire that hangs from the ceiling. When a second 100-N weight is added to the wire, it stretches 3.0 mm. The diameter and the length of the wire are 1.0 mm and 2.0 m, respectively. What is Young's modulus of the metal used to manufacture the wire?

61. The bulk modulus of a material is $1.0 \times 10^{11} \, \text{N/m}^2$. What fractional change in volume does a piece of this material undergo when it is subjected to a bulk stress increase of $10^7 \, \text{N/m}^2$? Assume that the force is applied uniformly over the surface.

62. Normal forces of magnitude $1.0 \times 10^6 \, \text{N}$ are applied uniformly to a spherical surface enclosing a volume of a liquid. This causes the radius of the surface to decrease from 50.000 cm to 49.995 cm. What is the bulk modulus of the liquid?

63. During a walk on a rope, a tightrope walker creates a tension of $3.94 \times 10^3 N$ in a wire that is stretched between two supporting poles that are 15.0 m apart. The wire has a diameter of 0.50 cm when it is not stretched. When the walker is on the wire in the middle between the poles the wire makes an angle of $5.0°$ below the horizontal. How much does this tension stretch the steel wire when the walker is this position?

64. When using a pencil eraser, you exert a vertical force of 6.00 N at a distance of 2.00 cm from the hardwood-eraser joint. The pencil is 6.00 mm in diameter and is held at an angle of $20.0°$ to the horizontal. (a) By how much does the wood flex perpendicular to its length? (b) How much is it compressed lengthwise?

65. Normal forces are applied uniformly over the surface of a spherical volume of water whose radius is 20.0 cm. If the pressure on the surface is increased by 200 MPa, by how much does the radius of the sphere decrease?

12.4 Elasticity and Plasticity

66. A uniform rope of cross-sectional area $0.50 \, \text{cm}^2$ breaks when the tensile stress in it reaches $6.00 \times 10^6 \, \text{N/m}^2$. (a) What is the maximum load that can be lifted slowly at a constant speed by the rope? (b) What is the maximum load that can be lifted by the rope with an acceleration of $4.00 \, \text{m/s}^2$?

67. One end of a vertical metallic wire of length 2.0 m and diameter 1.0 mm is attached to a ceiling, and the other end is attached to a 5.0-N weight pan, as shown below. The position of the pointer before the pan is 4.000 cm. Different weights are then added to the pan area, and the position of the pointer is recorded in the table shown. Plot stress versus strain for this wire, then use the resulting curve to determine Young's modulus and the proportionality limit of the metal. What metal is this most likely to be?

Added load (including pan) (N)	Scale reading (cm)
0	4.000

Added load (including pan) (N)	Scale reading (cm)
15	4.036
25	4.073
35	4.109
45	4.146
55	4.181
65	4.221
75	4.266
85	4.316

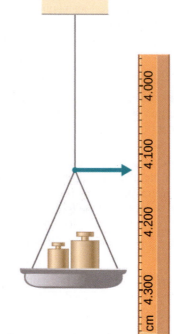

68. An aluminum $\left(\rho = 2.7 \, \text{g/cm}^3\right)$ wire is suspended from the ceiling and hangs vertically. How long must the wire be before the stress at its upper end reaches the proportionality limit, which is $8.0 \times 10^7 \, \text{N/m}^2$?

ADDITIONAL PROBLEMS

69. The coefficient of static friction between the rubber eraser of the pencil and the tabletop is $\mu_s = 0.80$. If the force $\vec{\textbf{F}}$ is applied along the axis of the pencil, as shown below, what is the minimum angle at which the pencil can stand without slipping? Ignore the weight of the pencil.

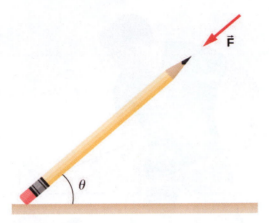

70. A pencil rests against a corner, as shown below. The sharpened end of the pencil touches a smooth vertical surface and the eraser end touches a rough horizontal floor. The coefficient of static friction between the eraser and the floor is $\mu_s = 0.80$. The center of mass of the pencil is located 9.0 cm from the tip of the eraser and 11.0 cm from the tip of the pencil lead. Find the minimum angle θ for which the pencil does not slip.

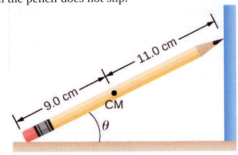

71. A uniform 4.0-m plank weighing 200.0 N rests against the corner of a wall, as shown below. There is no friction at the point where the plank meets the corner. (a) Find the forces that the corner and the floor exert on the plank. (b) What is the minimum coefficient of static friction between

the floor and the plank to prevent the plank from slipping?

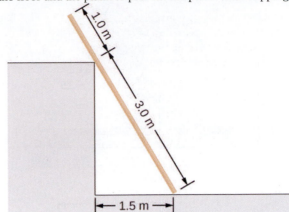

72. A 40-kg boy jumps from a height of 3.0 m, lands on one foot and comes to rest in 0.10 s after he hits the ground. Assume that he comes to rest with a constant deceleration. If the total cross-sectional area of the bones in his legs just above his ankles is $3.0 \, \text{cm}^2$, what is the compression stress in these bones? Leg bones can be fractured when they are subjected to stress greater than 1.7×10^8 Pa. Is the boy in danger of breaking his leg?

73. Two thin rods, one made of steel and the other of aluminum, are joined end to end. Each rod is 2.0 m long and has cross-sectional area $9.1 \, \text{mm}^2$. If a 10,000-N tensile force is applied at each end of the combination, find: (a) stress in each rod; (b) strain in each rod; and, (c) elongation of each rod.

74. Two rods, one made of copper and the other of steel, have the same dimensions. If the copper rod stretches by 0.15 mm under some stress, how much does the steel rod stretch under the same stress?

CHALLENGE PROBLEMS

75. A horizontal force $\vec{F}$ is applied to a uniform sphere in direction exact toward the center of the sphere, as shown below. Find the magnitude of this force so that the sphere remains in static equilibrium. What is the frictional force of the incline on the sphere?

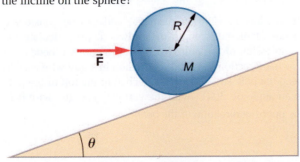

76. When a motor is set on a pivoted mount seen below, its weight can be used to maintain tension in the drive belt. When the motor is not running the tensions T_1 and T_2 are equal. The total mass of the platform and the motor is 100.0 kg, and the diameter of the drive belt pulley is 16.0 cm. when the motor is off, find: (a) the tension in the belt, and (b) the force at the hinged platform support at point C. Assume that the center of mass of the motor plus platform is at the center of the motor.

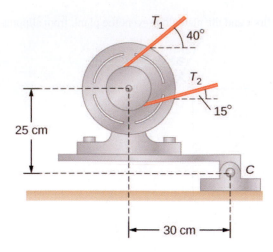

77. Two wheels A and B with weights w and $2w$, respectively, are connected by a uniform rod with weight $w/2$, as shown below. The wheels are free to roll on the sloped surfaces. Determine the angle that the rod forms with the horizontal when the system is in equilibrium. *Hint:* There are five forces acting on the rod, which is two weights of the wheels, two normal reaction forces at points where the wheels make contacts with the wedge, and the weight of the rod.

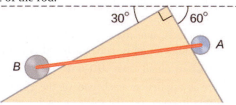

78. Weights are gradually added to a pan until a wheel of mass M and radius R is pulled over an obstacle of height d, as shown below. What is the minimum mass of the weights plus the pan needed to accomplish this?

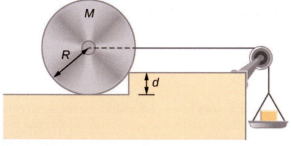

79. In order to lift a shovelful of dirt, a gardener pushes downward on the end of the shovel and pulls upward at distance l_2 from the end, as shown below. The weight of the shovel is $m\,\vec{g}$ and acts at the point of application of $\vec{F}_2$. Calculate the magnitudes of the forces $\vec{F}_1$ and $\vec{F}_2$ as functions of l_1, l_2, mg, and the weight W of the load. Why do your answers not depend on the angle θ that the shovel makes with the horizontal?

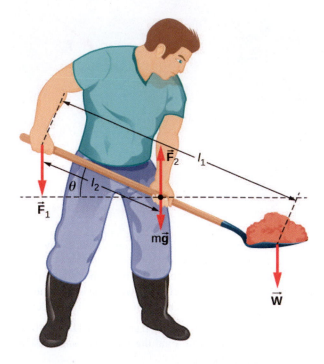

80. A uniform rod of length $2R$ and mass M is attached to a small collar C and rests on a cylindrical surface of radius R, as shown below. If the collar can slide without friction along the vertical guide, find the angle θ for which the rod is in static equilibrium.

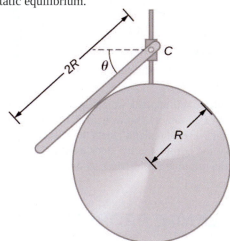

81. The pole shown below is at a $90.0°$ bend in a power line and is therefore subjected to more shear force than poles in straight parts of the line. The tension in each line is $4.00 \times 10^4\,\text{N}$, at the angles shown. The pole is 15.0 m tall, has an 18.0 cm diameter, and can be considered to have half the strength of hardwood. (a) Calculate the compression of the pole. (b) Find how much it bends and in what direction. (c) Find the tension in a guy wire used to keep the pole straight if it is attached to the top of the pole at an angle of $30.0°$ with the vertical. The guy wire is in the opposite direction of the bend.

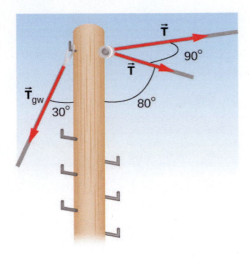

13 | GRAVITATION

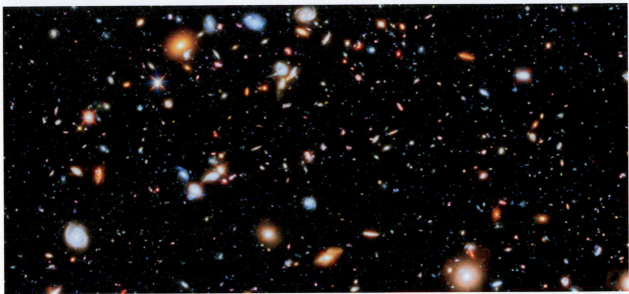

Figure 13.1 Our visible Universe contains billions of galaxies, whose very existence is due to the force of gravity. Gravity is ultimately responsible for the energy output of all stars—initiating thermonuclear reactions in stars, allowing the Sun to heat Earth, and making galaxies visible from unfathomable distances. Most of the dots you see in this image are not stars, but galaxies. (credit: modification of work by NASA/ESA)

Chapter Outline

13.1 Newton's Law of Universal Gravitation

13.2 Gravitation Near Earth's Surface

13.3 Gravitational Potential Energy and Total Energy

13.4 Satellite Orbits and Energy

13.5 Kepler's Laws of Planetary Motion

13.6 Tidal Forces

13.7 Einstein's Theory of Gravity

Introduction

In this chapter, we study the nature of the gravitational force for objects as small as ourselves and for systems as massive as entire galaxies. We show how the gravitational force affects objects on Earth and the motion of the Universe itself. Gravity is the first force to be postulated as an action-at-a-distance force, that is, objects exert a gravitational force on one another without physical contact and that force falls to zero only at an infinite distance. Earth exerts a gravitational force on you, but so do our Sun, the Milky Way galaxy, and the billions of galaxies, like those shown above, which are so distant that we cannot see them with the naked eye.

13.1 | Newton's Law of Universal Gravitation

We first review the history of the study of gravitation, with emphasis on those phenomena that for thousands of years have inspired philosophers and scientists to search for an explanation. Then we examine the simplest form of Newton's law of universal gravitation and how to apply it.

The History of Gravitation

The earliest philosophers wondered why objects naturally tend to fall toward the ground. Aristotle (384–322 BCE) believed that it was the nature of rocks to seek Earth and the nature of fire to seek the Heavens. Brahmagupta (598~665 CE) postulated that Earth was a sphere and that objects possessed a natural affinity for it, falling toward the center from wherever they were located.

The motions of the Sun, our Moon, and the planets have been studied for thousands of years as well. These motions were described with amazing accuracy by Ptolemy (90–168 CE), whose method of epicycles described the paths of the planets as circles within circles. However, there is little evidence that anyone connected the motion of astronomical bodies with the motion of objects falling to Earth—until the seventeenth century.

Nicolaus Copernicus (1473–1543) is generally credited as being the first to challenge Ptolemy's geocentric (Earth-centered) system and suggest a heliocentric system, in which the Sun is at the center of the solar system. This idea was supported by the incredibly precise naked-eye measurements of planetary motions by Tycho Brahe and their analysis by Johannes Kepler and Galileo Galilei. Kepler showed that the motion of each planet is an ellipse (the first of his three laws, discussed in **Kepler's Laws of Planetary Motion**), and Robert Hooke (the same Hooke who formulated Hooke's law for springs) intuitively suggested that these motions are due to the planets being attracted to the Sun. However, it was Isaac Newton who connected the acceleration of objects near Earth's surface with the centripetal acceleration of the Moon in its orbit about Earth.

Finally, in **Einstein's Theory of Gravity**, we look at the theory of general relativity proposed by Albert Einstein in 1916. His theory comes from a vastly different perspective, in which gravity is a manifestation of mass warping space and time. The consequences of his theory gave rise to many remarkable predictions, essentially all of which have been confirmed over the many decades following the publication of the theory (including the 2015 measurement of gravitational waves from the merger of two black holes).

Newton's Law of Universal Gravitation

Newton noted that objects at Earth's surface (hence at a distance of R_E from the center of Earth) have an acceleration of g, but the Moon, at a distance of about $60\,R_E$, has a centripetal acceleration about $(60)^2$ times smaller than g. He could explain this by postulating that a force exists between any two objects, whose magnitude is given by the product of the two masses divided by the square of the distance between them. We now know that this inverse square law is ubiquitous in nature, a function of geometry for point sources. The strength of any source at a distance r is spread over the surface of a sphere centered about the mass. The surface area of that sphere is proportional to r^2. In later chapters, we see this same form in the electromagnetic force.

Newton's Law of Gravitation

Newton's law of gravitation can be expressed as

$$\vec{\mathbf{F}}_{12} = G\frac{m_1 m_2}{r^2}\hat{\mathbf{r}}_{12}$$

(13.1)

where $\vec{\mathbf{F}}_{12}$ is the force on object 1 exerted by object 2 and $\hat{\mathbf{r}}_{12}$ is a unit vector that points from object 1 toward object 2.

As shown in **Figure 13.2**, the $\vec{\mathbf{F}}_{12}$ vector points from object 1 toward object 2, and hence represents an attractive force between the objects. The equal but opposite force $\vec{\mathbf{F}}_{21}$ is the force on object 2 exerted by object 1.

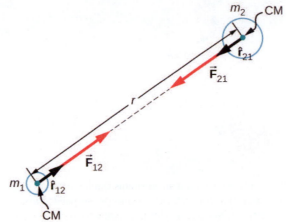

Figure 13.2 Gravitational force acts along a line joining the centers of mass of two objects.

These equal but opposite forces reflect Newton's third law, which we discussed earlier. Note that strictly speaking, **Equation 13.1** applies to point masses—all the mass is located at one point. But it applies equally to any spherically symmetric objects, where r is the distance between the centers of mass of those objects. In many cases, it works reasonably well for nonsymmetrical objects, if their separation is large compared to their size, and we take r to be the distance between the center of mass of each body.

The Cavendish Experiment

A century after Newton published his law of universal gravitation, Henry Cavendish determined the proportionality constant G by performing a painstaking experiment. He constructed a device similar to that shown in **Figure 13.3**, in which small masses are suspended from a wire. Once in equilibrium, two fixed, larger masses are placed symmetrically near the smaller ones. The gravitational attraction creates a torsion (twisting) in the supporting wire that can be measured.

The constant G is called the **universal gravitational constant** and Cavendish determined it to be $G = 6.67 \times 10^{-11} \ \text{N} \cdot \text{m}^2/\text{kg}^2$. The word 'universal' indicates that scientists think that this constant applies to masses of any composition and that it is the same throughout the Universe. The value of G is an incredibly small number, showing that the force of gravity is very weak. The attraction between masses as small as our bodies, or even objects the size of skyscrapers, is incredibly small. For example, two 1.0-kg masses located 1.0 meter apart exert a force of $6.7 \times 10^{-11} \ \text{N}$ on each other. This is the weight of a typical grain of pollen.

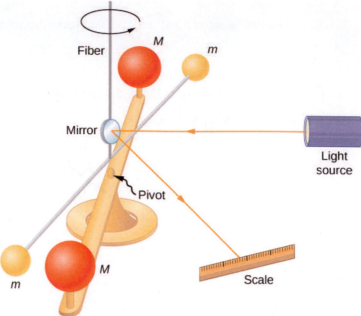

Figure 13.3 Cavendish used an apparatus similar to this to measure the gravitational attraction between two spheres (*m*) suspended from a wire and two stationary spheres (*M*). This is a common experiment performed in undergraduate laboratories, but it is quite challenging. Passing trucks outside the laboratory can create vibrations that overwhelm the gravitational forces.

Although gravity is the weakest of the four fundamental forces of nature, its attractive nature is what holds us to Earth, causes the planets to orbit the Sun and the Sun to orbit our galaxy, and binds galaxies into clusters, ranging from a few to millions. Gravity is the force that forms the Universe.

Problem-Solving Strategy: Newton's Law of Gravitation

To determine the motion caused by the gravitational force, follow these steps:

1. Identify the two masses, one or both, for which you wish to find the gravitational force.

2. Draw a free-body diagram, sketching the force acting on each mass and indicating the distance between their centers of mass.

3. Apply Newton's second law of motion to each mass to determine how it will move.

Example 13.1

A Collision in Orbit

Consider two nearly spherical *Soyuz* payload vehicles, in orbit about Earth, each with mass 9000 kg and diameter 4.0 m. They are initially at rest relative to each other, 10.0 m from center to center. (As we will see in **Kepler's Laws of Planetary Motion**, both orbit Earth at the same speed and interact nearly the same as if they were isolated in deep space.) Determine the gravitational force between them and their initial acceleration. Estimate how long it takes for them to drift together, and how fast they are moving upon impact.

Strategy

We use Newton's law of gravitation to determine the force between them and then use Newton's second law to find the acceleration of each. For the *estimate*, we assume this acceleration is constant, and we use the constant-acceleration equations from **Motion along a Straight Line** to find the time and speed of the collision.

Solution

The magnitude of the force is

$$\left| \overrightarrow{\mathbf{F}}_{12} \right| = F_{12} = G\frac{m_1 m_2}{r^2} = 6.67 \times 10^{-11} \ \text{N} \cdot \text{m}^2/\text{kg}^2 \frac{(9000 \ \text{kg})(9000 \ \text{kg})}{(10 \ \text{m})^2} = 5.4 \times 10^{-5} \ \text{N}.$$

The initial acceleration of each payload is

$$a = \frac{F}{m} = \frac{5.4 \times 10^{-5} \ \text{N}}{9000 \ \text{kg}} = 6.0 \times 10^{-9} \ \text{m/s}^2.$$

The vehicles are 4.0 m in diameter, so the vehicles move from 10.0 m to 4.0 m apart, or a distance of 3.0 m each. A similar calculation to that above, for when the vehicles are 4.0 m apart, yields an acceleration of 3.8×10^{-8} m/s^2, and the average of these two values is 2.2×10^{-8} m/s^2. If we assume a constant acceleration of this value and they start from rest, then the vehicles collide with speed given by

$$v^2 = v_0^2 + 2a(x - x_0), \ \text{where} \ v_0 = 0,$$

so

$$v = \sqrt{2(2.2 \times 10^{-9} \ \text{N})(3.0 \ \text{m})} = 3.6 \times 10^{-4} \ \text{m/s}.$$

We use $v = v_0 + at$ to find $t = v/a = 1.7 \times 10^4$ s or about 4.6 hours.

Significance

These calculations—including the initial force—are only estimates, as the vehicles are probably not spherically symmetrical. But you can see that the force is incredibly small. Astronauts must tether themselves when doing work outside even the massive International Space Station (ISS), as in **Figure 13.4**, because the gravitational attraction cannot save them from even the smallest push away from the station.

Figure 13.4 This photo shows Ed White tethered to the Space Shuttle during a spacewalk. (credit: NASA)

 13.1 Check Your Understanding What happens to force and acceleration as the vehicles fall together? What will our estimate of the velocity at a collision higher or lower than the speed actually be? And finally, what would happen if the masses were not identical? Would the force on each be the same or different? How about their accelerations?

The effect of gravity between two objects with masses on the order of these space vehicles is indeed small. Yet, the effect of gravity on you from Earth is significant enough that a fall into Earth of only a few feet can be dangerous. We examine the force of gravity near Earth's surface in the next section.

Example 13.2

Attraction between Galaxies

Find the acceleration of our galaxy, the Milky Way, due to the nearest comparably sized galaxy, the Andromeda galaxy (**Figure 13.5**). The approximate mass of each galaxy is 800 billion solar masses (a solar mass is the mass of our Sun), and they are separated by 2.5 million light-years. (Note that the mass of Andromeda is not so well known but is believed to be slightly larger than our galaxy.) Each galaxy has a diameter of roughly 100,000 light-years $(1 \text{ light-year} = 9.5 \times 10^{15} \text{ m})$.

Figure 13.5 Galaxies interact gravitationally over immense distances. The Andromeda galaxy is the nearest spiral galaxy to the Milky Way, and they will eventually collide. (credit: Boris Štromar)

Strategy

As in the preceding example, we use Newton's law of gravitation to determine the force between them and then use Newton's second law to find the acceleration of the Milky Way. We can consider the galaxies to be point masses, since their sizes are about 25 times smaller than their separation. The mass of the Sun (see **Appendix D**) is 2.0×10^{30} kg and a light-year is the distance light travels in one year, 9.5×10^{15} m.

Solution

The magnitude of the force is

$$F_{12} = G \frac{m_1 m_2}{r^2} = (6.67 \times 10^{-11} \text{ N} \cdot \text{m}^2 / \text{kg}^2) \frac{[(800 \times 10^9)(2.0 \times 10^{30} \text{ kg})]^2}{[(2.5 \times 10^6)(9.5 \times 10^{15} \text{ m})]^2} = 3.0 \times 10^{29} \text{ N}.$$

The acceleration of the Milky Way is

$$a = \frac{F}{m} = \frac{3.0 \times 10^{29} \text{ N}}{(800 \times 10^9)(2.0 \times 10^{30} \text{ kg})} = 1.9 \times 10^{-13} \text{ m/s}^2.$$

Significance

Does this value of acceleration seem astoundingly small? If they start from rest, then they would accelerate directly toward each other, "colliding" at their center of mass. Let's estimate the time for this to happen. The initial acceleration is $\sim 10^{-13}$ m/s^2, so using $v = at$, we see that it would take $\sim 10^{13}$ s for each galaxy to reach a speed of 1.0 m/s, and they would be only $\sim 0.5 \times 10^{13}$ m closer. That is nine orders of magnitude smaller than the initial distance between them. In reality, such motions are rarely simple. These two galaxies, along with about 50 other smaller galaxies, are all gravitationally bound into our local cluster. Our local cluster is gravitationally bound to other clusters in what is called a supercluster. All of this is part of the great cosmic dance that results from gravitation, as shown in **Figure 13.6**.

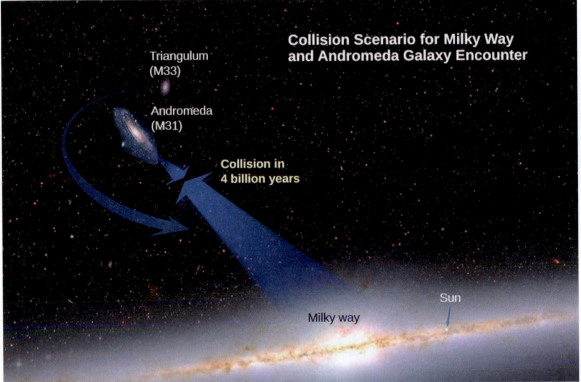

Figure 13.6 Based on the results of this example, plus what astronomers have observed elsewhere in the Universe, our galaxy will collide with the Andromeda Galaxy in about 4 billion years. (credit: modification of work by NASA; ESA; A. Feild and R. van der Marel, STScI)

13.2 | Gravitation Near Earth's Surface

Learning Objectives

By the end of this section, you will be able to:

- Explain the connection between the constants G and g
- Determine the mass of an astronomical body from free-fall acceleration at its surface
- Describe how the value of g varies due to location and Earth's rotation

In this section, we observe how Newton's law of gravitation applies at the surface of a planet and how it connects with what we learned earlier about free fall. We also examine the gravitational effects within spherical bodies.

Weight

Recall that the acceleration of a free-falling object near Earth's surface is approximately $g = 9.80 \text{ m/s}^2$. The force causing this acceleration is called the weight of the object, and from Newton's second law, it has the value mg. This weight is present regardless of whether the object is in free fall. We now know that this force is the gravitational force between the object and Earth. If we substitute mg for the magnitude of $\vec{F}_{12}$ in Newton's law of universal gravitation, m for m_1, and M_E for m_2, we obtain the scalar equation

$$mg = G\frac{mM_E}{r^2}$$

where r is the distance between the centers of mass of the object and Earth. The average radius of Earth is about 6370 km. Hence, for objects within a few kilometers of Earth's surface, we can take $r = R_E$ (**Figure 13.7**). The mass m of the object cancels, leaving

$$g = G\frac{M_E}{r^2}. \tag{13.2}$$

This explains why all masses free fall with the same acceleration. We have ignored the fact that Earth also accelerates toward the falling object, but that is acceptable as long as the mass of Earth is much larger than that of the object.

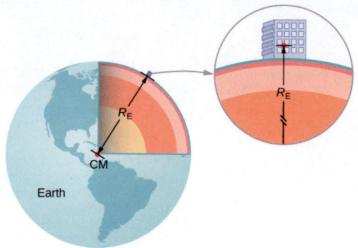

Figure 13.7 We can take the distance between the centers of mass of Earth and an object on its surface to be the radius of Earth, provided that its size is much less than the radius of Earth.

Example 13.3

Masses of Earth and Moon

Have you ever wondered how we know the mass of Earth? We certainly can't place it on a scale. The values of g and the radius of Earth were measured with reasonable accuracy centuries ago.

a. Use the standard values of g, R_E, and **Equation 13.2** to find the mass of Earth.

b. Estimate the value of g on the Moon. Use the fact that the Moon has a radius of about 1700 km (a value of this accuracy was determined many centuries ago) and assume it has the same average density as Earth, 5500 kg/m^3.

Strategy

With the known values of g and R_E, we can use **Equation 13.2** to find M_E. For the Moon, we use the assumption of equal average density to determine the mass from a ratio of the volumes of Earth and the Moon.

Solution

a. Rearranging **Equation 13.2**, we have

$$M_E = \frac{gR_E^2}{G} = \frac{9.80 \text{ m/s}^2 (6.37 \times 10^6 \text{ m})^2}{6.67 \times 10^{-11} \text{ N} \cdot \text{m}^2/\text{kg}^2} = 5.95 \times 10^{24} \text{ kg}.$$

b. The volume of a sphere is proportional to the radius cubed, so a simple ratio gives us

$$\frac{M_M}{M_E} = \frac{R_M^3}{R_E^3} \rightarrow M_M = \left(\frac{(1.7 \times 10^6 \text{ m})^3}{(6.37 \times 10^6 \text{ m})^3} \right) (5.95 \times 10^{24} \text{ kg}) = 1.1 \times 10^{23} \text{ kg}.$$

We now use **Equation 13.2**.

$$g_M = G\frac{M_M}{r_M^2} = (6.67 \times 10^{-11} \text{ N} \cdot \text{m}^2/\text{kg}^2)\frac{(1.1 \times 10^{23} \text{ kg})}{(1.7 \times 10^6 \text{ m})^2} = 2.5 \text{ m/s}^2$$

Significance

As soon as Cavendish determined the value of G in 1798, the mass of Earth could be calculated. (In fact, that was the ultimate purpose of Cavendish's experiment in the first place.) The value we calculated for g of the Moon is incorrect. The average density of the Moon is actually only 3340 kg/m^3 and $g = 1.6 \text{ m/s}^2$ at the surface.

Newton attempted to measure the mass of the Moon by comparing the effect of the Sun on Earth's ocean tides compared to that of the Moon. His value was a factor of two too small. The most accurate values for g and the mass of the Moon come from tracking the motion of spacecraft that have orbited the Moon. But the mass of the Moon can actually be determined accurately without going to the Moon. Earth and the Moon orbit about a common center of mass, and careful astronomical measurements can determine that location. The ratio of the Moon's mass to Earth's is the ratio of [the distance from the common center of mass to the Moon's center] to [the distance from the common center of mass to Earth's center].

Later in this chapter, we will see that the mass of other astronomical bodies also can be determined by the period of small satellites orbiting them. But until Cavendish determined the value of G, the masses of all these bodies were unknown.

Example 13.4

Gravity above Earth's Surface

What is the value of g 400 km above Earth's surface, where the International Space Station is in orbit?

Strategy

Using the value of M_E and noting the radius is $r = R_E + 400 \text{ km}$, we use **Equation 13.2** to find g.

From **Equation 13.2** we have

$$g = G\frac{M_E}{r^2} = 6.67 \times 10^{-11} \text{ N} \cdot \text{m}^2/\text{kg}^2 \frac{5.96 \times 10^{24} \text{ kg}}{(6.37 \times 10^6 + 400 \times 10^3 \text{ m})^2} = 8.67 \text{ m/s}^2.$$

Significance

We often see video of astronauts in space stations, apparently weightless. But clearly, the force of gravity is acting on them. Comparing the value of g we just calculated to that on Earth (9.80 m/s^2), we see that the astronauts in the International Space Station still have 88% of their weight. They only appear to be weightless because they are in free fall. We will come back to this in **Satellite Orbits and Energy**.

13.2 **Check Your Understanding** How does your weight at the top of a tall building compare with that on the first floor? Do you think engineers need to take into account the change in the value of *g* when designing structural support for a very tall building?

The Gravitational Field

Equation 13.2 is a scalar equation, giving the magnitude of the gravitational acceleration as a function of the distance from the center of the mass that causes the acceleration. But we could have retained the vector form for the force of gravity in **Equation 13.1**, and written the acceleration in vector form as

$$\vec{g} = G\frac{M}{r^2}\hat{r}.$$

We identify the vector field represented by $\vec{g}$ as the **gravitational field** caused by mass M. We can picture the field as shown **Figure 13.8**. The lines are directed radially inward and are symmetrically distributed about the mass.

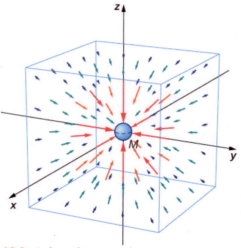

Figure 13.8 A three-dimensional representation of the gravitational field created by mass M. Note that the lines are uniformly distributed in all directions. (The box has been added only to aid in visualization.)

As is true for any vector field, the direction of $\vec{g}$ is parallel to the field lines at any point. The strength of $\vec{g}$ at any point is inversely proportional to the line spacing. Another way to state this is that the magnitude of the field in any region is proportional to the number of lines that pass through a unit surface area, effectively a density of lines. Since the lines are equally spaced in all directions, the number of lines per unit surface area at a distance r from the mass is the total number of lines divided by the surface area of a sphere of radius r, which is proportional to r^2. Hence, this picture perfectly represents the inverse square law, in addition to indicating the direction of the field. In the field picture, we say that a mass m interacts with the gravitational field of mass M. We will use the concept of fields to great advantage in the later chapters on electromagnetism.

Apparent Weight: Accounting for Earth's Rotation

As we saw in **Applications of Newton's Laws**, objects moving at constant speed in a circle have a centripetal acceleration directed toward the center of the circle, which means that there must be a net force directed toward the center of that circle. Since all objects on the surface of Earth move through a circle every 24 hours, there must be a net centripetal force on each object directed toward the center of that circle.

Let's first consider an object of mass m located at the equator, suspended from a scale (**Figure 13.9**). The scale exerts an upward force $\vec{F}_s$ away from Earth's center. This is the reading on the scale, and hence it is the **apparent weight** of the object. The weight (*mg*) points toward Earth's center. If Earth were not rotating, the acceleration would be zero and, consequently, the net force would be zero, resulting in $F_s = mg$. This would be the true reading of the weight.

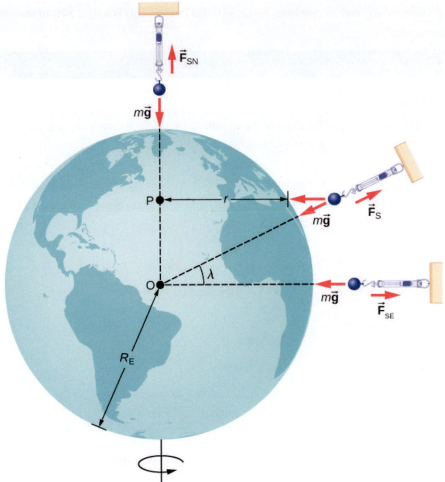

Figure 13.9 For a person standing at the equator, the centripetal acceleration (a_c) is in the same direction as the force of gravity. At latitude λ, the angle the between a_c and the force of gravity is λ and the magnitude of a_c decreases with $\cos\lambda$.

With rotation, the sum of these forces must provide the centripetal acceleration, a_c. Using Newton's second law, we have

$$\sum F = F_s - mg = ma_c \quad \text{where} \quad a_c = -\frac{v^2}{r}. \tag{13.3}$$

Note that a_c points in the same direction as the weight; hence, it is negative. The tangential speed v is the speed at the equator and r is R_E. We can calculate the speed simply by noting that objects on the equator travel the circumference of Earth in 24 hours. Instead, let's use the alternative expression for a_c from **Motion in Two and Three Dimensions**. Recall that the tangential speed is related to the angular speed (ω) by $v = r\omega$. Hence, we have $a_c = -r\omega^2$. By rearranging **Equation 13.3** and substituting $r = R_E$, the apparent weight at the equator is

$$F_s = m\left(g - R_E \omega^2\right).$$

The angular speed of Earth everywhere is

$$\omega = \frac{2\pi \, \text{rad}}{24 \, \text{hr} \times 3600 \, \text{s/hr}} = 7.27 \times 10^{-5} \, \text{rad/s}.$$

Substituting for the values or R_E and ω, we have $R_E \omega^2 = 0.0337 \text{ m/s}^2$. This is only 0.34% of the value of gravity, so it is clearly a small correction.

Example 13.5

Zero Apparent Weight

How fast would Earth need to spin for those at the equator to have zero apparent weight? How long would the length of the day be?

Strategy

Using **Equation 13.3**, we can set the apparent weight (F_s) to zero and determine the centripetal acceleration required. From that, we can find the speed at the equator. The length of day is the time required for one complete rotation.

Solution

From **Equation 13.2**, we have $\sum F = F_s - mg = ma_c$, so setting $F_s = 0$, we get $g = a_c$. Using the expression for a_c, substituting for Earth's radius and the standard value of gravity, we get

$$a_c = \frac{v^2}{r} = g$$
$$v = \sqrt{gr} = \sqrt{(9.80 \text{ m/s}^2)(6.37 \times 10^6 \text{ m})} = 7.91 \times 10^3 \text{ m/s}.$$

The period T is the time for one complete rotation. Therefore, the tangential speed is the circumference divided by T, so we have

$$v = \frac{2\pi r}{T}$$
$$T = \frac{2\pi r}{v} = \frac{2\pi(6.37 \times 10^6 \text{ m})}{7.91 \times 10^3 \text{ m/s}} = 5.06 \times 10^3 \text{ s}.$$

This is about 84 minutes.

Significance

We will see later in this chapter that this speed and length of day would also be the orbital speed and period of a satellite in orbit at Earth's surface. While such an orbit would not be possible near Earth's surface due to air resistance, it certainly is possible only a few hundred miles above Earth.

Results Away from the Equator

At the poles, $a_c \to 0$ and $F_s = mg$, just as is the case without rotation. At any other latitude λ, the situation is more complicated. The centripetal acceleration is directed toward point P in the figure, and the radius becomes $r = R_E \cos\lambda$.

The *vector* sum of the weight and $\vec{F}_s$ must point toward point P, hence $\vec{F}_s$ no longer points away from the center of Earth. (The difference is small and exaggerated in the figure.) A plumb bob will always point along this deviated direction. All buildings are built aligned along this deviated direction, not along a radius through the center of Earth. For the tallest buildings, this represents a deviation of a few feet at the top.

It is also worth noting that Earth is not a perfect sphere. The interior is partially liquid, and this enhances Earth bulging at the equator due to its rotation. The radius of Earth is about 30 km greater at the equator compared to the poles. It is left as an exercise to compare the strength of gravity at the poles to that at the equator using **Equation 13.2**. The difference is comparable to the difference due to rotation and is in the same direction. Apparently, you really can lose "weight" by moving to the tropics.

Gravity Away from the Surface

Earlier we stated without proof that the law of gravitation applies to spherically symmetrical objects, where the mass of each body acts as if it were at the center of the body. Since **Equation 13.2** is derived from **Equation 13.1**, it is also valid for symmetrical mass distributions, but both equations are valid only for values of $r \geq R_E$. As we saw in **Example 13.4**, at 400 km above Earth's surface, where the International Space Station orbits, the value of g is 8.67 m/s^2. (We will see later that this is also the centripetal acceleration of the ISS.)

For $r < R_E$, **Equation 13.1** and **Equation 13.2** are not valid. However, we can determine g for these cases using a principle that comes from Gauss's law, which is a powerful mathematical tool that we study in more detail later in the course. A consequence of Gauss's law, applied to gravitation, is that only the mass *within r* contributes to the gravitational force. Also, that mass, just as before, can be considered to be located at the center. The gravitational effect of the mass *outside r* has zero net effect.

Two very interesting special cases occur. For a spherical planet with constant density, the mass within r is the density times the volume within r. This mass can be considered located at the center. Replacing M_E with only the mass within r, $M = \rho \times (\text{volume of a sphere})$, and R_E with r, **Equation 13.2** becomes

$$g = G\frac{M_E}{R_E^2} = G\frac{\rho\left(4/3\pi r^3\right)}{r^2} = \frac{4}{3}G\rho\pi r.$$

The value of g, and hence your weight, decreases linearly as you descend down a hole to the center of the spherical planet. At the center, you are weightless, as the mass of the planet pulls equally in all directions. Actually, Earth's density is not constant, nor is Earth solid throughout. **Figure 13.10** shows the profile of g if Earth had constant density and the more likely profile based upon estimates of density derived from seismic data.

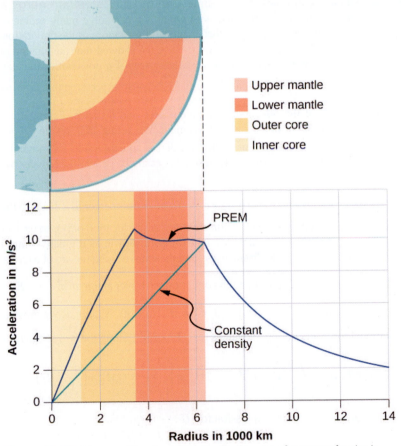

Figure 13.10 For $r < R_E$, the value of g for the case of constant density is the straight green line. The blue line from the PREM (Preliminary Reference Earth Model) is probably closer to the actual profile for g.

The second interesting case concerns living on a spherical shell planet. This scenario has been proposed in many science fiction stories. Ignoring significant engineering issues, the shell could be constructed with a desired radius and total mass, such that g at the surface is the same as Earth's. Can you guess what happens once you descend in an elevator to the inside of the shell, where there is no mass between you and the center? What benefits would this provide for traveling great distances from one point on the sphere to another? And finally, what effect would there be if the planet was spinning?

13.3 | Gravitational Potential Energy and Total Energy

Learning Objectives

By the end of this section, you will be able to:

*   Determine changes in gravitational potential energy over great distances
*   Apply conservation of energy to determine escape velocity
*   Determine whether astronomical bodies are gravitationally bound

We studied gravitational potential energy in **Potential Energy and Conservation of Energy**, where the value of g remained constant. We now develop an expression that works over distances such that g is not constant. This is necessary to correctly calculate the energy needed to place satellites in orbit or to send them on missions in space.

Gravitational Potential Energy beyond Earth

We defined work and potential energy in **Work and Kinetic Energy** and **Potential Energy and Conservation of Energy**. The usefulness of those definitions is the ease with which we can solve many problems using conservation of energy. Potential energy is particularly useful for forces that change with position, as the gravitational force does over large distances. In **Potential Energy and Conservation of Energy**, we showed that the change in gravitational potential energy near Earth's surface is $\Delta U = mg(y_2 - y_1)$. This works very well if g does not change significantly between y_1 and y_2. We return to the definition of work and potential energy to derive an expression that is correct over larger distances.

Recall that work (W) is the integral of the dot product between force and distance. Essentially, it is the product of the component of a force along a displacement times that displacement. We define ΔU as the *negative* of the work done by the force we associate with the potential energy. For clarity, we derive an expression for moving a mass m from distance r_1 from the center of Earth to distance r_2. However, the result can easily be generalized to any two objects changing their separation from one value to another.

Consider **Figure 13.11**, in which we take m from a distance r_1 from Earth's center to a distance that is r_2 from the center. Gravity is a conservative force (its magnitude and direction are functions of location only), so we can take any path we wish, and the result for the calculation of work is the same. We take the path shown, as it greatly simplifies the integration. We first move *radially* outward from distance r_1 to distance r_2, and then move along the arc of a circle until we reach the final position. During the radial portion, $\overrightarrow{\mathbf{F}}$ is opposite to the direction we travel along $d\overrightarrow{\mathbf{r}}$, so $E = K_1 + U_1 = K_2 + U_2$. Along the arc, $\overrightarrow{\mathbf{F}}$ is perpendicular to $d\overrightarrow{\mathbf{r}}$, so $\overrightarrow{\mathbf{F}} \cdot d\overrightarrow{\mathbf{r}} = 0$. No work is done as we move along the arc. Using the expression for the gravitational force and noting the values for $\overrightarrow{\mathbf{F}} \cdot d\overrightarrow{\mathbf{r}}$ along the two segments of our path, we have

$$\Delta U = -\int_{r_1}^{r_2} \overrightarrow{\mathbf{F}} \cdot d\overrightarrow{\mathbf{r}} = GM_E m \int_{r_1}^{r_2} \frac{dr}{r^2} = GM_E m\left(\frac{1}{r_1} - \frac{1}{r_2}\right).$$

Since $\Delta U = U_2 - U_1$, we can adopt a simple expression for U:

$$U = -\frac{GM_E m}{r}. \tag{13.4}$$

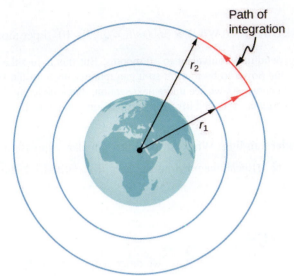

Figure 13.11 The work integral, which determines the change in potential energy, can be evaluated along the path shown in red.

Note two important items with this definition. First, $U \to 0$ as $r \to \infty$. The potential energy is zero when the two masses are infinitely far apart. Only the difference in U is important, so the choice of $U = 0$ for $r = \infty$ is merely one of convenience. (Recall that in earlier gravity problems, you were free to take $U = 0$ at the top or bottom of a building, or anywhere.) Second, note that U becomes increasingly more negative as the masses get closer. That is consistent with what you learned about potential energy in **Potential Energy and Conservation of Energy**. As the two masses are separated, positive work must be done against the force of gravity, and hence, U increases (becomes less negative). All masses naturally fall together under the influence of gravity, falling from a higher to a lower potential energy.

Example 13.6

Lifting a Payload

How much energy is required to lift the 9000-kg *Soyuz* vehicle from Earth's surface to the height of the ISS, 400 km above the surface?

Strategy

Use **Equation 13.2** to find the change in potential energy of the payload. That amount of work or energy must be supplied to lift the payload.

Solution

Paying attention to the fact that we start at Earth's surface and end at 400 km above the surface, the change in U is

$$\Delta U = U_{\text{orbit}} - U_{\text{Earth}} = -\frac{GM_{\text{E}}m}{R_{\text{E}} + 400\text{ km}} - \left(-\frac{GM_{\text{E}}m}{R_{\text{E}}}\right).$$

We insert the values

$$m = 9000\text{ kg}, \quad M_{\text{E}} = 5.96 \times 10^{24}\text{ kg}, \quad R_{\text{E}} = 6.37 \times 10^{6}\text{ m}$$

and convert 400 km into 4.00×10^{5} m. We find $\Delta U = 3.32 \times 10^{10}$ J. It is positive, indicating an increase in potential energy, as we would expect.

Significance

For perspective, consider that the average US household energy use in 2013 was 909 kWh per month. That is energy of

$$909 \text{ kWh} \times 1000 \text{ W/kW} \times 3600 \text{ s/h} = 3.27 \times 10^9 \text{ J per month.}$$

So our result is an energy expenditure equivalent to 10 months. But this is just the energy needed to raise the payload 400 km. If we want the *Soyuz* to be in orbit so it can rendezvous with the ISS and not just fall back to Earth, it needs a lot of kinetic energy. As we see in the next section, that kinetic energy is about five times that of ΔU. In addition, far more energy is expended lifting the propulsion system itself. Space travel is not cheap.

 13.3 Check Your Understanding Why not use the simpler expression $\Delta U = mg(y_2 - y_1)$? How significant would the error be? (Recall the previous result, in **Example 13.4**, that the value g at 400 km above the Earth is 8.67 m/s^2.)

Conservation of Energy

In **Potential Energy and Conservation of Energy**, we described how to apply conservation of energy for systems with conservative forces. We were able to solve many problems, particularly those involving gravity, more simply using conservation of energy. Those principles and problem-solving strategies apply equally well here. The only change is to place the new expression for potential energy into the conservation of energy equation, $E = K_1 + U_1 = K_2 + U_2$.

$$\frac{1}{2}mv_1^2 - \frac{GMm}{r_1} = \frac{1}{2}mv_2^2 - \frac{GMm}{r_2} \tag{13.5}$$

Note that we use M, rather than M_E, as a reminder that we are not restricted to problems involving Earth. However, we still assume that $m < < M$. (For problems in which this is not true, we need to include the kinetic energy of both masses and use conservation of momentum to relate the velocities to each other. But the principle remains the same.)

Escape velocity

Escape velocity is often defined to be the *minimum* initial velocity of an object that is required to escape the surface of a planet (or any large body like a moon) and never return. As usual, we assume no energy lost to an atmosphere, should there be any.

Consider the case where an object is launched from the surface of a planet with an initial velocity directed away from the planet. With the *minimum* velocity needed to escape, the object would *just* come to rest infinitely far away, that is, the object gives up the last of its kinetic energy just as it reaches infinity, where the force of gravity becomes zero. Since $U \to 0$ as $r \to \infty$, this means the total energy is zero. Thus, we find the escape velocity from the surface of an astronomical body of mass M and radius R by setting the total energy equal to zero. At the surface of the body, the object is located at $r_1 = R$ and it has escape velocity $v_1 = v_{esc}$. It reaches $r_2 = \infty$ with velocity $v_2 = 0$. Substituting into **Equation 13.5**, we have

$$\frac{1}{2}mv_{esc}^2 - \frac{GMm}{R} = \frac{1}{2}m0^2 - \frac{GMm}{\infty} = 0.$$

Solving for the escape velocity,

$$v_{esc} = \sqrt{\frac{2GM}{R}}. \tag{13.6}$$

Notice that m has canceled out of the equation. The escape velocity is the same for all objects, regardless of mass. Also, we are not restricted to the surface of the planet; R can be any starting point beyond the surface of the planet.

Example 13.7

Escape from Earth

What is the escape speed from the surface of Earth? Assume there is no energy loss from air resistance. Compare this to the escape speed from the Sun, starting from Earth's orbit.

Strategy

We use **Equation 13.6**, clearly defining the values of R and M. To escape Earth, we need the mass and radius of Earth. For escaping the Sun, we need the mass of the Sun, and the orbital distance between Earth and the Sun.

Solution

Substituting the values for Earth's mass and radius directly into **Equation 13.6**, we obtain

$$v_{esc} = \sqrt{\frac{2GM}{R}} = \sqrt{\frac{2(6.67 \times 10^{-11} \text{ N} \cdot \text{m}^2/\text{kg}^2)(5.96 \times 10^{24} \text{ kg})}{6.37 \times 10^6 \text{ m}}} = 1.12 \times 10^4 \text{ m/s}.$$

That is about 11 km/s or 25,000 mph. To escape the Sun, starting from Earth's orbit, we use $R = R_{ES} = 1.50 \times 10^{11}$ m and $M_{Sun} = 1.99 \times 10^{30}$ kg. The result is $v_{esc} = 4.21 \times 10^4$ m/s or about 42 km/s.

Significance

The speed needed to escape the Sun (leave the solar system) is nearly four times the escape speed from Earth's surface. But there is help in both cases. Earth is rotating, at a speed of nearly 1.7 km/s at the equator, and we can use that velocity to help escape, or to achieve orbit. For this reason, many commercial space companies maintain launch facilities near the equator. To escape the Sun, there is even more help. Earth revolves about the Sun at a speed of approximately 30 km/s. By launching in the direction that Earth is moving, we need only an additional 12 km/s. The use of gravitational assist from other planets, essentially a gravity slingshot technique, allows space probes to reach even greater speeds. In this slingshot technique, the vehicle approaches the planet and is accelerated by the planet's gravitational attraction. It has its greatest speed at the closest point of approach, although it decelerates in equal measure as it moves away. But relative to the planet, the vehicle's speed far before the approach, and long after, are the same. If the directions are chosen correctly, that can result in a significant increase (or decrease if needed) in the vehicle's speed relative to the rest of the solar system.

 Visit this **website (https://openstaxcollege.org/l/21escapevelocit)** to learn more about escape velocity.

 13.4   Check Your Understanding If we send a probe out of the solar system starting from Earth's surface, do we only have to escape the Sun?

Energy and gravitationally bound objects

As stated previously, escape velocity can be defined as the initial velocity of an object that can escape the surface of a moon or planet. More generally, it is the speed at *any* position such that the *total* energy is zero. If the total energy is zero or greater, the object escapes. If the total energy is negative, the object cannot escape. Let's see why that is the case.

As noted earlier, we see that $U \to 0$ as $r \to \infty$. If the total energy is zero, then as m reaches a value of r that approaches infinity, U becomes zero and so must the kinetic energy. Hence, m comes to rest infinitely far away from M. It has "just escaped" M. If the total energy is positive, then kinetic energy remains at $r = \infty$ and certainly m does not return. When the total energy is zero or greater, then we say that m is not gravitationally bound to M.

On the other hand, if the total energy is negative, then the kinetic energy must reach zero at some finite value of r, where U is negative and equal to the total energy. The object can never exceed this finite distance from M, since to do so would require the kinetic energy to become negative, which is not possible. We say m is **gravitationally bound** to M.

We have simplified this discussion by assuming that the object was headed directly away from the planet. What is remarkable is that the result applies for any velocity. Energy is a scalar quantity and hence **Equation 13.5** is a scalar equation—the direction of the velocity plays no role in conservation of energy. It is possible to have a gravitationally bound system where the masses do not "fall together," but maintain an orbital motion about each other.

We have one important final observation. Earlier we stated that if the total energy is zero or greater, the object escapes. Strictly speaking, **Equation 13.5** and **Equation 13.6** apply for point objects. They apply to finite-sized, spherically symmetric objects as well, provided that the value for r in **Equation 13.5** is always greater than the sum of the radii of the two objects. If r becomes less than this sum, then the objects collide. (Even for greater values of r, but near the sum of the radii, gravitational tidal forces could create significant effects if both objects are planet sized. We examine tidal effects in **Tidal Forces**.) Neither positive nor negative total energy precludes finite-sized masses from colliding. For real objects, direction is important.

Example 13.8

How Far Can an Object Escape?

Let's consider the preceding example again, where we calculated the escape speed from Earth and the Sun, starting from Earth's orbit. We noted that Earth already has an orbital speed of 30 km/s. As we see in the next section, that is the tangential speed needed to stay in circular orbit. If an object had this speed at the distance of Earth's orbit, but was headed directly away from the Sun, how far would it travel before coming to rest? Ignore the gravitational effects of any other bodies.

Strategy

The object has initial kinetic and potential energies that we can calculate. When its speed reaches zero, it is at its maximum distance from the Sun. We use **Equation 13.5**, conservation of energy, to find the distance at which kinetic energy is zero.

Solution

The initial position of the object is Earth's radius of orbit and the intial speed is given as 30 km/s. The final velocity is zero, so we can solve for the distance at that point from the conservation of energy equation. Using $R_{ES} = 1.50 \times 10^{11}$ m and $M_{Sun} = 1.99 \times 10^{30}$ kg , we have

$$\frac{1}{2}mv_1^2 - \frac{GMm}{r_1} = \frac{1}{2}mv_2^2 - \frac{GMm}{r_2}$$

$$\frac{1}{2}\cancel{m}(3.0 \times 10^3 \text{ m/s})^2 - \frac{(6.67 \times 10^{-11} \text{ N} \cdot \text{m/kg}^2)(1.99 \times 10^{30} \text{ kg})\cancel{m}}{1.50 \times 10^{11} \text{ m}}$$

$$= \frac{1}{2}\cancel{m}0^2 - \frac{(6.67 \times 10^{-11} \text{ N} \cdot \text{m/kg}^2)(1.99 \times 10^{30} \text{ kg})\cancel{m}}{r_2}$$

where the mass m cancels. Solving for r_2 we get $r_2 = 3.0 \times 10^{11}$ m . Note that this is twice the initial distance from the Sun and takes us past Mars's orbit, but not quite to the asteroid belt.

Significance

The object in this case reached a distance *exactly* twice the initial orbital distance. We will see the reason for this in the next section when we calculate the speed for circular orbits.

 13.5   Check Your Understanding Assume you are in a spacecraft in orbit about the Sun at Earth's orbit, but far away from Earth (so that it can be ignored). How could you redirect your tangential velocity to the radial direction such that you could then pass by Mars's orbit? What would be required to change just the direction of the velocity?

13.4 | Satellite Orbits and Energy

Learning Objectives

By the end of this section, you will be able to:

- Describe the mechanism for circular orbits
- Find the orbital periods and speeds of satellites
- Determine whether objects are gravitationally bound

The Moon orbits Earth. In turn, Earth and the other planets orbit the Sun. The space directly above our atmosphere is filled with artificial satellites in orbit. We examine the simplest of these orbits, the circular orbit, to understand the relationship between the speed and period of planets and satellites in relation to their positions and the bodies that they orbit.

Circular Orbits

As noted at the beginning of this chapter, Nicolaus Copernicus first suggested that Earth and all other planets orbit the Sun in circles. He further noted that orbital periods increased with distance from the Sun. Later analysis by Kepler showed that these orbits are actually ellipses, but the orbits of most planets in the solar system are nearly circular. Earth's orbital distance from the Sun varies a mere 2%. The exception is the eccentric orbit of Mercury, whose orbital distance varies nearly 40%.

Determining the **orbital speed** and **orbital period** of a satellite is much easier for circular orbits, so we make that assumption in the derivation that follows. As we described in the previous section, an object with negative total energy is gravitationally bound and therefore is in orbit. Our computation for the special case of circular orbits will confirm this. We focus on objects orbiting Earth, but our results can be generalized for other cases.

Consider a satellite of mass m in a circular orbit about Earth at distance r from the center of Earth (**Figure 13.12**). It has centripetal acceleration directed toward the center of Earth. Earth's gravity is the only force acting, so Newton's second law gives

$$\frac{GmM_{\text{E}}}{r^2} = ma_{\text{c}} = \frac{mv_{\text{orbit}}^2}{r}.$$

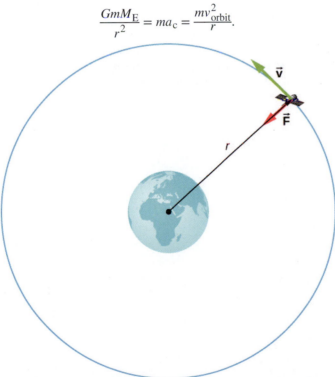

Figure 13.12 A satellite of mass m orbiting at radius r from the center of Earth. The gravitational force supplies the centripetal acceleration.

We solve for the speed of the orbit, noting that *m* cancels, to get the orbital speed

$$v_{\text{orbit}} = \sqrt{\frac{GM_{\text{E}}}{r}}. \tag{13.7}$$

Consistent with what we saw in **Equation 13.2** and **Equation 13.6**, *m* does not appear in **Equation 13.7**. The value of *g*, the escape velocity, and orbital velocity depend only upon the distance from the center of the planet, and *not* upon the mass of the object being acted upon. Notice the similarity in the equations for v_{orbit} and v_{esc}. The escape velocity is exactly $\sqrt{2}$ times greater, about 40%, than the orbital velocity. This comparison was noted in **Example 13.7**, and it is true for a satellite at any radius.

To find the period of a circular orbit, we note that the satellite travels the circumference of the orbit $2\pi r$ in one period *T*. Using the definition of speed, we have $v_{\text{orbit}} = 2\pi r/T$. We substitute this into **Equation 13.7** and rearrange to get

$$T = 2\pi\sqrt{\frac{r^3}{GM_{\text{E}}}}. \tag{13.8}$$

We see in the next section that this represents Kepler's third law for the case of circular orbits. It also confirms Copernicus's observation that the period of a planet increases with increasing distance from the Sun. We need only replace M_{E} with M_{Sun} in **Equation 13.8**.

We conclude this section by returning to our earlier discussion about astronauts in orbit appearing to be weightless, as if they were free-falling towards Earth. In fact, they are in free fall. Consider the trajectories shown in **Figure 13.13**. (This figure is based on a drawing by Newton in his *Principia* and also appeared earlier in **Motion in Two and Three Dimensions**.) All the trajectories shown that hit the surface of Earth have less than orbital velocity. The astronauts would accelerate toward Earth along the noncircular paths shown and feel weightless. (Astronauts actually train for life in orbit by riding in airplanes that free fall for 30 seconds at a time.) But with the correct orbital velocity, Earth's surface curves away from them at exactly the same rate as they fall toward Earth. Of course, staying the same distance from the surface is the point of a circular orbit.

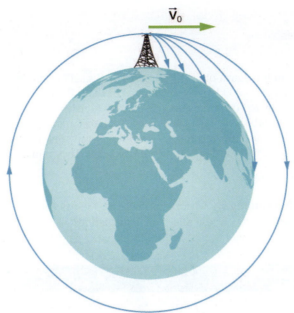

Figure 13.13 A circular orbit is the result of choosing a tangential velocity such that Earth's surface curves away at the same rate as the object falls toward Earth.

We can summarize our discussion of orbiting satellites in the following Problem-Solving Strategy.

Problem-Solving Strategy: Orbits and Conservation of Energy

1. Determine whether the equations for speed, energy, or period are valid for the problem at hand. If not, start with the first principles we used to derive those equations.

2. To start from first principles, draw a free-body diagram and apply Newton's law of gravitation and Newton's second law.

3. Along with the definitions for speed and energy, apply Newton's second law of motion to the bodies of interest.

Example 13.9

The International Space Station

Determine the orbital speed and period for the International Space Station (ISS).

Strategy

Since the ISS orbits 4.00×10^2 km above Earth's surface, the radius at which it orbits is $R_E + 4.00 \times 10^2$ km.

We use **Equation 13.7** and **Equation 13.8** to find the orbital speed and period, respectively.

Solution

Using **Equation 13.7**, the orbital velocity is

$$v_{\text{orbit}} = \sqrt{\frac{GM_E}{r}} = \sqrt{\frac{6.67 \times 10^{-11} \text{ N} \cdot \text{m}^2/\text{kg}^2 (5.96 \times 10^{24} \text{ kg})}{(6.36 \times 10^6 + 4.00 \times 10^5 \text{ m})}} = 7.67 \times 10^3 \text{ m/s}$$

which is about 17,000 mph. Using **Equation 13.8**, the period is

$$T = 2\pi \sqrt{\frac{r^3}{GM_E}} = 2\pi \sqrt{\frac{(6.37 \times 10^6 + 4.00 \times 10^5 \text{ m})^3}{(6.67 \times 10^{-11} \text{ N} \cdot \text{m}^2/\text{kg}^2)(5.96 \times 10^{24} \text{ kg})}} = 5.55 \times 10^3 \text{ s}$$

which is just over 90 minutes.

Significance

The ISS is considered to be in low Earth orbit (LEO). Nearly all satellites are in LEO, including most weather satellites. GPS satellites, at about 20,000 km, are considered medium Earth orbit. The higher the orbit, the more energy is required to put it there and the more energy is needed to reach it for repairs. Of particular interest are the satellites in geosynchronous orbit. All fixed satellite dishes on the ground pointing toward the sky, such as TV reception dishes, are pointed toward geosynchronous satellites. These satellites are placed at the exact distance, and just above the equator, such that their period of orbit is 1 day. They remain in a fixed position relative to Earth's surface.

 13.6   Check Your Understanding By what factor must the radius change to reduce the orbital velocity of a satellite by one-half? By what factor would this change the period?

Example 13.10

Determining the Mass of Earth

Determine the mass of Earth from the orbit of the Moon.

Strategy

We use **Equation 13.8**, solve for M_E, and substitute for the period and radius of the orbit. The radius and period of the Moon's orbit was measured with reasonable accuracy thousands of years ago. From the astronomical data in **Appendix D**, the period of the Moon is 27.3 days $= 2.36 \times 10^6$ s, and the *average* distance between the centers of Earth and the Moon is 384,000 km.

Solution

Solving for M_E,

$$T = 2\pi\sqrt{\frac{r^3}{GM_E}}$$

$$M_E = \frac{4\pi^2 r^3}{GT^2} = \frac{4\pi^2 (3.84 \times 10^8 \text{ m})^3}{(6.67 \times 10^{-11} \text{ N} \cdot \text{m}^2/\text{kg}^2)(2.36 \times 10^6 \text{ m})^2} = 6.01 \times 10^{24} \text{ kg.}$$

Significance

Compare this to the value of 5.96×10^{24} kg that we obtained in **Example 13.5**, using the value of g at the surface of Earth. Although these values are very close (~0.8%), both calculations use average values. The value of g varies from the equator to the poles by approximately 0.5%. But the Moon has an elliptical orbit in which the value of r varies just over 10%. (The apparent size of the full Moon actually varies by about this amount, but it is difficult to notice through casual observation as the time from one extreme to the other is many months.)

 13.7   Check Your Understanding There is another consideration to this last calculation of M_E. We derived **Equation 13.8** assuming that the satellite orbits around the center of the astronomical body at the same radius used in the expression for the gravitational force between them. What assumption is made to justify this? Earth is about 81 times more massive than the Moon. Does the Moon orbit about the exact center of Earth?

Example 13.11

Galactic Speed and Period

Let's revisit **Example 13.2**. Assume that the Milky Way and Andromeda galaxies are in a circular orbit about each other. What would be the velocity of each and how long would their orbital period be? Assume the mass of each is 800 billion solar masses and their centers are separated by 2.5 million light years.

Strategy

We cannot use **Equation 13.7** and **Equation 13.8** directly because they were derived assuming that the object of mass m orbited about the center of a much larger planet of mass M. We determined the gravitational force in **Example 13.2** using Newton's law of universal gravitation. We can use Newton's second law, applied to the centripetal acceleration of either galaxy, to determine their tangential speed. From that result we can determine the period of the orbit.

Solution

In **Example 13.2**, we found the force between the galaxies to be

$$F_{12} = G\frac{m_1 m_2}{r^2} = (6.67 \times 10^{-11} \text{ N} \cdot \text{m}^2/\text{kg}^2)\frac{[(800 \times 10^9)(2.0 \times 10^{30} \text{ kg})]^2}{[(2.5 \times 10^6)(9.5 \times 10^{15} \text{ m})]^2} = 3.0 \times 10^{29} \text{ N}$$

and that the acceleration of each galaxy is

$$a = \frac{F}{m} = \frac{3.0 \times 10^{29} \text{ N}}{(800 \times 10^9)(2.0 \times 10^{30} \text{ kg})} = 1.9 \times 10^{-13} \text{ m/s}^2.$$

Since the galaxies are in a circular orbit, they have centripetal acceleration. If we ignore the effect of other galaxies, then, as we learned in **Linear Momentum and Collisions** and **Fixed-Axis Rotation**, the centers of mass of the two galaxies remain fixed. Hence, the galaxies must orbit about this common center of mass. For equal masses, the center of mass is exactly half way between them. So the radius of the orbit, r_{orbit}, is not the same as the distance between the galaxies, but one-half that value, or 1.25 million light-years. These two different values are shown in **Figure 13.14**.

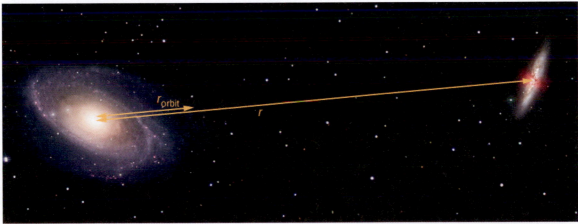

Figure 13.14 The distance between two galaxies, which determines the gravitational force between them, is r, and is different from r_{orbit}, which is the radius of orbit for each. For equal masses, $r_{\text{orbit}} = 1/2r$. (credit: modification of work by Marc Van Norden)

Using the expression for centripetal acceleration, we have

$$a_c = \frac{v_{\text{orbit}}^2}{r_{\text{orbit}}}$$

$$1.9 \times 10^{-13} \text{ m/s}^2 = \frac{v_{\text{orbit}}^2}{(1.25 \times 10^6)(9.5 \times 10^{15} \text{ m})}.$$

Solving for the orbit velocity, we have $v_{\text{orbit}} = 47 \text{ km/s}$. Finally, we can determine the period of the orbit directly from $T = 2\pi r/v_{\text{orbit}}$, to find that the period is $T = 1.6 \times 10^{18} \text{ s}$, about 50 billion years.

Significance

The orbital speed of 47 km/s might seem high at first. But this speed is comparable to the escape speed from the Sun, which we calculated in an earlier example. To give even more perspective, this period is nearly four times longer than the time that the Universe has been in existence.

In fact, the present relative motion of these two galaxies is such that they are expected to collide in about 4 billion years. Although the density of stars in each galaxy makes a direct collision of any two stars unlikely, such a collision will have a dramatic effect on the shape of the galaxies. Examples of such collisions are well known in astronomy.

13.8   Check Your Understanding Galaxies are not single objects. How does the gravitational force of one galaxy exerted on the "closer" stars of the other galaxy compare to those farther away? What effect would this have on the shape of the galaxies themselves?

See the **Sloan Digital Sky Survey page (https://openstaxcollege.org/l/21sloandigskysu)** for more information on colliding galaxies.

Energy in Circular Orbits

In **Gravitational Potential Energy and Total Energy**, we argued that objects are gravitationally bound if their total energy is negative. The argument was based on the simple case where the velocity was directly away or toward the planet. We now examine the total energy for a circular orbit and show that indeed, the total energy is negative. As we did earlier, we start with Newton's second law applied to a circular orbit,

$$\frac{GmM_{\text{E}}}{r^2} = ma_c = \frac{mv^2}{r}$$

$$\frac{GmM_{\text{E}}}{r} = mv^2.$$

In the last step, we multiplied by r on each side. The right side is just twice the kinetic energy, so we have

$$K = \frac{1}{2}mv^2 = \frac{GmM_{\text{E}}}{2r}.$$

The total energy is the sum of the kinetic and potential energies, so our final result is

$$E = K + U = \frac{GmM_{\text{E}}}{2r} - \frac{GmM_{\text{E}}}{r} = -\frac{GmM_{\text{E}}}{2r}. \tag{13.9}$$

We can see that the total energy is negative, with the same magnitude as the kinetic energy. For circular orbits, the magnitude of the kinetic energy is exactly one-half the magnitude of the potential energy. Remarkably, this result applies to any two masses in circular orbits about their common center of mass, at a distance r from each other. The proof of this is left as an exercise. We will see in the next section that a very similar expression applies in the case of elliptical orbits.

Example 13.12

Energy Required to Orbit

In **Example 13.8**, we calculated the energy required to simply lift the 9000-kg *Soyuz* vehicle from Earth's surface to the height of the ISS, 400 km above the surface. In other words, we found its *change* in potential energy.

We now ask, what total energy change in the *Soyuz* vehicle is required to take it from Earth's surface and put it in orbit with the ISS for a rendezvous (**Figure 13.15**)? How much of that total energy is kinetic energy?

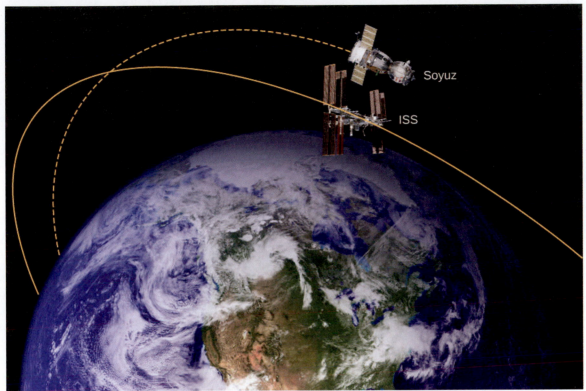

Figure 13.15 The *Soyuz* in a rendezvous with the ISS. Note that this diagram is not to scale; the *Soyuz* is very small compared to the ISS and its orbit is much closer to Earth. (credit: modification of works by NASA)

Strategy

The energy required is the difference in the *Soyuz*'s total energy in orbit and that at Earth's surface. We can use **Equation 13.9** to find the total energy of the *Soyuz* at the ISS orbit. But the total energy at the surface is simply the potential energy, since it starts from rest. [Note that we *do not* use **Equation 13.9** at the surface, since we are not in orbit at the surface.] The kinetic energy can then be found from the difference in the total energy change and the change in potential energy found in **Example 13.8**. Alternatively, we can use **Equation 13.7** to find v_{orbit} and calculate the kinetic energy directly from that. The total energy required is then the kinetic energy plus the change in potential energy found in **Example 13.8**.

Solution

From **Equation 13.9**, the total energy of the *Soyuz* in the same orbit as the ISS is

$$E_{\text{orbit}} = K_{\text{orbit}} + U_{\text{orbit}} = -\frac{GmM_{\text{E}}}{2r}$$

$$= \frac{(6.67 \times 10^{-11} \text{ N} \cdot \text{m}^2/\text{kg}^2)(9000 \text{ kg})(5.96 \times 10^{24} \text{ kg})}{2(6.36 \times 10^6 + 4.00 \times 10^5 \text{ m})} = -2.65 \times 10^{11} \text{ J.}$$

The total energy at Earth's surface is

$$E_{\text{surface}} = K_{\text{surface}} + U_{\text{surface}} = 0 - \frac{GmM_{\text{E}}}{r}$$

$$= -\frac{(6.67 \times 10^{-11} \text{ N} \cdot \text{m}^2/\text{kg}^2)(9000 \text{ kg})(5.96 \times 10^{24} \text{ kg})}{(6.36 \times 10^6 \text{ m})}$$

$$= -5.63 \times 10^{11} \text{ J.}$$

The change in energy is $\Delta E = E_{\text{orbit}} - E_{\text{surface}} = 2.98 \times 10^{11}$ J. To get the kinetic energy, we subtract the change in potential energy from **Example 13.6**, $\Delta U = 3.32 \times 10^{10}$ J. That gives us $K_{\text{orbit}} = 2.98 \times 10^{11} - 3.32 \times 10^{10} = 2.65 \times 10^{11}$ J. As stated earlier, the kinetic energy of a circular orbit is always one-half the magnitude of the potential energy, and the same as the magnitude of the total energy. Our result confirms this.

The second approach is to use **Equation 13.7** to find the orbital speed of the *Soyuz*, which we did for the ISS in **Example 13.9**.

$$v_{\text{orbit}} = \sqrt{\frac{GM_{\text{E}}}{r}} = \sqrt{\frac{(6.67 \times 10^{-11}\ \text{N} \cdot \text{m}^2/\text{kg}^2)(5.96 \times 10^{24}\ \text{kg})}{(6.36 \times 10^6 + 4.00 \times 10^5\ \text{m})}} = 7.67 \times 10^3\ \text{m/s}.$$

So the kinetic energy of the *Soyuz* in orbit is

$$K_{\text{orbit}} = \tfrac{1}{2}mv^2_{\text{orbit}} = \tfrac{1}{2}(9000\ \text{kg})(7.67 \times 10^3\ \text{m/s})^2 = 2.65 \times 10^{11}\ \text{J},$$

the same as in the previous method. The total energy is just

$$E_{\text{orbit}} = K_{\text{orbit}} + \Delta U = 2.65 \times 10^{11} + 3.32 \times 10^{10} = 2.95 \times 10^{11}\ \text{J}.$$

Significance

The kinetic energy of the *Soyuz* is nearly eight times the change in its potential energy, or 90% of the total energy needed for the rendezvous with the ISS. And it is important to remember that this energy represents only the energy that must be given to the *Soyuz*. With our present rocket technology, the mass of the propulsion system (the rocket fuel, its container and combustion system) far exceeds that of the payload, and a tremendous amount of kinetic energy must be given to that mass. So the actual cost in energy is many times that of the change in energy of the payload itself.

13.5 | Kepler's Laws of Planetary Motion

Learning Objectives

By the end of this section, you will be able to:

- Describe the conic sections and how they relate to orbital motion
- Describe how orbital velocity is related to conservation of angular momentum
- Determine the period of an elliptical orbit from its major axis

Using the precise data collected by Tycho Brahe, Johannes Kepler carefully analyzed the positions in the sky of all the known planets and the Moon, plotting their positions at regular intervals of time. From this analysis, he formulated three laws, which we address in this section.

Kepler's First Law

The prevailing view during the time of Kepler was that all planetary orbits were circular. The data for Mars presented the greatest challenge to this view and that eventually encouraged Kepler to give up the popular idea. **Kepler's first law** states that every planet moves along an ellipse, with the Sun located at a focus of the ellipse. An ellipse is defined as the set of all points such that the sum of the distance from each point to two foci is a constant. **Figure 13.16** shows an ellipse and describes a simple way to create it.

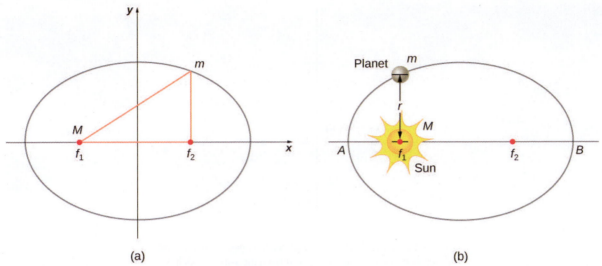

(a)   (b)

Figure 13.16 (a) An ellipse is a curve in which the sum of the distances from a point on the curve to two foci $(f_1$ and $f_2)$ is a constant. From this definition, you can see that an ellipse can be created in the following way. Place a pin at each focus, then place a loop of string around a pencil and the pins. Keeping the string taught, move the pencil around in a complete circuit. If the two foci occupy the same place, the result is a circle—a special case of an ellipse. (b) For an elliptical orbit, if $m \ll M$, then m follows an elliptical path with M at one focus. More exactly, both m and M move in their own ellipse about the common center of mass.

For elliptical orbits, the point of closest approach of a planet to the Sun is called the **perihelion**. It is labeled point A in **Figure 13.16**. The farthest point is the **aphelion** and is labeled point B in the figure. For the Moon's orbit about Earth, those points are called the perigee and apogee, respectively.

An ellipse has several mathematical forms, but all are a specific case of the more general equation for conic sections. There are four different conic sections, all given by the equation

$$\frac{\alpha}{r} = 1 + e\cos\theta. \tag{13.10}$$

The variables r and θ are shown in **Figure 13.17** in the case of an ellipse. The constants α and e are determined by the total energy and angular momentum of the satellite at a given point. The constant e is called the eccentricity. The values of α and e determine which of the four conic sections represents the path of the satellite.

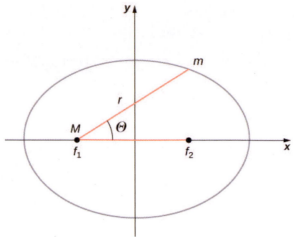

Figure 13.17 As before, the distance between the planet and the Sun is r, and the angle measured from the x-axis, which is along the major axis of the ellipse, is θ.

One of the real triumphs of Newton's law of universal gravitation, with the force proportional to the inverse of the distance squared, is that when it is combined with his second law, the solution for the path of any satellite is a conic section. Every path taken by m is one of the four conic sections: a circle or an ellipse for bound or closed orbits, or a parabola or hyperbola for unbounded or open orbits. These conic sections are shown in **Figure 13.18**.

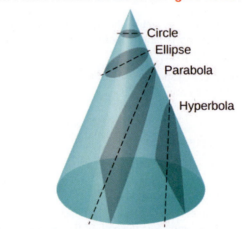

Figure 13.18 All motion caused by an inverse square force is one of the four conic sections and is determined by the energy and direction of the moving body.

If the total energy is negative, then $0 \le e < 1$, and **Equation 13.10** represents a bound or closed orbit of either an ellipse or a circle, where $e = 0$. [You can see from **Equation 13.10** that for $e = 0$, $r = \alpha$, and hence the radius is constant.] For ellipses, the eccentricity is related to how oblong the ellipse appears. A circle has zero eccentricity, whereas a very long, drawn-out ellipse has an eccentricity near one.

If the total energy is exactly zero, then $e = 1$ and the path is a parabola. Recall that a satellite with zero total energy has exactly the escape velocity. (The parabola is formed only by slicing the cone parallel to the tangent line along the surface.) Finally, if the total energy is positive, then $e > 1$ and the path is a hyperbola. These last two paths represent unbounded orbits, where m passes by M once and only once. This situation has been observed for several comets that approach the Sun and then travel away, never to return.

We have confined ourselves to the case in which the smaller mass (planet) orbits a much larger, and hence stationary, mass (Sun), but **Equation 13.10** also applies to any two gravitationally interacting masses. Each mass traces out the exact same-shaped conic section as the other. That shape is determined by the total energy and angular momentum of the system, with the center of mass of the system located at the focus. The ratio of the dimensions of the two paths is the inverse of the ratio of their masses.

 You can see an animation of two interacting objects at the *My Solar System* page at **Phet (https://openstaxcollege.org/l/21mysolarsys)** . Choose the Sun and Planet preset option. You can also view the more complicated multiple body problems as well. You may find the actual path of the Moon quite surprising, yet is obeying Newton's simple laws of motion.

Orbital Transfers

People have imagined traveling to the other planets of our solar system since they were discovered. But how can we best do this? The most efficient method was discovered in 1925 by Walter Hohmann, inspired by a popular science fiction novel of that time. The method is now called a Hohmann transfer. For the case of traveling between two circular orbits, the transfer is along a "transfer" ellipse that perfectly intercepts those orbits at the aphelion and perihelion of the ellipse. **Figure 13.19** shows the case for a trip from Earth's orbit to that of Mars. As before, the Sun is at the focus of the ellipse.

For any ellipse, the semi-major axis is defined as one-half the sum of the perihelion and the aphelion. In **Figure 13.17**, the semi-major axis is the distance from the origin to either side of the ellipse along the x-axis, or just one-half the longest axis (called the major axis). Hence, to travel from one circular orbit of radius r_1 to another circular orbit of radius r_2, the aphelion of the transfer ellipse will be equal to the value of the larger orbit, while the perihelion will be the smaller orbit. The semi-major axis, denoted a, is therefore given by $a = \frac{1}{2}(r_1 + r_2)$.

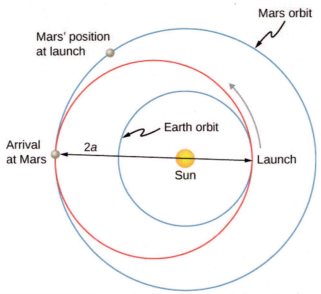

Figure 13.19 The transfer ellipse has its perihelion at Earth's orbit and aphelion at Mars' orbit.

Let's take the case of traveling from Earth to Mars. For the moment, we ignore the planets and assume we are alone in Earth's orbit and wish to move to Mars' orbit. From **Equation 13.9**, the expression for total energy, we can see that the total energy for a spacecraft in the larger orbit (Mars) is greater (less negative) than that for the smaller orbit (Earth). To move onto the transfer ellipse from Earth's orbit, we will need to increase our kinetic energy, that is, we need a velocity boost. The most efficient method is a very quick acceleration along the circular orbital path, which is also along the path of the ellipse at that point. (In fact, the acceleration should be instantaneous, such that the circular and elliptical orbits are congruent during the acceleration. In practice, the finite acceleration is short enough that the difference is not a significant consideration.) Once you have arrived at Mars orbit, you will need another velocity boost to move into that orbit, or you will stay on the elliptical orbit and simply fall back to perihelion where you started. For the return trip, you simply reverse the process with a retro-boost at each transfer point.

To make the move onto the transfer ellipse and then off again, we need to know each circular orbit velocity and the transfer orbit velocities at perihelion and aphelion. The velocity boost required is simply the difference between the circular orbit velocity and the elliptical orbit velocity at each point. We can find the circular orbital velocities from **Equation 13.7**. To determine the velocities for the ellipse, we state without proof (as it is beyond the scope of this course) that total energy for

an elliptical orbit is

$$E = -\frac{GmM_S}{2a}$$

where M_S is the mass of the Sun and a is the semi-major axis. Remarkably, this is the same as **Equation 13.9** for circular orbits, but with the value of the semi-major axis replacing the orbital radius. Since we know the potential energy from **Equation 13.4**, we can find the kinetic energy and hence the velocity needed for each point on the ellipse. We leave it as a challenge problem to find those transfer velocities for an Earth-to-Mars trip.

We end this discussion by pointing out a few important details. First, we have not accounted for the gravitational potential energy due to Earth and Mars, or the mechanics of landing on Mars. In practice, that must be part of the calculations. Second, timing is everything. You do not want to arrive at the orbit of Mars to find out it isn't there. We must leave Earth at precisely the correct time such that Mars will be at the aphelion of our transfer ellipse just as we arrive. That opportunity comes about every 2 years. And returning requires correct timing as well. The total trip would take just under 3 years! There are other options that provide for a faster transit, including a gravity assist flyby of Venus. But these other options come with an additional cost in energy and danger to the astronauts.

 Visit this **site (https://openstaxcollege.org/l/21plantripmars)** for more details about planning a trip to Mars.

Kepler's Second Law

Kepler's second law states that a planet sweeps out equal areas in equal times, that is, the area divided by time, called the areal velocity, is constant. Consider **Figure 13.20**. The time it takes a planet to move from position A to B, sweeping out area A_1, is exactly the time taken to move from position C to D, sweeping area A_2, and to move from E to F, sweeping out area A_3. These areas are the same: $A_1 = A_2 = A_3$.

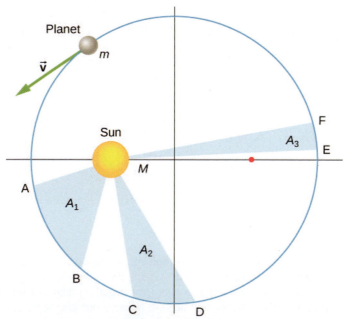

Figure 13.20 The shaded regions shown have equal areas and represent the same time interval.

Comparing the areas in the figure and the distance traveled along the ellipse in each case, we can see that in order for the areas to be equal, the planet must speed up as it gets closer to the Sun and slow down as it moves away. This behavior is completely consistent with our conservation equation, **Equation 13.5**. But we will show that Kepler's second law is actually a consequence of the conservation of angular momentum, which holds for any system with only radial forces.

Recall the definition of angular momentum from **Angular Momentum**, $\vec{L} = \vec{r} \times \vec{p}$. For the case of orbiting motion, $\overrightarrow{\textbf{L}}$ is the angular momentum of the planet about the Sun, $\overrightarrow{\textbf{r}}$ is the position vector of the planet measured from the Sun, and $\overrightarrow{\textbf{p}} = m\overrightarrow{\textbf{v}}$ is the instantaneous linear momentum at any point in the orbit. Since the planet moves along the ellipse, $\overrightarrow{\textbf{p}}$ is always tangent to the ellipse.

We can resolve the linear momentum into two components: a radial component $\overrightarrow{\textbf{p}}_{\text{rad}}$ along the line to the Sun, and a component $\overrightarrow{\textbf{p}}_{\text{perp}}$ perpendicular to $\overrightarrow{\textbf{r}}$. The cross product for angular momentum can then be written as

$$\overrightarrow{\textbf{L}} = \overrightarrow{\textbf{r}} \times \overrightarrow{\textbf{p}} = \overrightarrow{\textbf{r}} \times \left(\overrightarrow{\textbf{p}}_{\text{rad}} + \overrightarrow{\textbf{p}}_{\text{perp}} \right) = \overrightarrow{\textbf{r}} \times \overrightarrow{\textbf{p}}_{\text{rad}} + \overrightarrow{\textbf{r}} \times \overrightarrow{\textbf{p}}_{\text{perp}}.$$

The first term on the right is zero because $\overrightarrow{\textbf{r}}$ is parallel to $\overrightarrow{\textbf{p}}_{\text{rad}}$, and in the second term $\overrightarrow{\textbf{r}}$ is perpendicular to $\overrightarrow{\textbf{p}}_{\text{perp}}$, so the magnitude of the cross product reduces to $L = rp_{\text{perp}} = rmv_{\text{perp}}$. Note that the angular momentum does *not* depend upon p_{rad}. Since the gravitational force is only in the radial direction, it can change only p_{rad} and not p_{perp}; hence, the angular momentum must remain constant.

Now consider **Figure 13.21**. A small triangular area ΔA is swept out in time Δt. The velocity is along the path and it makes an angle θ with the radial direction. Hence, the perpendicular velocity is given by $v_{\text{perp}} = v\sin\theta$. The planet moves a distance $\Delta s = v\Delta t\sin\theta$ projected along the direction perpendicular to r. Since the area of a triangle is one-half the base (r) times the height (Δs), for a small displacement, the area is given by $\Delta A = \frac{1}{2}r\Delta s$. Substituting for Δs, multiplying by m in the numerator and denominator, and rearranging, we obtain

$$\Delta A = \frac{1}{2}r\Delta s = \frac{1}{2}r(v\Delta t\sin\theta) = \frac{1}{2m}r(mv\sin\theta\Delta t) = \frac{1}{2m}r(mv_{\text{perp}}\Delta t) = \frac{L}{2m}\Delta t.$$

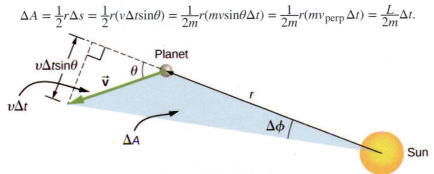

Figure 13.21 The element of area ΔA swept out in time Δt as the planet moves through angle $\Delta\phi$. The angle between the radial direction and $\overrightarrow{\textbf{v}}$ is θ.

The areal velocity is simply the rate of change of area with time, so we have

$$\text{areal velocity} = \frac{\Delta A}{\Delta t} = \frac{L}{2m}.$$

Since the angular momentum is constant, the areal velocity must also be constant. This is exactly Kepler's second law. As with Kepler's first law, Newton showed it was a natural consequence of his law of gravitation.

 You can view an **animated version (https://openstaxcollege.org/l/21animationgrav)** of **Figure 13.20**, and many other interesting animations as well, at the School of Physics (University of New South Wales) site.

Kepler's Third Law

Kepler's third law states that the square of the period is proportional to the cube of the semi-major axis of the orbit. In **Satellite Orbits and Energy**, we derived Kepler's third law for the special case of a circular orbit. **Equation 13.8** gives us the period of a circular orbit of radius r about Earth:

$$T = 2\pi\sqrt{\frac{r^3}{GM_{\text{E}}}}.$$

For an ellipse, recall that the semi-major axis is one-half the sum of the perihelion and the aphelion. For a circular orbit, the semi-major axis (*a*) is the same as the radius for the orbit. In fact, **Equation 13.8** gives us Kepler's third law if we simply replace *r* with *a* and square both sides.

$$T^2 = \frac{4\pi^2}{GM}a^3 \qquad\qquad\qquad\qquad \textbf{(13.11)}$$

We have changed the mass of Earth to the more general *M*, since this equation applies to satellites orbiting any large mass.

Example 13.13

Orbit of Halley's Comet

Determine the semi-major axis of the orbit of Halley's comet, given that it arrives at perihelion every 75.3 years. If the perihelion is 0.586 AU, what is the aphelion?

Strategy

We are given the period, so we can rearrange **Equation 13.11**, solving for the semi-major axis. Since we know the value for the perihelion, we can use the definition of the semi-major axis, given earlier in this section, to find the aphelion. We note that 1 Astronomical Unit (AU) is the average radius of Earth's orbit and is defined to be $1 \text{ AU} = 1.50 \times 10^{11} \text{ m}$.

Solution

Rearranging **Equation 13.11** and inserting the values of the period of Halley's comet and the mass of the Sun, we have

$$a = \left(\frac{GM}{4\pi^2}T^2\right)^{1/3}$$

$$= \left(\frac{(6.67 \times 10^{-11} \text{ N}\cdot\text{m}^2/\text{kg}^2)(2.00 \times 10^{30} \text{ kg})}{4\pi^2}(75.3 \text{ yr} \times 365 \text{ days/yr} \times 24 \text{ hr/day} \times 3600 \text{ s/hr})^2\right)^{1/3}.$$

This yields a value of 2.67×10^{12} m or 17.8 AU for the semi-major axis.

The semi-major axis is one-half the sum of the aphelion and perihelion, so we have

$$a = \frac{1}{2}(\text{aphelion} + \text{perihelion})$$
$$\text{aphelion} = 2a - \text{perihelion}.$$

Substituting for the values, we found for the semi-major axis and the value given for the perihelion, we find the value of the aphelion to be 35.0 AU.

Significance

Edmond Halley, a contemporary of Newton, first suspected that three comets, reported in 1531, 1607, and 1682, were actually the same comet. Before Tycho Brahe made measurements of comets, it was believed that they were one-time events, perhaps disturbances in the atmosphere, and that they were not affected by the Sun. Halley used Newton's new mechanics to predict his namesake comet's return in 1758.

 13.9 Check Your Understanding The nearly circular orbit of Saturn has an average radius of about 9.5 AU and has a period of 30 years, whereas Uranus averages about 19 AU and has a period of 84 years. Is this consistent with our results for Halley's comet?

13.6 | Tidal Forces

Learning Objectives
By the end of this section, you will be able to: • Explain the origins of Earth's ocean tides • Describe how neap and leap tides differ • Describe how tidal forces affect binary systems

The origin of Earth's ocean tides has been a subject of continuous investigation for over 2000 years. But the work of Newton is considered to be the beginning of the true understanding of the phenomenon. Ocean tides are the result of gravitational tidal forces. These same tidal forces are present in any astronomical body. They are responsible for the internal heat that creates the volcanic activity on Io, one of Jupiter's moons, and the breakup of stars that get too close to black holes.

Lunar Tides

If you live on an ocean shore almost anywhere in the world, you can observe the rising and falling of the sea level about twice per day. This is caused by a combination of Earth's rotation about its axis and the gravitational attraction of both the Moon and the Sun.

Let's consider the effect of the Moon first. In **Figure 13.22**, we are looking "down" onto Earth's North Pole. One side of Earth is closer to the Moon than the other side, by a distance equal to Earth's diameter. Hence, the gravitational force is greater on the near side than on the far side. The magnitude at the center of Earth is between these values. This is why a tidal bulge appears on both sides of Earth.

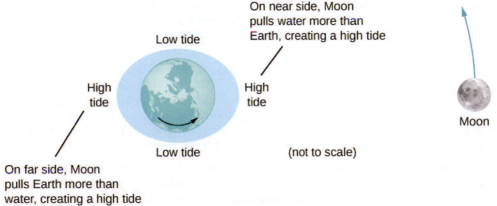

Figure 13.22 The tidal force stretches Earth along the line between Earth and the Moon. It is the *difference* between the gravitational force from the far side to the near side that creates the tidal bulge on both sides of the planet. Tidal variations of the oceans are on the order of few meters; hence, this diagram is greatly exaggerated.

The net force on Earth causes it to orbit about the Earth-Moon center of mass, located about 1600 km below Earth's surface along the line between Earth and the Moon. The **tidal force** can be viewed as the *difference* between the force at the center of Earth and that at any other location. In **Figure 13.23**, this difference is shown at sea level, where we observe the ocean tides. (Note that the change in sea level caused by these tidal forces is measured from the baseline sea level. We saw earlier that Earth bulges many kilometers at the equator due to its rotation. This defines the baseline sea level and here we consider only the much smaller tidal bulge measured from that baseline sea level.)

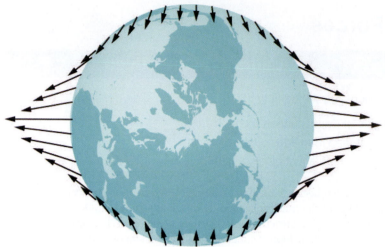

Figure 13.23 The tidal force is the *difference* between the gravitational force at the center and that elsewhere. In this figure, the tidal forces are shown at the ocean surface. These forces would diminish to zero as you approach Earth's center.

Why does the rise and fall of the tides occur twice per day? Look again at **Figure 13.22**. If Earth were not rotating and the Moon was fixed, then the bulges would remain in the same location on Earth. Relative to the Moon, the bulges stay fixed—along the line connecting Earth and the Moon. But Earth rotates (in the direction shown by the blue arrow) approximately every 24 hours. In 6 hours, the near and far locations of Earth move to where the low tides are occurring, and 6 hours later, those locations are back to the high-tide position. Since the Moon also orbits Earth approximately every 28 days, and in the same direction as Earth rotates, the time between high (and low) tides is actually about 12.5 hours. The actual timing of the tides is complicated by numerous factors, the most important of which is another astronomical body—the Sun.

The Effect of the Sun on Tides

In addition to the Moon's tidal forces on Earth's oceans, the Sun exerts a tidal force as well. The gravitational attraction of the Sun on any object on Earth is nearly 200 times that of the Moon. However, as we show later in an example, the *tidal* effect of the Sun is less than that of the Moon, but a significant effect nevertheless. Depending upon the positions of the Moon and Sun relative to Earth, the net tidal effect can be amplified or attenuated.

Figure 13.22 illustrates the relative positions of the Sun and the Moon that create the largest tides, called **spring tides** (or leap tides). During spring tides, Earth, the Moon, and the Sun are aligned and the tidal effects add. (Recall that the tidal forces cause bulges on both sides.) **Figure 13.22**(c) shows the relative positions for the smallest tides, called **neap tides**. The extremes of both high and low tides are affected. Spring tides occur during the new or full moon, and neap tides occur at half-moon.

 You can see **one (https://openstaxcollege.org/l/21tidesinmot01)** or **two (https://openstaxcollege.org/ l/21tidesinmot02)** animations of the tides in motion.

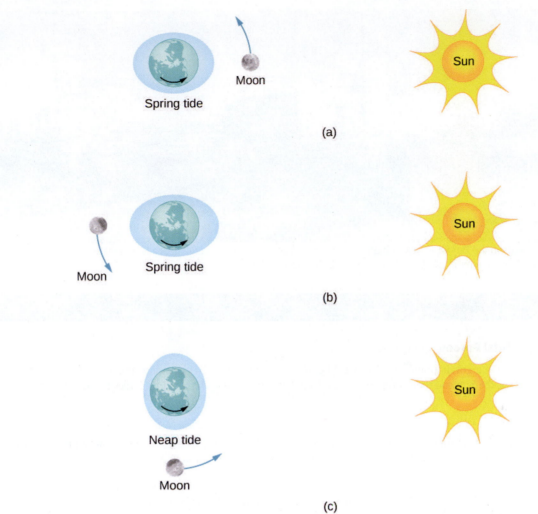

Figure 13.24 (a and b) The spring tides occur when the Sun and the Moon are aligned, whereas (c) the neap tides occur when the Sun and Moon make a right triangle with Earth. (Figure is not drawn to scale.)

The Magnitude of the Tides

With accurate data for the positions of the Moon and the Sun, the time of maximum and minimum tides at most locations on our planet can be predicted accurately.

 Visit **this site (https://openstaxcollege.org/l/21tidepredic)** to generate tide predictions for up to 2 years in the past or future, at more than 3000 locations around the United States.

The magnitude of the tides, however, is far more complicated. The relative angles of Earth and the Moon determine spring and neap tides, but the magnitudes of these tides are affected by the distances from Earth as well. Tidal forces are greater when the distances are smaller. Both the Moon's orbit about Earth and Earth's orbit about the Sun are elliptical, so a spring tide is exceptionally large if it occurs when the Moon is at perigee and Earth is at perihelion. Conversely, it is relatively small if it occurs when the Moon is at apogee and Earth is at aphelion.

The greatest causes of tide variation are the topography of the local shoreline and the bathymetry (the profile of the depth) of the ocean floor. The range of tides due to these effects is astounding. Although ocean tides are much smaller than a meter in many places around the globe, the tides at the Bay of Fundy (**Figure 13.25**), on the east coast of Canada, can be as much as 16.3 meters.

Figure 13.25 Boats in the Bay of Fundy at high and low tides. The twice-daily change in sea level creates a real challenge to the safe mooring of boats. (credit: modification of works by Dylan Kereluk)

Example 13.14

Comparing Tidal Forces

Compare the Moon's gravitational force on a 1.0-kg mass located on the near side and another on the far side of Earth. Repeat for the Sun and then compare the results to confirm that the Moon's tidal forces are about twice that of the Sun.

Strategy

We use Newton's law of gravitation given by **Equation 13.1**. We need the masses of the Moon and the Sun and their distances from Earth, as well as the radius of Earth. We use the astronomical data from **Appendix D**.

Solution

Substituting the mass of the Moon and mean distance from Earth to the Moon, we have

$$F_{12} = G\frac{m_1 m_2}{r^2} = (6.67 \times 10^{-11} \ \text{N} \cdot \text{m}^2/\text{kg}^2)\frac{(1.0 \ \text{kg})(7.35 \times 10^{22} \ \text{kg})}{(3.84 \times 10^8 \pm 6.37 \times 10^6 \ \text{m})^2}.$$

In the denominator, we use the minus sign for the near side and the plus sign for the far side. The results are

$$F_{\text{near}} = 3.44 \times 10^{-5} \ \text{N} \quad \text{and} \quad F_{\text{far}} = 3.22 \times 10^{-5} \ \text{N}.$$

The Moon's gravitational force is nearly 7% higher at the near side of Earth than at the far side, but both forces are much less than that of Earth itself on the 1.0-kg mass. Nevertheless, this small difference creates the tides. We now repeat the problem, but substitute the mass of the Sun and the mean distance between the Earth and Sun. The results are

$$F_{\text{near}} = 5.89975 \times 10^{-3} \ \text{N} \quad \text{and} \quad F_{\text{far}} = 5.89874 \times 10^{-3} \ \text{N}.$$

We have to keep six significant digits since we wish to compare the difference between them to the difference for the Moon. (Although we can't justify the absolute value to this accuracy, since all values in the calculation are the same except the distances, the accuracy in the difference is still valid to three digits.) The difference between the near and far forces on a 1.0-kg mass due to the Moon is

$$F_{\text{near}} = 3.44 \times 10^{-5} \ \text{N} - 3.22 \times 10^{-5} \ \text{N} = 0.22 \times 10^{-5} \ \text{N},$$

whereas the difference for the Sun is

$$F_{\text{near}} - F_{\text{far}} = 5.89975 \times 10^{-3} \ \text{N} - 5.89874 \times 10^{-3} \ \text{N} = 0.101 \times 10^{-5} \ \text{N}.$$

Note that a more proper approach is to write the difference in the two forces with the difference between the near and far distances explicitly expressed. With just a bit of algebra we can show that

$$F_{\text{tidal}} = \frac{GMm}{r_1^2} - \frac{GMm}{r_2^2} = GMm\left(\frac{(r_2 - r_1)(r_2 + r_1)}{r_1^2 r_2^2}\right),$$

where r_1 and r_2 are the same to three significant digits, but their difference $(r_2 - r_1)$, equal to the diameter of Earth, is also known to three significant digits. The results of the calculation are the same. This approach would be necessary if the number of significant digits needed exceeds that available on your calculator or computer.

Significance

Note that the forces exerted by the Sun are nearly 200 times greater than the forces exerted by the Moon. But the *difference* in those forces for the Sun is half that for the Moon. This is the nature of tidal forces. The Moon has a greater tidal effect because the fractional change in distance from the near side to the far side is so much greater for the Moon than it is for the Sun.

 13.10   Check Your Understanding Earth exerts a tidal force on the Moon. Is it greater than, the same as, or less than that of the Moon on Earth? Be careful in your response, as tidal forces arise from the *difference* in gravitational forces between one side and the other. Look at the calculations we performed for the tidal force on Earth and consider the values that would change significantly for the Moon. The diameter of the Moon is one-fourth that of Earth. Tidal forces on the Moon are not easy to detect, since there is no liquid on the surface.

Other Tidal Effects

Tidal forces exist between any two bodies. The effect stretches the bodies along the line between their centers. Although the tidal effect on Earth's seas is observable on a daily basis, long-term consequences cannot be observed so easily. One consequence is the dissipation of rotational energy due to friction during flexure of the bodies themselves. Earth's rotation rate is slowing down as the tidal forces transfer rotational energy into heat. The other effect, related to this dissipation and conservation of angular momentum, is called "locking" or tidal synchronization. It has already happened to most moons in our solar system, including Earth's Moon. The Moon keeps one face toward Earth—its rotation rate has locked into the orbital rate about Earth. The same process is happening to Earth, and eventually it will keep one face toward the Moon. If that does happen, we would no longer see tides, as the tidal bulge would remain in the same place on Earth, and half the planet would never see the Moon. However, this locking will take many billions of years, perhaps not before our Sun expires.

One of the more dramatic example of tidal effects is found on Io, one of Jupiter's moons. In 1979, the *Voyager* spacecraft sent back dramatic images of volcanic activity on Io. It is the only other astronomical body in our solar system on which we have found such activity. **Figure 13.26** shows a more recent picture of Io taken by the *New Horizons* spacecraft on its way to Pluto, while using a gravity assist from Jupiter.

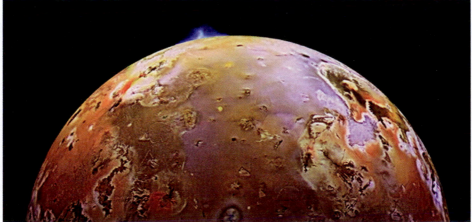

Figure 13.26   Dramatic evidence of tidal forces can be seen on Io. The eruption seen in blue is due to the internal heat created by the tidal forces exerted on Io by Jupiter. (credit: modification of work by NASA/JPL/University of Arizona)

For some stars, the effect of tidal forces can be catastrophic. The tidal forces in very close binary systems can be strong enough to rip matter from one star to the other, once the tidal forces exceed the cohesive self-gravitational forces that hold the stars together. This effect can be seen in normal stars that orbit nearby compact stars, such as neutron stars or black holes. **Figure 13.27** shows an artist's rendition of this process. As matter falls into the compact star, it forms an accretion disc that becomes super-heated and radiates in the X-ray spectrum.

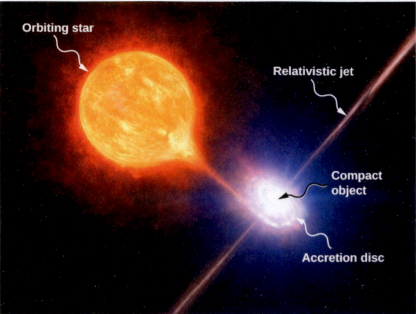

Figure 13.27 Tidal forces from a compact object can tear matter away from an orbiting star. In addition to the accretion disc orbiting the compact object, material is often ejected along relativistic jets as shown. (credit: modification of work by ESO/L. Calçada/M. Kornmesser)

The energy output of these binary systems can exceed the typical output of thousands of stars. Another example might be a quasar. Quasars are very distant and immensely bright objects, often exceeding the energy output of entire galaxies. It is the general consensus among astronomers that they are, in fact, massive black holes producing radiant energy as matter that has been tidally ripped from nearby stars falls into them.

13.7 | Einstein's Theory of Gravity

Learning Objectives

By the end of this section, you will be able to:

- Describe how the theory of general relativity approaches gravitation
- Explain the principle of equivalence
- Calculate the Schwarzschild radius of an object
- Summarize the evidence for black holes

Newton's law of universal gravitation accurately predicts much of what we see within our solar system. Indeed, only Newton's laws have been needed to accurately send every space vehicle on its journey. The paths of Earth-crossing asteroids, and most other celestial objects, can be accurately determined solely with Newton's laws. Nevertheless, many phenomena have shown a discrepancy from what Newton's laws predict, including the orbit of Mercury and the effect that gravity has on light. In this section, we examine a different way of envisioning gravitation.

A Revolution in Perspective

In 1905, Albert Einstein published his theory of special relativity. This theory is discussed in great detail in **Relativity** (http://cnx.org/content/m58555/latest/) , so we say only a few words here. In this theory, no motion can exceed the speed of light—it is the speed limit of the Universe. This simple fact has been verified in countless experiments. However, it has incredible consequences—space and time are no longer absolute. Two people moving relative to one another do not agree on the length of objects or the passage of time. Almost all of the mechanics you learned in previous chapters, while remarkably accurate even for speeds of many thousands of miles per second, begin to fail when approaching the speed of light.

This speed limit on the Universe was also a challenge to the inherent assumption in Newton's law of gravitation that gravity is an **action-at-a-distance force**. That is, without physical contact, any change in the position of one mass is instantly communicated to all other masses. This assumption does not come from any first principle, as Newton's theory simply does not address the question. (The same was believed of electromagnetic forces, as well. It is fair to say that most scientists were not completely comfortable with the action-at-a-distance concept.)

A second assumption also appears in Newton's law of gravitation **Equation 13.1**. The masses are assumed to be exactly the same as those used in Newton's second law, $\vec{F} = m\vec{a}$. We made that assumption in many of our derivations in this chapter. Again, there is no underlying principle that this must be, but experimental results are consistent with this assumption. In Einstein's subsequent **theory of general relativity** (1916), both of these issues were addressed. His theory was a theory of **space-time** geometry and how mass (and acceleration) distort and interact with that space-time. It was not a theory of gravitational forces. The mathematics of the general theory is beyond the scope of this text, but we can look at some underlying principles and their consequences.

The Principle of Equivalence

Einstein came to his general theory in part by wondering why someone who was free falling did not feel his or her weight. Indeed, it is common to speak of astronauts orbiting Earth as being weightless, despite the fact that Earth's gravity is still quite strong there. In Einstein's general theory, there is no difference between free fall and being weightless. This is called the **principle of equivalence**. The equally surprising corollary to this is that there is no difference between a uniform gravitational field and a uniform acceleration in the absence of gravity. Let's focus on this last statement. Although a perfectly uniform gravitational field is not feasible, we can approximate it very well.

Within a reasonably sized laboratory on Earth, the gravitational field $\vec{g}$ is essentially uniform. The corollary states that any physical experiments performed there have the identical results as those done in a laboratory accelerating at $\vec{a} = \vec{g}$ in deep space, well away from all other masses. **Figure 13.28** illustrates the concept.

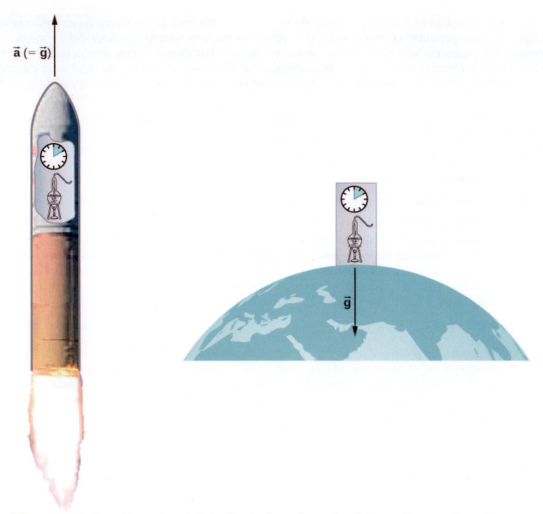

Figure 13.28 According to the principle of equivalence, the results of all experiments performed in a laboratory in a uniform gravitational field are identical to the results of the same experiments performed in a uniformly accelerating laboratory.

How can these two apparently fundamentally different situations be the same? The answer is that gravitation is not a force between two objects but is the result of each object responding to the effect that the other has on the space-time surrounding it. A uniform gravitational field and a uniform acceleration have exactly the same effect on space-time.

A Geometric Theory of Gravity

Euclidian geometry assumes a "flat" space in which, among the most commonly known attributes, a straight line is the shortest distance between two points, the sum of the angles of all triangles must be 180 degrees, and parallel lines never intersect. **Non-Euclidean geometry** was not seriously investigated until the nineteenth century, so it is not surprising that Euclidean space is inherently assumed in all of Newton's laws.

The general theory of relativity challenges this long-held assumption. Only empty space is flat. The presence of mass—or energy, since relativity does not distinguish between the two—distorts or curves space and time, or space-time, around it. The motion of any other mass is simply a response to this curved space-time. **Figure 13.29** is a two-dimensional representation of a smaller mass orbiting in response to the distorted space created by the presence of a larger mass. In a more precise but confusing picture, we would also see space distorted by the orbiting mass, and both masses would be in motion in response to the total distortion of space. Note that the figure is a representation to help visualize the concept. These are distortions in our three-dimensional space and time. We do not see them as we would a dimple on a ball. We see the distortion only by careful measurements of the motion of objects and light as they move through space.

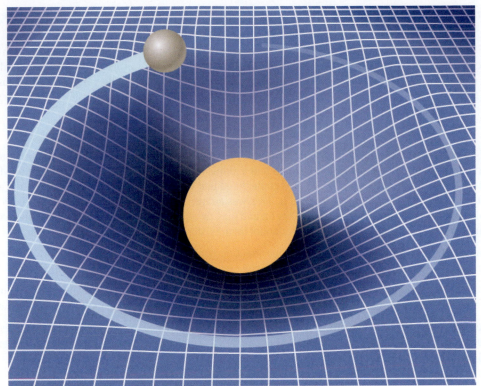

Figure 13.29 A smaller mass orbiting in the distorted space-time of a larger mass. In fact, all mass or energy distorts space-time.

For weak gravitational fields, the results of general relativity do not differ significantly from Newton's law of gravitation. But for intense gravitational fields, the results diverge, and general relativity has been shown to predict the correct results. Even in our Sun's relatively weak gravitational field at the distance of Mercury's orbit, we can observe the effect. Starting in the mid-1800s, Mercury's elliptical orbit has been carefully measured. However, although it is elliptical, its motion is complicated by the fact that the perihelion position of the ellipse slowly advances. Most of the advance is due to the gravitational pull of other planets, but a small portion of that advancement could not be accounted for by Newton's law. At one time, there was even a search for a "companion" planet that would explain the discrepancy. But general relativity correctly predicts the measurements. Since then, many measurements, such as the deflection of light of distant objects by the Sun, have verified that general relativity correctly predicts the observations.

We close this discussion with one final comment. We have often referred to distortions of space-time or distortions in both space and time. In both special and general relativity, the dimension of time has equal footing with each spatial dimension (differing in its place in both theories only by an ultimately unimportant scaling factor). Near a very large mass, not only is the nearby space "stretched out," but time is dilated or "slowed." We discuss these effects more in the next section.

Black Holes

Einstein's theory of gravitation is expressed in one deceptively simple-looking tensor equation (tensors are a generalization of scalars and vectors), which expresses how a mass determines the curvature of space-time around it. The solutions to that equation yield one of the most fascinating predictions: the **black hole**. The prediction is that if an object is sufficiently dense, it will collapse in upon itself and be surrounded by an **event horizon** from which nothing can escape. The name "black hole," which was coined by astronomer John Wheeler in 1969, refers to the fact that light cannot escape such an object. Karl Schwarzschild was the first person to note this phenomenon in 1916, but at that time, it was considered mostly to be a mathematical curiosity.

Surprisingly, the idea of a massive body from which light cannot escape dates back to the late 1700s. Independently, John Michell and Pierre Simon Laplace used Newton's law of gravitation to show that light leaving the surface of a star with enough mass could not escape. Their work was based on the fact that the speed of light had been measured by Ole Roemer in 1676. He noted discrepancies in the data for the orbital period of the moon Io about Jupiter. Roemer realized that the difference arose from the relative positions of Earth and Jupiter at different times and that he could find the speed of light from that difference. Michell and Laplace both realized that since light had a finite speed, there could be a star massive enough that the escape speed from its surface could exceed that speed. Hence, light always would fall back to the star.

Oddly, observers far enough away from the very largest stars would not be able see them, yet they could see a smaller star from the same distance.

Recall that in **Gravitational Potential Energy and Total Energy**, we found that the escape speed, given by **Equation 13.6**, is independent of the mass of the object escaping. Even though the nature of light was not fully understood at the time, the mass of light, if it had any, was not relevant. Hence, **Equation 13.6** should be valid for light. Substituting c, the speed of light, for the escape velocity, we have

$$v_{\text{esc}} = c = \sqrt{\frac{2GM}{R}}.$$

Thus, we only need values for R and M such that the escape velocity exceeds c, and then light will not be able to escape. Michell posited that if a star had the density of our Sun and a radius that extended just beyond the orbit of Mars, then light would not be able to escape from its surface. He also conjectured that we would still be able to detect such a star from the gravitational effect it would have on objects around it. This was an insightful conclusion, as this is precisely how we infer the existence of such objects today. While we have yet to, and may never, visit a black hole, the circumstantial evidence for them has become so compelling that few astronomers doubt their existence.

Before we examine some of that evidence, we turn our attention back to Schwarzschild's solution to the tensor equation from general relativity. In that solution arises a critical radius, now called the **Schwarzschild radius** (R_S). For any mass M, if that mass were compressed to the extent that its radius becomes less than the Schwarzschild radius, then the mass will collapse to a singularity, and anything that passes inside that radius cannot escape. Once inside R_S, the arrow of time takes all things to the singularity. (In a broad mathematical sense, a singularity is where the value of a function goes to infinity. In this case, it is a point in space of zero volume with a finite mass. Hence, the mass density and gravitational energy become infinite.) The Schwarzschild radius is given by

$$R_S = \frac{2GM}{c^2}. \tag{13.12}$$

If you look at our escape velocity equation with $v_{\text{esc}} = c$, you will notice that it gives precisely this result. But that is merely a fortuitous accident caused by several incorrect assumptions. One of these assumptions is the use of the *incorrect* classical expression for the kinetic energy for light. Just how dense does an object have to be in order to turn into a black hole?

Example 13.15

Calculating the Schwarzschild Radius

Calculate the Schwarzschild radius for both the Sun and Earth. Compare the density of the nucleus of an atom to the density required to compress Earth's mass uniformly to its Schwarzschild radius. The density of a nucleus is about 2.3×10^{17} kg/m^3.

Strategy

We use **Equation 13.12** for this calculation. We need only the masses of Earth and the Sun, which we obtain from the astronomical data given in **Appendix D**.

Solution

Substituting the mass of the Sun, we have

$$R_S = \frac{2GM}{c^2} = \frac{2(6.67 \times 10^{-11} \text{ N} \cdot \text{m}^2/\text{kg}^2)(1.99 \times 10^{30} \text{ kg})}{(3.0 \times 10^8 \text{ m/s})^2} = 2.95 \times 10^3 \text{ m}.$$

This is a diameter of only about 6 km. If we use the mass of Earth, we get $R_S = 8.85 \times 10^{-3}$ m. This is a diameter of less than 2 cm! If we pack Earth's mass into a sphere with the radius $R_S = 8.85 \times 10^{-3}$ m, we get a density of

$$\rho = \frac{\text{mass}}{\text{volume}} = \frac{5.97 \times 10^{24}\ \text{kg}}{(\frac{4}{3}\pi)\left(8.85 \times 10^{-3}\ \text{m}\right)^3} = 2.06 \times 10^{30}\ \text{kg/m}^3.$$

Significance

A **neutron star** is the most compact object known—outside of a black hole itself. The neutron star is composed of neutrons, with the density of an atomic nucleus, and, like many black holes, is believed to be the remnant of a supernova—a star that explodes at the end of its lifetime. To create a black hole from Earth, we would have to compress it to a density thirteen orders of magnitude greater than that of a neutron star. This process would require unimaginable force. There is no known mechanism that could cause an Earth-sized object to become a black hole. For the Sun, you should be able to show that it would have to be compressed to a density only about 80 times that of a nucleus. (Note: Once the mass is compressed within its Schwarzschild radius, general relativity dictates that it will collapse to a singularity. These calculations merely show the density we must achieve to initiate that collapse.)

 13.11   Check Your Understanding Consider the density required to make Earth a black hole compared to that required for the Sun. What conclusion can you draw from this comparison about what would be required to create a black hole? Would you expect the Universe to have many black holes with small mass?

The event horizon

The Schwarzschild radius is also called the event horizon of a black hole. We noted that both space and time are stretched near massive objects, such as black holes. **Figure 13.30** illustrates that effect on space. The distortion caused by our Sun is actually quite small, and the diagram is exaggerated for clarity. Consider the neutron star, described in **Example 13.15**. Although the distortion of space-time at the surface of a neutron star is very high, the radius is still larger than its Schwarzschild radius. Objects could still escape from its surface.

However, if a neutron star gains additional mass, it would eventually collapse, shrinking beyond the Schwarzschild radius. Once that happens, the entire mass would be pulled, inevitably, to a singularity. In the diagram, space is stretched to infinity. Time is also stretched to infinity. As objects fall toward the event horizon, we see them approaching ever more slowly, but never reaching the event horizon. As outside observers, we never see objects pass through the event horizon—effectively, time is stretched to a stop.

 Visit this **site (https://openstaxcollege.org/l/21spacetelescop)** to view an animated example of these spatial distortions.

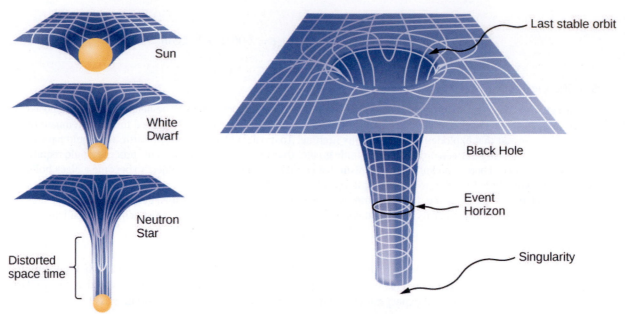

Figure 13.30 The space distortion becomes more noticeable around increasingly larger masses. Once the mass density reaches a critical level, a black hole forms and the fabric of space-time is torn. The curvature of space is greatest at the surface of each of the first three objects shown and is finite. The curvature then decreases (not shown) to zero as you move to the center of the object. But the black hole is different. The curvature becomes infinite: The surface has collapsed to a singularity, and the cone extends to infinity. (Note: These diagrams are not to any scale.) (credit: modification of work by NASA)

The evidence for black holes

Not until the 1960s, when the first neutron star was discovered, did interest in the existence of black holes become renewed. Evidence for black holes is based upon several types of observations, such as radiation analysis of X-ray binaries, gravitational lensing of the light from distant galaxies, and the motion of visible objects around invisible partners. We will focus on these later observations as they relate to what we have learned in this chapter. Although light cannot escape from a black hole for us to see, we can nevertheless see the gravitational effect of the black hole on surrounding masses.

The closest, and perhaps most dramatic, evidence for a black hole is at the center of our Milky Way galaxy. The UCLA Galactic Group, using data obtained by the W. M. Keck telescopes, has determined the orbits of several stars near the center of our galaxy. Some of that data is shown in **Figure 13.31**. The orbits of two stars are highlighted. From measurements of the periods and sizes of their orbits, it is estimated that they are orbiting a mass of approximately 4 million solar masses. Note that the mass must reside in the region created by the intersection of the ellipses of the stars. The region in which that mass must reside would fit inside the orbit of Mercury—yet nothing is seen there in the visible spectrum.

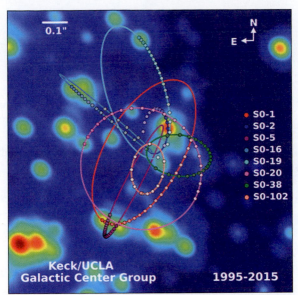

Figure 13.31 Paths of stars orbiting about a mass at the center of our Milky Way galaxy. From their motion, it is estimated that a black hole of about 4 million solar masses resides at the center. (credit: modification of work by UCLA Galactic Center Group – W.M. Keck Observatory Laser Team)

The physics of stellar creation and evolution is well established. The ultimate source of energy that makes stars shine is the self-gravitational energy that triggers fusion. The general behavior is that the more massive a star, the brighter it shines and the shorter it lives. The logical inference is that a mass that is 4 million times the mass of our Sun, confined to a very small region, and that cannot be seen, has no viable interpretation other than a black hole. Extragalactic observations strongly suggest that black holes are common at the center of galaxies.

 Visit the **UCLA Galactic Center Group main page (https://openstaxcollege.org/l/21galacenter01)** for information on X-ray binaries and gravitational lensing. Visit this **page (https://openstaxcollege.org/l/21galacenter02)** to view a three-dimensional visualization of the stars orbiting near the center of our galaxy, where the animation is near the bottom of the page.

Dark matter

Stars orbiting near the very heart of our galaxy provide strong evidence for a black hole there, but the orbits of stars far from the center suggest another intriguing phenomenon that is observed indirectly as well. Recall from **Gravitation Near Earth's Surface** that we can consider the mass for spherical objects to be located at a point at the center for calculating their gravitational effects on other masses. Similarly, we can treat the total mass that lies within the orbit of any star in our galaxy as being located at the center of the Milky Way disc. We can estimate that mass from counting the visible stars and include in our estimate the mass of the black hole at the center as well.

But when we do that, we find the orbital speed of the stars is far too fast to be caused by that amount of matter. **Figure 13.32** shows the orbital velocities of stars as a function of their distance from the center of the Milky Way. The blue line represents the velocities we would expect from our estimates of the mass, whereas the green curve is what we get from direct measurements. Apparently, there is a lot of matter we don't see, estimated to be about five times as much as what we do see, so it has been dubbed dark matter. Furthermore, the velocity profile does not follow what we expect from the observed distribution of visible stars. Not only is the estimate of the total mass inconsistent with the data, but the expected distribution is inconsistent as well. And this phenomenon is not restricted to our galaxy, but seems to be a feature of all galaxies. In fact, the issue was first noted in the 1930s when galaxies within clusters were measured to be orbiting about the center of mass of those clusters faster than they should based upon visible mass estimates.

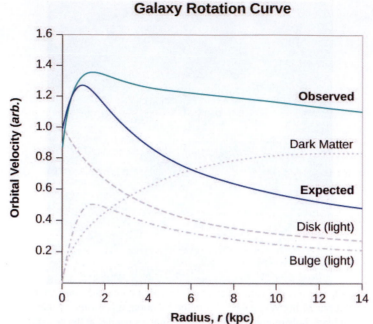

Figure 13.32 The blue curve shows the expected orbital velocity of stars in the Milky Way based upon the visible stars we can see. The green curve shows that the actually velocities are higher, suggesting additional matter that cannot be seen. (credit: modification of work by Matthew Newby)

There are two prevailing ideas of what this matter could be—WIMPs and MACHOs. WIMPs stands for weakly interacting massive particles. These particles (neutrinos are one example) interact very weakly with ordinary matter and, hence, are very difficult to detect directly. MACHOs stands for massive compact halo objects, which are composed of ordinary baryonic matter, such as neutrons and protons. There are unresolved issues with both of these ideas, and far more research will be needed to solve the mystery.

CHAPTER 13 REVIEW

KEY TERMS

action-at-a-distance force type of force exerted without physical contact

aphelion farthest point from the Sun of an orbiting body; the corresponding term for the Moon's farthest point from Earth is the apogee

apparent weight reading of the weight of an object on a scale that does not account for acceleration

black hole mass that becomes so dense, that it collapses in on itself, creating a singularity at the center surround by an event horizon

escape velocity initial velocity an object needs to escape the gravitational pull of another; it is more accurately defined as the velocity of an object with zero total mechanical energy

event horizon location of the Schwarzschild radius and is the location near a black hole from within which no object, even light, can escape

gravitational field vector field that surrounds the mass creating the field; the field is represented by field lines, in which the direction of the field is tangent to the lines, and the magnitude (or field strength) is inversely proportional to the spacing of the lines; other masses respond to this field

gravitationally bound two object are gravitationally bound if their orbits are closed; gravitationally bound systems have a negative total mechanical energy

Kepler's first law law stating that every planet moves along an ellipse, with the Sun located at a focus of the ellipse

Kepler's second law law stating that a planet sweeps out equal areas in equal times, meaning it has a constant areal velocity

Kepler's third law law stating that the square of the period is proportional to the cube of the semi-major axis of the orbit

neap tide low tide created when the Moon and the Sun form a right triangle with Earth

neutron star most compact object known—outside of a black hole itself

Newton's law of gravitation every mass attracts every other mass with a force proportional to the product of their masses, inversely proportional to the square of the distance between them, and with direction along the line connecting the center of mass of each

non-Euclidean geometry geometry of curved space, describing the relationships among angles and lines on the surface of a sphere, hyperboloid, etc.

orbital period time required for a satellite to complete one orbit

orbital speed speed of a satellite in a circular orbit; it can be also be used for the instantaneous speed for noncircular orbits in which the speed is not constant

perihelion point of closest approach to the Sun of an orbiting body; the corresponding term for the Moon's closest approach to Earth is the perigee

principle of equivalence part of the general theory of relativity, it states that there no difference between free fall and being weightless, or a uniform gravitational field and uniform acceleration

Schwarzschild radius critical radius (R_S) such that if a mass were compressed to the extent that its radius becomes less than the Schwarzschild radius, then the mass will collapse to a singularity, and anything that passes inside that radius cannot escape

space-time concept of space-time is that time is essentially another coordinate that is treated the same way as any individual spatial coordinate; in the equations that represent both special and general relativity, time appears in the same context as do the spatial coordinates

spring tide high tide created when the Moon, the Sun, and Earth are along one line

theory of general relativity Einstein's theory for gravitation and accelerated reference frames; in this theory, gravitation is the result of mass and energy distorting the space-time around it; it is also often referred to as Einstein's theory of gravity

tidal force *difference* between the gravitational force at the center of a body and that at any other location on the body; the tidal force stretches the body

universal gravitational constant constant representing the strength of the gravitational force, that is believed to be the same throughout the universe

KEY EQUATIONS

Newton's law of gravitation	$\vec{\mathbf{F}}_{12} = G\dfrac{m_1 m_2}{r^2} \hat{\mathbf{r}}_{12}$
Acceleration due to gravity at the surface of Earth	$g = G\dfrac{M_E}{r^2}$
Gravitational potential energy beyond Earth	$U = -\dfrac{GM_E m}{r}$
Conservation of energy	$\dfrac{1}{2}mv_1^2 - \dfrac{GMm}{r_1} = \dfrac{1}{2}mv_2^2 - \dfrac{GMm}{r_2}$
Escape velocity	$v_{esc} = \sqrt{\dfrac{2GM}{R}}$
Orbital speed	$v_{orbit} = \sqrt{\dfrac{GM_E}{r}}$
Orbital period	$T = 2\pi\sqrt{\dfrac{r^3}{GM_E}}$
Energy in circular orbit	$E = K + U = -\dfrac{GmM_E}{2r}$
Conic sections	$\dfrac{\alpha}{r} = 1 + e\cos\theta$
Kepler's third law	$T^2 = \dfrac{4\pi^2}{GM}a^3$
Schwarzschild radius	$R_S = \dfrac{2GM}{c^2}$

SUMMARY

13.1 Newton's Law of Universal Gravitation

- All masses attract one another with a gravitational force proportional to their masses and inversely proportional to the square of the distance between them.
- Spherically symmetrical masses can be treated as if all their mass were located at the center.
- Nonsymmetrical objects can be treated as if their mass were concentrated at their center of mass, provided their distance from other masses is large compared to their size.

13.2 Gravitation Near Earth's Surface

- The weight of an object is the gravitational attraction between Earth and the object.
- The gravitational field is represented as lines that indicate the direction of the gravitational force; the line spacing indicates the strength of the field.
- Apparent weight differs from actual weight due to the acceleration of the object.

13.3 Gravitational Potential Energy and Total Energy

- The acceleration due to gravity changes as we move away from Earth, and the expression for gravitational potential energy must reflect this change.

- The total energy of a system is the sum of kinetic and gravitational potential energy, and this total energy is conserved in orbital motion.

- Objects must have a minimum velocity, the escape velocity, to leave a planet and not return.

- Objects with total energy less than zero are bound; those with zero or greater are unbounded.

13.4 Satellite Orbits and Energy

- Orbital velocities are determined by the mass of the body being orbited and the distance from the center of that body, and not by the mass of a much smaller orbiting object.

- The period of the orbit is likewise independent of the orbiting object's mass.

- Bodies of comparable masses orbit about their common center of mass and their velocities and periods should be determined from Newton's second law and law of gravitation.

13.5 Kepler's Laws of Planetary Motion

- All orbital motion follows the path of a conic section. Bound or closed orbits are either a circle or an ellipse; unbounded or open orbits are either a parabola or a hyperbola.

- The areal velocity of any orbit is constant, a reflection of the conservation of angular momentum.

- The square of the period of an elliptical orbit is proportional to the cube of the semi-major axis of that orbit.

13.6 Tidal Forces

- Earth's tides are caused by the difference in gravitational forces from the Moon and the Sun on the different sides of Earth.

- Spring or neap (high) tides occur when Earth, the Moon, and the Sun are aligned, and neap or (low) tides occur when they form a right triangle.

- Tidal forces can create internal heating, changes in orbital motion, and even destruction of orbiting bodies.

13.7 Einstein's Theory of Gravity

- According to the theory of general relativity, gravity is the result of distortions in space-time created by mass and energy.

- The principle of equivalence states that that both mass and acceleration distort space-time and are indistinguishable in comparable circumstances.

- Black holes, the result of gravitational collapse, are singularities with an event horizon that is proportional to their mass.

- Evidence for the existence of black holes is still circumstantial, but the amount of that evidence is overwhelming.

CONCEPTUAL QUESTIONS

13.1 Newton's Law of Universal Gravitation

1. Action at a distance, such as is the case for gravity, was once thought to be illogical and therefore untrue. What is the ultimate determinant of the truth in science, and why was this action at a distance ultimately accepted?

2. In the law of universal gravitation, Newton assumed that the force was proportional to the product of the two masses ($\sim m_1 m_2$). While all scientific conjectures must be experimentally verified, can you provide arguments as to why this must be? (You may wish to consider simple examples in which any other form would lead to contradictory results.)

13.2 Gravitation Near Earth's Surface

3. Must engineers take Earth's rotation into account when constructing very tall buildings at any location other than the equator or very near the poles?

13.3 Gravitational Potential Energy and Total Energy

4. It was stated that a satellite with negative total energy is in a bound orbit, whereas one with zero or positive total energy is in an unbounded orbit. Why is this true? What choice for gravitational potential energy was made such that this is true?

5. It was shown that the energy required to lift a satellite into a *low* Earth orbit (the change in potential energy) is only a small fraction of the kinetic energy needed to keep it in orbit. Is this true for larger orbits? Is there a trend to the ratio of kinetic energy to change in potential energy as the size of the orbit increases?

13.4 Satellite Orbits and Energy

6. One student argues that a satellite in orbit is in free fall because the satellite keeps falling toward Earth. Another says a satellite in orbit is not in free fall because the acceleration due to gravity is not 9.80 m/s^2. With whom do you agree with and why?

7. Many satellites are placed in geosynchronous orbits. What is special about these orbits? For a global communication network, how many of these satellites would be needed?

13.5 Kepler's Laws of Planetary Motion

8. Are Kepler's laws purely descriptive, or do they contain causal information?

9. In the diagram below for a satellite in an elliptical orbit about a much larger mass, indicate where its speed is the greatest and where it is the least. What conservation law dictates this behavior? Indicate the directions of the force, acceleration, and velocity at these points. Draw vectors for these same three quantities at the two points where the *y*-axis intersects (along the semi-minor axis) and from this determine whether the speed is increasing decreasing, or at a max/min.

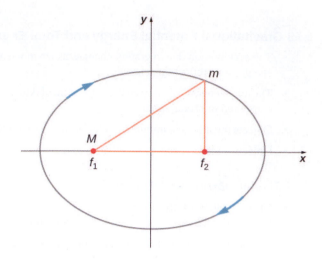

13.6 Tidal Forces

10. As an object falls into a black hole, tidal forces increase. Will these tidal forces always tear the object apart as it approaches the Schwarzschild radius? How does the mass of the black hole and size of the object affect your answer?

13.7 Einstein's Theory of Gravity

11. The principle of equivalence states that all experiments done in a lab in a uniform gravitational field cannot be distinguished from those done in a lab that is not in a gravitational field but is uniformly accelerating. For the latter case, consider what happens to a laser beam at some height shot perfectly horizontally to the floor, across the accelerating lab. (View this from a nonaccelerating frame outside the lab.) Relative to the height of the laser, where will the laser beam hit the far wall? What does this say about the effect of a gravitational field on light? Does the fact that light has no mass make any difference to the argument?

12. As a person approaches the Schwarzschild radius of a black hole, outside observers see all the processes of that person (their clocks, their heart rate, etc.) slowing down, and coming to a halt as they reach the Schwarzschild radius. (The person falling into the black hole sees their own processes unaffected.) But the speed of light is the same everywhere for all observers. What does this say about space as you approach the black hole?

PROBLEMS

13.1 Newton's Law of Universal Gravitation

13. Evaluate the magnitude of gravitational force between two 5-kg spherical steel balls separated by a center-to-center distance of 15 cm.

14. Estimate the gravitational force between two sumo wrestlers, with masses 220 kg and 240 kg, when they are embraced and their centers are 1.2 m apart.

15. Astrology makes much of the position of the planets

at the moment of one's birth. The only known force a planet exerts on Earth is gravitational. (a) Calculate the gravitational force exerted on a 4.20-kg baby by a 100-kg father 0.200 m away at birth (he is assisting, so he is close to the child). (b) Calculate the force on the baby due to Jupiter if it is at its closest distance to Earth, some 6.29×10^{11} m away. How does the force of Jupiter on the baby compare to the force of the father on the baby? Other objects in the room and the hospital building also exert similar gravitational forces. (Of course, there could be an unknown force acting, but scientists first need to be convinced that there is even an effect, much less that an unknown force causes it.)

16. A mountain 10.0 km from a person exerts a gravitational force on him equal to 2.00% of his weight. (a) Calculate the mass of the mountain. (b) Compare the mountain's mass with that of Earth. (c) What is unreasonable about these results? (d) Which premises are unreasonable or inconsistent? (Note that accurate gravitational measurements can easily detect the effect of nearby mountains and variations in local geology.)

17. The International Space Station has a mass of approximately 370,000 kg. (a) What is the force on a 150-kg suited astronaut if she is 20 m from the center of mass of the station? (b) How accurate do you think your answer would be?

Figure 13.33 (credit: ©ESA–David Ducros)

18. Asteroid Toutatis passed near Earth in 2006 at four times the distance to our Moon. This was the closest approach we will have until 2060. If it has mass of 5.0×10^{13} kg, what force did it exert on Earth at its closest approach?

19. (a) What was the acceleration of Earth caused by asteroid Toutatis (see previous problem) at its closest approach? (b) What was the acceleration of Toutatis at this point?

13.2 Gravitation Near Earth's Surface

20. (a) Calculate Earth's mass given the acceleration due to gravity at the North Pole is measured to be 9.832 m/s^2 and the radius of the Earth at the pole is 6356 km. (b) Compare this with the NASA's Earth Fact Sheet value of 5.9726×10^{24} kg .

21. (a) What is the acceleration due to gravity on the surface of the Moon? (b) On the surface of Mars? The mass of Mars is 6.418×10^{23} kg and its radius is 3.38×10^6 m .

22. (a) Calculate the acceleration due to gravity on the surface of the Sun. (b) By what factor would your weight increase if you could stand on the Sun? (Never mind that you cannot.)

23. The mass of a particle is 15 kg. (a) What is its weight on Earth? (b) What is its weight on the Moon? (c) What is its mass on the Moon? (d) What is its weight in outer space far from any celestial body? (e) What is its mass at this point?

24. On a planet whose radius is 1.2×10^7 m , the acceleration due to gravity is 18 m/s^2 . What is the mass of the planet?

25. The mean diameter of the planet Saturn is 1.2×10^8 m , and its mean mass density is 0.69 g/cm^3 . Find the acceleration due to gravity at Saturn's surface.

26. The mean diameter of the planet Mercury is 4.88×10^6 m , and the acceleration due to gravity at its surface is 3.78 m/s^2 . Estimate the mass of this planet.

27. The acceleration due to gravity on the surface of a planet is three times as large as it is on the surface of Earth. The mass density of the planet is known to be twice that of Earth. What is the radius of this planet in terms of Earth's radius?

28. A body on the surface of a planet with the same radius as Earth's weighs 10 times more than it does on Earth. What is the mass of this planet in terms of Earth's mass?

13.3 Gravitational Potential Energy and Total Energy

29. Find the escape speed of a projectile from the surface of Mars.

30. Find the escape speed of a projectile from the surface of Jupiter.

31. What is the escape speed of a satellite located at the Moon's orbit about Earth? Assume the Moon is not nearby.

32. (a) Evaluate the gravitational potential energy between two 5.00-kg spherical steel balls separated by a center-to-center distance of 15.0 cm. (b) Assuming that they are both initially at rest relative to each other in deep space, use conservation of energy to find how fast will they be traveling upon impact. Each sphere has a radius of 5.10 cm.

33. An average-sized asteroid located 5.0×10^7 km from Earth with mass 2.0×10^{13} kg is detected headed directly toward Earth with speed of 2.0 km/s. What will its speed be just before it hits our atmosphere? (You may ignore the size of the asteroid.)

34. (a) What will be the kinetic energy of the asteroid in the previous problem just before it hits Earth? b) Compare this energy to the output of the largest fission bomb, 2100 TJ. What impact would this have on Earth?

35. (a) What is the change in energy of a 1000-kg payload taken from rest at the surface of Earth and placed at rest on the surface of the Moon? (b) What would be the answer if the payload were taken from the Moon's surface to Earth? Is this a reasonable calculation of the energy needed to move a payload back and forth?

13.4 Satellite Orbits and Energy

36. If a planet with 1.5 times the mass of Earth was traveling in Earth's orbit, what would its period be?

37. Two planets in circular orbits around a star have speeds of v and $2v$. (a) What is the ratio of the orbital radii of the planets? (b) What is the ratio of their periods?

38. Using the average distance of Earth from the Sun, and the orbital period of Earth, (a) find the centripetal acceleration of Earth in its motion about the Sun. (b) Compare this value to that of the centripetal acceleration at the equator due to Earth's rotation.

39. What is the orbital radius of an Earth satellite having a period of 1.00 h? (b) What is unreasonable about this result?

40. Calculate the mass of the Sun based on data for Earth's orbit and compare the value obtained with the Sun's actual mass.

41. Find the mass of Jupiter based on the fact that Io, its innermost moon, has an average orbital radius of 421,700 km and a period of 1.77 days.

42. Astronomical observations of our Milky Way galaxy indicate that it has a mass of about 8.0×10^{11} solar masses. A star orbiting on the galaxy's periphery is about 6.0×10^4 light-years from its center. (a) What should the orbital period of that star be? (b) If its period is 6.0×10^7 years instead, what is the mass of the galaxy? Such calculations are used to imply the existence of other matter, such as a very massive black hole at the center of the Milky Way.

43. (a) In order to keep a small satellite from drifting into a nearby asteroid, it is placed in orbit with a period of 3.02 hours and radius of 2.0 km. What is the mass of the asteroid? (b) Does this mass seem reasonable for the size of the orbit?

44. The Moon and Earth rotate about their common center of mass, which is located about 4700 km from the center of Earth. (This is 1690 km below the surface.) (a) Calculate the acceleration due to the Moon's gravity at that point. (b) Calculate the centripetal acceleration of the center of Earth as it rotates about that point once each lunar month (about 27.3 d) and compare it with the acceleration found in part (a). Comment on whether or not they are equal and why they should or should not be.

45. The Sun orbits the Milky Way galaxy once each 2.60×10^8 years , with a roughly circular orbit averaging a radius of 3.00×10^4 light-years. (A light-year is the distance traveled by light in 1 year.) Calculate the centripetal acceleration of the Sun in its galactic orbit. Does your result support the contention that a nearly inertial frame of reference can be located at the Sun? (b) Calculate the average speed of the Sun in its galactic orbit. Does the answer surprise you?

46. A geosynchronous Earth satellite is one that has an orbital period of precisely 1 day. Such orbits are useful for communication and weather observation because the satellite remains above the same point on Earth (provided it orbits in the equatorial plane in the same direction as Earth's rotation). Calculate the radius of such an orbit based on the data for Earth in **Appendix D**.

13.5 Kepler's Laws of Planetary Motion

47. Calculate the mass of the Sun based on data for average Earth's orbit and compare the value obtained with the Sun's commonly listed value of 1.989×10^{30} kg .

48. Io orbits Jupiter with an average radius of 421,700 km and a period of 1.769 days. Based upon these data, what is the mass of Jupiter?

49. The "mean" orbital radius listed for astronomical objects orbiting the Sun is typically not an integrated average but is calculated such that it gives the correct period when applied to the equation for circular orbits. Given that, what is the mean orbital radius in terms of aphelion and perihelion?

50. The perihelion of Halley's comet is 0.586 AU and the aphelion is 17.8 AU. Given that its speed at perihelion is 55 km/s, what is the speed at aphelion ($1 \text{ AU} = 1.496 \times 10^{11} \text{ m}$)? (*Hint:* You may use either conservation of energy or angular momentum, but the latter is much easier.)

51. The perihelion of the comet Lagerkvist is 2.61 AU and it has a period of 7.36 years. Show that the aphelion for this comet is 4.95 AU.

52. What is the ratio of the speed at perihelion to that at aphelion for the comet Lagerkvist in the previous problem?

53. Eros has an elliptical orbit about the Sun, with a perihelion distance of 1.13 AU and aphelion distance of 1.78 AU. What is the period of its orbit?

13.6 Tidal Forces

54. (a) What is the difference between the forces on a 1.0-kg mass on the near side of Io and far side due to Jupiter? Io has a mean radius of 1821 km and a mean orbital radius about Jupiter of 421,700 km. (b) Compare this difference to that calculated for the difference for Earth due to the Moon calculated in **Example 13.14**. Tidal forces are the cause of Io's volcanic activity.

55. If the Sun were to collapse into a black hole, the point of no return for an investigator would be approximately 3 km from the center singularity. Would the investigator be able to survive visiting even 300 km from the center? Answer this by finding the difference in the gravitational attraction the black holes exerts on a 1.0-kg mass at the head and at the feet of the investigator.

56. Consider **Figure 13.23** in **Tidal Forces**. This diagram represents the tidal forces for spring tides. Sketch a similar diagram for neap tides. (*Hint:* For simplicity, imagine that the Sun and the Moon contribute equally. Your diagram would be the vector sum of two force fields (as in **Figure 13.23**), reduced by a factor of two, and superimposed at right angles.)

13.7 Einstein's Theory of Gravity

57. What is the Schwarzschild radius for the black hole at the center of our galaxy if it has the mass of 4 million solar masses?

58. What would be the Schwarzschild radius, in light years, if our Milky Way galaxy of 100 billion stars collapsed into a black hole? Compare this to our distance from the center, about 13,000 light years.

ADDITIONAL PROBLEMS

59. A neutron star is a cold, collapsed star with nuclear density. A particular neutron star has a mass twice that of our Sun with a radius of 12.0 km. (a) What would be the weight of a 100-kg astronaut on standing on its surface? (b) What does this tell us about landing on a neutron star?

60. (a) How far from the center of Earth would the net gravitational force of Earth and the Moon on an object be zero? (b) Setting the *magnitudes* of the forces equal should result in two answers from the quadratic. Do you understand why there are two positions, but only one where the net force is zero?

61. How far from the center of the Sun would the net gravitational force of Earth and the Sun on a spaceship be zero?

62. Calculate the values of g at Earth's surface for the following changes in Earth's properties: (a) its mass is doubled and its radius is halved; (b) its mass density is doubled and its radius is unchanged; (c) its mass density is halved and its mass is unchanged.

63. Suppose you can communicate with the inhabitants of a planet in another solar system. They tell you that on their planet, whose diameter and mass are 5.0×10^3 km and 3.6×10^{23} kg, respectively, the record for the high jump is 2.0 m. Given that this record is close to 2.4 m on Earth, what would you conclude about your extraterrestrial friends' jumping ability?

64. (a) Suppose that your measured weight at the equator is one-half your measured weight at the pole on a planet whose mass and diameter are equal to those of Earth. What is the rotational period of the planet? (b) Would you need to take the shape of this planet into account?

65. A body of mass 100 kg is weighed at the North Pole and at the equator with a spring scale. What is the scale reading at these two points? Assume that $g = 9.83 \text{ m/s}^2$ at the pole.

66. Find the speed needed to escape from the solar system starting from the surface of Earth. Assume there are no other bodies involved and do not account for the fact that Earth is moving in its orbit. [*Hint:* **Equation 13.6** does not apply. Use **Equation 13.5** and include the potential energy of both Earth and the Sun.

67. Consider the previous problem and include the fact that Earth has an orbital speed about the Sun of 29.8 km/s. (a) What speed relative to Earth would be needed and in what direction should you leave Earth? (b) What will be the shape of the trajectory?

68. A comet is observed 1.50 AU from the Sun with a speed of 24.3 km/s. Is this comet in a bound or unbound orbit?

69. An asteroid has speed 15.5 km/s when it is located 2.00 AU from the sun. At its closest approach, it is 0.400 AU from the Sun. What is its speed at that point?

70. Space debris left from old satellites and their launchers is becoming a hazard to other satellites. (a) Calculate the speed of a satellite in an orbit 900 km above Earth's surface. (b) Suppose a loose rivet is in an orbit of the same radius that intersects the satellite's orbit at an angle of $90°$. What is the velocity of the rivet relative to the satellite just before striking it? (c) If its mass is 0.500 g, and it comes to rest inside the satellite, how much energy in joules is generated by the collision? (Assume the satellite's velocity does not change appreciably, because its mass is much greater than the rivet's.)

71. A satellite of mass 1000 kg is in circular orbit about Earth. The radius of the orbit of the satellite is equal to two times the radius of Earth. (a) How far away is the satellite? (b) Find the kinetic, potential, and total energies of the satellite.

72. After Ceres was promoted to a dwarf planet, we now recognize the largest known asteroid to be Vesta, with a mass of $2.67 \times 10^{20} \text{ kg}$ and a diameter ranging from 578 km to 458 km. Assuming that Vesta is spherical with radius 520 km, find the approximate escape velocity from its surface.

73. (a) Using the data in the previous problem for the asteroid Vesta which has a diameter of 520 km and mass of $2.67 \times 10^{20} \text{ kg}$, what would be the orbital period for a space probe in a circular orbit of 10.0 km from its surface?

(b) Why is this calculation marginally useful at best?

74. What is the orbital velocity of our solar system about the center of the Milky Way? Assume that the mass within a sphere of radius equal to our distance away from the center is about a 100 billion solar masses. Our distance from the center is 27,000 light years.

75. (a) Using the information in the previous problem, what velocity do you need to escape the Milky Way galaxy from our present position? (b) Would you need to accelerate a spaceship to this speed relative to Earth?

76. Circular orbits in **Equation 13.10** for conic sections must have eccentricity zero. From this, and using Newton's second law applied to centripetal acceleration, show that the value of α in **Equation 13.10** is given by $\alpha = \dfrac{L^2}{GMm^2}$ where L is the angular momentum of the orbiting body. The value of α is constant and given by this expression regardless of the type of orbit.

77. Show that for eccentricity equal to one in **Equation 13.10** for conic sections, the path is a parabola. Do this by substituting Cartesian coordinates, x and y, for the polar coordinates, r and θ, and showing that it has the general form for a parabola, $x = ay^2 + by + c$.

78. Using the technique shown in **Satellite Orbits and Energy**, show that two masses m_1 and m_2 in circular orbits about their common center of mass, will have total energy $E = K + E = K_1 + K_2 - \dfrac{Gm_1m_2}{r} = -\dfrac{Gm_1m_2}{2r}$. We have shown the kinetic energy of both masses explicitly. (*Hint:* The masses orbit at radii r_1 and r_2, respectively, where $r = r_1 + r_2$. Be sure not to confuse the radius needed for centripetal acceleration with that for the gravitational force.)

79. Given the perihelion distance, p, and aphelion distance, q, for an elliptical orbit, show that the velocity at perihelion, v_p, is given by $v_p = \sqrt{\dfrac{2GM_{\text{Sun}}}{(q+p)}\dfrac{q}{p}}$. (*Hint:* Use conservation of angular momentum to relate v_p and v_q, and then substitute into the conservation of energy equation.)

80. Comet P/1999 R1 has a perihelion of 0.0570 AU and aphelion of 4.99 AU. Using the results of the previous problem, find its speed at aphelion. (*Hint:* The expression is for the perihelion. Use symmetry to rewrite the expression for aphelion.)

CHALLENGE PROBLEMS

81. A tunnel is dug through the center of a perfectly spherical and airless planet of radius R. Using the expression for g derived in **Gravitation Near Earth's Surface** for a uniform density, show that a particle of mass m dropped in the tunnel will execute simple harmonic motion. Deduce the period of oscillation of m and show that it has the same period as an orbit at the surface.

82. Following the technique used in **Gravitation Near Earth's Surface**, find the value of g as a function of the radius r from the center of a spherical shell planet of constant density ρ with inner and outer radii R_{in} and R_{out}. Find g for both $R_{in} < r < R_{out}$ and for $r < R_{in}$. Assuming the inside of the shell is kept airless, describe travel inside the spherical shell planet.

83. Show that the areal velocity for a circular orbit of radius r about a mass M is $\frac{\Delta A}{\Delta t} = \frac{1}{2}\sqrt{GMr}$. Does your expression give the correct value for Earth's areal velocity about the Sun?

84. Show that the period of orbit for two masses, m_1 and m_2, in circular orbits of radii r_1 and r_2, respectively, about their common center-of-mass, is given by $T = 2\pi\sqrt{\dfrac{r^3}{G(m_1 + m_2)}}$ where $r = r_1 + r_2$. (*Hint:* The masses orbit at radii r_1 and r_2, respectively where $r = r_1 + r_2$. Use the expression for the center-of-mass to relate the two radii and note that the two masses must have equal but opposite momenta. Start with the relationship of the period to the circumference and speed of orbit for one of the masses. Use the result of the previous problem using momenta in the expressions for the kinetic energy.)

85. Show that for small changes in height h, such that $h < < R_E$, **Equation 13.4** reduces to the expression $\Delta U = mgh$.

86. Using **Figure 13.9**, carefully sketch a free body diagram for the case of a simple pendulum hanging at latitude lambda, labeling all forces acting on the point mass, m. Set up the equations of motion for equilibrium, setting one coordinate in the direction of the centripetal acceleration (toward P in the diagram), the other perpendicular to that. Show that the deflection angle ε, defined as the angle between the pendulum string and the radial direction toward the center of Earth, is given by the expression below. What is the deflection angle at latitude 45 degrees? Assume that Earth is a perfect sphere.
$$\tan(\lambda + \varepsilon) = \frac{g}{(g - \omega^2 R_E)}\tan\lambda ,$$
where ω is the angular velocity of Earth.

87. (a) Show that tidal force on a small object of mass m, defined as the *difference* in the gravitational force that would be exerted on m at a distance at the near and the far side of the object, due to the gravitation at a distance R from M, is given by $F_{tidal} = \dfrac{2GMm}{R^3}\Delta r$ where Δr is the distance between the near and far side and $\Delta r < < R$. (b) Assume you are falling feet first into the black hole at the center of our galaxy. It has mass of 4 million solar masses. What would be the difference between the force at your head and your feet at the Schwarzschild radius (event horizon)? Assume your feet and head each have mass 5.0 kg and are 2.0 m apart. Would you survive passing through the event horizon?

88. Find the Hohmann transfer velocities, $\Delta v_{EllipseEarth}$ and $\Delta v_{EllipseMars}$, needed for a trip to Mars. Use **Equation 13.7** to find the circular orbital velocities for Earth and Mars. Using **Equation 13.4** and the total energy of the ellipse (with semi-major axis a), given by $E = -\dfrac{GmM_s}{2a}$, find the velocities at Earth (perihelion) and at Mars (aphelion) required to be on the transfer ellipse. The difference, Δv, at each point is the velocity boost or transfer velocity needed.

14 | FLUID MECHANICS

850hpa Heights (in Dyn. meters)
Forecast valid 19Z03Jul2014

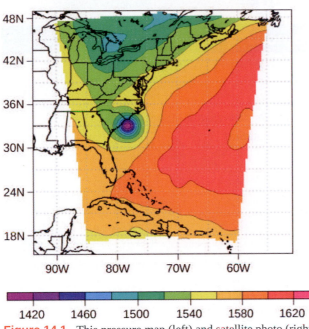

Figure 14.1 This pressure map (left) and satellite photo (right) were used to model the path and impact of Hurricane Arthur as it traveled up the East Coast of the United States in July 2014. Computer models use force and energy equations to predict developing weather patterns. Scientists numerically integrate these time-dependent equations, along with the energy budgets of long- and short-wave solar energy, to model changes in the atmosphere. The pressure map on the left was created using the Weather Research and Forecasting Model designed at the National Center for Atmospheric Research. The colors represent the height of the 850-mbar pressure surface. (credit left: modification of work by The National Center for Atmospheric Research; credit right: modification of work by NRL Monterey Marine Meteorology Division, The National Oceanic and Atmospheric Administration)

Chapter Outline

14.1 Fluids, Density, and Pressure

14.2 Measuring Pressure

14.3 Pascal's Principle and Hydraulics

14.4 Archimedes' Principle and Buoyancy

14.5 Fluid Dynamics

14.6 Bernoulli's Equation

14.7 Viscosity and Turbulence

Introduction

Picture yourself walking along a beach on the eastern shore of the United States. The air smells of sea salt and the sun warms your body. Suddenly, an alert appears on your cell phone. A tropical depression has formed into a hurricane. Atmospheric pressure has fallen to nearly 15% below average. As a result, forecasters expect torrential rainfall, winds in excess of 100 mph, and millions of dollars in damage. As you prepare to evacuate, you wonder: How can such a small drop in pressure lead to such a severe change in the weather?

Pressure is a physical phenomenon that is responsible for much more than just the weather. Changes in pressure cause ears to "pop" during takeoff in an airplane. Changes in pressure can also cause scuba divers to suffer a sometimes fatal disorder known as the "bends," which occurs when nitrogen dissolved in the water of the body at extreme depths returns to a gaseous state in the body as the diver surfaces. Pressure lies at the heart of the phenomena called buoyancy, which causes hot air balloons to rise and ships to float. Before we can fully understand the role that pressure plays in these phenomena, we need to discuss the states of matter and the concept of density.

14.1 | Fluids, Density, and Pressure

Learning Objectives

By the end of this section, you will be able to:

* State the different phases of matter
* Describe the characteristics of the phases of matter at the molecular or atomic level
* Distinguish between compressible and incompressible materials
* Define density and its related SI units
* Compare and contrast the densities of various substances
* Define pressure and its related SI units
* Explain the relationship between pressure and force
* Calculate force given pressure and area

Matter most commonly exists as a solid, liquid, or gas; these states are known as the three common phases of matter. We will look at each of these phases in detail in this section.

Characteristics of Solids

Solids are rigid and have specific shapes and definite volumes. The atoms or molecules in a solid are in close proximity to each other, and there is a significant force between these molecules. Solids will take a form determined by the nature of these forces between the molecules. Although true solids are not incompressible, it nevertheless requires a large force to change the shape of a solid. In some cases, the force between molecules can cause the molecules to organize into a lattice as shown in **Figure 14.2**. The structure of this three-dimensional lattice is represented as molecules connected by rigid bonds (modeled as stiff springs), which allow limited freedom for movement. Even a large force produces only small displacements in the atoms or molecules of the lattice, and the solid maintains its shape. Solids also resist shearing forces. (Shearing forces are forces applied tangentially to a surface, as described in **Static Equilibrium and Elasticity**.)

Characteristics of Fluids

Liquids and gases are considered to be **fluids** because they yield to shearing forces, whereas solids resist them. Like solids, the molecules in a liquid are bonded to neighboring molecules, but possess many fewer of these bonds. The molecules in a liquid are not locked in place and can move with respect to each other. The distance between molecules is similar to the distances in a solid, and so liquids have definite volumes, but the shape of a liquid changes, depending on the shape of its container. Gases are not bonded to neighboring atoms and can have large separations between molecules. Gases have neither specific shapes nor definite volumes, since their molecules move to fill the container in which they are held (**Figure 14.2**).

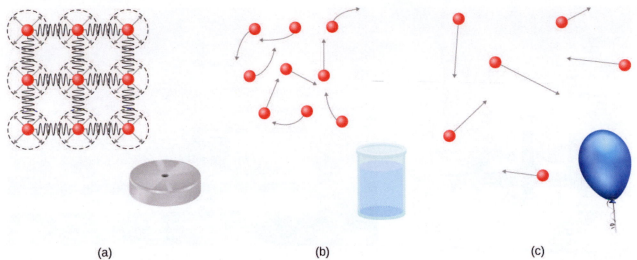

Figure 14.2 (a) Atoms in a solid are always in close contact with neighboring atoms, held in place by forces represented here by springs. (b) Atoms in a liquid are also in close contact but can slide over one another. Forces between the atoms strongly resist attempts to compress the atoms. (c) Atoms in a gas move about freely and are separated by large distances. A gas must be held in a closed container to prevent it from expanding freely and escaping.

Liquids deform easily when stressed and do not spring back to their original shape once a force is removed. This occurs because the atoms or molecules in a liquid are free to slide about and change neighbors. That is, liquids flow (so they are a type of fluid), with the molecules held together by mutual attraction. When a liquid is placed in a container with no lid, it remains in the container. Because the atoms are closely packed, liquids, like solids, resist compression; an extremely large force is necessary to change the volume of a liquid.

In contrast, atoms in gases are separated by large distances, and the forces between atoms in a gas are therefore very weak, except when the atoms collide with one another. This makes gases relatively easy to compress and allows them to flow (which makes them fluids). When placed in an open container, gases, unlike liquids, will escape.

In this chapter, we generally refer to both gases and liquids simply as fluids, making a distinction between them only when they behave differently. There exists one other phase of matter, plasma, which exists at very high temperatures. At high temperatures, molecules may disassociate into atoms, and atoms disassociate into electrons (with negative charges) and protons (with positive charges), forming a plasma. Plasma will not be discussed in depth in this chapter because plasma has very different properties from the three other common phases of matter, discussed in this chapter, due to the strong electrical forces between the charges.

Density

Suppose a block of brass and a block of wood have exactly the same mass. If both blocks are dropped in a tank of water, why does the wood float and the brass sink (**Figure 14.3**)? This occurs because the brass has a greater density than water, whereas the wood has a lower density than water.

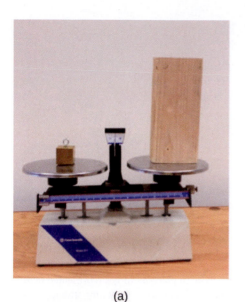
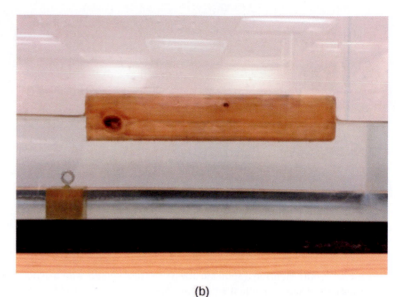

(a)   (b)

Figure 14.3 (a) A block of brass and a block of wood both have the same weight and mass, but the block of wood has a much greater volume. (b) When placed in a fish tank filled with water, the cube of brass sinks and the block of wood floats. (The block of wood is the same in both pictures; it was turned on its side to fit on the scale.) (credit: modification of works by Joseph J. Trout, Stockton University)

Density is an important characteristic of substances. It is crucial, for example, in determining whether an object sinks or floats in a fluid.

Density

The average density of a substance or object is defined as its mass per unit volume,

$$\rho = \frac{m}{V}$$ (14.1)

where the Greek letter ρ (rho) is the symbol for density, m is the mass, and V is the volume.

The SI unit of density is kg/m^3. **Table 14.1** lists some representative values. The cgs unit of density is the gram per cubic centimeter, g/cm^3, where

$$1 \, g/cm^3 = 1000 \, kg/m^3.$$

The metric system was originally devised so that water would have a density of $1 \, g/cm^3$, equivalent to $10^3 \, kg/m^3$. Thus, the basic mass unit, the kilogram, was first devised to be the mass of 1000 mL of water, which has a volume of $1000 \, cm^3$.

Solids (0.0°C)		Liquids (0.0°C)		Gases (0.0°C, 101.3 kPa)	
Substance	$\rho(kg/m^3)$	**Substance**	$\rho(kg/m^3)$	**Substance**	$\rho(kg/m^3)$
Aluminum	2.70×10^3	Benzene	8.79×10^2	Air	1.29×10^0
Bone	1.90×10^3	Blood	1.05×10^3	Carbon dioxide	1.98×10^0
Brass	8.44×10^3	Ethyl alcohol	8.06×10^2	Carbon monoxide	1.25×10^0

Table 14.1 Densities of Some Common Substances

Solids (0.0°C)		Liquids (0.0°C)		Gases (0.0°C, 101.3 kPa)	
Concrete	2.40×10^3	Gasoline	6.80×10^2	Helium	1.80×10^{-1}
Copper	8.92×10^3	Glycerin	1.26×10^3	Hydrogen	9.00×10^{-2}
Cork	2.40×10^2	Mercury	1.36×10^4	Methane	7.20×10^{-2}
Earth's crust	3.30×10^3	Olive oil	9.20×10^2	Nitrogen	1.25×10^0
Glass	2.60×10^3			Nitrous oxide	1.98×10^0
Gold	1.93×10^4			Oxygen	1.43×10^0
Granite	2.70×10^3				
Iron	7.86×10^3				
Lead	1.13×10^4				
Oak	7.10×10^2				
Pine	3.73×10^2				
Platinum	2.14×10^4				
Polystyrene	1.00×10^2				
Tungsten	1.93×10^4				
Uranium	1.87×10^3				

Table 14.1 Densities of Some Common Substances

As you can see by examining **Table 14.1**, the density of an object may help identify its composition. The density of gold, for example, is about 2.5 times the density of iron, which is about 2.5 times the density of aluminum. Density also reveals something about the phase of the matter and its substructure. Notice that the densities of liquids and solids are roughly comparable, consistent with the fact that their atoms are in close contact. The densities of gases are much less than those of liquids and solids, because the atoms in gases are separated by large amounts of empty space. The gases are displayed for a standard temperature of 0.0°C and a standard pressure of 101.3 kPa, and there is a strong dependence of the densities on temperature and pressure. The densities of the solids and liquids displayed are given for the standard temperature of 0.0°C and the densities of solids and liquids depend on the temperature. The density of solids and liquids normally increase with decreasing temperature.

Table 14.2 shows the density of water in various phases and temperature. The density of water increases with decreasing temperature, reaching a maximum at 4.0°C, and then decreases as the temperature falls below 4.0°C. This behavior of the density of water explains why ice forms at the top of a body of water.

Substance	$\rho(\text{kg/m}^3)$
Ice (0°C)	9.17×10^2
Water (0°C)	9.998×10^2
Water (4°C)	1.000×10^3

Table 14.2 Densities of Water

Substance	$\rho(\text{kg/m}^3)$
Water (20°C)	9.982×10^2
Water (100°C)	9.584×10^2
Steam (100°C, 101.3 kPa)	1.670×10^2
Sea water (0°C)	1.030×10^3

Table 14.2 Densities of Water

The density of a substance is not necessarily constant throughout the volume of a substance. If the density is constant throughout a substance, the substance is said to be a homogeneous substance. A solid iron bar is an example of a homogeneous substance. The density is constant throughout, and the density of any sample of the substance is the same as its average density. If the density of a substance were not constant, the substance is said to be a heterogeneous substance. A chunk of Swiss cheese is an example of a heterogeneous material containing both the solid cheese and gas-filled voids. The density at a specific location within a heterogeneous material is called *local density*, and is given as a function of location, $\rho = \rho(x, y, z)$ (**Figure 14.4**).

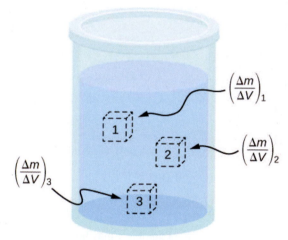

Figure 14.4 Density may vary throughout a heterogeneous mixture. Local density at a point is obtained from dividing mass by volume in a small volume around a given point.

Local density can be obtained by a limiting process, based on the average density in a small volume around the point in question, taking the limit where the size of the volume approaches zero,

$$\rho = \lim_{\Delta V \to 0} \frac{\Delta m}{\Delta V} \tag{14.2}$$

where ρ is the density, m is the mass, and V is the volume.

Since gases are free to expand and contract, the densities of the gases vary considerably with temperature, whereas the densities of liquids vary little with temperature. Therefore, the densities of liquids are often treated as constant, with the density equal to the average density.

Density is a dimensional property; therefore, when comparing the densities of two substances, the units must be taken into consideration. For this reason, a more convenient, dimensionless quantity called the **specific gravity** is often used to compare densities. Specific gravity is defined as the ratio of the density of the material to the density of water at 4.0 °C and one atmosphere of pressure, which is 1000 kg/m^3:

$$\text{Specific gravity} = \frac{\text{Density of material}}{\text{Density of water}}.$$

The comparison uses water because the density of water is 1 g/cm^3, which was originally used to define the kilogram. Specific gravity, being dimensionless, provides a ready comparison among materials without having to worry about the unit of density. For instance, the density of aluminum is 2.7 in g/cm^3 (2700 in kg/m^3), but its specific gravity is 2.7, regardless of the unit of density. Specific gravity is a particularly useful quantity with regard to buoyancy, which we will discuss later in this chapter.

Pressure

You have no doubt heard the word 'pressure' used in relation to blood (high or low blood pressure) and in relation to weather (high- and low-pressure weather systems). These are only two of many examples of pressure in fluids. (Recall that we introduced the idea of pressure in **Static Equilibrium and Elasticity**, in the context of bulk stress and strain.)

Pressure

Pressure (p) is defined as the normal force F per unit area A over which the force is applied, or

$$p = \frac{F}{A}.$$
(14.3)

To define the pressure at a specific point, the pressure is defined as the force dF exerted by a fluid over an infinitesimal element of area dA containing the point, resulting in $p = \frac{dF}{dA}$.

A given force can have a significantly different effect, depending on the area over which the force is exerted. For instance, a force applied to an area of 1 mm^2 has a pressure that is 100 times as great as the same force applied to an area of 1 cm^2. That is why a sharp needle is able to poke through skin when a small force is exerted, but applying the same force with a finger does not puncture the skin (**Figure 14.5**).

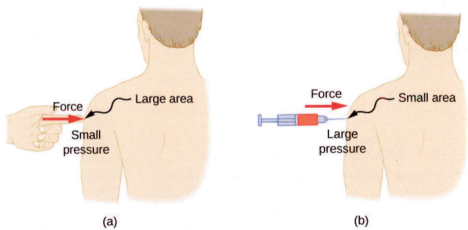

(a)   (b)

Figure 14.5 (a) A person being poked with a finger might be irritated, but the force has little lasting effect. (b) In contrast, the same force applied to an area the size of the sharp end of a needle is enough to break the skin.

Note that although force is a vector, pressure is a scalar. Pressure is a scalar quantity because it is defined to be proportional to the magnitude of the force acting perpendicular to the surface area. The SI unit for pressure is the *pascal* (Pa), named after the French mathematician and physicist Blaise Pascal (1623–1662), where

$$1 \text{ Pa} = 1 \text{ N/m}^2.$$

Several other units are used for pressure, which we discuss later in the chapter.

Variation of pressure with depth in a fluid of constant density

Pressure is defined for all states of matter, but it is particularly important when discussing fluids. An important characteristic of fluids is that there is no significant resistance to the component of a force applied parallel to the surface of a fluid. The molecules of the fluid simply flow to accommodate the horizontal force. A force applied perpendicular to the surface compresses or expands the fluid. If you try to compress a fluid, you find that a reaction force develops at each point inside the fluid in the outward direction, balancing the force applied on the molecules at the boundary.

Consider a fluid of constant density as shown in **Figure 14.6**. The pressure at the bottom of the container is due to the pressure of the atmosphere (p_0) plus the pressure due to the weight of the fluid. The pressure due to the fluid is equal to the weight of the fluid divided by the area. The weight of the fluid is equal to its mass times the acceleration due to gravity.

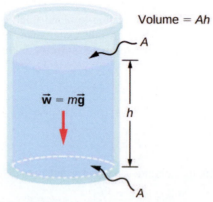

Figure 14.6 The bottom of this container supports the entire weight of the fluid in it. The vertical sides cannot exert an upward force on the fluid (since it cannot withstand a shearing force), so the bottom must support it all.

Since the density is constant, the weight can be calculated using the density:

$$w = mg = \rho V g = \rho A h g.$$

The pressure at the bottom of the container is therefore equal to atmospheric pressure added to the weight of the fluid divided by the area:

$$p = p_0 + \frac{\rho A h g}{A} = p_0 + \rho h g.$$

This equation is only good for pressure at a depth for a fluid of constant density.

Pressure at a Depth for a Fluid of Constant Density

The pressure at a depth in a fluid of constant density is equal to the pressure of the atmosphere plus the pressure due to the weight of the fluid, or

$$p = p_0 + \rho h g, \tag{14.4}$$

Where p is the pressure at a particular depth, p_0 is the pressure of the atmosphere, ρ is the density of the fluid, g is the acceleration due to gravity, and h is the depth.

Figure 14.7 The Three Gorges Dam, erected on the Yangtze River in central China in 2008, created a massive reservoir that displaced more than one million people. (credit: modification of work by "Le Grand Portage"/Flickr)

Example 14.1

What Force Must a Dam Withstand?

Consider the pressure and force acting on the dam retaining a reservoir of water (**Figure 14.7**). Suppose the dam is 500-m wide and the water is 80.0-m deep at the dam, as illustrated below. (a) What is the average pressure on the dam due to the water? (b) Calculate the force exerted against the dam.

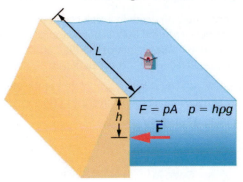

The average pressure p due to the weight of the water is the pressure at the average depth h of 40.0 m, since pressure increases linearly with depth. The force exerted on the dam by the water is the average pressure times the area of contact, $F = pA$.

solution

a. The average pressure due to the weight of a fluid is

$$p = h\rho g. \tag{14.5}$$

Entering the density of water from **Table 14.1** and taking h to be the average depth of 40.0 m, we obtain

$$p = (40.0 \text{ m})\left(10^3 \, \frac{\text{kg}}{\text{m}^3}\right)\left(9.80\frac{\text{m}}{\text{s}^2}\right)$$

$$= 3.92 \times 10^5 \, \frac{\text{N}}{\text{m}^2} = 392 \text{ kPa}.$$

b. We have already found the value for p. The area of the dam is

$$A = 80.0 \text{ m} \times 500 \text{ m} = 4.00 \times 10^4 \text{ m}^2,$$

so that

$$F = (3.92 \times 10^5 \text{ N/m}^2)(4.00 \times 10^4 \text{ m}^2)$$
$$= 1.57 \times 10^{10} \text{ N.}$$

Significance

Although this force seems large, it is small compared with the 1.96×10^{13} N weight of the water in the reservoir. In fact, it is only 0.0800% of the weight.

 14.1   Check Your Understanding If the reservoir in **Example 14.1** covered twice the area, but was kept to the same depth, would the dam need to be redesigned?

Pressure in a static fluid in a uniform gravitational field

A *static fluid* is a fluid that is not in motion. At any point within a static fluid, the pressure on all sides must be equal—otherwise, the fluid at that point would react to a net force and accelerate.

The pressure at any point in a static fluid depends only on the depth at that point. As discussed, pressure in a fluid near Earth varies with depth due to the weight of fluid above a particular level. In the above examples, we assumed density to be constant and the average density of the fluid to be a good representation of the density. This is a reasonable approximation for liquids like water, where large forces are required to compress the liquid or change the volume. In a swimming pool, for example, the density is approximately constant, and the water at the bottom is compressed very little by the weight of the water on top. Traveling up in the atmosphere is quite a different situation, however. The density of the air begins to change significantly just a short distance above Earth's surface.

To derive a formula for the variation of pressure with depth in a tank containing a fluid of density ρ on the surface of Earth, we must start with the assumption that the density of the fluid is not constant. Fluid located at deeper levels is subjected to more force than fluid nearer to the surface due to the weight of the fluid above it. Therefore, the pressure calculated at a given depth is different than the pressure calculated using a constant density.

Imagine a thin element of fluid at a depth h, as shown in **Figure 14.8**. Let the element have a cross-sectional area A and height Δy. The forces acting upon the element are due to the pressures $p(y)$ above and $p(y + \Delta y)$ below it. The weight of the element itself is also shown in the free-body diagram.

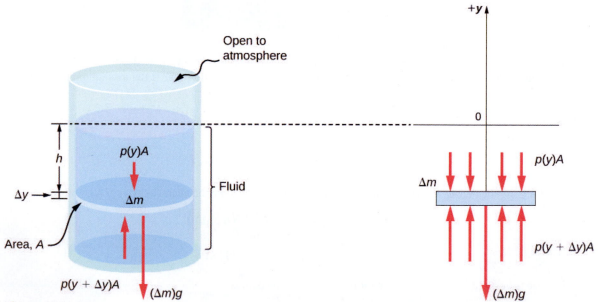

Figure 14.8   Forces on a mass element inside a fluid. The weight of the element itself is shown in the free-body diagram.

Since the element of fluid between y and $y + \Delta y$ is not accelerating, the forces are balanced. Using a Cartesian y-axis oriented up, we find the following equation for the y-component:

$$p(y + \Delta y)A - p(y)A - g\Delta m = 0(\Delta y > 0).$$ (14.6)

Note that if the element had a non-zero y-component of acceleration, the right-hand side would not be zero but would instead be the mass times the y-acceleration. The mass of the element can be written in terms of the density of the fluid and the volume of the elements:

$$\Delta m = |\rho A \Delta y| = -\rho A \Delta y \quad (\Delta y > 0).$$

Putting this expression for Δm into **Equation 14.6** and then dividing both sides by $A\Delta y$, we find

$$\frac{p(y + \Delta y) - p(y)}{\Delta y} = -\rho g.$$ (14.7)

Taking the limit of the infinitesimally thin element $\Delta y \to 0$, we obtain the following differential equation, which gives the variation of pressure in a fluid:

$$\frac{dp}{dy} = -\rho g.$$ (14.8)

This equation tells us that the rate of change of pressure in a fluid is proportional to the density of the fluid. The solution of this equation depends upon whether the density ρ is constant or changes with depth; that is, the function $\rho(y)$.

If the range of the depth being analyzed is not too great, we can assume the density to be constant. But if the range of depth is large enough for the density to vary appreciably, such as in the case of the atmosphere, there is significant change in density with depth. In that case, we cannot use the approximation of a constant density.

Pressure in a fluid with a constant density

Let's use **Equation 14.9** to work out a formula for the pressure at a depth h from the surface in a tank of a liquid such as water, where the density of the liquid can be taken to be constant.

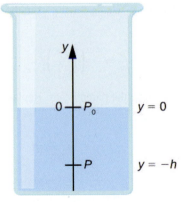

We need to integrate **Equation 14.9** from $y = 0$, where the pressure is atmospheric pressure (p_0), to $y = -h$, the y-coordinate of the depth:

$$\int_{p_0}^{p} dp = -\int_{0}^{-h} \rho g dy$$ (14.9)

$$p - p_0 = \rho g h$$

$$p = p_0 + \rho g h.$$

Hence, pressure at a depth of fluid on the surface of Earth is equal to the atmospheric pressure plus $\rho g h$ if the density of the fluid is constant over the height, as we found previously.

Note that the pressure in a fluid depends only on the depth from the surface and not on the shape of the container. Thus, in a container where a fluid can freely move in various parts, the liquid stays at the same level in every part, regardless of the shape, as shown in **Figure 14.9**.

Figure 14.9 If a fluid can flow freely between parts of a container, it rises to the same height in each part. In the container pictured, the pressure at the bottom of each column is the same; if it were not the same, the fluid would flow until the pressures became equal.

Variation of atmospheric pressure with height

The change in atmospheric pressure with height is of particular interest. Assuming the temperature of air to be constant, and that the ideal gas law of thermodynamics describes the atmosphere to a good approximation, we can find the variation of atmospheric pressure with height, when the temperature is constant. (We discuss the ideal gas law in a later chapter, but we assume you have some familiarity with it from high school and chemistry.) Let $p(y)$ be the atmospheric pressure at height y. The density ρ at y, the temperature T in the Kelvin scale (K), and the mass m of a molecule of air are related to the absolute pressure by the ideal gas law, in the form

$$p = \rho \frac{k_{\mathrm{B}} T}{m} \text{ (atmosphere)},$$
(14.10)

where k_{B} is Boltzmann's constant, which has a value of $1.38 \times 10^{-23} \, \mathrm{J/K}$.

You may have encountered the ideal gas law in the form $pV = nRT$, where n is the number of moles and R is the gas constant. Here, the same law has been written in a different form, using the density ρ instead of volume V. Therefore, if pressure p changes with height, so does the density ρ. Using density from the ideal gas law, the rate of variation of pressure with height is given as

$$\frac{dp}{dy} = -p \left(\frac{mg}{k_{\mathrm{B}} T} \right),$$

where constant quantities have been collected inside the parentheses. Replacing these constants with a single symbol α, the equation looks much simpler:

$$\begin{aligned}
\frac{dp}{dy} &= -\alpha p \\
\frac{dp}{p} &= -\alpha dy \\
\int_{p_0}^{p(y)} \frac{dp}{p} &= \int_0^y -\alpha dy \\
[\ln(p)]_{p_0}^{p(y)} &= [-\alpha y]_0^y \\
\ln(p) - \ln(p_0) &= -\alpha y \\
\ln\left(\frac{p}{p_0} \right) &= -\alpha y
\end{aligned}$$

This gives the solution

$$p(y) = p_0 \exp(-\alpha y).$$

Thus, atmospheric pressure drops exponentially with height, since the y-axis is pointed up from the ground and y has positive values in the atmosphere above sea level. The pressure drops by a factor of $\frac{1}{e}$ when the height is $\frac{1}{\alpha}$, which gives us a physical interpretation for α: The constant $\frac{1}{\alpha}$ is a length scale that characterizes how pressure varies with height and is often referred to as the pressure scale height.

We can obtain an approximate value of α by using the mass of a nitrogen molecule as a proxy for an air molecule. At temperature $27\,°\mathrm{C}$, or 300 K, we find

$$\alpha = -\frac{mg}{k_\mathrm{B} T} = \frac{4.8 \times 10^{-26}\ \mathrm{kg} \times 9.81\ \mathrm{m/s}^2}{1.38 \times 10^{-23}\ \mathrm{J/K} \times 300\ \mathrm{K}} = \frac{1}{8800\ \mathrm{m}}.$$

Therefore, for every 8800 meters, the air pressure drops by a factor $1/e$, or approximately one-third of its value. This gives us only a rough estimate of the actual situation, since we have assumed both a constant temperature and a constant g over such great distances from Earth, neither of which is correct in reality.

Direction of pressure in a fluid

Fluid pressure has no direction, being a scalar quantity, whereas the forces due to pressure have well-defined directions: They are always exerted perpendicular to any surface. The reason is that fluids cannot withstand or exert shearing forces. Thus, in a static fluid enclosed in a tank, the force exerted on the walls of the tank is exerted perpendicular to the inside surface. Likewise, pressure is exerted perpendicular to the surfaces of any object within the fluid. **Figure 14.10** illustrates the pressure exerted by air on the walls of a tire and by water on the body of a swimmer.

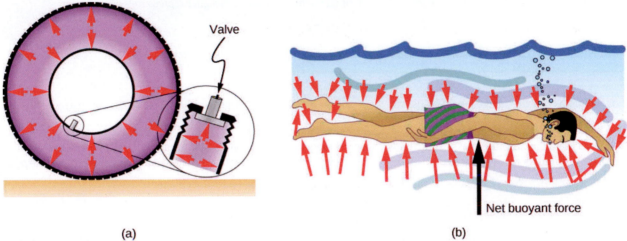

(a)   (b)

Figure 14.10 (a) Pressure inside this tire exerts forces perpendicular to all surfaces it contacts. The arrows represent directions and magnitudes of the forces exerted at various points. (b) Pressure is exerted perpendicular to all sides of this swimmer, since the water would flow into the space he occupies if he were not there. The arrows represent the directions and magnitudes of the forces exerted at various points on the swimmer. Note that the forces are larger underneath, due to greater depth, giving a net upward or buoyant force. The net vertical force on the swimmer is equal to the sum of the buoyant force and the weight of the swimmer.

14.2 | Measuring Pressure

Learning Objectives

By the end of this section, you will be able to:

- Define gauge pressure and absolute pressure
- Explain various methods for measuring pressure
- Understand the working of open-tube barometers
- Describe in detail how manometers and barometers operate

In the preceding section, we derived a formula for calculating the variation in pressure for a fluid in hydrostatic equilibrium. As it turns out, this is a very useful calculation. Measurements of pressure are important in daily life as well as in science and engineering applications. In this section, we discuss different ways that pressure can be reported and measured.

Gauge Pressure vs. Absolute Pressure

Suppose the pressure gauge on a full scuba tank reads 3000 psi, which is approximately 207 atmospheres. When the valve is opened, air begins to escape because the pressure inside the tank is greater than the atmospheric pressure outside the tank. Air continues to escape from the tank until the pressure inside the tank equals the pressure of the atmosphere outside the tank. At this point, the pressure gauge on the tank reads zero, even though the pressure inside the tank is actually 1 atmosphere—the same as the air pressure outside the tank.

Most pressure gauges, like the one on the scuba tank, are calibrated to read zero at atmospheric pressure. Pressure readings from such gauges are called **gauge pressure**, which is the pressure relative to the atmospheric pressure. When the pressure inside the tank is greater than atmospheric pressure, the gauge reports a positive value.

Some gauges are designed to measure negative pressure. For example, many physics experiments must take place in a vacuum chamber, a rigid chamber from which some of the air is pumped out. The pressure inside the vacuum chamber is less than atmospheric pressure, so the pressure gauge on the chamber reads a negative value.

Unlike gauge pressure, **absolute pressure** accounts for atmospheric pressure, which in effect adds to the pressure in any fluid not enclosed in a rigid container.

> ### Absolute Pressure
>
> The absolute pressure, or total pressure, is the sum of gauge pressure and atmospheric pressure:
>
> $$p_{abs} = p_g + p_{atm} \qquad (14.11)$$
>
> where p_{abs} is absolute pressure, p_g is gauge pressure, and p_{atm} is atmospheric pressure.

For example, if a tire gauge reads 34 psi, then the absolute pressure is 34 psi plus 14.7 psi (p_{atm} in psi), or 48.7 psi (equivalent to 336 kPa).

In most cases, the absolute pressure in fluids cannot be negative. Fluids push rather than pull, so the smallest absolute pressure in a fluid is zero (a negative absolute pressure is a pull). Thus, the smallest possible gauge pressure is $p_g = -p_{atm}$ (which makes p_{abs} zero). There is no theoretical limit to how large a gauge pressure can be.

Measuring Pressure

A host of devices are used for measuring pressure, ranging from tire gauges to blood pressure monitors. Many other types of pressure gauges are commonly used to test the pressure of fluids, such as mechanical pressure gauges. We will explore some of these in this section.

Any property that changes with pressure in a known way can be used to construct a pressure gauge. Some of the most common types include strain gauges, which use the change in the shape of a material with pressure; capacitance pressure gauges, which use the change in electric capacitance due to shape change with pressure; piezoelectric pressure gauges, which generate a voltage difference across a piezoelectric material under a pressure difference between the two sides; and ion gauges, which measure pressure by ionizing molecules in highly evacuated chambers. Different pressure gauges are useful in different pressure ranges and under different physical situations. Some examples are shown in **Figure 14.11**.

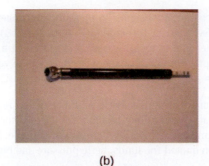
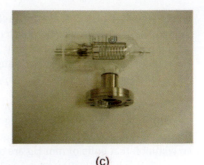

(a)	(b)	(c)

Figure 14.11 (a) Gauges are used to measure and monitor pressure in gas cylinders. Compressed gases are used in many industrial as well as medical applications. (b) Tire pressure gauges come in many different models, but all are meant for the same purpose: to measure the internal pressure of the tire. This enables the driver to keep the tires inflated at optimal pressure for load weight and driving conditions. (c) An ionization gauge is a high-sensitivity device used to monitor the pressure of gases in an enclosed system. Neutral gas molecules are ionized by the release of electrons, and the current is translated into a pressure reading. Ionization gauges are commonly used in industrial applications that rely on vacuum systems.

Manometers

One of the most important classes of pressure gauges applies the property that pressure due to the weight of a fluid of constant density is given by $p = h\rho g$. The U-shaped tube shown in **Figure 14.12** is an example of a *manometer*; in part (a), both sides of the tube are open to the atmosphere, allowing atmospheric pressure to push down on each side equally so that its effects cancel.

A manometer with only one side open to the atmosphere is an ideal device for measuring gauge pressures. The gauge pressure is $p_g = h\rho g$ and is found by measuring h. For example, suppose one side of the U-tube is connected to some source of pressure p_{abs}, such as the balloon in part (b) of the figure or the vacuum-packed peanut jar shown in part (c). Pressure is transmitted undiminished to the manometer, and the fluid levels are no longer equal. In part (b), p_{abs} is greater than atmospheric pressure, whereas in part (c), p_{abs} is less than atmospheric pressure. In both cases, p_{abs} differs from atmospheric pressure by an amount $h\rho g$, where ρ is the density of the fluid in the manometer. In part (b), p_{abs} can support a column of fluid of height h, so it must exert a pressure $h\rho g$ greater than atmospheric pressure (the gauge pressure p_g is positive). In part (c), atmospheric pressure can support a column of fluid of height h, so p_{abs} is less than atmospheric pressure by an amount $h\rho g$ (the gauge pressure p_g is negative).

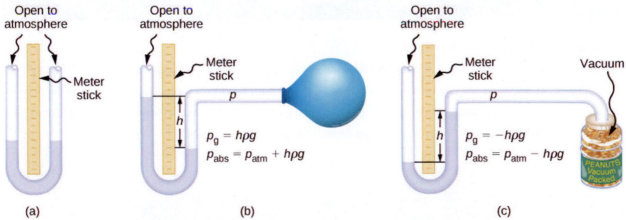

(a)	(b)	(c)

Figure 14.12 An open-tube manometer has one side open to the atmosphere. (a) Fluid depth must be the same on both sides, or the pressure each side exerts at the bottom will be unequal and liquid will flow from the deeper side. (b) A positive gauge pressure $p_g = h\rho g$ transmitted to one side of the manometer can support a column of fluid of height h. (c) Similarly, atmospheric pressure is greater than a negative gauge pressure p_g by an amount $h\rho g$. The jar's rigidity prevents atmospheric pressure from being transmitted to the peanuts.

Barometers

Manometers typically use a U-shaped tube of a fluid (often mercury) to measure pressure. A *barometer* (see **Figure 14.13**) is a device that typically uses a single column of mercury to measure atmospheric pressure. The barometer, invented by the Italian mathematician and physicist Evangelista Torricelli (1608–1647) in 1643, is constructed from a glass tube closed at one end and filled with mercury. The tube is then inverted and placed in a pool of mercury. This device measures atmospheric pressure, rather than gauge pressure, because there is a nearly pure vacuum above the mercury in the tube. The height of the mercury is such that $h\rho g = p_{\text{atm}}$. When atmospheric pressure varies, the mercury rises or falls.

Weather forecasters closely monitor changes in atmospheric pressure (often reported as barometric pressure), as rising mercury typically signals improving weather and falling mercury indicates deteriorating weather. The barometer can also be used as an altimeter, since average atmospheric pressure varies with altitude. Mercury barometers and manometers are so common that units of mm Hg are often quoted for atmospheric pressure and blood pressures.

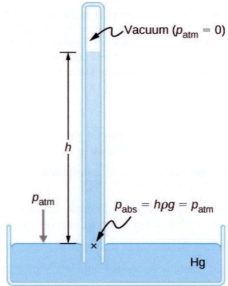

Figure 14.13 A mercury barometer measures atmospheric pressure. The pressure due to the mercury's weight, $h\rho g$, equals atmospheric pressure. The atmosphere is able to force mercury in the tube to a height h because the pressure above the mercury is zero.

Example 14.2

Fluid Heights in an Open U-Tube

A U-tube with both ends open is filled with a liquid of density ρ_1 to a height h on both sides (**Figure 14.14**). A liquid of density $\rho_2 < \rho_1$ is poured into one side and Liquid 2 settles on top of Liquid 1. The heights on the two sides are different. The height to the top of Liquid 2 from the interface is h_2 and the height to the top of Liquid 1 from the level of the interface is h_1. Derive a formula for the height difference.

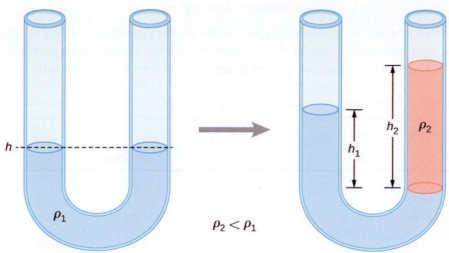

Figure 14.14 Two liquids of different densities are shown in a U-tube.

Strategy

The pressure at points at the same height on the two sides of a U-tube must be the same as long as the two points are in the same liquid. Therefore, we consider two points at the same level in the two arms of the tube: One point is the interface on the side of the Liquid 2 and the other is a point in the arm with Liquid 1 that is at the same level as the interface in the other arm. The pressure at each point is due to atmospheric pressure plus the weight of the liquid above it.

$$\text{Pressure on the side with Liquid 1} = p_0 + \rho_1 g h_1$$
$$\text{Pressure on the side with Liquid 2} = p_0 + \rho_2 g h_2$$

Solution

Since the two points are in Liquid 1 and are at the same height, the pressure at the two points must be the same. Therefore, we have

$$p_0 + \rho_1 g h_1 = p_0 + \rho_2 g h_2.$$

Hence,

$$\rho_1 h_1 = \rho_2 h_2.$$

This means that the difference in heights on the two sides of the U-tube is

$$h_2 - h_1 = \left(1 - \frac{p_1}{p_2}\right)h_2.$$

The result makes sense if we set $p_2 = p_1$, which gives $h_2 = h_1$. If the two sides have the same density, they have the same height.

 14.2 Check Your Understanding Mercury is a hazardous substance. Why do you suppose mercury is typically used in barometers instead of a safer fluid such as water?

Units of pressure

As stated earlier, the SI unit for pressure is the pascal (Pa), where

$$1 \, \text{Pa} = 1 \, \text{N/m}^2.$$

In addition to the pascal, many other units for pressure are in common use (**Table 14.3**). In meteorology, atmospheric pressure is often described in the unit of millibars (mbar), where

$$1000 \, \text{mbar} = 1 \times 10^5 \, \text{Pa}.$$

The millibar is a convenient unit for meteorologists because the average atmospheric pressure at sea level on Earth is $1.013 \times 10^5 \, \text{Pa} = 1013 \, \text{mbar} = 1 \, \text{atm}$. Using the equations derived when considering pressure at a depth in a fluid, pressure can also be measured as millimeters or inches of mercury. The pressure at the bottom of a 760-mm column of mercury at $0 \, °\text{C}$ in a container where the top part is evacuated is equal to the atmospheric pressure. Thus, 760 mm Hg is also used in place of 1 atmosphere of pressure. In vacuum physics labs, scientists often use another unit called the torr, named after Torricelli, who, as we have just seen, invented the mercury manometer for measuring pressure. One torr is equal to a pressure of 1 mm Hg.

Unit	Definition
SI unit: the Pascal	$1 \, \text{Pa} = 1 \, \text{N/m}^2$
English unit: pounds per square inch (lb/in.^2 or psi)	$1 \, \text{psi} = 6.895 \times 10^3 \, \text{Pa}$
Other units of pressure	$\begin{aligned} 1 \, \text{atm} \ &= 760 \, \text{mmHg} \\ &= 1.013 \times 10^5 \, \text{Pa} \\ &= 14.7 \, \text{psi} \\ &= 29.9 \, \text{inches of Hg} \\ &= 1013 \, \text{mbar} \end{aligned}$
	$1 \, \text{bar} = 10^5 \, \text{Pa}$
	$1 \, \text{torr} = 1 \, \text{mm Hg} = 133.3 \, \text{Pa}$

Table 14.3 Summary of the Units of Pressure

14.3 | Pascal's Principle and Hydraulics

Learning Objectives

By the end of this section, you will be able to:

- State Pascal's principle
- Describe applications of Pascal's principle
- Derive relationships between forces in a hydraulic system

In 1653, the French philosopher and scientist Blaise Pascal published his *Treatise on the Equilibrium of Liquids*, in which he discussed principles of static fluids. A static fluid is a fluid that is not in motion. When a fluid is not flowing, we say that the fluid is in static equilibrium. If the fluid is water, we say it is in **hydrostatic equilibrium**. For a fluid in static equilibrium, the net force on any part of the fluid must be zero; otherwise the fluid will start to flow.

Pascal's observations—since proven experimentally—provide the foundation for hydraulics, one of the most important developments in modern mechanical technology. Pascal observed that a change in pressure applied to an enclosed fluid is transmitted undiminished throughout the fluid and to the walls of its container. Because of this, we often know more about pressure than other physical quantities in fluids. Moreover, Pascal's principle implies that the total pressure in a fluid is the sum of the pressures from different sources. A good example is the fluid at a depth depends on the depth of the fluid and the pressure of the atmosphere.

Pascal's Principle

Pascal's principle (also known as Pascal's law) states that when a change in pressure is applied to an enclosed fluid, it is transmitted undiminished to all portions of the fluid and to the walls of its container. In an enclosed fluid, since atoms of the fluid are free to move about, they transmit pressure to all parts of the fluid *and* to the walls of the container. Any change in pressure is transmitted undiminished.

Note that this principle does not say that the pressure is the same at all points of a fluid—which is not true, since the pressure in a fluid near Earth varies with height. Rather, this principle applies to the *change* in pressure. Suppose you place some water in a cylindrical container of height H and cross-sectional area A that has a movable piston of mass m (**Figure 14.15**). Adding weight Mg at the top of the piston increases the pressure at the top by Mg/A, since the additional weight also acts over area A of the lid:

$$\Delta p_{top} = \frac{Mg}{A}.$$

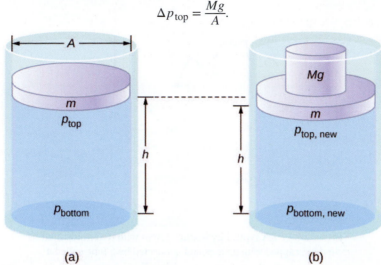

Figure 14.15 Pressure in a fluid changes when the fluid is compressed. (a) The pressure at the top layer of the fluid is different from pressure at the bottom layer. (b) The increase in pressure by adding weight to the piston is the same everywhere, for example, $p_{\text{top new}} - p_{\text{top}} = p_{\text{bottom new}} - p_{\text{bottom}}$.

According to Pascal's principle, the pressure at all points in the water changes by the same amount, Mg/A. Thus, the pressure at the bottom also increases by Mg/A. The pressure at the bottom of the container is equal to the sum of the atmospheric pressure, the pressure due the fluid, and the pressure supplied by the mass. The change in pressure at the bottom of the container due to the mass is

$$\Delta p_{bottom} = \frac{Mg}{A}.$$

Since the pressure changes are the same everywhere in the fluid, we no longer need subscripts to designate the pressure change for top or bottom:

$$\Delta p = \Delta p_{top} = \Delta p_{bottom} = \Delta p_{everywhere}.$$

 Pascal's Barrel is a great demonstration of Pascal's principle. Watch a **simulation (https://openstaxcollege.org/l/21pascalbarrel)** of Pascal's 1646 experiment, in which he demonstrated the effects of changing pressure in a fluid.

Applications of Pascal's Principle and Hydraulic Systems

Hydraulic systems are used to operate automotive brakes, hydraulic jacks, and numerous other mechanical systems (**Figure 14.16**).

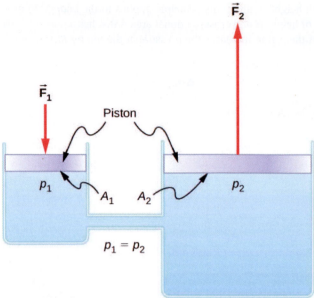

Figure 14.16 A typical hydraulic system with two fluid-filled cylinders, capped with pistons and connected by a tube called a hydraulic line. A downward force $\vec{\mathbf{F}}_1$ on the left piston creates a change in pressure that is transmitted undiminished to all parts of the enclosed fluid. This results in an upward force $\vec{\mathbf{F}}_2$ on the right piston that is larger than $\vec{\mathbf{F}}_1$ because the right piston has a larger surface area.

We can derive a relationship between the forces in this simple hydraulic system by applying Pascal's principle. Note first that the two pistons in the system are at the same height, so there is no difference in pressure due to a difference in depth. The pressure due to F_1 acting on area A_1 is simply

$$p_1 = \frac{F_1}{A_1}, \text{ as defined by } p = \frac{F}{A}.$$

According to Pascal's principle, this pressure is transmitted undiminished throughout the fluid and to all walls of the container. Thus, a pressure p_2 is felt at the other piston that is equal to p_1. That is, $p_1 = p_2$. However, since $p_2 = F_2/A_2$, we see that

$$\frac{F_1}{A_1} = \frac{F_2}{A_2}. \tag{14.12}$$

This equation relates the ratios of force to area in any hydraulic system, provided that the pistons are at the same vertical height and that friction in the system is negligible.

Hydraulic systems can increase or decrease the force applied to them. To make the force larger, the pressure is applied to a larger area. For example, if a 100-N force is applied to the left cylinder in **Figure 14.16** and the right cylinder has an area five times greater, then the output force is 500 N. Hydraulic systems are analogous to simple levers, but they have the advantage that pressure can be sent through tortuously curved lines to several places at once.

The **hydraulic jack** is such a hydraulic system. A hydraulic jack is used to lift heavy loads, such as the ones used by auto mechanics to raise an automobile. It consists of an incompressible fluid in a U-tube fitted with a movable piston on each side. One side of the U-tube is narrower than the other. A small force applied over a small area can balance a much larger force on the other side over a larger area (**Figure 14.17**).

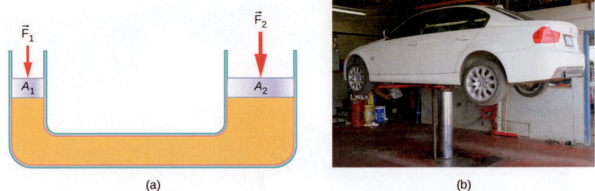

(a)   (b)

Figure 14.17 (a) A hydraulic jack operates by applying forces (F_1, F_2) to an incompressible fluid in a U-tube, using a movable piston (A_1, A_2) on each side of the tube. (b) Hydraulic jacks are commonly used by car mechanics to lift vehicles so that repairs and maintenance can be performed. (credit b: modification of work by Jane Whitney)

From Pascal's principle, it can be shown that the force needed to lift the car is less than the weight of the car:

$$F_1 = \frac{A_1}{A_2} F_2,$$

where F_1 is the force applied to lift the car, A_1 is the cross-sectional area of the smaller piston, A_2 is the cross sectional area of the larger piston, and F_2 is the weight of the car.

Example 14.3

Calculating Force on Wheel Cylinders: Pascal Puts on the Brakes

Consider the automobile hydraulic system shown in **Figure 14.18**. Suppose a force of 100 N is applied to the brake pedal, which acts on the pedal cylinder (acting as a "master" cylinder) through a lever. A force of 500 N is exerted on the pedal cylinder. Pressure created in the pedal cylinder is transmitted to the four wheel cylinders. The pedal cylinder has a diameter of 0.500 cm and each wheel cylinder has a diameter of 2.50 cm. Calculate the magnitude of the force F_2 created at each of the wheel cylinders.

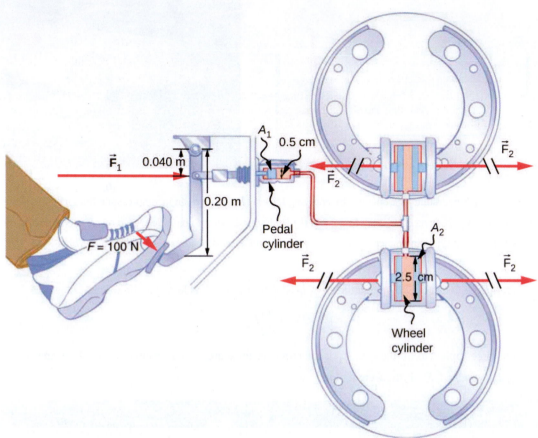

Figure 14.18 Hydraulic brakes use Pascal's principle. The driver pushes the brake pedal, exerting a force that is increased by the simple lever and again by the hydraulic system. Each of the identical wheel cylinders receives the same pressure and, therefore, creates the same force output F_2. The circular cross-sectional areas of the pedal and wheel cylinders are represented by A_1 and A_2, respectively.

Strategy

We are given the force F_1 applied to the pedal cylinder. The cross-sectional areas A_1 and A_2 can be calculated from their given diameters. Then we can use the following relationship to find the force F_2:

$$\frac{F_1}{A_1} = \frac{F_2}{A_2}.$$

Manipulate this algebraically to get F_2 on one side and substitute known values.

Solution

Pascal's principle applied to hydraulic systems is given by $\frac{F_1}{A_1} = \frac{F_2}{A_2}$:

$$F_2 = \frac{A_2}{A_1}F_1 = \frac{\pi r_2^2}{\pi r_1^2}F_1$$

$$= \frac{(1.25 \text{ cm})^2}{(0.250 \text{ cm})^2} \times 500 \text{ N} = 1.25 \times 10^4 \text{ N}.$$

Significance

This value is the force exerted by each of the four wheel cylinders. Note that we can add as many wheel cylinders as we wish. If each has a 2.50-cm diameter, each will exert 1.25×10^4 N. A simple hydraulic system, as an example of a simple machine, can increase force but cannot do more work than is done on it. Work is force times distance moved, and the wheel cylinder moves through a smaller distance than the pedal cylinder. Furthermore, the more wheels added, the smaller the distance each one moves. Many hydraulic systems—such as power brakes and those in bulldozers—have a motorized pump that actually does most of the work in the system.

 14.3   Check Your Understanding Would a hydraulic press still operate properly if a gas is used instead of a liquid?

14.4 | Archimedes' Principle and Buoyancy

Learning Objectives

By the end of this section, you will be able to:

- Define buoyant force
- State Archimedes' principle
- Describe the relationship between density and Archimedes' principle

When placed in a fluid, some objects float due to a buoyant force. Where does this buoyant force come from? Why is it that some things float and others do not? Do objects that sink get any support at all from the fluid? Is your body buoyed by the atmosphere, or are only helium balloons affected (**Figure 14.19**)?

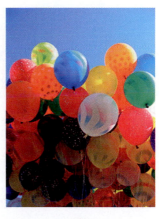

(a)   (b)   (c)

Figure 14.19   (a) Even objects that sink, like this anchor, are partly supported by water when submerged. (b) Submarines have adjustable density (ballast tanks) so that they may float or sink as desired. (c) Helium-filled balloons tug upward on their strings, demonstrating air's buoyant effect. (credit b: modification of work by Allied Navy; credit c: modification of work by "Crystl"/Flickr)

Answers to all these questions, and many others, are based on the fact that pressure increases with depth in a fluid. This means that the upward force on the bottom of an object in a fluid is greater than the downward force on top of the object. There is an upward force, or **buoyant force**, on any object in any fluid (**Figure 14.20**). If the buoyant force is greater than the object's weight, the object rises to the surface and floats. If the buoyant force is less than the object's weight, the object sinks. If the buoyant force equals the object's weight, the object can remain suspended at its present depth. The buoyant force is always present, whether the object floats, sinks, or is suspended in a fluid.

Buoyant Force

The buoyant force is the upward force on any object in any fluid.

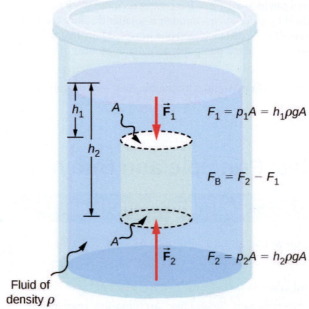

Figure 14.20 Pressure due to the weight of a fluid increases with depth because $p = h\rho g$. This change in pressure and associated upward force on the bottom of the cylinder are greater than the downward force on the top of the cylinder. The differences in the force results in the buoyant force F_B.

(Horizontal forces cancel.)

Archimedes' Principle

Just how large a force is buoyant force? To answer this question, think about what happens when a submerged object is removed from a fluid, as in **Figure 14.21**. If the object were not in the fluid, the space the object occupied would be filled by fluid having a weight w_{fl}. This weight is supported by the surrounding fluid, so the buoyant force must equal w_{fl}, the weight of the fluid displaced by the object.

Archimedes' Principle

The buoyant force on an object equals the weight of the fluid it displaces. In equation form, **Archimedes' principle** is

$$F_B = w_{fl},$$

where F_B is the buoyant force and w_{fl} is the weight of the fluid displaced by the object.

This principle is named after the Greek mathematician and inventor Archimedes (ca. 287–212 BCE), who stated this principle long before concepts of force were well established.

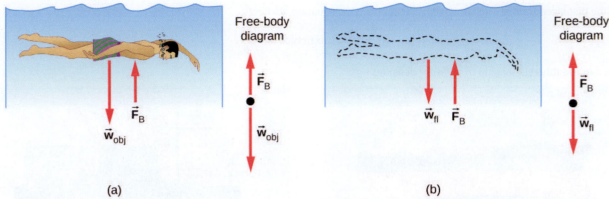

Figure 14.21 (a) An object submerged in a fluid experiences a buoyant force F_B. If F_B is greater than the weight of the object, the object rises. If F_B is less than the weight of the object, the object sinks. (b) If the object is removed, it is replaced by fluid having weight w_{fl}. Since this weight is supported by surrounding fluid, the buoyant force must equal the weight of the fluid displaced.

Archimedes' principle refers to the force of buoyancy that results when a body is submerged in a fluid, whether partially or wholly. The force that provides the pressure of a fluid acts on a body perpendicular to the surface of the body. In other words, the force due to the pressure at the bottom is pointed up, while at the top, the force due to the pressure is pointed down; the forces due to the pressures at the sides are pointing into the body.

Since the bottom of the body is at a greater depth than the top of the body, the pressure at the lower part of the body is higher than the pressure at the upper part, as shown in **Figure 14.20**. Therefore a net upward force acts on the body. This upward force is the force of buoyancy, or simply *buoyancy*.

 The exclamation "Eureka" (meaning "I found it") has often been credited to Archimedes as he made the discovery that would lead to Archimedes' principle. Some say it all started in a bathtub. To hear this story, watch this **video (https://openstaxcollege.org/l/21archNASA)** or explore **Scientific American (https://openstaxcollege.org/l/21archsciamer)** to learn more.

Density and Archimedes' Principle

If you drop a lump of clay in water, it will sink. But if you mold the same lump of clay into the shape of a boat, it will float. Because of its shape, the clay boat displaces more water than the lump and experiences a greater buoyant force, even though its mass is the same. The same is true of steel ships.

The average density of an object is what ultimately determines whether it floats. If an object's average density is less than that of the surrounding fluid, it will float. The reason is that the fluid, having a higher density, contains more mass and hence more weight in the same volume. The buoyant force, which equals the weight of the fluid displaced, is thus greater than the weight of the object. Likewise, an object denser than the fluid will sink.

The extent to which a floating object is submerged depends on how the object's density compares to the density of the fluid. In **Figure 14.22**, for example, the unloaded ship has a lower density and less of it is submerged compared with the same ship when loaded. We can derive a quantitative expression for the fraction submerged by considering density. The fraction submerged is the ratio of the volume submerged to the volume of the object, or

$$\text{fraction submerged} = \frac{V_{sub}}{V_{obj}} = \frac{V_{fl}}{V_{obj}}.$$

The volume submerged equals the volume of fluid displaced, which we call V_{fl}. Now we can obtain the relationship between the densities by substituting $\rho = \frac{m}{V}$ into the expression. This gives

$$\frac{V_{fl}}{V_{obj}} = \frac{m_{fl}/\rho_{fl}}{m_{obj}/\rho_{obj}},$$

where ρ_{obj} is the average density of the object and ρ_{fl} is the density of the fluid. Since the object floats, its mass and that of the displaced fluid are equal, so they cancel from the equation, leaving

$$\text{fraction submerged} = \frac{\rho_{\text{obj}}}{\rho_{\text{fl}}}.$$

We can use this relationship to measure densities.

 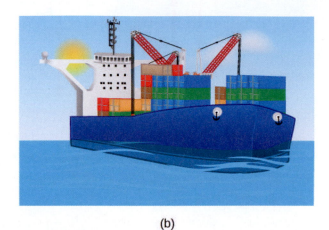

(a)   (b)

Figure 14.22 An unloaded ship (a) floats higher in the water than a loaded ship (b).

Example 14.4

Calculating Average Density

Suppose a 60.0-kg woman floats in fresh water with 97.0% of her volume submerged when her lungs are full of air. What is her average density?

Strategy

We can find the woman's density by solving the equation

$$\text{fraction submerged} = \frac{\rho_{\text{obj}}}{\rho_{\text{fl}}}$$

for the density of the object. This yields

$$\rho_{\text{obj}} = \rho_{\text{person}} = (\text{fraction submerged}) \cdot \rho_{\text{fl}}.$$

We know both the fraction submerged and the density of water, so we can calculate the woman's density.

Solution

Entering the known values into the expression for her density, we obtain

$$\rho_{\text{person}} = 0.970 \cdot \left(10^3 \; \frac{\text{kg}}{\text{m}^3}\right) = 970 \frac{\text{kg}}{\text{m}^3}.$$

Significance

The woman's density is less than the fluid density. We expect this because she floats.

Numerous lower-density objects or substances float in higher-density fluids: oil on water, a hot-air balloon in the atmosphere, a bit of cork in wine, an iceberg in salt water, and hot wax in a "lava lamp," to name a few. A less obvious example is mountain ranges floating on the higher-density crust and mantle beneath them. Even seemingly solid Earth has fluid characteristics.

Measuring Density

One of the most common techniques for determining density is shown in **Figure 14.23**.

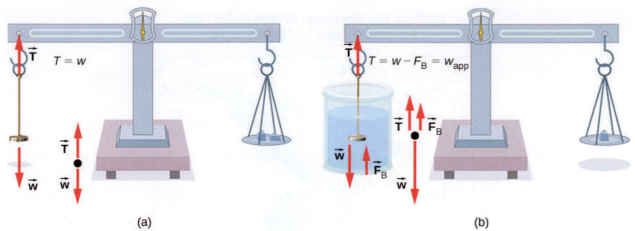

Figure 14.23 (a) A coin is weighed in air. (b) The apparent weight of the coin is determined while it is completely submerged in a fluid of known density. These two measurements are used to calculate the density of the coin.

An object, here a coin, is weighed in air and then weighed again while submerged in a liquid. The density of the coin, an indication of its authenticity, can be calculated if the fluid density is known. We can use this same technique to determine the density of the fluid if the density of the coin is known.

All of these calculations are based on Archimedes' principle, which states that the buoyant force on the object equals the weight of the fluid displaced. This, in turn, means that the object appears to weigh less when submerged; we call this measurement the object's apparent weight. The object suffers an apparent weight loss equal to the weight of the fluid displaced. Alternatively, on balances that measure mass, the object suffers an apparent mass loss equal to the mass of fluid displaced. That is, apparent weight loss equals weight of fluid displaced, or apparent mass loss equals mass of fluid displaced.

14.5 | Fluid Dynamics

Learning Objectives

By the end of this section, you will be able to:

- Describe the characteristics of flow
- Calculate flow rate
- Describe the relationship between flow rate and velocity
- Explain the consequences of the equation of continuity to the conservation of mass

The first part of this chapter dealt with fluid statics, the study of fluids at rest. The rest of this chapter deals with fluid dynamics, the study of fluids in motion. Even the most basic forms of fluid motion can be quite complex. For this reason, we limit our investigation to **ideal fluids** in many of the examples. An ideal fluid is a fluid with negligible **viscosity**. Viscosity is a measure of the internal friction in a fluid; we examine it in more detail in **Viscosity and Turbulence**. In a few examples, we examine an incompressible fluid—one for which an extremely large force is required to change the volume—since the density in an incompressible fluid is constant throughout.

Characteristics of Flow

Velocity vectors are often used to illustrate fluid motion in applications like meteorology. For example, wind—the fluid motion of air in the atmosphere—can be represented by vectors indicating the speed and direction of the wind at any given point on a map. **Figure 14.24** shows velocity vectors describing the winds during Hurricane Arthur in 2014.

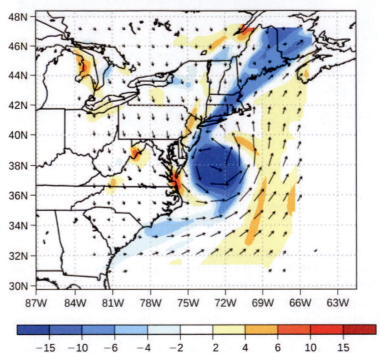

Figure 14.24 The velocity vectors show the flow of wind in Hurricane Arthur. Notice the circulation of the wind around the eye of the hurricane. Wind speeds are highest near the eye. The colors represent the relative vorticity, a measure of turning or spinning of the air. (credit: modification of work by Joseph Trout, Stockton University)

Another method for representing fluid motion is a *streamline*. A streamline represents the path of a small volume of fluid as it flows. The velocity is always tangential to the streamline. The diagrams in **Figure 14.25** use streamlines to illustrate two examples of fluids moving through a pipe. The first fluid exhibits a **laminar flow** (sometimes described as a steady flow), represented by smooth, parallel streamlines. Note that in the example shown in part (a), the velocity of the fluid is greatest in the center and decreases near the walls of the pipe due to the viscosity of the fluid and friction between the pipe walls and the fluid. This is a special case of laminar flow, where the friction between the pipe and the fluid is high, known as no slip boundary conditions. The second diagram represents **turbulent flow**, in which streamlines are irregular and change over time. In turbulent flow, the paths of the fluid flow are irregular as different parts of the fluid mix together or form small circular regions that resemble whirlpools. This can occur when the speed of the fluid reaches a certain critical speed.

(a) Laminar Flow   (b) Turbulent Flow

Figure 14.25 (a) Laminar flow can be thought of as layers of fluid moving in parallel, regular paths. (b) In turbulent flow, regions of fluid move in irregular, colliding paths, resulting in mixing and swirling.

Flow Rate and its Relation to Velocity

The volume of fluid passing by a given location through an area during a period of time is called **flow rate** Q, or more precisely, volume flow rate. In symbols, this is written as

$$Q = \frac{dV}{dt} \qquad\qquad \text{(14.13)}$$

where V is the volume and t is the elapsed time. In **Figure 14.26**, the volume of the cylinder is Ax, so the flow rate is

$$Q = \frac{dV}{dt} = \frac{d}{dt}(Ax) = A\frac{dx}{dt} = Av.$$

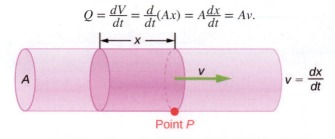

$$Q = \frac{dv}{dt} = \frac{d}{dt}(Ax) = A\frac{dx}{dt} = Av$$

Figure 14.26 Flow rate is the volume of fluid flowing past a point through the area A per unit time. Here, the shaded cylinder of fluid flows past point P in a uniform pipe in time t.

The SI unit for flow rate is m^3/s, but several other units for Q are in common use, such as liters per minute (L/min). Note that a liter (L) is 1/1000 of a cubic meter or 1000 cubic centimeters $(10^{-3} \text{ m}^3 \text{ or } 10^3 \text{ cm}^3)$.

Flow rate and velocity are related, but quite different, physical quantities. To make the distinction clear, consider the flow rate of a river. The greater the velocity of the water, the greater the flow rate of the river. But flow rate also depends on the size and shape of the river. A rapid mountain stream carries far less water than the Amazon River in Brazil, for example. **Figure 14.26** illustrates the volume flow rate. The volume flow rate is $Q = \frac{dV}{dt} = Av,$ where A is the cross-sectional area of the pipe and v is the magnitude of the velocity.

The precise relationship between flow rate Q and average speed v is

$$Q = Av,$$

where A is the cross-sectional area and v is the average speed. The relationship tells us that flow rate is directly proportional to both the average speed of the fluid and the cross-sectional area of a river, pipe, or other conduit. The larger the conduit, the greater its cross-sectional area. **Figure 14.26** illustrates how this relationship is obtained. The shaded cylinder has a volume $V = Ad$, which flows past the point P in a time t. Dividing both sides of this relationship by t gives

$$\frac{V}{t} = \frac{Ad}{t}.$$

We note that $Q = V/t$ and the average speed is $v = d/t$. Thus the equation becomes $Q = Av$.

Figure 14.27 shows an incompressible fluid flowing along a pipe of decreasing radius. Because the fluid is incompressible, the same amount of fluid must flow past any point in the tube in a given time to ensure continuity of flow. The flow is continuous because they are no sources or sinks that add or remove mass, so the mass flowing into the pipe must be equal the mass flowing out of the pipe. In this case, because the cross-sectional area of the pipe decreases, the velocity must necessarily increase. This logic can be extended to say that the flow rate must be the same at all points along the pipe. In particular, for arbitrary points 1 and 2,

$$\begin{aligned} Q_1 &= Q_2, \\ A_1 v_1 &= A_2 v_2. \end{aligned} \qquad\qquad \text{(14.14)}$$

This is called the *equation of continuity* and is valid for any incompressible fluid (with constant density). The consequences of the equation of continuity can be observed when water flows from a hose into a narrow spray nozzle: It emerges with a large speed—that is the purpose of the nozzle. Conversely, when a river empties into one end of a reservoir, the water slows considerably, perhaps picking up speed again when it leaves the other end of the reservoir. In other words, speed increases when cross-sectional area decreases, and speed decreases when cross-sectional area increases.

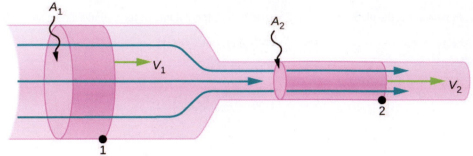

Figure 14.27 When a tube narrows, the same volume occupies a greater length. For the same volume to pass points 1 and 2 in a given time, the speed must be greater at point 2. The process is exactly reversible. If the fluid flows in the opposite direction, its speed decreases when the tube widens. (Note that the relative volumes of the two cylinders and the corresponding velocity vector arrows are not drawn to scale.)

Since liquids are essentially incompressible, the equation of continuity is valid for all liquids. However, gases are compressible, so the equation must be applied with caution to gases if they are subjected to compression or expansion.

Example 14.5

Calculating Fluid Speed through a Nozzle

A nozzle with a diameter of 0.500 cm is attached to a garden hose with a radius of 0.900 cm. The flow rate through hose and nozzle is 0.500 L/s. Calculate the speed of the water (a) in the hose and (b) in the nozzle.

Strategy

We can use the relationship between flow rate and speed to find both speeds. We use the subscript 1 for the hose and 2 for the nozzle.

Solution

a. We solve the flow rate equation for speed and use πr_1^2 for the cross-sectional area of the hose, obtaining

$$v = \frac{Q}{A} = \frac{Q}{\pi r_1^2}.$$

Substituting values and using appropriate unit conversions yields

$$v = \frac{(0.500 \text{ L/s})(10^{-3} \text{ m}^3/\text{L})}{3.14(9.00 \times 10^{-3} \text{ m})^2} = 1.96 \text{ m/s}.$$

b. We could repeat this calculation to find the speed in the nozzle v_2, but we use the equation of continuity to give a somewhat different insight. The equation states

$$A_1 v_1 = A_2 v_2.$$

Solving for v_2 and substituting πr^2 for the cross-sectional area yields

$$v_2 = \frac{A_1}{A_2} v_1 = \frac{\pi r_1^2}{\pi r_2^2} v_1 = \frac{r_1^2}{r_2^2} v_1.$$

Substituting known values,

$$v_2 = \frac{(0.900 \text{ cm})^2}{(0.250 \text{ cm})^2} 1.96 \text{ m/s} = 25.5 \text{ m/s}.$$

Significance

A speed of 1.96 m/s is about right for water emerging from a hose with no nozzle. The nozzle produces a considerably faster stream merely by constricting the flow to a narrower tube.

The solution to the last part of the example shows that speed is inversely proportional to the square of the radius of the tube, making for large effects when radius varies. We can blow out a candle at quite a distance, for example, by pursing our lips, whereas blowing on a candle with our mouth wide open is quite ineffective.

Mass Conservation

The rate of flow of a fluid can also be described by the *mass flow rate* or mass rate of flow. This is the rate at which a mass of the fluid moves past a point. Refer once again to **Figure 14.26**, but this time consider the mass in the shaded volume. The mass can be determined from the density and the volume:

$$m = \rho V = \rho A x.$$

The mass flow rate is then

$$\frac{dm}{dt} = \frac{d}{dt}(\rho A x) = \rho A \frac{dx}{dt} = \rho A v,$$

where ρ is the density, A is the cross-sectional area, and v is the magnitude of the velocity. The mass flow rate is an important quantity in fluid dynamics and can be used to solve many problems. Consider **Figure 14.28**. The pipe in the figure starts at the inlet with a cross sectional area of A_1 and constricts to an outlet with a smaller cross sectional area of A_2. The mass of fluid entering the pipe has to be equal to the mass of fluid leaving the pipe. For this reason the velocity at the outlet (v_2) is greater than the velocity of the inlet (v_1). Using the fact that the mass of fluid entering the pipe must be equal to the mass of fluid exiting the pipe, we can find a relationship between the velocity and the cross-sectional area by taking the rate of change of the mass in and the mass out:

$$\left(\frac{dm}{dt}\right)_1 = \left(\frac{dm}{dt}\right)_2 \tag{14.15}$$
$$\rho_1 A_1 v_1 = \rho_2 A_2 v_2.$$

Equation 14.15 is also known as the continuity equation in general form. If the density of the fluid remains constant through the constriction—that is, the fluid is incompressible—then the density cancels from the continuity equation,

$$A_1 v_1 = A_2 v_2.$$

The equation reduces to show that the volume flow rate into the pipe equals the volume flow rate out of the pipe.

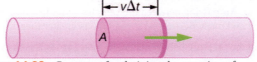

Figure 14.28 Geometry for deriving the equation of continuity. The amount of liquid entering the cross-sectional (shaded) area must equal the amount of liquid leaving the cross-sectional area if the liquid is incompressible.

14.6 | Bernoulli's Equation

As we showed in **Figure 14.27**, when a fluid flows into a narrower channel, its speed increases. That means its kinetic energy also increases. The increased kinetic energy comes from the net work done on the fluid to push it into the channel. Also, if the fluid changes vertical position, work is done on the fluid by the gravitational force.

A pressure difference occurs when the channel narrows. This pressure difference results in a net force on the fluid because the pressure times the area equals the force, and this net force does work. Recall the work-energy theorem,

$$W_{net} = \frac{1}{2}mv^2 - \frac{1}{2}mv_0^2.$$

The net work done increases the fluid's kinetic energy. As a result, the pressure drops in a rapidly moving fluid whether or not the fluid is confined to a tube.

There are many common examples of pressure dropping in rapidly moving fluids. For instance, shower curtains have a disagreeable habit of bulging into the shower stall when the shower is on. The reason is that the high-velocity stream of water and air creates a region of lower pressure inside the shower, whereas the pressure on the other side remains at the standard atmospheric pressure. This pressure difference results in a net force, pushing the curtain inward. Similarly, when a car passes a truck on the highway, the two vehicles seem to pull toward each other. The reason is the same: The high velocity of the air between the car and the truck creates a region of lower pressure between the vehicles, and they are pushed together by greater pressure on the outside (**Figure 14.29**). This effect was observed as far back as the mid-1800s, when it was found that trains passing in opposite directions tipped precariously toward one another.

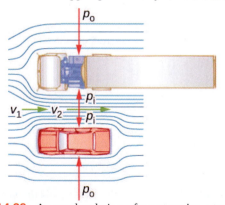

Figure 14.29 An overhead view of a car passing a truck on a highway. Air passing between the vehicles flows in a narrower channel and must increase its speed (v_2 is greater than v_1), causing the pressure between them to drop (p_i is less than p_o). Greater pressure on the outside pushes the car and truck together.

Energy Conservation and Bernoulli's Equation

The application of the principle of conservation of energy to frictionless laminar flow leads to a very useful relation between pressure and flow speed in a fluid. This relation is called **Bernoulli's equation**, named after Daniel Bernoulli (1700–1782), who published his studies on fluid motion in his book *Hydrodynamica* (1738).

Consider an incompressible fluid flowing through a pipe that has a varying diameter and height, as shown in **Figure 14.30**. Subscripts 1 and 2 in the figure denote two locations along the pipe and illustrate the relationships between the areas of the cross sections A, the speed of flow v, the height from ground y, and the pressure p at each point. We assume here that the density at the two points is the same—therefore, density is denoted by ρ without any subscripts—and since the fluid in incompressible, the shaded volumes must be equal.

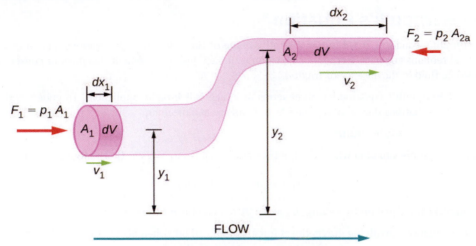

Figure 14.30 The geometry used for the derivation of Bernoulli's equation.

We also assume that there are no viscous forces in the fluid, so the energy of any part of the fluid will be conserved. To derive Bernoulli's equation, we first calculate the work that was done on the fluid:

$$dW = F_1\,dx_1 - F_2\,dx_2$$

$$dW = p_1 A_1\,dx_1 - p_2 A_2\,dx_2 = p_1\,dV - p_2\,dV = (p_1 - p_2)dV.$$

The work done was due to the conservative force of gravity and the change in the kinetic energy of the fluid. The change in the kinetic energy of the fluid is equal to

$$dK = \tfrac{1}{2}m_2 v_2^2 - \tfrac{1}{2}m_1 v_1^2 = \tfrac{1}{2}\rho dV\!\left(v_2^2 - v_1^2\right).$$

The change in potential energy is

$$dU = mgy_2 - mgy_1 = \rho dV g(y_2 - y_1).$$

The energy equation then becomes

$$
\begin{aligned}
dW &= dK + dU \\
(p_1 - p_2)dV &= \tfrac{1}{2}\rho dV\!\left(v_2^2 - v_1^2\right) + \rho dV g(y_2 - y_1) \\
(p_1 - p_2) &= \tfrac{1}{2}\rho\!\left(v_2^2 - v_1^2\right) + \rho g(y_2 - y_1).
\end{aligned}
$$

Rearranging the equation gives Bernoulli's equation:

$$p_1 + \tfrac{1}{2}\rho v_1^2 + \rho g y_1 = p_2 + \tfrac{1}{2}\rho v_2^2 + \rho g y_2.$$

This relation states that the mechanical energy of any part of the fluid changes as a result of the work done by the fluid external to that part, due to varying pressure along the way. Since the two points were chosen arbitrarily, we can write Bernoulli's equation more generally as a conservation principle along the flow.

Bernoulli's Equation

For an incompressible, frictionless fluid, the combination of pressure and the sum of kinetic and potential energy densities is constant not only over time, but also along a streamline:

$$p + \frac{1}{2}\rho v^2 + \rho g y = \text{constant}$$

(14.16)

A special note must be made here of the fact that in a dynamic situation, the pressures at the same height in different parts of the fluid may be different if they have different speeds of flow.

Analyzing Bernoulli's Equation

According to Bernoulli's equation, if we follow a small volume of fluid along its path, various quantities in the sum may change, but the total remains constant. Bernoulli's equation is, in fact, just a convenient statement of conservation of energy for an incompressible fluid in the absence of friction.

The general form of Bernoulli's equation has three terms in it, and it is broadly applicable. To understand it better, let us consider some specific situations that simplify and illustrate its use and meaning.

Bernoulli's equation for static fluids

First consider the very simple situation where the fluid is static—that is, $v_1 = v_2 = 0$. Bernoulli's equation in that case is

$$p_1 + \rho g h_1 = p_2 + \rho g h_2.$$

We can further simplify the equation by setting $h_2 = 0$. (Any height can be chosen for a reference height of zero, as is often done for other situations involving gravitational force, making all other heights relative.) In this case, we get

$$p_2 = p_1 + \rho g h_1.$$

This equation tells us that, in static fluids, pressure increases with depth. As we go from point 1 to point 2 in the fluid, the depth increases by h_1, and consequently, p_2 is greater than p_1 by an amount $\rho g h_1$. In the very simplest case, p_1 is zero at the top of the fluid, and we get the familiar relationship $p = \rho g h$. (Recall that $p = \rho g h$ and $\Delta U_g = -mgh$.) Thus, Bernoulli's equation confirms the fact that the pressure change due to the weight of a fluid is $\rho g h$. Although we introduce Bernoulli's equation for fluid motion, it includes much of what we studied for static fluids earlier.

Bernoulli's principle

Suppose a fluid is moving but its depth is constant—that is, $h_1 = h_2$. Under this condition, Bernoulli's equation becomes

$$p_1 + \frac{1}{2}\rho v_1^2 = p_2 + \frac{1}{2}\rho v_2^2.$$

Situations in which fluid flows at a constant depth are so common that this equation is often also called **Bernoulli's principle**, which is simply Bernoulli's equation for fluids at constant depth. (Note again that this applies to a small volume of fluid as we follow it along its path.) Bernoulli's principle reinforces the fact that pressure drops as speed increases in a moving fluid: If v_2 is greater than v_1 in the equation, then p_2 must be less than p_1 for the equality to hold.

Example 14.6

Calculating Pressure

In **Example 14.5**, we found that the speed of water in a hose increased from 1.96 m/s to 25.5 m/s going from the hose to the nozzle. Calculate the pressure in the hose, given that the absolute pressure in the nozzle is 1.01×10^5 N/m^2 (atmospheric, as it must be) and assuming level, frictionless flow.

Strategy

Level flow means constant depth, so Bernoulli's principle applies. We use the subscript 1 for values in the hose and 2 for those in the nozzle. We are thus asked to find $p1$.

Solution

Solving Bernoulli's principle for p_1 yields

$$p_1 = p_2 + \frac{1}{2}\rho v_2^2 - \frac{1}{2}\rho v_1^2 = p_2 + \frac{1}{2}\rho(v_2^2 - v_1^2).$$

Substituting known values,

$$p_1 = 1.01 \times 10^5 \ \text{N/m}^2 + \frac{1}{2}(10^3 \ \text{kg/m}^3)[(25.5 \ \text{m/s})^2 - (1.96 \ \text{m/s})^2]$$

$$= 4.24 \times 10^5 \ \text{N/m}^2.$$

Significance

This absolute pressure in the hose is greater than in the nozzle, as expected, since v is greater in the nozzle. The pressure p_2 in the nozzle must be atmospheric, because the water emerges into the atmosphere without other changes in conditions.

Applications of Bernoulli's Principle

Many devices and situations occur in which fluid flows at a constant height and thus can be analyzed with Bernoulli's principle.

Entrainment

People have long put the Bernoulli principle to work by using reduced pressure in high-velocity fluids to move things about. With a higher pressure on the outside, the high-velocity fluid forces other fluids into the stream. This process is called *entrainment*. Entrainment devices have been in use since ancient times as pumps to raise water to small heights, as is necessary for draining swamps, fields, or other low-lying areas. Some other devices that use the concept of entrainment are shown in **Figure 14.31**.

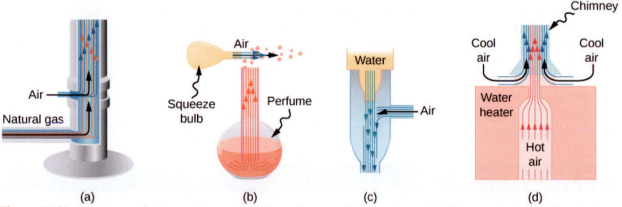

Figure 14.31 Entrainment devices use increased fluid speed to create low pressures, which then entrain one fluid into another. (a) A Bunsen burner uses an adjustable gas nozzle, entraining air for proper combustion. (b) An atomizer uses a squeeze bulb to create a jet of air that entrains drops of perfume. Paint sprayers and carburetors use very similar techniques to move their respective liquids. (c) A common aspirator uses a high-speed stream of water to create a region of lower pressure. Aspirators may be used as suction pumps in dental and surgical situations or for draining a flooded basement or producing a reduced pressure in a vessel. (d) The chimney of a water heater is designed to entrain air into the pipe leading through the ceiling.

Velocity measurement

Figure 14.32 shows two devices that apply Bernoulli's principle to measure fluid velocity. The manometer in part (a) is connected to two tubes that are small enough not to appreciably disturb the flow. The tube facing the oncoming fluid creates a dead spot having zero velocity ($v_1 = 0$) in front of it, while fluid passing the other tube has velocity v_2. This means that Bernoulli's principle as stated in

$$p_1 + \frac{1}{2}\rho v_1^2 = p_2 + \frac{1}{2}\rho v_2^2$$

becomes

$$p_1 = p_2 + \frac{1}{2}\rho v_2^2.$$

Thus pressure p_2 over the second opening is reduced by $\frac{1}{2}\rho v_2^2$, so the fluid in the manometer rises by h on the side connected to the second opening, where

$$h \propto \frac{1}{2}\rho v_2^2.$$

(Recall that the symbol $\propto$ means "proportional to.") Solving for v_2, we see that

$$v_2 \propto \sqrt{h}.$$

Part (b) shows a version of this device that is in common use for measuring various fluid velocities; such devices are frequently used as air-speed indicators in aircraft.

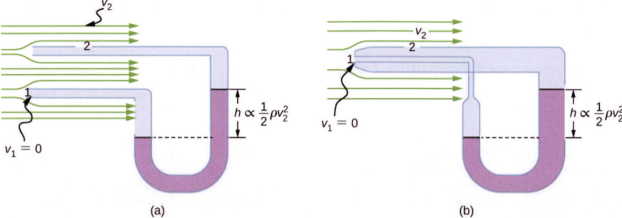

(a)   **(b)**

Figure 14.32 Measurement of fluid speed based on Bernoulli's principle. (a) A manometer is connected to two tubes that are close together and small enough not to disturb the flow. Tube 1 is open at the end facing the flow. A dead spot having zero speed is created there. Tube 2 has an opening on the side, so the fluid has a speed v across the opening; thus, pressure there drops. The difference in pressure at the manometer is $\frac{1}{2}\rho v_2^2$, so h is proportional to $\frac{1}{2}\rho v_2^2$. (b) This type of velocity measuring device is a Prandtl tube, also known as a pitot tube.

A fire hose

All preceding applications of Bernoulli's equation involved simplifying conditions, such as constant height or constant pressure. The next example is a more general application of Bernoulli's equation in which pressure, velocity, and height all change.

Example 14.7

Calculating Pressure: A Fire Hose Nozzle

Fire hoses used in major structural fires have an inside diameter of 6.40 cm (**Figure 14.33**). Suppose such a hose carries a flow of 40.0 L/s, starting at a gauge pressure of $1.62 \times 10^6 \, \text{N/m}^2$. The hose rises up 10.0 m along a ladder to a nozzle having an inside diameter of 3.00 cm. What is the pressure in the nozzle?

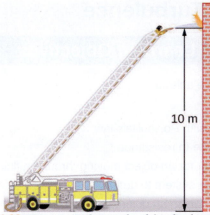

Figure 14.33 Pressure in the nozzle of this fire hose is less than at ground level for two reasons: The water has to go uphill to get to the nozzle, and speed increases in the nozzle. In spite of its lowered pressure, the water can exert a large force on anything it strikes by virtue of its kinetic energy. Pressure in the water stream becomes equal to atmospheric pressure once it emerges into the air.

Strategy

We must use Bernoulli's equation to solve for the pressure, since depth is not constant.

Solution

Bernoulli's equation is

$$p_1 + \frac{1}{2}\rho v_1^2 + \rho g h_1 = p_2 + \frac{1}{2}\rho v_2^2 + \rho g h_2$$

where subscripts 1 and 2 refer to the initial conditions at ground level and the final conditions inside the nozzle, respectively. We must first find the speeds v_1 and v_2. Since $Q = A_1 v_1$, we get

$$v_1 = \frac{Q}{A_1} = \frac{40.0 \times 10^{-3}\,\text{m}^3/\text{s}}{\pi (3.20 \times 10^{-2}\,\text{m})^2} = 12.4\,\text{m/s}.$$

Similarly, we find

$$v_2 = 56.6\ \text{m/s}.$$

This rather large speed is helpful in reaching the fire. Now, taking h_1 to be zero, we solve Bernoulli's equation for p_2:

$$p_2 = p_1 + \frac{1}{2}\rho(v_1^2 - v_2^2) - \rho g h_2.$$

Substituting known values yields

$$
\begin{aligned}
p_2 &= 1.62 \times 10^6\ \text{N/m}^2 + \tfrac{1}{2}(1000\ \text{kg/m}^3)[(12.4\ \text{m/s})^2 - (56.6\ \text{m/s})^2] \\
&\quad - (1000\ \text{kg/m}^3)(9.80\ \text{m/s}^2)(10.0\ \text{m}) \\
&= 0.
\end{aligned}
$$

Significance

This value is a gauge pressure, since the initial pressure was given as a gauge pressure. Thus, the nozzle pressure equals atmospheric pressure as it must, because the water exits into the atmosphere without changes in its conditions.

14.7 | Viscosity and Turbulence

Learning Objectives

By the end of this section, you will be able to:

- Explain what viscosity is
- Calculate flow and resistance with Poiseuille's law
- Explain how pressure drops due to resistance
- Calculate the Reynolds number for an object moving through a fluid
- Use the Reynolds number for a system to determine whether it is laminar or turbulent
- Describe the conditions under which an object has a terminal speed

In **Applications of Newton's Laws**, which introduced the concept of friction, we saw that an object sliding across the floor with an initial velocity and no applied force comes to rest due to the force of friction. Friction depends on the types of materials in contact and is proportional to the normal force. We also discussed drag and air resistance in that same chapter. We explained that at low speeds, the drag is proportional to the velocity, whereas at high speeds, drag is proportional to the velocity squared. In this section, we introduce the forces of friction that act on fluids in motion. For example, a fluid flowing through a pipe is subject to resistance, a type of friction, between the fluid and the walls. Friction also occurs between the different layers of fluid. These resistive forces affect the way the fluid flows through the pipe.

Viscosity and Laminar Flow

When you pour yourself a glass of juice, the liquid flows freely and quickly. But if you pour maple syrup on your pancakes, that liquid flows slowly and sticks to the pitcher. The difference is fluid friction, both within the fluid itself and between the fluid and its surroundings. We call this property of fluids *viscosity*. Juice has low viscosity, whereas syrup has high viscosity.

The precise definition of viscosity is based on laminar, or nonturbulent, flow. **Figure 14.34** shows schematically how laminar and turbulent flow differ. When flow is laminar, layers flow without mixing. When flow is turbulent, the layers mix, and significant velocities occur in directions other than the overall direction of flow.

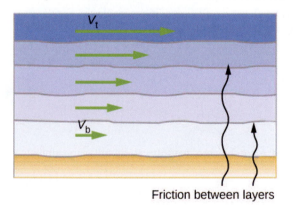
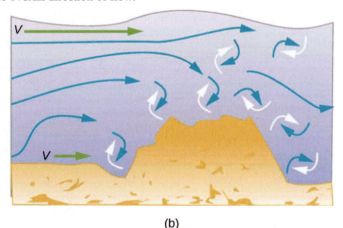

(a)   (b)

Figure 14.34 (a) Laminar flow occurs in layers without mixing. Notice that viscosity causes drag between layers as well as with the fixed surface. The speed near the bottom of the flow (v_b) is less than speed near the top (v_t) because in this case, the

surface of the containing vessel is at the bottom. (b) An obstruction in the vessel causes turbulent flow. Turbulent flow mixes the fluid. There is more interaction, greater heating, and more resistance than in laminar flow.

Turbulence is a fluid flow in which layers mix together via eddies and swirls. It has two main causes. First, any obstruction or sharp corner, such as in a faucet, creates turbulence by imparting velocities perpendicular to the flow. Second, high speeds cause turbulence. The drag between adjacent layers of fluid and between the fluid and its surroundings can form swirls and eddies if the speed is great enough. In **Figure 14.35**, the speed of the accelerating smoke reaches the point that it begins to swirl due to the drag between the smoke and the surrounding air.

Figure 14.35 Smoke rises smoothly for a while and then begins to form swirls and eddies. The smooth flow is called laminar flow, whereas the swirls and eddies typify turbulent flow. Smoke rises more rapidly when flowing smoothly than after it becomes turbulent, suggesting that turbulence poses more resistance to flow. (credit: "Creativity103"/Flickr)

Figure 14.36 shows how viscosity is measured for a fluid. The fluid to be measured is placed between two parallel plates. The bottom plate is held fixed, while the top plate is moved to the right, dragging fluid with it. The layer (or lamina) of fluid in contact with either plate does not move relative to the plate, so the top layer moves at speed v while the bottom layer remains at rest. Each successive layer from the top down exerts a force on the one below it, trying to drag it along, producing a continuous variation in speed from v to 0 as shown. Care is taken to ensure that the flow is laminar, that is, the layers do not mix. The motion in the figure is like a continuous shearing motion. Fluids have zero shear strength, but the rate at which they are sheared is related to the same geometrical factors A and L as is shear deformation for solids. In the diagram, the fluid is initially at rest. The layer of fluid in contact with the moving plate is accelerated and starts to move due to the internal friction between moving plate and the fluid. The next layer is in contact with the moving layer; since there is internal friction between the two layers, it also accelerates, and so on through the depth of the fluid. There is also internal friction between the stationary plate and the lowest layer of fluid, next to the station plate. The force is required to keep the plate moving at a constant velocity due to the internal friction.

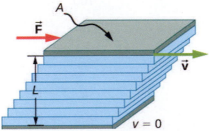

Figure 14.36 Measurement of viscosity for laminar flow of fluid between two plates of area A. The bottom plate is fixed. When the top plate is pushed to the right, it drags the fluid along with it.

A force F is required to keep the top plate in **Figure 14.36** moving at a constant velocity v, and experiments have shown that this force depends on four factors. First, F is directly proportional to v (until the speed is so high that turbulence occurs—then a much larger force is needed, and it has a more complicated dependence on v). Second, F is proportional to the area A of the plate. This relationship seems reasonable, since A is directly proportional to the amount of fluid being moved. Third, F is inversely proportional to the distance between the plates L. This relationship is also reasonable; L is like a lever arm, and the greater the lever arm, the less the force that is needed. Fourth, F is directly proportional to the coefficient of viscosity, η. The greater the viscosity, the greater the force required. These dependencies are combined into the equation

$$F = \eta \frac{vA}{L}.$$

This equation gives us a working definition of fluid viscosity η. Solving for η gives

$$\eta = \frac{FL}{vA} \qquad (14.17)$$

which defines viscosity in terms of how it is measured.

The SI unit of viscosity is $N \cdot m/\left[(m/s)m^2\right] = \left(N/m^2\right)s$ or $Pa \cdot s$. **Table 14.4** lists the coefficients of viscosity for various fluids. Viscosity varies from one fluid to another by several orders of magnitude. As you might expect, the viscosities of gases are much less than those of liquids, and these viscosities often depend on temperature.

Fluid	Temperature (°C)	Viscosity η (Pa · s)
Air	0	0.0171
	20	0.0181
	40	0.0190
	100	0.0218
Ammonia	20	0.00974
Carbon dioxide	20	0.0147
Helium	20	0.0196
Hydrogen	0	0.0090
Mercury	20	0.0450
Oxygen	20	0.0203
Steam	100	0.0130
Liquid water	0	1.792
	20	1.002
	37	0.6947
	40	0.653
	100	0.282
Whole blood	20	3.015
	37	2.084
Blood plasma	20	1.810
	37	1.257
Ethyl alcohol	20	1.20
Methanol	20	0.584

Table 14.4 Coefficients of Viscosity of Various Fluids

Fluid	Temperature (°C)	Viscosity η (Pa · s)
Oil (heavy machine)	20	660
Oil (motor, SAE 10)	30	200
Oil (olive)	20	138
Glycerin	20	1500
Honey	20	2000–10000
Maple syrup	20	2000–3000
Milk	20	3.0
Oil (corn)	20	65

Table 14.4 Coefficients of Viscosity of Various Fluids

Laminar Flow Confined to Tubes: Poiseuille's Law

What causes flow? The answer, not surprisingly, is a pressure difference. In fact, there is a very simple relationship between horizontal flow and pressure. Flow rate Q is in the direction from high to low pressure. The greater the pressure differential between two points, the greater the flow rate. This relationship can be stated as

$$Q = \frac{p_2 - p_1}{R}$$

where p_1 and p_2 are the pressures at two points, such as at either end of a tube, and R is the resistance to flow. The resistance R includes everything, except pressure, that affects flow rate. For example, R is greater for a long tube than for a short one. The greater the viscosity of a fluid, the greater the value of R. Turbulence greatly increases R, whereas increasing the diameter of a tube decreases R.

If viscosity is zero, the fluid is frictionless and the resistance to flow is also zero. Comparing frictionless flow in a tube to viscous flow, as in **Figure 14.37**, we see that for a viscous fluid, speed is greatest at midstream because of drag at the boundaries. We can see the effect of viscosity in a Bunsen burner flame [part (c)], even though the viscosity of natural gas is small.

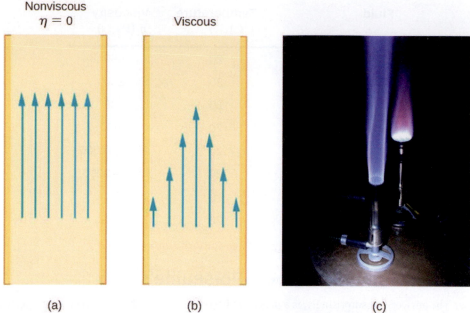

Figure 14.37 (a) If fluid flow in a tube has negligible resistance, the speed is the same all across the tube. (b) When a viscous fluid flows through a tube, its speed at the walls is zero, increasing steadily to its maximum at the center of the tube. (c) The shape of a Bunsen burner flame is due to the velocity profile across the tube. (credit c: modification of work by "jasonwoodhead23"/Flickr)

The resistance R to laminar flow of an incompressible fluid with viscosity η through a horizontal tube of uniform radius r and length l, is given by

$$R = \frac{8\eta l}{\pi r^4}.$$ (14.18)

This equation is called **Poiseuille's law for resistance**, named after the French scientist J. L. Poiseuille (1799–1869), who derived it in an attempt to understand the flow of blood through the body.

Let us examine Poiseuille's expression for R to see if it makes good intuitive sense. We see that resistance is directly proportional to both fluid viscosity η and the length l of a tube. After all, both of these directly affect the amount of friction encountered—the greater either is, the greater the resistance and the smaller the flow. The radius r of a tube affects the resistance, which again makes sense, because the greater the radius, the greater the flow (all other factors remaining the same). But it is surprising that r is raised to the fourth power in Poiseuille's law. This exponent means that any change in the radius of a tube has a very large effect on resistance. For example, doubling the radius of a tube decreases resistance by a factor of $2^4 = 16$.

Taken together, $Q = \frac{p_2 - p_1}{R}$ and $R = \frac{8\eta l}{\pi r^4}$ give the following expression for flow rate:

$$Q = \frac{(p_2 - p_1)\pi r^4}{8\eta l}.$$ (14.19)

This equation describes laminar flow through a tube. It is sometimes called Poiseuille's law for laminar flow, or simply **Poiseuille's law** (**Figure 14.38**).

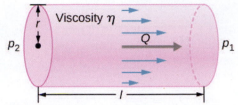

Figure 14.38 Poiseuille's law applies to laminar flow of an incompressible fluid of viscosity η through a tube of length l and radius r. The direction of flow is from greater to lower pressure. Flow rate Q is directly proportional to the pressure difference $p_2 - p_1$, and inversely proportional to the length l of the tube and viscosity η of the fluid. Flow rate increases with radius by a factor of r^4.

Example 14.8

Using Flow Rate: Air Conditioning Systems

An air conditioning system is being designed to supply air at a gauge pressure of 0.054 Pa at a temperature of 20 °C. The air is sent through an insulated, round conduit with a diameter of 18.00 cm. The conduit is 20-meters long and is open to a room at atmospheric pressure 101.30 kPa. The room has a length of 12 meters, a width of 6 meters, and a height of 3 meters. (a) What is the volume flow rate through the pipe, assuming laminar flow? (b) Estimate the length of time to completely replace the air in the room. (c) The builders decide to save money by using a conduit with a diameter of 9.00 cm. What is the new flow rate?

Strategy

Assuming laminar flow, Poiseuille's law states that

$$Q = \frac{(p_2 - p_1)\pi r^4}{8\eta l} = \frac{dV}{dt}.$$

We need to compare the artery radius before and after the flow rate reduction. Note that we are given the diameter of the conduit, so we must divide by two to get the radius.

Solution

a. Assuming a constant pressure difference and using the viscosity $\eta = 0.0181 \text{ mPa} \cdot \text{s}$,

$$Q = \frac{(0.054 \text{ Pa})(3.14)(0.09 \text{ m})^4}{8(0.0181 \times 10^{-3} \text{ Pa} \cdot \text{s})(20 \text{ m})} = 3.84 \times 10^{-3} \text{ } \frac{\text{m}^3}{\text{s}}.$$

b. Assuming constant flow $Q = \frac{dV}{dt} \approx \frac{\Delta V}{\Delta t}$

$$\Delta t = \frac{\Delta V}{Q} = \frac{(12 \text{ m})(6 \text{ m})(3 \text{ m})}{3.84 \times 10^{-3} \text{ } \frac{\text{m}^3}{\text{s}}} = 5.63 \times 10^4 \text{ s} = 15.63 \text{ hr}.$$

c. Using laminar flow, Poiseuille's law yields

$$Q = \frac{(0.054 \text{ Pa})(3.14)(0.045 \text{ m})^4}{8(0.0181 \times 10^{-3} \text{ Pa} \cdot \text{s})(20 \text{ m})} = 2.40 \times 10^{-4} \text{ } \frac{\text{m}^3}{\text{s}}.$$

Thus, the radius of the conduit decreases by half reduces the flow rate to 6.25% of the original value.

Significance

In general, assuming laminar flow, decreasing the radius has a more dramatic effect than changing the length. If the length is increased and all other variables remain constant, the flow rate is decreased:

$$\frac{Q_A}{Q_B} = \frac{\frac{(p_2 - p_1)\pi r_A^4}{8\eta l_A}}{\frac{(p_2 - p_1)\pi r_B^4}{8\eta l_B}} = \frac{l_B}{l_A}$$

$$Q_B = \frac{l_A}{l_B}Q_A.$$

Doubling the length cuts the flow rate to one-half the original flow rate.

If the radius is decreased and all other variables remain constant, the volume flow rate decreases by a much larger factor.

$$\frac{Q_A}{Q_B} = \frac{\frac{(p_2 - p_1)\pi r_A^4}{8\eta l_A}}{\frac{(p_2 - p_1)\pi r_B^4}{8\eta l_B}} = \left(\frac{r_A}{r_B}\right)^4$$

$$Q_B = \left(\frac{r_B}{r_A}\right)^4 Q_A$$

Cutting the radius in half decreases the flow rate to one-sixteenth the original flow rate.

Flow and Resistance as Causes of Pressure Drops

Water pressure in homes is sometimes lower than normal during times of heavy use, such as hot summer days. The drop in pressure occurs in the water main before it reaches individual homes. Let us consider flow through the water main as illustrated in **Figure 14.39**. We can understand why the pressure p_1 to the home drops during times of heavy use by rearranging the equation for flow rate:

$$Q = \frac{p_2 - p_1}{R}$$

$$p_2 - p_1 = RQ.$$

In this case, p_2 is the pressure at the water works and R is the resistance of the water main. During times of heavy use, the flow rate Q is large. This means that $p_2 - p_1$ must also be large. Thus p_1 must decrease. It is correct to think of flow and resistance as causing the pressure to drop from p_2 to p_1. The equation $p_2 - p_1 = RQ$ is valid for both laminar and turbulent flows.

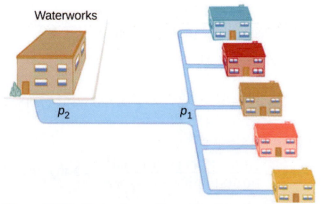

Figure 14.39 During times of heavy use, there is a significant pressure drop in a water main, and p_1 supplied to users is significantly less than p_2 created at the water works. If the flow is very small, then the pressure drop is negligible, and $p_2 \approx p_1$.

We can also use $p_2 - p_1 = RQ$ to analyze pressure drops occurring in more complex systems in which the tube radius is not the same everywhere. Resistance is much greater in narrow places, such as in an obstructed coronary artery. For a given flow rate Q, the pressure drop is greatest where the tube is most narrow. This is how water faucets control flow. Additionally, R is greatly increased by turbulence, and a constriction that creates turbulence greatly reduces the pressure downstream. Plaque in an artery reduces pressure and hence flow, both by its resistance and by the turbulence it creates.

Measuring Turbulence

An indicator called the **Reynolds number** N_R can reveal whether flow is laminar or turbulent. For flow in a tube of uniform diameter, the Reynolds number is defined as

$$N_R = \frac{2\rho vr}{\eta} \text{(flow in tube)}$$

(14.20)

where ρ is the fluid density, v its speed, η its viscosity, and r the tube radius. The Reynolds number is a dimensionless quantity. Experiments have revealed that N_R is related to the onset of turbulence. For N_R below about 2000, flow is laminar. For N_R above about 3000, flow is turbulent.

For values of N_R between about 2000 and 3000, flow is unstable—that is, it can be laminar, but small obstructions and surface roughness can make it turbulent, and it may oscillate randomly between being laminar and turbulent. In fact, the flow of a fluid with a Reynolds number between 2000 and 3000 is a good example of chaotic behavior. A system is defined to be chaotic when its behavior is so sensitive to some factor that it is extremely difficult to predict. It is difficult, but not impossible, to predict whether flow is turbulent or not when a fluid's Reynold's number falls in this range due to extremely sensitive dependence on factors like roughness and obstructions on the nature of the flow. A tiny variation in one factor has an exaggerated (or nonlinear) effect on the flow.

Example 14.9

Using Flow Rate: Turbulent Flow or Laminar Flow

In **Example 14.8**, we found the volume flow rate of an air conditioning system to be $Q = 3.84 \times 10^{-3}$ m^3/s. This calculation assumed laminar flow. (a) Was this a good assumption? (b) At what velocity would the flow become turbulent?

Strategy

To determine if the flow of air through the air conditioning system is laminar, we first need to find the velocity, which can be found by

$$Q = Av = \pi r^2 v.$$

Then we can calculate the Reynold's number, using the equation below, and determine if it falls in the range for laminar flow

$$R = \frac{2\rho vr}{\eta}.$$

Solution

a. Using the values given:

$$v = \frac{Q}{\pi r^2} = \frac{3.84 \times 10^{-3} \frac{\text{m}^3}{\text{s}}}{3.14(0.09 \text{ m})^2} = 0.15\frac{\text{m}}{\text{s}}$$

$$R = \frac{2\rho vr}{\eta} = \frac{2\left(1.23\frac{\text{kg}}{\text{m}^3}\right)(0.15 \frac{\text{m}}{\text{s}})(0.09 \text{ m})}{0.0181 \times 10^{-3} \text{ Pa} \cdot \text{s}} = 1835.$$

Since the Reynolds number is 1835 < 2000, the flow is laminar and not turbulent. The assumption that the flow was laminar is valid.

b. To find the maximum speed of the air to keep the flow laminar, consider the Reynold's number.

$$R = \frac{2\rho v r}{\eta} \leq 2000$$

$$v = \frac{2000\left(0.0181 \times 10^{-3}\,\text{Pa} \cdot \text{s}\right)}{2\left(1.23\frac{\text{kg}}{\text{m}^3}\right)(0.09\,\text{m})} = 0.16\frac{\text{m}}{\text{s}}.$$

Significance

When transferring a fluid from one point to another, it desirable to limit turbulence. Turbulence results in wasted energy, as some of the energy intended to move the fluid is dissipated when eddies are formed. In this case, the air conditioning system will become less efficient once the velocity exceeds 0.16 m/s, since this is the point at which turbulence will begin to occur.

CHAPTER 14 REVIEW

KEY TERMS

absolute pressure sum of gauge pressure and atmospheric pressure

Archimedes' principle buoyant force on an object equals the weight of the fluid it displaces

Bernoulli's equation equation resulting from applying conservation of energy to an incompressible frictionless fluid:
$p + \frac{1}{2}\rho v^2 + \rho g h = $ constant, throughout the fluid

Bernoulli's principle Bernoulli's equation applied at constant depth:
$$p_1 + \frac{1}{2}\rho v_1^2 = p_2 + \frac{1}{2}\rho v_2^2$$

buoyant force net upward force on any object in any fluid due to the pressure difference at different depths

density mass per unit volume of a substance or object

flow rate abbreviated Q, it is the volume V that flows past a particular point during a time t, or $Q = dV/dt$

fluids liquids and gases; a fluid is a state of matter that yields to shearing forces

gauge pressure pressure relative to atmospheric pressure

hydraulic jack simple machine that uses cylinders of different diameters to distribute force

hydrostatic equilibrium state at which water is not flowing, or is static

ideal fluid fluid with negligible viscosity

laminar flow type of fluid flow in which layers do not mix

Pascal's principle change in pressure applied to an enclosed fluid is transmitted undiminished to all portions of the fluid and to the walls of its container

Poiseuille's law rate of laminar flow of an incompressible fluid in a tube: $Q = \frac{(p_2 - p_1)\pi r^4}{8\eta l}$.

Poiseuille's law for resistance resistance to laminar flow of an incompressible fluid in a tube: $R = \frac{8\eta l}{\pi r^4}$

pressure force per unit area exerted perpendicular to the area over which the force acts

Reynolds number dimensionless parameter that can reveal whether a particular flow is laminar or turbulent

specific gravity ratio of the density of an object to a fluid (usually water)

turbulence fluid flow in which layers mix together via eddies and swirls

turbulent flow type of fluid flow in which layers mix together via eddies and swirls

viscosity measure of the internal friction in a fluid

KEY EQUATIONS

Density of a sample at constant density	$\rho = \frac{m}{V}$
Pressure	$p = \frac{F}{A}$
Pressure at a depth h in a fluid of constant density	$p = p_0 + \rho g h$
Change of pressure with height in a constant-density fluid	$\frac{dp}{dy} = -\rho g$

Absolute pressure	$p_{abs} = p_g + p_{atm}$
Pascal's principle	$\dfrac{F_1}{A_1} = \dfrac{F_2}{A_2}$
Volume flow rate	$Q = \dfrac{dV}{dt}$
Continuity equation (constant density)	$A_1 v_1 = A_2 v_2$
Continuity equation (general form)	$\rho_1 A_1 v_1 = \rho_2 A_2 v_2$
Bernoulli's equation	$p + \frac{1}{2}\rho v^2 + \rho g y = \text{constant}$
Viscosity	$\eta = \dfrac{FL}{vA}$
Poiseuille's law for resistance	$R = \dfrac{8\eta l}{\pi r^4}$
Poiseuille's law	$Q = \dfrac{(p_2 - p_1)\pi r^4}{8\eta l}$

SUMMARY

14.1 Fluids, Density, and Pressure

- A fluid is a state of matter that yields to sideways or shearing forces. Liquids and gases are both fluids. Fluid statics is the physics of stationary fluids.

- Density is the mass per unit volume of a substance or object, defined as $\rho = m/V$. The SI unit of density is kg/m^3.

- Pressure is the force per unit perpendicular area over which the force is applied, $p = F/A$. The SI unit of pressure is the pascal: $1\ Pa = 1\ N/m^2$.

- Pressure due to the weight of a liquid of constant density is given by $p = \rho g h$, where p is the pressure, h is the depth of the liquid, ρ is the density of the liquid, and g is the acceleration due to gravity.

14.2 Measuring Pressure

- Gauge pressure is the pressure relative to atmospheric pressure.

- Absolute pressure is the sum of gauge pressure and atmospheric pressure.

- Open-tube manometers have U-shaped tubes and one end is always open. They are used to measure pressure. A mercury barometer is a device that measures atmospheric pressure.

- The SI unit of pressure is the pascal (Pa), but several other units are commonly used.

14.3 Pascal's Principle and Hydraulics

- Pressure is force per unit area.

- A change in pressure applied to an enclosed fluid is transmitted undiminished to all portions of the fluid and to the walls of its container.

- A hydraulic system is an enclosed fluid system used to exert forces.

14.4 Archimedes' Principle and Buoyancy

- Buoyant force is the net upward force on any object in any fluid. If the buoyant force is greater than the object's weight, the object will rise to the surface and float. If the buoyant force is less than the object's weight, the object will sink. If the buoyant force equals the object's weight, the object can remain suspended at its present depth. The buoyant force is always present and acting on any object immersed either partially or entirely in a fluid.

- Archimedes' principle states that the buoyant force on an object equals the weight of the fluid it displaces.

14.5 Fluid Dynamics

- Flow rate Q is defined as the volume V flowing past a point in time t, or $Q = \frac{dV}{dt}$ where V is volume and t is time.

  The SI unit of flow rate is m^3/s, but other rates can be used, such as L/min.

- Flow rate and velocity are related by $Q = Av$ where A is the cross-sectional area of the flow and v is its average velocity.

- The equation of continuity states that for an incompressible fluid, the mass flowing into a pipe must equal the mass flowing out of the pipe.

14.6 Bernoulli's Equation

- Bernoulli's equation states that the sum on each side of the following equation is constant, or the same at any two points in an incompressible frictionless fluid:

$$p_1 + \frac{1}{2}\rho v_1^2 + \rho g h_1 = p_2 + \frac{1}{2}\rho v_2^2 + \rho g h_2.$$

- Bernoulli's principle is Bernoulli's equation applied to situations in which the height of the fluid is constant. The terms involving depth (or height h) subtract out, yielding

$$p_1 + \frac{1}{2}\rho v_1^2 = p_2 + \frac{1}{2}\rho v_2^2.$$

- Bernoulli's principle has many applications, including entrainment and velocity measurement.

14.7 Viscosity and Turbulence

- Laminar flow is characterized by smooth flow of the fluid in layers that do not mix.

- Turbulence is characterized by eddies and swirls that mix layers of fluid together.

- Fluid viscosity η is due to friction within a fluid.

- Flow is proportional to pressure difference and inversely proportional to resistance:

$$Q = \frac{p - 2p_1}{R}.$$

- The pressure drop caused by flow and resistance is given by $p_2 - p_1 = RQ$.

- The Reynolds number N_R can reveal whether flow is laminar or turbulent. It is $N_R = \frac{2\rho vr}{\eta}$.

- For N_R below about 2000, flow is laminar. For N_R above about 3000, flow is turbulent. For values of N_R between 2000 and 3000, it may be either or both.

CONCEPTUAL QUESTIONS

14.1 Fluids, Density, and Pressure

1. Which of the following substances are fluids at room temperature and atmospheric pressure: air, mercury, water, glass?

2. Why are gases easier to compress than liquids and solids?

3. Explain how the density of air varies with altitude.

4. The image shows a glass of ice water filled to the brim. Will the water overflow when the ice melts? Explain your answer.

5. How is pressure related to the sharpness of a knife and its ability to cut?

6. Why is a force exerted by a static fluid on a surface always perpendicular to the surface?

7. Imagine that in a remote location near the North Pole, a chunk of ice floats in a lake. Next to the lake, a glacier with the same volume as the floating ice sits on land. If both chunks of ice should melt due to rising global temperatures, and the melted ice all goes into the lake, which one would cause the level of the lake to rise the most? Explain.

8. In ballet, dancing *en pointe* (on the tips of the toes) is much harder on the toes than normal dancing or walking. Explain why, in terms of pressure.

9. Atmospheric pressure exerts a large force (equal to the weight of the atmosphere above your body—about 10 tons) on the top of your body when you are lying on the beach sunbathing. Why are you able to get up?

10. Why does atmospheric pressure decrease more rapidly than linearly with altitude?

11. The image shows how sandbags placed around a leak outside a river levee can effectively stop the flow of water under the levee. Explain how the small amount of water inside the column of sandbags is able to balance the much larger body of water behind the levee.

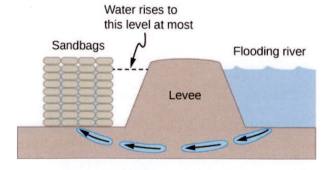

12. Is there a net force on a dam due to atmospheric pressure? Explain your answer.

13. Does atmospheric pressure add to the gas pressure in a rigid tank? In a toy balloon? When, in general, does atmospheric pressure not affect the total pressure in a fluid?

14. You can break a strong wine bottle by pounding a cork into it with your fist, but the cork must press directly against the liquid filling the bottle—there can be no air between the cork and liquid. Explain why the bottle breaks only if there is no air between the cork and liquid.

14.2 Measuring Pressure

15. Explain why the fluid reaches equal levels on either side of a manometer if both sides are open to the atmosphere, even if the tubes are of different diameters.

14.3 Pascal's Principle and Hydraulics

16. Suppose the master cylinder in a hydraulic system is at a greater height than the cylinder it is controlling. Explain how this will affect the force produced at the cylinder that is being controlled.

14.4 Archimedes' Principle and Buoyancy

17. More force is required to pull the plug in a full bathtub than when it is empty. Does this contradict Archimedes' principle? Explain your answer.

18. Do fluids exert buoyant forces in a "weightless" environment, such as in the space shuttle? Explain your answer.

19. Will the same ship float higher in salt water than in freshwater? Explain your answer.

20. Marbles dropped into a partially filled bathtub sink to the bottom. Part of their weight is supported by buoyant force, yet the downward force on the bottom of the tub increases by exactly the weight of the marbles. Explain why.

14.5 Fluid Dynamics

21. Many figures in the text show streamlines. Explain why fluid velocity is greatest where streamlines are closest together. (*Hint:* Consider the relationship between fluid velocity and the cross-sectional area through which the fluid flows.)

14.6 Bernoulli's Equation

22. You can squirt water from a garden hose a considerably greater distance by partially covering the opening with your thumb. Explain how this works.

23. Water is shot nearly vertically upward in a decorative fountain and the stream is observed to broaden as it rises. Conversely, a stream of water falling straight down from a faucet narrows. Explain why.

24. Look back to **Figure 14.29**. Answer the following two questions. Why is p_o less than atmospheric? Why is p_o greater than p_i?

25. A tube with a narrow segment designed to enhance entrainment is called a Venturi, such as shown below. Venturis are very commonly used in carburetors and aspirators. How does this structure bolster entrainment?

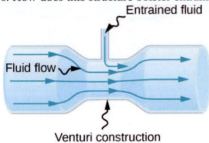

Venturi construction

26. Some chimney pipes have a T-shape, with a crosspiece on top that helps draw up gases whenever there is even a slight breeze. Explain how this works in terms of Bernoulli's principle.

27. Is there a limit to the height to which an entrainment device can raise a fluid? Explain your answer.

28. Why is it preferable for airplanes to take off into the wind rather than with the wind?

29. Roofs are sometimes pushed off vertically during a tropical cyclone, and buildings sometimes explode outward when hit by a tornado. Use Bernoulli's principle to explain these phenomena.

30. It is dangerous to stand close to railroad tracks when a rapidly moving commuter train passes. Explain why atmospheric pressure would push you toward the moving train.

31. Water pressure inside a hose nozzle can be less than atmospheric pressure due to the Bernoulli effect. Explain in terms of energy how the water can emerge from the nozzle against the opposing atmospheric pressure.

32. David rolled down the window on his car while driving on the freeway. An empty plastic bag on the floor promptly flew out the window. Explain why.

33. Based on Bernoulli's equation, what are three forms of energy in a fluid? (Note that these forms are conservative, unlike heat transfer and other dissipative forms not included in Bernoulli's equation.)

34. The old rubber boot shown below has two leaks. To what maximum height can the water squirt from Leak 1? How does the velocity of water emerging from Leak 2 differ from that of Leak 1? Explain your responses in terms of energy.

35. Water pressure inside a hose nozzle can be less than atmospheric pressure due to the Bernoulli effect. Explain in terms of energy how the water can emerge from the nozzle against the opposing atmospheric pressure.

14.7 Viscosity and Turbulence

36. Explain why the viscosity of a liquid decreases with temperature, that is, how might an increase in temperature reduce the effects of cohesive forces in a liquid? Also explain why the viscosity of a gas increases with temperature, that is, how does increased gas temperature create more collisions between atoms and molecules?

37. When paddling a canoe upstream, it is wisest to travel as near to the shore as possible. When canoeing downstream, it is generally better to stay near the middle. Explain why.

38. Plumbing usually includes air-filled tubes near water faucets (see the following figure). Explain why they are needed and how they work.

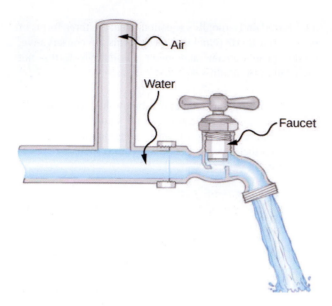

39. Doppler ultrasound can be used to measure the speed of blood in the body. If there is a partial constriction of an artery, where would you expect blood speed to be greatest: at or after the constriction? What are the two distinct causes of higher resistance in the constriction?

40. Sink drains often have a device such as that shown below to help speed the flow of water. How does this work?

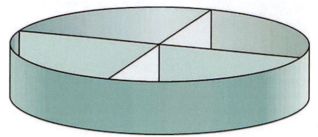

PROBLEMS

14.1 Fluids, Density, and Pressure

41. Gold is sold by the troy ounce (31.103 g). What is the volume of 1 troy ounce of pure gold?

42. Mercury is commonly supplied in flasks containing 34.5 kg (about 76 lb.). What is the volume in liters of this much mercury?

43. What is the mass of a deep breath of air having a volume of 2.00 L? Discuss the effect taking such a breath has on your body's volume and density.

44. A straightforward method of finding the density of an object is to measure its mass and then measure its volume by submerging it in a graduated cylinder. What is the density of a 240-g rock that displaces $89.0 \, \text{cm}^3$ of water? (Note that the accuracy and practical applications of this technique are more limited than a variety of others that are based on Archimedes' principle.)

45. Suppose you have a coffee mug with a circular cross-section and vertical sides (uniform radius). What is its inside radius if it holds 375 g of coffee when filled to a depth of 7.50 cm? Assume coffee has the same density as water.

46. A rectangular gasoline tank can hold 50.0 kg of gasoline when full. What is the depth of the tank if it is 0.500-m wide by 0.900-m long? (b) Discuss whether this gas tank has a reasonable volume for a passenger car.

47. A trash compactor can compress its contents to 0.350 times their original volume. Neglecting the mass of air

expelled, by what factor is the density of the rubbish increased?

48. A 2.50-kg steel gasoline can holds 20.0 L of gasoline when full. What is the average density of the full gas can, taking into account the volume occupied by steel as well as by gasoline?

49. What is the density of 18.0-karat gold that is a mixture of 18 parts gold, 5 parts silver, and 1 part copper? (These values are parts by mass, not volume.) Assume that this is a simple mixture having an average density equal to the weighted densities of its constituents.

50. The tip of a nail exerts tremendous pressure when hit by a hammer because it exerts a large force over a small area. What force must be exerted on a nail with a circular tip of 1.00-mm diameter to create a pressure of $3.00 \times 10^9 \, \text{N/m}^2$? (This high pressure is possible because the hammer striking the nail is brought to rest in such a short distance.)

51. A glass tube contains mercury. What would be the height of the column of mercury which would create pressure equal to 1.00 atm?

52. The greatest ocean depths on Earth are found in the Marianas Trench near the Philippines. Calculate the pressure due to the ocean at the bottom of this trench, given its depth is 11.0 km and assuming the density of seawater is constant all the way down.

53. Verify that the SI unit of $h\rho g$ is N/m^2.

54. What pressure is exerted on the bottom of a gas tank that is 0.500-m wide and 0.900-m long and can hold 50.0 kg of gasoline when full?

55. A dam is used to hold back a river. The dam has a height $H = 12\,\text{m}$ and a width $W = 10\,\text{m}$. Assume that the density of the water is $\rho = 1000\,\text{kg/m}^3$. (a) Determine the net force on the dam. (b) Why does the thickness of the dam increase with depth?

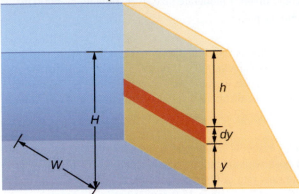

14.2 Measuring Pressure

56. Find the gauge and absolute pressures in the balloon and peanut jar shown in **Figure 14.12**, assuming the manometer connected to the balloon uses water and the manometer connected to the jar contains mercury. Express in units of centimeters of water for the balloon and millimeters of mercury for the jar, taking $h = 0.0500\text{m}$ for each.

57. How tall must a water-filled manometer be to measure blood pressure as high as 300 mm Hg?

58. Assuming bicycle tires are perfectly flexible and support the weight of bicycle and rider by pressure alone, calculate the total area of the tires in contact with the ground if a bicycle and rider have a total mass of 80.0 kg, and the gauge pressure in the tires is $3.50 \times 10^5\,\text{Pa}$.

14.3 Pascal's Principle and Hydraulics

59. How much pressure is transmitted in the hydraulic system considered in **Example 14.3**? Express your answer in atmospheres.

60. What force must be exerted on the master cylinder of a hydraulic lift to support the weight of a 2000-kg car (a large car) resting on a second cylinder? The master cylinder has a 2.00-cm diameter and the second cylinder has a 24.0-cm diameter.

61. A host pours the remnants of several bottles of wine into a jug after a party. The host then inserts a cork with a 2.00-cm diameter into the bottle, placing it in direct contact with the wine. The host is amazed when the host pounds the cork into place and the bottom of the jug (with a 14.0-cm diameter) breaks away. Calculate the extra force exerted against the bottom if he pounded the cork with a 120-N force.

62. A certain hydraulic system is designed to exert a force 100 times as large as the one put into it. (a) What must be the ratio of the area of the cylinder that is being controlled to the area of the master cylinder? (b) What must be the ratio of their diameters? (c) By what factor is the distance through which the output force moves reduced relative to the distance through which the input force moves? Assume no losses due to friction.

63. Verify that work input equals work output for a hydraulic system assuming no losses due to friction. Do this by showing that the distance the output force moves is reduced by the same factor that the output force is increased. Assume the volume of the fluid is constant. What effect would friction within the fluid and between components in the system have on the output force? How would this depend on whether or not the fluid is moving?

14.4 Archimedes' Principle and Buoyancy

64. What fraction of ice is submerged when it floats in freshwater, given the density of water at $0\,°C$ is very close to $1000\,\text{kg/m}^3$?

65. If a person's body has a density of $995\,\text{kg/m}^3$, what fraction of the body will be submerged when floating gently in (a) freshwater? (b) In salt water with a density of $1027\,\text{kg/m}^3$?

66. A rock with a mass of 540 g in air is found to have an apparent mass of 342 g when submerged in water. (a) What mass of water is displaced? (b) What is the volume of the rock? (c) What is its average density? Is this consistent with the value for granite?

67. Archimedes' principle can be used to calculate the density of a fluid as well as that of a solid. Suppose a chunk of iron with a mass of 390.0 g in air is found to have an apparent mass of 350.5 g when completely submerged in an unknown liquid. (a) What mass of fluid does the iron displace? (b) What is the volume of iron, using its density as given in **Table 14.1**? (c) Calculate the fluid's density and identify it.

68. Calculate the buoyant force on a 2.00-L helium balloon. (b) Given the mass of the rubber in the balloon is 1.50 g, what is the net vertical force on the balloon if it is let go? Neglect the volume of the rubber.

69. What is the density of a woman who floats in fresh water with 4.00% of her volume above the surface? (This could be measured by placing her in a tank with marks on the side to measure how much water she displaces when floating and when held under water.) (b) What percent of her volume is above the surface when she floats in seawater?

70. A man has a mass of 80 kg and a density of $955 kg/m^3$ (excluding the air in his lungs). (a) Calculate his volume. (b) Find the buoyant force air exerts on him. (c) What is the ratio of the buoyant force to his weight?

71. A simple compass can be made by placing a small bar magnet on a cork floating in water. (a) What fraction of a plain cork will be submerged when floating in water? (b) If the cork has a mass of 10.0 g and a 20.0-g magnet is placed on it, what fraction of the cork will be submerged? (c) Will the bar magnet and cork float in ethyl alcohol?

72. What percentage of an iron anchor's weight will be supported by buoyant force when submerged in salt water?

73. Referring to **Figure 14.20**, prove that the buoyant force on the cylinder is equal to the weight of the fluid displaced (Archimedes' principle). You may assume that the buoyant force is $F_2 - F_1$ and that the ends of the cylinder have equal areas A. Note that the volume of the cylinder (and that of the fluid it displaces) equals $(h_2 - h_1)A$.

74. A 75.0-kg man floats in freshwater with 3.00% of his volume above water when his lungs are empty, and 5.00% of his volume above water when his lungs are full. Calculate the volume of air he inhales—called his lung capacity—in liters. (b) Does this lung volume seem reasonable?

14.5 Fluid Dynamics

75. What is the average flow rate in cm^3/s of gasoline to the engine of a car traveling at 100 km/h if it averages 10.0 km/L?

76. The heart of a resting adult pumps blood at a rate of 5.00 L/min. (a) Convert this to cm^3/s. (b) What is this rate in m^3/s?

77. The Huka Falls on the Waikato River is one of New Zealand's most visited natural tourist attractions. On average, the river has a flow rate of about 300,000 L/s.

At the gorge, the river narrows to 20-m wide and averages 20-m deep. (a) What is the average speed of the river in the gorge? (b) What is the average speed of the water in the river downstream of the falls when it widens to 60 m and its depth increases to an average of 40 m?

78. (a) Estimate the time it would take to fill a private swimming pool with a capacity of 80,000 L using a garden hose delivering 60 L/min. (b) How long would it take if you could divert a moderate size river, flowing at $5000 m^3/s$ into the pool?

79. What is the fluid speed in a fire hose with a 9.00-cm diameter carrying 80.0 L of water per second? (b) What is the flow rate in cubic meters per second? (c) Would your answers be different if salt water replaced the fresh water in the fire hose?

80. Water is moving at a velocity of 2.00 m/s through a hose with an internal diameter of 1.60 cm. (a) What is the flow rate in liters per second? (b) The fluid velocity in this hose's nozzle is 15.0 m/s. What is the nozzle's inside diameter?

81. Prove that the speed of an incompressible fluid through a constriction, such as in a Venturi tube, increases by a factor equal to the square of the factor by which the diameter decreases. (The converse applies for flow out of a constriction into a larger-diameter region.)

82. Water emerges straight down from a faucet with a 1.80-cm diameter at a speed of 0.500 m/s. (Because of the construction of the faucet, there is no variation in speed across the stream.) (a) What is the flow rate in cm^3/s? (b) What is the diameter of the stream 0.200 m below the faucet? Neglect any effects due to surface tension.

14.6 Bernoulli's Equation

83. Verify that pressure has units of energy per unit volume.

84. Suppose you have a wind speed gauge like the pitot tube shown in **Figure 14.32**. By what factor must wind speed increase to double the value of h in the manometer? Is this independent of the moving fluid and the fluid in the manometer?

85. If the pressure reading of your pitot tube is 15.0 mm Hg at a speed of 200 km/h, what will it be at 700 km/h at the same altitude?

86. Every few years, winds in Boulder, Colorado, attain sustained speeds of 45.0 m/s (about 100 mph) when the jet stream descends during early spring. Approximately what is the force due to the Bernoulli equation on a roof having an area of $220 m^2$? Typical air density in Boulder is $1.14 kg/m^3$, and the corresponding atmospheric pressure is $8.89 \times 10^4 N/m^2$. (Bernoulli's principle as stated in the text assumes laminar flow. Using the principle here produces only an approximate result, because there is significant turbulence.)

87. What is the pressure drop due to the Bernoulli Effect as water goes into a 3.00-cm-diameter nozzle from a 9.00-cm-diameter fire hose while carrying a flow of 40.0 L/s? (b) To what maximum height above the nozzle can this water rise? (The actual height will be significantly smaller due to air resistance.)

88. (a) Using Bernoulli's equation, show that the measured fluid speed v for a pitot tube, like the one in **Figure 14.32**(b), is given by $v = \left(\frac{2\rho' gh}{\rho}\right)^{1/2}$, where h is the height of the manometer fluid, ρ' is the density of the manometer fluid, ρ is the density of the moving fluid, and g is the acceleration due to gravity. (Note that v is indeed proportional to the square root of h, as stated in the text.) (b) Calculate v for moving air if a mercury manometer's h is 0.200 m.

89. A container of water has a cross-sectional area of $A = 0.1 m^2$. A piston sits on top of the water (see the following figure). There is a spout located 0.15 m from the bottom of the tank, open to the atmosphere, and a stream of water exits the spout. The cross sectional area of the spout is $A_s = 7.0 \times 10^{-4} m^2$. (a) What is the velocity of the water as it leaves the spout? (b) If the opening of the spout is located 1.5 m above the ground, how far from the spout does the water hit the floor? Ignore all friction and dissipative forces.

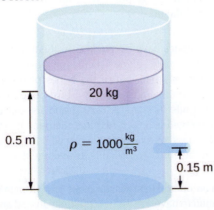

90. A fluid of a constant density flows through a reduction in a pipe. Find an equation for the change in pressure, in terms of v_1, A_1, A_2, and the density.

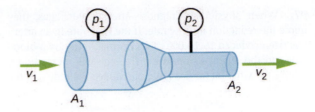

14.7 Viscosity and Turbulence

91. (a) Calculate the retarding force due to the viscosity of the air layer between a cart and a level air track given the following information: air temperature is $20\,°C$, the cart is moving at 0.400 m/s, its surface area is $2.50 \times 10^{-2} m^2$, and the thickness of the air layer is $6.00 \times 10^{-5} m$. (b) What is the ratio of this force to the weight of the 0.300-kg cart?

92. The arterioles (small arteries) leading to an organ constrict in order to decrease flow to the organ. To shut down an organ, blood flow is reduced naturally to 1.00% of its original value. By what factor do the radii of the arterioles constrict?

93. A spherical particle falling at a terminal speed in a liquid must have the gravitational force balanced by the drag force and the buoyant force. The buoyant force is equal to the weight of the displaced fluid, while the drag force is assumed to be given by Stokes Law, $F_s = 6\pi r\eta v$. Show that the terminal speed is given by

$$v = \frac{2R^2 g}{9\eta}(\rho_s - \rho_1),$$

where R is the radius of the sphere, ρ_s is its density, and ρ_1 is the density of the fluid, and η the coefficient of viscosity.

94. Using the equation of the previous problem, find the viscosity of motor oil in which a steel ball of radius 0.8 mm falls with a terminal speed of 4.32 cm/s. The densities of the ball and the oil are 7.86 and 0.88 g/mL, respectively.

95. A skydiver will reach a terminal velocity when the air drag equals his or her weight. For a skydiver with a large body, turbulence is a factor at high speeds. The drag force then is approximately proportional to the square of the velocity. Taking the drag force to be $F_D = \frac{1}{2}\rho A v^2$, and setting this equal to the skydiver's weight, find the terminal speed for a person falling "spread eagle."

96. (a) Verify that a 19.0% decrease in laminar flow through a tube is caused by a 5.00% decrease in radius, assuming that all other factors remain constant. (b) What increase in flow is obtained from a 5.00% increase in radius, again assuming all other factors remain constant?

97. When physicians diagnose arterial blockages, they quote the reduction in flow rate. If the flow rate in an artery has been reduced to 10.0% of its normal value by a blood clot and the average pressure difference has increased by 20.0%, by what factor has the clot reduced the radius of the artery?

98. An oil gusher shoots crude oil 25.0 m into the air through a pipe with a 0.100-m diameter. Neglecting air resistance but not the resistance of the pipe, and assuming laminar flow, calculate the pressure at the entrance of the 50.0-m-long vertical pipe. Take the density of the oil to be $900 \, \text{kg/m}^3$ and its viscosity to be $1.00 (\text{N/m}^2) \cdot \text{s}$ (or $1.00 \, \text{Pa} \cdot \text{s}$). Note that you must take into account the pressure due to the 50.0-m column of oil in the pipe.

99. Concrete is pumped from a cement mixer to the place it is being laid, instead of being carried in wheelbarrows. The flow rate is 200 L/min through a 50.0-m-long, 8.00-cm-diameter hose, and the pressure at the pump is $8.00 \times 10^6 \, \text{N/m}^2$. (a) Calculate the resistance of the hose. (b) What is the viscosity of the concrete, assuming the flow is laminar? (c) How much power is being supplied, assuming the point of use is at the same level as the pump? You may neglect the power supplied to increase the concrete's velocity.

100. Verify that the flow of oil is laminar for an oil gusher that shoots crude oil 25.0 m into the air through a pipe with a 0.100-m diameter. The vertical pipe is 50 m long. Take the density of the oil to be $900 \, \text{kg/m}^3$ and its viscosity to be $1.00 (\text{N/m}^2) \cdot \text{s}$ (or $1.00 \, \text{Pa} \cdot \text{s}$).

101. Calculate the Reynolds numbers for the flow of water through (a) a nozzle with a radius of 0.250 cm and (b) a garden hose with a radius of 0.900 cm, when the nozzle is attached to the hose. The flow rate through hose and nozzle is 0.500 L/s. Can the flow in either possibly be laminar?

102. A fire hose has an inside diameter of 6.40 cm. Suppose such a hose carries a flow of 40.0 L/s starting at a gauge pressure of $1.62 \times 10^6 \, \text{N/m}^2$. The hose goes 10.0 m up a ladder to a nozzle having an inside diameter of 3.00 cm. Calculate the Reynolds numbers for flow in the fire hose and nozzle to show that the flow in each must be turbulent.

103. At what flow rate might turbulence begin to develop in a water main with a 0.200-m diameter? Assume a $20 \, °\text{C}$ temperature.

ADDITIONAL PROBLEMS

104. Before digital storage devices, such as the memory in your cell phone, music was stored on vinyl disks with grooves with varying depths cut into the disk. A phonograph used a needle, which moved over the grooves, measuring the depth of the grooves. The pressure exerted by a phonograph needle on a record is surprisingly large. If the equivalent of 1.00 g is supported by a needle, the tip of which is a circle with a 0.200-mm radius, what pressure is exerted on the record in Pa?

105. Water towers store water above the level of consumers for times of heavy use, eliminating the need for high-speed pumps. How high above a user must the water level be to create a gauge pressure of $3.00 \times 10^5 \, \text{N/m}^2$?

106. The aqueous humor in a person's eye is exerting a force of 0.300 N on the 1.10-cm^2 area of the cornea. What pressure is this in mm Hg?

107. (a) Convert normal blood pressure readings of 120 over 80 mm Hg to newtons per meter squared using the relationship for pressure due to the weight of a fluid $(p = h\rho g)$ rather than a conversion factor. (b) Explain why the blood pressure of an infant would likely be smaller than that of an adult. Specifically, consider the smaller

height to which blood must be pumped.

108. Pressure cookers have been around for more than 300 years, although their use has greatly declined in recent years (early models had a nasty habit of exploding). How much force must the latches holding the lid onto a pressure cooker be able to withstand if the circular lid is 25.0 cm in diameter and the gauge pressure inside is 300 atm? Neglect the weight of the lid.

109. Bird bones have air pockets in them to reduce their weight—this also gives them an average density significantly less than that of the bones of other animals. Suppose an ornithologist weighs a bird bone in air and in water and finds its mass is 45.0 g and its apparent mass when submerged is 3.60 g (assume the bone is watertight). (a) What mass of water is displaced? (b) What is the volume of the bone? (c) What is its average density?

110. In an immersion measurement of a woman's density, she is found to have a mass of 62.0 kg in air and an apparent mass of 0.0850 kg when completely submerged with lungs empty. (a) What mass of water does she displace? (b) What is her volume? (c) Calculate her density. (d) If her lung capacity is 1.75 L, is she able to float without treading water with her lungs filled with air?

111. Some fish have a density slightly less than that of water and must exert a force (swim) to stay submerged. What force must an 85.0-kg grouper exert to stay submerged in salt water if its body density is 1015 kg/m^3?

112. The human circulation system has approximately 1×10^9 capillary vessels. Each vessel has a diameter of about $8\mu\text{m}$. Assuming cardiac output is 5 L/min, determine the average velocity of blood flow through each capillary vessel.

113. The flow rate of blood through a 2.00×10^{-6} m -radius capillary is $3.80 \times 10^9 \text{ cm}^3/\text{s}$. (a) What is the speed of the blood flow? (b) Assuming all the blood in the body passes through capillaries, how many of them must there be to carry a total flow of $90.0 \text{ cm}^3/\text{s}$?

114. The left ventricle of a resting adult's heart pumps blood at a flow rate of $83.0 \text{ cm}^3/\text{s}$, increasing its pressure by 110 mm Hg, its speed from zero to 30.0 cm/s, and its height by 5.00 cm. (All numbers are averaged over the entire heartbeat.) Calculate the total power output of the left ventricle. Note that most of the power is used to increase blood pressure.

115. A sump pump (used to drain water from the basement of houses built below the water table) is draining a flooded basement at the rate of 0.750 L/s, with an output pressure

of $3.00 \times 10^5 \text{ N/m}^2$. (a) The water enters a hose with a 3.00-cm inside diameter and rises 2.50 m above the pump. What is its pressure at this point? (b) The hose goes over the foundation wall, losing 0.500 m in height, and widens to 4.00 cm in diameter. What is the pressure now? You may neglect frictional losses in both parts of the problem.

116. A glucose solution being administered with an IV has a flow rate of $4.00 \text{ cm}^3/\text{min}$. What will the new flow rate be if the glucose is replaced by whole blood having the same density but a viscosity 2.50 times that of the glucose? All other factors remain constant.

117. A small artery has a length of 1.1×10^{-3} m and a radius of 2.5×10^{-5} m. If the pressure drop across the artery is 1.3 kPa, what is the flow rate through the artery? (Assume that the temperature is $37 \,°\text{C}$.)

118. Angioplasty is a technique in which arteries partially blocked with plaque are dilated to increase blood flow. By what factor must the radius of an artery be increased in order to increase blood flow by a factor of 10?

119. Suppose a blood vessel's radius is decreased to 90.0% of its original value by plaque deposits and the body compensates by increasing the pressure difference along the vessel to keep the flow rate constant. By what factor must the pressure difference increase? (b) If turbulence is created by the obstruction, what additional effect would it have on the flow rate?

CHALLENGE PROBLEMS

120. The pressure on the dam shown early in the problems section increases with depth. Therefore, there is a net torque on the dam. Find the net torque.

121. The temperature of the atmosphere is not always constant and can increase or decrease with height. In a neutral atmosphere, where there is not a significant amount of vertical mixing, the temperature decreases at a rate of approximately 6.5 K per km. The magnitude of the decrease in temperature as height increases is known as the lapse rate (Γ). (The symbol is the upper case Greek letter gamma.) Assume that the surface pressure is $p_0 = 1.013 \times 10^5$ Pa where $T = 293$ K and the lapse rate is $\left(-\Gamma = 6.5 \frac{\text{K}}{\text{km}}\right)$. Estimate the pressure 3.0 km above the surface of Earth.

122. A submarine is stranded on the bottom of the ocean with its hatch 25.0 m below the surface. Calculate the force needed to open the hatch from the inside, given it is circular

and 0.450 m in diameter. Air pressure inside the submarine is 1.00 atm.

123. Logs sometimes float vertically in a lake because one end has become water-logged and denser than the other. What is the average density of a uniform-diameter log that floats with 20.0% of its length above water?

124. Scurrilous con artists have been known to represent gold-plated tungsten ingots as pure gold and sell them at prices much below gold value but high above the cost of tungsten. With what accuracy must you be able to measure the mass of such an ingot in and out of water to tell that it is almost pure tungsten rather than pure gold?

125. The inside volume of a house is equivalent to that of a rectangular solid 13.0 m wide by 20.0 m long by 2.75 m high. The house is heated by a forced air gas heater. The main uptake air duct of the heater is 0.300 m in diameter. What is the average speed of air in the duct if it carries a volume equal to that of the house's interior every 15 minutes?

126. A garden hose with a diameter of 2.0 cm is used to fill a bucket, which has a volume of 0.10 cubic meters. It takes 1.2 minutes to fill. An adjustable nozzle is attached to the hose to decrease the diameter of the opening, which increases the speed of the water. The hose is held level to the ground at a height of 1.0 meters and the diameter is decreased until a flower bed 3.0 meters away is reached. (a) What is the volume flow rate of the water through the nozzle when the diameter is 2.0 cm? (b) What is the speed of the water coming out of the hose? (c) What does the speed of the water coming out of the hose need to be to reach the flower bed 3.0 meters away? (d) What is the diameter of the nozzle needed to reach the flower bed?

127. A frequently quoted rule of thumb in aircraft design is that wings should produce about 1000 N of lift per square meter of wing. (The fact that a wing has a top and bottom surface does not double its area.) (a) At takeoff, an aircraft travels at 60.0 m/s, so that the air speed relative to the bottom of the wing is 60.0 m/s. Given the sea level density of air as $1.29 \, \text{kg/m}^3$, how fast must it move over the upper surface to create the ideal lift? (b) How fast must air move over the upper surface at a cruising speed of 245 m/s and at an altitude where air density is one-fourth that at sea level? (Note that this is not all of the aircraft's lift—some comes from the body of the plane, some from engine thrust, and so on. Furthermore, Bernoulli's principle gives an approximate answer because flow over the wing creates turbulence.)

128. Two pipes of equal and constant diameter leave a water pumping station and dump water out of an open end that is open to the atmosphere (see the following figure). The water enters at a pressure of two atmospheres and a speed of $(v_1 = 1.0 \, \text{m/s})$. One pipe drops a height of 10 m. What is the velocity of the water as the water leaves each pipe?

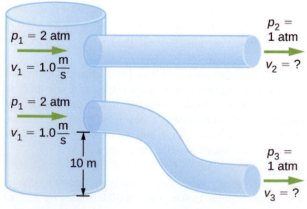

129. Fluid originally flows through a tube at a rate of $100 \, \text{cm}^3/\text{s}$. To illustrate the sensitivity of flow rate to various factors, calculate the new flow rate for the following changes with all other factors remaining the same as in the original conditions. (a) Pressure difference increases by a factor of 1.50. (b) A new fluid with 3.00 times greater viscosity is substituted. (c) The tube is replaced by one having 4.00 times the length. (d) Another tube is used with a radius 0.100 times the original. (e) Yet another tube is substituted with a radius 0.100 times the original and half the length, and the pressure difference is increased by a factor of 1.50.

130. During a marathon race, a runner's blood flow increases to 10.0 times her resting rate. Her blood's viscosity has dropped to 95.0% of its normal value, and the blood pressure difference across the circulatory system has increased by 50.0%. By what factor has the average radii of her blood vessels increased?

131. Water supplied to a house by a water main has a pressure of $3.00 \times 10^5 \, \text{N/m}^2$ early on a summer day when neighborhood use is low. This pressure produces a flow of 20.0 L/min through a garden hose. Later in the day, pressure at the exit of the water main and entrance to the house drops, and a flow of only 8.00 L/min is obtained through the same hose. (a) What pressure is now being supplied to the house, assuming resistance is constant? (b) By what factor did the flow rate in the water main increase in order to cause this decrease in delivered pressure? The pressure at the entrance of the water main is $5.00 \times 10^5 \, \text{N/m}^2$, and the original flow rate was 200 L/min. (c) How many more users are there, assuming each would consume 20.0 L/min in the morning?

132. Gasoline is piped underground from refineries to major users. The flow rate is $3.00 \times 10^{-2} \, \text{m}^3/\text{s}$ (about 500 gal/min), the viscosity of gasoline is $1.00 \times 10^{-3} \, (\text{N/m}^2) \cdot \text{s}$, and its density is $680 \, \text{kg/m}^3$. (a) What minimum diameter must the pipe have if the Reynolds number is to be less than 2000? (b) What pressure difference must be maintained along each kilometer of the pipe to maintain this flow rate?

15 | OSCILLATIONS

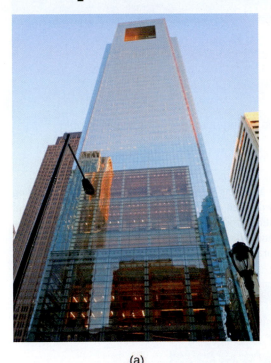

(a)

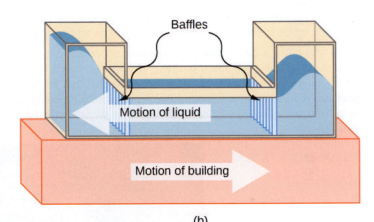

(b)

Figure 15.1 (a) The Comcast Building in Philadelphia, Pennsylvania, looming high above the skyline, is approximately 305 meters (1000 feet) tall. At this height, the top floors can oscillate back and forth due to seismic activity and fluctuating winds. (b) Shown above is a schematic drawing of a tuned, liquid-column mass damper, installed at the top of the Comcast, consisting of a 300,000-gallon reservoir of water to reduce oscillations.

Chapter Outline

Introduction

We begin the study of oscillations with simple systems of pendulums and springs. Although these systems may seem quite basic, the concepts involved have many real-life applications. For example, the Comcast Building in Philadelphia, Pennsylvania, stands approximately 305 meters (1000 feet) tall. As buildings are built taller, they can act as inverted, physical pendulums, with the top floors oscillating due to seismic activity and fluctuating winds. In the Comcast Building, a tuned-mass damper is used to reduce the oscillations. Installed at the top of the building is a tuned, liquid-column mass damper, consisting of a 300,000-gallon reservoir of water. This U-shaped tank allows the water to oscillate freely at a frequency that matches the natural frequency of the building. Damping is provided by tuning the turbulence levels in the moving water using baffles.

15.1 | Simple Harmonic Motion

When you pluck a guitar string, the resulting sound has a steady tone and lasts a long time (**Figure 15.2**). The string vibrates around an equilibrium position, and one oscillation is completed when the string starts from the initial position, travels to one of the extreme positions, then to the other extreme position, and returns to its initial position. We define **periodic motion** to be any motion that repeats itself at regular time intervals, such as exhibited by the guitar string or by a child swinging on a swing. In this section, we study the basic characteristics of oscillations and their mathematical description.

Figure 15.2 When a guitar string is plucked, the string oscillates up and down in periodic motion. The vibrating string causes the surrounding air molecules to oscillate, producing sound waves. (credit: Yutaka Tsutano)

Period and Frequency in Oscillations

In the absence of friction, the time to complete one oscillation remains constant and is called the **period (T)**. Its units are usually seconds, but may be any convenient unit of time. The word 'period' refers to the time for some event whether repetitive or not, but in this chapter, we shall deal primarily in periodic motion, which is by definition repetitive.

A concept closely related to period is the frequency of an event. **Frequency (f)** is defined to be the number of events per unit time. For periodic motion, frequency is the number of oscillations per unit time. The relationship between frequency and period is

$$f = \frac{1}{T}. \tag{15.1}$$

The SI unit for frequency is the *hertz* (Hz) and is defined as one *cycle per second*:

$$1 \text{ Hz} = 1\frac{\text{cycle}}{\text{sec}} \quad \text{or} \quad 1 \text{ Hz} = \frac{1}{\text{s}} = 1 \text{ s}^{-1}.$$

A cycle is one complete **oscillation**.

Example 15.1

Determining the Frequency of Medical Ultrasound

Ultrasound machines are used by medical professionals to make images for examining internal organs of the body. An ultrasound machine emits high-frequency sound waves, which reflect off the organs, and a computer receives the waves, using them to create a picture. We can use the formulas presented in this module to determine the frequency, based on what we know about oscillations. Consider a medical imaging device that produces ultrasound by oscillating with a period of $0.400\ \mu s$. What is the frequency of this oscillation?

Strategy

The period (T) is given and we are asked to find frequency (f).

Solution

Substitute $0.400\ \mu s$ for T in $f = \frac{1}{T}$:

$$f = \frac{1}{T} = \frac{1}{0.400 \times 10^{-6}\ \text{s}}.$$

Solve to find

$$f = 2.50 \times 10^6\ \text{Hz}.$$

Significance

This frequency of sound is much higher than the highest frequency that humans can hear (the range of human hearing is 20 Hz to 20,000 Hz); therefore, it is called ultrasound. Appropriate oscillations at this frequency generate ultrasound used for noninvasive medical diagnoses, such as observations of a fetus in the womb.

Characteristics of Simple Harmonic Motion

A very common type of periodic motion is called **simple harmonic motion (SHM)**. A system that oscillates with SHM is called a **simple harmonic oscillator**.

Simple Harmonic Motion

In simple harmonic motion, the acceleration of the system, and therefore the net force, is proportional to the displacement and acts in the opposite direction of the displacement.

A good example of SHM is an object with mass m attached to a spring on a frictionless surface, as shown in **Figure 15.3**. The object oscillates around the equilibrium position, and the net force on the object is equal to the force provided by the spring. This force obeys Hooke's law $F_s = -kx$, as discussed in a previous chapter.

If the net force can be described by Hooke's law and there is no *damping* (slowing down due to friction or other nonconservative forces), then a simple harmonic oscillator oscillates with equal displacement on either side of the equilibrium position, as shown for an object on a spring in **Figure 15.3**. The maximum displacement from equilibrium is called the **amplitude (A)**. The units for amplitude and displacement are the same but depend on the type of oscillation. For the object on the spring, the units of amplitude and displacement are meters.

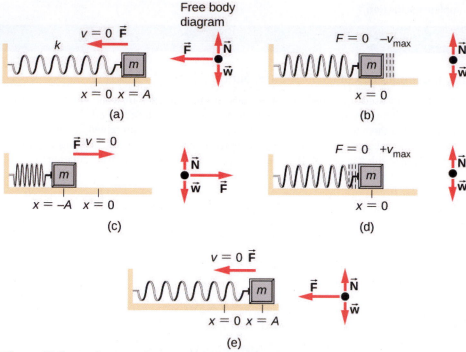

Figure 15.3 An object attached to a spring sliding on a frictionless surface is an uncomplicated simple harmonic oscillator. In the above set of figures, a mass is attached to a spring and placed on a frictionless table. The other end of the spring is attached to the wall. The position of the mass, when the spring is neither stretched nor compressed, is marked as $x = 0$ and is the equilibrium position. (a) The mass is displaced to a position $x = A$ and released from rest. (b) The mass accelerates as it moves in the negative x-direction, reaching a maximum negative velocity at $x = 0$. (c) The mass continues to move in the negative x-direction, slowing until it comes to a stop at $x = -A$. (d) The mass now begins to accelerate in the positive x-direction, reaching a positive maximum velocity at $x = 0$. (e) The mass then continues to move in the positive direction until it stops at $x = A$. The mass continues in SHM that has an amplitude A and a period T. The object's maximum speed occurs as it passes through equilibrium. The stiffer the spring is, the smaller the period T. The greater the mass of the object is, the greater the period T.

What is so significant about SHM? For one thing, the period T and frequency f of a simple harmonic oscillator are independent of amplitude. The string of a guitar, for example, oscillates with the same frequency whether plucked gently or hard.

Two important factors do affect the period of a simple harmonic oscillator. The period is related to how stiff the system is. A very stiff object has a large **force constant (k)**, which causes the system to have a smaller period. For example, you can adjust a diving board's stiffness—the stiffer it is, the faster it vibrates, and the shorter its period. Period also depends on the mass of the oscillating system. The more massive the system is, the longer the period. For example, a heavy person on a diving board bounces up and down more slowly than a light one. In fact, the mass m and the force constant k are the *only* factors that affect the period and frequency of SHM. To derive an equation for the period and the frequency, we must first define and analyze the equations of motion. Note that the force constant is sometimes referred to as the *spring constant*.

Equations of SHM

Consider a block attached to a spring on a frictionless table (**Figure 15.4**). The **equilibrium position** (the position where the spring is neither stretched nor compressed) is marked as $x = 0$. At the equilibrium position, the net force is zero.

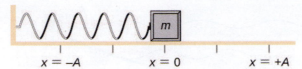

Figure 15.4 A block is attached to a spring and placed on a frictionless table. The equilibrium position, where the spring is neither extended nor compressed, is marked as $x = 0$.

Work is done on the block to pull it out to a position of $x = +A$, and it is then released from rest. The maximum x-position (A) is called the amplitude of the motion. The block begins to oscillate in SHM between $x = +A$ and $x = -A$, where A is the amplitude of the motion and T is the period of the oscillation. The period is the time for one oscillation. **Figure 15.5** shows the motion of the block as it completes one and a half oscillations after release. **Figure 15.6** shows a plot of the position of the block versus time. When the position is plotted versus time, it is clear that the data can be modeled by a cosine function with an amplitude A and a period T. The cosine function $\cos\theta$ repeats every multiple of 2π, whereas the motion of the block repeats every period T. However, the function $\cos\left(\frac{2\pi}{T}t\right)$ repeats every integer multiple of the period. The maximum of the cosine function is one, so it is necessary to multiply the cosine function by the amplitude A.

$$x(t) = A\cos\left(\frac{2\pi}{T}t\right) = A\cos(\omega t). \qquad\qquad (15.2)$$

Recall from the chapter on rotation that the angular frequency equals $\omega = \frac{d\theta}{dt}$. In this case, the period is constant, so the angular frequency is defined as 2π divided by the period, $\omega = \frac{2\pi}{T}$.

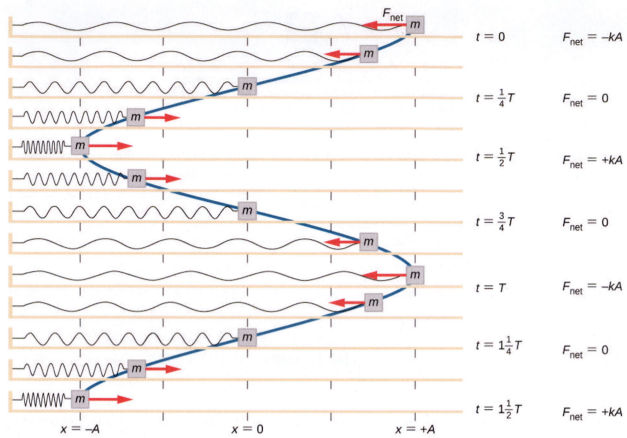

Figure 15.5 A block is attached to one end of a spring and placed on a frictionless table. The other end of the spring is anchored to the wall. The equilibrium position, where the net force equals zero, is marked as $x = 0$ m. Work is done on the block, pulling it out to $x = +A$, and the block is released from rest. The block oscillates between $x = +A$ and $x = -A$. The force is also shown as a vector.

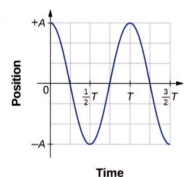

Figure 15.6 A graph of the position of the block shown in **Figure 15.5** as a function of time. The position can be modeled as a periodic function, such as a cosine or sine function.

The equation for the position as a function of time $x(t) = A\cos(\omega t)$ is good for modeling data, where the position of the block at the initial time $t = 0.00$ s is at the amplitude A and the initial velocity is zero. Often when taking experimental data, the position of the mass at the initial time $t = 0.00$ s is not equal to the amplitude and the initial velocity is not zero. Consider 10 seconds of data collected by a student in lab, shown in **Figure 15.7**.

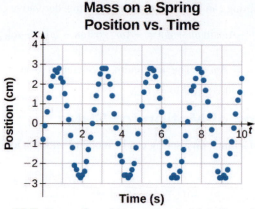

Figure 15.7 Data collected by a student in lab indicate the position of a block attached to a spring, measured with a sonic range finder. The data are collected starting at time $t = 0.00\text{s}$, but the initial position is near position $x \approx -0.80\text{ cm} \neq 3.00\text{ cm}$, so the initial position does not equal the amplitude $x_0 = +A$. The velocity is the time derivative of the position, which is the slope at a point on the graph of position versus time. The velocity is not $v = 0.00$ m/s at time $t = 0.00\text{ s}$, as evident by the slope of the graph of position versus time, which is not zero at the initial time.

The data in **Figure 15.7** can still be modeled with a periodic function, like a cosine function, but the function is shifted to the right. This shift is known as a **phase shift** and is usually represented by the Greek letter phi (ϕ). The equation of the position as a function of time for a block on a spring becomes

$$x(t) = A\cos(\omega t + \phi).$$

This is the generalized equation for SHM where t is the time measured in seconds, ω is the angular frequency with units of inverse seconds, A is the amplitude measured in meters or centimeters, and ϕ is the phase shift measured in radians (**Figure 15.8**). It should be noted that because sine and cosine functions differ only by a phase shift, this motion could be modeled using either the cosine or sine function.

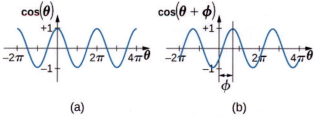

(a)   (b)

Figure 15.8 (a) A cosine function. (b) A cosine function shifted to the right by an angle ϕ. The angle ϕ is known as the phase shift of the function.

The velocity of the mass on a spring, oscillating in SHM, can be found by taking the derivative of the position equation:

$$v(t) = \frac{dx}{dt} = \frac{d}{dt}(A\cos(\omega t + \phi)) = -A\omega\sin(\omega t + \varphi) = -v_{\max}\sin(\omega t + \phi).$$

Because the sine function oscillates between –1 and +1, the maximum velocity is the amplitude times the angular frequency, $v_{\max} = A\omega$. The maximum velocity occurs at the equilibrium position $(x = 0)$ when the mass is moving toward $x = +A$. The maximum velocity in the negative direction is attained at the equilibrium position $(x = 0)$ when the mass is moving toward $x = -A$ and is equal to $-v_{\max}$.

The acceleration of the mass on the spring can be found by taking the time derivative of the velocity:

$$a(t) = \frac{dv}{dt} = \frac{d}{dt}(-A\omega\sin(\omega t + \phi)) = -A\omega^2\cos(\omega t + \varphi) = -a_{max}\cos(\omega t + \phi).$$

The maximum acceleration is $a_{max} = A\omega^2$. The maximum acceleration occurs at the position $(x = -A)$, and the acceleration at the position $(x = -A)$ and is equal to $-a_{max}$.

Summary of Equations of Motion for SHM

In summary, the oscillatory motion of a block on a spring can be modeled with the following equations of motion:

$$x(t) = A\cos(\omega t + \phi) \tag{15.3}$$
$$v(t) = -v_{max}\sin(\omega t + \phi) \tag{15.4}$$
$$a(t) = -a_{max}\cos(\omega t + \phi) \tag{15.5}$$
$$x_{max} = A \tag{15.6}$$
$$v_{max} = A\omega \tag{15.7}$$
$$a_{max} = A\omega^2. \tag{15.8}$$

Here, A is the amplitude of the motion, T is the period, ϕ is the phase shift, and $\omega = \frac{2\pi}{T} = 2\pi f$ is the angular frequency of the motion of the block.

Example 15.2

Determining the Equations of Motion for a Block and a Spring

A 2.00-kg block is placed on a frictionless surface. A spring with a force constant of $k = 32.00$ N/m is attached to the block, and the opposite end of the spring is attached to the wall. The spring can be compressed or extended. The equilibrium position is marked as $x = 0.00$ m.

Work is done on the block, pulling it out to $x = +0.02$ m. The block is released from rest and oscillates between $x = +0.02$ m and $x = -0.02$ m. The period of the motion is 1.57 s. Determine the equations of motion.

Strategy

We first find the angular frequency. The phase shift is zero, $\phi = 0.00$ rad, because the block is released from rest at $x = A = +0.02$ m. Once the angular frequency is found, we can determine the maximum velocity and maximum acceleration.

Solution

The angular frequency can be found and used to find the maximum velocity and maximum acceleration:

$$\omega = \frac{2\pi}{1.57\text{ s}} = 4.00\text{ s}^{-1};$$
$$v_{max} = A\omega = 0.02\text{m}(4.00\text{ s}^{-1}) = 0.08\text{ m/s};$$
$$a_{max} = A\omega^2 = 0.02\text{ m}(4.00\text{ s}^{-1})^2 = 0.32\text{ m/s}^2.$$

All that is left is to fill in the equations of motion:

$$x(t) = A\cos(\omega t + \phi) = (0.02\text{ m})\cos(4.00\text{ s}^{-1}t);$$
$$v(t) = -v_{max}\sin(\omega t + \phi) = (-0.08\text{ m/s})\sin(4.00\text{ s}^{-1}t);$$
$$a(t) = -a_{max}\cos(\omega t + \phi) = (-0.32\text{ m/s}^2)\cos(4.00\text{ s}^{-1}t).$$

The position, velocity, and acceleration can be found for any time. It is important to remember that when using these equations, your calculator must be in radians mode.

The Period and Frequency of a Mass on a Spring

One interesting characteristic of the SHM of an object attached to a spring is that the angular frequency, and therefore the period and frequency of the motion, depend on only the mass and the force constant, and not on other factors such as the amplitude of the motion. We can use the equations of motion and Newton's second law ($\overrightarrow{\mathbf{F}}_{net} = m \overrightarrow{\mathbf{a}}$) to find equations for the angular frequency, frequency, and period.

Consider the block on a spring on a frictionless surface. There are three forces on the mass: the weight, the normal force, and the force due to the spring. The only two forces that act perpendicular to the surface are the weight and the normal force, which have equal magnitudes and opposite directions, and thus sum to zero. The only force that acts parallel to the surface is the force due to the spring, so the net force must be equal to the force of the spring:

$$F_x = -kx;$$

$$ma = -kx;$$

$$m\frac{d^2x}{dt^2} = -kx;$$

$$\frac{d^2x}{dt^2} = -\frac{k}{m}x.$$

Substituting the equations of motion for x and a gives us

$$-A\omega^2 \cos(\omega t + \phi) = -\frac{k}{m}A\cos(\omega t + \phi).$$

Cancelling out like terms and solving for the angular frequency yields

$$\omega = \sqrt{\frac{k}{m}}. \tag{15.9}$$

The angular frequency depends only on the force constant and the mass, and not the amplitude. The angular frequency is defined as $\omega = 2\pi/T$, which yields an equation for the period of the motion:

$$T = 2\pi\sqrt{\frac{m}{k}}. \tag{15.10}$$

The period also depends only on the mass and the force constant. The greater the mass, the longer the period. The stiffer the spring, the shorter the period. The frequency is

$$f = \frac{1}{T} = \frac{1}{2\pi}\sqrt{\frac{k}{m}}. \tag{15.11}$$

Vertical Motion and a Horizontal Spring

When a spring is hung vertically and a block is attached and set in motion, the block oscillates in SHM. In this case, there is no normal force, and the net effect of the force of gravity is to change the equilibrium position. Consider **Figure 15.9**. Two forces act on the block: the weight and the force of the spring. The weight is constant and the force of the spring changes as the length of the spring changes.

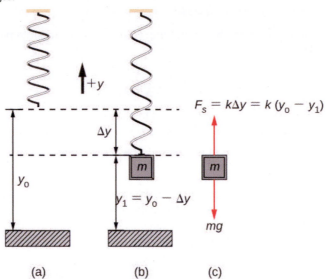

Figure 15.9 A spring is hung from the ceiling. When a block is attached, the block is at the equilibrium position where the weight of the block is equal to the force of the spring. (a) The spring is hung from the ceiling and the equilibrium position is marked as y_o. (b) A mass is attached to the spring and a new equilibrium position is reached ($y_1 = y_o - \Delta y$) when the force provided by the spring equals the weight of the mass. (c) The free-body diagram of the mass shows the two forces acting on the mass: the weight and the force of the spring.

When the block reaches the equilibrium position, as seen in **Figure 15.9**, the force of the spring equals the weight of the block, $F_{net} = F_s - mg = 0$, where

$$-k(-\Delta y) = mg.$$

From the figure, the change in the position is $\Delta y = y_0 - y_1$ and since $-k(-\Delta y) = mg$, we have

$$k(y_0 - y_1) - mg = 0.$$

If the block is displaced and released, it will oscillate around the new equilibrium position. As shown in **Figure 15.10**, if the position of the block is recorded as a function of time, the recording is a periodic function.

If the block is displaced to a position y, the net force becomes $F_{net} = k(y - y_0) - mg = 0$. But we found that at the equilibrium position, $mg = k\Delta y = ky_0 - ky_1$. Substituting for the weight in the equation yields

$$F_{net} = ky - ky_0 - (ky_0 - ky_1) = -k(y - y_1).$$

Recall that y_1 is just the equilibrium position and any position can be set to be the point $y = 0.00$m. So let's set y_1 to $y = 0.00$ m. The net force then becomes

$$
\begin{aligned}
F_{net} &= -ky; \\
m\frac{d^2 y}{dt^2} &= -ky.
\end{aligned}
$$

This is just what we found previously for a horizontally sliding mass on a spring. The constant force of gravity only served to shift the equilibrium location of the mass. Therefore, the solution should be the same form as for a block on a horizontal spring, $y(t) = A\cos(\omega t + \phi)$. The equations for the velocity and the acceleration also have the same form as for the horizontal case. Note that the inclusion of the phase shift means that the motion can actually be modeled using either a cosine or a sine function, since these two functions only differ by a phase shift.

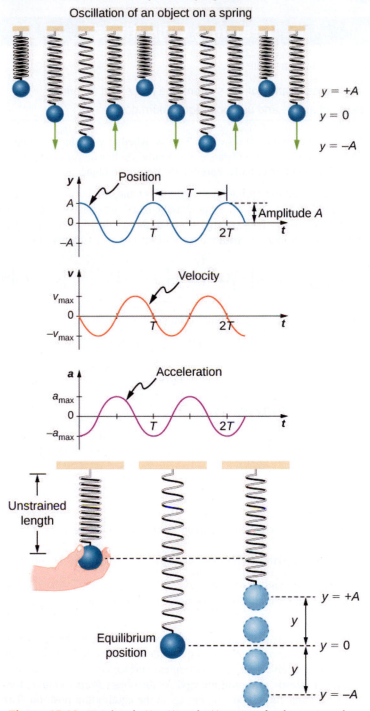

Figure 15.10 Graphs of $y(t)$, $v(t)$, and $a(t)$ versus t for the motion of an object on a vertical spring. The net force on the object can be described by Hooke's law, so the object undergoes SHM. Note that the initial position has the vertical displacement at its maximum value A; v is initially zero and then negative as the object moves down; the initial acceleration is negative, back toward the equilibrium position and becomes zero at that point.

15.2 | Energy in Simple Harmonic Motion

Learning Objectives

By the end of this section, you will be able to:

- Describe the energy conservation of the system of a mass and a spring
- Explain the concepts of stable and unstable equilibrium points

To produce a deformation in an object, we must do work. That is, whether you pluck a guitar string or compress a car's shock absorber, a force must be exerted through a distance. If the only result is deformation, and no work goes into thermal, sound, or kinetic energy, then all the work is initially stored in the deformed object as some form of potential energy.

Consider the example of a block attached to a spring on a frictionless table, oscillating in SHM. The force of the spring is a conservative force (which you studied in the chapter on potential energy and conservation of energy), and we can define a potential energy for it. This potential energy is the energy stored in the spring when the spring is extended or compressed. In this case, the block oscillates in one dimension with the force of the spring acting parallel to the motion:

$$W = \int_{x_i}^{x_f} F_x \, dx = \int_{x_i}^{x_f} -kx \, dx = \left[-\tfrac{1}{2}kx^2\right]_{x_i}^{x_f} = -\left[\tfrac{1}{2}kx_f^2 - \tfrac{1}{2}kx_i^2\right] = -\left[U_f - U_i\right] = -\Delta U.$$

When considering the energy stored in a spring, the equilibrium position, marked as $x_i = 0.00 \text{ m},$ is the position at which the energy stored in the spring is equal to zero. When the spring is stretched or compressed a distance x, the potential energy stored in the spring is

$$U = \tfrac{1}{2}kx^2.$$

Energy and the Simple Harmonic Oscillator

To study the energy of a simple harmonic oscillator, we need to consider all the forms of energy. Consider the example of a block attached to a spring, placed on a frictionless surface, oscillating in SHM. The potential energy stored in the deformation of the spring is

$$U = \tfrac{1}{2}kx^2.$$

In a simple harmonic oscillator, the energy oscillates between kinetic energy of the mass $K = \tfrac{1}{2}mv^2$ and potential energy $U = \tfrac{1}{2}kx^2$ stored in the spring. In the SHM of the mass and spring system, there are no dissipative forces, so the total energy is the sum of the potential energy and kinetic energy. In this section, we consider the conservation of energy of the system. The concepts examined are valid for all simple harmonic oscillators, including those where the gravitational force plays a role.

Consider Figure 15.11, which shows an oscillating block attached to a spring. In the case of undamped SHM, the energy oscillates back and forth between kinetic and potential, going completely from one form of energy to the other as the system oscillates. So for the simple example of an object on a frictionless surface attached to a spring, the motion starts with all of the energy stored in the spring as **elastic potential energy**. As the object starts to move, the elastic potential energy is converted into kinetic energy, becoming entirely kinetic energy at the equilibrium position. The energy is then converted back into elastic potential energy by the spring as it is stretched or compressed. The velocity becomes zero when the kinetic energy is completely converted, and this cycle then repeats. Understanding the conservation of energy in these cycles will provide extra insight here and in later applications of SHM, such as alternating circuits.

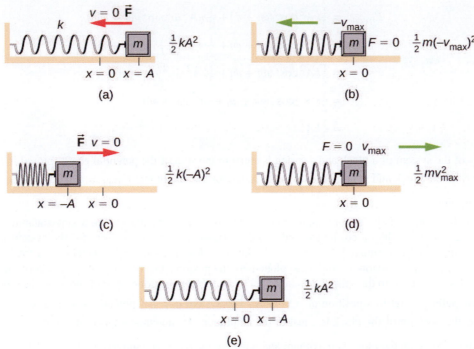

Figure 15.11 The transformation of energy in SHM for an object attached to a spring on a frictionless surface. (a) When the mass is at the position $x = +A$, all the energy is stored as potential energy in the spring $U = \frac{1}{2}kA^2$. The kinetic energy is equal to zero because the velocity of the mass is zero. (b) As the mass moves toward $x = -A$, the mass crosses the position $x = 0$. At this point, the spring is neither extended nor compressed, so the potential energy stored in the spring is zero. At $x = 0$, the total energy is all kinetic energy where $K = \frac{1}{2}m(-v_{max})^2$. (c) The mass continues to move until it reaches $x = -A$ where the mass stops and starts moving toward $x = +A$. At the position $x = -A$, the total energy is stored as potential energy in the compressed $U = \frac{1}{2}k(-A)^2$ and the kinetic energy is zero. (d) As the mass passes through the position $x = 0$, the kinetic energy is $K = \frac{1}{2}mv_{max}^2$ and the potential energy stored in the spring is zero. (e) The mass returns to the position $x = +A$, where $K = 0$ and $U = \frac{1}{2}kA^2$.

Consider **Figure 15.11**, which shows the energy at specific points on the periodic motion. While staying constant, the energy oscillates between the kinetic energy of the block and the potential energy stored in the spring:

$$E_{Total} = U + K = \frac{1}{2}kx^2 + \frac{1}{2}mv^2.$$

The motion of the block on a spring in SHM is defined by the position $x(t) = A\cos(\omega t + \phi)$ with a velocity of $v(t) = -A\omega\sin(\omega t + \phi)$. Using these equations, the trigonometric identity $\cos^2\theta + \sin^2\theta = 1$ and $\omega = \sqrt{\frac{k}{m}}$, we can find the total energy of the system:

$$E_{Total} = \frac{1}{2}kA^2\cos^2(\omega t + \phi) + \frac{1}{2}mA^2\omega^2\sin^2(\omega t + \phi)$$

$$= \frac{1}{2}kA^2\cos^2(\omega t + \phi) + \frac{1}{2}mA^2\left(\frac{k}{m}\right)\sin^2(\omega t + \phi)$$

$$= \frac{1}{2}kA^2\cos^2(\omega t + \phi) + \frac{1}{2}kA^2\sin^2(\omega t + \phi)$$

$$= \frac{1}{2}kA^2\left(\cos^2(\omega t + \phi) + \sin^2(\omega t + \phi)\right)$$

$$= \frac{1}{2}kA^2.$$

The total energy of the system of a block and a spring is equal to the sum of the potential energy stored in the spring plus the kinetic energy of the block and is proportional to the square of the amplitude $E_{Total} = (1/2)kA^2$. The total energy of the system is constant.

A closer look at the energy of the system shows that the kinetic energy oscillates like a sine-squared function, while the potential energy oscillates like a cosine-squared function. However, the total energy for the system is constant and is proportional to the amplitude squared. **Figure 15.12** shows a plot of the potential, kinetic, and total energies of the block and spring system as a function of time. Also plotted are the position and velocity as a function of time. Before time $t = 0.0\,\text{s},$ the block is attached to the spring and placed at the equilibrium position. Work is done on the block by applying an external force, pulling it out to a position of $x = +A$. The system now has potential energy stored in the spring. At time $t = 0.00\,\text{s},$ the position of the block is equal to the amplitude, the potential energy stored in the spring is equal to $U = \frac{1}{2}kA^2$, and the force on the block is maximum and points in the negative x-direction $(F_S = -kA)$. The velocity and kinetic energy of the block are zero at time $t = 0.00\,\text{s}.$ At time $t = 0.00\,\text{s},$ the block is released from rest.

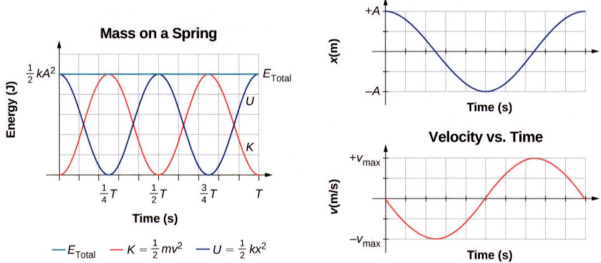

Figure 15.12 Graph of the kinetic energy, potential energy, and total energy of a block oscillating on a spring in SHM. Also shown are the graphs of position versus time and velocity versus time. The total energy remains constant, but the energy oscillates between kinetic energy and potential energy. When the kinetic energy is maximum, the potential energy is zero. This occurs when the velocity is maximum and the mass is at the equilibrium position. The potential energy is maximum when the speed is zero. The total energy is the sum of the kinetic energy plus the potential energy and it is constant.

Oscillations About an Equilibrium Position

We have just considered the energy of SHM as a function of time. Another interesting view of the simple harmonic oscillator is to consider the energy as a function of position. **Figure 15.13** shows a graph of the energy versus position of a system undergoing SHM.

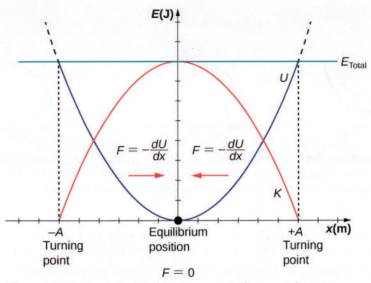

Figure 15.13 A graph of the kinetic energy (red), potential energy (blue), and total energy (green) of a simple harmonic oscillator. The force is equal to $F = -\dfrac{dU}{dx}$. The equilibrium position is shown as a black dot and is the point where the force is equal to zero. The force is positive when $x < 0$, negative when $x > 0$, and equal to zero when $x = 0$.

The potential energy curve in **Figure 15.13** resembles a bowl. When a marble is placed in a bowl, it settles to the equilibrium position at the lowest point of the bowl $(x = 0)$. This happens because a **restoring force** points toward the equilibrium point. This equilibrium point is sometimes referred to as a *fixed point*. When the marble is disturbed to a different position $(x = +A)$, the marble oscillates around the equilibrium position. Looking back at the graph of potential energy, the force can be found by looking at the slope of the potential energy graph $\left(F = -\dfrac{dU}{dx} \right)$. Since the force on either side of the fixed point points back toward the equilibrium point, the equilibrium point is called a **stable equilibrium point**. The points $x = A$ and $x = -A$ are called the turning points. (See **Potential Energy and Conservation of Energy**.)

Stability is an important concept. If an equilibrium point is stable, a slight disturbance of an object that is initially at the stable equilibrium point will cause the object to oscillate around that point. The stable equilibrium point occurs because the force on either side is directed toward it. For an unstable equilibrium point, if the object is disturbed slightly, it does not return to the equilibrium point.

Consider the marble in the bowl example. If the bowl is right-side up, the marble, if disturbed slightly, will oscillate around the stable equilibrium point. If the bowl is turned upside down, the marble can be balanced on the top, at the equilibrium point where the net force is zero. However, if the marble is disturbed slightly, it will not return to the equilibrium point, but will instead roll off the bowl. The reason is that the force on either side of the equilibrium point is directed away from that point. This point is an unstable equilibrium point.

Figure 15.14 shows three conditions. The first is a stable equilibrium point (a), the second is an unstable equilibrium point (b), and the last is also an unstable equilibrium point (c), because the force on only one side points toward the equilibrium point.

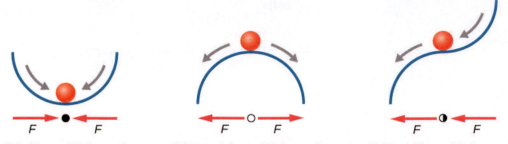

(a) Stable equilibrium point   (b) Unstable equilibrium point   (c) Unstable equilibrium point

Figure 15.14 Examples of equilibrium points. (a) Stable equilibrium point; (b) unstable equilibrium point; (c) unstable equilibrium point (sometimes referred to as a half-stable equilibrium point).

The process of determining whether an equilibrium point is stable or unstable can be formalized. Consider the potential energy curves shown in **Figure 15.15**. The force can be found by analyzing the slope of the graph. The force is $F = -\frac{dU}{dx}$. In (a), the fixed point is at $x = 0.00$ m. When $x < 0.00$ m, the force is positive. When $x > 0.00$ m, the force is negative. This is a stable point. In (b), the fixed point is at $x = 0.00$ m. When $x < 0.00$ m, the force is negative. When $x > 0.00$ m, the force is also negative. This is an unstable point.

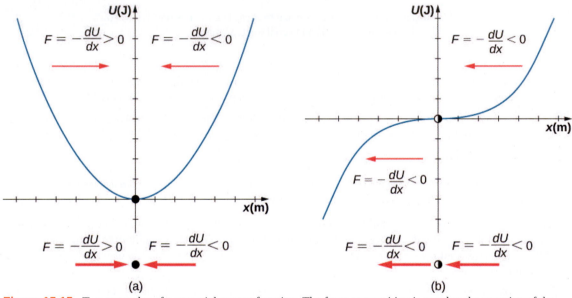

Figure 15.15 Two examples of a potential energy function. The force at a position is equal to the negative of the slope of the graph at that position. (a) A potential energy function with a stable equilibrium point. (b) A potential energy function with an unstable equilibrium point. This point is sometimes called half-stable because the force on one side points toward the fixed point.

A practical application of the concept of stable equilibrium points is the force between two neutral atoms in a molecule. If two molecules are in close proximity, separated by a few atomic diameters, they can experience an attractive force. If the molecules move close enough so that the electron shells of the other electrons overlap, the force between the molecules becomes repulsive. The attractive force between the two atoms may cause the atoms to form a molecule. The force between the two molecules is not a linear force and cannot be modeled simply as two masses separated by a spring, but the atoms of the molecule can oscillate around an equilibrium point when displaced a small amount from the equilibrium position. The atoms oscillate due the attractive force and repulsive force between the two atoms.

Consider one example of the interaction between two atoms known as the van Der Waals interaction. It is beyond the scope of this chapter to discuss in depth the interactions of the two atoms, but the oscillations of the atoms can be examined by considering one example of a model of the potential energy of the system. One suggestion to model the potential energy of this molecule is with the Lennard-Jones 6-12 potential:

$$U(x) = 4\varepsilon\left[\left(\frac{\sigma}{x}\right)^{12} - \left(\frac{\sigma}{x}\right)^{6}\right].$$

A graph of this function is shown in **Figure 15.16**. The two parameters ε and σ are found experimentally.

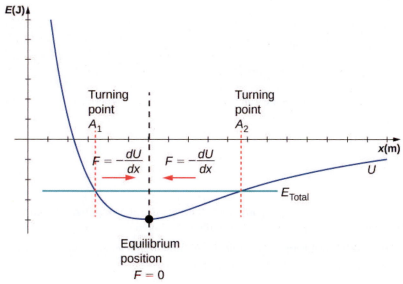

Figure 15.16 The Lennard-Jones potential energy function for a system of two neutral atoms. If the energy is below some maximum energy, the system oscillates near the equilibrium position between the two turning points.

From the graph, you can see that there is a potential energy well, which has some similarities to the potential energy well of the potential energy function of the simple harmonic oscillator discussed in **Figure 15.13**. The Lennard-Jones potential has a stable equilibrium point where the potential energy is minimum and the force on either side of the equilibrium point points toward equilibrium point. Note that unlike the simple harmonic oscillator, the potential well of the Lennard-Jones potential is not symmetric. This is due to the fact that the force between the atoms is not a Hooke's law force and is not linear. The atoms can still oscillate around the equilibrium position x_{min} because when $x < x_{min}$, the force is positive; when $x > x_{min}$, the force is negative. Notice that as x approaches zero, the slope is quite steep and negative, which means that the force is large and positive. This suggests that it takes a large force to try to push the atoms close together. As x becomes increasingly large, the slope becomes less steep and the force is smaller and negative. This suggests that if given a large enough energy, the atoms can be separated.

If you are interested in this interaction, find the force between the molecules by taking the derivative of the potential energy function. You will see immediately that the force does not resemble a Hooke's law force $(F = -kx)$, but if you are familiar with the binomial theorem:

$$(1 + x)^{n} = 1 + nx + \frac{n(n - 1)}{2\,!}x^{2} + \frac{n(n - 1)(n - 2)}{3\,!}x^{3} + \cdots,$$

the force can be approximated by a Hooke's law force.

Velocity and Energy Conservation

Getting back to the system of a block and a spring in **Figure 15.11**, once the block is released from rest, it begins to move in the negative direction toward the equilibrium position. The potential energy decreases and the magnitude of the velocity and the kinetic energy increase. At time $t = T/4$, the block reaches the equilibrium position $x = 0.00\ \mathrm{m}$, where the force on the block and the potential energy are zero. At the equilibrium position, the block reaches a negative velocity with a magnitude equal to the maximum velocity $v = -A\omega$. The kinetic energy is maximum and equal to $K = \frac{1}{2}mv^{2} = \frac{1}{2}mA^{2}\omega^{2} = \frac{1}{2}kA^{2}$. At this point, the force on the block is zero, but momentum carries the block, and it continues in the negative direction toward $x = -A$. As the block continues to move, the force on it acts in the positive direction and the magnitude of the velocity and kinetic energy decrease. The potential energy increases as the spring compresses. At time $t = T/2$, the block reaches $x = -A$. Here the velocity and kinetic energy are equal to zero. The force on the block is $F = +kA$ and the potential energy stored in the spring is $U = \frac{1}{2}kA^2$. During the oscillations, the total energy is constant and equal to the sum of the potential energy and the kinetic energy of the system,

$$E_{\text{Total}} = \frac{1}{2}kx^2 + \frac{1}{2}mv^2 = \frac{1}{2}kA^2.$$

(15.12)

The equation for the energy associated with SHM can be solved to find the magnitude of the velocity at any position:

$$|v| = \sqrt{\frac{k}{m}\left(A^2 - x^2\right)}.$$

(15.13)

The energy in a simple harmonic oscillator is proportional to the square of the amplitude. When considering many forms of oscillations, you will find the energy proportional to the amplitude squared.

 15.1 **Check Your Understanding** Why would it hurt more if you snapped your hand with a ruler than with a loose spring, even if the displacement of each system is equal?

 15.2 **Check Your Understanding** Identify one way you could decrease the maximum velocity of a simple harmonic oscillator.

15.3 | Comparing Simple Harmonic Motion and Circular Motion

Learning Objectives

By the end of this section, you will be able to:

- Describe how the sine and cosine functions relate to the concepts of circular motion
- Describe the connection between simple harmonic motion and circular motion

An easy way to model SHM is by considering uniform circular motion. **Figure 15.17** shows one way of using this method. A peg (a cylinder of wood) is attached to a vertical disk, rotating with a constant angular frequency. **Figure 15.18** shows a side view of the disk and peg. If a lamp is placed above the disk and peg, the peg produces a shadow. Let the disk have a radius of $r = A$ and define the position of the shadow that coincides with the center line of the disk to be $x = 0.00\, m$. As the disk rotates at a constant rate, the shadow oscillates between $x = +A$ and $x = -A$. Now imagine a block on a spring beneath the floor as shown in **Figure 15.18**.

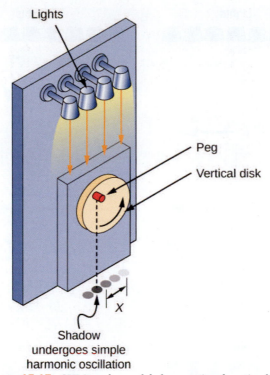

Figure 15.17 SHM can be modeled as rotational motion by looking at the shadow of a peg on a wheel rotating at a constant angular frequency.

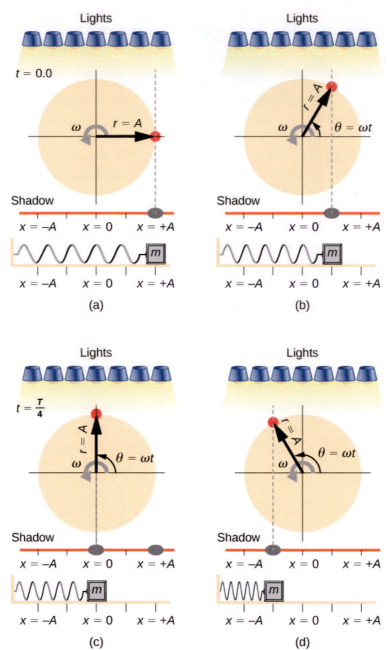

Figure 15.18 Light shines down on the disk so that the peg makes a shadow. If the disk rotates at just the right angular frequency, the shadow follows the motion of the block on a spring. If there is no energy dissipated due to nonconservative forces, the block and the shadow will oscillate back and forth in unison. In this figure, four snapshots are taken at four different times. (a) The wheel starts at $\theta = 0^o$ and the shadow of the peg is at $x = +A$, representing the mass at position $x = +A$. (b) As the disk rotates through an angle $\theta = \omega t$, the shadow of the peg is between $x = +A$ and $x = 0$. (c) The disk continues to rotate until $\theta = 90^0$, at which the shadow follows the mass to $x = 0$. (d) The disk continues to rotate, the shadow follows the position of the mass.

If the disk turns at the proper angular frequency, the shadow follows along with the block. The position of the shadow can be modeled with the equation

$$x(t) = A\cos(\omega t). \tag{15.14}$$

Recall that the block attached to the spring does not move at a constant velocity. How often does the wheel have to turn to have the peg's shadow always on the block? The disk must turn at a constant angular frequency equal to 2π times the frequency of oscillation $(\omega = 2\pi f)$.

Figure 15.19 shows the basic relationship between uniform circular motion and SHM. The peg lies at the tip of the radius, a distance A from the center of the disk. The x-axis is defined by a line drawn parallel to the ground, cutting the disk in half. The y-axis (not shown) is defined by a line perpendicular to the ground, cutting the disk into a left half and a right half. The center of the disk is the point $(x = 0, y = 0)$. The projection of the position of the peg onto the fixed x-axis gives the position of the shadow, which undergoes SHM analogous to the system of the block and spring. At the time shown in the figure, the projection has position x and moves to the left with velocity v. The tangential velocity of the peg around the circle equals $\bar{v}_{max}$ of the block on the spring. The x-component of the velocity is equal to the velocity of the block on the spring.

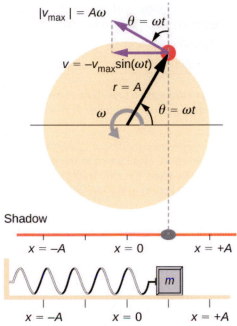

Figure 15.19 A peg moving on a circular path with a constant angular velocity ω is undergoing uniform circular motion. Its projection on the x-axis undergoes SHM. Also shown is the velocity of the peg around the circle, v_{max}, and its projection, which is v. Note that these velocities form a similar triangle to the displacement triangle.

We can use **Figure 15.19** to analyze the velocity of the shadow as the disk rotates. The peg moves in a circle with a speed of $v_{max} = A\omega$. The shadow moves with a velocity equal to the component of the peg's velocity that is parallel to the surface where the shadow is being produced:

$$v = -v_{max}\sin(\omega t). \tag{15.15}$$

It follows that the acceleration is

$$a = -a_{max}\cos(\omega t). \tag{15.16}$$

 15.3 Check Your Understanding Identify an object that undergoes uniform circular motion. Describe how you could trace the SHM of this object.

15.4 | Pendulums

Learning Objectives

By the end of this section, you will be able to:

- State the forces that act on a simple pendulum
- Determine the angular frequency, frequency, and period of a simple pendulum in terms of the length of the pendulum and the acceleration due to gravity
- Define the period for a physical pendulum
- Define the period for a torsional pendulum

Pendulums are in common usage. Grandfather clocks use a pendulum to keep time and a pendulum can be used to measure the acceleration due to gravity. For small displacements, a pendulum is a simple harmonic oscillator.

The Simple Pendulum

A **simple pendulum** is defined to have a point mass, also known as the pendulum bob, which is suspended from a string of length L with negligible mass (**Figure 15.20**). Here, the only forces acting on the bob are the force of gravity (i.e., the weight of the bob) and tension from the string. The mass of the string is assumed to be negligible as compared to the mass of the bob.

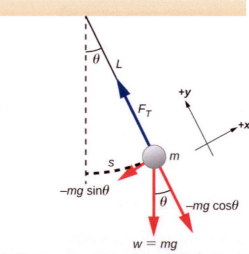

Figure 15.20 A simple pendulum has a small-diameter bob and a string that has a very small mass but is strong enough not to stretch appreciably. The linear displacement from equilibrium is s, the length of the arc. Also shown are the forces on the bob, which result in a net force of $-mg\sin\theta$ toward the equilibrium position—that is, a restoring force.

Consider the torque on the pendulum. The force providing the restoring torque is the component of the weight of the pendulum bob that acts along the arc length. The torque is the length of the string L times the component of the net force that is perpendicular to the radius of the arc. The minus sign indicates the torque acts in the opposite direction of the angular displacement:

$$\tau = -L(mg \sin \theta);$$
$$I\alpha = -L(mg \sin \theta);$$
$$I\frac{d^2\theta}{dt^2} = -L(mg \sin \theta);$$
$$mL^2\frac{d^2\theta}{dt^2} = -L(mg \sin \theta);$$
$$\frac{d^2\theta}{dt^2} = -\frac{g}{L}\sin \theta.$$

The solution to this differential equation involves advanced calculus, and is beyond the scope of this text. But note that for small angles (less than 15 degrees), $\sin \theta$ and θ differ by less than 1%, so we can use the small angle approximation $\sin \theta \approx \theta$. The angle θ describes the position of the pendulum. Using the small angle approximation gives an approximate solution for small angles,

$$\frac{d^2\theta}{dt^2} = -\frac{g}{L}\theta. \tag{15.17}$$

Because this equation has the same form as the equation for SHM, the solution is easy to find. The angular frequency is

$$\omega = \sqrt{\frac{g}{L}} \tag{15.18}$$

and the period is

$$T = 2\pi\sqrt{\frac{L}{g}}. \tag{15.19}$$

The period of a simple pendulum depends on its length and the acceleration due to gravity. The period is completely independent of other factors, such as mass and the maximum displacement. As with simple harmonic oscillators, the period T for a pendulum is nearly independent of amplitude, especially if θ is less than about $15°$. Even simple pendulum clocks can be finely adjusted and remain accurate.

Note the dependence of T on g. If the length of a pendulum is precisely known, it can actually be used to measure the acceleration due to gravity, as in the following example.

Example 15.3

Measuring Acceleration due to Gravity by the Period of a Pendulum

What is the acceleration due to gravity in a region where a simple pendulum having a length 75.000 cm has a period of 1.7357 s?

Strategy

We are asked to find g given the period T and the length L of a pendulum. We can solve $T = 2\pi\sqrt{\frac{L}{g}}$ for g, assuming only that the angle of deflection is less than $15°$.

Solution

1. Square $T = 2\pi\sqrt{\frac{L}{g}}$ and solve for g:

$$g = 4\pi^2 \frac{L}{T^2}.$$

2. Substitute known values into the new equation:

$$g = 4\pi^2 \frac{0.75000\,\text{m}}{(1.7357\,\text{s})^2}.$$

3. Calculate to find g:

$$g = 9.8281\,\text{m/s}^2.$$

Significance

This method for determining g can be very accurate, which is why length and period are given to five digits in this example. For the precision of the approximation $\sin\theta \approx \theta$ to be better than the precision of the pendulum length and period, the maximum displacement angle should be kept below about $0.5°$.

 15.4 Check Your Understanding An engineer builds two simple pendulums. Both are suspended from small wires secured to the ceiling of a room. Each pendulum hovers 2 cm above the floor. Pendulum 1 has a bob with a mass of 10 kg. Pendulum 2 has a bob with a mass of 100 kg. Describe how the motion of the pendulums will differ if the bobs are both displaced by $12°$.

Physical Pendulum

Any object can oscillate like a pendulum. Consider a coffee mug hanging on a hook in the pantry. If the mug gets knocked, it oscillates back and forth like a pendulum until the oscillations die out. We have described a simple pendulum as a point mass and a string. A **physical pendulum** is any object whose oscillations are similar to those of the simple pendulum, but cannot be modeled as a point mass on a string, and the mass distribution must be included into the equation of motion.

As for the simple pendulum, the restoring force of the physical pendulum is the force of gravity. With the simple pendulum, the force of gravity acts on the center of the pendulum bob. In the case of the physical pendulum, the force of gravity acts on the center of mass (CM) of an object. The object oscillates about a point O. Consider an object of a generic shape as shown in **Figure 15.21**.

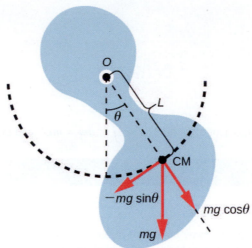

Figure 15.21 A physical pendulum is any object that oscillates as a pendulum, but cannot be modeled as a point mass on a string. The force of gravity acts on the center of mass (CM) and provides the restoring force that causes the object to oscillate. The minus sign on the component of the weight that provides the restoring force is present because the force acts in the opposite direction of the increasing angle θ.

When a physical pendulum is hanging from a point but is free to rotate, it rotates because of the torque applied at the CM, produced by the component of the object's weight that acts tangent to the motion of the CM. Taking the counterclockwise direction to be positive, the component of the gravitational force that acts tangent to the motion is $-mg \sin \theta$. The minus sign is the result of the restoring force acting in the opposite direction of the increasing angle. Recall that the torque is equal to $\vec{\tau} = \vec{r} \times \vec{F}$. The magnitude of the torque is equal to the length of the radius arm times the tangential component of the force applied, $|\tau| = rF\sin \theta$. Here, the length L of the radius arm is the distance between the point of rotation and the CM. To analyze the motion, start with the net torque. Like the simple pendulum, consider only small angles so that $\sin \theta \approx \theta$. Recall from **Fixed-Axis Rotation** on rotation that the net torque is equal to the moment of inertia $I = \int r^2 \, dm$ times the angular acceleration α, where $\alpha = \dfrac{d^2\theta}{dt^2}$:

$$I\alpha = \tau_{\text{net}} = L(-mg)\sin \theta.$$

Using the small angle approximation and rearranging:

$$I\alpha = -L(mg)\theta;$$
$$I\frac{d^2\theta}{dt^2} = -L(mg)\theta;$$
$$\frac{d^2\theta}{dt^2} = -\left(\frac{mgL}{I}\right)\theta.$$

Once again, the equation says that the second time derivative of the position (in this case, the angle) equals minus a constant $\left(-\dfrac{mgL}{I}\right)$ times the position. The solution is

$$\theta(t) = \Theta\cos(\omega t + \phi),$$

where Θ is the maximum angular displacement. The angular frequency is

$$\omega = \sqrt{\frac{mgL}{I}}. \tag{15.20}$$

The period is therefore

$$T = 2\pi\sqrt{\frac{I}{mgL}}. \qquad \qquad \textbf{(15.21)}$$

Note that for a simple pendulum, the moment of inertia is $I = \int r^2\, dm = mL^2$ and the period reduces to $T = 2\pi\sqrt{\frac{L}{g}}$.

Example 15.4

Reducing the Swaying of a Skyscraper

In extreme conditions, skyscrapers can sway up to two meters with a frequency of up to 20.00 Hz due to high winds or seismic activity. Several companies have developed physical pendulums that are placed on the top of the skyscrapers. As the skyscraper sways to the right, the pendulum swings to the left, reducing the sway. Assuming the oscillations have a frequency of 0.50 Hz, design a pendulum that consists of a long beam, of constant density, with a mass of 100 metric tons and a pivot point at one end of the beam. What should be the length of the beam?

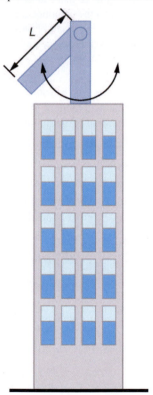

Strategy

We are asked to find the length of the physical pendulum with a known mass. We first need to find the moment of inertia of the beam. We can then use the equation for the period of a physical pendulum to find the length.

Solution

1. Find the moment of inertia for the CM:

2. Use the parallel axis theorem to find the moment of inertia about the point of rotation:

$$I = I_{\text{CM}} + \frac{L^2}{4}M = \frac{1}{12}ML^2 + \frac{1}{4}ML^2 = \frac{1}{3}ML^2.$$

3. The period of a physical pendulum has a period of $T = 2\pi\sqrt{\frac{I}{mgL}}$. Use the moment of inertia to solve for

the length L:

$$T = 2\pi\sqrt{\frac{I}{MgL}} = 2\pi\sqrt{\frac{\frac{1}{3}ML^2}{MgL}} = 2\pi\sqrt{\frac{L}{3g}};$$

$$L = 3g\left(\frac{T}{2\pi}\right)^2 = 3\left(9.8\frac{m}{s^2}\right)\left(\frac{2\,s}{2\pi}\right)^2 = 2.98\ m.$$

Significance

There are many ways to reduce the oscillations, including modifying the shape of the skyscrapers, using multiple physical pendulums, and using tuned-mass dampers.

Torsional Pendulum

A **torsional pendulum** consists of a rigid body suspended by a light wire or spring (**Figure 15.22**). When the body is twisted some small maximum angle (Θ) and released from rest, the body oscillates between $(\theta = +\Theta)$ and $(\theta = -\Theta)$. The restoring torque is supplied by the shearing of the string or wire.

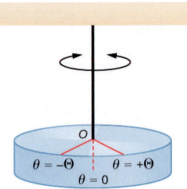

Figure 15.22 A torsional pendulum consists of a rigid body suspended by a string or wire. The rigid body oscillates between $\theta = +\Theta$ and $\theta = -\Theta$.

The restoring torque can be modeled as being proportional to the angle:

$$\tau = -\kappa\theta.$$

The variable kappa (κ) is known as the torsion constant of the wire or string. The minus sign shows that the restoring torque acts in the opposite direction to increasing angular displacement. The net torque is equal to the moment of inertia times the angular acceleration:

$$I\frac{d^2\theta}{dt^2} = -\kappa\theta;$$

$$\frac{d^2\theta}{dt^2} = -\frac{\kappa}{I}\theta.$$

This equation says that the second time derivative of the position (in this case, the angle) equals a negative constant times the position. This looks very similar to the equation of motion for the SHM $\frac{d^2x}{dt^2} = -\frac{k}{m}x$, where the period was found to be $T = 2\pi\sqrt{\frac{m}{k}}$. Therefore, the period of the torsional pendulum can be found using

$$T = 2\pi\sqrt{\frac{I}{\kappa}}. \tag{15.22}$$

The units for the torsion constant are $[\kappa] = \text{N-m} = \left(\text{kg}\frac{\text{m}}{\text{s}^2}\right)\text{m} = \text{kg}\,\frac{\text{m}^2}{\text{s}^2}$ and the units for the moment of inertial are

$[I] = \text{kg-m}^2$, which show that the unit for the period is the second.

Example 15.5

Measuring the Torsion Constant of a String

A rod has a length of $l = 0.30$ m and a mass of 4.00 kg. A string is attached to the CM of the rod and the system is hung from the ceiling (**Figure 15.23**). The rod is displaced 10 degrees from the equilibrium position and released from rest. The rod oscillates with a period of 0.5 s. What is the torsion constant κ?

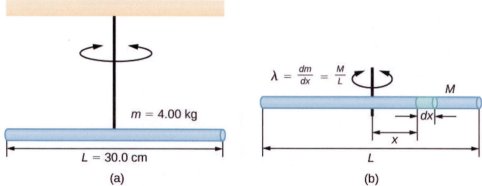

Figure 15.23 (a) A rod suspended by a string from the ceiling. (b) Finding the rod's moment of inertia.

Strategy

We are asked to find the torsion constant of the string. We first need to find the moment of inertia.

Solution

1. Find the moment of inertia for the CM:

$$I_{CM} = \int x^2 dm = \int_{-L/2}^{+L/2} x^2 \lambda dx = \lambda\left[\frac{x^3}{3}\right]_{-L/2}^{+L/2} = \lambda\frac{2L^3}{24} = \left(\frac{M}{L}\right)\frac{2L^3}{24} = \frac{1}{12}ML^2.$$

2. Calculate the torsion constant using the equation for the period:

$$T = 2\pi\sqrt{\frac{I}{\kappa}};$$

$$\kappa = I\left(\frac{2\pi}{T}\right)^2 = \left(\frac{1}{12}ML^2\right)\left(\frac{2\pi}{T}\right)^2;$$

$$= \left(\frac{1}{12}(4.00\,\text{kg})(0.30\,\text{m})^2\right)\left(\frac{2\pi}{0.50\,\text{s}}\right)^2 = 4.73\,\text{N}\cdot\text{m}.$$

Significance

Like the force constant of the system of a block and a spring, the larger the torsion constant, the shorter the period.

15.5 | Damped Oscillations

Learning Objectives

By the end of this section, you will be able to:

- Describe the motion of damped harmonic motion
- Write the equations of motion for damped harmonic oscillations
- Describe the motion of driven, or forced, damped harmonic motion
- Write the equations of motion for forced, damped harmonic motion

In the real world, oscillations seldom follow true SHM. Friction of some sort usually acts to dampen the motion so it dies away, or needs more force to continue. In this section, we examine some examples of damped harmonic motion and see how to modify the equations of motion to describe this more general case.

A guitar string stops oscillating a few seconds after being plucked. To keep swinging on a playground swing, you must keep pushing (**Figure 15.24**). Although we can often make friction and other nonconservative forces small or negligible, completely undamped motion is rare. In fact, we may even want to damp oscillations, such as with car shock absorbers.

Figure 15.24 To counteract dampening forces, you need to keep pumping a swing. (credit: Bob Mical)

Figure 15.25 shows a mass m attached to a spring with a force constant k. The mass is raised to a position A_0, the initial amplitude, and then released. The mass oscillates around the equilibrium position in a fluid with viscosity but the amplitude decreases for each oscillation. For a system that has a small amount of damping, the period and frequency are constant and are nearly the same as for SHM, but the amplitude gradually decreases as shown. This occurs because the non-conservative damping force removes energy from the system, usually in the form of thermal energy.

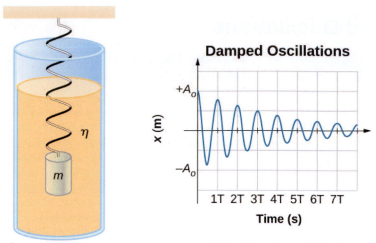

Figure 15.25 For a mass on a spring oscillating in a viscous fluid, the period remains constant, but the amplitudes of the oscillations decrease due to the damping caused by the fluid.

Consider the forces acting on the mass. Note that the only contribution of the weight is to change the equilibrium position, as discussed earlier in the chapter. Therefore, the net force is equal to the force of the spring and the damping force (F_D). If the magnitude of the velocity is small, meaning the mass oscillates slowly, the damping force is proportional to the velocity and acts against the direction of motion $(F_D = -bv)$. The net force on the mass is therefore

$$ma = -bv - kx.$$

Writing this as a differential equation in x, we obtain

$$m\frac{d^2x}{dt^2} + b\frac{dx}{dt} + kx = 0. \tag{15.23}$$

To determine the solution to this equation, consider the plot of position versus time shown in **Figure 15.26**. The curve resembles a cosine curve oscillating in the envelope of an exponential function $A_0 e^{-\alpha t}$ where $\alpha = \frac{b}{2m}$. The solution is

$$x(t) = A_0 e^{-\frac{b}{2m}t} \cos(\omega t + \phi). \tag{15.24}$$

It is left as an exercise to prove that this is, in fact, the solution. To prove that it is the right solution, take the first and second derivatives with respect to time and substitute them into **Equation 15.23**. It is found that **Equation 15.24** is the solution if

$$\omega = \sqrt{\frac{k}{m} - \left(\frac{b}{2m}\right)^2}.$$

Recall that the angular frequency of a mass undergoing SHM is equal to the square root of the force constant divided by the mass. This is often referred to as the **natural angular frequency**, which is represented as

$$\omega_0 = \sqrt{\frac{k}{m}}. \tag{15.25}$$

The angular frequency for damped harmonic motion becomes

$$\omega = \sqrt{\omega_0^2 - \left(\frac{b}{2m}\right)^2}.$$

(15.26)

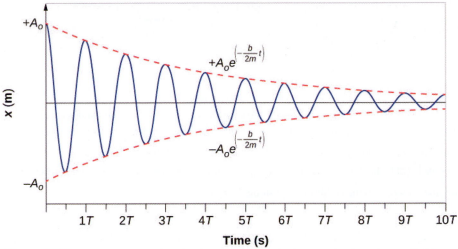

Figure 15.26 Position versus time for the mass oscillating on a spring in a viscous fluid. Notice that the curve appears to be a cosine function inside an exponential envelope.

Recall that when we began this description of damped harmonic motion, we stated that the damping must be small. Two questions come to mind. Why must the damping be small? And how small is small? If you gradually *increase* the amount of damping in a system, the period and frequency begin to be affected, because damping opposes and hence slows the back and forth motion. (The net force is smaller in both directions.) If there is very large damping, the system does not even oscillate—it slowly moves toward equilibrium. The angular frequency is equal to

$$\omega = \sqrt{\frac{k}{m} - \left(\frac{b}{2m}\right)^2}.$$

As b increases, $\frac{k}{m} - \left(\frac{b}{2m}\right)^2$ becomes smaller and eventually reaches zero when $b = \sqrt{4mk}$. If b becomes any larger, $\frac{k}{m} - \left(\frac{b}{2m}\right)^2$ becomes a negative number and $\sqrt{\frac{k}{m} - \left(\frac{b}{2m}\right)^2}$ is a complex number.

Figure 15.27 shows the displacement of a harmonic oscillator for different amounts of damping. When the damping constant is small, $b < \sqrt{4mk}$, the system oscillates while the amplitude of the motion decays exponentially. This system is said to be **underdamped**, as in curve (a). Many systems are underdamped, and oscillate while the amplitude decreases exponentially, such as the mass oscillating on a spring. The damping may be quite small, but eventually the mass comes to rest. If the damping constant is $b = \sqrt{4mk}$, the system is said to be **critically damped**, as in curve (b). An example of a critically damped system is the shock absorbers in a car. It is advantageous to have the oscillations decay as fast as possible. Here, the system does not oscillate, but asymptotically approaches the equilibrium condition as quickly as possible. Curve (c) in **Figure 15.27** represents an **overdamped** system where $b > \sqrt{4mk}$. An overdamped system will approach equilibrium over a longer period of time.

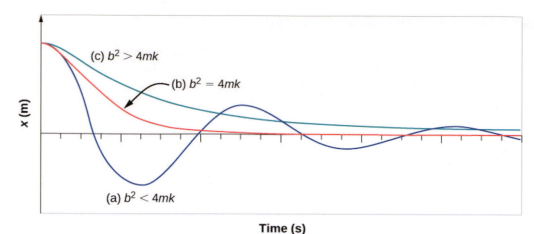

Time (s)

Figure 15.27 The position versus time for three systems consisting of a mass and a spring in a viscous fluid. (a) If the damping is small $\left(b < \sqrt{4mk}\right)$, the mass oscillates, slowly losing amplitude as the energy is dissipated by the non-conservative force(s). The limiting case is (b) where the damping is $\left(b = \sqrt{4mk}\right)$. (c) If the damping is very large $\left(b > \sqrt{4mk}\right)$, the mass does not oscillate when displaced, but attempts to return to the equilibrium position.

Critical damping is often desired, because such a system returns to equilibrium rapidly and remains at equilibrium as well. In addition, a constant force applied to a critically damped system moves the system to a new equilibrium position in the shortest time possible without overshooting or oscillating about the new position.

 15.5 Check Your Understanding Why are completely undamped harmonic oscillators so rare?

15.6 | Forced Oscillations

Learning Objectives
By the end of this section, you will be able to: • Define forced oscillations • List the equations of motion associated with forced oscillations • Explain the concept of resonance and its impact on the amplitude of an oscillator • List the characteristics of a system oscillating in resonance

Sit in front of a piano sometime and sing a loud brief note at it with the dampers off its strings (**Figure 15.28**). It will sing the same note back at you—the strings, having the same frequencies as your voice, are resonating in response to the forces from the sound waves that you sent to them. This is a good example of the fact that objects—in this case, piano strings—can be forced to oscillate, and oscillate most easily at their natural frequency. In this section, we briefly explore applying a periodic driving force acting on a simple harmonic oscillator. The driving force puts energy into the system at a certain frequency, not necessarily the same as the natural frequency of the system. Recall that the natural frequency is the frequency at which a system would oscillate if there were no driving and no damping force.

Figure 15.28 You can cause the strings in a piano to vibrate simply by producing sound waves from your voice. (credit: Matt Billings)

Most of us have played with toys involving an object supported on an elastic band, something like the paddle ball suspended from a finger in **Figure 15.29**. Imagine the finger in the figure is your finger. At first, you hold your finger steady, and the ball bounces up and down with a small amount of damping. If you move your finger up and down slowly, the ball follows along without bouncing much on its own. As you increase the frequency at which you move your finger up and down, the ball responds by oscillating with increasing amplitude. When you drive the ball at its natural frequency, the ball's oscillations increase in amplitude with each oscillation for as long as you drive it. The phenomenon of driving a system with a frequency equal to its natural frequency is called **resonance**. A system being driven at its natural frequency is said to *resonate*. As the driving frequency gets progressively higher than the resonant or natural frequency, the amplitude of the oscillations becomes smaller until the oscillations nearly disappear, and your finger simply moves up and down with little effect on the ball.

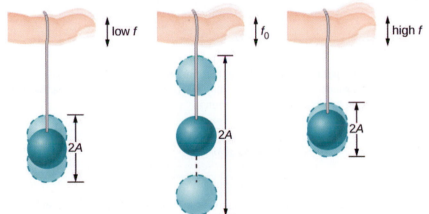

Figure 15.29 The paddle ball on its rubber band moves in response to the finger supporting it. If the finger moves with the natural frequency f_0 of the ball on the

rubber band, then a resonance is achieved, and the amplitude of the ball's oscillations increases dramatically. At higher and lower driving frequencies, energy is transferred to the ball less efficiently, and it responds with lower-amplitude oscillations.

Consider a simple experiment. Attach a mass m to a spring in a viscous fluid, similar to the apparatus discussed in the damped harmonic oscillator. This time, instead of fixing the free end of the spring, attach the free end to a disk that is driven by a variable-speed motor. The motor turns with an angular driving frequency of ω. The rotating disk provides energy to the system by the work done by the driving force $(F_d = F_0 \sin(\omega t))$. The experimental apparatus is shown in **Figure 15.30**.

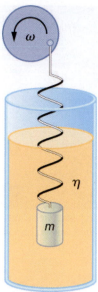

Figure 15.30 Forced, damped harmonic motion produced by driving a spring and mass with a disk driven by a variable-speed motor.

Using Newton's second law ($\overrightarrow{\mathbf{F}}_{net} = m \overrightarrow{\mathbf{a}}$), we can analyze the motion of the mass. The resulting equation is similar to the force equation for the damped harmonic oscillator, with the addition of the driving force:

$$-kx - b\frac{dx}{dt} + F_0 \sin(\omega t) = m\frac{d^2 x}{dt^2}.$$ (15.27)

When an oscillator is forced with a periodic driving force, the motion may seem chaotic. The motions of the oscillator is known as transients. After the transients die out, the oscillator reaches a steady state, where the motion is periodic. After some time, the steady state solution to this differential equation is

$$x(t) = A\cos(\omega t + \phi).$$ (15.28)

Once again, it is left as an exercise to prove that this equation is a solution. Taking the first and second time derivative of $x(t)$ and substituting them into the force equation shows that $x(t) = A\sin(\omega t + \phi)$ is a solution as long as the amplitude is equal to

$$A = \frac{F_0}{\sqrt{m^2\left(\omega^2 - \omega_0^2\right)^2 + b^2 \omega^2}}$$ (15.29)

where $\omega_0 = \sqrt{\frac{k}{m}}$ is the natural angular frequency of the system of the mass and spring. Recall that the angular frequency, and therefore the frequency, of the motor can be adjusted. Looking at the denominator of the equation for the amplitude, when the driving frequency is much smaller, or much larger, than the natural frequency, the square of the difference of the two angular frequencies $\left(\omega^2 - \omega_0^2\right)^2$ is positive and large, making the denominator large, and the result is a small amplitude for the oscillations of the mass. As the frequency of the driving force approaches the natural frequency of the system, the denominator becomes small and the amplitude of the oscillations becomes large. The maximum amplitude results when the frequency of the driving force equals the natural frequency of the system $\left(A_{\max} = \frac{F_0}{b\omega}\right)$.

Figure 15.31 shows a graph of the amplitude of a damped harmonic oscillator as a function of the frequency of the periodic force driving it. Each of the three curves on the graph represents a different amount of damping. All three curves peak at the point where the frequency of the driving force equals the natural frequency of the harmonic oscillator. The highest peak, or greatest response, is for the least amount of damping, because less energy is removed by the damping force. Note that since the amplitude grows as the damping decreases, taking this to the limit where there is no damping $(b = 0)$, the amplitude becomes infinite.

Note that a small-amplitude driving force can produce a large-amplitude response. This phenomenon is known as resonance. A common example of resonance is a parent pushing a small child on a swing. When the child wants to go higher, the parent does not move back and then, getting a running start, slam into the child, applying a great force in a short interval. Instead, the parent applies small pushes to the child at just the right frequency, and the amplitude of the child's swings increases.

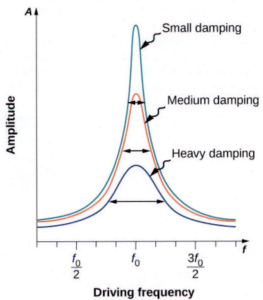

Figure 15.31 Amplitude of a harmonic oscillator as a function of the frequency of the driving force. The curves represent the same oscillator with the same natural frequency but with different amounts of damping. Resonance occurs when the driving frequency equals the natural frequency, and the greatest response is for the least amount of damping. The narrowest response is also for the least damping.

It is interesting to note that the widths of the resonance curves shown in **Figure 15.31** depend on damping: the less the damping, the narrower the resonance. The consequence is that if you want a driven oscillator to resonate at a very specific frequency, you need as little damping as possible. For instance, a radio has a circuit that is used to choose a particular radio station. In this case, the forced damped oscillator consists of a resistor, capacitor, and inductor, which will be discussed later in this course. The circuit is "tuned" to pick a particular radio station. Here it is desirable to have the resonance curve be very narrow, to pick out the exact frequency of the radio station chosen. The narrowness of the graph, and the ability to pick out a certain frequency, is known as the quality of the system. The quality is defined as the spread of the angular frequency, or equivalently, the spread in the frequency, at half the maximum amplitude, divided by the natural frequency $\left(Q = \frac{\Delta\omega}{\omega_0}\right)$

as shown in **Figure 15.32**. For a small damping, the quality is approximately equal to $Q \approx \frac{2b}{m}$.

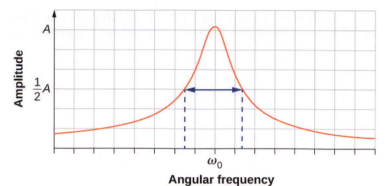

Figure 15.32 The quality of a system is defined as the spread in the frequencies at half the amplitude divided by the natural frequency.

These features of driven harmonic oscillators apply to a huge variety of systems. For instance, magnetic resonance imaging (MRI) is a widely used medical diagnostic tool in which atomic nuclei (mostly hydrogen nuclei or protons) are made to resonate by incoming radio waves (on the order of 100 MHz). In all of these cases, the efficiency of energy transfer from the driving force into the oscillator is best at resonance. **Figure 15.33** shows a photograph of a famous example (the Tacoma Narrows bridge) of the destructive effects of a driven harmonic oscillation. The Millennium bridge in London was closed for a short period of time for the same reason while inspections were carried out. Observations lead to modifications being made to the bridge prior to the reopening.

Figure 15.33 In 1940, the Tacoma Narrows bridge in the state of Washington collapsed. Moderately high, variable cross-winds (much slower than hurricane force winds) drove the bridge into oscillations at its resonant frequency. Damping decreased when support cables broke loose and started to slip over the towers, allowing increasingly greater amplitudes until the structure failed. (credit: "PRI's Studio 360"/Flickr)

 15.6 **Check Your Understanding** A famous magic trick involves a performer singing a note toward a crystal glass until the glass shatters. Explain why the trick works in terms of resonance and natural frequency.

CHAPTER 15 REVIEW

KEY TERMS

amplitude (A) maximum displacement from the equilibrium position of an object oscillating around the equilibrium position

critically damped condition in which the damping of an oscillator causes it to return as quickly as possible to its equilibrium position without oscillating back and forth about this position

elastic potential energy potential energy stored as a result of deformation of an elastic object, such as the stretching of a spring

equilibrium position position where the spring is neither stretched nor compressed

force constant (k) characteristic of a spring which is defined as the ratio of the force applied to the spring to the displacement caused by the force

frequency (f) number of events per unit of time

natural angular frequency angular frequency of a system oscillating in SHM

oscillation single fluctuation of a quantity, or repeated and regular fluctuations of a quantity, between two extreme values around an equilibrium or average value

overdamped condition in which damping of an oscillator causes it to return to equilibrium without oscillating; oscillator moves more slowly toward equilibrium than in the critically damped system

period (T) time taken to complete one oscillation

periodic motion motion that repeats itself at regular time intervals

phase shift angle, in radians, that is used in a cosine or sine function to shift the function left or right, used to match up the function with the initial conditions of data

physical pendulum any extended object that swings like a pendulum

resonance large amplitude oscillations in a system produced by a small amplitude driving force, which has a frequency equal to the natural frequency

restoring force force acting in opposition to the force caused by a deformation

simple harmonic motion (SHM) oscillatory motion in a system where the restoring force is proportional to the displacement, which acts in the direction opposite to the displacement

simple harmonic oscillator a device that oscillates in SHM where the restoring force is proportional to the displacement and acts in the direction opposite to the displacement

simple pendulum point mass, called a pendulum bob, attached to a near massless string

stable equilibrium point point where the net force on a system is zero, but a small displacement of the mass will cause a restoring force that points toward the equilibrium point

torsional pendulum any suspended object that oscillates by twisting its suspension

underdamped condition in which damping of an oscillator causes the amplitude of oscillations of a damped harmonic oscillator to decrease over time, eventually approaching zero

KEY EQUATIONS

Relationship between frequency and period	$f = \frac{1}{T}$
Position in SHM with $\phi = 0.00$	$x(t) = A\cos(\omega t)$
General position in SHM	$x(t) = A\cos(\omega t + \phi)$

General velocity in SHM	$v(t) = -A\omega\sin(\omega t + \phi)$
General acceleration in SHM	$a(t) = -A\omega^2\cos(\omega t + \phi)$
Maximum displacement (amplitude) of SHM	$x_{max} = A$
Maximum velocity of SHM	$|v_{max}| = A\omega$
Maximum acceleration of SHM	$|a_{max}| = A\omega^2$
Angular frequency of a mass-spring system in SHM	$\omega = \sqrt{\frac{k}{m}}$
Period of a mass-spring system in SHM	$T = 2\pi\sqrt{\frac{m}{k}}$
Frequency of a mass-spring system in SHM	$f = \frac{1}{2\pi}\sqrt{\frac{k}{m}}$
Energy in a mass-spring system in SHM	$E_{Total} = \frac{1}{2}kx^2 + \frac{1}{2}mv^2 = \frac{1}{2}kA^2$
The velocity of the mass in a spring-mass system in SHM	$v = \pm\sqrt{\frac{k}{m}(A^2 - x^2)}$
The x-component of the radius of a rotating disk	$x(t) = A\cos(\omega t + \phi)$
The x-component of the velocity of the edge of a rotating disk	$v(t) = -v_{max}\sin(\omega t + \phi)$
The x-component of the acceleration of the edge of a rotating disk	$a(t) = -a_{max}\cos(\omega t + \phi)$
Force equation for a simple pendulum	$\frac{d^2\theta}{dt^2} = -\frac{g}{L}\theta$
Angular frequency for a simple pendulum	$\omega = \sqrt{\frac{g}{L}}$
Period of a simple pendulum	$T = 2\pi\sqrt{\frac{L}{g}}$
Angular frequency of a physical pendulum	$\omega = \sqrt{\frac{mgL}{I}}$
Period of a physical pendulum	$T = 2\pi\sqrt{\frac{I}{mgL}}$
Period of a torsional pendulum	$T = 2\pi\sqrt{\frac{I}{\kappa}}$
Newton's second law for harmonic motion	$m\frac{d^2x}{dt^2} + b\frac{dx}{dt} + kx = 0$
Solution for underdamped harmonic motion	$x(t) = A_0 e^{-\frac{b}{2m}t}\cos(\omega t + \phi)$
Natural angular frequency of a mass-spring system	$\omega_0 = \sqrt{\frac{k}{m}}$
Angular frequency of underdamped harmonic motion	$\omega = \sqrt{\omega_0^2 - \left(\frac{b}{2m}\right)^2}$

Newton's second law for forced, damped oscillation	$-kx - b\frac{dx}{dt} + F_o \sin(\omega t) = m\frac{d^2 x}{dt^2}$
Solution to Newton's second law for forced, damped oscillations	$x(t) = A\cos(\omega t + \phi)$
Amplitude of system undergoing forced, damped oscillations	$A = \dfrac{F_o}{\sqrt{m^2\left(\omega^2 - \omega_o^2\right)^2 + b^2 \omega^2}}$

SUMMARY

15.1 Simple Harmonic Motion

- Periodic motion is a repeating oscillation. The time for one oscillation is the period T and the number of oscillations per unit time is the frequency f. These quantities are related by $f = \frac{1}{T}$.

- Simple harmonic motion (SHM) is oscillatory motion for a system where the restoring force is proportional to the displacement and acts in the direction opposite to the displacement.

- Maximum displacement is the amplitude A. The angular frequency ω, period T, and frequency f of a simple harmonic oscillator are given by $\omega = \sqrt{\frac{k}{m}}$, $T = 2\pi\sqrt{\frac{m}{k}}$, and $f = \frac{1}{2\pi}\sqrt{\frac{k}{m}}$, where m is the mass of the system and k is the force constant.

- Displacement as a function of time in SHM is given by $x(t) = A\cos\left(\frac{2\pi}{T}t + \phi\right) = A\cos(\omega t + \phi)$.

- The velocity is given by $v(t) = -A\omega\sin(\omega t + \phi) = -v_{\max}\sin(\omega t + \phi)$, where $v_{\max} = A\omega = A\sqrt{\frac{k}{m}}$.

- The acceleration is $a(t) = -A\omega^2\cos(\omega t + \phi) = -a_{\max}\cos(\omega t + \phi)$, where $a_{\max} = A\omega^2 = A\frac{k}{m}$.

15.2 Energy in Simple Harmonic Motion

- The simplest type of oscillations are related to systems that can be described by Hooke's law, $F = -kx$, where F is the restoring force, x is the displacement from equilibrium or deformation, and k is the force constant of the system.

- Elastic potential energy U stored in the deformation of a system that can be described by Hooke's law is given by $U = \frac{1}{2}kx^2$.

- Energy in the simple harmonic oscillator is shared between elastic potential energy and kinetic energy, with the total being constant:

$$E_{\text{Total}} = \frac{1}{2}mv^2 + \frac{1}{2}kx^2 = \frac{1}{2}kA^2 = \text{constant}.$$

- The magnitude of the velocity as a function of position for the simple harmonic oscillator can be found by using

$$|v| = \sqrt{\frac{k}{m}\left(A^2 - x^2\right)}.$$

15.3 Comparing Simple Harmonic Motion and Circular Motion

- A projection of uniform circular motion undergoes simple harmonic oscillation.

- Consider a circle with a radius A, moving at a constant angular speed ω. A point on the edge of the circle moves at a constant tangential speed of $v_{\max} = A\omega$. The projection of the radius onto the x-axis is $x(t) = A\cos(\omega t + \phi)$, where (ϕ) is the phase shift. The x-component of the tangential velocity is $v(t) = -A\omega\sin(\omega t + \phi)$.

15.4 Pendulums

- A mass m suspended by a wire of length L and negligible mass is a simple pendulum and undergoes SHM for amplitudes less than about $15°$. The period of a simple pendulum is $T = 2\pi\sqrt{\frac{L}{g}}$, where L is the length of the string and g is the acceleration due to gravity.

- The period of a physical pendulum $T = 2\pi\sqrt{\frac{I}{mgL}}$ can be found if the moment of inertia is known. The length between the point of rotation and the center of mass is L.

- The period of a torsional pendulum $T = 2\pi\sqrt{\frac{I}{\kappa}}$ can be found if the moment of inertia and torsion constant are known.

15.5 Damped Oscillations

- Damped harmonic oscillators have non-conservative forces that dissipate their energy.

- Critical damping returns the system to equilibrium as fast as possible without overshooting.

- An underdamped system will oscillate through the equilibrium position.

- An overdamped system moves more slowly toward equilibrium than one that is critically damped.

15.6 Forced Oscillations

- A system's natural frequency is the frequency at which the system oscillates if not affected by driving or damping forces.

- A periodic force driving a harmonic oscillator at its natural frequency produces resonance. The system is said to resonate.

- The less damping a system has, the higher the amplitude of the forced oscillations near resonance. The more damping a system has, the broader response it has to varying driving frequencies.

CONCEPTUAL QUESTIONS

15.1 Simple Harmonic Motion

1. What conditions must be met to produce SHM?

2. (a) If frequency is not constant for some oscillation, can the oscillation be SHM? (b) Can you think of any examples of harmonic motion where the frequency may depend on the amplitude?

3. Give an example of a simple harmonic oscillator, specifically noting how its frequency is independent of amplitude.

4. Explain why you expect an object made of a stiff material to vibrate at a higher frequency than a similar object made of a more pliable material.

5. As you pass a freight truck with a trailer on a highway, you notice that its trailer is bouncing up and down slowly. Is it more likely that the trailer is heavily loaded or nearly empty? Explain your answer.

6. Some people modify cars to be much closer to the ground than when manufactured. Should they install stiffer springs? Explain your answer.

15.2 Energy in Simple Harmonic Motion

7. Describe a system in which elastic potential energy is stored.

8. Explain in terms of energy how dissipative forces such as friction reduce the amplitude of a harmonic oscillator. Also explain how a driving mechanism can compensate. (A pendulum clock is such a system.)

9. The temperature of the atmosphere oscillates from a maximum near noontime and a minimum near sunrise. Would you consider the atmosphere to be in stable or unstable equilibrium?

15.3 Comparing Simple Harmonic Motion and Circular Motion

10. Can this analogy of SHM to circular motion be carried out with an object oscillating on a spring vertically hung from the ceiling? Why or why not? If given the choice, would you prefer to use a sine function or a cosine function to model the motion?

11. If the maximum speed of the mass attached to a spring, oscillating on a frictionless table, was increased, what characteristics of the rotating disk would need to be changed?

15.4 Pendulums

12. Pendulum clocks are made to run at the correct rate by adjusting the pendulum's length. Suppose you move from one city to another where the acceleration due to gravity is slightly greater, taking your pendulum clock with you, will you have to lengthen or shorten the pendulum to keep the correct time, other factors remaining constant? Explain your answer.

13. A pendulum clock works by measuring the period of a pendulum. In the springtime the clock runs with perfect time, but in the summer and winter the length of the pendulum changes. When most materials are heated, they expand. Does the clock run too fast or too slow in the summer? What about the winter?

14. With the use of a phase shift, the position of an object may be modeled as a cosine or sine function. If given the option, which function would you choose? Assuming that the phase shift is zero, what are the initial conditions of function; that is, the initial position, velocity, and acceleration, when using a sine function? How about when a cosine function is used?

15.5 Damped Oscillations

15. Give an example of a damped harmonic oscillator. (They are more common than undamped or simple harmonic oscillators.)

16. How would a car bounce after a bump under each of these conditions?

(a) overdamping

(b) underdamping

(c) critical damping

17. Most harmonic oscillators are damped and, if undriven, eventually come to a stop. Why?

15.6 Forced Oscillations

18. Why are soldiers in general ordered to "route step" (walk out of step) across a bridge?

19. Do you think there is any harmonic motion in the physical world that is not damped harmonic motion? Try to make a list of five examples of undamped harmonic motion and damped harmonic motion. Which list was easier to make?

20. Some engineers use sound to diagnose performance problems with car engines. Occasionally, a part of the engine is designed that resonates at the frequency of the engine. The unwanted oscillations can cause noise that irritates the driver or could lead to the part failing prematurely. In one case, a part was located that had a length L made of a material with a mass M. What can be done to correct this problem?

PROBLEMS

15.1 Simple Harmonic Motion

21. Prove that using $x(t) = A\sin(\omega t + \phi)$ will produce the same results for the period for the oscillations of a mass and a spring. Why do you think the cosine function was chosen?

22. What is the period of 60.0 Hz of electrical power?

23. If your heart rate is 150 beats per minute during strenuous exercise, what is the time per beat in units of seconds?

24. Find the frequency of a tuning fork that takes 2.50×10^{-3} s to complete one oscillation.

25. A stroboscope is set to flash every 8.00×10^{-5} s. What is the frequency of the flashes?

26. A tire has a tread pattern with a crevice every 2.00 cm. Each crevice makes a single vibration as the tire moves. What is the frequency of these vibrations if the car moves at 30.0 m/s?

27. Each piston of an engine makes a sharp sound every other revolution of the engine. (a) How fast is a race car going if its eight-cylinder engine emits a sound of frequency 750 Hz, given that the engine makes 2000 revolutions per kilometer? (b) At how many revolutions per minute is the engine rotating?

28. A type of cuckoo clock keeps time by having a mass

bouncing on a spring, usually something cute like a cherub in a chair. What force constant is needed to produce a period of 0.500 s for a 0.0150-kg mass?

29. A mass m_0 is attached to a spring and hung vertically. The mass is raised a short distance in the vertical direction and released. The mass oscillates with a frequency f_0. If the mass is replaced with a mass nine times as large, and the experiment was repeated, what would be the frequency of the oscillations in terms of f_0 ?

30. A 0.500-kg mass suspended from a spring oscillates with a period of 1.50 s. How much mass must be added to the object to change the period to 2.00 s?

31. By how much leeway (both percentage and mass) would you have in the selection of the mass of the object in the previous problem if you did not wish the new period to be greater than 2.01 s or less than 1.99 s?

15.2 Energy in Simple Harmonic Motion

32. Fish are hung on a spring scale to determine their mass. (a) What is the force constant of the spring in such a scale if it the spring stretches 8.00 cm for a 10.0 kg load? (b) What is the mass of a fish that stretches the spring 5.50 cm? (c) How far apart are the half-kilogram marks on the scale?

33. It is weigh-in time for the local under-85-kg rugby team. The bathroom scale used to assess eligibility can be described by Hooke's law and is depressed 0.75 cm by its maximum load of 120 kg. (a) What is the spring's effective force constant? (b) A player stands on the scales and depresses it by 0.48 cm. Is he eligible to play on this under-85-kg team?

34. One type of BB gun uses a spring-driven plunger to blow the BB from its barrel. (a) Calculate the force constant of its plunger's spring if you must compress it 0.150 m to drive the 0.0500-kg plunger to a top speed of 20.0 m/s. (b) What force must be exerted to compress the spring?

35. When an 80.0-kg man stands on a pogo stick, the spring is compressed 0.120 m. (a) What is the force constant of the spring? (b) Will the spring be compressed more when he hops down the road?

36. A spring has a length of 0.200 m when a 0.300-kg mass hangs from it, and a length of 0.750 m when a 1.95-kg mass hangs from it. (a) What is the force constant of the spring? (b) What is the unloaded length of the spring?

37. The length of nylon rope from which a mountain climber is suspended has an effective force constant of

1.40×10^4 N/m . (a) What is the frequency at which he bounces, given his mass plus and the mass of his equipment are 90.0 kg? (b) How much would this rope stretch to break the climber's fall if he free-falls 2.00 m before the rope runs out of slack? (*Hint:* Use conservation of energy.) (c) Repeat both parts of this problem in the situation where twice this length of nylon rope is used.

15.3 Comparing Simple Harmonic Motion and Circular Motion

38. The motion of a mass on a spring hung vertically, where the mass oscillates up and down, can also be modeled using the rotating disk. Instead of the lights being placed horizontally along the top and pointing down, place the lights vertically and have the lights shine on the side of the rotating disk. A shadow will be produced on a nearby wall, and will move up and down. Write the equations of motion for the shadow taking the position at $t = 0.0$ s to be $y = 0.0$ m with the mass moving in the positive y-direction.

39. (a) A novelty clock has a 0.0100-kg-mass object bouncing on a spring that has a force constant of 1.25 N/m. What is the maximum velocity of the object if the object bounces 3.00 cm above and below its equilibrium position? (b) How many joules of kinetic energy does the object have at its maximum velocity?

40. Reciprocating motion uses the rotation of a motor to produce linear motion up and down or back and forth. This is how a reciprocating saw operates, as shown below.

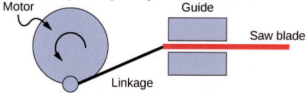

If the motor rotates at 60 Hz and has a radius of 3.0 cm, estimate the maximum speed of the saw blade as it moves up and down. This design is known as a scotch yoke.

41. A student stands on the edge of a merry-go-round which rotates five times a minute and has a radius of two meters one evening as the sun is setting. The student produces a shadow on the nearby building. (a) Write an equation for the position of the shadow. (b) Write an equation for the velocity of the shadow.

15.4 Pendulums

42. What is the length of a pendulum that has a period of 0.500 s?

43. Some people think a pendulum with a period of 1.00

s can be driven with "mental energy" or psycho kinetically, because its period is the same as an average heartbeat. True or not, what is the length of such a pendulum?

44. What is the period of a 1.00-m-long pendulum?

45. How long does it take a child on a swing to complete one swing if her center of gravity is 4.00 m below the pivot?

46. The pendulum on a cuckoo clock is 5.00-cm long. What is its frequency?

47. Two parakeets sit on a swing with their combined CMs 10.0 cm below the pivot. At what frequency do they swing?

48. (a) A pendulum that has a period of 3.00000 s and that is located where the acceleration due to gravity is 9.79 m/s^2 is moved to a location where the acceleration due to gravity is 9.82 m/s^2. What is its new period? (b) Explain why so many digits are needed in the value for the period, based on the relation between the period and the acceleration due to gravity.

49. A pendulum with a period of 2.00000 s in one location ($g = 9.80 \text{m/s}^2$) is moved to a new location where the period is now 1.99796 s. What is the acceleration due to gravity at its new location?

50. (a) What is the effect on the period of a pendulum if you double its length? (b) What is the effect on the period of a pendulum if you decrease its length by 5.00%?

15.5 Damped Oscillations

51. The amplitude of a lightly damped oscillator decreases by 3.0% during each cycle. What percentage of the mechanical energy of the oscillator is lost in each cycle?

15.6 Forced Oscillations

52. How much energy must the shock absorbers of a 1200-kg car dissipate in order to damp a bounce that initially has a velocity of 0.800 m/s at the equilibrium position? Assume the car returns to its original vertical position.

53. If a car has a suspension system with a force constant of $5.00 \times 10^4 \text{ N/m}$, how much energy must the car's shocks remove to dampen an oscillation starting with a maximum displacement of 0.0750 m?

54. (a) How much will a spring that has a force constant of 40.0 N/m be stretched by an object with a mass of 0.500 kg when hung motionless from the spring? (b) Calculate the decrease in gravitational potential energy of the 0.500-kg object when it descends this distance. (c) Part of this gravitational energy goes into the spring. Calculate the energy stored in the spring by this stretch, and compare it with the gravitational potential energy. Explain where the rest of the energy might go.

55. Suppose you have a 0.750-kg object on a horizontal surface connected to a spring that has a force constant of 150 N/m. There is simple friction between the object and surface with a static coefficient of friction $\mu_s = 0.100$.

(a) How far can the spring be stretched without moving the mass? (b) If the object is set into oscillation with an amplitude twice the distance found in part (a), and the kinetic coefficient of friction is $\mu_k = 0.0850$, what total distance does it travel before stopping? Assume it starts at the maximum amplitude.

ADDITIONAL PROBLEMS

56. Suppose you attach an object with mass m to a vertical spring originally at rest, and let it bounce up and down. You release the object from rest at the spring's original rest length, the length of the spring in equilibrium, without the mass attached. The amplitude of the motion is the distance between the equilibrium position of the spring without the mass attached and the equilibrium position of the spring with the mass attached. (a) Show that the spring exerts an upward force of $2.00mg$ on the object at its lowest point. (b) If the spring has a force constant of 10.0 N/m, is hung horizontally, and the position of the free end of the spring is marked as $y = 0.00 \text{ m}$, where is the new equilibrium position if a 0.25-kg-mass object is hung from the spring?

(c) If the spring has a force constant of 10.0 M/m and a 0.25-kg-mass object is set in motion as described, find the amplitude of the oscillations. (d) Find the maximum velocity.

57. A diver on a diving board is undergoing SHM. Her mass is 55.0 kg and the period of her motion is 0.800 s. The next diver is a male whose period of simple harmonic oscillation is 1.05 s. What is his mass if the mass of the board is negligible?

58. Suppose a diving board with no one on it bounces up and down in a SHM with a frequency of 4.00 Hz. The board has an effective mass of 10.0 kg. What is the frequency of the SHM of a 75.0-kg diver on the board?

59. The device pictured in the following figure entertains infants while keeping them from wandering. The child bounces in a harness suspended from a door frame by a spring. (a) If the spring stretches 0.250 m while supporting an 8.0-kg child, what is its force constant? (b) What is the time for one complete bounce of this child? (c) What is the child's maximum velocity if the amplitude of her bounce is 0.200 m?

Figure 15.34 (credit: Lisa Doehnert)

60. A mass is placed on a frictionless, horizontal table. A spring $(k = 100 \, \text{N/m})$, which can be stretched or compressed, is placed on the table. A 5.00-kg mass is attached to one end of the spring, the other end is anchored to the wall. The equilibrium position is marked at zero. A student moves the mass out to $x = 4.0 \text{cm}$ and releases it from rest. The mass oscillates in SHM. (a) Determine the equations of motion. (b) Find the position, velocity, and acceleration of the mass at time $t = 3.00 \, \text{s}$.

61. Find the ratio of the new/old periods of a pendulum if the pendulum were transported from Earth to the Moon, where the acceleration due to gravity is $1.63 \, \text{m/s}^2$.

62. At what rate will a pendulum clock run on the Moon, where the acceleration due to gravity is $1.63 \, \text{m/s}^2$, if it keeps time accurately on Earth? That is, find the time (in hours) it takes the clock's hour hand to make one revolution on the Moon.

63. If a pendulum-driven clock gains 5.00 s/day, what fractional change in pendulum length must be made for it to keep perfect time?

64. A 2.00-kg object hangs, at rest, on a 1.00-m-long string attached to the ceiling. A 100-g mass is fired with a speed of 20 m/s at the 2.00-kg mass, and the 100.00-g mass collides perfectly elastically with the 2.00-kg mass. Write an equation for the motion of the hanging mass after the collision. Assume air resistance is negligible.

65. A 2.00-kg object hangs, at rest, on a 1.00-m-long string attached to the ceiling. A 100-g object is fired with a speed of 20 m/s at the 2.00-kg object, and the two objects collide and stick together in a totally inelastic collision. Write an equation for the motion of the system after the collision. Assume air resistance is negligible.

66. Assume that a pendulum used to drive a grandfather clock has a length $L_0 = 1.00 \, \text{m}$ and a mass M at temperature $T = 20.00°\text{C}$. It can be modeled as a physical pendulum as a rod oscillating around one end. By what percentage will the period change if the temperature increases by $10°\text{C}$? Assume the length of the rod changes linearly with temperature, where $L = L_0(1 + \alpha \Delta T)$ and the rod is made of brass $\left(\alpha = 18 \times 10^{-6}°\text{C}^{-1}\right)$.

67. A 2.00-kg block lies at rest on a frictionless table. A spring, with a spring constant of 100 N/m is attached to the wall and to the block. A second block of 0.50 kg is placed on top of the first block. The 2.00-kg block is gently pulled to a position $x = +A$ and released from rest. There is a coefficient of friction of 0.45 between the two blocks. (a) What is the period of the oscillations? (b) What is the largest amplitude of motion that will allow the blocks to oscillate without the 0.50-kg block sliding off?

CHALLENGE PROBLEMS

68. A suspension bridge oscillates with an effective force constant of 1.00×10^8 N/m. (a) How much energy is needed to make it oscillate with an amplitude of 0.100 m? (b) If soldiers march across the bridge with a cadence equal to the bridge's natural frequency and impart 1.00×10^4 J of energy each second, how long does it take for the bridge's oscillations to go from 0.100 m to 0.500 m amplitude.

69. Near the top of the Citigroup Center building in New York City, there is an object with mass of 4.00×10^5 kg on springs that have adjustable force constants. Its function is to dampen wind-driven oscillations of the building by oscillating at the same frequency as the building is being driven—the driving force is transferred to the object, which oscillates instead of the entire building. (a) What effective force constant should the springs have to make the object oscillate with a period of 2.00 s? (b) What energy is stored in the springs for a 2.00-m displacement from equilibrium?

70. Parcels of air (small volumes of air) in a stable atmosphere (where the temperature increases with height) can oscillate up and down, due to the restoring force provided by the buoyancy of the air parcel. The frequency of the oscillations are a measure of the stability of the atmosphere. Assuming that the acceleration of an air parcel can be modeled as $\dfrac{\partial^2 z'}{\partial t^2} = \dfrac{g}{\rho_o}\dfrac{\partial \rho(z)}{\partial z}z'$, prove that $z' = z_0' e^{t\sqrt{-N^2}}$ is a solution, where N is known as the Brunt-Väisälä frequency. Note that in a stable atmosphere, the density decreases with height and parcel oscillates up and down.

71. Consider the van der Waals potential $U(r) = U_o\left[\left(\dfrac{R_o}{r}\right)^{12} - 2\left(\dfrac{R_o}{r}\right)^{6}\right]$, used to model the potential energy function of two molecules, where the minimum potential is at $r = R_o$. Find the force as a function of r. Consider a small displacement $r = R_o + r'$ and use the binomial theorem:

$$(1 + x)^n = 1 + nx + \frac{n(n-1)}{2!}x^2 + \frac{n(n-1)(n-2)}{3!}x^3 + \cdots$$

,

to show that the force does approximate a Hooke's law force.

72. Suppose the length of a clock's pendulum is changed by 1.000%, exactly at noon one day. What time will the clock read 24.00 hours later, assuming it the pendulum has kept perfect time before the change? Note that there are two answers, and perform the calculation to four-digit precision.

73. (a) The springs of a pickup truck act like a single spring with a force constant of 1.30×10^5 N/m. By how much will the truck be depressed by its maximum load of 1000 kg? (b) If the pickup truck has four identical springs, what is the force constant of each?

16 | WAVES

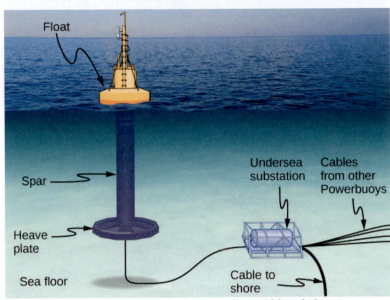

Figure 16.1 From the world of renewable energy sources comes the electric power-generating buoy. Although there are many versions, this one converts the up-and-down motion, as well as side-to-side motion, of the buoy into rotational motion in order to turn an electric generator, which stores the energy in batteries.

Chapter Outline

16.1 Traveling Waves

16.2 Mathematics of Waves

16.3 Wave Speed on a Stretched String

16.4 Energy and Power of a Wave

16.5 Interference of Waves

16.6 Standing Waves and Resonance

Introduction

In this chapter, we study the physics of wave motion. We concentrate on mechanical waves, which are disturbances that move through a medium such as air or water. Like simple harmonic motion studied in the preceding chapter, the energy transferred through the medium is proportional to the amplitude squared. Surface water waves in the ocean are transverse waves in which the energy of the wave travels horizontally while the water oscillates up and down due to some restoring force. In the picture above, a buoy is used to convert the awesome power of ocean waves into electricity. The up-and-down motion of the buoy generated as the waves pass is converted into rotational motion that turns a rotor in an electric generator. The generator charges batteries, which are in turn used to provide a consistent energy source for the end user. This model was successfully tested by the US Navy in a project to provide power to coastal security networks and was able to provide an average power of 350 W. The buoy survived the difficult ocean environment, including operation off the New Jersey coast through Hurricane Irene in 2011.

The concepts presented in this chapter will be the foundation for many interesting topics, from the transmission of information to the concepts of quantum mechanics.

16.1 | Traveling Waves

<table>
<tr><td>
Learning Objectives

By the end of this section, you will be able to:

* Describe the basic characteristics of wave motion
* Define the terms wavelength, amplitude, period, frequency, and wave speed
* Explain the difference between longitudinal and transverse waves, and give examples of each type
* List the different types of waves
</td></tr>
</table>

We saw in **Oscillations** that oscillatory motion is an important type of behavior that can be used to model a wide range of physical phenomena. Oscillatory motion is also important because oscillations can generate waves, which are of fundamental importance in physics. Many of the terms and equations we studied in the chapter on oscillations apply equally well to wave motion (**Figure 16.2**).

Figure 16.2 An ocean wave is probably the first picture that comes to mind when you hear the word "wave." Although this breaking wave, and ocean waves in general, have apparent similarities to the basic wave characteristics we will discuss, the mechanisms driving ocean waves are highly complex and beyond the scope of this chapter. It may seem natural, and even advantageous, to apply the concepts in this chapter to ocean waves, but ocean waves are nonlinear, and the simple models presented in this chapter do not fully explain them. (credit: Steve Jurvetson)

Types of Waves

A **wave** is a disturbance that propagates, or moves from the place it was created. There are three basic types of waves: mechanical waves, electromagnetic waves, and matter waves.

Basic **mechanical waves** are governed by Newton's laws and require a medium. A medium is the substance a mechanical waves propagates through, and the medium produces an elastic restoring force when it is deformed. Mechanical waves transfer energy and momentum, without transferring mass. Some examples of mechanical waves are water waves, sound waves, and seismic waves. The medium for water waves is water; for sound waves, the medium is usually air. (Sound waves can travel in other media as well; we will look at that in more detail in **Sound**.) For surface water waves, the disturbance occurs on the surface of the water, perhaps created by a rock thrown into a pond or by a swimmer splashing the surface repeatedly. For sound waves, the disturbance is a change in air pressure, perhaps created by the oscillating cone inside a speaker or a vibrating tuning fork. In both cases, the disturbance is the oscillation of the molecules of the fluid. In mechanical waves, energy and momentum transfer with the motion of the wave, whereas the mass oscillates around an equilibrium point. (We discuss this in **Energy and Power of a Wave**.) Earthquakes generate seismic waves from several types of disturbances, including the disturbance of Earth's surface and pressure disturbances under the surface. Seismic waves travel through the solids and liquids that form Earth. In this chapter, we focus on mechanical waves.

Electromagnetic waves are associated with oscillations in electric and magnetic fields and do not require a medium. Examples include gamma rays, X-rays, ultraviolet waves, visible light, infrared waves, microwaves, and radio waves. Electromagnetic waves can travel through a vacuum at the speed of light, $v = c = 2.99792458 \times 10^8$ m/s. For example, light from distant stars travels through the vacuum of space and reaches Earth. Electromagnetic waves have some characteristics that are similar to mechanical waves; they are covered in more detail in **Electromagnetic Waves (http://cnx.org/content/m58495/latest/)** .

Matter waves are a central part of the branch of physics known as quantum mechanics. These waves are associated with protons, electrons, neutrons, and other fundamental particles found in nature. The theory that all types of matter have wave-like properties was first proposed by Louis de Broglie in 1924. Matter waves are discussed in **Photons and Matter Waves (http://cnx.org/content/m58757/latest/)** .

Mechanical Waves

Mechanical waves exhibit characteristics common to all waves, such as amplitude, wavelength, period, frequency, and energy. All wave characteristics can be described by a small set of underlying principles.

The simplest mechanical waves repeat themselves for several cycles and are associated with simple harmonic motion. These simple harmonic waves can be modeled using some combination of sine and cosine functions. For example, consider the simplified surface water wave that moves across the surface of water as illustrated in **Figure 16.3**. Unlike complex ocean waves, in surface water waves, the medium, in this case water, moves vertically, oscillating up and down, whereas the disturbance of the wave moves horizontally through the medium. In **Figure 16.3**, the waves causes a seagull to move up and down in simple harmonic motion as the wave crests and troughs (peaks and valleys) pass under the bird. The crest is the highest point of the wave, and the trough is the lowest part of the wave. The time for one complete oscillation of the up-and-down motion is the wave's period T. The wave's frequency is the number of waves that pass through a point per unit time and is equal to $f = 1/T$. The period can be expressed using any convenient unit of time but is usually measured in seconds; frequency is usually measured in hertz (Hz), where $1 \text{ Hz} = 1 \text{ s}^{-1}$.

The length of the wave is called the **wavelength** and is represented by the Greek letter lambda (λ), which is measured in any convenient unit of length, such as a centimeter or meter. The wavelength can be measured between any two similar points along the medium that have the same height and the same slope. In **Figure 16.3**, the wavelength is shown measured between two crests. As stated above, the period of the wave is equal to the time for one oscillation, but it is also equal to the time for one wavelength to pass through a point along the wave's path.

The amplitude of the wave (A) is a measure of the maximum displacement of the medium from its equilibrium position. In the figure, the equilibrium position is indicated by the dotted line, which is the height of the water if there were no waves moving through it. In this case, the wave is symmetrical, the crest of the wave is a distance $+A$ above the equilibrium position, and the trough is a distance $-A$ below the equilibrium position. The units for the amplitude can be centimeters or meters, or any convenient unit of distance.

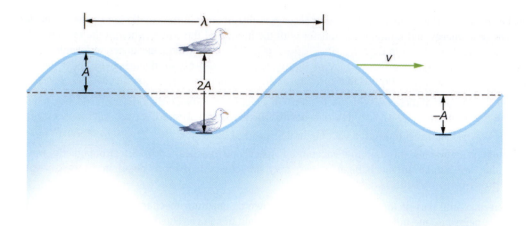

Figure 16.3 An idealized surface water wave passes under a seagull that bobs up and down in simple harmonic motion. The wave has a wavelength λ, which is the distance between adjacent identical parts of the wave. The amplitude A of the wave is the maximum displacement of the wave from the equilibrium position, which is indicated by the dotted line. In this example, the medium moves up and down, whereas the disturbance of the surface propagates parallel to the surface at a speed v.

The water wave in the figure moves through the medium with a propagation velocity $\vec{\mathbf{v}}$. The magnitude of the **wave velocity** is the distance the wave travels in a given time, which is one wavelength in the time of one period, and the **wave speed** is the magnitude of wave velocity. In equation form, this is

$$v = \frac{\lambda}{T} = \lambda f. \tag{16.1}$$

This fundamental relationship holds for all types of waves. For water waves, v is the speed of a surface wave; for sound, v is the speed of sound; and for visible light, v is the speed of light.

Transverse and Longitudinal Waves

We have seen that a simple mechanical wave consists of a periodic disturbance that propagates from one place to another through a medium. In **Figure 16.4**(a), the wave propagates in the horizontal direction, whereas the medium is disturbed in the vertical direction. Such a wave is called a **transverse wave**. In a transverse wave, the wave may propagate in any direction, but the disturbance of the medium is perpendicular to the direction of propagation. In contrast, in a **longitudinal wave** or compressional wave, the disturbance is parallel to the direction of propagation. **Figure 16.4**(b) shows an example of a longitudinal wave. The size of the disturbance is its amplitude A and is completely independent of the speed of propagation v.

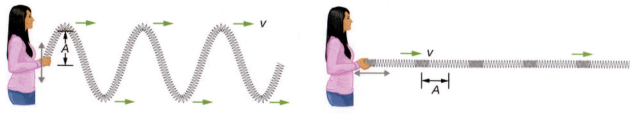

(a) Transverse wave   (b) Longitudinal wave

Figure 16.4 (a) In a transverse wave, the medium oscillates perpendicular to the wave velocity. Here, the spring moves vertically up and down, while the wave propagates horizontally to the right. (b) In a longitudinal wave, the medium oscillates parallel to the propagation of the wave. In this case, the spring oscillates back and forth, while the wave propagates to the right.

A simple graphical representation of a section of the spring shown in **Figure 16.4**(b) is shown in **Figure 16.5**. **Figure 16.5**(a) shows the equilibrium position of the spring before any waves move down it. A point on the spring is marked with a blue dot. **Figure 16.5**(b) through (g) show snapshots of the spring taken one-quarter of a period apart, sometime after the end of` the spring is oscillated back and forth in the x-direction at a constant frequency. The disturbance of the wave is seen as the compressions and the expansions of the spring. Note that the blue dot oscillates around its equilibrium position a distance A, as the longitudinal wave moves in the positive x-direction with a constant speed. The distance A is the amplitude of the wave. The y-position of the dot does not change as the wave moves through the spring. The wavelength of the wave is measured in part (d). The wavelength depends on the speed of the wave and the frequency of the driving force.

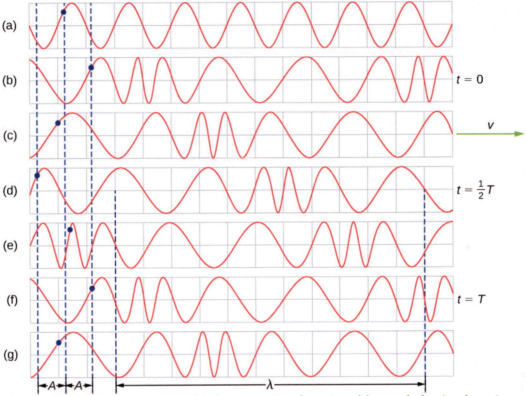

Figure 16.5 (a) This is a simple, graphical representation of a section of the stretched spring shown in **Figure 16.4**(b), representing the spring's equilibrium position before any waves are induced on the spring. A point on the spring is marked by a blue dot. (b–g) Longitudinal waves are created by oscillating the end of the spring (not shown) back and forth along the x-axis. The longitudinal wave, with a wavelength λ, moves along the spring in the $+x$-direction with a wave speed v. For convenience, the wavelength is measured in (d). Note that the point on the spring that was marked with the blue dot moves back and forth a distance A from the equilibrium position, oscillating around the equilibrium position of the point.

Waves may be transverse, longitudinal, or a combination of the two. Examples of transverse waves are the waves on stringed instruments or surface waves on water, such as ripples moving on a pond. Sound waves in air and water are longitudinal. With sound waves, the disturbances are periodic variations in pressure that are transmitted in fluids. Fluids do not have appreciable shear strength, and for this reason, the sound waves in them are longitudinal waves. Sound in solids can have both longitudinal and transverse components, such as those in a seismic wave. Earthquakes generate seismic waves under Earth's surface with both longitudinal and transverse components (called compressional or P-waves and shear or S-waves, respectively). The components of seismic waves have important individual characteristics—they propagate at different speeds, for example. Earthquakes also have surface waves that are similar to surface waves on water. Ocean waves also have both transverse and longitudinal components.

Example 16.1

Wave on a String

A student takes a 30.00-m-long string and attaches one end to the wall in the physics lab. The student then holds the free end of the rope, keeping the tension constant in the rope. The student then begins to send waves down the string by moving the end of the string up and down with a frequency of 2.00 Hz. The maximum displacement of the end of the string is 20.00 cm. The first wave hits the lab wall 6.00 s after it was created. (a) What is the speed of the wave? (b) What is the period of the wave? (c) What is the wavelength of the wave?

Strategy

a. The speed of the wave can be derived by dividing the distance traveled by the time.

b. The period of the wave is the inverse of the frequency of the driving force.

c. The wavelength can be found from the speed and the period $v = \lambda/T$.

Solution

a. The first wave traveled 30.00 m in 6.00 s:

$$v = \frac{30.00 \text{ m}}{6.00 \text{ s}} = 5.00 \frac{\text{m}}{\text{s}}.$$

b. The period is equal to the inverse of the frequency:

$$T = \frac{1}{f} = \frac{1}{2.00 \text{ s}^{-1}} = 0.50 \text{ s}.$$

c. The wavelength is equal to the velocity times the period:

$$\lambda = vT = 5.00 \frac{\text{m}}{\text{s}}(0.50 \text{ s}) = 2.50 \text{ m}.$$

Significance

The frequency of the wave produced by an oscillating driving force is equal to the frequency of the driving force.

 16.1   **Check Your Understanding** When a guitar string is plucked, the guitar string oscillates as a result of waves moving through the string. The vibrations of the string cause the air molecules to oscillate, forming sound waves. The frequency of the sound waves is equal to the frequency of the vibrating string. Is the wavelength of the sound wave always equal to the wavelength of the waves on the string?

Example 16.2

Characteristics of a Wave

A transverse mechanical wave propagates in the positive x-direction through a spring (as shown in **Figure 16.4**(a)) with a constant wave speed, and the medium oscillates between $+A$ and $-A$ around an equilibrium position. The graph in **Figure 16.6** shows the height of the spring (y) versus the position (x), where the x-axis points in the direction of propagation. The figure shows the height of the spring versus the x-position at $t = 0.00 \text{ s}$ as a dotted line and the wave at $t = 3.00 \text{ s}$ as a solid line. (a) Determine the wavelength and amplitude of the wave. (b) Find the propagation velocity of the wave. (c) Calculate the period and frequency of the wave.

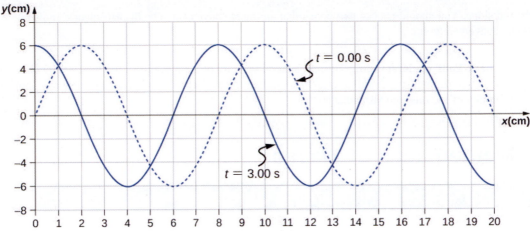

Figure 16.6 A transverse wave shown at two instants of time.

Strategy

a. The amplitude and wavelength can be determined from the graph.

b. Since the velocity is constant, the velocity of the wave can be found by dividing the distance traveled by the wave by the time it took the wave to travel the distance.

c. The period can be found from $v = \frac{\lambda}{T}$ and the frequency from $f = \frac{1}{T}$.

Solution

a. Read the wavelength from the graph, looking at the purple arrow in **Figure 16.7**. Read the amplitude by looking at the green arrow. The wavelength is $\lambda = 8.00 \, \text{cm}$ and the amplitude is $A = 6.00 \, \text{cm}$.

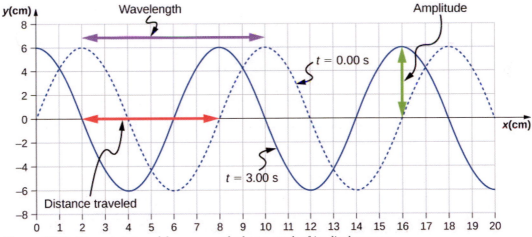

Figure 16.7 Characteristics of the wave marked on a graph of its displacement.

b. The distance the wave traveled from time $t = 0.00 \, \text{s}$ to time $t = 3.00 \, \text{s}$ can be seen in the graph. Consider the red arrow, which shows the distance the crest has moved in 3 s. The distance is $8.00 \, \text{cm} - 2.00 \, \text{cm} = 6.00 \, \text{cm}$. The velocity is

$$v = \frac{\Delta x}{\Delta t} = \frac{8.00 \, \text{cm} - 2.00 \, \text{cm}}{3.00 \, \text{s} - 0.00 \, \text{s}} = 2.00 \, \text{cm/s}.$$

c. The period is $T = \frac{\lambda}{v} = \frac{8.00 \, \text{cm}}{2.00 \, \text{cm/s}} = 4.00 \, \text{s}$ and the frequency is $f = \frac{1}{T} = \frac{1}{4.00 \, \text{s}} = 0.25 \, \text{Hz}$.

Significance

Note that the wavelength can be found using any two successive identical points that repeat, having the same height and slope. You should choose two points that are most convenient. The displacement can also be found using any convenient point.

 16.2   Check Your Understanding The propagation velocity of a transverse or longitudinal mechanical wave may be constant as the wave disturbance moves through the medium. Consider a transverse mechanical wave: Is the velocity of the medium also constant?

16.2 | Mathematics of Waves

Learning Objectives

By the end of this section, you will be able to:

- Model a wave, moving with a constant wave velocity, with a mathematical expression
- Calculate the velocity and acceleration of the medium
- Show how the velocity of the medium differs from the wave velocity (propagation velocity)

In the previous section, we described periodic waves by their characteristics of wavelength, period, amplitude, and wave speed of the wave. Waves can also be described by the motion of the particles of the medium through which the waves move. The position of particles of the medium can be mathematically modeled as **wave functions**, which can be used to find the position, velocity, and acceleration of the particles of the medium of the wave at any time.

Pulses

A **pulse** can be described as wave consisting of a single disturbance that moves through the medium with a constant amplitude. The pulse moves as a pattern that maintains its shape as it propagates with a constant wave speed. Because the wave speed is constant, the distance the pulse moves in a time Δt is equal to $\Delta x = v\Delta t$ (**Figure 16.8**).

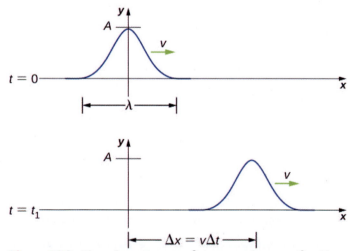

Figure 16.8 The pulse at time $t = 0$ is centered on $x = 0$ with amplitude A. The pulse moves as a pattern with a constant shape, with a constant maximum value A. The velocity is constant and the pulse moves a distance $\Delta x = v\Delta t$ in a time Δt. The distance traveled is measured with any convenient point on the pulse. In this figure, the crest is used.

Modeling a One-Dimensional Sinusoidal Wave using a Wave Function

Consider a string kept at a constant tension F_T where one end is fixed and the free end is oscillated between $y = +A$ and

$y = -A$ by a mechanical device at a constant frequency. **Figure 16.9** shows snapshots of the wave at an interval of an eighth of a period, beginning after one period ($t = T$).

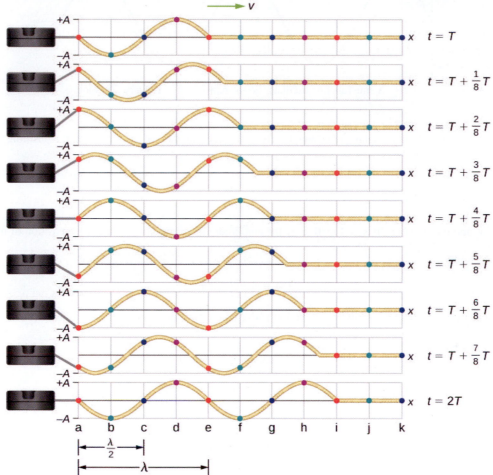

Figure 16.9 Snapshots of a transverse wave moving through a string under tension, beginning at time $t = T$ and taken at intervals of $\frac{1}{8}T$. Colored dots are used to highlight points on the string.

Points that are a wavelength apart in the x-direction are highlighted with the same color dots.

Notice that each select point on the string (marked by colored dots) oscillates up and down in simple harmonic motion, between $y = +A$ and $y = -A$, with a period T. The wave on the string is sinusoidal and is translating in the positive x-direction as time progresses.

At this point, it is useful to recall from your study of algebra that if $f(x)$ is some function, then $f(x - d)$ is the same function translated in the positive x-direction by a distance d. The function $f(x + d)$ is the same function translated in the negative x-direction by a distance d. We want to define a wave function that will give the y-position of each segment of the string for every position x along the string for every time t.

Looking at the first snapshot in **Figure 16.9**, the y-position of the string between $x = 0$ and $x = \lambda$ can be modeled as a sine function. This wave propagates down the string one wavelength in one period, as seen in the last snapshot. The wave therefore moves with a constant wave speed of $v = \lambda/T$.

Recall that a sine function is a function of the angle θ, oscillating between $+1$ and -1, and repeating every 2π radians (**Figure 16.10**). However, the y-position of the medium, or the wave function, oscillates between $+A$ and $-A$, and repeats every wavelength λ.

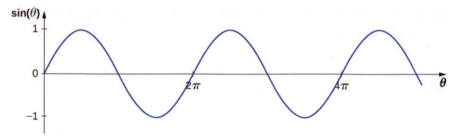

Figure 16.10 A sine function oscillates between $+1$ and -1 every 2π radians.

To construct our model of the wave using a periodic function, consider the ratio of the angle and the position,

$$\frac{\theta}{x} = \frac{2\pi}{\lambda},$$

$$\theta = \frac{2\pi}{\lambda}x.$$

Using $\theta = \frac{2\pi}{\lambda}x$ and multiplying the sine function by the amplitude A, we can now model the y-position of the string as a function of the position x:

$$y(x) = A \sin\left(\frac{2\pi}{\lambda}x\right).$$

The wave on the string travels in the positive x-direction with a constant velocity v, and moves a distance vt in a time t. The wave function can now be defined by

$$y(x, t) = A \sin\left(\frac{2\pi}{\lambda}(x - vt)\right).$$

It is often convenient to rewrite this wave function in a more compact form. Multiplying through by the ratio $\frac{2\pi}{\lambda}$ leads to the equation

$$y(x, t) = A \sin\left(\frac{2\pi}{\lambda}x - \frac{2\pi}{\lambda}vt\right).$$

The value $\frac{2\pi}{\lambda}$ is defined as the **wave number**. The symbol for the wave number is k and has units of inverse meters, m^{-1}:

$$k \equiv \frac{2\pi}{\lambda} \qquad\qquad (16.2)$$

Recall from **Oscillations** that the angular frequency is defined as $\omega \equiv \frac{2\pi}{T}$. The second term of the wave function becomes

$$\frac{2\pi}{\lambda}vt = \frac{2\pi}{\lambda}\left(\frac{\lambda}{T}\right)t = \frac{2\pi}{T}t = \omega t.$$

The wave function for a simple harmonic wave on a string reduces to

$$y(x, t) = A \sin(kx \mp \omega t),$$

where A is the amplitude, $k = \frac{2\pi}{\lambda}$ is the wave number, $\omega = \frac{2\pi}{T}$ is the angular frequency, the minus sign is for waves moving in the positive x-direction, and the plus sign is for waves moving in the negative x-direction. The velocity of the wave is equal to

$$v = \frac{\lambda}{T} = \frac{\lambda}{T}\left(\frac{2\pi}{2\pi}\right) = \frac{\omega}{k}. \qquad\qquad (16.3)$$

Think back to our discussion of a mass on a spring, when the position of the mass was modeled as $x(t) = A\cos(\omega t + \phi)$. The angle ϕ is a phase shift, added to allow for the fact that the mass may have initial conditions other than $x = +A$ and $v = 0$. For similar reasons, the initial phase is added to the wave function. The wave function modeling a sinusoidal wave, allowing for an initial phase shift ϕ, is

$$y(x,\ t) = A\sin(kx \mp \omega t + \phi) \qquad\qquad (16.4)$$

The value

$$(kx \mp \omega t + \phi) \qquad\qquad (16.5)$$

is known as the phase of the wave, where ϕ is the initial phase of the wave function. Whether the temporal term ωt is negative or positive depends on the direction of the wave. First consider the minus sign for a wave with an initial phase equal to zero $(\phi = 0)$. The phase of the wave would be $(kx - \omega t)$. Consider following a point on a wave, such as a crest. A crest will occur when $\sin(kx - \omega t) = 1.00$, that is, when $kx - \omega t = n\pi + \frac{\pi}{2}$, for any integral value of n. For instance, one particular crest occurs at $kx - \omega t = \frac{\pi}{2}$. As the wave moves, time increases and x must also increase to keep the phase equal to $\frac{\pi}{2}$. Therefore, the minus sign is for a wave moving in the positive x-direction. Using the plus sign, $kx + \omega t = \frac{\pi}{2}$. As time increases, x must decrease to keep the phase equal to $\frac{\pi}{2}$. The plus sign is used for waves moving in the negative x-direction. In summary, $y(x,\ t) = A\sin(kx - \omega t + \phi)$ models a wave moving in the positive x-direction and $y(x,\ t) = A\sin(kx + \omega t + \phi)$ models a wave moving in the negative x-direction.

Equation 16.4 is known as a simple harmonic wave function. A wave function is any function such that $f(x,\ t) = f(x - vt)$. Later in this chapter, we will see that it is a solution to the linear wave equation. Note that $y(x,\ t) = A\cos(kx + \omega t + \phi')$ works equally well because it corresponds to a different phase shift $\phi' = \phi - \frac{\pi}{2}$.

Problem-Solving Strategy: Finding the Characteristics of a Sinusoidal Wave

1. To find the amplitude, wavelength, period, and frequency of a sinusoidal wave, write down the wave function in the form $y(x,\ t) = A\sin(kx - \omega t + \phi)$.

2. The amplitude can be read straight from the equation and is equal to A.

3. The period of the wave can be derived from the angular frequency $\left(T = \frac{2\pi}{\omega}\right)$.

4. The frequency can be found using $f = \frac{1}{T}$.

5. The wavelength can be found using the wave number $\left(\lambda = \frac{2\pi}{k}\right)$.

Example 16.3

Characteristics of a Traveling Wave on a String

A transverse wave on a taut string is modeled with the wave function

$$y(x, t) = A \sin(kx - wt) = 0.2 \text{ m } \sin\left(6.28 \text{ m}^{-1}x - 1.57 \text{ s}^{-1}t\right).$$

Find the amplitude, wavelength, period, and speed of the wave.

Strategy

All these characteristics of the wave can be found from the constants included in the equation or from simple combinations of these constants.

Solution

1. The amplitude, wave number, and angular frequency can be read directly from the wave equation:

$$y(x, t) = A \sin(kx - wt) = 0.2 \text{ m } \sin\left(6.28 \text{ m}^{-1}x - 1.57 \text{ s}^{-1}t\right).$$

$$(A = 0.2 \text{ m}; k = 6.28 \text{ m}^{-1}; \omega = 1.57 \text{ s}^{-1})$$

2. The wave number can be used to find the wavelength:

$$k = \frac{2\pi}{\lambda}.$$

$$\lambda = \frac{2\pi}{k} = \frac{2\pi}{6.28 \text{ m}^{-1}} = 1.0 \text{ m}.$$

3. The period of the wave can be found using the angular frequency:

$$\omega = \frac{2\pi}{T}.$$

$$T = \frac{2\pi}{\omega} = \frac{2\pi}{1.57 \text{ s}^{-1}} = 4 \text{ s}.$$

4. The speed of the wave can be found using the wave number and the angular frequency. The direction of the wave can be determined by considering the sign of $kx \mp \omega t$: A negative sign suggests that the wave is moving in the positive x-direction:

$$|v| = \frac{\omega}{k} = \frac{1.57 \text{ s}^{-1}}{6.28 \text{ m}^{-1}} = 0.25 \text{ m/s}.$$

Significance

All of the characteristics of the wave are contained in the wave function. Note that the wave speed is the speed of the wave in the direction parallel to the motion of the wave. Plotting the height of the medium y versus the position x for two times $t = 0.00 \text{ s}$ and $t = 0.80 \text{ s}$ can provide a graphical visualization of the wave (**Figure 16.11**).

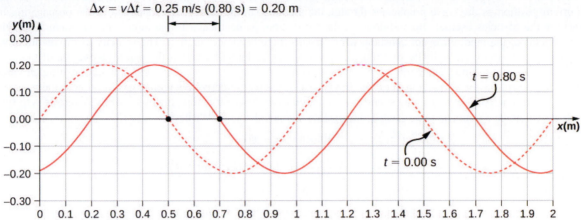

Figure 16.11 A graph of height of the wave y as a function of position x for snapshots of the wave at two times. The dotted line represents the wave at time $t = 0.00$ s and the solid line represents the wave at $t = 0.80$ s. Since the wave velocity is constant, the distance the wave travels is the wave velocity times the time interval. The black dots indicate the points used to measure the displacement of the wave. The medium moves up and down, whereas the wave moves to the right.

There is a second velocity to the motion. In this example, the wave is transverse, moving horizontally as the medium oscillates up and down perpendicular to the direction of motion. The graph in **Figure 16.12** shows the motion of the medium at point $x = 0.60$ m as a function of time. Notice that the medium of the wave oscillates up and down between $y = +0.20$ m and $y = -0.20$ m every period of 4.0 seconds.

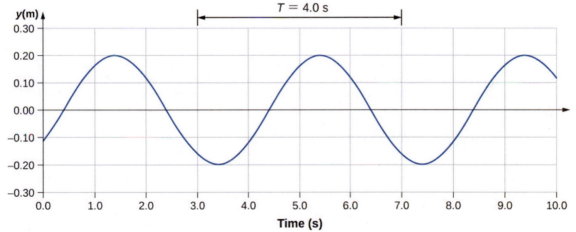

Figure 16.12 A graph of height of the wave y as a function of time t for the position $x = 0.6$ m. The medium oscillates between $y = +0.20$ m and $y = -0.20$ m every period. The period represented picks two convenient points in the oscillations to measure the period. The period can be measured between any two adjacent points with the same amplitude and the same velocity, $(\partial y/\partial t)$. The velocity can be found by looking at the slope tangent to the point on a y-versus-t plot. Notice that at times $t = 3.00$ s and $t = 7.00s$, the heights and the velocities are the same and the period of the oscillation is 4.00 s.

 16.3   Check Your Understanding The wave function above is derived using a sine function. Can a cosine function be used instead?

Velocity and Acceleration of the Medium

As seen in **Example 16.4**, the wave speed is constant and represents the speed of the wave as it propagates through the medium, not the speed of the particles that make up the medium. The particles of the medium oscillate around an equilibrium position as the wave propagates through the medium. In the case of the transverse wave propagating in the x-direction, the particles oscillate up and down in the y-direction, perpendicular to the motion of the wave. The velocity of the particles of the medium is not constant, which means there is an acceleration. The velocity of the medium, which is perpendicular to the wave velocity in a transverse wave, can be found by taking the partial derivative of the position equation with respect to time. The partial derivative is found by taking the derivative of the function, treating all variables as constants, except for the variable in question. In the case of the partial derivative with respect to time t, the position x is treated as a constant. Although this may sound strange if you haven't seen it before, the object of this exercise is to find the transverse velocity at a point, so in this sense, the x-position is not changing. We have

$$y(x, t) = A \sin(kx - \omega t + \phi)$$

$$v_y(x, t) = \frac{\partial y(x, t)}{\partial t} = \frac{\partial}{\partial t}(A \sin(kx - \omega t + \phi))$$

$$= -A\omega \cos(kx - \omega t + \phi)$$

$$= -v_{y\,max} \cos(kx - \omega t + \phi).$$

The magnitude of the maximum velocity of the medium is $\left|v_{y\,max}\right| = A\omega$. This may look familiar from the **Oscillations** and a mass on a spring.

We can find the acceleration of the medium by taking the partial derivative of the velocity equation with respect to time,

$$a_y(x, t) = \frac{\partial v_y}{\partial t} = \frac{\partial}{\partial t}(-A\omega \cos(kx - \omega t + \phi))$$

$$= -A\omega^2 \sin(kx - \omega t + \phi)$$

$$= -a_{y\,max} \sin(kx - \omega t + \phi).$$

The magnitude of the maximum acceleration is $\left|a_{y\,max}\right| = A\omega^2$. The particles of the medium, or the mass elements, oscillate in simple harmonic motion for a mechanical wave.

The Linear Wave Equation

We have just determined the velocity of the medium at a position x by taking the partial derivative, with respect to time, of the position y. For a transverse wave, this velocity is perpendicular to the direction of propagation of the wave. We found the acceleration by taking the partial derivative, with respect to time, of the velocity, which is the second time derivative of the position:

$$a_y(x, t) = \frac{\partial^2 y(x, t)}{\partial t^2} = \frac{\partial^2}{\partial t^2}(A \sin(kx - \omega t + \phi)) = -A\omega^2 \sin(kx - \omega t + \phi).$$

Now consider the partial derivatives with respect to the other variable, the position x, holding the time constant. The first derivative is the slope of the wave at a point x at a time t,

$$slope = \frac{\partial y(x, t)}{\partial x} = \frac{\partial}{\partial x}(A \sin(kx - \omega t + \phi)) = Ak \cos(kx - \omega t + \phi).$$

The second partial derivative expresses how the slope of the wave changes with respect to position—in other words, the curvature of the wave, where

$$curvature = \frac{\partial^2 y(x, t)}{\partial x^2} = \frac{\partial^2}{\partial^2 x}(A \sin(kx - \omega t + \phi)) = -Ak^2 \sin(kx - \omega t + \phi).$$

The ratio of the acceleration and the curvature leads to a very important relationship in physics known as the **linear wave equation**. Taking the ratio and using the equation $v = \omega/k$ yields the linear wave equation (also known simply as the wave equation or the equation of a vibrating string),

$$\frac{\frac{\partial^2 y(x,\, t)}{\partial t^2}}{\frac{\partial^2 y(x,\, t)}{\partial x^2}} = \frac{-A\omega^2 \sin(kx - \omega t + \phi)}{-Ak^2 \sin(kx - \omega t + \phi)}$$

$$= \frac{\omega^2}{k^2} = v^2,$$

$$\frac{\partial^2 y(x,\, t)}{\partial x^2} = \frac{1}{v^2}\frac{\partial^2 y(x,\, t)}{\partial t^2}. \qquad\qquad (16.6)$$

Equation 16.6 is the linear wave equation, which is one of the most important equations in physics and engineering. We derived it here for a transverse wave, but it is equally important when investigating longitudinal waves. This relationship was also derived using a sinusoidal wave, but it successfully describes any wave or pulse that has the form $y(x,\, t) = f(x \mp vt)$. These waves result due to a linear restoring force of the medium—thus, the name linear wave equation. Any wave function that satisfies this equation is a linear wave function.

An interesting aspect of the linear wave equation is that if two wave functions are individually solutions to the linear wave equation, then the sum of the two linear wave functions is also a solution to the wave equation. Consider two transverse waves that propagate along the x-axis, occupying the same medium. Assume that the individual waves can be modeled with the wave functions $y_1(x,\, t) = f(x \mp vt)$ and $y_2(x,\, t) = g(x \mp vt)$, which are solutions to the linear wave equations and are therefore linear wave functions. The sum of the wave functions is the wave function

$$y_1(x,\, t) + y_2(x,\, t) = f(x \mp vt) + g(x \mp vt).$$

Consider the linear wave equation:

$$\frac{\partial^2(f + g)}{\partial x^2} = \frac{1}{v^2}\frac{\partial^2(f + g)}{\partial t^2}$$

$$\frac{\partial^2 f}{\partial x^2} + \frac{\partial^2 g}{\partial x^2} = \frac{1}{v^2}\left[\frac{\partial^2 f}{\partial t^2} + \frac{\partial^2 g}{\partial t^2}\right].$$

This has shown that if two linear wave functions are added algebraically, the resulting wave function is also linear. This wave function models the displacement of the medium of the resulting wave at each position along the x-axis. If two linear waves occupy the same medium, they are said to interfere. If these waves can be modeled with a linear wave function, these wave functions add to form the wave equation of the wave resulting from the interference of the individual waves. The displacement of the medium at every point of the resulting wave is the algebraic sum of the displacements due to the individual waves.

Taking this analysis a step further, if wave functions $y_1(x,\, t) = f(x \mp vt)$ and $y_2(x,\, t) = g(x \mp vt)$ are solutions to the linear wave equation, then $Ay_1(x,\, t) + By_2(x,\, y)$, where A and B are constants, is also a solution to the linear wave equation. This property is known as the principle of superposition. Interference and superposition are covered in more detail in **Interference of Waves**.

Example 16.4

Interference of Waves on a String

Consider a very long string held taut by two students, one on each end. Student A oscillates the end of the string producing a wave modeled with the wave function $y_1(x,\, t) = A\sin(kx - \omega t)$ and student B oscillates the string producing at twice the frequency, moving in the opposite direction. Both waves move at the same speed $v = \frac{\omega}{k}$.

The two waves interfere to form a resulting wave whose wave function is $y_R(x,\, t) = y_1(x,\, t) + y_2(x,\, t)$. Find the velocity of the resulting wave using the linear wave equation $\dfrac{\partial^2 y(x,\,t)}{\partial x^2} = \dfrac{1}{v^2}\dfrac{\partial^2 y(x,\,t)}{\partial t^2}$.

Strategy

First, write the wave function for the wave created by the second student. Note that the angular frequency of the second wave is twice the frequency of the first wave (2ω), and since the velocity of the two waves are the same, the wave number of the second wave is twice that of the first wave $(2k)$. Next, write the wave equation for the resulting wave function, which is the sum of the two individual wave functions. Then find the second partial derivative with respect to position and the second partial derivative with respect to time. Use the linear wave equation to find the velocity of the resulting wave.

Solution

1. Write the wave function of the second wave: $y_2(x,\,t) = A\sin(2kx + 2\omega t)$.

2. Write the resulting wave function:

$$y_R(x,\,t) = y_1(x,\,t) + y(x,\,t) = A\sin(kx - \omega t) + A\sin(2kx + 2\omega t).$$

3. Find the partial derivatives:

$$\frac{\partial y_R(x,\,t)}{\partial x} = -Ak\cos(kx - \omega t) + 2Ak\cos(2kx + 2\omega t),$$

$$\frac{\partial^2 y_R(x,\,t)}{\partial^2 x} = -Ak^2\sin(kx - \omega t) - 4Ak^2\sin(2kx + 2\omega t),$$

$$\frac{\partial y_R(x,\,t)}{\partial t} = -A\omega\cos(kx - \omega t) + 2A\omega\cos(2kx + 2\omega t),$$

$$\frac{\partial^2 y_R(x,\,t)}{\partial^2 t} = -A\omega^2\sin(kx - \omega t) - 4A\omega^2\sin(2kx + 2\omega t).$$

4. Use the wave equation to find the velocity of the resulting wave:

$$\frac{\partial^2 y(x,\,t)}{\partial x^2} = \frac{1}{v^2}\frac{\partial^2 y(x,\,t)}{\partial t^2},$$

$$-Ak^2\sin(kx - \omega t) - 4Ak^2\sin(2kx + 2\omega t) = \frac{1}{v^2}\left(-A\omega^2\sin(kx - \omega t) - 4A\omega^2\sin(2kx + 2\omega t)\right),$$

$$k^2\left(-A\sin(kx - \omega t) - 4A\sin(2kx + 2\omega t)\right) = \frac{\omega^2}{v^2}\left(-A\sin(kx - \omega t) - 4A\sin(2kx + 2\omega t)\right),$$

$$k^2 = \frac{\omega^2}{v^2},\quad |v| = \frac{\omega}{k}.$$

Significance

The speed of the resulting wave is equal to the speed of the original waves $\left(v = \frac{\omega}{k}\right)$. We will show in the next section that the speed of a simple harmonic wave on a string depends on the tension in the string and the mass per length of the string. For this reason, it is not surprising that the component waves as well as the resultant wave all travel at the same speed.

16.4 **Check Your Understanding** The wave equation $\dfrac{\partial^2 y(x,\,t)}{\partial x^2} = \dfrac{1}{v^2}\dfrac{\partial^2 y(x,\,t)}{\partial t^2}$ works for any wave of the form $y(x,\,t) = f(x \mp vt)$. In the previous section, we stated that a cosine function could also be used to model a simple harmonic mechanical wave. Check if the wave

$$y(x,\,t) = 0.50\text{ m}\cos\left(0.20\pi\text{ m}^{-1}x - 4.00\pi\text{ s}^{-1}t + \frac{\pi}{10}\right)$$

is a solution to the wave equation.

Any disturbance that complies with the wave equation can propagate as a wave moving along the *x*-axis with a wave speed *v*. It works equally well for waves on a string, sound waves, and electromagnetic waves. This equation is extremely useful. For example, it can be used to show that electromagnetic waves move at the speed of light.

16.3 | Wave Speed on a Stretched String

The speed of a wave depends on the characteristics of the medium. For example, in the case of a guitar, the strings vibrate to produce the sound. The speed of the waves on the strings, and the wavelength, determine the frequency of the sound produced. The strings on a guitar have different thickness but may be made of similar material. They have different *linear densities*, where the linear density is defined as the mass per length,

$$\mu = \frac{\text{mass of string}}{\text{length of string}} = \frac{m}{l}. \tag{16.7}$$

In this chapter, we consider only string with a constant linear density. If the linear density is constant, then the mass (Δm) of a small length of string (Δx) is $\Delta m = \mu \Delta x$. For example, if the string has a length of 2.00 m and a mass of 0.06 kg, then the linear density is $\mu = \frac{0.06\,\text{kg}}{2.00\,\text{m}} = 0.03\frac{\text{kg}}{\text{m}}$. If a 1.00-mm section is cut from the string, the mass of the 1.00-mm length is $\Delta m = \mu \Delta x = \left(0.03\frac{\text{kg}}{\text{m}}\right)0.001\,\text{m} = 3.00 \times 10^{-5}\,\text{kg}$. The guitar also has a method to change the tension of the strings.

The tension of the strings is adjusted by turning spindles, called the tuning pegs, around which the strings are wrapped. For the guitar, the linear density of the string and the tension in the string determine the speed of the waves in the string and the frequency of the sound produced is proportional to the wave speed.

Wave Speed on a String under Tension

To see how the speed of a wave on a string depends on the tension and the linear density, consider a pulse sent down a taut string (**Figure 16.13**). When the taut string is at rest at the equilibrium position, the tension in the string F_T is constant. Consider a small element of the string with a mass equal to $\Delta m = \mu \Delta x$. The mass element is at rest and in equilibrium and the force of tension of either side of the mass element is equal and opposite.

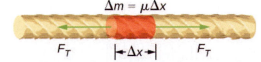

Figure 16.13 Mass element of a string kept taut with a tension F_T. The mass element is in static equilibrium, and the

force of tension acting on either side of the mass element is equal in magnitude and opposite in direction.

If you pluck a string under tension, a transverse wave moves in the positive *x*-direction, as shown in **Figure 16.14**. The mass element is small but is enlarged in the figure to make it visible. The small mass element oscillates perpendicular to the wave motion as a result of the restoring force provided by the string and does not move in the *x*-direction. The tension F_T in the string, which acts in the positive and negative *x*-direction, is approximately constant and is independent of position and time.

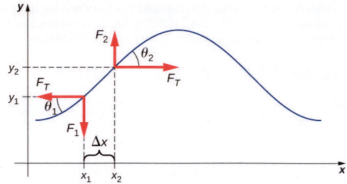

Figure 16.14 A string under tension is plucked, causing a pulse to move along the string in the positive x-direction.

Assume that the inclination of the displaced string with respect to the horizontal axis is small. The net force on the element of the string, acting parallel to the string, is the sum of the tension in the string and the restoring force. The x-components of the force of tension cancel, so the net force is equal to the sum of the y-components of the force. The magnitude of the x-component of the force is equal to the horizontal force of tension of the string F_T as shown in **Figure 16.14**. To obtain the y-components of the force, note that $\tan\theta_1 = \dfrac{-F_1}{F_T}$ and $\tan\theta_2 = \dfrac{F_2}{F_T}$. The $\tan\theta$ is equal to the slope of a function at a point, which is equal to the partial derivative of y with respect to x at that point. Therefore, $\dfrac{F_1}{F_T}$ is equal to the negative slope of the string at x_1 and $\dfrac{F_2}{F_T}$ is equal to the slope of the string at x_2 :

$$\frac{F_1}{F_T} = -\left(\frac{\partial y}{\partial x}\right)_{x_1} \text{ and } \frac{F_2}{F_T} = -\left(\frac{\partial y}{\partial x}\right)_{x_2}.$$

The net force is on the small mass element can be written as

$$F_{\text{net}} = F_1 + F_2 = F_T\left[\left(\frac{\partial y}{\partial x}\right)_{x_2} - \left(\frac{\partial y}{\partial x}\right)_{x_1}\right].$$

Using Newton's second law, the net force is equal to the mass times the acceleration. The linear density of the string μ is the mass per length of the string, and the mass of the portion of the string is $\mu\Delta x$,

$$F_T\left[\left(\frac{\partial y}{\partial x}\right)_{x_2} - \left(\frac{\partial y}{\partial x}\right)_{x_1}\right] = \Delta ma,$$

$$F_T\left[\left(\frac{\partial y}{\partial x}\right)_{x_2} - \left(\frac{\partial y}{\partial x}\right)_{x_1}\right] = \mu\Delta x\frac{\partial^2 y}{\partial t^2}.$$

Dividing by $F_T\Delta x$ and taking the limit as Δx approaches zero,

$$\frac{\left[\left(\frac{\partial y}{\partial x}\right)_{x_2} - \left(\frac{\partial y}{\partial x}\right)_{x_1}\right]}{\Delta x} = \frac{\mu}{F_T}\frac{\partial^2 y}{\partial t^2}$$

$$\lim_{\Delta x \to 0}\frac{\left[\left(\frac{\partial y}{\partial x}\right)_{x_2} - \left(\frac{\partial y}{\partial x}\right)_{x_1}\right]}{\Delta x} = \frac{\mu}{F_T}\frac{\partial^2 y}{\partial t^2}$$

$$\frac{\partial^2 y}{\partial x^2} = \frac{\mu}{F_T}\frac{\partial^2 y}{\partial t^2}.$$

Recall that the linear wave equation is

$$\frac{\partial^2 y(x,\ t)}{\partial x^2} = \frac{1}{v^2} \frac{\partial^2 y(x,\ t)}{\partial t^2}.$$

Therefore,

$$\frac{1}{v^2} = \frac{\mu}{F_T}.$$

Solving for v, we see that the speed of the wave on a string depends on the tension and the linear density.

Speed of a Wave on a String Under Tension

The speed of a pulse or wave on a string under tension can be found with the equation

$$|v| = \sqrt{\frac{F_T}{\mu}} \qquad (16.8)$$

where F_T is the tension in the string and μ is the mass per length of the string.

Example 16.5

The Wave Speed of a Guitar Spring

On a six-string guitar, the high E string has a linear density of $\mu_{\text{High E}} = 3.09 \times 10^{-4}$ kg/m and the low E string has a linear density of $\mu_{\text{Low E}} = 5.78 \times 10^{-3}$ kg/m. (a) If the high E string is plucked, producing a wave in the string, what is the speed of the wave if the tension of the string is 56.40 N? (b) The linear density of the low E string is approximately 20 times greater than that of the high E string. For waves to travel through the low E string at the same wave speed as the high E, would the tension need to be larger or smaller than the high E string? What would be the approximate tension? (c) Calculate the tension of the low E string needed for the same wave speed.

Strategy

a. The speed of the wave can be found from the linear density and the tension $v = \sqrt{\frac{F_T}{\mu}}$.

b. From the equation $v = \sqrt{\frac{F_T}{\mu}}$, if the linear density is increased by a factor of almost 20, the tension would need to be increased by a factor of 20.

c. Knowing the velocity and the linear density, the velocity equation can be solved for the force of tension $F_T = \mu v^2$.

Solution

a. Use the velocity equation to find the speed:

$$v = \sqrt{\frac{F_T}{\mu}} = \sqrt{\frac{56.40\ \text{N}}{3.09 \times 10^{-4}\ \text{kg/m}}} = 427.23\ \text{m/s}.$$

b. The tension would need to be increased by a factor of approximately 20. The tension would be slightly less than 1128 N.

c. Use the velocity equation to find the actual tension:

$$F_T = \mu v^2 = 5.78 \times 10^{-3}\ \text{kg/m}(427.23\ \text{m/s})^2 = 1055.00\ \text{N}.$$

This solution is within 7% of the approximation.

Significance

The standard notes of the six string (high E, B, G, D, A, low E) are tuned to vibrate at the fundamental frequencies (329.63 Hz, 246.94Hz, 196.00Hz, 146.83Hz, 110.00Hz, and 82.41Hz) when plucked. The frequencies depend on the speed of the waves on the string and the wavelength of the waves. The six strings have different linear densities and are "tuned" by changing the tensions in the strings. We will see in **Interference of Waves** that the wavelength depends on the length of the strings and the boundary conditions. To play notes other than the fundamental notes, the lengths of the strings are changed by pressing down on the strings.

 16.5 Check Your Understanding The wave speed of a wave on a string depends on the tension and the linear mass density. If the tension is doubled, what happens to the speed of the waves on the string?

Speed of Compression Waves in a Fluid

The speed of a wave on a string depends on the square root of the tension divided by the mass per length, the linear density. In general, the speed of a wave through a medium depends on the elastic property of the medium and the inertial property of the medium.

$$|v| = \sqrt{\frac{\text{elastic property}}{\text{inertial property}}}$$

The elastic property describes the tendency of the particles of the medium to return to their initial position when perturbed. The inertial property describes the tendency of the particle to resist changes in velocity.

The speed of a longitudinal wave through a liquid or gas depends on the density of the fluid and the bulk modulus of the fluid,

$$v = \sqrt{\frac{B}{\rho}}. \tag{16.9}$$

Here the bulk modulus is defined as $B = -\frac{\Delta P}{\frac{\Delta V}{V_0}}$, where ΔP is the change in the pressure and the denominator is the ratio of the change in volume to the initial volume, and $\rho \equiv \frac{m}{V}$ is the mass per unit volume. For example, sound is a mechanical wave that travels through a fluid or a solid. The speed of sound in air with an atmospheric pressure of 1.013×10^5 Pa and a temperature of $20°C$ is $v_s \approx 343.00$ m/s. Because the density depends on temperature, the speed of sound in air depends on the temperature of the air. This will be discussed in detail in **Sound**.

16.4 | Energy and Power of a Wave

Learning Objectives

By the end of this section, you will be able to:

- Explain how energy travels with a pulse or wave
- Describe, using a mathematical expression, how the energy in a wave depends on the amplitude of the wave

All waves carry energy, and sometimes this can be directly observed. Earthquakes can shake whole cities to the ground, performing the work of thousands of wrecking balls (**Figure 16.15**). Loud sounds can pulverize nerve cells in the inner ear, causing permanent hearing loss. Ultrasound is used for deep-heat treatment of muscle strains. A laser beam can burn away a malignancy. Water waves chew up beaches.

Figure 16.15 The destructive effect of an earthquake is observable evidence of the energy carried in these waves. The Richter scale rating of earthquakes is a logarithmic scale related to both their amplitude and the energy they carry.

In this section, we examine the quantitative expression of energy in waves. This will be of fundamental importance in later discussions of waves, from sound to light to quantum mechanics.

Energy in Waves

The amount of energy in a wave is related to its amplitude and its frequency. Large-amplitude earthquakes produce large ground displacements. Loud sounds have high-pressure amplitudes and come from larger-amplitude source vibrations than soft sounds. Large ocean breakers churn up the shore more than small ones. Consider the example of the seagull and the water wave earlier in the chapter (**Figure 16.3**). Work is done on the seagull by the wave as the seagull is moved up, changing its potential energy. The larger the amplitude, the higher the seagull is lifted by the wave and the larger the change in potential energy.

The energy of the wave depends on both the amplitude and the frequency. If the energy of each wavelength is considered to be a discrete packet of energy, a high-frequency wave will deliver more of these packets per unit time than a low-frequency wave. We will see that the average rate of energy transfer in mechanical waves is proportional to both the square of the amplitude and the square of the frequency. If two mechanical waves have equal amplitudes, but one wave has a frequency equal to twice the frequency of the other, the higher-frequency wave will have a rate of energy transfer a factor of four times as great as the rate of energy transfer of the lower-frequency wave. It should be noted that although the rate of energy transport is proportional to both the square of the amplitude and square of the frequency in mechanical waves, the rate of energy transfer in electromagnetic waves is proportional to the square of the amplitude, but independent of the frequency.

Power in Waves

Consider a sinusoidal wave on a string that is produced by a string vibrator, as shown in **Figure 16.16**. The string vibrator is a device that vibrates a rod up and down. A string of uniform linear mass density is attached to the rod, and the rod oscillates the string, producing a sinusoidal wave. The rod does work on the string, producing energy that propagates along the string. Consider a mass element of the string with a mass Δm, as seen in **Figure 16.16**. As the energy propagates along the string, each mass element of the string is driven up and down at the same frequency as the wave. Each mass element of the string can be modeled as a simple harmonic oscillator. Since the string has a constant linear density $\mu = \frac{\Delta m}{\Delta x}$, each mass element of the string has the mass $\Delta m = \mu \Delta x$.

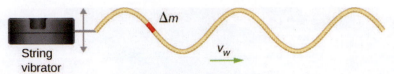

Figure 16.16 A string vibrator is a device that vibrates a rod. A string is attached to the rod, and the rod does work on the string, driving the string up and down. This produces a sinusoidal wave in the string, which moves with a wave velocity v. The wave speed depends on the tension in the string and the linear mass density of the string. A section of the string with mass Δm oscillates at the same frequency as the wave.

The total mechanical energy of the wave is the sum of its kinetic energy and potential energy. The kinetic energy $K = \frac{1}{2}mv^2$ of each mass element of the string of length Δx is $\Delta K = \frac{1}{2}(\Delta m)v_y^2$, as the mass element oscillates perpendicular to the direction of the motion of the wave. Using the constant linear mass density, the kinetic energy of each mass element of the string with length Δx is

$$\Delta K = \frac{1}{2}(\mu \Delta x)v_y^2.$$

A differential equation can be formed by letting the length of the mass element of the string approach zero,

$$dK = \lim_{\Delta x \to 0} \frac{1}{2}(\mu \Delta x)v_y^2 = \frac{1}{2}(\mu dx)v_y^2.$$

Since the wave is a sinusoidal wave with an angular frequency ω, the position of each mass element may be modeled as $y(x, t) = A\sin(kx - \omega t)$. Each mass element of the string oscillates with a velocity $v_y = \frac{\partial y(x, t)}{\partial t} = -A\omega\cos(kx - \omega t)$. The kinetic energy of each mass element of the string becomes

$$dK = \frac{1}{2}(\mu dx)(-A\omega\cos(kx - \omega t))^2,$$
$$= \frac{1}{2}(\mu dx)A^2\omega^2\cos^2(kx - \omega t).$$

The wave can be very long, consisting of many wavelengths. To standardize the energy, consider the kinetic energy associated with a wavelength of the wave. This kinetic energy can be integrated over the wavelength to find the energy associated with each wavelength of the wave:

$$dK = \frac{1}{2}(\mu dx)A^2\omega^2\cos^2(kx),$$
$$\int_0^{K_\lambda} dK = \int_0^\lambda \frac{1}{2}\mu A^2\omega^2\cos^2(kx)dx = \frac{1}{2}\mu A^2\omega^2\int_0^\lambda\cos^2(kx)dx,$$
$$K_\lambda = \frac{1}{2}\mu A^2\omega^2\left[\frac{1}{2}x + \frac{1}{4k}\sin(2kx)\right]_0^\lambda = \frac{1}{2}\mu A^2\omega^2\left[\frac{1}{2}\lambda + \frac{1}{4k}\sin(2k\lambda) - \frac{1}{4k}\sin(0)\right],$$
$$K_\lambda = \frac{1}{4}\mu A^2\omega^2\lambda.$$

There is also potential energy associated with the wave. Much like the mass oscillating on a spring, there is a conservative restoring force that, when the mass element is displaced from the equilibrium position, drives the mass element back to the equilibrium position. The potential energy of the mass element can be found by considering the linear restoring force of the string, In **Oscillations**, we saw that the potential energy stored in a spring with a linear restoring force is equal to $U = \frac{1}{2}k_s x^2$, where the equilibrium position is defined as $x = 0.00$ m. When a mass attached to the spring oscillates in simple harmonic motion, the angular frequency is equal to $\omega = \sqrt{\frac{k_s}{m}}$. As each mass element oscillates in simple harmonic motion, the spring constant is equal to $k_s = \Delta m\omega^2$. The potential energy of the mass element is equal to

$$\Delta U = \frac{1}{2}k_s x^2 = \frac{1}{2}\Delta m\omega^2 x^2.$$

Note that k_s is the spring constant and not the wave number $k = \frac{2\pi}{\lambda}$. This equation can be used to find the energy over a wavelength. Integrating over the wavelength, we can compute the potential energy over a wavelength:

$$dU = \frac{1}{2}k_s x^2 = \frac{1}{2}\mu\omega^2 x^2 dx,$$

$$U_\lambda = \frac{1}{2}\mu\omega^2 A^2 \int_0^\lambda \cos^2(kx)dx = \frac{1}{4}\mu A^2 \omega^2 \lambda.$$

The potential energy associated with a wavelength of the wave is equal to the kinetic energy associated with a wavelength.

The total energy associated with a wavelength is the sum of the potential energy and the kinetic energy:

$$E_\lambda = U_\lambda + K_\lambda,$$
$$E_\lambda = \frac{1}{4}\mu A^2 \omega^2 \lambda + \frac{1}{4}\mu A^2 \omega^2 \lambda = \frac{1}{2}\mu A^2 \omega^2 \lambda.$$

The time-averaged power of a sinusoidal mechanical wave, which is the average rate of energy transfer associated with a wave as it passes a point, can be found by taking the total energy associated with the wave divided by the time it takes to transfer the energy. If the velocity of the sinusoidal wave is constant, the time for one wavelength to pass by a point is equal to the period of the wave, which is also constant. For a sinusoidal mechanical wave, the time-averaged power is therefore the energy associated with a wavelength divided by the period of the wave. The wavelength of the wave divided by the period is equal to the velocity of the wave,

$$P_{ave} = \frac{E_\lambda}{T} = \frac{1}{2}\mu A^2 \omega^2 \frac{\lambda}{T} = \frac{1}{2}\mu A^2 \omega^2 v. \tag{16.10}$$

Note that this equation for the time-averaged power of a sinusoidal mechanical wave shows that the power is proportional to the square of the amplitude of the wave and to the square of the angular frequency of the wave. Recall that the angular frequency is equal to $\omega = 2\pi f$, so the power of a mechanical wave is equal to the square of the amplitude and the square of the frequency of the wave.

Example 16.6

Power Supplied by a String Vibrator

Consider a two-meter-long string with a mass of 70.00 g attached to a string vibrator as illustrated in **Figure 16.16**. The tension in the string is 90.0 N. When the string vibrator is turned on, it oscillates with a frequency of 60 Hz and produces a sinusoidal wave on the string with an amplitude of 4.00 cm and a constant wave speed. What is the time-averaged power supplied to the wave by the string vibrator?

Strategy

The power supplied to the wave should equal the time-averaged power of the wave on the string. We know the mass of the string (m_s), the length of the string (L_s), and the tension (F_T) in the string. The speed of the wave on the string can be derived from the linear mass density and the tension. The string oscillates with the same frequency as the string vibrator, from which we can find the angular frequency.

Solution

1. Begin with the equation of the time-averaged power of a sinusoidal wave on a string:

$$P = \frac{1}{2}\mu A^2 \omega^2 v.$$

   The amplitude is given, so we need to calculate the linear mass density of the string, the angular frequency of the wave on the string, and the speed of the wave on the string.

2. We need to calculate the linear density to find the wave speed:

$$\mu = \frac{m_s}{L_s} = \frac{0.070 \text{ kg}}{2.00 \text{ m}} = 0.035 \text{ kg/m}.$$

3. The wave speed can be found using the linear mass density and the tension of the string:

$$v = \sqrt{\frac{F_T}{\mu}} = \sqrt{\frac{90.00 \text{ N}}{0.035 \text{ kg/m}}} = 50.71 \text{ m/s}.$$

4. The angular frequency can be found from the frequency:

$$\omega = 2\pi f = 2\pi\left(60 \text{ s}^{-1}\right) = 376.80 \text{ s}^{-1}.$$

5. Calculate the time-averaged power:

$$P = \frac{1}{2}\mu A^2 \omega^2 v = \frac{1}{2}\left(0.035\frac{\text{kg}}{\text{m}}\right)(0.040 \text{ m})^2\left(376.80 \text{ s}^{-1}\right)^2\left(50.71\frac{\text{m}}{\text{s}}\right) = 201.59 \text{ W}.$$

Significance

The time-averaged power of a sinusoidal wave is proportional to the square of the amplitude of the wave and the square of the angular frequency of the wave. This is true for most mechanical waves. If either the angular frequency or the amplitude of the wave were doubled, the power would increase by a factor of four. The time-averaged power of the wave on a string is also proportional to the speed of the sinusoidal wave on the string. If the speed were doubled, by increasing the tension by a factor of four, the power would also be doubled.

 16.6 Check Your Understanding Is the time-averaged power of a sinusoidal wave on a string proportional to the linear density of the string?

The equations for the energy of the wave and the time-averaged power were derived for a sinusoidal wave on a string. In general, the energy of a mechanical wave and the power are proportional to the amplitude squared and to the angular frequency squared (and therefore the frequency squared).

Another important characteristic of waves is the intensity of the waves. Waves can also be concentrated or spread out. Waves from an earthquake, for example, spread out over a larger area as they move away from a source, so they do less damage the farther they get from the source. Changing the area the waves cover has important effects. All these pertinent factors are included in the definition of **intensity (*I*)** as power per unit area:

$$I = \frac{P}{A},$$
(16.11)

where *P* is the power carried by the wave through area *A*. The definition of intensity is valid for any energy in transit, including that carried by waves. The SI unit for intensity is watts per square meter (W/m^2). Many waves are spherical waves that move out from a source as a sphere. For example, a sound speaker mounted on a post above the ground may produce sound waves that move away from the source as a spherical wave. Sound waves are discussed in more detail in the next chapter, but in general, the farther you are from the speaker, the less intense the sound you hear. As a spherical wave moves out from a source, the surface area of the wave increases as the radius increases $\left(A = 4\pi r^2\right)$. The intensity for a spherical wave is therefore

$$I = \frac{P}{4\pi r^2}.$$
(16.12)

If there are no dissipative forces, the energy will remain constant as the spherical wave moves away from the source, but the intensity will decrease as the surface area increases.

In the case of the two-dimensional circular wave, the wave moves out, increasing the circumference of the wave as the radius of the circle increases. If you toss a pebble in a pond, the surface ripple moves out as a circular wave. As the ripple moves away from the source, the amplitude decreases. The energy of the wave spreads around a larger circumference and the amplitude decreases proportional to $\frac{1}{r}$, which is also the same in the case of a spherical wave, since intensity is proportional to the amplitude squared.

16.5 | Interference of Waves

Learning Objectives

By the end of this section, you will be able to:

- Explain how mechanical waves are reflected and transmitted at the boundaries of a medium
- Define the terms interference and superposition
- Find the resultant wave of two identical sinusoidal waves that differ only by a phase shift

Up to now, we have been studying mechanical waves that propagate continuously through a medium, but we have not discussed what happens when waves encounter the boundary of the medium or what happens when a wave encounters another wave propagating through the same medium. Waves do interact with boundaries of the medium, and all or part of the wave can be reflected. For example, when you stand some distance from a rigid cliff face and yell, you can hear the sound waves reflect off the rigid surface as an echo. Waves can also interact with other waves propagating in the same medium. If you throw two rocks into a pond some distance from one another, the circular ripples that result from the two stones seem to pass through one another as they propagate out from where the stones entered the water. This phenomenon is known as interference. In this section, we examine what happens to waves encountering a boundary of a medium or another wave propagating in the same medium. We will see that their behavior is quite different from the behavior of particles and rigid bodies. Later, when we study modern physics, we will see that only at the scale of atoms do we see similarities in the properties of waves and particles.

Reflection and Transmission

When a wave propagates through a medium, it reflects when it encounters the boundary of the medium. The wave before hitting the boundary is known as the incident wave. The wave after encountering the boundary is known as the reflected wave. How the wave is reflected at the boundary of the medium depends on the boundary conditions; waves will react differently if the boundary of the medium is fixed in place or free to move (**Figure 16.17**). A **fixed boundary condition** exists when the medium at a boundary is fixed in place so it cannot move. A **free boundary condition** exists when the medium at the boundary is free to move.

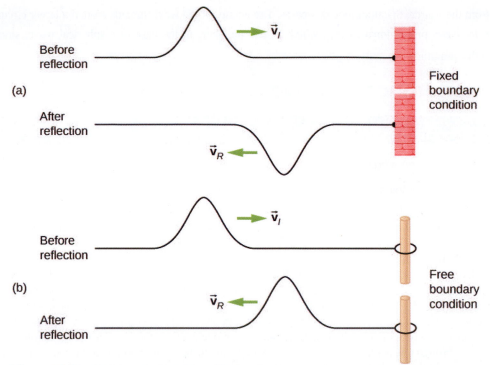

Figure 16.17 (a) One end of a string is fixed so that it cannot move. A wave propagating on the string, encountering this *fixed boundary condition*, is reflected $180°(\pi \text{ rad})$ out of phase with respect to the incident wave. (b) One end of a string is tied to a solid ring of negligible mass on a frictionless lab pole, where the ring is free to move. A wave propagating on the string, encountering this *free boundary condition*, is reflected in phase $0°(0 \text{ rad})$ with respect to the wave.

Part (a) of the **Figure 16.17** shows a fixed boundary condition. Here, one end of the string is fixed to a wall so the end of the string is fixed in place and the medium (the string) at the boundary cannot move. When the wave is reflected, the amplitude of the reflected way is exactly the same as the amplitude of the incident wave, but the reflected wave is reflected $180°(\pi \text{ rad})$ out of phase with respect to the incident wave. The phase change can be explained using Newton's third law: Recall that Newton's third law states that when object *A* exerts a force on object *B*, then object *B* exerts an equal and opposite force on object *A*. As the incident wave encounters the wall, the string exerts an upward force on the wall and the wall reacts by exerting an equal and opposite force on the string. The reflection at a fixed boundary is inverted. Note that the figure shows a crest of the incident wave reflected as a trough. If the incident wave were a trough, the reflected wave would be a crest.

Part (b) of the figure shows a free boundary condition. Here, one end of the string is tied to a solid ring of negligible mass on a frictionless pole, so the end of the string is free to move up and down. As the incident wave encounters the boundary of the medium, it is also reflected. In the case of a free boundary condition, the reflected wave is in phase with respect to the incident wave. In this case, the wave encounters the free boundary applying an upward force on the ring, accelerating the ring up. The ring travels up to the maximum height equal to the amplitude of the wave and then accelerates down towards the equilibrium position due to the tension in the string. The figure shows the crest of an incident wave being reflected in phase with respect to the incident wave as a crest. If the incident wave were a trough, the reflected wave would also be a trough. The amplitude of the reflected wave would be equal to the amplitude of the incident wave.

In some situations, the boundary of the medium is neither fixed nor free. Consider **Figure 16.18**(a), where a low-linear mass density string is attached to a string of a higher linear mass density. In this case, the reflected wave is out of phase with respect to the incident wave. There is also a transmitted wave that is in phase with respect to the incident wave. Both the incident and the reflected waves have amplitudes less than the amplitude of the incident wave. If the tension is the same in both strings, the wave speed is higher in the string with the lower linear mass density.

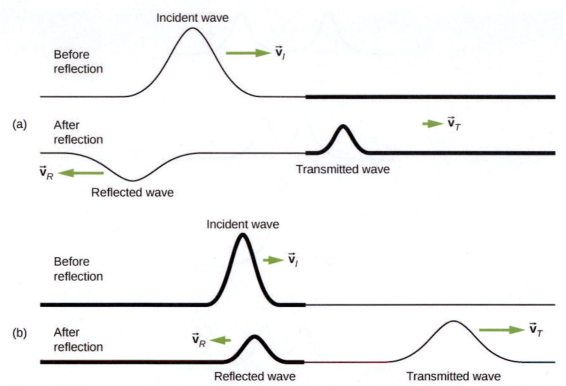

Figure 16.18 Waves traveling along two types of strings: a thick string with a high linear density and a thin string with a low linear density. Both strings are under the same tension, so a wave moves faster on the low-density string than on the high-density string. (a) A wave moving from a low-speed to a high-speed medium results in a reflected wave that is $180°(\pi \text{ rad})$ out of phase with respect to the incident pulse (or wave) and a transmitted wave that is in phase with the incident wave. (b) When a wave moves from a low-speed medium to a high-speed medium, both the reflected and transmitted wave are in phase with respect to the incident wave.

Part (b) of the figure shows a high-linear mass density string is attached to a string of a lower linear density. In this case, the reflected wave is in phase with respect to the incident wave. There is also a transmitted wave that is in phase with respect to the incident wave. Both the incident and the reflected waves have amplitudes less than the amplitude of the incident wave. Here you may notice that if the tension is the same in both strings, the wave speed is higher in the string with the lower linear mass density.

Superposition and Interference

Most waves do not look very simple. Complex waves are more interesting, even beautiful, but they look formidable. Most interesting mechanical waves consist of a combination of two or more traveling waves propagating in the same medium. The principle of superposition can be used to analyze the combination of waves.

Consider two simple pulses of the same amplitude moving toward one another in the same medium, as shown in **Figure 16.19**. Eventually, the waves overlap, producing a wave that has twice the amplitude, and then continue on unaffected by the encounter. The pulses are said to interfere, and this phenomenon is known as **interference**.

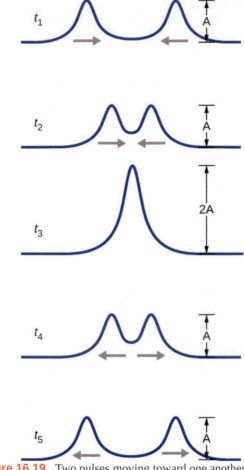

Figure 16.19 Two pulses moving toward one another experience interference. The term interference refers to what happens when two waves overlap.

To analyze the interference of two or more waves, we use the principle of superposition. For mechanical waves, the principle of **superposition** states that if two or more traveling waves combine at the same point, the resulting position of the mass element of the medium, at that point, is the algebraic sum of the position due to the individual waves. This property is exhibited by many waves observed, such as waves on a string, sound waves, and surface water waves. Electromagnetic waves also obey the superposition principle, but the electric and magnetic fields of the combined wave are added instead of the displacement of the medium. Waves that obey the superposition principle are linear waves; waves that do not obey the superposition principle are said to be nonlinear waves. In this chapter, we deal with linear waves, in particular, sinusoidal waves.

The superposition principle can be understood by considering the linear wave equation. In **Mathematics of a Wave**, we defined a linear wave as a wave whose mathematical representation obeys the linear wave equation. For a transverse wave on a string with an elastic restoring force, the linear wave equation is

$$\frac{\partial^2 y(x,\, t)}{\partial x^2} = \frac{1}{v^2} \frac{\partial^2 y(x,\, t)}{\partial t^2}.$$

Any wave function $y(x,\, t) = y(x \mp vt),$ where the argument of the function is linear $(x \mp vt)$ is a solution to the linear wave equation and is a linear wave function. If wave functions $y_1(x,\, t)$ and $y_2(x,\, t)$ are solutions to the linear wave equation, the sum of the two functions $y_1(x,\, t) + y_2(x,\, t)$ is also a solution to the linear wave equation. Mechanical waves that obey superposition are normally restricted to waves with amplitudes that are small with respect to their wavelengths. If the amplitude is too large, the medium is distorted past the region where the restoring force of the medium is linear.

Waves can interfere constructively or destructively. **Figure 16.20** shows two identical sinusoidal waves that arrive at the same point exactly in phase. **Figure 16.20**(a) and (b) show the two individual waves, **Figure 16.20**(c) shows the resultant wave that results from the algebraic sum of the two linear waves. The crests of the two waves are precisely aligned, as are the troughs. This superposition produces **constructive interference**. Because the disturbances add, constructive interference produces a wave that has twice the amplitude of the individual waves, but has the same wavelength.

Figure 16.21 shows two identical waves that arrive exactly $180°$ out of phase, producing **destructive interference**. **Figure 16.21**(a) and (b) show the individual waves, and **Figure 16.21**(c) shows the superposition of the two waves. Because the troughs of one wave add the crest of the other wave, the resulting amplitude is zero for destructive interference—the waves completely cancel.

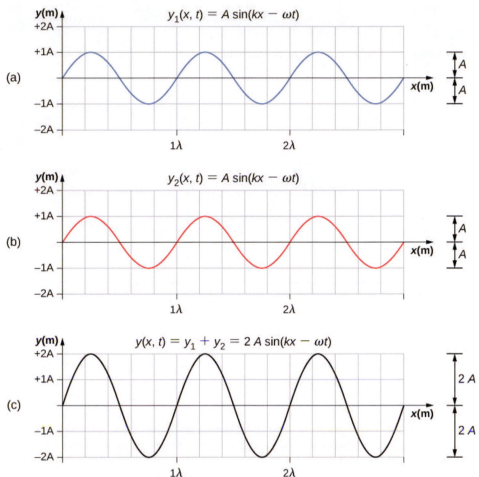

Figure 16.20 Constructive interference of two identical waves produces a wave with twice the amplitude, but the same wavelength.

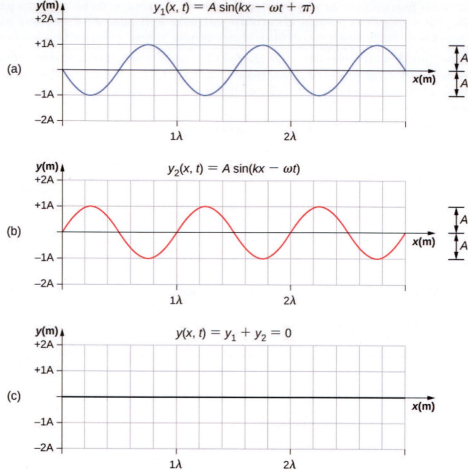

Figure 16.21 Destructive interference of two identical waves, one with a phase shift of $180°(\pi \text{ rad})$, produces zero amplitude, or complete cancellation.

When linear waves interfere, the resultant wave is just the algebraic sum of the individual waves as stated in the principle of superposition. **Figure 16.22** shows two waves (red and blue) and the resultant wave (black). The resultant wave is the algebraic sum of the two individual waves.

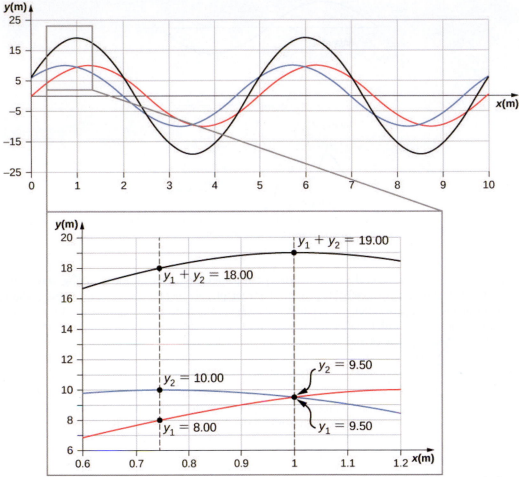

Figure 16.22 When two linear waves in the same medium interfere, the height of resulting wave is the sum of the heights of the individual waves, taken point by point. This plot shows two waves (red and blue) added together, along with the resulting wave (black). These graphs represent the height of the wave at each point. The waves may be any linear wave, including ripples on a pond, disturbances on a string, sound, or electromagnetic waves.

The superposition of most waves produces a combination of constructive and destructive interference, and can vary from place to place and time to time. Sound from a stereo, for example, can be loud in one spot and quiet in another. Varying loudness means the sound waves add partially constructively and partially destructively at different locations. A stereo has at least two speakers creating sound waves, and waves can reflect from walls. All these waves interfere, and the resulting wave is the superposition of the waves.

We have shown several examples of the superposition of waves that are similar. **Figure 16.23** illustrates an example of the superposition of two dissimilar waves. Here again, the disturbances add, producing a resultant wave.

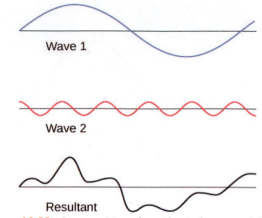

Figure 16.23 Superposition of nonidentical waves exhibits both constructive and destructive interference.

At times, when two or more mechanical waves interfere, the pattern produced by the resulting wave can be rich in complexity, some without any readily discernable patterns. For example, plotting the sound wave of your favorite music can look quite complex and is the superposition of the individual sound waves from many instruments; it is the complexity that makes the music interesting and worth listening to. At other times, waves can interfere and produce interesting phenomena, which are complex in their appearance and yet beautiful in simplicity of the physical principle of superposition, which formed the resulting wave. One example is the phenomenon known as standing waves, produced by two identical waves moving in different directions. We will look more closely at this phenomenon in the next section.

 Try this **simulation (https://openstaxcollege.org/l/21waveinterfer)** to make waves with a dripping faucet, audio speaker, or laser! Add a second source or a pair of slits to create an interference pattern. You can observe one source or two sources. Using two sources, you can observe the interference patterns that result from varying the frequencies and the amplitudes of the sources.

Superposition of Sinusoidal Waves that Differ by a Phase Shift

Many examples in physics consist of two sinusoidal waves that are identical in amplitude, wave number, and angular frequency, but differ by a phase shift:

$$y_1(x, t) = A \sin(kx - \omega t + \phi),$$
$$y_2(x, t) = A \sin(kx - \omega t).$$

When these two waves exist in the same medium, the resultant wave resulting from the superposition of the two individual waves is the sum of the two individual waves:

$$y_R(x, t) = y_1(x, t) + y_2(x, t) = A \sin(kx - \omega t + \phi) + A \sin(kx - \omega t).$$

The resultant wave can be better understood by using the trigonometric identity:

$$\sin u + \sin v = 2 \sin\left(\frac{u+v}{2}\right)\cos\left(\frac{u-v}{2}\right),$$

where $u = kx - \omega t + \phi$ and $v = kx - \omega t$. The resulting wave becomes

$$
\begin{aligned}
y_R(x, t) &= y_1(x, t) + y_2(x, t) = A \sin(kx - \omega t + \phi) + A \sin(kx - \omega t) \\
&= 2A \sin\left(\frac{(kx - \omega t + \phi) + (kx - \omega t)}{2}\right)\cos\left(\frac{(kx - \omega t + \phi) - (kx - \omega t)}{2}\right) \\
&= 2A \sin\left(kx - \omega t + \frac{\phi}{2}\right)\cos\left(\frac{\phi}{2}\right).
\end{aligned}
$$

This equation is usually written as

$$y_R(x, t) = \left[2A \cos\left(\frac{\phi}{2}\right)\right] \sin\left(kx - \omega t + \frac{\phi}{2}\right).$$

(16.13)

The resultant wave has the same wave number and angular frequency, an amplitude of $A_R = \left[2A \cos\left(\frac{\phi}{2}\right)\right]$, and a phase shift equal to half the original phase shift. Examples of waves that differ only in a phase shift are shown in **Figure 16.24**. The red and blue waves each have the same amplitude, wave number, and angular frequency, and differ only in a phase shift. They therefore have the same period, wavelength, and frequency. The green wave is the result of the superposition of the two waves. When the two waves have a phase difference of zero, the waves are in phase, and the resultant wave has the same wave number and angular frequency, and an amplitude equal to twice the individual amplitudes (part (a)). This is constructive interference. If the phase difference is $180°$, the waves interfere in destructive interference (part (c)).

The resultant wave has an amplitude of zero. Any other phase difference results in a wave with the same wave number and angular frequency as the two incident waves but with a phase shift of $\phi/2$ and an amplitude equal to $2A \cos(\phi/2)$.

Examples are shown in parts (b) and (d).

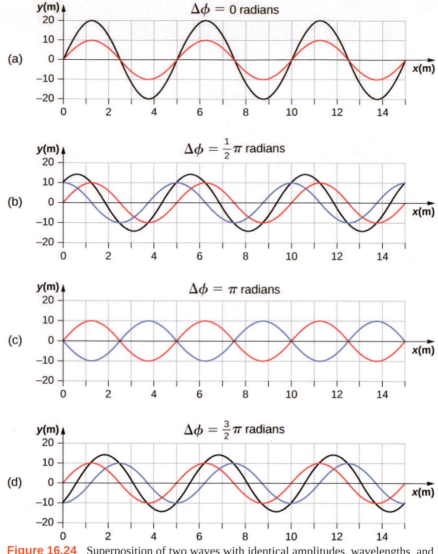

Figure 16.24 Superposition of two waves with identical amplitudes, wavelengths, and frequency, but that differ in a phase shift. The red wave is defined by the wave function $y_1(x, t) = A \sin(kx - \omega t)$ and the blue wave is defined by the wave function $y_2(x, t) = A \sin(kx - \omega t + \phi)$. The black line shows the result of adding the two waves. The phase difference between the two waves are (a) 0.00 rad, (b) $\pi/2$ rad, (c) π rad, and (d) $3\pi/2$ rad .

16.6 | Standing Waves and Resonance

Learning Objectives

By the end of this section, you will be able to:

- Describe standing waves and explain how they are produced
- Describe the modes of a standing wave on a string
- Provide examples of standing waves beyond the waves on a string

Throughout this chapter, we have been studying traveling waves, or waves that transport energy from one place to another. Under certain conditions, waves can bounce back and forth through a particular region, effectively becoming stationary. These are called **standing waves**.

Another related effect is known as resonance. In **Oscillations**, we defined resonance as a phenomenon in which a small-amplitude driving force could produce large-amplitude motion. Think of a child on a swing, which can be modeled as a physical pendulum. Relatively small-amplitude pushes by a parent can produce large-amplitude swings. Sometimes this resonance is good—for example, when producing music with a stringed instrument. At other times, the effects can be devastating, such as the collapse of a building during an earthquake. In the case of standing waves, the relatively large amplitude standing waves are produced by the superposition of smaller amplitude component waves.

Standing Waves

Sometimes waves do not seem to move; rather, they just vibrate in place. You can see unmoving waves on the surface of a glass of milk in a refrigerator, for example. Vibrations from the refrigerator motor create waves on the milk that oscillate up and down but do not seem to move across the surface. **Figure 16.25** shows an experiment you can try at home. Take a bowl of milk and place it on a common box fan. Vibrations from the fan will produce circular standing waves in the milk. The waves are visible in the photo due to the reflection from a lamp. These waves are formed by the superposition of two or more traveling waves, such as illustrated in **Figure 16.26** for two identical waves moving in opposite directions. The waves move through each other with their disturbances adding as they go by. If the two waves have the same amplitude and wavelength, then they alternate between constructive and destructive interference. The resultant looks like a wave standing in place and, thus, is called a standing wave.

Figure 16.25 Standing waves are formed on the surface of a bowl of milk sitting on a box fan. The vibrations from the fan causes the surface of the milk of oscillate. The waves are visible due to the reflection of light from a lamp. (credit: David Chelton)

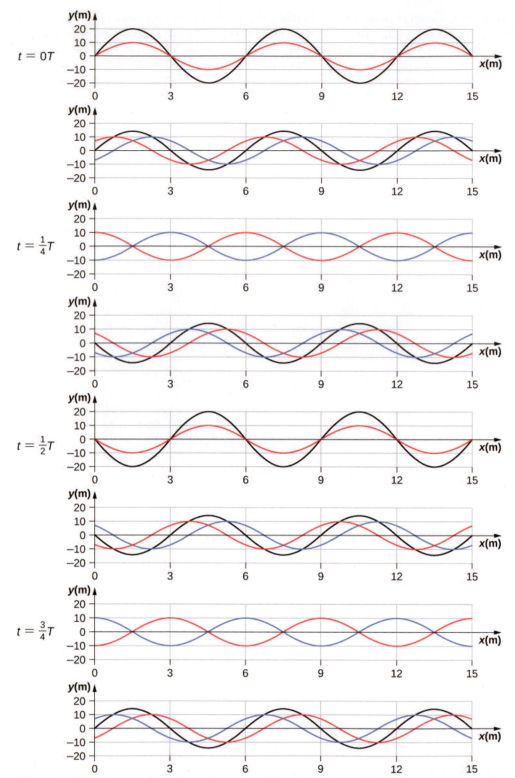

Figure 16.26 Time snapshots of two sine waves. The red wave is moving in the $-x$-direction and the blue wave is moving in the $+x$-direction. The resulting wave is shown in black. Consider the resultant wave at the points $x = 0\,\text{m}$, $3\,\text{m}$, $6\,\text{m}$, $9\,\text{m}$, $12\,\text{m}$, $15\,\text{m}$ and notice that the resultant wave always equals zero at these points, no matter what the time is. These points are known as fixed points (nodes). In between each two nodes is an antinode, a place where the medium oscillates with an amplitude equal to the sum of the amplitudes of the individual waves.

Consider two identical waves that move in opposite directions. The first wave has a wave function of

$y_1(x, t) = A\sin(kx - \omega t)$ and the second wave has a wave function $y_2(x, t) = A\sin(kx + \omega t)$. The waves interfere and form a resultant wave

$$y(x, t) = y_1(x, t) + y_2(x, t),$$
$$y(x, t) = A\sin(kx - \omega t) + A\sin(kx + \omega t).$$

This can be simplified using the trigonometric identity

$$\sin(\alpha \pm \beta) = \sin\alpha\cos\beta \pm \cos\alpha\sin\beta,$$

where $\alpha = kx$ and $\beta = \omega t$, giving us

$$y(x, t) = A[\sin(kx)\cos(\omega t) - \cos(kx)\sin(\omega t) + \sin(kx)\cos(\omega t) - \cos(kx)\sin(\omega t)],$$

which simplifies to

$$y(x, t) = [2A\sin(kx)]\cos(\omega t). \qquad\qquad (16.14)$$

Notice that the resultant wave is a sine wave that is a function only of position, multiplied by a cosine function that is a function only of time. Graphs of $y(x,t)$ as a function of x for various times are shown in **Figure 16.26**. The red wave moves in the negative x-direction, the blue wave moves in the positive x-direction, and the black wave is the sum of the two waves. As the red and blue waves move through each other, they move in and out of constructive interference and destructive interference.

Initially, at time $t = 0$, the two waves are in phase, and the result is a wave that is twice the amplitude of the individual waves. The waves are also in phase at the time $t = \frac{T}{2}$. In fact, the waves are in phase at any integer multiple of half of a period:

$$t = n\frac{T}{2} \text{ where } n = 0,\ 1,\ 2,\ 3.... \text{ (in phase)}.$$

At other times, the two waves are $180°(\pi \text{ radians})$ out of phase, and the resulting wave is equal to zero. This happens at

$$t = \tfrac{1}{4}T,\ \tfrac{3}{4}T,\ \tfrac{5}{4}T,...,\ \tfrac{n}{4}T \text{ where } n = 1,\ 3,\ 5.... \text{ (out of phase)}.$$

Notice that some x-positions of the resultant wave are always zero no matter what the phase relationship is. These positions are called **nodes**. Where do the nodes occur? Consider the solution to the sum of the two waves

$$y(x, t) = [2A\sin(kx)]\cos(\omega t).$$

Finding the positions where the sine function equals zero provides the positions of the nodes.

$$
\begin{aligned}
\sin(kx) &= 0 \\
kx &= 0,\ \pi,\ 2\pi,\ 3\pi,... \\
\tfrac{2\pi}{\lambda}x &= 0,\ \pi,\ 2\pi,\ 3\pi,... \\
x &= 0,\ \tfrac{\lambda}{2},\ \lambda,\ \tfrac{3\lambda}{2},... = n\tfrac{\lambda}{2} \quad n = 0,\ 1,\ 2,\ 3,....
\end{aligned}
$$

There are also positions where y oscillates between $y = \pm A$. These are the **antinodes**. We can find them by considering which values of x result in $\sin(kx) = \pm 1$.

$$
\begin{aligned}
\sin(kx) &= \pm 1 \\
kx &= \tfrac{\pi}{2},\ \tfrac{3\pi}{2},\ \tfrac{5\pi}{2},... \\
\tfrac{2\pi}{\lambda}x &= \tfrac{\pi}{2},\ \tfrac{3\pi}{2},\ \tfrac{5\pi}{2},... \\
x &= \tfrac{\lambda}{4},\ \tfrac{3\lambda}{4},\ \tfrac{5\lambda}{4},... = n\tfrac{\lambda}{4} \quad n = 1,\ 3,\ 5,....
\end{aligned}
$$

What results is a standing wave as shown in **Figure 16.27**, which shows snapshots of the resulting wave of two identical waves moving in opposite directions. The resulting wave appears to be a sine wave with nodes at integer multiples of half wavelengths. The antinodes oscillate between $y = \pm 2A$ due to the cosine term, $\cos(\omega t)$, which oscillates between ± 1.

The resultant wave appears to be standing still, with no apparent movement in the x-direction, although it is composed of one wave function moving in the positive, whereas the second wave is moving in the negative x-direction. **Figure 16.27** shows various snapshots of the resulting wave. The nodes are marked with red dots while the antinodes are marked with blue dots.

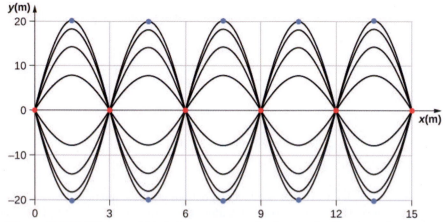

Figure 16.27 When two identical waves are moving in opposite directions, the resultant wave is a standing wave. Nodes appear at integer multiples of half wavelengths. Antinodes appear at odd multiples of quarter wavelengths, where they oscillate between $y = \pm A$. The nodes are marked with red dots and the antinodes are marked with blue dots.

A common example of standing waves are the waves produced by stringed musical instruments. When the string is plucked, pulses travel along the string in opposite directions. The ends of the strings are fixed in place, so nodes appear at the ends of the strings—the boundary conditions of the system, regulating the resonant frequencies in the strings. The resonance produced on a string instrument can be modeled in a physics lab using the apparatus shown in **Figure 16.28**.

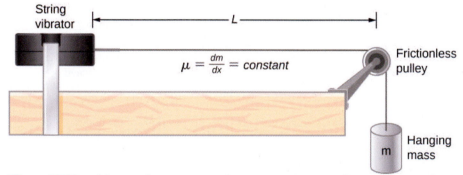

Figure 16.28 A lab setup for creating standing waves on a string. The string has a node on each end and a constant linear density. The length between the fixed boundary conditions is L. The hanging mass provides the tension in the string, and the speed of the waves on the string is proportional to the square root of the tension divided by the linear mass density.

The lab setup shows a string attached to a string vibrator, which oscillates the string with an adjustable frequency f. The other end of the string passes over a frictionless pulley and is tied to a hanging mass. The magnitude of the tension in the string is equal to the weight of the hanging mass. The string has a constant linear density (mass per length) μ and the speed at which a wave travels down the string equals $v = \sqrt{\frac{F_T}{\mu}} = \sqrt{\frac{mg}{\mu}}$ **Equation 16.7**. The symmetrical boundary conditions (a node at each end) dictate the possible frequencies that can excite standing waves. Starting from a frequency of zero and slowly increasing the frequency, the first mode $n = 1$ appears as shown in **Figure 16.29**. The first mode, also called the fundamental mode or the first harmonic, shows half of a wavelength has formed, so the wavelength is equal to twice the length between the nodes $\lambda_1 = 2L$. The **fundamental frequency**, or first harmonic frequency, that drives this mode is

$$f_1 = \frac{v}{\lambda_1} = \frac{v}{2L},$$

where the speed of the wave is $v = \sqrt{\frac{F_T}{\mu}}$. Keeping the tension constant and increasing the frequency leads to the second harmonic or the $n = 2$ mode. This mode is a full wavelength $\lambda_2 = L$ and the frequency is twice the fundamental frequency:

$$f_2 = \frac{v}{\lambda_2} = \frac{v}{L} = 2f_1.$$

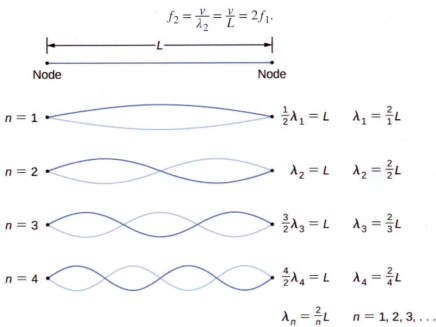

$$\lambda_n = \frac{2}{n}L \qquad n = 1, 2, 3, \ldots$$

Figure 16.29 Standing waves created on a string of length L. A node occurs at each end of the string. The nodes are boundary conditions that limit the possible frequencies that excite standing waves. (Note that the amplitudes of the oscillations have been kept constant for visualization. The standing wave patterns possible on the string are known as the normal modes. Conducting this experiment in the lab would result in a decrease in amplitude as the frequency increases.)

The next two modes, or the third and fourth harmonics, have wavelengths of $\lambda_3 = \frac{2}{3}L$ and $\lambda_4 = \frac{2}{4}L$, driven by frequencies of $f_3 = \frac{3v}{2L} = 3f_1$ and $f_4 = \frac{4v}{2L} = 4f_1$. All frequencies above the frequency f_1 are known as the **overtones**. The equations for the wavelength and the frequency can be summarized as:

$$\lambda_n = \frac{2}{n}L \quad n = 1, 2, 3, 4, 5\ldots \tag{16.15}$$

$$f_n = n\frac{v}{2L} = nf_1 \quad n = 1, 2, 3, 4, 5\ldots \tag{16.16}$$

The standing wave patterns that are possible for a string, the first four of which are shown in **Figure 16.29**, are known as the **normal modes**, with frequencies known as the normal frequencies. In summary, the first frequency to produce a normal mode is called the fundamental frequency (or first harmonic). Any frequencies above the fundamental frequency are overtones. The second frequency of the $n = 2$ normal mode of the string is the first overtone (or second harmonic). The frequency of the $n = 3$ normal mode is the second overtone (or third harmonic) and so on.

The solutions shown as **Equation 16.15** and **Equation 16.16** are for a string with the boundary condition of a node on each end. When the boundary condition on either side is the same, the system is said to have symmetric boundary conditions. **Equation 16.15** and **Equation 16.16** are good for any symmetric boundary conditions, that is, nodes at both ends or antinodes at both ends.

Example 16.7

Standing Waves on a String

Consider a string of $L = 2.00$ m. attached to an adjustable-frequency string vibrator as shown in **Figure 16.30**. The waves produced by the vibrator travel down the string and are reflected by the fixed boundary condition at the pulley. The string, which has a linear mass density of $\mu = 0.006$ kg/m, is passed over a frictionless pulley of a negligible mass, and the tension is provided by a 2.00-kg hanging mass. (a) What is the velocity of the waves on the string? (b) Draw a sketch of the first three normal modes of the standing waves that can be produced on the string and label each with the wavelength. (c) List the frequencies that the string vibrator must be tuned to in order to produce the first three normal modes of the standing waves.

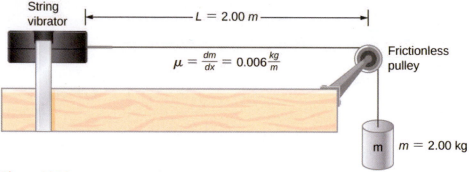

Figure 16.30   A string attached to an adjustable-frequency string vibrator.

Strategy

a. The velocity of the wave can be found using $v = \sqrt{\frac{F_T}{\mu}}$. The tension is provided by the weight of the hanging mass.

b. The standing waves will depend on the boundary conditions. There must be a node at each end. The first mode will be one half of a wave. The second can be found by adding a half wavelength. That is the shortest length that will result in a node at the boundaries. For example, adding one quarter of a wavelength will result in an antinode at the boundary and is not a mode which would satisfy the boundary conditions. This is shown in **Figure 16.31**.

c. Since the wave speed velocity is the wavelength times the frequency, the frequency is wave speed divided by the wavelength.

Figure 16.31   (a) The figure represents the second mode of the string that satisfies the boundary conditions of a node at each end of the string. (b)This figure could not possibly be a normal mode on the string because it does not satisfy the boundary conditions. There is a node on one end, but an antinode on the other.

Solution

a. Begin with the velocity of a wave on a string. The tension is equal to the weight of the hanging mass. The linear mass density and mass of the hanging mass are given:

$$v = \sqrt{\frac{F_T}{\mu}} = \sqrt{\frac{mg}{\mu}} = \sqrt{\frac{2\,\text{kg}(9.8\frac{m}{s})}{0.006\frac{kg}{m}}} = 57.15\ \text{m/s}.$$

b. The first normal mode that has a node on each end is a half wavelength. The next two modes are found by adding a half of a wavelength.

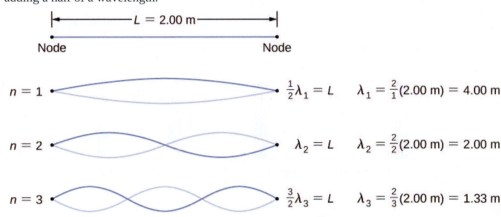

$n = 1$ $\frac{1}{2}\lambda_1 = L$ $\lambda_1 = \frac{2}{1}(2.00\ \text{m}) = 4.00\ \text{m}$

$n = 2$ $\lambda_2 = L$ $\lambda_2 = \frac{2}{2}(2.00\ \text{m}) = 2.00\ \text{m}$

$n = 3$ $\frac{3}{2}\lambda_3 = L$ $\lambda_3 = \frac{2}{3}(2.00\ \text{m}) = 1.33\ \text{m}$

c. The frequencies of the first three modes are found by using $f = \frac{v_w}{\lambda}$.

$$f_1 = \frac{v_w}{\lambda_1} = \frac{57.15\ \text{m/s}}{4.00\ \text{m}} = 14.29\ \text{Hz}$$

$$f_2 = \frac{v_w}{\lambda_2} = \frac{57.15\ \text{m/s}}{2.00\ \text{m}} = 28.58\ \text{Hz}$$

$$f_3 = \frac{v_w}{\lambda_3} = \frac{57.15\ \text{m/s}}{1.333\ \text{m}} = 42.87\ \text{Hz}$$

Significance

The three standing modes in this example were produced by maintaining the tension in the string and adjusting the driving frequency. Keeping the tension in the string constant results in a constant velocity. The same modes could have been produced by keeping the frequency constant and adjusting the speed of the wave in the string (by changing the hanging mass.)

 Visit this **simulation (https://openstaxcollege.org/l/21normalmodes)** to play with a 1D or 2D system of coupled mass-spring oscillators. Vary the number of masses, set the initial conditions, and watch the system evolve. See the spectrum of normal modes for arbitrary motion. See longitudinal or transverse modes in the 1D system.

 16.7 Check Your Understanding The equations for the wavelengths and the frequencies of the modes of a wave produced on a string:

$$\lambda_n = \frac{2}{n}L \quad n = 1,\ 2,\ 3,\ 4,\ 5... \text{ and}$$

$$f_n = n\frac{v}{2L} = nf_1 \quad n = 1,\ 2,\ 3,\ 4,\ 5...$$

were derived by considering a wave on a string where there were symmetric boundary conditions of a node at each end. These modes resulted from two sinusoidal waves with identical characteristics except they were moving in opposite directions, confined to a region L with nodes required at both ends. Will the same equations work if there were symmetric boundary conditions with antinodes at each end? What would the normal modes look like for a medium that was free to oscillate on each end? Don't worry for now if you cannot imagine such a medium, just consider two sinusoidal wave functions in a region of length L, with antinodes on each end.

The free boundary conditions shown in the last Check Your Understanding may seem hard to visualize. How can there be a system that is free to oscillate on each end? In **Figure 16.32** are shown two possible configuration of a metallic rods (shown in red) attached to two supports (shown in blue). In part (a), the rod is supported at the ends, and there are fixed boundary conditions at both ends. Given the proper frequency, the rod can be driven into resonance with a wavelength equal to length of the rod, with nodes at each end. In part (b), the rod is supported at positions one quarter of the length from each end of the rod, and there are free boundary conditions at both ends. Given the proper frequency, this rod can also be driven into resonance with a wavelength equal to the length of the rod, but there are antinodes at each end. If you are having trouble visualizing the wavelength in this figure, remember that the wavelength may be measured between any two nearest identical points and consider **Figure 16.33**.

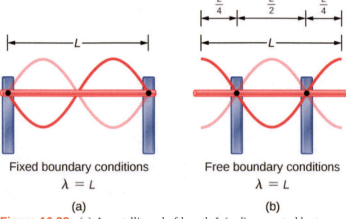

Figure 16.32 (a) A metallic rod of length L (red) supported by two supports (blue) on each end. When driven at the proper frequency, the rod can resonate with a wavelength equal to the length of the rod with a node on each end. (b) The same metallic rod of length L (red) supported by two supports (blue) at a position a quarter of the length of the rod from each end. When driven at the proper frequency, the rod can resonate with a wavelength equal to the length of the rod with an antinode on each end.

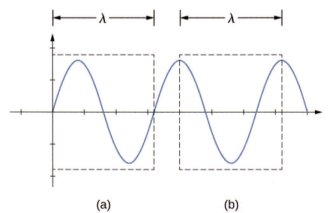

Figure 16.33 A wavelength may be measure between the nearest two repeating points. On the wave on a string, this means the same height and slope. (a) The wavelength is measured between the two nearest points where the height is zero and the slope is maximum and positive. (b) The wavelength is measured between two identical points where the height is maximum and the slope is zero.

Note that the study of standing waves can become quite complex. In **Figure 16.32**(a), the $n = 2$ mode of the standing wave is shown, and it results in a wavelength equal to L. In this configuration, the $n = 1$ mode would also have been possible with a standing wave equal to $2L$. Is it possible to get the $n = 1$ mode for the configuration shown in part (b)? The answer is no. In this configuration, there are additional conditions set beyond the boundary conditions. Since the rod is mounted at a point one quarter of the length from each side, a node must exist there, and this limits the possible modes of standing waves that can be created. We leave it as an exercise for the reader to consider if other modes of standing waves are possible. It should be noted that when a system is driven at a frequency that does not cause the system to resonate, vibrations may still occur, but the amplitude of the vibrations will be much smaller than the amplitude at resonance.

A field of mechanical engineering uses the sound produced by the vibrating parts of complex mechanical systems to troubleshoot problems with the systems. Suppose a part in an automobile is resonating at the frequency of the car's engine, causing unwanted vibrations in the automobile. This may cause the engine to fail prematurely. The engineers use microphones to record the sound produced by the engine, then use a technique called Fourier analysis to find frequencies of sound produced with large amplitudes and then look at the parts list of the automobile to find a part that would resonate at that frequency. The solution may be as simple as changing the composition of the material used or changing the length of the part in question.

There are other numerous examples of resonance in standing waves in the physical world. The air in a tube, such as found in a musical instrument like a flute, can be forced into resonance and produce a pleasant sound, as we discuss in **Sound**.

At other times, resonance can cause serious problems. A closer look at earthquakes provides evidence for conditions appropriate for resonance, standing waves, and constructive and destructive interference. A building may vibrate for several seconds with a driving frequency matching that of the natural frequency of vibration of the building—producing a resonance resulting in one building collapsing while neighboring buildings do not. Often, buildings of a certain height are devastated while other taller buildings remain intact. The building height matches the condition for setting up a standing wave for that particular height. The span of the roof is also important. Often it is seen that gymnasiums, supermarkets, and churches suffer damage when individual homes suffer far less damage. The roofs with large surface areas supported only at the edges resonate at the frequencies of the earthquakes, causing them to collapse. As the earthquake waves travel along the surface of Earth and reflect off denser rocks, constructive interference occurs at certain points. Often areas closer to the epicenter are not damaged, while areas farther away are damaged.

CHAPTER 16 REVIEW

KEY TERMS

antinode location of maximum amplitude in standing waves

constructive interference when two waves arrive at the same point exactly in phase; that is, the crests of the two waves are precisely aligned, as are the troughs

destructive interference when two identical waves arrive at the same point exactly out of phase; that is, precisely aligned crest to trough

fixed boundary condition when the medium at a boundary is fixed in place so it cannot move

free boundary condition exists when the medium at the boundary is free to move

fundamental frequency lowest frequency that will produce a standing wave

intensity (*I*) power per unit area

interference overlap of two or more waves at the same point and time

linear wave equation equation describing waves that result from a linear restoring force of the medium; any function that is a solution to the wave equation describes a wave moving in the positive *x*-direction or the negative *x*-direction with a constant wave speed *v*

longitudinal wave wave in which the disturbance is parallel to the direction of propagation

mechanical wave wave that is governed by Newton's laws and requires a medium

node point where the string does not move; more generally, nodes are where the wave disturbance is zero in a standing wave

normal mode possible standing wave pattern for a standing wave on a string

overtone frequency that produces standing waves and is higher than the fundamental frequency

pulse single disturbance that moves through a medium, transferring energy but not mass

standing wave wave that can bounce back and forth through a particular region, effectively becoming stationary

superposition phenomenon that occurs when two or more waves arrive at the same point

transverse wave wave in which the disturbance is perpendicular to the direction of propagation

wave disturbance that moves from its source and carries energy

wave function mathematical model of the position of particles of the medium

wave number $\frac{2\pi}{\lambda}$

wave speed magnitude of the wave velocity

wave velocity velocity at which the disturbance moves; also called the propagation velocity

wavelength distance between adjacent identical parts of a wave

KEY EQUATIONS

Wave speed	$v = \frac{\lambda}{T} = \lambda f$
Linear mass density	$\mu = \dfrac{\text{mass of the string}}{\text{length of the string}}$
Speed of a wave or pulse on a string under tension	$|v| = \sqrt{\dfrac{F_T}{\mu}}$

Speed of a compression wave in a fluid	$v = \sqrt{\dfrac{B}{\rho}}$
Resultant wave from superposition of two sinusoidal waves that are identical except for a phase shift	$y_R(x,\ t) = \left[2A\cos\left(\dfrac{\phi}{2}\right)\right]\sin\left(kx - \omega t + \dfrac{\phi}{2}\right)$
Wave number	$k \equiv \dfrac{2\pi}{\lambda}$
Wave speed	$v = \dfrac{\omega}{k}$
A periodic wave	$y(x,\ t) = A\sin(kx \mp \omega t + \phi)$
Phase of a wave	$kx \mp \omega t + \phi$
The linear wave equation	$\dfrac{\partial^2 y(x,\ t)}{\partial x^2} = \dfrac{1}{v_w^2}\dfrac{\partial^2 y(x,\ t)}{\partial t^2}$
Power averaged over a wavelength	$P_{ave} = \dfrac{E_\lambda}{T} = \dfrac{1}{2}\mu A^2 \omega^2 \dfrac{\lambda}{T} = \dfrac{1}{2}\mu A^2 \omega^2 v$
Intensity	$I = \dfrac{P}{A}$
Intensity for a spherical wave	$I = \dfrac{P}{4\pi r^2}$
Equation of a standing wave	$y(x,\ t) = [2A\sin(kx)]\cos(\omega t)$
Wavelength for symmetric boundary conditions	$\lambda_n = \dfrac{2}{n}L, \quad n = 1,\ 2,\ 3,\ 4,\ 5...$
Frequency for symmetric boundary conditions	$f_n = n\dfrac{v}{2L} = nf_1, \quad n = 1,\ 2,\ 3,\ 4,\ 5...$

SUMMARY

16.1 Traveling Waves

- A wave is a disturbance that moves from the point of origin with a wave velocity v.

- A wave has a wavelength λ, which is the distance between adjacent identical parts of the wave. Wave velocity and wavelength are related to the wave's frequency and period by $v = \dfrac{\lambda}{T} = \lambda f$.

- Mechanical waves are disturbances that move through a medium and are governed by Newton's laws.

- Electromagnetic waves are disturbances in the electric and magnetic fields, and do not require a medium.

- Matter waves are a central part of quantum mechanics and are associated with protons, electrons, neutrons, and other fundamental particles found in nature.

- A transverse wave has a disturbance perpendicular to the wave's direction of propagation, whereas a longitudinal wave has a disturbance parallel to its direction of propagation.

16.2 Mathematics of Waves

- A wave is an oscillation (of a physical quantity) that travels through a medium, accompanied by a transfer of energy. Energy transfers from one point to another in the direction of the wave motion. The particles of the medium oscillate up and down, back and forth, or both up and down and back and forth, around an equilibrium position.

- A snapshot of a sinusoidal wave at time $t = 0.00\ \text{s}$ can be modeled as a function of position. Two examples of such functions are $y(x) = A \sin(kx + \phi)$ and $y(x) = A \cos(kx + \phi)$.

- Given a function of a wave that is a snapshot of the wave, and is only a function of the position x, the motion of the pulse or wave moving at a constant velocity can be modeled with the function, replacing x with $x \mp vt$. The minus sign is for motion in the positive direction and the plus sign for the negative direction.

- The wave function is given by $y(x, t) = A \sin(kx - \omega t + \phi)$ where $k = 2\pi/\lambda$ is defined as the wave number, $\omega = 2\pi/T$ is the angular frequency, and ϕ is the phase shift.

- The wave moves with a constant velocity v_w, where the particles of the medium oscillate about an equilibrium position. The constant velocity of a wave can be found by $v = \frac{\lambda}{T} = \frac{\omega}{k}$.

16.3 Wave Speed on a Stretched String

- The speed of a wave on a string depends on the linear density of the string and the tension in the string. The linear density is mass per unit length of the string.

- In general, the speed of a wave depends on the square root of the ratio of the elastic property to the inertial property of the medium.

- The speed of a wave through a fluid is equal to the square root of the ratio of the bulk modulus of the fluid to the density of the fluid.

- The speed of sound through air at $T = 20°C$ is approximately $v_s = 343.00$ m/s.

16.4 Energy and Power of a Wave

- The energy and power of a wave are proportional to the square of the amplitude of the wave and the square of the angular frequency of the wave.

- The time-averaged power of a sinusoidal wave on a string is found by $P_{ave} = \frac{1}{2}\mu A^2 \omega^2 v$, where μ is the linear mass density of the string, A is the amplitude of the wave, ω is the angular frequency of the wave, and v is the speed of the wave.

- Intensity is defined as the power divided by the area. In a spherical wave, the area is $A = 4\pi r^2$ and the intensity is $I = \frac{P}{4\pi r^2}$. As the wave moves out from a source, the energy is conserved, but the intensity decreases as the area increases.

16.5 Interference of Waves

- Superposition is the combination of two waves at the same location.

- Constructive interference occurs from the superposition of two identical waves that are in phase.

- Destructive interference occurs from the superposition of two identical waves that are $180°(\pi \text{ radians})$ out of phase.

- The wave that results from the superposition of two sine waves that differ only by a phase shift is a wave with an amplitude that depends on the value of the phase difference.

16.6 Standing Waves and Resonance

- A standing wave is the superposition of two waves which produces a wave that varies in amplitude but does not propagate.

- Nodes are points of no motion in standing waves.

- An antinode is the location of maximum amplitude of a standing wave.

- Normal modes of a wave on a string are the possible standing wave patterns. The lowest frequency that will produce a standing wave is known as the fundamental frequency. The higher frequencies which produce standing waves are called overtones.

CONCEPTUAL QUESTIONS

16.1 Traveling Waves

1. Give one example of a transverse wave and one example of a longitudinal wave, being careful to note the relative directions of the disturbance and wave propagation in each.

2. A sinusoidal transverse wave has a wavelength of 2.80 m. It takes 0.10 s for a portion of the string at a position x to move from a maximum position of $y = 0.03$ m to the equilibrium position $y = 0$. What are the period, frequency, and wave speed of the wave?

3. What is the difference between propagation speed and the frequency of a mechanical wave? Does one or both affect wavelength? If so, how?

4. Consider a stretched spring, such as a slinky. The stretched spring can support longitudinal waves and transverse waves. How can you produce transverse waves on the spring? How can you produce longitudinal waves on the spring?

5. Consider a wave produced on a stretched spring by holding one end and shaking it up and down. Does the wavelength depend on the distance you move your hand up and down?

6. A sinusoidal, transverse wave is produced on a stretched spring, having a period T. Each section of the spring moves perpendicular to the direction of propagation of the wave, in simple harmonic motion with an amplitude A. Does each section oscillate with the same period as the wave or a different period? If the amplitude of the transverse wave were doubled but the period stays the same, would your answer be the same?

7. An electromagnetic wave, such as light, does not require a medium. Can you think of an example that would support this claim?

16.2 Mathematics of Waves

8. If you were to shake the end of a taut spring up and down 10 times a second, what would be the frequency and the period of the sinusoidal wave produced on the spring?

9. If you shake the end of a stretched spring up and down with a frequency f, you can produce a sinusoidal, transverse wave propagating down the spring. Does the wave number depend on the frequency you are shaking the spring?

10. Does the vertical speed of a segment of a horizontal taut string through which a sinusoidal, transverse wave is propagating depend on the wave speed of the transverse wave?

11. In this section, we have considered waves that move at a constant wave speed. Does the medium accelerate?

12. If you drop a pebble in a pond you may notice that several concentric ripples are produced, not just a single ripple. Why do you think that is?

16.3 Wave Speed on a Stretched String

13. If the tension in a string were increased by a factor of four, by what factor would the wave speed of a wave on the string increase?

14. Does a sound wave move faster in seawater or fresh water, if both the sea water and fresh water are at the same temperature and the sound wave moves near the surface? $\left(\rho_w \approx 1000 \dfrac{\text{kg}}{\text{m}^3}, \rho_s \approx 1030 \dfrac{\text{kg}}{\text{m}^3}, B_w = 2.15 \times 10^9 \text{ Pa}, \right.$

$\left. B_s = 2.34 \times 10^9 \text{ Pa} \right)$

15. Guitars have strings of different linear mass density. If the lowest density string and the highest density string are under the same tension, which string would support waves with the higher wave speed?

16. Shown below are three waves that were sent down a string at different times. The tension in the string remains constant. (a) Rank the waves from the smallest wavelength to the largest wavelength. (b) Rank the waves from the lowest frequency to the highest frequency.

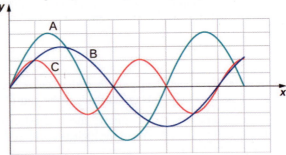

17. Electrical power lines connected by two utility poles are sometimes heard to hum when driven into oscillation by the wind. The speed of the waves on the power lines depend on the tension. What provides the tension in the power lines?

18. Two strings, one with a low mass density and one with a high linear density are spliced together. The higher density end is tied to a lab post and a student holds the free end of the low-mass density string. The student gives the string a flip and sends a pulse down the strings. If the tension is the same in both strings, does the pulse travel at the same wave velocity in both strings? If not, where does it travel faster, in the low density string or the high density string?

16.4 Energy and Power of a Wave

19. Consider a string with under tension with a constant linear mass density. A sinusoidal wave with an angular frequency and amplitude produced by some external driving force. If the frequency of the driving force is decreased to half of the original frequency, how is the time-averaged power of the wave affected? If the amplitude of the driving force is decreased by half, how is the time-averaged power affected? Explain your answer.

20. Circular water waves decrease in amplitude as they move away from where a rock is dropped. Explain why.

21. In a transverse wave on a string, the motion of the string is perpendicular to the motion of the wave. If this is so, how is possible to move energy along the length of the string?

22. The energy from the sun warms the portion of the earth facing the sun during the daylight hours. Why are the North and South Poles cold while the equator is quite warm?

23. The intensity of a spherical waves decreases as the wave moves away from the source. If the intensity of the wave at the source is I_0, how far from the source will the intensity decrease by a factor of nine?

16.5 Interference of Waves

24. An incident sinusoidal wave is sent along a string that is fixed to the wall with a wave speed of v. The wave reflects off the end of the string. Describe the reflected wave.

25. A string of a length of 2.00 m with a linear mass density of $\mu = 0.006$ kg/m is attached to the end of a 2.00-m-long string with a linear mass density of $\mu = 0.012$ kg/m. The free end of the higher-density string is fixed to the wall, and a student holds the free end of the low-density string, keeping the tension constant in both strings. The student sends a pulse down the string. Describe what happens at the interface between the two strings.

26. A long, tight spring is held by two students, one

student holding each end. Each student gives the end a flip sending one wavelength of a sinusoidal wave down the spring in opposite directions. When the waves meet in the middle, what does the wave look like?

27. Many of the topics discussed in this chapter are useful beyond the topics of mechanical waves. It is hard to conceive of a mechanical wave with sharp corners, but you could encounter such a wave form in your digital electronics class, as shown below. This could be a signal from a device known as an analog to digital converter, in which a continuous voltage signal is converted into a discrete signal or a digital recording of sound. What is the result of the superposition of the two signals?

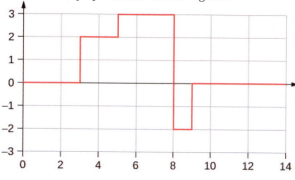

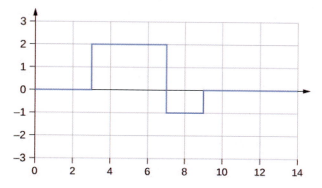

28. A string of a constant linear mass density is held taut by two students, each holding one end. The tension in the string is constant. The students each send waves down the string by wiggling the string. (a) Is it possible for the waves to have different wave speeds? (b) Is it possible for the waves to have different frequencies? (c) Is it possible for the waves to have different wavelengths?

16.6 Standing Waves and Resonance

29. A truck manufacturer finds that a strut in the engine is failing prematurely. A sound engineer determines that the strut resonates at the frequency of the engine and suspects that this could be the problem. What are two possible characteristics of the strut can be modified to correct the problem?

30. Why do roofs of gymnasiums and churches seem to fail more than family homes when an earthquake occurs?

31. Wine glasses can be set into resonance by moistening your finger and rubbing it around the rim of the glass. Why?

32. Air conditioning units are sometimes placed on the roof of homes in the city. Occasionally, the air conditioners cause an undesirable hum throughout the upper floors of the homes. Why does this happen? What can be done to reduce the hum?

33. Consider a standing wave modeled as $y(x, t) = 4.00\,\text{cm}\,\sin(3\,\text{m}^{-1}\,x)\cos(4\,\text{s}^{-1}\,t)$. Is there a node or an antinode at $x = 0.00\,\text{m}$? What about a standing wave modeled as $y(x, t) = 4.00\,\text{cm}\,\sin\left(3\,\text{m}^{-1}\,x + \frac{\pi}{2}\right)\cos(4\,\text{s}^{-1}\,t)$? Is there a node or an antinode at the $x = 0.00\,\text{m}$ position?

PROBLEMS

16.1 Traveling Waves

34. Storms in the South Pacific can create waves that travel all the way to the California coast, 12,000 km away. How long does it take them to travel this distance if they travel at 15.0 m/s?

35. Waves on a swimming pool propagate at 0.75 m/s. You splash the water at one end of the pool and observe the wave go to the opposite end, reflect, and return in 30.00 s. How far away is the other end of the pool?

36. Wind gusts create ripples on the ocean that have a wavelength of 5.00 cm and propagate at 2.00 m/s. What is their frequency?

37. How many times a minute does a boat bob up and down on ocean waves that have a wavelength of 40.0 m and a propagation speed of 5.00 m/s?

38. Scouts at a camp shake the rope bridge they have just crossed and observe the wave crests to be 8.00 m apart. If they shake the bridge twice per second, what is the propagation speed of the waves?

39. What is the wavelength of the waves you create in a swimming pool if you splash your hand at a rate of 2.00 Hz and the waves propagate at a wave speed of 0.800 m/s?

40. What is the wavelength of an earthquake that shakes you with a frequency of 10.0 Hz and gets to another city 84.0 km away in 12.0 s?

41. Radio waves transmitted through empty space at the speed of light $\left(v = c = 3.00 \times 10^8 \text{ m/s}\right)$ by the *Voyager* spacecraft have a wavelength of 0.120 m. What is their frequency?

42. Your ear is capable of differentiating sounds that arrive at each ear just 0.34 ms apart, which is useful in determining where low frequency sound is originating from. (a) Suppose a low-frequency sound source is placed to the right of a person, whose ears are approximately 18 cm apart, and the speed of sound generated is 340 m/s. How long is the interval between when the sound arrives at the right ear and the sound arrives at the left ear? (b) Assume the same person was scuba diving and a low-frequency sound source was to the right of the scuba diver. How long is the interval between when the sound arrives at the right ear and the sound arrives at the left ear, if the speed of sound in water is 1500 m/s? (c) What is significant about the time interval of the two situations?

43. (a) Seismographs measure the arrival times of earthquakes with a precision of 0.100 s. To get the distance to the epicenter of the quake, geologists compare the arrival times of S- and P-waves, which travel at different speeds. If S- and P-waves travel at 4.00 and 7.20 km/s, respectively, in the region considered, how precisely can the distance to the source of the earthquake be determined? (b) Seismic waves from underground detonations of nuclear bombs can be used to locate the test site and detect violations of test bans. Discuss whether your answer to (a) implies a serious limit to such detection. (Note also that the uncertainty is greater if there is an uncertainty in the propagation speeds of the S- and P-waves.)

44. A Girl Scout is taking a 10.00-km hike to earn a merit badge. While on the hike, she sees a cliff some distance away. She wishes to estimate the time required to walk to the cliff. She knows that the speed of sound is approximately 343 meters per second. She yells and finds that the echo returns after approximately 2.00 seconds. If she can hike 1.00 km in 10 minutes, how long would it take her to reach the cliff?

45. A quality assurance engineer at a frying pan company is asked to qualify a new line of nonstick-coated frying pans. The coating needs to be 1.00 mm thick. One method to test the thickness is for the engineer to pick a percentage of the pans manufactured, strip off the coating, and measure the thickness using a micrometer. This method is a destructive testing method. Instead, the engineer decides that every frying pan will be tested using a nondestructive method. An ultrasonic transducer is used that produces sound waves with a frequency of $f = 25\,\text{kHz}$. The sound waves are sent through the coating and are reflected by the interface between the coating and the metal pan, and the time is recorded. The wavelength of the ultrasonic waves in the coating is 0.076 m. What should be the time recorded if the coating is the correct thickness (1.00 mm)?

16.2 Mathematics of Waves

46. A pulse can be described as a single wave disturbance that moves through a medium. Consider a pulse that is defined at time $t = 0.00$ s by the equation

$$y(x) = \frac{6.00 \text{ m}^3}{x^2 + 2.00 \text{ m}^2}$$ centered around $x = 0.00$ m. The

pulse moves with a velocity of $v = 3.00$ m/s in the positive x-direction. (a) What is the amplitude of the pulse? (b) What is the equation of the pulse as a function of position and time? (c) Where is the pulse centered at time $t = 5.00$ s?

47. A transverse wave on a string is modeled with the wave function

$$y(x, t) = (0.20 \text{ cm})\sin\left(2.00 \text{ m}^{-1}x - 3.00 \text{ s}^{-1}t + \frac{\pi}{16}\right).$$

What is the height of the string with respect to the equilibrium position at a position $x = 4.00$ m and a time $t = 10.00$ s?

48. Consider the wave function

$$y(x, t) = (3.00 \text{ cm})\sin\left(0.4 \text{ m}^{-1}x + 2.00 \text{ s}^{-1}t + \frac{\pi}{10}\right).$$

What are the period, wavelength, speed, and initial phase shift of the wave modeled by the wave function?

49. A pulse is defined as

$$y(x, t) = e^{-2.77\left(\frac{2.00(x - 2.00 \text{ m/s}(t))}{5.00 \text{ m}}\right)^2}.$$ Use a spreadsheet,

or other computer program, to plot the pulse as the height of medium y as a function of position x. Plot the pulse at times $t = 0.00$ s and $t = 3.00$ s on the same graph. Where is the pulse centered at time $t = 3.00$ s? Use your spreadsheet to check your answer.

50. A wave is modeled at time $t = 0.00$ s with a wave function that depends on position. The equation is $y(x) = (0.30 \text{ m})\sin\left(6.28 \text{ m}^{-1}x\right)$. The wave travels a distance of 4.00 meters in 0.50 s in the positive x-direction. Write an equation for the wave as a function of position and time.

51. A wave is modeled with the function $y(x, t) = (0.25 \text{ m})\cos\left(0.30 \text{ m}^{-1}x - 0.90 \text{ s}^{-1}t + \frac{\pi}{3}\right).$ Find the (a) amplitude, (b) wave number, (c) angular

frequency, (d) wave speed, (e) phase shift, (f) wavelength, and (g) period of the wave.

52. A surface ocean wave has an amplitude of 0.60 m and the distance from trough to trough is 8.00 m. It moves at a constant wave speed of 1.50 m/s propagating in the positive x-direction. At $t = 0$, the water displacement at $x = 0$ is zero, and v_y is positive. (a) Assuming the wave can be modeled as a sine wave, write a wave function to model the wave. (b) Use a spreadsheet to plot the wave function at times $t = 0.00$ s and $t = 2.00$ s on the same graph. Verify that the wave moves 3.00 m in those 2.00 s.

53. A wave is modeled by the wave function $y(x, t) = (0.30 \text{ m})\sin\left[\frac{2\pi}{4.50 \text{ m}}\left(x - 18.00\frac{\text{m}}{\text{s}}t\right)\right].$ What are the amplitude, wavelength, wave speed, period, and frequency of the wave?

54. A transverse wave on a string is described with the wave function

$$y(x, t) = (0.50 \text{ cm})\sin\left(1.57 \text{ m}^{-1}x - 6.28 \text{ s}^{-1}t\right). \quad \text{(a)}$$

What is the wave velocity of the wave? (b) What is the magnitude of the maximum velocity of the string perpendicular to the direction of the motion?

55. A swimmer in the ocean observes one day that the ocean surface waves are periodic and resemble a sine wave. The swimmer estimates that the vertical distance between the crest and the trough of each wave is approximately 0.45 m, and the distance between each crest is approximately 1.8 m. The swimmer counts that 12 waves pass every two minutes. Determine the simple harmonic wave function that would describes these waves.

56. Consider a wave described by the wave function $y(x, t) = 0.3 \text{ m} \sin\left(2.00 \text{ m}^{-1}x - 628.00 \text{ s}^{-1}t\right).$ (a) How many crests pass by an observer at a fixed location in 2.00 minutes? (b) How far has the wave traveled in that time?

57. Consider two waves defined by the wave functions $y_1(x, t) = 0.50 \text{ m} \sin\left(\frac{2\pi}{3.00 \text{ m}}x + \frac{2\pi}{4.00 \text{ s}}t\right)$ and $y_2(x, t) = 0.50 \text{ m} \sin\left(\frac{2\pi}{6.00 \text{ m}}x - \frac{2\pi}{4.00 \text{ s}}t\right).$ What are the similarities and differences between the two waves?

58. Consider two waves defined by the wave functions $y_1(x, t) = 0.20 \text{ m} \sin\left(\frac{2\pi}{6.00 \text{ m}}x - \frac{2\pi}{4.00 \text{ s}}t\right)$ and $y_2(x, t) = 0.20 \text{ m} \cos\left(\frac{2\pi}{6.00 \text{ m}}x - \frac{2\pi}{4.00 \text{ s}}t\right).$ What are the similarities and differences between the two waves?

59. The speed of a transverse wave on a string is 300.00 m/s, its wavelength is 0.50 m, and the amplitude is 20.00 cm. How much time is required for a particle on the string to move through a distance of 5.00 km?

16.3 Wave Speed on a Stretched String

60. Transverse waves are sent along a 5.00-m-long string with a speed of 30.00 m/s. The string is under a tension of 10.00 N. What is the mass of the string?

61. A copper wire has a density of $\rho = 8920 \, \text{kg/m}^3$, a radius of 1.20 mm, and a length L. The wire is held under a tension of 10.00 N. Transverse waves are sent down the wire. (a) What is the linear mass density of the wire? (b) What is the speed of the waves through the wire?

62. A piano wire has a linear mass density of $\mu = 4.95 \times 10^{-3} \, \text{kg/m}$. Under what tension must the string be kept to produce waves with a wave speed of 500.00 m/s?

63. A string with a linear mass density of $\mu = 0.0060 \, \text{kg/m}$ is tied to the ceiling. A 20-kg mass is tied to the free end of the string. The string is plucked, sending a pulse down the string. Estimate the speed of the pulse as it moves down the string.

64. A cord has a linear mass density of $\mu = 0.0075 \, \text{kg/m}$ and a length of three meters. The cord is plucked and it takes 0.20 s for the pulse to reach the end of the string. What is the tension of the string?

65. A string is 3.00 m long with a mass of 5.00 g. The string is held taut with a tension of 500.00 N applied to the string. A pulse is sent down the string. How long does it take the pulse to travel the 3.00 m of the string?

66. Two strings are attached to poles, however the first string is twice as long as the second. If both strings have the same tension and mu, what is the ratio of the speed of the pulse of the wave from the first string to the second string?

67. Two strings are attached to poles, however the first string is twice the linear mass density mu of the second. If both strings have the same tension, what is the ratio of the speed of the pulse of the wave from the first string to the second string?

68. Transverse waves travel through a string where the tension equals 7.00 N with a speed of 20.00 m/s. What tension would be required for a wave speed of 25.00 m/s?

69. Two strings are attached between two poles separated by a distance of 2.00 m as shown below, both under the same tension of 600.00 N. String 1 has a linear density of $\mu_1 = 0.0025 \, \text{kg/m}$ and string 2 has a linear mass density of $\mu_2 = 0.0035 \, \text{kg/m}$. Transverse wave pulses are generated simultaneously at opposite ends of the strings. How much time passes before the pulses pass one another?

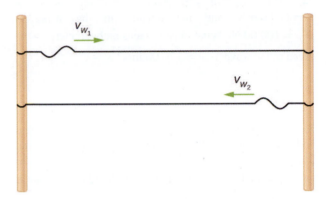

70. Two strings are attached between two poles separated by a distance of 2.00 meters as shown in the preceding figure, both strings have a linear density of $\mu_1 = 0.0025 \, \text{kg/m}$, the tension in string 1 is 600.00 N and the tension in string 2 is 700.00 N. Transverse wave pulses are generated simultaneously at opposite ends of the strings. How much time passes before the pulses pass one another?

71. The note E_4 is played on a piano and has a frequency of $f = 393.88$. If the linear mass density of this string of the piano is $\mu = 0.012 \, \text{kg/m}$ and the string is under a tension of 1000.00 N, what is the speed of the wave on the string and the wavelength of the wave?

72. Two transverse waves travel through a taut string. The speed of each wave is $v = 30.00 \, \text{m/s}$. A plot of the vertical position as a function of the horizontal position is shown below for the time $t = 0.00 \, \text{s}$. (a) What is the wavelength of each wave? (b) What is the frequency of each wave? (c) What is the maximum vertical speed of each string?

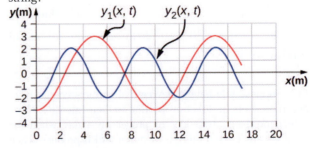

73. A sinusoidal wave travels down a taut, horizontal string with a linear mass density of $\mu = 0.060 \, \text{kg/m}$. The maximum vertical speed of the wave is

$v_{y\,max} = 0.30$ cm/s. The wave is modeled with the wave equation $y(x,\,t) = A\sin(6.00\text{ m}^{-1}x - 24.00\text{ s}^{-1}t)$. (a) What is the amplitude of the wave? (b) What is the tension in the string?

74. The speed of a transverse wave on a string is $v = 60.00$ m/s and the tension in the string is $F_T = 100.00$ N. What must the tension be to increase the speed of the wave to $v = 120.00$ m/s?

16.4 Energy and Power of a Wave

75. A string of length 5 m and a mass of 90 g is held under a tension of 100 N. A wave travels down the string that is modeled as $y(x,\,t) = 0.01\text{ m}\sin(0.40\text{ m}^{-1}x - 1170.12\text{ s}^{-1})$. What is the power over one wavelength?

76. Ultrasound of intensity 1.50×10^2 W/m^2 is produced by the rectangular head of a medical imaging device measuring 3.00 cm by 5.00 cm. What is its power output?

77. The low-frequency speaker of a stereo set has a surface area of $A = 0.05$ m^2 and produces 1 W of acoustical power. (a) What is the intensity at the speaker? (b) If the speaker projects sound uniformly in all directions, at what distance from the speaker is the intensity 0.1 W/m^2?

78. To increase the intensity of a wave by a factor of 50, by what factor should the amplitude be increased?

79. A device called an insolation meter is used to measure the intensity of sunlight. It has an area of 100 cm^2 and registers 6.50 W. What is the intensity in W/m^2?

80. Energy from the Sun arrives at the top of Earth's atmosphere with an intensity of 1400 W/m^2. How long does it take for 1.80×10^9 J to arrive on an area of 1.00 m^2?

81. Suppose you have a device that extracts energy from ocean breakers in direct proportion to their intensity. If the device produces 10.0 kW of power on a day when the breakers are 1.20 m high, how much will it produce when they are 0.600 m high?

82. A photovoltaic array of (solar cells) is 10.0% efficient in gathering solar energy and converting it to

electricity. If the average intensity of sunlight on one day is 70.00 W/m^2, what area should your array have to gather energy at the rate of 100 W? (b) What is the maximum cost of the array if it must pay for itself in two years of operation averaging 10.0 hours per day? Assume that it earns money at the rate of 9.00 cents per kilowatt-hour.

83. A microphone receiving a pure sound tone feeds an oscilloscope, producing a wave on its screen. If the sound intensity is originally 2.00×10^{-5} W/m^2, but is turned up until the amplitude increases by 30.0%, what is the new intensity?

84. A string with a mass of 0.30 kg has a length of 4.00 m. If the tension in the string is 50.00 N, and a sinusoidal wave with an amplitude of 2.00 cm is induced on the string, what must the frequency be for an average power of 100.00 W?

85. The power versus time for a point on a string $(\mu = 0.05$ kg/m) in which a sinusoidal traveling wave is induced is shown in the preceding figure. The wave is modeled with the wave equation $y(x,\,t) = A\sin(20.93\text{ m}^{-1}x - \omega t)$. What is the frequency and amplitude of the wave?

86. A string is under tension F_{T1}. Energy is transmitted by a wave on the string at rate P_1 by a wave of frequency f_1. What is the ratio of the new energy transmission rate P_2 to P_1 if the tension is doubled?

87. A 250-Hz tuning fork is struck and the intensity at the source is I_1 at a distance of one meter from the source. (a) What is the intensity at a distance of 4.00 m from the source? (b) How far from the tuning fork is the intensity a tenth of the intensity at the source?

88. A sound speaker is rated at a voltage of $P = 120.00$ V and a current of $I = 10.00$ A. Electrical power consumption is $P = IV$. To test the speaker, a signal of a sine wave is applied to the speaker. Assuming that the sound wave moves as a spherical wave and that all of the energy applied to the speaker is converted to sound energy, how far from the speaker is the intensity equal to 3.82 W/m^2?

89. The energy of a ripple on a pond is proportional to the amplitude squared. If the amplitude of the ripple is 0.1 cm at a distance from the source of 6.00 meters, what was the amplitude at a distance of 2.00 meters from the source?

16.5 Interference of Waves

90. Consider two sinusoidal waves traveling along a string, modeled as

$y_1(x, t) = 0.3\ \text{m}\ \sin\!\left(4\ \text{m}^{-1}x + 3\ \text{s}^{-1}t\right)$ and

$y_2(x, t) = 0.6\ \text{m}\ \sin\!\left(8\ \text{m}^{-1}x - 6\ \text{s}^{-1}t\right)$. What is the height of the resultant wave formed by the interference of the two waves at the position $x = 0.5\ \text{m}$ at time $t = 0.2\ \text{s}$?

91. Consider two sinusoidal sine waves traveling along a string, modeled as

$y_1(x, t) = 0.3\ \text{m}\ \sin\!\left(4\ \text{m}^{-1}x + 3\ \text{s}^{-1}t + \frac{\pi}{3}\right)$ and

$y_2(x, t) = 0.6\ \text{m}\ \sin\!\left(8\ \text{m}^{-1}x - 6\ \text{s}^{-1}t\right)$. What is the height of the resultant wave formed by the interference of the two waves at the position $x = 1.0\ \text{m}$ at time $t = 3.0\ \text{s}$?

92. Consider two sinusoidal sine waves traveling along a string, modeled as

$y_1(x, t) = 0.3\ \text{m}\ \sin\!\left(4\ \text{m}^{-1}x - 3\ \text{s}^{-1}t\right)$ and

$y_2(x, t) = 0.3\ \text{m}\ \sin\!\left(4\ \text{m}^{-1}x + 3\ \text{s}^{-1}t\right)$. What is the wave function of the resulting wave? [*Hint:* Use the trig identity $\sin(u \pm v) = \sin u \cos v \pm \cos u \sin v$

93. Two sinusoidal waves are moving through a medium in the same direction, both having amplitudes of 3.00 cm, a wavelength of 5.20 m, and a period of 6.52 s, but one has a phase shift of an angle ϕ. What is the phase shift if the resultant wave has an amplitude of 5.00 cm? [*Hint:* Use the trig identity $\sin u + \sin v = 2 \sin\!\left(\frac{u+v}{2}\right)\cos\!\left(\frac{u-v}{2}\right)$

94. Two sinusoidal waves are moving through a medium in the positive x-direction, both having amplitudes of 6.00 cm, a wavelength of 4.3 m, and a period of 6.00 s, but one has a phase shift of an angle $\phi = 0.50\ \text{rad}$. What is the height of the resultant wave at a time $t = 3.15\ \text{s}$ and a position $x = 0.45\ \text{m}$?

95. Two sinusoidal waves are moving through a medium in the positive x-direction, both having amplitudes of 7.00 cm, a wave number of $k = 3.00\ \text{m}^{-1}$, an angular frequency of $\omega = 2.50\ \text{s}^{-1}$, and a period of 6.00 s, but one has a phase shift of an angle $\phi = \frac{\pi}{12}\ \text{rad}$. What is the height of the resultant wave at a time $t = 2.00\ \text{s}$ and a position $x = 0.53\ \text{m}$?

96. Consider two waves $y_1(x, t)$ and $y_2(x, t)$ that are identical except for a phase shift propagating in the same medium. (a)What is the phase shift, in radians, if the amplitude of the resulting wave is 1.75 times the amplitude of the individual waves? (b) What is the phase shift in degrees? (c) What is the phase shift as a percentage of the individual wavelength?

97. Two sinusoidal waves, which are identical except for a phase shift, travel along in the same direction. The wave equation of the resultant wave is $y_R(x, t) = 0.70\ \text{m}\ \sin\!\left(3.00\ \text{m}^{-1}x - 6.28\ \text{s}^{-1}t + \pi/16\ \text{rad}\right)$. What are the angular frequency, wave number, amplitude, and phase shift of the individual waves?

98. Two sinusoidal waves, which are identical except for a phase shift, travel along in the same direction. The wave equation of the resultant wave is $y_R(x, t) = 0.35\ \text{cm}\ \sin\!\left(6.28\ \text{m}^{-1}x - 1.57\ \text{s}^{-1}t + \frac{\pi}{4}\right)$. What are the period, wavelength, amplitude, and phase shift of the individual waves?

99. Consider two wave functions, $y_1(x, t) = 4.00\ \text{m}\ \sin\!\left(\pi\ \text{m}^{-1}x - \pi\ \text{s}^{-1}t\right)$ and

$y_2(x, t) = 4.00\ \text{m}\ \sin\!\left(\pi\ \text{m}^{-1}x - \pi\ \text{s}^{-1}t + \frac{\pi}{3}\right)$. (a) Using a spreadsheet, plot the two wave functions and the wave that results from the superposition of the two wave functions as a function of position $(0.00 \le x \le 6.00\ \text{m})$ for the time $t = 0.00\ \text{s}$. (b) What are the wavelength and amplitude of the two original waves? (c) What are the wavelength and amplitude of the resulting wave?

100. Consider two wave functions, $y_2(x, t) = 2.00\ \text{m}\ \sin\!\left(\frac{\pi}{2}\text{m}^{-1}x - \frac{\pi}{3}\text{s}^{-1}t\right)$ and

$y_2(x, t) = 2.00\ \text{m}\ \sin\!\left(\frac{\pi}{2}\text{m}^{-1}x - \frac{\pi}{3}\text{s}^{-1}t + \frac{\pi}{6}\right)$. (a) Verify that $y_R = 2A\cos\!\left(\frac{\phi}{2}\right)\sin\!\left(kx - \omega t + \frac{\phi}{2}\right)$ is the solution for the wave that results from a superposition of the two waves. Make a column for x, y_1, y_2, $y_1 + y_2$, and $y_R = 2A\cos\!\left(\frac{\phi}{2}\right)\sin\!\left(kx - \omega t + \frac{\phi}{2}\right)$. Plot four waves as a function of position where the range of x is from 0 to 12 m.

101. Consider two wave functions that differ only by a phase shift, $y_1(x, t) = A\cos(kx - \omega t)$ and $y_2(x, t) = A\cos(kx - \omega t + \phi)$. Use the trigonometric identities $\cos u + \cos v = 2\cos\!\left(\frac{u-v}{2}\right)\cos\!\left(\frac{u+v}{2}\right)$ and $\cos(-\theta) = \cos(\theta)$ to find a wave equation for the wave resulting from the superposition of the two waves. Does the resulting wave function come as a surprise to you?

16.6 Standing Waves and Resonance

102. A wave traveling on a Slinky® that is stretched to 4 m takes 2.4 s to travel the length of the Slinky and back again. (a) What is the speed of the wave? (b) Using the same Slinky stretched to the same length, a standing wave is created which consists of three antinodes and four nodes. At what frequency must the Slinky be oscillating?

103. A 2-m long string is stretched between two supports with a tension that produces a wave speed equal to $v_w = 50.00$ m/s. What are the wavelength and frequency of the first three modes that resonate on the string?

104. Consider the experimental setup shown below. The length of the string between the string vibrator and the pulley is $L = 1.00$ m. The linear density of the string is $\mu = 0.006$ kg/m. The string vibrator can oscillate at any frequency. The hanging mass is 2.00 kg. (a)What are the wavelength and frequency of $n = 6$ mode? (b) The string oscillates the air around the string. What is the wavelength of the sound if the speed of the sound is $v_s = 343.00$ m/s?

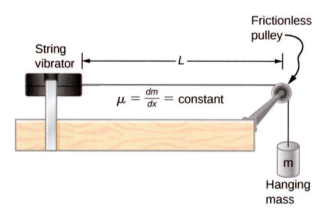

105. A cable with a linear density of $\mu = 0.2$ kg/m is hung from telephone poles. The tension in the cable is 500.00 N. The distance between poles is 20 meters. The wind blows across the line, causing the cable resonate. A standing waves pattern is produced that has 4.5 wavelengths between the two poles. The air temperature is $T = 20°C$. What are the frequency and wavelength of the hum?

106. Consider a rod of length L, mounted in the center to a support. A node must exist where the rod is mounted on a support, as shown below. Draw the first two normal modes of the rod as it is driven into resonance. Label the wavelength and the frequency required to drive the rod into resonance.

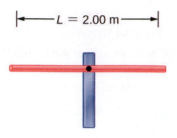

$$\vdash\!\!-\!\!-\!L = 2.00 \text{ m}\!-\!\!-\!\!\dashv$$

107. Consider two wave functions $y(x, t) = 0.30 \text{ cm} \sin(3 \text{ m}^{-1}x - 4 \text{ s}^{-1}t)$ and $y(x, t) = 0.30 \text{ cm} \sin(3 \text{ m}^{-1}x + 4 \text{ s}^{-1}t)$. Write a wave function for the resulting standing wave.

108. A 2.40-m wire has a mass of 7.50 g and is under a tension of 160 N. The wire is held rigidly at both ends and set into oscillation. (a) What is the speed of waves on the wire? The string is driven into resonance by a frequency that produces a standing wave with a wavelength equal to 1.20 m. (b) What is the frequency used to drive the string into resonance?

109. A string with a linear mass density of 0.0062 kg/m and a length of 3.00 m is set into the $n = 100$ mode of resonance. The tension in the string is 20.00 N. What is the wavelength and frequency of the wave?

110. A string with a linear mass density of 0.0075 kg/m and a length of 6.00 m is set into the $n = 4$ mode of resonance by driving with a frequency of 100.00 Hz. What is the tension in the string?

111. Two sinusoidal waves with identical wavelengths and amplitudes travel in opposite directions along a string producing a standing wave. The linear mass density of the string is $\mu = 0.075$ kg/m and the tension in the string is $F_T = 5.00$ N. The time interval between instances of total destructive interference is $\Delta t = 0.13$ s. What is the wavelength of the waves?

112. A string, fixed on both ends, is 5.00 m long and has a mass of 0.15 kg. The tension if the string is 90 N. The string is vibrating to produce a standing wave at the fundamental frequency of the string. (a) What is the speed of the waves on the string? (b) What is the wavelength of the standing wave produced? (c) What is the period of the standing wave?

113. A string is fixed at both end. The mass of the string is 0.0090 kg and the length is 3.00 m. The string is under a tension of 200.00 N. The string is driven by a variable frequency source to produce standing waves on the string. Find the wavelengths and frequency of the first four modes of standing waves.

114. The frequencies of two successive modes of standing waves on a string are 258.36 Hz and 301.42 Hz. What is the next frequency above 100.00 Hz that would produce a standing wave?

115. A string is fixed at both ends to supports 3.50 m apart and has a linear mass density of $\mu = 0.005$ kg/m.

The string is under a tension of 90.00 N. A standing wave is produced on the string with six nodes and five antinodes. What are the wave speed, wavelength, frequency, and period of the standing wave?

116. Sine waves are sent down a 1.5-m-long string fixed at both ends. The waves reflect back in the opposite direction. The amplitude of the wave is 4.00 cm. The propagation velocity of the waves is 175 m/s. The $n = 6$ resonance mode of the string is produced. Write an equation for the resulting standing wave.

ADDITIONAL PROBLEMS

117. Ultrasound equipment used in the medical profession uses sound waves of a frequency above the range of human hearing. If the frequency of the sound produced by the ultrasound machine is $f = 30$ kHz, what is the wavelength of the ultrasound in bone, if the speed of sound in bone is $v = 3000$ m/s?

118. Shown below is the plot of a wave function that models a wave at time $t = 0.00$ s and $t = 2.00$ s. The dotted line is the wave function at time $t = 0.00$ s and the solid line is the function at time $t = 2.00$ s. Estimate the amplitude, wavelength, velocity, and period of the wave.

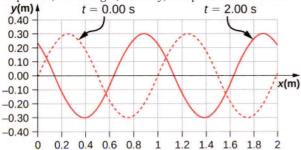

119. The speed of light in air is approximately $v = 3.00 \times 10^8$ m/s and the speed of light in glass is $v = 2.00 \times 10^8$ m/s. A red laser with a wavelength of $\lambda = 633.00$ nm shines light incident of the glass, and some of the red light is transmitted to the glass. The frequency of the light is the same for the air and the glass. (a) What is the frequency of the light? (b) What is the wavelength of the light in the glass?

120. A radio station broadcasts radio waves at a frequency of 101.7 MHz. The radio waves move through the air at approximately the speed of light in a vacuum. What is the wavelength of the radio waves?

121. A sunbather stands waist deep in the ocean and observes that six crests of periodic surface waves pass each minute. The crests are 16.00 meters apart. What is the wavelength, frequency, period, and speed of the waves?

122. A tuning fork vibrates producing sound at a frequency of 512 Hz. The speed of sound of sound in air is $v = 343.00$ m/s if the air is at a temperature of $20.00°C$. What is the wavelength of the sound?

123. A motorboat is traveling across a lake at a speed of $v_b = 15.00$ m/s. The boat bounces up and down every 0.50 s as it travels in the same direction as a wave. It bounces up and down every 0.30 s as it travels in a direction opposite the direction of the waves. What is the speed and wavelength of the wave?

124. Use the linear wave equation to show that the wave speed of a wave modeled with the wave function $y(x,\ t) = 0.20$ m $\sin(3.00$ m$^{-1}x + 6.00$ s$^{-1}t)$ is $v = 2.00$ m/s. What are the wavelength and the speed of the wave?

125. Given the wave functions $y_1(x,\ t) = A\ \sin(kx - \omega t)$ and $y_2(x,\ t) = A\ \sin(kx - \omega t + \phi)$ with $\phi \neq \frac{\pi}{2}$, show that $y_1(x,\ t) + y_2(x,\ t)$ is a solution to the linear wave equation with a wave velocity of $v = \sqrt{\frac{\omega}{k}}$.

126. A transverse wave on a string is modeled with the wave function $y(x,\ t) = 0.10$ m $\sin(0.15$ m$^{-1}x + 1.50$ s$^{-1}t + 0.20)$. (a) Find the wave velocity. (b) Find the position in the y-direction, the velocity perpendicular to the motion of the wave, and the acceleration perpendicular to the motion of the wave, of a small segment of the string centered at $x = 0.40$ m at time $t = 5.00$ s.

127. A sinusoidal wave travels down a taut, horizontal string with a linear mass density of $\mu = 0.060$ kg/m. The magnitude of maximum vertical acceleration of the wave is $a_{y\ max} = 0.90$ cm/s^2 and the amplitude of the wave is

0.40 m. The string is under a tension of $F_T = 600.00\,\text{N}$. The wave moves in the negative x-direction. Write an equation to model the wave.

128. A transverse wave on a string $(\mu = 0.0030\,\text{kg/m})$ is described with the equation $y(x, t) = 0.30\,\text{m}\,\sin\!\left(\frac{2\pi}{4.00\,\text{m}}(x - 16.00\frac{\text{m}}{\text{s}}t)\right)$. What is the tension under which the string is held taut?

129. A transverse wave on a horizontal string $(\mu = 0.0060\,\text{kg/m})$ is described with the equation $y(x, t) = 0.30\,\text{m}\,\sin\!\left(\frac{2\pi}{4.00\,\text{m}}(x - v_w t)\right)$. The string is under a tension of 300.00 N. What are the wave speed, wave number, and angular frequency of the wave?

130. A student holds an inexpensive sonic range finder and uses the range finder to find the distance to the wall. The sonic range finder emits a sound wave. The sound wave reflects off the wall and returns to the range finder. The round trip takes 0.012 s. The range finder was calibrated for use at room temperature $T = 20°C$, but the temperature in the room is actually $T = 23°C$. Assuming that the timing mechanism is perfect, what percentage of error can the student expect due to the calibration?

131. A wave on a string is driven by a string vibrator, which oscillates at a frequency of 100.00 Hz and an amplitude of 1.00 cm. The string vibrator operates at a voltage of 12.00 V and a current of 0.20 A. The power consumed by the string vibrator is $P = IV$. Assume that the string vibrator is 90% efficient at converting electrical energy into the energy associated with the vibrations of the string. The string is 3.00 m long, and is under a tension of 60.00 N. What is the linear mass density of the string?

132. A traveling wave on a string is modeled by the wave equation $y(x, t) = 3.00\,\text{cm}\,\sin\!\left(8.00\,\text{m}^{-1}x + 100.00\,\text{s}^{-1}t\right)$. The string is under a tension of 50.00 N and has a linear mass density of $\mu = 0.008\,\text{kg/m}$. What is the average power transferred by the wave on the string?

133. A transverse wave on a string has a wavelength of 5.0 m, a period of 0.02 s, and an amplitude of 1.5 cm. The average power transferred by the wave is 5.00 W. What is the tension in the string?

134. (a) What is the intensity of a laser beam used to burn away cancerous tissue that, when 90.0% absorbed, puts 500 J of energy into a circular spot 2.00 mm in diameter in 4.00 s? (b) Discuss how this intensity compares to the average intensity of sunlight (about) and the implications that would have if the laser beam entered your eye. Note how your answer depends on the time duration of the exposure.

135. Consider two periodic wave functions, $y_1(x, t) = A\sin(kx - \omega t)$ and $y_2(x, t) = A\sin(kx - \omega t + \phi)$. (a) For what values of ϕ will the wave that results from a superposition of the wave functions have an amplitude of $2A$? (b) For what values of ϕ will the wave that results from a superposition of the wave functions have an amplitude of zero?

136. Consider two periodic wave functions, $y_1(x, t) = A\sin(kx - \omega t)$ and $y_2(x, t) = A\cos(kx - \omega t + \phi)$. (a) For what values of ϕ will the wave that results from a superposition of the wave functions have an amplitude of $2A$? (b) For what values of ϕ will the wave that results from a superposition of the wave functions have an amplitude of zero?

137. A trough with dimensions 10.00 meters by 0.10 meters by 0.10 meters is partially filled with water. Small-amplitude surface water waves are produced from both ends of the trough by paddles oscillating in simple harmonic motion. The height of the water waves are modeled with two sinusoidal wave equations, $y_1(x, t) = 0.3\,\text{m}\,\sin\!\left(4\,\text{m}^{-1}x - 3\,\text{s}^{-1}t\right)$ and $y_2(x, t) = 0.3\,\text{m}\,\cos\!\left(4\,\text{m}^{-1}x + 3\,\text{s}^{-1}t - \frac{\pi}{2}\right)$. What is the wave function of the resulting wave after the waves reach one another and before they reach the end of the trough (i.e., assume that there are only two waves in the trough and ignore reflections)? Use a spreadsheet to check your results. (*Hint:* Use the trig identities $\sin(u \pm v) = \sin u \cos v \pm \cos u \sin v$ and $\cos(u \pm v) = \cos u \cos v \mp \sin u \sin v$)

138. A seismograph records the S- and P-waves from an earthquake 20.00 s apart. If they traveled the same path at constant wave speeds of $v_S = 4.00\,\text{km/s}$ and $v_P = 7.50\,\text{km/s}$, how far away is the epicenter of the earthquake?

139. Consider what is shown below. A 20.00-kg mass rests on a frictionless ramp inclined at $45°$. A string with a linear mass density of $\mu = 0.025\,\text{kg/m}$ is attached to the 20.00-kg mass. The string passes over a frictionless pulley of negligible mass and is attached to a hanging mass (m). The system is in static equilibrium. A wave is induced on the string and travels up the ramp. (a) What is the mass of the hanging mass (m)? (b) At what wave speed does the wave travel up the string?

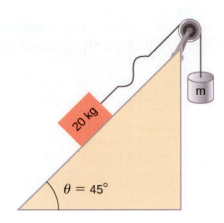

140. Consider the superposition of three wave functions

$y(x, t) = 3.00\,\text{cm} \sin\!\left(2\,\text{m}^{-1}x - 3\,\text{s}^{-1}t\right),$

$y(x, t) = 3.00\,\text{cm} \sin\!\left(6\,\text{m}^{-1}x + 3\,\text{s}^{-1}t\right),$ and

$y(x, t) = 3.00\,\text{cm} \sin\!\left(2\,\text{m}^{-1}x - 4\,\text{s}^{-1}t\right).$ What is the

height of the resulting wave at position $x = 3.00\,\text{m}$ at time $t = 10.0\,\text{s}$?

141. A string has a mass of 150 g and a length of 3.4 m. One end of the string is fixed to a lab stand and the other is attached to a spring with a spring constant of $k_s = 100\,\text{N/m}$. The free end of the spring is attached to another lab pole. The tension in the string is maintained by the spring. The lab poles are separated by a distance that stretches the spring 2.00 cm. The string is plucked and a pulse travels along the string. What is the propagation speed of the pulse?

142. A standing wave is produced on a string under a tension of 70.0 N by two sinusoidal transverse waves that are identical, but moving in opposite directions. The string is fixed at $x = 0.00\,\text{m}$ and $x = 10.00\,\text{m}$. Nodes appear at $x = 0.00\,\text{m},\ 2.00\,\text{m},\ 4.00\,\text{m},\ 6.00\,\text{m},\ 8.00\,\text{m},$ and 10.00 m. The amplitude of the standing wave is 3.00 cm. It takes 0.10 s for the antinodes to make one complete oscillation. (a) What are the wave functions of the two sine waves that produce the standing wave? (b) What are the maximum velocity and acceleration of the string, perpendicular to the direction of motion of the transverse waves, at the antinodes?

143. A string with a length of 4 m is held under a constant tension. The string has a linear mass density of $\mu = 0.006\,\text{kg/m}$. Two resonant frequencies of the string are 400 Hz and 480 Hz. There are no resonant frequencies between the two frequencies. (a) What are the wavelengths of the two resonant modes? (b) What is the tension in the string?

CHALLENGE PROBLEMS

144. A copper wire has a radius of $200\,\mu\text{m}$ and a length of 5.0 m. The wire is placed under a tension of 3000 N and the wire stretches by a small amount. The wire is plucked and a pulse travels down the wire. What is the propagation speed of the pulse? (Assume the temperature does not change: $\left(\rho = 8.96\dfrac{\text{g}}{\text{cm}^3},\ Y = 1.1 \times 10^{11}\dfrac{\text{N}}{\text{m}}\right).$)

145. A pulse moving along the x axis can be modeled as the wave function $y(x, t) = 4.00\,\text{m}\,e^{-\left(\frac{x + (2.00\ \text{m/s})t}{1.00\ \text{m}}\right)^2}.$

(a) What are the direction and propagation speed of the pulse? (b) How far has the wave moved in 3.00 s? (c) Plot the pulse using a spreadsheet at time $t = 0.00\,\text{s}$ and $t = 3.00\,\text{s}$ to verify your answer in part (b).

146. A string with a linear mass density of $\mu = 0.0085\,\text{kg/m}$ is fixed at both ends. A 5.0-kg mass is hung from the string, as shown below. If a pulse is sent along section A, what is the wave speed in section A and the wave speed in section B?

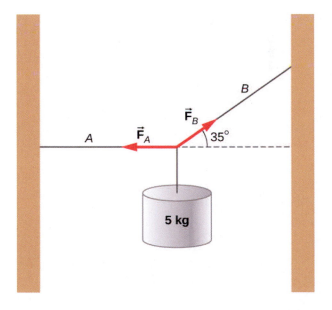

147. Consider two wave functions $y_1(x, t) = A \sin(kx - \omega t)$ and $y_2(x, t) = A \sin(kx + \omega t + \phi).$ What is the wave function resulting from the interference of the two wave? (*Hint:* $\sin(\alpha \pm \beta) = \sin\alpha\cos\beta \pm \cos\alpha\sin\beta$ and $\phi = \dfrac{\phi}{2} + \dfrac{\phi}{2}$.)

148. The wave function that models a standing wave is given as $y_R(x, t) = 6.00 \text{ cm } \sin(3.00 \text{ m}^{-1} x + 1.20 \text{ rad}) \cos(6.00 \text{ s}^{-1} t + 1.20 \text{ rad})$. What are two wave functions that interfere to form this wave function? Plot the two wave functions and the sum of the sum of the two wave functions at $t = 1.00 \text{ s}$ to verify your answer.

149. Consider two wave functions $y_1(x, t) = A \sin(kx - \omega t)$ and $y_2(x, t) = A \sin(kx + \omega t + \phi)$. The resultant wave form when you add the two functions is $y_R = 2A \sin\left(kx + \dfrac{\phi}{2}\right) \cos\left(\omega t + \dfrac{\phi}{2}\right)$. Consider the case where $A = 0.03 \text{ m}^{-1}$, $k = 1.26 \text{ m}^{-1}$, $\omega = \pi \text{ s}^{-1}$, and $\phi = \dfrac{\pi}{10}$. (a) Where are the first three nodes of the standing wave function starting at zero and moving in the positive x direction? (b) Using a spreadsheet, plot the two wave functions and the resulting function at time $t = 1.00 \text{ s}$ to verify your answer.

17 | SOUND

Figure 17.1 Hearing is an important human sense that can detect frequencies of sound, ranging between 20 Hz and 20 kHz. However, other species have very different ranges of hearing. Bats, for example, emit clicks in ultrasound, using frequencies beyond 20 kHz. They can detect nearby insects by hearing the echo of these ultrasonic clicks. Ultrasound is important in several human applications, including probing the interior structures of human bodies, Earth, and the Sun. Ultrasound is also useful in industry for nondestructive testing. (credit: modification of work by Angell Williams)

Chapter Outline

17.1 Sound Waves

17.2 Speed of Sound

17.3 Sound Intensity

17.4 Normal Modes of a Standing Sound Wave

17.5 Sources of Musical Sound

17.6 Beats

17.7 The Doppler Effect

17.8 Shock Waves

Introduction

Sound is an example of a mechanical wave, specifically, a pressure wave: Sound waves travel through the air and other media as oscillations of molecules. Normal human hearing encompasses an impressive range of frequencies from 20 Hz to 20 kHz. Sounds below 20 Hz are called infrasound, whereas those above 20 kHz are called ultrasound. Some animals, like the bat shown in **Figure 17.1**, can hear sounds in the ultrasonic range.

Many of the concepts covered in **Waves** also have applications in the study of sound. For example, when a sound wave encounters an interface between two media with different wave speeds, reflection and transmission of the wave occur.

Ultrasound has many uses in science, engineering, and medicine. Ultrasound is used for nondestructive testing in engineering, such as testing the thickness of coating on metal. In medicine, sound waves are far less destructive than X-rays and can be used to image the fetus in a mother's womb without danger to the fetus or the mother. Later in this chapter, we discuss the Doppler effect, which can be used to determine the velocity of blood in the arteries or wind speed in weather systems.

17.1 | Sound Waves

The physical phenomenon of **sound** is a disturbance of matter that is transmitted from its source outward. **Hearing** is the perception of sound, just as seeing is the perception of visible light. On the atomic scale, sound is a disturbance of atoms that is far more ordered than their thermal motions. In many instances, sound is a periodic wave, and the atoms undergo simple harmonic motion. Thus, sound waves can induce oscillations and resonance effects (**Figure 17.2**).

Figure 17.2 This glass has been shattered by a high-intensity sound wave of the same frequency as the resonant frequency of the glass. (credit: "‖read‖"/Flickr)

 This **video (https://openstaxcollege.org/l/21waveswineglas)** shows waves on the surface of a wine glass, being driven by sound waves from a speaker. As the frequency of the sound wave approaches the resonant frequency of the wine glass, the amplitude and frequency of the waves on the wine glass increase. When the resonant frequency is reached, the glass shatters.

A speaker produces a sound wave by oscillating a cone, causing vibrations of air molecules. In **Figure 17.3**, a speaker vibrates at a constant frequency and amplitude, producing vibrations in the surrounding air molecules. As the speaker oscillates back and forth, it transfers energy to the air, mostly as thermal energy. But a small part of the speaker's energy goes into compressing and expanding the surrounding air, creating slightly higher and lower local pressures. These compressions (high-pressure regions) and rarefactions (low-pressure regions) move out as longitudinal pressure waves having the same frequency as the speaker—they are the disturbance that is a sound wave. (Sound waves in air and most fluids are longitudinal, because fluids have almost no shear strength. In solids, sound waves can be both transverse and longitudinal.)

Figure 17.3(a) shows the compressions and rarefactions, and also shows a graph of gauge pressure versus distance from a speaker. As the speaker moves in the positive x-direction, it pushes air molecules, displacing them from their equilibrium positions. As the speaker moves in the negative x-direction, the air molecules move back toward their equilibrium positions due to a restoring force. The air molecules oscillate in simple harmonic motion about their equilibrium positions, as shown in part (b). Note that sound waves in air are longitudinal, and in the figure, the wave propagates in the positive x-direction and the molecules oscillate parallel to the direction in which the wave propagates.

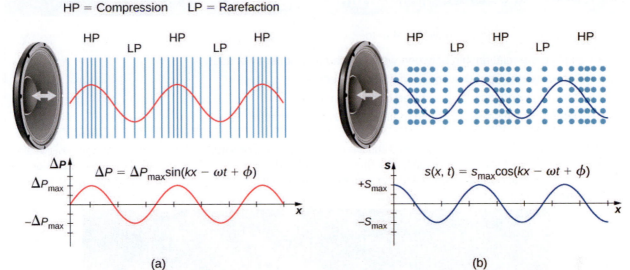

(a)   (b)

Figure 17.3   (a) A vibrating cone of a speaker, moving in the positive x-direction, compresses the air in front of it and expands the air behind it. As the speaker oscillates, it creates another compression and rarefaction as those on the right move away from the speaker. After many vibrations, a series of compressions and rarefactions moves out from the speaker as a sound wave. The red graph shows the gauge pressure of the air versus the distance from the speaker. Pressures vary only slightly from atmospheric pressure for ordinary sounds. Note that gauge pressure is modeled with a sine function, where the crests of the function line up with the compressions and the troughs line up with the rarefactions. (b) Sound waves can also be modeled using the displacement of the air molecules. The blue graph shows the displacement of the air molecules versus the position from the speaker and is modeled with a cosine function. Notice that the displacement is zero for the molecules in their equilibrium position and are centered at the compressions and rarefactions. Compressions are formed when molecules on either side of the equilibrium molecules are displaced toward the equilibrium position. Rarefactions are formed when the molecules are displaced away from the equilibrium position.

Models Describing Sound

Sound can be modeled as a pressure wave by considering the change in pressure from average pressure,

$$\Delta P = \Delta P_{max} \sin(kx \mp \omega t + \phi). \tag{17.1}$$

This equation is similar to the periodic wave equations seen in **Waves**, where ΔP is the change in pressure, ΔP_{max} is the maximum change in pressure, $k = \frac{2\pi}{\lambda}$ is the wave number, $\omega = \frac{2\pi}{T} = 2\pi f$ is the angular frequency, and ϕ is the initial phase. The wave speed can be determined from $v = \frac{\omega}{k} = \frac{\lambda}{T}$. Sound waves can also be modeled in terms of the displacement of the air molecules. The displacement of the air molecules can be modeled using a cosine function:

$$s(x, t) = s_{max} \cos(kx \mp \omega t + \phi). \tag{17.2}$$

In this equation, s is the displacement and s_{max} is the maximum displacement.

Not shown in the figure is the amplitude of a sound wave as it decreases with distance from its source, because the energy of the wave is spread over a larger and larger area. The intensity decreases as it moves away from the speaker, as discussed in **Waves**. The energy is also absorbed by objects and converted into thermal energy by the viscosity of the air. In addition, during each compression, a little heat transfers to the air; during each rarefaction, even less heat transfers from the air, and these heat transfers reduce the organized disturbance into random thermal motions. Whether the heat transfer from compression to rarefaction is significant depends on how far apart they are—that is, it depends on wavelength. Wavelength, frequency, amplitude, and speed of propagation are important characteristics for sound, as they are for all waves.

17.2 | Speed of Sound

Sound, like all waves, travels at a certain speed and has the properties of frequency and wavelength. You can observe direct evidence of the speed of sound while watching a fireworks display (**Figure 17.4**). You see the flash of an explosion well before you hear its sound and possibly feel the pressure wave, implying both that sound travels at a finite speed and that it is much slower than light.

Figure 17.4 When a firework shell explodes, we perceive the light energy before the sound energy because sound travels more slowly than light does.

The difference between the speed of light and the speed of sound can also be experienced during an electrical storm. The flash of lighting is often seen before the clap of thunder. You may have heard that if you count the number of seconds between the flash and the sound, you can estimate the distance to the source. Every five seconds converts to about one mile. The velocity of any wave is related to its frequency and wavelength by

$$v = f\lambda, \tag{17.3}$$

where v is the speed of the wave, f is its frequency, and λ is its wavelength. Recall from **Waves** that the wavelength is the length of the wave as measured between sequential identical points. For example, for a surface water wave or sinusoidal wave on a string, the wavelength can be measured between any two convenient sequential points with the same height and slope, such as between two sequential crests or two sequential troughs. Similarly, the wavelength of a sound wave is the distance between sequential identical parts of a wave—for example, between sequential compressions (**Figure 17.5**). The frequency is the same as that of the source and is the number of waves that pass a point per unit time.

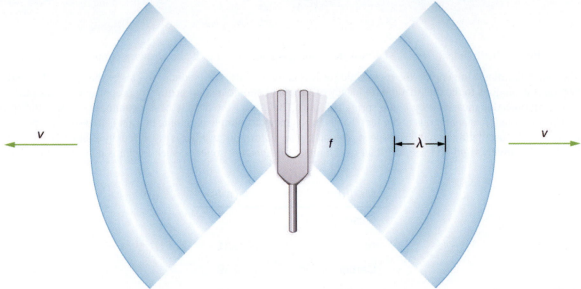

Figure 17.5 A sound wave emanates from a source, such as a tuning fork, vibrating at a frequency f. It propagates at speed v and has a wavelength λ.

Speed of Sound in Various Media

Table 17.1 shows that the speed of sound varies greatly in different media. The speed of sound in a medium depends on how quickly vibrational energy can be transferred through the medium. For this reason, the derivation of the speed of sound in a medium depends on the medium and on the state of the medium. In general, the equation for the speed of a mechanical wave in a medium depends on the square root of the restoring force, or the elastic property, divided by the inertial property,

$$v = \sqrt{\frac{\text{elastic property}}{\text{inertial property}}}.$$

Also, sound waves satisfy the wave equation derived in **Waves**,

$$\frac{\partial^2 y(x,\, t)}{\partial x^2} = \frac{1}{v^2} \frac{\partial^2 y(x,\, t)}{\partial t^2}.$$

Recall from **Waves** that the speed of a wave on a string is equal to $v = \sqrt{\frac{F_T}{\mu}}$, where the restoring force is the tension in the string F_T and the linear density μ is the inertial property. In a fluid, the speed of sound depends on the bulk modulus and the density,

$$v = \sqrt{\frac{B}{\rho}}. \tag{17.4}$$

The speed of sound in a solid the depends on the Young's modulus of the medium and the density,

$$v = \sqrt{\frac{Y}{\rho}}. \tag{17.5}$$

In an ideal gas (see **The Kinetic Theory of Gases (http://cnx.org/content/m58390/latest/)**), the equation for the speed of sound is

$$v = \sqrt{\frac{\gamma R T_K}{M}},$$

(17.6)

where γ is the adiabatic index, $R = 8.31 \, \text{J/mol} \cdot \text{K}$ is the gas constant, T_K is the absolute temperature in kelvins, and M is the molecular mass. In general, the more rigid (or less compressible) the medium, the faster the speed of sound. This observation is analogous to the fact that the frequency of simple harmonic motion is directly proportional to the stiffness of the oscillating object as measured by k, the spring constant. The greater the density of a medium, the slower the speed of sound. This observation is analogous to the fact that the frequency of a simple harmonic motion is inversely proportional to m, the mass of the oscillating object. The speed of sound in air is low, because air is easily compressible. Because liquids and solids are relatively rigid and very difficult to compress, the speed of sound in such media is generally greater than in gases.

Medium	v (m/s)
Gases at $0°C$	
Air	331
Carbon dioxide	259
Oxygen	316
Helium	965
Hydrogen	1290
Liquids at $20°C$	
Ethanol	1160
Mercury	1450
Water, fresh	1480
Sea Water	1540
Human tissue	1540
Solids (longitudinal or bulk)	
Vulcanized rubber	54
Polyethylene	920
Marble	3810
Glass, Pyrex	5640
Lead	1960
Aluminum	5120
Steel	5960

Table 17.1 Speed of Sound in Various Media

Because the speed of sound depends on the density of the material, and the density depends on the temperature, there is a relationship between the temperature in a given medium and the speed of sound in the medium. For air at sea level, the speed of sound is given by

$$v = 331\frac{m}{s}\sqrt{1 + \frac{T_C}{273°C}} = 331\frac{m}{s}\sqrt{\frac{T_K}{273\ K}} \qquad \text{(17.7)}$$

where the temperature in the first equation (denoted as T_C) is in degrees Celsius and the temperature in the second equation (denoted as T_K) is in kelvins. The speed of sound in gases is related to the average speed of particles in the gas, $v_{rms} = \sqrt{\frac{3k_B T}{m}}$, where k_B is the Boltzmann constant $(1.38 \times 10^{-23}$ J/K$)$ and m is the mass of each (identical) particle in the gas. Note that v refers to the speed of the coherent propagation of a disturbance (the wave), whereas v_{rms} describes the speeds of particles in random directions. Thus, it is reasonable that the speed of sound in air and other gases should depend on the square root of temperature. While not negligible, this is not a strong dependence. At $0°C$, the speed of sound is 331 m/s, whereas at $20.0°C$, it is 343 m/s, less than a 4% increase. **Figure 17.6** shows how a bat uses the speed of sound to sense distances.

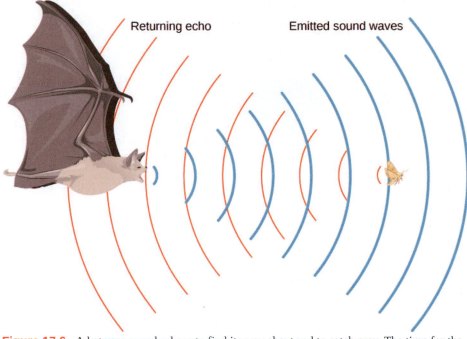

Figure 17.6 A bat uses sound echoes to find its way about and to catch prey. The time for the echo to return is directly proportional to the distance.

Derivation of the Speed of Sound in Air

As stated earlier, the speed of sound in a medium depends on the medium and the state of the medium. The derivation of the equation for the speed of sound in air starts with the mass flow rate and continuity equation discussed in **Fluid Mechanics**.

Consider fluid flow through a pipe with cross-sectional area A (**Figure 17.7**). The mass in a small volume of length x of the pipe is equal to the density times the volume, or $m = \rho V = \rho A x$. The mass flow rate is

$$\frac{dm}{dt} = \frac{d}{dt}(\rho V) = \frac{d}{dt}(\rho A x) = \rho A \frac{dx}{dt} = \rho A v.$$

The continuity equation from **Fluid Mechanics** states that the mass flow rate into a volume has to equal the mass flow rate out of the volume, $\rho_{in} A_{in} v_{in} = \rho_{out} A_{out} v_{out}$.

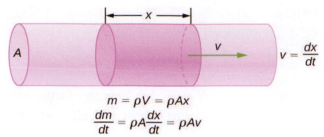

$$m = \rho V = \rho A x$$
$$\frac{dm}{dt} = \rho A \frac{dx}{dt} = \rho A v$$

Figure 17.7 The mass of a fluid in a volume is equal to the density times the volume, $m = \rho V = \rho A x$. The mass flow rate is the time derivative of the mass.

Now consider a sound wave moving through a parcel of air. A parcel of air is a small volume of air with imaginary boundaries (**Figure 17.8**). The density, temperature, and velocity on one side of the volume of the fluid are given as $\rho, T, v,$ and on the other side are $\rho + d\rho, T + dT, v + dv$.

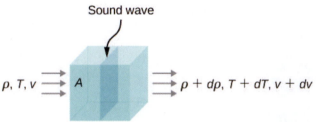

Figure 17.8 A sound wave moves through a volume of fluid. The density, temperature, and velocity of the fluid change from one side to the other.

The continuity equation states that the mass flow rate entering the volume is equal to the mass flow rate leaving the volume, so

$$\rho A v = (\rho + d\rho)A(v + dv).$$

This equation can be simplified, noting that the area cancels and considering that the multiplication of two infinitesimals is approximately equal to zero: $d\rho(dv) \approx 0,$

$$\begin{aligned} \rho v &= (\rho + d\rho)(v + dv) \\ \rho v &= \rho v + \rho(dv) + (d\rho)v + (d\rho)(dv) \\ 0 &= \rho(dv) + (d\rho)v \\ \rho\, dv &= -v d\rho. \end{aligned}$$

The net force on the volume of fluid (**Figure 17.9**) equals the sum of the forces on the left face and the right face:

$$\begin{aligned} F_{\text{net}} &= p\, dy\, dz - (p + dp)dy\, dz \\ &= p\, dy\, dz - p dy\, dz - dp\, dy\, dz \\ &= -dp\, dy\, dz \\ ma &= -dp\, dy\, dz. \end{aligned}$$

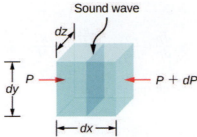

Figure 17.9 A sound wave moves through a volume of fluid. The force on each face can be found by the pressure times the area.

The acceleration is the force divided by the mass and the mass is equal to the density times the volume, $m = \rho V = \rho \, dx \, dy \, dz$. We have

$$ma = -dp \, dy \, dz$$

$$a = -\frac{dp \, dy \, dz}{m} = -\frac{dp \, dy \, dz}{\rho \, dx \, dy \, dz} = -\frac{dp}{(\rho \, dx)}$$

$$\frac{dv}{dt} = -\frac{dp}{(\rho \, dx)}$$

$$dv = -\frac{dp}{(\rho \, dx)} dt = -\frac{dp}{\rho} \frac{1}{v}$$

$$\rho v \, dv = -dp.$$

From the continuity equation $\rho \, dv = -v d\rho$, we obtain

$$\rho v dv = -dp$$

$$(-v d\rho)v = -dp$$

$$v = \sqrt{\frac{dp}{d\rho}}.$$

Consider a sound wave moving through air. During the process of compression and expansion of the gas, no heat is added or removed from the system. A process where heat is not added or removed from the system is known as an adiabatic system. Adiabatic processes are covered in detail in **The First Law of Thermodynamics (http://cnx.org/content/m58721/ latest/)** , but for now it is sufficient to say that for an adiabatic process, $pV^{\gamma} = $ constant, where p is the pressure, V is the volume, and gamma (γ) is a constant that depends on the gas. For air, $\gamma = 1.40$. The density equals the number of moles times the molar mass divided by the volume, so the volume is equal to $V = \frac{nM}{\rho}$. The number of moles and the molar mass are constant and can be absorbed into the constant $p\left(\frac{1}{\rho}\right)^{\gamma} = $ constant. Taking the natural logarithm of both sides yields $\ln p - \gamma \ln \rho = $ constant. Differentiating with respect to the density, the equation becomes

$$\ln p - \gamma \ln \rho = \text{constant}$$

$$\frac{d}{d\rho}(\ln p - \gamma \ln \rho) = \frac{d}{d\rho}(\text{constant})$$

$$\frac{1}{P}\frac{dp}{d\rho} - \frac{\gamma}{\rho} = 0$$

$$\frac{dp}{d\rho} = \frac{\gamma p}{\rho}.$$

If the air can be considered an ideal gas, we can use the ideal gas law:

$$pV = nRT = \frac{m}{M}RT$$

$$p = \frac{m}{V}\frac{RT}{M} = \rho\frac{RT}{M}.$$

Here M is the molar mass of air:

$$\frac{dp}{d\rho} = \frac{\gamma p}{\rho} = \frac{\gamma\left(\rho\frac{RT}{M}\right)}{\rho} = \frac{\gamma RT}{M}.$$

Since the speed of sound is equal to $v = \sqrt{\frac{dp}{d\rho}}$, the speed is equal to

$$v = \sqrt{\frac{\gamma\,RT}{M}}.$$

Note that the velocity is faster at higher temperatures and slower for heavier gases. For air, $\gamma = 1.4$, $M = 0.02897\frac{\text{kg}}{\text{mol}}$, and $R = 8.31\frac{\text{J}}{\text{mol}\cdot\text{K}}$. If the temperature is $T_C = 20°\text{C}(T = 293\text{ K})$, the speed of sound is $v = 343$ m/s.

The equation for the speed of sound in air $v = \sqrt{\frac{\gamma RT}{M}}$ can be simplified to give the equation for the speed of sound in air as a function of absolute temperature:

$$\begin{aligned} v &= \sqrt{\frac{\gamma RT}{M}} \\ &= \sqrt{\frac{\gamma RT}{M}\left(\frac{273\text{ K}}{273\text{ K}}\right)} = \sqrt{\frac{(273\text{ K})\gamma R}{M}}\sqrt{\frac{T}{273\text{ K}}} \\ &\approx 331\frac{\text{m}}{\text{s}}\sqrt{\frac{T}{273\text{ K}}}. \end{aligned}$$

One of the more important properties of sound is that its speed is nearly independent of the frequency. This independence is certainly true in open air for sounds in the audible range. If this independence were not true, you would certainly notice it for music played by a marching band in a football stadium, for example. Suppose that high-frequency sounds traveled faster—then the farther you were from the band, the more the sound from the low-pitch instruments would lag that from the high-pitch ones. But the music from all instruments arrives in cadence independent of distance, so all frequencies must travel at nearly the same speed. Recall that

$$v = f\lambda.$$

In a given medium under fixed conditions, v is constant, so there is a relationship between f and λ; the higher the frequency, the smaller the wavelength (**Figure 17.10**).

High f, small λ

Small f, large λ

Figure 17.10 Because they travel at the same speed in a given medium, low-frequency sounds must have a greater wavelength than high-frequency sounds. Here, the lower-frequency sounds are emitted by the large speaker, called a woofer, whereas the higher-frequency sounds are emitted by the small speaker, called a tweeter. (credit: modification of work by Jane Whitney)

Example 17.1

Calculating Wavelengths

Calculate the wavelengths of sounds at the extremes of the audible range, 20 and 20,000 Hz, in 30.0°C air. (Assume that the frequency values are accurate to two significant figures.)

Strategy

To find wavelength from frequency, we can use $v = f\lambda$.

Solution

1. Identify knowns. The value for v is given by

$$v = (331 \text{ m/s})\sqrt{\frac{T}{273 \text{ K}}}.$$

2. Convert the temperature into kelvins and then enter the temperature into the equation

$$v = (331 \text{ m/s})\sqrt{\frac{303 \text{ K}}{273 \text{ K}}} = 348.7 \text{ m/s}.$$

3. Solve the relationship between speed and wavelength for λ:

$$\lambda = \frac{v}{f}.$$

4. Enter the speed and the minimum frequency to give the maximum wavelength:

$$\lambda_{max} = \frac{348.7 \text{ m/s}}{20 \text{ Hz}} = 17 \text{ m}.$$

5. Enter the speed and the maximum frequency to give the minimum wavelength:

$$\lambda_{min} = \frac{348.7 \text{ m/s}}{20,000 \text{ Hz}} = 0.017 \text{ m} = 1.7 \text{ cm}.$$

Significance

Because the product of f multiplied by λ equals a constant, the smaller f is, the larger λ must be, and vice versa.

The speed of sound can change when sound travels from one medium to another, but the frequency usually remains the same. This is similar to the frequency of a wave on a string being equal to the frequency of the force oscillating the string. If v changes and f remains the same, then the wavelength λ must change. That is, because $v = f\lambda$, the higher the speed of a sound, the greater its wavelength for a given frequency.

 17.1   Check Your Understanding Imagine you observe two firework shells explode. You hear the explosion of one as soon as you see it. However, you see the other shell for several milliseconds before you hear the explosion. Explain why this is so.

Although sound waves in a fluid are longitudinal, sound waves in a solid travel both as longitudinal waves and transverse waves. Seismic waves, which are essentially sound waves in Earth's crust produced by earthquakes, are an interesting example of how the speed of sound depends on the rigidity of the medium. Earthquakes produce both longitudinal and transverse waves, and these travel at different speeds. The bulk modulus of granite is greater than its shear modulus. For that reason, the speed of longitudinal or pressure waves (P-waves) in earthquakes in granite is significantly higher than the speed of transverse or shear waves (S-waves). Both types of earthquake waves travel slower in less rigid material, such as sediments. P-waves have speeds of 4 to 7 km/s, and S-waves range in speed from 2 to 5 km/s, both being faster in more rigid material. The P-wave gets progressively farther ahead of the S-wave as they travel through Earth's crust. The time between the P- and S-waves is routinely used to determine the distance to their source, the epicenter of the earthquake. Because S-waves do not pass through the liquid core, two shadow regions are produced (**Figure 17.11**).

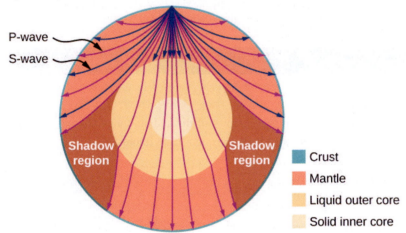

Figure 17.11   Earthquakes produce both longitudinal waves (P-waves) and transverse waves (S-waves), and these travel at different speeds. Both waves travel at different speeds in the different regions of Earth, but in general, P-waves travel faster than S-waves. S-waves cannot be supported by the liquid core, producing shadow regions.

As sound waves move away from a speaker, or away from the epicenter of an earthquake, their power per unit area decreases. This is why the sound is very loud near a speaker and becomes less loud as you move away from the speaker. This also explains why there can be an extreme amount of damage at the epicenter of an earthquake but only tremors are felt in areas far from the epicenter. The power per unit area is known as the intensity, and in the next section, we will discuss how the intensity depends on the distance from the source.

17.3 | Sound Intensity

In a quiet forest, you can sometimes hear a single leaf fall to the ground. But when a passing motorist has his stereo turned up, you cannot even hear what the person next to you in your car is saying (**Figure 17.12**). We are all very familiar with the loudness of sounds and are aware that loudness is related to how energetically the source is vibrating. High noise exposure is hazardous to hearing, which is why it is important for people working in industrial settings to wear ear protection. The relevant physical quantity is sound intensity, a concept that is valid for all sounds whether or not they are in the audible range.

Figure 17.12 Noise on crowded roadways, like this one in Delhi, makes it hard to hear others unless they shout. (credit: "Lingaraj G J"/Flickr)

In **Waves**, we defined intensity as the power per unit area carried by a wave. Power is the rate at which energy is transferred by the wave. In equation form, intensity I is

$$I = \frac{P}{A},$$ (17.8)

where P is the power through an area A. The SI unit for I is W/m^2. If we assume that the sound wave is spherical, and that no energy is lost to thermal processes, the energy of the sound wave is spread over a larger area as distance increases, so the intensity decreases. The area of a sphere is $A = 4\pi r^2$. As the wave spreads out from r_1 to r_2, the energy also spreads out over a larger area:

$$P_1 = P_2$$
$$I_1 4\pi r_1^2 = I_2 4\pi r_2^2;$$

$$I_2 = I_1 \left(\frac{r_1}{r_2}\right)^2 . \qquad\qquad (17.9)$$

The intensity decreases as the wave moves out from the source. In an inverse square relationship, such as the intensity, when you double the distance, the intensity decreases to one quarter,

$$I_2 = I_1 \left(\frac{r_1}{r_2}\right)^2 = I_1 \left(\frac{r_1}{2r_1}\right)^2 = \tfrac{1}{4} I_1 .$$

Generally, when considering the intensity of a sound wave, we take the intensity to be the time-averaged value of the power, denoted by $\langle P \rangle$, divided by the area,

$$I = \frac{\langle P \rangle}{A} . \qquad\qquad (17.10)$$

The intensity of a sound wave is proportional to the change in the pressure squared and inversely proportional to the density and the speed. Consider a parcel of a medium initially undisturbed and then influenced by a sound wave at time t, as shown in **Figure 17.13**.

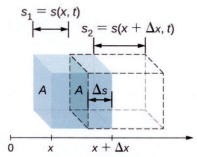

Figure 17.13 An undisturbed parcel of a medium with a volume $V = A\Delta x$ shown in blue. A sound wave moves through the medium at time t, and the parcel is displaced and expands, as shown by dotted lines. The change in volume is $\Delta V = A\Delta s = A(s_2 - s_1)$, where s_1 is the displacement of the leading edge of the parcel and s_2 is the displacement of the trailing edge of the parcel. In the figure, $s_2 > s_1$ and the parcel expands, but the parcel can either expand or compress $(s_2 < s_1)$, depending on which part of the sound wave (compression or rarefaction) is moving through the parcel.

As the sound wave moves through the parcel, the parcel is displaced and may expand or contract. If $s_2 > s_1$, the volume has increased and the pressure decreases. If $s_2 < s_1$, the volume has decreased and the pressure increases. The change in the volume is

$$\Delta V = A\Delta s = A(s_2 - s_1) = A(s(x + \Delta x, t) - s(x, t)).$$

The fractional change in the volume is the change in volume divided by the original volume:

$$\frac{dV}{V} = \lim_{\Delta x \to 0} \frac{A[s(x + \Delta x, t) - s(x, t)]}{A\Delta x} = \frac{\partial s(x, t)}{\partial x} .$$

The fractional change in volume is related to the pressure fluctuation by the bulk modulus $\beta = -\dfrac{\Delta p(x, t)}{dV/V}$. Recall that the minus sign is required because the volume is *inversely* related to the pressure. (We use lowercase p for pressure to distinguish it from power, denoted by P.) The change in pressure is therefore $\Delta p(x, t) = -\beta\frac{dV}{V} = -\beta\frac{\partial s(x, t)}{\partial x}$. If the sound wave is sinusoidal, then the displacement as shown in **Equation 17.2** is $s(x, t) = s_{max}\cos(kx \mp \omega t + \phi)$ and the pressure is found to be

$$\Delta p(x, t) = -\beta\frac{dV}{V} = -\beta\frac{\partial s(x, t)}{\partial x} = \beta k s_{max}\sin(kx - \omega t + \phi) = \Delta p_{max}\sin(kx - \omega t + \phi).$$

The intensity of the sound wave is the power per unit area, and the power is the force times the velocity, $I = \frac{P}{A} = \frac{Fv}{A} = pv$.

Here, the velocity is the velocity of the oscillations of the medium, and not the velocity of the sound wave. The velocity of the medium is the time rate of change in the displacement:

$$v(x, t) = \frac{\partial}{\partial y}s(x, t) = \frac{\partial}{\partial y}(s_{max}\cos(kx - \omega t + \phi)) = s_{max}\omega\sin(kx - \omega t + \phi).$$

Thus, the intensity becomes

$$\begin{aligned} I &= \Delta p(x, t)v(x, t) \\ &= \beta k s_{max}\sin(kx - \omega t + \phi)[s_{max}\omega\sin(kx - \omega t + \phi)] \\ &= \beta k \omega s_{max}^2\sin^2(kx - \omega t + \phi). \end{aligned}$$

To find the time-averaged intensity over one period $T = \frac{2\pi}{\omega}$ for a position x, we integrate over the period, $I = \frac{\beta k\omega s_{max}^2}{2}$.

Using $\Delta p_{max} = \beta k s_{max}$, $v = \sqrt{\frac{\beta}{\rho}}$, and $v = \frac{\omega}{k}$, we obtain

$$I = \frac{\beta k \omega s_{max}^2}{2} = \frac{\beta^2 k^2 \omega s_{max}^2}{2\beta k} = \frac{\omega(\Delta p_{max})^2}{2(\rho v^2)k} = \frac{v(\Delta p_{max})^2}{2(\rho v^2)} = \frac{(\Delta p_{max})^2}{2\rho v}.$$

That is, the intensity of a sound wave is related to its amplitude squared by

$$I = \frac{(\Delta p_{max})^2}{2\rho v}. \tag{17.11}$$

Here, Δp_{max} is the pressure variation or pressure amplitude in units of pascals (Pa) or N/m^2. The energy (as kinetic energy $\frac{1}{2}mv^2$) of an oscillating element of air due to a traveling sound wave is proportional to its amplitude squared. In this equation, ρ is the density of the material in which the sound wave travels, in units of kg/m^3, and v is the speed of sound in the medium, in units of m/s. The pressure variation is proportional to the amplitude of the oscillation, so I varies as $(\Delta p)^2$. This relationship is consistent with the fact that the sound wave is produced by some vibration; the greater its pressure amplitude, the more the air is compressed in the sound it creates.

Human Hearing and Sound Intensity Levels

As stated earlier in this chapter, hearing is the perception of sound. The hearing mechanism involves some interesting physics. The sound wave that impinges upon our ear is a pressure wave. The ear is a **transducer** that converts sound waves into electrical nerve impulses in a manner much more sophisticated than, but analogous to, a microphone. **Figure 17.14** shows the anatomy of the ear.

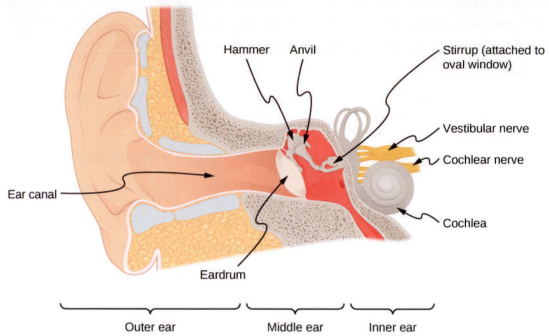

Figure 17.14 The anatomy of the human ear.

The outer ear, or ear canal, carries sound to the recessed, protected eardrum. The air column in the ear canal resonates and is partially responsible for the sensitivity of the ear to sounds in the 2000–5000-Hz range. The middle ear converts sound into mechanical vibrations and applies these vibrations to the cochlea.

 Watch this **video (https://openstaxcollege.org/l/21humanear)** for a more detailed discussion of the workings of the human ear.

The range of intensities that the human ear can hear depends on the frequency of the sound, but, in general, the range is quite large. The minimum threshold intensity that can be heard is $I_0 = 10^{-12}$ W/m^2. Pain is experienced at intensities of $I_{\text{pain}} = 1$ W/m^2. Measurements of sound intensity (in units of W/m^2) are very cumbersome due to this large range in values. For this reason, as well as for other reasons, the concept of sound intensity level was proposed.

The **sound intensity level** β of a sound, measured in decibels, having an intensity I in watts per meter squared, is defined as

$$\beta(\text{dB}) = 10 \log_{10}\left(\frac{I}{I_0}\right),$$

(17.12)

where $I_0 = 10^{-12}$ W/m^2 is a reference intensity, corresponding to the threshold intensity of sound that a person with normal hearing can perceive at a frequency of 1.00 kHz. It is more common to consider sound intensity levels in dB than in W/m^2. How human ears perceive sound can be more accurately described by the logarithm of the intensity rather than directly by the intensity. Because β is defined in terms of a ratio, it is a unitless quantity, telling you the *level* of the sound relative to a fixed standard (10^{-12} W/m^2). The units of decibels (dB) are used to indicate this ratio is multiplied by 10 in its definition. The bel, upon which the decibel is based, is named for Alexander Graham Bell, the inventor of the telephone.

The decibel level of a sound having the threshold intensity of 10^{-12} W/m^2 is $\beta = 0$ dB, because $\log_{10} 1 = 0$. **Table 17.2** gives levels in decibels and intensities in watts per meter squared for some familiar sounds. The ear is sensitive to as little as a trillionth of a watt per meter squared—even more impressive when you realize that the area of the eardrum is only about 1 cm^2, so that only 10^{-16} W falls on it at the threshold of hearing. Air molecules in a sound wave of this intensity vibrate over a distance of less than one molecular diameter, and the gauge pressures involved are less than 10^{-9} atm.

Sound intensity level β (dB)	Intensity I (W/m^2)	Example/effect
0	1×10^{-12}	Threshold of hearing at 1000 Hz
10	1×10^{-11}	Rustle of leaves
20	1×10^{-10}	Whisper at 1-m distance
30	1×10^{-9}	Quiet home
40	1×10^{-8}	Average home
50	1×10^{-7}	Average office, soft music
60	1×10^{-6}	Normal conversation
70	1×10^{-5}	Noisy office, busy traffic
80	1×10^{-4}	Loud radio, classroom lecture
90	1×10^{-3}	Inside a heavy truck; damage from prolonged exposure[1]
100	1×10^{-2}	Noisy factory, siren at 30 m; damage from 8 h per day exposure
110	1×10^{-1}	Damage from 30 min per day exposure
120	1	Loud rock concert; pneumatic chipper at 2 m; threshold of pain
140	1×10^{2}	Jet airplane at 30 m; severe pain, damage in seconds
160	1×10^{4}	Bursting of eardrums

Table 17.2 Sound Intensity Levels and Intensities [1] Several government agencies and health-related professional associations recommend that 85 dB not be exceeded for 8-hour daily exposures in the absence of hearing protection.

An observation readily verified by examining **Table 17.2** or by using **Equation 17.12** is that each factor of 10 in intensity corresponds to 10 dB. For example, a 90-dB sound compared with a 60-dB sound is 30 dB greater, or three factors of 10 (that is, 10^3 times) as intense. Another example is that if one sound is 10^7 as intense as another, it is 70 dB higher (**Table 17.3**).

---

1. Several government agencies and health-related professional associations recommend that 85 dB not be exceeded for 8-hour daily exposures in the absence of hearing protection.

I_2/I_1	$\beta_2 - \beta_1$
2.0	3.0 dB
5.0	7.0 dB
10.0	10.0 dB
100.0	20.0 dB
1000.0	30.0 dB

Table 17.3 Ratios of Intensities and Corresponding Differences in Sound Intensity Levels

Example 17.2

Calculating Sound Intensity Levels

Calculate the sound intensity level in decibels for a sound wave traveling in air at $0°C$ and having a pressure amplitude of 0.656 Pa.

Strategy

We are given Δp, so we can calculate I using the equation $I = \frac{(\Delta p)^2}{2\rho v_{\mathrm{w}}}$. Using I, we can calculate β straight from its definition in $\beta(dB) = 10 \log_{10}\left(\frac{I}{I_0}\right)$.

Solution

1. Identify knowns:
   Sound travels at 331 m/s in air at $0°C$.

   Air has a density of $1.29 \ \text{kg/m}^3$ at atmospheric pressure and $0°C$.

2. Enter these values and the pressure amplitude into $I = \frac{(\Delta p)^2}{2\rho v}$.

$$I = \frac{(\Delta p)^2}{2\rho v} = \frac{(0.656 \ \text{Pa})^2}{2(1.29 \ \text{kg/m}^3)(331 \ \text{m/s})} = 5.04 \times 10^{-4} \ \text{W/m}^2.$$

3. Enter the value for I and the known value for I_0 into $\beta(\text{dB}) = 10 \log_{10}(I/I_0)$. Calculate to find the sound intensity level in decibels:

$$10 \log_{10}(5.04 \times 10^8) = 10(8.70)\text{dB} = 87 \ \text{dB}.$$

Significance

This 87-dB sound has an intensity five times as great as an 80-dB sound. So a factor of five in intensity corresponds to a difference of 7 dB in sound intensity level. This value is true for any intensities differing by a factor of five.

Example 17.3

Changing Intensity Levels of a Sound

Show that if one sound is twice as intense as another, it has a sound level about 3 dB higher.

Strategy

We are given that the ratio of two intensities is 2 to 1, and are then asked to find the difference in their sound levels in decibels. We can solve this problem by using of the properties of logarithms.

Solution

1. Identify knowns:
    The ratio of the two intensities is 2 to 1, or

    $$\frac{I_2}{I_1} = 2.00.$$

    We wish to show that the difference in sound levels is about 3 dB. That is, we want to show:

    $$\beta_2 - \beta_1 = 3 \text{ dB}.$$

    Note that

    $$\log_{10} b - \log_{10} a = \log_{10}\left(\frac{b}{a}\right).$$

2. Use the definition of β to obtain

    $$\beta_2 - \beta_1 = 10 \log_{10}\left(\frac{I_2}{I_1}\right) = 10 \log_{10} 2.00 = 10(0.301) \text{ dB}.$$

    Thus,

    $$\beta_2 - \beta_1 = 3.01 \text{ dB}.$$

Significance

This means that the two sound intensity levels differ by 3.01 dB, or about 3 dB, as advertised. Note that because only the ratio I_2/I_1 is given (and not the actual intensities), this result is true for any intensities that differ by a factor of two. For example, a 56.0-dB sound is twice as intense as a 53.0-dB sound, a 97.0-dB sound is half as intense as a 100-dB sound, and so on.

 17.2 Check Your Understanding Identify common sounds at the levels of 10 dB, 50 dB, and 100 dB.

Another decibel scale is also in use, called the **sound pressure level**, based on the ratio of the pressure amplitude to a reference pressure. This scale is used particularly in applications where sound travels in water. It is beyond the scope of this text to treat this scale because it is not commonly used for sounds in air, but it is important to note that very different decibel levels may be encountered when sound pressure levels are quoted.

Hearing and Pitch

The human ear has a tremendous range and sensitivity. It can give us a wealth of simple information—such as pitch, loudness, and direction.

The perception of frequency is called **pitch**. Typically, humans have excellent relative pitch and can discriminate between two sounds if their frequencies differ by 0.3% or more. For example, 500.0 and 501.5 Hz are noticeably different. Musical **notes** are sounds of a particular frequency that can be produced by most instruments and in Western music have particular names, such as A-sharp, C, or E-flat.

The perception of intensity is called **loudness**. At a given frequency, it is possible to discern differences of about 1 dB, and a change of 3 dB is easily noticed. But loudness is not related to intensity alone. Frequency has a major effect on how loud a sound seems. Sounds near the high- and low-frequency extremes of the hearing range seem even less loud, because the ear is less sensitive at those frequencies. When a violin plays middle C, there is no mistaking it for a piano playing the same note. The reason is that each instrument produces a distinctive set of frequencies and intensities. We call our perception of these combinations of frequencies and intensities tone quality or, more commonly, the **timbre** of the sound. Timbre is the shape of the wave that arises from the many reflections, resonances, and superposition in an instrument.

A unit called a **phon** is used to express loudness numerically. Phons differ from decibels because the phon is a unit of loudness perception, whereas the decibel is a unit of physical intensity. **Figure 17.15** shows the relationship of loudness to intensity (or intensity level) and frequency for persons with normal hearing. The curved lines are equal-loudness curves. Each curve is labeled with its loudness in phons. Any sound along a given curve is perceived as equally loud by the average person. The curves were determined by having large numbers of people compare the loudness of sounds at different frequencies and sound intensity levels. At a frequency of 1000 Hz, phons are taken to be numerically equal to decibels.

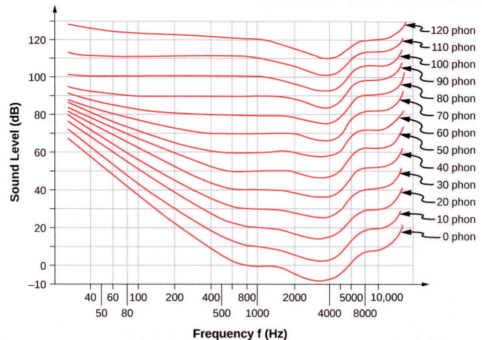

Figure 17.15 The relationship of loudness in phons to intensity level (in decibels) and intensity (in watts per meter squared) for persons with normal hearing. The curved lines are equal-loudness curves—all sounds on a given curve are perceived as equally loud. Phons and decibels are defined to be the same at 1000 Hz.

Example 17.4

Measuring Loudness

(a) What is the loudness in phons of a 100-Hz sound that has an intensity level of 80 dB? (b) What is the intensity level in decibels of a 4000-Hz sound having a loudness of 70 phons? (c) At what intensity level will an 8000-Hz sound have the same loudness as a 200-Hz sound at 60 dB?

Strategy

The graph in **Figure 17.15** should be referenced to solve this example. To find the loudness of a given sound, you must know its frequency and intensity level, locate that point on the square grid, and then interpolate between loudness curves to get the loudness in phons. Once that point is located, the intensity level can be determined from the vertical axis.

Solution

1. Identify knowns: The square grid of the graph relating phons and decibels is a plot of intensity level versus frequency—both physical quantities: 100 Hz at 80 dB lies halfway between the curves marked 70 and 80 phons.
   Find the loudness: 75 phons.

2. Identify knowns: Values are given to be 4000 Hz at 70 phons.
   Follow the 70-phon curve until it reaches 4000 Hz. At that point, it is below the 70 dB line at about 67 dB.
   Find the intensity level: 67 dB.

3. Locate the point for a 200 Hz and 60 dB sound.
    Find the loudness: This point lies just slightly above the 50-phon curve, and so its loudness is 51 phons.
    Look for the 51-phon level is at 8000 Hz: 63 dB.

Significance

These answers, like all information extracted from **Figure 17.15**, have uncertainties of several phons or several decibels, partly due to difficulties in interpolation, but mostly related to uncertainties in the equal-loudness curves.

 17.3   Check Your Understanding Describe how amplitude is related to the loudness of a sound.

In this section, we discussed the characteristics of sound and how we hear, but how are the sounds we hear produced? Interesting sources of sound are musical instruments and the human voice, and we will discuss these sources. But before we can understand how musical instruments produce sound, we need to look at the basic mechanisms behind these instruments. The theories behind the mechanisms used by musical instruments involve interference, superposition, and standing waves, which we discuss in the next section.

17.4 | Normal Modes of a Standing Sound Wave

Learning Objectives

By the end of this section, you will be able to:

- Explain the mechanism behind sound-reducing headphones
- Describe resonance in a tube closed at one end and open at the other end
- Describe resonance in a tube open at both ends

Interference is the hallmark of waves, all of which exhibit constructive and destructive interference exactly analogous to that seen for water waves. In fact, one way to prove something "is a wave" is to observe interference effects. Since sound is a wave, we expect it to exhibit interference.

Interference of Sound Waves

In **Waves**, we discussed the interference of wave functions that differ only in a phase shift. We found that the wave function resulting from the superposition of $y_1(x, t) = A \sin(kx - \omega t + \phi)$ and $y_2(x, t) = A \sin(kx - \omega t)$ is

$$y(x, t) = \left[2A \cos\left(\frac{\phi}{2}\right)\right] \sin\left(kx - \omega t + \frac{\phi}{2}\right).$$

One way for two identical waves that are initially in phase to become out of phase with one another is to have the waves travel different distances; that is, they have different path lengths. Sound waves provide an excellent example of a phase shift due to a path difference. As we have discussed, sound waves can basically be modeled as longitudinal waves, where the molecules of the medium oscillate around an equilibrium position, or as pressure waves.

When the waves leave the speakers, they move out as spherical waves (**Figure 17.16**). The waves interfere; constructive inference is produced by the combination of two crests or two troughs, as shown. Destructive interference is produced by the combination of a trough and a crest.

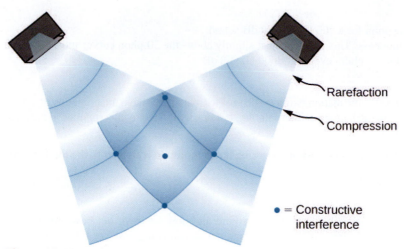

Rarefaction

Compression

● = Constructive
interference

Figure 17.16 When sound waves are produced by a speaker, they travel at the speed of sound and move out as spherical waves. Here, two speakers produce the same steady tone (frequency). The result is points of high-intensity sound (highlighted), which result from two crests (compression) or two troughs (rarefaction) overlapping. Destructive interference results from a crest and trough overlapping. The points where there is constructive interference in the figure occur because the two waves are in phase at those points. Points of destructive interference (**Figure 17.17**) are the result of the two waves being out of phase.

Chapter 17 | Sound

869

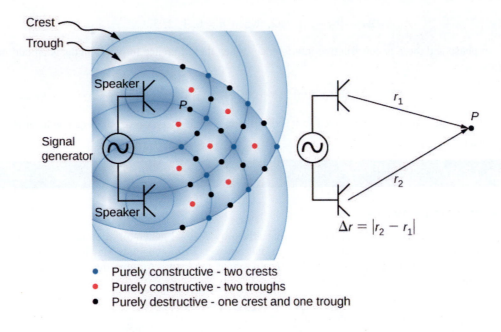

- (blue) Purely constructive - two crests
- (red) Purely constructive - two troughs
- (black) Purely destructive - one crest and one trough

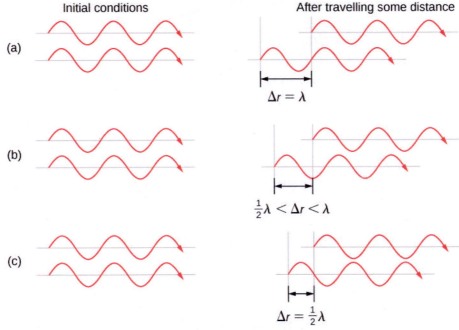

Figure 17.17 Two speakers being driven by a single signal generator. The sound waves produced by the speakers are in phase and are of a single frequency. The sound waves interfere with each other. When two crests or two troughs coincide, there is constructive interference, marked by the red and blue dots. When a trough and a crest coincide, destructive interference occurs, marked by black dots. The phase difference is due to the path lengths traveled by the individual waves. Two identical waves travel two different path lengths to a point P. (a) The difference in the path lengths is one wavelength, resulting in total constructive interference and a resulting amplitude equal to twice the original amplitude. (b) The difference in the path lengths is less than one wavelength but greater than one half a wavelength, resulting in an amplitude greater than zero and less than twice the original amplitude. (c) The difference in the path lengths is one half of a wavelength, resulting in total destructive interference and a resulting amplitude of zero.

The phase difference at each point is due to the different path lengths traveled by each wave. When the difference in the path lengths is an integer multiple of a wavelength,

$$\Delta r = |r_2 - r_1| = n\lambda, \text{ where } n = 0, 1, 2, 3,...,$$

the waves are in phase and there is constructive interference. When the difference in path lengths is an odd multiple of a half wavelength,

$$\Delta r = |r_2 - r_1| = n\frac{\lambda}{2}, \text{ where } n = 1, 3, 5,...,$$

the waves are $180°(\pi \text{ rad})$ out of phase and the result is destructive interference. These points can be located with a sound-level intensity meter.

Example 17.5

Interference of Sound Waves

Two speakers are separated by 5.00 m and are being driven by a signal generator at an unknown frequency. A student with a sound-level meter walks out 6.00 m and down 2.00 m, and finds the first minimum intensity, as shown below. What is the frequency supplied by the signal generator? Assume the wave speed of sound is $v = 343.00$ m/s.

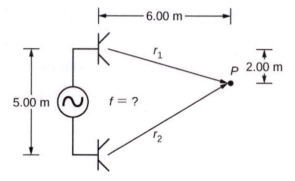

Strategy

The wave velocity is equal to $v = \frac{\lambda}{T} = \lambda f$. The frequency is then $f = \frac{v}{\lambda}$. A minimum intensity indicates destructive interference and the first such point occurs where there is path difference of $\Delta r = \lambda/2$, which can be found from the geometry.

Solution

1. Find the path length to the minimum point from each speaker.

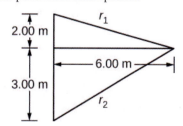

$$r_1 = \sqrt{(6.00 \text{ m})^2 + (2.00 \text{ m})^2} = 6.32 \text{ m}, \quad r_2 = \sqrt{(6.00 \text{ m})^2 + (3.00 \text{ m})^2} = 6.71 \text{ m}$$

2. Use the difference in the path length to find the wavelength.

$$\Delta r = |r_2 - r_1| = |6.71 \text{ m} - 6.32 \text{ m}| = 0.39 \text{ m}$$

$$\lambda = 2\Delta r = 2(0.39 \text{ m}) = 0.78 \text{ m}$$

3. Find the frequency.

$$f = \frac{v}{\lambda} = \frac{343.00 \text{ m/s}}{0.78 \text{ m}} = 439.74 \text{ Hz}$$

Significance

If point *P* were a point of maximum intensity, then the path length would be an integer multiple of the wavelength.

 17.4 Check Your Understanding If you walk around two speakers playing music, how come you do not notice places where the music is very loud or very soft, that is, where there is constructive and destructive interference?

The concept of a phase shift due to a difference in path length is very important. You will use this concept again in **Interference (http://cnx.org/content/m58536/latest/)** and **Photons and Matter Waves (http://cnx.org/content/m58757/latest/)** , where we discuss how Thomas Young used this method in his famous double-slit experiment to provide evidence that light has wavelike properties.

Noise Reduction through Destructive Interference

Figure 17.18 shows a clever use of sound interference to cancel noise. Larger-scale applications of active noise reduction by destructive interference have been proposed for entire passenger compartments in commercial aircraft. To obtain destructive interference, a fast electronic analysis is performed, and a second sound is introduced $180°$ out of phase with the original sound, with its maxima and minima exactly reversed from the incoming noise. Sound waves in fluids are pressure waves and are consistent with Pascal's principle; that is, pressures from two different sources add and subtract like simple numbers. Therefore, positive and negative gauge pressures add to a much smaller pressure, producing a lower-intensity sound. Although completely destructive interference is possible only under the simplest conditions, it is possible to reduce noise levels by 30 dB or more using this technique.

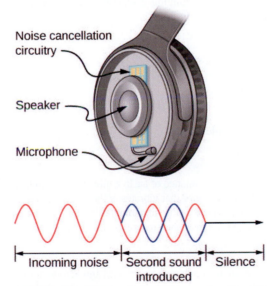

Figure 17.18 Headphones designed to cancel noise with destructive interference create a sound wave exactly opposite to the incoming sound. These headphones can be more effective than the simple passive attenuation used in most ear protection. Such headphones were used on the record-setting, around-the-world nonstop flight of the *Voyager* aircraft in 1986 to protect the pilots' hearing from engine noise.

 17.5 Check Your Understanding Describe how noise-canceling headphones differ from standard headphones used to block outside sounds.

Where else can we observe sound interference? All sound resonances, such as in musical instruments, are due to constructive and destructive interference. Only the resonant frequencies interfere constructively to form standing waves, whereas others interfere destructively and are absent.

Resonance in a Tube Closed at one End

As we discussed in **Waves**, *standing waves* are formed by two waves moving in opposite directions. When two identical sinusoidal waves move in opposite directions, the waves may be modeled as

$$y_1(x, t) = A \sin(kx - \omega t) \text{ and } y_2(x, t) = A \sin(kx + \omega t).$$

When these two waves interfere, the resultant wave is a standing wave:

$$y_R(x, t) = [2A \sin(kx)]\cos(\omega t).$$

Resonance can be produced due to the boundary conditions imposed on a wave. In **Waves**, we showed that resonance could be produced in a string under tension that had symmetrical boundary conditions, specifically, a node at each end. We defined a node as a fixed point where the string did not move. We found that the symmetrical boundary conditions resulted in some frequencies resonating and producing standing waves, while other frequencies interfere destructively. Sound waves can resonate in a hollow tube, and the frequencies of the sound waves that resonate depend on the boundary conditions.

Suppose we have a tube that is closed at one end and open at the other. If we hold a vibrating tuning fork near the open end of the tube, an incident sound wave travels through the tube and reflects off the closed end. The reflected sound has the same frequency and wavelength as the incident sound wave, but is traveling in the opposite direction. At the closed end of the tube, the molecules of air have very little freedom to oscillate, and a node arises. At the open end, the molecules are free to move, and at the right frequency, an antinode occurs. Unlike the symmetrical boundary conditions for the standing waves on the string, the boundary conditions for a tube open at one end and closed at the other end are anti-symmetrical: a node at the closed end and an antinode at the open end.

If the tuning fork has just the right frequency, the air column in the tube resonates loudly, but at most frequencies it vibrates very little. This observation just means that the air column has only certain natural frequencies. Consider the lowest frequency that will cause the tube to resonate, producing a loud sound. There will be a node at the closed end and an antinode at the open end, as shown in **Figure 17.19**.

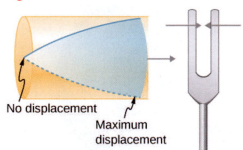

Figure 17.19 Resonance of air in a tube closed at one end, caused by a tuning fork that vibrates at the lowest frequency that can produce resonance (the fundamental frequency). A node exists at the closed end and an antinode at the open end.

The standing wave formed in the tube has an antinode at the open end and a node at the closed end. The distance from a node to an antinode is one-fourth of a wavelength, and this equals the length of the tube; thus, $\lambda_1 = 4L$. This same resonance can be produced by a vibration introduced at or near the closed end of the tube (**Figure 17.20**). It is best to consider this a natural vibration of the air column, independently of how it is induced.

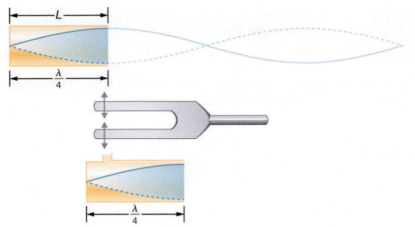

Figure 17.20 The same standing wave is created in the tube by a vibration introduced near its closed end.

Given that maximum air displacements are possible at the open end and none at the closed end, other shorter wavelengths can resonate in the tube, such as the one shown in **Figure 17.21**. Here the standing wave has three-fourths of its wavelength in the tube, or $\frac{3}{4}\lambda_3 = L$, so that $\lambda_3 = \frac{4}{3}L$. Continuing this process reveals a whole series of shorter-wavelength and higher-frequency sounds that resonate in the tube. We use specific terms for the resonances in any system. The lowest resonant frequency is called the **fundamental**, while all higher resonant frequencies are called **overtones**. The resonant frequencies that are integral multiples of the fundamental are collectively called **harmonics**. The fundamental is the first harmonic, the second harmonic is twice the frequency of the first harmonic, and so on. Some of these harmonics may not exist for a given scenario. **Figure 17.22** shows the fundamental and the first three overtones (or the first, third, fifth, and seventh harmonics) in a tube closed at one end.

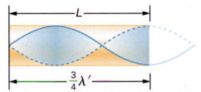

Figure 17.21 Another resonance for a tube closed at one end. This standing wave has maximum air displacement at the open end and none at the closed end. The wavelength is shorter, with three-fourths λ' equaling the length of the tube, so that $\lambda' = 4L/3$. This higher-frequency vibration is the first overtone.

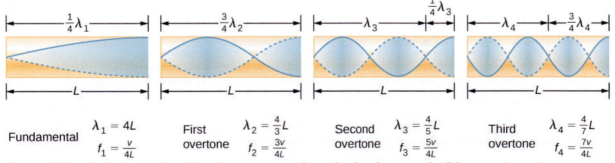

Figure 17.22 The fundamental and three lowest overtones for a tube closed at one end. All have maximum air displacements at the open end and none at the closed end.

The relationship for the resonant wavelengths of a tube closed at one end is

$$\lambda_n = \frac{4}{n}L \quad n = 1, 3, 5, \ldots \tag{17.13}$$

Now let us look for a pattern in the resonant frequencies for a simple tube that is closed at one end. The fundamental has $\lambda = 4L$, and frequency is related to wavelength and the speed of sound as given by

$v = f\lambda$.

Solving for f in this equation gives

$$f = \frac{v}{\lambda} = \frac{v}{4L},$$

where v is the speed of sound in air. Similarly, the first overtone has $\lambda = 4L/3$ (see **Figure 17.22**), so that

$$f_3 = 3\frac{v}{4L} = 3f_1.$$

Because $f_3 = 3f_1$, we call the first overtone the third harmonic. Continuing this process, we see a pattern that can be generalized in a single expression. The resonant frequencies of a tube closed at one end are

$$f_n = n\frac{v}{4L}, \quad n = 1, 3, 5, \ldots, \tag{17.14}$$

where f_1 is the fundamental, f_3 is the first overtone, and so on. It is interesting that the resonant frequencies depend on the speed of sound and, hence, on temperature. This dependence poses a noticeable problem for organs in old unheated cathedrals, and it is also the reason why musicians commonly bring their wind instruments to room temperature before playing them.

Resonance in a Tube Open at Both Ends

Another source of standing waves is a tube that is open at both ends. In this case, the boundary conditions are symmetrical: an antinode at each end. The resonances of tubes open at both ends can be analyzed in a very similar fashion to those for tubes closed at one end. The air columns in tubes open at both ends have maximum air displacements at both ends (**Figure 17.23**). Standing waves form as shown.

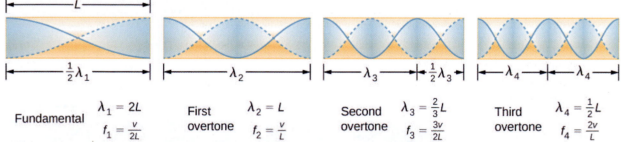

Figure 17.23 The resonant frequencies of a tube open at both ends, including the fundamental and the first three overtones. In all cases, the maximum air displacements occur at both ends of the tube, giving it different natural frequencies than a tube closed at one end.

The relationship for the resonant wavelengths of a tube open at both ends is

$$\lambda_n = \frac{2}{n}L, \quad n = 1, 2, 3, \ldots. \tag{17.15}$$

Based on the fact that a tube open at both ends has maximum air displacements at both ends, and using **Figure 17.23** as a guide, we can see that the resonant frequencies of a tube open at both ends are

$$f_n = n\frac{v}{2L}, \quad n = 1, 2, 3..., \tag{17.16}$$

where f_1 is the fundamental, f_2 is the first overtone, f_3 is the second overtone, and so on. Note that a tube open at both ends has a fundamental frequency twice what it would have if closed at one end. It also has a different spectrum of overtones than a tube closed at one end.

Note that a tube open at both ends has symmetrical boundary conditions, similar to the string fixed at both ends discussed in **Waves**. The relationships for the wavelengths and frequencies of a stringed instrument are the same as given in **Equation 17.15** and **Equation 17.16**. The speed of the wave on the string (from **Waves**) is $v = \sqrt{\frac{F_T}{\mu}}$. The air around the string vibrates at the same frequency as the string, producing sound of the same frequency. The sound wave moves at the speed of sound and the wavelength can be found using $v = \lambda f$.

 17.6   Check Your Understanding How is it possible to use a standing wave's node and antinode to determine the length of a closed-end tube?

 This **video (https://openstaxcollege.org/l/21soundwaves)** lets you visualize sound waves.

 17.7   Check Your Understanding You observe two musical instruments that you cannot identify. One plays high-pitched sounds and the other plays low-pitched sounds. How could you determine which is which without hearing either of them play?

17.5 | Sources of Musical Sound

Learning Objectives

By the end of this section, you will be able to:

* Describe the resonant frequencies in instruments that can be modeled as a tube with symmetrical boundary conditions
* Describe the resonant frequencies in instruments that can be modeled as a tube with anti-symmetrical boundary conditions

Some musical instruments, such as woodwinds, brass, and pipe organs, can be modeled as tubes with symmetrical boundary conditions, that is, either open at both ends or closed at both ends (**Figure 17.24**). Other instruments can be modeled as tubes with anti-symmetrical boundary conditions, such as a tube with one end open and the other end closed (**Figure 17.25**).

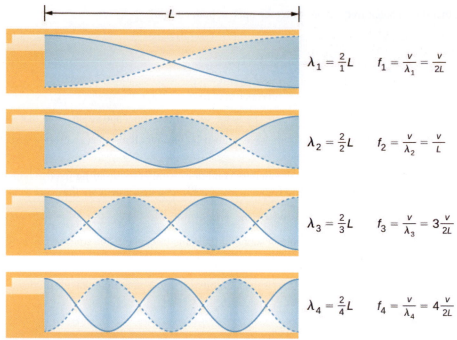

$$\lambda_1 = \frac{2}{1}L \qquad f_1 = \frac{v}{\lambda_1} = \frac{v}{2L}$$

$$\lambda_2 = \frac{2}{2}L \qquad f_2 = \frac{v}{\lambda_2} = \frac{v}{L}$$

$$\lambda_3 = \frac{2}{3}L \qquad f_3 = \frac{v}{\lambda_3} = 3\frac{v}{2L}$$

$$\lambda_4 = \frac{2}{4}L \qquad f_4 = \frac{v}{\lambda_4} = 4\frac{v}{2L}$$

Figure 17.24 Some musical instruments can be modeled as a pipe open at both ends.

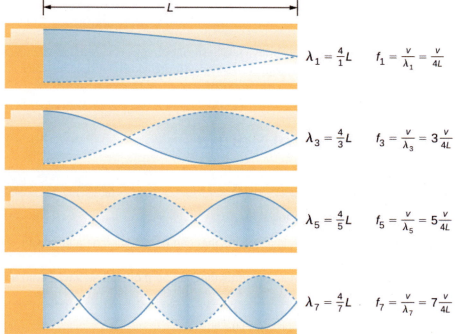

$$\lambda_1 = \frac{4}{1}L \qquad f_1 = \frac{v}{\lambda_1} = \frac{v}{4L}$$

$$\lambda_3 = \frac{4}{3}L \qquad f_3 = \frac{v}{\lambda_3} = 3\frac{v}{4L}$$

$$\lambda_5 = \frac{4}{5}L \qquad f_5 = \frac{v}{\lambda_5} = 5\frac{v}{4L}$$

$$\lambda_7 = \frac{4}{7}L \qquad f_7 = \frac{v}{\lambda_7} = 7\frac{v}{4L}$$

Figure 17.25 Some musical instruments can be modeled as a pipe closed at one end.

Resonant frequencies are produced by longitudinal waves that travel down the tubes and interfere with the reflected waves traveling in the opposite direction. A pipe organ is manufactured with various tubes of fixed lengths to produce different frequencies. The waves are the result of compressed air allowed to expand in the tubes. Even in open tubes, some reflection occurs due to the constraints of the sides of the tubes and the atmospheric pressure outside the open tube.

The antinodes do not occur at the opening of the tube, but rather depend on the radius of the tube. The waves do not fully expand until they are outside the open end of a tube, and for a thin-walled tube, an *end correction* should be added. This end correction is approximately 0.6 times the radius of the tube and should be added to the length of the tube.

Players of instruments such as the flute or oboe vary the length of the tube by opening and closing finger holes. On a trombone, you change the tube length by using a sliding tube. Bugles have a fixed length and can produce only a limited range of frequencies.

The fundamental and overtones can be present simultaneously in a variety of combinations. For example, middle C on a trumpet sounds distinctly different from middle C on a clarinet, although both instruments are modified versions of a tube closed at one end. The fundamental frequency is the same (and usually the most intense), but the overtones and their mix of intensities are different and subject to shading by the musician. This mix is what gives various musical instruments (and human voices) their distinctive characteristics, whether they have air columns, strings, sounding boxes, or drumheads. In fact, much of our speech is determined by shaping the cavity formed by the throat and mouth, and positioning the tongue to adjust the fundamental and combination of overtones. For example, simple resonant cavities can be made to resonate with the sound of the vowels (**Figure 17.26**). In boys at puberty, the larynx grows and the shape of the resonant cavity changes, giving rise to the difference in predominant frequencies in speech between men and women.

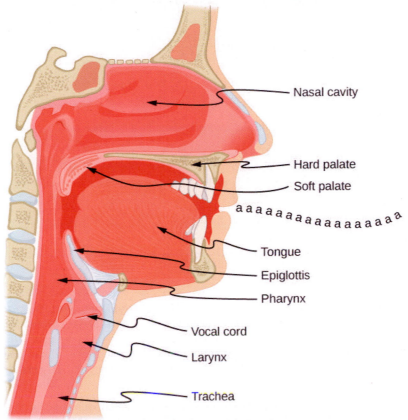

Figure 17.26 The throat and mouth form an air column closed at one end that resonates in response to vibrations in the voice box. The spectrum of overtones and their intensities vary with mouth shaping and tongue position to form different sounds. The voice box can be replaced with a mechanical vibrator, and understandable speech is still possible. Variations in basic shapes make different voices recognizable.

Example 17.6

Finding the Length of a Tube with a 128-Hz Fundamental

(a) What length should a tube closed at one end have on a day when the air temperature is 22.0°C if its fundamental frequency is to be 128 Hz (C below middle C)?

(b) What is the frequency of its fourth overtone?

Strategy

The length L can be found from the relationship $f_n = n\frac{v}{4L}$, but we first need to find the speed of sound v.

Solution

a. Identify knowns: The fundamental frequency is 128 Hz, and the air temperature is $22.0°C$.

Use $f_n = n\frac{v}{4L}$ to find the fundamental frequency ($n = 1$),

$$f_1 = \frac{v}{4L}.$$

Solve this equation for length,

$$L = \frac{v}{4f_1}.$$

Find the speed of sound using $v = (331 \text{ m/s})\sqrt{\frac{T}{273 \text{ K}}}$,

$$v = (331 \text{ m/s})\sqrt{\frac{295 \text{ K}}{273 \text{ K}}} = 344 \text{ m/s}.$$

Enter the values of the speed of sound and frequency into the expression for L.

$$L = \frac{v}{4f_1} = \frac{344 \text{ m/s}}{4(128 \text{ Hz})} = 0.672 \text{ m}$$

b. Identify knowns: The first overtone has $n = 3$, the second overtone has $n = 5$, the third overtone has $n = 7$, and the fourth overtone has $n = 9$.

Enter the value for the fourth overtone into $f_n = n\frac{v}{4L}$,

$$f_9 = 9\frac{v}{4L} = 9f_1 = 1.15 \text{ kHz}.$$

Significance

Many wind instruments are modified tubes that have finger holes, valves, and other devices for changing the length of the resonating air column and hence, the frequency of the note played. Horns producing very low frequencies require tubes so long that they are coiled into loops. An example is the tuba. Whether an overtone occurs in a simple tube or a musical instrument depends on how it is stimulated to vibrate and the details of its shape. The trombone, for example, does not produce its fundamental frequency and only makes overtones.

If you have two tubes with the same fundamental frequency, but one is open at both ends and the other is closed at one end, they would sound different when played because they have different overtones. Middle C, for example, would sound richer played on an open tube, because it has even multiples of the fundamental as well as odd. A closed tube has only odd multiples.

Resonance

Resonance occurs in many different systems, including strings, air columns, and atoms. As we discussed in earlier chapters, resonance is the driven or forced oscillation of a system at natural frequency. At resonance, energy is transferred rapidly to the oscillating system, and the amplitude of its oscillations grows until the system can no longer be described by Hooke's law. An example of this is the distorted sound intentionally produced in certain types of rock music.

Wind instruments use resonance in air columns to amplify tones made by lips or vibrating reeds. Other instruments also use air resonance in clever ways to amplify sound. **Figure 17.27** shows a violin and a guitar, both of which have sounding boxes but with different shapes, resulting in different overtone structures. The vibrating string creates a sound that resonates in the sounding box, greatly amplifying the sound and creating overtones that give the instrument its characteristic timbre. The more complex the shape of the sounding box, the greater its ability to resonate over a wide range of frequencies. The marimba, like the one shown in **Figure 17.28**, uses pots or gourds below the wooden slats to amplify their tones. The resonance of the pot can be adjusted by adding water.

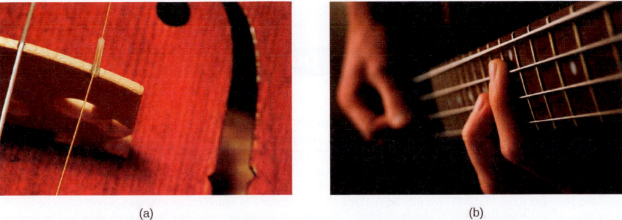

(a)   (b)

Figure 17.27 String instruments such as (a) violins and (b) guitars use resonance in their sounding boxes to amplify and enrich the sound created by their vibrating strings. The bridge and supports couple the string vibrations to the sounding boxes and air within. (credit a: modification of work by Feliciano Guimarães; credit b: modification of work by Steve Snodgrass)

Figure 17.28 Resonance has been used in musical instruments since prehistoric times. This marimba uses gourds as resonance chambers to amplify its sound. (credit: "APC Events"/Flickr)

We have emphasized sound applications in our discussions of resonance and standing waves, but these ideas apply to any system that has wave characteristics. Vibrating strings, for example, are actually resonating and have fundamentals and overtones similar to those for air columns. More subtle are the resonances in atoms due to the wave character of their electrons. Their orbitals can be viewed as standing waves, which have a fundamental (ground state) and overtones (excited states). It is fascinating that wave characteristics apply to such a wide range of physical systems.

17.6 | Beats

The study of music provides many examples of the superposition of waves and the constructive and destructive interference that occurs. Very few examples of music being performed consist of a single source playing a single frequency for an extended period of time. You will probably agree that a single frequency of sound for an extended period might be boring to the point of irritation, similar to the unwanted drone of an aircraft engine or a loud fan. Music is pleasant and interesting due to mixing the changing frequencies of various instruments and voices.

An interesting phenomenon that occurs due to the constructive and destructive interference of two or more frequencies of sound is the phenomenon of **beats**. If two sounds differ in frequencies, the sound waves can be modeled as

$$y_1 = A\cos(k_1 x - 2\pi f_1 t) \text{ and } y_2 = A\cos(k_2 x - 2\pi f_2 t).$$

Using the trigonometric identity $\cos u + \cos v = 2\cos\left(\dfrac{u+v}{2}\right)\cos\left(\dfrac{u-v}{2}\right)$ and considering the point in space as $x = 0.0 \text{ m}$, we find the resulting sound at a point in space, from the superposition of the two sound waves, is equal to **Figure 17.29**:

$$y(t) = 2A\cos(2\pi f_{avg} t)\cos\left(2\pi\left(\frac{|f_2 - f_1|}{2}\right)t\right),$$

where the **beat frequency** is

$$f_{\text{beat}} = |f_2 - f_1|. \qquad\qquad (17.17)$$

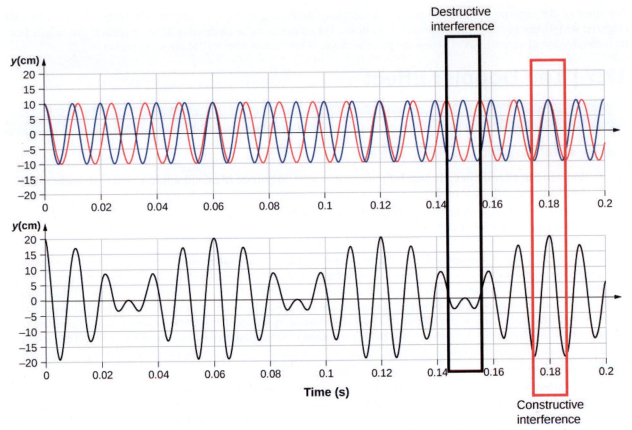

Figure 17.29 Beats produced by the constructive and destructive interference of two sound waves that differ in frequency.

These beats can be used by piano tuners to tune a piano. A tuning fork is struck and a note is played on the piano. As the piano tuner tunes the string, the beats have a lower frequency as the frequency of the note played approaches the frequency of the tuning fork.

Example 17.7

Find the Beat Frequency Between Two Tuning Forks

What is the beat frequency produced when a tuning fork of a frequency of 256 Hz and a tuning fork of a frequency of 512 Hz are struck simultaneously?

Strategy

The beat frequency is the difference of the two frequencies.

Solution

We use $f_{\text{beat}} = |f_2 - f_1|$:

$$|f_2 - f_1| = (512 - 256) \text{ Hz} = 256 \text{ Hz}.$$

Significance

The beat frequency is the absolute value of the difference between the two frequencies. A negative frequency would not make sense.

 17.8 Check Your Understanding What would happen if more than two frequencies interacted? Consider three frequencies.

The study of the superposition of various waves has many interesting applications beyond the study of sound. In later chapters, we will discuss the wave properties of particles. The particles can be modeled as a "wave packet" that results from the superposition of various waves, where the particle moves at the "group velocity" of the wave packet.

17.7 | The Doppler Effect

Learning Objectives

By the end of this section, you will be able to:

- Explain the change in observed frequency as a moving source of sound approaches or departs from a stationary observer
- Explain the change in observed frequency as an observer moves toward or away from a stationary source of sound

The characteristic sound of a motorcycle buzzing by is an example of the **Doppler effect**. Specifically, if you are standing on a street corner and observe an ambulance with a siren sounding passing at a constant speed, you notice two characteristic changes in the sound of the siren. First, the sound increases in loudness as the ambulance approaches and decreases in loudness as it moves away, which is expected. But in addition, the high-pitched siren shifts dramatically to a lower-pitched sound. As the ambulance passes, the frequency of the sound heard by a stationary observer changes from a constant high frequency to a constant lower frequency, even though the siren is producing a constant source frequency. The closer the ambulance brushes by, the more abrupt the shift. Also, the faster the ambulance moves, the greater the shift. We also hear this characteristic shift in frequency for passing cars, airplanes, and trains.

The Doppler effect is an alteration in the observed frequency of a sound due to motion of either the source or the observer. Although less familiar, this effect is easily noticed for a stationary source and moving observer. For example, if you ride a train past a stationary warning horn, you will hear the horn's frequency shift from high to low as you pass by. The actual change in frequency due to relative motion of source and observer is called a **Doppler shift**. The Doppler effect and Doppler shift are named for the Austrian physicist and mathematician Christian Johann Doppler (1803–1853), who did experiments with both moving sources and moving observers. Doppler, for example, had musicians play on a moving open train car and also play standing next to the train tracks as a train passed by. Their music was observed both on and off the train, and changes in frequency were measured.

What causes the Doppler shift? **Figure 17.30** illustrates sound waves emitted by stationary and moving sources in a stationary air mass. Each disturbance spreads out spherically from the point at which the sound is emitted. If the source is stationary, then all of the spheres representing the air compressions in the sound wave are centered on the same point, and the stationary observers on either side hear the same wavelength and frequency as emitted by the source (case a). If the source is moving, the situation is different. Each compression of the air moves out in a sphere from the point at which it was emitted, but the point of emission moves. This moving emission point causes the air compressions to be closer together on one side and farther apart on the other. Thus, the wavelength is shorter in the direction the source is moving (on the right in case b), and longer in the opposite direction (on the left in case b). Finally, if the observers move, as in case (c), the frequency at which they receive the compressions changes. The observer moving toward the source receives them at a higher frequency, and the person moving away from the source receives them at a lower frequency.

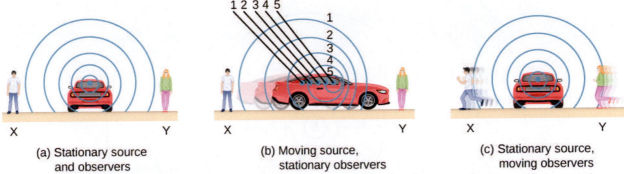

Figure 17.30 Sounds emitted by a source spread out in spherical waves. (a) When the source, observers, and air are stationary, the wavelength and frequency are the same in all directions and to all observers. (b) Sounds emitted by a source moving to the right spread out from the points at which they were emitted. The wavelength is reduced, and consequently, the frequency is increased in the direction of motion, so that the observer on the right hears a higher-pitched sound. The opposite is true for the observer on the left, where the wavelength is increased and the frequency is reduced. (c) The same effect is produced when the observers move relative to the source. Motion toward the source increases frequency as the observer on the right passes through more wave crests than she would if stationary. Motion away from the source decreases frequency as the observer on the left passes through fewer wave crests than he would if stationary.

We know that wavelength and frequency are related by $v = f\lambda$, where v is the fixed speed of sound. The sound moves in a medium and has the same speed v in that medium whether the source is moving or not. Thus, f multiplied by λ is a constant. Because the observer on the right in case (b) receives a shorter wavelength, the frequency she receives must be higher. Similarly, the observer on the left receives a longer wavelength, and hence he hears a lower frequency. The same thing happens in case (c). A higher frequency is received by the observer moving toward the source, and a lower frequency is received by an observer moving away from the source. In general, then, relative motion of source and observer toward one another increases the received frequency. Relative motion apart decreases frequency. The greater the relative speed, the greater the effect.

The Doppler effect occurs not only for sound, but for any wave when there is relative motion between the observer and the source. Doppler shifts occur in the frequency of sound, light, and water waves, for example. Doppler shifts can be used to determine velocity, such as when ultrasound is reflected from blood in a medical diagnostic. The relative velocities of stars and galaxies is determined by the shift in the frequencies of light received from them and has implied much about the origins of the universe. Modern physics has been profoundly affected by observations of Doppler shifts.

Derivation of the Observed Frequency due to the Doppler Shift

Consider two stationary observers X and Y in **Figure 17.31**, located on either side of a stationary source. Each observer hears the same frequency, and that frequency is the frequency produced by the stationary source.

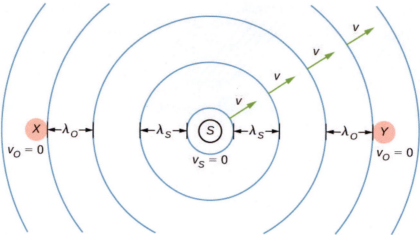

Figure 17.31 A stationary source sends out sound waves at a constant frequency f_s, with a constant wavelength λ_s, at the speed of sound v. Two stationary observers X and Y, on either side of the source, observe a frequency $f_o = f_s$, with a wavelength $\lambda_o = \lambda_s$.

Now consider a stationary observer X with a source moving away from the observer with a constant speed $v_s < v$ (**Figure 17.32**). At time $t = 0$, the source sends out a sound wave, indicated in black. This wave moves out at the speed of sound v. The position of the sound wave at each time interval of period T_s is shown as dotted lines. After one period, the source has moved $\Delta x = v_s T_s$ and emits a second sound wave, which moves out at the speed of sound. The source continues to move and produce sound waves, as indicated by the circles numbered 3 and 4. Notice that as the waves move out, they remained centered at their respective point of origin.

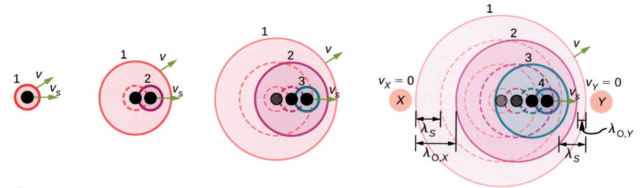

Figure 17.32 A source moving at a constant speed v_s away from an observer X. The moving source sends out sound waves at a constant frequency f_s, with a constant wavelength λ_s, at the speed of sound v. Snapshots of the source at an interval of T_s are shown as the source moves away from the stationary observer X. The solid lines represent the position of the sound waves after four periods from the initial time. The dotted lines are used to show the positions of the waves at each time period. The observer hears a wavelength of $\lambda_o = \lambda_s + \Delta x = \lambda_s + v_s T_s$.

Using the fact that the wavelength is equal to the speed times the period, and the period is the inverse of the frequency, we can derive the observed frequency:

$$\begin{aligned} \lambda_o &= \lambda_s + \Delta x \\ vT_o &= vT_s + v_s T_s \\ \frac{v}{f_o} &= \frac{v}{f_s} = \frac{v_s}{f_s} = \frac{v + v_s}{f_s} \\ f_o &= f_s\left(\frac{v}{v + v_s}\right). \end{aligned}$$

As the source moves away from the observer, the observed frequency is lower than the source frequency.

Now consider a source moving at a constant velocity v_S, moving toward a stationary observer Y, also shown in **Figure 17.32**. The wavelength is observed by Y as $\lambda_o = \lambda_s - \Delta x = \lambda_s - v_s T_s$. Once again, using the fact that the wavelength is equal to the speed times the period, and the period is the inverse of the frequency, we can derive the observed frequency:

$$
\begin{aligned}
\lambda_o &= \lambda_s - \Delta x \\
vT_o &= vT_s - v_s T_s \\
\frac{v}{f_o} &= \frac{v}{f_s} - \frac{v_s}{f_s} = \frac{v - v_s}{f_s} \\
f_o &= f_s\left(\frac{v}{v - v_s}\right).
\end{aligned}
$$

When a source is moving and the observer is stationary, the observed frequency is

$$
f_o = f_s\left(\frac{v}{v \mp v_s}\right), \tag{17.18}
$$

where f_o is the frequency observed by the stationary observer, f_s is the frequency produced by the moving source, v is the speed of sound, v_s is the constant speed of the source, and the top sign is for the source approaching the observer and the bottom sign is for the source departing from the observer.

What happens if the observer is moving and the source is stationary? If the observer moves toward the stationary source, the observed frequency is higher than the source frequency. If the observer is moving away from the stationary source, the observed frequency is lower than the source frequency. Consider observer X in **Figure 17.33** as the observer moves toward a stationary source with a speed v_o. The source emits a tone with a constant frequency f_s and constant period T_s. The observer hears the first wave emitted by the source. If the observer were stationary, the time for one wavelength of sound to pass should be equal to the period of the source T_s. Since the observer is moving toward the source, the time for one wavelength to pass is less than T_s and is equal to the observed period $T_o = T_s - \Delta t$. At time $t = 0$, the observer starts at the beginning of a wavelength and moves toward the second wavelength as the wavelength moves out from the source. The wavelength is equal to the distance the observer traveled plus the distance the sound wave traveled until it is met by the observer:

$$
\begin{aligned}
\lambda_s &= vT_o + v_o T_o \\
vT_s &= (v + v_o)T_o \\
v\left(\frac{1}{f_s}\right) &= (v + v_o)\left(\frac{1}{f_o}\right) \\
f_o &= f_s\left(\frac{v + v_o}{v}\right).
\end{aligned}
$$

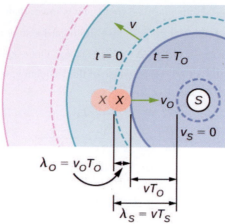

Figure 17.33 A stationary source emits a sound wave with a constant frequency f_s, with a constant wavelength λ_s moving at the speed of sound v. Observer X moves toward the source with a constant speed v_o, and the figure shows the initial and final position of observer X. Observer X observes a frequency higher than the source frequency. The solid lines show the position of the waves at $t = 0$. The dotted lines show the position of the waves at $t = T_o$.

If the observer is moving away from the source (**Figure 17.34**), the observed frequency can be found:

$$
\begin{aligned}
\lambda_s &= vT_o - v_o T_o \\
vT_s &= (v - v_o)T_o \\
v\left(\frac{1}{f_s}\right) &= (v - v_o)\left(\frac{1}{f_o}\right) \\
f_o &= f_s\left(\frac{v - v_o}{v}\right).
\end{aligned}
$$

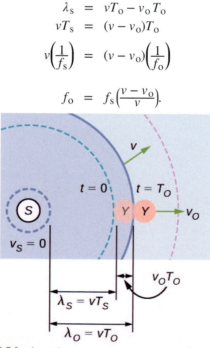

Figure 17.34 A stationary source emits a sound wave with a constant frequency f_s, with a constant wavelength λ_s moving at the speed of sound v. Observer Y moves away from the source with a constant speed v_o, and the figure shows initial and final position of the observer Y. Observer Y observes a frequency lower than the source frequency. The solid lines show the position of the waves at $t = 0$. The dotted lines show the position of the waves at $t = T_o$.

The equations for an observer moving toward or away from a stationary source can be combined into one equation:

$$f_o = f_s\left(\frac{v \pm v_o}{v}\right),$$ (17.19)

where f_o is the observed frequency, f_s is the source frequency, v_w is the speed of sound, v_o is the speed of the observer, the top sign is for the observer approaching the source and the bottom sign is for the observer departing from the source.

Equation 17.18 and Equation 17.19 can be summarized in one equation (the top sign is for approaching) and is further illustrated in Table 17.4:

$$f_o = f_s\left(\frac{v \pm v_o}{v \mp v_s}\right),$$ (17.20)

Doppler shift $f_o = f_s\left(\frac{v \pm v_o}{v \mp v_s}\right)$	Stationary observer	Observer moving towards source	Observer moving away from source
Stationary source	$f_o = f_s$	$f_o = f_s\left(\frac{v + v_o}{v}\right)$	$f_o = f_s\left(\frac{v - v_o}{v}\right)$
Source moving towards observer	$f_o = f_s\left(\frac{v}{v - v_s}\right)$	$f_o = f_s\left(\frac{v + v_o}{v - v_s}\right)$	$f_o = f_s\left(\frac{v - v_o}{v - v_s}\right)$
Source moving away from observer	$f_o = f_s\left(\frac{v}{v + v_s}\right)$	$f_o = f_s\left(\frac{v + v_o}{v + v_s}\right)$	$f_o = f_s\left(\frac{v - v_o}{v + v_s}\right)$

Table 17.4

where f_o is the observed frequency, f_s is the source frequency, v_w is the speed of sound, v_o is the speed of the observer, v_s is the speed of the source, the top sign is for approaching and the bottom sign is for departing.

 The Doppler effect involves motion and a **video (https://openstaxcollege.org/l/21doppler)** will help visualize the effects of a moving observer or source. This video shows a moving source and a stationary observer, and a moving observer and a stationary source. It also discusses the Doppler effect and its application to light.

Example 17.8

Calculating a Doppler Shift

Suppose a train that has a 150-Hz horn is moving at 35.0 m/s in still air on a day when the speed of sound is 340 m/s.

(a) What frequencies are observed by a stationary person at the side of the tracks as the train approaches and after it passes?

(b) What frequency is observed by the train's engineer traveling on the train?

Strategy

To find the observed frequency in (a), we must use $f_{obs} = f_s\left(\frac{v}{v \mp v_s}\right)$ because the source is moving. The minus sign is used for the approaching train, and the plus sign for the receding train. In (b), there are two Doppler shifts—one for a moving source and the other for a moving observer.

Solution

a. Enter known values into $f_o = f_s \left(\frac{v}{v - v_s} \right)$:

$$f_o = f_s \left(\frac{v}{v - v_s} \right) = (150 \text{ Hz}) \left(\frac{340 \text{ m/s}}{340 \text{ m/s} - 35.0 \text{ m/s}} \right).$$

Calculate the frequency observed by a stationary person as the train approaches:

$$f_o = (150 \text{ Hz})(1.11) = 167 \text{ Hz}.$$

Use the same equation with the plus sign to find the frequency heard by a stationary person as the train recedes:

$$f_o = f_s \left(\frac{v}{v + v_s} \right) = (150 \text{ Hz}) \left(\frac{340 \text{ m/s}}{340 \text{ m/s} + 35.0 \text{ m/s}} \right).$$

Calculate the second frequency:

$$f_o = (150 \text{ Hz})(0.907) = 136 \text{ Hz}.$$

b. Identify knowns:

   ◦ It seems reasonable that the engineer would receive the same frequency as emitted by the horn, because the relative velocity between them is zero.

   ◦ Relative to the medium (air), the speeds are $v_s = v_o = 35.0$ m/s.

   ◦ The first Doppler shift is for the moving observer; the second is for the moving source.

   Use the following equation:

   $$f_o = \left[f_s \left(\frac{v \pm v_o}{v} \right) \right] \left(\frac{v}{v \mp v_s} \right).$$

   The quantity in the square brackets is the Doppler-shifted frequency due to a moving observer. The factor on the right is the effect of the moving source.
   Because the train engineer is moving in the direction toward the horn, we must use the plus sign for v_{obs}; however, because the horn is also moving in the direction away from the engineer, we also use the plus sign for v_s. But the train is carrying both the engineer and the horn at the same velocity, so $v_s = v_o$. As a result, everything but f_s cancels, yielding

   $$f_o = f_s.$$

Significance

For the case where the source and the observer are not moving together, the numbers calculated are valid when the source (in this case, the train) is far enough away that the motion is nearly along the line joining source and observer. In both cases, the shift is significant and easily noticed. Note that the shift is 17.0 Hz for motion toward and 14.0 Hz for motion away. The shifts are not symmetric.

For the engineer riding in the train, we may expect that there is no change in frequency because the source and observer move together. This matches your experience. For example, there is no Doppler shift in the frequency of conversations between driver and passenger on a motorcycle. People talking when a wind moves the air between them also observe no Doppler shift in their conversation. The crucial point is that source and observer are not moving relative to each other.

 17.9 Check Your Understanding Describe a situation in your life when you might rely on the Doppler shift to help you either while driving a car or walking near traffic.

The Doppler effect and the Doppler shift have many important applications in science and engineering. For example, the Doppler shift in ultrasound can be used to measure blood velocity, and police use the Doppler shift in radar (a microwave) to measure car velocities. In meteorology, the Doppler shift is used to track the motion of storm clouds; such "Doppler Radar" can give the velocity and direction of rain or snow in weather fronts. In astronomy, we can examine the light emitted from distant galaxies and determine their speed relative to ours. As galaxies move away from us, their light is shifted to a lower frequency, and so to a longer wavelength—the so-called red shift. Such information from galaxies far, far away has allowed us to estimate the age of the universe (from the Big Bang) as about 14 billion years.

17.8 | Shock Waves

Learning Objectives

By the end of this section, you will be able to:

- Explain the mechanism behind sonic booms
- Describe the difference between sonic booms and shock waves
- Describe a bow wake

When discussing the Doppler effect of a moving source and a stationary observer, the only cases we considered were cases where the source was moving at speeds that were less than the speed of sound. Recall that the observed frequency for a moving source approaching a stationary observer is $f_o = f_s\left(\frac{v}{v - v_s}\right)$. As the source approaches the speed of sound, the observed frequency increases. According to the equation, if the source moves at the speed of sound, the denominator is equal to zero, implying the observed frequency is infinite. If the source moves at speeds greater than the speed of sound, the observed frequency is negative.

What could this mean? What happens when a source approaches the speed of sound? It was once argued by some scientists that such a large pressure wave would result from the constructive interference of the sound waves, that it would be impossible for a plane to exceed the speed of sound because the pressures would be great enough to destroy the airplane. But now planes routinely fly faster than the speed of sound. On July 28, 1976, Captain Eldon W. Joersz and Major George T. Morgan flew a Lockheed SR-71 Blackbird #61-7958 at 3529.60 km/h (2193.20 mi/h), which is Mach 2.85. The Mach number is the speed of the source divided by the speed of sound:

$$M = \frac{v_s}{v}. \qquad\qquad (17.21)$$

You will see that interesting phenomena occur when a source approaches and exceeds the speed of sound.

Doppler Effect and High Velocity

What happens to the sound produced by a moving source, such as a jet airplane, that approaches or even exceeds the speed of sound? The answer to this question applies not only to sound but to all other waves as well. Suppose a jet plane is coming nearly straight at you, emitting a sound of frequency f_s. The greater the plane's speed v_s, the greater the Doppler shift and the greater the value observed for f_o (**Figure 17.35**).

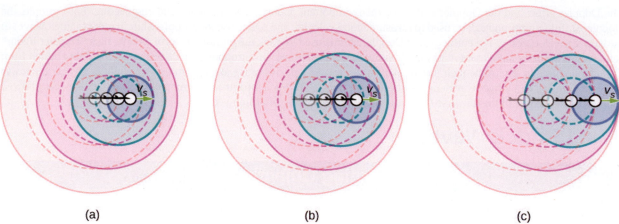

(a)   (b)   (c)

Figure 17.35 Because of the Doppler shift, as a moving source approaches a stationary observer, the observed frequency is higher than the source frequency. The faster the source is moving, the higher the observed frequency. In this figure, the source in (b) is moving faster than the source in (a). Shown are four time steps, the first three shown as dotted lines. (c) If a source moves at the speed of sound, each successive wave interfere with the previous one and the observer observes them all at the same instant.

Now, as v_s approaches the speed of sound, f_o approaches infinity, because the denominator in $f_o = f_s\left(\frac{v}{v \mp v_s}\right)$ approaches zero. At the speed of sound, this result means that in front of the source, each successive wave interferes with the previous one because the source moves forward at the speed of sound. The observer gets them all at the same instant, so the frequency is infinite [part (c) of the figure].

Shock Waves and Sonic Booms

If the source exceeds the speed of sound, no sound is received by the observer until the source has passed, so that the sounds from the approaching source are mixed with those from it when receding. This mixing appears messy, but something interesting happens—a shock wave is created (**Figure 17.36**).

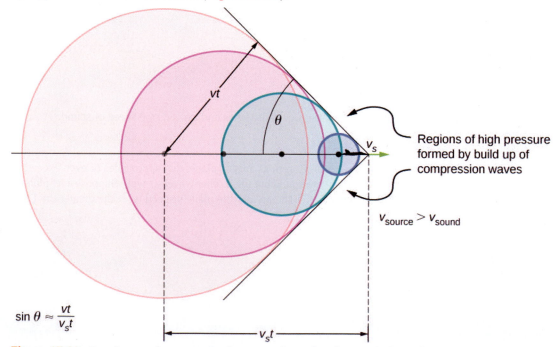

Figure 17.36 Sound waves from a source that moves faster than the speed of sound spread spherically from the point where they are emitted, but the source moves ahead of each wave. Constructive interference along the lines shown (actually a cone in three dimensions) creates a shock wave called a sonic boom. The faster the speed of the source, the smaller the angle θ.

Constructive interference along the lines shown (a cone in three dimensions) from similar sound waves arriving there simultaneously. This superposition forms a disturbance called a **shock wave**, a constructive interference of sound created by an object moving faster than sound. Inside the cone, the interference is mostly destructive, so the sound intensity there is much less than on the shock wave. The angle of the shock wave can be found from the geometry. In time t the source has moved $v_s t$ and the sound wave has moved a distance vt and the angle can be found using $\sin\theta = \frac{vt}{v_s t} = \frac{v}{v_s}$. Note that the

Mach number is defined as $\frac{v_s}{v}$ so the sine of the angle equals the inverse of the Mach number,

$$\sin\theta = \frac{v}{w_s} = \frac{1}{M}.$$

(17.22)

You may have heard of the common term '**sonic boom**.' A common misconception is that the sonic boom occurs as the plane breaks the sound barrier; that is, accelerates to a speed higher than the speed of sound. Actually, the sonic boom occurs as the shock wave sweeps along the ground.

An aircraft creates two shock waves, one from its nose and one from its tail (**Figure 17.37**). During television coverage of space shuttle landings, two distinct booms could often be heard. These were separated by exactly the time it would take the shuttle to pass by a point. Observers on the ground often do not see the aircraft creating the sonic boom, because it has passed by before the shock wave reaches them, as seen in the figure. If the aircraft flies close by at low altitude, pressures in the sonic boom can be destructive and break windows as well as rattle nerves. Because of how destructive sonic booms can be, supersonic flights are banned over populated areas.

Figure 17.37 Two sonic booms experienced by observers, created by the nose and tail of an aircraft as the shock wave sweeps along the ground, are observed on the ground after the plane has passed by.

Shock waves are one example of a broader phenomenon called bow wakes. A **bow wake**, such as the one in **Figure 17.38**, is created when the wave source moves faster than the wave propagation speed. Water waves spread out in circles from the point where created, and the bow wake is the familiar V-shaped wake, trailing the source. A more exotic bow wake is created when a subatomic particle travels through a medium faster than the speed of light travels in that medium. (In a vacuum, the maximum speed of light is $c = 3.00 \times 10^8$ m/s; in the medium of water, the speed of light is closer to $0.75c$.)

If the particle creates light in its passage, that light spreads on a cone with an angle indicative of the speed of the particle, as illustrated in **Figure 17.39**. Such a bow wake is called Cerenkov radiation and is commonly observed in particle physics.

Figure 17.38 Bow wake created by a duck. Constructive interference produces the rather structured wake, whereas relatively little wave action occurs inside the wake, where interference is mostly destructive. (credit: Horia Varlan)

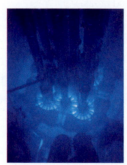

Figure 17.39 The blue glow in this research reactor pool is Cerenkov radiation caused by subatomic particles traveling faster than the speed of light in water. (credit: Idaho National Laboratory)

CHAPTER 17 REVIEW

KEY TERMS

beat frequency frequency of beats produced by sound waves that differ in frequency

beats constructive and destructive interference of two or more frequencies of sound

bow wake v-shaped disturbance created when the wave source moves faster than the wave propagation speed

Doppler effect alteration in the observed frequency of a sound due to motion of either the source or the observer

Doppler shift actual change in frequency due to relative motion of source and observer

fundamental the lowest-frequency resonance

harmonics the term used to refer collectively to the fundamental and its overtones

hearing perception of sound

loudness perception of sound intensity

notes basic unit of music with specific names, combined to generate tunes

overtones all resonant frequencies higher than the fundamental

phon numerical unit of loudness

pitch perception of the frequency of a sound

shock wave wave front that is produced when a sound source moves faster than the speed of sound

sonic boom loud noise that occurs as a shock wave as it sweeps along the ground

sound traveling pressure wave that may be periodic; the wave can be modeled as a pressure wave or as an oscillation of molecules

sound intensity level unitless quantity telling you the level of the sound relative to a fixed standard

sound pressure level ratio of the pressure amplitude to a reference pressure

timbre number and relative intensity of multiple sound frequencies

transducer device that converts energy of a signal into measurable energy form, for example, a microphone converts sound waves into an electrical signal

KEY EQUATIONS

Pressure of a sound wave	$\Delta P = \Delta P_{max} \sin(kx \mp \omega t + \phi)$
Displacement of the oscillating molecules of a sound wave	$s(x,\ t) = s_{max} \cos(kx \mp \omega t + \phi)$
Velocity of a wave	$v = f\lambda$
Speed of sound in a fluid	$v = \sqrt{\dfrac{\beta}{\rho}}$
Speed of sound in a solid	$v = \sqrt{\dfrac{Y}{\rho}}$
Speed of sound in an ideal gas	$v = \sqrt{\dfrac{\gamma R T}{M}}$
Speed of sound in air as a function of temperature	$v = 331\dfrac{m}{s}\sqrt{\dfrac{T_K}{273\ K}} = 331\dfrac{m}{s}\sqrt{1 + \dfrac{T_C}{273°C}}$

Decrease in intensity as a spherical wave expands	$I_2 = I_1 \left(\frac{r_1}{r_2}\right)^2$
Intensity averaged over a period	$I = \frac{\langle P \rangle}{A}$
Intensity of sound	$I = \frac{(\Delta p_{max})^2}{2\rho v}$
Sound intensity level	$\beta(dB) = 10 \log_{10}\left(\frac{I}{I_0}\right)$
Resonant wavelengths of a tube closed at one end	$\lambda_n = \frac{4}{n}L, \quad n = 1, 3, 5,\ldots$
Resonant frequencies of a tube closed at one end	$f_n = n\frac{v}{4L}, \quad n = 1, 3, 5,\ldots$
Resonant wavelengths of a tube open at both ends	$\lambda_n = \frac{2}{n}L, \quad n = 1, 2, 3,\ldots$
Resonant frequencies of a tube open at both ends	$f_n = n\frac{v}{2L}, \quad n = 1, 2, 3,\ldots$
Beat frequency produced by two waves that differ in frequency	$f_{beat} = |f_2 - f_1|$
Observed frequency for a stationary observer and a moving source	$f_o = f_s\left(\frac{v}{v \mp v_s}\right)$
Observed frequency for a moving observer and a stationary source	$f_o = f_s\left(\frac{v \pm v_o}{v}\right)$
Doppler shift for the observed frequency	$f_o = f_s\left(\frac{v \pm v_o}{v \mp v_s}\right)$
Mach number	$M = \frac{v_s}{v}$
Sine of angle formed by shock wave	$\sin\theta = \frac{v}{v_s} = \frac{1}{M}$

SUMMARY

17.1 Sound Waves

- Sound is a disturbance of matter (a pressure wave) that is transmitted from its source outward. Hearing is the perception of sound.

- Sound can be modeled in terms of pressure or in terms of displacement of molecules.

- The human ear is sensitive to frequencies between 20 Hz and 20 kHz.

17.2 Speed of Sound

- The speed of sound depends on the medium and the state of the medium.

- In a fluid, because the absence of shear forces, sound waves are longitudinal. A solid can support both longitudinal and transverse sound waves.

- In air, the speed of sound is related to air temperature T by $v = 331\frac{m}{s}\sqrt{\frac{T_K}{273\,K}} = 331\frac{m}{s}\sqrt{1 + \frac{T_C}{273°C}}$.

- v is the same for all frequencies and wavelengths of sound in air.

17.3 Sound Intensity

- Intensity $I = P/A$ is the same for a sound wave as was defined for all waves, where P is the power crossing area A. The SI unit for I is watts per meter squared. The intensity of a sound wave is also related to the pressure amplitude Δp:

$$I = \frac{(\Delta p)^2}{2\,\rho v},$$

  where ρ is the density of the medium in which the sound wave travels and v_w is the speed of sound in the medium.

- Sound intensity level in units of decibels (dB) is

$$\beta(\text{dB}) = 10 \log_{10}\left(\frac{I}{I_0}\right),$$

  where $I_0 = 10^{-12}\ \text{W/m}^2$ is the threshold intensity of hearing.

- The perception of frequency is pitch. The perception of intensity is loudness and loudness has units of phons.

17.4 Normal Modes of a Standing Sound Wave

- Unwanted sound can be reduced using destructive interference.

- Sound has the same properties of interference and resonance as defined for all waves.

- In air columns, the lowest-frequency resonance is called the fundamental, whereas all higher resonant frequencies are called overtones. Collectively, they are called harmonics.

17.5 Sources of Musical Sound

- Some musical instruments can be modeled as pipes that have symmetrical boundary conditions: open at both ends or closed at both ends. Other musical instruments can be modeled as pipes that have anti-symmetrical boundary conditions: closed at one end and open at the other.

- Some instruments, such as the pipe organ, have several tubes with different lengths. Instruments such as the flute vary the length of the tube by closing the holes along the tube. The trombone varies the length of the tube using a sliding bar.

- String instruments produce sound using a vibrating string with nodes at each end. The air around the string oscillates at the frequency of the string. The relationship for the frequencies for the string is the same as for the symmetrical boundary conditions of the pipe, with the length of the pipe replaced by the length of the string and the velocity replaced by $v = \sqrt{\frac{F_T}{\mu}}$.

17.6 Beats

- When two sound waves that differ in frequency interfere, beats are created with a beat frequency that is equal to the absolute value of the difference in the frequencies.

17.7 The Doppler Effect

- The Doppler effect is an alteration in the observed frequency of a sound due to motion of either the source or the observer.

- The actual change in frequency is called the Doppler shift.

17.8 Shock Waves

- The Mach number is the velocity of a source divided by the speed of sound, $M = \frac{v_s}{v}$.

- When a sound source moves faster than the speed of sound, a shock wave is produced as the sound waves interfere.
- A sonic boom is the intense sound that occurs as the shock wave moves along the ground.
- The angle the shock wave produces can be found as $\sin \theta = \frac{v}{v_s} = \frac{1}{M}$.
- A bow wake is produced when an object moves faster than the speed of a mechanical wave in the medium, such as a boat moving through the water.

CONCEPTUAL QUESTIONS

17.1 Sound Waves

1. What is the difference between sound and hearing?

2. You will learn that light is an electromagnetic wave that can travel through a vacuum. Can sound waves travel through a vacuum?

3. Sound waves can be modeled as a change in pressure. Why is the change in pressure used and not the actual pressure?

17.2 Speed of Sound

4. How do sound vibrations of atoms differ from thermal motion?

5. When sound passes from one medium to another where its propagation speed is different, does its frequency or wavelength change? Explain your answer briefly.

6. A popular party trick is to inhale helium and speak in a high-frequency, funny voice. Explain this phenomenon.

7. You may have used a sonic range finder in lab to measure the distance of an object using a clicking sound from a sound transducer. What is the principle used in this device?

8. The sonic range finder discussed in the preceding question often needs to be calibrated. During the calibration, the software asks for the room temperature. Why do you suppose the room temperature is required?

17.3 Sound Intensity

9. Six members of a synchronized swim team wear earplugs to protect themselves against water pressure at depths, but they can still hear the music and perform the combinations in the water perfectly. One day, they were asked to leave the pool so the dive team could practice a few dives, and they tried to practice on a mat, but seemed to have a lot more difficulty. Why might this be?

10. A community is concerned about a plan to bring train service to their downtown from the town's outskirts. The current sound intensity level, even though the rail yard is blocks away, is 70 dB downtown. The mayor assures the public that there will be a difference of only 30 dB in sound in the downtown area. Should the townspeople be concerned? Why?

17.4 Normal Modes of a Standing Sound Wave

11. You are given two wind instruments of identical length. One is open at both ends, whereas the other is closed at one end. Which is able to produce the lowest frequency?

12. What is the difference between an overtone and a harmonic? Are all harmonics overtones? Are all overtones harmonics?

13. Two identical columns, open at both ends, are in separate rooms. In room A, the temperature is $T = 20°C$ and in room B, the temperature is $T = 25°C$. A speaker is attached to the end of each tube, causing the tubes to resonate at the fundamental frequency. Is the frequency the same for both tubes? Which has the higher frequency?

17.5 Sources of Musical Sound

14. How does an unamplified guitar produce sounds so much more intense than those of a plucked string held taut by a simple stick?

15. Consider three pipes of the same length (L). Pipe A is open at both ends, pipe B is closed at both ends, and pipe C has one open end and one closed end. If the velocity of sound is the same in each of the three tubes, in which of the tubes could the lowest fundamental frequency be produced? In which of the tubes could the highest fundamental frequency be produced?

16. Pipe A has a length L and is open at both ends. Pipe B has a length $L/2$ and has one open end and one closed end. Assume the speed of sound to be the same in both tubes. Which of the harmonics in each tube would be equal?

17. A string is tied between two lab posts a distance L apart. The tension in the string and the linear mass density is such that the speed of a wave on the string is $v = 343$ m/s. A tube with symmetric boundary conditions has a length L and the speed of sound in the tube is $v = 343$ m/s. What could be said about the frequencies of the harmonics in the string and the tube? What if the velocity in the string were $v = 686$ m/s?

17.6 Beats

18. Two speakers are attached to variable-frequency signal generator. Speaker A produces a constant-frequency sound wave of 1.00 kHz, and speaker B produces a tone of 1.10 kHz. The beat frequency is 0.10 kHz. If the frequency of each speaker is doubled, what is the beat frequency produced?

19. The label has been scratched off a tuning fork and you need to know its frequency. From its size, you suspect that it is somewhere around 250 Hz. You find a 250-Hz tuning fork and a 270-Hz tuning fork. When you strike the 250-Hz fork and the fork of unknown frequency, a beat frequency of 5 Hz is produced. When you strike the unknown with the 270-Hz fork, the beat frequency is 15 Hz. What is the unknown frequency? Could you have deduced the frequency using just the 250-Hz fork?

20. Referring to the preceding question, if you had only the 250-Hz fork, could you come up with a solution to the problem of finding the unknown frequency?

21. A "showy" custom-built car has two brass horns that are supposed to produce the same frequency but actually emit 263.8 and 264.5 Hz. What beat frequency is produced?

17.7 The Doppler Effect

22. Is the Doppler shift real or just a sensory illusion?

23. Three stationary observers observe the Doppler shift from a source moving at a constant velocity. The observers are stationed as shown below. Which observer will observe

the highest frequency? Which observer will observe the lowest frequency? What can be said about the frequency observed by observer 3?

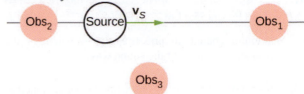

24. Shown below is a stationary source and moving observers. Describe the frequencies observed by the observers for this configuration.

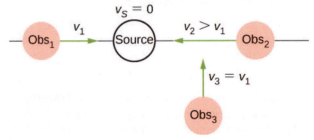

25. Prior to 1980, conventional radar was used by weather forecasters. In the 1960s, weather forecasters began to experiment with Doppler radar. What do you think is the advantage of using Doppler radar?

17.8 Shock Waves

26. What is the difference between a sonic boom and a shock wave?

27. Due to efficiency considerations related to its bow wake, the supersonic transport aircraft must maintain a cruising speed that is a constant ratio to the speed of sound (a constant Mach number). If the aircraft flies from warm air into colder air, should it increase or decrease its speed? Explain your answer.

28. When you hear a sonic boom, you often cannot see the plane that made it. Why is that?

PROBLEMS

17.1 Sound Waves

29. Consider a sound wave modeled with the equation $s(x, t) = 4.00 \text{ nm} \cos(3.66 \text{ m}^{-1} x - 1256 \text{ s}^{-1} t)$. What is the maximum displacement, the wavelength, the frequency, and the speed of the sound wave?

30. Consider a sound wave moving through the air modeled with the equation

$$s(x, t) = 6.00 \text{ nm} \cos(54.93 \text{ m}^{-1} x - 18.84 \times 10^3 \text{ s}^{-1} t).$$

What is the shortest time required for an air molecule to move between 3.00 nm and −3.00 nm?

31. Consider a diagnostic ultrasound of frequency 5.00 MHz that is used to examine an irregularity in soft tissue. (a) What is the wavelength in air of such a sound wave if the speed of sound is 343 m/s? (b) If the speed of sound in tissue is 1800 m/s, what is the wavelength of this wave in tissue?

32. A sound wave is modeled as $\Delta P = 1.80 \text{ Pa } \sin\left(55.41 \text{ m}^{-1} x - 18,840 \text{ s}^{-1} t\right)$. What is the maximum change in pressure, the wavelength, the frequency, and the speed of the sound wave?

33. A sound wave is modeled with the wave function $\Delta P = 1.20 \text{ Pa } \sin\left(kx - 6.28 \times 10^4 \text{ s}^{-1} t\right)$ and the sound wave travels in air at a speed of $v = 343.00$ m/s. (a) What is the wave number of the sound wave? (b) What is the value for $\Delta P(3.00 \text{ m}, 20.00 \text{ s})$?

34. The displacement of the air molecules in sound wave is modeled with the wave function $s(x, t) = 5.00 \text{ nm } \cos\left(91.54 \text{ m}^{-1} x - 3.14 \times 10^4 \text{ s}^{-1} t\right)$.
(a) What is the wave speed of the sound wave? (b) What is the maximum speed of the air molecules as they oscillate in simple harmonic motion? (c) What is the magnitude of the maximum acceleration of the air molecules as they oscillate in simple harmonic motion?

35. A speaker is placed at the opening of a long horizontal tube. The speaker oscillates at a frequency f, creating a sound wave that moves down the tube. The wave moves through the tube at a speed of $v = 340.00$ m/s. The sound wave is modeled with the wave function $s(x, t) = s_{max} \cos(kx - \omega t + \phi)$. At time $t = 0.00$ s, an air molecule at $x = 3.5$ m is at the maximum displacement of 7.00 nm. At the same time, another molecule at $x = 3.7$ m has a displacement of 3.00 nm. What is the frequency at which the speaker is oscillating?

36. A 250-Hz tuning fork is struck and begins to vibrate. A sound-level meter is located 34.00 m away. It takes the sound $\Delta t = 0.10$ s to reach the meter. The maximum displacement of the tuning fork is 1.00 mm. Write a wave function for the sound.

37. A sound wave produced by an ultrasonic transducer, moving in air, is modeled with the wave equation $s(x, t) = 4.50 \text{ nm } \cos\left(9.15 \times 10^4 \text{ m}^{-1} x - 2\pi(5.00 \text{ MHz})t\right)$. The transducer is to be used in nondestructive testing to test for fractures in steel beams. The speed of sound in the steel beam is $v = 5950$ m/s. Find the wave function for the sound wave in the steel beam.

38. Porpoises emit sound waves that they use for navigation. If the wavelength of the sound wave emitted is 4.5 cm, and the speed of sound in the water is $v = 1530$ m/s, what is the period of the sound?

39. Bats use sound waves to catch insects. Bats can detect frequencies up to 100 kHz. If the sound waves travel through air at a speed of $v = 343$ m/s, what is the wavelength of the sound waves?

40. A bat sends of a sound wave 100 kHz and the sound waves travel through air at a speed of $v = 343$ m/s. (a) If the maximum pressure difference is 1.30 Pa, what is a wave function that would model the sound wave, assuming the wave is sinusoidal? (Assume the phase shift is zero.) (b) What are the period and wavelength of the sound wave?

41. Consider the graph shown below of a compression wave. Shown are snapshots of the wave function for $t = 0.000$ s (blue) and $t = 0.005$ s (orange). What are the wavelength, maximum displacement, velocity, and period of the compression wave?

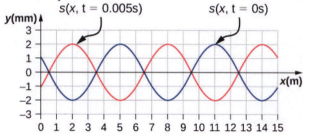

42. Consider the graph in the preceding problem of a compression wave. Shown are snapshots of the wave function for $t = 0.000$ s (blue) and $t = 0.005$ s (orange). Given that the displacement of the molecule at time $t = 0.00$ s and position $x = 0.00$ m is $s(0.00 \text{ m}, 0.00 \text{ s}) = 1.08$ mm, derive a wave function to model the compression wave.

43. A guitar string oscillates at a frequency of 100 Hz and produces a sound wave. (a) What do you think the frequency of the sound wave is that the vibrating string produces? (b) If the speed of the sound wave is $v = 343$ m/s, , what is the wavelength of the sound wave?

17.2 Speed of Sound

44. When poked by a spear, an operatic soprano lets out a 1200-Hz shriek. What is its wavelength if the speed of sound is 345 m/s?

45. What frequency sound has a 0.10-m wavelength when the speed of sound is 340 m/s?

46. Calculate the speed of sound on a day when a 1500-Hz frequency has a wavelength of 0.221 m.

47. (a) What is the speed of sound in a medium where a 100-kHz frequency produces a 5.96-cm wavelength? (b)

Which substance in **Table 17.1** is this likely to be?

48. Show that the speed of sound in 20.0°C air is 343 m/s, as claimed in the text.

49. Air temperature in the Sahara Desert can reach 56.0°C (about 134°F). What is the speed of sound in air at that temperature?

50. Dolphins make sounds in air and water. What is the ratio of the wavelength of a sound in air to its wavelength in seawater? Assume air temperature is 20.0°C.

51. A sonar echo returns to a submarine 1.20 s after being emitted. What is the distance to the object creating the echo? (Assume that the submarine is in the ocean, not in fresh water.)

52. (a) If a submarine's sonar can measure echo times with a precision of 0.0100 s, what is the smallest difference in distances it can detect? (Assume that the submarine is in the ocean, not in fresh water.) (b) Discuss the limits this time resolution imposes on the ability of the sonar system to detect the size and shape of the object creating the echo.

53. Ultrasonic sound waves are often used in methods of nondestructive testing. For example, this method can be used to find structural faults in a steel I-beams used in building. Consider a 10.00 meter long, steel I-beam with a cross-section shown below. The weight of the I-beam is 3846.50 N. What would be the speed of sound through in the I-beam? $\left(Y_{steel} = 200\,\text{GPa},\ \beta_{steel} = 159\,\text{GPa}\right)$.

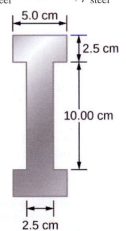

54. A physicist at a fireworks display times the lag between seeing an explosion and hearing its sound, and finds it to be 0.400 s. (a) How far away is the explosion if air temperature is 24.0°C and if you neglect the time taken for light to reach the physicist? (b) Calculate the distance to the explosion taking the speed of light into account. Note that this distance is negligibly greater.

55. During a 4th of July celebration, an M80 firework explodes on the ground, producing a bright flash and a loud bang. The air temperature of the night air is $T_F = 90.00°F$. Two observers see the flash and hear the bang. The first observer notes the time between the flash and the bang as 1.00 second. The second observer notes the difference as 3.00 seconds. The line of sight between the two observers meet at a right angle as shown below. What is the distance Δx between the two observers?

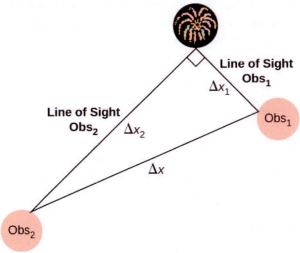

56. The density of a sample of water is $\rho = 998.00\,\text{kg/m}^3$ and the bulk modulus is $\beta = 2.15\,\text{GPa}$. What is the speed of sound through the sample?

57. Suppose a bat uses sound echoes to locate its insect prey, 3.00 m away. (See **Figure 17.6**.) (a) Calculate the echo times for temperatures of 5.00°C and 35.0°C. (b) What percent uncertainty does this cause for the bat in locating the insect? (c) Discuss the significance of this uncertainty and whether it could cause difficulties for the bat. (In practice, the bat continues to use sound as it closes in, eliminating most of any difficulties imposed by this and other effects, such as motion of the prey.)

17.3 Sound Intensity

58. What is the intensity in watts per meter squared of a 85.0-dB sound?

59. The warning tag on a lawn mower states that it produces noise at a level of 91.0 dB. What is this in watts per meter squared?

60. A sound wave traveling in air has a pressure amplitude of 0.5 Pa. What is the intensity of the wave?

61. What intensity level does the sound in the preceding problem correspond to?

62. What sound intensity level in dB is produced by earphones that create an intensity of 4.00×10^{-2} W/m^2 ?

63. What is the decibel level of a sound that is twice as intense as a 90.0-dB sound? (b) What is the decibel level of a sound that is one-fifth as intense as a 90.0-dB sound?

64. What is the intensity of a sound that has a level 7.00 dB lower than a 4.00×10^{-9}-W/m^2 sound? (b) What is the intensity of a sound that is 3.00 dB higher than a 4.00×10^{-9}-W/m^2 sound?

65. People with good hearing can perceive sounds as low as −8.00 dB at a frequency of 3000 Hz. What is the intensity of this sound in watts per meter squared?

66. If a large housefly 3.0 m away from you makes a noise of 40.0 dB, what is the noise level of 1000 flies at that distance, assuming interference has a negligible effect?

67. Ten cars in a circle at a boom box competition produce a 120-dB sound intensity level at the center of the circle. What is the average sound intensity level produced there by each stereo, assuming interference effects can be neglected?

68. The amplitude of a sound wave is measured in terms of its maximum gauge pressure. By what factor does the amplitude of a sound wave increase if the sound intensity level goes up by 40.0 dB?

69. If a sound intensity level of 0 dB at 1000 Hz corresponds to a maximum gauge pressure (sound amplitude) of 10^{-9} atm , what is the maximum gauge pressure in a 60-dB sound? What is the maximum gauge pressure in a 120-dB sound?

70. An 8-hour exposure to a sound intensity level of 90.0 dB may cause hearing damage. What energy in joules falls on a 0.800-cm-diameter eardrum so exposed?

71. Sound is more effectively transmitted into a stethoscope by direct contact rather than through the air, and it is further intensified by being concentrated on the smaller area of the eardrum. It is reasonable to assume that sound is transmitted into a stethoscope 100 times as effectively compared with transmission though the air. What, then, is the gain in decibels produced by a stethoscope that has a sound gathering area of $15.0\,\text{cm}^2$, and concentrates the sound onto two eardrums with a total area of $0.900\,\text{cm}^2$ with an efficiency of 40.0% ?

72. Loudspeakers can produce intense sounds with surprisingly small energy input in spite of their low efficiencies. Calculate the power input needed to produce a 90.0-dB sound intensity level for a 12.0-cm-diameter speaker that has an efficiency of 1.00% . (This value is the sound intensity level right at the speaker.)

73. The factor of 10^{-12} in the range of intensities to which the ear can respond, from threshold to that causing damage after brief exposure, is truly remarkable. If you could measure distances over the same range with a single instrument and the smallest distance you could measure was 1 mm, what would the largest be?

74. What are the closest frequencies to 500 Hz that an average person can clearly distinguish as being different in frequency from 500 Hz? The sounds are not present simultaneously.

75. Can you tell that your roommate turned up the sound on the TV if its average sound intensity level goes from 70 to 73 dB?

76. If a woman needs an amplification of 5.0×10^5 times the threshold intensity to enable her to hear at all frequencies, what is her overall hearing loss in dB? Note that smaller amplification is appropriate for more intense sounds to avoid further damage to her hearing from levels above 90 dB.

77. A person has a hearing threshold 10 dB above normal at 100 Hz and 50 dB above normal at 4000 Hz. How much more intense must a 100-Hz tone be than a 4000-Hz tone if they are both barely audible to this person?

17.4 Normal Modes of a Standing Sound Wave

78. (a) What is the fundamental frequency of a 0.672-m-long tube, open at both ends, on a day when the speed of sound is 344 m/s? (b) What is the frequency of its second harmonic?

79. What is the length of a tube that has a fundamental frequency of 176 Hz and a first overtone of 352 Hz if the speed of sound is 343 m/s?

80. The ear canal resonates like a tube closed at one end. (See [link]Figure 17_03_HumEar[/link].) If ear canals range in length from 1.80 to 2.60 cm in an average population, what is the range of fundamental resonant frequencies? Take air temperature to be 37.0°C, which is the same as body temperature.

81. Calculate the first overtone in an ear canal, which resonates like a 2.40-cm-long tube closed at one end, by taking air temperature to be 37.0°C . Is the ear particularly sensitive to such a frequency? (The resonances of the ear canal are complicated by its nonuniform shape, which we shall ignore.)

82. A crude approximation of voice production is to consider the breathing passages and mouth to be a resonating tube closed at one end. (a) What is the fundamental frequency if the tube is 0.240 m long, by taking air temperature to be $37.0°C$? (b) What would this frequency become if the person replaced the air with helium? Assume the same temperature dependence for helium as for air.

83. A 4.0-m-long pipe, open at one end and closed at one end, is in a room where the temperature is $T = 22°C$. A speaker capable of producing variable frequencies is placed at the open end and is used to cause the tube to resonate. (a) What is the wavelength and the frequency of the fundamental frequency? (b) What is the frequency and wavelength of the first overtone?

84. A 4.0-m-long pipe, open at both ends, is placed in a room where the temperature is $T = 25°C$. A speaker capable of producing variable frequencies is placed at the open end and is used to cause the tube to resonate. (a) What are the wavelength and the frequency of the fundamental frequency? (b) What are the frequency and wavelength of the first overtone?

85. A nylon guitar string is fixed between two lab posts 2.00 m apart. The string has a linear mass density of $\mu = 7.20$ g/m and is placed under a tension of 160.00 N. The string is placed next to a tube, open at both ends, of length L. The string is plucked and the tube resonates at the $n = 3$ mode. The speed of sound is 343 m/s. What is the length of the tube?

86. A 512-Hz tuning fork is struck and placed next to a tube with a movable piston, creating a tube with a variable length. The piston is slid down the pipe and resonance is reached when the piston is 115.50 cm from the open end. The next resonance is reached when the piston is 82.50 cm from the open end. (a) What is the speed of sound in the tube? (b) How far from the open end will the piston cause the next mode of resonance?

87. Students in a physics lab are asked to find the length of an air column in a tube closed at one end that has a fundamental frequency of 256 Hz. They hold the tube vertically and fill it with water to the top, then lower the water while a 256-Hz tuning fork is rung and listen for the first resonance. (a) What is the air temperature if the resonance occurs for a length of 0.336 m? (b) At what length will they observe the second resonance (first overtone)?

17.5 Sources of Musical Sound

88. If a wind instrument, such as a tuba, has a fundamental frequency of 32.0 Hz, what are its first three overtones? It is closed at one end. (The overtones of a real tuba are more complex than this example, because it is a tapered tube.)

89. What are the first three overtones of a bassoon that has a fundamental frequency of 90.0 Hz? It is open at both ends. (The overtones of a real bassoon are more complex than this example, because its double reed makes it act more like a tube closed at one end.)

90. How long must a flute be in order to have a fundamental frequency of 262 Hz (this frequency corresponds to middle C on the evenly tempered chromatic scale) on a day when air temperature is $20.0°C$? It is open at both ends.

91. What length should an oboe have to produce a fundamental frequency of 110 Hz on a day when the speed of sound is 343 m/s? It is open at both ends.

92. (a) Find the length of an organ pipe closed at one end that produces a fundamental frequency of 256 Hz when air temperature is $18.0°C$. (b) What is its fundamental frequency at $25.0°C$?

93. An organ pipe $(L = 3.00 \text{ m})$ is closed at both ends. Compute the wavelengths and frequencies of the first three modes of resonance. Assume the speed of sound is $v = 343.00$ m/s.

94. An organ pipe $(L = 3.00 \text{ m})$ is closed at one end. Compute the wavelengths and frequencies of the first three modes of resonance. Assume the speed of sound is $v = 343.00$ m/s.

95. A sound wave of a frequency of 2.00 kHz is produced by a string oscillating in the $n = 6$ mode. The linear mass density of the string is $\mu = 0.0065$ kg/m and the length of the string is 1.50 m. What is the tension in the string?

96. Consider the sound created by resonating the tube shown below. The air temperature is $T_C = 30.00°C$. What are the wavelength, wave speed, and frequency of the sound produced?

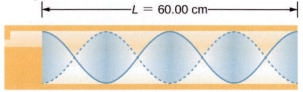

97. A student holds an 80.00-cm lab pole one quarter of the length from the end of the pole. The lab pole is made of aluminum. The student strikes the lab pole with a hammer. The pole resonates at the lowest possible frequency. What is that frequency?

98. A string on the violin has a length of 24.00 cm and a mass of 0.860 g. The fundamental frequency of the string is 1.00 kHz. (a) What is the speed of the wave on the string? (b) What is the tension in the string?

99. By what fraction will the frequencies produced by a wind instrument change when air temperature goes from $10.0°C$ to $30.0°C$? That is, find the ratio of the frequencies at those temperatures.

17.6 Beats

100. What beat frequencies are present: (a) If the musical notes A and C are played together (frequencies of 220 and 264 Hz)? (b) If D and F are played together (frequencies of 297 and 352 Hz)? (c) If all four are played together?

101. What beat frequencies result if a piano hammer hits three strings that emit frequencies of 127.8, 128.1, and 128.3 Hz?

102. A piano tuner hears a beat every 2.00 s when listening to a 264.0-Hz tuning fork and a single piano string. What are the two possible frequencies of the string?

103. Two identical strings, of identical lengths of 2.00 m and linear mass density of $\mu = 0.0065\,\text{kg/m}$, are fixed on both ends. String A is under a tension of 120.00 N. String B is under a tension of 130.00 N. They are each plucked and produce sound at the $n = 10$ mode. What is the beat frequency?

104. A piano tuner uses a 512-Hz tuning fork to tune a piano. He strikes the fork and hits a key on the piano and hears a beat frequency of 5 Hz. He tightens the string of the piano, and repeats the procedure. Once again he hears a beat frequency of 5 Hz. What happened?

105. A string with a linear mass density of $\mu = 0.0062\,\text{kg/m}$ is stretched between two posts 1.30 m apart. The tension in the string is 150.00 N. The string oscillates and produces a sound wave. A 1024-Hz tuning fork is struck and the beat frequency between the two sources is 52.83 Hz. What are the possible frequency and wavelength of the wave on the string?

106. A car has two horns, one emitting a frequency of 199 Hz and the other emitting a frequency of 203 Hz. What beat frequency do they produce?

107. The middle C hammer of a piano hits two strings, producing beats of 1.50 Hz. One of the strings is tuned to 260.00 Hz. What frequencies could the other string have?

108. Two tuning forks having frequencies of 460 and 464 Hz are struck simultaneously. What average frequency will you hear, and what will the beat frequency be?

109. Twin jet engines on an airplane are producing an average sound frequency of 4100 Hz with a beat frequency of 0.500 Hz. What are their individual frequencies?

110. Three adjacent keys on a piano (F, F-sharp, and G) are struck simultaneously, producing frequencies of 349, 370, and 392 Hz. What beat frequencies are produced by this discordant combination?

17.7 The Doppler Effect

111. (a) What frequency is received by a person watching an oncoming ambulance moving at 110 km/h and emitting a steady 800-Hz sound from its siren? The speed of sound on this day is 345 m/s. (b) What frequency does she receive after the ambulance has passed?

112. (a) At an air show a jet flies directly toward the stands at a speed of 1200 km/h, emitting a frequency of 3500 Hz, on a day when the speed of sound is 342 m/s. What frequency is received by the observers? (b) What frequency do they receive as the plane flies directly away from them?

113. What frequency is received by a mouse just before being dispatched by a hawk flying at it at 25.0 m/s and emitting a screech of frequency 3500 Hz? Take the speed of sound to be 331 m/s.

114. A spectator at a parade receives an 888-Hz tone from an oncoming trumpeter who is playing an 880-Hz note. At what speed is the musician approaching if the speed of sound is 338 m/s?

115. A commuter train blows its 200-Hz horn as it approaches a crossing. The speed of sound is 335 m/s. (a) An observer waiting at the crossing receives a frequency of 208 Hz. What is the speed of the train? (b) What frequency does the observer receive as the train moves away?

116. Can you perceive the shift in frequency produced when you pull a tuning fork toward you at 10.0 m/s on a day when the speed of sound is 344 m/s? To answer this question, calculate the factor by which the frequency shifts and see if it is greater than 0.300%.

117. Two eagles fly directly toward one another, the first at 15.0 m/s and the second at 20.0 m/s. Both screech, the first one emitting a frequency of 3200 Hz and the second one emitting a frequency of 3800 Hz. What frequencies do they receive if the speed of sound is 330 m/s?

118. Student A runs down the hallway of the school at a speed of $v_o = 5.00$ m/s, carrying a ringing 1024.00-Hz tuning fork toward a concrete wall. The speed of sound is $v = 343.00$ m/s. Student B stands at rest at the wall. (a) What is the frequency heard by student B? (b) What is the beat frequency heard by student A?

119. An ambulance with a siren $(f = 1.00\text{kHz})$ blaring is approaching an accident scene. The ambulance is moving at 70.00 mph. A nurse is approaching the scene from the opposite direction, running at $v_o = 7.00$ m/s. What frequency does the nurse observe? Assume the speed of sound is $v = 343.00$ m/s.

120. The frequency of the siren of an ambulance is 900 Hz and is approaching you. You are standing on a corner and observe a frequency of 960 Hz. What is the speed of the ambulance (in mph) if the speed of sound is $v = 340.00$ m/s?

121. What is the minimum speed at which a source must travel toward you for you to be able to hear that its frequency is Doppler shifted? That is, what speed produces a shift of 0.300% on a day when the speed of sound is 331 m/s?

17.8 Shock Waves

122. An airplane is flying at Mach 1.50 at an altitude of 7500.00 meters, where the speed of sound is $v = 343.00$ m/s. How far away from a stationary observer will the plane be when the observer hears the sonic boom?

123. A jet flying at an altitude of 8.50 km has a speed of Mach 2.00, where the speed of sound is $v = 340.00$ m/s. How long after the jet is directly overhead, will a stationary observer hear a sonic boom?

124. The shock wave off the front of a fighter jet has an angle of $\theta = 70.00°$. The jet is flying at 1200 km/h. What is the speed of sound?

125. A plane is flying at Mach 1.2, and an observer on the ground hears the sonic boom 15.00 seconds after the plane is directly overhead. What is the altitude of the plane? Assume the speed of sound is $v_w = 343.00$ m/s.

126. A bullet is fired and moves at a speed of 1342 mph. Assume the speed of sound is $v = 340.00$ m/s. What is the angle of the shock wave produced?

127. A speaker is placed at the opening of a long horizontal tube. The speaker oscillates at a frequency of f, creating a sound wave that moves down the tube. The wave moves through the tube at a speed of $v = 340.00$ m/s. The sound wave is modeled with the wave function $s(x, t) = s_{max} \cos(kx - \omega t + \phi)$. At time $t = 0.00$ s, an air molecule at $x = 2.3$ m is at the maximum displacement of 6.34 nm. At the same time, another molecule at $x = 2.7$ m has a displacement of 2.30 nm. What is the wave function of the sound wave, that is, find the wave number, angular frequency, and the initial phase shift?

128. An airplane moves at Mach 1.2 and produces a shock wave. (a) What is the speed of the plane in meters per second? (b) What is the angle that the shock wave moves?

ADDITIONAL PROBLEMS

129. A 0.80-m-long tube is opened at both ends. The air temperature is $26°C$. The air in the tube is oscillated using a speaker attached to a signal generator. What are the wavelengths and frequencies of first two modes of sound waves that resonate in the tube?

130. A tube filled with water has a valve at the bottom to allow the water to flow out of the tube. As the water is emptied from the tube, the length L of the air column changes. A 1024-Hz tuning fork is placed at the opening of the tube. Water is removed from the tube until the $n = 5$ mode of a sound wave resonates. What is the length of the air column if the temperature of the air in the room is $18°C$?

f = 1024 Hz

132. Early Doppler shift experiments were conducted using a band playing music on a train. A trumpet player on a moving railroad flatcar plays a 320-Hz note. The sound waves heard by a stationary observer on a train platform hears a frequency of 350 Hz. What is the flatcar's speed in mph? The temperature of the air is $T_C = 22°C$.

133. Two cars move toward one another, both sounding their horns $(f_s = 800 \text{ Hz})$. Car A is moving at 65 mph and Car B is at 75 mph. What is the beat frequency heard by each driver? The air temperature is $T_C = 22.00°C$.

134. Student A runs after Student B. Student A carries a tuning fork ringing at 1024 Hz, and student B carries a tuning fork ringing at 1000 Hz. Student A is running at a speed of $v_A = 5.00$ m/s and Student B is running at $v_B = 6.00$ m/s. What is the beat frequency heard by each student? The speed of sound is $v = 343.00$ m/s.

135. Suppose that the sound level from a source is 75 dB and then drops to 52 dB, with a frequency of 600 Hz. Determine the (a) initial and (b) final sound intensities and the (c) initial and (d) final sound wave amplitudes. The air temperature is $T_C = 24.00°C$ and the air density is $\rho = 1.184 \text{ kg/m}^3$.

131. Consider the following figure. The length of the string between the string vibrator and the pulley is $L = 1.00$ m. The linear density of the string is $\mu = 0.006$ kg/m. The string vibrator can oscillate at any frequency. The hanging mass is 2.00 kg. (a)What are the wavelength and frequency of $n = 6$ mode? (b) The string oscillates the air around the string. What is the wavelength of the sound if the speed of the sound is $v_s = 343.00$ m/s ?

136. The Doppler shift for a Doppler radar is found by $f = f_R \left(\frac{1 + \frac{v}{c}}{1 - \frac{v}{c}} \right)$, where f_R is the frequency of the radar, f is the frequency observed by the radar, c is the speed of light, and v is the speed of the target. What is the beat frequency observed at the radar, assuming the speed of the target is much slower than the speed of light?

137. A stationary observer hears a frequency of 1000.00 Hz as a source approaches and a frequency of 850.00 Hz as a source departs. The source moves at a constant velocity of 75 mph. What is the temperature of the air?

138. A flute plays a note with a frequency of 600 Hz. The flute can be modeled as a pipe open at both ends, where the flute player changes the length with his finger positions. What is the length of the tube if this is the fundamental frequency?

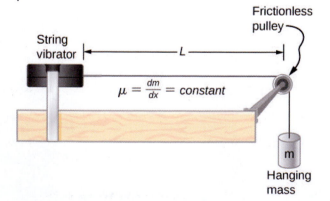

Frictionless pulley

String vibrator

$\mu = \frac{dm}{dx} = constant$

L

m

Hanging mass

CHALLENGE PROBLEMS

139. Two sound speakers are separated by a distance d,   each sounding a frequency f. An observer stands at one speaker and walks in a straight line a distance x, perpendicular to the the two speakers, until he comes to the first maximum intensity of sound. The speed of sound is v. How far is he from the speaker?

140. Consider the beats shown below. This is a graph of the gauge pressure versus time for the position $x = 0.00$ m. The wave moves with a speed of $v = 343.00$ m/s. (a) How many beats are there per second? (b) How many times does the wave oscillate per second? (c) Write a wave function for the gauge pressure as a function of time.

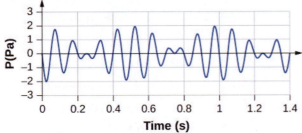

141. Two speakers producing the same frequency of sound are a distance of d apart. Consider an arc along a circle of radius R, centered at the midpoint of the speakers, as shown below. (a) At what angles will there be maxima? (b) At what angle will there be minima?

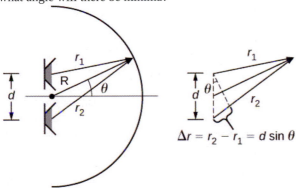

$$\Delta r = r_2 - r_1 = d \sin \theta$$

142. A string has a length of 1.5 m, a linear mass density $\mu = 0.008$ kg/m, , and a tension of 120 N. If the air temperature is $T = 22°C$, what should the length of a pipe open at both ends for it to have the same frequency for the $n = 3$ mode?

143. A string $\left(\mu = 0.006 \dfrac{\text{kg}}{\text{m}}, L = 1.50 \text{ m} \right)$ is fixed at both ends and is under a tension of 155 N. It oscillates in the $n = 10$ mode and produces sound. A tuning fork is ringing nearby, producing a beat frequency of 23.76 Hz. (a) What is the frequency of the sound from the string? (b) What is the frequency of the tuning fork if the tuning fork frequency is lower? (c) What should be the tension of the string for the beat frequency to be zero?

144. A string has a linear mass density μ, a length L, and a tension of F_T, and oscillates in a mode n at a frequency f. Find the ratio of $\dfrac{\Delta f}{f}$ for a small change in tension.

145. A string has a linear mass density $\mu = 0.007$ kg/m, a length $L = 0.70$ m, a tension of $F_T = 110$ N, and oscillates in a mode $n = 3$. (a) What is the frequency of the oscillations? (b) Use the result in the preceding problem to find the change in the frequency when the tension is increased by 1.00%.

146. A speaker powered by a signal generator is used to study resonance in a tube. The signal generator can be adjusted from a frequency of 1000 Hz to 1800 Hz. First, a 0.75-m-long tube, open at both ends, is studied. The temperature in the room is $T_F = 85.00°F$. (a) Which normal modes of the pipe can be studied? What are the frequencies and wavelengths? Next a cap is place on one end of the 0.75-meter-long pipe. (b) Which normal modes of the pipe can be studied? What are the frequencies and wavelengths?

147. A string on the violin has a length of 23.00 cm and a mass of 0.900 grams. The tension in the string 850.00 N. The temperature in the room is $T_C = 24.00°C$. The string is plucked and oscillates in the $n = 9$ mode. (a) What is the speed of the wave on the string? (b) What is the wavelength of the sounding wave produced? (c) What is the frequency of the oscillating string? (d) What is the frequency of the sound produced? (e) What is the wavelength of the sound produced?

APPENDIX A | UNITS

Quantity	Common Symbol	Unit	Unit in Terms of Base SI Units
Acceleration	$\vec{a}$	m/s²	m/s²
Amount of substance	n	**mole**	mol
Angle	θ, ϕ	radian (rad)	
Angular acceleration	$\vec{\alpha}$	rad/s²	s^{-2}
Angular frequency	ω	rad/s	s^{-1}
Angular momentum	$\vec{L}$	$kg \cdot m^2/s$	$kg \cdot m^2/s$
Angular velocity	$\vec{\omega}$	rad/s	s^{-1}
Area	A	m²	m²
Atomic number	Z		
Capacitance	C	farad (F)	$A^2 \cdot s^4/kg \cdot m^2$
Charge	q, Q, e	coulomb (C)	$A \cdot s$
Charge density:			
Line	λ	C/m	$A \cdot s/m$
Surface	σ	C/m²	$A \cdot s/m^2$
Volume	ρ	C/m³	$A \cdot s/m^3$
Conductivity	σ	$1/\Omega \cdot m$	$A^2 \cdot s^3/kg \cdot m^3$
Current	I	**ampere**	A
Current density	$\vec{J}$	A/m²	A/m²
Density	ρ	kg/m³	kg/m³
Dielectric constant	κ		
Electric dipole moment	$\vec{p}$	$C \cdot m$	$A \cdot s \cdot m$
Electric field	$\vec{E}$	N/C	$kg \cdot m/A \cdot s^3$
Electric flux	Φ	$N \cdot m^2/C$	$kg \cdot m^3/A \cdot s^3$
Electromotive force	ε	volt (V)	$kg \cdot m^2/A \cdot s^3$
Energy	E, U, K	joule (J)	$kg \cdot m^2/s^2$
Entropy	S	J/K	$kg \cdot m^2/s^2 \cdot K$

Table A1 Units Used in Physics (Fundamental units in bold)

Quantity	Common Symbol	Unit	Unit in Terms of Base SI Units
Force	$\vec{F}$	newton (N)	$kg \cdot m/s^2$
Frequency	f	hertz (Hz)	s^{-1}
Heat	Q	joule (J)	$kg \cdot m^2/s^2$
Inductance	L	henry (H)	$kg \cdot m^2/A^2 \cdot s^2$
Length:	ℓ, L	**meter**	m
Displacement	$\Delta x, \Delta \vec{r}$		
Distance	d, h		
Position	$x, y, z, \vec{r}$		
Magnetic dipole moment	$\vec{\mu}$	$N \cdot J/T$	$A \cdot m^2$
Magnetic field	$\vec{B}$	$tesla(T) = \left(Wb/m^2\right)$	$kg/A \cdot s^2$
Magnetic flux	Φ_m	weber (Wb)	$kg \cdot m^2/A \cdot s^2$
Mass	m, M	**kilogram**	kg
Molar specific heat	C	$J/mol \cdot K$	$kg \cdot m^2/s^2 \cdot mol \cdot K$
Moment of inertia	I	$kg \cdot m^2$	$kg \cdot m^2$
Momentum	$\vec{p}$	$kg \cdot m/s$	$kg \cdot m/s$
Period	T	s	s
Permeability of free space	μ_0	$N/A^2 = (H/m)$	$kg \cdot m/A^2 \cdot s^2$
Permittivity of free space	ε_0	$C^2/N \cdot m^2 = (F/m)$	$A^2 \cdot s^4/kg \cdot m^3$
Potential	V	$volt(V) = (J/C)$	$kg \cdot m^2/A \cdot s^3$
Power	P	$watt(W) = (J/s)$	$kg \cdot m^2/s^3$
Pressure	p	$pascal(Pa) = \left(N/m^2\right)$	$kg/m \cdot s^2$
Resistance	R	$ohm(\Omega) = (V/A)$	$kg \cdot m^2/A^2 \cdot s^3$
Specific heat	c	$J/kg \cdot K$	$m^2/s^2 \cdot K$
Speed	ν	m/s	m/s
Temperature	T	**kelvin**	K
Time	t	**second**	s
Torque	$\vec{\tau}$	$N \cdot m$	$kg \cdot m^2/s^2$

Table A1 Units Used in Physics (Fundamental units in bold)

Quantity	Common Symbol	Unit	Unit in Terms of Base SI Units
Velocity	$\vec{v}$	m/s	m/s
Volume	V	m^3	m^3
Wavelength	λ	m	m
Work	W	$\text{joule}(J) = (N \cdot m)$	$kg \cdot m^2/s^2$

Table A1 Units Used in Physics (Fundamental units in bold)

APPENDIX B | CONVERSION FACTORS

	m	cm	km
1 meter	1	10^2	10^{-3}
1 centimeter	10^{-2}	1	10^{-5}
1 kilometer	10^3	10^5	1
1 inch	2.540×10^{-2}	2.540	2.540×10^{-5}
1 foot	0.3048	30.48	3.048×10^{-4}
1 mile	1609	1.609×10^4	1.609
1 angstrom	10^{-10}		
1 fermi	10^{-15}		
1 light-year			9.460×10^{12}

	in.	ft	mi
1 meter	39.37	3.281	6.214×10^{-4}
1 centimeter	0.3937	3.281×10^{-2}	6.214×10^{-6}
1 kilometer	3.937×10^4	3.281×10^3	0.6214
1 inch	1	8.333×10^{-2}	1.578×10^{-5}
1 foot	12	1	1.894×10^{-4}
1 mile	6.336×10^4	5280	1

Table B1 Length

Area

$1\ cm^2 = 0.155\ in.^2$

$1\ m^2 = 10^4\ cm^2 = 10.76\ ft^2$

$1\ in.^2 = 6.452\ cm^2$

$1\ ft^2 = 144\ in.^2 = 0.0929\ m^2$

Volume

$1\ liter = 1000\ cm^3 = 10^{-3}\ m^3 = 0.03531\ ft^3 = 61.02\ in.^3$

$1\ ft^3 = 0.02832\ m^3 = 28.32\ liters = 7.477\ gallons$

$1\ gallon = 3.788\ liters$

	s	min	h	day	yr
1 second	1	1.667×10^{-2}	2.778×10^{-4}	1.157×10^{-5}	3.169×10^{-8}
1 minute	60	1	1.667×10^{-2}	6.944×10^{-4}	1.901×10^{-6}
1 hour	3600	60	1	4.167×10^{-2}	1.141×10^{-4}
1 day	8.640×10^4	1440	24	1	2.738×10^{-3}
1 year	3.156×10^7	5.259×10^5	8.766×10^3	365.25	1

Table B2 Time

	m/s	cm/s	ft/s	mi/h
1 meter/second	1	10^2	3.281	2.237
1 centimeter/second	10^{-2}	1	3.281×10^{-2}	2.237×10^{-2}
1 foot/second	0.3048	30.48	1	0.6818
1 mile/hour	0.4470	44.70	1.467	1

Table B3 Speed

Acceleration

$1 \text{ m/s}^2 = 100 \text{ cm/s}^2 = 3.281 \text{ ft/s}^2$

$1 \text{ cm/s}^2 = 0.01 \text{ m/s}^2 = 0.03281 \text{ ft/s}^2$

$1 \text{ ft/s}^2 = 0.3048 \text{ m/s}^2 = 30.48 \text{ cm/s}^2$

$1 \text{ mi/h} \cdot \text{s} = 1.467 \text{ ft/s}^2$

	kg	g	slug	u
1 kilogram	1	10^3	6.852×10^{-2}	6.024×10^{26}
1 gram	10^{-3}	1	6.852×10^{-5}	6.024×10^{23}
1 slug	14.59	1.459×10^4	1	8.789×10^{27}
1 atomic mass unit	1.661×10^{-27}	1.661×10^{-24}	1.138×10^{-28}	1
1 metric ton	1000			

Table B4 Mass

	N	dyne	lb
1 newton	1	10^5	0.2248
1 dyne	10^{-5}	1	2.248×10^{-6}
1 pound	4.448	4.448×10^5	1

Table B5 Force

	Pa	dyne/cm^2	atm	cmHg	lb/in.^{2}
1 pascal	1	10	9.869×10^{-6}	7.501×10^{-4}	1.450×10^{-4}
1 dyne/centimeter^2	10^{-1}	1	9.869×10^{-7}	7.501×10^{-5}	1.450×10^{-5}
1 atmosphere	1.013×10^5	1.013×10^6	1	76	14.70
1 centimeter mercury*	1.333×10^3	1.333×10^4	1.316×10^{-2}	1	0.1934
1 pound/inch^2	6.895×10^3	6.895×10^4	6.805×10^{-2}	5.171	1
1 bar	10^5				
1 torr				1 (mmHg)	

*Where the acceleration due to gravity is 9.80665 m/s^2 and the temperature is $0°C$

Table B6 Pressure

	J	erg	ft.lb
1 joule	1	10^7	0.7376
1 erg	10^{-7}	1	7.376×10^{-8}
1 foot-pound	1.356	1.356×10^7	1
1 electron-volt	1.602×10^{-19}	1.602×10^{-12}	1.182×10^{-19}
1 calorie	4.186	4.186×10^7	3.088
1 British thermal unit	1.055×10^3	1.055×10^{10}	7.779×10^2
1 kilowatt-hour	3.600×10^6		

	eV	cal	Btu
1 joule	6.242×10^{18}	0.2389	9.481×10^{-4}
1 erg	6.242×10^{11}	2.389×10^{-8}	9.481×10^{-11}
1 foot-pound	8.464×10^{18}	0.3239	1.285×10^{-3}
1 electron-volt	1	3.827×10^{-20}	1.519×10^{-22}
1 calorie	2.613×10^{19}	1	3.968×10^{-3}
1 British thermal unit	6.585×10^{21}	2.520×10^2	1

Table B7 Work, Energy, Heat

Power

1 W = 1 J/s

$1 \text{ hp} = 746 \text{ W} = 550 \text{ ft} \cdot \text{lb/s}$

1 Btu/h = 0.293 W

Angle

1 rad = 57.30° = 180°/π

1° = 0.01745 rad = π/180 rad

1 revolution = 360° = 2π rad

1 rev/min(rpm) = 0.1047 rad/s

APPENDIX C |
FUNDAMENTAL
CONSTANTS

Quantity	Symbol	Value
Atomic mass unit	u	$1.660\ 538\ 782\ (83) \times 10^{-27}$ kg $931.494\ 028\ (23)$ MeV/c^2
Avogadro's number	N_A	$6.022\ 141\ 79\ (30) \times 10^{23}$ particles/mol
Bohr magneton	$\mu_B = \frac{e\hbar}{2m_e}$	$9.274\ 009\ 15\ (23) \times 10^{-24}$ J/T
Bohr radius	$a_0 = \frac{\hbar^2}{m_e e^2 k_e}$	$5.291\ 772\ 085\ 9\ (36) \times 10^{-11}$ m
Boltzmann's constant	$k_B = \frac{R}{N_A}$	$1.380\ 650\ 4\ (24) \times 10^{-23}$ J/K
Compton wavelength	$\lambda_C = \frac{h}{m_e c}$	$2.426\ 310\ 217\ 5\ (33) \times 10^{-12}$ m
Coulomb constant	$k_e = \frac{1}{4\pi\varepsilon_0}$	$8.987\ 551\ 788... \times 10^9$ N·m^2/C^2 (exact)
Deuteron mass	m_d	$3.343\ 583\ 20\ (17) \times 10^{-27}$ kg $2.013\ 553\ 212\ 724(78)$ u $1875.612\ 859$ MeV/c^2
Electron mass	m_e	$9.109\ 382\ 15\ (45) \times 10^{-31}$ kg $5.485\ 799\ 094\ 3(23) \times 10^{-4}$ u $0.510\ 998\ 910\ (13)$ MeV/c^2
Electron volt	eV	$1.602\ 176\ 487\ (40) \times 10^{-19}$ J
Elementary charge	e	$1.602\ 176\ 487\ (40) \times 10^{-19}$ C
Gas constant	R	$8.314\ 472\ (15)$ J/mol·K
Gravitational constant	G	$6.674\ 28\ (67) \times 10^{-11}$ N·m^2/kg^2

Table C1 Fundamental Constants *Note:* These constants are the values recommended in 2006 by CODATA, based on a least-squares adjustment of data from different measurements. The numbers in parentheses for the values represent the uncertainties of the last two digits.

Quantity	Symbol	Value
Neutron mass	m_n	$1.674\ 927\ 211\ (84) \times 10^{-27}$ kg
		$1.008\ 664\ 915\ 97\ (43)$ u
		$939.565\ 346\ (23)$ MeV$/c^2$
Nuclear magneton	$\mu_n = \dfrac{e\hbar}{2m_p}$	$5.050\ 783\ 24\ (13) \times 10^{-27}$ J/T
Permeability of free space	μ_0	$4\pi \times 10^{-7}$ T$\cdot$m/A(exact)
Permittivity of free space	$\varepsilon_0 = \dfrac{1}{\mu_0 c^2}$	$8.854\ 187\ 817... \times 10^{-12}$ C^2/N$\cdot$m^2 (exact)
Planck's constant	h	$6.626\ 068\ 96\ (33) \times 10^{-34}$ J$\cdot$s
	$\hbar = \dfrac{h}{2\pi}$	$1.054\ 571\ 628\ (53) \times 10^{-34}$ J$\cdot$s
Proton mass	m_p	$1.672\ 621\ 637\ (83) \times 10^{-27}$ kg
		$1.007\ 276\ 466\ 77\ (10)$ u
		$938.272\ 013\ (23)$ MeV$/c^2$
Rydberg constant	R_H	$1.097\ 373\ 156\ 852\ 7\ (73) \times 10^7$ m^{-1}
Speed of light in vacuum	c	$2.997\ 924\ 58 \times 10^8$ m/s (exact)

Table C1 Fundamental Constants *Note:* These constants are the values recommended in 2006 by CODATA, based on a least-squares adjustment of data from different measurements. The numbers in parentheses for the values represent the uncertainties of the last two digits.

Useful combinations of constants for calculations:

$hc = 12{,}400$ eV$\cdot$Å $= 1240$ eV$\cdot$nm $= 1240$ MeV$\cdot$fm

$\hbar c = 1973$ eV$\cdot$Å $= 197.3$ eV$\cdot$nm $= 197.3$ MeV$\cdot$fm

$k_e e^2 = 14.40$ eV$\cdot$Å $= 1.440$ eV$\cdot$nm $= 1.440$ MeV$\cdot$fm

$k_B T = 0.02585$ eV at $T = 300$ K

APPENDIX D |
ASTRONOMICAL DATA

Celestial Object	Mean Distance from Sun (million km)	Period of Revolution (d = days) (y = years)	Period of Rotation at Equator	Eccentricity of Orbit
Sun	–	–	27 d	–
Mercury	57.9	88 d	59 d	0.206
Venus	108.2	224.7 d	243 d	0.007
Earth	149.6	365.26 d	23 h 56 min 4 s	0.017
Mars	227.9	687 d	24 h 37 min 23 s	0.093
Jupiter	778.4	11.9 y	9 h 50 min 30 s	0.048
Saturn	1426.7	29.5 6	10 h 14 min	0.054
Uranus	2871.0	84.0 y	17 h 14 min	0.047
Neptune	4498.3	164.8 y	16 h	0.009
Earth's Moon	149.6 (0.386 from Earth)	27.3 d	27.3 d	0.055

Celestial Object	Equatorial Diameter (km)	Mass (Earth = 1)	Density (g/cm^3)
Sun	1,392,000	333,000.00	1.4
Mercury	4879	0.06	5.4
Venus	12,104	0.82	5.2
Earth	12,756	1.00	5.5
Mars	6794	0.11	3.9
Jupiter	142,984	317.83	1.3
Saturn	120,536	95.16	0.7
Uranus	51,118	14.54	1.3
Neptune	49,528	17.15	1.6
Earth's Moon	3476	0.01	3.3

Table D1 Astronomical Data

Other Data:

Mass of Earth: 5.97×10^{24} kg

Mass of the Moon: 7.36×10^{22} kg

Mass of the Sun: 1.99×10^{30} kg

APPENDIX E | MATHEMATICAL FORMULAS

Quadratic formula

If $ax^2 + bx + c = 0$, then $x = \dfrac{-b \pm \sqrt{b^2 - 4ac}}{2a}$

Triangle of base b and height h	Area $= \frac{1}{2}bh$	
Circle of radius r	Circumference $= 2\pi r$	Area $= \pi r^2$
Sphere of radius r	Surface area $= 4\pi r^2$	Volume $= \frac{4}{3}\pi r^3$
Cylinder of radius r and height h	Area of curved surface $= 2\pi rh$	Volume $= \pi r^2 h$

Table E1 Geometry

Trigonometry

Trigonometric Identities

1. $\sin \theta = 1/\csc \theta$

2. $\cos \theta = 1/\sec \theta$

3. $\tan \theta = 1/\cot \theta$

4. $\sin(90^0 - \theta) = \cos \theta$

5. $\cos(90^0 - \theta) = \sin \theta$

6. $\tan(90^0 - \theta) = \cot \theta$

7. $\sin^2 \theta + \cos^2 \theta = 1$

8. $\sec^2 \theta - \tan^2 \theta = 1$

9. $\tan \theta = \sin \theta/\cos \theta$

10. $\sin(\alpha \pm \beta) = \sin \alpha \cos \beta \pm \cos \alpha \sin \beta$

11. $\cos(\alpha \pm \beta) = \cos \alpha \cos \beta \mp \sin \alpha \sin \beta$

12. $\tan(\alpha \pm \beta) = \dfrac{\tan \alpha \pm \tan \beta}{1 \mp \tan \alpha \tan \beta}$

13. $\sin 2\theta = 2\sin \theta \cos \theta$

14. $\cos 2\theta = \cos^2 \theta - \sin^2 \theta = 2\cos^2 \theta - 1 = 1 - 2\sin^2 \theta$

15. $\sin \alpha + \sin \beta = 2\sin\frac{1}{2}(\alpha + \beta)\cos\frac{1}{2}(\alpha - \beta)$

16. $\cos \alpha + \cos \beta = 2\cos\frac{1}{2}(\alpha + \beta)\cos\frac{1}{2}(\alpha - \beta)$

Triangles

1.   Law of sines: $\dfrac{a}{\sin \alpha} = \dfrac{b}{\sin \beta} = \dfrac{c}{\sin \gamma}$

2.   Law of cosines: $c^2 = a^2 + b^2 - 2ab\cos\gamma$

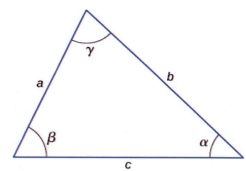

3.   Pythagorean theorem: $a^2 + b^2 = c^2$

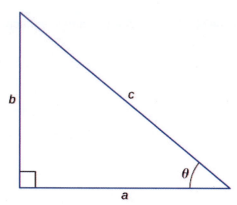

Series expansions

1.   Binomial theorem: $(a+b)^n = a^n + na^{n-1}b + \dfrac{n(n-1)a^{n-2}b^2}{2!} + \dfrac{n(n-1)(n-2)a^{n-3}b^3}{3!} + \cdots$

2. $(1 \pm x)^n = 1 \pm \dfrac{nx}{1!} + \dfrac{n(n-1)x^2}{2!} \pm \cdots \left(x^2 < 1\right)$

3. $(1 \pm x)^{-n} = 1 \mp \dfrac{nx}{1!} + \dfrac{n(n+1)x^2}{2!} \mp \cdots \left(x^2 < 1\right)$

4. $\sin x = x - \dfrac{x^3}{3!} + \dfrac{x^5}{5!} - \cdots$

5. $\cos x = 1 - \dfrac{x^2}{2!} + \dfrac{x^4}{4!} - \cdots$

6. $\tan x = x + \dfrac{x^3}{3} + \dfrac{2x^5}{15} + \cdots$

7. $e^x = 1 + x + \dfrac{x^2}{2!} + \cdots$

8. $\ln(1+x) = x - \dfrac{1}{2}x^2 + \dfrac{1}{3}x^3 - \cdots (|x| < 1)$

Derivatives

1. $\dfrac{d}{dx}[af(x)] = a\dfrac{d}{dx}f(x)$

2. $\dfrac{d}{dx}[f(x) + g(x)] = \dfrac{d}{dx}f(x) + \dfrac{d}{dx}g(x)$

3. $\dfrac{d}{dx}[f(x)g(x)] = f(x)\dfrac{d}{dx}g(x) + g(x)\dfrac{d}{dx}f(x)$

4. $\dfrac{d}{dx}f(u) = \left[\dfrac{d}{du}f(u)\right]\dfrac{du}{dx}$

5. $\dfrac{d}{dx}x^m = mx^{m-1}$

6. $\dfrac{d}{dx}\sin x = \cos x$

7. $\dfrac{d}{dx}\cos x = -\sin x$

8. $\dfrac{d}{dx}\tan x = \sec^2 x$

9. $\dfrac{d}{dx}\cot x = -\csc^2 x$

10. $\dfrac{d}{dx}\sec x = \tan x \sec x$

11. $\dfrac{d}{dx}\csc x = -\cot x \csc x$

12. $\dfrac{d}{dx}e^x = e^x$

13. $\dfrac{d}{dx}\ln x = \dfrac{1}{x}$

14. $\dfrac{d}{dx}\sin^{-1} x = \dfrac{1}{\sqrt{1-x^2}}$

15. $\dfrac{d}{dx}\cos^{-1} x = -\dfrac{1}{\sqrt{1-x^2}}$

16. $\dfrac{d}{dx}\tan^{-1} x = -\dfrac{1}{1+x^2}$

Integrals

1. $\displaystyle\int af(x)dx = a\int f(x)dx$

2. $\displaystyle\int [f(x) + g(x)]dx = \int f(x)dx + \int g(x)dx$

3. $\displaystyle\int x^m\,dx = \dfrac{x^{m+1}}{m+1}\ (m \neq -1)$
   $= \ln x (m = -1)$

4. $\displaystyle\int \sin x\,dx = -\cos x$

5. $\displaystyle\int \cos x\,dx = \sin x$

6. $\displaystyle\int \tan x\,dx = \ln|\sec x|$

7. $\int \sin^2 ax\, dx = \frac{x}{2} - \frac{\sin 2ax}{4a}$

8. $\int \cos^2 ax\, dx = \frac{x}{2} + \frac{\sin 2ax}{4a}$

9. $\int \sin ax \cos ax\, dx = -\frac{\cos 2ax}{4a}$

10. $\int e^{ax}\, dx = \frac{1}{a} e^{ax}$

11. $\int x e^{ax}\, dx = \frac{e^{ax}}{a^2}(ax - 1)$

12. $\int \ln ax\, dx = x \ln ax - x$

13. $\int \frac{dx}{a^2 + x^2} = \frac{1}{a} \tan^{-1} \frac{x}{a}$

14. $\int \frac{dx}{a^2 - x^2} = \frac{1}{2a} \ln\left|\frac{x + a}{x - a}\right|$

15. $\int \frac{dx}{\sqrt{a^2 + x^2}} = \sinh^{-1} \frac{x}{a}$

16. $\int \frac{dx}{\sqrt{a^2 - x^2}} = \sin^{-1} \frac{x}{a}$

17. $\int \sqrt{a^2 + x^2}\, dx = \frac{x}{2}\sqrt{a^2 + x^2} + \frac{a^2}{2} \sinh^{-1} \frac{x}{a}$

18. $\int \sqrt{a^2 - x^2}\, dx = \frac{x}{2}\sqrt{a^2 - x^2} + \frac{a^2}{2} \sin^{-1} \frac{x}{a}$

APPENDIX F | CHEMISTRY

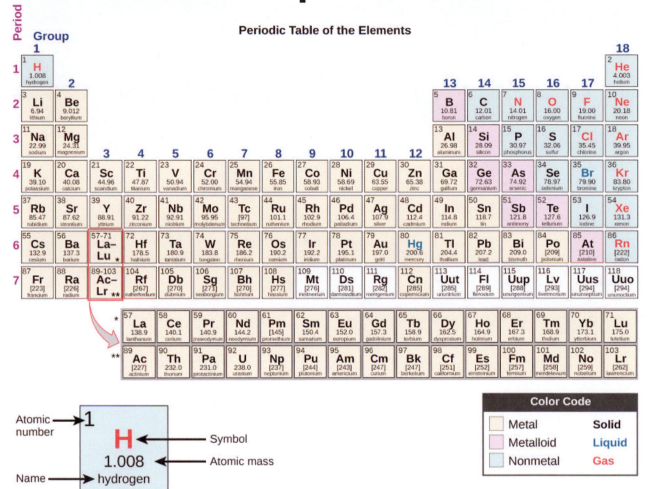

Periodic Table of the Elements

APPENDIX G | THE GREEK ALPHABET

Name	Capital	Lowercase	Name	Capital	Lowercase
Alpha	A	α	Nu	N	ν
Beta	B	β	Xi	Ξ	ξ
Gamma	Γ	γ	Omicron	O	o
Delta	Δ	δ	Pi	Π	π
Epsilon	E	ε	Rho	P	ρ
Zeta	Z	ζ	Sigma	Σ	σ
Eta	H	η	Tau	T	τ
Theta	Θ	θ	Upsilon	Υ	υ
Iota	I	ι	Phi	Φ	ϕ
Kappa	K	κ	Chi	X	χ
Lambda	Λ	λ	Psi	ψ	ψ
Mu	M	μ	Omega	Ω	ω

Table G1 The Greek Alphabet

ANSWER KEY

CHAPTER 1

CHECK YOUR UNDERSTANDING

1.1. 4.79×10^2 Mg or 479 Mg

1.2. 3×10^8 m/s

1.3. 10^8 km^2

1.4. The numbers were too small, by a factor of 4.45.

1.5. $4\pi r^3/3$

1.6. yes

1.7. 3×10^4 m or 30 km. It is probably an underestimate because the density of the atmosphere decreases with altitude. (In fact, 30 km does not even get us out of the stratosphere.)

1.8. No, the coach's new stopwatch will not be helpful. The uncertainty in the stopwatch is too great to differentiate between the sprint times effectively.

CONCEPTUAL QUESTIONS

1. Physics is the science concerned with describing the interactions of energy, matter, space, and time to uncover the fundamental mechanisms that underlie every phenomenon.

3. No, neither of these two theories is more valid than the other. Experimentation is the ultimate decider. If experimental evidence does not suggest one theory over the other, then both are equally valid. A given physicist might prefer one theory over another on the grounds that one seems more simple, more natural, or more beautiful than the other, but that physicist would quickly acknowledge that he or she cannot say the other theory is invalid. Rather, he or she would be honest about the fact that more experimental evidence is needed to determine which theory is a better description of nature.

5. Probably not. As the saying goes, "Extraordinary claims require extraordinary evidence."

7. Conversions between units require factors of 10 only, which simplifies calculations. Also, the same basic units can be scaled up or down using metric prefixes to sizes appropriate for the problem at hand.

9. a. Base units are defined by a particular process of measuring a base quantity whereas derived units are defined as algebraic combinations of base units. b. A base quantity is chosen by convention and practical considerations. Derived quantities are expressed as algebraic combinations of base quantities. c. A base unit is a standard for expressing the measurement of a base quantity within a particular system of units. So, a measurement of a base quantity could be expressed in terms of a base unit in any system of units using the same base quantities. For example, length is a base quantity in both SI and the English system, but the meter is a base unit in the SI system only.

11. a. Uncertainty is a quantitative measure of precision. b. Discrepancy is a quantitative measure of accuracy.

13. Check to make sure it makes sense and assess its significance.

PROBLEMS

15. a. 10^3; b. 10^5; c. 10^2; d. 10^{15}; e. 10^2; f. 10^{57}

17. 10^2 generations

19. 10^{11} atoms

21. 10^3 nerve impulses/s

23. 10^{26} floating-point operations per human lifetime

25. a. 957 ks; b. 4.5 cs or 45 ms; c. 550 ns; d. 31.6 Ms

27. a. 75.9 Mm; b. 7.4 mm; c. 88 pm; d. 16.3 Tm

29. a. 3.8 cg or 38 mg; b. 230 Eg; c. 24 ng; d. 8 Eg e. 4.2 g

31. a. 27.8 m/s; b. 62 mi/h

33. a. 3.6 km/h; b. 2.2 mi/h

35. 1.05×10^5 ft^2

37. 8.847 km

39. a. 1.3×10^{-9} m; b. 40 km/My

41. 10^6 Mg/μL

43. 62.4 lbm/ft^3

45. 0.017 rad

47. 1 light-nanosecond

49. 3.6×10^{-4} m^3

51. a. Yes, both terms have dimension L^{2}T^{-2} b. No. c. Yes, both terms have dimension LT^{-1} d. Yes, both terms have dimension

LT^{-2}

53. a. $[v] = LT^{-1}$; b. $[a] = LT^{-2}$; c. $\left[\int v\,dt\right] = L$; d. $\left[\int a\,dt\right] = LT^{-1}$; e. $\left[\frac{da}{dt}\right] = LT^{-3}$

55. a. L; b. L; c. $L^0 = 1$ (that is, it is dimensionless)

57. 10^{28} atoms

59. 10^{51} molecules

61. 10^{16} solar systems

63. a. Volume $= 10^{27}\ m^3$, diameter is 10^9 m.; b. 10^{11} m

65. a. A reasonable estimate might be one operation per second for a total of 10^9 in a lifetime.; b. about $(10^9)(10^{-17}\ s) = 10^{-8}$ s, or about 10 ns

67. 2 kg

69. 4%

71. 67 mL

73. a. The number 99 has 2 significant figures; 100. has 3 significant figures. b. 1.00%; c. percent uncertainties

75. a. 2%; b. 1 mm Hg

77. $7.557\ cm^2$

79. a. 37.2 lb; because the number of bags is an exact value, it is not considered in the significant figures; b. 1.4 N; because the value 55 kg has only two significant figures, the final value must also contain two significant figures

ADDITIONAL PROBLEMS

81. a. $[s_0] = L$ and units are meters (m); b. $[v_0] = LT^{-1}$ and units are meters per second (m/s); c. $[a_0] = LT^{-2}$ and units are meters per second squared (m/s²); d. $[j_0] = LT^{-3}$ and units are meters per second cubed (m/s³); e. $[S_0] = LT^{-4}$ and units are m/s⁴; f. $[c] = LT^{-5}$ and units are m/s⁵.

83. a. 0.059%; b. 0.01%; c. 4.681 m/s; d. 0.07%, 0.003 m/s

85. a. 0.02%; b. 1×10^4 lbm

87. a. $143.6\ cm^3$; b. $0.1\ cm^3$ or 0.084%

CHALLENGE PROBLEMS

89. Since each term in the power series involves the argument raised to a different power, the only way that every term in the power series can have the same dimension is if the argument is dimensionless. To see this explicitly, suppose $[x] = L^a M^b T^c$. Then, $[x^n] = [x]^n = L^{an} M^{bn} T^{cn}$. If we want $[x] = [x^n]$, then $an = a$, $bn = b$, and $cn = c$ for all n. The only way this can happen is if $a = b = c = 0$.

CHAPTER 2

CHECK YOUR UNDERSTANDING

2.1. a. not equal because they are orthogonal; b. not equal because they have different magnitudes; c. not equal because they have different magnitudes and directions; d. not equal because they are antiparallel; e. equal.

2.2. 16 m; $\vec{D} = -16\ m\,\hat{u}$

2.3. $G = 28.2$ cm, $\theta_G = 291°$

2.4. $\vec{D} = (-5.0\,\hat{i} - 3.0\,\hat{j})$cm; the fly moved 5.0 cm to the left and 3.0 cm down from its landing site.

2.5. 5.83 cm, 211°

2.6. $\vec{D} = (-20\ m)\,\hat{j}$

2.7. 35.1 m/s = 126.4 km/h

2.8. $\vec{G} = (10.25\,\hat{i} - 26.22\,\hat{j})$cm

2.9. $D = 55.7$ N; direction 65.7° north of east

2.10. $\hat{v} = 0.8\,\hat{i} + 0.6\,\hat{j}$, 36.87° north of east

2.11. $\vec{A} \cdot \vec{B} = -57.3$, $\vec{F} \cdot \vec{C} = 27.8$

2.13. 131.9°

2.14. $W_1 = 1.5\ J$, $W_2 = 0.3\ J$

2.15. $\vec{A} \times \vec{B} = -40.1\,\hat{k}$ or, equivalently, $\left|\vec{A} \times \vec{B}\right| = 40.1$, and the direction is into the page;

$\vec{C} \times \vec{F} = +157.6\hat{k}$ or, equivalently, $\left| \vec{C} \times \vec{F} \right| = 157.6$, and the direction is out of the page.

2.16. a. $-2\hat{k}$, b. 2, c. $153.4°$, d. $135°$

CONCEPTUAL QUESTIONS

1. scalar
3. answers may vary
5. parallel, sum of magnitudes, antiparallel, zero
7. no, yes
9. zero, yes
11. no
13. equal, equal, the same
15. a unit vector of the *x*-axis
17. They are equal.
19. yes
21. a. $C = \vec{A} \cdot \vec{B}$, b. $\vec{C} = \vec{A} \times \vec{B}$ or $\vec{C} = \vec{A} - \vec{B}$, c. $\vec{C} = \vec{A} \times \vec{B}$, d. $\vec{C} = A\vec{B}$, e. $\vec{C} + 2\vec{A} = \vec{B}$, f. $\vec{C} = \vec{A} \times \vec{B}$, g. left side is a scalar and right side is a vector, h. $\vec{C} = 2\vec{A} \times \vec{B}$, i. $\vec{C} = \vec{A}/B$, j. $\vec{C} = \vec{A}/B$
23. They are orthogonal.

PROBLEMS

25. $\vec{h} = -49 \text{ m}\hat{u}$, 49 m
27. 30.8 m, $35.7°$ west of north
29. 134 km, $80°$
31. 7.34 km, $63.5°$ south of east
33. 3.8 km east, 3.2 km north, 7.0 km
35. 14.3 km, $65°$
37. a. $\vec{A} = +8.66\hat{i} + 5.00\hat{j}$, b. $\vec{B} = +30.09\hat{i} + 39.93\hat{j}$, c. $\vec{C} = +6.00\hat{i} - 10.39\hat{j}$, d. $\vec{D} = -15.97\hat{i} + 12.04\hat{j}$, f. $\vec{F} = -17.32\hat{i} - 10.00\hat{j}$

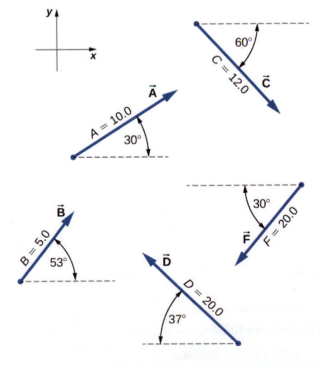

39. a. 1.94 km, 7.24 km; b. proof

41. 3.8 km east, 3.2 km north, 2.0 km, $\vec{D} = (3.8\,\hat{i} + 3.2\,\hat{j})$km

43. $P_1(2.165\,m,\ 1.250\,m)$, $P_2(-1.900\,m,\ 3.290\,m)$, 5.27 m

45. 8.60 m, $A(2\sqrt{5}\,m,\ 0.647\pi)$, $B(3\sqrt{2}\,m,\ 0.75\pi)$

47.   a. $\vec{A} + \vec{B} = -4\,\hat{i} - 6\,\hat{j}$, $\left|\vec{A} + \vec{B}\right| = 7.211,\ \theta = 213.7°$;   b. $\vec{A} - \vec{B} = 2\,\hat{i} - 2\,\hat{j}$, $\left|\vec{A} - \vec{B}\right| = 2\sqrt{2},\ \theta = -45°$

49. a. $\vec{C} = (5.0\,\hat{i} - 1.0\,\hat{j} - 3.0\,\hat{k})$m, $C = 5.92$ m;

b. $\vec{D} = (4.0\,\hat{i} - 11.0\,\hat{j} + 15.0\,\hat{k})$m, $D = 19.03$ m

51. $\vec{D} = (3.3\,\hat{i} - 6.6\,\hat{j})$km, $\hat{i}$ is to the east, 7.34 km, $-63.5°$

53. a. $\vec{R} = -1.35\,\hat{i} - 22.04\,\hat{j}$, b. $\vec{R} = -17.98\,\hat{i} + 0.89\,\hat{j}$

55. $\vec{D} = (200\,\hat{i} + 300\,\hat{j})$yd, $D = 360.5$ yd, $56.3°$ north of east; The numerical answers would stay the same but the physical unit would be meters. The physical meaning and distances would be about the same because 1 yd is comparable with 1 m.

57. $\vec{R} = -3\,\hat{i} - 16\,\hat{j}$

59. $\vec{E} = E\hat{E}$, $E_x = +178.9$V/m, $E_y = -357.8$V/m, $E_z = 0.0$V/m, $\theta_E = -\tan^{-1}(2)$

61.   a. $-34.290\ \vec{R}_B = (-12.278\,\hat{i} + 7.089\,\hat{j} + 2.500\,\hat{k})$km, $\vec{R}_D = (-34.290\,\hat{i} + 3.000\,\hat{k})$km;   b. $\left|\vec{R}_B - \vec{R}_D\right| = 23.131$ km

63. a. 0, b. 0, c. 0.866, d. 17.32

65. $\theta_i = 64.12°$, $\theta_j = 150.79°$, $\theta_k = 77.39°$

67. a. $-120\,\hat{k}$, b. $0\,\hat{k}$, c. $-94\,\hat{k}$, d. $-240\,\hat{k}$, e. $4.0\,\hat{k}$, f. $-3.0\,\hat{k}$, g. $15\,\hat{k}$, h. 0

69. a. 0, b. 0, c. $+-20{,}000\,\hat{k}$

ADDITIONAL PROBLEMS

71. a. 18.4 km and 26.2 km, b. 31.5 km and 5.56 km

73. a. $(r,\ \pi - \varphi)$, b. $(2r,\ \varphi + 2\pi)$, (c) $(3r,\ -\varphi)$

75. $d_{PM} = 6.2$ nmi $= 11.4$ km, $d_{NP} = 7.2$ nmi $= 13.3$ km

77. proof

79. a. 10.00 m, b. 5π m, c. 0

81. 22.2 km/h, $35.8°$ south of west

83. 270 m, $4.2°$ north of west

85. $\vec{B} = -4.0\,\hat{i} + 3.0\,\hat{j}$ or $\vec{B} = 4.0\,\hat{i} - 3.0\,\hat{j}$

87. proof

CHALLENGE PROBLEMS

89. $G_H = 19$ N $/ \sqrt{17} \approx 4.6$ N

91. proof

CHAPTER 3

CHECK YOUR UNDERSTANDING

3.1. (a) The rider's displacement is $\Delta x = x_f - x_0 = -1$ km. (The displacement is negative because we take east to be positive and west to be negative.) (b) The distance traveled is 3 km + 2 km = 5 km. (c) The magnitude of the displacement is 1 km.

3.2. (a) Taking the derivative of $x(t)$ gives $v(t) = -6t$ m/s. (b) No, because time can never be negative. (c) The velocity is $v(1.0\ s) =$

−6 m/s and the speed is $|v(1.0 \text{ s})| = 6$ m/s .

3.3. Inserting the knowns, we have

$$\bar{a} = \frac{\Delta v}{\Delta t} = \frac{2.0 \times 10^7 \text{ m/s} - 0}{10^{-4} \text{ s} - 0} = 2.0 \times 10^{11} \text{ m/s}^2.$$

3.4. If we take east to be positive, then the airplane has negative acceleration because it is accelerating toward the west. It is also decelerating; its acceleration is opposite in direction to its velocity.

3.5. To answer this, choose an equation that allows us to solve for time t, given only a, v_0, and v:

$$v = v_0 + at.$$

Rearrange to solve for t:

$$t = \frac{v - v_0}{a} = \frac{400 \text{ m/s} - 0 \text{ m/s}}{20 \text{ m/s}^2} = 20 \text{ s}.$$

3.6. $a = \frac{2}{3} \text{ m/s}^2$.

3.7. It takes 2.47 s to hit the water. The quantity distance traveled increases faster.

3.8.

    a. The velocity function is the integral of the acceleration function plus a constant of integration. By **Equation 3.91**,

$$v(t) = \int a(t)dt + C_1 = \int (5 - 10t)dt + C_1 = 5t - 5t^2 + C_1.$$

       Since $v(0) = 0$, we have $C_1 = 0$; so,

$$v(t) = 5t - 5t^2.$$

    b. By **Equation 3.93**,

$$x(t) = \int v(t)dt + C_2 = \int (5t - 5t^2)dt + C_2 = \frac{5}{2}t^2 - \frac{5}{3}t^3 + C_2.$$

       Since $x(0) = 0$, we have $C_2 = 0$, and

$$x(t) = \frac{5}{2}t^2 - \frac{5}{3}t^3.$$

    c. The velocity can be written as $v(t) = 5t(1 - t)$, which equals zero at $t = 0$, and $t = 1$ s.

CONCEPTUAL QUESTIONS

1. You drive your car into town and return to drive past your house to a friend's house.

3. If the bacteria are moving back and forth, then the displacements are canceling each other and the final displacement is small.

5. Distance traveled

7. Average speed is the total distance traveled divided by the elapsed time. If you go for a walk, leaving and returning to your home, your average speed is a positive number. Since Average velocity = Displacement/Elapsed time, your average velocity is zero.

9. Average speed. They are the same if the car doesn't reverse direction.

11. No, in one dimension constant speed requires zero acceleration.

13. A ball is thrown into the air and its velocity is zero at the apex of the throw, but acceleration is not zero.

15. Plus, minus

17. If the acceleration, time, and displacement are the knowns, and the initial and final velocities are the unknowns, then two kinematic equations must be solved simultaneously. Also if the final velocity, time, and displacement are the knowns then two kinematic equations must be solved for the initial velocity and acceleration.

19. a. at the top of its trajectory; b. yes, at the top of its trajectory; c. yes

21.   Earth $v = v_0 - gt = -gt$; Moon $v' = \frac{g}{6}t'$ $v = v'$ $-gt = -\frac{g}{6}t'$ $t' = 6t$; Earth $y = -\frac{1}{2}gt^2$ Moon

$$y' = -\frac{1}{2}\frac{g}{6}(6t)^2 = -\frac{1}{2}g6t^2 = -6\left(\frac{1}{2}gt^2\right) = -6y$$

PROBLEMS

25. a. $\vec{x}_1 = (-2.0 \text{ m})\hat{i}$, $\vec{x}_2 = (5.0 \text{ m})\hat{i}$; b. 7.0 m east

27. a. $t = 2.0$ s; b. $x(6.0) - x(3.0) = -8.0 - (-2.0) = -6.0$ m

29. a. 150.0 s, $\bar{v} = 156.7$ m/s ; b. 45.7% the speed of sound at sea level

Velocity vs. Time

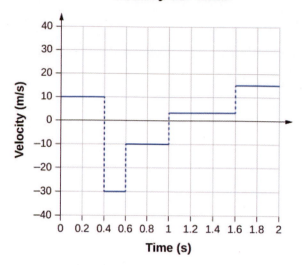

31.

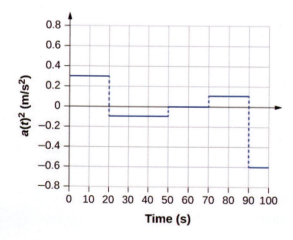

33.

35. a. $v(t) = (10 - 4t)$m/s ; $v(2$ s$) = 2$ m/s, $v(3$ s$) = -2$ m/s; b. $|v(2$ s$)| = 2$ m/s, $|v(3$ s$)| = 2$ m/s ; (c) $\bar{v} = 0$ m/s

37. $a = 4.29$m/s^2

Acceleration vs. Time

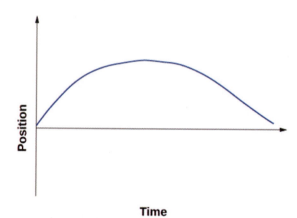

39.

41. $a = 11.1g$

43. 150 m

45. a. 525 m;

b. $v = 180$ m/s

47. a.

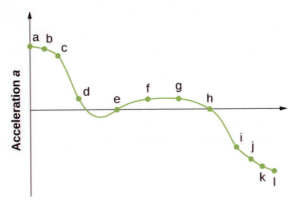

Time t

b. The acceleration has the greatest positive value at t_a

c. The acceleration is zero at t_e and t_h

d. The acceleration is negative at t_i, t_j, t_k, t_l

49. a. $a = -1.3$ m/s^2;

b. $v_0 = 18$ m/s;

c. $t = 13.8$ s

51. $v = 502.20$ m/s

53. a.

$v_0 = 0$ m/s $v_0 = ?$ m/s

$t_0 = 0$ s $t_0 = 12.0$ s
$x_0 = 0$ m $x_0 = ?$ m
$a = 2.40$ m/s^2 $a = 2.40$ m/s^2

b. Knowns: $a = 2.40$ m/s^2, $t = 12.0$ s, $v_0 = 0$ m/s, and $x_0 = 0$ m;

c. $x = x_0 + v_0 t + \frac{1}{2}at^2 = \frac{1}{2}at^2 = 2.40$ m/s^2 $(12.0$ s$)^2 = 172.80$ m, the answer seems reasonable at about 172.8 m; d. $v = 28.8$ m/s

55. a.

$t_0 = 0$ s $t_0 = ?$
$x_0 = 0$ m $x_0 = 1.80$ cm
$v_0 = 0$ m/s $v_0 = 30.0$ cm/s
$a = ?$ $a = ?$

b. Knowns: $v = 30.0$ cm/s, $x = 1.80$ cm;

c. $a = 250$ cm/s^2, $t = 0.12$ s;

d. yes

57. a. 6.87 m/s^2; b. $x = 52.26$ m

59. a. $a = 8450$ m/s^2;

b. $t = 0.0077$ s

61. a. $a = 9.18\, g$;

b. $t = 6.67 \times 10^{-3}$ s;

c. $a = -40.0 \text{ m/s}^2$
$a = 4.08 \, g$

63. Knowns: $x = 3$ m, $v = 0$ m/s, $v_0 = 54$ m/s. We want a, so we can use this equation: $a = -486 \text{ m/s}^2$.

65. a. $a = 32.58 \text{ m/s}^2$;

b. $v = 161.85$ m/s;

c. $v > v_{\max}$, because the assumption of constant acceleration is not valid for a dragster. A dragster changes gears and would have a greater acceleration in first gear than second gear than third gear, and so on. The acceleration would be greatest at the beginning, so it would not be accelerating at 32.6 m/s^2 during the last few meters, but substantially less, and the final velocity would be less than 162 m/s.

67. a. $y = -8.23$ m ; $v_1 = -18.9$ m/s

b. $y = -18.9$ m ; $v_2 = -23.8$ m/s

c. $y = -32.0$ m ; $v_3 = -28.7$ m/s

d. $y = -47.6$ m ; $v_4 = -33.6$ m/s

e. $y = -65.6$ m $v_5 = -38.5$ m/s

69. a. Knowns: $a = -9.8 \text{ m/s}^2$ $v_0 = -1.4$ m/s $t = 1.8$ s $y_0 = 0$ m;

b. $y = y_0 + v_0 t - \frac{1}{2}gt^2$ $y = v_0 t - \frac{1}{2}gt = -1.4 \text{ m/s}(1.8 \text{ sec}) - \frac{1}{2}(9.8)(1.8 \text{ s})^2 = -18.4$ m and the origin is at the rescuers, who are 18.4 m above the water.

71. a. $v^2 = v_0^2 - 2g(y - y_0)$ $y_0 = 0$ $v = 0$ $y = \frac{v_0^2}{2g} = \frac{(4.0 \text{ m/s})^2}{2(9.80)} = 0.82$ m ; b. to the apex $v = 0.41$ s times 2 to the board = 0.82 s from the board to the water $y = y_0 + v_0 t - \frac{1}{2}gt^2$ $y = -1.80$ m $y_0 = 0$ $v_0 = 4.0$ m/s $-1.8 = 4.0t - 4.9t^2$ $4.9t^2 - 4.0t - 1.80 = 0$, solution to quadratic equation gives 1.13 s; c. $v^2 = v_0^2 - 2g(y - y_0)$ $y_0 = 0$ $v_0 = 4.0$ m/s $y = -1.80$ m $v = 7.16$ m/s

73. Time to the apex: $t = 1.12$ s times 2 equals 2.24 s to a height of 2.20 m. To 1.80 m in height is an additional 0.40 m.

$y = y_0 + v_0 t - \frac{1}{2}gt^2$ $y = -0.40$ m $y_0 = 0$ $v_0 = -11.0$ m/s

$y = y_0 + v_0 t - \frac{1}{2}gt^2$ $y = -0.40$ m $y_0 = 0$ $v_0 = -11.0$ m/s .

$-0.40 = -11.0t - 4.9t^2$ or $4.9t^2 + 11.0t - 0.40 = 0$

Take the positive root, so the time to go the additional 0.4 m is 0.04 s. Total time is $2.24 \text{ s} + 0.04 \text{ s} = 2.28 \text{ s}$.

75. a. $v^2 = v_0^2 - 2g(y - y_0)$ $y_0 = 0$ $v = 0$ $y = 2.50$ m ; b. $t = 0.72$ s times 2 gives 1.44 s in the air
$v_0^2 = 2gy \Rightarrow v_0 = \sqrt{2(9.80)(2.50)} = 7.0$ m/s

77. a. $v = 70.0$ m/s ; b. time heard after rock begins to fall: 0.75 s, time to reach the ground: 6.09 s

79. a. $A = m/s^2 \quad B = m/s^{5/2}$;

b.
$$v(t) = \int a(t)dt + C_1 = \int \left(A - Bt^{1/2}\right)dt + C_1 = At - \frac{2}{3}Bt^{3/2} + C_1$$
$$v(0) = 0 = C_1 \quad \text{so} \quad v(t_0) = At_0 - \frac{2}{3}Bt_0^{3/2}$$

c.
$$x(t) = \int v(t)dt + C_2 = \int \left(At - \frac{2}{3}Bt^{3/2}\right)dt + C_2 = \frac{1}{2}At^2 - \frac{4}{15}Bt^{5/2} + C_2$$
$$x(0) = 0 = C_2 \quad \text{so} \quad x(t_0) = \frac{1}{2}At_0^2 - \frac{4}{15}Bt_0^{5/2}$$

81. a. $a(t) = 3.2 m/s^2 \quad t \le 5.0\,s$
$\quad a(t) = 1.5 m/s^2 \quad 5.0\,s \le t \le 11.0\,s$;
$\quad a(t) = 0 m/s^2 \quad t > 11.0\,s$

$$x(t) = \int v(t)dt + C_2 = \int 3.2t\, dt + C_2 = 1.6t^2 + C_2$$
$$t \le 5.0\,s$$
$$x(0) = 0 \Rightarrow C_2 = 0 \quad \text{therefore,} \quad x(2.0\,s) = 6.4\,m$$
$$x(t) = \int v(t)dt + C_2 = \int [16.0 - 1.5(t - 5.0)]dt + C_2 = 16t - 1.5\left(\frac{t^2}{2} - 5.0t\right) + C_2$$

$$5.0 \le t \le 11.0\,s$$
$$x(5\,s) = 1.6(5.0)^2 = 40\,m = 16(5.0\,s) - 1.5\left(\frac{5^2}{2} - 5.0(5.0)\right) + C_2$$

b. $\quad 40 = 98.75 + C_2 \Rightarrow C_2 = -58.75$
$$x(7.0\,s) = 16(7.0) - 1.5\left(\frac{7^2}{2} - 5.0(7)\right) - 58.75 = 69\,m$$
$$x(t) = \int 7.0\,dt + C_2 = 7t + C_2$$
$$t \ge 11.0\,s$$
$$x(11.0\,s) = 16(11) - 1.5\left(\frac{11^2}{2} - 5.0(11)\right) - 58.75 = 109 = 7(11.0\,s) + C_2 \Rightarrow C_2 = 32\,m$$
$$x(t) = 7t + 32\,m$$
$$x \ge 11.0\,s \Rightarrow x(12.0\,s) = 7(12) + 32 = 116\,m$$

ADDITIONAL PROBLEMS

83. Take west to be the positive direction.
1st plane: $\bar{v} = 600\,km/h$
2nd plane $\bar{v} = 667.0\,km/h$

85. $\quad a = \frac{v - v_0}{t - t_0}, \quad t = 0, \quad a = \frac{-3.4\,cm/s - v_0}{4\,s} = 1.2\,cm/s^2 \Rightarrow v_0 = -8.2\,cm/s \quad v = v_0 + at = -8.2 + 1.2\,t$;
$v = -7.0\,cm/s \quad v = -1.0\,cm/s$

87. $a = -3\,m/s^2$

89. a.
$v = 8.7 \times 10^5\,m/s$;
b. $t = 7.8 \times 10^{-8}\,s$

91. $1\,km = v_0(80.0\,s) + \frac{1}{2}a(80.0)^2$; $2\,km = v_0(200.0) + \frac{1}{2}a(200.0)^2$ solve simultaneously to get $a = -\frac{0.1}{2400.0}km/s^2$
and $v_0 = 0.014167\,km/s$, which is $51.0\,km/h$. Velocity at the end of the trip is $v = 21.0\,km/h$.

93. $a = -0.9\,m/s^2$

95. Equation for the speeding car: This car has a constant velocity, which is the average velocity, and is not accelerating, so use the equation for displacement with $x_0 = 0$: $x = x_0 + \bar{v}t = \bar{v}t$; Equation for the police car: This car is accelerating, so use the equation for displacement with $x_0 = 0$ and $v_0 = 0$, since the police car starts from rest: $x = x_0 + v_0 t + \frac{1}{2}at^2 = \frac{1}{2}at^2$; Now we have an equation of motion for each car with a common parameter, which can be eliminated to find the solution. In this case, we solve for t. Step 1, eliminating x: $x = \bar{v}t = \frac{1}{2}at^2$; Step 2, solving for t: $t = \frac{2\bar{v}}{a}$. The speeding car has a constant velocity of 40 m/s, which is its average velocity. The acceleration of the police car is 4 m/s^2. Evaluating t, the time for the police car to reach the speeding car, we have $t = \frac{2\bar{v}}{a} = \frac{2(40)}{4} = 20$ s.

97. At this acceleration she comes to a full stop in $t = \frac{-v_0}{a} = \frac{8}{0.5} = 16$ s, but the distance covered is $x = 8$ m/s$(16$ s$) - \frac{1}{2}(0.5)(16$ s$)^2 = 64$ m, which is less than the distance she is away from the finish line, so she never finishes the race.

99. $x_1 = \frac{3}{2}v_0 t$

$x_2 = \frac{5}{3}x_1$

101. $v_0 = 7.9$ m/s velocity at the bottom of the window.

$v = 7.9$ m/s

$v_0 = 14.1$ m/s

103. a. $v = 5.42$ m/s;

b. $v = 4.64$ m/s;

c. $a = 2874.28$ m/s^2;

d. $(x - x_0) = 5.11 \times 10^{-3}$ m

105. Consider the players fall from rest at the height 1.0 m and 0.3 m.
0.9 s
0.5 s

107. a. $t = 6.37$ s taking the positive root;

b. $v = 59.5$ m/s

109. a. $y = 4.9$ m;

b. $v = 38.3$ m/s;

c. -33.3 m

111. $h = \frac{1}{2}gt^2$, h = total height and time to drop to ground

$\frac{2}{3}h = \frac{1}{2}g(t-1)^2$ in $t - 1$ seconds it drops $2/3h$

$\frac{2}{3}\left(\frac{1}{2}gt^2\right) = \frac{1}{2}g(t-1)^2$ or $\frac{t^2}{3} = \frac{1}{2}(t-1)^2$

$0 = t^2 - 6t + 3$ $t = \frac{6 \pm \sqrt{6^2 - 4\cdot 3}}{2} = 3 \pm \frac{\sqrt{24}}{2}$

$t = 5.45$ s and $h = 145.5$ m. Other root is less than 1 s. Check for $t = 4.45$ s $h = \frac{1}{2}gt^2 = 97.0$ m $= \frac{2}{3}(145.5)$

CHALLENGE PROBLEMS

113. a. $v(t) = 10t - 12t^2$ m/s, $a(t) = 10 - 24t$ m/s^2;

b. $v(2$ s$) = -28$ m/s, $a(2$ s$) = -38$m/s^2; c. The slope of the position function is zero or the velocity is zero. There are two possible solutions: $t = 0$, which gives $x = 0$, or $t = 10.0/12.0 = 0.83$ s, which gives $x = 1.16$ m. The second answer is the correct choice; d. 0.83 s (e) 1.16 m

115. 96 km/h $= 26.67$ m/s, $a = \frac{26.67 \text{ m/s}}{4.0 \text{ s}} = 6.67$m/s^2, 295.38 km/h $= 82.05$ m/s, $t = 12.3$ s time to accelerate to maximum speed

$x = 504.55$ m distance covered during acceleration

7495.44 m at a constant speed

$\dfrac{7495.44 \text{ m}}{82.05 \text{ m/s}} = 91.35 \text{ s}$ so total time is $91.35 \text{ s} + 12.3 \text{ s} = 103.65 \text{ s}$.

CHAPTER 4

CHECK YOUR UNDERSTANDING

4.1. (a) Taking the derivative with respect to time of the position function, we have $\vec{v}(t) = 9.0t^2\,\hat{i}$ and $\vec{v}(3.0\text{s}) = 81.0\,\hat{i}$ m/s. (b) Since the velocity function is nonlinear, we suspect the average velocity is not equal to the instantaneous velocity. We check this and find

$$\vec{v}_{avg} = \frac{\vec{r}(t_2) - \vec{r}(t_1)}{t_2 - t_1} = \frac{\vec{r}(4.0\text{ s}) - \vec{r}(2.0\text{ s})}{4.0\text{ s} - 2.0\text{ s}} = \frac{(144.0\,\hat{i} - 36.0\,\hat{i})\text{ m}}{2.0\text{ s}} = 54.0\,\hat{i}\text{ m/s},$$

which is different from $\vec{v}(3.0\text{s}) = 81.0\,\hat{i}$ m/s.

4.2. The acceleration vector is constant and doesn't change with time. If a, b, and c are not zero, then the velocity function must be linear in time. We have $\vec{v}(t) = \int \vec{a}\, dt = \int (a\,\hat{i} + b\,\hat{j} + c\,\hat{k})dt = (a\,\hat{i} + b\,\hat{j} + c\,\hat{k})t$ m/s, since taking the derivative of the velocity function produces $\vec{a}(t)$. If any of the components of the acceleration are zero, then that component of the velocity would be a constant.

4.3. (a) Choose the top of the cliff where the rock is thrown from the origin of the coordinate system. Although it is arbitrary, we typically choose time $t = 0$ to correspond to the origin. (b) The equation that describes the horizontal motion is $x = x_0 + v_x t$. With $x_0 = 0$, this equation becomes $x = v_x t$. (c) **Equation 4.27** through **Equation 4.29** and **Equation 4.46** describe the vertical motion, but since $y_0 = 0$ and $v_{0y} = 0$, these equations simplify greatly to become $y = \frac{1}{2}(v_{0y} + v_y)t = \frac{1}{2}v_y t$, $v_y = -gt$, $y = -\frac{1}{2}gt^2$, and $v_y^2 = -2gy$. (d) We use the kinematic equations to find the x and y components of the velocity at the point of impact. Using $v_y^2 = -2gy$ and noting the point of impact is -100.0 m, we find the y component of the velocity at impact is $v_y = 44.3$ m/s. We are given the x component, $v_x = 15.0$ m/s, so we can calculate the total velocity at impact: $v = 46.8$ m/s and $\theta = 71.3°$ below the horizontal.

4.4. The golf shot at $30°$.

4.5. 134.0 cm/s

4.6. Labeling subscripts for the vector equation, we have B = boat, R = river, and E = Earth. The vector equation becomes $\vec{v}_{BE} = \vec{v}_{BR} + \vec{v}_{RE}$. We have right triangle geometry shown in Figure 04_05_BoatRiv_img. Solving for $\vec{v}_{BE}$, we have

$$v_{BE} = \sqrt{v_{BR}^2 + v_{RE}^2} = \sqrt{4.5^2 + 3.0^2}$$

$$v_{BE} = 5.4 \text{ m/s}, \quad \theta = \tan^{-1}\left(\frac{3.0}{4.5}\right) = 33.7°.$$

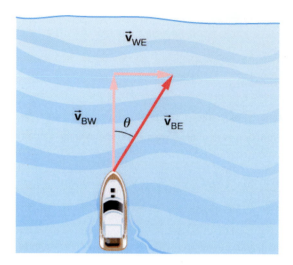

CONCEPTUAL QUESTIONS

1. straight line

3. The slope must be zero because the velocity vector is tangent to the graph of the position function.

5. No, motions in perpendicular directions are independent.

7. a. no; b. minimum at apex of trajectory and maximum at launch and impact; c. no, velocity is a vector; d. yes, where it lands

9. They both hit the ground at the same time.

11. yes

13. If he is going to pass the ball to another player, he needs to keep his eyes on the reference frame in which the other players on the team are located.

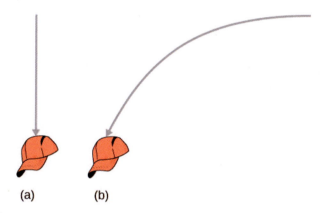

15.

PROBLEMS

17. $\vec{r} = 1.0\,\hat{i} - 4.0\,\hat{j} + 6.0\hat{k}$

19. $\Delta \vec{r}_{Total} = 472.0\text{ m }\hat{i} + 80.3\text{ m }\hat{j}$

21. Sum of displacements $= -6.4\text{ km }\hat{i} + 9.4\text{ km }\hat{j}$

23. a. $\vec{v}(t) = 8.0t\,\hat{i} + 6.0t^2\,\hat{k}, \quad \vec{v}(0) = 0, \quad \vec{v}(1.0) = 8.0\,\hat{i} + 6.0\hat{k}\text{ m/s},$

b. $\vec{v}_{avg} = 4.0\,\hat{i} + 2.0\hat{k}\text{ m/s}$

25. $\Delta \vec{r}_1 = 20.00\text{ m }\hat{j}, \Delta \vec{r}_2 = (2.000 \times 10^4\text{ m})\,(\cos 30°\,\hat{i} + \sin 30°\,\hat{j})$

$\Delta \vec{r} = 1.700 \times 10^4\text{ m }\hat{i} + 1.002 \times 10^4\text{ m }\hat{j}$

27. a. $\vec{\mathbf{v}}(t) = (4.0t\,\hat{\mathbf{i}} + 3.0t\,\hat{\mathbf{j}})$m/s, $\quad \vec{\mathbf{r}}(t) = (2.0t^2\,\hat{\mathbf{i}} + \frac{3}{2}t^2\,\hat{\mathbf{j}})$ m,

b. $x(t) = 2.0t^2$ m, $y(t) = \frac{3}{2}t^2$ m, $t^2 = \frac{x}{2} \Rightarrow y = \frac{3}{4}x$

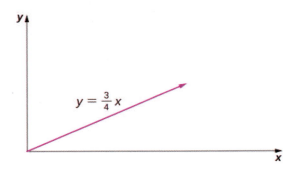

29. a. $\vec{\mathbf{v}}(t) = (6.0t\,\hat{\mathbf{i}} - 21.0t^2\,\hat{\mathbf{j}} + 10.0t^{-3}\,\hat{\mathbf{k}})$m/s,

b. $\vec{\mathbf{a}}(t) = (6.0\,\hat{\mathbf{i}} - 42.0t\,\hat{\mathbf{j}} - 30t^{-4}\,\hat{\mathbf{k}})$m/s^2,

c. $\vec{\mathbf{v}}(2.0s) = (12.0\,\hat{\mathbf{i}} - 84.0\,\hat{\mathbf{j}} + 1.25\,\hat{\mathbf{k}})$m/s,

d. $\vec{\mathbf{v}}(1.0\,\text{s}) = 6.0\,\hat{\mathbf{i}} - 21.0\,\hat{\mathbf{j}} + 10.0\,\hat{\mathbf{k}}$m/s, $\left|\vec{\mathbf{v}}(1.0\,\text{s})\right| = 24.0$ m/s

$\vec{\mathbf{v}}(3.0\,\text{s}) = 18.0\,\hat{\mathbf{i}} - 189.0\,\hat{\mathbf{j}} + 0.37\,\hat{\mathbf{k}}$m/s, $\left|\vec{\mathbf{v}}(3.0\,\text{s})\right| = 199.0$ m/s,

e. $\vec{\mathbf{r}}(t) = (3.0t^2\,\hat{\mathbf{i}} - 7.0t^3\,\hat{\mathbf{j}} - 5.0t^{-2}\,\hat{\mathbf{k}})$m

$\vec{\mathbf{v}}_{\text{avg}} = 9.0\,\hat{\mathbf{i}} - 49.0\,\hat{\mathbf{j}} + 3.75\,\hat{\mathbf{k}}$m/s

31. a. $\vec{\mathbf{v}}(t) = -\sin(1.0t)\,\hat{\mathbf{i}} + \cos(1.0t)\,\hat{\mathbf{j}} + \hat{\mathbf{k}}$, b. $\vec{\mathbf{a}}(t) = -\cos(1.0t)\,\hat{\mathbf{i}} - \sin(1.0t)\,\hat{\mathbf{j}}$

33. a. $t = 0.55$ s, b. $x = 110$ m

35. a. $t = 0.24$s, $d = 0.28$ m, b. They aim high.

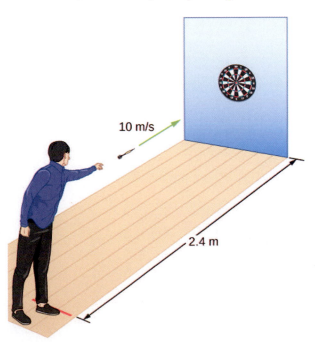

37. a., $t = 12.8$ s, $x = 5619$ m b. $v_y = 125.0$ m/s, $v_x = 439.0$ m/s, $|\vec{v}| = 456.0$ m/s

39. a. $v_y = v_{0y} - gt$, $t = 10$s, $v_y = 0$, $v_{0y} = 98.0$ m/s, $v_0 = 196.0$ m/s , b. $h = 490.0$ m,

c. $v_{0x} = 169.7$ m/s, $x = 3394.0$ m,

$\quad x = 2545.5$ m

d. $y = 465.5$ m

$\quad \vec{s} = 2545.5$ m $\hat{i}$ + 465.5 m $\hat{j}$

41. -100 m $= (-2.0$ m/s$)t - (4.9$ m/s$^2)t^2$, $t = 4.3$ s, $x = 86.0$ m

43. $R_{Moon} = 48$ m

45. a. $v_{0y} = 24$ m/s $v_y^2 = v_{0y}^2 - 2gy \Rightarrow h = 23.4$ m,

b. $t = 3$ s $v_{0x} = 18$ m/s $x = 54$ m,

c. $y = -100$ m $y_0 = 0$ $y - y_0 = v_{0y}t - \frac{1}{2}gt^2$ $-100 = 24t - 4.9t^2$ $\Rightarrow t = 7.58$ s,

d. $x = 136.44$ m,

e. $t = 2.0$ s $y = 28.4$ m $x = 36$ m

$\quad t = 4.0$ s $y = 17.6$ m $x = 22.4$ m

$\quad t = 6.0$ s $y = -32.4$ m $x = 108$ m

47. $v_{0y} = 12.9$ m/s $y - y_0 = v_{0y}t - \frac{1}{2}gt^2$ $-20.0 = 12.9t - 4.9t^2$

$t = 3.7$ s $v_{0x} = 15.3$ m/s $\Rightarrow x = 56.7$ m

So the golfer's shot lands 13.3 m short of the green.

49. a. $R = 60.8$ m,

b. $R = 137.8$ m

51. a. $v_y^2 = v_{0y}^2 - 2gy \Rightarrow y = 2.9$ m/s

$y = 3.3$ m/s

$y = \frac{v_{0y}^2}{2g} = \frac{(v_0 \sin\theta)^2}{2g} \Rightarrow \sin\theta = 0.91 \Rightarrow \theta = 65.5°$

53. $R = 18.5$ m

55. $y = (\tan\theta_0)x - \left[\frac{g}{2(v_0 \cos\theta_0)^2}\right]x^2 \Rightarrow v_0 = 16.4$ m/s

57. $R = \frac{v_0^2 \sin 2\theta_0}{g} \Rightarrow \theta_0 = 15.9°$

59. It takes the wide receiver 1.1 s to cover the last 10 m of his run.

$T_{tof} = \frac{2(v_0 \sin\theta)}{g} \Rightarrow \sin\theta = 0.27 \Rightarrow \theta = 15.6°$

61. $a_C = 40$ m/s^2

63. $a_C = \frac{v^2}{r} \Rightarrow v^2 = r\, a_C = 78.4$, $v = 8.85$ m/s

$T = 5.68$ s, which is 0.176 rev/s $= 10.6$ rev/min

65. Venus is 108.2 million km from the Sun and has an orbital period of 0.6152 y.

$r = 1.082 \times 10^{11}$ m $T = 1.94 \times 10^7$ s

$v = 3.5 \times 10^4$ m/s, $a_C = 1.135 \times 10^{-2}$ m/s^2

67. 360 rev/min $= 6$ rev/s

$v = 3.8$ m/s $a_C = 144.$ m/s^2

69. a. $O'(t) = (4.0\,\hat{i} + 3.0\,\hat{j} + 5.0\hat{k})t$ m ,

b. $\vec{r}_{PS} = \vec{r}_{PS'} + \vec{r}_{S'S}$, $\vec{r}(t) = \vec{r}'(t) + (4.0\,\hat{i} + 3.0\,\hat{j} + 5.0\,\hat{k})t$ m,

c. $\vec{v}(t) = \vec{v}'(t) + (4.0\,\hat{i} + 3.0\,\hat{j} + 5.0\,\hat{k})$ m/s, d. The accelerations are the same.

71. $\vec{v}_{PC} = (2.0\,\hat{i} + 5.0\,\hat{j} + 4.0\,\hat{k})$m/s

73. a. A = air, S = seagull, G = ground

$\vec{v}_{SA} = 9.0$ m/s velocity of seagull with respect to still air

$\vec{v}_{AG} = ?$ $\vec{v}_{SG} = 5$ m/s $\vec{v}_{SG} = \vec{v}_{SA} + \vec{v}_{AG} \Rightarrow \vec{v}_{AG} = \vec{v}_{SG} - \vec{v}_{SA}$

$\vec{v}_{AG} = -4.0$ m/s

b. $\vec{v}_{SG} = \vec{v}_{SA} + \vec{v}_{AG} \Rightarrow \vec{v}_{SG} = -13.0$ m/s

$\dfrac{-6000 \text{ m}}{-13.0 \text{ m/s}} = 7$ min 42 s

75. Take the positive direction to be the same direction that the river is flowing, which is east. S = shore/Earth, W = water, and B = boat.

a. $\vec{v}_{BS} = 11$ km/h

$t = 8.2$ min

b. $\vec{v}_{BS} = -5$ km/h

$t = 18$ min

c. $\vec{v}_{BS} = \vec{v}_{BW} + \vec{v}_{WS}$ $\theta = 22°$ west of north

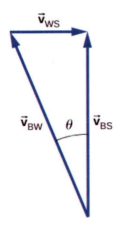

d. $|\vec{v}_{BS}| = 7.4$ km/h $t = 6.5$ min

e. $\vec{v}_{BS} = 8.54$ km/h, but only the component of the velocity straight across the river is used to get the time

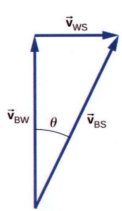

$t = 6.0$ min

Downstream = 0.3 km

77. $\vec{\mathbf{v}}_{AG} = \vec{\mathbf{v}}_{AC} + \vec{\mathbf{v}}_{CG}$

$\left| \vec{\mathbf{v}}_{AC} \right| = 25 \text{ km/h}$ $\left| \vec{\mathbf{v}}_{CG} \right| = 15 \text{ km/h}$ $\left| \vec{\mathbf{v}}_{AG} \right| = 29.15 \text{ km/h}$ $\vec{\mathbf{v}}_{AG} = \vec{\mathbf{v}}_{AC} + \vec{\mathbf{v}}_{CG}$

The angle between $\vec{\mathbf{v}}_{AC}$ and $\vec{\mathbf{v}}_{AG}$ is $31°$, so the direction of the wind is $14°$ north of east.

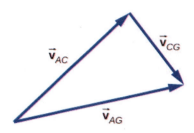

ADDITIONAL PROBLEMS

79. $a_C = 39.6 \text{ m/s}^2$

81. $90.0 \text{ km/h} = 25.0 \text{ m/s},\ 9.0 \text{ km/h} = 2.5 \text{ m/s},\ 60.0 \text{ km/h} = 16.7 \text{ m/s}$

$a_T = -2.5 \text{ m/s}^2, a_C = 1.86 \text{ m/s}^2, a = 3.1 \text{ m/s}^2$

83. The radius of the circle of revolution at latitude λ is $R_E \cos \lambda$. The velocity of the body is $\frac{2\pi r}{T}$. $a_C = \frac{4\pi^2 R_E \cos \lambda}{T^2}$ for

$\lambda = 40°,\ a_C = 0.26\% \, g$

85. $a_T = 3.00 \text{ m/s}^2$

$v(5 \text{ s}) = 15.00 \text{ m/s}$ $a_C = 150.00 \text{ m/s}^2$ $\theta = 88.8°$ with respect to the tangent to the circle of revolution directed inward.

$\left| \vec{\mathbf{a}} \right| = 150.03 \text{ m/s}^2$

87. $\vec{\mathbf{a}}(t) = -A\omega^2 \cos \omega t \, \hat{\mathbf{i}} - A\omega^2 \sin \omega t \, \hat{\mathbf{j}}$

$a_C = 5.0 \, m\omega^2$ $\omega = 0.89 \text{ rad/s}$

$\vec{\mathbf{v}}(t) = -2.24 \text{ m/s} \, \hat{\mathbf{i}} - 3.87 \text{ m/s} \, \hat{\mathbf{j}}$

89. $\vec{\mathbf{r}}_1 = 1.5 \, \hat{\mathbf{j}} + 4.0 \, \hat{\mathbf{k}}$ $\vec{\mathbf{r}}_2 = \Delta \vec{\mathbf{r}} + \vec{\mathbf{r}}_1 = 2.5 \, \hat{\mathbf{i}} + 4.7 \, \hat{\mathbf{j}} + 2.8 \, \hat{\mathbf{k}}$

91. $v_x(t) = 265.0 \text{ m/s}$

$v_y(t) = 20.0 \text{ m/s}$

$\vec{\mathbf{v}}(5.0 \text{ s}) = (265.0 \, \hat{\mathbf{i}} + 20.0 \, \hat{\mathbf{j}}) \text{m/s}$

93. $R = 1.07 \text{ m}$

95. $v_0 = 20.1 \text{ m/s}$

97. $v = 3072.5 \text{ m/s}$

$a_C = 0.223 \text{ m/s}^2$

CHALLENGE PROBLEMS

99. a. $-400.0 \text{ m} = v_{0y} t - 4.9 t^2$ $359.0 \text{ m} = v_{0x} t$ $t = \frac{359.0}{v_{0x}}$ $-400.0 = 359.0 \frac{v_{0y}}{v_{0x}} - 4.9 (\frac{359.0}{v_{0x}})^2$

$-400.0 = 359.0 \tan 40 - \frac{631,516.9}{v_{0x}^2} \Rightarrow v_{0x}^2 = 900.6$ $v_{0x} = 30.0 \text{ m/s}$ $v_{0y} = v_{0x} \tan 40 = 25.2 \text{ m/s}$

$v = 39.2 \text{ m/s}$, b. $t = 12.0 \text{ s}$

101. a. $\vec{\mathbf{r}}_{TC} = (-32 + 80t) \, \hat{\mathbf{i}} + 50t \, \hat{\mathbf{j}}$, $\left| \vec{\mathbf{r}}_{TC} \right|^2 = (-32 + 80t)^2 + (50t)^2$

$$2r\frac{dr}{dt} = 2(-32 + 80t)(80) + 5000t \quad \frac{dr}{dt} = \frac{160(-32 + 80t) + 5000t}{2r} = 0$$

$17800t = 5184 \Rightarrow t = 0.29 \text{ hr},$

b. $\left|\vec{\mathbf{r}}_{TC}\right| = 17 \text{ km}$

CHAPTER 5

CHECK YOUR UNDERSTANDING

5.1. 14 N, $56°$ measured from the positive x-axis

5.2. a. His weight acts downward, and the force of air resistance with the parachute acts upward. b. neither; the forces are equal in magnitude

5.3. 0.1 m/s^2

5.4. 40 m/s^2

5.5. a. $159.0 \hat{\mathbf{i}} + 770.0 \hat{\mathbf{j}} \text{ N}$; b. $0.1590 \hat{\mathbf{i}} + 0.7700 \hat{\mathbf{j}} \text{ N}$

5.6. $a = 2.78 \text{ m/s}^2$

5.7. a. 3.0 m/s^2; b. 18 N

5.8. a. 1.7 m/s^2; b. 1.3 m/s^2

5.9. $6.0 \times 10^2 \text{ N}$

5.10.

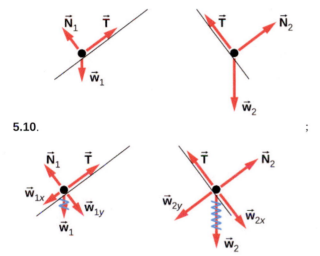

;

CONCEPTUAL QUESTIONS

1. Forces are directional and have magnitude.

3. The cupcake velocity before the braking action was the same as that of the car. Therefore, the cupcakes were unrestricted bodies in motion, and when the car suddenly stopped, the cupcakes kept moving forward according to Newton's first law.

5. No. If the force were zero at this point, then there would be nothing to change the object's momentary zero velocity. Since we do not observe the object hanging motionless in the air, the force could not be zero.

7. The astronaut is truly weightless in the location described, because there is no large body (planet or star) nearby to exert a gravitational force. Her mass is 70 kg regardless of where she is located.

9. The force you exert (a contact force equal in magnitude to your weight) is small. Earth is extremely massive by comparison. Thus, the acceleration of Earth would be incredibly small. To see this, use Newton's second law to calculate the acceleration you would cause if your weight is 600.0 N and the mass of Earth is $6.00 \times 10^{24} \text{ kg}$.

11. a. action: Earth pulls on the Moon, reaction: Moon pulls on Earth; b. action: foot applies force to ball, reaction: ball applies force to foot; c. action: rocket pushes on gas, reaction: gas pushes back on rocket; d. action: car tires push backward on road, reaction: road pushes forward on tires; e. action: jumper pushes down on ground, reaction: ground pushes up on jumper; f. action: gun pushes forward on bullet, reaction: bullet pushes backward on gun.

13. a. The rifle (the shell supported by the rifle) exerts a force to expel the bullet; the reaction to this force is the force that the bullet exerts on the rifle (shell) in opposite direction. b. In a recoilless rifle, the shell is not secured in the rifle; hence, as the bullet is pushed to move forward, the shell is pushed to eject from the opposite end of the barrel. c. It is not safe to stand behind a recoilless rifle.

15. a. Yes, the force can be acting to the left; the particle would experience deceleration and lose speed. B. Yes, the force can be acting downward because its weight acts downward even as it moves to the right.

17. two forces of different types: weight acting downward and normal force acting upward

PROBLEMS

19. a. $\overrightarrow{\textbf{F}}_{net} = 5.0\,\hat{\textbf{i}} + 10.0\,\hat{\textbf{j}}$ N; b. the magnitude is $F_{net} = 11$ N , and the direction is $\theta = 63°$

21. a. $\overrightarrow{\textbf{F}}_{net} = 660.0\,\hat{\textbf{i}} + 150.0\,\hat{\textbf{j}}$ N; b. $F_{net} = 676.6$ N at $\theta = 12.8°$ from David's rope

23. a. $\overrightarrow{\textbf{F}}_{net} = 95.0\,\hat{\textbf{i}} + 283\,\hat{\textbf{j}}$ N; b. 299 N at $71°$ north of east; c. $\overrightarrow{\textbf{F}}_{DS} = -\left(95.0\,\hat{\textbf{i}} + 283\,\hat{\textbf{j}}\right)$N

25. Running from rest, the sprinter attains a velocity of $v = 12.96$ m/s , at end of acceleration. We find the time for acceleration using $x = 20.00$ m $= 0 + 0.5at_1^2$, or $t_1 = 3.086$ s. For maintained velocity, $x_2 = vt_2$, or $t_2 = x_2/v = 80.00$ m/12.96 m/s $= 6.173$ s . Total time $= 9.259$ s .

27. a. $m = 56.0$ kg ; b. $a_{meas} = a_{astro} + a_{ship}$, where $a_{ship} = \dfrac{m_{astro}\,a_{astro}}{m_{ship}}$; c. If the force could be exerted on the astronaut by another source (other than the spaceship), then the spaceship would not experience a recoil.

29. $F_{net} = 4.12 \times 10^5$ N

31. $a = 253$ m/s^2

33. $F_{net} = F - f = ma \Rightarrow F = 1.26 \times 10^3$ N

35. $v^2 = v_0^2 + 2ax \Rightarrow a = -7.80$ m/s^2
$F_{net} = -7.80 \times 10^3$ N

37. a. $\overrightarrow{\textbf{F}}_{net} = m\,\overrightarrow{\textbf{a}} \Rightarrow \overrightarrow{\textbf{a}} = 9.0\,\hat{\textbf{i}}$ m/s^2; b. The acceleration has magnitude 9.0 m/s^2, so $x = 110$ m .

39. $1.6\,\hat{\textbf{i}} - 0.8\,\hat{\textbf{j}}$ m/s^2

41. a. $\begin{aligned} w_{Moon} &= mg_{Moon} \\ m &= 150\text{ kg} \\ w_{Earth} &= 1.5 \times 10^3 \text{ N} \end{aligned}$; b. Mass does not change, so the suited astronaut's mass on both Earth and the Moon is 150 kg.

43. a. $\begin{aligned} F_h &= 3.68 \times 10^3 \text{ N and} \\ w &= 7.35 \times 10^2 \text{ N} \end{aligned}$;
$\dfrac{F_h}{w} = 5.00$ times greater than weight
b. $\begin{aligned} F_{net} &= 3750\text{ N} \\ \theta &= 11.3° \text{ from horizontal} \end{aligned}$

45. $\begin{aligned} w &= 19.6\text{ N} \\ F_{net} &= 5.40\text{ N} \\ F_{net} &= ma \Rightarrow a = 2.70 \text{ m/s}^2 \end{aligned}$

47. $0.60\,\hat{\textbf{i}} - 8.4\,\hat{\textbf{j}}$ m/s^2

49. 497 N

51. a. $F_{net} = 2.64 \times 10^7$ N; b. The force exerted on the ship is also 2.64×10^7 N because it is opposite the shell's direction of motion.

53. Because the weight of the history book is the force exerted by Earth on the history book, we represent it as $\overrightarrow{\textbf{F}}_{EH} = -14\,\hat{\textbf{j}}$ N. Aside from this, the history book interacts only with the physics book. Because the acceleration of the history book is zero, the net force on it is zero by Newton's second law: $\overrightarrow{\textbf{F}}_{PH} + \overrightarrow{\textbf{F}}_{EH} = \overrightarrow{\textbf{0}}$, where $\overrightarrow{\textbf{F}}_{PH}$ is the force exerted by the physics book on the history book. Thus, $\vec{F}_{PH} = -\vec{F}_{EH} = -\left(-14\,\hat{j}\right)N = 14\,\hat{j}$ N. We find that the physics

book exerts an upward force of magnitude 14 N on the history book. The physics book has three forces exerted on it: $\vec{F}_{EP}$ due to

Earth, $\vec{F}_{HP}$ due to the history book, and $\vec{F}_{DP}$ due to the desktop. Since the physics book weighs 18 N, $\vec{F}_{EP} = -18\,\hat{j}$ N.

From Newton's third law, $\vec{F}_{HP} = -\vec{F}_{PH}$, so $\vec{F}_{HP} = -14\,\hat{j}$ N. Newton's second law applied to the physics book gives

$$\sum \vec{F} = \vec{0}, \quad \text{or} \quad \vec{F}_{DP} + \vec{F}_{EP} + \vec{F}_{HP} = \vec{0}, \quad \text{so} \quad \vec{F}_{DP} = -\left(-18\,\hat{j}\right) - \left(-14\,\hat{j}\right) = 32\,\hat{j}\ \text{N}.$$ The desk exerts an

upward force of 32 N on the physics book. To arrive at this solution, we apply Newton's second law twice and Newton's third law once.

55. a. The free-body diagram of pulley 4:

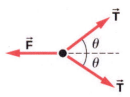

b. $T = mg$, $F = 2T\cos\theta = 2mg\cos\theta$

57. a.

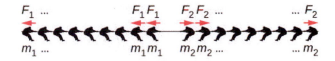

$$F_{net} = Ma; F_1 = 1350\ \text{N}; F_2 = 1365\ \text{N}$$
$$9(F_2 - F_1) = 9(m_1 + m_2)a; \ m_1 = 68\ \text{kg}; m_2 = 73\ \text{kg}$$
$$a = 0.11\ \text{m/s}^2;$$

Thus, the heavy team wins.

b.

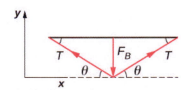

$$T - 9F_1 = 9m_1 a \Rightarrow T = 9m_1 a + 9F_1$$
$$= 1.2 \times 10^4\ \text{N}$$

59. a. $T = 1.96 \times 10^{-4}$ N;

$T' = 4.71 \times 10^{-4}$ N

b.

$\dfrac{T'}{T} = 2.40$ times the tension in the vertical strand

61.

$$F_{y\,net} = F_\perp - 2T\sin\theta = 0$$
$$F_\perp = 2T\sin\theta$$
$$T = \frac{F_\perp}{2\sin\theta}$$

63. a. see **Example 5.13**; b. 1.5 N; c. 15 N

65. a. 5.6 kg; b. 55 N; c. $T_2 = 60\,\text{N}$;

d.

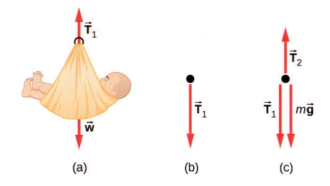

(a)   (b)   (c)

67. a. $4.9\,\text{m/s}^2$, 17 N; b. 9.8 N

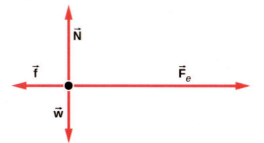

69.

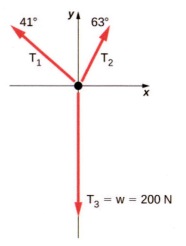

71.

ADDITIONAL PROBLEMS

73. 5.90 kg

75.

77. a. $F_{net} = \dfrac{m(v^2 - v_0{}^2)}{2x}$; b. 2590 N

79. $\vec{F}_{net} = 4.05\,\hat{i} + 12.0\,\hat{j}$ N

$\vec{F}_{net} = m\,\vec{a} \Rightarrow \vec{a} = 0.405\,\hat{i} + 1.20\,\hat{j}$ m/s^2

81. $\vec{F}_{net} = \vec{F}_A + \vec{F}_B$

$\vec{F}_{net} = A\,\hat{i} + \left(-1.41A\,\hat{i} - 1.41A\,\hat{j}\right)$

$\vec{F}_{net} = A\left(-0.41\,\hat{i} - 1.41\,\hat{j}\right)$

$\theta = 254°$

(We add $180°$, because the angle is in quadrant IV.)

83. $F = 2mk^2x^2$; First, take the derivative of the velocity function to obtain $a = 2kxv = 2kx\left(kx^2\right) = 2k^2x^3$. Then apply Newton's second law $F = ma = 2mk^2x^2$.

85. a. For box A, $N_A = mg$ and $N_B = mg\cos\theta$; b. $N_A > N_B$ because for $\theta < 90°$, $\cos\theta < 1$; c. $N_A > N_B$ when $\theta = 10°$

87. a. 8.66 N; b. 0.433 m

89. 0.40 or 40%

91. 16 N

CHALLENGE PROBLEMS

93. a.

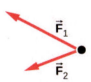

; b. No; $\vec{F}_R$ is not shown, because it would replace $\vec{F}_1$ and $\vec{F}_2$. (If we want to show it, we could draw it and then place squiggly lines on $\vec{F}_1$ and $\vec{F}_2$ to show that they are no longer considered.

95. a. 14.1 m/s; b. 601 N

97. $\dfrac{F}{m}t^2$

99. 936 N

101. $\vec{a} = -248\,\hat{i} - 433\,\hat{j}$ m/s^2

103. 0.548 m/s^2

105. a. $T_1 = \dfrac{2mg}{\sin\theta}$, $T_2 = \dfrac{mg}{\sin\left(\arctan\left(\frac{1}{2}\tan\theta\right)\right)}$, $T_3 = \dfrac{2mg}{\tan\theta}$; b. $\phi = \arctan\left(\frac{1}{2}\tan\theta\right)$; c. $2.56°$; (d)

$x = d\left(2\cos\theta + 2\cos\left(\arctan\left(\frac{1}{2}\tan\theta\right)\right) + 1\right)$

107. a. $\vec{a} = \left(\frac{5.00}{m}\hat{i} + \frac{3.00}{m}\hat{j}\right)$ m/s^2; b. 1.38 kg; c. 21.2 m/s; d. $\vec{v} = \left(18.1\hat{i} + 10.9\hat{j}\right)$ m/s^2

109. a. $0.900\hat{i} + 0.600\hat{j}$ N ; b. 1.08 N

CHAPTER 6

CHECK YOUR UNDERSTANDING

6.1. $F_s = 645$ N

6.2. $a = 3.68$ m/s^2, $T = 18.4$ N

6.3. $T = \frac{2m_1 m_2}{m_1 + m_2}g$ (This is found by substituting the equation for acceleration in **Figure 6.7** (a), into the equation for tension in **Figure 6.7** (b).)

6.4. 1.49 s

6.5. 49.4 degrees

6.6. 128 m; no

6.7. a. 4.9 N; b. 0.98 m/s^2

6.8. -0.23 m/s^2; the negative sign indicates that the snowboarder is slowing down.

6.9. 0.40

6.10. 34 m/s

6.11. 0.27 kg/m

CONCEPTUAL QUESTIONS

1. The scale is in free fall along with the astronauts, so the reading on the scale would be 0. There is no difference in the apparent weightlessness; in the aircraft and in orbit, free fall is occurring.

3. If you do not let up on the brake pedal, the car's wheels will lock so that they are not rolling; sliding friction is now involved and the sudden change (due to the larger force of static friction) causes the jerk.

5. 5.00 N

7. Centripetal force is defined as any net force causing uniform circular motion. The centripetal force is not a new kind of force. The label "centripetal" refers to *any* force that keeps something turning in a circle. That force could be tension, gravity, friction, electrical attraction, the normal force, or any other force. Any combination of these could be the source of centripetal force, for example, the centripetal force at the top of the path of a tetherball swung through a vertical circle is the result of both tension and gravity.

9. The driver who cuts the corner (on Path 2) has a more gradual curve, with a larger radius. That one will be the better racing line. If the driver goes too fast around a corner using a racing line, he will still slide off the track; the key is to stay at the maximum value of static friction. So, the driver wants maximum possible speed and maximum friction. Consider the equation for centripetal force: $F_c = m\frac{v^2}{r}$ where v is speed and r is the radius of curvature. So by decreasing the curvature ($1/r$) of the path that the car takes, we reduce the amount of force the tires have to exert on the road, meaning we can now increase the speed, v. Looking at this from the point of view of the driver on Path 1, we can reason this way: the sharper the turn, the smaller the turning circle; the smaller the turning circle, the larger is the required centripetal force. If this centripetal force is not exerted, the result is a skid.

11. The barrel of the dryer provides a centripetal force on the clothes (including the water droplets) to keep them moving in a circular path. As a water droplet comes to one of the holes in the barrel, it will move in a path tangent to the circle.

13. If there is no friction, then there is no centripetal force. This means that the lunch box will move along a path tangent to the circle, and thus follows path B. The dust trail will be straight. This is a result of Newton's first law of motion.

15. There must be a centripetal force to maintain the circular motion; this is provided by the nail at the center. Newton's third law explains the phenomenon. The action force is the force of the string on the mass; the reaction force is the force of the mass on the string. This reaction force causes the string to stretch.

17. Since the radial friction with the tires supplies the centripetal force, and friction is nearly 0 when the car encounters the ice, the car will obey Newton's first law and go off the road in a straight line path, tangent to the curve. A common misconception is that the car will follow a curved path off the road.

19. Anna is correct. The satellite is freely falling toward Earth due to gravity, even though gravity is weaker at the altitude of the satellite, and g is not 9.80 m/s^2. Free fall does not depend on the value of g; that is, you could experience free fall on Mars if you jumped off Olympus Mons (the tallest volcano in the solar system).

21. The pros of wearing body suits include: (1) the body suit reduces the drag force on the swimmer and the athlete can move more easily; (2) the tightness of the suit reduces the surface area of the athlete, and even though this is a small amount, it can make a difference in performance time. The cons of wearing body suits are: (1) The tightness of the suits can induce cramping and breathing problems. (2) Heat will be retained and thus the athlete could overheat during a long period of use.

23. The oil is less dense than the water and so rises to the top when a light rain falls and collects on the road. This creates a dangerous situation in which friction is greatly lowered, and so a car can lose control. In a heavy rain, the oil is dispersed and does not affect the motion of cars as much.

PROBLEMS

25. a. 170 N; b. 170 N

27. $\vec{F}_3 = (-7\hat{i} + 2\hat{j} + 4\hat{k})\,N$

29. 376 N pointing up (along the dashed line in the figure); the force is used to raise the heel of the foot.

31. −68.5 N

33. a. $7.70\ m/s^2$; b. 4.33 s

35. a. 46.4 m/s; b. $2.40 \times 10^3\ m/s^2$; c. 5.99×10^3 N; ratio of 245

37. a. 1.87×10^4 N; b. 1.67×10^4 N; c. 1.56×10^4 N; d. 19.4 m, 0 m/s

39. a. 10 kg; b. 140 N; c. 98 N; d. 0

41. a. $3.35\ m/s^2$; b. 4.2 s

43. a. $2.0\ m/s^2$; b. 7.8 N; c. 2.0 m/s

45. a. $4.43\ m/s^2$ (mass 1 accelerates up the ramp as mass 2 falls with the same acceleration); b. 21.5 N

47. a. 10.0 N; b. 97.0 N

49. a. $4.9\ m/s^2$; b. The cabinet will not slip. c. The cabinet will slip.

51. a. 32.3 N, $35.2°$; b. 0; c. $0.301\ m/s^2$ in the direction of $\vec{F}_{tot}$

53.
$$\begin{aligned}
net\,F_y &= 0 \Rightarrow N = mg\cos\theta \\
net\,F_x &= ma \\
a &= g(\sin\theta - \mu_k\cos\theta)
\end{aligned}$$

55. a. $0.737\ m/s^2$; b. $5.71°$

57. a. $10.8\ m/s^2$; b. $7.85\ m/s^2$; c. $2.00\ m/s^2$

59. a. $9.09\ m/s^2$; b. $6.16\ m/s^2$; c. $0.294\ m/s^2$

61. a. 272 N, 512 N; b. 0.268

63. a. 46.5 N; b. $0.629\ m/s^2$

65. a. 483 N; b. 17.4 N; c. 2.24, 0.0807

67. $4.14°$

69. a. 24.6 m; b. $36.6\ m/s^2$; c. 3.73 times g

71. a. 16.2 m/s; b. 0.234

73. a. 179 N; b. 290 N; c. 8.3 m/s

75. 20.7 m/s

77. 21 m/s

79. 115 m/s or 414 km/h

81. $v_T = 11.8\ m/s; v_2 = 9.9\ m/s$

83. $\left(\dfrac{110}{65}\right)^2 = 2.86$ times

85. Stokes' law is $F_s = 6\pi r\eta v$. Solving for the viscosity, $\eta = \dfrac{F_s}{6\pi r v}$. Considering only the units, this becomes $[\eta] = \dfrac{kg}{m \cdot s}$.

87. $0.76\ kg/m \cdot s$

89. a. 0.049 kg/s; b. 0.57 m

91. a. 1860 N, 2.53; b. The value (1860 N) is more force than you expect to experience on an elevator. The force of 1860 N is 418 pounds, compared to the force on a typical elevator of 904 N (which is about 203 pounds); this is calculated for a speed from 0 to 10 miles per hour, which is about 4.5 m/s, in 2.00 s). c. The acceleration $a = 1.53 \times g$ is much higher than any standard elevator.

The final speed is too large (30.0 m/s is VERY fast)! The time of 2.00 s is not unreasonable for an elevator.

93. 199 N

95. 15 N

97. 12 N

ADDITIONAL PROBLEMS

99. $a_x = 0.40 \text{ m/s}^2$ and $T = 11.2 \times 10^3 \text{ N}$

101. $m(6pt + 2q)$

103. $\vec{\mathbf{v}}(t) = \left(\frac{pt}{m} + \frac{nt^2}{2m}\right)\hat{\mathbf{i}} + \left(\frac{qt^2}{2m}\right)\hat{\mathbf{j}}$ and $\vec{\mathbf{r}}(t) = \left(\frac{pt^2}{2m} + \frac{nt^3}{6m}\right)\hat{\mathbf{i}} + \left(\frac{qt^3}{6m}\right)\hat{\mathbf{j}}$

105. 9.2 m/s

107. 1.3 s

109. 3.5 m/s^2

111. a. 0.75; b. 1200 N; c. 1.2 m/s^2 and 1080 N; d. -1.2 m/s^2; e. 120 N

113. 0.789

115. a. 0.186 N; b. 0.774 N; c. 0.48 N

117. 13 m/s

119. 0.21

121. a. 28,300 N; b. 2540 m

123. 25 N

125. $a = \frac{F}{4} - \mu_k g$

127. 11 m

CHALLENGE PROBLEMS

129. $v = \sqrt{v_0{}^2 - 2gr_0\left(1 - \frac{r_0}{r}\right)}$

131. 78.7 m

133. a. 98 m/s; b. 490 m; c. 107 m/s; d. 9.6 s

135. a. $v = 20.0(1 - e^{-0.01t})$; b. $v_{\text{limiting}} = 20 \text{ m/s}$

CHAPTER 7

CHECK YOUR UNDERSTANDING

7.1. No, only its magnitude can be constant; its direction must change, to be always opposite the relative displacement along the surface.

7.2. No, it's only approximately constant near Earth's surface.

7.3. $W = 35 \text{ J}$

7.4. a. The spring force is the opposite direction to a compression (as it is for an extension), so the work it does is negative. b. The work done depends on the square of the displacement, which is the same for $x = \pm 6 \text{ cm}$, so the magnitude is 0.54 J.

7.5. a. the car; b. the truck

7.6. against

7.7. $\sqrt{3} \text{ m/s}$

7.8. 980 W

CONCEPTUAL QUESTIONS

1. When you push on the wall, this "feels" like work; however, there is no displacement so there is no physical work. Energy is consumed, but no energy is transferred.

3. If you continue to push on a wall without breaking through the wall, you continue to exert a force with no displacement, so no work is done.

5. The total displacement of the ball is zero, so no work is done.

7. Both require the same gravitational work, but the stairs allow Tarzan to take this work over a longer time interval and hence gradually exert his energy, rather than dramatically by climbing a vine.

9. The first particle has a kinetic energy of $4(\frac{1}{2}mv^2)$ whereas the second particle has a kinetic energy of $2(\frac{1}{2}mv^2)$, so the first particle has twice the kinetic energy of the second particle.

11. The mower would gain energy if $-90° < \theta < 90°$. It would lose energy if $90° < \theta < 270°$. The mower may also lose energy due to friction with the grass while pushing; however, we are not concerned with that energy loss for this problem.

13. The second marble has twice the kinetic energy of the first because kinetic energy is directly proportional to mass, like the work done by gravity.

15. Unless the environment is nearly frictionless, you are doing some positive work on the environment to cancel out the frictional work against you, resulting in zero total work producing a constant velocity.

17. Appliances are rated in terms of the energy consumed in a relatively small time interval. It does not matter how long the appliance is on, only the rate of change of energy per unit time.

19. The spark occurs over a relatively short time span, thereby delivering a very low amount of energy to your body.

21. If the force is antiparallel or points in an opposite direction to the velocity, the power expended can be negative.

PROBLEMS

23. 3.00 J

25. a. 593 kJ; b. −589 kJ; c. 0

27. 3.14 kJ

29. a. −700 J; b. 0; c. 700 J; d. 38.6 N; e. 0

31. 100 J

33. a. 2.45 J; b. − 2.45 J; c. 0

35. a. 2.22 kJ; b. −2.22 kJ; c. 0

37. 18.6 kJ

39. a. 2.32 kN; b. 22.0 kJ

41. 835 N

43. 257 J

45. a. 1.47 m/s; b. answers may vary

47. a. 772 kJ; b. 4.0 kJ; c. 1.8×10^{-16} J

49. a. 2.6 kJ; b. 640 J

51. 2.72 kN

53. 102 N

55. 2.8 m/s

57. $W(\text{bullet}) = 20 \times W(\text{crate})$

59. 12.8 kN

61. 0.25

63. a. 24 m/s, −4.8 m/s^2; b. 29.4 m

65. 310 m/s

67. a. 40; b. 8 million

69. $149

71. a. 208 W; b. 141 s

73. a. 3.20 s; b. 4.04 s

75. a. 224 s; b. 24.8 MW; c. 49.7 kN

77. a. 1.57 kW; b. 6.28 kW

79. 6.83μW

81. a. 8.51 J; b. 8.51 W

83. 1.7 kW

ADDITIONAL PROBLEMS

85. $15 \, \text{N} \cdot \text{m}$

87. $39 \, \text{N} \cdot \text{m}$

89. a. $208 \, \text{N} \cdot \text{m}$; b. $240 \, \text{N} \cdot \text{m}$

91. a. $-0.9 \, \text{N} \cdot \text{m}$; b. $-0.83 \, \text{N} \cdot \text{m}$

93. a. 10. J; b. 10. J; c. 380 N/m

95. 160 J/s

97. a. 10 N; b. 20 W

CHALLENGE PROBLEMS

99. If crate goes up: a. 3.46 kJ; b. −1.89 kJ; c. −1.57 kJ; d. 0; If crate goes down: a. −0.39 kJ; b. −1.18 kJ; c. 1.57 kJ; d. 0

101. 8.0 J

103. 35.7 J

105. 24.3 J

107. a. 40 hp; b. 39.8 MJ, independent of speed; c. 80 hp, 79.6 MJ at 30 m/s; d. If air resistance is proportional to speed, the car gets about 22 mpg at 34 mph and half that at twice the speed, closer to actual driving experience.

CHAPTER 8

CHECK YOUR UNDERSTANDING

8.1. $(4.63\text{ J}) - (-2.38\text{ J}) = 7.00\text{ J}$

8.2. 35.3 kJ, 143 kJ, 0

8.3. 22.8 cm. Using 0.02 m for the initial displacement of the spring (see above), we calculate the final displacement of the spring to be 0.028 m; therefore the length of the spring is the unstretched length plus the displacement, or 22.8 cm.

8.4. It increases because you had to exert a downward force, doing positive work, to pull the mass down, and that's equal to the change in the total potential energy.

8.5. 2.83 N

8.6. $F = 4.8\text{ N},$ directed toward the origin

8.7. 0.033 m

8.8. b. At any given height, the gravitational potential energy is the same going up or down, but the kinetic energy is less going down than going up, since air resistance is dissipative and does negative work. Therefore, at any height, the speed going down is less than the speed going up, so it must take a longer time to go down than to go up.

8.9. constant $U(x) = -1\text{ J}$

8.10. a. yes, motion confined to $-1.055\text{ m} \le x \le 1.055\text{ m}$; b. same equilibrium points and types as in example

8.11. $x(t) = \pm\sqrt{(2E/k)}\,\sin\!\big[(\sqrt{k/m})t\big]$ and $v_0 = \pm\sqrt{(2E/m)}$

CONCEPTUAL QUESTIONS

1. The potential energy of a system can be negative because its value is relative to a defined point.

3. If the reference point of the ground is zero gravitational potential energy, the javelin first increases its gravitational potential energy, followed by a decrease in its gravitational potential energy as it is thrown until it hits the ground. The overall change in gravitational potential energy of the javelin is zero unless the center of mass of the javelin is lower than from where it is initially thrown, and therefore would have slightly less gravitational potential energy.

5. the vertical height from the ground to the object

7. A force that takes energy away from the system that can't be recovered if we were to reverse the action.

9. The change in kinetic energy is the net work. Since conservative forces are path independent, when you are back to the same point the kinetic and potential energies are exactly the same as the beginning. During the trip the total energy is conserved, but both the potential and kinetic energy change.

11. The car experiences a change in gravitational potential energy as it goes down the hills because the vertical distance is decreasing. Some of this change of gravitational potential energy will be taken away by work done by friction. The rest of the energy results in a kinetic energy increase, making the car go faster. Lastly, the car brakes and will lose its kinetic energy to the work done by braking to a stop.

13. It states that total energy of the system E is conserved as long as there are no non-conservative forces acting on the object.

15. He puts energy into the system through his legs compressing and expanding.

17. Four times the original height would double the impact speed.

PROBLEMS

19. 40,000

21. a. -200 J; b. -200 J; c. -100 J; d. -300 J

23. a. 0.068 J; b. -0.068 J; c. 0.068 J; d. 0.068 J; e. -0.068 J; f. 46 cm

25. a. -120 J; b. 120 J

27. a. $\left(\dfrac{-2a}{b}\right)^{1/6}$; b. 0 ; c. $\sim x^6$

29. 14 m/s

31. 14 J

33. proof

35. 9.7 m/s

37. 39 m/s

39. 1900 J

41. -137 J

43. 3.5 cm

45. 10x with x-axis pointed away from the wall and origin at the wall

47. 4.6 m/s

49. a. 5.6 m/s; b. 5.2 m/s; c. 6.4 m/s; d. no; e. yes

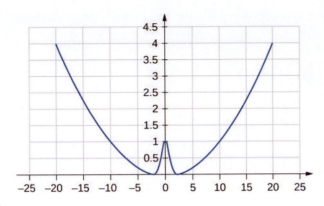

51. a.

where $k = 0.02$, $A = 1$, $\alpha = 1$; b. $F = kx - \alpha x A e^{-\alpha x^2}$; c. The potential energy at $x = 0$ must be less than the kinetic plus

potential energy at $x = a$ or $A \leq \frac{1}{2}mv^2 + \frac{1}{2}ka^2 + Ae^{-\alpha a^2}$. Solving this for A matches results in the problem.

53. 8700 N/m
55. a. 70.6 m/s; b. 69.9 m/s
57. a. 180 N/m; b. 11 m
59. a. 9.8×10^3 J; b. 1.4×10^3 J; c. 14 m/s

61. a. 47.6 m; b. 1.88×10^5 J; c. 373 N

63. 33.9 cm
65. a. Zero, since the total energy of the system is zero and the kinetic energy at the lowest point is zero; b. –0.038 J; c. 0.62 m/s
67. 42 cm

ADDITIONAL PROBLEMS

69. –0.44 J
71. 3.6 m/s
73. $bD^4/4$
75. proof

77. a. $\sqrt{\dfrac{2m^2 gh}{k(m+M)}}$; b. $\dfrac{mMgh}{m+M}$

79. a. 2.24 m/s; b. 1.94 m/s; c. 1.94 m/s

81. 18 m/s
83. $v_A = 24$ m/s; $v_B = 14$ m/s; $v_C = 31$ m/s

85. a. Loss of energy is $240 \, \text{N} \cdot \text{m}$; b. $F = 8 \, \text{N}$
87. 89.7 m/s
89. 32 J

CHAPTER 9

CHECK YOUR UNDERSTANDING

9.1. To reach a final speed of $v_f = \frac{1}{4}\left(3.0 \times 10^8 \text{ m/s}\right)$ at an acceleration of $10g$, the time

required is

$$10g = \frac{v_f}{\Delta t}$$

$$\Delta t = \frac{v_f}{10g} \frac{\frac{1}{4}\left(3.0 \times 10^8 \text{ m/s}\right)}{10g} = 7.7 \times 10^5 \text{ s} = 8.9 \text{ d}$$

9.2. If the phone bounces up with approximately the same initial speed as its impact speed, the change in momentum of the phone
will be $\Delta \vec{p} = m\Delta \vec{v} - \left(-m\Delta \vec{v}\right) = 2m\Delta \vec{v}$. This is twice the momentum change than when the phone does not bounce,

so the impulse-momentum theorem tells us that more force must be applied to the phone.
9.3. If the smaller cart were rolling at 1.33 m/s to the left, then conservation of momentum gives

$$(m_1 + m_2)\, \vec{v}_f = m_1 v_1\, \hat{i} - m_2 v_2\, \hat{i}$$

$$\vec{v}_f = \left(\frac{m_1 v_1 - m_2 v_2}{m_1 + m_2}\right)\hat{i}$$

$$= \left[\frac{(0.675\ \text{kg})(0.75\ \text{m/s}) - (0.500\ \text{kg})(1.33\ \text{m/s})}{1.175\ \text{kg}}\right]\hat{i}$$

$$= -(0.135\ \text{m/s})\,\hat{i}$$

Thus, the final velocity is 0.135 m/s to the left.

9.4. If the ball does not bounce, its final momentum $\vec{p}_2$ is zero, so

$$\Delta \vec{p} = \vec{p}_2 - \vec{p}_1$$

$$= (0)\,\hat{j} - (-1.4\ \text{kg}\cdot\text{m/s})\,\hat{j}$$

$$= +(1.4\ \text{kg}\cdot\text{m/s})\,\hat{j}$$

9.5. Consider the impulse momentum theory, which is $\vec{J} = \Delta\vec{p}$. If $\vec{J} = 0$, we have the situation described in the example. If a force acts on the system, then $\vec{J} = \vec{F}_{ave}\Delta t$. Thus, instead of $\vec{p}_f = \vec{p}_i$, we have

$$\vec{F}_{ave}\Delta t = \Delta\vec{p} = \vec{p}_f - \vec{p}_i$$

where $\vec{F}_{ave}$ is the force due to friction.

9.6. The impulse is the change in momentum multiplied by the time required for the change to occur. By conservation of momentum, the changes in momentum of the probe and the comment are of the same magnitude, but in opposite directions, and the interaction time for each is also the same. Therefore, the impulse each receives is of the same magnitude, but in opposite directions. Because they act in opposite directions, the impulses are not the same. As for the impulse, the force on each body acts in opposite directions, so the forces on each are not equal. However, the change in kinetic energy differs for each, because the collision is not elastic.

9.7. This solution represents the case in which no interaction takes place: the first puck misses the second puck and continues on with a velocity of 2.5 m/s to the left. This case offers no meaningful physical insights.

9.8. If zero friction acts on the car, then it will continue to slide indefinitely ($d \to \infty$), so we cannot use the work-kinetic-energy theorem as is done in the example. Thus, we could not solve the problem from the information given.

9.9. Were the initial velocities not at right angles, then one or both of the velocities would have to be expressed in component form. The mathematical analysis of the problem would be slightly more involved, but the physical result would not change.

9.10. The volume of a scuba tank is about 11 L. Assuming air is an ideal gas, the number of gas molecules in the tank is

$$PV = NRT$$

$$N = \frac{PV}{RT} = \frac{(2500\ \text{psi})(0.011\ \text{m}^3)}{(8.31\ \text{J/mol}\cdot\text{K})(300\ \text{K})}\left(\frac{6894.8\ \text{Pa}}{1\ \text{psi}}\right)$$

$$= 7.59 \times 10^1\ \text{mol}$$

The average molecular mass of air is 29 g/mol, so the mass of air contained in the tank is about 2.2 kg. This is about 10 times less than the mass of the tank, so it is safe to neglect it. Also, the initial force of the air pressure is roughly proportional to the surface area of each piece, which is in turn proportional to the mass of each piece (assuming uniform thickness). Thus, the initial acceleration of each piece would change very little if we explicitly consider the air.

9.11. The average radius of Earth's orbit around the Sun is 1.496×10^9 m. Taking the Sun to be the origin, and noting that the mass of the Sun is approximately the same as the masses of the Sun, Earth, and Moon combined, the center of mass of the Earth + Moon system and the Sun is

$$R_{CM} = \frac{m_{Sun}R_{Sun} + m_{em}R_{em}}{m_{Sun}}$$

$$= \frac{(1.989 \times 10^{30}\ \text{kg})(0) + (5.97 \times 10^{24}\ \text{kg} + 7.36 \times 10^{22}\ \text{kg})(1.496 \times 10^9\ \text{m})}{1.989 \times 10^{30}\ \text{kg}}$$

$$= 4.6\ \text{km}$$

Thus, the center of mass of the Sun, Earth, Moon system is 4.6 km from the center of the Sun.

9.12. On a macroscopic scale, the size of a unit cell is negligible and the crystal mass may be considered to be distributed homogeneously throughout the crystal. Thus,

$$\vec{r}_{CM} = \frac{1}{M}\sum_{j=1}^{N} m_j \,\vec{r}_j = \frac{1}{M}\sum_{j=1}^{N} m\,\vec{r}_j = \frac{m}{M}\sum_{j=1}^{N}\vec{r}_j = \frac{Nm}{M}\frac{\sum_{j=1}^{N}\vec{r}_j}{N}$$

where we sum over the number N of unit cells in the crystal and m is the mass of a unit cell. Because $Nm = M$, we can write

$$\vec{r}_{CM} = \frac{m}{M}\sum_{j=1}^{N}\vec{r}_j = \frac{Nm}{M}\frac{\sum_{j=1}^{N}\vec{r}_j}{N} = \frac{1}{N}\sum_{j=1}^{N}\vec{r}_j.$$

This is the definition of the geometric center of the crystal, so the center of mass is at the same point as the geometric center.
9.13. The explosions would essentially be spherically symmetric, because gravity would not act to distort the trajectories of the expanding projectiles.
9.14. The notation m_g stands for the mass of the fuel and m stands for the mass of the rocket plus the initial mass of the fuel. Note that m_g changes with time, so we write it as $m_g(t)$. Using m_R as the mass of the rocket with no fuel, the total mass of the rocket plus fuel is $m = m_R + m_g(t)$. Differentiation with respect to time gives

$$\frac{dm}{dt} = \frac{dm_R}{dt} + \frac{dm_g(t)}{dt} = \frac{dm_g(t)}{dt}$$

where we used $\frac{dm_R}{dt} = 0$ because the mass of the rocket does not change. Thus, time rate of change of the mass of the rocket is the same as that of the fuel.

CONCEPTUAL QUESTIONS

1. Since $K = p^2/2m$, then if the momentum is fixed, the object with smaller mass has more kinetic energy.

3. Yes; impulse is the force applied multiplied by the time during which it is applied ($J = F\Delta t$), so if a small force acts for a long time, it may result in a larger impulse than a large force acting for a small time.
5. By friction, the road exerts a horizontal force on the tires of the car, which changes the momentum of the car.
7. Momentum is conserved when the mass of the system of interest remains constant during the interaction in question and when no *net* external force acts on the system during the interaction.
9. To accelerate air molecules in the direction of motion of the car, the car must exert a force on these molecules by Newton's second law $\vec{F} = d\vec{p}/dt$. By Newton's third law, the air molecules exert a force of equal magnitude but in the opposite direction on the car. This force acts in the direction opposite the motion of the car and constitutes the force due to air resistance.
11. No, he is not a closed system because a net nonzero external force acts on him in the form of the starting blocks pushing on his feet.
13. Yes, all the kinetic energy can be lost if the two masses come to rest due to the collision (i.e., they stick together).
15. The angle between the directions must be 90°. Any system that has zero net external force in one direction and nonzero net external force in a perpendicular direction will satisfy these conditions.
17. Yes, the rocket speed can exceed the exhaust speed of the gases it ejects. The thrust of the rocket does not depend on the relative speeds of the gases and rocket, it simply depends on conservation of momentum.

PROBLEMS

19. a. magnitude: $25\ \text{kg}\cdot\text{m/s}$; b. same as a.

21. $1.78 \times 10^{29}\ \text{kg}\cdot\text{m/s}$

23. $1.3 \times 10^{9}\ \text{kg}\cdot\text{m/s}$

25. a. $1.50 \times 10^{6}\ \text{N}$; b. $1.00 \times 10^{5}\ \text{N}$

27. $4.69 \times 10^{5}\ \text{N}$

29. $2.10 \times 10^{3}\ \text{N}$

31. $\vec{p}(t) = \left(10\,\hat{i} + 20t\,\hat{j}\right)\text{kg}\cdot\text{m/s}$; $\vec{F} = (20\,\text{N})\,\hat{j}$

33. Let the positive x-axis be in the direction of the original momentum. Then $p_x = 1.5\ \text{kg}\cdot\text{m/s}$ and $p_y = 7.5\ \text{kg}\cdot\text{m/s}$

35. $(0.122\ \text{m/s})\,\hat{i}$

37. a. 47 m/s in the bullet to block direction; b. $70.6\ \text{N}\cdot\text{s}$, toward the bullet; c. $70.6\ \text{N}\cdot\text{s}$, toward the block; d. magnitude is

2.35×10^4 N

39. 2:5

41. 5.9 m/s

43. a. 6.80 m/s, 5.33°; b. yes (calculate the ratio of the initial and final kinetic energies)

45. 2.5 cm

47. the speed of the leading bumper car is 6.00 m/s and that of the trailing bumper car is 5.60 m/s

49. 6.6%

51. 1.8 m/s

53. 22.1 m/s at $32.2°$ below the horizontal

55. a. 33 m/s and 110 m/s; b. 57 m; c. 480 m

57. $(732 \text{ m/s}) \hat{\mathbf{i}} + (-79.6 \text{ m/s}) \hat{\mathbf{j}}$

59. $-(0.21 \text{ m/s}) \hat{\mathbf{i}} + (0.25 \text{ m/s}) \hat{\mathbf{j}}$

61. 341 m/s at 86.8° with respect to the $\hat{\mathbf{i}}$ axis.

63. With the origin defined to be at the position of the 150-g mass, $x_{CM} = -1.23$cm and $y_{CM} = 0.69$cm

65. $y_{CM} = \begin{cases} \frac{h}{2} - \frac{1}{4}gt^2, & t < T \\ h - \frac{1}{2}gt^2 - \frac{1}{4}gT^2 + \frac{1}{2}gtT, & t \geq T \end{cases}$

67. a. $R_1 = 4$ m, $R_2 = 2$ m; b. $X_{CM} = \frac{m_1 x_1 + m_2 x_2}{m_1 + m_2}$, $Y_{CM} = \frac{m_1 y_1 + m_2 y_2}{m_1 + m_2}$; c. yes, with $R = \frac{1}{m_1 + m_2}\sqrt{16m_1^2 + 4m_2^2}$

69. $x_{cm} = \frac{3}{4}L\left(\frac{\rho_1 + \rho_0}{\rho_1 + 2\rho_0}\right)$

71. $\left(\frac{2a}{3}, \frac{2b}{3}\right)$

73. $(x_{CM}, y_{CM}, z_{CM}) = (0, 0, h/4)$

75. $(x_{CM}, y_{CM}, z_{CM}) = (0, 4R/(3\pi), 0)$

77. (a) 0.413 m/s, (b) about 0.2 J

79. 1551 kg

81. 4.9 km/s

ADDITIONAL PROBLEMS

84. the elephant has a higher momentum

86. Answers may vary. The first clause is true, but the second clause is not true in general because the velocity of an object with small mass may be large enough so that the momentum of the object is greater than that of a larger-mass object with a smaller velocity.

88. 4.5×10^3 N

90. $\vec{\mathbf{J}} = \int_0^\tau \left[m \vec{\mathbf{g}} - m \vec{\mathbf{g}} \left(1 - e^{-bt/m}\right)\right]dt = \frac{m^2}{b} \vec{\mathbf{g}} \left(e^{-b\tau/m} - 1\right)$

92. a. $-\left(2.1 \times 10^3 \text{ kg} \cdot \text{m/s}\right)\hat{\mathbf{i}}$, b. $-\left(24 \times 10^3 \text{ N}\right)\hat{\mathbf{i}}$

94. a. $\left(1.1 \times 10^3 \text{ kg} \cdot \text{m/s}\right)\hat{\mathbf{i}}$, b. $(0.010 \text{ kg} \cdot \text{m/s})\hat{\mathbf{i}}$, c. $-(0.00093 \text{ m/s})\hat{\mathbf{i}}$, d. $-(0.0012 \text{ m/s})\hat{\mathbf{i}}$

96. 0.10 kg, $-(130 \text{ m/s})\hat{\mathbf{i}}$

98. $v_{1,f} = v_{1,i}\frac{m_1 - m_2}{m_1 + m_2}$, $v_{2,f} = v_{1,i}\frac{2m_1}{m_1 + m_2}$

100. 2.8 m/s

102. 0.094 m/s

104. final velocity of cue ball is $-(0.76 \text{ m/s})\hat{\mathbf{i}}$, final velocities of the other two balls are 2.6 m/s at ±30° with respect to the initial velocity of the cue ball

106. ball 1: $-(1.4 \text{ m/s}) \hat{i} - (0.4 \text{ m/s}) \hat{j}$, ball 2: $(2.2 \text{ m/s}) \hat{i} + (2.4 \text{ m/s}) \hat{j}$

108. ball 1: $(1.4 \text{ m/s}) \hat{i} - (1.7 \text{ m/s}) \hat{j}$, ball 2: $-(2.8 \text{ m/s}) \hat{i} + (0.012 \text{ m/s}) \hat{j}$

110. $(r, \theta) = (2R/3, \pi/8)$

112. Answers may vary. The rocket is propelled forward not by the gasses pushing against the surface of Earth, but by conservation of momentum. The momentum of the gas being expelled out the back of the rocket must be compensated by an increase in the forward momentum of the rocket.

CHALLENGE PROBLEMS

114. a. $617 \text{ N} \cdot \text{s}$, $108°$; b. $F_x = 2.91 \times 10^4 \text{ N}$, $F_y = 2.6 \times 10^5 \text{ N}$; c. $F_x = 5265 \text{ N}$, $F_y = 5850 \text{ N}$

116. Conservation of momentum demands $m_1 v_{1,i} + m_2 v_{2,i} = m_1 v_{1,f} + m_2 v_{2,f}$. We are given that $m_1 = m_2$, $v_{1,i} = v_{2,f}$, and $v_{2,i} = v_{1,f} = 0$. Combining these equations with the equation given by conservation of momentum gives $v_{1,i} = v_{1,i}$, which is true, so conservation of momentum is satisfied. Conservation of energy demands $\frac{1}{2} m_1 v_{1,i}^2 + \frac{1}{2} m_2 v_{2,i}^2 = \frac{1}{2} m_1 v_{1,f}^2 + \frac{1}{2} m_2 v_{2,f}^2$. Again combining this equation with the conditions given above give $v_{1,i} = v_{1,i}$, so conservation of energy is satisfied.

118. Assume origin on centerline and at floor, then $(x_{CM}, y_{CM}) = (0, 86 \text{ cm})$

CHAPTER 10

CHECK YOUR UNDERSTANDING

10.1. a. $40.0 \text{ rev/s} = 2\pi(40.0) \text{ rad/s}$, $\bar{\alpha} = \dfrac{\Delta\omega}{\Delta t} = \dfrac{2\pi(40.0) - 0 \text{ rad/s}}{20.0 \text{ s}} = 2\pi(2.0) = 4.0\pi \text{ rad/s}^2$; b. Since the angular velocity increases linearly, there has to be a constant acceleration throughout the indicated time. Therefore, the instantaneous angular acceleration at any time is the solution to $4.0\pi \text{ rad/s}^2$.

10.2. a. Using **Equation 10.24**, we have $7000 \text{ rpm} = \dfrac{7000.0(2\pi \text{ rad})}{60.0 \text{ s}} = 733.0 \text{ rad/s}$,

$\alpha = \dfrac{\omega - \omega_0}{t} = \dfrac{733.0 \text{ rad/s}}{10.0 \text{ s}} = 73.3 \text{ rad/s}^2$;

b. Using **Equation 10.28**, we have

$\omega^2 = \omega_0^2 + 2\alpha\Delta\theta \Rightarrow \Delta\theta = \dfrac{\omega^2 - \omega_0^2}{2\alpha} = \dfrac{0 - (733.0 \text{ rad/s})^2}{2(73.3 \text{ rad/s}^2)} = 3665.2 \text{ rad}$

10.3. The angular acceleration is $\alpha = \dfrac{(5.0 - 0)\text{rad/s}}{20.0 \text{ s}} = 0.25 \text{ rad/s}^2$. Therefore, the total angle that the boy passes through is

$\Delta\theta = \dfrac{\omega^2 - \omega_0^2}{2\alpha} = \dfrac{(5.0)^2 - 0}{2(0.25)} = 50 \text{ rad}$.

Thus, we calculate
$s = r\theta = 5.0 \text{ m}(50.0 \text{ rad}) = 250.0 \text{ m}$.

10.4. The initial rotational kinetic energy of the propeller is
$K_0 = \frac{1}{2}I\omega^2 = \frac{1}{2}(800.0 \text{ kg-m}^2)(4.0 \times 2\pi \text{ rad/s})^2 = 2.53 \times 10^5 \text{ J}$.

At 5.0 s the new rotational kinetic energy of the propeller is
$K_f = 2.03 \times 10^5 \text{ J}$.

and the new angular velocity is

$\omega = \sqrt{\dfrac{2(2.03 \times 10^5 \text{ J})}{800.0 \text{ kg-m}^2}} = 22.53 \text{ rad/s}$

which is 3.58 rev/s.

10.5. $I_{\text{parallel-axis}} = I_{\text{center of mass}} + md^2 = mR^2 + mR^2 = 2mR^2$

10.6. The angle between the lever arm and the force vector is $80°$; therefore, $r_\perp = 100\text{m}(\sin 80°) = 98.5 \text{ m}$. The cross product $\vec{\tau} = \vec{r} \times \vec{F}$ gives a negative or clockwise torque. The torque is then $\tau = -r_\perp F = -98.5 \text{ m}(5.0 \times 10^5 \text{ N}) = -4.9 \times 10^7 \text{ N} \cdot \text{m}$.

10.7. a. The angular acceleration is $\alpha = \dfrac{20.0(2\pi)\text{rad/s} - 0}{10.0 \text{ s}} = 12.56 \text{ rad/s}^2$. Solving for the torque, we have $\sum_i \tau_i = I\alpha = (30.0 \text{ kg} \cdot \text{m}^2)(12.56 \text{ rad/s}^2) = 376.80 \text{ N} \cdot \text{m}$; b. The angular acceleration is $\alpha = \dfrac{0 - 20.0(2\pi)\text{rad/s}}{20.0 \text{ s}} = -6.28 \text{ rad/s}^2$. Solving for the torque, we have $\sum_i \tau_i = I\alpha = (30.0 \text{ kg-m}^2)(-6.28 \text{ rad/s}^2) = -188.50 \text{ N} \cdot \text{m}$

10.8. 3 MW

CONCEPTUAL QUESTIONS

1. The second hand rotates clockwise, so by the right-hand rule, the angular velocity vector is into the wall.
3. They have the same angular velocity. Points further out on the bat have greater tangential speeds.
5. straight line, linear in time variable
7. constant
9. The centripetal acceleration vector is perpendicular to the velocity vector.
11. a. both; b. nonzero centripetal acceleration; c. both
13. The hollow sphere, since the mass is distributed further away from the rotation axis.
15. a. It decreases. b. The arms could be approximated with rods and the discus with a disk. The torso is near the axis of rotation so it doesn't contribute much to the moment of inertia.
17. Because the moment of inertia varies as the square of the distance to the axis of rotation. The mass of the rod located at distances greater than $L/2$ would provide the larger contribution to make its moment of inertia greater than the point mass at $L/2$.
19. magnitude of the force, length of the lever arm, and angle of the lever arm and force vector
21. The moment of inertia of the wheels is reduced, so a smaller torque is needed to accelerate them.
23. yes
25. $\left|\overrightarrow{\mathbf{r}}\right|$ can be equal to the lever arm but never less than the lever arm
27. If the forces are along the axis of rotation, or if they have the same lever arm and are applied at a point on the rod.

PROBLEMS

29. $\omega = \dfrac{2\pi \text{ rad}}{45.0 \text{ s}} = 0.14 \text{ rad/s}$

31. a. $\theta = \dfrac{s}{r} = \dfrac{3.0 \text{ m}}{1.5 \text{ m}} = 2.0 \text{ rad}$; b. $\omega = \dfrac{2.0 \text{ rad}}{1.0 \text{ s}} = 2.0 \text{ rad/s}$; c. $\dfrac{v^2}{r} = \dfrac{(3.0 \text{ m/s})^2}{1.5 \text{ m}} = 6.0 \text{ m/s}^2$.

33. The propeller takes only $\Delta t = \dfrac{\Delta \omega}{\alpha} = \dfrac{0 \text{ rad/s} - 10.0(2\pi) \text{ rad/s}}{-2.0 \text{ rad/s}^2} = 31.4 \text{ s}$ to come to rest, when the propeller is at 0 rad/s, it would start rotating in the opposite direction. This would be impossible due to the magnitude of forces involved in getting the propeller to stop and start rotating in the opposite direction.

35. a. $\omega = 25.0(2.0 \text{ s}) = 50.0 \text{ rad/s}$; b. $\alpha = \dfrac{d\omega}{dt} = 25.0 \text{ rad/s}^2$

37. a. $\omega = 54.8 \text{ rad/s}$;
b. $t = 11.0 \text{ s}$

39. a. 0.87 rad/s^2;
b. $\theta = 12{,}600 \text{ rad}$

41. a. $\omega = 42.0 \text{ rad/s}$;
b. $\theta = 200 \text{ rad}$; c. $\begin{aligned} v_t &= 42 \text{ m/s} \\ a_t &= 4.0 \text{ m/s}^2 \end{aligned}$

43. a. $\omega = 7.0 \text{ rad/s}$;
b. $\theta = 22.5 \text{ rad}$; c. $a_t = 0.1 \text{ m/s}$

45. $\alpha = 28.6 \text{ rad/s}^2$.

47. $r = 0.78 \text{ m}$

49. a. $\alpha = -0.314 \text{ rad/s}^2$,
b. $a_c = 197.4 \text{ m/s}^2$; c. $a = \sqrt{a_c^2 + a_t^2} = \sqrt{197.4^2 + (-6.28)^2} = 197.5 \text{ m/s}^2$

$\theta = \tan^{-1} \dfrac{-6.28}{197.4} = -1.8°$ in the clockwise direction from the centripetal acceleration vector

51. $ma = 40.0\,\text{kg}(5.1\,\text{m/s}^2) = 204.0\,\text{N}$

The maximum friction force is $\mu_S N = 0.6(40.0\,\text{kg})(9.8\,\text{m/s}^2) = 235.2\,\text{N}$ so the child does not fall off yet.

53.
$$
\begin{aligned}
v_t &= r\omega = 1.0(2.0t)\,\text{m/s} \\
a_c &= \frac{v_t^2}{r} = \frac{(2.0t)^2}{1.0\,\text{m}} = 4.0t^2\,\text{m/s}^2 \\
a_t(t) &= r\alpha(t) = r\frac{d\omega}{dt} = 1.0\,\text{m}(2.0) = 2.0\,\text{m/s}^2.
\end{aligned}
$$

Plotting both accelerations gives

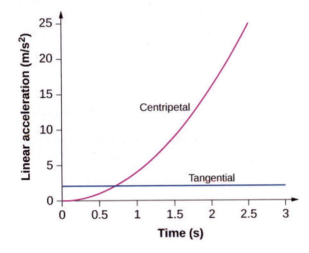

The tangential acceleration is constant, while the centripetal acceleration is time dependent, and increases with time to values much greater than the tangential acceleration after $t = 1s$. For times less than 0.7 s and approaching zero the centripetal acceleration is much less than the tangential acceleration.

55. a. $K = 2.56 \times 10^{29}\,\text{J}$;

b. $K = 2.68 \times 10^{33}\,\text{J}$

57. $K = 434.0\,\text{J}$

59. a. $v_f = 86.5\,\text{m/s}$;

b. The rotational rate of the propeller stays the same at 20 rev/s.

61. $K = 3.95 \times 10^{42}\,\text{J}$

63. a. $I = 0.315\,\text{kg} \cdot \text{m}^2$;

b. $K = 621.8\,\text{J}$

65. $I = \frac{7}{36}mL^2$

67. $v = 7.14\,\text{m/s}.$

69. $\theta = 10.2°$

71. $F = 30\,\text{N}$

73. a. $0.85\,\text{m}(55.0\,\text{N}) = 46.75\,\text{N} \cdot \text{m}$; b. It does not matter at what height you push.

75. $m_2 = \dfrac{4.9\,\text{N} \cdot \text{m}}{9.8(0.3\,\text{m})} = 1.67\,\text{kg}$

77. $\tau_{net} = -9.0\,\text{N} \cdot \text{m} + 3.46\,\text{N} \cdot \text{m} + 0 - 3.38\,\text{N} \cdot \text{m} = -8.92\,\text{N} \cdot \text{m}$

79. $\tau = 5.66\,\text{N} \cdot \text{m}$

81. $\sum \tau = 57.82\,\text{N} \cdot \text{m}$

83. $\vec{r} \times \vec{F} = 4.0\,\hat{i} + 2.0\,\hat{j} - 16.0\hat{k}\,\text{N} \cdot \text{m}$

85. a. $\tau = (0.280\,\text{m})(180.0\,\text{N}) = 50.4\,\text{N} \cdot \text{m}$; b. $\alpha = 17.14\,\text{rad/s}^2$;

c. $\alpha = 17.04 \text{ rad/s}^2$

87. $\tau = 8.0 \text{ N} \cdot \text{m}$

89. $\tau = -43.6 \text{ N} \cdot \text{m}$

91. a. $\alpha = 1.4 \times 10^{-10} \text{ rad/s}^2$;

b. $\tau = 1.36 \times 10^{28} \text{ N-m}$; c. $F = 2.1 \times 10^{21} \text{ N}$

93. $a = 3.6 \text{ m/s}^2$

95. a. $a = r\alpha = 14.7 \text{ m/s}^2$; b. $a = \frac{L}{2}\alpha = \frac{3}{4}g$

97. $\tau = \frac{P}{\omega} = \frac{2.0 \times 10^6 \text{ W}}{2.1 \text{ rad/s}} = 9.5 \times 10^5 \text{ N} \cdot \text{m}$

99. a. $K = 888.50 \text{ J}$;

b. $\Delta\theta = 294.6 \text{ rev}$

101. a. $I = 114.6 \text{ kg} \cdot \text{m}^2$;

b. $P = 104{,}700 \text{ W}$

103. $v = L\omega = \sqrt{3Lg}$

105. a. $a = 5.0 \text{ m/s}^2$; b. $W = 1.25 \text{ N} \cdot \text{m}$

ADDITIONAL PROBLEMS

107. $\Delta t = 10.0 \text{ s}$

109. a. 0.06 rad/s^2; b. $\theta = 105.0 \text{ rad}$

111. $s = 405.26 \text{ m}$

113. a. $I = 0.363 \text{ kg} \cdot \text{m}^2$;

b. $I = 2.34 \text{ kg} \cdot \text{m}^2$

115. $\omega = \sqrt{\dfrac{6.68 \text{ J}}{4.4 \text{ kgm}^2}} = 1.23 \text{ rad/s}$

117. $F = 23.3 \text{ N}$

119. $\alpha = \dfrac{190.0 \text{ N-m}}{2.94 \text{ kg-m}^2} = 64.4 \text{ rad/s}^2$

CHALLENGE PROBLEMS

121. a. $\omega = 2.0t - 1.5t^2$; b. $\theta = t^2 - 0.5t^3$; c. $\theta = -400.0 \text{ rad}$; d. the vector is at $-0.66(360°) = -237.6°$

123. $I = \frac{2}{5}mR^2$

125. a. $\omega = 8.2 \text{ rad/s}$; b. $\omega = 8.0 \text{ rad/s}$

CHAPTER 11

CHECK YOUR UNDERSTANDING

11.1. a. $\mu_S \geq \dfrac{\tan\theta}{1 + (mr^2/I_{CM})}$; inserting the angle and noting that for a hollow cylinder $I_{CM} = mr^2$, we have

$\mu_S \geq \dfrac{\tan 60°}{1 + (mr^2/mr^2)} = \frac{1}{2}\tan 60° = 0.87$; we are given a value of 0.6 for the coefficient of static friction, which is less than 0.87, so the condition isn't satisfied and the hollow cylinder will slip; b. The solid cylinder obeys the condition $\mu_S \geq \frac{1}{3}\tan\theta = \frac{1}{3}\tan 60° = 0.58$. The value of 0.6 for μ_S satisfies this condition, so the solid cylinder will not slip.

11.2. From the figure, we see that the cross product of the radius vector with the momentum vector gives a vector directed out of the page. Inserting the radius and momentum into the expression for the angular momentum, we have

$$\vec{l} = \vec{r} \times \vec{p} = (0.4 \text{ m } \hat{i}) \times (1.67 \times 10^{-27} \text{ kg}(4.0 \times 10^6 \text{ m/s}) \hat{j}) = 2.7 \times 10^{-21} \text{ kg} \cdot \text{m}^2/\text{s} \hat{k}$$

11.3. $I_{sphere} = \frac{2}{5}mr^2$, $I_{cylinder} = \frac{1}{2}mr^2$; Taking the ratio of the angular momenta, we have:

$\frac{L_{cylinder}}{L_{sphere}} = \frac{I_{cylinder}\,\omega_0}{I_{sphere}\,\omega_0} = \frac{\frac{1}{2}mr^2}{\frac{2}{5}mr^2} = \frac{5}{4}$. Thus, the cylinder has 25% more angular momentum. This is because the cylinder has

more mass distributed farther from the axis of rotation.

11.4. Using conservation of angular momentum, we have

$I(4.0\text{ rev/min}) = 1.25 I \omega_f, \quad \omega_f = \frac{1.0}{1.25}(4.0\text{ rev/min}) = 3.2\text{ rev/min}$

11.5. The Moon's gravity is 1/6 that of Earth's. By examining **Equation 11.83**, we see that the top's precession frequency is linearly proportional to the acceleration of gravity. All other quantities, mass, moment of inertia, and spin rate are the same on the Moon. Thus, the precession frequency on the Moon is

$\omega_P(\text{Moon}) = \frac{1}{6}\omega_P(\text{Earth}) = \frac{1}{6}(5.0\text{ rad/s}) = 0.83\text{ rad/s}.$

CONCEPTUAL QUESTIONS

1. No, the static friction force is zero.

3. The wheel is more likely to slip on a steep incline since the coefficient of static friction must increase with the angle to keep rolling motion without slipping.

5. The cylinder reaches a greater height. By **Equation 11.20**, its acceleration in the direction down the incline would be less.

7. All points on the straight line will give zero angular momentum, because a vector crossed into a parallel vector is zero.

9. The particle must be moving on a straight line that passes through the chosen origin.

11. Without the small propeller, the body of the helicopter would rotate in the opposite sense to the large propeller in order to conserve angular momentum. The small propeller exerts a thrust at a distance R from the center of mass of the aircraft to prevent this from happening.

13. The angular velocity increases because the moment of inertia is decreasing.

15. More mass is concentrated near the rotational axis, which decreases the moment of inertia causing the star to increase its angular velocity.

17. A torque is needed in the direction perpendicular to the angular momentum vector in order to change its direction. These forces on the space vehicle are external to the container in which the gyroscope is mounted and do not impart torques to the gyroscope's rotating disk.

PROBLEMS

19. $v_{CM} = R\omega \Rightarrow \omega = 66.7\text{ rad/s}$

21. $\alpha = 3.3\text{ rad/s}^2$

23. $I_{CM} = \frac{2}{5}mr^2$, $a_{CM} = 3.5\text{ m/s}^2$; $x = 15.75\text{ m}$

25. positive is down the incline plane;

$a_{CM} = \frac{mg\sin\theta}{m + (I_{CM}/r^2)} \Rightarrow I_{CM} = r^2\left[\frac{mg\sin 30}{a_{CM}} - m\right],$

$x - x_0 = v_0 t - \frac{1}{2}a_{CM}t^2 \Rightarrow a_{CM} = 2.96\text{ m/s}^2,$

$I_{CM} = 0.66\,mr^2$

27. $\alpha = 67.9\text{ rad/s}^2$,

$(a_{CM})_x = 1.5\text{ m/s}^2$

29. $W = -1080.0\text{ J}$

31. Mechanical energy at the bottom equals mechanical energy at the top;

$\frac{1}{2}mv_0^2 + \frac{1}{2}\left(\frac{1}{2}mr^2\right)\left(\frac{v_0}{r}\right)^2 = mgh \Rightarrow h = \frac{1}{g}\left(\frac{1}{2} + \frac{1}{4}\right)v_0^2,$

$h = 7.7\text{ m}$, so the distance up the incline is 22.5 m .

33. Use energy conservation

$\frac{1}{2}mv_0^2 + \frac{1}{2}I_{Cyl}\omega_0^2 = mgh_{Cyl},$

$\frac{1}{2}mv_0^2 + \frac{1}{2}I_{Sph}\omega_0^2 = mgh_{Sph}.$

Subtracting the two equations, eliminating the initial translational energy, we have

$$\frac{1}{2}I_{Cyl}\omega_0^2 - \frac{1}{2}I_{Sph}\omega_0^2 = mg(h_{Cyl} - h_{Sph}),$$

$$\frac{1}{2}mr^2(\frac{v_0}{r})^2 - \frac{1}{2}\frac{2}{3}mr^2(\frac{v_0}{r})^2 = mg(h_{Cyl} - h_{Sph}),$$

$$\frac{1}{2}v_0^2 - \frac{1}{2}\frac{2}{3}v_0^2 = g(h_{Cyl} - h_{Sph}),$$

$$h_{Cyl} - h_{Sph} = \frac{1}{g}\left(\frac{1}{2} - \frac{1}{3}\right)v_0^2 = \frac{1}{9.8\text{ m/s}^2}\left(\frac{1}{6}\right)(5.0\text{ m/s})^2 = 0.43\text{ m}.$$

Thus, the hollow sphere, with the smaller moment of inertia, rolls up to a lower height of $1.0 - 0.43 = 0.57$ m.

35. The magnitude of the cross product of the radius to the bird and its momentum vector yields $rp\sin\theta$, which gives $r\sin\theta$ as the altitude of the bird h. The direction of the angular momentum is perpendicular to the radius and momentum vectors, which we choose arbitrarily as $\hat{\mathbf{k}}$, which is in the plane of the ground:

$$\vec{\mathbf{L}} = \vec{\mathbf{r}} \times \vec{\mathbf{p}} = hmv\hat{\mathbf{k}} = (300.0\text{ m})(2.0\text{ kg})(20.0\text{ m/s})\hat{\mathbf{k}} = 12,000.0\text{ kg}\cdot\text{m}^2/\text{s}\,\hat{\mathbf{k}}$$

37. a. $\vec{\mathbf{l}} = 45.0\text{ kg}\cdot\text{m}^2/\text{s}\,\hat{\mathbf{k}}$;

b. $\vec{\tau} = 10.0\text{ N}\cdot\text{m}\,\hat{\mathbf{k}}$

39. a. $\vec{\mathbf{l}}_1 = -0.4\text{ kg}\cdot\text{m}^2/\text{s}\,\hat{\mathbf{k}}$, $\vec{\mathbf{l}}_2 = \vec{\mathbf{l}}_4 = 0$,

$\vec{\mathbf{l}}_3 = 1.35\text{ kg}\cdot\text{m}^2/\text{s}\,\hat{\mathbf{k}}$; b. $\vec{\mathbf{L}} = 0.95\text{ kg}\cdot\text{m}^2/\text{s}\,\hat{\mathbf{k}}$

41. a. $L = 1.0 \times 10^{11}\text{ kg}\cdot\text{m}^2/\text{s}$; b. No, the angular momentum stays the same since the cross-product involves only the perpendicular distance from the plane to the ground no matter where it is along its path.

43. a. $\vec{\mathbf{v}} = -gt\hat{\mathbf{j}}$, $\vec{\mathbf{r}}_\perp = -d\hat{\mathbf{i}}$, $\vec{\mathbf{l}} = mdgt\hat{\mathbf{k}}$;

b. $\vec{\mathbf{F}} = -mg\hat{\mathbf{j}}$, $\sum\vec{\tau} = dmg\hat{\mathbf{k}}$; c. yes

45. a. $mgh = \frac{1}{2}m(r\omega)^2 + \frac{1}{2}\frac{2}{5}mr^2\omega^2$;

$\omega = 51.2\text{ rad/s}$;

$L = 16.4\text{ kg}\cdot\text{m}^2/\text{s}$;

b. $\omega = 72.5\text{ rad/s}$;

$L = 23.2\text{ kg}\cdot\text{m}^2/\text{s}$

47. a. $I = 720.0\text{ kg}\cdot\text{m}^2$; $\alpha = 4.20\text{ rad/s}^2$;

$\omega(10\text{ s}) = 42.0\text{ rad/s}$; $L = 3.02 \times 10^4\text{ kg}\cdot\text{m}^2/\text{s}$;

$\omega(20\text{ s}) = 84.0\text{ rad/s}$;

b. $\tau = 3.03 \times 10^3\text{ N}\cdot\text{m}$

49. a. $L = 1.131 \times 10^7\text{ kg}\cdot\text{m}^2/\text{s}$;

b. $\tau = 3.77 \times 10^4\text{ N}\cdot\text{m}$

51. $\omega = 28.6\text{ rad/s} \Rightarrow L = 2.6\text{ kg}\cdot\text{m}^2/\text{s}$

53. $L_f = \frac{2}{5}M_S(3.5 \times 10^3\text{ km})^2\frac{2\pi}{T_f}$,

$$(7.0 \times 10^5\text{ km})^2\frac{2\pi}{28\text{ days}} = (3.5 \times 10^3\text{ km})^2\frac{2\pi}{T_f} \quad T_f$$

$$= 28\text{ days}\frac{(3.5 \times 10^3\text{ km})^2}{(7.0 \times 10^5\text{ km})^2} = 7.0 \times 10^{-4}\text{ day} = 60.5\text{ s}$$

55. $f_f = 2.1\text{ rev/s} \Rightarrow f_0 = 0.5\text{ rev/s}$

57. $r_P m v_P = r_A m v_A \Rightarrow v_P = 18.3 \, \text{km/s}$

59. a. $I_{\text{disk}} = 5.0 \times 10^{-4} \, \text{kg} \cdot \text{m}^2$,

$I_{\text{bug}} = 2.0 \times 10^{-4} \, \text{kg} \cdot \text{m}^2$,

$(I_{\text{disk}} + I_{\text{bug}}) \omega_1 = I_{\text{disk}} \omega_2$,

$\omega_2 = 14.0 \, \text{rad/s}$

b. $\Delta K = 0.014 \, \text{J}$;

c. $\omega_3 = 10.0 \, \text{rad/s}$ back to the original value;

d. $\frac{1}{2} (I_{\text{disk}} + I_{\text{bug}}) \omega_3^2 = 0.035 \, \text{J}$ back to the original value;

e. work of the bug crawling on the disk

61. $L_i = 400.0 \, \text{kg} \cdot \text{m}^2/\text{s}$,

$L_f = 500.0 \, \text{kg} \cdot \text{m}^2 \omega$,

$\omega = 0.80 \, \text{rad/s}$

63. $I_0 = 340.48 \, \text{kg} \cdot \text{m}^2$,

$I_f = 268.8 \, \text{kg} \cdot \text{m}^2$,

$\omega_f = 25.33 \, \text{rpm}$

65. a. $L = 280 \, \text{kg} \cdot \text{m}^2/\text{s}$,

$I_f = 89.6 \, \text{kg} \cdot \text{m}^2$,

$\omega_f = 3.125 \, \text{rad/s}$; b. $K_i = 437.5 \, \text{J}$,

$K_f = 437.5 \, \text{J}$

67. Moment of inertia in the record spin: $I_0 = 0.5 \, \text{kg} \cdot \text{m}^2$,

$I_f = 1.1 \, \text{kg} \cdot \text{m}^2$,

$\omega_f = \frac{I_0}{I_f} \omega_0 \Rightarrow f_f = 155.5 \, \text{rev/min}$

69. Her spin rate in the air is: $f_f = 2.0 \, \text{rev/s}$;

She can do four flips in the air.

71. Moment of inertia with all children aboard:

$I_0 = 2.4 \times 10^5 \, \text{kg} \cdot \text{m}^2$;

$I_f = 1.5 \times 10^5 \, \text{kg} \cdot \text{m}^2$;

$f_f = 0.3 \text{rev/s}$

73. $I_0 = 1.00 \times 10^{10} \, \text{kg} \cdot \text{m}^2$,

$I_f = 9.94 \times 10^9 \, \text{kg} \cdot \text{m}^2$,

$f_f = 3.32 \, \text{rev/min}$

75. $I = 2.5 \times 10^{-3} \, \text{kg} \cdot \text{m}^2$,

$\omega_P = 0.78 \, \text{rad/s}$

77. a. $L_{\text{Earth}} = 7.06 \times 10^{33} \, \text{kg} \cdot \text{m}^2/\text{s}$,

$\Delta L = 5.63 \times 10^{33} \, \text{kg} \cdot \text{m}^2/\text{s}$;

b. $\tau = 1.4 \times 10^{22} \, \text{N} \cdot \text{m}$;

c. The two forces at the equator would have the same magnitude but different directions, one in the north direction and the other in the south direction on the opposite side of Earth. The angle between the forces and the lever arms to the center of Earth is $90°$

, so a given torque would have magnitude $\tau = F R_E \sin 90° = F R_E$. Both would provide a torque in the same direction:

$\tau = 2FR_E \Rightarrow F = 1.3 \times 10^{15}\,\text{N}$

ADDITIONAL PROBLEMS

79. $a_{CM} = -\dfrac{3}{10}g$,

$v^2 = v_0^2 + 2a_{CM}x \Rightarrow v^2 = (7.0\,\text{m/s})^2 - 2\left(\dfrac{3}{10}g\right)x, \quad v^2 = 0 \Rightarrow x = 8.34\,\text{m};$

b. $t = \dfrac{v - v_0}{a_{CM}}, \quad v = v_0 + a_{CM}t \Rightarrow t = 2.38\,\text{s};$

The hollow sphere has a larger moment of inertia, and therefore is harder to bring to a rest than the marble, or solid sphere. The distance travelled is larger and the time elapsed is longer.

81. a. $W = -500.0\,\text{J};$

b. $K + U_{grav} = \text{constant}$,

$500\,\text{J} + 0 = 0 + (6.0\,\text{kg})(9.8\,\text{m/s}^2)h$,

$h = 8.5\,\text{m}, \; d = 17.0\,\text{m};$

The moment of inertia is less for the hollow sphere, therefore less work is required to stop it. Likewise it rolls up the incline a shorter distance than the hoop.

83. a. $\tau = 34.0\,\text{N}\cdot\text{m};$

b. $l = mr^2\omega \Rightarrow \omega = 3.6\,\text{rad/s}$

85. a. $d_M = 3.85 \times 10^8\,\text{m}$ average distance to the Moon; orbital period $27.32\text{d} = 2.36 \times 10^6\,\text{s}$; speed of the Moon

$\dfrac{2\pi 3.85 \times 10^8\,\text{m}}{2.36 \times 10^6\,\text{s}} = 1.0 \times 10^3\,\text{m/s}$; mass of the Moon $7.35 \times 10^{22}\,\text{kg}$,

$L = 2.90 \times 10^{34}\,\text{kgm}^2/\text{s};$

b. radius of the Moon $1.74 \times 10^6\,\text{m}$; the orbital period is the same as (a): $\omega = 2.66 \times 10^{-6}\,\text{rad/s}$,

$L = 2.37 \times 10^{29}\,\text{kg}\cdot\text{m}^2/\text{s};$

The orbital angular momentum is 1.22×10^5 times larger than the rotational angular momentum for the Moon.

87. $I = 0.135\,\text{kg}\cdot\text{m}^2$,

$\alpha = 4.19\,\text{rad/s}^2, \quad \omega = \omega_0 + \alpha t$,

$\omega(5\,\text{s}) = 21.0\,\text{rad/s}, \quad L = 2.84\,\text{kg}\cdot\text{m}^2/\text{s}$,

$\omega(10\,\text{s}) = 41.9\,\text{rad/s}, \quad L = 5.66\,\text{kg}\cdot\text{m/s}^2$

89. In the conservation of angular momentum equation, the rotation rate appears on both sides so we keep the (rev/min) notation as the angular velocity can be multiplied by a constant to get (rev/min):

$L_i = -0.04\,\text{kg}\cdot\text{m}^2(300.0\,\text{rev/min}),$

$L_f = 0.08\,\text{kg}\cdot\text{m}^2 f_f \Rightarrow f_f = -150.0\,\text{rev/min clockwise}$

91. $I_0\omega_0 = I_f\omega_f$,

$I_0 = 6120.0\,\text{kg}\cdot\text{m}^2$,

$I_f = 1180.0\,\text{kg}\cdot\text{m}^2$,

$\omega_f = 31.1\,\text{rev/min}$

93. $L_i = 1.00 \times 10^7\,\text{kg}\cdot\text{m}^2/\text{s}$,

$I_f = 2.025 \times 10^5\,\text{kg}\cdot\text{m}^2$,

$\omega_f = 7.86\,\text{rev/s}$

CHALLENGE PROBLEMS

95. Assume the roll accelerates forward with respect to the ground with an acceleration a'. Then it accelerates backwards relative to the truck with an acceleration $(a - a')$.

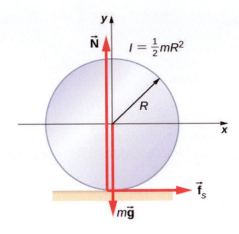

Also, $R\alpha = a - a'$ $I = \frac{1}{2}mR^2$ $\sum F_x = f_s = ma'$,

$\sum \tau = f_s R = I\alpha = I\frac{a - a'}{R}$ $f_s = \frac{I}{R^2}(a - a') = \frac{1}{2}m(a - a')$,

Solving for a' : $f_s = \frac{1}{2}m(a - a')$; $a' = \frac{a}{3}$,

$x - x_0 = v_0 t + \frac{1}{2}at^2$; $d = \frac{1}{3}at^2$; $t = \sqrt{\frac{3d}{a}}$;

therefore, $s = 1.5d$

97. a. The tension in the string provides the centripetal force such that $T \sin\theta = mr_\perp \omega^2$. The component of the tension that is vertical opposes the gravitational force such that $T \cos\theta = mg$. This gives $T = 5.7\,\text{N}$. We solve for $r_\perp = 0.16\,\text{m}$. This gives the length of the string as $r = 0.32\,\text{m}$.

At $\omega = 10.0\,\text{rad/s}$, there is a new angle, tension, and perpendicular radius to the rod. Dividing the two equations involving the tension to eliminate it, we have $\dfrac{\sin\theta}{\cos\theta} = \dfrac{(0.32\,\text{m}\sin\theta)\omega^2}{g} \Rightarrow \dfrac{1}{\cos\theta} = \dfrac{0.32\,m\omega^2}{g}$;

$\cos\theta = 0.31 \Rightarrow \theta = 72.2°$; b. $l_{\text{initial}} = 0.08\,\text{kg} \cdot \text{m}^2/\text{s}$,

$l_{\text{final}} = 0.46\,\text{kg} \cdot \text{m}^2/\text{s}$; c. No, the cosine of the angle is inversely proportional to the square of the angular velocity, therefore in order for $\theta \to 90°$, $\omega \to \infty$. The rod would have to spin infinitely fast.

CHAPTER 12

CHECK YOUR UNDERSTANDING

12.1. $x = 1.3\,\text{m}$

12.2. (b), (c)

12.3. 316.7 g; 5.8 N

12.4. $T = 1963\,\text{N};\ F = 1732\,\text{N}$

12.5. $\mu_s < 0.5 \cot\beta$

12.6. $\vec{\mathbf{F}}_{\text{door on } A} = 100.0\,\text{N}\,\hat{\mathbf{i}} - 200.0\,\text{N}\,\hat{\mathbf{j}}$; $\vec{\mathbf{F}}_{\text{door on } B} = -100.0\,\text{N}\,\hat{\mathbf{i}} - 200.0\,\text{N}\,\hat{\mathbf{j}}$

12.7. 711.0 N; 466.0 N

12.8. 1167 N; 980 N directed upward at $18°$ above the horizontal

12.9. 206.8 kPa; 4.6×10^{-5}

12.10. 5.0×10^{-4}

12.11. 63 mL

12.12. Fluids have different mechanical properties than those of solids; fluids flow.

CONCEPTUAL QUESTIONS

1. constant

3. magnitude and direction of the force, and its lever arm

5. True, as the sum of forces cannot be zero in this case unless the force itself is zero.

7. False, provided forces add to zero as vectors then equilibrium can be achieved.

9. It helps a wire-walker to maintain equilibrium.

11. (Proof)

13. In contact with the ground, stress in squirrel's limbs is smaller than stress in human's limbs.

15. tightly

17. compressive; tensile

19. no

23. It acts as "reinforcement," increasing a range of strain values before the structure reaches its breaking point.

PROBLEMS

25. $46.8 \, \text{N} \cdot \text{m}$

27. $153.4°$

29. $23.3 \, \text{N}$

31. $80.0 \, \text{kg}$

33. $40 \, \text{kg}$

35. right cable, 444.3 N; left cable, 888.5 N; weight of equipment 156.8 N; 16.0 kg

37. 784 N, 376 N

39. a. 539 N; b. 461 N; c. do not depend on the angle

41. tension 778 N; at hinge 778 N at $45°$ above the horizontal; no

43. 1500 N; 1620 N at $30°$

45. 0.3 mm

47. 9.0 cm

49. $4.0 \times 10^2 \, \text{N/cm}^2$

51. $0.149 \, \mu\text{m}$

53. 0.57 mm

55. 8.59 mm

57. $1.35 \times 10^9 \, \text{Pa}$

59. 259.0 N

61. 0.01%

63. 1.44 cm

65. 0.63 cm

ADDITIONAL PROBLEMS

69. $\tan^{-1}(1/\mu_s) = 51.3°$

71. a. at corner 66.7 N at $30°$ with the horizontal; at floor 192.4 N at $60°$ with the horizontal; b. $\mu_s = 0.577$

73. a. $1.10 \times 10^9 \, \text{N/m}^2$; b. 5.5×10^{-3}; c. 11.0 mm, 31.4 mm

CHALLENGE PROBLEMS

75. $F = Mg \tan\theta; \; f = 0$

77. with the horizontal, $\theta = 42.2°$; $\alpha = 17.8°$ with the steeper side of the wedge

79. $W(l_1/l_2 - 1); \; Wl_1/l_2 + mg$

81. a. 1.1 mm; b. 6.6 mm to the right; c. $1.11 \times 10^5 \, \text{N}$

CHAPTER 13

CHECK YOUR UNDERSTANDING

13.1. The force of gravity on each object increases with the square of the inverse distance as they fall together, and hence so does the acceleration. For example, if the distance is halved, the force and acceleration are quadrupled. Our average is accurate only for a linearly increasing acceleration, whereas the acceleration actually increases at a greater rate. So our calculated speed is too small. From Newton's third law (action-reaction forces), the force of gravity between any two objects must be the same. But the accelerations will not be if they have different masses.

13.2. The tallest buildings in the world are all less than 1 km. Since g is proportional to the distance squared from Earth's center, a simple ratio shows that the change in g at 1 km above Earth's surface is less than 0.0001%. There would be no need to consider this in structural design.

13.3. The value of g drops by about 10% over this change in height. So $\Delta U = mg(y_2 - y_1)$ will give too large a value. If we use $g = 9.80 \, \text{m/s}$, then we get $\Delta U = mg(y_2 - y_1) = 3.53 \times 10^{10}$ J which is about 6% greater than that found with the correct method.

13.4. The probe must overcome both the gravitational pull of Earth and the Sun. In the second calculation of our example, we found the speed necessary to escape the Sun from a distance of Earth's orbit, not from Earth itself. The proper way to find this value is to start with the energy equation, **Equation 13.26**, in which you would include a potential energy term for both Earth and the Sun.

13.5. You change the direction of your velocity with a force that is perpendicular to the velocity at all points. In effect, you must constantly adjust the thrusters, creating a centripetal force until your momentum changes from tangential to radial. A simple momentum vector diagram shows that the net *change* in momentum is $\sqrt{2}$ times the magnitude of momentum itself. This turns out to be a very inefficient way to reach Mars. We discuss the most efficient way in **Kepler's Laws of Planetary Motion**.

13.6. In **Equation 13.32**, the radius appears in the denominator inside the square root. So the radius must increase by a factor of 4, to decrease the orbital velocity by a factor of 2. The circumference of the orbit has also increased by this factor of 4, and so with half the orbital velocity, the period must be 8 times longer. That can also be seen directly from **Equation 13.33**.

13.7. The assumption is that orbiting object is much less massive than the body it is orbiting. This is not really justified in the case of the Moon and Earth. Both Earth and the Moon orbit about their common center of mass. We tackle this issue in the next example.

13.8. The stars on the "inside" of each galaxy will be closer to the other galaxy and hence will feel a greater gravitational force than those on the outside. Consequently, they will have a greater acceleration. Even without this force difference, the inside stars would be orbiting at a smaller radius, and, hence, there would develop an elongation or stretching of each galaxy. The force difference only increases this effect.

13.9. The semi-major axis for the highly elliptical orbit of Halley's comet is 17.8 AU and is the average of the perihelion and aphelion. This lies between the 9.5 AU and 19 AU orbital radii for Saturn and Uranus, respectively. The radius for a circular orbit is the same as the semi-major axis, and since the period increases with an increase of the semi-major axis, the fact that Halley's period is between the periods of Saturn and Uranus is expected.

13.10. Consider the last equation above. The values of r_1 and r_2 remain nearly the same, but the diameter of the Moon, $(r_2 - r_1)$, is one-fourth that of Earth. So the tidal forces on the Moon are about one-fourth as great as on Earth.

13.11. Given the incredible density required to force an Earth-sized body to become a black hole, we do not expect to see such small black holes. Even a body with the mass of our Sun would have to be compressed by a factor of 80 beyond that of a neutron star. It is believed that stars of this size cannot become black holes. However, for stars with a few solar masses, it is believed that gravitational collapse at the end of a star's life could form a black hole. As we will discuss later, it is now believed that black holes are common at the center of galaxies. These galactic black holes typically contain the mass of many millions of stars.

CONCEPTUAL QUESTIONS

1. The ultimate truth is experimental verification. Field theory was developed to help explain how force is exerted without objects being in contact for both gravity and electromagnetic forces that act at the speed of light. It has only been since the twentieth century that we have been able to measure that the force is not conveyed immediately.

3. The centripetal acceleration is not directed along the gravitational force and therefore the correct line of the building (i.e., the plumb bob line) is not directed towards the center of Earth. But engineers use either a plumb bob or a transit, both of which respond to both the direction of gravity and acceleration. No special consideration for their location on Earth need be made.

5. As we move to larger orbits, the change in potential energy increases, whereas the orbital velocity decreases. Hence, the ratio is highest near Earth's surface (technically infinite if we orbit at Earth's surface with no elevation change), moving to zero as we reach infinitely far away.

7. The period of the orbit must be 24 hours. But in addition, the satellite must be located in an equatorial orbit and orbiting in the same direction as Earth's rotation. All three criteria must be met for the satellite to remain in one position relative to Earth's surface. At least three satellites are needed, as two on opposite sides of Earth cannot communicate with each other. (This is not technically true, as a wavelength could be chosen that provides sufficient diffraction. But it would be totally impractical.)

9. The speed is greatest where the satellite is closest to the large mass and least where farther away—at the periapsis and apoapsis, respectively. It is conservation of angular momentum that governs this relationship. But it can also be gleaned from conservation of energy, the kinetic energy must be greatest where the gravitational potential energy is the least (most negative). The force, and hence acceleration, is always directed towards M in the diagram, and the velocity is always tangent to the path at all points. The acceleration vector has a tangential component along the direction of the velocity at the upper location on the y-axis; hence, the satellite is speeding up. Just the opposite is true at the lower position.

11. The laser beam will hit the far wall at a lower elevation than it left, as the floor is accelerating upward. Relative to the lab, the laser beam "falls." So we would expect this to happen in a gravitational field. The mass of light, or even an object with mass, is not relevant.

PROBLEMS

13. 7.4×10^{-8} N

15. a. 7.01×10^{-7} N ; b. The mass of Jupiter is

$m_J = 1.90 \times 10^{27}$ kg

$F_J = 1.35 \times 10^{-6}$ N

$\dfrac{F_f}{F_J} = 0.521$

17. a. 9.25×10^{-6} N ; b. Not very, as the ISS is not even symmetrical, much less spherically symmetrical.

19. a. 1.41×10^{-15} m/s^2 ; b. 1.69×10^{-4} m/s^2

21. a. 1.62 m/s^2 ; b. 3.75 m/s^2

23. a. 147 N; b. 25.5 N; c. 15 kg; d. 0; e. 15 kg

25. 12 m/s^2

27. $(3/2)R_E$

29. 5000 m/s

31. 1440 m/s

33. 11 km/s

35. a. 5.85×10^{10} J ; b. -5.85×10^{10} J ; No. It assumes the kinetic energy is recoverable. This would not even be reasonable if we had an elevator between Earth and the Moon.

37. a. 0.25; b. 0.125

39. a. 5.08×10^3 km ; b. This less than the radius of Earth.

41. 1.89×10^{27} kg

43. a. 4.01×10^{13} kg ; b. The satellite must be outside the radius of the asteroid, so it can't be larger than this. If it were this size, then its density would be about 1200 kg/m^3 . This is just above that of water, so this seems quite reasonable.

45. a. 1.66×10^{-10} m/s^2 ; Yes, the centripetal acceleration is so small it supports the contention that a nearly inertial frame of reference can be located at the Sun. b. 2.17×10^5 m/s

47. 1.98×10^{30} kg ; The values are the same within 0.05%.

49. Compare Equation 13.33 and Equation 13.53 to see that they differ only in that the circular radius, r, is replaced by the semi-major axis, a. Therefore, the mean radius is one-half the sum of the aphelion and perihelion, the same as the semi-major axis.

51. The semi-major axis, 3.78 AU is found from the equation for the period. This is one-half the sum of the aphelion and perihelion, giving an aphelion distance of 4.95 AU.

53. 1.75 years

55. 19,800 N; this is clearly not survivable

57. 1.19×10^7 km

ADDITIONAL PROBLEMS

59. a. 1.85×10^{14} N ; b. Don't do it!

61. 1.49×10^8 km

63. The value of g for this planet is 3.8 m/s^2, which is about one-fourth that of Earth. So they are weak high jumpers.

65. At the North Pole, 983 N; at the equator, 980 N

67. a. The escape velocity is still 43.6 km/s. By launching from Earth in the direction of Earth's tangential velocity, you need $43.4 - 29.8 = 13.8$ km/s relative to Earth. b. The total energy is zero and the trajectory is a parabola.

69. 61.5 km/s

71. a. 1.3×10^7 m ; b. 1.56×10^{10} J ; -3.12×10^{10} J ; -1.56×10^{10} J

73. a. 6.24×10^3 s or about 1.8 hours. This was using the 520 km average diameter. b. Vesta is clearly not very spherical, so you would need to be above the largest dimension, nearly 580 km. More importantly, the nonspherical nature would disturb the orbit very quickly, so this calculation would not be very accurate even for one orbit.

75. a. 323 km/s; b. No, you need only the difference between the solar system's orbital speed and escape speed, so about $323 - 228 = 95$ km/s .

77. Setting $e = 1$, we have $\frac{\alpha}{r} = 1 + \cos\theta \rightarrow \alpha = r + r\cos\theta = r + x$; hence, $r^2 = x^2 + y^2 = (\alpha - x)^2$. Expand and collect to show $x = \frac{1}{-2\alpha}y^2 + \frac{\alpha}{2}$.

79. Substitute directly into the energy equation using $pv_p = qv_q$ from conservation of angular momentum, and solve for v_p.

CHALLENGE PROBLEMS

81. $g = \frac{4}{3}G\rho\pi r \rightarrow F = mg = \left[\frac{4}{3}Gm\rho\pi\right]r$, and from $F = m\frac{d^2r}{dt^2}$, we get $\frac{d^2r}{dt^2} = \left[\frac{4}{3}G\rho\pi\right]r$ where the first term is ω^2.

Then $T = \frac{2\pi}{\omega} = 2\pi\sqrt{\frac{3}{4G\rho\pi}}$ and if we substitute $\rho = \frac{M}{4/3\pi R^3}$, we get the same expression as for the period of orbit R.

83. Using the mass of the Sun and Earth's orbital radius, the equation gives $2.24 \times 10^{15}\, m^2/s$. The value of $\pi R_{ES}^2/(1\text{ year})$ gives the same value.

85. $\Delta U = U_f - U_i = -\frac{GM_E m}{r_f} + \frac{GM_E m}{r_i} = GM_E m\left(\frac{r_f - r_i}{r_f r_i}\right)$ where $h = r_f - r_i$. If $h << R_E$, then $r_f r_i \approx R_E^2$, and

upon substitution, we have $\Delta U = GM_E m\left(\frac{h}{R_E^2}\right) = m\left(\frac{GM_E}{R_E^2}\right)h$ where we recognize the expression with the parenthesis as the definition of g.

87. a. Find the difference in force,

$F_{tidal} = = \frac{2GMm}{R^3}\Delta r$;

b. For the case given, using the Schwarzschild radius from a previous problem, we have a tidal force of $9.5 \times 10^{-3}\, N$. This won't even be noticed!

CHAPTER 14

CHECK YOUR UNDERSTANDING

14.1. The pressure found in part (a) of the example is completely independent of the width and length of the lake; it depends only on its average depth at the dam. Thus, the force depends only on the water's average depth and the dimensions of the dam, not on the horizontal extent of the reservoir. In the diagram, note that the thickness of the dam increases with depth to balance the increasing force due to the increasing pressure.

14.2. The density of mercury is 13.6 times greater than the density of water. It takes approximately 76 cm (29.9 in.) of mercury to measure the pressure of the atmosphere, whereas it would take approximately 10 m (34 ft.) of water.

14.3. Yes, it would still work, but since a gas is compressible, it would not operate as efficiently. When the force is applied, the gas would first compress and warm. Hence, the air in the brake lines must be bled out in order for the brakes to work properly.

CONCEPTUAL QUESTIONS

1. Mercury and water are liquid at room temperature and atmospheric pressure. Air is a gas at room temperature and atmospheric pressure. Glass is an amorphous solid (non-crystalline) material at room temperature and atmospheric pressure. At one time, it was thought that glass flowed, but flowed very slowly. This theory came from the observation that old glass planes were thicker at the bottom. It is now thought unlikely that this theory is accurate.

3. The density of air decreases with altitude. For a column of air of a constant temperature, the density decreases exponentially with altitude. This is a fair approximation, but since the temperature does change with altitude, it is only an approximation.

5. Pressure is force divided by area. If a knife is sharp, the force applied to the cutting surface is divided over a smaller area than the same force applied with a dull knife. This means that the pressure would be greater for the sharper knife, increasing its ability to cut.

7. If the two chunks of ice had the same volume, they would produce the same volume of water. The glacier would cause the greatest rise in the lake, however, because part of the floating chunk of ice is already submerged in the lake, and is thus already contributing to the lake's level.

9. The pressure is acting all around your body, assuming you are not in a vacuum.

11. Because the river level is very high, it has started to leak under the levee. Sandbags are placed around the leak, and the water held by them rises until it is the same level as the river, at which point the water there stops rising. The sandbags will absorb water until the water reaches the height of the water in the levee.

13. Atmospheric pressure does not affect the gas pressure in a rigid tank, but it does affect the pressure inside a balloon. In general, atmospheric pressure affects fluid pressure unless the fluid is enclosed in a rigid container.

15. The pressure of the atmosphere is due to the weight of the air above. The pressure, force per area, on the manometer will be the same at the same depth of the atmosphere.

17. Not at all. Pascal's principle says that the change in the pressure is exerted through the fluid. The reason that the full tub requires more force to pull the plug is because of the weight of the water above the plug.

19. The buoyant force is equal to the weight of the fluid displaced. The greater the density of the fluid, the less fluid that is needed to be displaced to have the weight of the object be supported and to float. Since the density of salt water is higher than that of fresh water, less salt water will be displaced, and the ship will float higher.

21. Consider two different pipes connected to a single pipe of a smaller diameter, with fluid flowing from the two pipes into the smaller pipe. Since the fluid is forced through a smaller cross-sectional area, it must move faster as the flow lines become closer together. Likewise, if a pipe with a large radius feeds into a pipe with a small radius, the stream lines will become closer together and the fluid will move faster.

23. The mass of water that enters a cross-sectional area must equal the amount that leaves. From the continuity equation, we know that the density times the area times the velocity must remain constant. Since the density of the water does not change, the velocity times the cross-sectional area entering a region must equal the cross-sectional area times the velocity leaving the region. Since the velocity of the fountain stream decreases as it rises due to gravity, the area must increase. Since the velocity of the faucet stream speeds up as it falls, the area must decrease.

25. When the tube narrows, the fluid is forced to speed up, thanks to the continuity equation and the work done on the fluid. Where the tube is narrow, the pressure decreases. This means that the entrained fluid will be pushed into the narrow area.

27. The work done by pressure can be used to increase the kinetic energy and to gain potential energy. As the height becomes larger, there is less energy left to give to kinetic energy. Eventually, there will be a maximum height that cannot be overcome.

29. Because of the speed of the air outside the building, the pressure outside the house decreases. The greater pressure inside the building can essentially blow off the roof or cause the building to explode.

31. The air inside the hose has kinetic energy due to its motion. The kinetic energy can be used to do work against the pressure difference.

33. Potential energy due to position, kinetic energy due to velocity, and the work done by a pressure difference.

35. The water has kinetic energy due to its motion. This energy can be converted into work against the difference in pressure.

37. The water in the center of the stream is moving faster than the water near the shore due to resistance between the water and the shore and between the layers of fluid. There is also probably more turbulence near the shore, which will also slow the water down. When paddling up stream, the water pushes against the canoe, so it is better to stay near the shore to minimize the force pushing against the canoe. When moving downstream, the water pushes the canoe, increasing its velocity, so it is better to stay in the middle of the stream to maximize this effect.

39. You would expect the speed to be slower after the obstruction. Resistance is increased due to the reduction in size of the opening, and turbulence will be created because of the obstruction, both of which will clause the fluid to slow down.

PROBLEMS

41. 1.610 cm^3

43. The mass is 2.58 g. The volume of your body increases by the volume of air you inhale. The average density of your body decreases when you take a deep breath because the density of air is substantially smaller than the average density of the body.

45. 3.99 cm

47. 2.86 times denser

49. 15.6 g/cm^3

51. $0.760 \text{ m} = 76.0 \text{ cm} = 760 \text{ mm}$

53. proof

55. a. Pressure at $h = 7.06 \times 10^6 \text{N}$;

b. The pressure increases as the depth increases, so the dam must be built thicker toward the bottom to withstand the greater pressure.

57. 4.08 m

59. 251 atm

61. $5.76 \times 10^3 \text{ N}$ extra force

63. If the system is not moving, the friction would not play a role. With friction, we know there are losses, so that $W_o = W_i - W_f$; therefore, the work output is less than the work input. In other words, to account for friction, you would need to push harder on the input piston than was calculated.

65. a. 99.5% submerged; b. 96.9% submerged

67. a. 39.5 g; b. 50 cm^3; c. 0.79 g/cm^3; ethyl alcohol

69. a. 960 kg/m^3; b. 6.34%; She floats higher in seawater.

71. a. 0.24; b. 0.72; c. Yes, the cork will float in ethyl alcohol.

73.
$$\text{net } F = F_2 - F_1 = p_2 A - p_1 A = (p_2 - p_1)A = (h_2 \rho_{fl} g - h_1 \rho_{fl} g)A$$
$$= (h_2 - h_1)\rho_{fl} gA, \text{ where } \rho_{fl} = \text{density of fluid.}$$
$$\text{net } F = (h_2 - h_1)A\rho_{fl} g = V_{fl} \rho_{fl} g = m_{fl} g = w_{fl}$$

75. $2.77 \text{ cm}^3/\text{s}$

77. a. 0.75 m/s; b. 0.13 m/s

79. a. 12.6 m/s; b. $0.0800\ \mathrm{m^3/s}$; c. No, the flow rate and the velocity are independent of the density of the fluid.

81. If the fluid is incompressible, the flow rate through both sides will be equal: $Q = A_1 \bar{v}_1 = A_2 \bar{v}_2$,   or

$$\pi \frac{d_1^2}{4} \bar{v}_1 = \pi \frac{d_2^2}{4} \bar{v}_2 \Rightarrow \bar{v}_2 = \bar{v}_1 (d_1^2 / d_2^2) = \bar{v}_1 (d_1 / d_2)^2$$

83. $F = pA \Rightarrow p = \frac{F}{A},$

$[p] = \mathrm{N/m^2} = \mathrm{N \cdot m/m^3} = \mathrm{J/m^3} = \text{energy/volume}$

85. −135 mm Hg

87. a. $1.58 \times 10^6\ \mathrm{N/m^2}$; b. 163 m

89. a. $v_2 = 3.28 \frac{\mathrm{m}}{\mathrm{s}}$;

b. $t = 0.55$ s

$x = vt = 1.81$ m

91. a. 3.02×10^{-3} N ; b. 1.03×10^{-3}

93. proof

95. 40 m/s

97. $0.537 r$; The radius is reduced to 53.7% of its normal value.

99. a. $2.40 \times 10^9\ \mathrm{N \cdot s/m^5}$; b. $48.3\ \mathrm{(N/m^2) \cdot s}$; c. 2.67×10^4 W

101. a. Nozzle: $v = 25.5 \frac{\mathrm{m}}{\mathrm{s}}$

$N_R = 1.27 \times 10^5 > 2000 \Rightarrow$

Flow is not laminar.

b. Hose: $v = 1.96 \frac{\mathrm{m}}{\mathrm{s}}$

$N_R = 35,100 > 2000 \Rightarrow$

Flow is not laminar.

103. $3.16 \times 10^{-4}\ \mathrm{m^3/s}$

ADDITIONAL PROBLEMS

105. 30.6 m

107. a. $p_{120} = 1.60 \times 10^4\ \mathrm{N/m^2}$;
$p_{80} = 1.07 \times 10^4\ \mathrm{N/m^2}$

b. Since an infant is only approximately 20 inches tall, while an adult is approximately 70 inches tall, the blood pressure for an infant would be expected to be smaller than that of an adult. The blood only feels a pressure of 20 inches rather than 70 inches, so the pressure should be smaller.

109. a. 41.4 g; b. 41.4 cm³; c. 1.09 g/cm³. This is clearly not the density of the bone everywhere. The air pockets will have a density of approximately 1.29×10^{-3} g/cm³ , while the bone will be substantially denser.

111. 8.21 N

113. a. 3.02×10^{-2} cm/s . (This small speed allows time for diffusion of materials to and from the blood.) b. 2.37×10^{10} capillaries. (This large number is an overestimate, but it is still reasonable.)

115. a. $2.76 \times 10^5\ \mathrm{N/m^2}$; b. $P_2 = 2.81 \times 10^5\ \mathrm{N/m^2}$

117. $8.7 \times 10^{-2}\ \mathrm{mm^3/s}$

119. a. 1.52; b. Turbulence would decrease the flow rate of the blood, which would require an even larger increase in the pressure difference, leading to higher blood pressure.

CHALLENGE PROBLEMS

121. $p = 0.99 \times 10^5$ Pa

123. $800\ \mathrm{kg/m^3}$

125. 11.2 m/s

127. a. 71.8 m/s; b. 257 m/s

129. a. $150 \, \text{cm}^3/\text{s}$; b. $33.3 \, \text{cm}^3/\text{s}$; c. $25.0 \, \text{cm}^3/\text{s}$; d. $0.0100 \, \text{cm}^3/\text{s}$; e. $0.0300 \, \text{cm}^3/\text{s}$

131. a. $1.20 \times 10^5 \, \text{N/m}^2$; b. The flow rate in the main increases by 90%. c. There are approximately 38 more users in the afternoon.

CHAPTER 15

CHECK YOUR UNDERSTANDING

15.1. The ruler is a stiffer system, which carries greater force for the same amount of displacement. The ruler snaps your hand with greater force, which hurts more.

15.2. You could increase the mass of the object that is oscillating. Other options would be to reduce the amplitude, or use a less stiff spring.

15.3. A ketchup bottle sits on a lazy Susan in the center of the dinner table. You set it rotating in uniform circular motion. A set of lights shine on the bottle, producing a shadow on the wall.

15.4. The movement of the pendulums will not differ at all because the mass of the bob has no effect on the motion of a simple pendulum. The pendulums are only affected by the period (which is related to the pendulum's length) and by the acceleration due to gravity.

15.5. Friction often comes into play whenever an object is moving. Friction causes damping in a harmonic oscillator.

15.6. The performer must be singing a note that corresponds to the natural frequency of the glass. As the sound wave is directed at the glass, the glass responds by resonating at the same frequency as the sound wave. With enough energy introduced into the system, the glass begins to vibrate and eventually shatters.

CONCEPTUAL QUESTIONS

1. The restoring force must be proportional to the displacement and act opposite to the direction of motion with no drag forces or friction. The frequency of oscillation does not depend on the amplitude.

3. Examples: Mass attached to a spring on a frictionless table, a mass hanging from a string, a simple pendulum with a small amplitude of motion. All of these examples have frequencies of oscillation that are independent of amplitude.

5. Since the frequency is proportional to the square root of the force constant and inversely proportional to the square root of the mass, it is likely that the truck is heavily loaded, since the force constant would be the same whether the truck is empty or heavily loaded.

7. In a car, elastic potential energy is stored when the shock is extended or compressed. In some running shoes elastic potential energy is stored in the compression of the material of the soles of the running shoes. In pole vaulting, elastic potential energy is stored in the bending of the pole.

9. The overall system is stable. There may be times when the stability is interrupted by a storm, but the driving force provided by the sun bring the atmosphere back into a stable pattern.

11. The maximum speed is equal to $v_{\text{max}} = A\omega$ and the angular frequency is independent of the amplitude, so the amplitude would be affected. The radius of the circle represents the amplitude of the circle, so make the amplitude larger.

13. The period of the pendulum is $T = 2\pi\sqrt{L/g}$. In summer, the length increases, and the period increases. If the period should be one second, but period is longer than one second in the summer, it will oscillate fewer than 60 times a minute and clock will run slow. In the winter it will run fast.

15. A car shock absorber.

17. The second law of thermodynamics states that perpetual motion machines are impossible. Eventually the ordered motion of the system decreases and returns to equilibrium.

19. All harmonic motion is damped harmonic motion, but the damping may be negligible. This is due to friction and drag forces. It is easy to come up with five examples of damped motion: (1) A mass oscillating on a hanging on a spring (it eventually comes to rest). (2) Shock absorbers in a car (thankfully they also come to rest). (3) A pendulum is a grandfather clock (weights are added to add energy to the oscillations). (4) A child on a swing (eventually comes to rest unless energy is added by pushing the child). (5) A marble rolling in a bowl (eventually comes to rest). As for the undamped motion, even a mass on a spring in a vacuum will eventually come to rest due to internal forces in the spring. Damping may be negligible, but cannot be eliminated.

PROBLEMS

21. Proof

23. 0.400 s/beat

25. 12,500 Hz

27. a. 340 km/hr; b. 11.3×10^3 rev/min

29. $f = \frac{1}{3}f_0$

31. 0.009 kg; 2%

33. a. $1.57 \times 10^5 \, \text{N/m}$; b. 77 kg, yes, he is eligible to play

35. a. 6.53×10^3 N/m ; b. yes, when the man is at his lowest point in his hopping the spring will be compressed the most

37. a. 1.99 Hz; b. 50.2 cm; c. 0.710 m

39. a. 0.335 m/s; b. 5.61×10^{-4} J

41. a. $x(t) = 2\,\text{mcos}\left(0.52\text{s}^{-1}t\right)$; b. $v(t) = (-1.05 \text{ m/s})\sin\left(0.52\text{s}^{-1}t\right)$

43. 24.8 cm

45. 4.01 s

47. 1.58 s

49. 9.82002 m/s^2

51. 9%

53. 141 J

55. a. 4.90×10^{-3} m ; b. 1.15×10^{-2} m

ADDITIONAL PROBLEMS

57. 94.7 kg

59. a. 314 N/m; b. 1.00 s; c. 1.25 m/s

61. ratio of 2.45

63. The length must increase by 0.0116%.

65. $\theta = (0.31 \text{ rad})\sin\left(3.13 \text{ s}^{-1}t\right)$

67. a. 0.99 s; b. 0.11 m

CHALLENGE PROBLEMS

69. a. 3.95×10^6 N/m ; b. 7.90×10^6 J

71. $F \approx -\text{constant } r'$

73. a. 7.54 cm; b. 3.25×10^4 N/m

CHAPTER 16

CHECK YOUR UNDERSTANDING

16.1. The wavelength of the waves depends on the frequency and the velocity of the wave. The frequency of the sound wave is equal to the frequency of the wave on the string. The wavelengths of the sound waves and the waves on the string are equal only if the velocities of the waves are the same, which is not always the case. If the speed of the sound wave is different from the speed of the wave on the string, the wavelengths are different. This velocity of sound waves will be discussed in **Sound**.

16.2. In a transverse wave, the wave may move at a constant propagation velocity through the medium, but the medium oscillates perpendicular to the motion of the wave. If the wave moves in the positive x-direction, the medium oscillates up and down in the y-direction. The velocity of the medium is therefore not constant, but the medium's velocity and acceleration are similar to that of the simple harmonic motion of a mass on a spring.

16.3. Yes, a cosine function is equal to a sine function with a phase shift, and either function can be used in a wave function. Which function is more convenient to use depends on the initial conditions. In **Figure 16.11** , the wave has an initial height of $y(0.00, 0.00) = 0$ and then the wave height increases to the maximum height at the crest. If the initial height at the initial time was equal to the amplitude of the wave $y(0.00, 0.00) = +A,$ then it might be more convenient to model the wave with a cosine function.

16.4. This wave, with amplitude $A = 0.5$ m, wavelength $\lambda = 10.00$ m, period $T = 0.50$ s, is a solution to the wave equation with a wave velocity $v = 20.00$ m/s.

16.5. Since the speed of a wave on a taunt string is proportional to the square root of the tension divided by the linear density, the wave speed would increase by $\sqrt{2}$.

16.6. At first glance, the time-averaged power of a sinusoidal wave on a string may look proportional to the linear density of the string because $P = \frac{1}{2}\mu A^2 \omega^2 v;$ however, the speed of the wave depends on the linear density. Replacing the wave speed with $\sqrt{\frac{F_T}{\mu}}$ shows that the power is proportional to the square root of tension and proportional to the square root of the linear mass density:

$$P = \frac{1}{2}\mu A^2 \omega^2 v = \frac{1}{2}\mu A^2 \omega^2 \sqrt{\frac{F_T}{\mu}} = \frac{1}{2}A^2 \omega^2 \sqrt{\mu F_T}.$$

16.7. Yes, the equations would work equally well for symmetric boundary conditions of a medium free to oscillate on each end where there was an antinode on each end. The normal modes of the first three modes are shown below. The dotted line shows the equilibrium position of the medium.

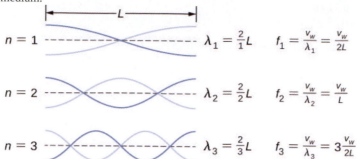

$$n = 1 \quad\quad\quad \lambda_1 = \frac{2}{1}L \quad f_1 = \frac{v_w}{\lambda_1} = \frac{v_w}{2L}$$

$$n = 2 \quad\quad\quad \lambda_2 = \frac{2}{2}L \quad f_2 = \frac{v_w}{\lambda_2} = \frac{v_w}{L}$$

$$n = 3 \quad\quad\quad \lambda_3 = \frac{2}{3}L \quad f_3 = \frac{v_w}{\lambda_3} = 3\frac{v_w}{2L}$$

Note that the first mode is two quarters, or one half, of a wavelength. The second mode is one quarter of a wavelength, followed by one half of a wavelength, followed by one quarter of a wavelength, or one full wavelength. The third mode is one and a half wavelengths. These are the same result as the string with a node on each end. The equations for symmetrical boundary conditions work equally well for fixed boundary conditions and free boundary conditions. These results will be revisited in the next chapter when discussing sound wave in an open tube.

CONCEPTUAL QUESTIONS

1. A wave on a guitar string is an example of a transverse wave. The disturbance of the string moves perpendicular to the propagation of the wave. The sound produced by the string is a longitudinal wave where the disturbance of the air moves parallel to the propagation of the wave.

3. Propagation speed is the speed of the wave propagating through the medium. If the wave speed is constant, the speed can be found by $v = \frac{\lambda}{T} = \lambda f$. The frequency is the number of wave that pass a point per unit time. The wavelength is directly proportional to the wave speed and inversely proportional to the frequency.

5. No, the distance you move your hand up and down will determine the amplitude of the wave. The wavelength will depend on the frequency you move your hand up and down, and the speed of the wave through the spring.

7. Light from the Sun and stars reach Earth through empty space where there is no medium present.

9. The wavelength is equal to the velocity of the wave times the frequency and the wave number is equal to $k = \frac{2\pi}{\lambda}$, so yes, the wave number will depend on the frequency and also depend on the velocity of the wave propagating through the spring.

11. The medium moves in simple harmonic motion as the wave propagates through the medium, continuously changing speed, therefore it accelerates. The acceleration of the medium is due to the restoring force of the medium, which acts in the opposite direction of the displacement.

13. The wave speed is proportional to the square root of the tension, so the speed is doubled.

15. Since the speed of a wave on a string is inversely proportional to the square root of the linear mass density, the speed would be higher in the low linear mass density of the string.

17. The tension in the wire is due to the weight of the electrical power cable.

19. The time averaged power is $P = \frac{E_\lambda}{T} = \frac{1}{2}\mu A^2 \omega^2 \frac{\lambda}{T} = \frac{1}{2}\mu A^2 \omega^2 v$. If the frequency or amplitude is halved, the power decreases by a factor of 4.

21. As a portion on the string moves vertically, it exerts a force on the neighboring portion of the string, doing work on the portion and transferring the energy.

23. The intensity of a spherical wave is $I = \frac{P}{4\pi r^2}$, if no energy is dissipated the intensity will decrease by a factor of nine at three meters.

25. At the interface, the incident pulse produces a reflected pulse and a transmitted pulse. The reflected pulse would be out of phase with respect to the incident pulse, and would move at the same propagation speed as the incident pulse, but would move in the opposite direction. The transmitted pulse would travel in the same direction as the incident pulse, but at half the speed. The transmitted pulse would be in phase with the incident pulse. Both the reflected pulse and the transmitted pulse would have amplitudes less than the amplitude of the incident pulse.

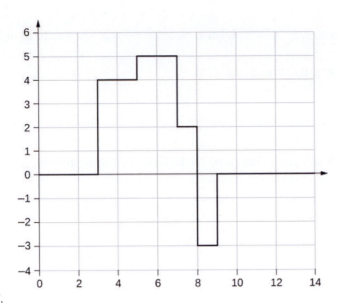

27.

29. It may be as easy as changing the length and/or the density a small amount so that the parts do not resonate at the frequency of the motor.

31. Energy is supplied to the glass by the work done by the force of your finger on the glass. When supplied at the right frequency, standing waves form. The glass resonates and the vibrations produce sound.

33. For the equation $y(x, t) = 4.00 \text{ cm} \sin(3 \text{ m}^{-1} x)\cos(4 \text{ s}^{-1} t)$, there is a node because when $x = 0.00 \text{ m}$,

$\sin(3 \text{ m}^{-1} (0.00 \text{ m})) = 0.00$, so $y(0.00 \text{ m}, t) = 0.00 \text{ m}$ for all time. For the equation

$y(x, t) = 4.00 \text{ cm} \sin\left(3 \text{ m}^{-1} x + \frac{\pi}{2}\right)\cos(4 \text{ s}^{-1} t)$, there is an antinode because when $x = 0.00 \text{ m}$,

$\sin\left(3 \text{ m}^{-1} (0.00 \text{ m}) + \frac{\pi}{2}\right) = +1.00$, so $y(0.00 \text{ m}, t)$ oscillates between +A and −A as the cosine term oscillates between +1 and -1.

PROBLEMS

35. $2d = vt \Rightarrow d = 11.25 \text{ m}$

37. $v = f\lambda$, so that $f = 0.125 \text{ Hz}$, so that
$N = 7.50 \text{ times}$

39. $v = f\lambda \Rightarrow \lambda = 0.400 \text{ m}$

41. $v = f\lambda \Rightarrow f = 2.50 \times 10^9 \text{ Hz}$

43. a. The P-waves outrun the S-waves by a speed of $v = 3.20 \text{ km/s}$; therefore, $\Delta d = 0.320 \text{ km}$. b. Since the uncertainty in the distance is less than a kilometer, our answer to part (a) does not seem to limit the detection of nuclear bomb detonations. However, if the velocities are uncertain, then the uncertainty in the distance would increase and could then make it difficult to identify the source of the seismic waves.

45. $v = 1900 \text{ m/s}$
$\Delta t = 1.05 \mu s$

47. $y(x, t) = -0.037 \text{ cm}$

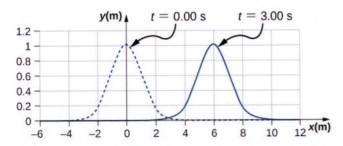

49.

The pulse will move $\Delta x = 6.00 \, \text{m}$.

51. a. $A = 0.25 \, \text{m}$; b. $k = 0.30 \, \text{m}^{-1}$; c. $\omega = 0.90 \, \text{s}^{-1}$; d. $v = 3.0 \, \text{m/s}$; e. $\phi = \pi/3 \, \text{rad}$; f. $\lambda = 20.93 \, \text{m}$; g. $T = 6.98 \, \text{s}$

53. $A = 0.30 \, \text{m}$, $\lambda = 4.50 \, \text{m}$, $v = 18.00 \, \text{m/s}$, $f = 4.00 \, \text{Hz}$, $T = 0.25 \, \text{s}$

55. $y(x, t) = 0.23 \, \text{m} \sin\left(3.49 \, \text{m}^{-1} x - 0.63 \, \text{s}^{-1} t\right)$

57. They have the same angular frequency, frequency, and period. They are traveling in opposite directions and $y_2(x, t)$ has twice the wavelength as $y_1(x, t)$ and is moving at half the wave speed.

59. Each particle of the medium moves a distance of $4A$ each period. The period can be found by dividing the velocity by the wavelength: $t = 10.42 \, \text{s}$

61. a. $\mu = 0.040 \, \text{kg/m}$; b. $v = 15.75 \, \text{m/s}$

63. $v = 180 \, \text{m/s}$

65. $v = 547.723 \, \text{m/s}$, $\Delta t = 5.48 \, \text{ms}$

67. 0.707

69. $v_1 t + v_2 t = 2.00 \, \text{m}$, $t = 1.69 \, \text{ms}$

71. $v = 288.68 \, \text{m/s}$, $\lambda = 0.73 \, \text{m}$

73. a. $A = 0.0125 \, \text{cm}$; b. $F_T = 0.96 \, \text{N}$

75. $v = 74.54 \, \text{m/s}$, $P_\lambda = 91.85 \, \text{W}$

77. a. $I = 20.0 \, \text{W/m}^2$; b. $\quad I = \dfrac{P}{A}, \ A = 10.0 \, \text{m}^2$
$$A = 4\pi r^2, \ r = 0.892 \, \text{m}$$

79. $I = 650 \, \text{W/m}^2$

81.
$$P \propto E \propto I \propto X^2 \Rightarrow \frac{P_2}{P_1} = \left(\frac{X_2}{X_1}\right)^2$$
$$P_2 = 2.50 \, \text{kW}$$

83.
$$I \propto X^2 \Rightarrow \frac{I_1}{I_2} = \left(\frac{X_1}{X_2}\right)^2 \Rightarrow$$
$$I_2 = 3.38 \times 10^{-5} \, \text{W/m}^2$$

85. $f = 100.00 \, \text{Hz}$, $A = 1.10 \, \text{cm}$

87. a. $I_2 = 0.063 I_1$; b. $\quad I_1 4\pi r_1^2 = I_2 4\pi r_2^2$
$$r_2 = 3.16 \, \text{m}$$

89. $2\pi r_1 A_1^2 = 2\pi r_2 A_2^2$, $A_1 = \left(\dfrac{r_2}{r_1}\right)^{1/2} A_1 = 0.17 \, \text{m}$

91. $y(x, t) = 0.63 \, \text{m}$

93. $A_R = 2A \cos\left(\dfrac{\phi}{2}\right)$, $\phi = 1.17 \, \text{rad}$

95. $y_R = 1.90 \, \text{cm}$

97.
$$\omega = 6.28 \, \text{s}^{-1}, \ k = 3.00 \, \text{m}^{-1}, \ \phi = \frac{\pi}{8} \, \text{rad},$$
$$A_R = 2A \cos\left(\frac{\phi}{2}\right), \ A = 0.37 \, \text{m}$$

99. a.

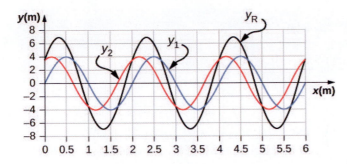

;

b. $\lambda = 2.0\,\text{m}$, $A = 4\,\text{m}$; c. $\lambda_R = 2.0\,\text{m}$, $A_R = 6.93\,\text{m}$

101. $y_R(x,\ t) = 2A\cos\!\left(\dfrac{\phi}{2}\right)\cos\!\left(kx - \omega t + \dfrac{\phi}{2}\right)$; The result is not surprising because $\cos(\theta) = \sin\!\left(\theta + \dfrac{\pi}{2}\right)$.

$\lambda_n = \dfrac{2.00}{n}L, \qquad f_n = \dfrac{v}{\lambda_n}$

103. $\lambda_1 = 4.00\,\text{m}, \qquad f_1 = 12.5\,\text{Hz}$
$\lambda_2 = 2.00\,\text{m}, \qquad f_2 = 25.00\,\text{Hz}$
$\lambda_3 = 1.33\,\text{m}, \qquad f_3 = 37.59\,\text{Hz}$

105. $v = 158.11\,\text{m/s}, \quad \lambda = 4.44\,\text{m}, \quad f = 35.61\,\text{Hz}$
$\lambda_s = 9.63\,\text{m}$

107. $y(x,\ t) = \left[0.60\,\text{cm}\,\sin\!\left(3\,\text{m}^{-1}\,x\right)\right]\cos\!\left(4\,\text{s}^{-1}\,t\right)$

109. $\lambda_{100} = 0.06\,\text{m}$
$v = 56.8\,\text{m/s}, \quad f_n = n f_1, \quad n = 1,\ 2,\ 3,\ 4,\ 5...$
$f_{100} = 947\,\text{Hz}$

111. $T = 2\Delta t, \qquad v = \dfrac{\lambda}{T}, \qquad \lambda = 2.12\,\text{m}$

113. $\lambda_1 = 6.00\,\text{m}, \qquad \lambda_2 = 3.00\,\text{m}, \qquad \lambda_3 = 2.00\,\text{m}, \qquad \lambda_4 = 1.50\,\text{m}$
$v = 258.20\,\text{m/s} = \lambda f$
$f_1 = 43.03\,\text{Hz}, \qquad f_2 = 86.07\,\text{Hz}, \qquad f_3 = 129.10\,\text{Hz}, \qquad f_4 = 172.13\,\text{Hz}$

115. $v = 134.16\,\text{ms}$, $\lambda = 1.4\,\text{m}$, $f = 95.83\,\text{Hz}$, $T = 0.0104\,\text{s}$

ADDITIONAL PROBLEMS

117. $\lambda = 0.10\,\text{m}$

119. a. $f = 4.74 \times 10^{14}\,\text{Hz}$; b. $\lambda = 422\,\text{nm}$

121. $\lambda = 16.00\,\text{m}, \qquad f = 0.10\,\text{Hz}, \qquad T = 10.00\,\text{s}, \qquad v = 1.6\,\text{m/s}$

123. $\lambda = (v_b + v)t_b, \qquad v = 3.75\,\text{m/s}, \qquad \lambda = 3.00\,\text{m}$

$$\frac{\partial^2 (y_1 + y_2)}{\partial t^2} = -A\omega^2 \sin(kx - \omega t) - A\omega^2 \sin(kx - \omega t + \phi)$$

$$\frac{\partial^2 (y_1 + y_2)}{\partial x^2} = -Ak^2 \sin(kx - \omega t) - Ak^2 \sin(kx - \omega t + \phi)$$

125. $\frac{\partial^2 y(x,\ t)}{\partial x^2} = \frac{1}{v^2}\frac{\partial^2 y(x,\ t)}{\partial t^2}$

$-A\omega^2 \sin(kx - \omega t) - A\omega^2 \sin(kx - \omega t + \phi) = \left(\frac{1}{v^2}\right)\left(-Ak^2 \sin(kx - \omega t) - Ak^2 \sin(kx - \omega t + \phi)\right)$

$v = \frac{\omega}{k}$

127. $y(x,\ t) = 0.40\text{ m} \sin\left(0.015\text{ m}^{-1} x + 1.5\text{ s}^{-1} t\right)$

129. $v = 223.61\text{ m/s},\ k = 1.57\text{ m}^{-1},\ \omega = 142.43\text{ s}^{-1}$

131. $P = \frac{1}{2}A^2 (2\pi f)^2 \sqrt{\mu F_T}$
$\mu = 2.00 \times 10^{-4}\text{ kg/m}$

133. $P = \frac{1}{2}\mu A^2 \omega^2 \frac{\lambda}{T},\ \mu = 0.0018\text{ kg/m}$

135. a. $A_R = 2A\cos\left(\frac{\phi}{2}\right),\ \cos\left(\frac{\phi}{2}\right) = 1,\ \phi = 0,\ 2\pi,\ 4\pi,...$; b. $A_R = 2A\cos\left(\frac{\phi}{2}\right),\ \cos\left(\frac{\phi}{2}\right) = 0,\ \phi = 0,\ \pi,\ 3\pi,\ 5\pi...$

137. $y_R(x,\ t) = 0.6\text{ m} \sin\left(4\text{ m}^{-1} x\right)\cos\left(3\text{ s}^{-1} t\right)$

139. a. $\begin{aligned}&(1) F_T - 20.00\text{ kg}\left(9.80\text{ m/s}^2\right)\cos 45° = 0\\&(2) m\left(9.80\text{ m/s}^2\right) - F_T = 0\\&m = 14.14\text{ kg}\end{aligned}$; b. $\begin{aligned}F_T &= 138.57\text{ N}\\v &= 67.96\text{ m/s}\end{aligned}$

141. $F_T = 12\text{ N},\ v = 16.49\text{ m/s}$

143. a. $\begin{aligned}&f_n = \frac{nv}{2L},\ v = \frac{2Lf_{n+1}}{n+1},\ \frac{n+1}{n} = \frac{2Lf_{n+1}}{2Lf_n},\ 1 + \frac{1}{n} = 1.2,\ n = 5\\&\lambda_n = \frac{2}{n}L,\ \lambda_5 = 1.6\text{ m},\ \lambda_6 = 1.33\text{ m}\end{aligned}$; b. $F_T = 245.76\text{ N}$

CHALLENGE PROBLEMS

145. a. Moves in the negative x direction at a propagation speed of $v = 2.00\text{ m/s}$. b. $\Delta x = -6.00\text{ m}$; c.

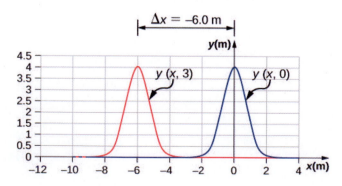

$$\sin(kx - \omega t) = \sin\left(kx + \frac{\phi}{2}\right)\cos\left(\omega t + \frac{\phi}{2}\right) - \cos\left(kx + \frac{\phi}{2}\right)\sin\left(\omega t + \frac{\phi}{2}\right)$$

$$\sin(kx - \omega t + \phi) = \sin\left(kx + \frac{\phi}{2}\right)\cos\left(\omega t + \frac{\phi}{2}\right) + \cos\left(kx + \frac{\phi}{2}\right)\sin\left(\omega t + \frac{\phi}{2}\right)$$

147.

$$\sin(kx - \omega t) + \sin(kx + \omega t + \phi) = 2\sin\left(kx + \frac{\phi}{2}\right)\cos\left(\omega t + \frac{\phi}{2}\right)$$

$$y_R = 2A\sin\left(kx + \frac{\phi}{2}\right)\cos\left(\omega t + \frac{\phi}{2}\right)$$

149.

$$\sin\left(kx + \frac{\phi}{2}\right) = 0, \ kx + \frac{\phi}{2} = 0, \ \pi, \ 2\pi, \ 1.26\,\text{m}^{-1}\,x + \frac{\pi}{20} = \pi, \ 2\pi, \ 3\pi \ ;$$

$$x = 2.37\,\text{m}, \ 4.86\,\text{m}, \ 7.35\,\text{m}$$

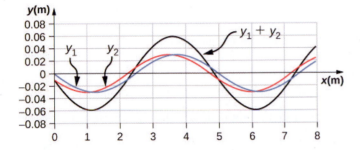

CHAPTER 17

CHECK YOUR UNDERSTANDING

17.1. Sound and light both travel at definite speeds, and the speed of sound is slower than the speed of light. The first shell is probably very close by, so the speed difference is not noticeable. The second shell is farther away, so the light arrives at your eyes noticeably sooner than the sound wave arrives at your ears.

17.2. 10 dB: rustle of leaves; 50 dB: average office; 100 dB: noisy factory

17.3. Amplitude is directly proportional to the experience of loudness. As amplitude increases, loudness increases.

17.4. In the example, the two speakers were producing sound at a single frequency. Music has various frequencies and wavelengths.

17.5. Regular headphones only block sound waves with a physical barrier. Noise-canceling headphones use destructive interference to reduce the loudness of outside sounds.

17.6. When the tube resonates at its natural frequency, the wave's node is located at the closed end of the tube, and the antinode is located at the open end. The length of the tube is equal to one-fourth of the wavelength of this wave. Thus, if we know the wavelength of the wave, we can determine the length of the tube.

17.7. Compare their sizes. High-pitch instruments are generally smaller than low-pitch instruments because they generate a smaller wavelength.

17.8. An easy way to understand this event is to use a graph, as shown below. It appears that beats are produced, but with a more complex pattern of interference.

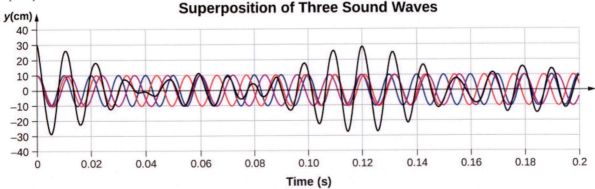

17.9. If I am driving and I hear Doppler shift in an ambulance siren, I would be able to tell when it was getting closer and also if it has passed by. This would help me to know whether I needed to pull over and let the ambulance through.

CONCEPTUAL QUESTIONS

1. Sound is a disturbance of matter (a pressure wave) that is transmitted from its source outward. Hearing is the human perception of sound.

3. Consider a sound wave moving through air. The pressure of the air is the equilibrium condition, it is the change in pressure that produces the sound wave.

5. The frequency does not change as the sound wave moves from one medium to another. Since the speed changes and the frequency does not, the wavelength must change. This is similar to the driving force of a harmonic oscillator or a wave on the string.

7. The transducer sends out a sound wave, which reflects off the object in question and measures the time it takes for the sound wave to return. Since the speed of sound is constant, the distance to the object can found by multiplying the velocity of sound by half the time interval measured.

9. The ear plugs reduce the intensity of the sound both in water and on land, but Navy researchers have found that sound under water is heard through vibrations mastoid, which is the bone behind the ear.

11. The fundamental wavelength of a tube open at each end is 2L, where the wavelength of a tube open at one end and closed at one end is 4L. The tube open at one end has the lower fundamental frequency, assuming the speed of sound is the same in both tubes.

13. The wavelength in each is twice the length of the tube. The frequency depends on the wavelength and the speed of the sound waves. The frequency in room B is higher because the speed of sound is higher where the temperature is higher.

15. When resonating at the fundamental frequency, the wavelength for pipe C is 4L, and for pipes A and B is 2L. The frequency is equal to $f = v/\lambda$. Pipe C has the lowest frequency and pipes A and B have equal frequencies, higher than the one in pipe C.

17. Since the boundary conditions are both symmetric, the frequencies are $f_n = \frac{nv}{2L}$. Since the speed is the same in each, the frequencies are the same. If the wave speed were doubled in the string, the frequencies in the string would be twice the frequencies in the tube.

19. The frequency of the unknown fork is 255 Hz. No, if only the 250 Hz fork is used, listening to the beat frequency could only limit the possible frequencies to 245 Hz or 255 Hz.

21. The beat frequency is 0.7 Hz.

23. Observer 1 will observe the highest frequency. Observer 2 will observe the lowest frequency. Observer 3 will hear a higher frequency than the source frequency, but lower than the frequency observed by observer 1, as the source approaches and a lower frequency than the source frequency, but higher than the frequency observed by observer 1, as the source moves away from observer 3.

25. Doppler radar can not only detect the distance to a storm, but also the speed and direction at which the storm is traveling.

27. The speed of sound decreases as the temperature decreases. The Mach number is equal to $M = \frac{v_s}{v}$, so the plane should slow down.

PROBLEMS

29. $s_{max} = 4.00$ nm, $\lambda = 1.72$ m, $f = 200$ Hz, $v = 343.17$ m/s

31. a. $\lambda = 68.60 \ \mu$m; b. $\lambda = 360.00 \ \mu$m

33. a. $k = 183.09$m^{-1};

b. $\Delta P = -1.11$ Pa

35.
$s_1 = 7.00$ nm, $s_2 = 3.00$ nm, $kx_1 + \phi = 0$ rad

$kx_2 + \phi = 1.128$ rad

$k(x_2 - x_1) = 1.128$ rad, $k = 5.64$ m^{-1}

$\lambda = 1.11$ m, $f = 306.31$ Hz

37.
$k = 5.28 \times 10^3$ m

$s(x, t) = 4.50$ nm $\cos\left(5.28 \times 10^3 \text{ m}^{-1} x - 2\pi(5.00 \text{ MHz})t\right)$

39. $\lambda = 3.43$ mm

41.
$$\lambda = 6.00 \text{ m}$$
$$s_{max} = 2.00 \text{ mm}$$
$$v = 600 \text{ m/s}$$
$$T = 0.01 \text{ s}$$

43. (a) $f = 100 \text{ Hz}$, (b) $\lambda = 3.43 \text{ m}$

45. $f = 3400 \text{ Hz}$

47. a. $v = 5.96 \times 10^3 \text{ m/s}$; b. steel (from value in **Table 17.1**)

49. $v = 363 \frac{\text{m}}{\text{s}}$

51. $\Delta x = 924 \text{ m}$

53.
$$V = 0.05 \text{ m}^3$$
$$m = 392.5 \text{ kg}$$
$$\rho = 7850 \text{ kg/m}^3$$
$$v = 5047.54 \text{ m/s}$$

55.
$$T_C = 35°C, \; v = 351.58 \text{ m/s}$$
$$\Delta x_1 = 35.16 \text{ m}, \; \Delta x_2 = 52.74 \text{ m}$$
$$\Delta x = 63.39 \text{ m}$$

57. a. $t_{5.00°C} = 0.0180 \text{ s}$, $t_{35.0°C} = 0.0171 \text{ s}$; b. % uncertainty = 5.00%; c. This uncertainty could definitely cause difficulties for the bat, if it didn't continue to use sound as it closed in on its prey. A 5% uncertainty could be the difference between catching the prey around the neck or around the chest, which means that it could miss grabbing its prey.

59. $1.26 \times 10^{-3} \text{ W/m}^2$

61. 85 dB

63. a. 93 dB; b. 83 dB

65. $1.58 \times 10^{-13} \text{ W/m}^2$

67. A decrease of a factor of 10 in intensity corresponds to a reduction of 10 dB in sound level: $120 \text{ dB} - 10 \text{ dB} = 110 \text{ dB}$.

69. We know that 60 dB corresponds to a factor of 10^6 increase in intensity. Therefore,

$$I \propto X^2 \Rightarrow \frac{I_2}{I_1} = \left(\frac{X_2}{X_1}\right)^2, \text{ so that } X_2 = 10^{-6} \text{ atm.}$$

120 dB corresponds to a factor of 10^{12} increase $\Rightarrow 10^{-9} \text{ atm}(10^{12})^{1/2} = 10^{-3} \text{ atm.}$

71. 28.2 dB

73. $1 \times 10^6 \text{ km}$

75. $73 \text{ dB} - 70 \text{ dB} = 3 \text{ dB}$; Such a change in sound level is easily noticed.

77. 2.5; The 100-Hz tone must be 2.5 times more intense than the 4000-Hz sound to be audible by this person.

79. 0.974 m

81. 11.0 kHz; The ear is not particularly sensitive to this frequency, so we don't hear overtones due to the ear canal.

83. a. $v = 344.08 \text{ m/s}$, $\lambda_1 = 16.00 \text{ m}$, $f_1 = 21.51 \text{ Hz}$;

b. $\lambda_3 = 5.33 \text{ m}$, $f_3 = 64.56 \text{ Hz}$

85.
$$v_{string} = 149.07 \text{ m/s}, \; \lambda_3 = 1.33 \text{ m}, \; f_3 = 112.08 \text{ Hz}$$
$$\lambda_1 = \frac{v}{f_1}, \quad L = 1.53 \text{ m}$$

87. a. $22.0°C$; b. 1.01 m

89.
first overtone = 180 Hz;
second overtone = 270 Hz;
third overtone = 360 Hz

91. 1.56 m

93. The pipe has symmetrical boundary conditions;

$\lambda_n = \frac{2}{n}L, \quad f_n = \frac{nv}{2L}, \quad n = 1, 2, 3$

$\lambda_1 = 6.00 \, \text{m}, \quad \lambda_2 = 3.00 \, \text{m}, \quad \lambda_3 = 2.00 \, \text{m}$

$f_1 = 57.17 \, \text{Hz}, \quad f_2 = 114.33 \, \text{Hz}, \quad f_3 = 171.50 \, \text{Hz}$

95. $\begin{aligned} &\lambda_6 = 0.5 \, \text{m} \\ &v = 1000 \, \text{m/s} \\ &F_T = 6500 \, \text{N} \end{aligned}$

97. $f = 6.40 \, \text{kHz}$

99. 1.03 or 3%

101. $\begin{aligned} f_B &= |f_1 - f_2| \\ |128.3 \, \text{Hz} - 128.1 \, \text{Hz}| &= 0.2 \, \text{Hz}; \\ |128.3 \, \text{Hz} - 127.8 \, \text{Hz}| &= 0.5 \, \text{Hz}; \\ |128.1 \, \text{Hz} - 127.8 \, \text{Hz}| &= 0.3 \, \text{Hz} \end{aligned}$

103. $\begin{aligned} &v_A = 135.87 \, \text{m/s}, \; v_B = 141.42 \, \text{m/s}, \\ &\lambda_A = \lambda_B = 0.40 \, \text{m} \\ &\Delta f = 15.00 \, \text{Hz} \end{aligned}$

105. $\begin{aligned} &v = 155.54 \, \text{m/s}, \\ &f_{\text{string}} = 971.17 \, \text{Hz}, \; n = 16.23 \\ &f_{\text{string}} = 1076.83 \, \text{Hz}, \; n = 18.00 \end{aligned}$

The frequency is 1076.83 Hz and the wavelength is 0.14 m.

107. $f_2 = f_1 \pm f_B = 260.00 \, \text{Hz} \pm 1.50 \, \text{Hz}$,

so that $f_2 = 261.50 \, \text{Hz}$ or $f_2 = 258.50 \, \text{Hz}$

109. $\begin{aligned} &f_{\text{ace}} = \frac{f_1 + f_2}{2}; \; f_B = f_1 - f_2 (\text{assume} f_1 > f_2) \\ &f_{\text{ace}} = \frac{(f_B + f_2) + f_2}{2} \Rightarrow \\ &f_2 = 4099.750 \, \text{Hz} \\ &f_1 = 4100.250 \, \text{Hz} \end{aligned}$

111. a. 878 Hz; b. 735 Hz

113. $3.79 \times 10^3 \, \text{Hz}$

115. a. 12.9 m/s; b. 193 Hz

117. The first eagle hears $4.23 \times 10^3 \, \text{Hz}$. The second eagle hears $3.56 \times 10^3 \, \text{Hz}$.

119. $\begin{aligned} &v_s = 31.29 \, \text{m/s} \\ &f_o = 1.12 \, \text{kHz} \end{aligned}$

121. An audible shift occurs when $\frac{f_{\text{obs}}}{f_s} \geq 1.003$; $f_{\text{obs}} = f_s \frac{v}{v - v_s} \Rightarrow \frac{f_{\text{obs}}}{f_s} = \frac{v}{v - v_s} \Rightarrow$

$v_s = 0.990 \, \text{m/s}$

$\theta = 30.02°$

123. $v_s = 680.00$ m/s

$\tan\theta = \dfrac{y}{v_s t}, \quad t = 21.65$ s

125. $\sin\theta = \dfrac{1}{M}, \quad \theta = 56.47°$

$y = 9.31$ km

127.
$s_1 = 6.34$ nm
$s_2 = 2.30$ nm
$kx_1 + \phi = 0$ rad
$kx_2 + \phi = 1.20$ rad
$k(x_2 - x_1) = 1.20$ rad
$k = 3.00 \text{ m}^{-1}$
$\omega = 1019.62 \text{ s}^{-1}$
$s_1 = s_{max}\cos(kx_1 - \phi)$

$\phi = 5.66$ rad
$s(x, t) = 6.30 \text{ nmcos}\left(3.00 \text{ m}^{-1}x - 1019.62 \text{ s}^{-1}t + 5.66\right)$

ADDITIONAL PROBLEMS

129. $v_s = 346.40$ m/s ;

$\lambda_n = \dfrac{2}{n}L \qquad f_n = \dfrac{v_s}{\lambda_n}$

$\lambda_1 = 1.60$ m $\quad f_1 = 216.50$ Hz
$\lambda_2 = 0.80$ m $\quad f_1 = 433.00$ Hz

$\lambda_6 = 0.40$ m

131. a. $v = 57.15\dfrac{m}{s}$ $\quad$; b. $\lambda_s = 2.40$ m

$f_6 = 142.89$ Hz

133.
$v = 344.08\dfrac{m}{s}$
$v_A = 29.05\dfrac{m}{s}, \quad v_B = 33.52$ m/s
$f_A = 961.18$ Hz,
$f_B = 958.89$ Hz
$f_{A,\,beat} = 161.18$ Hz, $f_{B,\,beat} = 158.89$ Hz

135. $v = 345.24\dfrac{m}{s}$; a. $I = 31.62\dfrac{\mu W}{m^2}$; b. $I = 0.16\dfrac{\mu W}{m^2}$; c. $s_{max} = 104.39$ μm ; d. $s_{max} = 7.43$ μm

137. $\dfrac{f_A}{f_D} = \dfrac{v + v_s}{v - v_s}, \quad (v - v_s)\dfrac{f_A}{f_D} = v + v_s, \quad v = 347.39\dfrac{m}{s}$

$T_C = 27.70°$

CHALLENGE PROBLEMS

$$\sqrt{x^2 + d^2} - x = \lambda, \quad x^2 + d^2 = (\lambda + x)^2$$

$$x^2 + d^2 = \lambda^2 + 2x\lambda + x^2, \quad d^2 = \lambda^2 + 2x\lambda$$

139.
$$x = \frac{d^2 - \left(\frac{v}{f}\right)^2}{2\frac{v}{f}}$$

141. a. For maxima
$$\Delta r = d \sin \theta$$
$$d \sin \theta = n\lambda \quad n = 0, \ \pm 1, \ \pm 2...., \quad \theta = \sin^{-1}\left(n\frac{\lambda}{d}\right) \ n = 0, \ \pm 1, \ \pm 2....$$

b. For minima,
$$\Delta r = d \sin \theta$$
$$d \sin \theta = \left(n + \frac{1}{2}\right)\lambda \quad n = 0, \ \pm 1, \ \pm 2....$$
$$\theta = \sin^{-1}\left(\left(n + \frac{1}{2}\right)\frac{\lambda}{d}\right) \ n = 0, \ \pm 1, \ \pm 2....$$

143. a. $v_{string} = 160.73\frac{m}{s}, \quad f_{string} = 535.77 \text{ Hz}$; b. $f_{fork} = 512 \text{ Hz}$; c. $f_{fork} = \frac{n\sqrt{\frac{F_T}{\mu}}}{2L}, \quad F_T = 141.56 \text{ N}$

145. a. $f = 268.62 \text{ Hz}$; b. $\Delta f \approx \frac{1}{2}\frac{\Delta F_T}{F_T}f = 1.34 \text{ Hz}$

147. a. $v = 466.07\frac{m}{s}$; b. $\lambda_9 = 51.11 \text{ mm}$; c. $f_9 = 9.12 \text{ kHz}$;
d. $f_{sound} = 9.12 \text{ kHz}$; e. $\lambda_{air} = 37.86 \text{ mm}$

INDEX